Principles of Gene Manipulation and Genomics

Principles of Gene Manipulation and Genomics

SEVENTH EDITION

S.B. Primrose and R.M. Twyman

Blackwell Publishing

© 2006 Blackwell Publishing

BLACKWELL PUBLISHING
350 Main Street, Malden, MA 02148-5020, USA
9600 Garsington Road, Oxford OX4 2DQ, UK
550 Swanston Street, Carlton, Victoria 3053, Australia

This material was originally published in two separate volumes: *Principles of Gene Manipulation*, 6th edition (2001) and *Principles of Genetic Analysis and Genomics*, 3rd edition (2003).

First published 1980
Second edition published 1981
Third edition published 1985
Fourth edition published 1989
Fifth edition published 1994
Sixth edition published 2001
Seventh edition published 2006

1   2006

Library of Congress Cataloging-in-Publication Data

Primrose, S.B.
    Principles of gene manipulation and genomics / S.B. Primrose and R.M. Twyman.—7th ed.
       p. ; cm.
    Rev. ed. of: Principles of gene manipulation. 6th ed. 2001 and: Principles of genome analysis and genomics / Sandy B. Primrose, Richard M. Twyman. 3rd ed. 2003.
    Includes bibliographical references and index.
    ISBN 1-4051-3544-1 (pbk. : alk. paper)   1. Genetic engineering.   2. Genomics.   3. Gene mapping.   4. Nucleotide sequence.
    [DNLM: 1. Genetic Engineering.   2. Base Sequence.   3. Chromosome Mapping.   4. DNA, Recombinant.   5. Genomics. QH 442 P952pa 2006]   I. Twyman, Richard M.   II. Primrose, S.B. Principles of gene manipulation.   III. Primrose, S. B. Principles of genome analysis and genomics.   IV. Title.

QH442.O42 2006
660.6′5—dc22

                                                                                          2005018202

A catalogue record for this title is available from the British Library.

Set in 10/12.5pt Photina
by Graphicraft Limited, Hong Kong
Printed and bound in the United Kingdom
by TJ International, Padstow, Cornwall, UK

The publisher's policy is to use permanent paper from mills that operate a sustainable forestry policy, and which has been manufactured from pulp processed using acid-free and elementary chlorine-free practices. Furthermore, the publisher ensures that the text paper and cover board used have met acceptable environmental accreditation standards.

For further information on
Blackwell Publishing, visit our website:
www.blackwellpublishing.com

Contents

Preface

The first edition of *Principles of Gene Manipulation* was published over 25 years ago when the recombinant DNA era was in its infancy and the idea of sequencing the entire human genome was inconceivable. In writing the first edition, the aim was to explain a new and rapidly growing technology. The basic philosophy was to present the principles of gene manipulation, and its associated techniques, in sufficient detail to enable the non-specialist reader to understand them. However, as the techniques became more sophisticated and advanced, so the book grew in size and complexity. Eventually, recombinant DNA technology advanced to the stage where the sequencing and analysis of entire genomes became possible. This gave rise to a whole new biological discipline, known as genomics, with its own principles and associated techniques. From this emerged the first edition of another book, *Principles of Genome Analysis*, whose title changed to *Principles of Genome Analysis and Genomics* in its third edition to reflect the rapid growth of post-sequencing technologies aiming at the large-scale analysis of gene function. It is now five years since the draft human genome sequence was published and we are reaching the stage where the technologies of gene manipulation and genomics are becoming increasingly integrated. Genome mapping and sequencing technologies borrow extensively from the early recombinant DNA technologies of library construction, cloning, and amplification using the polymerase chain reaction; gene transfer to microbes, animals, and plants is now widely used for the functional analysis of genomes; and the applications of genomics and recombinant DNA are becoming difficult to separate.

This new edition, entitled *Principles of Gene Manipulation and Genomics*, therefore unites the themes covered formerly by the two separate books and provides for the first time a fully integrated approach to the principles and practice of gene manipulation in the context of the genomics era. As in previous editions of the two books, we have written the text at an advanced undergraduate level, assuming a basic knowledge of molecular biology and genetics but no knowledge of recombinant DNA technology or genomics. However, we are aware that the book is favored not only by newcomers to the field but also by experts, and we have tried to remain faithful to both audiences with our coverage. As before we have not changed the level at which the book is written nor the general style, but we have divided the book into sections to enable the book to be used in different ways by different readers.

The basic methodologies are presented in the first part of the book, which is devoted to cloning in *Escherichia coli*, while more advanced gene-transfer techniques (applying to other microbes and to animals and plants) are presented in the second part. The reader who has read and understood the material in the first part, or already knows it, should have no difficulty in understanding any of the material in the second part of the book. The third part moves from the basic gene-manipulation technologies to genomics, transcriptomics, proteomics, and metabolomics, the major branches of the high-throughput, large-scale biology that has become synonymous with the new millennium. Finally, the fourth part of the book contains two chapters that discuss how recombinant DNA technology and genomics are being applied in the fields of medicine, agriculture, diagnostics, forensics, and biotechnology.

In writing the first part of the book, we thought carefully about the inclusion of early "historical" information. Although older readers may feel that some of this material is dated, we elected to leave much of it in place because it has an important bearing on today's methods and an understanding of it is incorrectly assumed in many of today's publications. We have included such information where it illustrates how modern techniques and procedures have evolved, but we have tried not to catalog outmoded or redundant methods that are no longer used. This is particularly the case in the genomics section where new technologies seem to come and go every day, and few stand the test of time or become truly indispensable. We have aimed to avoid as much jargon as possible, and to explain it clearly where it is absolutely necessary. As is common in all areas of science, the principles of gene manipulation and genomics abound with acronyms and synonyms which are often confusing particularly now molecular biology is becoming increasingly commercial in both basic research and its applications. Where appropriate, we have provided lists of definitions as boxes set aside from the text. Boxes are also used to illustrate key experiments or principles, historical information,

and applications. While the text is fully referenced throughout, we have also provided a list of classic papers and reviews at the end of each chapter to ease the wary reader into the scientific literature.

This book would not have been possible without the help and advice of many colleagues. Particular thanks are due to Sue Goddard and her library staff at HPA Porton for assistance with many literature searches. Sandy Primrose would like to dedicate this book to his wife Jill and Richard Twyman would like to dedicate this book to his parents, Irene and Peter, to his children Emily and Lucy, and to Liz for her endless support and encouragement.

Abbreviations

2DE	two-dimensional gel electrophoresis
Ac	*Activator*
ADME	adsorption, distribution, metabolism and excretion
AFBAC	affected family-based control
AFLP	amplified fragment length polymorphism
ALL	acute lymphoblastic leukemia
AML	acute myeloid leukemia
AMV	avian myeloblastosis virus
APL	acute promyelocytic leukemia
ARS	autonomously replicating sequence
ATRA	all-*trans*-retinoic acid
BAC	bacterial artificial chromosome
BCG	Bacille Calmette–Guérin
bFGF	basic fibroblast growth factor
BIND	Biomolecular Interaction Network Database
BLAST	Basic Local Alignment Search Tool
BLOSUM	Blocks Substitution Matrix
BMP	bone morphogenetic protein
bp	base pair
BRET	bioluminescence resonance energy transfer
CAPS	cleavable amplified polymorphic sequences
CASP	Critical Assessment of Structural Prediction
CATH	Class, Architecture, Topology and Homologous superfamily (database)
ccc DNA	covalently closed circular DNA
CCD	charge couple device
CD	circular dichroism
cDNA	complementary DNA
CEPH	Centre d'Etude du Polymorphisme Humain
cfu	commonly forming unit
CHEF	contour-clamped homogeneous electrical field
CID	chemically induced dimerization Also: collision-induced dissociation

cM	centimorgan
COG	cluster of orthologous groups
cR	centiRay
cRNA	complementary RNA
CSSL	chromosome segment substitution line
ct	chloroplast
DALPC	direct analysis of large protein complexes
DAS	distributed annotation system
DAS	downstream activation site
DBM	diazobenzyloxymethyl
DDBJ	DNA Databank of Japan
DIP	Database of Interacting Proteins
DMD	Duchenne muscular dystrophy
DNA	deoxyribonucleic acid
dNTP	deoxynucleoside triphosphate
Ds	*Dissociation*
dsDNA	double-stranded DNA
dsRNA	double-stranded RNA
EGF	epidermal growth factor
ELISA	enzyme-linked immunosorbent sandwich assay
EMBL	European Molecular Biology Laboratory
ENU	ethylnitrosourea
EOP	efficiency of plating
ES	embryonic stem (cells)
ESI	electrospray ionization
EST	expressed sequence tag
EUROFAN	European Functional Analysis Network (consortium)
FACS	fluorescence-activated cell sorting
FEN	flap endonuclease
FIAU	Fialuridine (1–2′-deoxy-2′-fluoro-β-D-arabinofuranosyl-5-iodouracil)
FIGE	field-inversion gel electrophoresis
FISH	fluorescence *in situ* hybridization
FPC	fingerprinted contigs
FRET	fluorescence resonance energy

	transfer	MCS	multiple cloning site
FSSP	Fold classification based on Structure–Structure alignment of Proteins (database)	MDA	multiple displacement amplification
		MGED	Microarray Gene Expression Database
		MHC	major histocompatibility complex
GASP	Genome Annotation aSsessment Project	MIAME	minimum information about a microarray experiment
G-CSF	granulocyte colony stimulating factor	MIP	molecularly imprinted polymer
GeneEMAC	gene external marker-based automatic congruencing	MIPS	Munich Information Center for Protein Sequences
GGTC	German Gene Trap Consortium	MM	'mismatch' oligonucleotide
GST	gene trap sequence tag	MMTV	mouse mammary tumor virus
GST	glutathione-*S*-transferase	MPSS	massively parallel signature sequencing
HAT	hypoxanthine, aminopterin and thymidine	mRNA	messenger RNA
HDL	high-density lipoprotein	MS	mass spectrometry
HERV	human endogenous retrovirus	MS/MS	tandem mass spectroscopy
HGP	Human Genome Project	mt	mitochondrial
HLA	human leukocyte antigen	MTM	Maize Targeted Mutagenesis project
HPRT	hypoxanthine phosphoribosyl-transferase	*Mu*	*Mutator*
HTF	*Hpa*II tiny fragment	MudPIT	multidimensional protein identification technology
htSNP	haplotype tag single nucleotide polymorphism	MuLV	Moloney murine leukemia virus
ibd	identical by descent	NCBI	National Center for Biotechnology Information
ICAT	isotope-coded affinity tag	NDB	Nucleic Acid Databank
IDA	interaction defective allele	NGF	nerve growth factor
IEF	isoelectric focusing	NIGMS	National Institute of General Medical Sciences
Ihh	Indian hedgehog		
IPTG	isopropylthio-β-D-galactopyranoside	NIL	near isogenic line
IST	interaction sequence tag	NMR	nuclear magnetic resonance
ITCHY	incremental truncation for the creation of hybrid enzymes	NOE	nuclear Overhauser effect
		NOESY	NOE spectroscopy
IVET	*in vivo* expression technology	nt	nucleotide
kb	kilobase	oc DNA	open circular DNA
LCR	low complexity region	OFAGE	orthogonal-field-alternation gel electrophoresis
LD	linkage disequilibrium		
LINE	long interspersed nuclear element	OMIM	on-line Mendelian inheritance in man
LOD	logarithm$_{10}$ of odds	ORF	open-reading frame
LTR	long terminal repeat	ORFan	orphan open-reading frame
m : z	mass : charge ratio	P/A	presence/absence polymorphism
MAD	multiwavelength anomalous diffraction	PAC	P1-derived artificial chromosome
		PAGE	polyacrylaminde gel electrophoresis
MAGE	microarray and gene expression	PAI	pathogenicity island
MAGE-ML	microarray and gene expression mark-up language	PAM	percentage of accepted point mutations
MAGE-OM	microarray and gene expression object model	PCR	polymerase chain reaction
		PDB	Protein Databank (database)
MALDI	matrix assisted laser desorption ionization	Pfam	Protein families database of alignments
MAR	matrix attachment region	PFGE	pulsed field gel electrophoresis
Mb	megabase	PM	'perfect match' oligonucleotide
MCAT	mass coded abundance tag	poly(A)$^+$	polyadenylated

PQL   protein quantity loci
PRINS   primed *in situ*
PS   position shift polymorphism
PSI-BLAST Position-Specific Iterated BLAST
           (software)
PTGS   post-transcriptional gene silencing
PVDF   polyvinylidine difluoride
QTL   quantitative trait loci
RACE   rapid amplification of cDNA ends
RAGE   recombinase-activated gene
           expression
RAPD   randomly amplified polymorphic DNA
RARE   RecA-assisted restriction
           endonuclease
RC   recombinant congenic (strains)
RCA   rolling circle amplification
RCSB   Research Collaboratory for Structural
           Bioinformatics
rDNA/RNA   ribosomal DNA/RNA
REMI   restriction enzyme-mediated
           integration
RFLP   restriction fragment length
           polymorphism
RIL   recombinant inbred line
R-M   restriction-modification
RNA   ribonucleic acid
RNAi   RNA interference
RNase   ribonuclease
RPMLC   reverse phase microcapillary liquid
           chromatography
RRS   Ras recruitment system
RT-PCR   reverse transcriptase polymerase
           chain reaction
RTX   repeats in toxins
SAGE   serial analysis of gene expression
SCOP   Structural Classification of Proteins
           (database)
SCOPE   structure-based combinatorial
           protein engineering
SDS   sodium dodecyl sulfate
SELDI   surface-enhanced laser desorption
           and ionization
SGA   synthetic genetic array
SGDP   *Saccharomyces* Gene Deletion Project
Shh   sonic hedgehog
SILAC   stable-isotope labeling with amino
           acids in cell culture

SINE   short interspersed nuclear element
SINS   sequenced insertion sites
SISDC   sequence-independent site-directed
           chimeragenesis
SNP   single nucleotide polymorphism
SPIN   Surface Properties of protein–protein
           Interfaces (database)
Spm   *Suppressor–mutator*
SPR   surface plasmon resonance
SRCD   synchrotron radiation circular
           dichroism
SRS   sequence retrieval system
SRS   SOS recruitment system
SSLP   simple sequence length
           polymorphism
SSR   simple sequence repeat
STC   sequence-tagged connector
STM   signature-tagged mutagenesis
STS   sequence-tagged site
TAC   transformation-competent artificial
           chromosome
TAFE   transversely alternating-field
           electrophoresis
TAP   tandem affinity purification
TAR   transformation-associated
           recombination
T-DNA   *Agrobacterium* transfer DNA
TIGR   The Institute for Genomic Research
TIM   triose phosphate isomerase
TOF   time of flight
tRNA   transfer RNA
TUSC   Trait Utility System for Corn
UAS   upstream activation site
UPA   universal protein array
URS   upstream repression site
USPS   ubiquitin-based split protein sensor
UTR   untranslated region
VDA   variant detector array
VIGS   virus-induced gene silencing
WGA   whole-genome amplification
Y2H   yeast two-hybrid
YAC   yeast artificial chromosome
YCp   yeast centromere plasmid
YEp   yeast episomal plasmid
YIp   yeast integrating plasmid
YRp   yeast replicating plasmid

CHAPTER 1

Gene manipulation in the post-genomics era

Introduction

Since the beginning of the last century, scientists have been interested in *genes*. First, they wanted to find out what genes were made of, how they worked, and how they were transmitted from generation to generation with the seemingly mythic ability to control both heredity and variation. Genes were initially thought of in functional terms as hereditary units responsible for the appearance of particular biological characteristics, such as eye or hair color in human beings, but their physical properties were unclear. It was not until the 1940s that genes were shown to be made of DNA, and that a workable physical and functional definition of the gene – a length of DNA encoding a particular protein – was achieved (Box 1.1). Next, scientists wanted to find ways to study the structure, behavior, and activity of genes in more detail. This required the simultaneous development of novel techniques for DNA analysis and manipulation. These developments began in the early 1970s with the first experiments involving the creation and manipulation of recombinant DNA. Thus began the *recombinant DNA revolution*.

Gene manipulation involves the creation and cloning of recombinant DNA

The definition of *recombinant DNA* is any artificially created DNA molecule which brings together DNA sequences that are not usually found together in nature. *Gene manipulation* refers to any of a variety of sophisticated techniques for the creation of recombinant DNA and, in many cases, its subsequent introduction into living cells. In the developed world there is a precise legal definition of gene manipulation as a result of government legislation to control it. In the UK, for example, gene manipulation is defined as: ". . . the formation of new combinations of heritable material by the insertion of nucleic acid molecules,

produced by whatever means outside the cell, into any virus, bacterial plasmid or other vector system so as to allow their incorporation into a host organism in which they do not naturally occur but in which they are capable of continued propagation." The propagation of recombinant DNA inside a particular host cell so that many copies of the same sequence are produced is known as *cloning*.

Cloning was a significant breakthrough in molecular biology because it became possible to obtain homogeneous preparations of any desired DNA molecule in amounts suitable for laboratory-scale experiments. A single organism, the bacterium *Escherichia coli*, played the dominant role in the early years of the recombinant DNA era. This bacterium had always been a popular model system for molecular geneticists and, prior to the development of recombinant DNA technology, there were already a large number of well-characterized mutants, gene regulation was understood, and many plasmids had been isolated. It is not surprising that the first cloning experiments were undertaken in *E. coli* and that this organism became the primary cloning host. Subsequently, cloning techniques were extended to a range of other microorganisms, such as *Bacillus subtilis*, *Pseudomonas* spp., yeasts, and filamentous fungi, and then to higher eukaryotes. Despite these advances, *E. coli* remains the most widely used cloning host even today because gene manipulation in this bacterium is technically easier than in any other organism. As a result, it is unusual for researchers to clone DNA directly in other organisms. Rather, DNA from the organism of choice is first manipulated in *E. coli* and subsequently transferred back to the original host or another organism, as appropriate. Without the ability to clone and manipulate DNA in *E. coli*, the application of recombinant DNA technology to other organisms would be greatly hindered.

Until the mid-1980s, all cloning was cell-based (i.e. the DNA molecule of interest had to be introduced into *E. coli* or another host for amplification).

Box 1.1 What is a gene?

The concept of the gene as a unit of hereditary information was introduced by the Austrian monk Gregor Mendel in an 1866 paper entitled 'Experiments in plant hybridization'. In this paper, he detailed the results of numerous crosses between pea plants of different characteristics, and from these data put forward a number of postulates concerning the principles of heredity. Although Mendel introduced the concept, the word *gene* was not used until 25 years after his death. It was coined by Wilhelm Johansen in 1909 to describe a heritable factor responsible for the transmission and expression of a given biological trait. In Mendel's work, published over 40 years earlier, these hereditary factors were given the rather less catchy name *Formbildungelementen* (form-building elements).

Mendel had no clear idea what his hereditary elements consisted of in a physical sense, and described them as purely mathematical entities. The first evidence as to the physical and functional nature of genes emerged in 1902. In this year, the chromosome theory of inheritance was put forward by William Sutton, after he noticed that chromosomes during meiosis behaved in the same way as Mendel's elements. Also in 1902, Archibald Garrod showed that the metabolic disorder alkaptonurea resulted from the failure of a specific enzyme and could be transmitted in an autosomal recessive fashion. This he called an inborn error of metabolism. This was the first evidence that genes were necessary to make proteins. In 1911, Thomas Hunt Morgan and colleagues performed the first genetic linkage experiments in the fruit fly *Drosophila melanogaster*, and hence showed that genes were located on chromosomes and were physically linked together.

A more precise idea of the physical and functional basis for the gene emerged during the Second World War. In 1942, George Beadle and Edward Tatum found that X-ray-induced mutations in fungi often caused specific biochemical defects, reflecting the absence or malfunction of a single enzyme. This led to the *one gene one enzyme* model of gene function. In 1944, Oswald Avery and colleagues showed that DNA was the genetic material. Thus evolved a simple picture of the gene – a length of DNA in a chromosome which encoded the information required to produce a single enzyme.

This definition had to be expanded in the following years to encompass new discoveries. For example, not all genes encode enzymes: many encode proteins with other functions, and some do not encode proteins at all, but produce functional RNA molecules. Further complexity results from the selective use of information in the gene to generate multiple products. In eukaryotes, this often reflects alternative splicing, but in both prokaryotes and eukaryotes multiple gene products can be generated by alternative promoter or polyadenylation site usage. In more obscure cases, two or more genes may be required to generate a single polypeptide, e.g. the rare phenomenon of trans-splicing.

In 1983, there was a further mini-revolution in molecular biology with the invention of the *polymerase chain reaction (PCR)*. This technique allowed DNA sequences to be amplified *in vitro* using pure enzymes. The great sensitivity and robustness of the PCR allows DNA to be prepared rapidly from very small amounts of starting material and material of very poor quality, but it is not as accurate as cell-based cloning and only works on relatively short DNA sequences. Therefore cell-based cloning and the PCR have complementary but overlapping uses in gene manipulation.

Although the initial cloning experiments generated a great deal of excitement, it is unlikely that any of the early workers in this field could have predicted the immense impact recombinant DNA technology would have on the progress of scientific understanding and indeed on society as a whole, particularly in the fields of medicine and agriculture. Today, gene manipulation underlies a multi-billion dollar industry, employing hundreds of thousands of people worldwide and offering solutions to some of mankind's most intractable problems. The ability to insert new combinations of genetic material into microbes, animals, and plants offers novel ways to produce valuable small molecules and proteins; provides the means

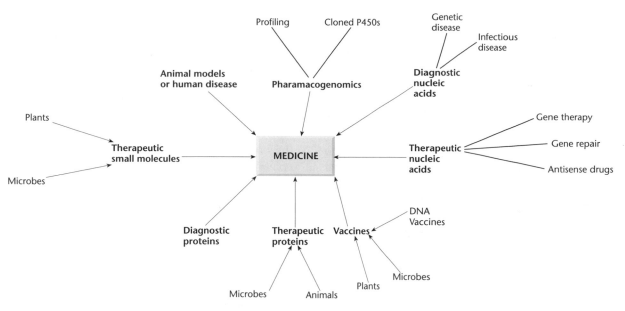

Fig. 1.1 The impact of gene manipulation on the practice of medicine.

to produce plants and animals that are disease-resistant, tolerant of harsh environments, and have higher yields of useful products; and provides new methods to treat and prevent human disease.

Recombinant DNA has opened new horizons in medicine

The developments in gene manipulation that have taken place in the last 30 years have revolutionized medicine by increasing our understanding of the basis of disease, providing new tools for disease diagnosis, and opening the way to the discovery or development of new drugs, treatments, and vaccines.

The first medical benefit to arise from recombinant DNA technology was the availability of significant quantities of therapeutic proteins, such as human growth hormone (HGH), which is used to treat growth defects. Originally HGH was purified from pituitary glands removed from cadavers. However, many pituitary glands are required to produce enough HGH to treat just one child. Furthermore, some children treated with pituitary-derived HGH have developed Creutzfeld–Jakob syndrome originating from cadavers. Following the cloning and expression of the HGH gene in *E. coli*, it became possible to produce enough HGH in a 10-liter fermenter to treat hundreds of children. Since then, many different therapeutic proteins have become available for the first time. Many of these proteins are also manufactured in *E. coli* but others are made in yeast or animal cells and some in plants or the milk of genetically modified animals. The only common factor is that the relevant gene has been cloned and overexpressed using the techniques of gene manipulation.

Medicine has benefited from recombinant DNA technology in other ways (Fig. 1.1). For example, novel routes to vaccines have been developed: the current hepatitis B vaccine is produced by the expression of a viral antigen on the surface of yeast cells, and a recombinant vaccine has been used to eliminate rabies from foxes in a large part of Europe. Gene manipulation can also be used to increase the levels of small molecules within microbial or plant cells. This can be done by cloning all the genes for a particular biosynthetic pathway and overexpressing them. Alternatively, it is possible to shut down particular metabolic pathways and thus redirect intermediates towards the desired end product. This approach has been used to facilitate production of chiral intermediates, antibiotics, and novel therapeutic entities. New antibiotics can also be created by mixing and matching genes from organisms producing different but related molecules in a technique known as combinatorial biosynthesis.

Gene cloning enables nucleic acid probes to be produced readily, and such probes have many uses in medicine. For example, they can be used to determine or confirm the identity of a microbial pathogen or to carry out pre- or peri-natal diagnosis of an inherited genetic disease. Increasingly, probes are being used to determine the likelihood of adverse reactions to drugs or to select the best class of drug to treat a particular illness in different groups of patients. Nucleic acids are also being used as therapeutic entities in their own right. For example, antisense nucleic acids are being used to downregulate gene expression in certain diseases, and the relatively new phenomenon of RNA interference is poised to become a breakthrough technology for the development of new therapeutic approaches. In other cases, nucleic acids are being administered to correct or repair inherited gene defects (gene therapy, gene repair) or as vaccines. In the reverse of gene repair, animals are being generated that have mutations identical to those found in human disease. These are being used as models to learn more about disease pathology and to test novel therapies.

Mapping and sequencing technologies formed a crucial link between gene manipulation and genomics

As well as techniques for DNA cloning and transfer to new host cells, the recombinant DNA revolution spawned new technologies for gene mapping (ordering genes on chromosomes) and DNA sequencing (determining the order of bases, identified by the letters A, C, G, and T, along the DNA molecule). Within the gene itself, the order of bases determines the protein encoded by the gene by specifying the order of amino acids. Thus, DNA sequencing made it possible to work out the amino acid sequence of the encoded protein without the direct analysis of the protein itself. This was extremely useful because, at the time DNA sequencing was first developed, only the most abundant proteins in the cell could be purified in sufficient quantities to facilitate direct analysis. Further elements surrounding the coding region of the gene were identified as control regions, specifying each gene's expression profile. As more sequence data accumulated, it became possible to identify common features in related genes, both in the coding region and the regulatory regions. This type of sequence analysis was greatly facilitated by the foundation of sequence databases, and the development of computer-aided techniques for sequence analysis and comparison, a field now known as *bioinformatics*. Today, DNA molecules can be scanned quickly for a whole series of structural features, e.g. restriction enzyme recognition sites, matches or overlaps with other sequences, start and stop signals for transcription and translation, and sequence repeats, using programs available on the Internet.

The original goal of sequencing was to determine the precise order of nucleotides in a gene, but soon the goal became the sequence of a small genome. A *genome* is the complete content of genetic information in an organism, i.e. all the genes and other sequences it contains. The first target was the genome of a small virus called φX174, then larger plasmid and viral genomes, then chromosomes and microbial genomes until ultimately the complete genomes of higher eukaryotes were sequenced (Table 1.1). In the mid-1980s, scientists began to discuss seriously how the entire human genome might be sequenced. To put these discussions in context, the largest stretch of DNA that can be sequenced in a single pass

Table 1.1 Timeline of genome sequencing, showing the increasing genome sizes that have been achieved.

Genome sequenced	Year	Genome size	Comment
Bacteriophage φX174	1977	5.38 kb	First genome sequenced
Plasmid pBR322	1979	4.3 kb	First plasmid sequenced
Bacteriophage λ	1982	48.5 kb	
Epstein–Barr virus	1984	172 kb	
Yeast chromosome III	1992	315 kb	First chromosome sequenced
Hemophilus influenzae	1995	1.8 Mb	First genome of cellular organism to be sequenced
Saccharomyces cerevisiae	1996	12 Mb	First eukaryotic genome to be sequenced
Ceanorhabditis elegans	1998	97 Mb	First genome of multicellular organism to be sequenced
Drosophila melanogaster	2000	165 Mb	
Arabidopsis thaliana	2000	125 Mb	First plant genome to be sequenced
Homo sapiens	2001	3000 Mb	First mammalian genome to be sequenced
Rice (*Oryza sativa*)	2002	430 Mb	First crop plant to be sequenced
Pufferfish (*Fugu rubripes*)	2002	400 Mb	Smallest known vertebrate genome
Mouse (*Mus musculis*)	2002/3	2700 Mb	Widely used model organism
Chimpanzee (*Pan troglodytes*)	2005	3000 Mb	Closest to human genome

(even today) is 600–800 nucleotides and the largest genome that had been sequenced in 1985 was that of the 172-kb Epstein–Barr virus (Baer *et al.* 1984). By comparison, the human genome is 3000 Mb in size, over 17,000 times bigger! One school of thought was that a completely new sequencing methodology would be required, and a number of different technologies were explored but with little success. Early on, however, it was realized that existing sequencing technology could be used if a large genome could be broken down into more manageable pieces for sequencing in a highly parallel fashion, and then the pieces could be joined together again. A strategy was agreed upon in which a map of the human genome would be used as a scaffold to assemble the sequence.

The problem here was that in 1985 there were not enough markers, or points of reference, on the human genome map to produce a physical scaffold on which to assemble the complete sequence. Genetic maps are based on recombination frequencies, and in model organisms they are constructed by carrying out large-scale crosses between different mutant strains. The principle of a genetic map is that the further apart two loci are on a chromosome, the more likely that a crossover will occur between them during meiosis. Recombination events resulting from crossovers can be scored in genetically amenable organisms such as the fruit fly *Drosophila melanogaster* and yeast by looking for new combinations of the mutant phenotypes in the offspring of the cross. This approach cannot be used in human populations because it would involve setting up large-scale matings between people with different inherited diseases. Instead, human genetic maps rely on the analysis of DNA sequence polymorphisms, i.e. naturally occurring DNA sequence differences in the population which do not have an overt, debilitating effect. A major breakthrough was the development of methods for using DNA probes to identify polymorphic sequences (Botstein *et al.* 1980).

Prior to the Human Genome Project (HGP), low-resolution genetic maps had been constructed using restriction fragment length polymorphisms (RFLPs). These are naturally occurring variations that create or destroy sites for restriction enzymes and therefore generate different sized bands on Southern blots (Fig. 1.2). The Southern blot is a technique for separating DNA fragments by size, see Fig. 2.6, p. 23. The problem with RFLPs was that they were too few and too widely spaced to be of much use for constructing a framework for physical mapping – the first RFLP map had just over 400 markers and a resolution of 10 cM, equivalent to one marker for

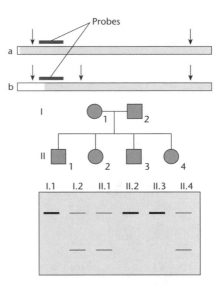

Fig. 1.2 Restriction fragment length polymorphisms (RFLPs) are sequence variants that create or destroy a restriction site in DNA therefore altering the length of the restriction fragment that is detected. The top panel shows two alternative alleles, in which the restriction fragment detected by a specific probe differs in length due to the presence or absence of the middle of three restriction sites (represented by vertical arrows). Alleles a and b therefore produce hybridizing bands of different sizes in Southern blots (lower panel). This allows the alleles to be traced through a family pedigree. For example child II.2 has inherited two copies of allele a, one from each parent, while child II.4 has inherited one copy of allele a and one copy of allele b.

every 10 Mb of DNA (Donis-Keller *et al.* 1987). The necessary breakthrough came with the discovery of new polymorphic markers, known as microsatellites, which were abundant and widely dispersed in the genome (Fig. 1.3). By 1992, a genetic map based on microsatellites had been constructed with a resolution of 1 cM (equivalent to one marker for every 1 Mb of DNA) which was a suitable template for physical mapping.

Unlike genetic maps, physical maps are based on real units of DNA and therefore provide a basis for sequence assembly. The physical mapping phase of the HGP involved the creation of genomic DNA libraries and the identification and assembly of overlapping clones to form contigs (unbroken series of clones representing contiguous segments of the genome). When the HGP was initiated, the highest-capacity vectors available for cloning were cosmids, with a maximum insert size of 40 kb. Because hundreds of thousands of cosmid clones would have to be screened to assemble a physical map, the HGP would not have progressed very quickly without the development of novel high-capacity vectors and methods to find overlaps between them so that clone contigs could be assembled on the genomic scaffold.

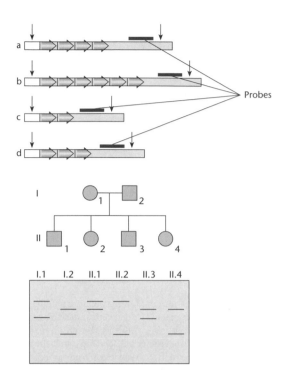

Fig. 1.3 Microsatellites are sequence variants that cause restriction fragments or PCR products to differ in length due to the number of copies of a short tandem repeat sequence, 1–12 nt in length. The top panel shows four alternative alleles, in which the restriction fragment detected by a specific probe differs in length due to a variable number of tandem repeats. All four alleles produce bands of different sizes on Southern blots (lower panel) or different sized PCR products (not shown). Unlike RFLPs, multiple allelism is common for microsatellites so the precise inheritance pattern in a family pedigree can be tracked. For example, the mother and father in the pedigree have alleles b/d and a/c, respectively (the smaller DNA fragments move further during electrophoresis). The first child, II.1, has inherited allele b from his mother and allele a from his father.

The genomics era began in earnest in 1995 with the complete sequencing of a bacterial genome

The late 1980s and early 1990s saw much debate about the desirability of sequencing the human genome. This debate often strayed from rational scientific debate into the realms of politics, personalities, and egos. Among the genuine issues raised were questions such as:

- Is the sequencing of the human genome an intellectually appropriate project for biologists?
- Is sequencing the human genome feasible?
- What benefits might arise from the project?
- Will these benefits justify the cost and are there alternative ways of achieving the same benefits?

- Will the project compete with other areas of biology for funding and intellectual resources?

Behind the debate was a fear that sequencing the human genome was an end in itself, much like a mountaineer who climbs a new peak just because it is there.

The publicly funded Human Genome Project was officially launched in 1990, and the scientific community began to develop new strategies to enable the large-scale mapping and sequencing that were required to complete the project, strategies which centered around high-throughput, highly parallel automated sequencing. One of the benefits of this new technology development was the completion of several pilot genome projects, beginning with that of the bacterium *Hemophilus influenzae* (Fleischmann *et al.* 1995). The net effect was that by the time the human genome had been sequenced (International Human Genome Sequencing Consortium 2001, Venter *et al.* 2001), the complete sequence was already known for over 30 bacterial genomes plus that of a yeast (*Saccharomyces cerevisiae*), the fruit fly, a nematode (*Caenorhabditis elegans*), and a plant (*Arabidopsis thaliana*).

Parallel developments in the field of bioinformatics were required to handle and analyze the exponentially increasing amounts of sequence data arising from the genome projects, but bioinformatics also facilitated the development of new sequencing strategies. For example, when a European consortium set itself the goal of sequencing the entire genome of the budding yeast *S. cerevisiae* (15 Mb), they segmented the task by allocating the sequencing of each chromosome to different groups. That is, they subdivided the genome into more manageable parts. At the time this project was initiated there was no other way of achieving the objective and when the resulting genomic sequence was published (Goffeau *et al.* 1996), it was the result of a unique multi-institution collaboration. While the *S. cerevisiae* sequencing project was underway, a new genomic sequencing strategy was unveiled: shotgun sequencing. In this approach, large numbers of genomic fragments are sequenced and sophisticated bioinformatics algorithms used to construct the finished sequence. In contrast to the consortium approach used with *S. cerevisiae*, a single laboratory set up as a sequencing factory undertook shotgun sequencing.

The first success with shotgun sequencing was the complete sequence of the bacterium *H. influenzae* (Fleischmann *et al.* 1995) and this was quickly followed with the sequences of *Mycoplasma genitalium* (Fraser *et al.* 1995), *Mycoplasma pneumoniae* (Himmelreich *et al.* 1996) and *Methanococcus jannaschii* (Bult *et al.* 1996). It should be noted that *H. influenzae* was selected for sequencing because so little was known about it: there was no genetic map and not much biochemical data either. By contrast, *S. cerevisiae* was a well-mapped and well-characterized organism. As will be seen in Chapter 17, the relative merits of shotgun sequencing vs. ordered, map-based sequencing are still being debated today. Nevertheless, the fact that a major sequencing laboratory can turn out the entire sequence of a bacterium in 1–2 months shows the power of shotgun sequencing.

Genome sequencing greatly increases our understanding of basic biology

Fears that sequencing the human genome would be an end in itself have proved groundless. Because so many different genomes have been sequenced it is now possible to undertake comparative analyses of genomes, a topic known as *comparative genomics*. By comparing genomes from distantly related species we can begin to decipher the major stages in evolution. By comparing more closely related species we can begin to uncover more recent events such as genome rearrangement which have facilitated speciation (see e.g. Murphy *et al.* 2004). Currently, the most fertile area of comparative genomics is the analysis of bacterial genomes because so many have been sequenced. Already this analysis is throwing up some interesting questions. For example, over 25% of the genes in any one bacterial genome have no equivalents in any other sequenced genome. Is this an artifact resulting from limited sequence data or does it reflect the unique evolutionary events that have shaped the genomes of these organisms? Similarly, comparative analysis of the genomes of a wide range of thermophiles has revealed numerous interesting features, including strong evidence of extensive horizontal gene transfer. However, what is the genomic basis for thermophily? We still do not know.

One of the fascinating aspects of the classic paper by Fleischmann *et al.* (1995) was their analysis of the metabolic capabilities of *H. influenzae*, which they deduced from sequence information alone. This analysis has been extended to every other sequenced genome and is providing tremendous insight into the physiology and ecological adaptability of different organisms. For example, obligate parasitism in bacteria is linked to the absence of genes for certain enzymes involved in central metabolic pathways. Another example is the correlation between genome size and the diversity of ecological niches that can be colonized. The larger the bacterial genome, the greater are the metabolic capabilities of the host organism and this means that the organism can be found in a greater number of habitats.

Another benefit of genome mapping and sequencing that deserves mention is the proliferation of international scientific collaborations. In magnitude, the goal of sequencing the human genome was equivalent to putting a man on the moon. However, putting a man on the moon was a race between two nations and was driven by global political ambitions as much as by scientific challenge. By contrast, genome sequencing truly has been an international effort requiring laboratories in Europe, North America, and Japan to collaborate in a way never seen before. The extent of this collaboration can be seen by looking at the affiliations of the authors on many of the classic genome papers (e.g. The *Arabidopsis* Genome Initiative 2000, International Human Genome Sequencing Consortium 2001). The fact that one US company, Celera Genomics Inc., has successfully undertaken many sequencing projects in no way diminishes this collaborative effort. Rather, they have constantly challenged the accepted way of doing things and have increased the efficiency with which key tasks have been undertaken.

Three other aspects of genome sequencing and genomics deserve mention. First, in other branches of science such as nuclear physics and space exploration, the concept of "superfacilities" is well established. With the advent of whole genome sequencing, biology is moving into the superfacility league and a number of sequencing "factories" have been established. Secondly, high throughput methodologies have become commonplace and this has meant a partnering of biology with automation, instrumentation, and data management. Thirdly, many biologists have eschewed chemistry, physics, and mathematics but progress in genomics demands that biologists have a much greater understanding of these subjects. For example, methodologies such as mass spectrometry, X-ray crystallography, and protein structure modeling are now fundamental to the identification of gene function. The impact that this has on undergraduate recruitment in the sciences remains to be seen.

The post-genomics era aims at the complete characterization of cells at all levels

Knowing the complete genome sequence of any organism is very useful, but more important is finding the genes and determining their functions. One of the most surprising results from the early genome projects was the discovery of how little was known about even the best-characterized organisms. In the case of the bakers' yeast (*S. cerevisiae*), which was considered a very well-characterized model species, only one-third of the genes identified in the sequencing project had been identified before. Over 4000 genes were discovered with no known function. Some of these could be assigned tentative functions on the basis of similarity to known genes either in the yeast or in other organisms, but this still left over 2000 genes whose function could only be established by direct experiments.

Following sequencing and annotation (gene finding) scientists then turned their attention to the functional characterization of newly identified genes. This has given rise to two new branches of biology, completely unheard of before 1995. These are *transcriptomics* (the large-scale study of mRNA expression) and *proteomics* (the large-scale study of proteins). While mRNA can yield useful information in terms of sequence, expression profile, and abundance, direct analysis of proteins is much more informative, since proteins can be analyzed not only in terms of sequence and abundance but also in terms of structure, post-translational modification, localization, and interactions with other molecules. No-one working in the 1970s, when recombinant DNA was a novel technology and protein analysis was laborious, could have imagined today's large-scale experiments, where thousands of proteins can be separated on a high-resolution gel, digested into peptides, and identified rapidly by mass spectrometry. In the post-genomics era, it is becoming possible to carry out complete characterizations of cells, at the level of the genome, the transcriptome, the proteome, and now even the metabolome (the global profile of small-molecule metabolites in the cell).

Recombinant DNA technology and genomics form the foundation of the biotechnology industry

The early successes in overproducing mammalian proteins in *E. coli* suggested to a few entrepreneurial individuals that a new company should be formed to exploit the potential of recombinant DNA technology. Thus was Genentech Inc. born (Box 1.2). Since then, thousands of biotechnology companies have been formed worldwide. As soon as major new developments in the science of gene manipulation are reported, a rash of new companies is formed to commercialize the new technology. For example, many recently formed companies are hoping the data from the Human Genome Project will result in the identification of a large number of new proteins with potential for human therapy. Other companies have been founded to exploit novel technologies for recombinant protein expression or the applications of therapeutic nucleic acids.

Although there are thousands of biotechnology companies, fewer than 100 have sales of their products and even fewer are profitable. Already many biotechnology companies have failed, but the technology advances at such a rate that there is no shortage of new company start-ups to take their place. One group of biotechnology companies that has prospered is those supplying specialist reagents to laboratory workers engaged in gene manipulation, genomics, and proteomics. In the very beginning, researchers had to make their own restriction enzymes and this limited the technology to those with protein chemistry skills. Soon a number of companies were formed which catered to the needs of researchers by supplying high-quality enzymes for DNA manipulation. Despite the availability of these enzymes, many people had great difficulty in cloning DNA. The reason for this was the need for careful quality control of all the components used in the preparation of reagents, something researchers are not good at! The supply companies responded by making easy-to-use cloning kits in addition to enzymes. Today, these supply companies can provide almost everything that is needed to clone, express, and analyze DNA and have thereby accelerated the use of recombinant DNA technology in all biological disciplines. In the early days of recombinant DNA technology, the development of methodology was an end in itself for many academic researchers. This is no longer true. The researchers have gone back to using the tools to further our knowledge of biology, and the development of new methodologies has largely fallen to the supply companies.

Outline of the rest of the book

The remainder of this book is divided into four parts. Part I is devoted to the basic methodology for manipulating genes, and covers techniques for cloning and gene manipulation in *E. coli* as well as *in vitro* methods

Box 1.2 The birth of an industry

Biotechnology is not new. Cheese, bread, and yogurt are products of biotechnology and have been known for centuries. However, the stock-market excitement about biotechnology stems from the potential of gene manipulation, which is the subject of this book. The birth of this modern version of biotechnology can be traced to the founding of the company Genentech.

In 1976, a 27-year-old venture capitalist called Robert Swanson had a discussion over a few beers with a University of California professor, Herb Boyer. The discussion centered on the commercial potential of gene manipulation. Swanson's enthusiasm for the technology and his faith in it were contagious. By the close of the meeting the decision was taken to found Genentech (Genetic Engineering Technology). Although Swanson and Boyer faced skepticism from both the academic and business communities they forged ahead with their idea. Successes came thick and fast (see Table B1.1) and within a few years they had proved their detractors wrong. Over 1000 biotechnology companies have been set up in the USA alone since the founding of Genentech but very, very few have been as successful.

Table B1.1 Key events at Genentech.

1976	Genentech founded
1977	Genentech produced first human protein (somatostatin) in a microorganism
1978	Human insulin cloned by Genentech scientists
1979	Human growth hormone cloned by Genentech scientists
1980	Genentech went public, raising $35 million
1982	First recombinant DNA drug (human insulin) marketed (Genentech product licensed to Eli Lilly & Co.)
1984	First laboratory production of factor VIII for therapy of hemophilia. License granted to Cutter Biological
1985	Genentech launched its first product, Protropin (human growth hormone), for growth hormone deficiency in children
1987	Genentech launched Activase (tissue plasminogen activator) for dissolving blood clots in heart-attack patients
1990	Genentech launched Actimmune (interferon-$\gamma_{1\beta}$) for treatment of chronic granulomatous disease
1990	Genentech and the Swiss pharmaceutical company Roche complete a $2.1 billion merger

such as the PCR (Fig. 1.4). Basic techniques for gene and protein analysis are also described. Chapter 2 covers many of the techniques that are common to all cloning experiments and are fundamental to the success of the technology. Chapter 3 is devoted to methods for selectively cutting DNA molecules into fragments that can be readily joined together again. Without the ability to do this, there would be no recombinant DNA technology. If fragments of DNA are inserted into cells, they fail to replicate except in those rare cases where they integrate into the chromosome. To enable such fragments to be propagated, they are inserted into DNA molecules (vectors) that are capable of extrachromosomal replication. These vectors are derived from plasmids and bacteriophages and their basic properties are described in Chapter 4.

Originally, the purpose of vectors was the propagation of cloned DNA but today vectors fulfil many other roles, such as facilitating DNA sequencing, promoting expression of cloned genes, facilitating purification of cloned gene products, and reporting the activity and localization of proteins. The specialist vectors for these tasks are described in Chapter 5. With this background in place it is possible to describe in detail how to clone the particular DNA sequences that one wants. There are two basic strategies. Either one clones all the DNA from an organism and then selects the very small number of clones of interest or one amplifies the DNA sequences

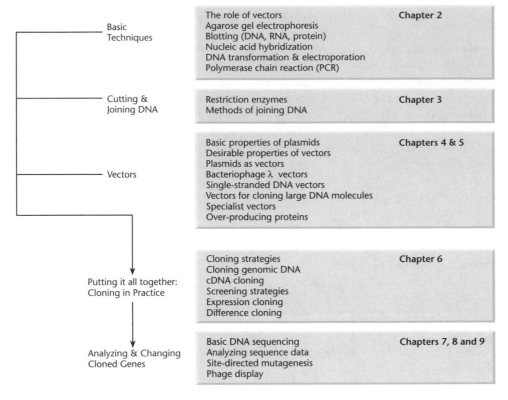

Fig. 1.4 Roadmap outlining the first section of the book, which covers basic techniques in gene manipulation and their relationships.

of interest and then clones these. Both these strategies are described in Chapter 6, which focuses on methods for cloning individual genes. Once the DNA of interest has been cloned, it can be sequenced and this will yield information on the proteins that are encoded and any regulatory signals that are present (Chapter 7). There might also be a wish to modify the DNA and/or protein sequence and determine the biological effects of such changes. The techniques for sequencing and changing cloned genes and the properties of the encoded protein are described in Chapter 8. Finally, Chapter 9 provides an overview of bioinformatics, the essential computer-based methods for the analysis of genes and their products.

Part II of the book describes the specialist techniques for cloning in organisms other than *E. coli* (Fig. 1.5). Each of these chapters can be read in isolation from the other chapters in this section provided that there is a thorough understanding of the material from the first part of the book. Chapter 10 details the methods for cloning in other bacteria. Originally it was thought that some of these bacteria, e.g. *B. subtilis*, would usurp the position of *E. coli*. This has not happened and gene manipulation techniques are used simply to better understand the biology of these bacteria. Chapter 11 focuses on cloning in fungi, although the emphasis is on the yeast *S. cerevisiae*.

Fungi are eukaryotes and are useful model systems for investigating topics such as meiosis, mitosis, and the control of cell division. Animal cells can be cultured like microorganisms and the techniques for introducing genes into them are described in Chapter 12. Chapters 13 and 14 describe basic procedures for the introduction of genes into animals and plants, respectively, while Chapter 15 covers some of the more cutting-edge techniques for these same systems.

Part III of the book moves from gene manipulation to genomics (Fig. 1.6). Chapter 16 introduces the topic of genomics by providing a biological survey of genomes. The genomes of free-living cellular organisms range in size from less than 1 Mb for some bacteria to millions, or tens of millions, of megabases for some plants. The sheer size of the genome of even a simple bacterium is such that to handle it in the laboratory we need to break it down into smaller pieces that are propagated as clones. As stated above, one way to approach this problem is to create a genome map, which can then be populated with physical landmarks onto which the smaller DNA fragments can be assembled. Another approach is to dispense with the map and break the entire genome into pieces, sequence them, and reassemble them. The methods for mapping genomes and

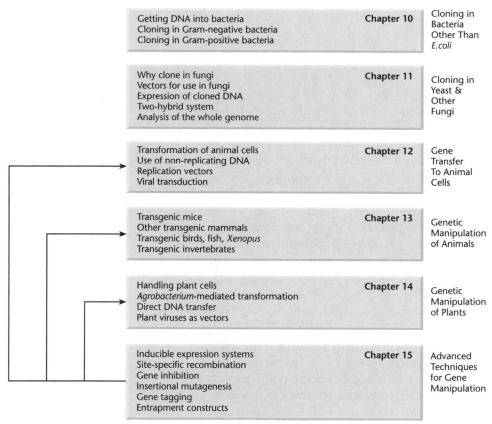

Fig. 1.5 Roadmap outlining the second section of the book, which covers advanced techniques in gene manipulation and their application to organisms other than *E. coli*.

Getting DNA into bacteria Cloning in Gram-negative bacteria Cloning in Gram-positive bacteria	**Chapter 10**	Cloning in Bacteria Other Than *E.coli*
Why clone in fungi Vectors for use in fungi Expression of cloned DNA Two-hybrid system Analysis of the whole genome	**Chapter 11**	Cloning in Yeast & Other Fungi
Transformation of animal cells Use of non-replicating DNA Replication vectors Viral transduction	**Chapter 12**	Gene Transfer To Animal Cells
Transgenic mice Other transgenic mammals Transgenic birds, fish, *Xenopus* Transgenic invertebrates	**Chapter 13**	Genetic Manipulation of Animals
Handling plant cells *Agrobacterium*-mediated transformation Direct DNA transfer Plant viruses as vectors	**Chapter 14**	Genetic Manipulation of Plants
Inducible expression systems Site-specific recombination Gene inhibition Insertional mutagenesis Gene tagging Entrapment constructs	**Chapter 15**	Advanced Techniques for Gene Manipulation

Fig. 1.6 Roadmap covering the early chapters of Part III, which discuss different methodologies for mapping and sequencing genomes.

Genome

Genome size Sequence complexity Introns and exons Genome structure Repetitive DNA	**Chapter 16**

Chromosome

Fragmentation with endonucleases Separation of large DNA fragments Isolation of chromosomes Chromosome microdissection Vectors for cloning	**Chapter 17**

Library

Restriction fingerprinting STSs, ESTs, SSLPs and SNPs RAPDs, CAPs and AFLPs Hybridization mapping Optical mapping, radiation hybrids and HAPPY mapping Integration of mapping methods	**Chapter 17**

Map

Sequencing methodology Automation and high throughput sequencing Sequencing strategies Sequencing large genomes Pyrosequencing Sequencing by hybridization	**Chapters 7 and 17**

Sequence

Databases and software Finding genes Identifying gene function Genome annotation Molecular phylogenetics	**Chapters 9 and 18**

Gene

assembling physical clone maps are discussed in Chapter 17.

Sequencing a genome is not an end in itself. Rather, it is just the first stage in a long journey whose goal is a detailed understanding of all the biological functions encoded in that genome and their evolution. To achieve this goal it is necessary to define all the genes in the genome and the functions that they encode. There are a number of different ways of doing this, one of which is comparative genomics (Chapter 18). The premise here is that DNA sequences encoding important cellular functions are likely to be conserved whereas dispensable or non-coding sequences will not. However, comparative genomics only gives a broad overview of the capabilities of different organisms. For a more detailed view one needs to identify each gene in the genome and determine its function. Over the last few years, technology developments in this new discipline of *functional genomics* have been nothing short of breathtaking. The final six chapters in this section look at ways in which large-scale functional analysis can be carried out (Fig. 1.7).

Chapter 19 explores the idea of determining gene function by inactivation. Whereas this is carried out on a gene-by-gene basis in classical genetics, in genomics it is performed on a genome-wide scale. Traditionally, this has involved the generation of populations of random mutants or the deliberate and systematic inactivation of every gene in the genome. More recently, the technique of RNA interference has risen to a dominant position, heralded by experiments in which up to 18,000 genes can be inactivated systematically to investigate their functions. Chapter 20 moves onto the next stage, the analysis of the transcriptome, focusing on sequence-based techniques such as serial analysis of gene expression (SAGE) and the use of DNA microarrays. Chapters 21–23 explore the burgeoning field of proteomics, which involves the large-scale analysis of many dif-

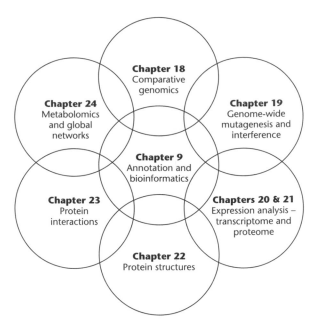

Fig. 1.7 Roadmap covering the later chapters of Part III, which discuss the 'omic' disciplines for determining gene and protein functions, scaling to the level of the complete cell or organism.

ferent properties of proteins – expression, abundance, physico-chemical properties, localization in the cell, interaction with other molecules, structure, state of modification – to create a robust definition of function. Finally, Chapter 24 explores the relatively new field of metabolomics, the systematic analysis of all small molecules (or metabolites) produced in the cell.

Part IV of the book provides some examples of how the techniques of gene manipulation and genomics are being applied in healthcare, agriculture, and industry. While some applications have been mentioned in boxes throughout the book, the final chapters concentrate on major applications, such as pharmacogenomics, the analysis of quantitative traits, biopharmaceutical production, gene therapy, and modern agriculture, which really emphasize the incredible potential of this technology.

Part I

Fundamental Techniques of Gene Manipulation

CHAPTER 2

Basic techniques

Introduction

The initial impetus for gene manipulation *in vitro* came about in the early 1970s with the simultaneous development of techniques for:

- genetic transformation of *Escherichia coli*;
- cutting and joining DNA molecules;
- monitoring the cutting and joining reactions.

In order to explain the significance of these developments we must first consider the essential requirements of a successful gene-manipulation procedure.

Three technical problems had to be solved before *in vitro* gene manipulation was possible on a routine basis

Before the advent of modern gene-manipulation methods there had been many early attempts at transforming pro- and eukaryotic cells with foreign DNA. But, in general, little progress could be made. The reasons for this are as follows. Let us assume that the exogenous DNA is taken up by the recipient cells. There are then two basic difficulties. First, where detection of uptake is dependent on gene expression, failure could be due to lack of accurate transcription or translation. Secondly, and more importantly, the exogenous DNA may not be maintained in the transformed cells. If the exogenous DNA is integrated into the host genome, there is no problem. The exact mechanism whereby this integration occurs is not clear and it is usually a rare event. However this occurs, the result is that the foreign DNA sequence becomes incorporated into the host cell's genetic material and will subsequently be propagated as part of that genome. If, however, the exogenous DNA fails to be integrated, it will probably be lost during subsequent multiplication of the host cells. The reason for this is simple. In order to be replicated, DNA molecules must contain an *origin of replication*, and in bacteria and viruses there is usually only one

per genome. Such molecules are called *replicons*. Fragments of DNA are not replicons and in the absence of replication will be diluted out of their host cells. It should be noted that, even if a DNA molecule contains an origin of replication, this may not function in a foreign host cell.

There is an additional, subsequent problem. If the early experiments were to proceed, a method was required for assessing the fate of the donor DNA. In particular, in circumstances where the foreign DNA was maintained because it had become integrated in the host DNA, a method was required for mapping the foreign DNA and the surrounding host sequences.

A number of basic techniques are common to most gene-cloning experiments

If fragments of DNA arc not replicated, the obvious solution is to attach them to a suitable replicon. Such replicons are known as *vectors* or *cloning vehicles.* Small plasmids and bacteriophages are the most suitable vectors for they are replicons in their own right, their maintenance does not necessarily require integration into the host genome and their DNA can be readily isolated in an intact form. The different plasmids and phages which are used as vectors are described in detail in Chapters 4 and 5. Suffice it to say at this point that initially plasmids and phages suitable as vectors were only found in *E. coli.* An important consequence follows from the use of a vector to carry the foreign DNA: simple methods become available for purifying the vector molecule, complete with its foreign DNA insert, from transformed host cells. Thus not only does the vector provide the replicon function, but it also permits the easy bulk preparation of the foreign DNA sequence free from host-cell DNA.

Composite molecules in which foreign DNA has been inserted into a vector molecule are sometimes called DNA *chimeras* because of their analogy with the Chimaera of mythology – a creature with the head of a lion, body of a goat, and tail of a serpent. The construction of such composite or *artificial recombinant* molecules has also been termed *genetic engineering* or *gene manipulation* because of the potential for creating novel genetic combinations by biochemical means. The process has also been termed *molecular cloning* or *gene cloning* because a line of genetically identical organisms, all of which contain the composite molecule, can be propagated and grown in bulk, hence *amplifying* the composite molecule and *any gene product whose synthesis it directs*.

Although conceptually very simple, cloning of a fragment of foreign, or *passenger*, or *target* DNA in a vector demands that the following can be accomplished:

* The vector DNA must be purified and cut open.
* The passenger DNA must be inserted into the vector molecule to create the artificial recombinant. DNA joining reactions must therefore be performed. Methods for cutting and joining DNA molecules are now so sophisticated that they warrant a chapter of their own (Chapter 3).
* The cutting and joining reactions must be readily monitored. This is achieved by the use of gel electrophoresis.
* Finally, the artificial recombinant must be introduced into *E. coli* or another host cell (transformation).

Further details on the use of gel electrophoresis and transformation of *E. coli* are given in the next section. As we have noted, the necessary techniques became available at about the same time and quickly led to many cloning experiments, the first of which were reported in 1972 (Jackson *et al.* 1972, Lobban & Kaiser 1973).

Gel electrophoresis is used to separate different nucleic acid molecules on the basis of their size

The progress of the first experiments on cutting and joining of DNA molecules was monitored by velocity sedimentation in sucrose gradients. However, this has been entirely superseded by gel electrophoresis. Gel electrophoresis is not only used as an analytical method, it is also routinely used preparatively for the purification of specific DNA fragments. The gel is composed of polyacrylamide or agarose. Agarose is convenient for separating DNA fragments ranging in size from a few hundred base pairs to about 20 kb

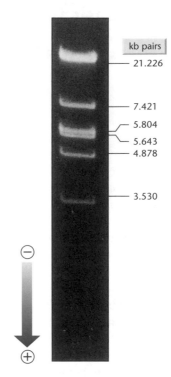

Fig. 2.1 Electrophoresis of DNA in agarose gels. The direction of migration is indicated by the arrow. DNA bands have been visualized by soaking the gel in a solution of ethidium bromide (see Fig. 2.3), which complexes with DNA by intercalating between stacked base pairs, and photographing the orange fluorescence which results upon ultraviolet irradiation.

(Fig. 2.1). Polyacrylamide is preferred for smaller DNA fragments.

The mechanism responsible for the separation of DNA molecules by molecular weight during gel electrophoresis is not well understood (Holmes & Stellwagen 1990). The migration of the DNA molecules through the pores of the matrix must play an important role in molecular-weight separations since the electrophoretic mobility of DNA in free solution is independent of molecular weight. An agarose gel is a complex network of polymeric molecules whose average pore size depends on the buffer composition and the type and concentration of agarose used. DNA movement through the gel was originally thought to resemble the motion of a snake (reptation). However, real-time fluorescence microscopy of stained molecules undergoing electrophoresis has revealed more subtle dynamics (Schwartz & Koval 1989, Smith *et al.* 1989). DNA molecules display elastic behavior by stretching in the direction of the applied field and then contracting into dense balls. The larger the pore size of the gel, the greater the ball of DNA which can pass through and hence the larger the molecules which can be separated. Once the globular volume of the DNA molecule exceeds the pore size, the DNA molecule can only pass through by reptation. This occurs with molecules about 20 kb in size and it is difficult to separate molecules larger than this without recourse to pulsed electrical fields.

In pulsed-field gel electrophoresis (PFGE) (Schwartz & Cantor 1984) molecules as large as 10 Mb can be separated in agarose gels. This is achieved by causing the DNA to periodically alter its direction of migration by regular changes in the orientation of the electric field with respect to the gel. With each change in the electric-field orientation, the DNA must realign its axis prior to migrating in the new direction. Electric-field parameters, such as the direction, intensity, and duration of the electric field, are set independently for each of the different fields and are chosen so that the net migration of the DNA is down the gel. The difference between the direction of migration induced by each of the electric fields is the *reorientation angle* and corresponds to the angle that the DNA must turn as it changes its direction of migration each time the fields are switched.

A major disadvantage of PFGE, as originally described, is that the samples do not run in straight lines. This makes subsequent analysis difficult. This problem has been overcome by the development of improved methods for alternating the electrical field. The most popular of these is contour-clamped homogeneous electrical-field (CHEF) electrophoresis (Chu *et al.* 1986). In early CHEF-type systems (Fig. 2.2) the reorientation angle was fixed at 120°. However, in newer systems, the reorientation angle can be varied and it has been found that for whole-yeast chromosomes the migration rate is much faster with an angle of 106° (Birren *et al.* 1988). Fragments of

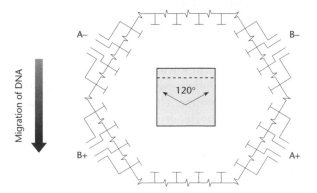

Fig. 2.2 Schematic representation of CHEF (contour-clamped homogeneous electrical field) pulsed-field gel electrophoresis.

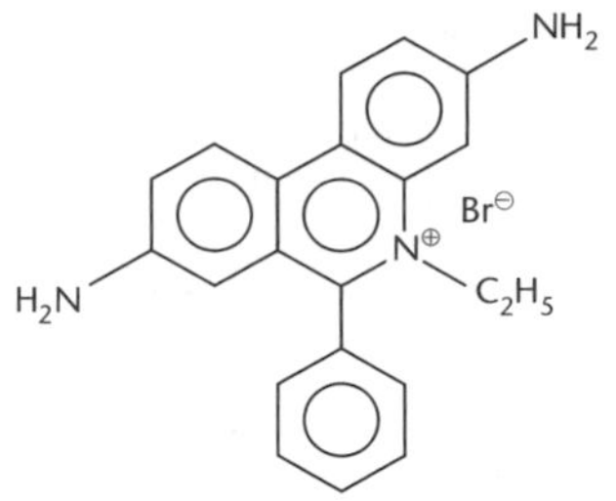

Fig. 2.3 Ethidium bromide.

DNA as large as 200–300 kb are routinely handled in genomics work and these can be separated in a matter of hours using CHEF systems with a reorientation angle of 90° or less (Birren & Lai 1994).

Aaij and Borst (1972) showed that the migration rates of DNA molecules were inversely proportional to the logarithms of their molecular weights. Subsequently, Southern (1979a,b) showed that plotting fragment length or molecular weight against the reciprocal of mobility gives a straight line over a wider range than the semilogarithmic plot. In any event, gel electrophoresis is frequently performed with marker DNA fragments of known size, which allows accurate size determination of an unknown DNA molecule by interpolation. A particular advantage of gel electrophoresis is that the DNA bands can be readily detected at high sensitivity. Traditionally, the bands of DNA have been stained with the intercalating dye ethidium bromide (Fig. 2.3) and as little as 0.05 μg of DNA can be detected as visible fluorescence when the gel is illuminated with ultraviolet light. A major disadvantage of ethidium bromide is that it is mutagenic in various laboratory tests and by inference is a potential carcinogen. To overcome this problem a new fluorescent DNA stain called SYBR Safe™ has been developed.

In addition to resolving DNA fragments of different lengths, gel electrophoresis can be used to separate different molecular configurations of a DNA molecule. Examples of this are given in Chapter 4 (see p. 56). Gel electrophoresis can also be used for investigating protein–nucleic acid interactions in the so-called *gel retardation* or *band shift* assay. It is based on the observation that binding of a protein to DNA fragments usually leads to a reduction in electrophoretic mobility. The assay typically involves the addition of protein to linear double-stranded DNA fragments, separation of complex and naked DNA by gel electrophoresis and visualization. A review of the physical basis of electrophoretic mobility shifts and their application is provided by Lane *et al.* (1992).

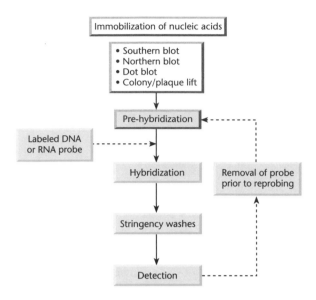

Fig. 2.4 Overview of nucleic acid blotting and hybridization (reproduced courtesy of Amersham Pharmacia Biotech).

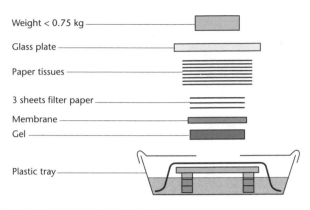

Fig. 2.5 A typical capillary blotting apparatus.

Blotting is used to transfer nucleic acids from gels to membranes for further analysis

Nucleic acid labeling and hybridization on membranes have formed the basis for a range of experimental techniques central to recent advances in our understanding of the organization and expression of the genetic material. These techniques may be applied in the isolation and quantification of specific nucleic acid sequences and in the study of their organization, intracellular localization, expression, and regulation. A variety of specific applications includes the diagnosis of infectious and inherited disease. Each of these topics is covered in depth in subsequent chapters.

An overview of the steps involved in nucleic acid blotting and membrane hybridization procedures is shown in Fig. 2.4. *Blotting* describes the immobilization of sample nucleic acids on to a solid support, generally nylon or nitrocellulose membranes. The blotted nucleic acids are then used as "targets" in subsequent hybridization experiments. The main blotting procedures are:

* blotting of nucleic acids from gels;
* dot and slot blotting;
* colony and plaque blotting.

Colony and plaque blotting are described in detail on p. 111 and dot and slot blotting in Chapter 6.

Southern blotting is the method used to transfer DNA from agarose gels to membranes so that the compositional properties of the DNA can be analyzed

The original method of blotting was developed by Southern (1975, 1979b) for detecting fragments in an agarose gel that are complementary to a given RNA or DNA sequence. In this procedure, referred to as Southern blotting, the agarose gel is mounted on a filter-paper wick which dips into a reservoir containing transfer buffer (Fig. 2.5). The hybridization membrane is sandwiched between the gel and a stack of paper towels (or other absorbent material), which serves to draw the transfer buffer through the gel by capillary action. The DNA molecules are carried out of the gel by the buffer flow and immobilized on the membrane. Initially, the membrane material used was nitrocellulose. The main drawback with this membrane is its fragile nature. Supported nylon membranes have since been developed which have greater binding capacity for nucleic acids in addition to high tensile strength.

For efficient Southern blotting, gel pretreatment is important. Large DNA fragments (>10 kb) require a longer transfer time than short fragments. To allow uniform transfer of a wide range of DNA fragment sizes, the electrophoresed DNA is exposed to a short depurination treatment (0.25 mol/l HCl) followed by alkali. This shortens the DNA fragments by alkaline hydrolysis at depurinated sites. It also denatures the fragments prior to transfer, ensuring that they are in the single-stranded state and accessible for probing. Finally, the gel is equilibrated in neutralizing solution prior to blotting. An alternative method uses positively charged nylon membranes, which remove the need for extended gel pretreatment. In this case, the DNA is transferred in native (non-denatured) form and then alkali-denatured *in situ* on the membrane.

After transfer, the nucleic acid needs to be fixed to the membrane and a number of methods are available. Oven baking at 80°C is the recommended method for nitrocellulose membranes and this can also be used with nylon membranes. Due to the flammable nature of nitrocellulose, it is important that it is baked in a vacuum oven. An alternative fixation method utilizes ultraviolet cross-linking. It is based on the formation of cross-links between a small fraction of the thymine residues in the DNA and positively charged amino groups on the surface of nylon membranes. A calibration experiment must be performed to determine the optimal fixation period.

Following the fixation step, the membrane is placed in a solution of labeled (radioactive or non-radioactive) RNA, single-stranded DNA, or oligodeoxynucleotide which is complementary in sequence to the blot-transferred DNA band or bands to be detected. Conditions are chosen so that the labeled nucleic acid hybridizes with the DNA on the membrane. Since this labeled nucleic acid is used to detect and locate the complementary sequence, it is called the *probe*. Conditions are chosen which maximize the rate of hybridization, compatible with a low background of non-specific binding on the membrane (see Box 2.1). After the hybridization reaction has been carried out, the membrane is washed to remove unbound radioactivity and regions of hybridization are detected autoradiographically by placing the membrane in contact with X-ray film (see Box 2.2). A common approach is to carry out the hybridization under conditions of relatively low stringency which permit a high rate of hybridization, followed by a series of post-hybridization washes of increasing stringency (i.e. higher temperature or, more commonly, lower ionic strength). Autoradiography following each washing stage will reveal any DNA bands that are related to, but not perfectly complementary with, the probe and will also permit an estimate of the degree of mismatching to be made.

The Southern blotting methodology can be extremely sensitive. It can be applied to mapping restriction sites around a single-copy gene sequence in a complex genome such as that of humans (Fig. 2.6), and when a "mini-satellite" probe is used it can be applied forensically to minute amounts of DNA (see p. 335).

Northern blotting is a variant of Southern blotting that is used for RNA analysis

Southern's technique has been of enormous value, but it was thought that it could not be applied directly to the blot-transfer of RNAs separated by gel electrophoresis, since RNA was found not to bind to nitrocellulose. Alwine *et al.* (1979) therefore devised a procedure in which RNA bands are blot-transferred from the gel on to chemically reactive paper, where they are bound covalently. The reactive paper is prepared by diazotization of aminobenzyloxymethyl paper (creating diazobenzyloxymethyl (DBM) paper), which itself can be prepared from Whatman 540 paper by a series of uncomplicated reactions. Once covalently bound, the RNA is available for hybridization with radiolabeled DNA probes. As before, hybridizing bands are located by autoradiography. Alwine *et al.*'s method thus extends that of Southern and for this reason it has acquired the jargon term *northern* blotting.

Subsequently it was found that RNA bands can indeed be blotted on to nitrocellulose membranes under appropriate conditions (Thomas 1980) and suitable nylon membranes have been developed. Because of the convenience of these more recent methods, which do not require freshly activated paper, the use of DBM paper has been superseded.

Western blotting is used to transfer proteins from acrylamide gels to membranes

The term "western" blotting (Burnette 1981) refers to a procedure which does not directly involve nucleic acids, but which is of importance in gene manipulation. It involves the transfer of electrophoresed protein bands from a polyacrylamide gel on to a membrane of nitrocellulose or nylon, to which they bind strongly (Gershoni & Palade 1982, Renart & Sandoval 1984). The bound proteins are then available for analysis by a variety of specific protein–ligand interactions. Most commonly, antibodies are used to detect specific antigens. Lectins have been used to identify glycoproteins. In these cases the probe may itself be labeled with radioactivity, or some other "tag" may be employed. Often, however, the probe is unlabeled and is itself detected in a "sandwich" reaction, using a second molecule which is labeled, for instance a species-specific second antibody, or protein A of *Staphylococcus aureus* (which binds to certain subclasses of IgG antibodies), or

Box 2.1   Hybridization of nucleic acids on membranes

The hybridization of nucleic acids on membranes is a widely used technique in gene manipulation and analysis. Unlike solution hybridizations, membrane hybridizations tend not to proceed to completion. One reason for this is that some of the bound nucleic acid is embedded in the membrane and is inaccessible to the probe. Prolonged incubations may not generate any significant increase in detection sensitivity.

The composition of the hybridization buffer can greatly affect the speed of the reaction and the sensitivity of detection. The key components of these buffers are shown below:

Rate enhancers	Dextran sulfate and other polymers act as volume excluders to increase both the rate and the extent of hybridization
Detergents and blocking agents	Dried milk, heparin, and detergents such as sodium dodecylsulfate (SDS) have been used to depress non-specific binding of the probe to the membrane. Denhardt's solution (Denhardt 1966) uses Ficoll, polyvinylpyrrolidone, and bovine serum albumin
Denaturants	Urea or formamide can be used to depress the melting temperature of the hybrid so that reduced temperatures of hybridization can be used
Heterologous DNA	This can reduce non-specific binding of probes to non-homologous DNA on the blot

Stringency control

Stringency can be regarded as the specificity with which a particular target sequence is detected by hybridization to a probe. Thus, at high stringency, only completely complementary sequences will be bound, whereas low-stringency conditions will allow hybridization to partially matched sequences. Stringency is most commonly controlled by the temperature and salt concentration in the post-hybridization washes, although these parameters can also be utilized in the hybridization step. In practice, the stringency washes are performed under successively more stringent conditions (lower salt or higher temperature) until the desired result is obtained.

The melting temperature (T_m) of a probe–target hybrid can be calculated to provide a starting point for the determination of correct stringency. The T_m is the temperature at which the probe and target are 50% dissociated. For probes longer than 100 base pairs:

$$T_m = 81.5°C + 16.6 \log M + 0.41 \,(\% \,G + C)$$

where M = ionic strength of buffer in moles/liter. With long probes, the hybridization is usually carried out at $T_m - 25°C$. When the probe is used to detect partially matched sequences, the hybridization temperature is reduced by 1°C for every 1% sequence divergence between probe and target.

Oligonucleotides can give a more rapid hybridization rate than long probes as they can be used at a higher molarity. Also, in situations where target is in excess to the probe, for example dot blots, the hybridization rate is diffusion-limited and longer probes diffuse more slowly than oligonucleotides. It is standard practice to use oligonucleotides to analyze putative mutants following a site-directed mutagenesis experiment where the difference between parental and mutant progeny is often only a single base-pair change.

The availability of the exact sequence of oligonucleotides allows conditions for hybridization and stringency washing to be tightly controlled so that the probe will only remain hybridized when it is 100% homologous to the target. Stringency is commonly controlled by adjusting the

continued

Box 2.1 *continued*

temperature of the wash buffer. The "Wallace rule" (Lay Thein & Wallace 1986) is used to determine the appropriate stringency wash temperature:

$$T_m = 4 \times (\text{number of GC base pairs}) + 2 \times (\text{number of AT base pairs})$$

In filter hybridizations with oligonucleotide probes, the hybridization step is usually performed at 5°C below T_m for perfectly matched sequences. For every mismatched base pair, a further 5°C reduction is necessary to maintain hybrid stability.

The design of oligonucleotides for hybridization experiments is critical to maximize hybridization specificity. Consideration should be given to:

- probe length – the longer the oligonucleotide, the less chance there is of it binding to sequences other than the desired target sequence under conditions of high stringency;
- oligonucleotide composition – the GC content will influence the stability of the resultant hybrid and hence the determination of the appropriate stringency washing conditions. Also the presence of any non-complementary bases will have an effect on the hybridization conditions.

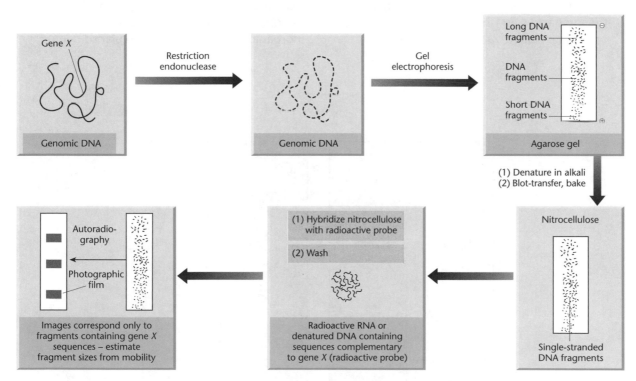

Fig. 2.6 Mapping restriction sites around a hypothetical gene sequence in total genomic DNA by the Southern blot method. Genomic DNA is cleaved with a restriction endonuclease into hundreds of thousands of fragments of various sizes. The fragments are separated according to size by gel electrophoresis and blot-transferred on to nitrocellulose paper. Highly radioactive RNA or denatured DNA complementary in sequence to gene *X* is applied to the nitrocellulose paper bearing the blotted DNA. The radiolabeled RNA or DNA will hybridize with gene *X* sequences and can be detected subsequently by autoradiography, so enabling the sizes of restriction fragments containing gene *X* sequences to be estimated from their electrophoretic mobility. By using several restriction endonucleases singly and in combination, a map of restriction sites in and around gene *X* can be built up.

Box 2.2   The principles of autoradiography

The localization and recording of a radiolabel within a solid specimen is known as autoradiography and involves the production of an image in a photographic emulsion. Such emulsions consist of silver halide crystals suspended in a clear phase composed mainly of gelatin. When a β-particle or γ-ray from a radionuclide passes through the emulsion, the silver ions are converted to silver atoms. This results in a latent image being produced, which is converted to a visible image when the image is developed. Development is a system of amplification in which the silver atoms cause the entire silver halide crystal to be reduced to metallic silver. Unexposed crystals are removed by dissolution in fixer, giving an autoradiographic image which represents the distribution of radiolabel in the original sample.

In direct autoradiography, the sample is placed in intimate contact with the film and the radioactive emissions produce black areas on the developed autoradiograph. It is best suited to detection of weak- to medium-strength β-emitting radionuclides (^{3}H, ^{14}C, ^{35}S). Direct autoradiography is not suited to the detection of highly energetic β-particles, such as those from ^{32}P, or for γ-rays emitted from isotopes like ^{125}I. These emissions pass through and beyond the film, with the majority of the energy being wasted. Both ^{32}P and ^{125}I are best detected by indirect autoradiography.

Indirect autoradiography describes the technique by which emitted energy is converted to light by means of a scintillator, using fluorography or intensifying screens. In fluorography the sample is impregnated with a liquid scintillator. The radioactive emissions transfer their energy to the scintillator molecules, which then emit photons which expose the photographic emulsion. Fluorography is mostly used to improve the detection of weak β-emitters (Fig. B2.1). Intensifying screens are sheets of a solid inorganic scintillator which are placed behind

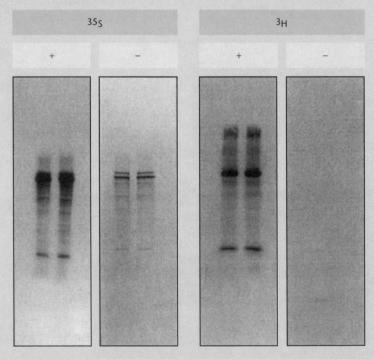

Fig. B2.1 Autoradiographs showing the detection of ^{35}S- and ^{3}H-labeled proteins in acrylamide gels with (+) and without (−) fluorography. (Photo courtesy of Amersham Pharmacia Biotech.)

continued

Box 2.2   *continued*

the film. Any emissions passing through the photographic emulsion are absorbed by the screen and converted to light, effectively superimposing a photographic image upon the direct autoradiographic image.

The gain in sensitivity which is achieved by use of indirect autoradiography is offset by nonlinearity of film response. A single hit by a β-particle or γ-ray can produce hundreds of silver atoms, but a single hit by a photon of light produces only a single silver atom. Although two or more silver atoms in a silver halide crystal are stable, a single silver atom is unstable and reverts to a silver ion very rapidly. This means that the probability of a second

photon being captured before the first silver atom has reverted is greater for large amounts of radioactivity than for small amounts. Hence small amounts of radioactivity are under-represented with the use of fluorography and intensifying screens. This problem can be overcome by a combination of pre-exposing a film to an instantaneous flash of light (pre-flashing) and exposing the autoradiograph at −70°C. Pre-flashing provides many of the silver halide crystals of the film with a stable pair of silver atoms. Lowering the temperature to −70°C increases the stability of a single silver atom, increasing the time available to capture a second photon (Fig. B2.2).

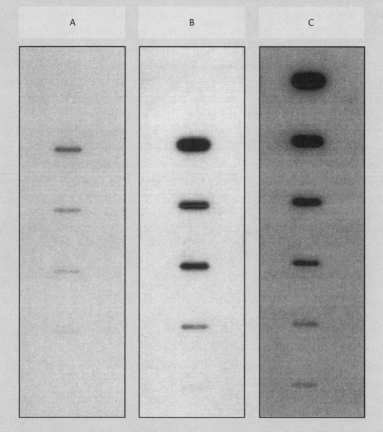

Fig. B2.2   The improvement in sensitivity of detection of ^{125}I-labeled IgG by autoradiography obtained by using an intensifying screen and pre-flashed film. A, no screen and no pre-flashing; B, screen present but film not pre-flashed; C, use of screen and pre-flashed film. (Photo courtesy of Amersham Pharmacia Biotech.)

streptavidin (which binds to antibody probes that have been biotinylated). These second molecules may be labeled in a variety of ways with radioactive, enzyme, or fluorescent tags. An advantage of the sandwich approach is that a single preparation of labeled second molecule can be employed as a general detector for different probes. For example, an antiserum may be raised in rabbits which reacts with a range of mouse immunoglobins. Such a rabbit anti-mouse (RAM) antiserum may be radio-labeled and used in a number of different applications to identify polypeptide bands probed with different, specific, monoclonal antibodies, each monoclonal antibody being of mouse origin. The sandwich method may also give a substantial increase in sensitivity, owing to the multivalent binding of antibody molecules.

A number of techniques have been devised to speed up and simplify the blotting process

The original blotting technique employed capillary blotting but nowadays the blotting is usually accomplished by electrophoretic transfer of polypeptides from an SDS-polyacrylamide gel on to the membrane (Towbin *et al.* 1979). Electrophoretic transfer is also the method of choice for transferring DNA or RNA from low-pore-size polyacrylamide gels. It can also be used with agarose gels. However, in this case, the rapid electrophoretic transfer process requires high currents, which can lead to extensive heating effects, resulting in distortion of agarose gels. The use of an external cooling system is necessary to prevent this.

Another alternative to capillary blotting is vacuum-driven blotting (Olszewska & Jones 1988), for which several devices are commercially available. Vacuum blotting has several advantages over capillary or electrophoretic transfer methods: transfer is very rapid and gel treatment can be performed *in situ* on the vacuum apparatus. This ensures minimal gel handling and, together with the rapid transfer, prevents significant DNA diffusion.

The ability to transform *E. coli* with DNA is an essential prerequisite for most experiments on gene manipulation

Early attempts to achieve transformation of *E. coli* were unsuccessful and it was generally believed that *E. coli* was refractory to transformation. However, Mandel and Higa (1970) found that treatment with CaCl$_2$ allowed *E. coli* cells to take up DNA from bacteriophage λ. A few years later Cohen *et al.* (1972) showed that CaCl$_2$-treated *E. coli* cells are also effective recipients for plasmid DNA. Almost any strain of *E. coli* can be transformed with plasmid DNA, albeit with varying efficiency, whereas it was thought that only *rec*BC$^-$ mutants could be transformed with linear bacterial DNA (Cosloy & Oishi 1973). Later, Hoekstra *et al.* (1980) showed that *rec*BC$^+$ cells can be transformed with linear DNA, but the efficiency is only 10% of that in otherwise isogenic *rec*BC$^-$ cells. Transformation of *rec*BC$^-$ cells with linear DNA is only possible if the cells are rendered recombination-proficient by the addition of a *sbc*A or *sbc*B mutation. The fact that the *rec*BC gene product is an exonuclease explains the difference in transformation efficiency of circular and linear DNA in *rec*BC$^+$ cells.

As will be seen from the next chapter, many bacteria contain restriction systems which can influence the efficiency of transformation. Although the complete function of these restriction systems is not yet known, one role they do play is the recognition and degradation of foreign DNA. For this reason it is usual to use a restriction-deficient strain of *E. coli* as a transformable host.

Since transformation of *E. coli* is an essential step in many cloning experiments, it is desirable that it be as efficient as possible. Several groups of workers have examined the factors affecting the efficiency of transformation. It has been found that *E. coli* cells and plasmid DNA interact productively in an environment of calcium ions and low temperature (0–5°C), and that a subsequent heat shock (37–45°C) is important, but not strictly required. Several other factors, especially the inclusion of metal ions in addition to calcium, have been shown to stimulate the process.

A very simple, moderately efficient transformation procedure for use with *E. coli* involves resuspending log-phase cells in ice-cold 50 mmol/l calcium chloride at about 10^{10} cells/ml and keeping them on ice for about 30 min. Plasmid DNA (0.1 μg) is then added to a small aliquot (0.2 ml) of these now *competent* (i.e. competent for transformation) cells, and the incubation on ice continued for a further 30 min, followed by a heat shock of 2 min at 42°C. The cells are then usually transferred to nutrient medium and incubated for some time (30 min to 1 h) to allow phenotypic properties conferred by the plasmid to be expressed, e.g. antibiotic resistance commonly used as a selectable marker for plasmid-containing cells. (This so-called *phenotypic lag* may not need to be taken into consideration with high-level ampicillin resistance. With this marker, significant resistance builds up very rapidly, and ampicillin exerts its effect on cell-wall biosynthesis only in cells which have progressed into active growth.) Finally the cells are plated out on selective medium. Just why such a transformation procedure is effective is not fully understood (Huang & Reusch 1995). The calcium chloride affects the cell wall and may also be responsible for binding DNA to the cell surface. The actual uptake of DNA is stimulated by the brief heat shock.

Hanahan (1983) re-examined the factors that affect the efficiency of transformation, and devised a set of conditions for optimal efficiency (expressed as transformants per μg plasmid DNA) applicable to most *E. coli* K12 strains. Typically, efficiencies of 10^7 to 10^9 transformants/μg can be achieved depending on the strain of *E. coli* and the method used (Liu & Rashidbaigi 1990). Ideally, one wishes to make a large batch of competent cells and store them frozen for future use. Unfortunately, competent cells made by the Hanahan procedure rapidly lose their competence on storage. Inoue *et al.* (1990) have optimized the conditions for the preparation of competent cells. Not only could they store cells for up to 40 days at −70°C while retaining efficiencies of 1–5×10^9 cfu/μg, but competence was affected only minimally by salts in the DNA preparation.

There are many enzymic activities in *E. coli* which can destroy incoming DNA from non-homologous sources (see Chapter 3) and reduce the transformation efficiency. Large DNAs transform less efficiently, on a molar basis, than small DNAs. Even with such improved transformation procedures, certain potential gene-cloning experiments requiring large numbers of clones are not reliable. One approach which can be used to circumvent the problem of low transformation efficiencies is to package recombinant DNA into virus particles *in vitro*. A particular form of this approach, the use of cosmids, is described in detail in Chapter 5. Another approach is electroporation, which is described below.

Electroporation is a means of introducing DNA into cells without making them competent for transformation

A rapid and simple technique for introducing cloned genes into a wide variety of microbial, plant, and animal cells, including *E. coli*, is electroporation. This technique depends on the original observation by Zimmerman & Vienken (1983) that high-voltage electric pulses can induce cell plasma membranes to fuse. Subsequently it was found that, when subjected to electric shock, the cells take up exogenous DNA from the suspending solution. A proportion of these cells become stably transformed and can be selected if a suitable marker gene is carried on the transforming DNA. Many different factors affect the efficiency of electroporation, including temperature, various electric-field parameters (voltage, resistance, and capacitance), topological form of the DNA, and various host-cell factors (genetic background, growth conditions, and post-pulse treatment). Some of these factors have been reviewed by Hanahan *et al.* (1991).

With *E. coli*, electroporation has been found to give plasmid transformation efficiencies (10^9 cfu/μg DNA) comparable with the best CaCl$_2$ methods (Dower *et al.* 1988). More recently, Zhu and Dean (1999) have reported 10-fold higher transformation efficiencies with plasmids (9×10^9 transformants/μg) by co-precipitating the DNA with transfer RNA (tRNA) prior to electroporation. With conventional CaCl$_2$-mediated transformation, the efficiency falls off rapidly as the size of the DNA molecule increases and is almost negligible when the size exceeds 50 kb. While size also affects the efficiency of electroporation (Sheng *et al.* 1995), it is possible to get transformation efficiencies of 10^6 cfu/μg DNA with molecules as big as 240 kb. Molecules three to four times this size also can be electroporated successfully. This is important because much of the work on mapping and sequencing of genomes demands the ability to handle large fragments of DNA (see Chapter 17).

The ability to transform organisms other than *E. coli* with recombinant DNA enables genes to be studied in different host backgrounds

Although *E. coli* often remains the host organism of choice for cloning experiments, many other hosts are now used, and with them transformation may still be a critical step. In the case of Gram-positive bacteria, the two most important groups of organisms are *Bacillus* spp. and actinomycetes. That *B. subtilis* is naturally competent for transformation has been known for a long time and hence the genetics of this organism are fairly advanced. For this reason *B. subtilis* is a particularly attractive alternative prokaryotic cloning host. The significant features of transformation with this organism are detailed in

Chapter 10. Of particular relevance here is that it is possible to transform protoplasts of *B. subtilis*, a technique which leads to improved transformation frequencies. A similar technique is used to transform actinomycetes, and recently it has been shown that the frequency can be increased considerably by first entrapping the DNA in liposomes, which then fuse with the host-cell membrane.

In later chapters we discuss ways, including electroporation, in which cloned DNA can be introduced into eukaryotic cells. With animal cells there is no great problem as only the membrane has to be crossed. In the case of yeast, protoplasts are required (Hinnen *et al.* 1978). With higher plants one strategy that has been adopted is either to package the DNA in a plant virus or to use a bacterial plant pathogen as the donor. It has also been shown that protoplasts prepared from plant cells are competent for transformation. A further remarkable approach that has been demonstrated with plants and animals (Klein & Fitzpatrick-McElligott 1993) is the use of microprojectiles shot from a gun (p. 291).

Animal cells and protoplasts of yeast, plant, and bacterial cells are susceptible to transformation by liposomes (Deshayes *et al.* 1985). A simple transformation system has been developed which makes use of liposomes prepared from a cationic lipid (Felgner *et al.* 1987). Small unilamellar (single-bilayer) vesicles are produced. DNA in solution spontaneously and efficiently complexes with these liposomes (in contrast to previously employed liposome encapsidation procedures involving non-ionic lipids). The positively charged liposomes not only complex with DNA, but also bind to cultured animal cells and are efficient in transforming them, probably by fusion with the plasma membrane. The use of liposomes as a transformation or transfection system is called *lipofection*.

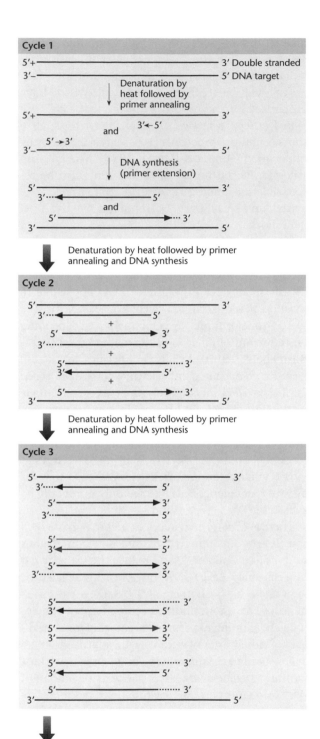

Repeated cycles lead to exponential doubling of the target sequence

Fig. 2.7 (*right*) The polymerase chain reaction. In cycle 1 two primers anneal to denatured DNA at opposite sides of the target region, and are extended by DNA polymerase to give new strands of variable length. In cycle 2, the original strands and the new strands from cycle 1 are separated, yielding a total of four primer sites with which primers anneal. The primers that are hybridized to the new strands from cycle 1 are extended by polymerase as far as the end of the template, leading to a precise copy of the target region. In cycle 3, double-stranded DNA molecules are produced (highlighted in color) that are precisely identical to the target region. Further cycles lead to exponential doubling of the target region. The original DNA strands and the variably extended strands become negligible after the exponential increase of target fragments.

The polymerase chain reaction (PCR) has revolutionized the way that biologists manipulate and analyze DNA

The impact of the PCR upon molecular biology has been profound. The reaction is easily performed, and leads to the amplification of specific DNA sequences by an enormous factor. From a simple basic principle, many variations have been developed with applications throughout gene technology (Erlich 1989, Innis *et al.* 1990). Very importantly, the PCR has revolutionized prenatal diagnosis by allowing tests to be performed using small samples of fetal tissue. In forensic science, the enormous sensitivity of PCR-based procedures is exploited in DNA profiling; following the publicity surrounding *Jurassic Park*, virtually everyone is aware of potential applications in paleontology and archeology. Many other processes have been described which should produce equivalent results to a PCR (for review, see Landegran 1996) but as yet none has found widespread use.

In many applications of the PCR to gene manipulation, the enormous amplification is secondary to the aim of altering the amplified sequence. This often involves incorporating extra sequences at the ends of the amplified DNA. In this section we shall consider only the amplification process. The applications of the PCR will be described in appropriate places later in the book.

The principle of the PCR is exceedingly simple

First we need to consider the basic PCR. The principle is illustrated in Fig. 2.7. The PCR involves two oligonucleotide primers, 17–30 nucleotides in length, which flank the DNA sequence that is to be amplified. The primers hybridize to opposite strands of the DNA after it has been denatured, and are orientated so that DNA synthesis by the polymerase proceeds through the region between the two primers. The extension reactions create two double-stranded target regions, each of which can again be denatured ready for a second cycle of hybridization and extension. The third cycle produces two double-stranded molecules that comprise precisely the target region in double-stranded form. By repeated cycles of heat denaturation, primer hybridization and extension, there follows a rapid exponential accumulation of the specific target fragment of DNA. After 22 cycles, an amplification of about 10^6-fold is expected (Fig. 2.8), and amplifications of this order are actually attained in practice.

In the original description of the PCR method (Mullis & Faloona 1987, Saiki *et al.* 1988, Mullis 1990), Klenow DNA polymerase was used and, because of the heat-denaturation step, fresh enzyme had to be added during each cycle. A breakthrough came with the introduction of *Taq* DNA polymerase (Lawyer *et al.* 1989) from the thermophilic bacterium *Thermus aquaticus*. The *Taq* DNA polymerase

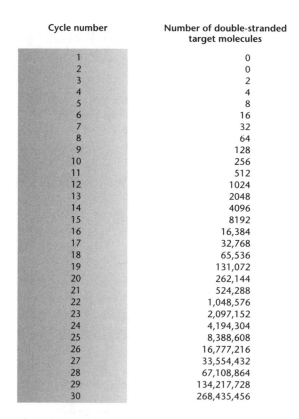

Cycle number	Number of double-stranded target molecules
1	0
2	0
3	2
4	4
5	8
6	16
7	32
8	64
9	128
10	256
11	512
12	1024
13	2048
14	4096
15	8192
16	16,384
17	32,768
18	65,536
19	131,072
20	262,144
21	524,288
22	1,048,576
23	2,097,152
24	4,194,304
25	8,388,608
26	16,777,216
27	33,554,432
28	67,108,864
29	134,217,728
30	268,435,456

Fig. 2.8 Theoretical PCR amplification of a target fragment with increasing number of cycles.

is resistant to high temperatures and so does not need to be replenished during the PCR (Erlich *et al.* 1988, Sakai *et al.* 1988). Furthermore, by enabling the extension reaction to be performed at higher temperatures, the specificity of the primer annealing is not compromised. As a consequence of employing the heat-resistant enzyme, the PCR could be automated very simply by placing the assembled reaction in a heating block with a suitable thermal cycling program (see Box 2.3).

Recent developments have sought to minimize amplification times. Such systems have used small reaction volumes in glass capillaries to give large surface area-to-volume ratios. This results in almost instantaneous temperature equilibration and minimal annealing and denaturation times. This, accompanied by temperature ramp rates of 10–20°C/s, made possible by the use of turbulent forced hot-air systems to heat the sample, results in an amplification reaction completed in tens of minutes.

While the PCR is simple in concept, practically there are a large number of variables which can influence the outcome of the reaction. This is especially important when the method is being used with rare samples of starting material or if the end result has diagnostic or forensic implications. For a

Box 2.3 The PCR achieves enormous amplifications of specific target sequence, very simply

The reaction is assembled in a single tube, and then placed in a thermal cycler (a programmable heating/cooling block), as described below.

A typical PCR for amplifying a human genomic DNA sequence has the following composition. The reaction volume is 100 µl.

Input genomic DNA, 0.1–1 µg
Primer 1, 20 pmol
Primer 2, 20 pmol
20 mmol/l Tris-HCl, pH 8.3 (at 20°C)
1.5 mmol/l magnesium chloride
25 mmol/l potassium chloride
50 mmol/l each deoxynucleoside triphosphate (dATP, dCTP, dGTP, dTTP)
2 units *Taq* DNA polymerase

A layer of mineral oil is placed over the reaction mix to prevent evaporation.

The reaction is cycled 25–35 times, with the following temperature program:

Denaturation	94°C, 0.5 min
Primer annealing	55°C, 1.5 min
Extension	72°C, 1 min

Typically, the reaction takes some 2–3 h overall.

Notes:

- The optimal temperature for the annealing step will depend upon the primers used.
- The pH of the Tris-HCl buffer decreases markedly with increasing temperature. The actual pH varies between about 6.8 and 7.8 during the thermal cycle.
- The time taken for each cycle is considerably longer than 3 min (0.5 + 1.5 + 1 min), depending upon the rates of heating and cooling between steps, but can be reduced considerably by using turbo systems (p. 27).
- The standard PCR does not efficiently amplify sequences much longer than about 3 kb.

detailed analysis of the factors affecting the PCR, the reader should consult Pavlov *et al.* (2004). There are many substances present in natural samples (e.g. blood, feces, environmental materials) which can interfere with the PCR, and ways of eliminating them have been reviewed by Bickley and Hopkins (1999).

RT-PCR enables the sequences on a mRNA molecule to be amplified as DNA

The thermostable polymerase used in the basic PCR requires a DNA template and hence is limited to the amplification of DNA samples. There are numerous instances in which the amplification of RNA would be preferred. For example, in analyses involving the differential expression of genes in tissues during development or the cloning of DNA derived from an mRNA (complementary DNA or *cDNA*), particularly a rare mRNA. In order to apply PCR methodology to the study of RNA, the RNA sample must first be reverse-transcribed to cDNA to provide the necessary DNA template for the thermostable polymerase. This

process is called reverse transcription (RT), hence the name RT-PCR.

Avian myeloblastosis virus (AMV) or Moloney murine leukemia virus (MuLV) reverse transcriptases are generally used to produce a DNA copy of the RNA template. Various strategies can be adopted for first-strand cDNA synthesis (Fig. 2.9), and more are described in Chapter 6.

The basic PCR is not efficient at amplifying long DNA fragments

Amplification of long DNA fragments is desirable for numerous applications of gene manipulation. The basic PCR works well when small fragments are amplified. The efficiency of amplification and therefore the yield of amplified fragments decrease significantly as the size of the amplicon increases over 5 kb. This decrease in yield of longer amplified fragments is attributable to partial synthesis across the desired sequence, which is not a suitable substrate for the subsequent cycles. This is demonstrated by the presence of smeared, as opposed to discrete, bands on a gel.

Fig. 2.9 Three strategies for synthesis of first-strand cDNA. (a) Random primer; (b) oligo (dT) primer; (c) sequence-specific primer.

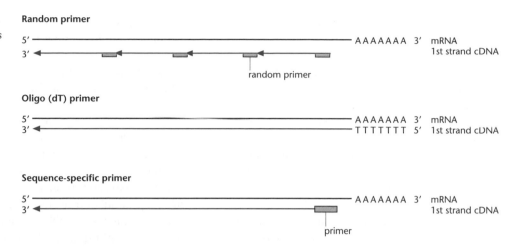

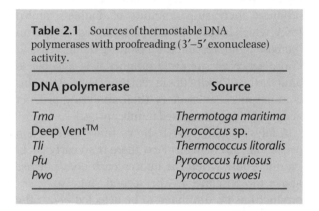

Table 2.1 Sources of thermostable DNA polymerases with proofreading (3′–5′ exonuclease) activity.

DNA polymerase	Source
Tma	*Thermotoga maritima*
Deep Vent™	*Pyrococcus* sp.
Tli	*Thermococcus litoralis*
Pfu	*Pyrococcus furiosus*
Pwo	*Pyrococcus woesi*

Barnes (1994) and Cheng *et al.* (1994a,b) examined the factors affecting the thermostable polymerization across larger regions of DNA and identified key variables affecting the yield of longer PCR fragments. Most significant of these was the absence of a 3′–5′ exonuclease (proofreading) activity in *Taq* polymerase. Presumably, when the *Taq* polymerase misincorporates a deoxynucleoside triphosphate (dNTP), subsequent extension of the strand either proceeds very slowly or stops completely. To overcome this problem, a second thermostable polymerase with proofreading capability is added. Thermostable DNA polymerases with proofreading capabilities are listed in Table 2.1.

The success of a PCR experiment is very dependent on the choice of experimental variables

The specificity of the PCR depends crucially upon the primers. The following factors are important in choosing effective primers.

- Primers should be 17 to 30 nucleotides in length.
- A GC content of about 50% is ideal. For primers with a low GC content it is desirable to choose a long primer so as to avoid a low melting temperature.
- Sequences with long runs (i.e. more than three or four) of a single nucleotide should be avoided.
- Primers with significant secondary structure are undesirable.
- There should be no complementarity between the two primers.

The great majority of primers which conform with these guidelines can be made to work, although not all comparable primer sets are equally effective even under optimized conditions.

In carrying out a PCR it is usual to employ a *hot-start* protocol. This entails adding the DNA polymerase after the heat-denaturation step of the first cycle, the addition taking place at a temperature at or above the annealing temperature and just prior to the annealing step of the first cycle. The hot start overcomes the problem that would arise if the DNA polymerase were added to complete the assembly of the PCR reaction mixture at a relatively low temperature. At low temperature, below the desired hybridization temperature for the primer (typically in the region 45–60°C), mismatched primers will form and may be extended somewhat by the polymerase. Once extended, the mismatched primer is stabilized at the unintended position. Having been incorporated into the extended DNA during the first cycle, the primer will hybridize efficiently in subsequent cycles and hence may cause the amplification of a spurious product.

Alternatives to the hot-start protocol include the use of *Taq* polymerase antibodies, which are inactivated as the temperature rises (Taylor & Logan 1995), and AmpliTaq Gold™, a modified *Taq* polymerase that is inactive until heated to 95°C (Birch 1996). Yet another means of inactivating *Taq* DNA polymerase at ambient temperatures is the SELEX method (systematic evolution of ligands by exponential enrichment). Here the polymerase is reversibly inactivated by the binding of nanomolar amounts of a 70-mer, which is itself a poor polymerase substrate and should not interfere with the amplification primers (Dang & Jayasena 1996).

In order to minimize further the amplification of spurious products, the strategy of *nested primers* may be deployed. Here the products of an initial PCR amplification are used to seed a second PCR amplification, in which one or both primers are located internally with respect to the primers of the first PCR. Since there is little chance of the spurious products containing sequences capable of hybridizing with the second primer set, the PCR with these nested primers selectively amplifies the sought-after DNA.

As noted above, the *Taq* DNA polymerase lacks a 3′–5′ proofreading exonuclease. This lack appears to contribute to errors during PCR amplification due to misincorporation of nucleotides (Eckert & Kunkel 1990). Partly to overcome this problem, other thermostable DNA polymerases with improved fidelity have been sought although the *Taq* DNA polymerase remains the most widely used enzyme for PCR. In certain applications, especially where amplified DNA is cloned, it is important to check the nucleotide sequence of the cloned product to reveal any mutations that may have occurred during the PCR. The fidelity of the amplification reaction can be assessed by cloning, sequencing, and comparing several independently amplified molecules.

By using special instrumentation it is possible to make the PCR quantitative

There are many applications of the PCR where it would be advantageous to be able to quantify the amount of starting material. Theoretically, there is a quantitative relationship between the amount of starting material (target sequence) and the amount of PCR product at any given cycle. In practice, replicate reactions yield different amounts of product, making quantitation unreliable. Higuchi *et al.* (1992, 1993) pioneered the use of ethidium bromide to quantify PCR products as they accumulate. Amplification produces increasing amounts of double-stranded DNA, which binds ethidium bromide, resulting in an increase in fluorescence. By plotting the increase in fluorescence versus cycle number it is possible to analyze the PCR kinetics in real time (Fig. 2.10). This is much more satisfactory than analyzing product accumulation after a fixed number of cycles.

Analysis of Fig. 2.10 shows that the reaction profile has three stages. First there is an early background phase when the fluorescence does not rise above the baseline. In the second stage sufficient product has accumulated to be detected above the background and the fluorescence increases exponentially. In practice, a fixed fluorescence threshold is set above the baseline and the parameter C_T (threshold cycle) is defined as the fractional cycle number at which the fluorescence passes the fixed threshold.

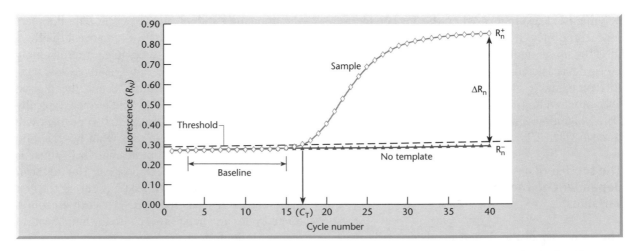

Fig. 2.10   Schematic representation of a typical DNA amplification plot obtained using real-time PCR. Figure reproduced courtesy of Applied Biosystems Inc.

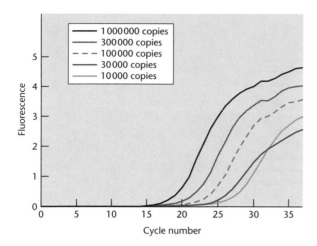

Fig. 2.11 The effect of the initial DNA concentration on the kinetics of PCR amplification as monitored by an increase in fluorescence.

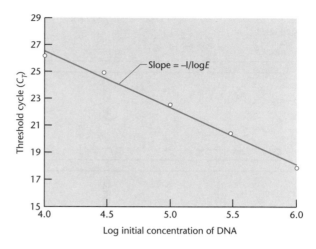

Fig. 2.12 A standard curve of C_T value versus the logarithm of the initial DNA concentration calculated from the data shown in Fig. 2.11. In this example the calculated value of E is 1.73.

Once C_T is reached, the amplification reaction is described by the equation:

$$T_n = T_0(E)^n$$

where T_n is the amount of target sequence at cycle n, T_0 is the initial amount of target, and E is the efficiency of amplification. In the final stage the reaction efficiency declines until no more product accumulates.

The value of C_T will depend on the starting concentration of target DNA: the lower the initial concentration the more cycles will be required to reach the threshold (Fig. 2.11). With the current generation of real-time PCR instruments about 10^{10} copies of PCR product are required to produce a signal above background. If you start with one million copies of target DNA and have an amplification efficiency of 1.9 then using the above equation it can be calculated that a signal will first be seen at cycle 14. On the other hand, if you start with 1000 copies you will first see a signal at cycle 25. A plot of the logarithm of the initial target copy number versus C_T gives a straight line (Fig. 2.12) and has a slope of $-1/\log E$. This graph has two functions. It permits the calculation of the initial target copy number from experimental C_T values and determination of the reaction efficiency. The maximum efficiency possible for a PCR reaction is two and is achieved if every PCR product is replicated in every cycle. In practice, the efficiency may be much lower and the reasons for this reduced efficiency include:

- the presence of PCR inhibitors;
- the presence of DNA fragments that cause non-specific priming events;
- inappropriate choice of primers, probes, and amplicons.

For a more detailed discussion of quantitative PCR the reader should consult www.idahotech.com/lightcycler_u/lectures/quantification_on_lc.htm

There are a number of different ways of generating fluorescence in quantitative PCR reactions

The original work on the development of quantitative PCR made use of the enhanced fluorescence of ethidium bromide when it intercalates into double-stranded DNA. An alternative to ethidium bromide is SYBR Green I. This dye binds in the minor groove of double-stranded DNA and on binding its fluorescence increases over 100-fold. The problem with both of these dyes is that they bind equally to the specific product of the amplification reaction and to any non-specific products and primer dimers as well. Although there are ways of handling this problem an alternative is to use specific amplification probes (Livak *et al.* 1995).

The most widely used probe system is TaqMan™. In this system (Holland *et al.* 1991, Woo *et al.* 1998), the probes used are oligonucleotides with a reporter fluorescent dye attached at the 5′ end and a quencher dye at the 3′ end. While the probe is intact,

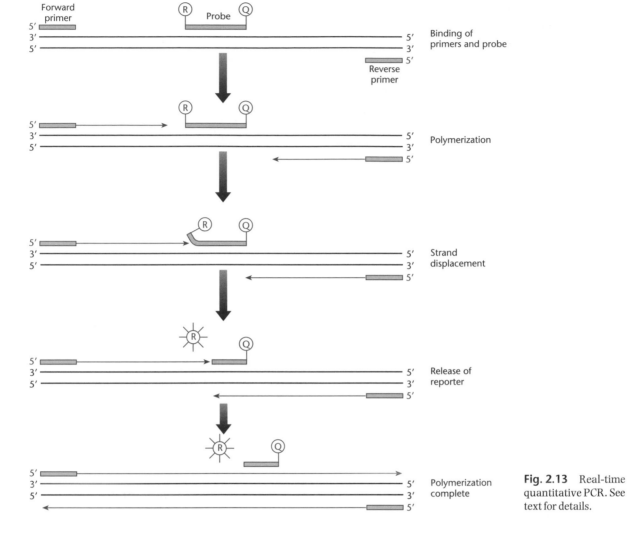

Fig. 2.13 Real-time quantitative PCR. See text for details.

the proximity of the quencher reduces the fluorescence emitted by the reporter dye. If the target sequence is present, the probe anneals downstream from one of the primer sites. As the primer is extended, the probe is cleaved by the 5′ nuclease activity of the *Taq* polymerase (Fig. 2.13). This cleavage of the probe separates the reporter dye from the quencher dye, thereby increasing the reporter-dye signal. Cleavage removes the probe from the target strand, allowing primer extension to continue to the end of the template strand. Additional reporter-dye molecules are cleaved from their respective probes with each cycle, effecting an increase in fluorescence intensity proportional to the amount of amplicon produced.

Instrumentation has been developed which combines thermal cycling with measurement of fluorescence, thereby enabling the progress of the PCR to be monitored in real time. This revolutionizes the way one approaches PCR-based quantitation of DNA. Reactions are characterized by the point in time during cycling when amplification of a product is first detected, rather than by the amount of PCR product accumulated after a fixed number of cycles. The higher the starting copy number of the target, the sooner a significant increase in fluorescence is noted. Quantitation of the amount of target in unknown samples is achieved by preparing a standard curve, using different starting copy numbers of the target sequence.

Two alternatives to the TaqMan system are molecular beacons (Tyagi & Kramer 1996, Tyagi *et al.* 1998) and Scorpion probes (Thelwell *et al.* 2000). These differ from TaqMan probes in that fluorescence occurs by hybridization of the probe rather than cleavage of the probe during amplification. Molecular

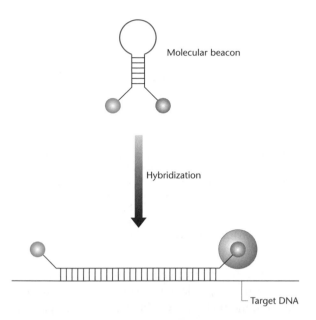

Fig. 2.14 The principle of molecular beacons. In the absence of a DNA target the fluorophore (purple) is held close to the quencher (gray) and fluorescence cannot occur. When the probe binds to its target, the rigidity of the probe-target helix forces the stem of the probe to unwind. This results in sufficient spatial separation of the fluorophore and quencher for fluorescence to occur.

beacons are hairpin-shaped oligonucleotides with a fluorophore at one end and a fluorescence quencher at the other end. While in the hairpin structure the fluorescence is quenched but when the probe binds to the target it undergoes a conformational reorganization that restores fluorescence (Fig. 2.14). A Scorpion probe also has a hairpin loop configuration but differs from a molecular beacon in that it has a 3′ complementary sequence that acts as a primer and is extended in the PCR reaction (Fig. 2.15). After denaturation, the probe sequence binds to the target sequence and this restores fluorescence. Scorpions perform better under rapid cycling conditions because binding obeys first-order kinetics whereas molecular beacons obey second-order kinetics.

A major advantage of fluorescent probes is that they can be constructed using a number of different fluorophores each of which has a characteristic color. This means that it is possible to determine the relative amounts of two targets in a mixture provided that they have the same amplification efficiency (Fig. 2.16).

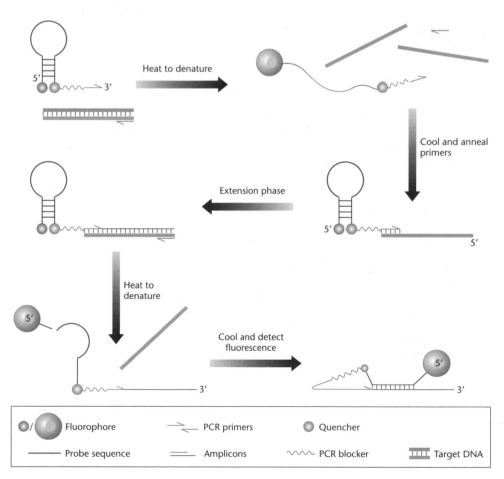

Fig. 2.15 The principle of Scorpion probes. (Redrawn from Thelwell *et al.*, 2000)

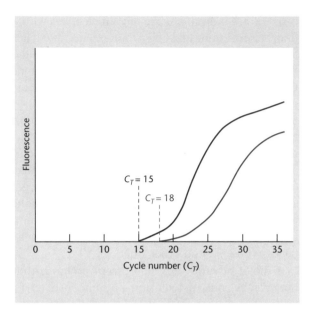

Fig. 2.16 The relative concentrations of two DNA samples as determined using specific probes labeled with different fluorescent dyes. The difference in C_T values for the two DNA samples is three and providing that they amplify with the same efficiency the difference in starting concentration is approximately a factor of 10.

It is now possible to amplify whole genomes as well as gene segments

The PCR is a technique that amplifies specific target sequences but recently it has been extended to whole-genome amplification (WGA). Unlike traditional PCR, the objective of WGA is to represent the entire genome with minimal amplification bias. In the case of the human genome this means the

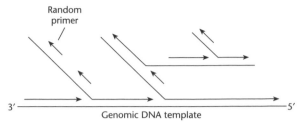

Fig. 2.17 Multiple displacement amplification reaction. DNA synthesis is primed by random hexamers. Exponential amplification occurs by a "hyperbranching" mechanism. Unlike PCR, which requires thermal cycling to repeatedly melt template and anneal primers, the φ29 DNA polymerase acts at 30°C to concurrently extend primers as it displaces downstream DNA products.

amplification of three billion bases without the loss or distortion of particular loci or alleles. To date, five different methods of WGA have been developed and the characteristics of each of them are compared in Table 2.2. Four of them are direct variants of PCR that generate only products of short length (<1000 nucleotides) and with poor fidelity (Lasken & Egholm 2003). For this reason they have not been widely adopted. An alternative method of WGA that does not involve PCR has been developed and is called multiple displacement amplification (MDA) (Dean *et al.* 2002; Hosono *et al.* 2003). In MDA the template is replicated again and again by a hyperbranching mechanism of strand displacement synthesis (Fig. 2.17). That is, the polymerase lays down new copies of the template concurrently with the displacement of new copies. MDA is an isothermal method that takes advantage of two key properties of the DNA polymerase from bacteriophage φ29. First, this

Table 2.2 Characteristics of commonly used whole genome amplification methods.

Method	MDA	PEP	iPEP	DOP	LL-DOP-PCR
DNA yield (per 100 μl reaction)	80 μg	40 ng	ND	1–6 μg	ND
Reactions scalable to any volume	Yes	No	No	No	No
DNA product length (bp)	2000–≥100,000	100–1000	100–1000	100–1000	500–>10,000
Amplification bias range between loci	<6-fold	10^3	ND	10^6	10^3
DNA polymerase error rate	$<10^{-6}$	3×10^{-4} to 1×10^{-5}	$\sim10^{-5}$	3×10^{-4}	-10^{-5}
Amplification from single cells demonstrated	ND	+	+	+	ND
Amplification of fixed, paraffin-embedded tissue	−	+	+	+	ND

ND, not determined; MDA, multiple displacement amplification; PEP, primer extension preamplification; iPEP, improved primer extension preamplification; DOP, degenerate-oligonucleotide-primed PCR; LL-DOP-PCR, long products from low DNA quantities degenerate-oligonucleotide-primed PCR.

polymerase adds ~70,000 nucleotides every time it binds to the primer and explains why MDA generates long DNA products. Second, it is an extremely high-fidelity polymerase with an error rate of only 1 in 10^6–10^7.

Suggested reading

Ashton R., Padala C. & Kane R.S. (2003) Microfluidic separation of DNA. *Current Opinion in Biotechnology* **14**, 497–504.

Barbier V. & Viovy J.L. (2003) Advanced polymers for DNA separation. *Current Opinion in Biotechnology* **14**, 51–7.

Slater G.W., Kenward M., McCormick L.C. & Gauthier M.G. (2003) The theory of DNA separation by capillary electrophoresis. *Current Opinion in Biotechnology* **14**, 58–64.

Three reviews from the same journal issue on the growing trend to separate DNA species by capillary electrophoresis rather than in slab gels.

Arezi B., Xing W., Sorge J.A. & Hogrefe H.H. (2003) Amplification efficiency of thermostable DNA polymerases. *Analytical Biochemistry* **321**, 226–35.

Pavlov A.R., Pavlova N.V., Kozyavkin S.A. & Slesarev A.I. (2004) Recent developments in the optimization of thermostable DNA polymerases for efficient applications. *Trends in Biotechnology* **22**, 253–60.

Two excellent papers on the factors affecting enzyme selection for use in the PCR.

Ding C. & Cantor C.R. (2004) Quantitative analysis of nucleic acids – the last few years of progress. *Journal of Biochemistry and Molecular Biology* **37**, 1–10.

Pardigol A., Guillet S. & Popping B. (2003) A simple procedure for quantification of genetically modified organisms using hybrid amplicon standards. *European Food Research Technology* **216**, 412–20.

These two papers discuss different aspects of the problems associated with quantitative PCR.

Barker D.L., Hansen M.S., Faruqi A.F., *et al.* (2004) Two methods of whole-genome amplification enable accurate genotyping across a 2320-SNP linkage panel. *Genome Research* **14**, 901–7.

Holbrook J.F., Stabley D. & Sol-Church K. (2005) Exploring whole genome amplification as a DNA recovery tool for molecular genetic analysis. *Journal of Biomolecular Techniques* **16**, 125–33.

Paez J.G., Lin M., Beroukhim R., *et al.* (2004) Genome coverage and sequence fidelity of the φ29 polymerase-based multiple strand displacement whole genome amplification. *Nucleic Acids Research* **32**, 71.

These three papers are good examples of the progress being made in the highly topical subject of whole genome amplification.

Useful website

http://www.idahotech.com/lightcycler_u/lectures/
This website contains everything you wanted to know about quantitative PCR but were afraid to ask about.

CHAPTER 3

Cutting and joining DNA molecules

Cutting DNA molecules

Before 1970 there was no method for cleaving DNA at discrete points. All the available methods for fragmenting DNA were non-specific. The available endonucleases had little site specificity and chemical methods produced very small fragments of DNA. The only method where any degree of control could be exercised was the use of mechanical shearing. The long, thin threads which constitute duplex DNA molecules are sufficiently rigid to be very easily broken by shear forces in solution. Intense sonication with ultrasound can reduce the length to about 300 nucleotide pairs. More controlled shearing can be achieved by high-speed stirring in a blender. Typically, high-molecular-weight DNA is sheared to a population of molecules with a mean size of about 8 kb by stirring at 1500 rev/min for 30 min (Wensink *et al.* 1974). Breakage occurs essentially at random with respect to DNA sequence. The termini consist of short, single-stranded regions which may have to be taken into account in subsequent joining procedures.

During the 1960s, phage biologists elucidated the biochemical basis of the phenomenon of host restriction and modification. The culmination of this work was the purification of the restriction endonuclease of *Escherichia coli* K12 by Meselson and Yuan (1968). Since this endonuclease cuts unmodified DNA into large discrete fragments, it was reasoned that it must recognize a target sequence. This in turn raised the prospect of controlled manipulation of DNA. Unfortunately, the K12 endonuclease turned out to be perverse in its properties. While the enzyme does bind to a defined recognition sequence, cleavage occurs at a "random" site several kilobases away (Yuan *et al.* 1980). The much sought-after breakthrough finally came in 1970 with the discovery in *Hemophilus influenzae* (Kelly & Smith 1970, Smith & Wilcox 1970) of an enzyme that behaves more simply. That is, the enzyme recognizes a particular target sequence in a duplex DNA molecule and breaks the polynucleotide chain within that sequence to give rise to discrete fragments of defined length and sequence.

The presence of restriction and modification systems is a double-edged sword. On the one hand, they provide a rich source of useful enzymes for DNA manipulation. On the other, these systems can significantly affect the recovery of recombinant DNA in cloning hosts. For this reason, some knowledge of restriction and modification is essential.

Understanding the biological basis of host-controlled restriction and modification of bacteriophage DNA led to the identification of restriction endonucleases

Restriction systems allow bacteria to monitor the origin of incoming DNA and to destroy it if it is recognized as foreign. Restriction endonucleases recognize specific sequences in the incoming DNA and cleave the DNA into fragments, either at specific sites or more randomly. When the incoming DNA is a bacteriophage genome, the effect is to reduce the efficiency of plating, i.e. to reduce the number of plaques formed in plating tests. The phenomena of restriction and modification were well illustrated and studied by the behavior of phage λ on two *E. coli* host strains.

If a stock preparation of phage λ, for example, is made by growth upon *E. coli* strain C and this stock is then titered upon *E. coli* C and *E. coli* K, the titers observed on these two strains will differ by several orders of magnitude, the titer on *E. coli* K being the lower. The phage are said to be *restricted* by the second host strain (*E. coli* K). When those phage that do result from the infection of *E. coli* K are now replated on *E. coli* K they are no longer restricted; but if they are first cycled through *E. coli* C they are once again restricted when plated upon *E. coli* K (Fig. 3.1). Thus the efficiency with which phage λ plates upon a particular host strain depends upon the strain on which it was last propagated. This non-heritable change conferred upon the phage by the second host strain

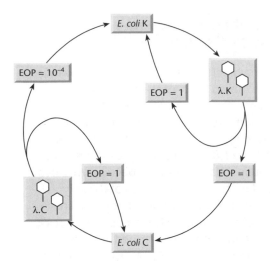

Fig. 3.1 Host-controlled restriction and modification of phage λ in *E. coli* strain K, analyzed by efficiency of plating (EOP). Phage propagated by growth on strains K or C (i.e. λ.K or λ.C) have EOPs on the two strains, as indicated by arrows.

(*E. coli* K) that allows it to be replated on that strain without further restriction is called *modification*.

The restricted phages adsorb to restrictive hosts and inject their DNA normally. When the phage are labeled with ^{32}P, it is apparent that their DNA is degraded soon after injection (Dussoix & Arber 1962) and the endonuclease that is primarily responsible for this degradation is called a *restriction endonuclease* or restriction enzyme (Lederberg & Meselson 1964). The restrictive host must, of course, protect its own DNA from the potentially lethal effects of the restriction endonuclease and so its DNA must be appropriately modified. Modification involves methylation of certain bases at a very limited number of sequences within DNA, which constitute the recognition sequences for the restriction endonuclease. This explains why phage that survive one cycle of growth upon the restrictive host can subsequently reinfect that host efficiently; their DNA has been replicated in the presence of the modifying methylase and so it, like the host DNA, becomes methylated and protected from the restriction system. Although phage infection has been chosen as our example to illustrate restriction and modification, these processes can occur whenever DNA is transferred from one bacterial strain to another.

Four different types of restriction and modification (R-M) system have been recognized but only one is widely used in gene manipulation

At least four different kinds of R-M system are known: type I, type II, type III, and type IIs. The essential differences between them are summarized in Table 3.1.

The type I systems were the first to be characterized and a typical example is that from *E. coli* K12. The active enzyme consists of two restriction subunits, two modification (methylation) subunits, and one recognition subunit. These subunits are the products of the *hsd*R, *hsd*M, and *hsd*S genes. The methylation and cutting reactions both require ATP and *S*-adenosylmethionine as cofactors. The recognition sequences are quite long with no recognizable features such as symmetry. The enzyme cuts unmodified DNA at some distance from the recognition sequence. However, because the methylation reaction is performed by the same enzyme which mediates cleavage, the target DNA may be modified before it is cut. These features mean that type I systems are of little value for gene manipulation (see also Box 3.1). However, their presence in *E. coli*

Table 3.1 Characteristics of the different types of endonucleases.

System	Key features
Type I	One enzyme with different subunits for recognition, cleavage, and methylation. Recognizes and methylates a single sequence but cleaves DNA up to 1000 bp away
Type II	Two different enzymes which both recognize the same target sequence, which is symmetrical. The two enzymes either cleave or modify the recognition sequence
Type III	One enzyme with two different subunits, one for recognition and modification and one for cleavage. Recognizes and methylates same sequence but cleaves 24–26 bp away
Type IIs	Two different enzymes but recognition sequence is asymmetric. Cleavage occurs on one side of recognition sequence up to 20 bp away

Box 3.1   Restriction: from a phenomenon in bacterial genetics to a biological revolution

In the two related phenomena of host-controlled restriction and modification, a single cycle of phage growth in a particular host bacterium, alters the host range of the progeny virus. The virus may fail to plate efficiently on a second host; its host range is restricted. This modification of the virus differs fundamentally from mutation because it is imposed by the host cell on which the virus has been grown but it is not inherited; when the phage is grown in some other host, the modification may be lost. In the 1950s, restriction and modification were recognized as common phenomena, affecting many virulent and temperate (i.e. capable of forming lysogens) phages and involving various bacterial species (Luria 1953, Lederberg 1957).

Arber and Dussoix clarified the molecular basis of the phenomena. They showed that restriction of phage λ is associated with rapid breakdown of the phage DNA in the host bacterium. They also showed that modification results from an alteration of the phage DNA which renders the DNA insensitive to restriction. They deduced that a single modified strand in the DNA duplex is sufficient to prevent restriction (Arber & Dussoix 1962, Dussoix & Arber 1962). Subsequent experiments implicated methylation of the DNA in the modification process (Arber 1965).

Detailed genetic analysis, in the 1960s, of the bacterial genes (in *E. coli* K and *E. coli* B) responsible for restriction and modification supported the duality of the two phenomena. Mutants of the bacteria could be isolated that were both restriction-deficient and modification-deficient (R^-M^-), or that were R^-M^+. The failure to recover R^+M^- mutants was correctly ascribed to the suicidal failure to confer protective modification upon the host's own DNA.

The biochemistry of restriction advanced with the isolation of the restriction endonuclease from *E. coli* K (Meselson & Yuan 1968). It was evident that the restriction endonucleases from *E. coli* K and *E. coli* B were important examples of proteins that recognize specific structures in DNA, but the properties of these type I enzymes as they are now known, were complex. Although the recognition sites in the phage could be mapped genetically (Murray *et al.* 1973a), determined efforts to define the DNA sequences cleaved were unsuccessful (Eskin & Linn 1972, Murray *et al.* 1973b).

The breakthrough came with Hamilton Smith's discovery of a restriction endonuclease from *Hemophilus influenzae* strain Rd (Smith & Wilcox 1970) and the elucidation of the nucleotide sequence at its cleavage sites in phage T7 DNA (Kelly & Smith 1970). This enzyme is now known as *Hind*II. The choice of T7 DNA as the substrate for cleavage was a good one, because the bacterium also contains another type II restriction enzyme, *Hind*III, in abundance. Fortunately, *Hind*III does not cleave T7 DNA, and so any contaminating *Hind*III in the *Hind*II preparation could not be problematical (Old *et al.* 1975). Shortly after the discovery of *Hind*II, several other type II restriction endonucleases were isolated and characterized. *Eco*RI was foremost among these (Hedgepeth *et al.* 1972). They were rapidly exploited in the first recombinant DNA experiments.

By the mid-1960s, restriction and modification had been recognized as important and interesting phenomena within the field of bacterial genetics (see, for example, Hayes 1968), but who could have foreseen the astonishing impact of restriction enzymes upon biology?

strains can affect recovery of recombinants (see p. 24). Type III enzymes have symmetrical recognition sequences but otherwise resemble type I systems and are of little value.

Most of the useful R-M systems are of type II. They have a number of advantages over type I and III systems. First, restriction and modification are mediated by separate enzymes so it is possible to cleave DNA in the absence of modification. Secondly, the restriction activities do not require cofactors such as ATP or *S*-adenosylmethionine, making them easier to use. Most important of all, type II enzymes recognize a defined, usually symmetrical, sequence *and cut within it*. Many of them also make a staggered break in the DNA and the usefulness of this will become apparent. Although type IIs systems have similar cofactors and macromolecular structure to those of type II systems, the fact that restriction occurs at a distance from the recognition site limits their usefulness.

The classification of R-M systems into types I to III is convenient but may require modification as new discoveries are made. For example, the *Eco*571 system comprises a single polypeptide which has both restriction and modification activities (Petrusyte *et al.* 1988). Other restriction systems are known which fall outside the above classification. Examples include the *mcr* and *mrr* systems (see p. 43) and homing endonucleases. The latter are double-stranded deoxyribonucleases (DNases) derived from introns or inteins (Belfort & Roberts 1997). They have large, asymmetric recognition sequences and, unlike standard restriction endonucleases, tolerate some sequence degeneracy within their recognition sequence.

The naming of restriction endonucleases provides information about their source

The discovery of a large number of restriction and modification systems called for a uniform system of nomenclature. A suitable system was proposed by Smith and Nathans (1973) and a simplified version of this is in use today. The key features are:

- The species name of the host organism is identified by the first letter of the genus name and the first two letters of the specific epithet to generate a three-letter abbreviation. This abbreviation is always written in italics.
- Where a particular strain has been the source then this is identified.
- When a particular host strain has several different R-M systems, these are identified by roman numerals.

Some examples are given in Table 3.2.

Homing endonucleases are named in a similar fashion except that intron-encoded endonucleases are given the prefix "I-" (e.g. I-*Ceu*I) and intein endonucleases have the prefix "PI-" (e.g. Pl-*Psp*I). Where it is necessary to distinguish between the restriction and methylating activities, they are given the prefixes "R" and "M", respectively, e.g. R.*Sma*I and M.*Sma*I.

Restriction enzymes cut DNA at sites of rotational symmetry and different enzymes recognize different sequences

Most, but not all, type II restriction endonucleases recognize and cleave DNA within particular sequences of four to eight nucleotides which have a twofold axis of *rotational symmetry*. Such sequences are often referred to as *palindromes* because of their similarity to words that read the same backwards as forwards. For example, the restriction and modification enzymes R.*Eco*RI and M.*Eco*RI recognize the sequence:

$$5'\text{-G A A} \mid \text{T T C-}3'$$
$$3'\text{-C T T} \mid \text{A A G-}5'$$

Axis of symmetry

Table 3.2 Examples of restriction endonuclease nomenclature.

Enzyme	Enzyme source	Recognition sequence
*Sma*I	*Serratia marcescens*, 1st enzyme	CCCGGG
*Hae*III	*Hemophilus aegyptius*, 3rd enzyme	GGCC
*Hind*II	*Hemophilus influenzae*, strain d, 2nd enzyme	GTPyPuAC
*Hind*III	*Hemophilus influenzae*, strain d, 3rd enzyme	AAGCTT
*Bam*HI	*Bacillus amyloliquefaciens*, strain H, 1st enzyme	GGATCC

The position at which the restricting enzyme cuts is usually shown by the symbol "/" and the nucleotides methylated by the modification enzyme are usually marked with an asterisk. For *Eco*RI these would be represented thus:

5'-G/AA*T T C-3'
3'-C T T A*A/G-5'

For convenience it is usual practice to simplify the description of recognition sequences by showing only one strand of DNA, that which runs in the 5' to 3' direction. Thus the *Eco*RI recognition sequence would be shown as G/AATTC.

From the information shown above we can see that *Eco*RI makes single-stranded breaks four bases apart in the opposite strands of its target sequence so generating fragments with protruding 5' termini:

5'-G   5'-AATTC-3'
3'-CTTAA-5'   G-5'

These DNA fragments can associate by hydrogen bonding between overlapping 5' termini, or the fragments can circularize by intramolecular reaction (Fig. 3.2). For this reason the fragments are said to have *sticky* or *cohesive* ends. In principle, DNA fragments from diverse sources can be joined by means of the cohesive ends and, as we shall see later, the nicks in the molecules can be sealed to form an intact *artificially recombinant* DNA molecule.

Not all type II enzymes cleave their target sites like *Eco*RI. Some, such as *Pst*I (CTGCA/G), produce fragments bearing 3' overhangs, while others, such as *Sma*I (CCC/GGG), produce *blunt* or *flush* ends.

Table 3.3 Some restriction endonucleases and their recognition sites.

Enzyme	Recognition sequence
4-base cutters	
*Mbo*I, *Dpn*I, *Sau*3AI	/GATC
*Msp*I, *Hpa*II	C/CGG
*Alu*I	AG/CT
*Hae*III	GG/CC
*Tai*I	ACGT/
6-base cutters	
*Bgl*II	A/GATCT
*Cla*I	AT/CGAT
*Pvu*II	CAG/CTG
*Pvu*I	CGAT/CG
*Kpn*I	GGTAC/C
8-base cutters	
*Not*I	GC/GGCCGC
*Sbf*I	CCTGCA/GG

To date, over 10,000 microbes from around the world have been screened for restriction enzymes. From these, over 3000 enzymes have been found representing approximately 200 different sequence specificities. Some representative examples are shown in Table 3.3. For a comprehensive database of information on restriction endonucleases and their associated methylases, including cleavage sites, commercial availability, and literature references, the reader should consult the website maintained by New England Biolabs (www.rebase.neb.com).

Occasionally enzymes with novel DNA sequence specificities are found but most prove to have the same specificity as enzymes already known. Restriction enzymes with the same sequence specificity and cut site are known as *isoschizomers*. Enzymes that recognize the same sequence but cleave at different points, for example *Sma*I (CCC/GGG) and *Xma*I (C/CCGGG), are sometimes known as *neoschizomers*.

Under extreme conditions, such as elevated pH or low ionic strength, restriction endonucleases are capable of cleaving sequences which are similar but not identical to their defined recognition sequence. This altered specificity is known as *star* activity. The most common types of altered activity are acceptance of base substitutions and truncation of the number of bases in the recognition sequence. For example, *Eco*RI* (*Eco*RI star activity) cleaves the sequence N/AATTN, where N is any base, whereas *Eco*RI cleaves the sequence GAATTC.

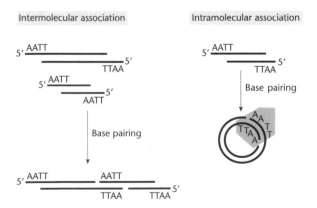

Fig. 3.2 Cohesive ends of DNA fragments produced by digestion with *Eco*RI.

The G+C content of a DNA molecule affects its susceptibility to different restriction endonucleases

The number and size of the fragments generated by a restriction enzyme depend on the frequency of occurrence of the target site in the DNA to be cut. Assuming a DNA molecule with a 50% G+C content and a random distribution of the four bases, a four-base recognition site occurs every 4^4 (256) bp. Similarly, a six-base recognition site occurs every 4^6 (4096) bp and an eight-base recognition sequence every 4^8 (65,536) bp. In practice, there is not a random distribution of the four bases and many organisms can be AT- or GC-rich, e.g. the nuclear genome of mammals is 40% G+C and the dinucleotide CG is fivefold less common than statistically expected. Similarly, CCG and CGG are the rarest trinucleotides in most A+T-rich bacterial genomes and CTAG is the rarest tetranucleotide in G+C-rich bacterial genomes. Thus different restriction endonucleases with six-base recognition sites can produce average fragment sizes significantly different from the expected 4096 bp (Table 3.4).

Certain restriction endonucleases show preferential cleavage of some sites in the same DNA molecule. For example, phage λ DNA has five sites for *Eco*RI but the different sites are cleaved non-randomly (Thomas & Davis 1975). The site nearest the right terminus is cleaved 10 times faster than the sites in the middle of the molecule. There are four sites for *Sac*II in λ DNA but the three sites in the center of the molecule are cleaved 50 times faster than the remaining site. There is a group of three restriction enzymes which show an even more dramatic site preference. These are *Nar*I, *Nae*I, and *Sac*II and they require simultaneous interaction with two copies of their recognition sequence before they will cleave DNA (Kruger 1988, Conrad & Topal, 1989). Thus *Nar*I will rapidly cleave two of the four recognition sites on plasmid pBR322 DNA but will seldom cleave the remaining two sites.

Simple DNA manipulations can convert a site for one restriction enzyme into a site for another enzyme

In order to join two fragments of DNA together, it is not essential that they are produced by the same restriction endonuclease. Many different restriction endonucleases produce compatible cohesive ends. For example, *Age*I (A/CCGGT) and *Ava*I (C/CCGGG) produce molecules with identical 5' overhangs and so can be ligated together. There are many other examples of compatible cohesive ends. What is more, if the cohesive ends were produced by six-base cutters, the ligation products are often recleavable by four-base cutters. Thus, in the example cited above, the hybrid site ACCGGG can be cleaved by *Hpa*II (C/CGG), *Nci*I (CC/GGG), or *Scr*FI (CC/NGG).

New restriction sites can be generated by filling in the overhangs generated by restriction endonucleases and ligating the products together. Figure 3.3 shows that after filling in the cohesive ends produced by *Eco*RI, ligation produces restriction sites recognized by four other enzymes. Many other examples of creating new target sites by filling and ligation are known.

There are also many examples of combinations of blunt-end restriction endonucleases that produce recleavable ligation products. For example, when molecules generated by cleavage with *Alu*I (AG/CT) are joined to ones produced by *Eco*RV (GAT/ATC), some of the ligation sites will have the sequence GATCT and others will have the sequence AGATC. Both can be cleaved by *Mbo*I (GATC).

A methyltransferase, M.*Sss*I, that methylates the dinucleotide CpG (Nur *et al.* 1985) has been isolated from *Spiroplasma*. This enzyme can be used to modify

Table 3.4 Average fragment size (bp) produced by different enzymes with DNA from different sources.

Enzyme	Target	*Arabidopsis*	Nematode	*Drosophila*	*E. coli*	Human
*Apa*I	GGGCCC	25,000	40,000	6,000	15,000	2,000
*Avr*II	CCTAGG	15,000	20,000	20,000	150,000	8,000
*Bam*HI	GGATCC	6,000	9,000	4,000	5,000	5,000
*Dra*I	TTTAAA	2,000	1,000	1,000	2,000	2,000
*Spe*I	ACTAGT	8,000	8,000	9,000	60,000	10,000

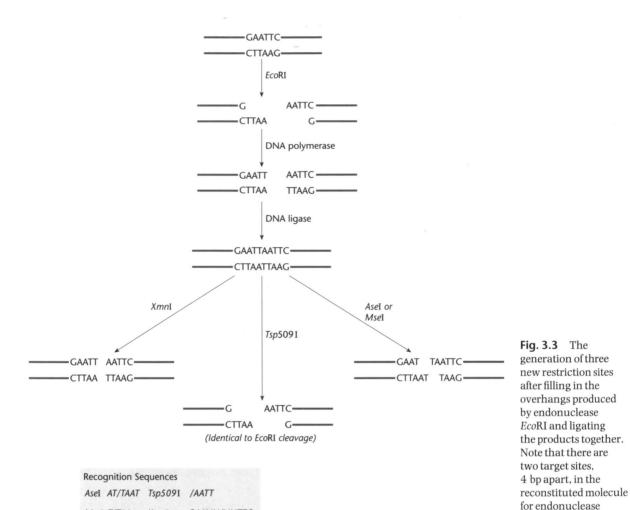

Fig. 3.3 The generation of three new restriction sites after filling in the overhangs produced by endonuclease *Eco*RI and ligating the products together. Note that there are two target sites, 4 bp apart, in the reconstituted molecule for endonuclease *Tsp*509I.

in vitro restriction endonuclease target sites which contain the CG sequence. Some of the target sequences modified in this way will be resistant to endonuclease cleavage, while others will remain sensitive. For example, if the sequence CCGG is modified with *Sss*I, it will be resistant to *Hpa*II but sensitive to *Msp*I. Since 90% of the methyl groups in the genomic DNA of many animals, including vertebrates and echinoderms, occur as 5-methylcytosine in the sequence CG, M.*Sss* can be used to imprint DNA from other sources with a vertebrate pattern.

Methylation can reduce the susceptibility of DNA to cleavage by restriction endonucleases and the efficiency of DNA transformation

Most laboratory strains of *E. coli* contain three site-specific DNA methylases. The methylase encoded by the *dam* gene transfers a methyl group from *S*-adenosylmethionine to the N^6 position of the adenine residue in the sequence GATC. The methylase encoded by the *dcm* gene (the Dcm methylase, previously called the Mec methylase) modifies the internal cytosine residues in the sequences CCAGG and CCTGG at the C^5 position (Marinus *et al.* 1983). In DNA in which the GC content is 50%, the sites for these two methylases occur, on average, every 256–512 bp. The third methylase is the enzyme M.*Eco*KI but the sites for this enzyme are much rarer and occur about once every 8 kb.

These enzymes are of interest for two reasons. First, some or all of the sites for a restriction endonuclease may be resistant to cleavage when isolated from strains expressing the Dcm or Dam methylases. This occurs when a particular base in the recognition site of a restriction endonuclease is methylated. The relevant base may be methylated by one of the *E. coli* methylases if the methylase recognition site overlaps the endonuclease recognition site. For example, DNA isolated from Dam⁺ *E. coli* is completely resistant to cleavage by *Mbo*I, but not *Sau*3AI, both of which recognize the sequence GATC. Similarly, DNA from a Dcm⁺ strain will be cleaved by *Bst*NI but not by *Eco*RII, even though both recognize the sequence CCATGG. It is worth noting that most cloning strains of *E. coli* are Dam⁺ Dcm⁺ but double mutants are available (Marinus *et al.* 1983).

The second reason these methylases are of interest is that the modification state of plasmid DNA can affect the frequency of transformation in special situations. Transformation efficiency will be reduced when Dam-modified plasmid DNA is introduced into Dam⁻ *E. coli* or Dam- or Dcm-modified DNA is introduced into other species (Russell & Zinder 1987). When DNA is to be moved from *E. coli* to another species it is best to use a strain lacking the Dam and Dcm methylases.

As will be seen later, it is difficult to clone DNA that contains short, direct repetitive sequences stably. Deletion of the repeat units occurs quickly, even when the host strain is deficient in recombination. However, the deletion mechanism appears to involve Dam methylation, for it does not occur in *dam* mutants (Troester *et al.* 2000).

It is important to eliminate restriction systems in *E. coli* strains used as hosts for recombinant DNA

If foreign DNA is introduced into an *E. coli* host it may be attacked by restriction systems active in the host cell. An important feature of these systems is that the fate of the incoming DNA in the restrictive host depends not only on the sequence of the DNA but also upon its history: the DNA sequence may or may not be restricted, depending upon its source immediately prior to transforming the *E. coli* host strain. As we have seen, post-replication modifications of the DNA, usually in the form of methylation of particular adenine or cytosine residues in the target sequence, protect against cognate restriction systems but not, in general, against different restriction systems.

Because restriction provides a natural defense against invasion by foreign DNA, it is usual to employ a K restriction-deficient *E. coli* K12 strain as a host in transformation with newly created recombinant molecules. Thus where, for example, mammalian DNA has been ligated into a plasmid vector, transformation of the *Eco*K restriction-deficient host eliminates the possibility that the incoming sequence will be restricted, even if the mammalian sequence

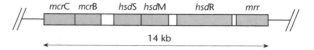

Fig. 3.4 The immigration control region of *E. coli* strain K12.

contains an unmodified *Eco*K target site. If the host happens to be *Eco*K restriction-deficient but *Eco*K modification-*proficient*, propagation on the host will confer modification methylation and hence allow subsequent propagation of the recombinant in *Eco*K restriction-proficient strains, if desired.

Whereas the *Eco*KI restriction system, encoded by the *hsd*RMS genes, cleaves DNA that is not protected by methylation at the target site, the McrA, McrBC, and Mrr endonucleases cut DNA that is methylated at specific positions. All three endonucleases restrict DNA modified by CpC methylase (M.*Sss*I) and the Mrr endonuclease will attack DNA with methyladenine in specific sequences. The significance of these restriction enzymes is that DNA from many bacteria, and from all plants and higher animals, is extensively methylated and its recovery in cloning experiments will be greatly reduced if the restriction activity is not eliminated. There is no problem with DNA from *Saccharomyces cerevisiae* or *Drosophila melanogaster* since there is little methylation of their DNA.

All the restriction systems in *E. coli* are clustered together in an "immigration control region" about 14 kb in length (Fig. 3.4). Some strains carry mutations in one of the genes. For example, strains DH1 and DH5 have a mutation in the *hsd*R gene and so are defective for the *Eco*KI endonuclease but still mediate the *Eco*KI modification of DNA. Strain DP50 has a mutation in the *hsd*S gene and so lacks both the *Eco*KI restriction and modification activities. Other strains, such as *E. coli* C and the widely used cloning strain HB101, have a deletion of the entire *mcr*C–*mrr* region and hence lack all restriction activities.

The success of a cloning experiment is critically dependent on the quality of any restriction enzymes that are used

Restriction enzymes are available from many different commercial sources. In choosing a source of enzyme, it is important to consider the quality of the enzyme supplied. High-quality enzymes are purified extensively to remove contaminating exonucleases and endonucleases and tests for the absence of such contaminants form part of routine quality control

(QC) on the finished product (see below). The absence of exonucleases is particularly important. If they are present, they can nibble away the overhangs of cohesive ends, thereby eliminating or reducing the production of subsequent recombinants. Contaminating phosphatases can remove the terminal phosphate residues, thereby preventing ligation. Even where subsequent ligation is achieved, the resulting product may contain small deletions. The message is clear: cheap restriction enzymes are in reality poor value for money!

A typical QC procedure is as follows. DNA fragments are produced by an excessive overdigestion of substrate DNA with each restriction endonuclease. These fragments are then ligated and recut with the same restriction endonuclease. Ligation can occur only if the 3′ and 5′ termini are left intact, and only those molecules with a perfectly restored recognition site can be recleaved. A normal banding pattern after cleavage indicates that both the 3′ and 5′ termini are intact and the enzyme preparation is free of detectable exonucleases and phosphatases (Fig. 3.5).

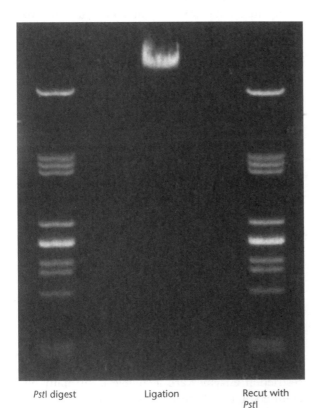

Fig. 3.5 Quality control of the enzyme *Pst*I. DNA was overdigested with the endonuclease and the fragments were ligated together and then recut. Note that the two digests give an identical banding pattern upon agarose gel electrophoresis.

An additional QC test is the *blue/white screening* assay. In this, the starting material is a plasmid carrying the *E. coli lacZ′* gene in which there is a single recognition site for the enzyme under test. The plasmid is overdigested with the restriction enzyme, religated and transformed into a lacZ⁻ strain of *E. coli*. The transformants are plated on media containing the β-galactosidase substrate Xgal. If the *lacZ* gene remains intact after digestion and ligation, it will give rise to a blue colony. If any degradation of the cut ends occurred, then a white colony will be produced (Box 3.2).

Joining DNA molecules

Having described the methods available for cutting DNA molecules, we must consider the ways in which DNA fragments can be joined to create artificially recombinant molecules. There are currently three methods for joining DNA fragments *in vitro*. The first of these capitalizes on the ability of DNA ligase to join covalently the annealed cohesive ends produced by certain restriction enzymes. The second depends upon the ability of DNA ligase from phage T4-infected *E. coli* to catalyze the formation of phosphodiester bonds between blunt-ended fragments. The third utilizes the enzyme terminal deoxynucleotidyltransferase to synthesize homopolymeric 3′ single-stranded tails at the ends of fragments. We can now look at these three methods a little more deeply.

The enzyme DNA ligase is the key to joining DNA molecules *in vitro*

E. coli and phage T4 encode an enzyme, DNA ligase, which seals single-stranded nicks between adjacent nucleotides in a duplex DNA chain (Olivera *et al.* 1968, Gumport & Lehman 1971). Although the reactions catalyzed by the enzymes of *E. coli* and T4-infected *E. coli* are very similar, they differ in their cofactor requirements. The T4 enzyme requires ATP, while the *E. coli* enzyme requires NAD⁺. In each case the cofactor is split and forms an enzyme–AMP complex. The complex binds to the nick, which must expose a 5′ phosphate and 3′ OH group, and makes a covalent bond in the phosphodiester chain, as shown in Fig. 3.6.

When termini created by a restriction endonuclease that creates cohesive ends associate, the joint has nicks a few base pairs apart in opposite strands. DNA ligase can then repair these nicks to form an intact duplex. This reaction, performed *in vitro* with purified

Box 3.2 α-Complementation of β-galactosidase and the use of Xgal

The activity of the enzyme β-galactosidase is easily monitored by including in the growth medium the chromogenic substrate 5-bromo-4-chloro-3-indolyl-β-D-galactoside (Xgal). This compound is colorless but on cleavage releases a blue indolyl derivative. On solid medium, colonies that are expressing active β-galactosidase are blue in color while those without the activity are white in color. This is often referred to as blue/white screening. Since Xgal is not an inducer of β-galactosidase, the non-substrate (*gratuitous*) inducer isopropyl-β-D-thiogalactoside (IPTG) is also added to the medium.

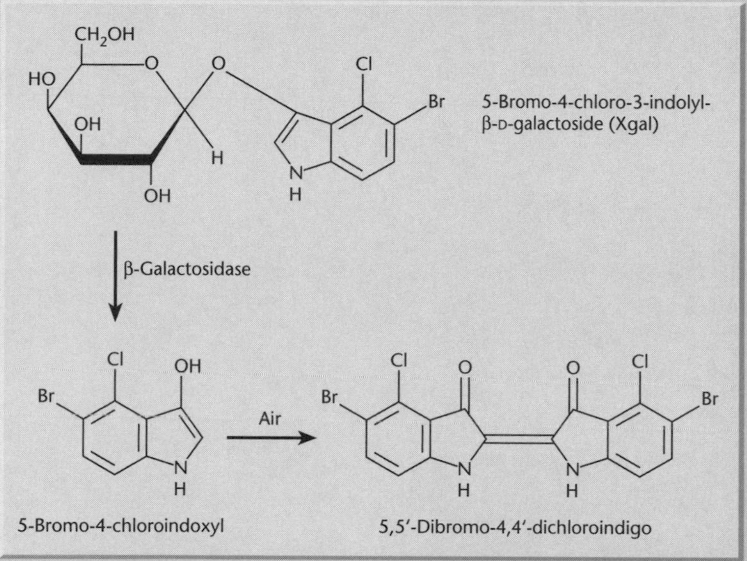

5-Bromo-4-chloro-3-indolyl-β-D-galactoside (Xgal)

β-Galactosidase

5-Bromo-4-chloroindoxyl

5,5'-Dibromo-4,4'-dichloroindigo

The phenomenon of α-complementation of β-galactosidase is widely used in molecular genetics. The starting-point for α-complementation is the M15 mutant of *E. coli*. This has a deletion of residues 11–41 in the *lacZ* gene and shows no β-galactosidase activity. Enzyme activity can be restored to the mutant enzyme *in vitro* by adding a cyanogen bromide peptide derived from amino acid residues 3–92 (Langley *et al.* 1975, Langley & Zabin 1976). Complementation can also be shown *in vivo*. If a plasmid carrying the N-terminal fragment of the *lacZ* gene encompassing the missing region is introduced into the M15 mutant, then β-galactosidase is produced, as demonstrated by the production of a blue color on medium containing Xgal. In practice, the plasmid usually carries the *lacI* gene and the first 146 codons of the *lacZ* gene, because in the early days of genetic engineering this was a convenient fragment of DNA to manipulate.

Since wild-type β-galactosidase has 1021 amino acids, it is encoded by a gene 3.1 kb in length. While a gene of this length is easily manipulated *in vitro*, there are practical disadvantages to using the whole gene. As will be seen later, it is preferable to keep cloning vectors and their inserts as small as possible. The phenomenon of α-complementation allows genetic engineers to take advantage of the *lac* system without having to have the entire Z gene on the vector.

DNA ligase, is fundamental to many gene-manipulation procedures, such as that shown in Fig. 3.7.

The optimum temperature for ligation of nicked DNA is 37°C, but at this temperature the hydrogen-bonded join between the sticky ends is unstable. *Eco*RI-generated termini associate through only four AT base pairs and these are not sufficient to resist thermal disruption at such a high temperature. The optimum

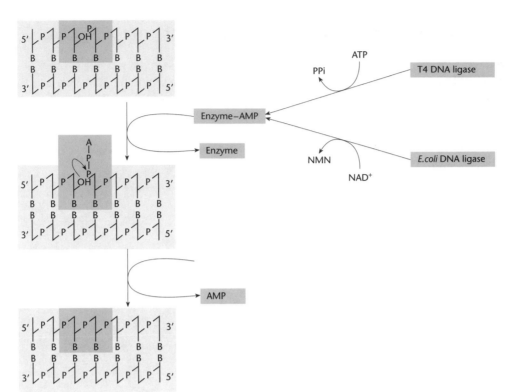

Fig. 3.6 Action of DNA ligase. An enzyme–AMP complex binds to a nick bearing 3′ OH and 5′ P groups. The AMP reacts with the phosphate group. Attack by the 3′ OH group on this moiety generates a new phosphodiester bond, which seals the nick.

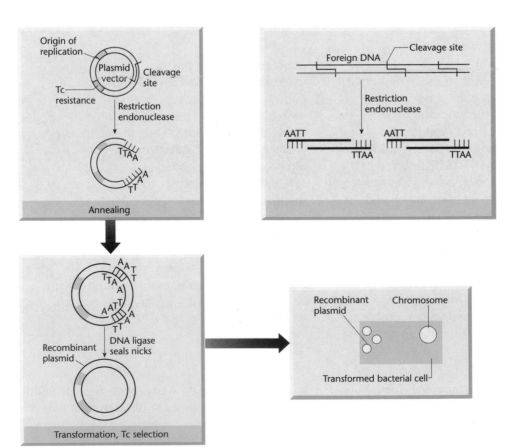

Fig. 3.7 Use of DNA ligase to create a covalent DNA recombinant joined through association of termini generated by *Eco*RI.

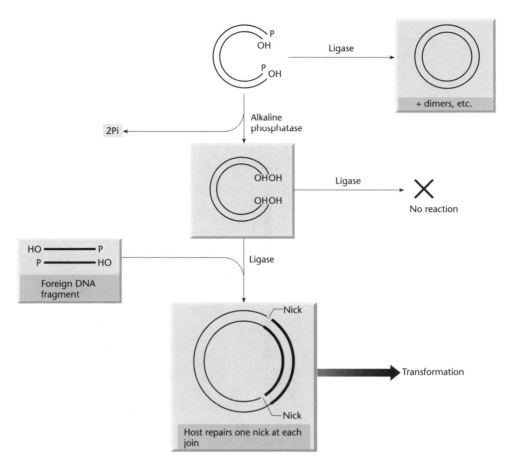

Fig. 3.8 Application of alkaline phosphatase treatment to prevent recircularization of vector plasmid without insertion of foreign DNA.

temperature for ligating the cohesive termini is therefore a compromise between the rate of enzyme action and association of the termini, and has been found experimentally to be in the range 4–15°C (Dugaiczyk *et al.* 1975, Ferretti & Sgaramella 1981).

The ligation reaction can be performed so as to favor the formation of recombinants. First, the population of recombinants can be increased by performing the reaction at a high DNA concentration; in dilute solutions *circularization* of linear fragments is relatively favored because of the reduced frequency of intermolecular reactions. Secondly, by treating linearized plasmid vector DNA with alkaline phosphatase to remove 5′-terminal phosphate groups, both recircularization and plasmid dimer formation are prevented (Fig. 3.8). In this case, circularization of the vector can occur only by insertion of non-phosphatase-treated foreign DNA which provides one 5′-terminal phosphate at each join. One nick at each join remains unligated, but, after transformation of host bacteria, cellular repair mechanisms reconstitute the intact duplex.

Joining DNA fragments with cohesive ends by DNA ligase is a relatively efficient process which has been used extensively to create artificial recombinants. A modification of this procedure depends upon the ability of T4 DNA ligase to join blunt-ended DNA molecules (Sgaramella 1972). The *E. coli* DNA ligase will not catalyze blunt ligation except under special reaction conditions of macromolecular crowding (Zimmerman & Pheiffer 1983). Blunt ligation is most usefully applied to joining blunt-ended fragments via *linker* molecules; in an early example of this, Scheller *et al.* (1977) synthesized self-complementary decameric oligonucleotides, which contain sites for one or more restriction endonucleases. One such molecule is shown in Fig. 3.9. The molecule can be ligated to both ends of the foreign DNA to be cloned, and then treated with restriction endonuclease to produce a sticky-ended fragment, which can be incorporated into a vector molecule that has been cut with the same restriction endonuclease. Insertion by means of the linker creates restriction-enzyme target sites at each end of the foreign DNA and so enables the foreign DNA to be excised and recovered after cloning and amplification in the host bacterium.

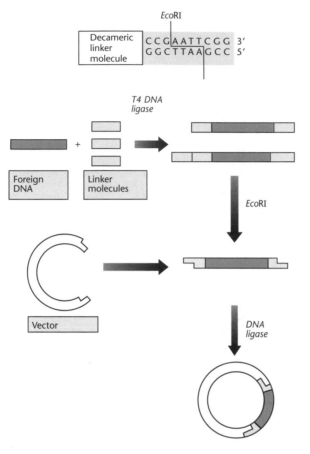

Fig. 3.9 A decameric linker molecule containing an *Eco*RI target site is joined by T4 DNA ligase to both ends of flush-ended foreign DNA. Cohesive ends are then generated by *Eco*RI. This DNA can then be incorporated into a vector that has been treated with the same restriction endonuclease.

Adaptors and linkers are short double-stranded DNA molecules that permit different cleavage sites to be interconnected

It may be the case that the restriction enzyme used to generate the cohesive ends in the linker will also cut the foreign DNA at internal sites. In this situation, the foreign DNA will be cloned as two or more sub-fragments. One solution to this problem is to choose another restriction enzyme, but there may not be a suitable choice if the foreign DNA is large and has sites for several restriction enzymes. Another solution is to methylate internal restriction sites with the appropriate modification methylase. An example of this is described in Fig. 6.2 (p. 99). Alternatively, a general solution to the problem is provided by chemically synthesized adaptor molecules which have a *preformed* cohesive end (Wu *et al.* 1978). Consider a blunt-ended foreign DNA containing an internal *Bam*HI site (Fig. 3.10), which is to be cloned in a *Bam*HI-cut vector. The *Bam*HI adaptor molecule has one blunt end bearing a 5′ phosphate group and a *Bam*HI cohesive end which is not phosphorylated. The adaptor can be ligated to the foreign DNA ends. The foreign DNA plus added adaptors is then phosphorylated at the 5′ termini and ligated into the *Bam*HI site of the vector. If the foreign DNA were to be recovered from the recombinant with *Bam*HI, it would be obtained in two fragments. However, the adaptor is designed to contain two other restriction sites (*Sma*I, *Hpa*II), which may enable the foreign DNA to be recovered intact.

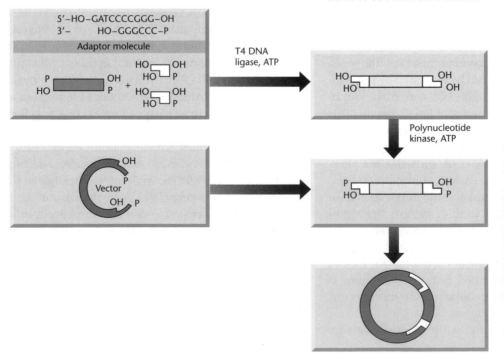

Fig. 3.10 Use of a *Bam*HI adaptor molecule. A synthetic adaptor molecule is ligated to the foreign DNA. The adaptor is used in the 5′-hydroxyl form to prevent self-polymerization. The foreign DNA plus ligated adaptors is phosphorylated at the 5′-termini and ligated into the vector previously cut with *Bam*HI.

Note that the only difference between an adaptor and a linker is that the former has cohesive ends and the latter has blunt ends. A wide range of adaptors is available commercially.

Homopolymer tailing is a general method for joining DNA molecules that has special uses

A general method for joining DNA molecules makes use of the annealing of complementary homopolymer sequences. Thus, by adding oligo(dA) sequences to the 3′ ends of one population of DNA molecules and oligo(dT) blocks to the 3′ ends of another population, the two types of molecule can anneal to form mixed dimeric circles (Fig. 3.11).

An enzyme purified from calf thymus, terminal deoxynucleotidyltransferase, provides the means by which the homopolymeric extensions can be synthesized, for if presented with a single deoxynucleotide triphosphate it will repeatedly add nucleotides to the 3′ OH termini of a population of DNA molecules (Chang & Bollum 1971). DNA with exposed 3′ OH groups, such as arise from pretreatment with phage λ exonuclease or restriction with an enzyme such as *Pst*I, is a very good substrate for the transferase. However, conditions have been found in which the enzyme will extend even the shielded 3′ OH of 5′ cohesive termini generated by *Eco*RI (Roychoudhury *et al.* 1976, Humphries *et al.* 1978).

The terminal transferase reactions have been characterized in detail with regard to their use in gene manipulation (Deng & Wu 1981, Michelson & Orkin 1982). Typically, 10–40 homopolymeric residues are added to each end.

One of the earliest examples of the construction of recombinant molecules, the insertion of a piece of λ DNA into SV40 viral DNA, made use of homopolymer tailing (Jackson *et al.* 1972). In their experiments, the single-stranded gaps which remained in the two strands at each join were repaired *in vitro* with DNA polymerase and DNA ligase so as to produce covalently closed circular molecules. The recombinants were then transfected into susceptible mammalian cells (see Chapter 12). Subsequently, the homopolymer method, using either dA.dT or dG.dC homopolymers was used extensively to construct recombinant plasmids for cloning in *E. coli*. In recent years, homopolymer tailing has been largely replaced as a result of the availability of a much wider range of restriction endonucleases and other DNA-modifying enzymes. However, it is still important for cDNA cloning (see p. 102 *et seq.*).

Special methods are often required if DNA produced by PCR amplification is to be cloned

Many of the strategies for cloning DNA fragments do not work well with PCR products. The reason for this is that the polymerases used in the PCR have a terminal transferase activity. For example, the *Taq*

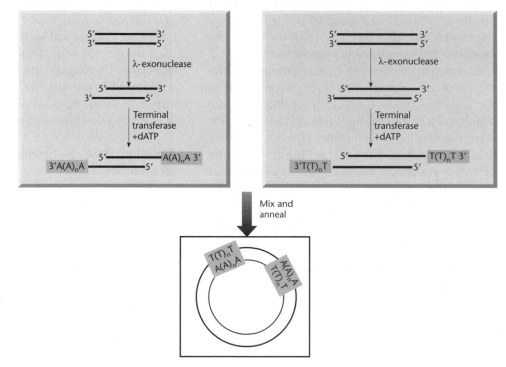

Fig. 3.11 Use of calf-thymus terminal deoxynucleotidyltransferase to add complementary homopolymer tails to two DNA molecules.

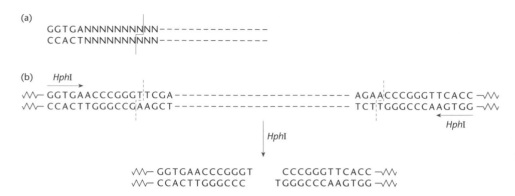

Fig. 3.12 Cleavage of a vector DNA molecule to generate single thymidylate overhangs. (a) The recognition sequence and cleavage point for the restriction endonuclease *Hph*I. (b) Sequences in the vector DNA which result in desired overhangs after cleavage with *Hph*I.

polymerase adds a single 3′ A overhang to each end of the PCR product. Thus PCR products cannot be blunt-end-ligated unless the ends are first *polished* (blunted). A DNA polymerase like Klenow can be used to fill in the ends. Alternatively, *Pfu* DNA polymerase can be used to remove extended bases with its 3′ to 5′ exonuclease activity. However, even when the PCR fragments are polished, blunt-end-ligating them into a vector still may be very inefficient. One solution to this problem is to use T/A cloning (Mead *et al.* 1991). In this method, the PCR fragment is ligated to a vector DNA molecule with a single 3′ deoxythymidylate extension (Fig. 3.12).

A PCR primer may be designed which, in addition to the sequence required for hybridization with the input DNA, includes an extra sequence at its 5′ end. The extra sequence does not participate in the first hybridization step – only the 3′ portion of the primer hybridizes – but it subsequently becomes incorporated into the amplified DNA fragment (Fig. 3.13). Because the extra sequence can be chosen at the will of the experimenter, great flexibility is available here.

A common application of this principle is the incorporation of restriction sites at each end of the amplified product. Figure 3.13 illustrates the addition of a *Hin*dIII site and an *Eco*RI site to the ends of an amplified DNA fragment. In order to ensure that the restriction sites are good substrates for the restriction endonucleases, four nucleotides are placed between the hexanucleotide restriction sites and the extreme ends of the DNA. The incorporation of these restriction sites provides one method for cloning amplified DNA fragments (see below).

DNA molecules can be joined without DNA ligase

In all the cutting and joining reactions described above, two separate protein components were required: a site-specific endonuclease and a DNA ligase. Shuman (1994) has described a novel approach to the synthesis of recombinant molecules in which a single enzyme, vaccinia DNA topoisomerase, both cleaves and rejoins DNA molecules. Placement of the CCCTT cleavage motif for vaccinia topoisomerase near the end of a duplex DNA permits efficient generation of a stable, highly recombinogenic protein DNA adduct that can only religate to acceptor DNAs that contain complementary single-strand extensions. Linear DNAs containing CCCTT cleavage sites at both ends can be activated by topoisomerase and inserted into a plasmid vector.

Heyman *et al.* (1999) have used the properties of vaccinia topoisomerase to develop a ligase-free technology for the covalent joining of DNA fragments to plasmid vectors. Whereas joining molecules with DNA ligase requires an overnight incubation, topoisomerase-mediated ligation occurs in 5 min. The method is particularly suited to cloning PCR fragments. A linearized vector with single 3′ T extensions is activated with the topoisomerase. On addition of the PCR product with 3′ A overhangs, ligation is very rapid. In addition, the high substrate specificity of the enzyme means that there is a low rate of formation of vectors without inserts.

Amplified DNA can be cloned using *in vitro* recombination

Conservative site-specific recombinases (see Box 3.3) are topoisomerases that catalyze rearrangements of DNA at specific sequences that are considerably longer than the cleavage sequence favored by vaccinia topoisomerase. These sites can be incorporated into PCR-amplified fragments by including them at the 5′ end of both of the amplification primers in exactly the same way as shown in Fig. 3.13 for restriction sites. All that is required to clone such

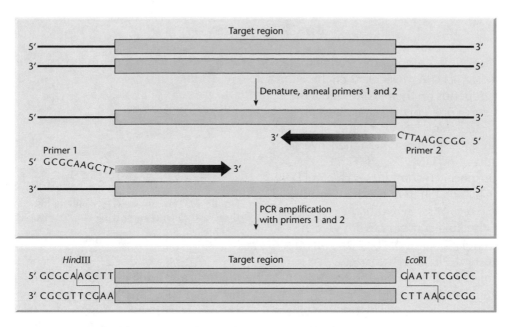

Fig. 3.13 Incorporation of extra sequence at the 5′ end of a primer. Two primers have sequences designed to hybridize at the ends of the target region. Primer 1 has an extra sequence near its 5′ end which forms a *Hind*III site (AAGCTT), and primer 2 has an extra sequence near its 5′ end which forms an *Eco*RI (GAATTC) site. Each primer has an additional 5′-terminal sequence of four nucleotides so that the hexanucleotide restriction sites are placed within the extreme ends of the amplified DNA, and so present good substrates for endonuclease cleavage.

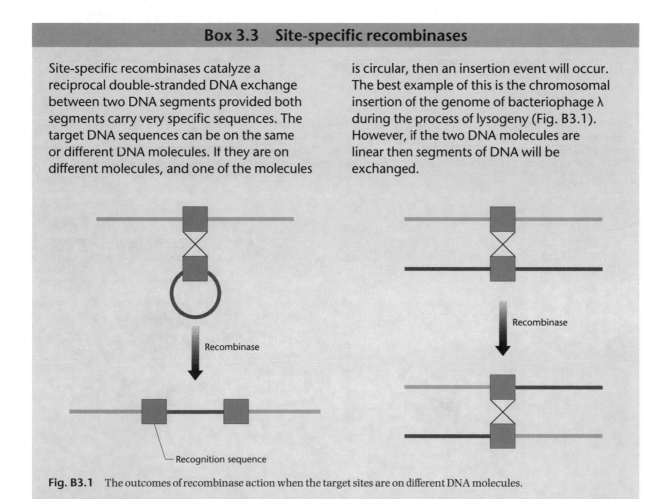

Box 3.3   Site-specific recombinases

Site-specific recombinases catalyze a reciprocal double-stranded DNA exchange between two DNA segments provided both segments carry very specific sequences. The target DNA sequences can be on the same or different DNA molecules. If they are on different molecules, and one of the molecules is circular, then an insertion event will occur. The best example of this is the chromosomal insertion of the genome of bacteriophage λ during the process of lysogeny (Fig. B3.1). However, if the two DNA molecules are linear then segments of DNA will be exchanged.

Fig. B3.1 The outcomes of recombinase action when the target sites are on different DNA molecules.

continued

Box 3.3 *continued*

If the recognition sites for the recombinase are on the same DNA molecule then the outcome depends on their alignment. If the two sites are aligned in the same direction then the intervening DNA is deleted. If the two sites are in opposite orientation then inversion of the intervening DNA occurs (Fig. B3.2).

The site-specific recombinases catalyze DNA exchange by two different mechanisms: the Int–Flp and resolvase–invertase mechanisms. Only the Int–Flp recombinases are used in gene manipulation and the best-known ones are phage λ Int, Flp from the yeast *Saccharomyces cerevisiae* and the Cre protein from bacteriophage P1. With some recombinases, like Cre and Flp, the sites where recombination takes place (Fig. B3.3) are identical and the enzyme can function without any accessory proteins.

With others, such as Int, the sites of recombination exhibit homology but are not

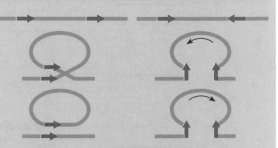

Fig. B3.2 The outcomes of recombinase action when the target sequences are on the same DNA molecule.

identical and additional proteins are required for activity (Fig. B3.4). All of them share a conserved carboxyterminal region that includes the active site tyrosine residue that forms a covalent protein–DNA intermediate during the recombination step. Since recombination does not involve the gain or loss of nucleotides it is said to be conservative.

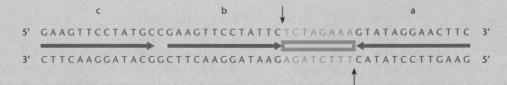

Fig. B3.3 The recognition site (*FRT*) for the Flp recombinases. The site consists of three 13 bp symmetry elements (shown by arrows), one of which (a) is in a different orientation to the other two (b and c). The "a" and "b" elements are separated by an 8-bp asymmetric sequence across which recombination takes place. The "c" element is dispensable. The Cre recognition site (*lox*P) is similar to *FRT* in consisting of two 13-bp inverted repeats separated by an asymmetric 8-bp region (Fig. 15.7).

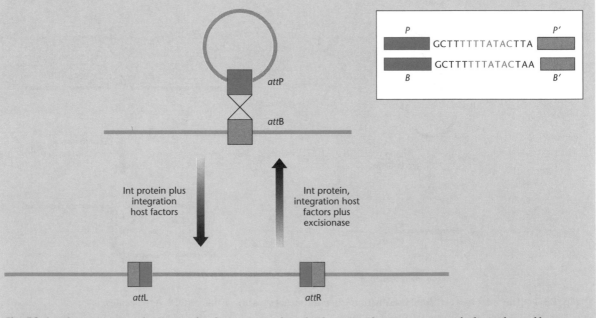

Fig. B3.4 The integration of DNA into the chromosome of *E. coli*. The *att*P and *att*B sites are mostly dissimilar and have only the central 15 bases in common (inset). Recombination occurs between the bases shown in red.

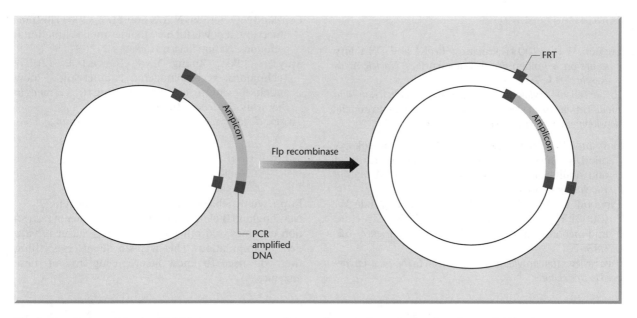

Fig. 3.14 Cloning PCR-amplified DNA using a site-specific recombinase. The DNA to be cloned is amplified using primers containing the recognition site for a site-specific recombinase (in this case, *FRT*, the site for the Flp recombinase). The amplified DNA is mixed with a vector containing two *FRT* sites and the Flp recombinase. The recombinase mediates the replacement of the vector section between the two *FRT* sites with the amplified DNA.

amplified DNA is to mix it with a vector containing two copies of the homologous recognition site (Fig. 3.14). Two variations of this method have been developed. In the original version, developed by Hartley *et al.* (2000), the λ integrase (Int) is used to catalyze the recombination event. This results in a recombinant molecule in which the insert is flanked by different sites than were present on the original amplified DNA. In the variation described by Sadowski (2003), the Flp recombinase is used and this generates recombinants in which the flanking sites remain unchanged.

There are two major applications of cloning with recombinases: recloning and recombineering. Once a PCR amplicon has been cloned it may be necessary to reclone it in another vector and the presence of recombinase sites makes this recloning easy (Fig. 3.15). A series of specialist vectors (the Gateway vectors) have been designed for this purpose and these are described on p. 94. In certain experiments it is necessary to clone very large fragments of DNA (>50 kb). It is difficult to manipulate *in vitro* vectors containing such large inserts and usually it is preferable to do the manipulations *in vivo*. This is known as recombineering and it is facilitated by the use of the Cre and Flp recombinases in conjunction with specialist vectors known as BACs and PACs (see p. 76). Other applications of recombinases are described in the review of Schweizer (2003).

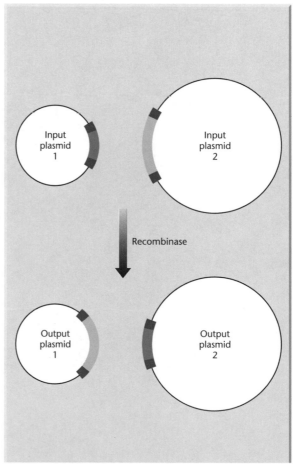

Fig. 3.15 Using recombinases to move cloned DNA from one vector to another.

Suggested reading

Loenen W.A. (2003) Tracking *Eco*KI and DNA fifty years on: a golden era full of surprises. *Nucleic Acids Research* **31**, 7059–69.
An excellent review of host-controlled restriction and modification which provides an historical perspective on the development of restriction enzymes.

Buryanov Y. & Shevchuk T. (2005) The use of prokaryotic DNA methylation transferases as experimental and analytical tools in modern biology. *Analytical Biochemistry* **338**, 1–11.

Pingoud A., Fuxreiter M., Pingoud V. & Wende W. (2005) Type II restriction endonucleases: structure and mechanism. *Cell and Molecular Life Sciences* **62**, 685–707.
Two reviews that provide additional detail to the material covered in this chapter.

Copeland N.G., Jenkins N.A. & Court D.L. (2001) Recombineering: a powerful new tool for mouse functional genomics. *Nature Reviews Genetics* **2**, 769–79.

Muyrers J.P.P., Zhang Y. & Stewart F. (2001) Techniques: recombinogenic engineering – new options for cloning and manipulating DNA. *Trends in Biochemical Sciences* **26**, 325–31.
Two excellent reviews on the topic of recombineering.

Useful website

http://www.neb.com/nebecomm/tech_reference/
New England Biolabs are the premier supplier of restriction enzymes and other enzymes involved in *in vitro* gene manipulation. This website details everything that you need to know before using any of these enzymes.

CHAPTER 4

Basic biology of plasmid and phage vectors

Plasmid biology and simple plasmid vectors

Plasmids are widely used as cloning vehicles but, before discussing their use in this context, it is appropriate to review some of their basic properties. Plasmids are replicons which are stably inherited in an extrachromosomal state. Most plasmids exist as double-stranded circular DNA molecules. If both strands of DNA are intact circles the molecules are described as covalently closed circles or CCC DNA (Fig. 4.1). If only one strand is intact, then the molecules are described as open circles or OC DNA. When isolated from cells, covalently closed circles often have a deficiency of turns in the double helix, such that they have a supercoiled configuration. The enzymatic interconversion of supercoiled, relaxed CCC DNA[1] and OC DNA is shown in Fig. 4.1. Because of their different structural configurations, super-coiled and OC DNA separate upon electrophoresis in agarose gels (Fig. 4.2). Addition of an intercalating agent, such as ethidium bromide, to supercoiled DNA causes the plasmid to unwind. If excess ethidium bromide is added, the plasmid will rewind in the opposite direction (Fig. 4.3). Use of this fact is made in the isolation of plasmid DNA (see p. 59).

Not all plasmids exist as circular molecules. Linear plasmids have been found in a variety of bacteria, e.g. *Streptomyces* sp. and *Borrelia burgdorferi*. To prevent nuclease digestion, the ends of linear plasmids need to be protected, and two general mechanisms have evolved. Either there are repeated sequences ending in a terminal DNA hairpin loop (*Borrelia*) or the ends are protected by covalent attachment of a protein (*Streptomyces*). For more details of linear plasmids the reader should consult Hinnebusch and Tilly (1993).

Plasmids are widely distributed throughout the prokaryotes, vary in size from less than 1×10^6 daltons to greater than 200×10^6, and are generally dispensable. Some of the phenotypes which these plasmids confer on their host cells are listed in Table 4.1. Plasmids to which phenotypic traits have not yet been ascribed are called *cryptic* plasmids.

Plasmids can be categorized into one of two major types – conjugative or non-conjugative – depending upon whether or not they carry a set of transfer genes, called the *tra* genes, which promote bacterial conjugation. Plasmids can also be categorized on the basis of their being maintained as multiple copies per cell (*relaxed* plasmids) or as a limited number of copies per cell (*stringent* plasmids). Generally, conjugative plasmids are of relatively high molecular weight and are present as one to three copies per chromosome, whereas non-conjugative plasmids

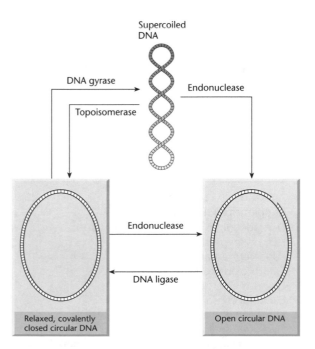

Fig. 4.1 The interconversion of supercoiled, relaxed covalently closed circular DNA and open circular DNA.

---

[1] The reader should not be confused by the terms *relaxed circle* and *relaxed plasmid*. Relaxed circles are CCC DNA that does not have a supercoiled configuration. Relaxed plasmids are plasmids with multiple copies per cell.

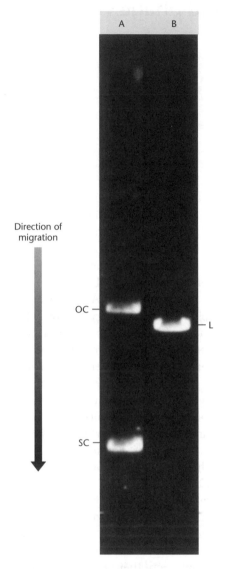

Direction of
migration

Fig. 4.2 Electrophoresis of DNA in agarose gels. The direction of migration is indicated by the arrow. DNA bands have been visualized by soaking the gel in a solution of ethidium bromide (which complexes with DNA by intercalating between stacked base pairs) and photographing the orange fluorescence which results upon ultraviolet irradiation. (A) Open circular (OC) and supercoiled (SC) forms of a plasmid of 6.4 kb pairs. Note that the compact supercoils migrate considerably faster than open circles. (B) Linear plasmid (L) DNA is produced by treatment of the preparation shown in lane (A) with *Eco*RI, for which there is a single target site. Under the conditions of electrophoresis employed here, the linear form migrates just ahead of the open circular form.

are of low molecular weight and are present as multiple copies per chromosome (Table 4.2). An exception is the conjugative plasmid R6K, which has a molecular weight of 25×10^6 daltons and is maintained as a relaxed plasmid.

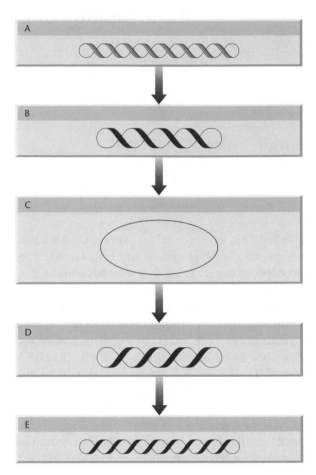

Fig. 4.3 Effect of intercalation of ethidium bromide on supercoiling of DNA. As the amount of intercalated ethidium bromide increases, the double helix untwists, with the result that the supercoiling decreases until the relaxed form of the circular molecule is produced. Further intercalation introduces excess turns in the double helix, resulting in supercoiling in the opposite sense (note the direction of coiling at B and D). For clarity, only a single line represents the double helix.

Table 4.1 Some phenotypic traits exhibited by plasmid-carried genes.

Antibiotic resistance
Antibiotic production
Degradation of aromatic compounds
Hemolysin production
Sugar fermentation
Enterotoxin production
Heavy-metal resistance
Bacteriocin production
Induction of plant tumors
Hydrogen sulfide production
Host-controled restriction and modification

Table 4.2 Properties of some conjugative and non-conjugative plasmids of Gram-negative organisms.

Plasmid	Size (MDa)	Conjugative	No. of plasmid copies/ chromosome equivalent	Phenotype
Col E1	4.2	No	10–15	Col E1 production
RSF1030	5.6	No	20–40	Ampicillin resistance
clo DF13	6	No	10	Cloacin production
R6K	25	Yes	13–38	Ampicillin and streptomycin resistance
F	62	Yes	1–2	–
RI	62.5	Yes	3–6	Multiple drug resistance
Ent P 307	65	Yes	1–3	Enterotoxin production

The host range of plasmids is determined by the replication proteins that they encode

Plasmids encode only a few of the proteins required for their own replication and in many cases encode only one of them. All the other proteins required for replication, e.g. DNA polymerases, DNA ligase, helicases, etc., are provided by the host cell. Those replication proteins that are plasmid-encoded are located very close to the *ori* (origin of replication) sequences at which they act. Thus, only a small region surrounding the *ori* site is required for replication. Other parts of the plasmid can be deleted and foreign sequences can be added to the plasmid and replication will still occur. This feature of plasmids has greatly simplified the construction of versatile cloning vectors.

The host range of a plasmid is determined by its *ori* region. Plasmids whose *ori* region is derived from plasmid Col E1 have a restricted host range: they only replicate in enteric bacteria, such as *E. coli*, *Salmonella*, etc. Other *promiscuous* plasmids have a broad host range and these include RP4 and RSF1010. Plasmids of the RP4 type will replicate in most Gram-negative bacteria, to which they are readily transmitted by conjugation. Such promiscuous plasmids offer the potential of readily transferring cloned DNA molecules into a wide range of genetic backgrounds. Plasmids like RSF1010 are not conjugative but can be transformed into a wide range of Gram-negative and Gram-positive bacteria, where they are stably maintained. Many of the plasmids isolated from *Staphylococcus aureus* also have a broad host range and can replicate in many other Gram-positive bacteria. Plasmids with a broad host range encode most, if not all, of the proteins required for replication. They must also be able to express these genes and thus their promoters and ribosome binding sites must have evolved such that they can be recognized in a diversity of bacterial families.

The number of copies of a plasmid in a cell varies between plasmids and is determined by the regulatory mechanisms controlling replication

The copy number of a plasmid is determined by regulating the initiation of plasmid replication. Two major mechanisms of control of initiation have been recognized: regulation by antisense RNA and regulation by binding of essential proteins to repeated sequences called iterons (for reviews, see Del Solar *et al.* 1998 and Grabherr & Bayer 2002). Most of the cloning vectors in current use carry an *ori* region derived from plasmid Col E1 and copy-number control is mediated by antisense RNA. In this type of plasmid, the primer for DNA replication is a 555-base ribonucleotide molecule called RNA II, which forms an RNA–DNA hybrid at the replication origin (Tomizawa & Itoh 1982). RNA II can only act as a primer if it is cleaved by RNase H to leave a free 3′ hydroxyl group. Unless RNA II is processed in this way, it will not function as a primer and replication will not ensue. Replication control is mediated by another small (108-base) RNA molecule called RNA I (Tomizawa & Itoh 1981), which is encoded by the same region of DNA as RNA II but by the complementary strand. Thus RNA I and RNA II are complementary to each other and can hybridize to form a double-stranded RNA helix. The formation of this duplex interferes with the processing of RNA II by RNase H and hence replication does not ensue (Fig. 4.4). Since RNA I is encoded by the plasmid, more of it will be synthesized when the copy number

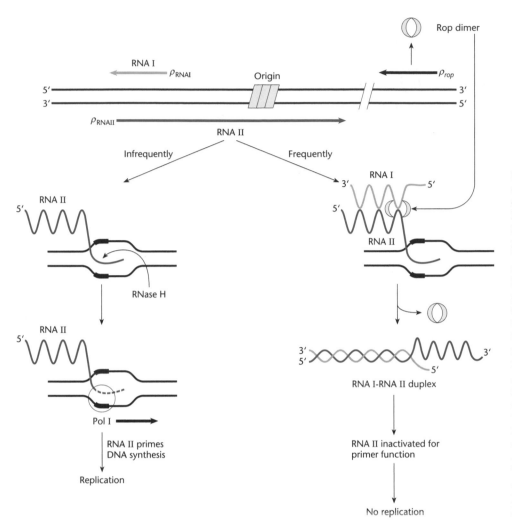

Fig. 4.4 Regulation of replication of Col E1-derived plasmids. RNA II must be processed by RNase H before it can prime replication. "Origin" indicates the transition point between the RNA primer and DNA. Most of the time, RNA I binds to RNA II and inhibits the processing, thereby regulating the copy number. ρ_{RNAI} and ρ_{RNAII} are the promoters for RNA I and RNA II transcription, respectively. RNA I is colored pale purple and RNA II dark purple. The Rop protein dimer enhances the initial pairing of RNA I and RNA II.

of the plasmid is high. As the host cell grows and divides, so the concentration of RNA I will fall and the plasmid will begin to replicate again (Cesarini *et al.* 1991, Eguchi *et al.* 1991).

In addition to RNA I, a plasmid-encoded protein called Rop helps maintain the copy number (Cesarini *et al.* 1982). This protein, which forms a dimer, enhances the pairing between RNA I and RNA II so that processing of the primer can be inhibited even at relatively low concentrations of RNA I. Deletion of the *ROP* gene (Twigg & Sherratt 1980) or mutations in RNA I (Muesing *et al.* 1981) result in increased copy numbers.

In plasmid pSC101 and many of the broad-host-range plasmids, the *ori* region contains three to seven copies of an iteron sequence which is 17 to 22 bp long. Close to the *ori* region there is a gene, called *repA* in pSC101, which encodes the RepA protein. This protein, which is the only plasmid-encoded pro-

tein required for replication, binds to the iterons and initiates DNA synthesis (Fig. 4.5).

Copy-number control is exerted by two superimposed mechanisms. First, the RepA protein represses its own synthesis by binding to its own promoter region and blocking transcription of its own gene (Ingmer & Cohen 1993). If the copy number is high, synthesis of RepA will be repressed. After cell division, the copy number and concentration of RepA will drop and replication will be initiated. Mutations in the RepA protein can lead to increased copy number (Ingmer & Cohen 1993, Cereghino *et al.* 1994). Secondly, the RepA protein can link two plasmids together, by binding to their iteron sequences, thereby preventing them from initiating replication. By this mechanism, known as *handcuffing* (McEachern *et al.* 1989), the replication of iteron plasmids will depend both on the concentration of RepA protein and on the concentration of the plasmids themselves.

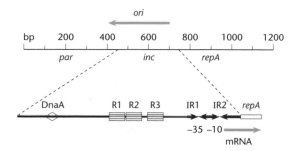

Fig. 4.5 The *ori* region of pSC101. R1, R2, and R3 are the three iteron sequences (CAAAGGTCTAGCAGCAGAATTTACAGA for R3) to which RepA binds to handcuff two plasmids. RepA autoregulates its own synthesis by binding to the inverted repeats IR1 and IR2. The location of the partitioning site *par* and the binding sites for the host protein DnaA are also shown.

The stable maintenance of plasmids in cells requires a specific partitioning mechanism

The loss of plasmids due to defective partitioning (segregation when the cell divides) is called segregative instability. Naturally occurring plasmids are stably maintained because they contain a partitioning function, *par*, which ensures that they are stably maintained at each cell division. Such *par* regions are essential for stability of low-copy-number plasmids (for review, see Bingle & Thomas 2001). The higher-copy-number plasmid Col E1 also contains a *par* region but this is deleted in many Col E1-derived cloning vectors, e.g. pBR322. Although the copy number of vectors such as pBR322 is usually high, plasmid-free cells arise under nutrient limitation or other stress conditions (Jones *et al.* 1980, Nugent *et al.* 1983). The *par* region from a plasmid such as pSC101 can be cloned into pBR322, thereby stabilizing the plasmid (Primrose *et al.* 1983).

DNA superhelicity is involved in the partitioning mechanism (Miller *et al.* 1990). pSC101 derivatives lacking the *par* locus show decreased overall superhelical density as compared with wild-type pSC101. Partition-defective mutants of pSC101 and similar mutants of unrelated plasmids are stabilized in *Escherichia coli* by *top*A mutations, which increase negative DNA supercoiling. Conversely, DNA gyrase inhibitors and mutations in DNA gyrase increase the rate of loss of *par*-defective pSC101 derivatives.

Plasmid instability may also arise due to the formation of multimeric forms of a plasmid. The mechanism that controls the copy number of a plasmid ensures a fixed number of plasmid origins per bacterium. Cells containing multimeric plasmids have the same number of plasmid origins but fewer plasmid molecules, which leads to segregative instability if they lack a partitioning function. These multimeric forms are not seen with Col E1, which has a natural method of resolving multimers back to monomers. It contains a highly recombinogenic site (*cer*). If the *cer* sequence occurs more than once in a plasmid, as in a multimer, the host-cell Xer protein promotes recombination, thereby regenerating monomers (Summers & Sherratt 1984, Guhathakurta *et al.* 1996; for review, see Summers 1998).

Plasmids with similar replication and partitioning systems cannot be maintained in the same cell

Plasmid *incompatibility* is the inability of two different plasmids to coexist in the same cell in the absence of selection pressure. The term incompatibility can only be used when it is certain that entry of the second plasmid has taken place and that DNA restriction is not involved. Groups of plasmids which are mutually incompatible are considered to belong to the same incompatibility (Inc) group. Over 30 incompatibility groups have been defined in *E. coli* and 13 for plasmids of *S. aureus*. Plasmids will be incompatible if they have the same mechanism of replication control. Not surprisingly, by changing the sequence of the RNA I/RNA II region of plasmids with antisense control of copy number, it is possible to change their incompatibility group. Alternatively, they will be incompatible if they share the same *par* region (Austin & Nordstrom 1990, Firsheim & Kim 1997).

The purification of plasmid DNA

An obvious prerequisite for cloning in plasmids is the purification of the plasmid DNA. Although a wide range of plasmid DNAs are now routinely purified, the methods used are not without their problems. Undoubtedly the trickiest stage is the lysis of the host cells; both incomplete lysis and total dissolution of the cells result in greatly reduced recoveries of plasmid DNA. The ideal situation occurs when each cell is just sufficiently broken to permit the plasmid DNA to escape without too much contaminating chromosomal DNA. Provided the lysis is done gently, most of the chromosomal DNA released will be of high molecular weight and can be removed, along with cell debris, by high-speed centrifugation to yield a *cleared lysate*. The production of satisfactory cleared lysates from bacteria other than *E. coli*, particularly if large plasmids are to be isolated, is frequently a combination of skill, luck, and patience.

Many methods are available for isolating pure plasmid DNA from cleared lysates but only two will be described here. The first of these is the "classical" method and is due to Vinograd (Radloff *et al.* 1967). This method involves isopycnic centrifugation of cleared lysates in a solution of CsCl containing ethidium bromide (EtBr). EtBr binds by intercalating between the DNA base pairs, and in so doing causes the DNA to unwind. A CCC DNA molecule, such as a plasmid, has no free ends and can only unwind to a limited extent, thus limiting the amount of EtBr bound. A linear DNA molecule, such as fragmented chromosomal DNA, has no such topological constraints and can therefore bind more of the EtBr molecules. Because the density of the DNA–EtBr complex decreases as more EtBr is bound, and because more EtBr can be bound to a linear molecule than to a covalent circle, the covalent circle has a higher density at saturating concentrations of EtBr. Thus covalent circles (i.e. plasmids) can be separated from linear chromosomal DNA (Fig. 4.6).

Currently the most popular method of extracting and purifying plasmid DNA is that of Birnboim and Doly (1979). This method makes use of the observation that there is a narrow range of pH (12.0–12.5)

within which denaturation of linear DNA, but not covalently closed circular DNA, occurs. Plasmid-containing cells are treated with lysozyme to weaken the cell wall and then lysed with sodium hydroxide and sodium dodecyl sulfate (SDS). Chromosomal DNA remains in a high-molecular-weight form but is denatured. Upon neutralization with acidic sodium acetate, the chromosomal DNA renatures and aggregates to form an insoluble network. Simultaneously, the high concentration of sodium acetate causes precipitation of protein–SDS complexes and of high-molecular-weight RNA. Provided the pH of the alkaline denaturation step has been carefully controlled, the CCC plasmid DNA molecules will remain in a native state and in solution, while the contaminating macromolecules co-precipitate. The precipitate can be removed by centrifugation and the plasmid concentrated by ethanol precipitation. If necessary, the plasmid DNA can be purified further by gel filtration.

A number of commercial suppliers of convenience molecular biology products have developed kits to improve the yield and purity of plasmid DNA. All of them take advantage of the benefits of alkaline lysis and have as their starting point the cleared lysate. With most of these kits, the plasmid DNA is selectively bound to an ion-exchange resin in the presence of a chaotropic agent (e.g. guanidinium isothiocyanate). The purified plasmid DNA is eluted in a small volume after washing away the contaminants. If traces of guanidinium isothiocyanate, or any other solvents that are used, contaminate the plasmid DNA then they can be inhibitory to the PCR and other enzymic reactions. This problem can be eliminated by the use of Charge Switch™ technology (European Patent EP98957019.7) that involves materials that behave as pH-dependent ionic switches. At low pH values, these materials are positively charged and bind DNA selectively. On raising the pH, the charge is switched off and the DNA is eluted free of all other reagents.

The yield of plasmid is affected by a number of factors. The first of these is the actual copy number inside the cells at the time of harvest. The copy-number control systems described earlier are not the only factors affecting yield. The copy number is also affected by the growth medium, the stage of growth and the genotype of the host cell (Nugent *et al.* 1983, Seelke *et al.* 1987, Duttweiler & Gross 1998). The second and most important factor is the care taken in making the cleared lysate. Unfortunately, the commercially available kits have not removed the vagaries of this procedure. Finally, the presence in the host

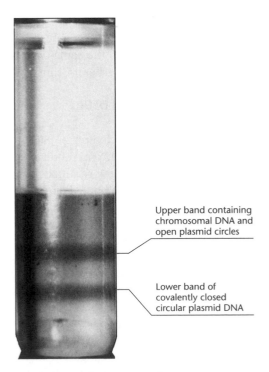

Upper band containing chromosomal DNA and open plasmid circles

Lower band of covalently closed circular plasmid DNA

Fig. 4.6 Purification of Col E1 *Kan*^R plasmid DNA by isopycnic centrifugation in a CsCl–EtBr gradient. (Photograph by courtesy of Dr. G. Birnie.)

cell of a wild-type *endA* gene can affect the recovery of plasmid. The product of the *endA* gene is endonuclease I, a periplasmic protein whose substrate is double-stranded DNA. The function of endonuclease I is not fully understood. Strains bearing *endA* mutations have no obvious phenotype other than improved stability and yield of plasmid obtained from them.

Although most cloning vehicles are of low molecular weight (see next section), it is sometimes necessary to use the much larger conjugative plasmids. Although these high-molecular-weight plasmids can be isolated by the methods just described, the yields are often very low. Either there is inefficient release of the plasmids from the cells as a consequence of their size or there is physical destruction caused by shear forces during the various manipulative steps. A number of alternative procedures have been described (Gowland & Hardmann 1986), many of which are a variation on that of Eckhardt (1978). Bacteria are suspended in a mixture of Ficoll and lysozyme and this results in a weakening of the cell walls. The samples are then placed in the slots of an agarose gel, where the cells are lysed by the addition

of detergent. The plasmids are subsequently extracted from the gel following electrophoresis. The use of agarose, which melts at low temperature, facilitates extraction of the plasmid from the gel.

Good plasmid cloning vehicles share a number of desirable features

An ideal cloning vehicle would have the following three properties:

* low molecular weight;
* ability to confer readily selectable phenotypic traits on host cells;
* single sites for a large number of restriction endonucleases, preferably in genes with a readily scorable phenotype (see Box 4.1).

The advantages of a low molecular weight are several. First, the plasmid is much easier to handle, i.e. it is more resistant to damage by shearing, and is readily isolated from host cells. Secondly, low-molecular-weight plasmids are usually present as

---

Box 4.1 Selectable markers and reporter genes for use with bacterial vectors

The judicious choice of markers on cloning vectors can greatly simplify the selection and analysis of recombinant clones. A key step in any cloning procedure is the selection of transformants carrying the desired recombinant plasmid. Because transformation efficiencies are so low it is essential to be able to select positively those rare cells that have been transformed. The commonest selectable markers are ones encoding resistance to antibiotics such as ampicillin (Ap^R), chloramphenicol (Cm^R), tetracycline (Tc^R), streptomycin (Sm^R), and kanamycin (Km^R). Another type of positive selection is reversal of auxotrophy. For example, if the *hisB*$^+$ gene is cloned in a vector then it is easy to select recombinants by transforming a *hisB* auxotroph and growing it in a medium lacking histidine. The *supE* (also known as *glnV*) and *supF* (also known as *tyrT*) genes encode tRNA molecules that suppress amber (UAG) mutations in essential genes of the vector host cells. Some other markers are useful because their *inactivation* can be positively selected. Two such markers are:

sacB The *sacB* gene encodes levansucrase and its activity is lethal to cells growing on medium containing 7% sucrose.

ccdB The *ccdB* gene is lethal to host cells unless they carry the DNA gyrase mutation *gyrA*.

Reporter genes are ones whose phenotype can be discerned by visual examination of colonies growing on a plate and/or ones that can be used to measure levels of gene expression. In terms of analysis of recombinants, the most widely used reporter gene is the *lacZ* gene encoding β-galactosidase. As noted in Box 3.2, the presence or absence of β-galactosidase activity is easily detected by growing cells on medium containing the chromogenic substrate Xgal. Another gene whose activity can be detected in a similar way is the *gusA* gene encoding β-glucuronidase. This enzyme cleaves 4-methylumbelliferyl-β-D-glucuronide (MUG) to release a blue pigment. Further reporter genes are discussed in Chapters 12, 14, and 15.

multiple copies (see Table 4.2), and this not only facilitates their isolation but leads to gene dosage effects for all cloned genes. Finally, with a low molecular weight there is less chance that the vector will have multiple substrate sites for any restriction endonuclease (see below).

After a piece of foreign DNA is inserted into a vector, the resulting chimeric molecules have to be transformed into a suitable recipient. Since the efficiency of transformation is so low, it is essential that the chimeras have some readily scorable phenotype. Usually this results from some gene, e.g. antibiotic resistance, carried on the vector, but could also be produced by a gene carried on the inserted DNA.

One of the first steps in cloning is to cut the vector DNA and the DNA to be inserted with either the same endonuclease or ones producing the same ends. If the vector has more than one site for the endonuclease, more than one fragment will be produced. When the two samples of cleaved DNA are subsequently mixed and ligated, the resulting chimeras will, in all probability, lack one of the vector fragments. It is advantageous if insertion of foreign DNA at endonuclease-sensitive sites inactivates a gene whose phenotype is readily scorable, for in this way it is

possible to distinguish chimeras from cleaved plasmid molecules which have self-annealed. Of course, readily detectable insertional inactivation is not essential if the vector and insert are to be joined by the homopolymer tailing method (see p. 49) or if the insert confers a new phenotype on host cells.

pBR322 is an early example of a widely used, purpose-built cloning vector

In early cloning experiments, the cloning vectors used were natural plasmid such as Col E1 and pSC101. While these plasmids are small and have single sites for the common restriction endonucleases, they have limited genetic markers for selecting transformants. For this reason, considerable effort was expended on constructing, *in vitro*, superior cloning vectors. The best, and most widely used of these early purpose-built vectors is pBR322. Plasmid pBR322 contains the Ap^R and Tc^R genes of RSF2124 and pSC101, respectively, combined with replication elements of pMB1, a Col E1-like plasmid (Fig. 4.7a). The origins of pBR322 and its progenitor, pBR313, are shown in Fig. 4.7b, and details of its construction can be found in the papers of Bolivar *et al.* (1977a,b).

(a)

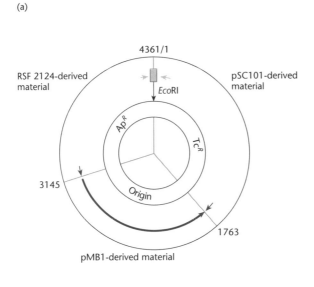

(b)

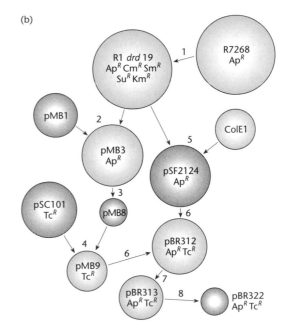

Fig. 4.7 The origins of plasmid pBR322. (a) The boundaries between the pSC101, pMB1, and RSF2124-derived material. The numbers indicate the positions of the junctions in base pairs from the unique *Eco*RI site. (b) The molecular origins of plasmid pBR322. R7268 was isolated in London in 1963 and later renamed R1. 1, A variant, R1*drd*19, which was derepressed for mating transfer, was isolated. 2, The Ap^R transposon, Tn3, from this plasmid was transposed on to pMB1 to form pMB3. 3, This plasmid was reduced in size by *Eco*RI* rearrangement to form a tiny plasmid, pMB8, which carries only colicin immunity. 4, *Eco*RI* fragments from pSC101 were combined with pMB8 opened at its unique *Eco*RI site and the resulting chimeric molecule rearranged by *Eco*RI* activity to generate pMB9. 5, In a separate event, the Tn3 of R1*drd*19 was transposed to Col E1 to form pSF2124. 6, The Tn3 element was then transposed to pMB9 to form pBR312. 7, *Eco*RI* rearrangement of pBR312 led to the formation of pBR313, from which (8) two separate fragments were isolated and ligated together to form pBR322. During this series of constructions, R1 and Col E1 served only as carries for Tn3. (Reproduced by courtesy of Dr. G. Sutcliffe and Cold Spring Harbor Laboratory.)

Plasmid pBR322 has been completely sequenced. The original published sequence (Sutcliffe 1979) was 4362 bp long. Position O of the sequence was arbitrarily set between the A and T residues of the *Eco*RI recognition sequence (GAATTC). The sequence was revised by the inclusion of an additional CG base pair at position 526, thus increasing the size of the plasmid to 4363 bp (Backman & Boyer 1983, Peden 1983). Watson (1988) later revised the size yet again, this time to 4361 bp, by eliminating base

pairs at coordinates 1893 and 1915. The most useful aspect of the DNA sequence is that it totally characterizes pBR322 in terms of its restriction sites, such that the exact length of every fragment can be calculated. These fragments can serve as DNA markers for sizing any other DNA fragment in the range of several base pairs up to the entire length of the plasmid.

There are over 40 enzymes with unique cleavage sites on the pBR322 genome (Fig. 4.8). The target

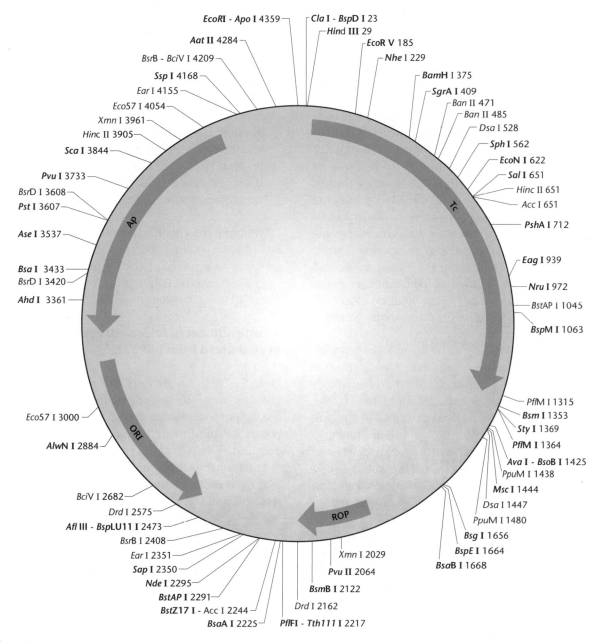

Fig. 4.8   Restriction map of plasmid pBR322 showing the location and direction of transcription of the ampicillin (Ap) and tetracycline (Tc) resistance loci, the origin of replication (*ori*), and the Col E1-derived *Rop* gene. The map shows the restriction sites of those enzymes that cut the molecule once or twice. The unique sites are shown in bold type. The coordinates refer to the position of the 5′ base in each recognition sequence with the first T in the *Eco*RI site being designated as nucleotide number 1. The exact positions of the loci are: Tc, 86–1268; Ap, 4084–3296; Rop, 1918–2105 and the origin of replication, 2535.

sites of 11 of these enzymes lie within the tetracycline resistant (Tc^R) gene, and there are sites for a further two (ClaI and HindIII) within the promoter of that gene. There are unique sites for six enzymes within the ampicillin resistant (Ap^R) gene. Thus, cloning in pBR322 with the aid of any one of those 19 enzymes will result in insertional inactivation of either the Ap^R or the Tc^R markers. However, cloning in the other unique sites does not permit the easy selection of recombinants, because neither of the antibiotic resistance determinants is inactivated.

Following manipulation *in vitro*, *E. coli* cells transformed with plasmids with inserts in the Tc^R gene can be distinguished from those cells transformed with recircularized vector. The former are Ap^R and tetracycline sensitive (Tc^S), whereas the latter are both Ap^R and Tc^R. In practice, transformants are selected on the basis of their Ap resistance and then replica-plated onto Tc-containing media to identify those that are Tc^S. Cells transformed with pBR322 derivatives carrying inserts in the Ap^R gene can be identified more readily (Boyko & Ganschow 1982). Detection is based upon the ability of the β-lactamase produced by Ap^R cells to convert penicillin to penicilloic acid, which in turn binds iodine. Transformants are selected on rich medium containing soluble starch and Tc. When colonized plates are flooded with an indicator solution of iodine and penicillin, β-lactamase-producing (Ap^R) colonies clear the indicator solution whereas ampicillin sensitive (Ap^S) colonies do not.

The PstI site in the Ap^R gene is particularly useful, because the 3′ tetranucleotide extensions formed on digestion are ideal substrates for terminal transferase. Thus this site is excellent for cloning by the homopolymer tailing method described in the previous chapter (see p. 49). If oligo-(dG.dC) tailing is used, the PstI site is regenerated (see Fig. 3.11) and the insert may be cut out with that enzyme.

Plasmid pBR322 has been a widely used cloning vehicle. In addition, it has been widely used as a model system for the study of prokaryotic transcription and translation, as well as investigation of the effects of topological changes on DNA conformation. The popularity of pBR322 is a direct result of the availability of an extensive body of information on its structure and function. This in turn is increased with each new study. The reader wishing more detail on the structural features, transcriptional signals, replication, amplification, stability, and conjugal mobility of pBR322 should consult the review of Balbás *et al.* (1986).

Example of the use of plasmid pBR322 as a vector: isolation of DNA fragments which carry promoters

Cloning into the HindIII site of pBR322 generally results in loss of tetracycline resistance. However, in some recombinants, Tc^R is retained or even increased. This is because the HindIII site lies within the promoter rather than the coding sequence. Thus whether or not insertional inactivation occurs depends on whether the cloned DNA carries a promoter-like sequence able to initiate transcription of the Tc^R gene. Widera *et al.* (1978) have used this technique to search for promoter-containing fragments.

Four structural domains can be recognized within *E. coli* promoters. These are:

* position 1, the purine initiation nucleotide from which RNA synthesis begins;
* position -6 to -12, the Pribnow box;
* the region around base pair -35;
* the sequence between base pairs -12 and -35.

Although the HindIII site lies within the Pribnow box (Rodriguez *et al.* 1979) the box is recreated on insertion of a foreign DNA fragment. Thus when insertional inactivation occurs it must be the region from -13 to -40 which is modified.

A large number of improved vectors have been derived from pBR322

Over the years, numerous different derivatives of pBR322 have been constructed, many to fulfil special-purpose cloning needs. A compilation of the properties of some of these plasmids has been provided by Balbás *et al.* (1986).

Much of the early work on the improvement of pBR322 centered on the insertion of additional unique restriction sites and selectable markers, e.g. pBR325 encodes chloramphenicol resistance in addition to ampicillin and tetracycline resistance and has a unique EcoRI site in the Cm^R gene. Initially, each new vector was constructed in a series of steps analogous to those used in the generation of pBR322 itself (Fig. 4.7). Then the construction of improved vectors was simplified (Vieira & Messing 1982, 1987, Yanisch-Perron *et al.* 1985) by the use of *polylinkers* or *multiple cloning sites* (MCSs), as exemplified by the pUC vectors (Fig. 4.9). An MCS is a short DNA sequence, 2.8 kb in the case of pUC19,

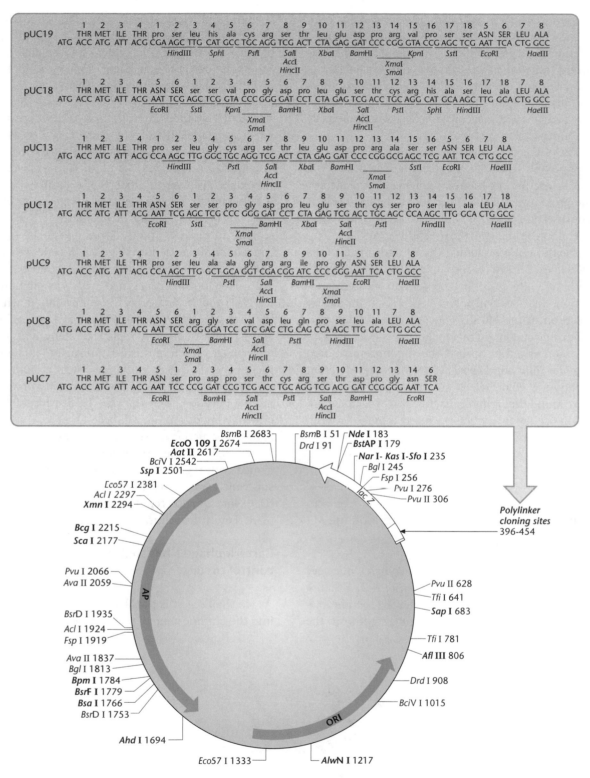

Fig. 4.9 Genetic maps of some pUC plasmids. The multiple cloning site (MCS) is inserted into the *lacZ'* gene but does not interfere with gene function. The additional codons present in the *lacZ'* gene as a result of the polylinker are labeled with lower-case letters. These polylinker regions (MCS) are identical to those of the M13 mp series of vectors (see p. 73).

carrying sites for many different restriction endonucleases. An MCS increases the number of potential cloning strategies available by extending the range of enzymes that can be used to generate a restriction fragment suitable for cloning. By combining them within an MCS, the sites are made contiguous, so that any two sites within it can be cleaved simultaneously without excising vector sequences.

The pUC vectors also incorporate a DNA sequence that permits rapid visual detection of an insert. The MCS is inserted into the *lacZ′* sequence, which encodes the promoter and the α-peptide of β-galactosidase. The insertion of the MCS into the *lacZ′* fragment does not affect the ability of the α-peptide to mediate complementation, but cloning DNA fragments into the MCS does. Therefore, recombinants can be detected by blue/white screening on growth medium containing Xgal (see Box 3.2 on p. 45). The usual site for insertion of the MCS is between the initiator ATG codon and codon 7, a region that encodes a functionally non-essential part of the α-complementation peptide. Slilaty and Lebel (1998) have reported that blue/white color selection can be variable. They have found that reliable inactivation of complementation occurs only when the insert is made between codons 11 and 36.

Bacteriophage λ

The genetic organization of bacteriophage λ favors its subjugation as a vector

Bacteriophage λ is a genetically complex but very extensively studied virus of *E. coli* (Box 4.2). Because it has been the object of so much research in molecular genetics, it was natural that, right from the beginnings of gene manipulation, it should have been investigated and developed as a vector. The DNA of phage λ, in the form in which it is isolated from the phage particle, is a linear duplex molecule of about 48.5 kbp. The entire DNA sequence has been determined (Sanger *et al.* 1982). At each end are short single-stranded 5′ projections of 12 nucleotides, which are complementary in sequence and by which the DNA adopts a circular structure when it is injected into its host cell, i.e. λ DNA naturally has cohesive termini, which associate to form the *cos* site.

Functionally related genes of phage λ are clustered together on the map, except for the two positive regulatory genes *N* and *Q*. Genes on the left of the conventional linear map (Fig. 4.10) code for head and tail proteins of the phage particle. Genes of the central region are concerned with recombination (e.g. *red*) and the process of lysogenization, in which the circularized chromosome is inserted into its host chromosome and stably replicated along with it as a prophage. Much of this central region, including these genes, is not essential for phage growth and can be deleted or replaced without seriously impairing the infectious growth cycle. Its dispensability is crucially important, as will become apparent later, in the construction of vector derivatives of the phage. To the right of the central region are genes concerned with regulation and prophage immunity to superinfection (*N*, *cro*, *cI*), followed by DNA synthesis (*O*, *P*), late function regulation (*Q*), and host cell lysis (*S*, *R*). Figure 4.11 illustrates the λ life cycle.

Bacteriophage λ has sophisticated control circuits

As we shall see, it is possible to insert foreign DNA into the chromosome of phage-λ derivative and, in

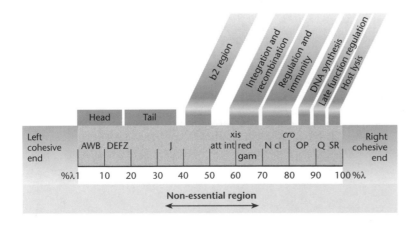

Fig. 4.10 Map of the λ chromosome, showing the physical position of some genes on the full-length DNA of wild-type bacteriophage λ. Clusters of functionally related genes are indicated.

Box 4.2 Bacteriophage λ: its important place in molecular biology and recombinant DNA technology

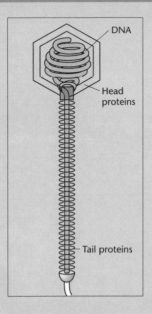

In the early 1950s, following some initial studies on *Bacillus megaterium*, André Lwoff and his colleagues at the Institut Pasteur described the phenomenon of lysogeny in *E. coli*. It became clear that certain strains of *E. coli* were lysogenized by phage, that is to say, these bacteria harbored phage λ in a dormant form, called a prophage. The lysogenic bacteria grew normally and might easily not have been recognized as lysogenic. However, when Lwoff exposed the bacteria to a moderate dose of ultraviolet light, the bacteria stopped growing, and after about 90 min of incubation the bacteria lysed, releasing a crop of phage into the medium.

The released phage were incapable of infecting more *E. coli* that had been lysogenized by phage λ (this is called immunity to superinfection), but non-lysogenic bacteria could be infected to yield another crop of virus. Not every non-lysogenic bacterium yielded virus; some bacteria were converted into lysogens because the phage switched to the dormant lifestyle – becoming prophage – rather than causing a lytic infection.

By the mid-1950s it was realized that the prophage consisted of a phage λ genome that had become integrated into the *E. coli* chromosome. It was also apparent to Lwoff's colleagues, Jacob and Monod, that the switching between the two states of the virus – the lytic and lysogenic lifestyles – was an example of a fundamental aspect of genetics that was gaining increasing attention, gene regulation.

Intensive genetic and molecular biological analysis of the phage, mainly in the 1960s and 1970s, led to a good understanding of the virus. The key molecule in maintaining the dormancy of the prophage and in conferring immunity to superinfection is the phage repressor, which is the product of the phage *cI* gene. In 1967 the phage repressor was isolated by Mark Ptashne (Ptashne 1967a,b). The advanced molecular genetics of the phage made it a good candidate for development as a vector, beginning in the 1970s and continuing to the present day, as described in the text. The development of vectors exploited the fact that a considerable portion of the phage genome encodes functions that are not needed for the infectious cycle. The ability to package recombinant phage DNA into virus particles *in vitro* was an important development for library construction (Hohn & Murray 1977).

A landmark in molecular biology was reached when the entire sequence of the phage λ genome, 48,502 nucleotide pairs, was determined by Fred Sanger and his colleagues (Sanger *et al.* 1982).

some cases, foreign genes can be expressed efficiently via λ promoters. We must therefore briefly consider the promoters and control circuits affecting λ gene expression (see Ptashne (1992) for an excellent monograph on phage-λ control circuits).

In the lytic cycle, λ transcription occurs in three temporal stages: early, middle, and late. Basically, early gene transcription establishes the lytic cycle (in competition with lysogeny), middle gene products replicate and recombine the DNA, and late gene

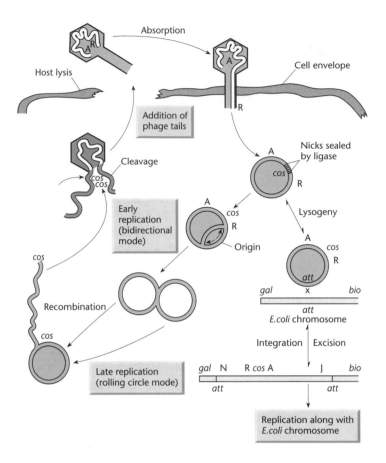

Fig. 4.11 Replication of phage-λ DNA in lytic and lysogenic cycles.

products package this DNA into mature phage particles. Following infection of a sensitive host, early transcription proceeds from major promoters situated immediately to the left (P_L) and right (P_R) of the repressor gene (*cI*) (Fig. 4.12). This transcription is subject to repression by the product of the *cI* gene and in a lysogen this repression is the basis of immunity to superinfecting λ. Early in infection, transcripts from P_L and P_R stop at termination sites t_L and t_{R1}. The site t_{R2} stops any transcripts that escape beyond t_{R1}. Lambda switches from early- to middle-stage transcription by anti-termination. The *N* gene product, expressed from P_L, directs this switch. It interacts with RNA polymerase and, antagonizing the action of host termination protein ρ, permits it to ignore the stop signals so that P_L and P_R transcripts extend into genes such as *red*, *O*, and *P* necessary for the middle stage. The early and middle transcripts and patterns of expression therefore overlap. The *cro* product, when sufficient has accumulated, prevents transcription from P_L and P_R. The gene *Q* is expressed from the distal portion of the extended P_R transcript and is responsible for the middle-to-late switch. This

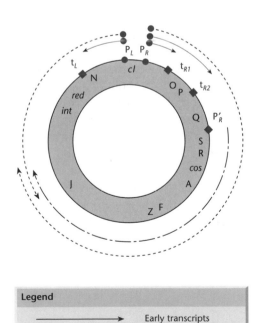

Legend

⟶	Early transcripts
⇢	Middle transcripts
⟶·⟶	Late transcripts

Fig. 4.12 Major promoters and transcriptional termination sites of phage λ. (See text for details.)

also operates by anti-termination. The *Q* product specifically anti-terminates the short P_R transcript, extending it into the late genes, across the cohered *cos* region, so that many mature phage particles are ultimately produced.

Both *N* and *Q* play positive regulatory roles essential for phage growth and plaque formation; but an N^- phage can produce a small plaque if the termination site t_{R2} is removed by a small deletion termed *nin* (*N*-independent) as in λN^- *nin*.

There are two basic types of phage λ vectors: insertional vectors and replacement vectors

Wild-type λ DNA contains several target sites for most of the commonly used restriction endonucleases and so is not itself suitable as a vector. Derivatives of the wild-type phage have therefore been produced that either have a single target site at which foreign DNA can be inserted (*insertional* vectors) or have a pair of sites defining a fragment that can be removed (*stuffer*) and replaced by foreign DNA (*replacement* vectors). Since phage λ can accommodate only about 5% more than its normal complement of DNA, vector derivatives are constructed with deletions to increase the space within the genome. The shortest λ DNA molecules that produce plaques of nearly normal size are 25% deleted. Apparently, if too much non-essential DNA is deleted from the genome, it cannot be packaged into phage particles efficiently. This can be turned to advantage for, if the replaceable fragment of a replacement-type vector is either removed by physical separation or effectively destroyed by treatment with a second restriction endonuclease that cuts it alone, then the deleted vector genome can give rise to plaques only if a new DNA segment is inserted into it. This amounts to positive selection for recombinant phage carrying foreign DNA.

Many vector derivatives of both the insertional and replacement types were produced by several groups of researchers early in the development of recombinant DNA technology (e.g. Thomas *et al.* 1974, Murray & Murray 1975, Blattner *et al.* 1977, Leder *et al.* 1977). Most of these vectors were constructed for use with *Eco*RI, *Bam*HI, or *Hind*III, but their application could be extended to other endonucleases by the use of linker molecules. These early vectors have been largely superseded by improved vectors for rapid and efficient genomic or complementary DNA (cDNA) library construction (see Chapter 6).

A number of phage λ vectors with improved properties have been described

As with plasmid vectors, improved phage-vector derivatives have been developed. There have been several aims, among which are the following.

* To increase the capacity for foreign DNA fragments, preferably for fragments generated by any one of several restriction enzymes (reviewed by Murray 1983).
* To devise methods for positively selecting recombinant formation.
* To allow RNA probes to be conveniently prepared by transcription of the foreign DNA insert; this facilitates the screening of libraries in chromosome walking procedures. An example of a vector with this property is λZAP (see p. 104).
* To develop vectors for the insertion of eukaryotic cDNA (p. 104) such that expression of the cDNA, in the form of a fusion polypeptide with β-galactosidase, is driven in *E. coli*; this form of expression vector is useful in antibody screening. An example of such a vector is λgt11.

The first two points will be discussed here. The discussion of improved vectors in library construction and screening is deferred until Chapter 6.

The maximum capacity of phage-λ derivatives can only be attained with vectors of the replacement type, so that there has also been an accompanying incentive to devise methods for positively selecting recombinant formation without the need for prior removal of the stuffer fragment. Even when steps are taken to remove the stuffer fragment by physical purification of vector arms, small contaminating amounts may remain, so that genetic selection for recombinant formation remains desirable. The usual method of achieving this is to exploit the Spi⁻ phenotype.

Wild-type λ cannot grow on *E. coli* strains lysogenic for phage P2; in other words, the λ phage is Spi⁺ (sensitive to P2 inhibition). It has been shown that the products of λ genes *red* and *gam*, which lie in the region 64–69% on the physical map, are responsible for the inhibition of growth in a P2 lysogen (Herskowitz 1974, Sprague *et al.* 1978, Murray 1983). Hence vectors have been derived in which the stuffer fragment includes the region 64–69%, so that recombinants in which this has been replaced by foreign DNA are phenotypically Spi⁻ and can be positively selected by plating on a P2 lysogen (Karn *et al.* 1984, Loenen & Brammar 1980).

Deletion of the *gam* gene has other consequences. The *gam* product is necessary for the normal switch in λ DNA replication from the bidirectional mode to the rolling-circle mode (see Fig. 4.11). Gam⁻ phage cannot generate the concatemeric linear DNA which is normally the substrate for packaging into phage heads. However, *gam*⁻ phage do form plaques because the *rec* and *red* recombination systems act on circular DNA molecules to form multimers, which can be packaged. *gam*⁻ *red*⁻ phage are totally dependent upon *rec*-mediated exchange for plaque formation on *rec*⁺ bacteria. λ DNA is a poor substrate for this *rec*-mediated exchange. Therefore, such phage make vanishingly small plaques unless they contain one or more short DNA sequences called *chi* (crossover hot-spot instigator) sites, which stimulate *rec*-mediated exchange. Many of the current replacement vectors generate *red*⁻ *gam*⁻ clones and so have been constructed with a *chi* site within the non-replaceable part of the phage.

The most recent generation of λ vectors, which are based on EMBL3 and EMBL4 (Frischauf *et al.* 1983; Fig. 4.13), have a capacity for DNA of size 9–23 kb. As well as being *chi*⁺, they have polylinkers flanking the stuffer fragment to facilitate library construction. Phages with inserts can be selected on the basis of their Spi⁻ phenotype, but there is an alternative. The vector can be digested with *Bam*HI and *Eco*RI prior to ligation with foreign DNA fragments produced with *Bam*HI. If the small *Bam*HI–*Eco*RI fragments from the polylinkers are removed, the stuffer fragment will not be reincorporated.

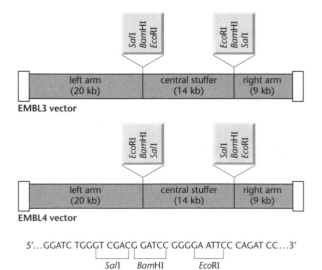

Fig. 4.13 The structure of bacteriophage λ cloning vectors EMBL3 and EMBL4. The polylinker sequence is present in opposite orientation in the two vectors.

By packaging DNA into phage λ *in vitro* it is possible to eliminate the need for competent cells of *E. coli*

So far, we have considered only one way of introducing manipulated phage DNA into the host bacterium, i.e. by transfection of competent bacteria (see Chapter 2). Transfection is the introduction of naked phage DNA into a bacterial cell. Using freshly prepared λ DNA that has not been subjected to any gene-manipulation procedures, transfection will result typically in about 10^5 plaques/µg of DNA. In a gene-manipulation experiment in which the vector DNA is restricted and then ligated with foreign DNA, this figure is reduced to about 10^4–10^3 plaques/µg of vector DNA. Even with perfectly efficient nucleic acid biochemistry, some of this reduction is inevitable. It is a consequence of the random association of fragments in the ligation reaction, which produces molecules with a variety of fragment combinations, many of which are inviable. Yet, in some contexts, 10^6 or more recombinants are required. The scale of such experiments can be kept within a reasonable limit by packaging the recombinant DNA into mature phage particles *in vitro*.

Placing the recombinant DNA in a phage coat allows it to be introduced into the host bacteria by the normal processes of phage infection, i.e. phage adsorption followed by DNA injection. This is known as transduction. Depending upon the details of the experimental design, packaging *in vitro* yields about 10^6 plaques/µg of vector DNA after the ligation reaction.

Figure 4.14 shows some of the events occurring during the packaging process that take place within the host during normal phage growth and which we now require to perform *in vitro*. Phage DNA in concatemeric form, produced by a rolling-circle replication mechanism (see Fig. 4.11), is the substrate for the packaging reaction. In the presence of phage head precursor (the product of gene *E* is the major capsid protein) and the product of gene *A*, the concatemeric DNA is cleaved into monomers and encapsidated. Nicks are introduced in opposite strands of the DNA, 12 nucleotide pairs apart at each *cos* site, to produce the linear monomer with its cohesive termini. The product of gene *D* is then incorporated into what now becomes a completed phage head. The products of genes *W* and *FII*, among others, then unite the head with a separately assembled tail structure to form the mature particle.

The principle of packaging *in vitro* is to supply the ligated recombinant DNA with high concentrations

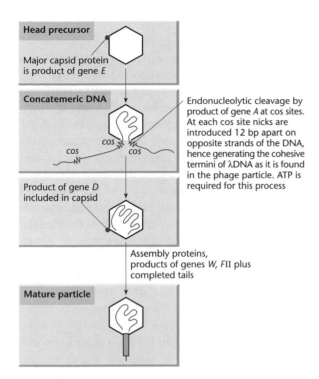

Fig. 4.14 Simplified scheme showing packaging of phage-λ DNA into phage particles.

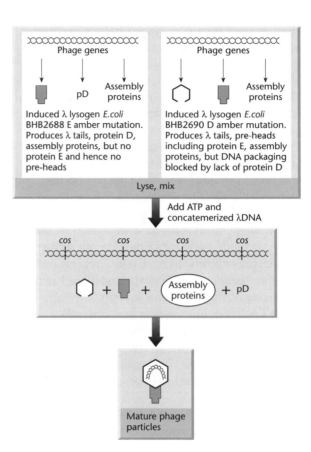

Fig. 4.15 *In vitro* packaging of concatemerized phage-λ DNA in a mixed lysate.

of phage-head precursor, packaging proteins, and phage tails. Practically, this is most efficiently performed in a very concentrated mixed lysate of two induced lysogens, one of which is blocked at the pre-head stage by an amber mutation in gene *D* and therefore accumulates this precursor, while the other is prevented from forming any head structure by an amber mutation in gene *E* (Hohn & Murray 1977). In the mixed lysate, genetic complementation occurs and exogenous DNA is packaged (Fig. 4.15). Although concatemeric DNA is the substrate for packaging (covalently joined concatemers are, of course, produced in the ligation reaction by association of the natural cohesive ends of λ), the *in vitro* system will package added monomeric DNA, which presumably first concatemerizes non-covalently.

There are two potential problems associated with packaging *in vitro*. First, endogenous DNA derived from the induced prophages of the lysogens used to prepare the packaging lysate can itself be packaged. This can be overcome by choosing the appropriate genotype for these prophages, i.e. excision upon induction is inhibited by the *b2* deletion (Gottesmann & Yarmolinsky 1968) and *imm 434* immunity will prevent plaque formation if an *imm 434* lysogenic bacterium is used for plating the complex reaction mixture. Additionally, if the vector does not contain

any amber mutation a non-suppressing host bacterium can be used so that endogenous DNA will not give rise to plaques. The second potential problem arises from recombination in the lysate between exogenous DNA and induced prophage markers. If troublesome, this can be overcome by using recombination-deficient (i.e. *red⁻ rec⁻*) lysogens and by UV-irradiating the cells used to prepare the lysate, so eliminating the biological activity of the endogenous DNA (Hohn & Murray 1977).

DNA cloning with single-stranded DNA vectors

M13, f1, and fd are filamentous coliphages containing a circular single-stranded DNA molecule. These coliphages have been developed as cloning vectors, for they have a number of advantages over other vectors, including the other two classes of vector for *E. coli*: plasmids and phage λ. However, in order to appreciate their advantages, it is essential to have a basic understanding of the biology of filamentous phages.

Filamentous bacteriophages have a number of unique properties that make them suitable as vectors

The phage particles have dimensions 900 nm × 9 nm and contain a single-stranded circular DNA molecule, which is 6407 (M13) or 6408 (fd) nucleotides long. The complete nucleotide sequences of fd and M13 are available and they are 97% identical. The differences consist mainly of isolated nucleotides here and there, mostly affecting the redundant bases of codons, with no blocks of sequence divergence.

The filamentous phages only infect strains of enteric bacteria harboring F pili. The adsorption site appears to be the end of the F pilus, but exactly how the phage genome gets from the end of the F pilus to the inside of the cell is not known. Replication of phage DNA does not result in host-cell lysis. Rather, infected cells continue to grow and divide, albeit at a slower rate than uninfected cells, and extrude virus particles. Up to 1000 phage particles may be released into the medium per cell per generation (Fig. 4.16).

The single-stranded phage DNA enters the cell by a process in which decapsidation and replication are tightly coupled. The capsid proteins enter the cytoplasmic membrane as the viral DNA passes into the cell while being converted to a double-stranded replicative form (RF). The RF multiplies rapidly until about 100 RF molecules are formed inside the cell. Replication of the RF then becomes asymmetric, due to the accumulation of a viral-encoded single-stranded specific DNA-binding protein. This protein binds to the viral strand and prevents synthesis of the complementary strand. From this point on, only viral single strands are synthesized. These progeny single strands are released from the cell as filamentous particles following morphogenesis at the cell membrane. As the DNA passes through the membrane, the DNA-binding protein is stripped off and replaced with capsid protein.

Vectors with single-stranded DNA genomes have specialist uses

Single-stranded DNA is required for several applications of cloned DNA. Sequencing by the original dideoxy method required single-stranded DNA, as do techniques for oligonucleotide-directed mutagenesis and certain methods of probe preparation. The use of vectors that occur in single-stranded form is an attractive means of combining the cloning, amplification, and strand separation of an originally double-stranded DNA fragment.

As single-stranded vectors, the filamentous phages have a number of advantages. First, the phage DNA is replicated via a double-stranded circular DNA (RF) intermediate. This RF can be purified and manipulated *in vitro* just like a plasmid. Secondly, both RF and single-stranded DNA will transfect competent *E. coli* cells to yield either plaques or infected colonies, depending on the assay method. Thirdly, the size of the phage particle is governed by the size of the viral DNA and therefore there are no packaging constraints. Indeed, viral DNA up to six times the length of M13 DNA has been packaged (Messing *et al.* 1981). Finally, with these phages it is very easy to determine the orientation of an insert. Although the relative orientation can be determined from restriction analysis of RF, there is an easier method (Barnes 1980). If two clones carry the insert in opposite directions, the single-stranded DNA from them will hybridize and this can be detected by agarose gel electrophoresis. Phage from as little as 0.1 ml of culture can be used in assays of this sort, making mass screening of cultures very easy.

In summary, as vectors, filamentous phages possess all the advantages of plasmids while producing particles containing single-stranded DNA in an easily obtainable form.

Phage M13 has been modified to make it a better vector

Unlike λ, the filamentous coliphages do not have any non-essential genes which can be used as cloning sites. However, in M13 there is a 507 bp intergenic region, from position 5498 to 6005 of the DNA sequence, which contains the origins of DNA replication for both the viral and the complementary strands. In most of the vectors developed so far, foreign DNA has been inserted at this site, although it is possible to clone at the carboxy-terminal end of gene IV (Boeke *et al.* 1979). The wild-type phages are not very promising as vectors because they contain very few unique sites within the intergenic region: *Asu*I in the case of fd, and *Asu*I and *Ava*I in the case of M13.

The first example of M13 cloning made use of one of 10 *Bsu*I sites in the genome, two of which are in the intergenic region (Messing *et al.* 1977). For cloning, M13 RF was partially digested with *Bsu*I and linear full-length molecules isolated by agarose gel electrophoresis. These linear monomers were

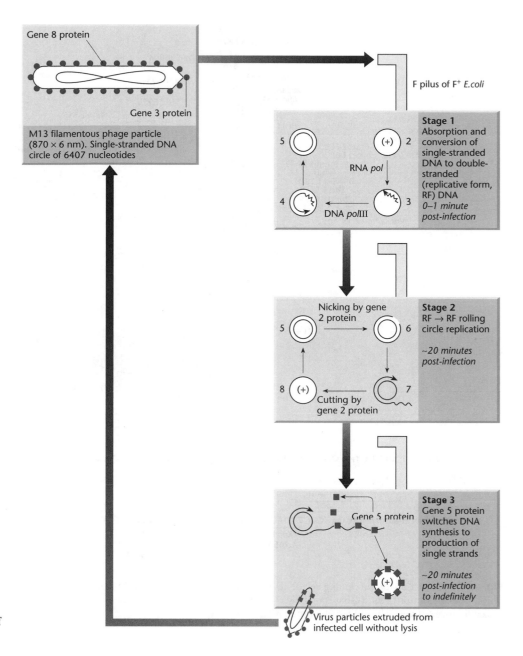

Fig. 4.16 Life cycle and DNA replication of phage M13.

blunt-end-ligated to a *Hind*II restriction fragment comprising the *E. coli lac* regulatory region and the genetic information for the α-peptide of β-galactosidase. The complete ligation mixture was used to transform a strain of *E. coli* with a deletion of the β-galactosidase α-fragment and recombinant phage detected by intragenic complementation on media containing IPTG and Xgal (see Box 3.2 on p. 45). One of the blue plaques was selected and the virus in it designated M13 mp1.

Insertion of DNA fragments into the *lac* region of M13 mp1 destroys its ability to form blue plaques, making detection of recombinants easy. However,

the *lac* region only contains unique sites for *Ava*II, *Bgl*I, and *Pvu*I and three sites for *Pvu*II, and there are no sites anywhere on the complete genome for the commonly used enzymes such as *Eco*RI or *Hind*III. To remedy this defect, Gronenborn and Messing (1978) used *in vitro* mutagenesis to change a single base pair, thereby creating a unique *Eco*RI site within the *lac* fragment. This variant was designated M13 mp2. This phage derivative was further modified to generate derivatives with polylinkers upstream of the lac α-fragment (Fig. 4.17). These derivatives (mp7–mp11, mp18, mp19) are the exact M13 counterparts of the pUC plasmids shown in Fig. 4.9.

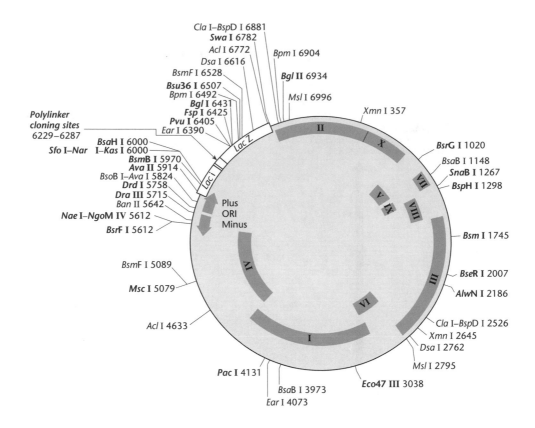

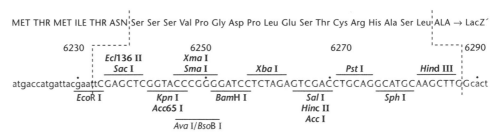

Fig. 4.17 Restriction map of cloning vector M13mp18. Phages M13mp18 and M13mp19 are 7249 bases in length and differ only in the orientation of the 54-base polylinker that they carry. The map shows the restriction sites of enzymes that cut the molecule once or twice. The unique sites are shown in bold type.

Suggested reading

Balbas P. & Bolivar F. (2004) Back to basics: PBR322 and protein expression on systems in *E. coli*. *Methods in Molecular Biology* **267**, 77–90.

Christensen A.C. (2001) Bacteriophage lambda-based expression vectors. *Molecular Biotechnology* **17**, 219–24.

A short review that covers much of the same ground as discussed in this book.

Grabherr R. & Bayer K. (2002) Impact of targeted vector design on ColE1 plasmid replication. *Trends in Biotechnology* **20**, 257–60.

Most of the vectors used in E. coli *are derived from plasmid ColE1 and this review discusses the factors that influence the plasmid copy number.*

Stadler J., Lemmens R. & Nyhammar T. (2004) Plasmid DNA purification. *Journal of Gene Medicine* **6**, Supplement 1, 554–66.

This review discusses plasmid purification from the perspective of large-scale manufacture for gene therapy purposes.

CHAPTER 5

Cosmids, phasmids, and other advanced vectors

Introduction

In the 1970s, when recombinant DNA technology was first being developed, only a limited number of vectors were available and these were based on either high-copy-number plasmids or phage λ. Later, phage M13 was developed as a specialist vector to facilitate DNA sequencing. Gradually, a number of purpose-built vectors were developed, of which pBR322 is probably the best example, but the creation of each one was a major task. Over time, a series of specialist vectors was constructed, each for a particular purpose. During this period, there were many arguments about the relative benefits of plasmid and phage vectors. Today, the molecular biologist has available an enormous range of vectors and these are notable for three reasons. First, many of them combine elements from both plasmids and phages and are known as *phasmids* or, if they contain an M13 *ori* region, *phagemids*. One group of phasmids that is widely used is the λZAP series of vectors used for cDNA cloning and these are described on p. 104. Secondly, many different features that facilitate cloning and expression can be found combined in a single vector. Thirdly, purified vector DNA plus associated reagents can be purchased from molecular-biology suppliers. The hapless scientist who opens a molecular-biology catalog is faced with a bewildering selection of vectors and each vender promotes different ones. Although the benefits of using each vector may be clear, the disadvantages are seldom obvious. The aim of this chapter is to provide the reader with a detailed explanation of the biological basis for the different designs of vector.

There are two general uses for cloning vectors: cloning large pieces of DNA and manipulating genes. When mapping and sequencing genomes, the first step is to subdivide the genome into manageable pieces. The larger these pieces, the easier it is to construct the final picture (see Chapter 17); hence the need to clone large fragments of DNA. Large fragments are also needed if it is necessary to "walk" along the genome to isolate a gene, and this topic is covered in Chapter 6. In many instances, the desired gene will be relatively easy to isolate and a simpler cloning vector can be used. Once isolated, the cloned gene may be expressed as a probe sequence or as a protein, it may be sequenced or it may be mutated *in vitro*. For all these applications, small specialist vectors are used.

Vectors for cloning large fragments of DNA

Cosmids are plasmids that can be packaged into bacteriophage λ particles

As we have seen, concatemers of unit-length λ DNA molecules can be efficiently packaged if the *cos* sites, substrates for the packaging-dependent cleavage, are 37–52 kb apart (75–105% the size of λ⁺ DNA). In fact, only a small region in the proximity of the *cos* site is required for recognition by the packaging system (Hohn 1975).

Plasmids have been constructed which contain a fragment of λ DNA including the *cos* site (Collins & Brüning 1978, Collins & Hohn 1979, Wahl *et al.* 1987, Evans *et al.* 1989). These plasmids have been termed *cosmids* and can be used as gene-cloning vectors in conjunction with the *in vitro* packaging system. Figure 5.1 shows a gene-cloning scheme employing a cosmid. Packaging the cosmid recombinants into phage coats imposes a desirable selection upon their size. With a cosmid vector of 5 kb, we demand the insertion of 32–47 kb of foreign DNA – much more than a phage-λ vector can accommodate. Note that, after packaging *in vitro*, the particle is used to infect a suitable host. The recombinant cosmid DNA is injected and circularizes like phage DNA but replicates as a normal plasmid without the expression of any phage functions. Transformed cells are selected on the basis of a vector drug-resistance marker.

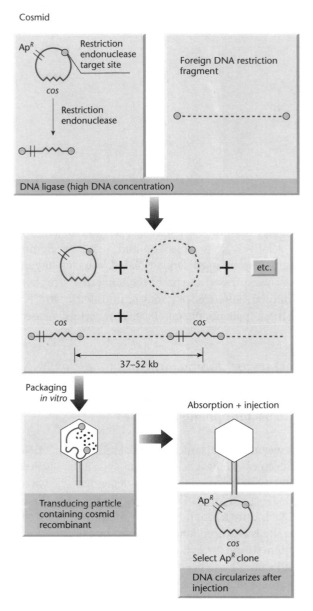

Fig. 5.1 Simple scheme for cloning in a cosmid vector. (See text for details.)

Cosmids provide an efficient means of cloning large pieces of foreign DNA. Because of their capacity for large fragments of DNA, cosmids are particularly attractive vectors for constructing libraries of eukaryotic genome fragments. Partial digestion with a restriction endonuclease provides suitably large fragments. However, there is a potential problem associated with the use of partial digests in this way. This is due to the possibility of two or more genome fragments joining together in the ligation reaction, hence creating a clone containing fragments that were not initially adjacent in the genome. This would give an incorrect picture of their chromosomal organ-

ization. The problem can be overcome by size fractionation of the partial digest.

Even with sized foreign DNA, it is possible for cosmid clones to be produced that contain non-contiguous DNA fragments ligated to form a single insert. The problem can be solved by dephosphorylating the foreign DNA fragments so as to prevent their ligation together. This method is very sensitive to the exact ratio of target-to-vector DNAs (Collins & Brüning 1978) because vector-to-vector ligation can occur. Furthermore, recombinants with a duplicated vector are unstable and break down in the host by recombination, resulting in the propagation of a non-recombinant cosmid vector.

Such difficulties have been overcome in a cosmid-cloning procedure devised by Ish-Horowicz and Burke (1981). By appropriate treatment of the cosmid vector pJB8 (Fig. 5.2), left-hand and right-hand vector ends are purified which are incapable of self-ligation but which accept dephosphorylated foreign DNA. Thus the method eliminates the need to size the foreign DNA fragments and prevents formation of clones containing short foreign DNA or multiple vector sequences.

An alternative solution to these problems has been devised by Bates and Swift (1983) who have constructed cosmid c2XB. This cosmid carries a *Bam*HI insertion site and two *cos* sites separated by a blunt-end restriction site (Fig. 5.3). The creation of these blunt ends, which ligate only very inefficiently under the conditions used, effectively prevents vector self-ligation in the ligation reaction.

Modern cosmids of the pWE and sCos series (Wahl *et al.* 1987, Evans *et al.* 1989) contain features such as: (i) multiple cloning sites (Bates & Swift 1983, Pirrotta *et al.* 1983, Breter *et al.* 1987) for simple cloning using non-size-selected DNA; (ii) phage promoters flanking the cloning site; and (iii) unique *Not*I, *Sac*II, or *Sfi*I sites (rare cutters, see Chapter 17) flanking the cloning site to permit removal of the insert from the vector as single fragments. Mammalian expression modules encoding dominant selectable markers (Chapter 12) may also be present, for gene transfer to mammalian cells if required.

BACs and PACs are vectors that can carry much larger fragments of DNA than cosmids because they do not have packaging constraints

Phage P1 is a temperate bacteriophage which has been extensively used for genetic analysis of *Escherichia*

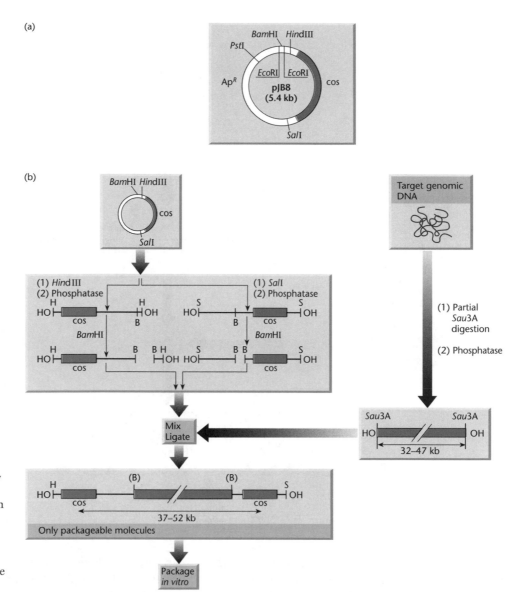

Fig. 5.2 Cosmid-cloning scheme of Ish-Horowicz and Burke (1981). (a) Map of cosmid pJB8. (b) Application to the construction of a genomic library of fragments obtained by partial digestion with *Sau3*A. This restriction endonuclease has a tetranucleotide recognition site and generates fragments with the same cohesive termini as *Bam*HI.

coli because it can mediate generalized transduction. Sternberg and co-workers have developed a P1 vector system which has a capacity for DNA fragments as large as 100 kb (Sternberg 1990, Pierce *et al.* 1992). Thus the capacity is about twice that of cosmid clones but less than that of yeast artificial chromosome (YAC) clones (see p. 213). The P1 vector contains a packaging site (*pac*) which is necessary for *in vitro* packaging of recombinant molecules into phage particles. The vectors contain two *lox*P sites. These are the sites recognized by the phage recombinase, the product of the phage *cre* gene, and which lead to circularization of the packaged DNA after it has been injected into an *E. coli* host expressing the recombinase (Fig. 5.4). Clones are maintained in *E. coli* as low-copy-number plasmids by selection for

a vector kanamycin-resistance marker. A high copy number can be induced by exploitation of the P1 lytic replicon (Sternberg 1990). This P1 system has been used to construct genomic libraries of mouse, human, fission yeast, and *Drosophila* DNA (Hoheisel *et al.* 1993, Hartl *et al.* 1994).

Shizuya *et al.* (1992) have developed a bacterial cloning system for mapping and analysis of complex genomes. This BAC system (bacterial artificial chromosome) is based on the single-copy sex factor F of *E. coli*. This vector (Fig. 5.5) includes the λ *cos* N and P1 *lox*P sites, two cloning sites (*Hind*III and *Bam*HI), and several G+C restriction enzyme sites (e.g. *Sfi*I, *Not*I, etc.) for potential excision of the inserts. The cloning site is also flanked by T7 and SP6 promoters for generating RNA probes. This BAC can be

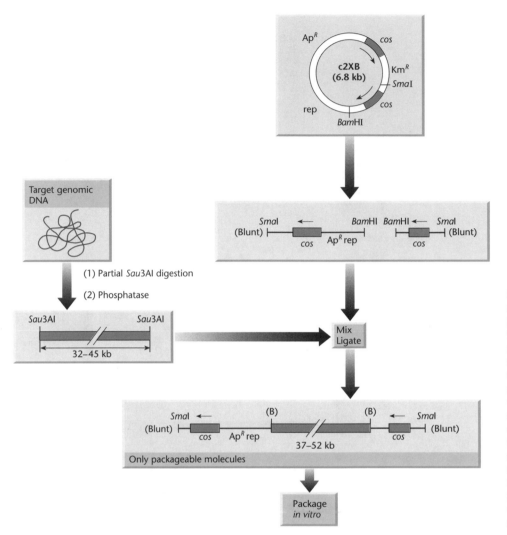

Fig. 5.3 Cosmid-cloning scheme of Bates and Swift (1983). The cosmid c2XB contains two *cos* sites, separated by a site for the restriction endonuclease *Sma*I which creates blunt ends. These blunt ends ligate only very inefficiently under the conditions used and effectively prevent the formation of recombinants containing multiple copies of the vector.

transformed into *E. coli* very efficiently, thus avoiding the packaging extracts that are required with the P1 system. BACs are capable of maintaining human and plant genomic fragments of greater than 300 kb for over 100 generations with a high degree of stability (Woo *et al.* 1994) and have been used to construct genome libraries with an average insert size of 125 kb (Wang *et al.* 1995a). Subsequently, Ioannou *et al.* (1994) have developed a P1-derived artificial chromosome (PAC), by combining features of both the P1 and the F-factor systems. Such PAC vectors are able to handle inserts in the 100–300 kb range.

The first BAC vector, pBAC108L, lacked a selectable marker for recombinants. Thus, clones with inserts had to be identified by colony hybridization. While this once was standard practice in gene manipulation work, today it is considered to be inconvenient! Two widely used BAC vectors, pBeloBAC11 and pECBAC1, are derivatives of pBAC108L in which the original cloning site is replaced with a *lac*Z gene

carrying a multiple cloning site (Kim *et al.* 1996, Frijters *et al.* 1997). pBeloBAC11 has two *Eco*RI sites, one in the *lac*Z gene and one in the CM^R gene, whereas pECBAC1 has only the *Eco*RI site in the *lac*Z gene. Further improvements to BACs have been made by replacing the *lac*Z gene with the *sac*B gene (Hamilton *et al.* 1996). Insertional inactivation of *sac*B permits growth of the host cell on sucrose-containing media, i.e. positive selection for inserts. Frengen *et al.* (1999) have further improved these BACs by including a site for the insertion of a transposon. This enables genomic inserts to be modified after cloning in bacteria, a procedure known as *retrofitting*. The principal uses of retrofitting are the simplified introduction of deletions (Chatterjee & Coren 1997) and the introduction of reporter genes for use in the original host of the genomic DNA (Wang *et al.* 2001). For example, Al-Hasani *et al.* (2003) and Magin-Lachmann *et al.* (2003) have used retrofitting to develop BACs that facilitate transfection, episomal maintenance,

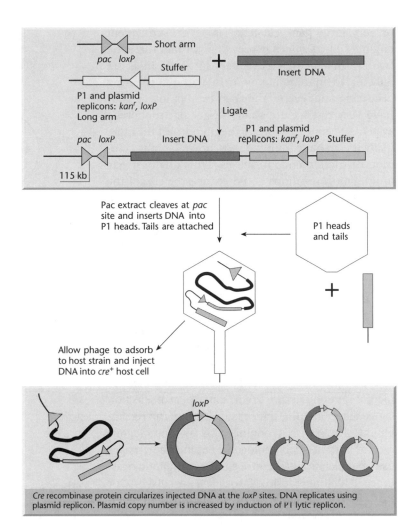

Fig. 5.4 The phage P1 vector system. The P1 vector Ad10 (Sternberg 1990) is digested to generate short and long vector arms. These are dephosphorylated to prevent self-ligation. Size-selected insert DNA (85–100 kb) is ligated with vector arms, ready for a two-stage processing by packaging extracts. First, the recombinant DNA is cleaved at the *pac* site by pacase in the packaging extract. Then the pacase works in concert with head/tail extract to insert DNA into phage heads, *pac* site first, cleaving off a headful of DNA at 115 kb. Heads and tails then unite. The resulting phage particle can inject recombinant DNA into host *E. coli*. The host is *cre*+. The *cre* recombinase acts on *lox*P sites to produce a circular plasmid. The plasmid is maintained at low copy number, but can be amplified by inducing the P1 lytic operon.

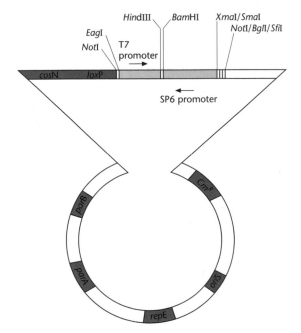

and functional analysis of large genomic fragments in eukaryotic cells.

Recombinogenic engineering (recombineering) simplifies the cloning of DNA, particularly with high-molecular-weight constructs

BACs and PACs are the vectors of choice if one wishes to clone a large fragment of DNA (e.g. >100 kb). They are especially useful for cloning complete operons

Fig. 5.5 (*left*) Structure of a BAC vector derived from a mini-F plasmid. The *ori*S and *rep*E genes mediate the unidirectional replication of the F factor, while *par*A and *par*B maintain the copy number at a level of one or two per genome. Cm^R is a chloramphenicol-resistance marker. *cos*N and *lox*P are the cleavage sites for λ terminase and P1 *cre* protein, respectively. *Hind*III and *Bam*HI are unique cleavage sites for inserting foreign DNA. (Adapted from Shizuya *et al.* 1992.)

or very large gene clusters. However, subsequent engineering of clones is very difficult and there are two reasons for this. First, very large plasmids are difficult to manipulate *in vitro* without shearing occurring and, even if they can be kept intact, they have very restricted mobility in gel electrophoresis systems. Second, the longer the DNA insert the more likely it is to have multiple sites for each of the common restriction endonucleases. Thus any cleavage and ligation reactions could reduce the size of the DNA insert and result in scrambling of fragments. To avoid these problems it is necessary to carry out the manipulations *in vivo* instead of *in vitro* by making use of homologous recombination.

Homologous recombination allows the exchange of genetic information between two DNA molecules in a precise, specific, and faithful manner. It occurs through homology regions which are stretches of DNA shared by the two molecules that recombine. Because the sequence of the homology regions can be chosen freely, any position on a target molecule can be specifically altered. Because homologous recombination is a rare event, some form of selection is needed to identify the cells that carry the recombinant. Hence DNA engineering by homologous recombination (recombinogenic engineering) makes use of a selection procedure such as antibiotic resistance. If all that is desired is to delete part of a BAC or PAC then only a single round of homologous recombination is required (Fig. 5.6a).

In many cases, the persistence of the selectable gene at the site of recombination in the product is undesirable and two rounds of recombinogenic engineering are required. In the first round, homologous recombination is used to generate an initial product by integration of the selectable gene, together with additional functional elements, at the intended site. In the second round, the extra functional elements are used to remove the selectable gene, thereby generating the final product. Different kinds of functional elements have been used, and Fig. 5.6b shows one of these, target sites for site-specific recombinases such as Cre or FLP (as described in Box 3.3). In this case, the first product is exposed to the relevant site-specific recombinase, which deletes the selectable cassette.

There are a number of ways of undertaking recombinogenic engineering (see reviews of Muyrers *et al.* (2001) and Copeland *et al.* (2001)) but only the most widely-used one, known as ET recombination, will be described here. In this method recombination is mediated by either RecE/RecT from the *Rac* phage or by Redα/Redβ from phage λ. These two systems are

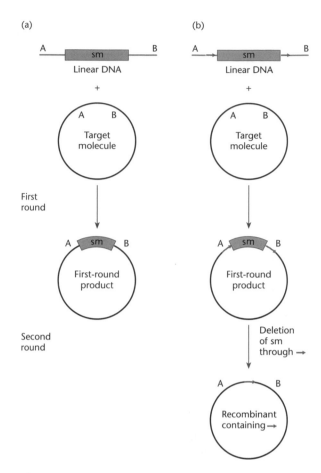

Fig. 5.6 Two variations of *in vivo* recombinogenic engineering. In both variations the target molecule remains intact and could be a BAC, PAC, any other plasmid, or the *E. coli* chromosome. (a) Replacement of a segment of a target molecule with a linear DNA molecule that is introduced to the host cell of the plasmid by electroporation. (b) Removal of the selectable gene by recombination at recombinase target sites (e.g. FRT or *loxP*). sm, selectable marker; arrowheads, recombinase target sites; A and B, regions of homology.

functionally equivalent. RecE and Redα are 5′–3′ exonucleases and RecT and Redβ are DNA annealing proteins. A functional interaction between RecE and RecT or Redα and Redβ also is required for homologous recombination. The easiest way of supplying these recombination functions is to provide them via plasmid-encoded genes, e.g. the pBAD-ETγ plasmid (Zhang *et al.* 1998). In addition to the *RecE* and *RecT* genes, this plasmid encodes the phage *gam* gene. The benefit of this is that the *gam* gene product inhibits any RecBCD-dependent destruction of the targeting cassette.

With ET recombination, as illustrated in Fig. 5.6, a linear targeting DNA carrying short homology regions flanking a selectable gene is introduced by electroporation into cells carrying the BAC or PAC to be modified. The length of homology required for efficient recombination is 35–60 nucleotides and these can be incorporated into PCR primers for amplification of the selectable marker (p. 9). It should be noted that expression of the *gam* gene prevents normal replication of plasmids with ColEl replication functions. Loss of pBAD-ETγ can be prevented until after the recombinogenic engineering step by selecting for the ampicillin resistance marker that it also carries.

A number of factors govern the choice of vector for cloning large fragments of DNA

The maximum size of insert that the different vectors will accommodate is shown in Table 5.1. The size of insert is not the only feature of importance. The absence of chimeras and deletions is even more important. In practice, some 50% of YACs (yeast artificial chromosomes) show structural instability of inserts or are chimeras in which two or more DNA fragments have become incorporated into one clone. These defective YACs are unsuitable for use as mapping and sequencing reagents and a great deal of effort is required to identify them. Cosmid inserts sometimes contain the same aberrations, and the greatest problem with them arises when the DNA being cloned contains tandem arrays of repeated sequences. The problem is particularly acute when the tandem array is several times larger than the allowable size of a cosmid insert. Potential advantages of the BAC and PAC systems over YACs include lower levels of chimerism (Hartl *et al.* 1994, Sternberg 1994), ease of library generation, and ease of manipulation and isolation of insert DNA. BAC clones seem to represent human DNA far more faithfully than their YAC or cosmid counterparts and appear to be excellent substrates for shotgun sequence analysis,

Table 5.1 Maximum DNA insert possible with different cloning vectors. YACs are discussed on p. 213.

Vector	Host	Insert size
λ phage	*E. coli*	5–25 kb
λ cosmids	*E. coli*	35–45 kb
P1 phage	*E. coli*	70–100 kb
PACs	*E. coli*	100–300 kb
BACs	*E. coli*	≤300 kb
YACs	*Saccharomyces cerevisiae*	200–2000 kb

resulting in accurate contiguous sequence data (Venter *et al.* 1998).

Specialist-purpose vectors

M13-based vectors can be used to make single-stranded DNA suitable for sequencing

Whenever a new gene is cloned or a novel genetic construct is made, it is usual practice to sequence all or part of the chimeric molecule. As will be seen later (p. 126), the Sanger method of sequencing requires single-stranded DNA as the starting material. Originally, single-stranded DNA was obtained by cloning the sequence of interest in an M13 vector (see p. 72). Today, it is more usual to clone the sequence into a pUC-based phagemid vector which contains the M13 *ori* region as well as the pUC (Col E1) origin of replication. Such vectors normally replicate inside the cell as double-stranded molecules. Single-stranded DNA for sequencing can be produced by infecting cultures with a helper phage such as M13K07. This helper phage has the origin of replication of P15A and a kanamycin-resistance gene inserted into the M13 *ori* region and carries a mutation in the *gII* gene (Vieira & Messing 1987). M13K07 can replicate on its own. However, in the presence of a phagemid bearing a wild-type origin of replication, single-stranded phagemid is packaged preferentially and secreted into the culture medium. DNA purified from the phagemids can be used directly for sequencing.

Expression vectors enable a cloned gene to be placed under the control of a promoter that functions in *E. coli*

Expression vectors are required if one wants to prepare RNA probes from the cloned gene or to purify large amounts of the gene product. In either case, transcription of the cloned gene is required. Although it is possible to have the cloned gene under the control of its own promoter, it is more usual to utilize a promoter specific to the vector. Such vector-carried promoters have been optimized for binding of the *E. coli* RNA polymerase and many of them can be regulated easily by changes in the growth conditions of the host cell.

E. coli RNA polymerase is a multi-subunit enzyme. The core enzyme consists of two identical α subunits and one each of the β and β′ subunits. The core enzyme is not active unless an additional subunit,

		-35 Region		-10 Region	
			1 2 3 4 5 6 7 8 9 10 11 12 13 14 15 16 17		
CONSENSUS	• • •	TTGACA	• • • • • • • • • • • • • • • • •	TATAAT	• •
lac	GGC	TTTACAC	TTTATGCTTCCGGCTCG	TATATT	GT
trp	CTG	TTGACAA	TTAATCAT CGAACTAG	TTAACT	AG
λP$_L$	GTG	TTGACAT	AAATACCA CTGGCGGT	GATACT	GA
rec A	CAC	TTGATAC	TGTATGAA GCATACAG	TATAAT	TG
tacI	CTG	TTGACAA	TTAATCAT CGGCTCG	TATAAT	GT
tacII	CTG	TTGACAA	TTAATCAT CGAACTAG	TTTAAT	GT

Fig. 5.7 The base sequence of the −10 and −35 regions of four natural promoters, two hybrid promoters, and the consensus promoter.

the σ factor, is present. RNA polymerase recognizes different types of promoters depending on which type of σ factor is attached. The most common promoters are those recognized by the RNA polymerase with σ^{70}. A large number of σ^{70} promoters from *E. coli* have been analyzed and a compilation of over 300 of them can be found in Lisser and Margalit (1993). A comparison of these promoters has led to the formulation of a consensus sequence (Fig. 5.7). If the transcription start point is assigned the position +1 then this consensus sequence consists of the −35 region (5′-TTGACA-) and the −10 region, or Pribnow box (5′-TATAAT). RNA polymerase must bind to both sequences to initiate transcription. The strength of a promoter, i.e. how many RNA copies are synthesized per unit time per enzyme molecule, depends on how close its sequence is to the consensus. While the −35 and −10 regions are the sites of nearly all mutations affecting promoter strength, other bases flanking these regions can affect promoter activity (Hawley & McClure 1983, Dueschle *et al.* 1986, Keilty & Rosenberg 1987). The distance between the −35 and −10 regions is also important. In all cases examined, the promoter was weaker when the spacing was increased or decreased from 17 bp.

Upstream (UP) elements located 5′ of the −35 hexamer in certain bacterial promoters are A+T-rich sequences that increase transcription by interacting with the α subunit of RNA polymerase. Gourse *et al.* (1998) have identified UP sequences conferring increased activity to the *rrn* core promoter. The best UP sequence was portable and increased heterologous protein expression from the *lac* promoter by a factor of 100.

Once RNA polymerase has initiated transcription at a promoter, it will polymerize ribonucleotides until it encounters a transcription-termination site in the DNA. Bacterial DNA has two types of transcription-termination site: factor-independent and factor-dependent. As their names imply, these types are distinguished by whether they work with just RNA

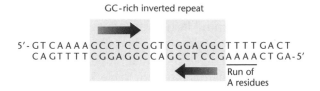

Fig. 5.8 Structure of a factor-independent transcriptional terminator.

polymerase and DNA alone or need other factors before they can terminate transcription. The factor-independent transcription terminators are easy to recognize because they have similar sequences: an inverted repeat followed by a string of A residues (Fig. 5.8). Transcription is terminated in the string of A residues, resulting in a string of U residues at the 3′ end of the mRNA. The factor-dependent transcription terminators have very little sequence in common with each other. Rather, termination involves interaction with one of the three known *E. coli* termination factors, Rho (ρ), Tau (τ), and NusA. Most expression vectors incorporate a factor-independent termination sequence downstream from the site of insertion of the cloned gene.

Specialist vectors have been developed that facilitate the production of RNA probes and interfering RNA

Although single-stranded DNA can be used as a sequence probe in hybridization experiments, RNA probes are preferred. The reasons for this are that the rate of hybridization and the stability are far greater for RNA–DNA hybrids compared with DNA–DNA hybrids. To make an RNA probe, the relevant gene sequence is cloned in a plasmid vector such that it is under the control of a phage promoter. After purification, the plasmid is linearized with a suitable restriction enzyme and then incubated with the phage RNA polymerase and the four ribonucleoside triphosphates (Fig. 5.9). No transcription terminator

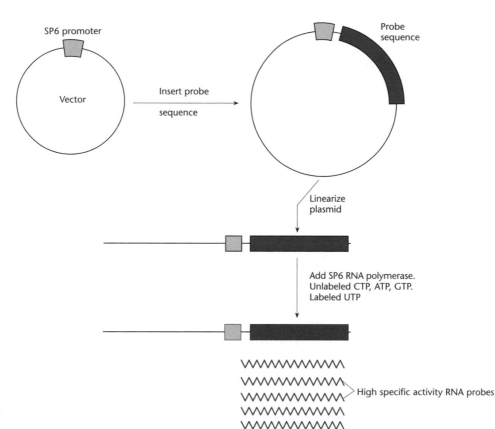

Fig. 5.9 Method for preparing RNA probes from a cloned DNA molecule using a phage SP6 promoter and SP6 RNA polymerase.

is required because the RNA polymerase will fall off the end of the linearized plasmid.

There are three reasons for using a phage promoter. First, such promoters are very strong, enabling large amounts of RNA to be made *in vitro*. Secondly, the phage promoter is not recognized by the *E. coli* RNA polymerase and so no transcription will occur inside the cell. This minimizes any selection of variant inserts. Thirdly, the RNA polymerases encoded by phages such as SP6, T7, and T3 are much simpler molecules to handle than the *E. coli* enzyme, since the active enzyme is a single polypeptide.

If it is planned to probe RNA or single-stranded DNA sequences, then it is essential to prepare RNA probes corresponding to both strands of the insert. One way of doing this is to have two different clones corresponding to the two orientations of the insert. An alternative method is to use a cloning vector in which the insert is placed between two different, opposing phage promoters (e.g. T7/T3 or T7/SP6) that flank a multiple cloning sequence (see Fig. 5.5). Since each of the two promoters is recognized by a different RNA polymerase, the direction of transcription is determined by which polymerase is used.

A further improvement has been introduced by Evans *et al.* (1995). In their LITMUS vectors, the polylinker regions are flanked by two modified T7 RNA polymerase promoters. Each contains a unique restriction site (*Spe*I or *Afl*II) that has been engineered into the T7 promoter consensus sequence such that cleavage with the corresponding endonuclease inactivates that promoter. Both promoters are active despite the presence of engineered sites. Selective unidirectional transcription is achieved by simply inactivating the other promoter by digestion with *Spe*I or *Afl*II prior to *in vitro* transcription (Fig. 5.10). Since efficient labeling of RNA probes demands that the template be linearized prior to transcription, at a site downstream from the insert, cutting at the site within the undesired promoter performs both functions in one step. Should the cloned insert contain either an *Spe*I or an *Afl*II site, the unwanted promoter can be inactivated by cutting at one of the unique sites within the polylinker.

RNA interference (RNAi) is a mechanism of post-transcriptional gene silencing in which double-stranded RNA corresponding to a gene of interest is introduced into an organism, thereby causing

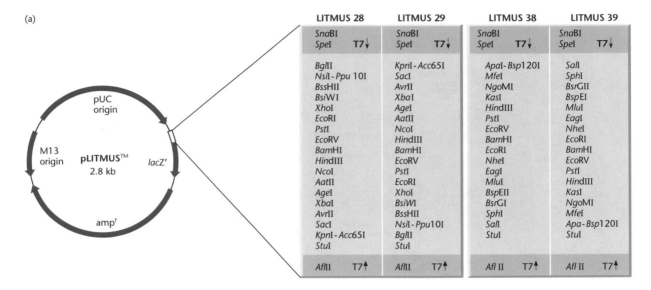

(a)

LITMUS 28	LITMUS 29	LITMUS 38	LITMUS 39
*Sna*BI	*Sna*BI	*Sna*BI	*Sna*BI
*Spe*I **T7↓**	*Spe*I **T7↓**	*Spe*I **T7↓**	*Spe*I **T7↓**
*Bgl*II	*Kpn*I - *Acc*65I	*Apa*I - *Bsp*120I	*Sal*I
*Nsi*I - *Ppu* 10I	*Sac*I	*Mfe*I	*Sph*I
*Bss*HII	*Avr*II	*Ngo*MI	*Bsr*GII
*Bsi*WI	*Xba*I	*Kas*I	*Bsp*EI
*Xho*I	*Age*I	*Hind*III	*Mlu*I
*Eco*RI	*Aat*II	*Pst*I	*Eag*I
*Pst*I	*Nco*I	*Eco*RV	*Nhe*I
*Eco*RV	*Hind*III	*Bam*HI	*Eco*RI
*Bam*HI	*Bam*HI	*Eco*RI	*Bam*HI
*Hind*III	*Eco*RV	*Nhe*I	*Eco*RV
*Nco*I	*Pst*I	*Eag*I	*Pst*I
*Aat*II	*Eco*RI	*Mlu*I	*Hind*III
*Age*I	*Xho*I	*Bsp*EII	*Kas*I
*Xba*I	*Bsi*WI	*Bsr*GI	*Ngo*MI
*Avr*II	*Bss*HII	*Sph*I	*Mfe*I
*Sac*I	*Nsi*I - *Ppu*10I	*Sal*I	*Apa* - *Bsp*120I
*Kpn*I - *Acc*65I	*Bgl*II	*Stu*I	*Stu*I
*Stu*I	*Stu*I		
*Afl*II **T7↑**	*Afl*II **T7↑**	*Afl* II **T7↑**	*Afl* II **T7↑**

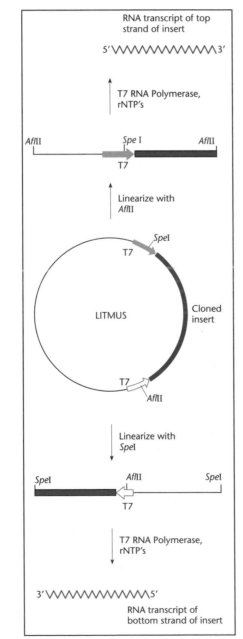

(b)

RNA transcript of top strand of insert

5′ VVVVVVVVVV 3′

↑ T7 RNA Polymerase, rNTP's

*Afl*II ———— *Spe* I ▬▬▬ *Afl*II
T7

↑ Linearize with *Afl*II

LITMUS
Cloned insert
T7 — *Spe*I
T7 — *Afl*II

↓ Linearize with *Spe*I

*Spe*I ▬▬▬ *Afl*II *Spe*I
T7

↓ T7 RNA Polymerase, rNTP's

3′ VVVVVVVVVV 5′

RNA transcript of bottom strand of insert

Fig. 5.10 Structure and use of the LITMUS vectors for making RNA probes. (a) Structure of the LITMUS vectors showing the orientation and restriction sites of the four polylinkers. (b) Method of using the LITMUS vectors to selectively synthesize RNA probes from each strand of a cloned insert. (Figure reproduced courtesy of New England Biolabs.)

degradation of the matching mRNA. The applications of this technique are discussed in detail on p. 318. The easiest way to make double-stranded RNA is to use vectors like the LITMUS ones just described. In this case the plasmid DNA containing the cloned target of interest is digested, in separate reactions, with *Spe*I and *Afl*II. This will generate a template for each RNA strand. If the templates are mixed and used for *in vitro* transcription then double-stranded RNA will be produced.

Vectors with strong, controllable promoters are used to maximize synthesis of cloned gene products

Provided that a cloned gene is preceded by a promoter recognized by the host cell, then there is a high probability that there will be *detectable* synthesis of the cloned gene product. However, much of the interest in the application of recombinant DNA technology lies in the possibility of facile synthesis of large quantities of protein, either to study its properties or because it has commercial value. In such instances, detectable synthesis is not sufficient: rather, it must be maximized. The factors affecting the level of expression of a cloned gene are shown in Table 5.2 and are reviewed by Baneyx (1999). Of these factors, only promoter strength is considered here.

When maximizing gene expression it is not enough to select the strongest promoter possible: the effects of overexpression on the host cell also need to be considered. Many gene products can be toxic to the host cell even when synthesized in small amounts. Examples include surface structural proteins (Beck & Bremer 1980), proteins such as the *PolA* gene product that regulate basic cellular metabolism (Murray & Kelley 1979), the cystic fibrosis transmembrane conductance regulator (Gregory *et al.* 1990), and lentivirus envelope sequences (Cunningham *et al.* 1993). If such cloned genes are allowed to be expressed there will be a rapid selection for mutants that no longer synthesize the toxic protein. Even

Table 5.2 Factors affecting the expression of cloned genes.

Factor	Text
Promoter strength	This page
Transcriptional termination	Page 82
Plasmid copy number	Page 57, Chapter 4
Plasmid stability	Page 59, Chapter 4
Host-cell physiology	Chapters 4 & 5
Translational initiation sequences	Box 5.1, page 88
Codon choice	Box 5.1, page 88
mRNA structure	Box 5.1, page 88

when overexpression of a protein is not toxic to the host cell, high-level synthesis exerts a metabolic drain on the cell. This leads to slower growth and hence in culture there is selection for variants with lower or no expression of the cloned gene because these will grow faster. To minimize the problems associated with high-level expression, it is usual to use a vector in which the cloned gene is under the control of a *regulated* promoter.

Many different vectors have been constructed for regulated expression of gene inserts but most of those in current use contain one of the following controllable promoters: λ P_L, T7, *trc* (*tac*), or BAD. Table 5.3 shows the different levels of expression that can be achieved when the gene for chloramphenicol transacetylase (CAT) is placed under the control of three of these promoters.

The *trc* and *tac* promoters are hybrid promoters derived from the *lac* and *trp* promoters (Brosius 1984). They are stronger than either of the two parental promoters because their sequences are more like the consensus sequence. Like *lac*, the *trc* and *tac* promoters are inducibile by lactose and isopropyl-β-D-thiogalactoside (IPTG). Vectors using these promoters also carry the *lac*O operator and the *lac*I gene, which encodes the repressor.

Table 5.3 Control of expression of chloramphenicol acetyltransferase (CAT) in *E. coli* by three different promoters. The levels of CAT are expressed as μg/mg total protein.

Promoter	Uninduced level of CAT	Induced level of CAT	Ratio
λP_L	0.0275	28.18	1025
trc	1.10	5.15	4.7
T7	1.14	15.40	13.5

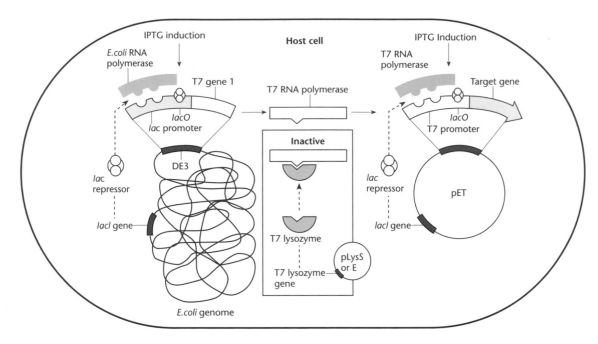

Fig. 5.11 Strategy for regulating the expression of genes cloned into a pET vector. The gene for T7 RNA polymerase (gene 1) is inserted into the chromosome of *E. coli* and transcribed from the *lac* promoter; therefore, it will be expressed only if the inducer IPTG is added. The T7 RNA polymerase will then transcribe the gene cloned into the pET vector. If the protein product of the cloned gene is toxic, it may be necessary to further reduce the transcription of the cloned gene before induction. The T7 lysozyme encoded by a compatible plasmid, pLysS, will bind to any residual T7 RNA polymerase made in the absence of induction and inactivate it. Also, the presence of *lac* operators between the T7 promoter and the cloned gene will further reduce transcription of the cloned gene in the absence of the inducer IPTG. (Reprinted with permission from the *Novagen Catalog*, Novagen, Madison, Wisconsin, 1995.)

The pET vectors are a family of expression vectors that utilize phage T7 promoters to regulate synthesis of cloned gene products (Studier *et al.* 1990). The general strategy for using a pET vector is shown in Fig. 5.11. To provide a source of phage-T7 RNA polymerase, *E. coli* strains that contain gene 1 of the phage have been constructed. This gene is cloned downstream of the *lac* promoter, in the chromosome, so that the phage polymerase will only be synthesized following IPTG induction. The newly synthesized T7 RNA polymerase will then transcribe the foreign gene in the pET plasmid. It is possible to minimize the uninduced level of T7 RNA polymerase, if the protein product of the cloned gene is toxic. First, a plasmid compatible with pET vectors is selected and the T7 *lys*S gene is cloned in it. When introduced into a host cell carrying a pET plasmid, the *lys*S gene will bind any residual T7 RNA polymerase (Studier 1991, Zhang & Studier 1997). Also, if a *lac* operator is placed between the T7 promoter and the cloned gene, this will further reduce transcription of the insert in the absence of IPTG (Dubendorff & Studier 1991). Improvements in the yield of heterologous proteins can sometimes be achieved by use of selected host cells (Miroux & Walker 1996).

The λ P_L promoter system combines very tight transcriptional control with high levels of gene expression. This is achieved by putting the cloned gene under the control of the P_L promoter carried on a vector, while the P_L promoter is controlled by a *c*I repressor gene in the *E. coli* host. This *c*I gene is itself under the control of the tryptophan (*trp*) promoter (Fig. 5.12). In the absence of exogenous tryptophan, the *c*I gene is transcribed and the *c*I repressor binds to the P_L promoter, preventing expression of the cloned gene. Upon addition of tryptophan, the *trp* repressor binds to the *c*I gene, preventing synthesis of the *c*I repressor. In the absence of *c*I repressor, there is a high level of expression from the very strong P_L promoter.

The pBAD vectors, like the ones based on P_L promoter, offer extremely tight control of expression of cloned genes (Guzman *et al.* 1995). The pBAD vectors carry the promoter of the *ara*BAD (arabinose) operon and the gene encoding the positive and negative regulator of this promoter, *ara*C. AraC is a transcriptional regulator that forms a complex with L-arabinose. In the absence of arabinose, AraC binds to the O_2 and I_1 half-sites of the *ara*BAD operon, forming a 210 bp DNA loop and thereby blocking transcription (Fig. 5.13). As arabinose is added to

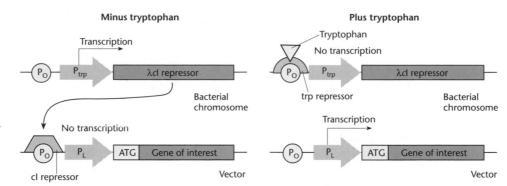

Fig. 5.12 Control of cloned gene expression using the λcI promoter. See text for details. (Diagram reproduced courtesy of InVitrogen Corporation.)

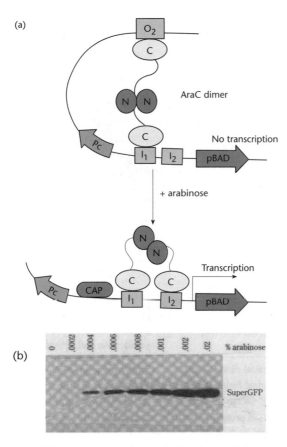

Fig. 5.13 Regulation of the pBAD promoter. (a) The conformational changes that take place on addition of arabinose. (b) Western blot showing the increase in synthesis of a cloned gene product when different levels of arabinose are added to a culture of the host cell.

cAMP synthesis, thereby decreasing expression from the *ara*BAD promoter. Thus one can titrate the level of cloned gene product by varying the glucose and arabinose content of the growth medium (Fig. 5.13). According to Guzman *et al.* (1995), the pBAD vectors permit fine-tuning of gene expression. All that is required is to change the sugar composition of the medium. However, this is disputed by others (Siegele & Hu 1997, Hashemzadeh-Bonehi *et al.* 1998).

Many of the vectors designed for high-level expression also contain translation-initiation signals optimized for *E. coli* expression (see Box 5.1).

Purification of a cloned gene product can be facilitated by use of purification tags

Many cloning vectors have been engineered so that the protein being expressed will be fused to another protein, called a *tag*, which can be used to facilitate protein purification. Examples of tags include glutathione-*S*-transferase, the MalE (maltose-binding) protein, and multiple histidine residues, which can easily be purified by affinity chromatography. The tag vectors are usually constructed so that the coding sequence for an amino acid sequence cleaved by a specific protease is inserted between the coding sequence for the tag and the gene being expressed. After purification, the tag protein can be cleaved off with the specific protease to leave a normal or nearly normal protein. It is also possible to include in the tag a protein sequence that can be assayed easily. This permits assay of the cloned gene product when its activity is not known or when the usual assay is inconvenient. Three different examples of tags are given below. The reader requiring a more detailed insight should consult the review by La Vallie and McCoy (1995).

To use a polyhistidine fusion for purification, the gene of interest is first engineered into a vector in which there is a polylinker downstream of six

the growth medium, it binds to AraC, thereby releasing the O_2 site. This in turn causes AraC to bind to the I_2 site adjacent to the I_1 site. This releases the DNA loop and allows transcription to begin. Binding of AraC to I_1 and I_2 is activated in the presence of cAMP activator protein (CAP) plus cyclic adenosine monophosphate (cAMP). If glucose is added to the growth medium, this will lead to a repression of

Box 5.1 Optimizing translation

High-level expression of a cloned gene requires more than a strong promoter. The mRNA produced during transcription needs to be effectively translated into protein. Although many factors can influence the rate of translation, the most important is the interaction of the ribosome with the bases immediately upstream from the initiation codon of the gene. In bacteria, a key sequence is the ribosome-binding site or Shine–Dalgarno (S–D) sequence. The degree of complementarity of this sequence with the 16S rRNA can affect the rate of translation (De Boer & Hui 1990). Maximum complementarity occurs with the sequence 5′-UAAGGAGG-3′ (Ringquist *et al.* 1992). The spacing between the S-D sequence and the initiation codon is also important. Usually there are five to 10 bases, with eight being optimal. Decreasing the distance below 4 bp or increasing it beyond 14 bp can reduce translation by several orders of magnitude.

Translation is affected by the sequence of bases that follow the S-D site (De Boer *et al.* 1983b). The presence of four A residues or four T residues in this position gave the highest translational efficiency. Translational efficiency was 50% or 25% of maximum when the region contained, respectively, four C residues or four G residues.

The composition of the triplet immediately preceding the AUG start codon also affects the efficiency of translation. For translation of β-galactosidase mRNA, the most favorable combinations of bases in this triplet are UAU and CUU. If UUC, UCA, or AGG replace UAU or CUU, the level of expression is 20-fold less (Hui *et al.* 1984).

The codon composition following the AUG start codon can also affect the rate of translation. For example, a change in the third position of the fourth codon of a human γ-interferon gene resulted in a 30-fold change in the level of expression (De Boer & Hui 1990). Also, there is a strong bias in the second codon of many natural mRNAs, which is quite different from the general bias in codon usage. Highly expressed genes have AAA (Lys) or GCU (Ala) as the second codon. Devlin *et al.* (1988) changed all the G and C nucleotides for the first four codons of a

granulocyte colony-stimulating factor gene and expression increased from undetectable to 17% of total cell protein.

Sequences upstream from the S-D site can affect the efficiency of translation of certain genes. In the *E. coli rnd* gene there is a run of eight uracil residues. Changing two to five of these residues has no effect on mRNA levels but reduces translation by up to 95% (Zhang & Deutscher 1992). Etchegaray and Inouye (1999) have identified an element downstream of the initiation codon, the downstream box, which facilitates formation of the translation-initiation complex. The sequence of the 3′ untranslated region of the mRNA can also be important. If this region is complementary to sequences within the gene, hairpin loops can form and hinder ribosome movement along the messenger.

The genetic code is degenerate, and hence for most of the amino acids there is more than one codon. However, in all genes, whatever their origin, the selection of synonymous codons is distinctly non-random (for reviews, see Kurland 1987, Ernst 1988 and McPherson 1988). The bias in codon usage has two components: correlation with tRNA availability in the cell, and non-random choices between pyrimidine-ending codons. Ikemura (1981a) measured the relative abundance of the 26 known tRNAs of *E. coli* and found a strong positive correlation between tRNA abundance and codon choice. Later, Ikemura (1981b) noted that the most highly expressed genes in *E. coli* contain mostly those codons corresponding to major tRNAs but few codons of minor tRNAs. In contrast, genes that are expressed less well use more suboptimal codons. Forman *et al.* (1998) noted significant misincorporation of lysine, in place of arginine, when the rare AGA codon was included in a gene overexpressed in *E. coli*. It should be noted that the bias in codon usage even extends to the stop codons (Sharp & Bulmer 1988). UAA is favored in genes expressed at high levels, whereas UAG and UGA are used more frequently in genes expressed at a lower level.

For a review of translation the reader should consult Kozak (1999).

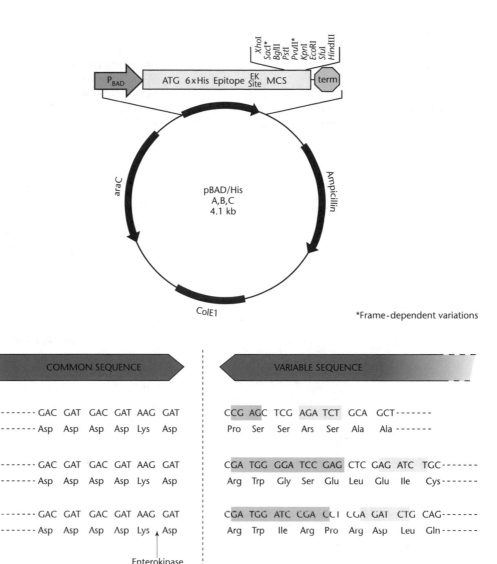

Fig. 5.14 Structure of a vector (pBAD/His, Invitrogen Corporation) designed for the expression of a cloned gene as a fusion protein containing a polyhistidine sequence. Three different variants (A, B, C) allow the insert to be placed in each of the three translational reading frames. The sequence shaded purple shows the base sequence which is altered in each of the three vectors. The lightly-shaded box (AGATCT) is the *Bgl* II site of the polylinker. Note that the initial A residue of the restriction site is at a different point in the triplet codon in each of the three sequences.

histidine residues and a proteolytic cleavage site. In the example shown in Fig. 5.14, the cleavage site is that for enterokinase. After induction of synthesis of the fusion protein, the cells are lysed and the viscosity of the lysate is reduced by nuclease treatment. The lysate is then applied to a column containing immobilized divalent nickel, which selectively binds the polyhistidine tag. After washing away any contaminating proteins, the fusion protein is eluted from the column and treated with enterokinase to release the cloned gene product.

For the cloned gene to be expressed correctly, it has to be in the correct translational reading frame. This is achieved by having three different vectors, each with a polylinker in a different reading frame (see Fig. 5.14). Enterokinase recognizes the sequence (Asp)$_4$Lys and cleaves immediately after the lysine

residue. Therefore, after enterokinase cleavage, the cloned gene protein will have a few extra amino acids at the N terminus. If desired, the cleavage site and polyhistidines can be synthesized at the C terminus. If the cloned gene product itself contains an enterokinase cleavage site, then an alternative protease, such as thrombin or factor Xa, with a different cleavage site can be used.

To facilitate assay of the fusion proteins, short antibody recognition sequences can be incorporated into the tag between the affinity label and the protease cleavage site. Some examples of recognizable epitopes are given in Table 5.4. These antibodies can be used to detect, by western blotting, fusion proteins carrying the appropriate epitope. Note that a polyhistidine tag at the C terminus can function for both assay and purification.

Peptide sequence	Antibody recognition
-Glu-Gln-Lys-Leu-Ile Ser-Glu-Glu-Asp-Leu- -His-His-His-His-His-His-COOH -Gly-Lys-Pro-Ile-Pro-Asn-Pro-Leu-Leu-Gly-Leu- Asp-Ser-Thr-	Anti-*myc* antibody Anti-His (C-terminal) antibody Anti-V5 antibody

Table 5.4 Peptide epitopes, and the antibodies that recognize them, for use in assaying fusion proteins.

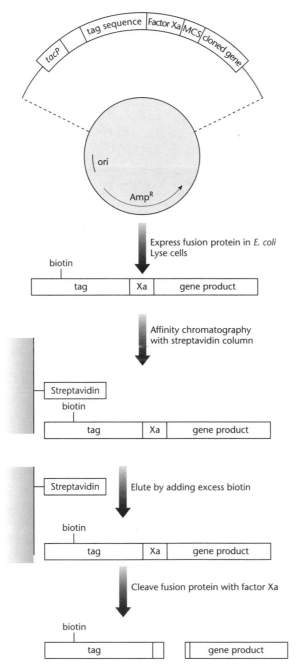

Fig. 5.15 Purification of a cloned gene product synthesized as a fusion to the biotin carboxylase carrier protein (tag). See text for details.

Biotin is an essential cofactor for a number of carboxylases important in cell metabolism. The biotin in these enzyme complexes is covalently attached at a specific lysine residue of the biotin carboxylase carrier protein. Fusions made to a segment of the carrier protein are recognized in *E. coli* by biotin ligase, the product of the *bir*A gene, and biotin is covalently attached in an ATP-dependent reaction. The expressed fusion protein can be purified using streptavidin affinity chromatography (Fig. 5.15). *E. coli* expresses a single endogenous biotinylated protein, but it does not bind to streptavidin in its native configuration, making the affinity purification highly specific for the recombinant fusion protein. The presence of biotin on the fusion protein has an additional advantage: its presence can be detected with enzymes coupled to streptavidin.

The affinity purification systems described above suffer from the disadvantage that a protease is required to separate the target protein from the affinity tag. Also, the protease has to be separated from the protein of interest. Chong *et al.* (1997, 1998) have described a unique purification system that has neither of these disadvantages. The system utilizes a protein splicing element, an intein, from the *Saccharomyces cerevisiae VMA1* gene (see Box 5.2). The intein is modified such that it undergoes a self-cleavage reaction at its N terminus at low temperatures in the presence of thiols, such as cysteine, dithiothreitol, or β-mercaptoethanol. The gene encoding the target protein is inserted into a multiple cloning site (MCS) of a vector to create a fusion between the C terminus of the target gene and the N terminus of the gene encoding the intein. DNA encoding a small (5 kDa) chitin-binding domain from *Bacillus circulans* was added to the C terminus of the intein for affinity purification (Fig. 5.16).

The above construct is placed under the control of an IPTG-inducible T7 promoter. When crude extracts from induced cells are passed through a chitin column, the fusion protein binds and all contaminating proteins are washed through. The fusion is then

Box 5.2 Inteins, exteins, and protein splicing

Protein splicing is defined as the excision of an intervening protein sequence (the *intein*) from a protein precursor. Splicing involves ligation of the flanking protein fragments (the *exteins*) to form a mature extein protein and the free intein. Protein splicing results in a native peptide bond between the ligated exteins and this differentiates it from other forms of autoproteolysis.

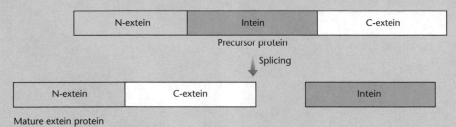

Sequence comparison and structural analysis have indicated that the residues responsible for splicing are ~100 amino acids at the N terminus of the intein and ~50 amino acids at the C terminus. These two splicing regions are separated by a linker or a gene encoding a *homing endonuclease*. If present in a cell, the homing endonuclease makes a double-stranded break in the DNA at or near the insertion site (the home) of the intein encoding it. This endonuclease activity initiates the movement of the intein into another allele of the same gene if that allele lacks the intein.

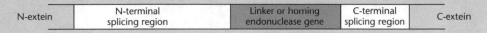

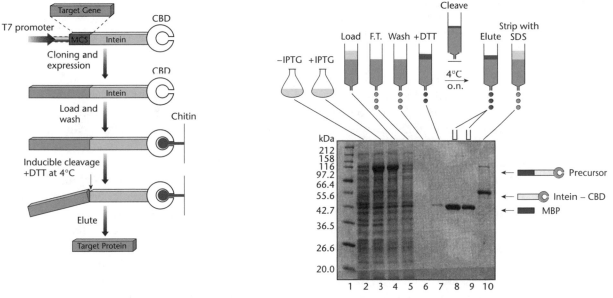

Lane 1: Protein Marker.
Lane 2: Crude extract from uninduced cells.
Lane 3: Crude extract from cells, induced at 15°C for 16 hours.
Lane 4: Clarified crude extract from induced cells.
Lane 5: Chitin column flow through (F.T.).
Lane 6: Chitin column wash.
Lane 7: Quick DTT wash to distribute DTT evenly throughout the chitin column.
Lanes 8-9: Fraction of eluted MBP after stopping column flow and inducing a self-cleavage reaction at 4°C overnight.
Lane 10: SDS stripping of remaining proteins bound to chitin column (mostly the cleaved intein-CBD fusion).

Fig. 5.16 Purification of a cloned gene product synthesized as a fusion with an intein protein. (Figure reproduced courtesy of New England Biolabs.)

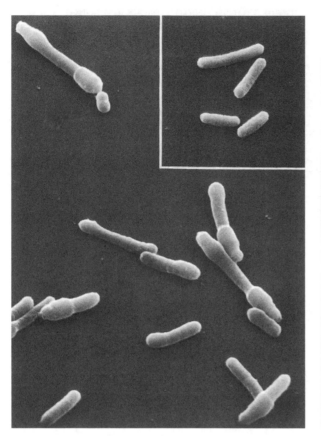
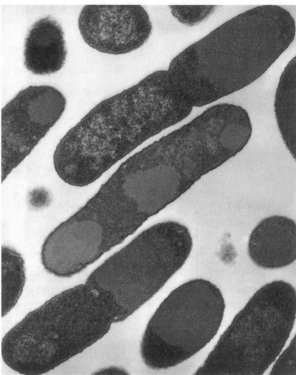

Fig. 5.17 Inclusions of Trp polypeptide–proinsulin fusion protein in *E. coli*. (*Left*) Scanning electron micrograph of cells fixed in the late logarithmic phase of growth; the inset shows normal *E. coli* cells. (*Right*) Thin section through *E. coli* cells producing Trp polypeptide–insulin A chain fusion protein. (Photographs reproduced from *Science* courtesy of Dr. D.C. Williams (Eli Lilly & Co.) and the American Association for the Advancement of Science.)

induced to undergo intein-mediated self-cleavage on the column by incubation with a thiol. This releases the target protein, while the intein chitin-binding domain remains bound to the column.

Vectors are available that promote solubilization of expressed proteins

One of the problems associated with the overproduction of proteins in *E. coli* is the sequestration of the product into insoluble aggregates or "inclusion bodies" (Fig. 5.17). They were first reported in strains overproducing insulin A and B chains (Williams *et al.* 1982). At first, their formation was thought to be restricted to the overexpression of heterologous proteins in *E. coli*, but they can form in the presence of high levels of normal *E. coli* proteins, e.g. subunits of RNA polymerase (Gribskov & Burgess 1983). Two parameters that can be manipulated to reduce inclusion-body formation are temperature and growth rate. There are a number of reports which show that lowering the temperature of growth increases the yield of correctly folded, soluble protein (Schein &

Noteborn 1988, Takagi *et al.* 1988, Schein 1991). Media compositions and pH values that reduce the growth rate also reduce inclusion-body formation. Renaturation of misfolded proteins can sometimes be achieved following solubilization in guanidinium hydrochloride (Lilie *et al.* 1998).

Three "genetic" methods of preventing inclusion-body formation have been described. In the first of these, the host cell is engineered to overproduce a chaperon (e.g. DnaK, GroEL, or GroES proteins) in addition to the protein of interest (Van Dyk *et al.* 1989, Blum *et al.* 1992, Thomas *et al.* 1997). Castanie *et al.* (1997) have developed a series of vectors which are compatible with pBR322-type plasmids and which encode the overproduction of chaperons (proteins whose function is to assist with the folding and refolding of other proteins). These vectors can be used to test the effect of chaperons on the solubilization of heterologous gene products. Even with excess chaperon there is no guarantee of proper folding. The second method involves making minor changes to the amino acid sequence of the target protein. For example, cysteine-to-serine changes in fibroblast growth factor minimized inclusion-body formation (Rinas *et al.* 1992). The third method is derived from the observation that many proteins produced as insoluble aggregates in their native state are synthesized in soluble form as thioredoxin fusion proteins (La Vallie *et al.* 1993). More recently, Davis *et al.* (1999) have shown that the NusA and GrpE proteins, as well as bacterioferritin, are even better than thioredoxin at solubilizing proteins expressed at a high level. Kapust and Waugh (1999) have reported that the maltose-binding protein is also much better than thioredoxin.

Building on the work of La Vallie *et al.* (1993), a series of vectors has been developed in which the gene of interest is cloned into an MCS and the gene product is produced as a thioredoxin fusion protein with an enterokinase cleavage site at the fusion point. After synthesis, the fusion protein is released from the producing cells by osmotic shock and purified. The desired protein is then released by enterokinase cleavage. To simplify the purification of thioredoxin fusion proteins, Lu *et al.* (1996) systematically mutated a cluster of surface amino acid residues. Residues 30 and 62 were converted to histidine and the modified ("histidine patch") thioredoxin could now be purified by affinity chromatography on immobilized divalent nickel. An alternative purification method was developed by Smith *et al.* (1998). They synthesized a gene in which a short biotinylation peptide is fused to the N terminus of the thioredoxin gene to generate a new protein called BIOTRX. They constructed a vector carrying the BIOTRX gene, with an MCS at the C terminus, and the *birA* gene. After cloning a gene in the MCS, a fused protein is produced which can be purified by affinity chromatography on streptavidin columns.

An alternative way of keeping recombinant proteins soluble is to export them to the periplasmic space (see next section). However, even here they may still be insoluble. Barth *et al.* (2000) solved this problem by growing the producing bacteria under osmotic stress (4% NaCl plus 0.5 mol/l sorbitol) in the presence of compatible solutes. Compatible solutes are low-molecular-weight osmolytes, such as glycine betaine, that occur naturally in halophilic bacteria and are known to protect proteins at high salt concentrations. Adding glycine betaine for the cultivation of *E. coli* under osmotic stress not only allowed the bacteria to grow under these otherwise inhibitory conditions but also produced a periplasmic environment for the generation of correctly folded recombinant proteins.

Proteins that are synthesized with signal sequences are exported from the cell

Gram-negative bacteria such as *E. coli* have a complex wall–membrane structure comprising an inner, cytoplasmic membrane separated from an outer membrane by a cell wall and periplasmic space. Secreted proteins may be released into the periplasm or integrated into or transported across the outer membrane. In *E. coli* it has been established that protein export through the inner membrane to the periplasm or to the outer membrane is achieved by a universal mechanism known as the general export pathway (GEP). This involves the *sec* gene products (for review see Lory 1998). Proteins that enter the GEP are synthesized in the cytoplasm with a signal sequence at the N terminus. This sequence is cleaved by a signal or leader peptidase during transport. A signal sequence has three domains: a positively charged amino-terminal region, a hydrophobic core, consisting of five to 15 hydrophobic amino acids, and a leader peptidase cleavage site. A signal sequence attached to a normally cytoplasmic protein will direct it to the export pathway.

Many signal sequences derived from naturally occurring secretory proteins (e.g. OmpA, OmpT, PelB, β-lactamase, alkaline phosphatase, and phage M13 gIII) support the efficient translocation of heterologous peptides across the inner membrane when fused to their amino termini. In some cases, however, the preproteins are not readily exported and either become "jammed" in the inner membrane, accumulate in precursor inclusion bodies or are rapidly degraded within the cytoplasm. In practice, it may be necessary to try several signal sequences (Berges *et al.* 1996) and/or overproduce different chaperons to optimize the translocation of a particular heterologous protein. A first step would be to try the secretion vectors offered by a number of molecular-biology suppliers and which are variants of the vectors described above.

It is possible to engineer proteins such that they are transported through the outer membrane and are secreted into the growth medium. This is achieved by making use of the type I, Sec-independent secretion system. The prototype type I system is the hemolysin transport system, which requires a short carboxy-terminal secretion signal, two translocators (HlyB and D), and the outer-membrane protein TolC. If the protein of interest is fused to the carboxy-terminal secretion signal of a type I-secreted protein, it will be secreted into the medium provided HlyB, HlyD, and TolC are expressed as well (Spreng *et al.*

2000). An alternative presentation of recombinant proteins is to express them on the surface of the bacterial cell using any one of a number of carrier proteins (for review, see Cornelis 2000).

The Gateway® system is a highly efficient method for transferring DNA fragments to a large number of different vectors

The traditional method for moving a gene fragment from one vector to another would involve restriction enzyme digestion and would include some or all of the following steps:

- restriction endonuclease digestion of the donor plasmid;
- purification of the gene insert;
- restriction digestion of the target vector;
- ligation of the gene insert with the digested target vector;
- transformation of *E. coli* and selection of the new recombinant plasmid;
- isolation of the new plasmid and confirmation by endonuclease digestion and gel electrophoresis that it has the correct properties.

The Gateway system is designed to replace all these steps by using the phage λ site-specific recombinase (Box 3.3, p. 51) in a simple two-step procedure (Fig. 5.18).

To use the Gateway system, the gene of interest is cloned by conventional means in a Gateway entry vector. This vector carries two *att* sites that are recognized by the λ site-specific recombinase and the cloned gene should lie between them. Moving this cloned gene to another vector (destination vector) is very simple. The entry vector containing the cloned gene is mixed with the destination vector and λ recombinase *in vitro* and after a short incubation period the desired recombinant is selected by transformation (Fig. 5.18). The beauty of the Gateway system is that after the initial entry clone is made the gene of interest can be transferred to many other vectors (Fig. 5.19) while maintaining orientation and reading frame with high efficiencies (>99%).

Putting it all together: vectors with combinations of features

Many of the vectors in current use, particularly those that are commercially available, have combinations of the features described in previous sections. Two

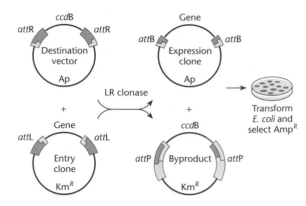

Fig. 5.18 The principle of the Gateway system. Following transformation, the only cells that can form colonies are ones carrying a plasmid encoding Ap^R and lacking *ccdB*. Ap^R, ampicillin resistance; Km^R, kanamycin resistance; *ccdB*, a counterselectable gene (see Box 4.1); *att*, sites for the λ recombinase (see Box 3.3). (Figure reproduced courtesy of the InVitrogen Life Technologies.)

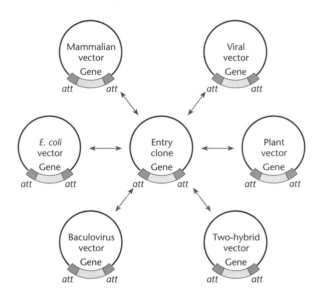

Fig. 5.19 The use of the Gateway system to move a gene unchanged from an entry clone to many different vectors. (Figure reproduced courtesy of the InVitrogen Life Technologies.)

examples are described here to show the connection between the different features. The first example is the LITMUS vectors that were described earlier (p. 83) and which are used for the generation of RNA probes. They exhibit the following features:

- The polylinkers are located in the *lacZ'* gene and inserts in the polylinker prevent α-complementation. Thus blue/white screening (see Box 3.2 on p. 45) can be used to distinguish clones with inserts from those containing vector only.

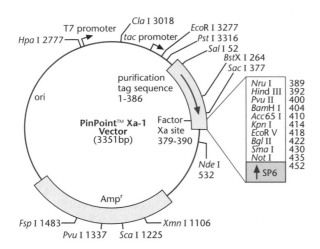

Fig. 5.20 Structural features of a PinPoint™ vector. (Figure reproduced courtesy of Promega Corporation.)

- The LITMUS polylinkers contain a total of 32 unique restriction sites. Twenty-nine of these enzymes leave four-base overhangs and three leave blunt ends. The three blunt cutting enzymes have been placed at either end of the polylinker and in the middle of it.
- The vectors carry both the pUC and the M13 *ori* regions. Under normal conditions the vector replicates as a double-stranded plasmid but, on infection with helper phage (M13KO7), single-stranded molecules are produced and packaged in phage protein.
- The single-stranded molecules produced on helper phage addition have all the features necessary for DNA sequencing (see p. 81).
- The vectors are small (<3 kb) and with a pUC *ori* have a high copy number.

The second example is the PinPoint series of expression vectors (Fig. 5.20). These vectors have the following features:

- Expression is under the control of either the T7 or the *tac* promoter, allowing the user great flexibility of control over the synthesis of the cloned gene product.
- Some of them carry a DNA sequence specifying synthesis of a signal peptide.
- Presence of an MCS adjacent to a factor-Xa cleavage site.
- Synthesis of an N-terminal biotinylated sequence to facilitate purification.

- Three different vectors of each type permitting translation of the cloned gene insert in each of the three reading frames.
- Presence of a phage SP6 promoter distal to the MCS to permit the synthesis of RNA probes complementary to the cloned gene. Note that the orientation of the cloned gene is known and so the RNA probe need only be synthesized from one strand.

What is absent from these vectors is an M13 origin of replication to facilitate synthesis of single strands for DNA sequencing.

Suggested reading

Goldberg A.L. (2003) Protein degradation and protection against misfolded or damaged proteins. *Nature* **426**, 895–9.
This review provides useful insight into the problems associated with overexpression of proteins.

Hurst M. & Dubel S. (2005) Phage display vectors for the *in vitro* generation of antibody fragments. *Methods in Molecular Biology* **295**, 71–96.

Lee S.Y., Choi J.H. & Xu Z. (2003) Microbial cell surface display. *Trends in Biotechnology* **21**, 45–52.

Mergulhao F.J., Summers D.K. & Monteiro G.A. (2005) Recombinant protein secretion in *Escherichia coli*. *Biotechnology Advances* **23**, 177–202.

Pugsley A.P., Francetic O., Driessen A.J. & de Lorenzo V. (2004) Getting out: protein traffic in prokaryotes. *Molecular Microbiology* **52**, 3–11.
These four papers present different aspects of protein export in E.coli and other bacteria.

Schweizer H.P. (2003) Applications of the *Saccharomyces cerevisiae* Flp-FRT system in bacterial genetics. *Journal of Molecular Microbiology and Biotechnology* **5**, 67–77.
A very useful review of methodology for allele replacement in strain construction.

Gustafsson C., Govinddarajan S. & Minshull J. (2004) Codon bias and heterologous protein expression. *Trends in Biotechnology* **22**, 346–53.

Swartz J.R. (2001) Advances in *Escherichia coli* production of therapeutic proteins. *Current Opinion in Biotechnology* **12**, 195–201.
Two short but excellent reviews of the factors that affect overexpression of proteins.

Jansson J.K. (2003) Marker and reporter genes: illuminating tools for environmental microbiologists. *Current Opinion in Microbiology* **6**, 310–16.
This review focuses on a topic not discussed here: the labeling of strains and plasmids with non-selective markers.

CHAPTER 6

Gene-cloning strategies

Introduction

Papers reporting the results of genome mapping and sequencing projects now appear in the scientific literature at the rate of about one every fortnight, and by the time this book is published it is likely that over 200 genomes will have been completely sequenced and annotated. Already, it is possible to obtain "off-the-shelf" cloned genes and cDNAs for many of our most important organisms, and even some very obscure ones. As an example, Table 6.1 lists the mammalian BAC libraries that are currently available. In addition to humans, this includes all the expected laboratory model species, many domestic mammals, and a diverse collection of other mammals which are being studied to facilitate phylogenetic analysis. Is it necessary or worth the effort to clone single genes any more?

There are several reasons why single-gene cloning is still an important part of molecular biology. One rather prosaic reason is that there remain many genomes that have yet to be mapped or sequenced – an investigator wishing to clone a specific gene from the polar bear, for example, would have no choice other than to work at the single gene level. More importantly, however, genome sequences reveal only part of the information available for a given gene. In contrast, cDNA sequences, which are reverse transcribed from mRNA, reveal expression profiles in different cell types, developmental stages, and in response to natural or experimentally simulated external stimuli. Additionally, for higher organisms, cDNA sequences provide useful information about splice isoforms and their abundance in different tissues and developmental stages. A further reason is that many cloning strategies reveal extra functional information about genes, e.g. expression profiles or biochemical functions. The task of functionally annotating genomes always lags way behind the structural annotation phase, and gene-cloning strategies therefore remain of value for the elucidation of gene function.

In Chapters 2–5 we discussed DNA cutting and joining techniques and the different types of vectors that are available for cloning DNA molecules. One question that was overlooked in these earlier chapters was how these segments of DNA for cloning are obtained in the first place. In simple subcloning procedures, where DNA fragments are removed from one vector and inserted into another, the target DNA is available in a pure form as part of the source vector. We now need to consider what happens when the source of donor DNA is very complex. We may wish, for example, to isolate a single gene from the human genome or from the maize genome. In such cases, the target sequence could be diluted over a million-fold by other genes and unwanted genomic DNA. We need to sift rapidly through large numbers of unwanted sequences to identify our particular target.

There are two major approaches for isolating sequences from complex sources such as genomic DNA or cDNA, but in each case the cloning strategy is divided into four stages as shown in Fig. 6.1. The first, a cell-based cloning strategy, is to divide the source DNA into manageable fragments and *clone everything*. Such a collection of clones, representative of the entire starting population, is known as a *library*. We must then *screen the library* to identify our clone of interest using a procedure that discriminates between the desired clone and all the others. A number of such procedures are discussed later in the chapter. The second strategy is to selectively amplify the target sequence directly from the source DNA using the polymerase chain reaction (PCR), and then clone this individual fragment. Each strategy has its advantages and disadvantages. In the library approach, screening is carried out after the entire source DNA population has been cloned indiscriminately. Conversely, in the PCR approach, the screening step is built into the first stage of the procedure, so that only these selected fragments are actually cloned. In this chapter we consider principles for the construction and screening of genomic

Table 6.1 Mammalian BAC libraries currently available or in production.

MONOTREMES Echidna, *Tachyglossus aculeatus* Platypus, *Ornithorhynchus anatinus* **MARSUPIALS** American opossum, *Didelphis virginianus* Laboratory opossum, *Monodelphis domestica* Tammar wallaby, *Macropus eugenii* **PLACENTALS** <u>Afrotheria</u> Proboscidea African savanna elephant, *Loxodonta africana* <u>Xenarthra</u> Nine-banded armadillo, *Dasypus novemcinctus* <u>Laurasiatheria</u> Carnivora Domestic dog, *Canis familiaris* Domestic cat, *Felis catus* Clouded leopard, *Neofelis nebulosa* Perissodactyla Domestic horse, *Equus cabailus* Cetartiodactyla Formosan muntjac, *Muntiacus reevesi* Indian muntjac, *Muntiacus muntjac* Domestic cattle, *Bos taurus* Domestic sheep, *Ovis aries* Domestic pig, *Sus scrofa* Domestic goat, *Capra hircus* Chiroptera Horseshoe bat, *Rhinolophus ferrumequinum* Brown bat, *Myotis lucifugus* Flying fox, *Pteropus livingstoni* Eulipotyphla Hedgehog, *Atilerex albiventris* Shrew, *Sorex araneus*	<u>Euarchontoglires</u> Lagomorpha Rabbit, *Oryctolagus cuniculus* Rodentia Deer mouse, *Peromyscus maniculatus* Hamster, *Cricetulus griseus* Mouse, *Mus musculus* Rat, *Rattus norvegicus* Ground squirrel, *Spermophilus tridecemlineatus* Primates Baboon, *Papio hamadryas* Black lemur, *Eulemur macaco* Chimpanzee, *Pan troglodytes* Colobus monkey, *Colobus guereza* Dusky titi, *Callicebus moloch* Galago, *Otolemur gametti* Gibbon, *Hylobates concolor* Gorilla, *Gorilla gorilla* Ring-tailed lemur, *Lemur catta* Macaque, *Macaca mulatta* Marmoset, *Callithrix jacchus* Mouse lemur, *Microcebus murinus* Sumatran orangutan, *Pongo pygmaeus* Owl Monkey, *Aotus trivirgatus/nancymai* Squirrel monkey, *Saimiri boliviensis* Tarsier, *Tarsius bancanus* Vervet monkey, *Cercopithacus aethiops* Scandentia Tree shrew, *Tupaia minor*

For more information, see http://www.genome.gov/10001852 and http://bacpac.chori.org/.

and cDNA libraries, and compare the library-based route for gene isolation to equivalent PCR-based techniques.

Genomic DNA libraries are generated by fragmenting the genome and cloning overlapping fragments in vectors

The first genomic libraries were cloned in simple plasmid and phage vectors

Although the human genome has been mapped, cloned, and sequenced in its entirety, it is still useful to examine how we might go about generating a genomic library and isolating a given gene, because

the principles apply to all genomes. We could simply digest total genomic DNA with a restriction endonuclease, such as *Eco*RI, insert the fragments into a suitable phage λ vector, and then attempt to isolate the desired clone. How many recombinants would we have to screen in order to isolate the right one? Assuming *Eco*RI gives, on average, fragments of about 4 kb, and given that the size of the human haploid genome is in the region of 2.8×10^6 kb, it is clear that over 7×10^5 independent recombinants must be prepared and screened in order to have a reasonable chance of including the desired sequence. In other words we have to obtain a very large number of recombinants, which together contain a complete collection of all of the DNA sequences in the entire genome, a *genomic library*.

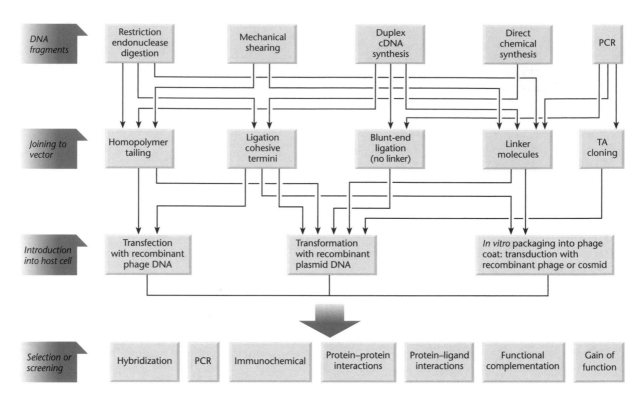

Fig. 6.1 Generalized overview of cloning strategies, with favored routes shown by arrows. Note that in cell-based cloning strategies, DNA fragments are initially generated and cloned in a non-specific manner, so that screening for the desired clone is carried out at the end of the process. Conversely, when specific DNA fragments are obtained by PCR or direct chemical synthesis, there is no need for subsequent screening.

There are two problems with the above approach. First, the gene may be cut internally one or more times by *Eco*RI so that it is not obtained as a single fragment. This is likely if the gene is large. Also, it may be desirable to obtain extensive regions flanking the gene or whole gene clusters. Fragments averaging about 4 kb are likely to be inconveniently short. Alternatively, the gene may be contained on an *Eco*RI fragment that is larger than the vector can accept. In this case the appropriate gene would not be cloned at all.

These problems can be overcome by cloning *random* DNA fragments of a large size (for λ replacement vectors, approximately 20 kb). Since the DNA is randomly fragmented, there will be no systematic exclusion of any sequence. Furthermore, clones will overlap one another, allowing the sequence of very large genes to be assembled. Because of the larger size of each cloned DNA fragment, fewer clones are required for a complete or nearly complete library. How many clones are required? Let n be the size of the genome relative to a single cloned fragment. Thus for the human genome (2.8×10^6 kb) and an average cloned fragment size of 20 kb, $n = 1.4 \times 10^5$. The number of independent recombinants required

in the library must be greater than n, because sampling variation will lead to the inclusion of some sequences several times, and the exclusion of other sequences in a library of just n recombinants. Clarke & Carbon (1976) derived a formula that relates the probability (P) of including any DNA sequence in a random library of N independent recombinants:

$$ N = \frac{\ln(1-P)}{\ln\left(1-\dfrac{1}{n}\right)} $$

Therefore, to achieve a 95% probability ($P = 0.95$) of including any particular sequence in a random human genomic DNA library of 20 kb fragment size:

$$ N = \frac{\ln(1-0.95)}{\ln\left(1-\dfrac{1}{1.4 \times 10^5}\right)} = 4.2 \times 10^5 $$

Notice that a considerably higher number of recombinants is required to achieve a 99% probability, for here $N = 6.5 \times 10^5$.

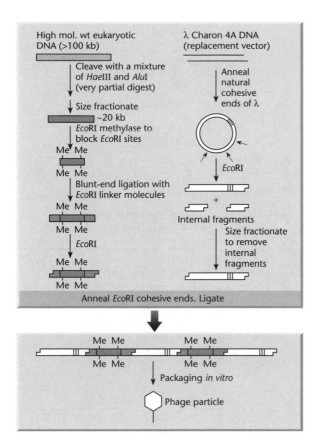

Fig. 6.2 Maniatis' strategy for producing a representative gene library.

How can appropriately sized random fragments be produced? Various methods are available. Random breakage by mechanical shearing is appropriate because the average fragment size can be controlled, but insertion of the resulting fragments into vectors requires additional modification steps. The more commonly used procedure involves restriction endonucleases. In the strategy devised by Maniatis *et al.* (1978) (Fig. 6.2) the target DNA is digested with a mixture of *two* restriction enzymes. These enzymes have tetranucleotide recognition sites, which therefore occur frequently in the target DNA and in a complete double-digest would produce fragments averaging less than 1 kb. However, only a partial restriction digest is carried out, and therefore the majority of the fragments are large (in the range 10–30 kb). Given that the chances of cutting at each of the available restriction sites are more or less equivalent, such a reaction effectively produces a random set of overlapping fragments. These can be size fractionated, e.g. by gel electrophoresis, so as to give a random population of fragments of about 20 kb, which are suitable for insertion into a λ

replacement vector. Packaging *in vitro* (p. 70) ensures that an appropriately large number of independent recombinants can be recovered, which will give an almost completely representative library.

More sophisticated vectors have been developed to facilitate genomic library construction

In the Maniatis strategy, the use of two different restriction endonucleases with completely unrelated recognition sites, *Hae*III and *Alu*I, assists in obtaining fragmentation that is nearly random. These enzymes both produce blunt ends, and the cloning strategy requires linkers (see Fig. 6.2). Therefore, in the early days of vector development, a large number of different vectors became available with alternative restriction sites and genetic markers suitable for varied cloning strategies. A good example of this diversity is the Charon series, which included both insertion and replacement type vectors (Blattner *et al.* 1977, Williams & Blattner 1979).

A convenient simplification can be achieved by using a *single* restriction endonuclease that cuts frequently, such as *Sau*3AI. This will create a partial digest that is slightly less random than that achieved with a pair of enzymes. However, it has the great advantage that the *Sau*3AI fragments can be readily inserted into λ replacement vectors, such as λEMBL3 (Frischauf *et al.* 1983), which have been digested with *Bam*HI (Fig. 6.3). This is because *Sau*3AI and *Bam*HI create the same cohesive ends (see p. 41). Due to the convenience and efficiency of this strategy, the λEMBL series of vectors have been very widely used for genomic library construction. Note that λEMBL vectors also carry the *red* and *gam* genes on the stuffer fragment and a *chi* site on one of the vector arms, allowing convenient positive selection on the basis of the Spi phenotype (see p. 67). Most λ vectors currently used for genomic library construction are positively selected on this basis, including λ2001 (Karn *et al.* 1984), λDASH, and λFIX (Sorge 1988). λDASH and λFIX and their derivatives are particularly versatile because the multiple cloning sites flanking the stuffer fragment contain opposed promoters for the T3 and T7 RNA polymerases. If the recombinant vector is digested with a restriction endonuclease that cuts frequently, only short fragments of insert DNA are left attached to these promoters. This allows RNA probes to be generated corresponding to the *ends* of any genomic insert. These are ideal for probing the library to identify

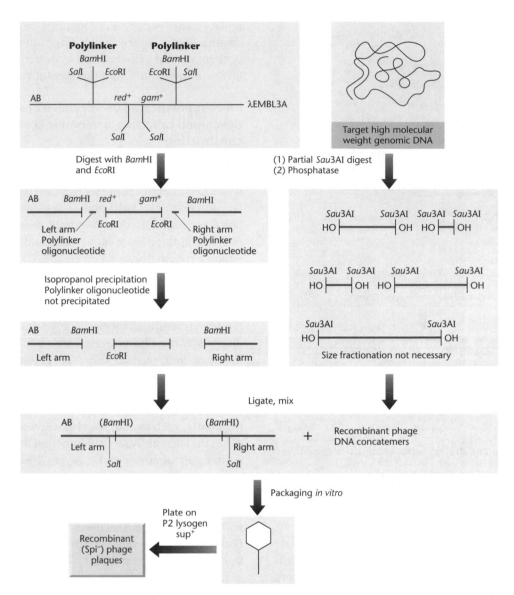

Fig. 6.3 Creation of a genomic DNA library using the phage-λ vector EMBL3A. High-molecular-weight genomic DNA is partially digested with *Sau*3AI. The fragments are treated with phosphatase to remove their 5′ phosphate groups. The vector is digested with *Bam*HI and *Eco*RI, which cut within the polylinker sites. The tiny *Bam*HI/*Eco*RI polylinker fragments are discarded in the isopropanol precipitation, or alternatively the vector arms may be purified by preparative agarose gel electrophoresis. The vector arms are then ligated with the partially digested genomic DNA. The phosphatase treatment prevents the genomic DNA fragments from ligating together. Non-recombinant vector cannot reform because the small polylinker fragments have been discarded. The only packageable molecules are recombinant phages. These are obtained as plaques on a P2 lysogen of *sup*+ *E. coli*. The Spi⁻ selection ensures recovery of phage lacking *red* and *gam* genes. A *sup*+ host is necessary because, in this example, the vector carries amber mutations in genes *A* and *B*. These mutations increase biological containment, and can be applied to selection procedures, such as recombinational selection, or tagging DNA with a *sup*+ gene. Ultimately, the foreign DNA can be excised from the vector by virtue of the *Sal*I sites in the polylinker. (*Note*: Rogers *et al.* (1988) have shown that the EMBL3 polylinker sequence is not exactly as originally described. It contains an extra sequence with a previously unreported *Pst*I site. This does not affect most applications as a vector.)

overlapping clones and have the great advantage that they can be made conveniently, directly from the vector, without recourse to subcloning. Vector maps of λDASH and λFIX are shown in Fig. 6.4. λFIX is similar to λDASH, except that it incorporates

additional *Xho*I sites flanking the stuffer fragment. Digestion of the vector with *Xho*I followed by partial filling of the sticky ends prevents vector re-ligation. However, partially filled *Sau*3AI sticky ends are compatible with the partially filled *Xho*I ends, although

(a)

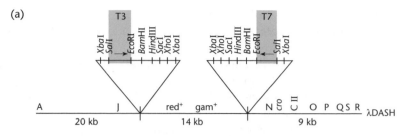

(b)

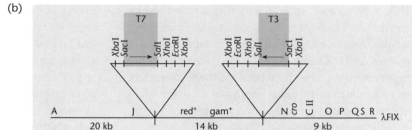

Fig. 6.4 The replacement vectors λDASH and λFIX. Promoters specific for the bacteriophage T3 and T7 RNA polymerases are located adjacent to the cloning sites, allowing RNA probes to be generated that correspond to each end of the insert.

not with each other. This strategy prevents the ligation of vector arms without genomic DNA, and also prevents the insertion of multiple fragments.

Genomic libraries for higher eukaryotes are usually constructed using high-capacity vectors

In place of phage λ derivatives, a number of higher capacity cloning vectors such as cosmids, BACs, PACs, and YACs are now available for the construction of genomic libraries. The advantage of such vectors is that the average insert size is much larger than for λ replacement vectors. Thus, the number of recombinants that need to be screened to identify a particular gene of interest is correspondingly lower, large genes are more likely to be contained within a single clone, and fewer clones are required to assemble a contig. Such vectors are therefore widely used for the construction of libraries representing large genomes.

Generally, strategies similar to the Maniatis method discussed above are used to construct such libraries, except that the partial restriction digest conditions are optimized for larger fragment sizes, and size fractionation must be preformed by specialized electrophoresis methods that can separate fragments over 30 kb in length. The development of modern vectors and cloning strategies has simplified library construction to the point where many workers now prefer to create a new library for each screening, but pre-made libraries are available from many commercial sources and the same companies often offer custom library services. These libraries are

often of high quality and such services are becoming increasingly popular. The highest capacity vectors – BACs, PACs, and YACs – would seem to be ideal for library construction because of the very large insert sizes. However, such libraries are generally more difficult to prepare, and the larger inserts can be less than straightforward to work with. The main application of BAC, PAC, and YAC libraries is for genome mapping, sequencing, and the assembly of clone contigs.

The PCR can be used as an alternative to genomic DNA cloning

The PCR is a robust technique for amplifying specific DNA sequences from complex sources. In principle, therefore, PCR with specific primers could be used to isolate genes directly from genomic DNA, obviating the need for the production of genomic libraries. However, a serious limitation is that standard PCR conditions are suitable only for the amplification of short products. The maximum product size that can be obtained is about 5 kb, although the typical size is more likely to be 1–2 kb. This reflects the poor processivity of PCR enzymes such as *Taq* polymerase, and their lack of proofreading activity. Both of these deficiencies increase the likelihood of the enzyme detaching from the template, especially if the template is long. The extreme reaction conditions required for the PCR are also thought to cause damage to bases and generate nicks in DNA strands, increasing the probability of premature termination on long templates.

Long PCR uses a mixture of enzymes to amplify long DNA templates

Modifications to reaction conditions can improve polymerase processivity by lowering the reaction temperature and increasing the pH, thereby protecting the template from damage (Foord & Rose 1994). The use of such conditions in combination with *two* DNA polymerases, one of which is a proofreading enzyme, has been shown to dramatically improve the performance of the PCR using long templates (Barnes 1994, Cheng *et al.* 1994a). Essentially, the improvements come about because the proofreading enzyme removes mismatched bases that are often incorporated into growing DNA strands by enzymes such as *Taq* polymerase. Under normal conditions, *Taq* polymerase would stall at these obstructions, and lacking the intrinsic proofreading activity to correct them, the reaction would most likely be aborted.

Using such polymerase mixtures, it has been possible to amplify DNA fragments of up to 22 kb directly from human genomic DNA, almost the entire 16.6-kb human mitochondrial genome and the complete or near complete genomes of several viruses, including 42 kb of the 45-kb phage λ genome (Cheng *et al.* 1994a, b). Several commercial companies now provide cocktails of enzymes suitable for long PCR, e.g. TaqPlus Long PCR system, marketed by Stratagene, which is essentially a mixture of *Taq* polymerase and the thermostable proofreading enzyme *Pfu* polymerase. The technique has been applied to the structural analysis of human genes (e.g. Ruzzo *et al.* 1998, Bochmann *et al.* 1999), and viral genomes, including HIV (Dittmar *et al.* 1997). Long PCR has particular diagnostic value for the analysis of human triplet repeat disorders, such as Friedreich's ataxia (Lamont *et al.* 1997). However, while long PCR is useful for the isolation of genes where sequence information is already available, it is unlikely to replace the use of genomic libraries since the latter represent a permanent, full-genome resource that can be shared by numerous laboratories. Indeed, genomic libraries may be used in preference to total genomic DNA as the starting point for gene isolation by long PCR.

Fragment libraries can be prepared from material that is unsuitable for conventional library cloning

Traditional genomic libraries cannot be prepared from small amounts of starting material, e.g. single cells, or from problematical sources such as fixed tissue. In these cases, the PCR is the only available strategy for gene isolation. However, as well as being useful for the isolation of specific fragments, the PCR can be used to generate libraries, i.e. by amplifying a representative collection of random genomic fragments. This can be achieved either using random primers followed by size selection for suitable PCR products, or a strategy in which genomic DNA is digested with restriction enzymes and then linkers are ligated to the ends of the DNA fragments providing annealing sites for one specific type of primer (e.g. Cheung & Nelson 1996, Zhang *et al.* 1992). These techniques are powerful because they allow genomic fragment libraries to be prepared from material that could not yield DNA of suitable quality or quantity for conventional library construction, but until recently competition among the templates generally has not allowed the production of truly representative libraries.

This problem has been addressed in a strategy called whole-genome amplification, in which the entire genome is amplified by PCR without any bias towards particular sequences (Lasken & Egholm 2003). Although several PCR-based techniques have been developed (p. 34) they produce short fragments which have limited usefulness. A more recent development called multiple displacement amplification (MDA) involves a branching reaction and utilizes the high-fidelity and highly processive DNA polymerase from bacteriophage φ29. The product length is usually between 20 kb and 200 kb, which is suitable for genomic library construction, and the bias between loci is less than one order of magnitude, while those of PCR-based techniques range from 10^3 to 10^6. The principle of MDA is explained on p. 34.

Complementary DNA (cDNA) libraries are generated by the reverse transcription of mRNA

cDNA is representative of the mRNA population, and therefore reflects mRNA levels and the diversity of splice isoforms in particular tissues

Complementary DNA (cDNA) is prepared by reverse transcribing cellular mRNA. Cloned eukaryotic cDNAs have their own special uses, which derive from the fact that they lack introns and other non-coding sequences present in the corresponding genomic

DNA. Introns are rare in bacteria but occur in most genes of higher eukaryotes. They can be situated within the coding sequence itself, where they then interrupt the colinear relationship of the gene and its encoded polypeptide, or they may occur in the 5′ or 3′ untranslated regions. In any event, they are copied by RNA polymerase when the gene is transcribed. The primary transcript goes through a series of processing events in the nucleus before appearing in the cytoplasm as mature mRNA. These events include the removal of intron sequences by a process called *splicing*. In mammals, some genes contain numerous large introns that represent the vast majority of the sequence. For example, the human dystrophin gene contains 79 introns, representing over 99% of the sequence. The gene is nearly 2.5 Mb in length yet the corresponding cDNA is only just over 11 kb. Thus, one advantage of cDNA cloning is that in many cases the size of the cDNA clone is significantly lower than that of the corresponding genomic clone. Since removal of eukaryotic intron transcripts by splicing does not occur in bacteria, eukaryotic cDNA clones find application where bacterial expression of the foreign DNA is necessary, either as a prerequisite for detecting the clone (see p. 116), or because expression of the polypeptide product is the primary objective. Also, where the sequence of the genomic DNA is available, the position of intron/exon boundaries can be assigned by comparison with the cDNA sequence.

Under some circumstances, it may be possible to prepare cDNA directly from a purified mRNA species. Much more commonly a *cDNA library* is prepared by reverse transcribing a population of mRNAs, and then screening for particular clones. An important concept is that the cDNA library is representative of the RNA population from which it was derived. Thus, whereas genomic libraries are

essentially the same regardless of the cell type or developmental stage from which the DNA was isolated, the contents of cDNA libraries will vary widely according to these parameters. A given cDNA library will also be enriched for abundant mRNAs but may contain only a few clones representing rare mRNAs. Furthermore, where a gene is differentially spliced, a cDNA library will contain different clones representing alternative splice variants.

Table 6.2 shows the abundances of different classes of mRNAs in two representative tissues. Generally, mRNAs can be described as abundant, moderately abundant, or rare. Notice that in the chicken oviduct, one mRNA type is superabundant. This encodes ovalbumin, the major egg-white protein. Therefore, the starting population is naturally so enriched in ovalbumin mRNA that isolating the ovalbumin cDNA can be achieved without the use of a library. An appropriate strategy for obtaining such abundant cDNAs is to clone them directly in an M13 vector such as M13mp8. A set of clones can then be sequenced immediately and identified on the basis of the polypeptide that each encodes. A successful demonstration of this strategy was reported by Putney *et al.* (1983), who determined DNA sequences of 178 randomly chosen muscle cDNA clones. Based on the amino acid sequences available for 19 abundant muscle-specific proteins, they were able to identify clones corresponding to 13 of these 19 proteins, including several protein variants.

For the isolation of cDNA clones in the moderate and low abundance classes it is usually necessary to construct a cDNA library. Once again the high efficiency obtained by packaging *in vitro* makes phage λ vectors attractive for obtaining large numbers of cDNA clones. λ insertion vectors are particularly well suited for cDNA cloning and some of the most widely used vectors are discussed in Box 6.1.

Table 6.2
Abundance classes of typical mRNA populations.

Source	Number of different mRNAs	Abundance (molecules/cell)
Mouse liver cytoplasmic poly(A)⁺	9	12,000
	700	300
	11,500	15
Chick oviduct polysomal poly(A)⁺	1	100,000
	7	4,000
	12,500	5

References: mouse (Young *et al.* 1976); chick oviduct (Axel *et al.* 1976).

Box 6.1 Phage-λ vectors for cDNA cloning and expression

λgt10 and λgt11

Most early cDNA libraries were constructed using plasmid vectors, and were difficult to store and maintain for long periods. They were largely replaced by phage-λ libraries, which can be stored indefinitely and can also be prepared to much higher titers. λgt10 and λgt11 were the standard vectors for cDNA cloning until about 1990. Both λgt10 and λgt11 are insertion vectors, and they can accept approximately 7.6 kb and 7.2 kb of foreign DNA, respectively. In each case, the foreign DNA is introduced at a unique *EcoRI* cloning site. λgt10 is used to make libraries that are screened by hybridization. The *EcoRI* site interrupts the phage *cI* gene, allowing selection on the basis of plaque morphology. λgt11 contains an *E. coli lacZ* gene driven by the *lac* promoter. If inserted

in the correct orientation and reading frame, cDNA sequences cloned in this vector can be expressed as β-galactosidase fusion proteins, and can be detected by immunological screening or screening with other ligands (see p. 117). λgt11 libraries can also be screened by hybridization, although λgt10 is more appropriate for this screening strategy because higher titers are possible.

λZAP series

While phage-λ vectors generate better libraries, they cannot be manipulated *in vitro* with the convenience of plasmid vectors. Therefore, phage clones have to be laboriously subcloned back into plasmids for further analysis. This limitation of conventional phage-λ vectors has been addressed by the development of hybrids, sometimes called

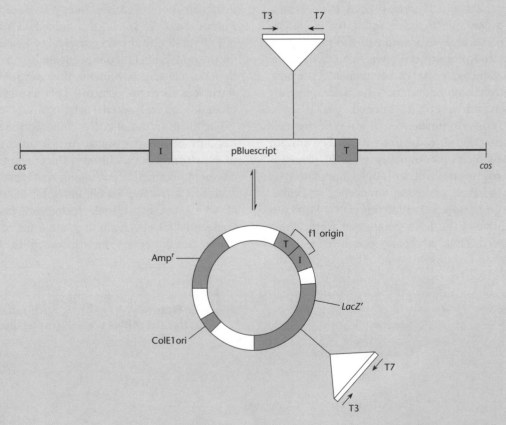

Fig. B6.1 Linear phage map of the prototype λZAP vector with the circular map of the excised pBluescript plasmid shown below it.

continued

Box 6.1 *continued*

phasmids, which possess the most attractive features of both bacteriophage λ and plasmids (see Chapter 5). The most popular current vectors for cDNA cloning are undoubtedly those of the λZAP series marketed by Stratagene (Short *et al.* 1988). A map of the original λZAP vector is shown opposite. The advantageous features of this vector are: (i) the high capacity – up to 10 kb of foreign DNA can be cloned, which is large enough to encompass most cDNAs; (ii) the presence of a polylinker with six unique restriction sites, which increases cloning versatility and also allows directional cloning; and (iii) the availability of T3 and T7 RNA polymerase sites flanking the polylinker, allowing sense and antisense RNA to be prepared from the insert. Most importantly, all these features

are included within a plasmid vector called pBluescript, which is itself inserted into the phage genome. Thus the cDNA clone can be recovered from the phage and propagated as a high-copy-number plasmid without any subcloning, simply by coinfecting the bacteria with a helper f1 phage that nicks the λZAP vector at the flanks of the plasmid and facilitates excision. Another member of this series, λZAP Express, also includes the human cytomegalovirus promoter and SV40 terminator, so that fusion proteins can be expressed in mammalian cells as well as bacteria. Thus, cDNA libraries can be cloned in the phage vector in *E. coli*, rescued as plasmids and then transfected into mammalian cells for expression cloning.

Typically, 10^5 clones is sufficient for the isolation of low-abundance mRNAs from most cell types, i.e. those present at 15 molecules per cell or above. However, some mRNAs are even less abundant than this, and may be further diluted if they are expressed in only a few specific cells in a particular tissue. Under these circumstances it may be worth enriching the mRNA preparation prior to library construction, e.g. by size fractionation, and testing the fractions for the presence of the desired molecule. One way in which this can be achieved is to inject mRNA fractions into *Xenopus* oocytes (p. 266) and test them for production of the corresponding protein (Melton 1987). See also the discussion of subtraction cloning on p. 122.

The first stage of cDNA library construction is the synthesis of double-stranded DNA using mRNA as the template

The synthesis of double-stranded cDNA suitable for insertion into a cloning vector involves three major steps: (1) first-strand DNA synthesis on the mRNA template, carried out with a reverse transcriptase; (2) removal of the RNA template; and (3) second-strand DNA synthesis using the first DNA strand as a template, carried out with a DNA-dependent DNA polymerase such as *E. coli* DNA polymerase I. All DNA polymerases, whether they use RNA or DNA as the template, require a primer to initiate strand synthesis.

The first reports of cDNA cloning were published in the mid-1970s, and were all based on the homopolymer tailing technique, which is described briefly in Chapter 3. Of several alternative methods, the one that became the most popular was that of Maniatis *et al.* (1976). This involved the use of an oligo-dT primer annealing at the polyadenylate tail of the mRNA to prime first-strand cDNA synthesis, and took advantage of the fact that the first cDNA strand has the tendency to transiently fold back on itself, forming a hairpin loop, resulting in self-priming of the second strand (Efstratiadis *et al.* 1976). After the synthesis of the second DNA strand, this loop must be cleaved with a single-strand-specific nuclease, e.g. S1 nuclease, to allow insertion into the cloning vector (Fig. 6.5).

A serious disadvantage of the hairpin method is that cleavage with S1 nuclease results in the loss of a certain amount of sequence at the 5′ end of the clone. This strategy has therefore been superseded by other methods in which the second strand is primed in a separate reaction. One of the simplest strategies is shown in Fig. 6.6 (Land *et al.* 1981). After first-strand synthesis, which is primed with an oligo-dT primer as usual, the cDNA is tailed with a string of cytidine residues using the enzyme terminal transferase. This artificial oligo-dC tail is then used as an annealing site for a synthetic oligo-dG primer, allowing synthesis of the second strand. Using this method, Land *et al.* (1981) were able to isolate a full-length cDNA

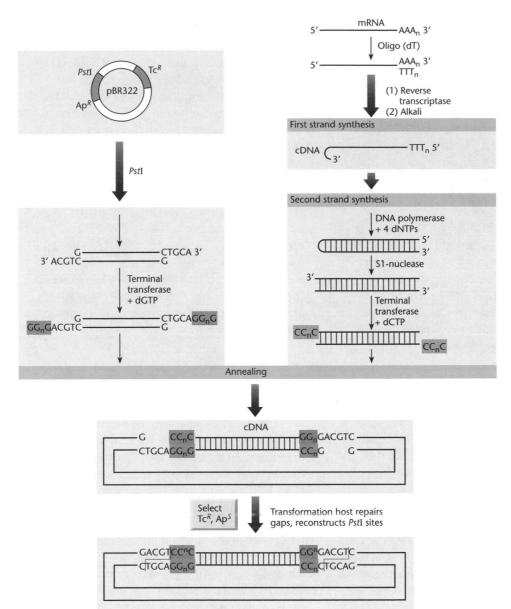

Fig. 6.5 An early cDNA cloning strategy, involving hairpin-primed second-strand DNA synthesis and homopolymer tailing to insert the cDNA into the vector.

corresponding to the chicken lysozyme gene. However, the efficiency can be lower for other cDNAs (e.g. Cooke *et al.* 1980).

For cDNA expression libraries, it is advantageous if the cDNA can be inserted into the vector in the correct orientation. With the self-priming method, this can be achieved by adding a synthetic linker to the double-stranded cDNA molecule before the hairpin loop is cleaved (e.g. Kurtz & Nicodemus 1981; Fig. 6.7a). Where the second strand is primed separately, direction cloning can be achieved using a oligo-dT primer containing a linker sequence (e.g. Coleclough & Erlitz; Fig. 6.7b). An alternative is to use primers for cDNA synthesis that are already

linked to a plasmid (Fig. 6.7c). This strategy was devised by Okayama & Berg (1982) and has two further notable characteristics. First, full-length cDNAs are *preferentially obtained* because an RNA–DNA hybrid molecule, the result of first-strand synthesis, is the substrate for a terminal transferase reaction. A cDNA that does not extend to the end of the mRNA will present a shielded 3-hydroxyl group, which is a poor substrate for tailing. Second, the second-strand synthesis step is primed by nicking the RNA at multiple sites with RNase H. Second-strand synthesis therefore occurs by a nick-translation type of reaction, which is highly efficient. Simpler cDNA cloning strategies incorporating replacement synthesis of the

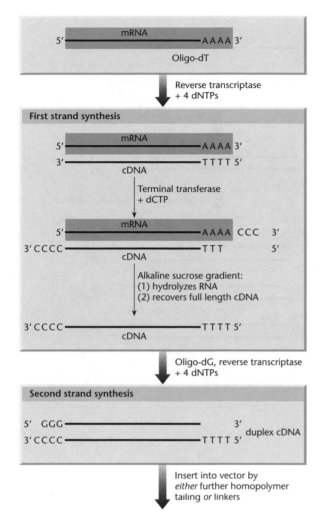

Fig. 6.6 Improved method for cDNA cloning. The first strand is tailed with oligo(dC) allowing the second strand to be initiated using an oligo(dG) primer.

second strand are widely used (e.g. Gubbler & Hoffman 1983, Lapeyre & Amalric 1985). The Gubbler–Hoffman reaction, as it is commonly known, is show in Fig. 6.8.

Obtaining full-length cDNA for cloning can be a challenge

Conventional approaches to the production of cDNA libraries have two major drawbacks. First, where oligo-dT primers are used to initiate first-strand synthesis, there is generally a 3′-end bias (preferential recovery of clones representing the 3′ end of cDNA sequences) in the resulting library. This can be addressed through the use of random oligonucleotide primers, usually hexamers, for both first- and second-

strand cDNA synthesis. However, while this eliminates 3′ end bias in library construction, the resulting clones are much smaller, such that full-length cDNAs must be assembled from several shorter fragments. Second, as the size of a cDNA increases, it becomes progressively more difficult to isolate full-length clones. This is partly due to deficiencies in the reverse transcriptase enzymes used for first-strand cDNA synthesis. The enzymes are usually purified from avian myelobastosis virus (AMV) or produced from a cloned Moloney murine leukemia virus (MuLV) gene in *E. coli*. Native enzymes have poor processivity and intrinsic RNase activity, which leads to degradation of the RNA template (Champoux 1995). Several companies produce engineered murine reverse transcriptases that lack RNase H activity, and these are more efficient in the production of full-length cDNAs (Gerard & D'Allesio 1993). An example is the enzyme SuperScript II, marketed by Life Technologies (Kotewicz *et al.* 1988). This enzyme can also carry out reverse transcription at temperatures of up to 50°C. The native enzymes function optimally at 37°C, and therefore tend to stall at sequences that are rich in secondary structure, as often found in 5′ and 3′ untranslated regions.

Despite improvements in reverse transcriptases, the generation of full-length clones corresponding to large mRNAs remains a problem. This has been addressed by the development of cDNA cloning strategies involving the selection of mRNAs with intact 5′ ends. Nearly all eukaryotic mRNAs have a 5′ end cap, a specialized, methylated guanidine residue that is inverted with respect to the rest of the strand and is recognized by the ribosome prior to the initiation of protein synthesis. Using a combination of cap selection and nuclease treatment, it is possible to select for full-length first-strand cDNAs, and thus generate libraries highly enriched in full-length clones.

An example of the above is the method described by Edery *et al.* (1995) (Fig. 6.9). In this strategy, first-strand cDNA synthesis is initiated as usual, using an oligo-dT primer. Following the synthesis reaction, the hybrid molecules are treated with RNase A, which only digests single-stranded RNA. DNA–RNA hybrids therefore remain intact. If the first-strand cDNA is full length, it reaches all the way to the 5′ cap of the mRNA, which is therefore protected from cleavage by RNase A. However, part-length cDNAs will leave a stretch of unprotected single-stranded RNA between the end of the double-stranded region and the cap, which is digested away with the

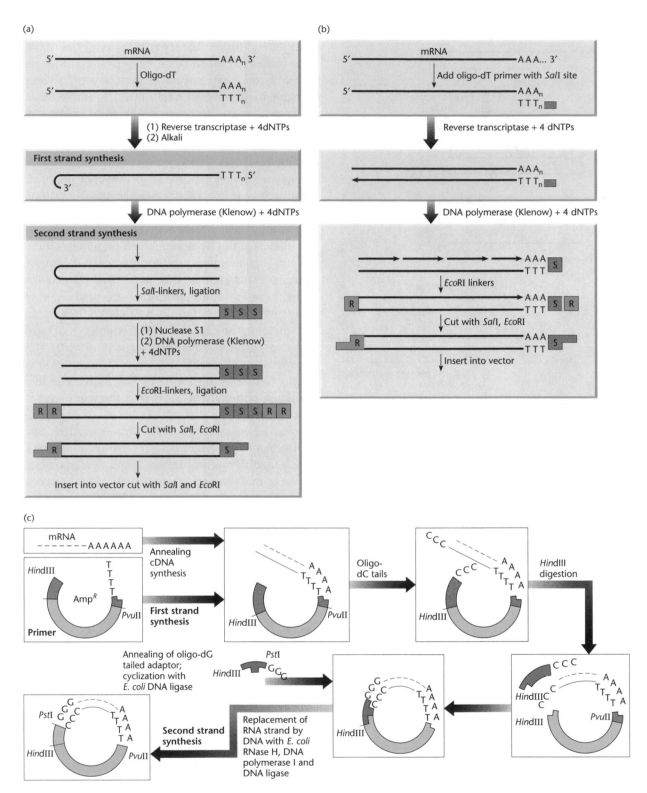

Fig. 6.7 Methods for directional cDNA cloning. (a) An early strategy in which the formation of a loop is exploited to place a specific linker (in this example, for *Sal*I) at the open end of the duplex cDNA. Following this ligation, the loop is cleaved and trimmed with S1 nuclease and *Eco*R1 linkers are added to both ends. Cleavage with *Eco*R1 and *Sal*I generates a restriction fragment that can be unidirectionally inserted into a vector cleaved with the same enzymes. (b) A similar strategy, but second-strand cDNA synthesis is random-primed. The oligo(dT) primer carries an extension forming a *Sal*I site. During second-strand synthesis, this forms a double-stranded *Sal*I linker. The addition of further *Eco*RI linkers to both ends allows the cDNA to be unidirectionally cloned, as above. (c) The strategy of Okayama & Berg (1982), where the mRNA is linked unidirectionally to the plasmid cloning vector prior to cDNA synthesis, by virtue of a cDNA tail.

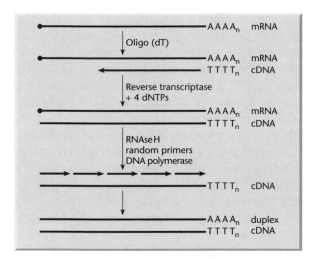

Fig. 6.8 The Gubbler–Hoffman method, a simple and general method for non-directional cDNA cloning. First-strand synthesis is primed using an oligo(dT) primer. When the first strand is complete, the RNA is removed with RNase H and the second strand is random-primed and synthesized with DNA polymerase I. T4 DNA polymerase is used to ensure that the molecule is blunt-ended prior to insertion into the vector.

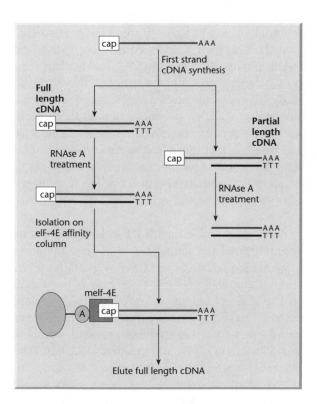

Fig. 6.9 The CAPture method of full-length cDNA cloning, using the eukaryotic initiation factor eIF-4E to select mRNAs with caps protected from RNase digestion by a complementary DNA strand.

enzyme. In the next stage of the procedure, the eukaryotic translational initiation factor eIF-4E is used to isolate full-length molecules by affinity capture. Incomplete cDNAs and cDNAs synthesized on broken templates will lack the cap, and will not be retained. A similar method based on the biotinylation of mRNA has also been reported (Caminci *et al.* 1996). Both methods, however, also co-purify cDNAs resulting from the mispriming of first-strand synthesis, which can account for up to 10% of the clones in a library. An alternative method, *oligo-capping*, addresses this problem by performing selection at the RNA stage (Maruyama & Sugano 1994, Suzuki *et al.* 1997, Suzuki *et al.* 2000; Fig. 6.10). The basis of the method is that RNA is sequentially treated with the enzymes alkaline phosphatase and

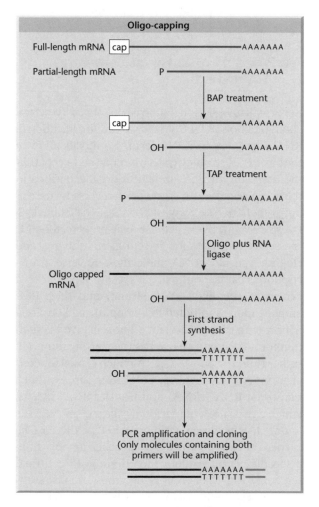

Fig. 6.10 Oligo-capping, the addition of specific oligonucleotide primers to full-length RNAs by sequential treatment with alkaline phosphatase and acid pyrophosphatase. Once the oligo cap has annealed to the 5′ end of the mRNA, it can serve as a primer binding site for PCR amplification.

acid pyrophosphatase. The first enzyme removes phosphate groups from the 5' ends of uncapped RNA molecules, but does not affect full-length molecules with a 5' cap. The second treatment removes the cap from full-length RNAs leaving a 5' terminal residue with a phosphate group. Full-length molecules can be ligated to a specific oligo-nucleotide, while broken and degraded molecules cannot. The result is an oligo-capped population of full-length mRNAs. This selected population is then reverse transcribed using an oligo-dT primer. Second-strand synthesis and cloning is then carried out by the PCR using the oligo-dT primer and a primer annealing to the oligonucleotide cap. Only full-length cDNAs annealing to both primers will be amplified, thus eliminating broken or degraded RNAs, incomplete first cDNA strands (which lack a 5' primer annealing site), and misprimed cDNAs (which lack a 3' primer annealing site).

The PCR can be used as an alternative to cDNA cloning

Reverse transcription followed by the polymerase chain reaction (RT-PCR) leads to the amplification of RNA sequences in cDNA form. No modification to the basic PCR strategy (p. 26) is required, except that the template for PCR amplification is generated in the same reaction tube in a prior reverse transcription reaction (see Kawasaki 1990, Dieffenbach & Dvesler 1995). Using gene-specific primers, RT-PCR is a sensitive means for detecting, quantifying, and cloning specific cDNA molecules. Reverse transcription is carried out using a specific 3' primer that generates the first cDNA strand, and then PCR amplification is initiated following the addition of a 5' primer to the reaction mix. The sensitivity is such that total RNA can be used as the starting material, rather than the poly(A)$^+$ RNA which is used for conventional cDNA cloning. Total RNA also contains ribosomal RNA (rRNA) and transfer RNA (tRNA), which can be present in a great excess to mRNA.

Due to the speed with which RT-PCR can be carried out, it is an attractive approach for obtaining a specific cDNA sequence for cloning. In contrast, screening a cDNA library is laborious, even presuming that a suitable cDNA library is already available and does not have to be constructed for the purpose. Quite apart from the labor involved, a cDNA library may not yield a cDNA clone with a full-length coding region because, as described above, generating a full-length cDNA clone may be technically challenging,

particularly with respect to long mRNAs. Furthermore, the sought-after cDNA may be very rare even in specialized libraries. Does this mean that cDNA libraries have been superseded? Despite the advantages of RT-PCR, there are still reasons for constructing cDNA libraries. The first reflects the availability of starting material, and the permanence of the library. A sought-after mRNA may occur in a source that is not readily available, perhaps a small number of cells in a particular human tissue. A good-quality cDNA library has only to be constructed once from this tissue to give a virtually infinite resource for future use. The specialized library is permanently available for screening. Indeed, the library may be used as a source from which a specific cDNA can be obtained by PCR amplification. The second reason concerns screening strategies. The PCR-based approaches are dependent upon specific primers. However, with cDNA libraries, screening strategies are possible that are based upon expression, e.g. immunochemical screening, rather than nucleic acid hybridization (see below).

As discussed above for genomic libraries, PCR can be used to provide the DNA for library construction when the source is unsuitable for conventional approaches, e.g. a very small amount of starting material or fixed tissue. Instead of gene-specific primers, universal primers can be used that lead to the amplification of all mRNAs, which can then be subcloned into suitable vectors. A disadvantage of PCR-based strategies for cDNA library construction is that the DNA polymerases used for PCR are more error-prone than those used conventionally for second-strand synthesis, so the library may contain a large number of mutations. There is also likely to be a certain amount of distortion due to competition among templates, and a bias towards shorter cDNAs.

A potential problem with RT-PCR is false results resulting from the amplification of contaminating genomic sequences in the RNA preparation. Even trace amounts of genomic DNA may be amplified. In the study of eukaryotic mRNAs, it is therefore desirable to choose primers that anneal in different exons such that the products expected from the amplification of cDNA and genomic DNA would be different sizes, or if the intron is suitably large, so that genomic DNA would not be amplified at all. Where this is not possible (e.g. when bacterial RNA is used as the template), the RNA can be treated with DNase prior to amplification to destroy any contaminating DNA.

Full-length cDNA cloning is facilitated by the rapid amplification of cDNA ends (RACE)

Another way to address the problem of incomplete cDNA sequences in libraries is to use a PCR-based technique for the *rapid amplification of cDNA ends* (RACE) (Frohman *et al.* 1988). Both 5′ RACE and 3′ RACE protocols are available, although 3′ RACE is usually only required if cDNAs have been generated using random primers. In each case only limited knowledge of the mRNA sequence is required. A single stretch of sequence within the mRNA is sufficient, so an incomplete clone from a cDNA library is a good starting point. From this sequence, specific primers are chosen which face *outwards*, and which produce overlapping cDNA fragments. In the two RACE protocols, extension of the cDNAs from the ends of the transcript to the specific primers is accomplished by using primers that hybridize either at the natural 3′ poly(A) tail of the mRNA, or at a synthetic poly(dA) tract added to the 5′ end of the first-strand cDNA (Fig. 6.11). Finally, after amplification, the overlapping RACE products can be combined if desired, to produce an intact full-length cDNA.

Although simple in principle, RACE suffers from the same limitations that affect conventional cDNA cloning procedures. In 5′ RACE, for example, the reverse transcriptase may not, in many cases, reach the authentic 5′ end of the mRNA, but all first-strand cDNAs whether full length or truncated, are tailed in the subsequent reaction, leading to the amplification of a population of variable-length products. Furthermore, as might be anticipated, since only a single *specific* primer is used in each of the RACE protocols, the specificity of amplification may not be very high. This is especially problematical where the specific primer is degenerate. In order to overcome this problem, a modification of the RACE method has been devised which is based on using nested primers to increase specificity (Frohman & Martin 1989). Strategies for improving the specificity of RACE have been reviewed (Schaefer 1995, Chen 1996).

Many different strategies are available for library screening

The identification of a specific clone from a DNA library can be carried out by exploiting either the sequence of the clone or the structure/function of its expressed product. The former applies to any type of library, genomic or cDNA, and can involve either nucleic acid hybridization or the PCR. In each case, the design of the probe or primers can be used to home in on one specific clone or a group of structurally related clones. Note that PCR screening can also be used to isolate DNA sequences from uncloned genomic DNA and cDNA. Screening the product of a clone applies only to expression libraries, i.e. libraries where the DNA fragment is expressed to yield a protein. In this case, the clone can be identified because its product is recognized by an antibody, or a ligand of some nature, or because the biological activity of the protein is preserved and can be assayed in an appropriate test system.

Both genomic and cDNA libraries can be screened by hybridization

Nucleic acid hybridization is the most commonly used method of library screening because it is rapid, it can be applied to very large numbers of clones, and in the case of cDNA libraries, can be used to identify clones that are not full length (and therefore cannot be expressed).

Grunstein & Hogness (1975) developed a screening procedure to detect DNA sequences in transformed colonies by hybridization *in situ* with radioactive RNA probes. Their procedure can rapidly determine which colony among thousands contains the target sequence. A modification of the method allows screening of colonies plated at a very high density (Hanahan & Meselson 1980). The colonies to be screened are first replica plated onto a nitrocellulose filter disk that has been placed on the surface of an agar plate prior to inoculation (Fig. 6.12). A reference set of these colonies on the master plate is retained. The filter bearing the colonies is removed and treated with alkali so that the bacterial colonies are lysed and the DNA they contain is denatured. The filter is then treated with proteinase K to remove protein and leave denatured DNA bound to the nitrocellulose, for which it has a high affinity, in the form of a "DNA-print" of the colonies. The DNA is fixed firmly by baking the filter at 80°C. The defining, labeled RNA is hybridized to this DNA and the result of this hybridization is monitored by autoradiography. A colony whose DNA print gives a positive autoradiographic result can then be picked from the reference plate.

Variations of this procedure can be applied to phage plaques (Jones & Murray 1975, Kramer *et al.* 1976). Benton & Davis (1977) devised a method called *plaque lift* in which the nitrocellulose filter is

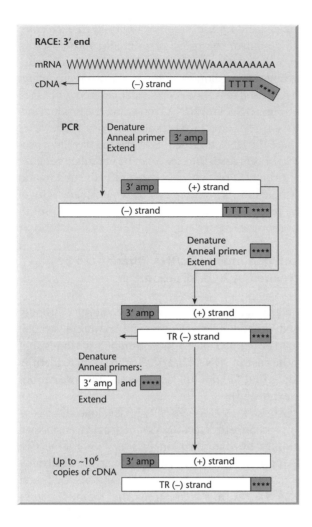

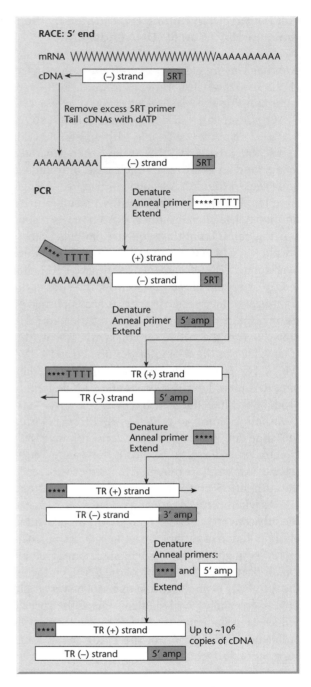

Fig. 6.11 Rapid amplification of cDNA ends (RACE) (Frohman *et al.* 1988). *3′ Protocol.* The mRNA is reverse transcribed using an oligo(dT$_{17}$) primer which has a 17 nucleotide extension at its 5′ end. This extension, the anchor sequence, is designed to contain restriction sites for subsequent cloning. Amplification is performed using the anchor 17-mer (which has a T_{m} higher than oligo(dT$_{17}$)) and a primer specific for the sought-after cDNA. *5′ Protocol.* The mRNA is reverse transcribed from a specific primer. The resultant cDNA is then extended by terminal transferase to create a poly(dA) tail at the 3′ end of the cDNA. Amplification is performed with the oligo(dT$_{17}$)/anchor system as used for the 3′ protocol, and the specific primer. Open boxes represent DNA strands being synthesized; colored boxes represent DNA from a previous step. The diagram is simplified to show only how the *new* product from a previous step is used. Molecules designated TR, truncated, are shorter than full-length (+) or (−) strands.

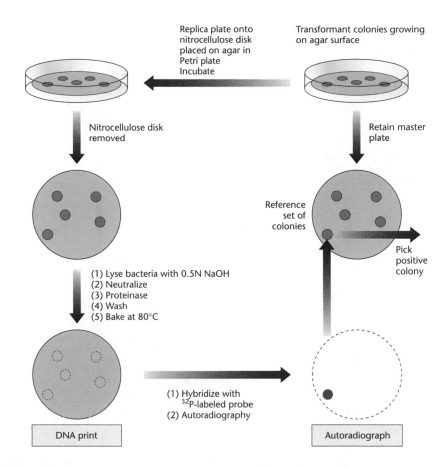

Replica plate onto
nitrocellulose disk
placed on agar in
Petri plate
Incubate

Transformant colonies growing
on agar surface

Nitrocellulose disk
removed

Retain master
plate

(1) Lyse bacteria with 0.5N NaOH
(2) Neutralize
(3) Proteinase
(4) Wash
(5) Bake at 80°C

Reference
set of
colonies

Pick
positive
colony

(1) Hybridize with
    ^{32}P-labeled probe
(2) Autoradiography

DNA print

Autoradiograph

Fig. 6.12 Grunstein–Hogness method for detection of recombinant clones by colony hybridization.

applied to the upper surface of agar plates, making direct contact between plaques and filter. The plaques contain phage particles as well as a considerable amount of unpackaged recombinant DNA. Both phage and unpackaged DNA bind to the filter and can be denatured, fixed, and hybridized. This method has the advantage that several identical DNA prints can easily be made from a single-phage plate: this allows the screening to be performed in duplicate, and hence with increased reliability, and also allows a single set of recombinants to be screened with two or more probes. The Benton and Davis procedure is probably the most widely applied method of library screening, successfully applied in thousands of laboratories to the isolation of recombinant phage by nucleic acid hybridization (Fig. 6.13). More recently, however, library presentation and screening have become increasingly automated. Box 6.2 considers the advantages of gridded reference libraries.

In place of RNA probes, DNA or synthetic oligonucleotide probes can be used. A number of alternative labeling methods are also available that avoid the use of radioactivity. These methods involve the incorporation of chemical labels into the probe, such

as digoxigenin or biotin, which can be detected with a specific antibody or the ligand streptavidin respectively.

Probes are designed to maximize the chances of recovering the desired clone

A great advantage of hybridization for library screening is that it is extremely versatile. Conditions can be used in which hybridization is very stringent, so that only sequences identical to the probe are identified. This is necessary, for example, to identify genomic clones corresponding to a specific cDNA, or to identify overlapping clones in a contig (Chapter 17). Alternatively, less stringent conditions can be used to identify both identical and related sequences. This is appropriate where a probe from one species is being used to isolate a homologous clone from another species (e.g. see Old *et al.* 1982). Probes corresponding to a conserved functional domain of a gene may also cross-hybridize with several different clones in the same species at lower stringency, and this can be exploited to identify members of a gene family. The identification of the vertebrate *Hox* genes

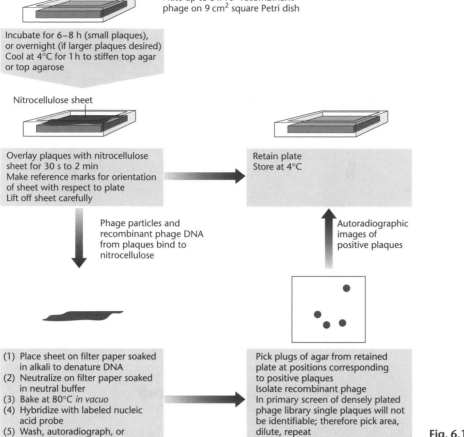

Plate up to 5×10^4 recombinant phage on 9 cm^2 square Petri dish

Incubate for 6–8 h (small plaques), or overnight (if larger plaques desired) Cool at 4°C for 1 h to stiffen top agar or top agarose

Nitrocellulose sheet

Overlay plaques with nitrocellulose sheet for 30 s to 2 min
Make reference marks for orientation of sheet with respect to plate
Lift off sheet carefully

Retain plate
Store at 4°C

Phage particles and recombinant phage DNA from plaques bind to nitrocellulose

Autoradiographic images of positive plaques

(1) Place sheet on filter paper soaked in alkali to denature DNA
(2) Neutralize on filter paper soaked in neutral buffer
(3) Bake at 80°C *in vacuo*
(4) Hybridize with labeled nucleic acid probe
(5) Wash, autoradiograph, or otherwise detect label

Pick plugs of agar from retained plate at positions corresponding to positive plaques
Isolate recombinant phage
In primary screen of densely plated phage library single plaques will not be identifiable; therefore pick area, dilute, repeat

Fig. 6.13 Benton and Davis' plaque-lift procedure.

Box 6.2   Gridded (arrayed) hybridization reference libraries

Traditionally, library screening by hybridization involves taking a plaque lift or colony blot, which generates a replica of the distribution of clones on an agar plate. However, an alternative is to individually pick clones and arrange them on the membrane in the form of a regular grid. Once a laborious process, gridding or arraying has been considerably simplified through the use of robotics. Machines can be programmed to pick clones from microtiter dishes and spot them onto membranes at a high density; then the membrane can be hybridized with a radioactive probe as normal. Using traditional libraries, positive clones are detected by autoradiography and the X-ray film must be aligned with the original plates in order to identify the corresponding plaques. With gridded libraries, however, positive signals can be used to obtain sets of coordinates, which then identify the corresponding clone from the original microtiter dishes. Since identical sets of membranes can be easily prepared,

duplicates can be distributed to other laboratories for screening. These laboratories can then determine the coordinates of their positive signals and order the corresponding clone from the source laboratory. Thus, one library can serve a number of different users and all data can be centralized (Zehetner & Lehrach 1994). Gridded libraries, while convenient for screening and data sharing, are more expensive to prepare than traditional libraries. Therefore, they are often prepared for high-value libraries with wide applications, such as genomic libraries cloned in high-capacity P1, BAC, or YAC vectors (Bentley *et al.* 1992) and also for valuable cDNA libraries (Lennon & Lehrach 1991). It is possible to plate libraries at a density of one clone per well, although for PACs and BACs it is more common to pool clones in a hierarchical manner, so that individual clones may be identified by successive rounds of screening on smaller subpools (e.g. Shepherd *et al.* 1994, Shepherd & Smoller 1994).

provides an example in which cross-species hybridization was used to identify a family of related clones (Levine *et al.* 1984). In this case a DNA sequence was identified that was conserved between the *Drosophila* developmental genes *fushi tarazu* and *Antennapedia*. When this sequence, the homeobox, was used to screen a Southern blot of DNA from other species, including frogs and humans, several hybridizing bands were revealed. This led to the isolation of a number of clones from vertebrate cDNA libraries representing the large family of *Hox* genes that play a central role in animal development.

Hybridization thus has the potential to isolate any sequence from any library *if a probe is available*. If a suitable DNA or RNA probe cannot be obtained from an existing cloned DNA, an alternative strategy is to make an oligonucleotide probe by chemical synthesis. This requires some knowledge of the amino acid sequence of the protein encoded by the target clone. However, since the genetic code is degenerate (i.e. most amino acids are specified by more than one codon) degeneracy must be incorporated into probe design so that a mixture of probes is made, at least one variant of which will specifically match the target clone. Amino acid sequences known to include methionine and tryptophan are particularly valuable because these amino acids are each specified by a single codon, hence reducing the degeneracy of the resulting probe. Thus, for example, the oligopeptide His-Phe-Pro-Phe-Met may be identified and chosen to provide a probe sequence, in which case 32 different oligonucleotides would be required:

$$5' \quad CA\,^{T}_{C}TT\,^{T}_{C}CCCTT\,^{T}_{C}ATG \quad 3'$$
$$\phantom{5' \quad CA\,^{T}_{C}TT\,^{T}_{C}CC}^{A}_{G}$$

These 32 different sequences do not have to be synthesized individually because it is possible to perform a mixed addition reaction for each polymerization step. This mixture is then end labeled with a single isotopic or alternatively labeled nucleotide using an exchange reaction. This mixed-probe method was originally devised by Wallace and co-workers (Suggs *et al.* 1981). To cover all codon possibilities, degeneracies of 64-fold (Orkin *et al.* 1983) or even 256-fold (Bell *et al.* 1984) have been employed successfully. What length of oligonucleotide is required for reliable hybridization? Even though 11-mers can be adequate for Southern blot hybridization (Singer-Sam *et al.* 1983) longer probes are necessary for good

colony and plaque hybridization. Mixed probes of 14 nucleotides have been successful, although 16-mers are typical (Singer-Sam *et al.* 1983).

An alternative strategy is to use a single longer probe of 40–60 nucleotides. Here the uncertainty at each codon is largely ignored and instead increased probe length confers specificity. Such probes are usually designed to incorporate the most commonly used codons in the target species, and they may include the nonstandard base inosine at positions of high uncertainty because this can pair with all four conventional bases. Such probes are sometimes termed *guessmers*. Hybridization is carried out under low stringency to allow for the presence of mismatches. This strategy is examined theoretically by Lathe (1985), and has been applied to sequences coding for human coagulation factor VIII (Wood *et al.* 1984, Toole *et al.* 1984) and the human insulin receptor (Ullrich *et al.* 1985).

The PCR can be used as an alternative to hybridization for the screening of genomic and cDNA libraries

The PCR is widely used to isolate specific DNA sequences from uncloned genomic DNA or cDNA, but it is also a useful technique for library screening (Takumi 1997). As a screening method, the PCR has the same versatility as hybridization, and the same limitations. It is possible to identify any clone by the PCR but only if there is sufficient information about its sequence to make suitable primers.[1]

To isolate a specific clone, the PCR is carried out with gene-specific primers that flank a unique sequence in the target. A typical strategy for library screening by the PCR is demonstrated by Takumi & Lodish (1994). Instead of plating the library out on agar as would be necessary for screening by hybridization, pools of clones are maintained in multiwell plates. Each well is screened by the PCR and positive wells are identified. The clones in each positive well are then diluted into a series in a secondary set of plates and screened again. The process is repeated

---

[1] Note that in certain situations, clever experimental design can allow the PCR to be used to isolate specific but unknown DNA sequences. One example of this is 5′ RACE, which is discussed on p. 111. Another is inverse PCR (p. 397) which can be used to isolate unknown flanking DNA surrounding the insertion site of an integrating vector. In each case, primers are designed to bind to known sequences that are joined to the DNA fragment of interest, e.g. synthetic homopolymer tails, linkers, or parts of the cloning vector.

until wells carrying homogeneous clones corresponding to the gene of interest have been identified.

There are also several applications where the use of *degenerate primers* is favorable. A degenerate primer is a mixture of primers, all of similar sequence but with variations at one or more positions. This is analogous to the use of degenerate oligonucleotides as hybridization probes, and the primers are synthesized in the same way. A common circumstance requiring the use of degenerate primers is when the primer sequences have to be deduced from amino acid sequences (Lee *et al.* 1988). Degenerate primers may also be employed to search for novel members of a known family of genes (Wilks 1989), or to search for homologous genes between species (Nunberg *et al.* 1989). As with oligonucleotide probes, the selection of amino acids with low codon degeneracy is desirable. However, a 128-fold degeneracy in each primer can be successful in amplifying a single copy target from the human genome (Girgis *et al.* 1988). Under such circumstances the concentration of any individual primer sequence is very low, so mismatching between primer and template must occur under the annealing conditions chosen. Since mismatching of the 3′-terminal nucleotide of the primer may prevent efficient extension, degeneracy at this position is to be avoided.

More diverse strategies are available for the screening of expression libraries

If a DNA library is established using expression vectors, each individual clone can be expressed to yield a polypeptide. While all libraries can be screened by hybridization or the PCR as discussed above, expression libraries are useful because they allow a range of alternative techniques to be employed, each of which exploits some structural or functional property of the gene product. This can be important in cases where the DNA sequence of the target clone is completely unknown and there is no strategy available to design a suitable probe or set of primers.

For higher eukaryotes, all expression libraries are cDNA libraries, since these lack introns and the clones are in most cases of a reasonable size. Generally, a random primer method is used for cDNA synthesis, so there is a greater representation of 5′ sequences. As discussed above, such libraries are representative of their source, so certain cDNAs are abundant and others rare. However, it should be noted that bacterial expression libraries and many yeast expression libraries are usually genomic, since there are few

introns in bacteria and some yeast, and very little intergenic DNA. Efficient expression libraries can be generated by cloning randomly sheared genomic DNA or partially digested DNA, and therefore all genes are potentially represented at the same frequency (Young *et al.* 1985). A problem with such libraries is that clones corresponding to a specific gene may carry termination sequences from the gene lying immediately upstream, which can prevent efficient expression. For this reason, conditions are imposed so that the size of the fragments for cloning are smaller than the target gene, and enough recombinants are generated so that there is a reasonable chance that each gene fragment will be cloned in all six possible reading frames (three in each orientation).

Immunological screening uses specific antibodies to detect expressed gene products

Immunological screening involves the use of antibodies that specifically recognize antigenic determinants on the polypeptide synthesized by a target clone. This is one of the most versatile expression-cloning strategies because it can be applied to any protein for which an antibody is available. Unlike the screening strategies discussed below, there is also no need for that protein to be functional. The molecular target for recognition is generally an *epitope*, a short sequence of amino acids that folds into a particular three-dimensional conformation on the surface of the protein. Epitopes can fold independently of the rest of the protein and therefore often form even when the polypeptide chain is incomplete, or when expressed as a fusion with another protein. Importantly, many epitopes can form under denaturing conditions when the overall conformation of the protein is abnormal.

The first immunological screening techniques were developed in the late 1970s, when expression libraries were generally constructed using plasmid vectors. The method of Broome & Gilbert (1978) was widely used at the time. This method exploited the facts that antibodies adsorb very strongly to certain types of plastic, such as polyvinyl, and that IgG antibodies can be readily labeled with ^{125}I by iodination *in vitro*. As usual, transformed cells were plated out on Petri dishes and allowed to form colonies. In order to release the antigen from positive clones, the colonies were lysed, e.g. using chloroform vapor or by spraying with an aerosol of virulent phage (a replica plate is required because this procedure kills the bacteria). A sheet of polyvinyl that had been

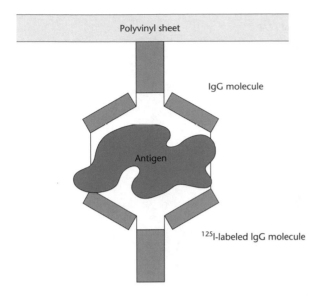

Polyvinyl sheet

IgG molecule

Antigen

^{125}I-labeled IgG molecule

Fig. 6.14 Antigen–antibody complex formation in the immunochemical detection method of Broome and Gilbert. (See text for details.)

coated with the appropriate antibody was then applied to the surface of the plate, allowing antigen–antibody complexes to form. The sheet was then removed and exposed to ^{125}I-labeled IgG specific to a *different* determinant on the surface of the antigen (i.e. a determinant not involved in the initial binding of the antigen to the antibody-coated sheet; Fig. 6.14). The sheet was then washed and exposed to X-ray film. The clones identified by this procedure could then be isolated from the replica plate. Note that this "sandwich" technique is applicable only where two antibodies recognizing different determinants of the same protein are available. However, if the protein is expressed as a fusion, antibodies that bind to each component of the fusion can be used, efficiently selecting for recombinant molecules.

While plasmid libraries have been useful for expression screening (Helfman *et al.* 1983, Helfman & Hughes 1987) it is much more convenient to use bacteriophage λ insertion vectors because these have a higher capacity and the efficiency of *in vitro* packaging allows large numbers of recombinants to be prepared and screened. Immunological screening with phage λ cDNA libraries was introduced by Young & Davies (1983) using the expression vector λgt11 which generates fusion proteins with β-galactosidase under the control of the *lac* promoter (see Box 6.1 for a discussion of λgt11 and similar fusion vectors such as λZAP). In the original technique, screening was carried out using colonies of induced lysogenic bacteria, which required the pro-

duction of replica plates as above. A simplification of the method is possible by directly screening plaques of recombinant phage. In this procedure (Fig. 6.15) the library is plated out at moderately high density (up to 5×10^4 plaques per 9 cm^2 plate), with *E. coli* strain Y1090 as the host. This *E. coli* strain overproduces the *lac* repressor and ensures that no expression of cloned sequences (which may be deleterious to the host) takes place until the inducer IPTG is presented to the infected cells. Y1090 is also deficient in the *lon* protease, hence increasing the stability of recombinant fusion proteins. Fusion proteins expressed in plaques are absorbed onto a nitrocellulose membrane overlay and this membrane is processed for antibody screening. When a positive signal is identified on the membrane, the positive plaque can be picked from the original agar plate (a replica is not necessary) and the recombinant phage can be isolated.

The original detection method using iodinated antibodies has been superseded by more convenient methods using non-isotopic labels, which are also more sensitive and have a lower background of non-specific signal. Generally, these involve the use of unlabeled primary antibodies directed against the polypeptide of interest, which are in turn recognized by secondary antibodies carrying an enzymatic label. As well as eliminating the need for isotopes, such methods also incorporate an amplification step, since two or more secondary antibodies bind to the primary antibody. Typically, the secondary antibody recognizes the species-specific constant region of the primary antibody, and is conjugated to either horseradish peroxidase (de Wet *et al.* 1984) or alkaline phosphatase (Mierendorf *et al.* 1987), each of which can in turn be detected using a simple colorimetric assay carried out directly on the nitrocellulose filter. Polyclonal antibodies, which recognize many different epitopes, provide a very sensitive probe for immunological screening, although they may also cross-react to proteins in the expression host. Monoclonal antibodies and cloned antibody fragments can also be used, although the sensitivity of such reagents is reduced because only a single epitope is recognized.

Southwestern and northwestern screening are used to detect clones encoding nucleic acid binding proteins

We have seen how fusion proteins expressed in plaques produced by recombinant λgt11 or λZAP vectors may be detected by immunochemical screening.

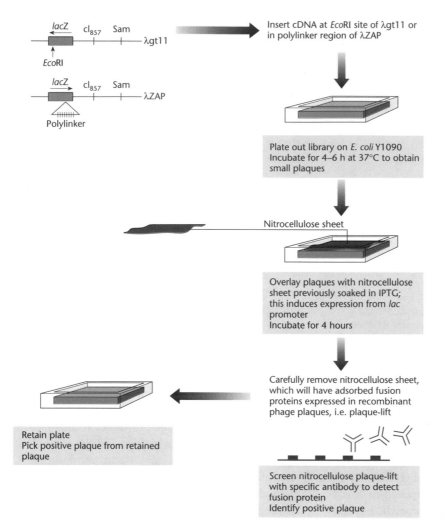

Fig. 6.15 Immunochemical screening of λgt11 or λZAP recombinant plaques.

A closely related approach has been used for the screening and isolation of clones expressing sequence-specific DNA-binding proteins. As above, a plaque lift is carried out to transfer a print of the library onto nitrocellulose membranes. However, the screening is carried out not using an antibody, by incubating the membranes with a radiolabeled double-stranded DNA oligonucleotide probe, containing the recognition sequence for the target DNA-binding protein. This technique is called southwestern screening because it combines the principles of Southern and western blots. It has been particularly successful in the isolation of clones expressing cDNA sequences corresponding to certain mammalian transcription factors (Vinson *et al.* 1988, Staudt *et al.* 1988, Singh *et al.* 1988, Katagiri *et al.* 1989, Xiao *et al.* 1991, Williams *et al.* 1991). A limitation of this technique is that, since individual plaques contain only single cDNA clones, transcription factors that function only in the form of heterodimers or as part of a multi-

meric complex do not recognize the DNA probe, and the corresponding cDNAs cannot be isolated. Clearly the procedure can also be successful only in cases where the transcription factor remains functional when expressed as a fusion polypeptide. It is also clear that the affinity of the polypeptide for the specific DNA sequence must be high, and this has led to the preferential isolation of certain types of transcription factor (reviewed by Singh 1993). More recently, a similar technique has been used to isolate sequence-specific RNA-binding proteins, in this case using a single-stranded RNA probe. By analogy to the above, this is termed northwestern screening and has been successful in a number of cases (e.g. see Qian & Wilusz 1993; reviewed by Bagga & Wilusz 1999). Both southwestern and northwestern screening are most efficient when the oligonucleotide contains the binding sequence in multimeric form. This may mean that several fusion polypeptides on the filter bind to each probe, hence greatly increasing the average dissociation time. To minimize non-specific binding, a large excess of unlabeled double-stranded DNA (or single-stranded RNA) is mixed with the specific probe. However, it is usually necessary to confirm the specificity of binding in a second round of screening using the specific oligonucleotide probe and one or more alternative probes containing similar sequences that are not expected to be recognized.

As well as DNA and RNA, a whole range of alternative "ligands" can be used to identify polypeptides that specifically bind certain molecules. Such techniques are not widely used because they generally have a low sensitivity and their success depends on the preservation of the appropriate interacting domain of the protein when exposed on the surface of a nitrocellulose filter. Furthermore, as discussed in Chapter 23, the yeast two-hybrid system and its derivatives now provide versatile assay formats for many specific types of protein–protein interaction, with the advantage that such interactions are tested in living cells so the proteins involved are more likely to retain their functional interacting domains.

Functional cloning exploits the biochemical or physiological activity of the gene product

Finally, we consider screening methods that depend on the full biological activity of the protein. This is often termed *functional cloning* or *functional complementation*. In this strategy, a particular DNA sequence compensates for a missing function in a mutant cell, and thus restores the wild-type phenotype. This can be a very powerful method of expression cloning, because if the mutant cells are non-viable under particular growth conditions, cells carrying the clone of interest can be positively selected, allowing the corresponding clones to be isolated.

Ratzkin & Carbon (1977) provide an early example of how certain eukaryotic genes can be cloned on the basis of their ability to complement auxotrophic mutations in *E. coli*. These investigators inserted fragments of yeast DNA, obtained by mechanical shearing, into the plasmid ColEl using a homopolymer tailing procedure. They transformed *E. coli his*B mutants, which are unable to synthesize histidine, with the recombinant plasmids and plated the bacteria on minimal medium. In this way, they selected for complementation of the mutation, and isolated clones carrying an expressed yeast *his* gene. If the function of the gene is highly conserved, it is quite possible to carry out functional cloning of, for example, mammalian proteins in bacteria and yeast.

Thus, complementation in yeast has been used to isolate cDNAs for a number of mammalian metabolic enzymes (e.g. Botstein & Fink 1988) and certain highly conserved transcription factors (e.g. Becker *et al.* 1991) as well as regulators of meiosis in plants (Hirayama *et al.* 1997). This approach can also be used in mammalian cells, as demonstrated by Strathdee *et al.* (1992), who succeeded in isolating the *FACC* gene, corresponding to complementation group C of Fanconi's anemia. Generally a pool system is employed, where cells are transfected with a complex mix of up to 100,000 clones. Pools which successfully complement the mutant phenotype are then subdivided for a further round of transfection, and the procedure repeated until the individual cDNA responsible is isolated.

Functional complementation is also possible in transgenic animals and plants. In this way, Probst *et al.* (1998) were able to clone the mouse deafness-associated gene *Shaker-2*, and from there its human homolog, *DFNB3* (Fig. 6.16). The *Shaker-2* mutation had previously been mapped to a region of the mouse genome that is syntenic (has a similar order) to the region involved in a human deafness disorder. BAC clones corresponding to this region were therefore prepared from wild-type mice and microinjected into the eggs of *Shaker-2* mutants. The resulting transgenic mice were screened for restoration of a normal hearing phenotype, allowing a BAC clone corresponding to the functional *Shaker-2* gene to be identified. The gene was shown to encode a cytoskeletal myosin protein. This was then used to screen a human genomic library, resulting in the identification of the equivalent human gene. Note that no sequence information was required for this screening procedure, and without the functional assay there would have been no way to identify either the mouse or human gene except through a laborious chromosome walk from a linked marker. The development of high-capacity transformation vectors for plants (p. 289) has allowed similar methods to be used to identify plant genes (e.g. Sawa *et al.* 1999, Kubo & Kakimoto 2001).

Complementation analysis can be used only if an appropriate mutant expression host is available. In many cases, however, the function of the target gene is too specialized for such a technique to work in a bacterial or yeast expression host, and even in a higher eukaryotic system, loss of function in the host may be fully or partially compensated by one or more other genes. As an alternative, it may be possible to identify clones on the basis that they confer a gain of

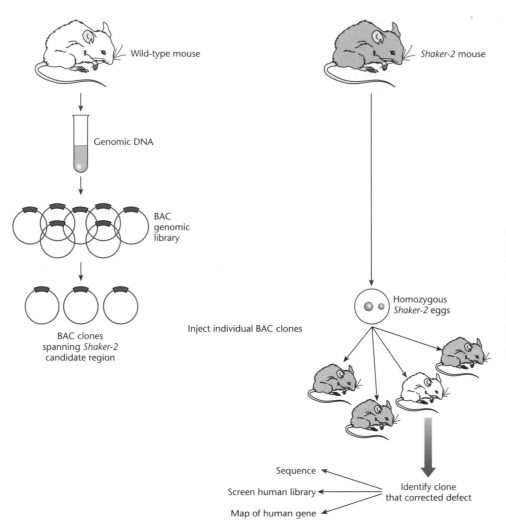

Fig. 6.16 Functional complementation in transgenic mice to isolate the *Shaker-2* gene. Homozygous *Shaker-2* fertilized mouse eggs were injected with BAC clones derived from the *Shaker-2* candidate region of a wild-type mouse. Progeny were screened for restoration of the wild-type phenotype, thus identifying the BAC clone corresponding to the *Shaker-2* gene. This clone is then sequenced and used to isolate and map the corresponding human disease gene *DFNB3*.

function on the host cell. In some cases, this gain of function is a selectable phenotype that allows cells containing the corresponding clone to be positively selected. For example, in an early example of the expression of a mammalian gene in *E. coli*, Chang *et al.* (1978) constructed a population of recombinant plasmids containing cDNA derived from unfractionated mouse mRNA. This population of mRNA molecules was expected to contain the transcript for dihydrofolate reductase (DHFR). Mouse DHFR is much less sensitive to inhibition by the drug trimethoprim than *E. coli* DHFR, so growing transformants in medium containing the drug allowed selection for those cells containing the mouse *Dhfr* cDNA.

In other cases, the phenotype conferred by the clone of interest is not selectable, but can be detected because it causes a visible change in phenotype. In mammalian cells, for example, clones corresponding to cellular oncogenes have been identified on the

basis of their ability to stimulate the proliferation of quiescent mouse 3T3 fibroblast cells either in culture or when transplanted into "nude mice" (e.g. Brady *et al.* 1985). Many different specific assays have also been developed for the functional cloning of cDNAs encoding particular types of gene product. For example, *Xenopus* melanophores have been used for the functional cloning of G-protein-coupled receptors. Melanophores are dark cells containing many pigment organelles called melanosomes. A useful characteristic of these organelles is that they disperse when adenyl cyclase or phospholipase C is active, and aggregate when these enzymes are inhibited. Therefore, the expression of cDNAs encoding G-protein-coupled receptors and many types of receptor tyrosine kinases leads to redistribution of pigmentation within the cell, which can be used as an assay for the identification of receptor cDNAs (reviewed by Lerner 1994).

Positional cloning is used when there is no biological information about a gene, but its position can be mapped relative to other genes or markers

Where there is no available biological information about a gene product (often the case for human disease genes) an approach known as *positional cloning* can be used, which requires as input only the mapped position of the gene (Parimoo *et al.* 1995). Armed with this information, researchers can locate the nearest physical markers and then institute what is known as a *chromosome walk*. In this technique, overlapping clones are obtained spanning the region of interest by using each successive clone as a probe to detect the next. Thus, one can "walk" from one flanking marker to the next, knowing that at least one of the clones will contain the gene of interest. Chromosome walking is simple in principle, but technically demanding. For large distances, it is advisable to use libraries based on high-capacity vectors, such as BACs and YACs, to reduce the number of steps involved. Before such libraries were available, some ingenious strategies were used to reduce the number of steps needed in a walk. In one of the first applications of this technology, Hogness and his coworkers (Bender *el al.* 1983) cloned DNA from the *Ace* and *rosy* loci and the homeotic *Bithorax* gene complex in *Drosophila*. The number of steps was minimized by exploiting the numerous strains carrying well-characterized inversions and translocations of specific chromosome regions. An example of chromosome walking in humans is discussed in Box 6.3. We return to the topic in Chapter 25, which discusses some of the applications of genomics.

Difference cloning exploits differences in the abundance of particular DNA fragments

Difference cloning refers to a range of techniques used to isolate sequences that are represented in one source of DNA but not another. Normally this means differentially expressed cDNAs, representing genes that are active in one tissue but inactive in another, but the technique can also be applied to genomic

---

Box 6.3   A landmark publication. Identification of the cystic fibrosis gene by chromosome walking and jumping

Cystic fibrosis (CF) is a relatively common severe autosomal recessive disorder. Until the CF gene was cloned, there was little definite information about the primary genetic defect. The cloning of the CF gene was a breakthrough for studying the biochemistry of the disorder (abnormal chloride channel function), for providing probes for pre-natal diagnosis, and for potential treatment by somatic gene therapy or other means. The publication discussed here (Rommens *et al.* 1989) is especially notable for the generality of the cloning strategy. In the absence of any direct functional information about the CF gene, the chromosomal location of the gene was used as the basis of the cloning strategy. Starting from markers identified by linkage analysis as being close to the CF locus on chromosome 7, a total of about 500 kb was encompassed by a combination of chromosome walking and jumping (a variation of walking using larger vectors to bridge unclonable gaps; Collins *et al.* 1987). In this work, large numbers of clones were involved, obtained from several different phage and cosmid genomic libraries. Among these libraries, one was prepared using the Maniatis strategy using the λCharon 4A vector, and several prepared using the λDASH and λFIX vectors after partial digestion of human genomic DNA with *Sau*3AI. Cloned regions were aligned with a map of the genome in the CF region, obtained by long-range restriction mapping using rare-cutting enzymes such as *Not*I in combination with pulsed-field gel electrophoresis. The actual CF gene was detected in this cloned region by a number of criteria, such as the identification of open reading frames, the detection of cDNAs hybridizing to the genomic clones, the detection of cross-hybridizing sequences in other species, and the presence of CpG islands, which are known to be associated with the 5′ ends of many genes in mammals.

DNA to identify genes corresponding to deletion mutants. There are a number of cell- based differential cloning methods and also a range of PCR techniques. Each method follows one of two principles: either the differences between two sources are displayed, allowing differentially expressed clones to be visually identified, or the differences are exploited to generate a collection of clones that are enriched for differentially expressed sequences. The analysis of differential gene expression has taken on new importance recently with the advent of high-throughput techniques allowing the monitoring of many, and in some cases all, genes simultaneously.

Library-based approaches may involve differential screening or the creation of subtracted libraries enriched for differentially represented clones

An early approach to difference cloning was *differential screening*, a simple variation on normal hybridization-based library screening protocols that is useful for the identification of differentially expressed cDNAs that are also moderately abundant (e.g. Dworkin & Dawid 1980). Let us consider, for example, the isolation of cDNAs derived from mRNAs which are abundant in the gastrula embryo of the frog *Xenopus* but which are absent, or present at low abundance, in the egg. A cDNA library is prepared from gastrula mRNA. Replica filters carrying identical sets of recombinant clones are then prepared. One of these filters is then probed with ^{32}P-labeled mRNA (or cDNA) from gastrula embryos and one with ^{32}P-labeled mRNA (or cDNA) from the egg. Some colonies will give a positive signal with both probes; these represent cDNAs derived from mRNA types that are abundant at both stages of development. Some colonies will not give a positive signal with either probe; these correspond to mRNA types present at undetectably low abundance in both tissues. This is a feature of using *complex probes*, which are derived from mRNA populations rather than single molecules: only abundant or moderately abundant sequences in the probe carry a significant proportion of the label and are effective in hybridization. Importantly, some colonies give a positive signal with the gastrula probe, but not with the egg probe. These can be visually identified and should correspond to differentially expressed sequences.

A recent resurgence in the popularity of differential screening has come about through the development of DNA microarrays (Schena *et al.* 1995). In this technique, cDNA clones are transferred to a miniature solid support in a dense grid pattern, and screened simultaneously with complex probes from two sources, which are labeled with different fluorochromes. Clones that are expressed in both tissues will fluoresce in a color that represents a mixture of fluorochromes, while differentially expressed clones will fluoresce in a color closer to the pure signal of one or other of the probes. A similar technique involves the use of DNA chips containing densely arrayed oligonucleotides. The development and application of DNA microarray technology are discussed in Chapter 20.

An alternative to differential screening is to generate a library that is enriched in differentially expressed clones by removing sequences that are common to two sources. This is called a *subtracted cDNA library* and should greatly assist the isolation of rare cDNAs. If we use the same example as above, the aim of the experiment would be to generate a library enriched for cDNAs derived from gastrula-specific mRNAs. This would be achieved by hybridizing first-strand cDNAs prepared from gastrula mRNA with a large excess of mRNA from *Xenopus* oocytes. If this driver population is labeled in some way allowing it to be removed from the mixed population, only gastrula-specific cDNAs would remain behind. A suitable labeling method would be to add biotin to all the oocyte mRNA, allowing oocyte/gastrula RNA/cDNA hybrids as well as excess oocyte mRNA to be subtracted by binding to streptavidin, for which biotin has great affinity (Duguid *et al.* 1988, Rubinstein *et al.* 1990). Libraries can be prepared by several rounds of extraction with driver mRNA, resulting in highly enriched subtracted libraries (reviewed by Sagerstrom *et al.* 1997). There are also a few examples where subtractive cloning has been used with genomic libraries to identify differentially represented sequences. Perhaps the most remarkable example of this approach is the cloning of the human gene for muscular dystrophy by subtractive cloning using normal DNA and DNA from an individual with a deletion spanning the *DMD* gene. This is discussed in Box 6.4.

Differentially expressed genes can also be identified using PCR-based methods

As expected, PCR-based methods for difference cloning are more sensitive and rapid than library-based methods, and can be applied to small amounts of starting material. Two similar methods have been described which use pairs of short arbitrary primers

Box 6.4   A landmark publication. Subtraction cloning of the human Duchenne muscular dystrophy (*DMD*) gene

While most subtractive-cloning experiments involve cDNAs, this publication reports one of the few successful attempts to isolate a gene using a subtracted *genomic* library. The study began with the identification of a young boy, known as "BB", who suffered from four X-linked disorders, including DMD. Cytogenetic analysis showed that the boy had a chromosome deletion in the region Xp21, which was known to be the DMD locus. A subtraction-cloning procedure was then devised to isolate the DNA sequences that were deleted in BB (Kunkel 1986). Genomic DNA was isolated from BB and randomly sheared, generating fragments with blunt ends and non-specific overhangs. DNA was also isolated from an aneuploid cell line with four (normal) X chromosomes. This DNA was digested with the restriction enzyme *Mbo*I, generating sticky ends suitable for cloning. The *Mbo*I fragments were mixed with a large excess of the randomly sheared DNA from BB, and the mixture was denatured and then persuaded to reanneal extensively, using phenol enhancement. The principle behind the strategy was that, since the randomly fragmented DNA was present in a vast excess, most of the DNA from the cell line would be sequestered into hybrid DNA molecules that would be unclonable. However, those sequences present among the *Mbo*I fragments but absent from BB's DNA due to the deletion would only be able to reanneal to complementary strands from the cell line. Such strands would have intact *Mbo*I sticky ends and could therefore be ligated into an appropriate cloning vector. Using this strategy, Kunkel and colleagues generated a genomic library that was highly enriched for fragments corresponding to the deletion in BB. Subclones from the library were tested by hybridization against normal DNA and DNA from BB to confirm that they mapped to the deletion. To confirm that the genuine *DMD* gene had been isolated, the positive subclones were then tested against DNA from many other patients with DMD, revealing similar deletions in 6.5% of cases.

From Kunkel (1986) *Nature* **322**: 73–77.

to amplify pools of partial cDNA sequences. If the same primer combinations are used to amplify cDNAs from two different tissues, the products can be fractionated side by side on a sequencing gel, and differences in the pattern of bands generated, the *mRNA fingerprint*, therefore reveal differentially expressed genes (Fig. 6.17). Essentially the distinction between the two techniques concerns the primer used for first-strand cDNA synthesis. In the differential display PCR technique (Liang & Pardee 1992) the antisense primer is an oligo-dT primer with a specific two-base extension, which thus binds at the 3′ end of the mRNA. Conversely, in the arbitrarily primed PCR method (Welsh *et al.* 1992), the antisense primer is arbitrary and can in principle anneal anywhere in the message. In each case, an arbitrary sense primer is used, allowing the amplification of partial cDNAs from pools of several hundred mRNA molecules. Following electrophoresis, differentially expressed cDNAs can be excised from

the gel and characterized further, usually to confirm their differential expression.

Despite the fact that these display techniques are problematical and appear to generate a large number of false positive results, there have been remarkable successes. In the original report by Liang & Pardee, the technique was used to study differences between tumor cells and normal cells, resulting in the identification of a number of genes associated with the onset of cancer (Liang *et al.* 1992). Further cancer-related gene products have been discovered by other groups using differential display (Sager *et al.* 1993, Okamato & Beach 1994). The technique has also been used successfully to identify developmentally regulated genes (e.g. Adati *et al.* 1995) and genes that are induced by hormone treatment (Nitsche *et al.* 1996). An advantage of display techniques over subtracted libraries is that changes can be detected in related mRNAs representing the same gene family. In subtractive-cloning procedures, such differences

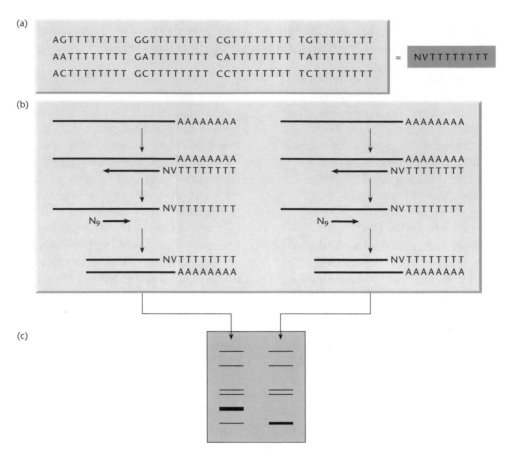

Fig. 6.17 Summary of the differential mRNA display technique, after Liang & Pardee (1992). (a) A set of 12 oligo(dT) primers is synthesized, each with a different two-base extension; the generic designation for this primer set is NVTTTTTTTTT, where N is any nucleotide and V is any nucleotide except T. (b) Messenger RNA from two sources is then converted into cDNA using these primers, generating 12 non-overlapping pools of first-strand cDNA molecules for each source. The PCR is then carried out using the appropriate oligo(dT) primer and a set of arbitrary 9-mers (N_9), which may anneal anywhere within the cDNA sequence. This facilitates the amplification of pools of cDNA fragments, essentially the same as expressed sequence tages (ESTs). (c) Pools of PCR products, derived from alternative mRNA sources but amplified with the same pair of primers, are then compared side by side on a sequencing gel. Bands present in one lane but absent from the other are likely to represent differentially expressed genes. The corresponding bands can be excised from the sequencing gel and the PCR products subcloned, allowing sequence annotation and expression analysis, e.g. by northern blot or *in situ* hybridization, to confirm differential expression.

are often overlooked because the excess of driver DNA can eliminate such sequences (see review by McClelland *et al.* 1995).

Representational difference analysis is a PCR-based subtractive-cloning procedure

Representational difference analysis is a PCR subtraction technique, i.e. common sequences between two sources are eliminated prior to amplification. The method was developed for the comparative analysis of genomes (Lisitsyn *et al.* 1993) but has been modified for cloning differentially expressed genes (Hubank & Schatz 1994). Essentially the technique involves the same principle as subtraction

hybridization in that a large excess of a DNA from one source, the driver, is used to make common sequences in the other source, the tester, unclonable (in this case unamplifiable). The general scheme is shown in Fig. 6.18. cDNA is prepared from two sources, digested with restriction enzymes and amplified. The amplified products from one source are then annealed to specific linkers that provide annealing sites for a unique pair of PCR primers. These linkers are not added to the driver cDNA. A large excess of driver cDNA is then added to the tester cDNA and the populations are mixed. Driver/driver fragments possess no linkers and cannot be amplified, while driver/tester fragments possess only one primer annealing site and will only be amplified in a

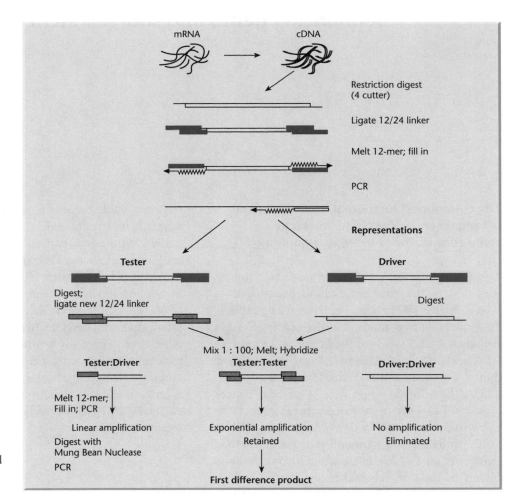

Fig. 6.18 Basic strategy for cDNA-based representational difference analysis. See text for details.

linear fashion. However, cDNAs that are present only in the tester will possess linkers on both strands and will be amplified exponentially, and can therefore be isolated and cloned.

Suggested reading

Bashiardes S. & Lovett M. (2000) cDNA detection and analysis. *Current Opinion in Chemical Biology* **5**, 15–20.

Glover D.M. & Hames B.D. (eds.) (1995) *DNA Cloning 1: A Practical Approach. Core Techniques*, 2nd edn. IRL Press, Oxford.

Glover D.M. & Hames B.D. (eds.) (1995) *DNA Cloning 3: A Practical Approach. Complex Systems*, 2nd edn. IRL Press, Oxford.

Hawkins T.K., Detter J.C. & Richardson P.M. (2002) Whole genome amplification – applications and advances. *Current Opinion in Biotechnology* **13**, 65–7.

McPherson M.J. & Muller S.G. (2000) *PCR: The Basics*. BIOS Scientific Publishers, Oxford.

Sambrook J. & Russel D. (2001) *Molecular Cloning: A Laboratory Manual*, 3rd edn. Cold Spring Harbor Laboratory Press, Cold Spring Harbor, New York.

Soares M.B. (1997) Identification and cloning of differentially expressed genes. *Current Opinion in Biotechnology* **8**, 542–6.

CHAPTER 7

Sequencing genes and short stretches of DNA

The commonest method of DNA sequencing is Sanger sequencing (also known as chain-terminator or dideoxy sequencing)

The first significant DNA sequence to be obtained was that of the cohesive ends of phage λ DNA (Wu & Taylor 1971) which are only 12 bases long. The methodology used was derived from RNA sequencing and was not applicable to large-scale DNA sequencing. An improved method, plus and minus sequencing, was developed and used to sequence the 5386 bp phage ΦX174 genome (Sanger *et al.* 1977a). This method was superseded in 1977 by two different methods, that of Maxam and Gilbert (1977) and the chain-termination or dideoxy method (Sanger *et al.* 1977b). For a while the Maxam and Gilbert method, which makes use of chemical reagents to bring about base-specific cleavage of DNA, was the favored procedure. However, refinements to the chain-termination method meant that by the early 1980s it became the preferred procedure. To date, most large sequences have been determined using this technology, with the notable exception of bacteriophage T7 (Dunn & Studier 1983). For this reason, only the chain-termination method will be described here.

The chain-terminator or dideoxy procedure for DNA sequencing capitalizes on two properties of DNA polymerases: (i) their ability to synthesize faithfully a complementary copy of a single-stranded DNA template; and (ii) their ability to use 2′, 3′-dideoxynucleotides as substrates (Fig. 7.1). Once the analog is incorporated at the growing point of the DNA chain, the 3′ end lacks a hydroxyl group and no longer is a substrate for chain elongation. Thus, the growing DNA chain is terminated, i.e. dideoxynucleotides act as chain terminators. In practice, the Klenow fragment of DNA polymerase is used because this lacks the 5′ → 3′ exonuclease activity associated with the intact enzyme. Initiation of DNA synthesis requires a primer and usually this is a chemically synthesized oligonucleotide which is annealed close to the sequence being analyzed.

DNA synthesis is carried out in the presence of the four deoxynucleoside triphosphates, one or more of which is labeled with ^{32}P, and in four separate incubation mixes containing a low concentration of one each of the four dideoxynucleoside triphosphate analogs. Therefore, in each reaction there is a population of partially synthesized radioactive DNA molecules, each having a common 5′ end, but each varying in length to a base-specific 3′ end (Fig. 7.2). After a suitable incubation period, the DNA in each mixture is denatured and electrophoresed in a sequencing gel.

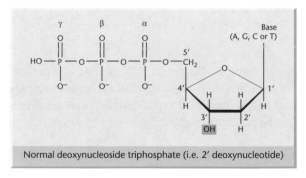

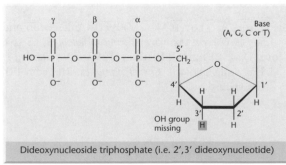

Fig. 7.1 Dideoxynucleoside triphosphates act as chain terminators because they lack a 3′-OH group. Numbering of the carbon atoms of the pentose is shown (primes distinguish these from atoms in the bases). The α, β, and γ phosphorus atoms are indicated.

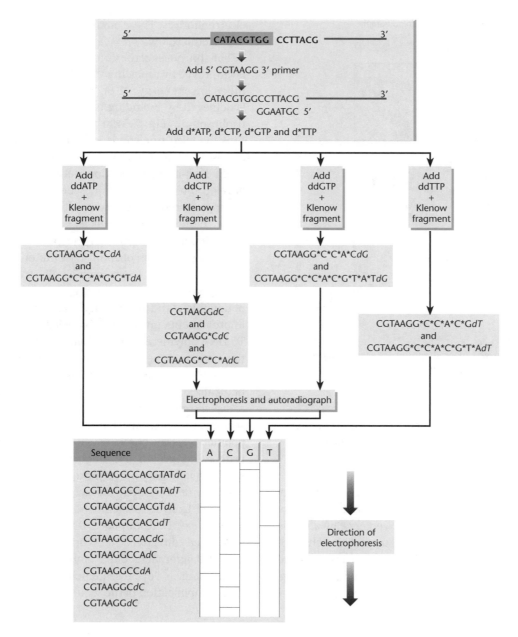

Fig. 7.2 DNA sequencing with dideoxynucleoside triphosphates as chain terminators. In this figure asterisks indicate the presence of ^{32}P and the prefix "d" indicates the presence of a dideoxynucleotide. At the top of the figure the DNA to be sequenced is enclosed within the box. Note that unless the primer is also labeled with a radioisotope the smallest band with the sequence CGTAAGG*dC* will not be detected by autoradiography as no labeled bases were incorporated.

A sequencing gel is a high-resolution gel designed to fractionate single-stranded (denatured) DNA fragments on the basis of their size and which is capable of resolving fragments differing in length by a single base. They routinely contain 6–20% polyacrylamide and 7 M urea. The function of the urea is to minimize DNA secondary structure which affects electrophoretic mobility. The gel is run at sufficient power to heat up to about 70°C. This also minimizes DNA secondary structure. The labeled DNA bands obtained after such electrophoresis are revealed by autoradiography on large sheets of X-ray film and from these the sequence can be read (Fig. 7.3).

To facilitate the isolation of single strands, the DNA to be sequenced may be cloned into one of the clustered cloning sites in the *lac* region of the M13 mp series of vectors (Fig. 7.4). A feature of these vectors is that cloning into the same region can be mediated by any one of a large selection of restriction enzymes but still permits the use of a single sequencing primer.

Sequencing gel autoradiograph

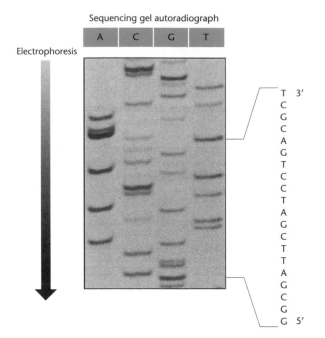

Fig. 7.3 Enlarged autoradiograph of a sequencing gel obtained with the chain-terminator DNA sequencing method.

The original Sanger method has been greatly improved by a number of experimental modifications

Since its first description, the Sanger method of sequencing has been greatly improved in terms of read length, accuracy, and convenience by a whole series of changes to the basic protocol. These changes include the choice of label, enzymes (Box 7.1) and template, the development of automated sequencing, and a move from slab gel electrophoresis to capillary

electrophoresis. These modifications are discussed in more detail below.

The sharpness of the autoradiographic images can be improved by replacing the ^{32}P-radiolabel with the much lower energy ^{33}P or ^{35}S. In the case of ^{35}S, this is achieved by including an α-^{35}S-deoxynucleoside triphosphate (Fig. 7.5) in the sequencing reaction. This modified nucleotide is accepted by DNA polymerase and incorporated into the growing DNA chain. Non-isotopic detection methods also have been developed with chemiluminescent, chromogenic, or fluorogenic reporter systems. Although the sensitivity of these methods is not as great as with radiolabels, it is adequate for many purposes.

The combination of chain-terminator sequencing and M13 vectors to produce single-stranded DNA is very powerful. Very good quality sequencing is obtainable with this technique, especially when the improvements given by ^{35}S-labeled precursors and T7 DNA polymerase are exploited. Further modifications allow sequencing of "double-stranded" DNA, i.e. double-stranded input DNA is denatured by alkali, neutralized, and one strand then is annealed with a specific primer for the actual chain-terminator sequencing reactions. This approach has gained in popularity as the convenience of having a universal primer has grown less important with the widespread availability of oligonucleotide synthesizers. With this development, Sanger sequencing has been liberated from its attachment to the M13 cloning system; e.g. PCR-amplified DNA segments can be sequenced directly. One variant of the double-stranded approach, often employed in automated sequencing, is "cycle sequencing". This involves a *linear* amplification of the sequencing reaction using

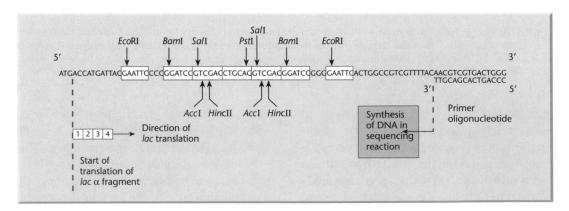

Fig. 7.4 Sequence of M13 mp7 DNA in the vicinity of the multipurpose cloning region. The upper sequence is that of M13 mp7 from the ATG start codon of the β-galactosidase α-fragment, through the multipurpose cloning region, and back into the β-galactosidase gene. The short sequence at the right-hand side is that of the primer used to initiate DNA synthesis across the cloned insert. The numbered boxes correspond to the amino acids of the β-galactosidase fragment.

Box 7.1   DNA polymerases used for Sanger sequencing

Many DNA polymerases have 5'→3' and 3'→5' exonuclease activities in addition to their polymerase activity. The 5'→3' nuclease activity is detrimental to sequencing regardless of whether the label for detection is on the 5' end of the sequencing fragment, incorporated into the fragment as an internal label, or is on the terminator. In some cases, the domain of the polymerase that has the 5'→3' exonuclease activity is absent, as in the case of T7 DNA polymerase. In others, the 5'→3' nuclease domain can be removed by protease cleavage as first demonstrated by Klenow & Henningsen (1970) with *E. coli* DNA polymerase I. Alternatively, the domain can be removed by deletion, although some enzymes with deletions lose processivity. By contrast, *Taq* DNA polymerases with such deletions have greater thermostability and greater fidelity than full-length enzymes.

The 3'→5' exonuclease activity is undesirable for sequencing applications because it hydrolyzes the single-stranded sequencing primers. In most DNA polymerases this activity can be destroyed by point mutations or small deletions but some DNA polymerases, like those from *Thermus* species, naturally lack the activity.

When Sanger sequencing was first developed the enzyme used was the Klenow fragment of DNA polymerase I. A disadvantage of this enzyme is a sequence-dependent discrimination of dideoxy nucleotides (ddNTPs) for ordinary nucleotides (dNTPs). This leads to a variation of the amount of fragments for each base in the sequencing reaction and hence uneven band intensities in autoradiographs or uneven peak heights if fluorescent labels are used (see p. 130). This leads to a requirement for high concentrations of ddNTPs, which are expensive, and increases the background noise with fluorescent labels. Problems also are encountered when using native *Taq* DNA polymerase (from *Thermus aquaticus*).

The discrimination of ddNTPs can be greatly reduced by replacing the Klenow polymerase with T7 DNA polymerase. The reason for this is that native *E. coli* DNA polymerase and *Taq* DNA polymerase have a phenylalanine in their active sites as compared with tyrosine in T7 DNA polymerase. Exchanging the phenylalanine residue with a tyrosine residue in the *E. coli* and *Taq* enzymes decreases the discrimination of ddNTPs by a factor of 250–8000 (Tabor & Richardson 1995).

Native *Taq* DNA polymerase exhibits another undesirable characteristic: uneven incorporation of ddNTPs with a strong bias in favor of ddGTP incorporation. This bias results from a strong interaction between ddGTP and an arginine residue at position 660. When this arginine residue is replaced with aspartate, serine, leucine, phenylalanine, or tyrosine, the bias is eliminated. This results in more even band intensities, greater accuracy, and longer read lengths (Li *et al.* 1999a).

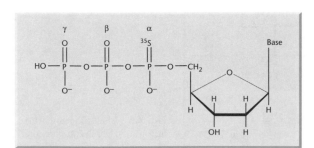

Fig. 7.5   Structure of an α-³⁵S-deoxynucleoside triphosphate.

25 cycles of denaturation, annealing of a specific primer to one strand only, and extension in the presence of *Taq* DNA polymerase plus labeled dideoxynucleotides. Alternatively, labeled primers can be used with unlabeled dideoxynucleotides.

There are some important differences between M13 phages, plasmids, and PCR products as templates in sequencing reactions. The single-stranded M13 phages can be sequenced only on one strand and share a limitation in size with PCR products to a practical maximum of 2–3 kb. Longer PCR products can be obtained under certain conditions but have not been used routinely for sequencing. Plasmids can harbor up to 10 kb fragments if low-copy-number vectors are used and this is a major advantage for linking sequences when very long stretches of DNA need to be sequenced. They also give better representation of DNA sequences from higher organisms with less tendency to eliminate repeated sequences (Chissoe *et al.* 1997, Elkin *et al.* 2001). A disadvantage of PCR sequencing is "polymerase slippage", a term that refers to the inability of the *Taq* DNA polymerase to incorporate the correct number of bases when copying runs of 12 or more identical bases.

It is possible to automate DNA sequencing by replacing radioactive labels with fluorescent labels

In manual sequencing, the DNA fragments are radiolabeled in four chain-termination reactions, separated on the sequencing gel in four lanes, and detected by autoradiography. This approach is not well suited to automation. To automate the process it is desirable to acquire sequence data in real time by detecting the DNA bands within the gel during the

electrophoretic separation. However, this is not trivial as there are only about 10^{-15}–10^{-16} moles of DNA per band. The solution to the detection problem is to use fluorescence methods. In practice, the fluorescent tags are attached to the chain-terminating nucleotides. Each of the four dideoxynucleotides carries a spectrally different fluorophore. The tag is incorporated into the DNA molecule by the DNA polymerase and accomplishes two operations in one step: it terminates synthesis and it attaches the fluorophore to the end of the molecule. Alternatively, fluorescent primers can be used with nonlabeled dideoxynucleotides. By using four different fluorescent dyes it is possible to electrophorese all four chain-terminating reactions together in one lane of a sequencing gel. The DNA bands are detected by their fluorescence as they electrophorese past a detector (Fig. 7.6). If the detector is made to scan horizontally across the base of a slab gel, many separate sequences can be scanned, one sequence per lane. Because the different fluorophores affect the mobility of fragments to different extents, sophisticated software is incorporated into the scanning step to ensure that bands are read in the correct order. This

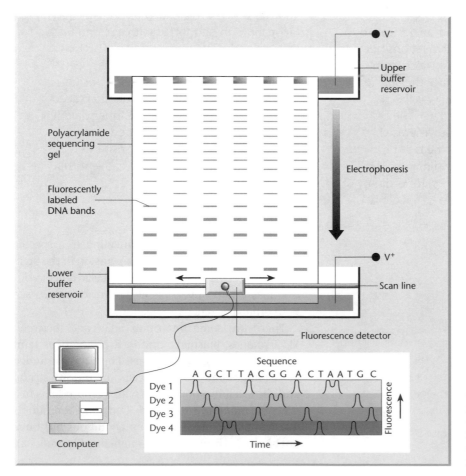

Fig. 7.6 Block diagram of an automated DNA sequencer and idealized representation of the correspondence between fluorescence in a single electrophoresis lane and nucleotide sequence.

is the principle on which the original Applied Biosystems (ABI) instruments operate.

An alternative to the four-dye system is to start with a single fluorescent-labeled primer which is used in all four sequencing reactions. The resulting fluorescent-labeled DNA strands are separated in four different lanes in the electrophoresis system. This is the basis of the Amersham Pharmacia Biotech ALF sequencer. It has a fixed argon laser which emits light that passes through the width of the gel and is sensed by detectors in each of the lanes.

Another variation is provided by the LI-COR two-dye near-infrared DNA analysis system. It can detect the products of two different sequencing reactions in parallel, enabling pooling reactions and simultaneous bidirectional sequencing. Sequencing both directions on a template by combining forward and reverse primers in the same direction produces twice the data from each reaction prepared.

Automated DNA sequencers offer a number of advantages that are not particularly obvious. First, manual sequencing can generate excellent data but even in the best sequencing laboratories poor autoradiographs frequently are produced that make sequence reading difficult or impossible. Usually the problem is related to the need to run different termination reactions in different tracks of the gel. Skilled DNA sequencers ignore bad sequencing tracks but many laboratories do not. This leads to poor quality sequence data. The use of a single-gel track for all four dideoxy reactions means that this problem is less acute in automated sequencing. Nevertheless, it is desirable to sequence a piece of DNA several times, and on both strands, to eliminate errors caused by technical problems. It should be noted that long runs of the same nucleotide or a high G + C content can cause compression of the bands on a gel, necessitating manual reading of the data, even with an automated system. Note also that multiple tandem short-repeats, which are common in the DNA of higher eukaryotes, can reduce the fidelity of DNA copying, particularly with *Taq* DNA polymerase. The second advantage of automated DNA sequencers is that the output from them is in machine-readable form. This eliminates the errors that arise when DNA sequences are read and transcribed manually.

DNA sequencing throughput can be greatly increased by replacing slab gels with capillary array electrophoresis

There are two problems with high-volume sequencing in slab gels. First, the preparation of the gels is very labor-intensive and is a significant cost in large-scale sequencing centers. Secondly, DNA sequencing speed is related to the electric field applied to separate the fragments. To a first approximation, the sequencing speed increases linearly with the applied electric field but application of a voltage across a material with high conductivity results in the generation of heat ("Joule heating"). Because of their thickness, slab gels cannot efficiently radiate heat, and Joule heating limits the maximum electric field that can be applied.

Capillary electrophoresis is undertaken in high-purity fused silica capillaries with an internal diameter of 50 μm. Because of their small diameter, these capillaries are not prone to Joule heating even when high electric fields are applied in order to obtain rapid separation of DNA fragments. In addition, silica capillaries are very flexible, are easily incorporated into automated instruments, and can be supplied prefilled with a gel matrix. This reduces the hands-on operator time for 1000 samples per day from 8 h with a slab gel system to 15 min with a capillary gel system (Venter *et al.* 1998). For this reason, capillary gels are very rapidly replacing slab gels in automated sequencing instruments. The current generation of capillary sequencers has arrays of up to 384 capillaries.

In practice, DNA sequencing by capillary electrophoresis is very simple (Dovichi & Zhang 2001). Instead of placing the product of a sequencing reaction in the well of a slab gel it is applied to the top of a capillary gel. A number of different materials can be used for the gel matrix but it usually is linear (i.e. non-crosslinked) polyacrylamide. The labeled DNA fragments migrate through the capillary and emerge at the end in a vertical stream where they are detected (Fig. 7.7). Consideration of the detection method shown in Fig. 7.7 reveals another advantage of capillary gels over slab gels. With slab gels, the DNA fragments do not always run in a straight line and this makes automated reading more complicated but with capillaries this problem does not occur.

The accuracy of automated DNA sequencing can be determined with basecalling algorithms

As DNA fragments pass the detector of a DNA sequencer they generate a signal. Information about the identity of the nucleotide bases is provided by the base-specific dye attached to the primer or dideoxy chain-terminating nucleotide. Additional steps

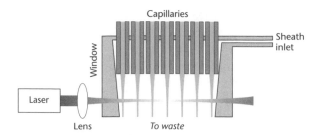

Fig. 7.7 Capillary linear-array sheath-flow cuvet instrument. Sheath fluid draws the analyte into thin streams in the center of the flow chamber, with a single stream produced downstream from each capillary. A laser beam is focused beneath the capillary tips on the sample streams. A lens collects the fluorescence emission signal, which is then spectrally filtered and detected with either an array of photodiodes or with a charged couple device (CCD) camera. (Redrawn with permission from Dovichi & Zhang 2001.)

include lane tracking and profiling, if slab gels have been used, and trace processing. The latter involves the production of a set of four traces of signal intensities corresponding to each of the four bases over the length of the sequencing run. Using algorithms the four traces then are converted into the actual sequence of nucleotides. This process is known as *basecalling*.

The accuracy of the basecalling algorithms directly impacts the quality of the resulting sequence. In an ideal world the traces would be free of noise and peaks would be evenly spaced, of equal height, and have a Gaussian shape. In reality, the peaks have variable spacing and height and there can be secondary peaks underneath the primary peaks. As a consequence, basecalling is error prone and for accurate sequence assembly it is essential to provide an estimate of quality for each assigned base (Buetow *et al.* 1999, Altshuler *et al.* 2000b). The algorithm that gives the best estimate of error rates for basecalling with slab-gel based sequencing is *PHRED* (Ewing & Green 1998, Ewing *et al.* 1998). A PHRED quality score of X corresponds to an error probability of approximately $10^{-X/10}$. Thus a PHRED score of 30 corresponds to a 99.9% accuracy for the basecall. An improved basecalling algorithm, *Life Trace*, which is particularly suitable for capillary sequencing, has been described by Walther *et al.* (2001).

Different strategies are required depending on the complexity of the DNA to be sequenced

Basically, there are three applications of sequencing:

1 sequencing of short regions of DNA to identify mutations of interest or single nucleotide polymorphisms (SNPs);
2 sequencing of complete genes and associated upstream and downstream control regions;
3 sequencing of complete genomes.

For each of these applications there are different strategies and different resource requirements. When the objective is the identification of mutations in a particular stretch of DNA, e.g. following site-directed mutagenesis of a cloned gene, the requirement is to analyze the same short sequence (10–20 nucleotides) from a large number of different samples. In essence, one *resequences* the same stretch of DNA many times. This is easily done with Sanger sequencing but the method generates many more data (500 bases of sequence) than are required. In such instances it may be more convenient to use one of the alternative sequencing methods described later in this chapter.

The strategy for sequencing a gene depends on its size. In prokaryotes and lower eukaryotes genes are seldom more than a few kilobases in size. Thus, all that is required is to sequence a series of overlapping gene fragments. By contrast, in higher eukaryotes a gene may span several hundred kilobases or even several megabases. Sequencing DNA of this size requires a different approach because 10 Mb of data are required to generate 1 Mb of confirmed sequence. This is the equivalent of sequencing a small bacterial genome and the strategy for this is discussed in Chapter 17. An alternative approach is to sequence only the cDNA since this might be only 2 kb in length compared to 1 Mb for the intact gene. There are two disadvantages with this approach. First, control elements such as enhancers and promoters as well as splice sites would not be sequenced by this approach. Second, many genes are expressed at very low levels or for very short periods and so may not be represented in cDNA libraries.

A third approach to sequencing the genes of higher eukaryotes is to use feature mapping (Krishnan *et al.* 1995) to help guide the choice of regions to be subjected to detailed analyses. In feature mapping, large DNA fragments are cloned into a transposon-based cosmid vector designed for generating nested deletions by *in vivo* transposition and simple bacteriological selection. These deletions place primer sites throughout the DNA of interest at locations that are easily determined by plasmid size (Fig. 7.8). An alternative way of using transposons to facilitate gene sequencing is to insert them at random into the gene

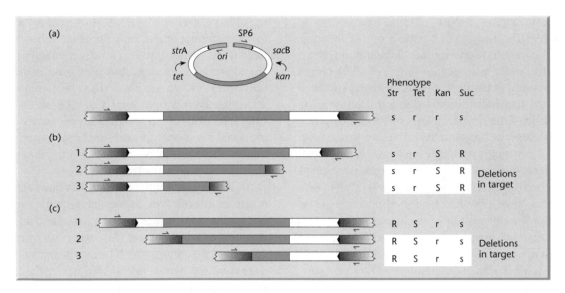

Fig. 7.8 Schematic representation of hypothetical nested deletions obtained using a transposon-based cosmid vector. (a) Depiction of a hypothetical DNA clone. The purple portion represents cloned DNA, the filled arrowheads the transposon sequences, and the purple half-arrows represent locations of primer binding sites. (b) Deletions extending into various sites within the cloned fragment resulting from selection for resistance to sucrose, caused by loss of *sac*B, and tetracycline. (c) Deletions resulting from selection for resistance to streptomycin, caused by loss of *str*A, and kanamycin. R/r indicates resistance and S/s indicates sensitivity. (Redrawn from Krishnan *et al.* 1995, by permission of Oxford University Press.)

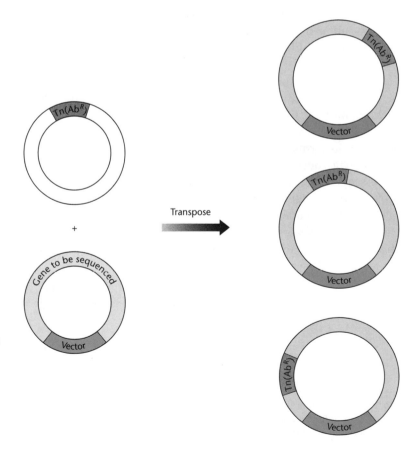

Fig. 7.9 Random insertion of a transposon into a DNA sequence cloned in a vector. A plasmid containing the transposon is mixed with the vector containing the target sequence and incubated with purified transposase. This results in *in vitro* transposition with the transposon inserting at different sites in the target sequence.

of interest (Biery *et al.* 2000). This is done *in vitro* by incubating the target DNA with a non-replicating plasmid carrying a transposon encoding antibiotic resistance and a transposase (Fig. 7.9). The target DNA may be a plasmid, a cosmid, or a BAC and after the *in vitro* transposition selection is made *in vivo* for acquisition of the new antibiotic resistance determinant. Because of target immunity there is only one insertion per target molecule and these occur at different positions. Unique priming sites are included at the ends of the transposon and this permits DNA sequences to be obtained from both strands of the target DNA at the position of the insertion.

The sequencing of genomes requires a completely different approach from that used to sequence genes. The reason for this is quite simple. The maximum length of continuous DNA sequence that can be determined with Sanger sequencing is 500–1000 bases whereas genomes range in size from 0.5–1 Mb to thousands of megabases. Consequently, the genome must first be fragmented into thousands of pieces of a size suitable for sequencing in such a way that the order of the fragments can be reconstructed. This requires special techniques and these are described in Chapter 17.

Alternatives to Sanger sequencing have been developed and are particularly useful for resequencing of DNA

Over the last 20 years many research groups have developed alternatives to the Sanger sequencing method. Initially, the impetus for developing alternative methods was to increase the throughput of sequence in terms of bases read per day per individual. The advent of multichannel automated DNA sequencers, particularly the newer capillary sequencers, obviated the need for alternative methods for *de novo* sequencing of DNA molecules larger than a few megabases. Today, the rationale for using these alternative methodologies is not the sequenc-

ing of large stretches of DNA. Rather, they are used for applications where only short stretches of DNA need to be sequenced, e.g. analysis of SNPs, mutation detection, resequencing of disease genes, partial expressed sequence tag (EST) sequencing, and microbial typing by analysis of 16S rRNA genes (Ronaghi 2001). Although Sanger sequencing is the "gold standard", it generates much more information than is necessary and is time consuming because it involves electrophoresis.

Although many different methods have been proposed as alternatives to Sanger sequencing, most have generated much initial excitement and then disappeared without trace. The exception is sequencing by hybridization which now is performed using microarrays ("gene chips"). Two newer methods of sequencing short stretches of DNA that may be widely adopted are pyrosequencing and massively parallel signature sequencing (MPSS). Pyrosequencing most closely resembles Sanger sequencing in that it involves DNA synthesis, whereas MPSS has similarities to sequencing by hybridization. The continued refinement of two methods for producing gas-phase ions, electrospray ionization, and matrix-assisted laser desorption ionization, has resulted in methods for the rapid characterization of oligonucleotides by mass spectrometry (Limbach *et al.* 1995). This methodology has not been widely adopted for sequencing and will not be discussed further here.

Pyrosequencing permits sequence analysis in real time

Pyrosequencing is a DNA sequencing method that involves determining which of the four bases is incorporated at each step in the copying of a DNA template. As DNA polymerase moves along a single-stranded template, each of the four nucleoside triphosphates is fed sequentially and then removed. If one of the four bases is incorporated then pyrophosphate is released and this is detected in an

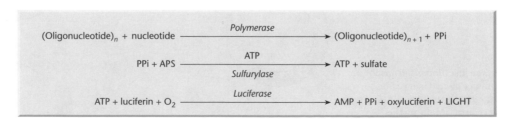

Fig. 7.10 The general principle of pyrosequencing. A polymerase catalyzes incorporation of nucleotides into a nucleic acid chain. As each nucleotide is incorporated a pyrophosphate (PPi) molecule is released and incorporated into ATP by ATP sulfurylase. On addition of luciferin and the enzyme luciferase, this ATP is degraded to AMP with the production of light.

(a)

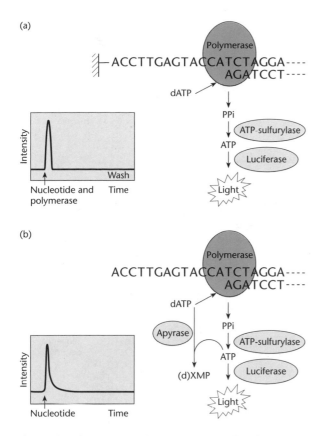

(b)

Fig. 7.11 The two types of pyrosequencing. (a) Schematic representation of the progress of the enzyme reaction in solid-phase pyrosequencing. The four different nucleotides are added stepwise to the immobilized primed DNA template and the incorporation event is followed using the enzymes ATP sulfurylase and luciferase. After each nucleotide addition, a washing step is performed to allow iterative addition. (b) Schematic representation of the progress of the enzyme reaction in liquid-phase pyrosequencing. The primed DNA template and four enzymes involved in liquid-phase pyrosequencing are placed in the well of a microtiter plate. The four different nucleotides are added stepwise and incorporation is followed using the enzymes ATP sulfurylase and luciferase. The nucleotides are continuously degraded by an enzyme allowing the addition of the subsequent nucleotide. dXTP indicates one of the four nucleotides. (Redrawn with permission from Ronaghi 2001.)

enzyme cascade that emits light (Fig. 7.10). There are two variants of the pyrosequencing technique (Fig. 7.11). In solid-phase pyrosequencing (Ronaghi *et al.* 1996), the DNA to be sequenced is immobilized and a washing step is used to remove the excess substrate after each nucleotide addition. In liquid-phase sequencing (Ronaghi *et al.* 1998b) a nucleotide degrading enzyme (apyrase) is introduced to make a four-enzyme system. Addition of this enzyme has eliminated the need for a solid support and intermediate washing thereby enabling the pyrosequencing reaction to be performed in a single tube. However, without the washing step, inhibitory substances can accumulate. The output from a typical pyrogram is shown in Fig. 7.12. It is worth noting that because the light emitted by the enzyme cascade is directly proportional to the amount of pyrophosphate released, it is easy to detect runs of 5–6 identical bases. For longer runs of a single base it may be necessary to use software algorithms to determine the exact number of incorporated nucleotides.

Template preparation for pyrosequencing is very easy. After generation of the template by the PCR, unincorporated nucleotides and PCR primers are removed. Two methods have been developed for this purification step. In the first, biotinylated PCR product is captured on magnetic beads, washed, and denatured with alkali (Ronaghi *et al.* 1998a). In the second method, akaline phosphatase or apyrase and exonuclease I are added to the PCR product to destroy the nucleotides and primers, respectively. The sequencing primer is then added and the mixture rapidly heated and cooled. This inactivates the enzymes, denatures the DNA and enables the primers to anneal to the templates (Nordstrom *et al.* 2000a,b).

The acceptable read length of pyrosequencing currently is about 200 nucleotides, i.e. much less

Fig. 7.12 Pyrogram of the raw data obtained from liquid-phase pyrosequencing. Proportional signals are obtained for one, two, three, and four base incorporations. Nucleotide addition, according to the order of nucleotides, is indicated below the pyrogram and the obtained sequence is indicated above the pyrogram. (Redrawn with permission from Ronaghi 2001.)

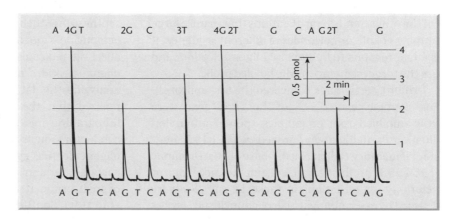

than is achieved with Sanger sequencing. However, many modifications are being made to the reaction conditions to extend the read length. For example, the addition of ssDNA-binding protein to the reaction mixture increases read length, facilitates sequencing of difficult templates, and provides flexibility in primer design. The availability of automated systems for pyrosequencing (Ronaghi 2001) is facilitating the use of the technique for high-throughput analyses. A key benefit of the technique is the absence of a requirement to label the test sequence.

It is possible to sequence DNA by hybridization using microarrays

The principle of sequencing by hybridization can best be explained by starting with a simple example. Consider the tetranucleotide CTCA, whose complementary strand is TGAG, and a matrix of the whole set of $4^3 = 64$ trinucleotides. This tetranucleotide will specifically hybridize only with complementary trinucleotides TGA and GAG, revealing the presence of these blocks in the complementary sequence. From this the sequence TGAG can be reconstructed. If instead of using trinucleotides, $4^8 = 65,536$ octanucleotides were used, it should be possible to sequence DNA fragments up to 200 bases long (Bains & Smith 1988, Lysov et al. 1988, Southern 1988, Drmanac et al. 1989). The length of the target that can be analyzed is approximately equal to the square root of the number of oligonucleotides in the array (Southern et al. 1992). Two different experimental configurations have been developed for the hybridization reaction. Either the target sequence may be immobilized and oligonucleotides labeled or the oligonucleotides may be immobilized and the target sequence labeled. Each method has advantages over the other for particular applications. It is an advantage to label the oligonucleotides to analyze a large number of target sequences for fingerprinting. On the other hand, for applications that require large numbers of oligonucleotides of different sequence, it is advantageous to immobilize oligonucleotides and use the target sequence as the labeled probe.

Drmanac et al. have pioneered the first approach. For example, exons 5–8 of the *TP53* gene were PCR-amplified from 12 samples, spotted onto nylon filters and individually probed with 8192 non-complementary radiolabeled 7-base oligos (Drmanac et al. 1998). In this way, 13 distinct homozygous or heterozygous mutations in these 12 samples were detected by determining which oligonucleotide probes

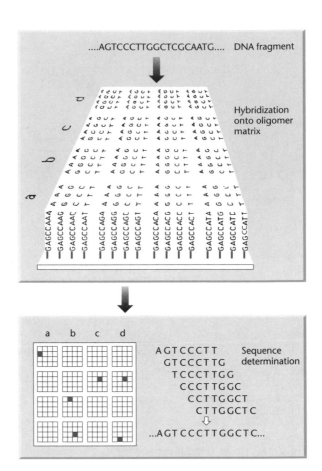

Fig. 7.13 Schematic representation of sequencing by hybridization where the sequence being examined is labeled and hybridized with immobilized oligomers. (Reprinted from Hoheisel 1994 by permission of Elsevier Science.)

hybridized to the immobilized targets. The downside of this approach is the need to perform thousands of separate hybridization reactions.

The feasibility of the second approach (Fig. 7.13) was first demonstrated by Southern et al. (1992). They developed technology for making complete sets of oligonucleotides of defined length and covalently attaching them to the surface of a glass plate by synthesizing them *in situ*. A device carrying all octapurine sequences was used to explore factors affecting molecular hybridization of the tethered oligonucleotides and computer-aided methods for analyzing the data were developed. It was quickly realized that the light-directed synthetic (photolithography) methods routinely used in the semiconductor industry could be combined with standard oligonucleotide synthesis to prepare microarrays carrying hundreds of thousands of oligonucleotides of predetermined sequence (Fodor et al. 1993, Pease et al. 1994, Southern 1996a).

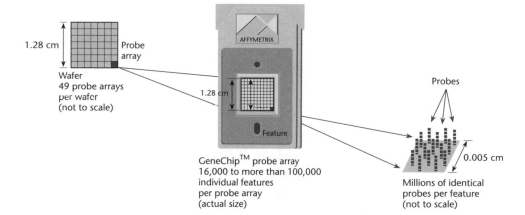

Fig. 7.14 Structure of a sequencing microchip. (Courtesy of Dr. M. Chee, Affymetrix.)

Manufacturing a microarray is deceptively simple. In the first step a mercury lamp is shone through a standard computer-industry photolithographic mask onto the synthesis surface. This activates specific areas for chemical coupling with a nucleoside which itself contains a $5'$ protecting group. Further exposure to light removes this group, leaving a $5'$-hydroxyl group capable of reacting with another nucleoside in the subsequent cycle. The choice of which nucleoside to activate is thus controlled by the composition of the mask. Successive rounds of deprotection and chemistry can result in an exponential increase in oligonucleotide complexity on a chip for a linear number of steps. For example, it requires only $4 \times 15 = 60$ cycles to synthesize a complete set of ~1 billion different 15 mers. The space occupied by each specific oligonucleotide is termed a feature and may house at least 1 million identical molecules (Fig. 7.14). A standard 1.6 cm square chip can house more than 1 million different oligos with a spacing of 1 µm, cf. the computer industry which routinely achieves resolution at 0.3 µm!

Although microchips have not been used for *de novo* sequencing of a genome, they have been used for resequencing genomes as exemplified by human mitochondrial DNA (Chee *et al.* 1996). For this purpose a 4L tiled array is set up in which L corresponds to the length of the sequence to be analyzed. The sequence is probed with a series of oligomers of length P which exactly match the target sequence except for one position which is systematically substituted with each of the four bases A, T, G, or C. In the example shown in Fig. 7.15, a tiled array of 15 mers varied at position 7 from the $3'$ end is used. This

is known as a $P^{15,7}$ array. To use such an array, the DNA to be analyzed is amplified by long-range PCR. Fluorescently labeled RNA is then prepared by *in vitro* transcription and this is hybridized to the array. The hybridization patterns are imaged with a high-resolution confocal scanner and a typical result is shown in Fig. 7.15. For the purpose of resequencing the human mitochondrial genome (L = 16,569 bp) with a tiled array of $P^{15,7}$ probes then a total of 66,276 probes ($4 \times 16,569$) of the possible ~10^9 15 mers would be required.

Microchips are particularly useful for detecting mutations and polymorphisms, particularly SNPs, (Chee *et al.* 1996, Hacia *et al.* 1996, 1998, Kozal *et al.* 1996). As an example, consider again the mitochondrial genome. Chee *et al.* (1996) prepared a $P^{25,13}$ tiling array consisting of 136,528 synthesis cells, each 35 µm square in size. In addition to a 4L tiling across the genome, the array contained a set of probes representing a single-base deletion at every position across the genome and sets of probes designed to match a range of specific mtDNA haplotypes. After hybridization of fluorescently labeled target RNA, 99% of the sequence could be read correctly simply by identifying the highest intensity in each column of four substitution probes. The array also was used to detect three disease-causing mutations in a patient with hereditary optic neuropathy (Fig. 7.16). A refinement to this method also has been developed. Ideally, the hybridization signals from the reference and test DNAs should be compared by hybridization to the same array. This can be carried out by using a two-color labeling and detection scheme in which the reference is labeled with phycoerythrin (red) and the target

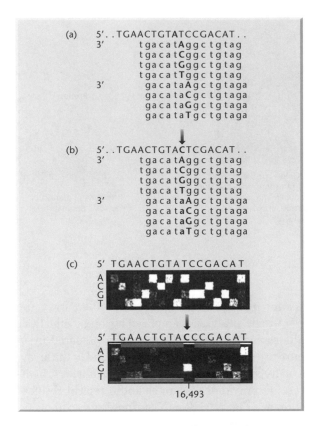

Fig. 7.15 Design and use of a 4L tiled array. Each position in the target sequence (uppercase letters) is queried by a set of four probes on the chip (lowercase letters), identical except at a single position, termed the substitution position, which is either A, C, G, or T (bold black indicates complementarity, green a mismatch). Two sets of probes are shown, querying adjacent positions in target. (b) Effect of a change in the target sequence. The probes are the same as in (a), but the target now contains a single-base substitution (base C, shown in green and arrowed). The probe set querying the changed base still has a perfect match (the G probe). However, probes in adjacent sets that overlap the altered target position now have either one or two mismatches (green) instead of zero or one, because they were designed to match the target shown in (a). (c) Hybridization to a 4L tiled array and detection of a base change in the target. The array shown was designed to the mt1 sequence. (Top) Hybridization to mt1. The substitution used in each row of probes is indicated to the left of the image. The target sequence can be read 5′ to 3′ from left to right as the complement of the substitution base with the brightest signal. With hybridization to mt2 (bottom), which differs from mt1 in this region by a T → C transition, the G probe at position 16,493 is now a perfect match, with the other three probes having single-base mismatches (**A** 5, **C** 3, **G** 37, **T** 4 counts). However, at flanking positions, the probes have either single- or double-base mismatches, because the mt2 transition now occurs away from the query position. (Reprinted from Chee *et al.* 1996 by permission of the American Association for the Advancement of Science.)

with fluorescein (green). By processing the reference and target together, experimental variability during the fragmentation, hybridization, washing, and detection steps is eliminated.

Variations of the microarray method

It should be noted that there are two variants of the hybridization-based method of DNA sequencing. The first approach is exemplified by the mitochondrial analysis described above. That is, oligos are used that are complementary to a significant subset of sequence changes of interest and one measures *gain of hybridization* signal to these oligos in test samples relative to reference samples. Relative "gain" of signal by these oligos indicates a sequence change. In this respect, the microarray is being used in an analogous fashion to a conventional dot blot. This "gain-of-signal" approach allows for a partial scan of a DNA segment for all possible sequence variations. An array designed to interrogate both strands of a target of length N for all possible single nucleotide substitutions would consist of 8N oligos of length 20–25 nucleotides, i.e. 80,000 oligos for a 10 kb sequence. To interrogate the same DNA target for all deletions of a particular length would require an additional 2N oligos, i.e. 100,000 oligos if screening for deletions of 1–5 bp. However, interrogating both target strands for insertions of length X requires $2(4^X)N$ oligos, i.e. 27,280,000 oligos if looking for all 1–5 bp insertions!

In the loss-of-signal approach, sequence variations are scored by quantitating relative losses of hybridization signal to perfect match oligonucleotide probes in test samples relative to wild-type reference targets. Ideally, a homozygous sequence change results in a complete loss of hybridization signal to perfect match probes interrogating the region surrounding the sequence change (Fig. 7.17). For heterozygous sequence variations the signal loss theoretically is 50%. With this approach, an array designed to interrogate both strands of a 10 kb sequence for all possible sequences requires just 20,000 oligos. An added advantage is that multiple probes contribute to the detection of a sequence variation (see Fig. 7.17) thereby minimizing random sources of error caused by hybridization signal fluctuations – a problem with the gain-of-signal approach. The disadvantage of the loss-of-signal method is that the identity of the mutation cannot be determined without subsequent conventional sequencing.

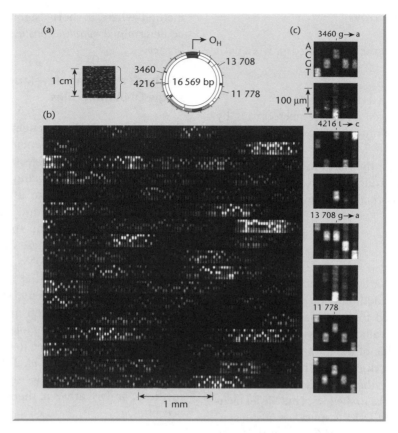

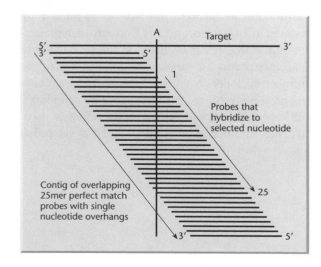

Fig. 7.16 Use of a sequencing microchip to analyze the human mitochondrial genome. (a) An image of the array hybridized to 16.6 kb of mitochondrial target RNA (L strand). The 16,569 bp map of the genome is shown, and the H strand origin of replication (O_H), located in the control region, is indicated. (b) A portion of the hybridization pattern magnified. In each column there are five probes. A, C, G, T, and Δ, from top to bottom. The Δ probe has a single base deletion instead of a substitution and hence is 24 instead of 25 bases in length. The scale is indicated by the bar beneath the image. Although there is considerable sequence-dependent intensity variation, most of the array can be read directly. The image was collected at a resolution of ~100 pixels per probe cell. (c) The ability of the array to detect and read single base differences in a 16.6 kb sample is illustrated. Two different target sequences were hybridized in parallel to different chips. The hybridization patterns are compared for four positions in the sequence. Only the $P^{25,13}$ probes are shown. The top panel of each pair shows the hybridization of the mt3 target, which matches the chip P^0 sequence at these positions. The lower panel shows the pattern generated by a sample from a patient with Leber's hereditary optic neuropathy (LHON). Three known pathogenic mutations, LHON3460, LHON4216, and LHON13,708, are clearly detected. For comparison, the fourth panel in the set shows a region around position 11,778 that is identical in both samples. (Reprinted from Chee *et al.* 1996 by permission of the American Association for the Advancement of Science.)

Fig. 7.17 (*right*) Oligonucleotide probes used in loss of hybridization signal sequence analysis. Ideally each target nucleotide position contributes to hybridization to a set of N overlapping N-base perfect match probes in an oligonucleotide array. In this example, hybridization to 25 overlapping 25-base probes is affected by changes in a single target nucleotide. (Redrawn with permission from Hacia 1999.)

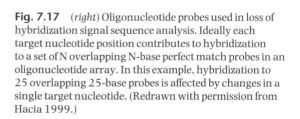

Availability of microarrays

Five different components are required for work with microarrays.

1 The chip itself with its special surface.
2 The device for producing the microarrays by spotting the oligos onto the chip or for their *in situ* synthesis.
3 A fluidic system for hybridization of the test DNA to the immobilized oligos.
4 A scanner to read the micoarrays after hybridization.
5 Software programs to quantify and interpret the results.

All of the above are commercially available (Meldrum 2000b). In addition, micoarrays carrying sets of oligos matched to particular sequences are also commercially available. However, it must be stressed that the value of the data obtained with a microarray depends critically on the quality of the arraying. Fortunately the laying down of arrays is becoming easier thanks to the use of bubble jet technology (Okamoto *et al.* 2000), maskless *in situ* synthesis of oligonucleotides (Singh-Gasson *et al.* 1999), and improvements in microarray surface chemistry (Beier & Hoheisel 2000). This technology is discussed in more detail in Chapter 20.

Massively parallel signature sequencing can be used to monitor RNA abundance

This sequencing method was developed by Brenner *et al.* (2000) to enable them to undertake a global analysis of gene regulation. This topic is covered in more detail later (p. 410) but essentially the relative abundance of different mRNAs is estimated by sequencing a large number of clones from representational cDNA libraries. Because the probability of finding a particular message is proportional to the number of clones sequenced, a large-scale sequencing project is required to detect very low abundance mRNAs if traditional Sanger sequencing is used. Instead, Brenner *et al.* (2000) have developed a method where a short "signature" sequence can be determined *simultaneously* for every cDNA in a library.

Methods are being developed for sequencing single DNA molecules

Of the various DNA sequencing methods described above, only Sanger sequencing has been used in large-scale sequencing projects. The other methods are restricted to resequencing. Two disadvantages of the Sanger method are a need to amplify the DNA before sequencing and the requirement for electrophoresis to separate the DNA ladders that are generated. Both of these disadvantages would be eliminated if single DNA molecules could be sequenced directly but this in turn generates a need for exquisite sensitivity. In recent years considerable progress has been made in detecting the fluorescence from single molecules and the technology now is being applied to single-molecule DNA sequencing (Braslavsky *et al.* 2003, Ramanathan *et al.* 2004). It remains to be seen if either of these methods, or derivatives of them, achieve wide-scale acceptance.

Suggested reading

Chee M. *et al.* (1996) Accessing genetic information with high-density arrays. *Science* **274**, 610–14.
This is a classic paper that describes in detail the use of DNA arrays to resequence the human mitochondrial genome and the identification of base changes in certain diseases.

Drmanac R. *et al.* (2002) Sequencing by hybridization (SBH): advantages, achievements and opportunities. *Advances in Biochemical Engineering and Biotechnology* **77**, 75–101.
This review describes sequencing by hybridization in much greater detail than is possible in this chapter.

Paegel B.M., Blazej R.G. & Mathies R.A. (2003) Microfluidic devices for DNA sequencing: sample preparation and electrophoretic analysis. *Current Opinion in Biotechnology* **14**, 42–50.
This slightly futuristic review describes how all the different stages of DNA sequencing from clone isolation to electrophoretic analysis could be combined into a single device.

CHAPTER 8

Changing genes: site-directed mutagenesis and protein engineering

Introduction

The generation and characterization of mutants is an essential component of any study on structure–function relationships. Knowledge of the three-dimensional structure of a protein, RNA species, or DNA regulatory element (e.g. a promoter) can provide clues to the way in which they function but proof that the correct mechanism has been elucidated requires the analysis of mutants that have amino acid or nucleotide changes at key residues (see Box 8.2).

Classically, mutants are generated by treating the test organism with chemical or physical agents that modify DNA (mutagens). This method of mutagenesis has been extremely successful, as witnessed by the growth of molecular biology and functional genomics, but suffers from a number of disadvantages. First, any gene in the organism can be mutated and the frequency with which mutants occur in the gene of interest can be very low. This means that selection strategies have to be developed. Second, even when mutants with the desired phenotype are isolated, there is no guarantee that the mutation has occurred in the gene of interest. Third, prior to the development of gene-cloning and sequencing techniques, there was no way of knowing where in the gene the mutation had occurred and whether it arose by a single base change, an insertion of DNA, or a deletion.

As techniques in molecular biology have developed, so that the isolation and study of a single gene is not just possible but routine, so mutagenesis has also been refined. Instead of crudely mutagenizing many cells or organisms and then analyzing many thousands or millions of offspring to isolate a desired mutant, it is now possible to change specifically any given base in a cloned DNA sequence. This technique is known as *site-directed mutagenesis*. It has become a basic tool of gene manipulation, for it simplifies DNA manipulations that in the past required a great deal of ingenuity and hard work, e.g. the creation or elimination of cleavage sites for restriction endonucleases. The importance of site-directed mutagenesis goes beyond gene structure–function relationships for the technique enables mutant proteins with novel properties of value to be created (protein engineering). Such mutant proteins may have only minor changes but it is not uncommon for entire domains to be deleted or new domains added.

Primer extension (the single-primer method) is a simple method for site-directed mutation

The first method of site-directed mutagenesis to be developed was the single-primer method (Gillam *et al.* 1980, Zoller & Smith 1983). As originally described the method involves *in vitro* DNA synthesis with a chemically synthesized oligonucleotide (7–20 nucleotides long) that carries a base mismatch with the complementary sequence. As shown in Fig. 8.1, the method requires that the DNA to be mutated is available in single-stranded form, and cloning the gene in M13-based vectors makes this easy. However, DNA cloned in a plasmid and obtained in duplex form can also be converted to a partially single-stranded molecule that is suitable (Dalbadie-McFarland *et al.* 1982).

The synthetic oligonucleotide primes DNA synthesis and is itself incorporated into the resulting heteroduplex molecule. After transformation of the host *E. coli*, this heteroduplex gives rise to homoduplexes whose sequences are either that of the original wild-type DNA or that containing the mutated base. The frequency with which mutated clones arise, compared with wild-type clones, may be low. In order to pick out mutants, the clones can be screened by nucleic acid hybridization with ^{32}P-labeled oligonucleotide as probe. Under suitable conditions of stringency, i.e. temperature and cation concentration, a positive signal will be obtained only with mutant clones. This allows ready detection of the

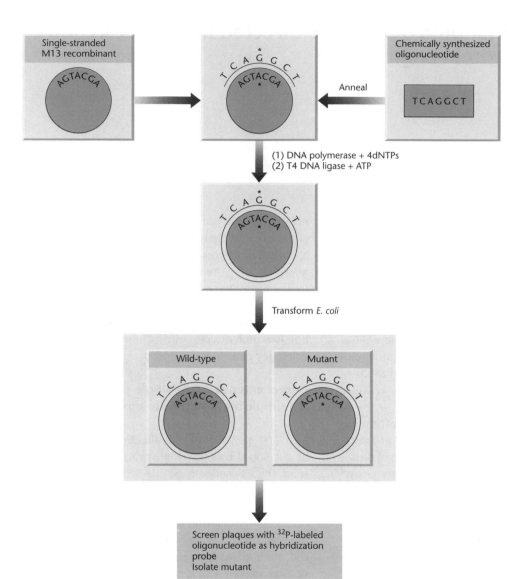

Fig. 8.1
Oligonucleotide-directed mutagenesis. Asterisks indicate mismatched bases. Originally the Klenow fragment of DNA polymerase was used, but now this has been largely replaced with T7 polymerase.

desired mutant (Wallace *et al.* 1981, Traboni *et al.* 1983). It is prudent to check the sequence of the mutant directly by DNA sequencing, in order to check that the procedure has not introduced other adventitious changes. This was a particular necessity with early versions of the technique which made use of *E. coli* DNA polymerase. The more recent use of the high-fidelity DNA polymerases has minimized the problem of extraneous mutations as well as shortening the time for copying the second strand. Also, these polymerases do not "strand-displace" the oligomer, a process which would eliminate the original mutant oligonucleotide.

A variation of the procedure (Fig. 8.2) outlined above involves oligonucleotides containing inserted or deleted sequences. As long as stable hybrids are formed with single-stranded wild-type DNA, prim-

ing of *in vitro* DNA synthesis can occur, ultimately giving rise to clones corresponding to the inserted or deleted sequence (Wallace *et al.* 1980, Norrander *et al.* 1983).

The single-primer method has a number of deficiencies

The efficiency with which the single-primer method yields mutants is dependent upon several factors. The double-stranded heteroduplex molecules that are generated will be contaminated both by any single-stranded non-mutant template DNA that has remained uncopied and by partially double-stranded molecules. The presence of these species considerably reduces the proportion of mutant progeny. They can be removed by sucrose gradient centrifugation

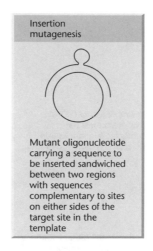

Fig. 8.2
Oligonucleotide-directed mutagenesis used for multiple point mutation, insertion mutagenesis, and deletion mutagenesis.

or by agarose gel electrophoresis, but this is time-consuming and inconvenient.

Following transformation and *in vivo* DNA synthesis, segregation of the two strands of the heteroduplex molecule can occur, yielding a mixed population of mutant and non-mutant progeny. Mutant progeny have to be purified away from parental molecules, and this process is complicated by the cell's mismatch repair system. In theory, the mismatch repair system should yield equal numbers of mutant and non-mutant progeny, but in practice mutants are counterselected. The major reason for this low yield of mutant progeny is that the methyl-directed mismatch repair system of *E. coli* favors the repair of non-methylated DNA. In the cell, newly synthesized DNA strands that have not yet been methylated are preferentially repaired at the position of the mismatch, thereby eliminating a mutation. In a similar way, the non-methylated *in vitro*-generated mutant strand is repaired by the cell so that the majority of progeny are wild type (Kramer *et al.* 1984). The problems associated with the mismatch repair system can be overcome by using host strains carrying the *mut*L, *mut*S, or *mut*H mutations, which prevent the methyl-directed repair of mismatches.

A heteroduplex molecule with one mutant and one non-mutant strand must inevitably give rise to both mutant and non-mutant progeny upon replication. It would be desirable to suppress the growth of non-mutants, and various strategies have been developed with this in mind (Kramer, B. 1984, Carter *et al.* 1985, Kunkel 1985, Sayers & Eckstein 1991).

Another disadvantage of all of the primer extension methods is that they require a single-stranded template. In contrast, with PCR-based mutagenesis

(see below) the template can be single-stranded or double-stranded, circular or linear. In comparison with single-stranded DNAs, double-stranded DNAs are much easier to prepare. Also, gene inserts are in general more stable with double-stranded DNAs.

The issues raised above account for the fact that most of the mutagenesis kits that are available commercially make use of multiple primers and double-stranded templates. For example, in the GeneEditor™ system (Fig. 8.3), two primers are used. One of these primers encodes the mutation to be inserted into the target gene. The second encodes a mutation that enhances the antibiotic resistance properties of the ampicillin-resistance determinant on the vector by conferring resistance to ceftazidime as well. After extending the two primers to yield an intact circular DNA molecule, the mutated plasmid is transformed into *E. coli* and selection made for the enhanced antibiotic resistance. Plasmids encoding the enhanced antibiotic resistance also should carry the mutated target gene. In a variant of this procedure, the vector has two antibiotic resistance determinants (ampicillin and tetracycline) but one of these (Amp^R) carries a mutation. Again, two primers are used: one carrying the mutation to be introduced to the target gene and the other restores ampicillin resistance. After the *in vitro* mutagenesis steps, the plasmid is transformed into *E. coli* and selection made for ampicillin resistance.

Methods have been developed that simplify the process of making all possible amino acid substitutions at a selected site

Using site-directed mutagenesis it is possible to change two or three adjacent nucleotides so that

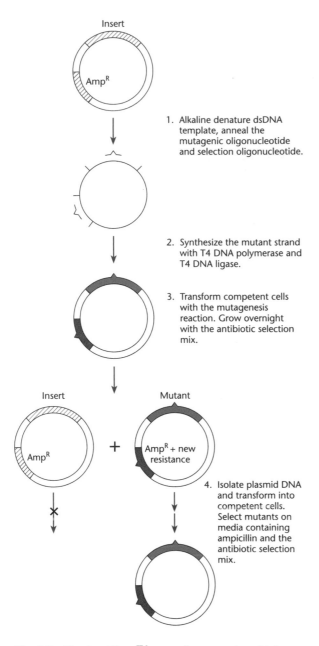

1. Alkaline denature dsDNA template, anneal the mutagenic oligonucleotide and selection oligonucleotide.

2. Synthesize the mutant strand with T4 DNA polymerase and T4 DNA ligase.

3. Transform competent cells with the mutagenesis reaction. Grow overnight with the antibiotic selection mix.

4. Isolate plasmid DNA and transform into competent cells. Select mutants on media containing ampicillin and the antibiotic selection mix.

Fig. 8.3 The GeneEditor™ system for generating a high frequency of mutations using site-directed mutagenesis.

every possible amino acid substitution is made at a site of interest. This generates a requirement for 19 different mutagenic oligonucleotides assuming only one codon will be used for each substitution. An alternative way of changing one amino acid to all the alternatives is cassette mutagenesis. This involves replacing a fragment of the gene with different fragments containing the desired codon changes. It is a simple method for which the efficiency of mutagenesis is close to 100%. However, if it is desired to change the amino acids at two sites to all the possible alternatives then 400 different oligos or fragments would be required and the practicality of the method becomes questionable. One solution to this problem is to use doped oligonucleotides (Fig. 8.4). Many different variations of this technique have been developed and the interested reader should consult the review of Neylon (2004).

The PCR can be used for site-directed mutagenesis

Early work on the development of the PCR method of DNA amplification showed its potential for mutagenesis (Scharf *et al.* 1986). Single bases mismatched between the amplification primer and the template become incorporated into the template sequence as a result of amplification (Fig. 8.5). Higuchi *et al.* (1988) have described a variation of the basic method which enables a mutation in a PCR-produced DNA fragment to be introduced anywhere along its length. Two primary PCR reactions produce two overlapping DNA fragments, both bearing the same mutation in the overlap region. The overlap in sequence allows the fragments to hybridize (Fig. 8.5). One of the two possible hybrids is extended by DNA polymerase to produce a duplex fragment. The other hybrid has recessed 5′ ends and, since it is not a substrate for the polymerase, is effectively lost from the reaction

Fig. 8.4 Mutagenesis by means of doped oligonucleotides. During synthesis of the upper strand of the oligonucleotide, a mixture of all four nucleotides is used at the positions indicated by the letter N. When the lower strand is synthesized, inosine (I) is inserted at the positions shown. The double-stranded oligonucleotide is inserted into the relevant position of the vector.

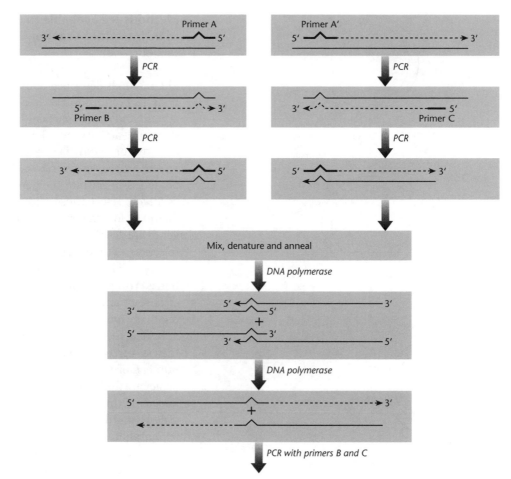

Fig. 8.5 Site-directed mutagenesis by means of the PCR. The steps shown in the top-left corner of the diagram show the basic PCR method of mutagenesis. The bottom half of the figure shows how the mutation can be moved to the middle of a DNA molecule. Primers are shown in bold and primers A and A′ are complementary.

mixture. As with conventional primer-extension mutagenesis, deletions and insertions can also be created.

The method of Higuchi *et al.* (1988) is rather complicated in that it requires four primers and three PCRs (a pair of PCRs to amplify the overlapping segments and a third PCR to fuse the two segments). Commercial suppliers of reagents have developed simpler methods and two of these methods are described below. Two features of PCR mutagenesis should be noted. First, the procedure is not restricted to single base changes: by selecting appropriate primers it is possible to make insertions and deletions as well. Second, *Taq* polymerase copies DNA with low fidelity (see p. 29) and there is a significant risk of extraneous mutations being introduced during the amplification reaction. This problem can be minimized by using a high fidelity thermostable polymerase, and a high template concentration, and fewer than 10 cycles of amplification.

In the Exsite™ method (Fig. 8.6), both strands of the vector carrying the target gene are amplified using the PCR but one of the primers carries the

desired mutation. This results in the production of a population of linear duplexes carrying the mutated gene that is contaminated with a low level of the original circular template DNA. If the template DNA was derived from an *E. coli* cell with an intact restriction modification system then it will be methylated and will be sensitive to restriction by the *Dpn*I endonuclease. The linear DNA produced by amplification will be resistant to *Dpn*I cleavage and after circularization by blunt-end ligation can be recovered by transformation into *E. coli*. Any hybrid molecules consisting of a single strand of the methylated template DNA and unmethylated amplicon also will be destroyed by the endonuclease.

In the GeneTailor™ method (Fig. 8.7), the target DNA is methylated *in vitro* before the mutagenesis step and overlapping primers are used. Once again, linear amplicons are produced that carry the desired mutation but in this case they are transformed directly into *E. coli*. The host-cell repair enzymes circularize the linear mutated DNA while the *Mcr*BC endonuclease digests the methylated template DNA leaving only unmethylated, mutated product.

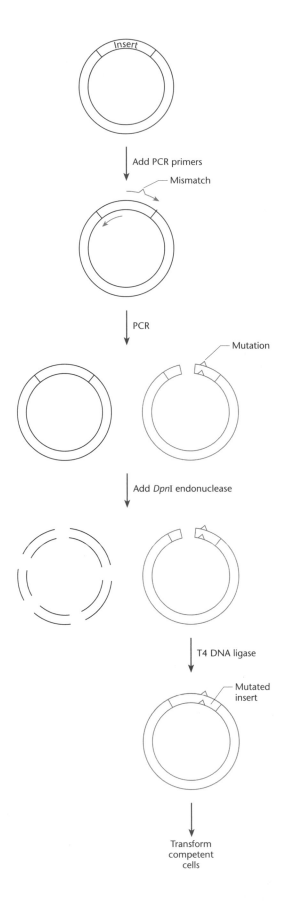

Methods are available to enable mutations to be introduced randomly throughout a target gene

The methods described above enable defined mutations to be introduced at defined locations within a gene and are of particular value in determining structure–activity relationships. However, if the objective of a study is to select mutants with altered and/or improved characteristics then a better approach is to mutate the gene at random and then positively select those with the desired properties. Methods for the random mutagenesis of cloned genes are described in this section and the next while selection methods are described later (p. 148).

It is well known that the polymerase chain reaction is error prone and that there is a high probability of base changes in amplicons. However, even the relatively low fidelity *Taq* polymerase is too accurate to be of value in generating mutant libraries. Nevertheless, increases in error rates can be obtained in a number of ways. One of the commonest ways of achieving this is to introduce a small amount of Mn^{2+}, in place of the normal Mg^{2+}, and to include an excess of dGTP and dTTP relative to the other two nucleotide triphosphates. With this protocol it is possible to achieve error rates of one nucleotide per kilobase (Caldwell & Joyce 1994, Cirino *et al.* 2003). Even higher rates of mutagenesis can be achieved by using nucleoside triphosphate analogs (Zaccolo *et al.* 1996).

The methodologies for error-prone PCR all involve either a misincorporation process in which the polymerase adds an incorrect base to the growing daughter strand or a lack of proofreading ability on the part of the polymerase. It might be expected that they generate a completely random set of mutants but in reality the mutant libraries produced are heavily biased. There are three sources of bias. First, the inherent characteristics of the DNA polymerase used mean that some types of errors are more common than others (Cirino *et al.* 2003). The second source of bias arises because of the nature of the genetic code. For example, a single point mutation in a valine codon can change it to one encoding phenylalanine, leucine, isoleucine, alanine, aspartate, or glycine but

Fig. 8.6   (*left*) The Exsite™ method for generating mutants using the PCR. The parental plasmid (shown in blue) carrying the target gene is derived from a restriction-proficient strain of *E. coli* and so is methylated. This makes it sensitive to the *Dpn*I endonuclease and hence it can be eliminated selectively from the final PCR mixture.

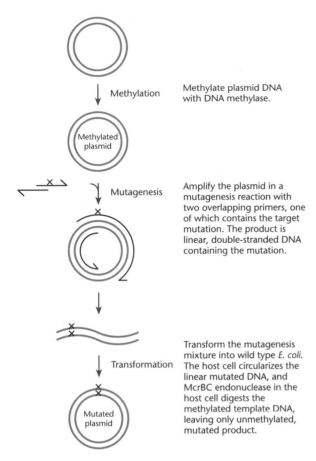

Methylation — Methylate plasmid DNA with DNA methylase.

Methylated plasmid

Mutagenesis — Amplify the plasmid in a mutagenesis reaction with two overlapping primers, one of which contains the target mutation. The product is linear, double-stranded DNA containing the mutation.

Transformation — Transform the mutagenesis mixture into wild type *E. coli*. The host cell circularizes the linear mutated DNA, and McrBC endonuclease in the host cell digests the methylated template DNA, leaving only unmethylated, mutated product.

Mutated plasmid

Fig. 8.7 The GeneTailor™ method for generating mutants using the PCR.

two or three adjacent point mutations are required to change it to one encoding all the other amino acids. The final source of bias arises from the process of amplification. A mutant that is generated early in the amplification process will be over-represented in the final library compared to one that arises in later rounds of amplification.

Error-prone PCR protocols are effective as a means of randomly changing one amino acid into another in the final protein. However, sometimes it might be desirable to explore the effect of randomly deleting or inserting amino acids and this is possible using the random insertion/deletion (RID) process devised by Murakami *et al.* (2002, 2003). The method is based on ligating an insertion or deletion cassette at nearly random locations within the gene.

Altered proteins can be produced by inserting unusual amino acids during protein synthesis

All the mutation methods described above result in the replacement of one or more amino acid residues

in a protein with other *natural* amino acids, e.g. the replacement of a phenylalanine residue with tyrosine, tryptophan, histidine, etc. The ability to incorporate unnatural amino acids into proteins *in vivo* would permit the production of large quantities of proteins with novel properties. For example, the replacement of methionine with selenomethionine facilitates the determination of the three-dimensional structure of proteins (Hendrickson *et al.* 1990). While it is possible to "force" bacteria to incorporate unnatural amino acids into proteins (for review, see Link *et al.* 2003) a better method is to engineer the translational apparatus. This is achieved by generating an aminoacyl-tRNA synthetase and tRNA pair that function independently of the synthetases and tRNAs endogenous to *E. coli* (Wang *et al.* 2001a, Santoro *et al.* 2003). Such a pair are said to be *orthogonal* and satisfy a number of criteria:

- The tRNA is not a substrate for any of the endogenous *E. coli* synthetases but functions efficiently in protein translation.
- The orthogonal synthetase efficiently aminoacylates the orthogonal tRNA whose anticodon has been modified to recognize an amber (UAG) or opal (UGA) stop codon.
- The synthetase does not aminoacylate any of the endogenous *E. coli* tRNAs.

Archaebacteria appear to be an especially good source of orthogonal pairs for use in *E. coli*.

Modifying the anticodon on the tRNA such that it recognizes amber and opal codons is relatively easy. However, the synthetase also needs to be modified such that it charges the cognate tRNA with unusual amino acids more efficiently than the normal amino acid. To do this a library of synthetase mutants is generated and subjected to positive selection based on suppression of an amber codon located in a plasmid-borne gene encoding chloramphenicol acetyltransferase (Wang *et al.* 2001a). Using this approach the tyrosyl-tRNA synthetase of *Methanococcus jannaschii* was modified to permit the site-specific incorporation into proteins of phenylalanine and tyrosine derivatives such as *O*-allyltyrosine, *p*-acetyl-phenylalanine, and *p*-benzoyl-phenylalanine. These modified amino acids can be used as sites for chemical modification of the protein *in vitro* after purification, e.g. the attachment of fluorescent labels (Chin *et al.* 2003, Link *et al.* 2003).

There have been two significant developments of the above technique. In the first of these, Zhang *et al.* (2003) have shown that chemical modification of proteins can occur *in vivo* as well as *in vitro*. For example, *m*-acetylphenylalanine was substituted for Lys7 of the cytoplasmic domain of protein Z and for Arg200 of the outer membrane protein LamB. On addition of a membrane-permeable dye (fluorescein hydrazide) to intact cells, these modified proteins were selectively labeled. In the case of cells expressing the modified LamB derivative, labeling was possible with a range of fluorescein derivatives that are not membrane permeable. The second development is the ability to charge the orthogonal tRNA with glycosylated amino acids. For example, Zhang *et al.* (2004) were able to synthesize in *E. coli* a myoglobin derivative containing β–N-acetylglucosamine (GlcNAc) at a defined position. This GlcNAc moiety was recognized by a saccharide-binding protein and could be modified by a galactosyltransferase.

Phage display can be used to facilitate the selection of mutant peptides

In phage display, a segment of foreign DNA is inserted into either a phagemid or an infectious filamentous phage genome and expressed as a fusion product with a phage coat protein. It is a very powerful technique for selecting and engineering polypeptides with novel functions. The technique was developed first for the *E. coli* phage M13 (Parmley & Smith 1988), but has since been extended to other phages such as T4 and λ (Ren & Black 1998, Santini *et al.* 1998).

The M13 phage particle consists of a single-stranded DNA molecule surrounded by a coat consisting of several thousand copies of the major coat protein, P8. At one end of the particle are five copies each of the two minor coat proteins P9 and P7 and at the other end five copies each of P3 and P6. In early examples of phage display, a random DNA cassette (see above) was inserted into either the P3 or the P8 gene at the junction between the signal sequence and the native peptide. *E. coli* transfected with the recombinant DNA molecules secreted phage particles that displayed on their surface the amino acids encoded by the foreign DNA. Particular phage displaying peptide motifs with, for example, antibody-binding properties were isolated by affinity chromatography (Fig. 8.8). Several rounds of affinity chromatography and phage propagation can be

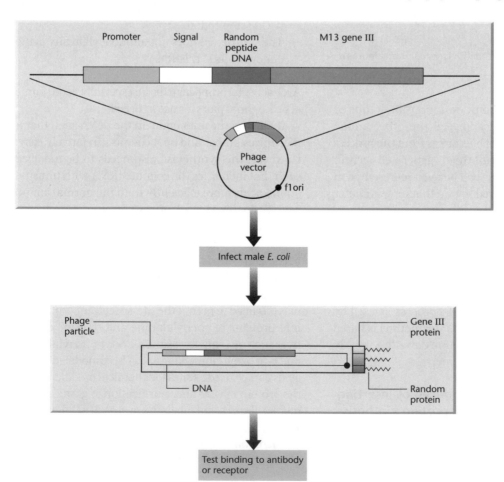

Fig. 8.8 The principle of phage display of random peptides.

used to further enrich for phage with the desired binding characteristics. In this way, millions of random peptides have been screened for their ability to bind to an anti-peptide antibody or to streptavidin (Cwirla *et al.* 1990, Devlin *et al.* 1990, Scott & Smith 1990), and variants of human growth hormone with improved affinity and receptor specificity have been isolated (Lowman *et al.* 1991).

One disadvantage of the original method of phage display is that polypeptide inserts greater than 10 residues compromise coat-protein function and so cannot be efficiently displayed. This problem can be solved by the use of phagemid display (Bass *et al.* 1990). In this system, the starting-point is a plasmid carrying a single copy of the P3 or P8 gene from M13 plus the M13 *ori* sequence (i.e. a phagemid, see p. 75). As before, the random DNA sequence is inserted into the P3 or P8 gene downstream from the signal peptide-cleavage site and the construct transformed into *E. coli*. Phage particles displaying the amino acid sequences encoded by the DNA insert are obtained by superinfecting the transformed cells with helper phage. The resulting phage particles are phenotypically mixed and their surfaces are a mosaic of normal coat protein and fusion protein.

Specialized phagemid display vectors have been developed for particular purposes. For example, phagemids have been constructed that have an amber (chain-terminating) codon immediately downstream

from the foreign DNA insert and upstream from the body of P3 or P8. When the recombinant phagemid is transformed into non-suppressing strains of *E. coli*, the protein encoded by the foreign DNA terminates at the amber codon and is secreted into the medium. However, if the phagemid is transformed into cells carrying an amber suppressor, the entire fusion protein is synthesized and displayed on the surface of the secreted phage particles (Winter *et al.* 1994). Other studies (Jespers *et al.* 1995, Fuh & Sidhu 2000, Fuh *et al.* 2000) have shown that proteins can be displayed as fusions to the carboxy terminus of P3, P6, and P8. Although amino-terminal display formats are likely to dominate established applications, carboxy-terminal display permits constructs that are unsuited to amino-terminal display.

For a detailed review of phage and phagemid display, the reader should consult Sidhu (2000) and Sidhu *et al.* (2000).

Cell-surface display is a more versatile alternative to phage display

As noted in the previous section, the size of foreign protein that can be expressed by phage display is rather limited. Microbial cell-surface display systems were developed to solve this problem (for review, see Lee *et al.* 2003) and these systems also have far more applications (Box 8.1). These display systems

Box 8.1 Applications of cell-surface display

There are many different biotechnological and industrial applications of the cell-surface display technology (Fig. B8.1). For example, key proteins from microbial pathogens can be displayed on the surface of bacteria and their ability to elicit antigen-specific responses determined as a major step towards the development of live vaccines. Proteins that bind heavy metals or specific organic pollutants can be expressed on the surface of cells and these cells can be used as specific bioadsorbents for environmental remediation. Alternatively, new enzyme activities can be expressed on the cell surface to promote environmental degradation of pollutants or for use in industrial biocatalysis. Finally, by anchoring enzymes, receptors, or other signal-sensitive components to the cell's surface new biosensors could be developed. For a review of this topic the reader should consult Benhar (2001).

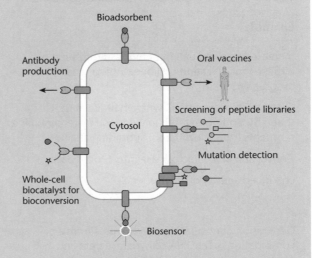

Fig. B8.1 Applications of microbial cell-surface display. Reproduced from Lee *et al.* (2003), with permission from Elsevier.

involve expressing a heterologous peptide or protein of interest (the passenger or target protein) as a fusion protein with various cell-surface proteins (carrier proteins). Depending on the properties of the passenger and carrier proteins, the passenger protein is expressed as an N-terminal, a C-terminal or a sandwich fusion.

For a cell-surface protein to be a successful carrier it should satisfy four requirements. First, it should have an efficient signal peptide to permit the fusion protein to pass through the inner membrane. Second, it should have a strong anchoring structure to keep fusion proteins on the cell surface without detachment. Third, it should be compatible with the passenger protein such that the fusion is not unstable. Finally, it should be resistant to attack by proteases present in the periplasmic space or the growth medium. In Gram-negative bacteria such as *E. coli* many different proteins have been subjugated as carriers. Basically, these proteins fall into two classes: outer membrane proteins (e.g. the adhesin protein, peptidiglycan-associated lipoprotein, and the OmpC and TraT proteins) and protein components of appendages such as pili and flagella. Where outer membrane proteins are used as the carrier it is important to know which part of them is exposed on the outer surface of the cell since this needs to be the site of insertion of the passenger protein.

The passenger protein to be displayed is selected by the required application but its properties influence the translocation process and the effectiveness of the display procedure. For example, the formation of disulfide bridges at the periplasmic side of the outer membrane can affect the efficiency of translocation. Also, the presence of many charged or hydrophobic residues can result in inefficient secretion. Thus, if display technology is used to screen variants produced by random mutagenesis, there may be negative or positive selection for those mutants that affect the efficiency of translocation.

Protein engineering

One of the most exciting aspects of recombinant DNA technology is that it permits the design, development, and isolation of proteins with improved operating characteristics and even completely novel proteins (Table 8.1). The principle of the methods described so far in this chapter is that the gene is mutated, either at a discrete site or at random, and then selection made for a protein variant with the desired property. The improved variant can be subjected to further rounds of mutagenesis and selection, a process known as *directed evolution*. The paradigm for this approach is the enzyme subtilisin. Every

Table 8.1 Some examples of protein engineering.

Example	Method	Reference
Increased rate and extent of biodesulfurization of diesel by modification of dibenzothiophene mono-oxygenase	RACHITT	Coco *et al.* 2001
Generation of a subtilisin with a half-life at 65°C that is 50 times greater than wild type by recombining segments from five different subtilisin variants	StEP	Zhao *et al.* 1998
Conversion of a galactosidase into a fucosidase	Shuffling	Zhang *et al.* 1997
Enhanced activity of amylosucrase	Random mutagenesis plus shuffling	Van der Veen *et al.* 2004
Generation of novel DNA polymerases from a combination of rat DNA polymerase beta and African swine fever virus DNA polymerase X	SCOPE	O'Maille *et al.* 2002
Generation of novel β-lactamase by recombining two genes with 40% amino acid identity and 49% nucleotide sequence identity	SISDC	Hiraga & Arnold 2003

property of this serine protease has been altered including its rate of catalysis, substrate specificity, pH-rate profile, and stability to oxidative, thermal, and alkaline inactivation (for review, see Bryan 2000). Variants also have been produced that favor

aminolysis (synthesis) over hydrolysis in aqueous solvents (see Box 8.2).

An alternative approach to directed evolution is *gene shuffling*. The principle of this method is that many protein variants with desirable characteristics

---

Box 8.2   Improving enzymes

Oxidation-resistant variants of α_1-antitrypsin (AAT)

Cumulative damage to lung tissue is thought to be responsible for the development of emphysema, an irreversible disease characterized by loss of lung elasticity. The primary defense against elastase damage is AAT, a glycosylated serum protein of 394 amino acids. The function of AAT is known because its genetic deficiency leads to a premature breakdown of connective tissue. In healthy individuals there is an association between AAT and neutrophil elastase followed by cleavage of AAT between methionine residue 358 and serine residue 359 (see Fig. B8.2). After cleavage, there is negligible dissociation of the complex.

Smokers are more prone to emphysema, because smoking results in an increased concentration of leucocytes in the lung and consequently increased exposure to neutrophil elastase. In addition, leucocytes liberate oxygen free radicals and these can oxidize methionine-358 to methionine sulfoxide. Since methionine sulfoxide is much bulkier than methionine, it does not fit into the active site of elastase. Hence oxidized AAT is a poor inhibitor. By means of site-directed mutagenesis, an oxidation-resistant mutant of AAT has been constructed by replacing methionine-358 with valine (Courtney *et al.*

1985). In a laboratory model of inflammation, the modified AAT was an effective inhibitor of elastase and was not inactivated by oxidation. Clinically, this could be important, since intravenous replacement therapy with plasma concentrates of AAT is used with patients with a genetic deficiency in AAT production.

Improving the performance of subtilisin

Proof of the power of gene manipulation coupled with the techniques of *in vitro* (random and site-directed) mutagenesis as a means of generating improved enzymes is provided by the work done on subtilisin over the past 15 years (for review, see Bryan 2000). Every property of this serine protease has been altered, including its rate of catalysis, substrate specificity, pH-rate profile, and stability to oxidative, thermal, and alkaline inactivation. In the process, well over 50% of the 275 amino acids of subtilisin have been changed. At some positions in the molecule, the effects of replacing the usual amino acid with all the other 19 natural amino acids have been evaluated.

Many of the changes described above were made to improve the ability of subtilisin to hydrolyze protein when incorporated into detergents. However, serine proteases can be used to synthesize peptides and this approach has a number of advantages over conventional methods (Abrahmsen *et al.* 1991). A problem

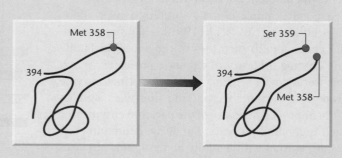

Fig. B8.2   The cleavage of α_1-antitrypsin on binding to neutrophil elastase.

continued

Box 8.2   *continued*

with the use of subtilisin for peptide synthesis is that hydrolysis is strongly favored over aminolysis, unless the reaction is undertaken in organic solvents. Solvents, in turn, reduce the half-life of subtilisin. Using site-directed mutagenesis, a number of variants of subtilisin have been isolated with greatly enhanced solvent stability (Wong *et al.* 1990, Zhong *et al.* 1991). Changes introduced included the minimization of surface changes to reduce solvation energy, the enhancement of internal polar and hydrophobic interactions, and the introduction of conformational restrictions to reduce the tendency of the protein to denature. Designing these changes requires an extensive knowledge of the enzyme's structure and function. Chen and Arnold (1991, 1993) have provided an alternative solution. They utilized random mutagenesis combined with screening for enhanced proteolysis in the presence of solvent (dimethyl formamide) and substrate (casein).

The engineering of subtilisin has now gone one step further, in that it has been modified

such that aminolysis (synthesis) is favored over hydrolysis, even in aqueous solvents. This was achieved by changing a serine residue in the active site to cysteine (Abrahmsen *et al.* 1991). The reasons for this enhancement derive mainly from the increased affinity and reactivity of the acyl intermediate for the amino nucleophile (Fig. B8.3). These engineered "peptide ligases" are in turn being used to synthesize novel glycopeptides. A glycosyl amino acid is used in peptide synthesis to form a glycosyl peptide ester, which will react with another *C*-protected peptide in the presence of the peptide ligase to form a larger glycosyl peptide.

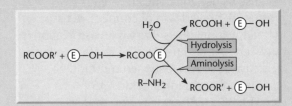

Fig. B8.3   The aminolysis (synthetic) and hydrolysis reactions mediated by an acylated protease.

already exist in nature and novel combinations of these variants may have even more desirable properties (Fig. 8.9). There are three sources of variants for gene shuffling. First, different polymorphisms of the gene of interest might exist naturally in a single organism or might have been created by random *in vitro* mutagenesis (as described on p. 146). Second, the same protein with the same activity may be found in other organisms but the gene and protein sequences will be different. Third, the protein of interest might belong to a protein family where the different members have different but related activities.

A good example of gene shuffling is work done on subtilisin by Ness *et al.* (1999). They started with the genes for 26 members of the subtilisin family and created a library of chimeric proteases. When this library was screened for four distinct enzyme properties, variants were found that were significantly improved over any of the parental enzymes for each individual property. Similarly, Lehmann *et al.* (2000) started with a family of mesophilic phytases whose amino acid sequence had been determined. Using these data they constructed a "consensus"

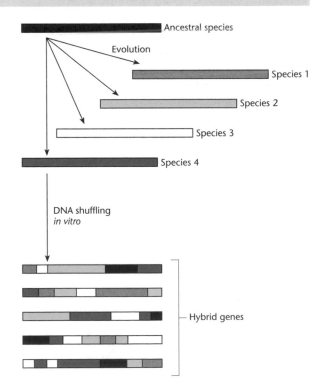

Fig. 8.9   Schematic representation of gene shuffling.

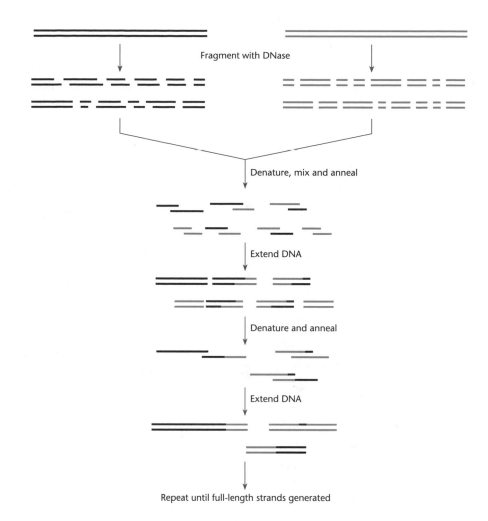

Fragment with DNase

Denature, mix and anneal

Extend DNA

Denature and anneal

Extend DNA

Repeat until full-length strands generated

Fig. 8.10 The original method of gene shuffling. After fragmentation of the two homologous genes, the cycles of denaturation, annealing, and extension are continued until full-length genes can be detected by gel electrophoresis.

phytase sequence and found that an enzyme with this sequence was much more thermostable than any of the parent enzymes.

A number of different methods of gene shuffling have been developed

In the original method of gene shuffling (Stemmer 1993, 2004), one starts by purifying the different genes that will provide the source of variation. These genes are digested with DNase to generate the fragments that will be recombined. The fragments from the different sources are mixed together and subjected to repeated rounds of melting, annealing, and extension (Fig. 8.10). Eventually a full-length gene should be synthesized and this can be amplified by the PCR and cloned. The smaller the fragments that are produced in the initial step the greater the number of single site variations that can be incorporated in the final product. However, the smaller the fragments the greater the number of cycles needed to reassemble a complete gene.

An alternative method is the staggered extension process (StEP, Zhao *et al.* 1998). This also relies on repeated cycles of melting, annealing, and extension to build the variant genes. However, in the StEP process one starts with a mixture of full-length genes, denatures them, and then primes the synthesis of complementary strands (Fig. 8.11). After a short period of primer extension, the DNA is subjected to a round of melting, annealing, and extension. Some of the extended primers will anneal to templates with a different base sequence and on further extension will generate chimeras. The more cycles of extension, melting, and annealing the greater the variability that can be produced.

RACHITT (random chimeragenesis on transient templates) is conceptually similar to the original DNA-shuffling method but is designed to produce chimeras with a much larger number of crossovers (Coco *et al.* 2001, Coco 2003). In this method the gene fragments are generated from one strand of all but one of the parental DNAs (Fig. 8.12). These fragments then are reassembled on the full-length

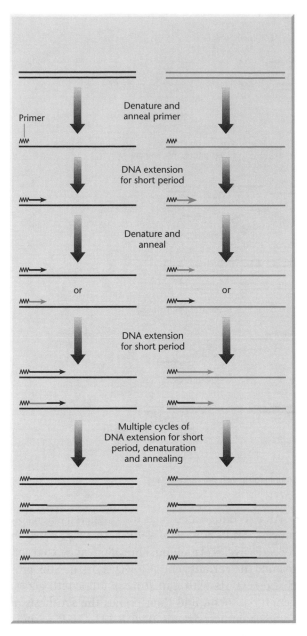

Fig. 8.11 The StEP method for generating hybrid proteins. In the example shown, a hybrid gene will be constructed from two homologous genes (shown in purple and black). Cloning of the hybrid gene will result in the production of a hybrid protein. For clarity, only one strand of each gene is shown after the initial denaturation step.

opposite strand of the remaining parent (the transient template). The fragments are cut back to remove mismatched sections, extended, and then ligated to generate full-length genes. Finally, the template strand is destroyed to leave only the ligated gene fragments to be converted to double-stranded DNA.

Each of the methods described above has its advantages and disadvantages and all of them rely

to a greater or lesser extent on the annealing of mismatched DNA sequences. Thus there is always a chance that the parental molecules will be recreated preferentially or that the degree of variation generated will not be as great as expected. However, methods for "forcing" the generation of recombinants have been developed (for review, see Neylon 2004).

Chimeric proteins can be produced in the absence of gene homology

The gene-shuffling methods described above have an absolute requirement for significant homology between the parental sequences. However, there may be a wish to create hybrids between proteins with functional similarities but whose sequence homology is less than 50%. Achieving this requires methods for combining non-homologous sequences and the first one to be developed (Ostermeier *et al.* 1999) was ITCHY (incremental truncation for the creation of hybrid enzymes). This method is based on the direct ligation of libraries of fragments generated by the truncation of two template sequences, each template being truncated from opposite ends (Fig. 8.13). This ligation procedure removes any need for homology at the point of crossover but the downside is that the DNA fragments may be reconnected in a way that is not at all analogous to their position in the template gene.

In the original ITCHY process the incremental truncation was performed using timed exonuclease digestions. In practice, these digestions are difficult to control. An improved process was developed where the initial templates are generated with phosphorothioate linkages incorporated at random along the length of the gene (Lutz *et al.* 2001a). Complete exonuclease digestion then generates fragments with lengths determined by the position of the nuclease-resistant phosphorothioate linkage. This method is known as thio-ITCHY and is much simpler to perform. One drawback of ITCHY libraries is that they contain only one crossover per gene. However, by combining ITCHY libraries with DNA-shuffling methods, a process known as SCRATCHY, it is possible to generate additional variation (Lutz *et al.* 2001b).

A major problem with methods such as ITCHY is that they generate large numbers of non-functional sequences due to mutations, insertions, and deletions. Furthermore, when one examines the three-dimensional structure of proteins it is clear that they are organized into domains and motifs. Therefore, a more attractive way of generating chimeric

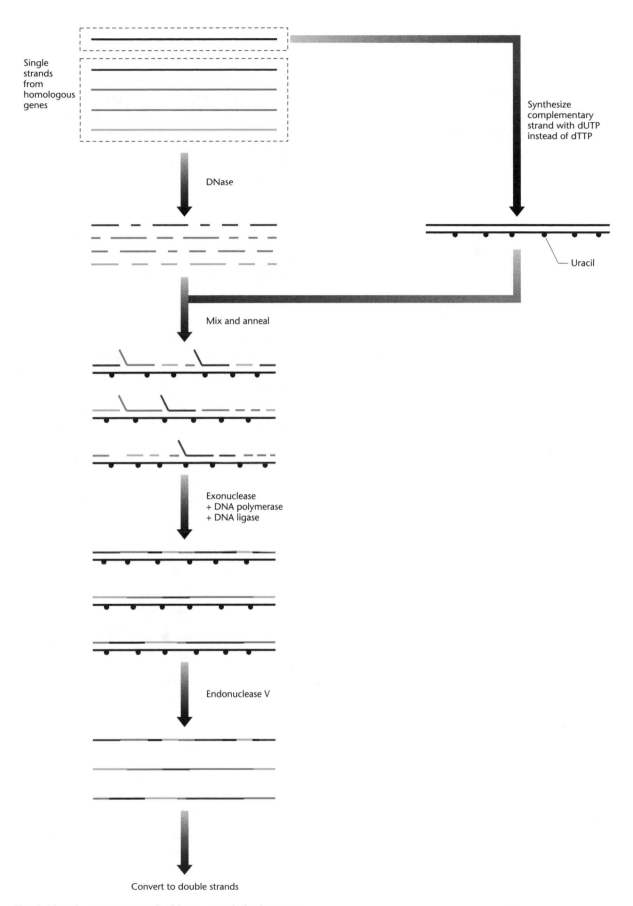

Fig. 8.12 The RACHITT method for creating hybrid proteins.

proteins might be to recombine these domains and motifs in novel ways. Two general methods of doing this have been developed (O'Maille *et al.* 2002, Hiraga & Arnold 2003) and these are SCOPE (structure-based combinatorial protein engineering) and SISDC (sequence-independent site-directed chimeragenesis).

Suggested reading

Brannigan J.A. & Wilkinson A.J. (2002) Protein engineering 20 years on. *Nature Reviews Molecular Cell Biology* **3**, 964–70.
A short but excellent review of the development and pharmaceutical applications of protein engineering.

Collins C.H., Yokobayashi Y., Umeno D. & Arnold F.H. (2003) Engineering proteins that bind, move, make and break DNA. *Current Opinion in Biotechnology* **14**, 371–8.
Another short but excellent review that focuses on what can be achieved with protein engineering rather than on the methods themselves.

Link A.J., Mock M.L. & Tirrell D.A. (2003) Non-canonical amino acids in protein engineering. *Current Opinion in Biotechnology* **14**, 603–9.

Lu Y. (2005) Design and engineering of metalloproteins containing unnatural amino acids as non-native metal-containing cofactors. *Current Opinion in Chemical Biology* **9**, 118–26.
These two papers provide short reviews of the novel chemistries that are possible once unusual amino acids are introduced to proteins.

Lutz S. & Patrick W.M. (2004) Novel methods for directed evolution of enzymes: quality not quantity. *Current Opinion in Biotechnology* **15**, 291–7.

Neylon C. (2004) Chemical and biochemical strategies for the randomisation of protein encoding DNA sequences: library construction methods for directed evolution. *Nucleic Acids Research* **32**, 1448–59.
Each of the methods for generating gene libraries is reviewed in these papers with particular attention being given to the practicality of the methods and the characteristics of the libraries that are produced.

Roodveldt C., Aharoni A. & Tawfik D.S. (2005) Directed evolution of proteins for heterologous expression and stability. *Current Opinion in Structural Biology* **15**, 50–6.
A short review of the application of protein engineering for overproduction of commercial proteins.

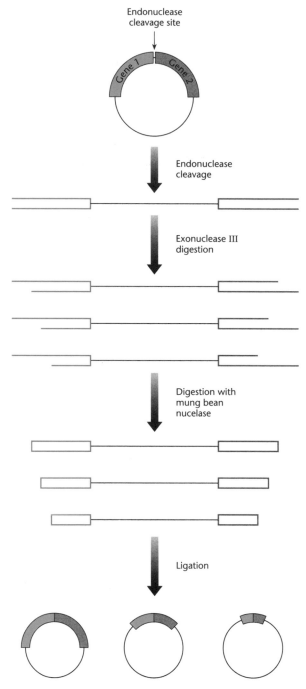

Fig. 8.13 The ITCHY method for creating hybrids of two related proteins. In the figure, the two related proteins are encoded by genes 1 (shown in purple) and 2 (shown in gray). The end result is a hybrid gene comprising the 5' end of gene 1 and the 3' end of gene 2.

CHAPTER 9

Bioinformatics

Introduction

Bioinformatics is the branch of biology that deals with the storage, analysis, and interpretation of experimental data. In today's climate of large-scale biology, the datasets can be very large and computers are essential. The instruments of bioinformatics are computers, databases, and the statistical tools and algorithms that are used for data analysis. The purpose of bioinformatics is to extract information and identify relationships between datasets. The datasets often comprise nucleotide or protein sequences, protein structures, gene-expression profiles, molecular weights, digitized images of gels, and biochemical or metabolic pathways.

Bioinformatics as a defined scientific discipline emerged in the mid-1990s as the genomic revolution geared up and the amount of sequence, structural, and biochemical data began to accumulate. However, the roots of bioinformatics can be traced back to the 1960s, when Margaret Dayhoff established the first database of protein sequences, a database that was published annually as a series of volumes entitled *Atlas of Protein Sequence and Structure* (Dayhoff *et al.* 1965). It could therefore be argued that bioinformatics was born when the first complete protein sequence was determined. This was bovine insulin, sequenced between 1951 and 1955 by Frederick Sanger and colleagues (Ryle *et al.* 1955). By 1965, when the *Atlas of Protein Sequence and Structure* was first published, there were more than 100 protein sequences in the scientific literature. However, the *Atlas* contained very few uncharacterized proteins; most of the sequences were redundant and were used to investigate sequence diversity between homologous proteins in large families such as the globins. It was at this time that the foundations of bioinformatics were laid with the development of mathematical tools for sequence comparison.

It is now much easier to obtain a DNA sequence than a protein sequence, but before 1977, when reliable methods for DNA sequencing became available, this was not the case. The first nucleotide sequence to be determined was that of a yeast tRNA (Madison *et al.* 1966), and most reported nucleotide sequences prior to about 1975 were from RNA molecules. During the late 1970s and early 1980s, DNA sequences began to accumulate slowly in the literature and it became more common to predict protein sequences by translating sequenced genes than by direct analysis of the proteins themselves. Thus the number of uncharacterized protein sequences began to increase. In 1982 there were enough DNA sequences to justify the establishment of the first nucleotide sequence database, GenBank (Benson *et al.* 2004). By the end of 1982, GenBank contained a grand total of 606 sequences. The database grew steadily until about 1994, when the genomics era really kicked in, and then the number of sequences began to grow exponentially as large numbers of genomic clones and expressed sequence tags (ESTs) were deposited. In 1994 the number of sequences in GenBank was just over 200,000. Ten years later, the figure stands at 30 million and shows no sign of slowing down (Fig. 9.1).

As the number of sequences has grown, so has the necessity to use computer-based algorithms to analyze them. In this chapter, we focus on the use of algorithms for the analysis and interpretation of sequence data, and the development and implementation of databases that permit efficient access and management of different types of information. A detailed analysis of the algorithms and statistics used in bioinformatics is beyond the scope of this book, and the interested reader should consult a specialist text such as Attwood & Parry-Smith (1999). Later in the book, we consider how bioinformatics is used to assist with the analysis, modeling, and prediction of protein structures and interactions, and how bioinformatics forms an essential component of genomics, transcriptomics, and proteomics.

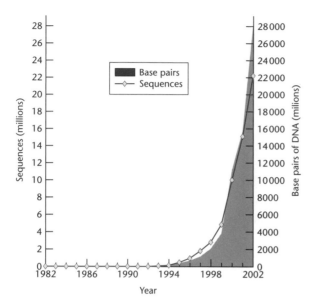

Fig. 9.1 Growth of the GenBank sequence database from 1982–2002. Reproduced with kind permission from GenBank.

Databases are required to store and cross-reference large biological datasets

Databases are at the heart of bioinformatics. Essentially, they are electronic filing cabinets that offer a convenient and efficient method of storing vast amounts of information. There are many different database types, depending both on the nature of the information being stored (e.g. sequences, structures, gel images) and on the manner of data storage (e.g. flat files, tables in relational databases, object-oriented databases). The number of different databases is growing very rapidly. During the year 2005, 171 new molecular biology databases were added to the collection listed by the journal *Nucleic Acids Research*, bringing the total to 719 (Galperin *et al.* 2005). An electronic version of the database list is available at http://nar.oxfordjournals.org/.

The primary nucleotide sequence databases are repositories for annotated nucleotide sequence data

Perhaps the most important databases in molecular biology are the three so-called primary sequence databases. These are GenBank, maintained by the US National Center for Biotechnology Information (NCBI); the DNA Databank of Japan (DDBJ); and the Nucleotide Sequence Database maintained by the

European Molecular Biology Laboratory (EMBL). New sequence data can be deposited with any one of these three groups since they automatically share the data on a daily basis. Further data is collected into the databases by automatically scanning the scientific literature.

The primary sequence databases are repositories for raw sequence data derived directly from experiments and sequencing projects, but the entries are extensively annotated. A typical DNA sequence record held by EMBL is shown in Fig. 9.2. The left-hand column is a list of headings and data is entered in the right-hand column. The file contains not only the sequence, but also information on literature references, and a features table which shows start and stop sites and, if appropriate, the predicted protein sequence. The main sequence databases have subsidiaries with specific types of sequence data in them. For example, GenBank has a subsidiary called dbEST, which is a database of ESTs (see Box 9.1). ESTs have been instrumental in generating gene maps both by hands-on experimentation and pure *in silico* analysis. For example, the first-generation human gene map was generated by mapping ESTs onto radiation hybrid panels and YAC clones using PCR assays. Now genomic sequences are compared directly to the contents of dbEST in order to identify potential open reading frames (ORFs).

SWISS-PROT and TrEMBL are databases of annotated protein sequences

The SWISS-PROT database is not just a repository for protein sequences. Rather, it is a collection of confirmed protein sequences that is extensively annotated with information about structure and biological function, protein family assignments, and bibliographic references (Fig. 9.3). The quality of the data in SWISS-PROT is very high because the content is actively managed (*curated*). Since verification and complete annotation are time consuming, the content of SWISS-PROT is always slightly out of date. Therefore, the less robust TrEMBL database has been developed consisting of entries in the same format as those in SWISS-PROT, derived from the translation of all coding sequences in the EMBL nucleotide sequence database that are not in SWISS-PROT already. Unlike SWISS-PROT entries, those in TrEMBL are awaiting manual annotation. However, they are given a potential functional annotation by similarity to homologous proteins in SWISS-PROT.

```
EMBL (Release):RC22378
ID   RC22378  standard; RNA; PLN; 1440 BP.
XX
AC   U22378;
XX
SV   U22378.1
XX
DT   11-APR-1995 (Rel. 43, Created)
DT   04-MAR-2000 (Rel. 63, Last updated, Version 5)
XX
DE   Ricinus communis oleate 12-hydroxylase mRNA, complete cds.
XX
KW   .
XX
OS   Ricinus communis (castor bean)
OC   Eukaryota; Viridiplantae; Streptophyta; Embryophyta; Tracheophyta;
OC   Spermatophyta; Magnoliophyta; eudicotyledons; core eudicots; Rosidae;
OC   eurosids I; Malpighiales; Euphorbiaceae; Ricinus.
XX
RN   [1]
RP   1-1440
RX   MEDLINE; 95350145.
RA   van de Loo F.J., Broun P., Turner S., Somerville C.;
RT   "An oleate 12-hydroxylase from Ricinus communis L. is a fatty acyl
RT   desaturase homolog";
RL   Proc. Natl. Acad. Sci. U.S.A. 92(15):6743-6747(1995).
XX
RN   [2]
RP   1-1440
RA   Somerville C.R.;
RT   ;
RL   Submitted (08-MAR-1995) to the EMBL/GenBank/DDBJ databases.
RL   Chris R. Somerville, Plant Biology, Carnegie Institution of Washington, 290
RL   Panama Street, Stanford, CA 94305-4101, USA
XX
DR   AGDR; U22378; U22378.
DR   MENDEL; 10454; Ricco;1207;10454.
DR   SPTREMBL; Q41131; Q41131.
XX
FH   Key             Location/Qualifiers
FH
FT   source          1..1440
FT                   /db_xref="taxon:3988"
FT                   /organism="Ricinus communis"
FT                   /strain="Baker 296"
FT                   /tissue_type="developing endosperm"
FT   CDS             187..1350
FT                   /codon_start=1
FT                   /db_xref="SPTREMBL:Q41131"
FT                   /note="expressed only in developing endosperm of castor;
FT                   possible integral membrane protein of endoplasmic
FT                   reticulum; uses cytochrome b5 as intermediate electron
FT                   donor; fatty acid hydroxylase"
FT                   /product="oleate 12-hydroxylase"
FT                   /protein_id="AAC49010.1"
FT                   /translation="MGGGGRMSTVITSNNSEKKGGSSHLKRAPHTKPPFTLGDLKRAIP
FT                   PHCFERSFVRSFSYVAYDVCLSFLFYSIATNFFPYISSPLSYVAWLVYWLFQGCILTGL
.
.
.
FT                   KPIMGEYYRYDGTPFYKALWREAKECLFVEPDEGAPTQGVFWYRNKY"
FT   polyA site      1440
FT                   /note="8 A nucleotides"
XX
SQ   Sequence 1440 BP; 367 A; 340 C; 321 G; 412 T; 0 other;
     gccaccttaa  gcgagcgccg  cacacgaagc  ctcctttcac  acttggtgac  ctcaaatcaa        60
     acaccacacc  ttataactta  gtcttaagag  agagagagag  agagaggaga  catttctctt       120
                                                                                    .
                                                                                    .
     ggcgttttct  ggtaccggaa  caagtattaa  aaaagtgtca  tgtagcctgt  ttctttaaga      1380
     gaagtaatta  gaacaagaag  gaatgtgtgt  gtagtgtaat  gtgttctaat  aaagaaggca      1440
//
```

Fig. 9.2   A typical DNA sequence database entry (dotted lines denote points at which, for convenience, material has been excised).

Box 9.1   Expressed sequence tags (ESTs) provide a rich resource for bioinformatics

The gold standard in the analysis of individual genes is a full-length cDNA clone that has been independently sequenced several times to ensure accuracy. Such clones are desirable for accurate archiving and for the detailed mapping of genomic transcription units, i.e. to determine the transcriptional start and stop sites, and all intron/exon boundaries. However, as discussed in Chapter 6, such clones can be difficult and expensive to obtain. Technology has not yet advanced to the stage where full-length cDNAs can be produced and sequenced in a high-throughput manner.

Fortunately, full-length cDNA clones are not required for many types of analysis. Even short cDNA sequence fragments can be used to identify specific genes unambiguously, and therefore map them onto physical gene maps or provide information about their expression patterns. The development of high-throughput sequencing technology has allowed thousands of clones to be picked randomly from cDNA libraries and subjected to single-pass sequencing to generate 2–300-bp cDNA signatures called *expressed sequence tags* (ESTs) (Wilcox *et al.* 1991, Okubo *et al.* 1992). Although short and somewhat inaccurate, very large numbers of sequences can be collected rapidly and inexpensively, and deposited into public databases that can

be searched using the Internet. The vast majority of database sequences are now ESTs rather than full cDNA or genomic clones. ESTs have been used for gene discovery, as physical markers on genomic maps, and for the identification of genes in genomic clones (e.g. Adams *et al.* 1991, 1992, Banfi *et al.* 1996). Nearly 30 million ESTs from numerous species are currently searchable using the major public EST database, dbEST. The development of EST informatics has been reviewed (Boguski 1995, Gerhold & Caskey 1996, Hard 1996, Okubo & Matsubara 1997).

As well as their use for mapping, ESTs are also useful for expression analysis. PCR primers designed around ESTs have been used to generate large numbers of target sequences for cDNA microarrays. (see Chapter 20), and the partial cDNA fragments used for techniques such as differential display PCR are also essentially ESTs (Chapter 6). The ultimate EST approach to expression analysis is *serial analysis of gene expression* (SAGE). In this technique, the size of the sequence tag is only 9–10 bp (the minimum that is sufficient to identify specific transcripts uniquely) and multiple tags are ligated into a large concatemer allowing expressed genes to be "read" by cloning and sequencing the tags serially arranged in each clone (Chapter 20).

The Protein Databank is the main repository for protein structural information

While SWISS-PROT is the major database of protein sequences, three-dimensional structures are stored in the Protein Databank (PDB). This is the single world-wide archive of structural data derived by X-ray crystallography, nuclear magnetic resonance spectroscopy, and other techniques, as well as structural models (Chapter 22). The database is maintained by the Research Collaboratory for Structural Bioinformatics (RCSB), at Rutgers University. The associated Nucleic Acid Databank (NDB) which shows three-dimensional structures of nucleic acids (e.g. tRNAs) is also maintained there. Like SWISS-

PROT, data in the PDB are very high quality and are extensively curated. There are also other structural databases such as the NCBI's Molecular Modeling Database (MMDB) which aims to provide information on sequence and structure neighbors, links between the scientific literature and 3D structures, and sequence and structure visualization.

Secondary sequence databases pull out common features of protein sequences and structures

Secondary sequence databases take data from the primary databases and use them to classify genes and proteins into different families using the techniques

```
SWALL:PAGT HUMAN
ID   PAGT     HUMAN     STANDARD;     PRT;      559 AA.
AC   Q10472;
DT   01-OCT-1996 (Rel. 34, Created)
DT   01-OCT-1996 (Rel. 34, Last sequence update)
DT   01-MAR-2002 (Rel. 41, Last annotation update)
DE   Polypeptide N-acetylgalactosaminyltransferase (EC 2.4.1.41) (Protein-
DE   UDP acetylgalactosaminyltransferase) (UDP-GalNAc:polypeptide, N-
DE   acetylgalactosaminyltransferase) (GalNAc-T1).
GN   GALNT1.
OS   Homo sapiens (Human).
OC   Eukaryota; Metazoa; Chordata; Craniata; Vertebrata; Euteleostomi;
OC   Mammalia; Eutheria; Primates; Catarrhini; Hominidae; Homo.
OX   NCBI_TaxID=9606;
RN   [1]
RP   SEQUENCE FROM N.A.
RC   TISSUE=Salivary gland;
RX   MEDLINE=96115928; PubMed=8690719;
RA   Meurer J.A., Naylor J.M., Baker C.A., Thomsen D.R., Homa F.L.,
RA   Elhammer A.P.;
RT   "cDNA cloning, expression, and chromosomal localization of a human
RT   UDP-GalNAc:polypeptide, N-acetylgalactosaminyltransferase.";
RL   J. Biochem. 118:568-574(1995).
RN   [2]
RP   SEQUENCE FROM N.A.
RX   MEDLINE=96025800; PubMed=7592619;
RA   White T., Bennett E.P., Takio K., Soerensen T., Bonding N.,
RA   Clausen H.;
RT   "Purification and cDNA cloning of a human UDP-N-acetyl-alpha-D-
RT   galactosamine:polypeptide N-acetylgalactosaminyltransferase.";
RL   J. Biol. Chem. 270:24156-24165(1995).
CC   -!- FUNCTION: THIS PROTEIN CATALYZES THE INITIAL REACTION IN O-LINKED
CC       OLIGOSACCHARIDE BIOSYNTHESIS, THE TRANSFER OF AN N-ACETYL-D-
CC       GALACTOSAMINE RESIDUE TO A SERINE OR THREONINE RESIDUE ON THE
CC       PROTEIN RECEPTOR.
CC   -!- CATALYTIC ACTIVITY: UDP-N-acetyl-D-galactosamine + polypeptide =
CC       UDP + N-acetyl-D-galactosaminyl-polypeptide.
CC   -!- COFACTOR: MANGANESE AND CALCIUM.
CC   -!- PATHWAY: GLYCOSYLATION.
CC   -!- SUBCELLULAR LOCATION: Type II membrane protein. Golgi.
CC   -!- SIMILARITY: BELONGS TO THE GLYCOSYLTRANSFERASE FAMILY 2.
CC   -!- SIMILARITY: CONTAINS 1 RICIN B-TYPE LECTIN DOMAIN.
CC   -----------------------------------------------------------------------
CC   This SWISS-PROT entry is copyright. It is produced through a collaboration
CC   between the Swiss Institute of Bioinformatics and the EMBL outstation -
CC   the European Bioinformatics Institute. There are no restrictions on its
CC   use by non-profit institutions as long as its content is in no way
CC   modified and this statement is not removed. Usage by and for commercial
CC   entities requires a license agreement (See http://www.isb-sib.ch/announce/
CC   or send an email to license@isb-sib.ch).
CC   -----------------------------------------------------------------------
DR   EMBL; U41514; AAC50327.1; -.
DR   EMBL; X85018; CAA59380.1; -.
DR   MIM; 602273; -.
DR   InterPro; IPR001173; Glycos_transf_2.
DR   InterPro; IPR000772; Ricin_B_lectin.
DR   Pfam; PF00535; Glycos_transf_2; 1.
DR   Pfam; PF00652; Ricin_B_lectin; 3.
DR   SMART; SM00458; RICIN; 1.
DR   PROSITE; PS50231; RICIN_B_LECTIN; 1.
KW   Transferase; Glycosyltransferase; Transmembrane; Signal-anchor;
KW   Golgi stack; Glycoprotein; Manganese; Calcium; Lectin.
FT   PROPEP        1     40       REMOVED IN SOLUBLE POLYPEPTIDE
FT                                N-ACETYLGALACTOSAMINYLTRANSFERASE
FT                                (BY SIMILARITY).
FT   CHAIN        41    559       POLYPEPTIDE N-
FT                                ACETYLGALACTOSAMINYLTRANSFERASE, SOLUBLE
FT                                FORM.
FT   DOMAIN        1      8       CYTOPLASMIC (POTENTIAL).
FT   TRANSMEM      9     28       SIGNAL-ANCHOR (TYPE-II MEMBRANE PROTEIN)
FT                                (POTENTIAL).
FT   DOMAIN       29    559       LUMENAL, CATALYTIC (POTENTIAL).
FT   DOMAIN      439    559       RICIN B-TYPE LECTIN.
FT   CARBOHYD     95     95       N-LINKED (GLCNAC . . .) (POTENTIAL).
FT   CARBOHYD    117    117       O-LINKED (POTENTIAL).
FT   CARBOHYD    118    118       O-LINKED (POTENTIAL).
FT   CARBOHYD    119    119       O-LINKED (POTENTIAL).
FT   CARBOHYD    141    141       N-LINKED (GLCNAC . . .) (POTENTIAL).
FT   CARBOHYD    288    288       O-LINKED (POTENTIAL).
FT   CARBOHYD    541    541       N-LINKED (GLCNAC . . .) (POTENTIAL).
FT   CARBOHYD    552    552       N-LINKED (GLCNAC . . .) (POTENTIAL).
SQ   SEQUENCE    559 AA;     64219 MW;    CD68118CB201EE5B    CRC64;
     MRKFAYCKVV  LATSLIWVLL  DMFLLLYFSE  CNKCDEKKER  GLPAGDVLEP  VQKPHEGPGE
     .
     .
     .
     LCLDVSKLNG  PVTMLKCHHL  KGNQLWEYDP  VKLTLQHVNS  NQCLDKATEE  DSQVPSIRDC
     NGSRSQQWLL  RNVTLPEIF
//
```

Fig. 9.3 An abbreviated version of a typical entry in the SWISS-PROT database (dotted lines denote points at which, for convenience, material has been excised).

Table 9.1   A selection of widely used secondary sequence and structural databases.

Database	Contents	URL
PROSITE	*Sequence patterns* associated with protein families and longer *sequence profiles* representing full protein domains	http://ca.expasy.org/prosite
PRINTS, BLOCKS	Highly conserved regions in multiple alignments of protein families. These are called *motifs* in PRINTS and *blocks* in BLOCKS	http://bioinf.man.ac.uk/dbbrowser/PRINTS http://www.blocks.fhcrc.org
Pfam, SMART, ProDom	Collections of protein domains	http://www.sanger.ac.uk/Software/Pfam http://smart.embl-heidelberg.de/ http://prodes.toulouse.inra.fr/prodom/current/html/home.php
Superfamily	HMM library and genome assignments	http://supfam.org/SUPERFAMILY/
PROT-FAM	Protein sequence homology database	http://www.mips.biochem.mpg.de/desc/protfam/
ProClass and iProclass	Protein classifications based on PROSITE patterns and PIR superfamilies	http://pir.georgetown.edu/iproclass/ http://pir.georgetown.edu/gfserver/proclass.html
ProtoMap	Automatic hierarchical classification of all SWISS-PROT and TrEMBL sequences	http://protomap.cornell.edu/
SYSTERS	Protein families database	http://systers.molgen.mpg.de/
CATH	Hierarchical classification of protein structures	http://www.biochem.ucl.ac.uk/bsm/cath_new/
DDD	Structural classification of recurring protein domains	http://www2.ebi.ac.uk/dali/domain
FSSP	Fold classification based on structural alignments	http://www.embl.ebi.ac.uk/dali/fssp
SCOP	Manually curated structural classification of proteins	http://scop.mrc-lmb.cam.ac.uk/scop
Interpro	A search facility that integrates the information from other secondary databases	http://www.ebi.ac.uk/interpro/

of pattern matching. Similarly, there are various structural databases which classify proteins into families based on the possession of particular structural motifs. These databases work by extracting information from the primary databases and using algorithms sometimes with additional manual curation in an attempt to classify sequences and structures into different families. As more sequence and structural data accumulate, such classification becomes more refined, and it becomes easier to assign functions to newly discovered genes and proteins based on sequence or structural similarity to known genes and proteins. Some of the more widely used secondary databases are listed in Table 9.1. The entries in these databases are extensively cross-references with GenBank, SWISS-PROT, the PDB, and additional databases dealing with functional classifications and biochemical pathways (see below). The secondary databases have been constructed using different analytical methods such as motif

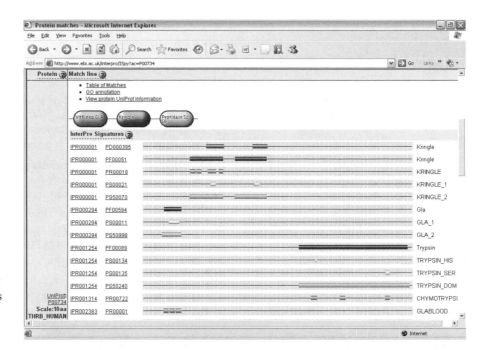

Fig. 9.4 InterPro analysis of human prothrombin. The bars represent matches to different pattern, profile, and structure databases listed in the key.

recognition, fingerprints of collections of motifs, domain profiles, and hidden Markov models. Consequently, each has different strengths and weaknesses and hence different areas of optimum application. This makes it difficult to interpret the results when a predicted protein hits entries in several of the databases. To resolve this issue, a cross-referencing system called InterPro has been developed (Apweiler *et al.* 2001a,b). InterPro permits a protein sequence to be screened against each of the secondary databases and then extracts all the relevant information (Fig. 9.4). An example of the use of InterPro can be found in the analysis of the proteome of *Drosophila* (Rubin *et al.* 2000). We discuss the role of structural databases further in Chapter 22.

Other databases cover a variety of useful topics

A whole book would be required to describe all the databases and their salient features available to biology researchers. However, there are a number of integrated data-retrieval tools which can be accessed over the Internet, and these can be used to search many databases at once. Perhaps the best known is Entrez, a data-retrieval tool developed by the NCBI, which links together all the NCBI databases including GenBank and MMDB as well as the literature database MEDLINE and OMIM (on-line Mendelian inheritance in man), a database of gene products and human phenotypes. Similar tools include SRS

(sequence retrieval system) and DBGet (Kaneisha 1998, Lewitter 1998).

The more notable types of database available on the Internet include those dedicated to specific organisms or genome-sequencing projects, and databases for displaying genome-sequence data complete with gene annotations. There are several so-called gateway sites that contain compendiums of such databases, and these provide a useful first port of call for the uninitiated. Some examples are listed in Table 9.2. The more recently established databases deal with protein functions and attempt to link all proteins into functional networks. Some of these databases focus on specific product types (e.g. receptors, transcription factors) while others have a more general remit. A good example of the latter is the *Kyoto Encyclopedia of Genes and Genomes* (http://www.genome.ad.jp/kegg/), which integrates molecular pathways with gene and protein functions and metabolism. Databases such as DIP and BIND focus on cataloging protein–protein and protein–small molecule interactions, as discussed in greater detail in Chapter 23.

Sequence analysis is based on alignment scores

The basis of sequence comparison is the ability to align two sequences and determine the number of shared residues. The result is an alignment score,

Table 9.2   A selection of gateway sites for bioinformatics on the Internet.

URL	Description
http://www.ncbi.nlm.nih.gov	National Center for Biotechnology Information (NCBI) Homepage, linking to many useful databases, bioinformatics tools, and applications. Home of Entrez, PubMed, GenBank, UniGene, Genome databases and Online Mendelian Inheritance in Man
http://www.ebi.ac.uk	The European Bioinformatics Institute (EBI) is a non-profit academic organization that forms part of the European Molecular Biology Laboratory (EMBL). Another good starting point to find biological databases and bioinformatics software, with good tutorial support
http://www.expasy.ch	The ExPASy (Expert Protein Analysis System) proteomics server of the Swiss Institute of Bioinformatics (SIB). Provides links to many resources, databases and bioinformatics tools relating to the analysis of proteins
http://www.ornl.gov/sci/techresources/Human_Genome/links.shtml	Human Genome Project information "links to the genetic world" hosted by Oak Ridge National Laboratory, a massive collection of links to all manner of useful sites
http://www.highveld.com/pages/molbiol.html	Molecular Biology Jumpstation. Another very extensive and useful collection of links to molecular biology websites. An excellent place to start looking for information
http://wit.integratedgenomics.com/GOLD/	Genomes On Line Database, with links to genomic databases and progress reports on genome projects
http://www.genome.ad.jp/kegg/	Kyoto Encyclopedia of Genes and Genomes. A very comprehensive Japanese site including metabolic maps
http://bioinformatics.ubc.ca/resources/links_directory/	An impressively comprehensive site maintained by *Nucleic Acids Research* and the UBC Bioinformatics Centre at the University of British Columbia (UBC) listing freely available tools, databases, and resources for bioinformatics research organized within categories familiar to biologists.

which represents the quality of the alignment and, at the same time, the closeness of the evolutionary relationship between the two sequences. For nucleotide sequences, comparisons are always made on the basis of sequence *identity*, which is the percentage of identical residues in the alignment. For protein sequences, identity can be suitable for the comparison of very closely related sequences but a more useful measure is sequence *similarity*, which takes into account conservative substitutions between chemically or physically similar amino acids (e.g. valine and isoleucine). When evolutionary changes occur in protein sequences, they tend to involve substitutions between amino acids with similar properties because such changes are less likely to affect the structure and function of the protein. Tables known as substitution score matrices are used to

assign weightings or probabilities to particular substitutions. Several of these matrices are in common use, such as the PAM_{250} matrix and the $BLOSUM_{80}$ matrix. The weightings have been derived by looking at the type of substitutions that have occurred over an evolutionary time scale.

Algorithms for pairwise similarity searching find the best alignment between pairs of sequences

Very short nucleotide or protein sequences can be compared manually, but computer algorithms are required to find the best alignments when the sequences are longer than about 10–15 residues. There are two algorithms in common use, known as the Needleman–Wunsch and Smith–Waterman algorithms, and both of them use dynamic programming to achieve the best alignment scores. Although the algorithms work on similar principles, the Needleman–Wunsch algorithm looks for global similarity between sequences while the Smith–Waterman algorithm focuses on shorter regions of local similarity. Gap penalties are usually applied so that indiscriminate sequence gaps cannot be introduced into the alignment to force sequences to match. Usually, an affine gap penalty is employed, where the alignment score suffers a penalty when a gap is opened, and a smaller penalty when an existing gap is extended.

Dynamic programming algorithms are guaranteed to find the best alignment between two sequences for a given substitution matrix and gap penalty system but they are slow and resource hungry. Therefore, if they are applied to large sequence databases, the searches could take many hours to perform. To allow more rapid searches, alternative methods have been developed which are not based on dynamic programming, and which are faster but less accurate. These have been important in the development of Internet-based database search facilities which otherwise could be rapidly saturated by researchers carrying out similarity searches. The two principal algorithms are BLAST and FASTA. There are several variants of each algorithm that are adapted for different types of searches depending on the nature of the query sequence and the database (Table 9.3). Both BLAST and FASTA take into account the fact that high-scoring alignments are likely to contain short stretches of identical or near identical letters, which are sometimes termed words. In the case of BLAST, the first step is to look for words of a certain fixed word length (W, which is usually equivalent to three amino acids) that score above a given threshold level, T, set by the user. In FASTA, this word length is two amino acids and there is no T value because the match must be perfect. Both programs then attempt to extend their matching segments to produce longer alignments, which in BLAST terminology are called high-scoring segment pairs. FASTA is slower than BLAST because the final stage of the alignment process involves alignment of the high-scoring regions using full dynamic programming.

The significance of a sequence-identity or sequence-similarity score depends on the length of the sequence over which the alignment takes place. The difference between chance similarity and alignments that have real biological significance is determined by the statistical analysis of search scores, particularly the calculation of p values and E values. The p value of a similarity score S is the probability that a score of at least S would have been obtained in a match between any two unrelated protein sequences of similar composition and length. Significant matches are therefore identified by low p values (e.g. $p = 0.01$), which indicate that it is very unlikely that the similarity score was obtained by chance, and probably indicates a real evolutionary relationship. The E value is related to p and is the expected frequency of similarity scores of at least S that would occur by chance. E increases in proportion to the size of the

Table 9.3 Variants of the BLAST and FASTA algorithms.

Program	Compares
FASTA	A nucleotide sequence against a nucleotide sequence database, or an amino acid sequence against a protein sequence database
TFASTA	An amino acid sequence against a nucleotide sequence database translated in all six reading frames
BLASTN	A nucleotide sequence against a nucleotide sequence database
BLASTX	A nucleotide sequence translated in all six reading frames against a protein sequence database
EST BLAST	A cDNA/EST sequence against cDNA/EST sequence databases
BLASTP	An amino acid sequence against a protein sequence database
TBLASTN	An amino acid sequence against a nucleotide sequence database translated in all six reading frames

database that is searched, so even searches with low *p* values (e.g. *p* = 0.0001) might uncover some spurious matches in a database containing 100,000 sequences ($E = 0.001 \times 100,000 = 100$).

Multiple alignments allow important features of gene and protein families to be identified

While pairwise alignments can be used to search for related proteins and provide identification and an initial classification of a newly determined protein sequence, the inter-relationships between members of a protein family are better illustrated by multiple alignments. This is because the conservation of any two amino acid residues between two protein sequences could occur by chance, but if that same residue is found in five or 10 proteins in the family it may play a key functional role.

There are several software packages that can be used for multiple sequence alignment, perhaps the most commonly used of which is ClustalW/X. These programs use progressive alignment strategies in which pairwise alignments are carried out first to assess the degree of similarity between each sequence and to produce a dendrogram of these relationships, which is similar to a phylogenetic tree. The two most similar sequences are aligned first and the others are added in order of similarity. The advantage of this method is its speed, but a disadvantage is that information in distant sequence alignments that could improve the overall alignment is lost. In many cases, the multiple alignments have to be adjusted manually, e.g. to bring conserved cysteine residues into register when it is known that such residues are involved in disulfide bonds.

Sequence analysis of genomic DNA involves the *de novo* identification of genes and other features

When a long piece of DNA has been sequenced, the first task is to identify any genes that are present. In prokaryotes, gene density is generally high and most protein-coding genes lack introns, so the task is relatively straightforward. However, if the DNA in question came from a higher eukaryote a gene is much harder to recognize because it may be divided into many small exons and a similar number of larger introns. For example, in the human genome a typical exon is 150 bp and a typical intron is several kilobases so a complete gene can be hundreds of kilobases in length. Also, the mRNA from some genes can be edited in a number of different ways, resulting in the generation of splice variants. That is, different polypeptides are synthesized from a single gene. An added problem in gene finding is the signal-to-noise ratio. In bacterial genomes, genes make up 80–85% of the DNA. In the yeast *Saccharomyces cerevisiae* this figure drops to 70%. In the fruit fly and nematode this figure drops to 25% and in the human genome genes account for only 3% of the DNA. Thus, defining the precise start and stop positions of a gene and the splicing pattern of its exons among all the non-coding sequence is exceedingly difficult.

Once a gene has been identified the nucleic acid sequence is converted into a protein sequence. The question then is, what is the function of this protein? By searching all the information contained in the various databases it may be possible to identify other proteins with a similar sequence and this may help to identify its function. Sequence comparisons also can be used to identify particular motifs in a protein such as ATP-binding or DNA-binding structures and these too can give information about function.

Genes in prokaryotic DNA can often be found by six-frame translation

If one starts with a length of DNA sequence and wants to identify possible genes then the first task is to identify the correct reading frame. Since there are three possible reading frames on each strand of a DNA molecule this is done by carrying out a *six-frame translation*. The result is six potential protein sequences (Fig. 9.5). The correct reading frame is assumed to be the longest frame uninterrupted by a stop codon (TGA, TAA, or TAG). The longer this open reading frame (ORF), the more likely it is to be a gene since long ORFs are unlikely to occur by chance. Finding the end of such an ORF is easier than finding its beginning. The N-terminal amino acid of a protein usually is methionine so the presence of an ATG codon might indicate the 5′ end of a gene. However, methionine is not always the first amino acid in a protein sequence and it can occur at other positions. Consequently, additional techniques are required to identify the start of an ORF.

In a mRNA molecule, the start codon may be flanked by a Kozak sequence (CCGCCAUGG) and finding this sequence helps to identify the 5′ end of the gene. Analysis of the codon usage can also be helpful since there are marked differences between

```
Query Sequence:
                 10         20         30         40         50
   0  TCCATTGAGC CTTATACCAG TAACATCTAC ACTCGAAGAT CTTGTCAGGG
  50  GAATTTCAGA TTGTGAATCC TCACTTACTG AAAGATCTTA CTGAGCGGGG
 100  CTTGTGGAAT GAAGAGATGA AAAATCAGAT TATTGCATGC AATGGCTCCA
 150  TTCAGTTTTC CTTTTTCAGA GCATACCAGA AATTCCTGAT GACCTGAAGC
 200  AACTCTATAA GACCGTGTGG GAAATCTCTC AGAAGACTGT TCTCAAGATG

Six-Frame Amino Acid Translation:

Forward 0
                 10         20         30         40         50
   0  SIEPYTSNIY TRRSCQGNFR L!ILTY!KIL LSGACGMKR! KIRLLHAMAP
  50  FSFPFSEHTR NS!!PEATL! DRVGNLSEDC SQD

Forward 1
                 10         20         30         40         50
   0  PLSLIPVTST LEDLVRGISD CESSLTERSY !AGLVE!RDE KSDYCMQWLH
  50  SVFLFQSIPE IPDDLKQLYK TVWEISQKTV LKM

Forward 2
                 10         20         30         40         50
   0  H!ALYQ!HLH SKILSGEFQI VNPHLLKDLT ERGLWNEEMK NQIIACNGSI
  50  QFSFFRAYQK FLMT!SNSIR PCGKSLRRLF SR

Reverse 0
                 10         20         30         40         50
   0  HLENSLLRDF PHGLIELLQV IRNFWYALKK EN!MEPLHAI I!FFISSFHK
  50  PRSVRSFSK! GFTI!NSPDK IFECRCYWYK AQW

Reverse 1
                 10         20         30         40         50
   0  ILRTVF!EIS HTVL!SCFRS SGISGML!KR KTEWSHCMQ! SDFSSLHSTS
  50  PAQ!DLSVSE DSQSEIPLTR SSSVDVTGIR LNG

Reverse 2
                 10         20         30         40         50
   0  S!EQSSERFP TRSYRVASGH QEFLVCSEKG KLNGATACNN LIFHLFIPQA
  50  PLSKIFQ!VR IHNLKFP!QD LRV!MLLV!G SM
```

Fig. 9.5 A six-frame translation of an arbitrary DNA sequence. ! denotes a stop codon. (From Attwood & Parry-Smith 1999 © Pearson Education Limited 1999, reprinted by permission of Pearson Education Ltd.)

coding and non-coding regions. Also, the use of codons for particular amino acids varies according to species (Table 9.4). These codon-use rules break down in sequences that are not destined to be translated. Such untranslated regions often have an uncharacteristically high representation of rarely used codons. The identification of segments with a much higher than average GC content, and a higher than average frequency of the CpG dinucleotide, could be indicative of a CpG island. Such islands are found at the 5′ end of many vertebrate genes (Ioshikhes & Zhang 2000). As a final aid to finding ORFs, use can be made of the bias towards G or C as the third base in a codon. Although all of these tools can be applied manually to a DNA sequence, sophisticated computer programs are available (see below) that will do this for you.

There is an important caveat associated with the searching of nucleotide sequences for ORFs. If care is not taken at the sequencing stage then errors can creep into the finished sequence (see p. 131). Although incorrect base calling is undesirable, its effects are fairly minimal unless it results in the erroneous creation or elimination of a termination codon. More important, an erroneous single-base addition or deletion ("phantom indels") will disturb the reading frame and make correct identification of the ORF much more difficult.

Table 9.4
Percentage use of the different serine codons in different organisms. (Reproduced with permission from Attwood & Parry-Smith 1999.)

Codon	*Escherichia coli*	*Drosophila*	Human	Maize	Yeast
AGT	3	1	10	3	5
AGC	20	23	34	30	4
TCG	4	17	9	22	1
TCA	2	2	5	4	6
TCT	34	9	13	4	52
TCC	37	48	28	37	33

Algorithms have been developed that find genes automatically

The most important single development in genome annotation is the use of computers to predict the existence of genes in unprocessed genome-sequence data (see reviews by Fickett 1996, Claverie 1997, Burge & Karlin 1998, Lewis *et al.* 2000 and Gaasterland & Oprea 2001). Before the advent of such algorithms, genes in large genomic constructs had to be identified by painstaking experiments, and several years could be spent finding genes in a section of genomic DNA that had been identified by linkage analysis (Box 9.2). The advantage of computer-based prediction is its speed – annotation can be carried out concurrently with sequencing itself – but a disadvantage is its accuracy, particularly in the complex genome of higher eukaryotes. Essentially, two strategies are used for gene prediction: homology searching and *ab initio* prediction. Homology-searching programs compare genomic-sequence data to gene, cDNA, EST, and protein sequences already present in databases, and are based on the BLAST or FASTA algorithms discussed above. *Ab initio* prediction algorithms search for gene-specific features such as promoters, splice sites, and polyadenylation sites or for pertinent gene content, such as ORFs. Many of the currently available programs combine different search criteria, and their sensitivities vary widely (e.g. see Burset & Guigó 1996).

The identification of ORFs exceeding a certain length (usually about 300 nucleotides, equivalent to 100 amino acids) is sufficient to find most genes in prokaryote genomes. Genuine genes that are smaller than this will be missed if such a threshold is rigorously applied, and there are also difficulties in identifying so-called *shadow genes*, i.e. overlapping ORFs on opposite strands. However, ambiguities arising from this type of genome organization can be resolved using algorithms that incorporate Markov models to highlight differences in base composition between genes and non-coding DNA, e.g. GENMARK (Borodovsky & McIninch 1993), a modified GeneScan algorithm (Ramakrishna & Srinivasan 1999), and Glimmer (Salzberg *et al.* 1999). As a result, it is now possible to identify all genes with near certainty in bacterial genomes. For example, Ramakrishna & Srinivasan (1999), using the improved GeneScan algorithm, reported a near 100% success rate in three microbial genomes – *Hemophilus influenze*, *Plasmodium falciparum*, and *Mycoplasma genitalium*.

Box 9.2   Traditional methods for finding genes

It is hard to imagine in these days of sequencing projects and computer-based annotation the difficulty faced by researchers in the 1970s and 1980s in tracking down genes responsible for particular traits in higher eukaryotes. If some biochemical information was available about the gene or its product, it was sometimes possible to devise a cloning strategy that allowed direct isolation of the gene from a suitable cDNA or genomic library (*functional cloning*; p. 119). However, for most traits (including thousands of inherited human diseases), no relevant biochemical information was available. In such cases, *positional cloning* (p. 120) strategies were developed in which the gene was first mapped to a particular candidate region, and this was progressively narrowed down to a small number of clones spanning the disease locus.

At this stage, various transcript-mapping strategies could be employed:

- *Hybridization of genomic clones to zoo blots.* This approach is based on the observation that coding sequences are strongly conserved during evolution, whereas non-coding DNA generally is not. DNA from a genomic clone that may contain a gene is hybridized to a Southern blot containing whole-genomic DNA from a variety of species (a zoo blot). At reduced stringency, probes containing human genes, for example, will generate strong hybridization signals on genomic DNA from other animals (Monaco *et al.* 1986).
- *Hybridization of genomic clones to northern/reverse northern blots and cDNA libraries.* The major defining feature of a

continued

Box 9.2 *continued*

gene is that it is expressed, producing an RNA transcript. Therefore, if a genomic clone hybridizes to a northern blot (a blot containing only RNA) or a reverse northern blot (a blot containing cDNA, which is derived from RNA) or a cDNA library, it is likely that the genomic clone contains a gene. Unfortunately, this technique relies on the gene being expressed at a significant level in the tissue used to prepare the RNA or cDNA since the sensitivity is low. Part of the sensitivity problem reflects the small exons and large introns characteristic of higher eukaryote genes, which means that large genomic clones may only contain a few hundred base pairs of expressed DNA.

- *Identification of CpG islands.* CpG islands are short stretches of hypomethylated GC-rich DNA often found associated with vertebrate genes. Their function is unclear, but about 50% of human genes have associated CpG islands and these motifs can be exploited for gene identification. One approach is to search raw sequence data for CpG islands by computer. However, the sequences found in CpG islands are scarce in bulk genomic DNA so certain rare-cutter restriction enzymes, such as *Sac*II (which recognizes the site CCGCGG), generate small fragments which can indicate the presence of a gene (Cross & Bird 1995). An alternative PCR-based technique has also been used to identify CpG islands (Valdes *et al.* 1994).

- *cDNA selection and cDNA capture.* In the cDNA selection approach (Lovett *et al.* 1991, Parimoo *et al.* 1991) an amplified cDNA library is hybridized to immobilized genomic clones covering the candidate genomic region. Of the cDNAs selected by this procedure, at least one should correspond to the desired gene. More recently, hybridization has been carried out in solution (cDNA capture) to enrich for cDNAs corresponding to a genomic clone. As above, this technique relies on adequate

expression of the target gene in the tissue used for cDNA preparation, but its sensitivity is much greater than other blot-based techniques. One further problem that has been encountered with this approach is the hybridization of cDNAs to (non-expressed) repetitive DNA elements and pseudogenes (see Lovett 1994).

- *Exon trapping (exon amplification).* This technique was independently devised by a number of research groups and involves an artificial splicing assay (Auch & Reth 1990, Duyk *et al.* 1990, Buckler *et al.* 1991, Hamaguchi *et al.* 1992). The advantage of this over the RNA/cDNA methods discussed above is that there is no need for the gene to be expressed. The general principle is that a genomic clone is inserted into an "intron" flanked by two artificial exons within an expression vector. The vector is then introduced into mammalian cells by transfection and the recombinant expression cassette is transcribed and spliced to yield an artificial mRNA that can be amplified by RT-PCR. If the genomic clone does not contain an exon, the RT-PCR product will contain the two artificial exons in the vector and will be of a defined size. If the genomic clone does contain an exon, it will be spliced into the mature transcript and the RT-PCR product will be larger than expected (Church & Buckler 1999). Cosmid-based exon trap vectors can be used to trap multiple exons in one experiment (Datson *et al.* 1996, den Dunnen 1999).

Powerful as these methods are, they are limited by the fact that experiments have to be carried out at the bench on individual DNA clones. This is suitable where the goal is to identify individual genes, but for the high-throughput annotation of entire genomes, this is simply not fast enough to keep up with the rate at which sequence data accumulate. For predominantly this reason, bioinformatics has largely replaced experimental approaches to gene identification.

Table 9.5 Types of algorithms used for searching for genes in DNA sequences.

Type of algorithm	Principle	Examples
Neural network (Uberacher & Mural 1991)	These are analytical techniques modeled on the processes of learning in cognitive systems. They use a data-training set to build rules that can make predictions or classifications on datasets	GRAIL
Rule-based system	Uses an explicit set of rules to make decisions	GeneFinder
Hidden Markov model (Burge & Karlin 1997)	Represents a system as a set of discrete states and transitions between those states. Markov models are "hidden" when one or more of the states cannot be directly observed. Each transition has an associated probability. Has the advantage of explicitly modeling how the individual probabilities of a sequence of features are combined into a probability estimate for the whole gene	GENSCAN GENIE HMMGene GeneMarkHMM FGENEH

Several sophisticated software algorithms have been devised to handle gene prediction in eukaryotic genomes. Some of these gene predictors only predict a single feature, e.g. the exon predictors HEXON and MZEF. Most, however, attempt to use the output of several algorithms to generate a whole-gene model in which a gene is defined as a series of exons that are coordinately transcribed. The principles used in these algorithms are summarized in Table 9.5. Most of these algorithms are available free of charge over the Internet, as listed in Box 9.3.

What features of eukaryotic genes are recognized by gene-prediction programs? All protein-coding genes are transcribed and translated, so transcriptional and translational control signals, such as the TATA box, cap site, Kozak consensus, and polyadenylation site would seem useful targets. Unfortunately, the diversity of eukaryotic promoters in combination with the small size of these target motifs detracts from their usefulness. For example, a TATA box is found in only about 70% of human genes (Fickett & Hatzigeorgiou 1997) while

---

Box 9.3 Internet resources for genome annotation

Gene prediction software

http://genomic.sanger.ac.uk/gf/gfs.shtml FGENEH

http://www1.imim.es/geneid.html GENEID

http://www.fruitfly.org/seq_tools/genie.html GENIE

http://CCR-081.mit.edu/GENSCAN.html GENSCAN

http://www.cbs.dtu.dk/services/HMMgene HMMGene

http://genemark.biology.gatech.edu/ GeneMark GeneMarkHMM

http://compbio.ornl.gov GRAIL

http://www.tigr.org/softlab/glimmer/ glimmer.html GlimmerM

http://www.itba.mi.cnr.it/webgene GeneBuilder

http://www.sanger.ac.uk/Software/Wise2 Wise2/Genewise

http://blocks.fhcrc.org BLOCKS

Sites providing information on annotation

http://www.fruitfly.org/GASP1 Genome Annotation Assessment Project

http://www.geneontology.org Gene Ontology project

http://www.ebi.ac.uk/interpro InterPro

polyadenylation signals can differ considerably from the consensus sequence AATAAA. Additionally, such signals identify only the first and last exons of a gene. Splice signals are much more useful because they define each exon and are almost invariant. Early gene-finding models assumed independence between positions in the 5′ and 3′ splice sites. More recently, however, dependencies between positions have been identified and have been built into gene-prediction algorithms (e.g. Burge & Karlin 1993). As well as these feature-dependent methods, differences in base composition between coding and non-coding DNA play an important role in gene prediction. Fickett and Tung (1992) compared a large number of base-composition para-meters in coding and non-coding DNA, and reached the conclusion that comparisons of hexamer base composition gave the best discrimination. Many of the currently used gene-prediction programs incorporate Markov models to distinguish hexamer usage between coding and non-coding DNA.

It should be stressed that each of these gene-prediction algorithms needs to be "trained" with data or else implanted with a set of rules. These activities are essential so that the algorithm can recognize key features that distinguish a gene or exon from non-coding DNA. These features are not identical in all organisms. Thus an algorithm trained with nematode DNA will not perform satisfactorily with plant DNA without being retrained. For example, GRAIL is one of the oldest gene-prediction programs and can be used with human, mouse, *Arabidopsis*, *Drosophila*, and *E. coli* sequences but GENIE has been trained only on human and *Drosophila* sequences.

Even when an algorithm has been trained with data from a particular species it is not 100% accurate at identifying genes. This has been addressed in an ongoing international collaborative venture called the Genome Annotation aSsessment Project (GASP). For example, Reese *et al.* (2000) selected two well-characterized regions of the *Drosophila* genome and presented the nucleotide sequence to the authors of the various algorithms for analysis. The best gene predictor had a sensitivity (detection of true positives) of 40% and a specificity (elimination of false positives) of 30% when required to predict entire gene structures. The errors generated included incorrect calling of exon boundaries, missed or phantom exons, or failure to detect entire genes. In a similar study using human DNA, Fortna & Gardiner (2001) found that the best results were obtained

by running five different programs and counting consensus exons obtained from any two or more programs (Table 9.6).

The algorithms described in Table 9.5 are known as *ab initio* programs since they attempt to predict genes from sequence data without the use of prior knowledge about similarities to other genes. Finding genes in long sequences can be facilitated by looking for matches with sequences that are known to be transcribed, e.g. a cDNA, an EST, or even a gene in another species. The pace of genome annotation changed radically with the growth of EST data, since genomic sequences could be rapidly screened for EST hits to identify potential genes. EST clones are derived from the 3′ ends of polyA⁺ transcripts and contain 3′ untranslated sequences (Box 9.1). However, they often extend far enough towards the 5′ end to reach the coding sequence and thus overlap with predicted exons but they cannot be expected to identify all coding exons. It should be noted that not all ESTs can be assumed to be reliable indicators of a gene or a mature mRNA. In some cases they can be derived from unprocessed intronic sequences, primed from the genomic polyA tract, or from processed pseudogenes.

The current trend in gene prediction is to make as much use of sequence-similarity data as possible. The latest generation of gene-prediction algorithms, such as Grail/Exp, GenicEST and GenomeScan (Yeh *et al.* 2001), combine *ab initio* predictions with similarity data into a single probability model. Both Reese *et al.* (2000) and Fortna & Gardiner (2001) found that algorithms that take similarity data into account are better at predicting gene structure.

Additional algorithms are necessary to find non-coding RNA genes and regulatory elements

In all cells, but particularly multicellular eukaryotes, there is much more to the genome than coding regions. For example, analysis of non-coding RNAs and regulatory regions can provide much useful information. Of the non-coding RNAs, rRNAs are the easiest to find and this usually is done by similarity searching. tRNAs can be found by using tRNAScan-SE, a program that includes searching for characteristic structural features such as the ability to form hairpins (Eddy 1999). More recently there has been intense interest in the detection and identification of genes for short regulatory RNA molecules variously known as small temporal RNAs (stRNAs)

Table 9.6 Exon predictions using different exon search programs with and without expressed sequence tags (ESTs). For each gene the actual number of exons has been experimentally verified. (Reprinted from Fortna & Gardiner 2001 by permission of Elsevier Science.)

Accession no.	Gene	No. of exons	GRAIL			Genscan			MZEF			FGENES			FGENE H			≥2 Exons*			ESTs†			Exon + EST‡		
			TP	FP	ME	TP	FP	ME	TP	FP	ME	TP	FP	ME	TP	FP	ME	TP	FP	ME	TP	FP	ME	TP	FP	ME
AP001715	B3/GCFC	17	14	5	3	17	2	0	14	1	3	16	4	1	17	2	0	17	4	0	9	19	8	9	2	8
	ORF4	3	1	2	2	0	1	3	1	4	2	0	0	3	0	0	3	1	1	2	2	9	1	1	2	2
	B37	3	2	1	1	0	0	3	3	1	0	2	1	1	1	0	2	3	0	0	1	4	2	1	0	2
AP001717	PRKCBP2	1	1	0	0	1	0	0	0	0	1	1	0	0	1	0	0	1	0	0	1	0	0	1	0	0
	IFNGR2	7	6	4	1	4	1	3	2	3	5	4	3	3	3	3	4	5	1	2	7	8	0	6	2	1
	C21orf4	7	2	0	5	3	0	4	4	1	3	4	1	3	0	0	7	4	0	3	7	5	0	4	1	3
	C21orf55	2	0	0	2	0	0	2	0	1	2	0	0	2	0	0	2	0	0	2	1	0	1	0	0	2
	GART	22	19	2	3	15	0	7	19	4	3	19	0	3	14	1	8	20	0	0	22	9	0	21	2	1
	SON	11	9	2	2	10	1	1	9	4	2	7	1	4	10	1	1	11	1	0	11	23	0	10	11	1
	CRYZL1	13	7	6	6	7	1	6	7	8	6	8	1	5	7	0	6	8	1	5	12	14	1	9	2	4
	B17	9	4	0	5	4	0	5	7	2	2	5	1	4	8	1	1	7	1	2	9	6	0	8	1	1
AP001753	AGPAT3	8	7	8	1	7	6	1	6	1	2	7	5	1	7	7	1	7	5	1	8	5	0	7	1	1
	TMEM1	23	17	12	6	20	6	3	13	1	10	18	7	5	14	7	9	19	6	4	16	25	7	15	12	8
	PWP2H	21	17	2	4	20	4	1	14	1	7	19	1	2	19	6	2	21	3	0	18	4	3	18	3	3
	C21orf33	7	4	0	3	7	4	0	3	0	4	5	1	2	6	0	1	7	0	0	7	4	0	7	0	0
	KIAA0653	6	4	2	2	5	2	1	6	2	0	6	1	0	5	1	1	5	0	1	4	0	2	4	0	2
	DNMT3L	12	10	1	2	11	0	1	6	0	6	6	4	6	11	3	1	11	1	1	10	3	2	10	1	2
Total		172	124	47	48	131	28	41	110	32	62	127	31	45	123	32	49	147	24	25	145	138	27	131	41	41

* Consistent exon predictions from any two or more programs.
† Exon predictions from ESTs alone.
‡ Consistent exon prediction from EST match and any one or more exon programs.
TP, True positives; FP, false positives; ME, missed exons.

or micro-RNAs (miRNAs) (Kim 2005). These are discussed in more detail in Chapter 15.

Regulatory regions are particularly important sequence features of genomes. To date, a relatively small number of transcriptional-factor binding sites have been identified by classical experimental methods. The existence of such sites in a query sequence is suggestive of a regulatory region. However, these sequences typically are fairly short and could occur by chance. Better evidence for such regulatory regions is their conservation in the sequences upstream from the same gene in two related species, e.g. mouse and human.

Several *in silico* methods are available for the functional annotation of genes

The simplest way to identify the function of new genes, or putative genes, is to search for sequence homologs that have functions assigned already. The principle upon which bioinformatics is built is that similar sequences yield similar structures, and similar structures have similar functions. If a putative protein encoded by an uncharacterized ORF shows statistically significant similarity to another protein of known function, this strongly indicates that the ORF in question is a *bona fide* new gene and identifies its function. A search of this kind is initially undertaken with BLAST or FASTA to identify close-sequence relatives.

Significant matches of a novel gene to another sequence may be in any of four classes (Oliver 1996). First, a match may predict both the biochemical and physiological function of the novel gene. An example is ORF YCR24c, identified during the whole-genome sequencing of *S. cerevisiae*, which has a close-sequence similarity to an Asn-tRNA synthetase from *E. coli*. Second, a match may define the biochemical function of a gene product without revealing its cellular function. An example of this is five protein kinase genes found on yeast chromosome III whose biochemical function is clear (they phosphorylate proteins) but whose particular physiological function in yeast is unknown. Third, a match may be to a gene from another organism whose function in that organism is unknown. For example, ORF YCR63w from yeast matched protein G10 from *Xenopus* and novel genes from *Caenorhabditis elegans* and humans but at the time the function of all of them was unknown. Finally, a match may occur to a gene of known function that merely reveals that our understanding of that function is superficial, e.g. yeast ORF YCL17c and the NifS protein of nitrogen-fixing bacteria. After similar sequences were found in a number of bacteria that do not fix nitrogen it was shown that the NifS protein is a pyridoxal phosphate-dependent aminotransferase.

If a standard BLAST or FASTA search fails to identify a homolog that has already been annotated, more advanced search tools can be employed. One example is PSI-BLAST, an extension of the basic BLAST program which can identify three times as many related sequences. The principle of PSI-BLAST is iterated database searching, where the results of a standard BLAST search are collected into a profile, which is then used for a second round of searching. The process can be repeated for a defined number of cycles as determined by the user, or it can be repeated indefinitely until no more hits are obtained.

Even more distant relationships can be found by pattern matching, in which large collections of sequences are aligned and screened for conserved features. The simplest example of pattern matching is the consensus sequence, which is a representation of the most common residue found at each position of a multiple alignment. More complex sequence patterns and profiles are based on similar principles, but allow variation at each position to be represented. Pattern matching is implemented in the secondary-sequence databases discussed earlier in the chapter (Eddy 1998).

Structures can be used in addition to sequences for functional annotation. This is because protein structures are far better conserved than sequences over evolutionary time. Proteins that show less than 10% sequence identity can have very similar structures, and structural comparisons can therefore reveal more distant evolutionary relationships than any of the sequence-based methods. We return to this topic in Chapter 22.

As is the case for sequences and structures, the large amount of information now available about protein functions makes it necessary to develop systems for functional classification. Several such systems have been devised. One of the oldest and best established, but which only applies to enzymes, is the Enzyme Commission hierarchical system for enzyme classification. Other more general approaches are used in the *Kyoto Encyclopedia of Genes and Genomes* (http://www.genome.ad.jp/kegg/), and most recently, the Gene Ontology system (http://www.geneontology.org/). Gene ontology (Ashburner *et al.* 2000) is a classification system

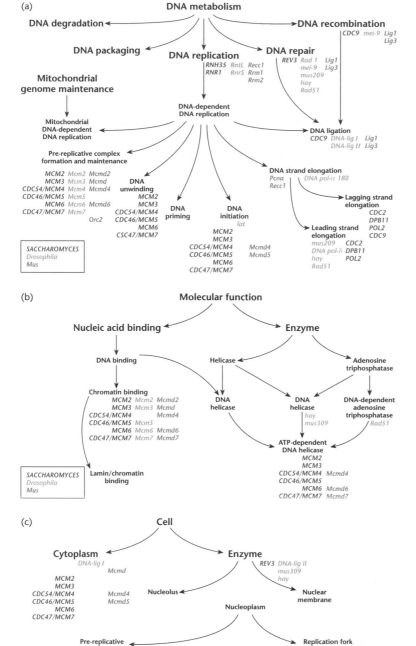

Fig. 9.6 Examples of gene ontology. Three examples illustrate the structure and style used by gene ontology to represent the gene ontologies and to associate genes with nodes within an ontology. The ontologies are built from a structured, controlled vocabulary. The illustrations are the products of work in progress and are subject to change when new evidence becomes available. For simplicity, not all known gene annotations have been included in the figures. (a) Biological process ontology. This section illustrates a portion of the biological process ontology describing DNA metabolism. Note that a node may have more than one parent. For example, "DNA ligation" has three parents: "DNA-dependent DNA replication"; "DNA repair"; and "DNA recombination". (b) Molecular function ontology. The ontology is not intended to represent a reaction pathway, but instead reflects conceptual categories of gene-product function. A gene product can be associated with more than one node within an ontology, as illustrated by the MCM proteins. These proteins have been shown to bind chromatin and to possess ATP-dependent DNA helicase activity, and are annotated to both nodes. (c) Cellular component ontology. The ontologies are designed for a generic eukaryotic cell, and are flexible enough to represent the known differences between diverse organisms. (Reprinted from Ashburner *et al.* 2000 by permission of Nature Publishing Group, New York.)

that enables protein function to be related to gross cellular or whole-organism functions such as central metabolism, nucleic acid replication, cell division, pathogenesis, etc. (Fig. 9.6). It combines the breadth required to describe biological functions among very diverse species with the specificity and depth needed to distinguish a particular protein from another member of the same family. All the organism-specific databases are using the standard gene ontology vocabulary to annotate genes and proteins (Stein 2001).

Caution must be exercised when using purely *in silico* methods to annotate genomes

Many different genomes have been completely sequenced and as the sequence data accumulate, attempts are made, using the tools described above, to identify all the genes and ascribe functions to them. However, a predicted gene or predicted protein function is exactly that – a prediction. The reliability of that prediction depends on the experimental data to support it. Therefore, in annotating genes and genomes, biologists must use their own knowledge and intuition plus information from the literature to design experiments to support their interpretations. An example is provided by Pollack (2001) who found that examination of genomic or enzymatic data alone provided an incomplete picture of metabolic function in the bacterium *Ureaplasma urealyticum*.

The genome of *U. urealyticum* is 752 kb in length and has 613 protein-coding genes and 39 RNA-coding genes. Biological roles have been ascribed to 53% of the protein-coding genes, 19% have no known function although they are similar to other orphan genes, and 28% are genes with no significant relationship to any other gene in any database (Glass *et al.* 2000). The metabolism of *Ureaplasma* has been studied for a long time and hence one would expect a good fit between annotated genes and known enzyme activities. Although in the majority of cases there is such agreement, a considerable number of examples were found where there were annotated genes without detectable activity (e.g. deoxyguanosine kinase, hypoxanthine-guanine phosphoribosyl transferase) or reported enzyme activity without annotation (e.g. pyruvate carboxylase, aspartate aminotransferase).

There are a number of explanations for the apparent disagreement between biological function and gene annotation. These include inaccurate sequencing, inaccurate annotation, mistakes in carrying out enzymatic assays, mutations in crucial residues that eliminate enzymatic activity, and differences between strains. It could be that the annotation is correct but that post-translational modification renders the enzyme undetectable or inactive or the enzyme only is synthesized in certain media or environmental conditions. For genome annotation it is therefore essential to bring biologists and bioinformaticians together and get them working in teams where they can both stimulate and challenge each other. Stein (2001) has reviewed the different organizational models that have been used for these annotation teams. Given the importance of genome annotation, it is essential that different teams use similar methods. Fortunately, standard genome annotation languages are gaining general acceptance. GAME is particularly valuable for describing experimental evidence that supports an annotation and DAS (distributed annotation system) is particularly useful for indexing and visualization. These are accompanied by a number of software tools (e.g. BioPerl, BioPython, BioJava, and BioCORBA) for storing, manipulating, and visualizing these genome annotations.

Sequencing also provides new data for molecular phylogenetics

The major use of DNA and protein sequence data is the analysis of cellular function. However, the data also can be used to investigate the evolution of genes and their protein products and this is known as molecular phylogeny. There are two approaches to building models of sequence evolution (Whelan *et al.* 2001). Empirical models are built through comparisons of large numbers of observed sequences whereas parametric models are built on the basis of the chemical or biological properties of DNA and amino acids.

The parametric approach is favored for studies on DNA evolution and the parameters used are base frequency, base exchangeability, and rate heterogeneity. The base frequency parameter describes the frequency of the bases A, G, C, and T averaged over all sequence sites and is influenced by the overall GC content. Base exchangeability describes the relative tendencies for one base to be substituted for another and base transitions (purine/purine and pyrimidine/pyrimidine) are expected to be more frequent than base transversions (purine/pyrimidine). Rate heterogeneity is the variation in mutation rates along a stretch of DNA because of biochemical constraints, structural features, etc.

In contrast to phylogenetic studies with DNA, those on proteins use the empirical approach in which the number of amino acid substitutions is computed. On its own, the number of changes is not very revealing and so phylogenists use sophisticated statistical methods such as maximum likelihood to make inferences about the patterns and processes of evolution. Since proteins are encoded by DNA it might be assumed that protein and DNA phylogenies would be identical but this is not necessarily so. The reason for this is that silent mutations occur in DNA because of the redundancy of the genetic code. Mutations at the DNA level that do not result in amino acid substitutions only get incorporated into DNA phylogenies.

Suggested reading

Atwood T.K. & Parry-Smith D.J. (1999) *Introduction to Bioinformatics*. Addison Wesley Longman, Harlow.

Benoit G. (2005) Bioinformatics. *Annual Review of Information Science* **39**, 179–218.

Goodman N. (2002) Biological data becomes computer literate: new advances in bioinformatics. *Current Opinion in Biotechnology* **13**, 68–71.

Orengo C.A., Jones D.T. & Thornton J.M. (eds) (2003) *Advanced Text. Bioinformatics: Genes, Proteins and Computers*. BIOS Scientific Publishers, Oxford.

Patterson M. & Handel M. (eds.) (1998) Trends guide to bioinformatics. A supplement to *Trends in Genetics*, 1998.

Reffern O., Grant A., Maibaum M., *et al.* (2005) Survey of current protein family databases and their application in comparative, structural and functional genomics. *Journal of Chromatography* **815**, 97–107.

Stern L. (2001) Genome annotation: from sequence to biology. *Nature Review Genetics* **2**, 493–503.

Part II

Manipulating DNA in Microbes, Plants, and Animals

CHAPTER 10

Cloning in bacteria other than *Escherichia coli*

Introduction

For many experiments it is convenient to use *E. coli* as a recipient for genes cloned from eukaryotes or other prokaryotes. Transformation is easy and there is available a wide range of easy-to-use vectors with specialist properties, e.g. regulatable high-level gene expression. However, use of *E. coli* is not always practicable because it lacks some auxiliary biochemical pathways that are essential for the phenotypic expression of certain functions, e.g. degradation of aromatic compounds, antibiotic synthesis, pathogenicity, sporulation, etc. In such circumstances, the genes have to be cloned back into species similar to those from which they were derived.

There are three prerequisites for cloning genes in a new host. First, there needs to be a method for introducing the DNA of interest into the potential recipient. The methods available include transformation, conjugation, and electroporation, and these will be discussed in more detail later. Secondly, the introduced DNA needs to be maintained in the new host. Either it must function as a replicon in its new environment or it has to be integrated into the chromosome or a pre-existing plasmid. Finally, the uptake and maintenance of the cloned genes will only be detected if they are expressed. Thus the inability to detect a cloned gene in a new bacterial host could be due to failure to introduce the gene, to maintain it or to express it, or to a combination of these factors. Another cause of failure could be restriction. For example, the frequency of transformation of *Pseudomonas putida* with plasmid RSF1010 is 10^5 transformants/µg DNA, but only if the plasmid is prepared from another *P. putida* strain. Otherwise, no transformants are obtained (Bagdasarian *et al.* 1979). Similarly, the frequency of transformation of *Pseudomonas stutzeri* with plasmid DNA that has been prepared from *E. coli* is 43 times higher if the recepient strain is defective for host restriction (Berndt *et al.* 2003). Wilkins *et al.* (1996) have noted

that conjugative transfer of promiscuous IncP plasmids is unusually sensitive to restriction.

Many bacteria are naturally competent for transformation

DNA can be transferred between different strains of *E. coli* by the three classical methods of conjugation, transduction, and transformation, as well as by the newer method of electroporation. For gene-manipulation work, transformation is nearly always used. The reasons for this are threefold. First, it is relatively simple to do, particularly now that competent cells are commercially available. Secondly, it can be very efficient. Efficiencies of 10^8–10^9 transformants/µg plasmid DNA are readily achievable and are more than adequate for most applications. Thirdly, self-transmissible cloning vectors are much larger than their non-transmissible counterparts because they have to carry all the genes required for conjugal transfer.

The strains of *E. coli* used in gene manipulation are not naturally transformable. Rather, competence is induced by chemically treating the cells (see p. 24). By contrast, over 40 species of bacteria are known to be naturally transformable (Lorenz & Wackernagel 1994). Transformation in these organisms differs in a number of respects from chemically induced competence in *E. coli*. First, with the exception of *Neisseria gonorrhoeae*, competence for transformation is a transient phenomenon. Second, transformation can be sequence independent, as in *Bacillus subtilis* and *Acinetobacter calcoaceticus*, but in other species (*Hemophilus influenzae*, *N. gonorrhoeae*) it is dependent on the presence of specific uptake sequences. Third, the mechanism of natural transformation involves nuclease cleavage of the DNA duplex and degradation of one of the two strands so that a linear single strand can enter the cell. This mechanism of DNA uptake is not compatible with efficient plasmid transformation and geneticists working with *B. subtilis* have developed specialized methods for overcoming this problem (see Box. 10.1). For work with other species, electroporation (see p. 25) offers a much simpler alternative although the efficiency may be very low. For example, in *Methanococcus voltae* electroporation yielded only 10^2 transformants/µg plasmid DNA (Tumbula & Whitman 1999).

Genetic analysis of a number of model organisms has shown that the mechanism of DNA uptake has been conserved in Gram-positive and Gram-negative bacteria (Fig. 10.1) and shares a number of features with Type II secretion systems. By contrast, the regulation of competence is not conserved although a number of distinct mechanisms can be recognized. Based on such studies with model organisms it is possible to define a set of genes encoding homologous proteins potentially related to DNA uptake. Using various bioinformatics tools it is possible to search for similar genes in those organisms whose genomes have been completely sequenced. If these genes are found in a particular bacterium then it suggests that that bacterium could be naturally transformable (Claverys & Martin 2003). Such analysis indicated that *Lactococcus lactis*, *Listeria monocytogenes*, *Streptococcus pyogenes*, and even *E. coli* could be naturally transformable.

Given that plasmid transformation is difficult in many non-enteric bacteria, conjugation represents

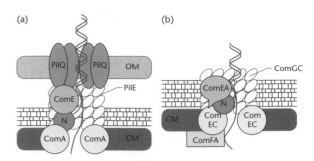

Fig. 10.1 Organization of the DNA uptake machinery in a Gram-negative bacterium (*Neisseria gonorrhoeae*) and a Gram-positive bacterium (*Bacillus subtilis*). (a) In *N. gonorrhoeae*, exogenous DNA binds to the bacterial surface and crosses the outer membrane (OM) through PilQ. A pilin complex made of PilE is required for DNA to traverse the periplasm and the peptidoglycan layer aided by the DNA-binding protein ComE. (b) In *B. subtilis*, ComGC is required for DNA to access the DNA-binding protein ComEA. DNA then is delivered to a nuclease which degrades one strand. The DNA-translocase ComFA drives the complementary strand into the cytosol through a channel in the cytoplasmic membrane formed by ComEC. Reprinted from Claverys *et al.* (2003), with permission from Elsevier.

an acceptable alternative. The term *promiscuous plasmids* was originally coined for those plasmids which are self-transmissible to a wide range of other Gram-negative bacteria (see p. 185), where they are stably maintained. The best examples are the IncP alpha plasmids RP4, RP1, RK2, etc., which

---

Box 10.1 Transforming *Bacillus subtilis* with plasmid DNA

Although it is very easy to transform *B. subtilis* with fragments of chromosomal DNA, there are problems associated with transformation by plasmid molecules. Ehrlich (1977) first reported that competent cultures of *B. subtilis* can be transformed with covalently closed circular (CCC) plasmid DNA from *Staphylococcus aureus* and that this plasmid DNA is capable of autonomous replication and expression in its new host. The development of competence for transformation by plasmid and chromosomal DNA follows a similar time course and in both cases transformation is first-order with respect to DNA concentration, suggesting that a single DNA molecule is sufficient for successful transformation (Contente & Dubnau 1979). However, transformation of *B. subtilis* with plasmid

DNA is very inefficient in comparison with chromosomal transformation, for only one transformant is obtained per 10^3–10^4 plasmid molecules.

An explanation for the poor transformability of plasmid DNA molecules was provided by Canosi *et al.* (1978). They found that the specific activity of plasmid DNA in the transformation of *B. subtilis* was dependent on the degree of oligomerization of the plasmid genome. Purified monomeric CCC forms of plasmids transform *B. subtilis* several orders of magnitude less efficiently than do unfractionated plasmid preparations or multimers. Furthermore, the low residual transforming activity of monomeric CCC DNA molecules can be attributed to low-level contamination with multimers (Mottes *et al.*

continued

Box 10.1 *continued*

1979). Using a recombinant plasmid capable of replication in both *E. coli* and *B. subtilis* (pHV14) (see p. 192), Mottes *et al.* (1979) were able to show that plasmid transformation of *E. coli* occurs regardless of the degree of oligomerization, in contrast to the situation with *B. subtilis*. Oligomerization of linearized plasmid DNA by DNA ligase resulted in a substantial increase of specific transforming activity when assayed with *B. subtilis* and caused a decrease when used to transform *E. coli*. An explanation of the molecular events in transformation which generate the requirement for oligomers has been presented by De Vos *et al.* (1981). Basically, the plasmids are cleaved into linear molecules upon contact with competent cells, just as chromosomal DNA is cleaved during transformation of *Bacillus*. Once the linear single-stranded form of the plasmid enters the cell, it is not reproduced unless it can circularize; hence the need for multimers to provide regions of homology that can recombine. Michel *et al.* (1982) have shown that multimers, or even dimers, are not required, provided that part of the plasmid genome is duplicated. They constructed plasmids carrying direct internal repeats 260–2000 bp long and found that circular or linear monomers of such plasmids were active in transformation.

Canosi *et al.* (1981) have shown that plasmid monomers will transform recombination-proficient *B. subtilis* if they contain an insert of *B. subtilis* DNA. However, the transformation efficiency of such monomers is still considerably less than that of oligomers. One consequence of the requirement for plasmid oligomers for efficient transformation of *B. subtilis* is that there have been very few successes in obtaining large numbers of clones in *B. subtilis* recipients (Keggins *et al.* 1978, Michel *et al.* 1980). The potential for generating multimers during ligation of vector and foreign DNA is limited.

Transformation by plasmid rescue

An alternative strategy for transforming

B. subtilis has been suggested by Gryczan *et al.* (1980). If plasmid DNA is linearized by restriction-endonuclease cleavage, no transformation of *B. subtilis* results. However, if the recipient carries a homologous plasmid and if the restriction cut occurs within a homologous marker, then this same marker transforms efficiently. Since this rescue of donor plasmid markers by a homologous resident plasmid requires the *B. subtilis recE* gene product, it must be due to recombination between the linear donor DNA and the resident plasmid. Since DNA linearized by restriction-endonuclease cleavage at a unique site is monomeric, this rescue system (*plasmid rescue*) bypasses the requirement for a multimeric vector. The model presented by De Vos *et al.* (1981) to explain the requirement for oligomers (see above) can be adapted to account for transformation by monomers by means of plasmid rescue. In practice, foreign DNA is ligated to monomeric vector DNA and the *in vitro* recombinants are used to transform *B. subtilis* cells carrying a homologous plasmid. Using such a "plasmid-rescue" system, Gryczan *et al.* (1980) were able to clone various genes from *B. licheniformis* in *B. subtilis*.

One disadvantage of the plasmid-rescue method is that transformants contain both the recombinant molecule and the resident plasmid. Incompatibility will result in segregation of the two plasmids. This may require several subculture steps, although Haima *et al.* (1990) observed very rapid segregation. Alternatively, the recombinant plasmids can be transformed into plasmid-free cells.

Transformation of protoplasts

A third method for plasmid DNA transformation in *B. subtilis* involves polyethylene glycol (PEG) induction of DNA uptake in protoplasts and subsequent regeneration of the bacterial cell wall (Chang & Cohen 1979). The procedure is highly efficient and yields up to 80% transformants, making the method suitable

continued

Box 10.1 *continued*

for the introduction even of cryptic plasmids. In addition to its much higher yield of plasmid-containing transformants, the protoplast transformation system differs in two respects from the "traditional" system using physiologically competent cells. First, linear plasmid DNA and non-supercoiled circular plasmid DNA molecules constructed by ligation *in vitro* can be introduced at high efficiency into *B. subtilis* by the protoplast transformation system, albeit at a frequency 10–1000 lower than the frequency observed for CCC plasmid DNA. However, the efficiency of shotgun cloning is much lower

with protoplasts than with competent cells (Haima *et al.* 1988). Secondly, while competent cells can be transformed easily for genetic determinants located on the *B. subtilis* chromosome, no detectable transformation with chromosomal DNA is seen using the protoplast assay. Until recently, a disadvantage of the protoplast system was that the regeneration medium was nutritionally complex, necessitating a two-step selection procedure for auxotrophic markers. Details have been presented of a defined regeneration medium by Puyet *et al.* (1987).

Table B10.1 Comparison of the different methods of transforming *B. subtilis*.

System	Efficiency (transformants/μg DNA)		Advantages	Disadvantages
Competent cells	Unfractionated plasmid	2×10^4	Competent cells readily prepared	Requires plasmid oligomers or internally duplicated plasmids, which makes shotgun experiments difficult unless high DNA concentrations and high vector/donor DNA ratios are used
	Linear	0	Transformants can be selected readily on any medium	
	CCC monomer	4×10^4	Recipient can be Rec$^-$	
	CCC dimer	8×10^3		
	CCC multimer	2.6×10^5		Not possible to use phosphatase-treated vector
Plasmid rescue	Unfractionated plasmid	2×10^6	Oligomers not required Can transform with linear DNA Transformants can be selected on any medium	Transformants contain resident plasmid and incoming plasmid and these have to be separated by segregation or retransformation Recipient must be Rec$^+$
Protoplasts	Unfractionated plasmid	3.8×10^6	Most efficient system Gives up to 80% transformants Does not require competent cells Can transform with linear DNA and can use phosphatase-treated vector	Efficiency lower with molecules which have been cut and religated Efficiency also very size-dependent, and declines steeply as size increases
	Linear	2×10^4		
	CCC monomer	3×10^6		
	CCC dimer	2×10^6		
	CCC multimer	2×10^6		

are about 60 kb in size, but the IncW plasmid Sa (29.6 kb) has been used extensively in *Agrobacterium tumefaciens*. Self-transmissible plasmids carry *tra* (transfer) genes encoding the conjugative apparatus. IncP plasmids are able to promote the transfer of other compatible plasmids, as well as themselves. For example, they can mobilize IncQ plasmids, such as RSF1010. More important, transfer can be mediated between *E. coli* and Gram-positive bacteria (Trieu-Cuot *et al.* 1987, Gormley & Davies 1991), as well as to yeasts and fungi (Heinemann & Sprague 1989, Hayman & Bolen 1993, Bates *et al.* 1998). However, the transfer range of a plasmid may be greater than its replication maintenance or host range (Mazodier & Davies 1991).

Self-transmissible plasmids have been identified in many different Gram-positive genera. The transfer regions of these plasmids are much smaller than those for plasmids from Gram-negative bacteria. One reason for this difference may be the much simpler cell-wall structure in Gram-positive bacteria. As well as being self-transmissible, many of these Gram-positive plasmids are promiscuous. For example, plasmid pAMβ1 was originally isolated from *Enterococcus faecalis* but can transfer to staphylococci, streptococci, bacilli, and lactic acid bacteria (De Vos *et al.* 1997), as well as to Gram-negative bacteria, such as *E. coli* (Trieu-Cuot *et al.* 1988). The self-transmissible plasmids of Gram-positive bacteria, like their counterparts in the Gram-negative bacteria, can also mobilize other plasmids between different genera (Projan & Archer 1989, Charpentier *et al.* 1999). Non-self-transmissible plasmids can also be mobilized within and between genera by conjugative transposons (Salyers *et al.* 1995, Charpentier *et al.* 1999).

It should be noted that conjugation is not a replacement for transformation or electroporation. If DNA is manipulated *in vitro*, then it has to be transferred into a host cell at some stage. In many cases, this will be *E. coli* and transformation is a suitable procedure. Once in *E. coli*, or any other organism for that matter, it may be moved to other bacteria directly by conjugation, as an alternative to purifying the DNA and moving it by transformation or electroporation.

Recombinant DNA needs to replicate or be integrated into the chromosome in new hosts

For recombinant DNA to be maintained in a new host cell, either it must be capable of replication or it must integrate into the chromosome or a plasmid. In most instances, the recombinant will be introduced as a covalently closed circle (CCC) plasmid and maintenance will depend on the host range of the plasmid. As noted in Chapter 4 (p. 57), the host range of a plasmid is dependent on the number of encoded proteins required for its replication. Some plasmids have a very narrow host range, whereas others can be maintained in a wide range of Gram-negative or Gram-positive genera. Some, such as the plasmids from *Staphylococcus aureus* and RSF1010, can replicate in both Gram-negative and Grampositive species (Lacks *et al.* 1986, Gormley & Davies 1991, Leenhouts *et al.* 1991).

As noted in Chapter 5, there is a very wide range of specialist vectors for use in *E. coli*. However, most of these vectors have a very narrow host range and can be maintained only in enteric bacteria. A common way of extending the host range of these vectors is to form hybrids with plasmids from the target species. The first such *shuttle* vectors to be described were fusions between the *E. coli* vector pBR322 and the *S. aureus*/*B. subtilis* plasmids pC194 and pUB110 (Ehrlich 1978). The advantage of shuttle plasmids is that *E. coli* can be used as an efficient *intermediate* host for cloning. This is particularly important if transformation or electroporation of the alternative host is very inefficient.

Recombinant DNA can integrate into the chromosome in different ways

When a recombinant plasmid is transferred to an unrelated host cell, there are a number of possible outcomes:

- It may be stably maintained as a plasmid.
- It may be lost.
- It may integrate into another replicon, usually the chromosome.
- A gene on the plasmid may recombine with a homologous gene elsewhere in the cell.

Under normal circumstances, a plasmid will be maintained if it can replicate in the new host and will be lost if it cannot. Plasmids which will be lost in their new host are particularly useful for delivering transposons (Saint *et al.* 1995, Maguin *et al.* 1996). If the plasmid carries a cloned insert with homology to a region of the chromosome, then the outcome is quite different. The homologous region may be excised and incorporated into the chromosome by

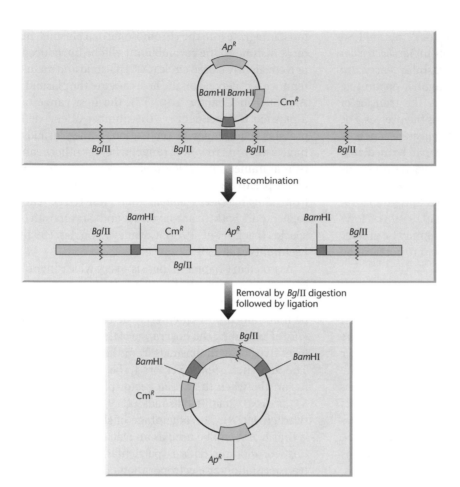

Fig. 10.2 Cloning DNA sequences flanking the site of insertion. The purple bar on the plasmid represents a *Bam*HI fragment of *B. subtilis* chromosomal DNA carrying the *amyE* gene. Note that the plasmid has no *Bgl*II sites. (See text for details.)

the normal recombination process, i.e. substitution occurs via a double crossover event. If only a single crossover occurs, the entire recombinant plasmid is integrated into the chromosome.

It is possible to favor integration by transferring a plasmid into a host in which it cannot replicate and selecting for a plasmid-borne marker. For example, Stoss *et al.* (1997) have constructed an integrative vector by cloning a neomycin resistance gene and part of the *amyE* (alpha amylase) gene of *B. subtilis* in plasmid pBR322. When this plasmid is transferred to *B. subtilis*, it is unable to replicate, but, if selection is made for neomycin resistance, then integration of the plasmid occurs at the *amyE* locus (Fig. 10.2). This technique is particularly useful if one wishes to construct a recombinant carrying a single copy of a foreign gene. However, integrated plasmids tend to be unstable unless selective pressure is maintained.

Once a recombinant plasmid has integrated into the chromosome, it is relatively easy to clone adjacent sequences. Suppose, for example, that a vector carrying *B. subtilis* DNA in the *Bam*HI site (Fig. 10.2) has recombined into the chromosome. If the recombinant plasmid has no *Bgl*II sites, it can be recovered by digesting the chromosomal DNA with *Bgl*II, ligating the resulting fragments, and transforming *E. coli* to Ap^R. However, the plasmid which is isolated will be larger than the original one, because DNA flanking the site of insertion will also have been cloned. In this way, Niaudet *et al.* (1982) used a plasmid carrying a portion of the *B. subtilis ilvA* gene to clone the adjacent *thyA* gene.

Genes cloned into a plasmid and flanked by regions homologous to the chromosome can also integrate without tandem duplication of the chromosomal segments. In this case, the plasmid DNA is linearized before transformation, as shown in Fig. 10.3. The same technique can be used to generate deletions. The gene of interest is cloned, a portion of the gene replaced *in vitro* with a fragment bearing an antibiotic marker and the linearized plasmid transformed into *B. subtilis*, with selection made for antibiotic resistance.

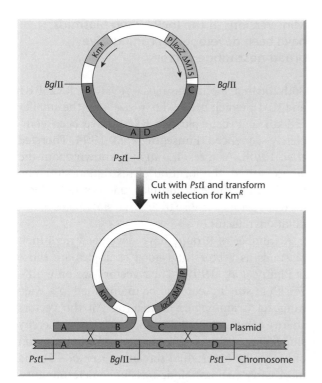

Cut with *Pst*I and transform with selection for Km^R

Fig. 10.3 Insertion of plasmid DNA into the chromosome by a double crossover event. The *B. subtilis* DNA is shown in gray and the letters A to D represent different chromosomal sequences. Vector DNA is shown in white and other vector-borne genes in purple.

Cloning in Gram-negative bacteria other than *E. coli*

To clone DNA in non-enteric bacteria, a plasmid cloning vehicle is required which can replicate in the selected organism(s). Under normal circumstances, *E. coli* will be used as an intermediate host for transformation of the ligation mix and screening for recombinant plasmids. Therefore, the vector must be able to replicate in *E. coli* as well. The options which are available are to generate a shuttle vector or to use a broad-host-range plasmid as a vector. If a small plasmid can be isolated from the bacterium of interest, then it is easy to splice it into an existing *E. coli* vector to generate a shuttle vector. Recent examples of this approach are the construction of vectors for use in *Pasteurella* (Bills *et al.* 1993), *Desulfovibrio* (Rousset *et al.* 1998), and *Thermus* (De Grado *et al.* 1999). This approach is particularly useful if the selectable markers used in *E. coli* also function in the new host. Then one can take advantage of the many different specialist vectors

(see Chapter 5) which already exist, e.g. expression vectors, secretion vectors, etc. (for reviews, see Schweizer 2001 and Davison 2002). If the selectable markers are not expressed in the new host, then extensive manipulations may be necessary just to enable transformants to be detected.

With broad-host-range plasmids, there is a high probability that the selectable markers will be expressed in the new host and confirming that this is indeed the case is easy to do. However, the naturally occurring plasmids do not fulfil all the criteria for an ideal vector, which are:

* small size;
* having multiple selectable markers;
* having unique sites for a large number of restriction enzymes, preferably in genes with readily scorable phenotypes.

Consequently the natural plasmids have been extensively modified, but few approach the degree of sophistication found in the standard *E. coli* vectors.

Vectors derived from the IncQ-group plasmid RSF1010 are not self-transmissible

Plasmid RSF1010 is a multicopy replicon which specifies resistance to two antimicrobial agents, sulfonamide and streptomycin. The plasmid DNA, which is 8684 bp long, has been completely sequenced (Scholz *et al.* 1989). A detailed physical and functional map has been constructed (Bagdasarian *et al.* 1981, Scherzinger *et al.* 1984). The features mapped are the restriction-endonuclease recognition sites, RNA polymerase binding sites, resistance determinants, genes for plasmid mobilization (*mob*), three replication proteins (Rep A, B, and C), and the origins of vegetative (*ori*) and transfer (*nic*) replication.

Plasmid RSF1010 has unique cleavage sites for *Eco*RI, *Bst*EII, *Hpa*I, *Dra*II, *Nsi*I, and *Sac*I and, from the nucleotide sequence data, is predicted to have unique sites for *Afl*III, *Ban*II, *Not*I, *Sac*II, *Sfi*I, and *Spl*I. There are two *Pst*I sites, about 750 bp apart, which flank the sulfonamide-resistance determinant (Fig. 10.4). None of the unique cleavage sites is located within the antibiotic-resistance determinants and none is particularly useful for cloning. Before the *Bst*EII, *Eco*RI, and *Pst*I sites can be used, another selective marker must be introduced into the RSF1010 genome. This need arises because the *Sm*^R and *Su*^R genes are transcribed from the same promoter (Bagdasarian *et al.* 1981). Insertion of a DNA fragment between the *Pst*I

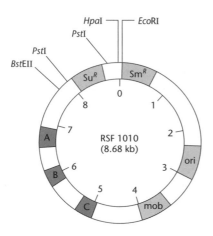

Fig. 10.4 The structure of plasmid RSFI010. The gray tinted areas show the positions of the *Sm*^R and *Su*^R genes. The region marked *ori* indicates the location of the origin of replication. The *mob* function is required for conjugal mobilization by a compatible self-transmissible plasmid. A, B, and C are the regions encoding the three replication proteins.

sites inactivates both resistance determinants. Although the *Eco*RI and *Bst*EII sites lie outside the coding regions of the *Sm*^R gene, streptomycin resistance is lost if a DNA fragment is inserted at these sites unless the fragment provides a new promoter. Furthermore, the *Su*^R determinant which remains is a poor selective marker.

A whole series of improved vectors has been derived from RSF1010 but only a few are mentioned here. The earliest vectors contained additional unique cleavage sites and more useful antibiotic-resistance determinants. For example, plasmids KT230 and KT231 encode *Km*^R and *Sm*^R and have unique sites for *Hind*III, *Xma*I, *Xho*RI, and *Sst*I which can be used for insertional inactivation. These two vectors have been used to clone (in *P. putida*) genes involved in the catabolism of aromatic compounds (Franklin *et al.* 1981). Vectors for the regulated expression of cloned genes have also been described. Some of these make use of the *tac* promoter (Bagdasarian *et al.* 1983, Deretic *et al.* 1987) or the phage T7 promoter (Davison *et al.* 1989), which will function in *P. putida* as well as *E. coli*. Another makes use of positively activated twin promoters from a plasmid specifying catabolism of toluene and xylenes (Mermod *et al.* 1986). Expression of cloned genes can be obtained in a wide range of Gram-negative bacteria following induction with micromolar quantities of benzoate, and the product of the cloned gene can account for 5% of total cell protein.

Mini-versions of the IncP-group plasmids have been developed as conjugative broad-host-range vectors

Both the IncP alpha plasmids (R18, R68, RK2, RP1 and RP4), which are 60 kb in size, and the smaller (52 kb) IncP beta plasmid R751 have been completely sequenced (Pansegrau *et al.* 1994, Thorsted *et al.* 1998). As a result, much is known about the genes carried, the location of restriction sites, etc. Despite this, the P-group plasmids are not widely used as vectors because their large size makes manipulations difficult.

A number of groups have developed mini-IncP plasmids as vectors and good examples are those of Blatny *et al.* (1997). Their vectors are only 4.8–7.1 kb in size but can still be maintained in a wide range of Gram-negative bacteria. All the vectors share a common polylinker and *lacZ'* region, thereby simplifying cloning procedures and identification of inserts by blue/white screening (see p. 45) and most carry two antibiotic-resistance determinants. All the vectors retain the *ori*T (origin of transfer) locus, enabling them to be conjugally transferred in those cases where the recipient cannot be successfully transformed or electroporated. Two other features of these vectors deserve mention. First, the *parDE* and *parCBA* operons from the parent plasmid have been incorporated in some of the vectors since these greatly enhance segregative stability in many hosts. Secondly, the *trfA* locus on the vector contains unique sites for the restriction enzymes *Nde*I and *Sfi*I. Removal of the *Nde*I–*Sfi*I fragment results in an increased copy number. Expression vectors have also been developed by the inclusion of controllable promoters. Representative examples of these vectors are shown in Fig. 10.5.

An alternative way of using P-group plasmids as cloning vectors has been described by Kok *et al.* (1994). Their method combines the advantages of high-copy-number pBR322 vectors with the convenience of conjugative plasmids. This is achieved by converting the pBR322 vector into a transposable element. Most pBR322 derivatives contain the β-lactamase gene and one of two 38 bp inverted repeats of transposon Tn2. By adding a second inverted repeat, a transposable element is created (Fig. 10.6). All that is missing is transposase activity and this is provided by another plasmid, which is a pSC101 derivative carrying the *tnpA* gene. To use this system, the desired DNA sequence is cloned into the transposition vector. The recombinant molecules

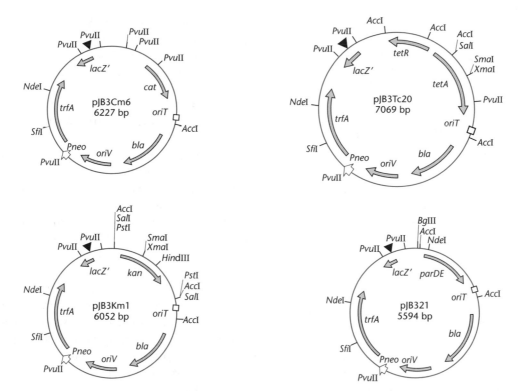

Fig. 10.5 Map and construction of general-purpose broad-host-range cloning vectors derived from plasmid RP4. The restriction sites in the polylinker downstream of the *lacZ* promoter are marked (▼), and the sites are, in the counterclockwise direction, *Hind*III, *Sph*I, *Pst*I, *Sal*I/*Hinc*II/*Acc*I, *Xba*I, *Bam*HI, *Xma*I/*Sma*I, *Kpn*I, *Sac*I, and *Eco*RI. Sites in the polylinker that are not unique are indicated elsewhere on each vector. Note that the sites for *Nde*I and *Sfi*I are unique for all of the vectors except pJB321. *Pneo*, promoter from the neomycin resistance gene; *bla*, *kan*, *tet*, and *cat*, genes encoding ampicillin, kanamycin, tetracycline, and chloramphenicol resistance, respectively. (Figure modified from Blatny *et al.* 1997.)

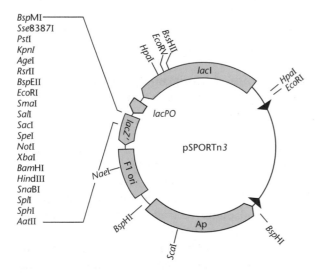

Fig. 10.6 A transposable vector derived from pBR322. The solid arrowheads indicate the inverted repeats required for transposition. Insertion of a DNA fragment in the multiple-cloning site results in inactivation of the *lacZ'* gene and regulated gene expression from the wild-type *lac* promoter.

are transformed into an *E. coli* strain carrying the P-group plasmid (e.g. R751) and the pSC101 *tnp*A derivative and selection is made for the desired characteristics. Once a suitable transformant has been selected, it is conjugated with other Gram-negative bacteria and selection made for the ampicillin-resistance marker carried on the transposon.

Vectors derived from the broad-host-range plasmid Sa are used mostly with *Agrobacterium tumefaciens*

Although a group-W plasmid, such as plasmid pSa (Fig. 10.7) can infect a wide range of Gram-negative bacteria, it has been developed mainly as a vector for use with the oncogenic bacterium *A. tumefaciens* (see p. 277). Two regions have been identified as involved in conjugal transfer of the plasmid and one of them has the unexpected property of suppressing oncogenicity by *A. tumefaciens* (Tait *et al.* 1982). Information encoding the replication of the plasmid

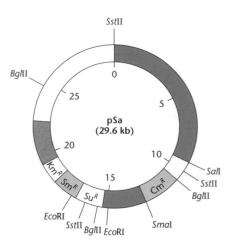

Fig. 10.7 The structure of plasmid Sa. The gray area encodes the functions essential for plasmid replication. The dark purple areas represent the regions containing functions essential for self-transmission, the one between the SstII and SalI sites being responsible for suppression of tumor induction by Agrobacterium tumefaciens.

in *E. coli* and *A. tumefaciens* is contained within a 4 kb DNA fragment. Leemans *et al.* (1982b) have described four small (5.6–7.2 MDa), multiply marked derivatives of pSa. The derivatives contain single target sites for a number of the common restriction endonucleases and at least one marker in each is subject to insertional inactivation. Although these Sa derivatives are non-conjugative, they can be mobilized by other conjugative plasmids. Tait *et al.* (1983) have also constructed a set of broad-host-range vectors from pSa. The properties of their derivatives are similar to those of Leemans *et al.* (1982b), but one of them also contains the bacteriophage λ *cos* sequence and hence functions as a cosmid. Specialist vectors for use in *Agrobacterium* and which are derived from a natural *Agrobacterium* plasmid have been described by Gallie *et al.* (1988).

pBBR1 is another plasmid that has been used to develop broad-host-range cloning vectors

Plasmid BBR1 is a broad-host-range plasmid originally isolated from *Bordatella bronchiseptica* that is compatible with IncP, IncQ, and IncW plasmids and replicates in a wide range of bacteria. Kovach *et al.* (1995) have developed a series of vectors from pBBR1 that are relatively small (<5.3 kb), possess multiple cloning sites, allow direct selection of recombinants in *E. coli* by blue/white screening, and are mobilizable by IncP plasmids. Newman & Fuqua (1999) have developed an expression vector from one of these pBBR1 derivatives by incorporating the *araBAD/araC*

cassette from *E. coli* (see p. 86) and shown that the promoter is controllable in *Agrobacterium*. Sukchawalit *et al.* (1999), using a similar vector, have shown that the promoter is also controllable in *Xanthomonas*.

Cloned DNA can be shuttled between high-copy-number and low-copy-number vectors

Ouimet & Marczynski (2000) have described an elegant method for the transfer of cloned DNA from a high-copy-number *E. coli* vector to a low-copy-number vector by *in vivo* DNA shuttling (Fig. 10.8). A transcriptional gene fusion to a promoterless β-galactosidase gene ('lacZ) is constructed using an *E. coli* vector derived from ColE1 that encodes ampicillin resistance. This recombinant plasmid is then transferred to a second, recombination-proficient strain containing a promiscuous low-copy-number vector (e.g. one derived from RK2) and a helper plasmid. The promiscuous vector carries sequences homologous to the ColE1 vector, a gene encoding tetracycline resistance, and the *sacB-npt* gene cluster that confers sensitivity to sucrose and resistance to kanamycin. The helper plasmid encodes spectinomycin and streptomycin resistance as well as the RNA1 antisense RNA that inhibits ColE1 replication (see p. 57). When the recipient is plated on medium containing sucrose and tetracycline selection is made for cells in which a double crossover has occurred. The *sacB-nrt* gene cluster is removed and replaced with the original gene fusion. Such cells should be both ampicillin and kanamycin sensitive and the plasmid that they contain can be conjugated into other recipients, e.g. *Pseudomonas* species.

Proper transcriptional analysis of a cloned gene requires that it is present on the chromosome

Whole-genome sequencing has identified many genes of unknown function. Fusion of such genes to reporter genes such as β-galactosidase and luciferase can be very useful for initial transcriptional studies and many specialist vectors are available for this purpose. However, such plasmid systems suffer from a number of inherent problems. For example, high-copy-number vectors may be useful for identification of weak promoters but do not reflect the natural situation, and titration effects involving single-copy, chromosomally encoded transcriptional regulators may lead to improperly regulated gene expression. Furthermore, genomic DNA and plasmid DNA differ in their extent of supercoiling, which

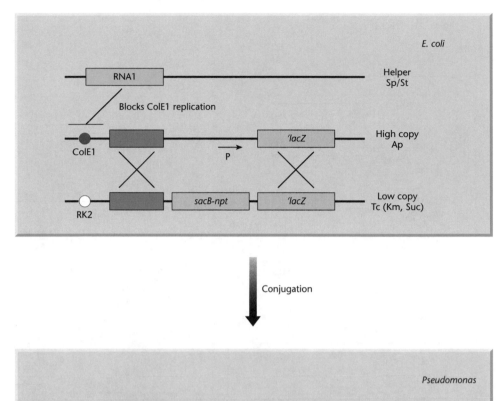

Fig. 10.8 *In vivo* transfer of cloned DNA from a high-copy vector to a low-copy vector. See text for details. Abbreviations: Sp, spectinomycin resistance; St, streptomycin resistance; Ap, ampicillin resistance; Tc, tetracycline resistance; Km, kanamycin resistance; Suc, sensitivity to sucrose; P, promoter under investigation. Reprinted from Schweizer (2001), with permission from Elsevier.

is known to play a major role in regulating gene expression. Chromosomal integration is the best way of circumventing such problems. One way of achieving this is integration of non-replicative plasmids but, as noted earlier (p. 184), selective pressure needs to be maintained. An alternative strategy is to replace the existing gene with the desired gene fusion but this is not suitable for the analysis of essential genes. Ideally, the reporter gene fusion should be integrated at a neutral site and this has been achieved in *Pseudomonas aeruginosa* using novel integrative vectors (Hoang *et al.* 2000, Becher & Schweizer 2000).

The integrative vectors for *P. aeruginosa* consist of a number of key elements (Fig. 10.9). These are the pMB1-derived origin of replication, a selectable marker (tetracycline resistance), an *ori*T for conjugation-mediated plasmid transfer from *E. coli*, the attachment site for *Pseudomonas* phage ΦCTX, the ΦCTX integrase gene, and a multiple cloning site (MCS) flanked by phage T4 transcription termination sequences. Following transfer of the integrative vector from *E. coli* to *P. aeruginosa* the plasmid-encoded integrase facilitates insertion at the chromosomal attachment site for ΦCTX. If a gene was inserted at the MCS in the vector then a single copy now will be present at a neutral site (attΦCTX) on the *P. aeruginosa* chromosome. As an additional feature, the MCS and termination sequences in the vector are flanked by Flp recombinase target sites (*FRT*). These enable unwanted plasmid backbone sequences to be removed *in vivo* (see Box 3.3). To achieve this it is necessary to transfer a second plasmid to the *P. aeruginosa* recepient that encodes the gene for Flp recombinase.

Cloning in Gram-positive bacteria

In Gram-positive bacteria, the base composition of the different genomes ranges from <30% GC to >70% GC. Given this disparity in GC content, the preferred codons and regulatory signals used by organisms at one end of the % GC spectrum will not be recognized by organisms at the other end. This in turn means that there are no universal cloning vehicles for use with all Gram-positive bacteria. Rather, one set of systems has been developed for

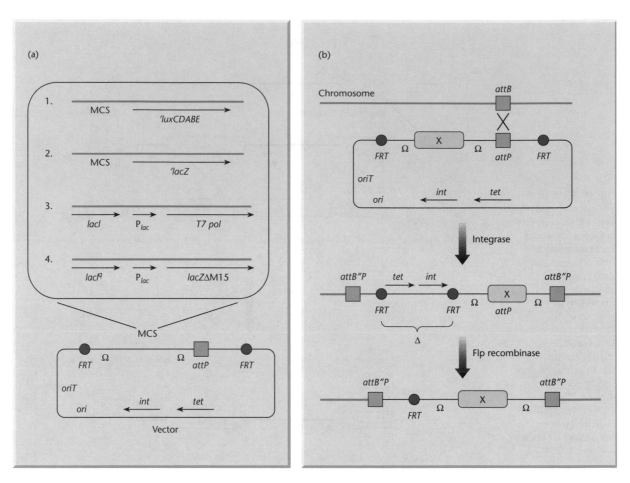

Fig. 10.9 Integration vectors for use in *Pseudomonas aeruginosa*. (a) Maps of various integration vectors containing the indicated genetic elements and (b) an illustration of the integrase-mediated chromosomal integration of genes. The four cassettes allow construction of host strains for regulated expression from the *lac*-based and T7 promoters and enable the construction of transcriptional fusions to a promoterless *lacZ* gene or *lux* operon. The locations of genes and their transcriptional orientations are shown, including the T4 transcription terminators (Ω). Abbreviations: *attB*, chromosomal φCTX attachment site; *attP*, φCTX attachment site; *FRT*, Flp recombinase target site; *int*, φCTX integrase-encoding gene; *lacA*, *E. coli lacI* gene with promoter-up mutation; P$_{lac}$, *E. coli lac* operon promoter; *lacZΔM15*, *E. coli* β-galactosidase gene containing the in-frame M15 deletion; MCS, multiple cloning site; *ori*, ColE1 origin of replication; *oriT*, origin of transfer; *T7 pol*, phage T7 polymerase-encoding gene; *tet*, tetracycline resistance encoding gene. Integration of a genetic element (X) at the chromosomal *attB* locus and removal of unwanted plasmid sequences involves two steps: Integrase-mediated integration of the non-replicative plasmid, which is selected by tetracycline resistance, and introduction of an Flp-recombinase-encoding plasmid, followed by screening for loss of tetracycline resistance and curing of the Flp-encoding plasmid. Reprinted from Schweizer (2001), with permission from Elsevier.

high-GC organisms (e.g. streptomycetes) and another for low-GC organisms. This latter group comprises bacteria from the unrelated genera *Bacillus*, *Clostridium*, and *Staphylococcus* and the lactic acid bacteria *Streptococcus*, *Lactococcus*, and *Lactobacillus*.

Many of the cloning vectors used with *Bacillus subtilis* and other low-GC bacteria are derived from plasmids found in *Staphylococcus aureus*

The development of *B. subtilis* vectors began with the observation (Ehrlich 1977) that plasmids from

S. aureus (Table 10.1) can be transformed into *B. subtilis*, where they replicate and express antibiotic resistance normally.

As can be seen from Table 10.1, none of the natural *S. aureus* plasmids carries more than one selectable marker and so improved vectors have been constructed by gene manipulation, e.g. pHV11 is pC194 carrying the *Tc*^{R} gene of pT127 (Ehrlich 1978). In general, these plasmids are stable in *B. subtilis*, but segregative stability is greatly reduced following insertion of exogenous DNA (Bron & Luxen 1985). Reasoning that stable host–vector systems in *B. subtilis* are more likely if endogenous plasmids

Table 10.1 Properties of some *S. aureus* plasmids used as vectors in *B. subtilis*.

Plasmid	Phenotype conferred on host cell	Size	Copy no.	Other comments
pC194	Chloramphenicol resistance	2906 bp	15	Generates large amount of high-molecular-weight DNA when carrying heterologous inserts
pE194	Erythromycin resistance	3728 bp	10	*cop*-6 derivative has copy number of 100 Plasmid is naturally temperature-sensitive for replication
pUB110	Kanamycin resistance	4548 bp	50	Virtually the complete sequence is involved in replication maintenance, site-specific plasmid recombination or conjugal transfer

are used, Bron and colleagues have developed the cryptic *Bacillus* plasmid pTA1060 as a vector (Haima *et al.* 1987, Bron *et al.* 1989).

Because of the difficulties experienced in direct cloning in *B. subtilis*, hybrid plasmids were constructed which can replicate in both *E. coli* and *B. subtilis*. Originally most of these were constructed as fusions between pBR322 and pC194 or pUB110. With such plasmids, *E. coli* can be used as an efficient intermediate host for cloning. Plasmid preparations extracted from *E. coli* clones are subsequently used to transform competent *B. subtilis* cells. Such preparations contain sufficient amounts of multimeric plasmid molecules to be efficient in *B. subtilis*-competent cell transformation (see p. 180).

Table 10.2 lists some of the commonly used shuttle plasmids. Note that some of them carry some of the features described earlier for *E. coli* plasmids, e.g. the *E. coli lacZα*-complementation fragment, multiple cloning sites (MCS) (see p. 65), and the phage f1 origin for subsequent production of single-stranded DNA in a suitable *E. coli* host (see p. 81).

The mode of plasmid replication can affect the stability of cloning vectors in *B. subtilis*

Early in the development of *B. subtilis* cloning vectors, it was noted that only short DNA fragments could be efficiently cloned (Michel *et al.* 1980) and that longer DNA segments often undergo rearrangements (Ehrlich *et al.* 1986). This structural instability is independent of the host recombination systems, for it still occurs in Rec⁻ strains (Peijnenburg *et al.* 1987).

A major contributing factor to structural instability of recombinant DNA in *B. subtilis* appears

to be the mode of replication of the plasmid vector (Gruss & Ehrlich 1989, Jannière *et al.* 1990). All the *B. subtilis* vectors described above replicate by a rolling-circle mechanism (see Box 10.2). Nearly every step in the process digresses or could digress from its usual function, thus effecting rearrangements. Also, single-stranded DNA is known to be a reactive intermediate in every recombination process, and single-stranded DNA is generated during rolling-circle replication.

If structural instability is a consequence of rolling-circle replication, then vectors which replicate by the alternative theta mechanism could be more stable. Jannière *et al.* (1990) have studied two potentially useful plasmids, pAMβ1 and pTB19, which are large (26.5 kb) natural plasmids derived from *Streptococcus (Enterococcus) faecalis* and *B. subtilis*, respectively. Replication of these plasmids does not lead to accumulation of detectable amounts of single-stranded DNA, whereas the rolling-circle mode of replication does. Also, the replication regions of these two large plasmids share no sequence homology with the corresponding highly conserved regions of the rolling-circle-type plasmids. It is worth noting that the classical *E. coli* vectors, which are derived from plasmid ColE1, all replicate via theta-like structures.

Renault *et al.* (1996) have developed a series of cloning vectors from pAMβ1. All the vectors carry a gene essential for replication, *repE*, and its regulator, *copF*. The latter gene can be inactivated by inserting a linker into its unique *Kpn*I site. Since *copF* downregulates the expression of *repE*, its inactivation leads to an increase in the plasmid copy number per cell. The original low-copy-number state can be restored by removal of the linker by cleavage and

Table 10.2 *B. subtilis–E. coli* shuttle plasmids.

Plasmid	Size (kbp)	Replicon		Markers		Comments
		E. coli	*B. subtilis*	*E. coli*	*B. subtilis*	
pHV14	4.6	pBR322	pC194	Ap, Cm	Cm	pBR322/pC194 fusion. Sites: *Pst*I, *Bam*HI, *Sal*I, *Nco*I (Ehrlich 1978)
pHV15	4.6	pBR322	pC194	Ap, Cm	Cm	pHV14, reversed orientation of pC194 relative to pBR322
pHV33	4.6	pBR322	pC194	Ap, Tc, Cm	Cm	Revertant of pHV14 (Primrose & Ehrlich 1981)
pEB10	8.9	pBR322	pUB110	Ap, Km	Km	pBR322/pUB110 fusion (Bron *et al.* 1988)
pLB5	5.8	pBR322	pUB110	Ap, Cm, Km	Cm, Km	Deletion of pBR322/pUB110 fusion, *Cm*^R gene of pCl94 Segregationally unstable (Bron & Luxen 1985). Sites: *Bam*HI, *Eco*RI, *Bgl*III (in *Km*^R gene), *Nco*I (in *Cm*^R gene)
pHP3	4.8	pBR322	pTA1060	Em, Cm	Em, Cm	Segregationally stable pTA1060 replicon (Peeters *et al.* 1988). Copy number *c.* 5. Sites: *Nco*I (*Cm*^R gene), *Bcl*I and *Hpa*I (both *Em*^R gene)
pHP3Ff	5.3	pBR322	pTA1060	Em, Cm	Em, Cm	Like pHP3; phage f1 replication origin and packaging signal
pGPA14	5.8	pBR322	pTA1060	Em	Em	Stable pTA1060 replicon. Copy number *c.* 5. α-Amylase-based selection vector for protein export functions (Smith *et al.* 1987). MCS of M13*mp*11 in *lacZ*α
pGPB14	5.7	pBR322	pTA1060	Em	Em	As pGPA14, probe gene TEM-β-lactamase
pHP13	4.9	pBR322	pTA1060	Em, Cm	Em, Cm	Stable pTA1060 replicon. Copy number *c.* 5. Efficient (shotgun) cloning vector (Haima *et al.* 1987). MCS of M13*mp*9 in *lacZ*α *LacZ*α not expressed in *B. subtilis*. Additional sites: *Bcl*I and *Hpa*I (both *Em*^R gene)
pHV1431	10.9	pBR322	pAMβ1	Ap, Tc, Cm	Cm	Efficient cloning vector based on segregationally stable pAMβ1 (Jannière *et al.* 1990). Copy number *c.* 200. Sites: *Bgm*HI, *Sal*I, *Pst*I, *Nco*I. Structurally unstable in *E. coli*
pHV1432	8.8	pBR322	pAMβ1	Ap, Tc, Cm	Cm	pHV1431 lacking stability fragment orfH. Structurally stable in *E. coli*
pHV1436	8.2	pBR322	pTB19	Ap, Tc, Cm	Cm	Low-copy-number cloning vector (Jannière *et al.* 1990) Structurally stable

Box 10.2   The two modes of replication of circular DNA molecules

There are two modes of replication of circular DNA molecules: via theta-like structures or by a rolling-circle type of mechanism. Visualization by electron microscopy of the replicating intermediates of many circular DNA molecules reveals that they retain a ring structure throughout replication. They always possess a theta-like shape that comes into existence by the initiation of a replicating bubble at the origin of replication (Fig. B10.1). Replication can proceed either uni- or bidirectionally. As long as each chain remains intact, even minor untwisting of a section of the circular double helix results in the creation of positive supercoils in the other direction. This supercoiling is relaxed by the action of topoisomerases (see Fig. 4.1), which create single-stranded breaks (relaxed molecules) and then reseal them.

An alternative way to replicate circular DNA is the rolling-circle mechanism (Fig. B10.2). DNA synthesis starts with a cut in one strand at the origin of replication. The 5′ end of the cut strand is displaced from the duplex, permitting the addition of deoxyribonucleotides at the free 3′ end. As replication proceeds, the 5′ end of the cut strand is rolled out as a free tail of increasing length. When a full-length tail is produced, the replicating machinery cuts it off and ligates the two ends together. The double-stranded progeny can reinitiate replication, whereas the single-stranded progeny must first be converted to a double-stranded form. Gruss & Ehrlich (1989) have suggested how deletants and defective molecules can be produced at each step in the rolling-circle process.

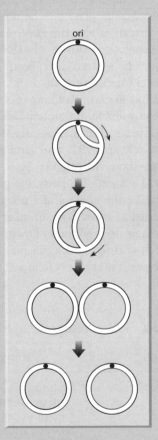

Fig. B10.1   Theta replication of a circular DNA molecule. The original DNA is shown in black and the newly synthesized DNA in purple. • represents the origin of replication and the arrow shows the direction of leading strand replication.

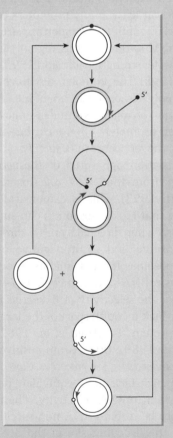

Fig. B10.2   Rolling-circle replication of a circular DNA molecule. The original DNA is shown in black and the newly synthesized DNA in purple. The solid and open circles represent the positions of the replication origins of the outer (+) and inner (−) circles, respectively.

religation. This new replicon has been used to build vectors for making transcriptional and translational fusions and for expression of native proteins. Poyart and Trieu-Cuot (1997) have constructed a shuttle vector based on pAMβ1 for the construction of transcriptional fusions; it can be conjugally transferred between *E. coli* and a wide range of Gram-positive bacteria.

Compared with *E. coli, B. subtilis* has additional requirements for efficient transcription and translation and this can prevent the expression of genes from Gram-negative organisms in ones that are Gram-positive

The composition of the core RNA polymerase in *B. subtilis* and other low-GC hosts resembles that of *E. coli*. The number of sigma factors is different in each of the various genera but the principal sigma factor is sigma A. Analysis of many sigma A-dependent *Bacillus* promoters shows that they contain the canonical −35 and −10 sequences found in *E. coli* promoters. In *B. subtilis*, at least, many promoters contain an essential TGTG motif (−16 region) upstream of the −10 region. Mutations of this region significantly reduce promoter strength (Helmann 1995, Voskuil & Chambliss 1998). The promoters also have conserved polyA and polyT tracts upstream of the −35 region. Although the −16 region is found in some *E. coli* promoters, such promoters often lack the −35 region, whereas this never occurs in *B. subtilis*.

The translation apparatus of *B. subtilis* differs significantly from that of *E. coli* (for review, see Vellanoweth 1993). This is demonstrated by the observation that *E. coli* ribosomes can support protein synthesis when directed by mRNA from a range of Gram-positive and Gram-negative organisms, whereas ribosomes from *B. subtilis* recognize only homologous mRNA (Stallcup *et al.* 1974). The explanation for the selectivity of *B. subtilis* ribosomes is that they lack a counterpart of the largest *E. coli* ribosomal protein, S1 (Higo *et al.* 1982, Roberts & Rabinowitz 1989). Other Gram-positive bacteria, such as *Staphylococcus, Streptococcus, Clostridium*, and *Lactobacillus*, also lack an S1-equivalent protein and they too exhibit mRNA selectivity. The role of S1 is believed to be to bind RNA non-specifically and bring it to the decoding site of the 30S subunit, where proper positioning of the Shine–Dalgarno (S-D) sequence and initiation codon signals can take place. This is reflected in a more extensive complementarity between the S-D sequences and the 3′

end of the 16S ribosomal RNA (rRNA) than found in bacteria which have ribosomal protein S1.

The additional sequence requirements for efficient transcription and translation in *B. subtilis* and other low-GC organisms probably explain why many *E. coli* genes are not expressed in these hosts.

Specialist vectors have been developed that permit controlled expression in *B. subtilis* and other low-GC hosts

The first controlled expression system to be used in *B. subtilis* was the *Spac* system (Yansura & Henner 1984). This consists of the *E. coli lacI* gene and the promoter of phage SPO-1 coupled to the *lac* operator. More recently, the *E. coli* T7 system has been successfully implemented in *B. subtilis* (Conrad *et al.* 1996). This was achieved by inserting the T7 RNA polymerase gene (*rpoT7*) into the chromosome under the control of a xylose-inducible promoter and cloning the gene of interest, coupled to a T7 promoter, on a *B. subtilis* vector. Of course, expression of the heterologous gene can be made simpler by putting it directly under the control of the xylose-inducible promoter (Kim *et al.* 1996). A similar xylose-inducible system has been developed in staphylococci (Sizemore *et al.* 1991, Peschel *et al.* 1996) and *Lactobacillus* (Lokman *et al.* 1997). Many different controllable promoters are available in *Lactococcus lactis* (for reviews see Kuipers *et al.* 1995, De Vos *et al.* 1997) and some representative examples are shown in Table 10.3.

In the φ31 system of *L. lactis*, the gene of interest is placed under the control of a phage middle promoter inserted in a low-copy-number vector carrying the phage *ori* region. Following infection of the host cell with φ31, the plasmid copy number rapidly increases and this is followed by expression

Table 10.3 Some inducible systems in *L. lactis*.

Promoter	Inducer
lacA or *lacR*	Lactose
dnaJ	High temperature
sodA	Aeration
PA170	Low pH, low temperature
trpE	Absence of tryptophan
φ31 and *ori*	φ31 infection
nisA or *nisF*	Nisin

from the phage promoter. Following induction in this manner, the level of expression of the cloned gene can increase over 1000-fold (O'Sullivan *et al.* 1996). Similar levels of expression can be achieved by using the *nisA* and *nisF* systems but with the added advantage that the exact level of expression depends on the amount of nisin added to the medium (Kuipers *et al.* 1995, De Ruyter *et al.* 1996).

Vectors have been developed that facilitate secretion of foreign proteins from *B. subtilis*

The export mechanism in *Bacillus* and other low-GC bacteria resembles that of *E. coli* (for review, see Tjalsma *et al.* 2000). However, there are differences in the signal peptides compared with those found in *E. coli* and eukaryotes. The NH$_2$ termini are more positively charged. The signal peptides are also larger and the extra length is distributed among all three regions of the signal peptide. Hols *et al.* (1992) developed two probe vectors for the identification of Gram-positive secretion signals. These vectors made use of a silent reporter gene encoding the mature α-amylase from *Bacillus licheniformis*. The disadvantage of this system is that detection of secreted amylase involves flooding starch-containing media with iodine and this kills the bacteria in the colonies. Consequently, replica plates must be made before iodine addition.

Poquet *et al.* (1998) have developed an alternative probe system which uses the *S. aureus*-secreted nuclease as a reporter. This nuclease is a small (168 amino acid), stable, monomeric enzyme that is devoid of cysteine residues and the enzymatic test is nontoxic to bacterial colonies. The probe vectors have the nuclease gene, lacking its signal sequence, located downstream from an MCS. Cloning DNA in the vectors results in the synthesis of fusion proteins and those containing signal sequences are detected by nuclease activity in the growth medium. Le Loir *et al.* (1998) have noted that inclusion of a nine-residue synthetic propeptide immediately downstream of the signal-peptide cleavage site significantly enhances secretion.

As an aid to understanding gene function in *B. subtilis*, vectors have been developed for directed gene inactivation

With the advent of mass sequencing of genomes (see Chapter 7), many genes have been discovered whose function is unknown. One way of determin-

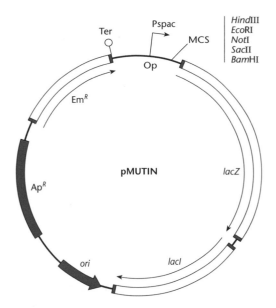

Fig. 10.10 A typical pMUTIN vector. The Em^R, *lacI*, and *lacZ* genes are expressed in *B. subtilis* and the Ap^R gene is expressed in *E. coli*. Ter indicates the presence of a terminator to prevent run-through transcription from the Em^R gene. Op represents the LacI operator.

ing function is to inactivate the gene and then monitor the effect of this on cell fitness under different growth conditions. To study the functions of uncharacterized open reading frames in *B. subtilis*, Vagner *et al.* (1998) constructed a series of vectors to perform directed insertional mutagenesis in the chromosome. These vectors, which have been given the designation pMUTIN, have the following properties:

- an inability to replicate in *B. subtilis*, which allows insertional mutagenesis;
- a reporter *lacZ* gene to facilitate the measurement of expression of the target gene;
- the inducible Pspac promoter to allow controlled expression of genes downstream of and found in the same operon as the target gene.

A typical pMUTIN vector is shown in Fig. 10.10 and their mode of use is as follows. An internal fragment of the target gene is amplified by the PCR and cloned in a pMUTIN vector and the resulting plasmid is used to transform *B. subtilis*. Upon integration, the target gene is interrupted and a transcriptional fusion is generated between its promoter and the reporter *lacZ* gene (Fig. 10.11). If the targeted gene is part of an operon, then any genes downstream of it are placed under the control of the Pspac promoter.

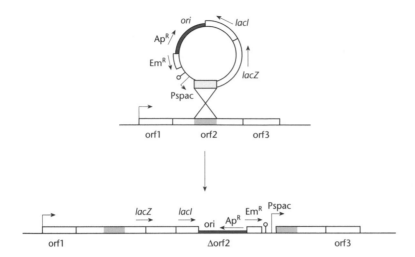

Fig. 10.11 Integration of pMUTIN into a target gene. Genes of the *orf1–orf3* operon are indicated as white boxes. Purple box corresponds to the internal segment of the target gene. The vector is integrated in *orf2* by a single crossing-over event. (Figure reproduced from *Microbiology* courtesy of Dr. S.D. Ehrlich and the Society for General Microbiology.)

It should be noted that the procedure shown in Fig. 10.11 simultaneously generates two types of mutants: an absolute (null) mutation in orf2 through gene inactivation, and a conditional mutation in orf3, which can be relieved by induction with isopropyl-β-D-thiogalactoside (IPTG).

The mechanism whereby *B. subtilis* is transformed with plasmid DNA facilitates the ordered assembly of dispersed genes

There are many economically important biomolecules, such as antibiotics, that are the end product of long metabolic pathways. The easiest way to manipulate the synthesis of these molecules is to clone all the relevant genes and coordinate their expression. Such a cloning exercise is easy if all the genes are clustered in one or two locations in their natural host. However, if the genes are dispersed throughout the chromosome then cloning them all becomes a major exercise. For example, when Szczebara *et al.* (2003) generated a microbial strain that could synthesize hydrocortisone it was necessary to introduce progressively eight different mammalian genes.

A major problem in the assembly of dispersed genes is the requirement for circular DNA molecules for plasmid transformation of *E. coli*. When different gene segments are assembled in a linear order then it is necessary to undertake intermolecule ligation. To form a circular plasmid requires intramolecule ligation. The experimental conditions that favor the two types of ligation are quite different and the greater the number of fragments to be ligated the less likely it will be that a circular plasmid carrying all the fragments will be formed. By contrast, *B. subtilis* can be transformed with linear DNA molecules provided that they carry partial duplications (see

Box 10.1). Tsuge *et al.* (2003) have made use of this fact to develop a method for the one-step assembly of multiple DNA fragments with a desired order and orientation.

Suppose we wish to construct a single plasmid carrying five genes (A, B, C, D, E) that are dispersed in their natural host. The first step is to isolate each of the genes by PCR-mediated amplification using primers with sites for endonuclease *Sfi*I at their 5′ ends. This results in the isolation of duplex DNA molecules carrying the desired genes flanked by sequences carrying sites for *Sfi*I. By carefully selecting the 5′ primer extensions it is possible to generate different three-base overhangs after digestion with *Sfi*I (Fig. 10.12) so that no individual duplex can self-anneal. In this way, each of the five gene fragments is bounded by a unique pair of overhangs and these control the sequence in which the fragments can be assembled. The cloning vector that will carry the gene assembly has two sites for *Sfi*I and after digestion a linear molecule is produced that also has unique three-base overhangs (Fig. 10.12). Multimers are produced when the gene fragments and the linearized vector are ligated together (Fig. 10.13) and following transformation of *B. subtilis* circular plasmid monomers are formed with the different genes in the correct order.

A variety of different methods can be used to transform high-GC organisms such as the streptomycetes

Cloning in *Streptomyces* has attracted a lot of interest because of the large number of antibiotics that are made by members of this genus. Although *Streptomyces coelicolor* is the model species for genetic studies (Hopwood 1999), many other species are

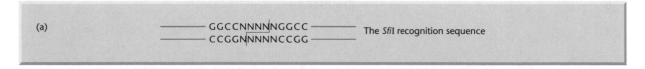

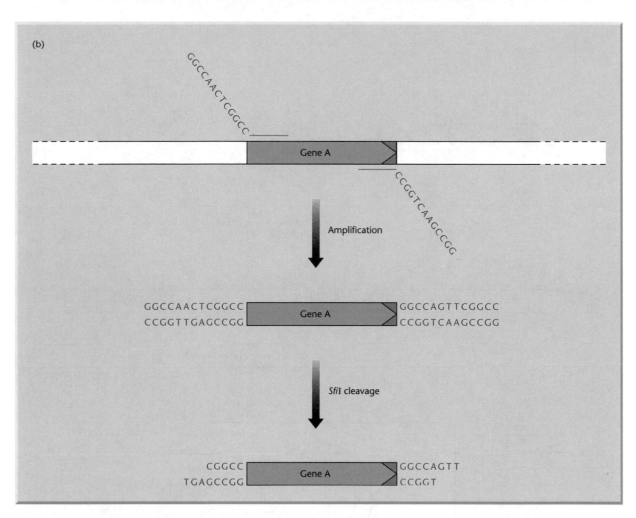

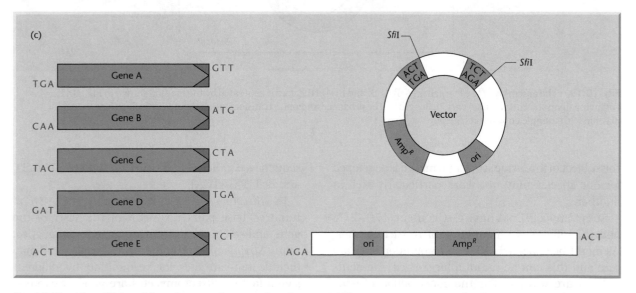

Fig. 10.12 The addition of three-base overhangs to cloned genes. (a) The recognition sequence for the restriction endonuclease *Sfi*I. Note that the base sequence in the middle of the recognition sequence can be varied at will. (b) The use of primers with 5′ extensions encoding an *Sfi*I recognition site to amplify a gene (gene A). Note that the nucleotides in the center of the two *Sfi*I sites are not the same. (c) The overhangs generated on the ends of each amplified gene and the vector.

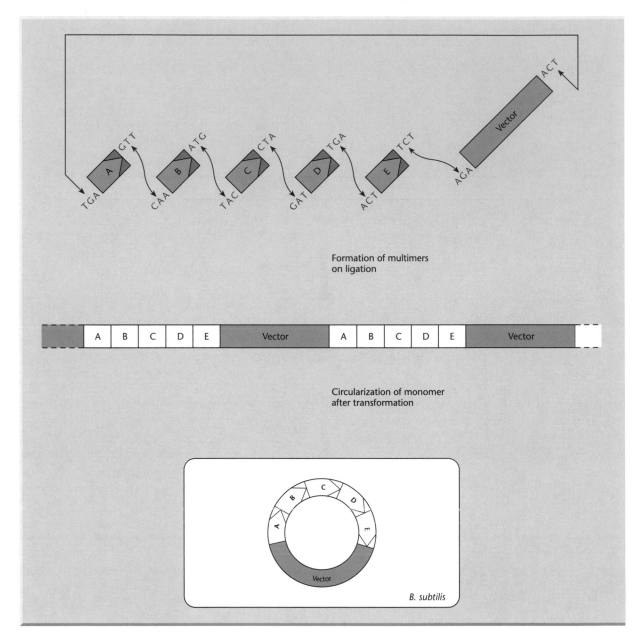

Fig. 10.13 The assembly of the five genes (A, B, C, D, and E) and the vector generated by the method shown in Fig. 10.12. Following ligation, multimers of varying length will be produced and upon transformation these are resolved into a circular plasmid with a single copy of each gene.

the subject of intensive study and methods developed for one species may not work particularly well in another.

Streptomycete DNA has a G+C content of 70–75% and this affects the frequency of restriction sites. As might be expected, AT-rich recognition sites are rare and this can be useful if large-sized fragments of DNA are wanted. For the construction of gene libraries, the most commonly used enzymes are ones with a high GC content in their recognition

sequence, e.g. *Bam*H1 (G′GATCC), *Bgl*II (A′GATCT), and *Bcl*I (T′GATCA).

In *Streptomyces*, promoters may be several hundred base pairs upstream of the start of the gene and so can be lost during gene cloning. Also, many *Streptomyces* promoters are complex and may include tandem sites for recognition by different sigma factors. Streptomycetes are good at expressing genes (promoters, ribosome binding sites, etc.) from low-G+C organisms, but *Streptomyces* genes are usually difficult to express in *E. coli* because most promoters do not function, and translation may be inefficient unless the initial amino acid codons are changed to lower-G+C alternatives.

There are several ways in which DNA can be introduced into streptomycetes, including transformation, transfection, and conjugation. Transformation is achieved by using protoplasts, rather than competent cells, and high frequencies of plasmid DNA uptake can be achieved in the presence of polyethylene glycol (Bibb *et al.* 1978). Plasmid monomers that are covalently closed will yield 10^6–10^7 transformants/μg of DNA, even with plasmids up to 60 kb in size. Open circular and linearized molecules with sticky ends transform with 10–100-fold lower efficiency (Bibb *et al.* 1980). The number of transformants obtained with non-replicating plasmids that integrate by homologous recombination into the recipient chromosome is greatly stimulated by simple denaturation of the donor DNA (Oh & Chater 1997). This stimulation reflects an increased frequency of recombination rather than an increased frequency of DNA uptake. Electroporation has been used to transform streptomycetes, since it bypasses the need to develop protoplast regeneration procedures (Pigac & Schrempf 1995, Tyurin *et al.* 1995). For electroporation, limited non-protoplasting lysozyme treatment is used to weaken the cell wall and improve DNA uptake. Intergeneric conjugation of mobilizable plasmids from *E. coli* into streptomycetes (see p. 186) is increasingly being used, because the required constructs can be made easily in *E. coli* and the conjugation protocols are simple. For intergeneric conjugation to occur, the vectors have to carry the *oriT* locus from RP4 and the *E. coli* strain needs to supply the transfer functions *in trans* (Mazodier *et al.* 1989).

Transformants are generally identified by the selection of appropriate phenotypes. However, antibiotic resistance has much less utility than in other organisms, because many streptomycetes produce antibiotics and hence have innate resistance to them. One particularly useful phenomenon is that clones harboring conjugative plasmids can be detected by the visualization of pocks. The property of pock formation, also known as *lethal zygosis*, is exhibited if a strain containing a conjugative plasmid is replica-plated on to a lawn of the corresponding plasmid-free strain. Under these conditions, clones containing plasmids are surrounded by a narrow zone in which the growth of the plasmid-free strain is retarded (Chater & Hopwood 1983).

Most of the vectors used with streptomycetes are derivatives of endogenous plasmids and bacteriophages

With the exception of RSF1010 (see p. 185), no plasmid from any other organism has been found to replicate in *Streptomyces*. For this reason, all the cloning vectors used in streptomycetes are derived from plasmids and phages that occur naturally in them. The different replicons that have been subjugated as vectors are listed in Table 10.4. Nearly all *Streptomyces* plasmids carry transfer functions that permit conjugative plasmid transfer and provide different levels of chromosome-mobilizing activity. These transfer functions are very simple, consisting of a single transfer gene and a repressor of gene function.

Plasmid SCP2* is a deriviative of the sex plasmid SCP2 (Lydiate *et al.* 1985). Both plasmids have a size of 31.4 kb and are physically indistinguishable, although SCP2* exhibits a much more pronounced lethal zygosis reaction. SCP2* is important because it is the progenitor of many very low-copy-number, stable vectors. High-copy-number derivatives have also been isolated with the exact copy number (10 or 1000) being dependent on the sequences from the replication region

Table 10.4

Streptomyces plasmids that have been used in the development of vectors.

Plasmid	Size	Mode of replication	Copy number	Host range
pIJ101	8.8 kb	Rolling circle	300	
pJV1	11.1 kb	Rolling circle		Broad
pSG5	12.2 kb	Rolling circle	20–50	Broad
SCP2*	31 kb	Theta	1–4	
SLP1	17.2 kb	Rolling circle	Integrating	Limited
pSAM2	10.9 kb	Rolling circle	Integrating	Broad

that are present. SLP1 and pSAM2 are examples of *Streptomyces* plasmids that normally reside integrated into a specific highly conserved chromosomal transfer RNA (tRNA) sequence (Kieser & Hopwood 1991). Many different specialist vectors have been derived from these plasmids, including cosmids, expression vectors, vectors with promoterless reporter genes, positive-selection vectors, temperature-sensitive vectors, etc., and full details can be found in Kieser *et al.* (2000).

The temperate phage φC31 is the streptomycete equivalent of phage λ and has been subjugated as a vector. φC31-derived vectors have upper and lower size limits for clonable fragments with an average insert size of 8 kb. In contrast, there are no such size constraints on plasmid cloning, although recombinant plasmids of a size greater than 35 kb are rare with the usual vectors. However, phage vectors do have one important advantage: plaques can be obtained overnight, whereas plasmid transformants can take up to 1 week to sporulate. Plasmid-integrating vectors can be generated by incorporating the integration functions of φC31.

As noted earlier, a major reason for cloning in streptomycetes is to analyze the genetics and regulation of antibiotic synthesis. Although all the genes for a few complete biosynthetic pathways have been cloned (Malpartida & Hopwood 1985, Kao *et al.* 1994, Schwecke *et al.* 1995), some gene clusters may be too large to be cloned in the standard vectors. For this reason, Sosio *et al.* (2000) generated bacterial artificial chromosomes (BACs) that can accommodate up to 100 kb of streptomycete DNA. These vectors can be shuttled between *E. coli*, where they replicate autonomously, and *Streptomyces*, where they integrate site-specifically into the chromosome.

Cloning in Archaea

Organisms are divided into three domains: *Bacteria* (eubacteria), *Archaea* (archaebacteria), and *Eucarya* (eukaryotes). The Archaea comprises at least three major groups of prokaryotic organisms with unusual phenotypes when compared with their eubacterial counterparts. For example, many of them thrive in extreme environments such as ones with very high temperatures, high pH, or high salt concentrations. Others are strictly anaerobic methanogens that live in the rumen of herbivores. Because many Archaea exhibit interesting physiological properties,

coupled with a growing number of complete genome sequences, there has been considerable interest in developing suitable gene-cloning procedures.

As noted earlier (p. 179), there are three basic requirements for gene cloning in any organism: a means of introducing DNA into recipients (usually transformation); suitable plasmid vectors; and selectable markers. All three requirements have been met (for review, see Luo & Wasserfallen 2001) but progress lags far behind that made with the major eubacterial groups. In most Archaea investigated so far, electroporation is the only method of getting DNA into cells and the observed efficiencies are low (~10^2 transformants/µg). Several species appear to be naturally transformable but, again, efficiencies are as low as with electroporation. Much higher efficiencies (e.g. 10^7 transformants/g) have been observed with some methanogens using polyethylene glycol-mediated transformation or liposome-mediated transformation.

A number of plasmids and viruses have been identified and a number of these have been converted into shuttle vectors and used to develop gene-transfer systems (Sowers & Schreier 1999). Most of the selective markers used are based on resistance to antibiotics such as puromycin, novobiocin, thiostreptin, and mevinolin. More advanced vectors such as integrating vectors and expression vectors also have been developed. The usefulness of integration vectors has been demonstrated by insertional mutagenesis in methanogens. Three reporter genes for expression studies also have been used in methanogens: β-glucuronidase and β-galactosidase from *E. coli* and trehalase from *B. subtilis*.

Suggested reading

Chen I. & Dubnau D. (2004) DNA uptake during bacterial transformation. *Nature Reviews Microbiology* **2**, 241–9.
This paper reviews the current state of knowledge about transformation in bacteria that naturally are competent for DNA uptake.

Davison J. (2002) Genetic tools for Pseudomonads, Rhizobia, and other Gram-negative bacteria. *BioTechniques* **32**, 386–401.
A review giving an additional perspective on the material presented in this chapter.

Luo Y. & Wasserfallen A. (2001) Gene transfer systems and their applications in Archaea. *Systematic and Applied Microbiology* **24**, 15–25.

A review of gene manipulation in a group of "exotic" bacteria that are attracting much interest because of the novel properties of their proteins.

Martinez A., Kolvek S.J., Yip C.L., *et al.* (2004) Genetically modified bacterial strains and novel bacterial artificial chromosome shuttle vectors for constructing environmental libraries and detecting heterologous natural products in multiple expression hosts. *Applied and Environmental Microbiology* **70**, 2452–63.

Middleton R. & Hofmeister A. (2004) New shuttle vectors for ectopic insertion of genes into *Bacillus subtilis*. *Plasmid* **51**, 238–45.

Quandt J., Clark R.G., Venter A.P., *et al.* (2004) Modified RP4 and Tn5-Mob derivatives for facilitated manipulation of large plasmids in Gram-negative bacteria. *Plasmid* **52**, 1–12.

Takken F.L., Van Wijk R., Michielse C.B., *et al.* (2004) A one-step method to convert vectors into binary vectors suited for *Agrobacterium*-mediated transformation. *Current Genetics* **45**, 242–8.

The four papers listed above represent a good selection of current work on the development of improved vectors for use in bacteria other than Escherichia coli.

CHAPTER 11

Cloning in *Saccharomyces cerevisiae* and other fungi

There are a number of reasons for cloning DNA in *S. cerevisiae*

When recombinant DNA technology was first applied to fungi in the late 1970s the organism of choice was the yeast *Saccharomyces cerevisiae*. At that time the primary purpose of cloning was to understand what particular genes do *in vivo* and the concomitant development of DNA sequencing methodology facilitated the identification of the different elements that control gene expression in fungi. A secondary reason for cloning in yeast was to understand those cellular functions unique to eukaryotes such as mitosis, meiosis, signal transduction, obligate cellular differentiation, etc. Today, just over a quarter of a century later, there are different reasons for cloning in *S. cerevisiae* and other fungi.

In 1996 the sequencing of the entire 12 Mb genome of *S. cerevisiae* was completed and most, if not all, of the genes have been identified. The key biological questions now are what products do each of these genes encode, how do the different gene products interact, and under what circumstances is each gene expressed? Answering these questions still requires gene manipulation but the emphasis has switched from analysis of individual genes to that of the whole genome. That is, there has been a switch from the reductive approach to a holistic one.

A second reason for the current interest in gene manipulation in fungi is the overproduction of proteins of commercial value. Fungi offer a number of advantages, such as the ability to glycosylate protein, the absence of pyrogenic toxins, and in the case of the methylotrophic yeast *Pichia pastoris*, the ability to get very high yields of recombinant proteins. Yeasts and other fungi are widely used in the production of food and beverages and recombinant DNA technology can be used to enhance their desirable properties.

The third reason for the current interest in the application of recombinant DNA technology in yeast is the ability to clone very large pieces of DNA, an essential requirement for many whole-genome studies. Although there is no theoretical limit to the size of DNA that can be cloned in a bacterial plasmid in practice it is found that large inserts cause structural and segregative instability. Certain yeast vectors (YACs, p. 213) can accept inserts greater than 1 Mb, much greater than those found in BACs and PACs. Although the early YACs had a tendency to be unstable this problem has largely been eliminated in the new generation of YACs.

Fungi are not naturally transformable and special methods are required to introduce exogenous DNA

Like *E. coli*, fungi are not naturally transformable and artificial means have to be used for introducing foreign DNA. One method involves the use of spheroplasts (i.e. wall-less cells) and was first developed for *S. cerevisiae* (Hinnen *et al.* 1978). In this method, the cell wall is removed enzymically and the resulting spheroplasts are fused with ethylene glycol in the presence of DNA and $CaCl_2$. The spheroplasts are then allowed to generate new cell walls in a stabilizing medium containing 3% agar. This latter step makes subsequent retrieval of cells inconvenient. Electroporation provides a simpler and more convenient alternative to the use of spheroplasts. Cells transformed by electroporation can be selected on the surface of solid media, thus facilitating subsequent manipulation. Both the spheroplast technique and electroporation have been applied to a wide range of yeasts and filamentous fungi.

DNA can also be introduced into yeasts and filamentous fungi by conjugation. Heinemann & Sprague (1989) and Sikorski *et al.* (1990) found that enterobacterial plasmids, such as R751 (IncPβ) and F (IncF), could facilitate plasmid transfer from *E. coli* to *S. cerevisiae* and *Schizosaccharomyces pombe*. The bacterial plant pathogen *Agrobacterium tumefaciens* contains a large plasmid, the Ti plasmid, and part of this plasmid (the transferred DNA (T-DNA)) can be conjugally transferred to protoplasts of *S. cerevisiae* (Bundock *et al.* 1995) and a range of filamentous fungi (De Groot *et al.* 1998). T-DNA can also be transferred to hyphae and conidia.

Exogenous DNA that is not carried on a vector can only be maintained by integration into a chromosome

In the original experiments on transformation of *S. cerevisiae*, Hinnen *et al.* (1978) transformed a leucine auxotroph with the plasmid pYeLeu 10. This plasmid is a hybrid composed of the enterobacterial plasmid ColE1 and a segment of yeast DNA containing the *LEU2⁺* gene and is unable to replicate in yeast. Analysis of the transformants showed that in some of them there had been reciprocal recombination between the incoming *LEU2⁺* and the recipient *Leu2⁻* alleles. In the majority of the transformants, ColE1 DNA was also present and genetic analysis showed that in some of them the *LEU2⁺* allele was closely linked to the original *Leu2⁻* allele, whereas in the remaining ones the *LEU2⁺* allele was on a different chromosome.

The results described above can be confirmed by restriction-endonuclease analysis, since pYeLeu 10 contains no cleavage sites for *Hind*III. When DNA from the *Leu2⁻* parent was digested with endonuclease *Hind*III and electrophoresed in agarose, multiple DNA fragments were observed but only one of these hybridized with DNA from pYeLeu 10. With the transformants in which the *Leu2⁻* and *LEU2⁺*

alleles were linked, only a single fragment of DNA hybridized to pYeLeu 10, but this had an increased size, consistent with the insertion of a complete pYeLeu 10 molecule into the original fragment. These data are consistent with there being a tandem duplication of the *Leu2* region of the chromosome (Fig. 11.1). With the remaining transformants, two DNA fragments that hybridized to pYeLeu 10 could be found on electrophoresis. One fragment corresponded to the fragment seen with DNA from the recipient cells, the other to the plasmid genome which had been inserted in another chromosome (see Fig. 11.1). These results represented the first unambiguous demonstration that foreign DNA, in this case cloned ColE1 DNA, can integrate into the genome of a eukaryote. A plasmid such as pYeLeu 10 which can do this is known as a yeast integrating plasmid (YIp).

During transformation, the integration of exogenous DNA can occur by recombination with a homologous or an unrelated sequence. In most cases, non-homologous integration is more common than homologous recombination (Fincham 1989), but this is not so in *S. cerevisiae* (Schiestl & Petes 1991). In the experiments of Hinnen *et al.* (1978) described above, sequences of the yeast retrotransposon Ty2 were probably responsible for the integration of the plasmid in novel locations of the genome, i.e. the "illegitimate" recombinants were the result of homologous crossovers within a repeated element (Kudla & Nicolas 1992). Based on a similar principle, a novel vector has been constructed by Kudla and Nicolas (1992) which allows integration of a cloned DNA sequence at different sites in the genome. This feature

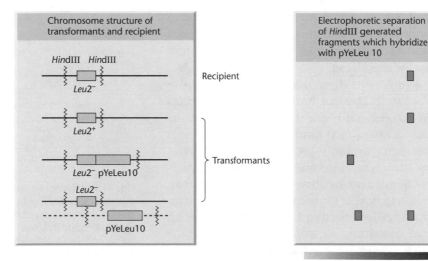

Fig. 11.1 Analysis of yeast transformants. (See text for details.)

is provided by the inclusion in the vector of a repeated yeast *sigma* sequence present in approximately 20–30 copies per genome and spread over most or all of the 16 chromosomes.

When T-DNA from the Ti plasmid of *Agrobacterium* is transferred to yeast, it too will insert in different parts of the genome by illegitimate recombination (Bundock & Hooykaas 1996).

Schiestl & Petes (1991) developed a method for forcing illegitimate recombination by transforming yeast with *Bam*HI-generated fragments in the presence of the *Bam*HI enzyme. Not only did this increase the frequency of transformants but the transformants which were obtained had the exogenous DNA integrated into genomic *Bam*HI sites. This technique, which is sometimes referred to as restriction-enzyme-mediated integration (REMI), has been extended to other fungi, such as *Cochliobolus* (Lu *et al.* 1994), *Ustilago* (Bolker *et al.* 1995), and *Aspergillus* (Sanchez *et al.* 1998).

Different kinds of vectors have been developed for use in *S. cerevisiae*

If the heterologous DNA introduced into fungi is to be maintained in an extrachromosomal state then plasmid vectors are required which are capable of replicating in the fungal host. Four types of plasmid vector have been developed: yeast episomal plasmids (YEps), yeast replicating plasmids (YRps), yeast centromere plasmids (YCps), and yeast artificial chromosomes (YACs). All of them have features in common. First, they all contain unique target sites for a number of restriction endonucleases. Secondly, they can all replicate in *E. coli*, often at high copy number. This is important, because for many experiments it is necessary to amplify the vector DNA in *E. coli* before transformation of the ultimate yeast recipient. Finally, they all employ markers that can be selected readily in yeast and which will often complement the corresponding mutations in *E. coli* as well. The four most widely used markers are *His*3, *Leu*2, *Trp*1, and *Ura*3. Mutations in the cognate chromosomal markers are recessive, and non-reverting mutants are available. Two yeast selectable markers, *Ura*3 and *Lys*2, have the advantage of offering both positive and negative selection. Positive selection is for complementation of auxotrophy. Negative selection is for ability to grow on medium containing a compound that inhibits the growth of cells expressing the wild-type function. In the case of *Ura*3, it is 5-fluoro-orotic acid (Boeke *et al.*

1984) and for *Lys*2 it is α-aminoadipate (Chatoo *et al.* 1979). These inhibitors permit the ready selection of those rare cells that have undergone a recombination or loss event to remove the plasmid DNA sequences. The *Lys*2 gene is not utilized frequently because it is large and contains sites within the coding sequence for many of the commonly used restriction sites.

Yeast episomal plasmids

YEps were first constructed by Beggs (1978) by recombining an *E. coli* cloning vector with the naturally occurring yeast 2 μm plasmid. This plasmid is 6.3 kb in size, has a copy number of 50–100 per haploid cell and has no known function. A representative YEp is shown in Fig. 11.2.

Yeast replicating plasmids

YRps were initially constructed by Struhl *et al.* (1979). They isolated chromosomal fragments of DNA which carry sequences that enable *E. coli* vectors to replicate in yeast cells. Such sequences are known as *ars* (autonomously replicating sequences). An *ars* is quite different from a centromere: the former acts as an origin of replication (Palzkill & Newlon 1988, Huang & Kowalski 1993), whereas the latter is involved in chromosome segregation.

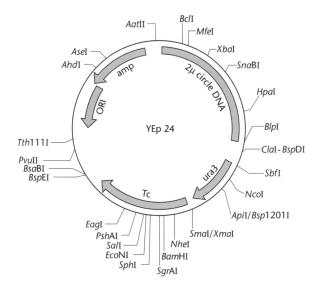

Fig. 11.2 Schematic representation of a typical yeast episomal plasmid (YEp 24). The plasmid can replicate both in *E. coli* (due to the presence of the pBR 322 origin of replication) and in *S. cerevisiae* (due to the presence of the yeast 2 μm origin of replication). The ampicillin and tetracycline determinants are derived from pBR 322 and the *URA*3 gene from yeast.

The structure of *ars* elements has been elucidated and shown to be a short (~100 bp) AT-rich sequence with the 17-bp core consensus WWWWTTTAYRTTTWGTT where W = A or T, Y = T or C, and R = A or G (Theis & Newlon 1997).

Although plasmids containing an *ars* transform yeast very efficiently, the resulting transformants are exceedingly unstable. For unknown reasons, YRps tend to remain associated with the mother cell and are not efficiently distributed to the daughter cell. (Note: *S. cerevisiae* does not undergo binary fission but buds off daughter cells instead.) Occasional stable transformants are found and these appear to be cases in which the entire YRp has integrated into a homologous region on a chromosome in a manner identical to that of YIps (Stinchcomb *et al.* 1979, Nasmyth & Reed 1980).

Yeast centromere plasmids

Using a YRp vector, Clarke and Carbon (1980) isolated a number of hybrid plasmids containing DNA segments from around the centromere-linked *leu2*, *cdc*10, and *pgk* loci on chromosome III of yeast. As expected for plasmids carrying an *ars*, most of the recombinants were unstable in yeast. However, one of them was maintained stably through mitosis and meiosis. The stability segment was confined to a 1.6 kb region lying between the *leu2* and *cdc*10 loci and its presence on plasmids carrying either of two *ars* tested resulted in those plasmids behaving like minichromosomes (Clarke & Carbon 1980, Hsiao & Carbon 1981). Genetic markers on the minichromosomes acted as linked markers segregating in the first meiotic division as centromere-linked genes and were unlinked to genes on other chromosomes.

Structurally, plasmid-borne centromere sequences have the same distinctive chromatin structure that occurs in the centromere region of yeast chromosomes (Bloom & Carbon 1982). Functionally YCps exhibit three characteristics of chromosomes in yeast cells. First, they are mitotically stable in the absence of selective pressure. Secondly, they segregate during meiosis in a Mendelian manner. Finally, they are found at low copy number in the host cell.

Yeast artificial chromosomes

All three autonomous plasmid vectors described above are maintained in yeast as circular DNA molecules – even the YCp vectors, which possess yeast centromeres. Thus, none of these vectors resembles the normal yeast chromosomes, which have a linear structure. The ends of all yeast chromosomes, like those of all other linear eukaryotic chromosomes, have unique structures that are called *telomeres*. Telomere structure has evolved as a device to preserve the integrity of the ends of DNA molecules, which often cannot be finished by the conventional mechanisms of DNA replication (for detailed discussion see Watson 1972). Szostak & Blackburn (1982) developed the first vector which could be maintained as a linear molecule, thereby mimicking a chromosome, by cloning yeast telomeres into a YRp. Such vectors are known as yeast artificial chromosomes (YACs).

The method for cloning large DNA sequences in YACs developed by Burke *et al.* (1987) is shown in Fig. 11.3.

The availability of different kinds of vector offers yeast geneticists great flexibility

The analysis of yeast DNA sequences has been facilitated by the ease with which DNA from eukaryotes can be cloned in *E. coli* using the vectors described in earlier chapters. Such cloned sequences can be obtained easily in large amounts and can be altered *in vivo* by bacterial genetic techniques and *in vitro* by specific enzyme modifications. To determine the effects of these experimentally induced changes on the function and expression of yeast genes they need to be taken out of bacteria and returned to the yeast cell. The availability of different kinds of vectors with different properties (Table 11.1) enables yeast geneticists to do this with relative ease and to perform subsequent manipulations in yeast itself. Thus cloned genes can be used in conventional genetic analysis by means of recombination using YIp vectors or linearized YRp vectors (Orr-Weaver *et al.* 1981). Complementation can be carried out using YEp, YRp, YCp, or YAC vectors, but there are a number of factors which make YCps the vectors of choice (Rose *et al.* 1987). For example, YEps and YRps exist at high copy number in yeast and this can prevent the isolation of genes whose products are toxic when overexpressed, e.g. the genes for actin and tubulin. In other cases, the overexpression of genes other than the gene of interest can suppress the mutation used for selection (Kuo & Campbell 1983). All the yeast vectors can be used to create partial diploids or partial polyploids and the extra gene sequences can be integrated or extrachromosomal. Deletions, point mutations, and frame-shift mutations can be

Table 11.1 Properties of the different yeast vectors.

Vector	Transformation frequency	Copy no./cell	Loss in non-selective medium	Disadvantages	Advantages
YIp	10^2 transformants per µg DNA	1	Much less than 1% per generation	1 Low transformation frequency 2 Can only be recovered from yeast by cutting chromosomal DNA with restriction endonuclease which does not cleave original vector containing cloned gene	1 Of all vectors, this kind give most stable maintenance of cloned genes 2 An integrated YIp plasmid behaves as an ordinary genetic marker, e.g. a diploid heterozygous for an integrated plasmid segregates the plasmid in a Mendelian fashion 3 Most useful for surrogate genetics of yeast, e.g. can be used to introduce deletions, inversions, and transpositions (see Botstein & Davis 1982)
YEp	10^3–10^5 transformants per µg DNA	25–200	1% per generation	Novel recombinants generated *in vivo* by recombination with endogenous 2 µm plasmid	1 Readily recovered from yeast 2 High copy number 3 High transformation frequency 4 Very useful for complementation studies
YRp	10^4 transformants per µg DNA	1–20	Much greater than 1% per generation but can get chromosomal integration	Instability of transformants	1 Readily recovered from yeast 2 High copy number. Note that the copy number is usually less than that of YEp vectors but this may be useful if cloning gene whose product is deleterious to the cell if produced in excess 3 High transformation frequency 4 Very useful for complementation studies 5 Can integrate into the chromosome
YCp	10^4 transformants per µg DNA	1–2	Less than 1% per generation	Low copy number makes recovery from yeast more difficult than that with YEp or YRp vectors	1 Low copy number is useful if product of cloned gene is deleterious to cell 2 High transformation frequency 3 Very useful for complementation studies 4 At meiosis generally shows Mendelian segregation
YAC		1–2	Depends on length: the longer the YAC the more stable it is	Difficult to map by standard techniques	1 High-capacity cloning system permitting DNA molecules greater than 40 kb to be cloned 2 Can amplify large DNA molecules in a simple genetic background

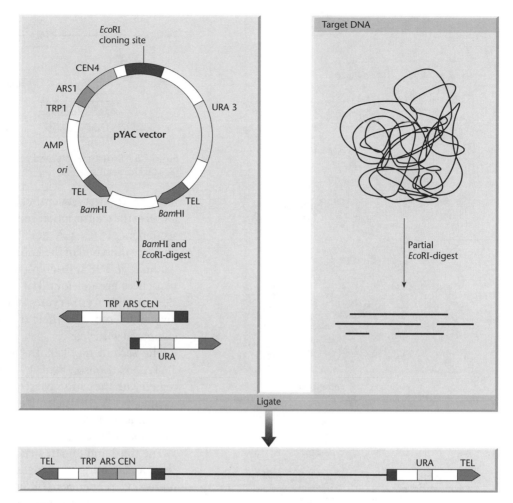

Fig. 11.3
Construction of a yeast artificial chromosome containing large pieces of cloned DNA. Key regions of the pYAC vector are as follows: TEL, yeast telomeres; ARS1, autonomously replicating sequence; CEN4, centromere from chromosome 4; *URA3* and *TRP1*, yeast marker genes; Amp, ampicillin-resistance determinant of pBR322; *ori*, origin of replication of pBR322.

introduced *in vitro* into cloned genes and the altered genes returned to yeast and used to replace the wild-type allele. Excellent reviews of these techniques have been presented by Botstein and Davis (1982), Hicks *et al.* (1982), Struhl (1983) and Stearns *et al.* (1990).

Recombinogenic engineering can be used to move genes from one vector to another

During the process of analyzing a particular cloned gene it is often necessary to change the plasmid's selective marker. Alternatively, it may be desired to move the cloned gene to a different plasmid, e.g. from a YCp to a YEp. Again, genetic analysis may require many different alleles of a cloned gene to be introduced to a particular plasmid for subsequent functional studies. All these objectives can be achieved by standard *in vitro* techniques, but Ma *et al.* (1987) showed that methods based on recombination *in vivo* are much quicker. The underlying principle is that linearized plasmids are repaired during yeast transformation by recombination with

a homologous DNA restriction fragment. It should be noted that this was the first demonstration of recombinogenic engineering (see p. 79) and significantly predated the development of this technique in *E. coli.*

Suppose we wish to move the *HIS3* gene from pBR328, which cannot replicate in yeast, to YEp420 (see Fig. 11.4). Plasmid pRB328 is cut with *Pvu*I and *Pvu*II and the *HIS3* fragment selected. The *HIS3* fragment is mixed with YEp420 which has been linearized with *Eco*RI and the mixture transformed into yeast. Two crossover events occurring between homologous regions flanking the *Eco*RI site of YEp420 will result in the generation of a recombinant YEp containing both the *HIS3* and *URA3* genes. The *HIS3* gene can be selected directly. If this were not possible, selection could be made for the *URA3* gene, for a very high proportion of the clones will also carry the *HIS3* gene. Many other variations of the above method have been described by Ma *et al.* (1987), to whom the interested reader is referred for details.

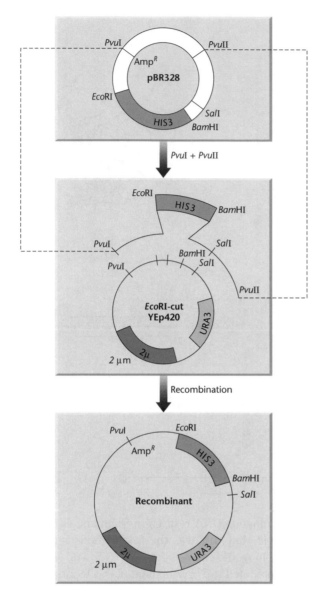

Fig. 11.4 Plasmid construction by homologous recombination in yeast. pRB328 is digested with *Pvu*I and *Pvu*II and the *HIS3*-containing fragment is transformed into yeast along with the *Eco*RI-cut YEp420. Homologous recombination occurs between pBR322 sequences, shown as thin lines, to generate a new plasmid carrying both *HIS3* and *URA3*.

Yeast promoters are more complex than bacterial promoters

When gene manipulation in fungi first became possible, there were many unsuccessful attempts to express heterologous genes from bacteria or higher eukaryotes. This suggested that fungal promoters have a unique structure, a feature first shown for *S. cerevisiae* (Guarente 1987). Four structural elements can be recognized in the average yeast promoter

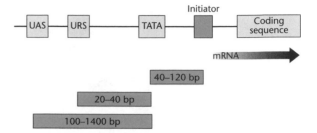

Fig. 11.5 Structure of typical yeast promoters. (See text for details.)

(Fig. 11.5). First, several consensus sequences are found at the transcription-initiation site. Two of these sequences, TC(G/A)A and PuPuPyPuPu, account for more than half of the known yeast initiation sites (Hahn *et al.* 1985, Rudolph & Hinnen 1987). These sequences are not found at transcription-initiation sites in higher eukaryotes, which implies a mechanistic difference in their transcription machinery compared with yeast.

The second motif in the yeast promoter is the TATA box (Dobson *et al.* 1982). This is an AT-rich region with the canonical sequence TATAT/AAT/A, located 60–120 nucleotides before the initiation site. Functionally, it can be considered equivalent to the Pribnow box of *E. coli* promoters (see p. 82).

The third and fourth structural elements are upstream activating sequences (UASs) and upstream repressing sequences (URSs). These are found in genes whose transcription is regulated. Binding of positive-control proteins to UASs increases the rate of transcription and deletion of the UASs abolishes transcription. An important structural feature of UASs is the presence of one or more regions of dyad symmetry (Rudolph & Hinnen 1987). Binding of negative-control proteins to URSs reduces the transcription rate of those genes that need to be negatively regulated.

The level of transcription can be affected by sequences located within the gene itself and which are referred to as downstream activating sequences (DASs). Chen *et al.* (1984) noted that, when using the phosphoglycerate kinase (*PGK*) promoter, several heterologous proteins accumulate to 1–2% of total cell protein, whereas phosphoglycerate kinase itself accumulates to over 50%. These disappointing amounts of heterologous protein reflect the levels of mRNA which were due to a lower level of initiation rather than a reduced mRNA half-life (Mellor *et al.* 1987). Addition of downstream PGK sequences restored the rate of mRNA transcription, indicating the presence of a DAS. Evidence for these DASs has been found in a number of other genes.

Promoter systems have been developed to facilitate overexpression of recombinant proteins in yeast

The first overexpression systems developed were for *S. cerevisiae* and used promoters from genes encoding abundant glycolytic enzymes, e.g. alcohol dehydrogenase (ADH1), PGK or glyceraldehyde-3-phosphate dehydrogenase (GAP). These are strong promoters and mRNA transcribed from them can accumulate up to 5% of total. They were at first thought to be constitutive but later were shown to be induced by glucose (Tuite *et al*. 1982). Now there is a large variety of native and engineered promoters available (Table 11.2), differing in strength, regulation, and induction ratio. These have been reviewed in detail by Romanos *et al*. (1992).

The ideal promoter is one that is tightly regulated so that the growth phase can be separated from the induction phase. This minimizes the selection of non-expressing cells and can permit the expression of proteins normally toxic to the cell. The ideal promoter will also have a high induction ratio. One promoter which has these characteristics and which is now the most widely used is that from the *GAL1* gene. Galactose regulation in yeast is now extremely well studied and has become a model system for eukaryotic transcriptional regulation (see Box 11.1).

Table 11.2 Common fungal promoters used for manipulation of gene expression.

Species	Promoter	Gene	Regulation
General			
S. cerevisiae	PGK	Phosphoglycerate kinase	Glucose-induced
	GAL1	Galactokinase	Galactose-induced
	PHO5	Acid phosphatase	Phosphate-repressed
	ADH2	Alcohol dehydrogenase II	Glucose-repressed
	CUP1	Copper metallothionein	Copper-induced
	MFα1	Mating factor α1	Constitutive but temperature-induced variant available
Candida albicans	MET3	ATP sulfur lyase	Repressed by methionine and cysteine
Methanol utilizers			
Candida boidnii	AOD1	Alcohol oxidase	Methanol-induced
Hansenula polymorpha	MOX	Alcohol oxidase	Methanol-induced
Pichia methanolica	AUG1	Alcohol oxidase	Methanol-induced
Pichia pastoris	AOX1	Alcohol oxidase	Methanol-induced
	GAP	Glyceraldehyde-3-phosphate dehydrogenase	Strong constitutive
	FLD1	Formaldehyde dehydrogenase	Methanol- or methylamine-induced
	PEX8	Peroxin	Methanol-induced
	YPT1	Secretory GTPase	Medium constitutive
Lactose utilizer			
Kluyveromyces lactis	LAC4	β-Galactosidase	Lactose-induced
	PGK	Phosphoglycerate kinase	Strong constitutive
	ADH4	Alcohol dehydrogenase	Ethanol-induced
Starch utilizer			
Schwanniomyces occidentalis	AMY1	α-Amylase	Maltose- or starch-induced
	GAM1	Glucoamylase	Maltose- or starch-induced
Xylose utilizer			
Pichia stipitis	XYL1		Xylose-induced
Alkane utilizer			
Yarrowia lipolytica	XPR2	Extracellular protease	Peptone-induced
	TEF	Translation elongation factor	Strong constitutive
	RPS7	Ribosomal protein S7	Strong constitutive

Box 11.1   Galactose metabolism and its control in *Saccharomyces cerevisiae*

Galactose is metabolized to glucose-6-phosphate in yeast by an identical pathway to that operating in other organisms (Fig. B11.1). The key enzymes and their corresponding genes are a kinase (*GAL1*), a transferase (*GAL7*), an epimerase (*GAL10*), and a mutase (*GAL5*). Melibiose (galactosyl-glucose) is metabolized by the same enzymes after cleavage by an α-galactosidase encoded by the *MEL1* gene. Galactose uptake by yeast cells is via a permease encoded by the *GAL2* gene. The *GAL5* gene is constitutively expressed. All the others are induced by growth on galactose and repressed during growth on glucose.

The *GAL1*, *GAL7*, and *GAL10* genes are clustered on chromosome II but transcribed separately from individual promoters. The *GAL2* and *MEL1* genes are on other chromosomes. The *GAL4* gene encodes a protein that activates transcription of the catabolic genes by binding UAS 5′ to each gene. The *GAL80* gene encodes a repressor that binds directly to the *GAL4* gene product, thus preventing it from activating transcription. The *GAL3* gene product catalyzes the conversion of galactose to an inducer, which combines with the *GAL80* gene product, preventing it from inhibiting the *GAL4* protein from binding to DNA (Fig. B11.2).

The expression of the *GAL* genes is repressed during growth on glucose. The regulatory circuit responsible for this phenomenon, termed catabolite repression, is superimposed upon the circuit responsible for induction of *GAL* gene expression. Very little is known about its mechanism.

For a review of galactose metabolism in *S. cerevisiae*, the reader should consult Johnston (1987).

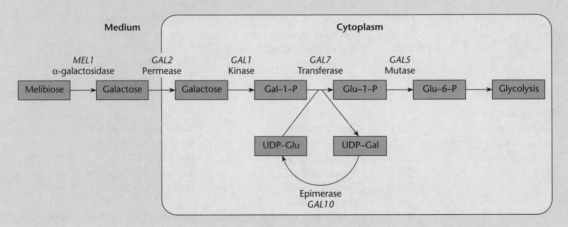

Fig. B11.1   The genes and enzymes associated with the metabolism of galactose by yeast.

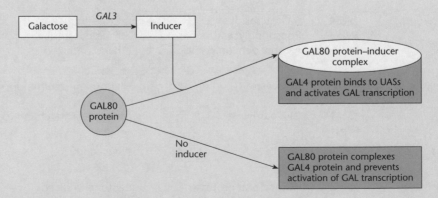

Fig. B11.2   The regulation of transcription of the yeast galactose genes.

Following addition of galactose, *GAL1* mRNA is rapidly induced over 1000-fold and can reach 1% of total mRNA. However, the promoter is strongly repressed by glucose and so in glucose-grown cultures this induction only occurs following depletion of glucose. To facilitate galactose induction in the presence of glucose, mutants have been isolated which are insensitive to glucose repression (Matsumoto *et al.* 1983, Horland *et al.* 1989). The *trans*-activator GAL4 protein is present in only one or two molecules per cell and so *GAL1* transcription is limited. With multicopy expression vectors, *GAL4* limitation is exacerbated. However, *GAL4* expression can be made autocatalytic by fusing the GAL4 gene to a *GAL10* promoter (Schultz *et al.* 1987), i.e. *GAL4* expression is now regulated (induced) by galactose.

In recent years, methylotrophic yeasts, such as *Pichia pastoris*, have proved extremely popular as hosts for the overexpression of heterologous proteins. There are a number of reasons for this. First, the alcohol oxidase (AOX1) promoter is one of the strongest and most regulatable promoters known. Second, it is possible to stably integrate expression plasmids at specific sites in the genome in either single or multiple copies. Third, the strains can be

cultivated to very high density. To date, over 300 foreign proteins have been produced in *P. pastoris* (Cereghino & Cregg 1999, 2000). Promoters for use in other yeasts are shown in Table 11.2.

A number of specialist multi-purpose vectors have been developed for use in yeast

Many different specialist yeast vectors have been developed which incorporate the useful features found in the corresponding *E. coli* vectors (see p. 94), e.g. an f1 origin to permit sequencing of inserts, production of the cloned gene product as a purification fusion, etc. Some representative examples are shown in Fig. 11.6. Two aspects of these vectors warrant further discussion: secretion and surface display.

In yeast, proteins destined for the cell surface or for export from the cell are synthesized on and translocated into the endoplasmic reticulum. From there they are transported to the Golgi body for processing and packaging into secretory vesicles. Fusion of the secretory vesicles with the plasma membrane then occurs constitutively or in response to an external signal (reviewed by Rothman & Orci 1992). Of the proteins naturally synthesized and secreted by yeast,

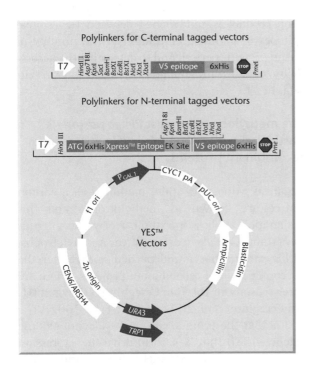
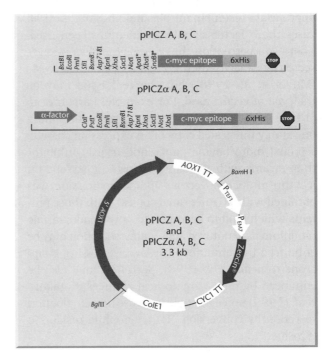

Fig. 11.6 Two specialized vectors for use in *Saccharomyces* (YES vectors) and *Pichia* (pPICZ). V5, Express, 6XHis, and c-myc encode epitopes which can be readily detected and purified by affinity chromatography. The YES vectors offer a choice of 2 μm origin for high copy or CEN6/ARSH4 origin for low copy in yeast, a choice of *URA3* or *TRP1* genes for auxotrophic selection on minimal medium, blasticidin resistance for dominant selection in any strain, and an f1 origin to facilitate DNA sequencing. In the pPICZ vectors, the zeocin-resistance gene is driven by the EM-7 promoter for selection in *E. coli* and the TEF1 promoter for selection in *Pichia*. More details of these features can be found in Chapter 5. (Figure reproduced courtesy of Invitrogen Corporation.)

only a few end up in the growth medium, e.g. the mating pheromone α factor and the killer toxin. The remainder, such as invertase and acid phosphatase, cross the plasma membrane but remain within the periplasmic space or become associated with the cell wall.

Polypeptides destined for secretion have a hydrophobic amino-terminal extension, which is responsible for translocation to the endoplasmic reticulum (Blobel & Dobberstein 1975). The extension is usually composed of about 20 amino acids and is cleaved from the mature protein within the endoplasmic reticulum. Such signal sequences precede the mature yeast invertase and acid phosphatase sequences. Rather longer leader sequences precede the mature forms of the α mating factor and the killer toxin (Kurjan & Herskowitz 1982, Bostian *et al.* 1984). The initial 20 amino acids or so are similar to the conventional hydrophobic signal sequences, but cleavage does not occur in the endoplasmic reticulum. In the case of the α factor, which has an 89 amino acid leader sequence, the first cleavage occurs after a Lys–Arg sequence at positions 84 and 85 and happens in the Golgi body (Julius *et al.* 1983, 1984).

To date, a large number of non-yeast polypeptides have been secreted from yeast cells containing the appropriate recombinant plasmid and in almost all cases the α-factor signal sequence has been used. There is a perception that *S. cerevisiae* has a lower secretory capacity than *P. pastoris* and other yeasts (Muller *et al.* 1998), but the real issue may be the type of vector used. For example, Parekh *et al.* (1996) found that *S. cerevisiae* strains containing one stably integrated copy of an expression cassette secreted more bovine pancreatic trypsin inhibitor than strains with the same expression cassette on a 2 μm multicopy vector. Optimal expression was obtained with 10 integrated copies. With those proteins which tend to accumulate in the endoplasmic reticulum as denatured aggregates, secretion may be enhanced by simultaneously overexpressing chaperons (Shusta *et al.* 1998). Secretion may also be enhanced by minor amino acid changes. Katakura *et al.* (1999) noted a sixfold increase in lactoglobulin secretion by conversion of a tryptophan residue to tyrosine.

Heterologous proteins can be synthesized as fusions for display on the cell surface of yeast

S. cerevisiae can be used to elucidate and dissect the function of a protein in a manner similar to

phage-display systems. Either can be used to detect protein–ligand interactions and to select mutant proteins with altered binding capacity (Shusta *et al.* 1999). However, phage-display systems often cannot display secreted eukaryotic proteins in their native functional conformation, whereas yeast surface display can.

Yeast surface display makes use of the cell surface receptor α-agglutinin (Aga), which is a two-subunit glycoprotein. The 725-residue Aga1p subunit anchors the assembly to the cell wall via a covalent linkage. The 69-residue binding subunit (Aga2p) is linked to Aga1p by two disulfide bonds. To achieve surface display, the appropriate gene is inserted at the C terminus of a vector-borne Aga2p gene under the control of the GAL1 promoter. The construct is then transformed into a yeast strain carrying a chromosomal copy of the Aga1p gene, also under the control of the GAL1 promoter. If the cloned gene has been inserted in the correct translational reading frame, its gene product will be synthesized as a fusion with the Aga2p subunit. The fusion product will associate with the Aga1p subunit within the secretory pathway and be exported to the cell surface (Boder & Wittrup 1997). In practice, the gene fusion is somewhat more complicated. Usually the cloned gene product is sandwiched between two simple epitopes to permit quantitation of the number of fusion proteins per cell and to determine the accessibility of different domains of the fusion protein (Fig. 11.7).

The methylotrophic yeast *Pichia pastoris* is particularly suited to high-level expression of recombinant proteins

Organisms that can utilize methanol as a carbon source are known as methylotrophs and this designation applies to a number of yeasts such as *Pichia* and *Candida* species. Of the methylotrophs, *Pichia pastoris* has found particular favor for the overproduction of recombinant proteins. There are three reasons for this (Lin Cereghino & Cregg 2001, Lin Cereghino *et al.* 2002, Houard *et al.* 2002). The first is that the promoter from the alcohol oxidase I gene (*AOX1*) that is used to drive the expression of foreign genes is extremely efficient and tightly regulated. The *AOX1* promoter is strongly repressed in cells grown on glucose and most other carbon sources but is induced over 1000-fold if the cells are shifted to a medium with methanol as the sole carbon source. Second, the *S. cerevisiae MATα* signal

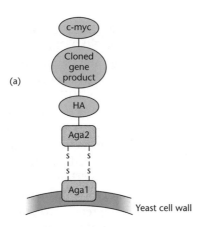

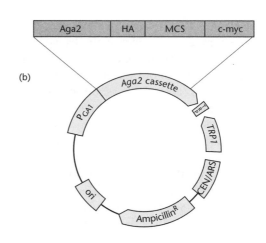

Fig. 11.7 Yeast surface display of heterologous proteins. (a) Schematic representation of surface display. (b) A vector used for facilitating surface display. MCS, multiple cloning site; HA, hemagglutinin epitope; c-myc, c-myc epitope. (See text for details.)

peptide can be attached to the protein of interest, causing it to be exported out of the cell. Interestingly, this signal sequence may be much more efficient in *P. pastoris* than in its native *S. cerevisiae*. Third, when organisms like *E. coli* and *S. cerevisiae* are grown to high cell density they have a tendency to switch to a fermentative mode of growth rather than a respiratory mode. This results in the accumulation of ethanol and acetic acid and these quickly become toxic. The rate of accumulation can be reduced by various cultural manipulations, such as batch feeding of glucose, but the basic problem still persists. By contrast, *P. pastoris* can only grow in respiratory mode on methanol and as a consequence it is possible to achieve cell densities of >100 g/l dry cell weight (>400 g/l wet cell weight). At these cell densities it is possible to get yields of recombinant proteins of 10–12 g/l, much higher than can be achieved with *S. cerevisiae* or *E. coli*.

Cloning and manipulating large fragments of DNA

Yeast artificial chromosomes can be used to clone very large fragments of DNA

When the first YACs were developed it was found that their segregative stability was determined by their size. If the size of the YAC was less than 20 kb then centromere function was impaired whereas much larger YACs segregated normally. Burke *et al.* (1987) made use of this fact in developing a vector (Fig. 11.3) for cloning large DNA molecules. One problem with large DNA molecules is that they are difficult to manipulate in the liquid phase prior to transformation and keeping them intact is very

difficult. Thus, many of the early YAC libraries had average insert sizes of only 50–100 kb. By removing small DNA fragments by PFGE fractionation prior to cloning, Anand *et al.* (1990) were able to increase the average insert size to 350 kb. By including polyamines to prevent DNA degradation, Larin *et al.* (1991) were able to construct YAC libraries from mouse and human DNA with average insert sizes of 700 and 620 kb, respectively. Later, Bellanné-Chantelot *et al.* (1992) constructed a human library with an average insert size of 810 kb and with some inserts as large as 1800 kb. The ability of YACs to accept such large inserts means that, in principle, it should be possible for them to carry the very large genes (>1 Mb) found in higher eukaryotes.

Classical YACs have a number of deficiencies as vectors

There are a number of operational problems associated with the use of YACs (Kouprina *et al.* 1994, Monaco & Larin 1994). The first of these is that it is estimated that 10–60% of clones in existing libraries represent chimeric DNA sequences: i.e. sequences from different regions of the genome cloned into a single YAC. Chimeras may arise by co-ligation of DNA inserts *in vitro* prior to yeast transformation, or by recombination between two DNA molecules that were introduced into the same yeast cell. It is possible to detect chimeras by *in situ* hybridization of the YAC to metaphase chromosomes: hybridization to two or more chromosomes or to geographically disparate regions of the same chromosome is indicative of a chimera.

A second problem with YACs is that many clones are unstable and tend to delete internal regions from their inserts. Using a model system, Kouprina *et al.* (1994) were able to show that deletions can be generated both during the transformation process and during mitotic growth of transformants and that the size of the deletions varied from 20 to 260 kb. Ling *et al.* (1993) showed that the frequency of deletion formation could be reduced by use of a strain rendered recombination-deficient as a result of a *rad52* mutation. However, such strains grow more slowly and transform less efficiently than RAD strains and therefore are not ideal hosts for YAC library construction. Le & Dobson (1997) have shown that the *rad54-3* allele significantly stabilizes YAC clones containing human satellite DNA sequences. Strains carrying this allele can undergo meiosis and have growth and transformation rates comparable with wild-type strains. Heale *et al.* (1994) have shown that chimera formation results from the yeast's mitotic recombination system, which is stimulated by the spheroplasting step of the standard YAC transformation system. Transformation of intact yeast cells is much less recombinogenic. An additional limitation on the use of YACs is the high rate of loss of some YACs during mitotic growth.

The third major problem with YAC clones is that the 15 Mb yeast host chromosome background cannot be separated from the YACs by simple methods, nor is the yield of DNA very high. Unlike plasmid vectors in bacteria, YACs have a structure very similar to natural yeast chromosomes. Thus, purifying YAC DNA from the yeast chromosomes usually requires separation by PFGE. Alternatively, the entire yeast genome is subcloned in bacteriophage or cosmid vectors followed by identification of those clones carrying the original YAC insert.

Circular YACs have a number of advantages over classical YACs

To overcome the disadvantages of classical YACs, Cocchia *et al.* (2000) developed a method to convert linear YACs into circular chromosomes that also can be propagated in *E. coli* as bacterial artificial chromosomes (BACs). The method involves a specialised BAC that contains a yeast centromeric sequence, a marker (G418^R) for selection in yeast, and two sequences homologous to those flanking the *Eco*RI site in a standard YAC (pYAC4). These two sequences are known as hooks and they are separated by a *Sal*I sequence thereby enabling the BAC to be linearized (Fig. 11.8). The linearized BAC is electroporated into a yeast cell carrying a standard linear YAC and recombination results in the generation

of a circular BAC/YAC and loss of the telomeres of the original YAC. Note that the use of recombination in this application is identical to the principle of recombinogenic engineering described earlier in this chapter (p. 207) and in chapter 3 (p. 79).

Circular YACs have a number of advantages over classical linear YACs. First, they can be separated easily from linear yeast chromosomes using standard alkaline lysis methods and are much less susceptible to shear forces. Thus molecules greater than 250 kb in size can be isolated intact. Second, circular YACs exhibit far greater structural stability than linear YACs and in the size range 100–200 kb have comparable stability to BACs. In the larger size range, where only YACs provide coverage, up to 40% of the circular YACs are chimeric but it is possible to isolate stable clones. Finally, circular YACs up to 250 kb can be electroporated into *E. coli*, where they behave as BACs.

Transformation-associated recombination (TAR) cloning in yeast permits selective isolation of large chromosomal fragments

The traditional method for isolating a specific eukaryotic gene, particularly if it is very large like the cystic fibrosis gene (230 kb) or the human dystrophin gene (2 Mb), has two disadvantages. First, it is a time-consuming process to analyze hundreds of thousands of BAC or YAC clones. Second, since restriction digestion is used for library construction, a gene may be available in a library only as a *collection* of several overlapping fragments that need to be correctly assembled by recombination between different YACs or BACs. Transformation-associated recombination (TAR) cloning (Noskov *et al.* 2003, Kouprina & Larionov 2003) is an alternative method that enables very large genes or chromosomal segments to be selectively and accurately isolated from genomic DNA.

TAR cloning makes use of circular YACs carrying the *S. cerevisiae URA3* gene that is under the control of the promoter from the *Schizosaccharomyces pombe* alcohol dehydrogenase (*ADH1*) gene. This promoter tolerates the insertion of up to a 130 bp sequence between the TATA box and the transcription initiation site. In TAR vectors, an insertion is made that carries hooks that are homologous to the 3′ and 5′ regions of the gene of interest and which are separated by a unique restriction site. These hooks can be as short as 60 bp and since their combined length is less than 130 bp they do not interfere

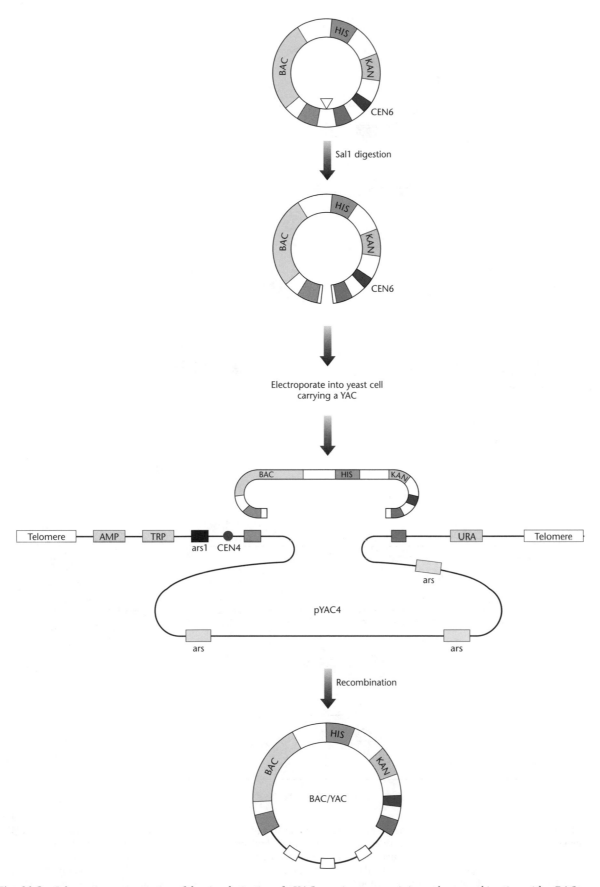

Fig. 11.8 Schematic representation of the circularization of a YAC carrying a genomic insert by recombination with a BAC carrying yeast functional elements. The hooks are shown in purple and gray. The BAC carries the *HIS* and *KAN* (G418^R) markers that are selectable in yeast plus the centromeric sequence CEN6 but has no *ars* sequence. The BAC also has a single recognition site (shown by arrowhead) for the restriction endonuclease *Sal*I. The BAC cannot replicate in yeast, even if it recyclizes, unless it acquires an *ars*.

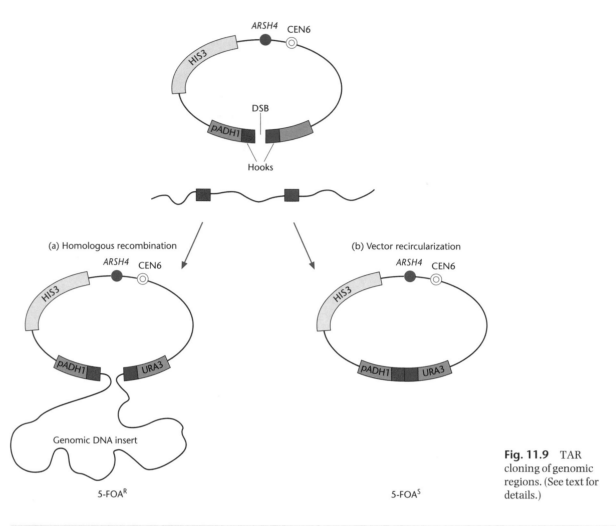

Fig. 11.9 TAR cloning of genomic regions. (See text for details.)

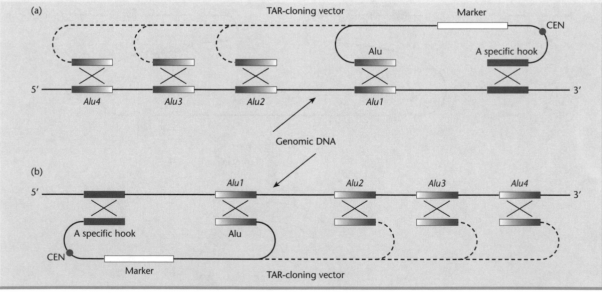

Fig. 11.10 Isolation of the entire copy gene from a complex genome by radial TAR cloning. The TAR vector contains either a 3′ or 5′ gene-specific sequence as one hook and a common repeat as the second hook; hooks are placed at the ends of the linearized TAR vector. Recombination between the sequences in the vector and the gene-containing fragment(s) leads to isolation of the gene as a series of circular, overlapping YACs that extend from the unique targeting sequence to various *Alu* (or *Bl*) positions. Simply by changing the orientation of the unique targeting hook, it is possible to clone large regions that extend in both directions from the specific targeting sequence. Propagation of the YAC in yeast cells depends on acquisition of *ars* sequences in the cloned genomic segment. CEN corresponds to the yeast centromere, and Marker is a marker for selection in yeast. Reprinted from Kouprina *et al.* Copyright (1998) National Academy of Sciences, USA.

with *URA3* expression. To clone the gene of interest, the circular YAC is linearized and co-transformed with genomic DNA into yeast spheroplasts (Fig. 11.9). Clones carrying a genomic insert will have their *URA3* gene disrupted and hence will be resistant to 5-fluoroorotic acid (5-FOA) whereas cells carrying recircularized vector will be FOA-sensitive.

When DNA sequence information is available from only one of the flanking regions then a modified version of TAR cloning, known as radial TAR cloning, can be used. In this variant, one hook is a unique sequence from the chromosomal region of interest and the other is a repeated sequence that occurs frequently and randomly throughout the genomic DNA. These repeated sequences could be *Alu* repeats if human DNA is being cloned, *B1* repeats for mouse DNA cloning, or any other similar repeat motif. When used in TAR cloning such vectors generate a series of nested overlapping fragments that extend from the gene-specific targeting hook to upstream or downstream *Alu* or *Bl* positions in the gene of interest (Fig. 11.10).

The development of TAR cloning has greatly facilitated the study of genes and genome structure in complex genomes (for review, see Kouprina & Larionov 2003). For example, the technique has been used to isolate dozens of unique regions and full-size genes from the DNA of humans, mice, and other organisms. It also has been used (Kouprina *et al.* 2003a) to clone centromeric DNA from mammals which previously had proved refractory to cloning because of its high content of repetitive DNA (see p. 344). Yet another application has been the cloning of human DNA that is unclonable in bacterial vectors (Kouprina *et al.* 2003b).

Suggested reading

Houard S., Heinderyckx M. & Bollen A. (2002) Engineering of non-conventional yeasts for efficient synthesis of macromolecules: the methylotrophic genera. *Biochimie* **84**, 1089–93.

Olsen D., Yang C., Bodo M., *et al.* Recombinant collagen and gelatin for drug delivery. *Advances in Drug Delivery* **55**, 1547–67.

Yokoyama S. (2003) Protein expression systems for structural genomics and proteomics. *Current Opinion in Chemical Biology* **7**, 39–43.

Three quite different perspectives on the utility of Pichia pastoris *and its relative merits over other yeasts and* Escherichia coli.

Klinner U. & Schafer B. (2004) Genetic aspects of targeted insertion mutagenesis in yeasts. *FEMS Microbiology Reviews* **28**, 201–23.

A comprehensive review of a topic that was only discussed briefly in this chapter.

Kouprina N. & Larionov V. (2003) Exploiting the yeast *Saccharomyces cerevisiae* for the study of the organization and evolution of complex genomes. *FEMS Microbiology Reviews* **27**, 629–49.

An excellent review that focuses on the use of recombinational cloning techniques to analyze genome structure and gene expression.

CHAPTER 12

Gene transfer to animal cells

Introduction

The genetic modification of animal cells underlies many aspects of basic research and applied biotechnology. Techniques are available for the introduction of DNA or RNA into hundreds of different cell types in culture, and such experiments can be used to study gene function and regulation, to produce recombinant proteins and to manipulate the endogenous genome. Mammalian cells are the most widely used hosts for gene delivery, since they allow the production of recombinant human proteins with authentic post-translational modifications that are not carried out by bacteria, yeast, or plants. Indeed, mammalian cells are cultured on a commercial scale for the synthesis of many valuable products, including antibodies, hormones, growth factors, cytokines, and vaccines. There has been intense research into the development of efficient vector systems and transformation methods for animal cells, based on viral, bacterial, and synthetic delivery vectors. Although this research has concentrated mainly on mammalian cell lines, other systems have also become popular, such as the baculovirus expression system which is used in insects.

Research has focused not only on gene-delivery systems for cultured cells, but also on systems for gene transfer to animal cells *in vivo*. The most important application of this technology is *in vivo* gene therapy, i.e. the introduction of DNA into living human beings in order to treat disease. Viral gene-delivery vectors are favored for therapeutic applications because of their efficiency, but safety concerns have prompted research into alternative non-viral transfer procedures such as liposome-mediated delivery, particle bombardment, and ultrasound.

This chapter considers methods for the introduction of DNA into *somatic cells*, i.e. cells that do not contribute to the animal germline. Unlike the situation in plants, animal somatic cells are restricted in their developmental potential and cannot under normal circumstances be used to generate transgenic animals. Transformation of the animal germline requires gene transfer to pluripotent or totipotent cells, such as eggs, early embryos, isolated germ cells, or gametes. Exceptionally, germline transformation can be achieved by the transformation of cultured cells derived from pluripotent cell lines, such as murine embryonic stem (ES) cells, or the transfer of somatic nuclei into enucleated eggs. We consider these methods for animal germline transformation in Chapter 13.

There are four major strategies for gene transfer to animal cells

Gene transfer to animal cells can be achieved using four broad types of delivery mechanism (Fig. 12.1). Two of these are described as *biological* mechanisms because the target cells need to be infected with a biological delivery vector, such as a virus or bacterium, which carries the exogenous genetic material. Delivery using a viral vector is known as *transduction*, and many different viruses have been adapted as gene-delivery vectors. The exogenous genetic material, or transgene, is either added to the complete viral genome or used to replace one or more viral genes. The transgene is therefore delivered as part of a recombinant viral genome, exploiting the virus's natural ability to infect and transfer nucleic acid into animal cells. Delivery using bacterial vectors is a more recent development, which in most cases relies on bacteria which also invade animal cells. In this case, however, the transgene is delivered not as part of the bacterial genome, but on a plasmid which is carried by the bacterium. Bacterial gene delivery is sometimes termed *bactofection*. The other two delivery mechanisms are described as non-biological because biological delivery vectors are not required. To distinguish such methods from those involving *infection* with a bacterium or virus, the term transfection is used. In

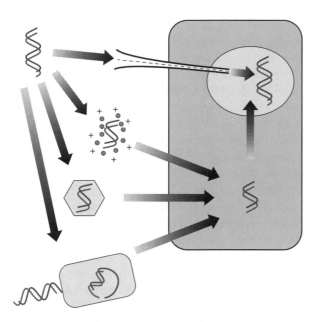

Fig. 12.1   The four principal mechanisms by which DNA can be introduced into animal cells – direct transfer by physical transfection, chemical-mediated transfection in which the cell is persuaded to take up DNA by endocytosis, introduction of DNA packaged inside a virus (transduction), and introduction of DNA packaged inside a bacterium (bactofection).

chemical transfection methods, cells are persuaded to take up DNA from their surroundings when the DNA is presented as a synthetic complex – either a complex with an overall positive charge, allowing it to interact with the negatively charged cell membrane and promote uptake by endocytosis; or a lipophilic complex which fuses with the membrane and deposits the transgene directly into the cytoplasm (Fig. 12.2). In *physical transfection* methods, naked DNA is introduced directly into the cell by exploiting a physical force. Examples of physical transfection

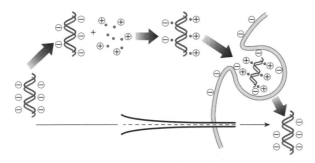

Fig. 12.2   The introduction of DNA into animal cells by chemical transfection involves the formation of a complex that facilitates interactions with the plasma membrane of the cell. This can be via electrostatic attraction or the fusion of lipophilic vesicles. In contrast, physical transfection involves direct transfer through the cell membrane, although the DNA may be complexed to protect it from shear forces.

methods include microinjection, particle bombardment, ultrasound, and electroporation. Whichever gene-transfer strategy is chosen, the result is *transformation*, i.e. a change of the recipient cell's genotype caused by the acquired transgene. This change in genotype may be transient or stable, depending on the type of vector and the fate of the introduced genetic material.

There are several chemical transfection techniques for animal cells but all are based on similar principles

The calcium phosphate method involves the formation of a co-precipitate which is taken up by endocytosis

The ability of mammalian cells to take up exogenous DNA from the culture medium was first reported by Szybalska & Szybalski (1962). They used total uncloned genomic DNA to transfect human cells deficient for the enzyme hypoxanthine-guanine phosphoribosyltransferase (HPRT). Rare HPRT-positive cells, which had presumably taken up fragments of DNA containing the functional gene, were identified by selection on HAT medium (p. 225). At this time, the actual mechanism of DNA uptake was not understood. Much later, it was appreciated that successful DNA transfer in such experiments was dependent on the formation of a fine DNA/calcium phosphate co-precipitate, which first settles onto the cells and is then internalized. The precipitate must be formed freshly at the time of transfection. It is thought that small granules of calcium phosphate associated with DNA are taken up by endocytosis. The DNA then escapes and some reaches the nucleus and can be expressed (Orrantia & Cheng 1990, Jordan & Wurm 2004). The technique became generally accepted after Graham & van der Erb (1973) used calcium phosphate to increase the infectivity of adenovirus DNA. It is now established as a general method for the introduction of DNA into a wide range of cell types in culture. However, since the precipitate must coat the cells, this method is suitable for cells growing in monolayers or in suspension, but not those growing as clumps.

As originally described, calcium phosphate transfection was limited by the variable and rather low proportion of cells that took up DNA (1–2%). Only a small number of these would be stably transformed. Improvements to the method have increased the transfection frequency up to 20% for some cell lines (Chu & Sharp 1981). A variant of the technique, using a different buffer system, allows the precipitate to form slowly over a number of hours, and this can increase stable transformation efficiency by up to 100-fold when using high-quality plasmid DNA (Chen & Okayama 1987, 1988). Even so, particle size is the most important parameter for successful transfection and is difficult to control, since it is dependent on the concentrations of DNA, calcium and phosphate, the pH of the medium, the presence of medium additives, and the mixing technique (Jordan & Wurm 2004).

Transfection with polyplexes is more efficient because of the uniform particle size

The calcium phosphate method is applicable to many cell types, but some cell lines are adversely affected by the co-precipitate due to its toxicity and are hence difficult to transfect. Alternative chemical transfection methods have been developed to address this problem using polycationic compounds that form soluble complexes (*polyplexes*) through spontaneous electrostatic interactions with DNA. The earliest of these methods utilized DEAE-dextran (diethylamino-ethyl dextran), a soluble polycationic carbohydrate. This technique was initially devised to introduce viral RNA and DNA into cells (Pagano & Vaheri 1965, McCutchan & Pagano 1968) but was later adapted as a method for plasmid DNA transfer (Milman & Herzberg 1981). The efficiency of the original procedure was improved by Lopata *et al.* (1984) and Sussman & Milman (1984) by adding after-treatments such as osmotic shock[1] or exposure to chloroquine, the latter having been shown to inhibit the acidification of endosomal vesicles (Luthmann & Magnusson 1983). Although efficient for the transient transfection of many cell types, DEAE-dextran cannot be used to generate stably transformed cell lines and has thus fallen out of favor.

A new generation of polycationic compounds has been developed more recently, which are more versatile and can be used for the stable transformation not only of proliferating cell lines but also of primary cells and post-mitotic cells such as neurons and myocytes (Davis 2002). Simple, linear compounds such as poly-L-lysine and spermine have been

successful with some cell types, but more progress has been made with synthetic polyamines, polyethylenimines (PEIs), and dendrimers (Boussif *et al.* 1995, Abdallah *et al.* 1996, Tang *et al.* 1996, Kukowska-Latallo *et al.* 1996). Polyamines are organic molecules carrying multiple amine groups, which increase the number of positive charges available for interactions with DNA. PEIs are linear or branched hydrocarbon chains with nitrogens at every third position, maximizing the available positive charges. Dendrimers are highly complex molecules, built in layers from a central initiator such as ammonia or ethylenediamine to adopt an overall spherical or star-like geometry with multiple internal and external amine groups. Their shape, size, and chemical properties can be controlled precisely, facilitating the formation of homogeneous DNA complexes. Many proprietary polyplex-forming agents are available commercially and a small selection is shown in Table 12.1.

As well as increasing the efficiency of gene transfer in cultured cells, polyplexes can also be used for *in vivo* gene transfer. For this to be successful, precise control over transgene expression is required. Recent developments in polymer technology provide a novel solution – controlling the physical and chemical properties of the gene-delivery vehicle itself to regulate transgene expression (Yokoyama 2002). The basis of these developments is that some polymers change their properties in response to a physical stimulus. For example, Nagasaki *et al.* (2000) have described a synthetic polymer whose affinity to DNA changes upon exposure to UV light. The polymer is based on azobenzene, and the azo moiety undergoes *trans*-to-*cis* isomerism following UV irradiation, causing DNA to be released from the complex. In cultured cells, exposure to UV light increases transfection efficiency by up to 50%. Polymeric complexes responsive to UV light have limited use for gene therapy because it would be difficult to illuminate internal organs. However, another synthetic complex has been described that is responsive to heat (Kurisawa *et al.* 2000a,b, Yokoyama *et al.* 2001). This polymer is based on N-isopropylacrylamide (PIPAAm) and undergoes a phase transition at a lower critical solution temperature of 32°C. Below 32°C, the polymer is hydrophilic and soluble, and forms a loose complex with DNA. Above 32°C, the polymer is hydrophobic and compact, and forms a tight complex by aggregation. A tight complex is preferable for DNA delivery because it is suitable for uptake and resistant to nucleases, whereas a loose complex is better for transcription because it

---

[1] Osmotic shock is also used with calcium phosphate transfection, lipofection, and electroporation to increase transfection efficiencies.

Table 12.1 Some cationic lipid and cationic polymer transfection reagents.

Reagent or reagent family	Composition, if known	Supplier
CellFectin	Cationic lipopolyamine	Invitrogen
CLONfectin	Cationic lipid	BD Biosciences/CLONTECH
CytoFectene	Cationic lipid	Amersham
Cytofectin GS	Cationic lipid	Glen Research
CytoPure	Cationic polymer, biodegradable	Qbiogene
DOSPAR	Polycationic lipid	Roche Applied Science
DOTAP	Cationic lipid	Roche Applied Science
Effectegene	Nonliposomal lipid	Qiagen
Escort family	Cationic lipids	Sigma Aldritch
ExGen family	Polyethylenimine	MBI Fermentas
FluoroFectin family	DOTAP, neutral lipid and fluorescent tracer	Qbiogene
FuGENE 6	Nonliposomal lipid	Roche Applied Science
GeneJammer	Polyamine	Strategene
GeneJuice	Polyamine	Novagen
GeneLimo family	Polycationic lipid and helper	CPG
GenePORTER family		Gene Therapy Systems
GeneSHUTTLE family	Polycationic lipids	Qbiogene
Genetransfer	Cationic liposome	Wako
GenFect	Cationic lipid	MoleculA
jetPEI	Polyethylenimine	Qbiogene
LipofectAMINE family	Polycationic lipid	Invitrogen
Lipofectin	DOTMA and DOPE	Invitrogen
LipoGen	Nonliposomal lipid	InvivoGen
LipoTaxi	Cationic lipid	Strategene
LipoVec	Cationic phosphonolipid mix	InvivoGen
MaxFect	Cationic lipid	MoleculA
Metafectene	Cationic lipid	Biontex
Polyethylenimine-Transferrinfection	Polyethylenimine	Bender MedSystems
PolyFect	Dendrimer	Qiagen
SuperFect	Dendrimer	Qiagen
SureFECTOR	Cationic lipid and neutral lipid	B-Bridge International
Targetfect F1	Cationic lipid	Targeting Systems
Targetfect F2 and F4	Cationic polymer	Targeting Systems
Tfx family	Cationic lipids	Promega
TransFAST	Cationic lipid	Promega
Transfectam	Cationic lipids	Promega
TransFectin	Cationic lipid	Bio-Rad
TransIT family*	Polyamine	Pan Vera Corp
TransIT-Insecta	Cationic lipid	Pan Vera Corp
TransMessenger		Qiagen
UniFECTOR	Polyationic lipid and neutral lipid	B-Bridge International
X-tremeGENE family		Roche Applied Science

facilitates access by transcription factors. Therefore, at normal body temperatures the DNA complex may be taken up efficiently by all cells, but poorly expressed. Local cooling, e.g. through the application of ice to the body surface or the use of catheters, can then induce gene expression in specific tissues or organs. The properties of the polymer can be modified by increasing the proportion of hydrophobic or hydrophilic chemical groups, thus lowering or raising the transition temperature, respectively.

Transfection can also be achieved using liposomes and lipoplexes

An alternative chemical transfection procedure is to package DNA inside a *fusogenic* phosopholipid vesicle, which interacts with the target cell membrane and facilitates DNA uptake. The first example of this approach was provided by Schaffner (1980), who used bacterial protoplasts containing plasmids to transfer DNA into mammalian cells. Briefly, bacterial cells were transformed with a suitable plasmid vector and then treated with chloramphenicol to amplify the plasmid copy number. Lysozyme was used to remove the cell walls, and the resulting protoplasts were gently centrifuged onto a monolayer of mammalian cells and induced to fuse with them using polyethylene glycol. A similar strategy was employed by Wiberg *et al.* (1987), who used the hemoglobin-free ghosts of erythrocytes as delivery vehicles. The procedures are very efficient in terms of the number of transformants obtained, but they are also labor-intensive and so have not been widely adopted as a general transfection method. However, an important advantage is that they are gentle, allowing the transfer of large DNA fragments without shearing. Yeast cells with the cell wall removed (spheroplasts) have therefore been used to introduce YAC DNA into mouse ES cells by this method, for the production of YAC transgenic mice (see Chapter 13).

More widespread use has been made of artificial phospholipid vesicles, which are called *liposomes* (Schaefer-Ridder *et al.* 1982). Initial liposome-based procedures were hampered by the difficulty encountered in encapsulating the DNA, and the transfection efficiency was no better than that of the calcium phosphate method. However, a breakthrough came with the discovery that cationic/neutral lipid mixtures can spontaneously form stable complexes with DNA (*lipoplexes*) that interact productively with the cell membrane, resulting in DNA uptake by endocytosis (Felgner *et al.* 1987, 1994). This low-toxicity transfection method, commonly known as *lipofection*, is one of the simplest to perform and is applicable to many cell types that are difficult to transfect by other means, including primary cells and cells growing in suspension (e.g. Ruysscharet *et al.* 1994). The technique facilitates transient and stable transformation, and is sufficiently gentle to be used with YACs and other large DNA fragments. The efficiency is also much higher than that of other chemical transfection methods – up to 90% of cells in a culture dish can be transfected. A large number of proprietary lipid mixtures is available, varying in efficiency for different cell lines (Table 12.1). A useful feature of liposome and lipoplex gene-delivery vehicles is their ability to transform cells in live animals following injection into target tissues or even the bloodstream. Transfection efficiency has been improved and targeting to specific cell types achieved by combining liposomes and lipoplexes with viral proteins that promote cell fusion, nuclear targeting signals, and various molecular conjugates that recognize specific cell-surface molecules. The development of liposomes for gene therapy and the determination of important parameters for efficient transfection have been comprehensively reviewed (Scheule & Cheng 1996, Tseng & Huang 1998, Rose & Hui 1999).

Physical transfection techniques have diverse mechanisms

Electroporation and ultrasound create transient pores in the cell

Electroporation is a physical transfection technique which involves the generation of transient, nanometer-sized pores in the cell membrane, by exposing cells to a brief pulse of electricity. DNA enters the cell through these pores, and is transported to the nucleus. This technique was first applied to animal cells by Wong & Neumann (1982), who successfully introduced plasmid DNA into mouse fibroblasts. The electroporation technique has been adapted to many other cell types (Potter *et al.* 1984). The most critical parameters are the intensity and duration of the electric pulse, and these must be determined empirically for different cell types. However, once optimal electroporation parameters have been established, the method is simple to carry out and highly reproducible. The technique has high input costs because a specialized capacitor discharge machine is required that can accurately control pulse length and amplitude (Potter 1988). Additionally, larger numbers of cells may be required than for other methods because in many cases, the most efficient electroporation occurs when there is up to 50% cell death.

Electroporation can also be used as a method for *in vivo* gene transfer, particularly for surface or near-surface tissues such as skin, muscle, and certain tumors (Bigley *et al.* 2002). This can be achieved by the direct application of electrodes to the skin following shaving and mild abrasion (Dujardin *et al.* 2001), although DNA is generally injected into the skin or muscle before the application of the electric field (Aihara *et al.* 1998, Rols *et al.* 1998). In combination with conventional needle injection, electroporation can also be used to introduce DNA into internal organs, such as the liver. In this case, gene transfer is achieved through the use of needle electrodes (Heller *et al.* 1996).

Ultrasound transfection involves the exposure of cells to a rapidly oscillating probe, such as the tip of a sonicator (Mitragotri *et al.* 1996, Wyber *et al.* 1997). The transfection mechanism is similar in some ways to electroporation in that the application of ultrasound waves to a dish of cells or a particular tissue results in the formation and collapse of bubbles in the liquid, including the cell membrane, a process known as cavitation. The transient appearance of such cavities allows DNA to cross the membrane into the cytoplasm. It has been shown that the application of low-frequency ultrasound allows the efficient delivery of nucleic acids into mammalian cells both *in vitro* and *in vivo*, because the plasmid DNA is left structurally intact. Furthermore, ultrasound-mediated gene delivery raises no safety concerns because the ultrasound waves appear to have no adverse effects when focused on different anatomic locations in the human body. Gene transfer *in vivo* is generally achieved by injection followed by the application of a focused ultrasound device.

Other physical transfection methods pierce the cell membrane and introduce DNA directly into the cell

Another group of transfection methods involves the direct transfer of DNA into the cell, without a synthetic carrier. One such procedure is microinjection, a technique that is guaranteed to generate successful hits on target cells but that can only be applied to a few cells in any one experiment. This technique has been applied to cultured cells that are recalcitrant to other transfection methods (Capecchi 1980, Greassmann & Graessmann 1983), but its principle use is to introduce DNA and other molecules into large cells, such as oocytes, eggs, and the cells of early embryos, as discussed in Chapter 13. For *in vivo* applications, conventional needle injection appears to be an efficient way to transfer DNA into target cells. The DNA can be injected directly into tissues such as skin, muscle, or internal organs (Wolff *et al.*

1990), or it can be injected into the blood. In the latter case, because the DNA is rapidly degraded, a hydrodynamic gene-transfer method has been devised which involves the transfer of a large amount of DNA into the bloodstream in a small volume of saline (Liu & Knapp 2002).

Particle bombardment is another direct delivery method, initially developed for the transformation of plants (Chapter 14). This involves coating small metal particles with DNA and then accelerating them into target tissues using a powerful force, such as a blast of high-pressure gas or an electric discharge through a water droplet (O'Brien & Lummis 2004). In animals, this technique is most often used to transfect multiple cells in tissue slices rather than cultured cells (e.g. Lo *et al.* 1994, Arnold *et al.* 1994) but it has also been used to transfect cells in culture (e.g. Burkholder *et al.* 1993). Particle bombardment has most recently been adapted as a method for the transfer of DNA into skin cells *in vivo* (Haynes *et al.* 1996, Roy *et al.* 2001).

Cells can be transfected with either replicating or non-replicating DNA

The transformation of animal cells occurs in two stages, the first involving the introduction of DNA into the cell (the transfection stage), and the second involving its incorporation into the nucleus, often by integration into the host chromosome. Transfection is much more efficient than integration, hence a large proportion of transfected cells never integrate the foreign DNA they contain. The DNA is maintained in the nucleus in an extrachromosomal state and, assuming it does not contain an origin of replication that functions in the host cell, it persists for just a short time before it is diluted and degraded. This is known as *transient transformation* (the term *transient transfection* is also used) reflecting the fact that the properties of the cell are changed by the introduced transgene, but only for a short duration. In a small proportion of transfected cells, the DNA will integrate into the genome, forming a new genetic locus that will be inherited by all clonal descendants. This is known as stable transformation, and results in the formation of a "cell line" carrying and expressing the transgene. Since integration is such an inefficient process, the rare stably transformed cells must be isolated from the large background of non-transformed and transiently transformed cells by selection.

Three types of selectable marker have been developed for animal cells

Endogenous selectable markers are already present in the cellular genome, and mutant cell lines are required when they are used

Following the general acceptance of the calcium phosphate transfection method, it was shown that mouse cells deficient for the enzyme thymidine kinase (TK) could be stably transformed to a wild-type phenotype by transfecting them with the herpes simplex virus (HSV) *Tk* gene (Wigler *et al.* 1977). As for the HPRT⁺ transformants discussed earlier in the chapter, cells positive for TK can be selected on HAT medium. This is because both enzymes are required

for nucleotide biosynthesis via the *salvage pathway* (Fig. 12.3). In mammals, nucleotides are produced via two alternative routes, the *de novo* and salvage pathways. In the *de novo* pathway, nucleotides are synthesized from basic precursors such as sugars and amino acids, while the salvage pathway recycles nucleotides from DNA and RNA. If the *de novo* pathway is blocked, nucleotide synthesis becomes dependent on the salvage pathway, and this can be exploited for the selection of cells carrying functional *Hprt* and *Tk* genes. The drug aminopterin blocks the *de novo* synthesis of both inosine monophosphate (IMP) and thymidine monophosphate (TMP) by inhibiting key enzymes in the *de novo* pathway. Cells exposed to aminopterin can thus survive only if they have functional *Hprt* and *Tk* genes and a source of

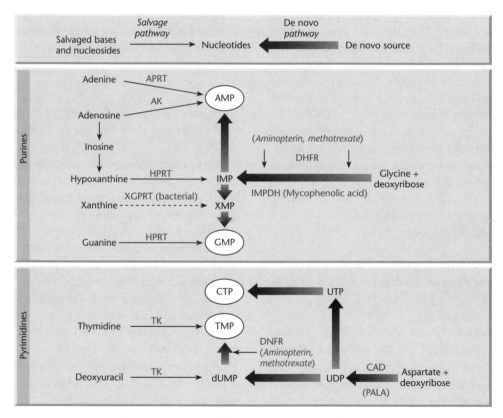

Fig. 12.3 Simplified representation of the *de novo* and salvage nucleotide synthesis pathways. Top panel: purine synthesis. *De novo* purine nucleotide synthesis (shown on the right) initially involves the formation of inosine monophosphate (IMP) which is then converted into either adenosine monophosphate (AMP) or, via xanthine monophosphate (XMP), guanosine monophosphate (GMP). The *de novo* synthesis of IMP requires the enzyme dihydrofolate reductase (DHFR), whose activity can be blocked by aminopterin or methotrexate. In the presence of such inhibitors, cell survival depends on nucleotide salvage, as shown on the left. Cells lacking one of the essential salvage enzymes, such as HPRT or APRT, therefore cannot survive in the presence of aminopterin or methotrexate unless they are transformed with a functional copy of the corresponding gene. Thus, the genes encoding salvage-pathway enzymes can be used as selectable markers. Note that the enzyme XGPTR, which converts xanthine to XMP, is found only in bacterial cells and not in animals. Bottom panel: pyrimidine synthesis. *De novo* pyrimidine nucleotide synthesis (shown on the right) initially involves the formation of uridine diphosphate (UDP). This step requires a multifunctional enzyme, CAD, whose activity can be blocked by N-phosphonacetyl-L-aspartate (PALA). UDP is then converted into either thymidine monophosphate (TMP) or cytidine triphosphate (CTP), the latter via uridine triphospate (UTP). *De novo* TMP synthesis requires DHFR, so the reaction can be blocked in the same way as in *de novo* purine synthesis, making cell survival dependent on the salvage enzyme thymidine kinase (TK). Thus, the *Tk* gene can be used as a selectable marker. There is no salvage pathway for CTP.

Table 12.2 Commonly used endogenous selectable marker genes. Most of these markers are involved in the redundant endogenous nucleotide biosynthetic pathways (Fig. 12.3). They can also be used as counterselectable markers. For example, negative selection for TK activity is achieved using toxic thymidine analogs (e.g. *5-bromodeoxyuridine, ganciclovir*), which are incorporated into DNA only if there is TK activity in the cell.

Marker	Product	Selection	References
Ada	Adenosine deaminase	Xyl-A (9-β-D-xylofuranosyl adenosine) and 2′-deoxycoformycin	Kaufman *et al.* 1986
Aprt	Adenine phosphoribosyltransferase	Adenine plus *azaserine*, to block *de novo* dATP synthesis	Lester *et al.* 1980
Cad	Multifunctional enzyme	PALA (*N*-phosphonacetyl-L-aspartate) inhibits the aspartate transcarbamylase activity of CAD*	De Saint-Vincent *et al.* 1981
Hprt	Hypoxanthine-guanine phosphoribosyltransferase	Hypoxanthine and *aminopterin*, to block *de novo* IMP synthesis. Selected on HAT medium	Lester *et al.* 1980
Tk	Thymidine kinase	Thymidine and *aminopterin* to block *de novo* dTTP synthesis. Selected on HAT medium	Wigler *et al.* 1977

*CAD: carbamyl phosphate synthetase/aspartate transcarbamylase/dihydroorotase.

hypoxanthine and thymidine. *Hprt+* and *Tk+* transformants can therefore both be selected using HAT medium, which contains hypoxanthine, aminopterin, and thymidine.

In these early experiments, the transgene of interest conferred a selectable phenotype on the cell. However, most genes do not generate a conveniently selectable phenotype, and the isolation of transformants in such experiments was initially problematic. A breakthrough was made when it was discovered that transfection with two physically unlinked DNAs resulted in *co-transformation*, i.e. the integration of both transgenes into the genome (Wigler *et al.* 1979). To obtain co-transformants, cultured cells were transfected with the HSV *Tk* gene and a vast excess of well-defined DNA, such as the plasmid pBR322. Cells selected on HAT medium were then tested by Southern blot hybridization for the presence of the non-selected plasmid DNA. Wigler and colleagues found evidence for the presence of non-selected DNA in nearly 90% of the TK$^+$ cells, indicating that the HSV *Tk* gene could be used as a *selectable marker*. In subsequent experiments, it was shown that the initially unlinked donor DNA fragments were incorporated into large concatemeric structures up to 2 Mbp in length prior to integration (Perucho *et al.* 1980).

The phenomenon of co-transformation allows the stable introduction of any foreign DNA sequence into mammalian cells as long as a selectable marker is introduced at the same time. The HSV *Tk* gene is representative of a class of genes known as *endogenous markers*, because they confer a property that is already present in wild-type cells. A number of such markers have been used, all of which act in redundant metabolic pathways (Table 12.2).

There is no competing activity for dominant selectable markers

The major disadvantage of endogenous markers is that they can only be used with mutant cell lines in which the corresponding host gene is non-functional. This restricts the range of cells that can be transfected. Endogenous markers have therefore been largely superseded by so-called *dominant selectable markers*, which confer a phenotype that is entirely novel to the cell and can hence be used in any cell type. Such markers are usually drug-resistance genes of bacterial origin, and transformed cells are selected on a medium that contains the drug at an appropriate concentration. For example, *E. coli* transposons Tn5 and Tn601 contain distinct genes encoding neomycin phosphotransferase, whose expression confers resistance to aminoglycoside antibiotics (kanamycin, neomycin, and G418). These are protein synthesis inhibitors, active against bacterial and eukaryotic cells, and can therefore be used for selection in either bacteria or animals. By attaching the selectable marker to the SV40 early promoter (Berg 1981) or the HSV *Tk* promoter (Colbère-Garapin *et al.* 1981), which function

Table 12.3 Commonly used dominant selectable marker genes in animals (see Box 14.2 for selectable markers used in plants).

Marker	Product (and source)	Principles of selection	References
as	Asparagine synthase (*Escherichia coli*)	Toxic glutamine analog *albizziin*	Andrulis & Siminovitch 1981
ble	Glycopeptide-binding protein (*Streptoalloteichus hindustantus*)	Confers resistance to glycopeptide antibiotics *bleomycin, pheomycin, Zeocin*™	Genilloud *et al.* 1984
bsd	Blasticidin deaminase (*Aspergillus terreus*)	Confers resistance to *basticidin S*	Izumi *et al.* 1991
gpt	Guanine-xanthine phosphoribosyltransferase (*E. coli*)	Analogous to *Hprt* in mammals, but possesses additional xanthine phosphoribosyltransferase activity, allowing survival in medium containing *aminopterin* and *mycophenolic acid* (Fig. 12.1)	Mulligan & Berg 1981b
hisD	Histidinol dehydrogenase (*Salmonella typhimurium*)	Confers resistance to *histidinol*	Mantei *et al.* 1979
hpt	Hygromycin phosphotransferase (*E. coli*)	Confers resistance to *hygromycin-B*	Blochlinger & Diggelmann 1984
neo (nptII)	Neomycin phosphotransferase (*E. coli*)	Confers resistance to *aminoglycoside antibiotics* (e.g. neomycin, kanamycin, G418)	Colbère-Garapin *et al.* 1981
pac	Puromycin N-acetyltransferase (*Streptomyces alboniger*)	Confers resistance to *puromycin*	Vara *et al.* 1986
trpB	Tryptophan synthesis (*E. coli*)	Confers resistance to *indole*	Hartman & Mulligan 1988

in many cell types, neomycin phosphotransferase was shown to confer antibiotic resistance in a variety of non-mutant cell lines. This marker continues to be used in many contemporary expression vectors. The power of aminoglycoside antibiotic resistance as a selective system in eukaryotes is now very evident. It also has applications in yeast (Chapter 11) and plants (Chapter 14). Other commonly used dominant selectable markers are listed in Table 12.3.

Some marker genes facilitate stepwise transgene amplification

If animal cells are exposed to toxic concentrations of certain drugs, rare individual cells can survive because they have spontaneously undergone a mutation that confers resistance to that drug. The first such compound to be investigated in this manner was the folic acid analog methotrexate, which is a competitive inhibitor of the enzyme dihydrofolate

reductase (DHFR). The analysis of surviving cells showed that some had undergone point mutations in the *Dhfr* locus, producing an enzyme with resistance to the inhibitor. Others had undergone mutations in other loci, for example preventing the uptake of the drug. The most interesting group comprised those cells that had survived by amplifying the *Dhfr* locus, therefore providing enough enzyme to out-compete the inhibitor (Schimke *et al.* 1978). This type of amplification mutation is useful because the drug dose can be progressively increased, resulting in the stepwise selection of cells with massively amplified *Dhfr* gene arrays. Such cells can survive in media containing methotrexate concentrations up to 10,000 times higher than the nominal dose lethal to wild-type cells. The amplified loci are often maintained within the chromosome as extended homogeneously staining regions, or alternatively as small extra chromosomes called double minutes.

Cells with high copy numbers of the *Dhfr* locus

are not generated in response to methotrexate exposure, but arise randomly in the population and are selected on the basis of their resistance to the drug. The random nature of the amplification is confirmed by the fact that, as well as the *Dhfr* gene itself, extensive regions of flanking DNA are also amplified, even though they confer no advantage on the cell. This phenomenon can be exploited to co-amplify transgenes introduced along with a *Dhfr* marker gene, resulting in high-level expression. Wigler *et al.* (1980) demonstrated this principle by transfecting methotrexate-sensitive mouse cells with genomic DNA from the methotrexate-resistant cell line A29, which contains multiple copies of an altered *Dhfr* gene. They linearized the A29 genomic DNA with the restriction enzyme *Sal*I, and ligated it to *Sal*I-linearized pBR322 DNA prior to transfection. Following stepwise drug selection of the transformed cells, Southern blot hybridization showed that the amount of pBR322 DNA had increased more than 50-fold.

Methotrexate selection has been used for the large-scale expression of many recombinant proteins, including tissue plasminogen activator (Kaufman *et al.* 1985), hepatitis B surface antigen (Patzer *et al.* 1986), and clotting factor VIII (Kaufman *et al.* 1988). CHO cells are preferred as hosts for this expression system because of the availability of a number of *dhfr* mutants (Urlaub *et al.* 1983). In non-mutant cell lines, non-transformed cells can survive selection by amplifying the endogenous *Dhfr* genes, generating a background of "false positives". Alternative strategies have been developed that allow DHFR selection to be used in wild-type cells. Expressing the *Dhfr* marker using a strong constitutive promoter (Murray *et al.* 1983) or using a methotrexate-resistant allele of the mouse gene (Simonsen & Levinson 1983) allows selection at methotrexate concentrations much higher than wild-type cells can tolerate. The *E. coli dhfr* gene is also naturally resistant to methotrexate, although for this reason it cannot be used for amplifiable selection (O'Hare *et al.* 1981). Another useful strategy is to employ a second marker gene, such as *neo*, allowing non-transformed cells to be eliminated using G418 (Kim & Wold 1985). Although *dhfr* is the most widely used amplifiable marker, many others have been evaluated, as shown in Table 12.4.

Table 12.4 Common markers used for *in situ* gene amplification. Many amplifiable markers can also be used as endogenous or dominant selectable markers, but, in some cases, the drug used for amplification may not be the same as that used for standard selection.

Marker	Product	Amplifying selective drug	References
Ada	Adenosine deaminase	Deoxycoformycin	Kaufman *et al.* 1986
as	Asparagine synthase	β-Aspartylhydroxamate	Cartier *et al.* 1987
Cad	Aspartate transcarbamylase	*N*-Phosphonacetyl-L-aspartate	Wahl *et al.* 1984
Dhfr	Dihydrofolate reductase	Methotrexate	Kaufman *et al.* 1985
gpt	Xanthine-guanine phosphoribosyltransferase	Mycophenolic acid	Chapman *et al.* 1983
GS	Glutamine synthase	Methionine sulphoxamine	Cockett *et al.* 1990
Hprt	Hypoxanthine-guanine phosphoribosyltransferase	Aminopterin	Kanalas & Suttle 1984
Impdh	Inosine monophosphate dehydrogenase	Mycophenolic acid	Collart & Huberman 1987
Mt-1	Metallothionein 1	Cd^{2+}	Beach & Palmiter 1981
M[res]	Multidrug resistance: P-glycoprotein 170 gene	Adriamycin, colchicine, others	Kane *et al.* 1988
Odc	Ornithine decarboxylase	Difluoromethylornithine	Chiang & McConlogue 1988
Tk	Thymidine kinase	Aminopterin	Roberts & Axel 1982
Umps	Uridine monophosphate synthases	Pyrazofurin	Kanalas & Suttle 1984

Plasmid vectors for the transfection of animal cells contain modules from bacterial and animal genes

Stable transformation by integration can be achieved using any source of DNA. The early gene-transfer experiments discussed above were carried out using complex DNA mixtures, e.g. genomic DNA, bacterial plasmids, and phage. Calcium phosphate transfection was used in most of these experiments, and the specific donor DNA was often bulked up with a non-specific carrier such as cleaved salmon sperm DNA. However, it is generally more beneficial to use a purified source of the donor transgene. This principle was originally demonstrated by Wigler *et al.* (1977) who transfected cultured mouse cells with a homogeneous preparation of the HSV *Tk* gene. Later, this gene was cloned in *E. coli* plasmids to provide a more convenient source. The use of plasmid vectors for transfection provides numerous other advantages, depending on the modular elements included on the plasmid backbone. (i) The convenience of bacterial plasmid vectors can be extended to animal cells, in terms of the ease of subcloning, *in vitro* manipulation, and purification of recombinant proteins (Chapter 5). (ii) More importantly, modular elements can be included to drive transgene expression, and these can be used with any transgene of interest. The pSV and pRSV plasmids are examples of early expression vectors for use in animal cells, containing transcriptional control sequences from SV40 and Rous sarcoma virus which are functional in a wide range of cell types. The incorporation of these sequences into pBR322 generated convenient expression vectors in which any transgene could be controlled by these promoters when integrated into the genome of a transfected cell (Fig. 12.4). (iii) The inclusion of a selectable marker gene obviates the need for co-transformation, since the transgene

and marker remain linked when they co-integrate into the recipient cell's genome. A range of pSV and pRSV vectors was developed containing alternative selectable marker genes, e.g. pSV2-neo (Southern & Berg 1982), pSV2-gpt (Mulligan & Berg 1980), and pSV2-dhfr (Subramani *et al.* 1981). (iv) As discussed below, some plasmid vectors for gene transfer to animal cells are designed to be shuttle vectors, i.e. they contain origins of replication functional in animal cells allowing the vector to be maintained as an episomal replicon.

Non-replicating plasmid vectors persist for a short time in an extrachromosomal state

One application in which the use of plasmid vectors is critical is transient transformation. Here, the goal is to exploit the short-term persistence of extrachromosomal DNA. Such experiments have a variety of uses, including transient assays of gene expression and the recovery of moderate amounts of recombinant protein. Generally, transient transformation is used as a test system, e.g. to assay regulatory elements using reporter genes (Box 12.1), to check the correct function of an expression construct before going to the expense of generating stable cell lines, or to recover moderate amounts of recombinant protein for verification purposes. Transient transformation is particularly useful for testing large numbers of alternative constructs in parallel. No regime of selection is required because stable cell lines are not recovered – the cells are generally transfected, assayed after one or two days, and then discarded. The simplest way to achieve the transient transformation of animal cells is to use a plasmid vector lacking an origin of replication functional in the host. Although the vector cannot replicate, gene expression from a mammalian transcription unit is possible for as long as the plasmid remains stable, which

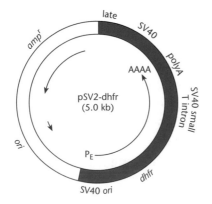
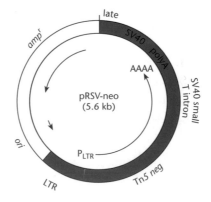

Fig. 12.4 The mammalian plasmid expression vectors pSV2-dhfr and pRSV-neo.

Box 12.1 Reporter genes and promoter analysis

Reporter genes are also known as screenable marker genes. These differ from selectable marker genes in that they do not confer a property that allows transformed cells to survive under selective conditions. Instead, they encode a product that can be detected using a simple and inexpensive assay.

When controlled by a strong constitutive promoter, reporter genes are often used as markers to confirm transient or stable transformation, since only cells containing the reporter-gene construct can express the corresponding protein. Importantly, the assays used to detect reporter-gene activity are *quantitative*, so they can also be used to *measure* transformation efficiency. If attached to a cloned promoter, reporter genes can therefore be used to determine transcriptional activity in different cell types and under different conditions. Transient reporter assays have been widely used to characterize and dissect the regulatory elements driving eukaryotic genes, as shown in the example below. The use of reporters is advantageous, because it circumvents the necessity to derive different assays for individual genes and also allows the activities of transgenes and homologous endogenous genes to be distinguished in the same cell.

An example of *in vitro* promoter analysis using chloramphenicol acetyltransferase

The first reporter gene to be used in animal cells was *cat*, derived from *E. coli* transposon Tn*9* (Gorman *et al.* 1982b); it has also been used to a certain extent in plants (Herrera-Estrella *et al.* 1983a,b). This gene encodes the enzyme chloramphenicol acetyltransferase (CAT), which confers resistance to the antibiotic chloramphenicol by transferring acetyl groups on to the chloramphenicol molecule from acetyl-CoA. If ^{14}C-labeled chloramphenicol is used as the substrate, CAT activity produces a mixture of labeled monoacetylated and diacetylated forms,

which can be separated by thin-layer chromatography and detected by autoradiography. The higher the CAT activity, the more acetylated forms of chloramphenicol are produced. These can be quantified in a scintillation counter or using a phosphorimager. Gorman *et al.* (1982a,b) placed the *cat* gene downstream of the SV40 and Rous sarcoma virus (RSV) promoters in the expression vectors pSV2 and pRSV2, to create the pSV-CAT and pRSV-CAT constructs, respectively. These vectors, and derivatives thereof, have been widely used to analyze transient-transfection efficiency, because the promoters are active in many animal cells and CAT activity can be assayed rapidly in cell homogenates.

The *cat* gene has also been used to test regulatory elements by attaching it to a "minimal promoter", typically a simple TATA box. This basic construct generates only low-level background transcription in transiently transfected cells. The activity of other regulatory elements, such as promoters and enhancers, and *response elements*, which activate transcription in response to external signals, can be tested by subcloning them upstream of the minimal promoter and testing their activity in appropriate cell types. In an early example of this type of experiment, Walker *et al.* (1986) attached the promoter and 5′-flanking sequences of human and rat insulin and rat chymotrypsin genes, which are expressed at a high level only in the pancreas, to the *cat* gene. Each gene is expressed in clearly distinct cell types: insulin is synthesized in endocrine β cells and chymotrypsin in exocrine cells. Plasmid DNA was introduced into either pancreatic endocrine or pancreatic exocrine cell lines in culture and, after a subsequent 44 h incubation, cell extracts were assayed for CAT activity. It was found that the constructs retained their preferential expression in the appropriate cell type. The insulin 5′-flanking DNA conferred a high level of CAT expression in the endocrine but not the exocrine cell line, with the

continued

Box 12.1   *continued*

converse being the case for the chymotrypsin 5'-flanking DNA. The analysis was extended by creating deletions in the 5'-flanking sequences and testing their effects on expression. From such experiments it could be concluded that sequences located 150–300 bp upstream of the transcription start are essential for appropriate cell-specific transcription.

Other reporter genes

The *cat* reporter gene has been widely used for *in vitro* assays but has a generally low sensitivity and is dependent on a rather cumbersome isotopic assay format. An alternative reporter gene, *SeAP* (secreted alkaline phosphatase), has been useful in many cases because various sensitive colorimetric,

fluorometric, and chemiluminescent assay formats are available. Also, since the reporter protein is secreted, it can be assayed in the growth medium, so there is no need to kill the cells (Berger *et al.* 1988, Cullen & Malim 1992). The bacterial genes *lacZ* and *gusA* have been used as reporters for *in vitro* assays using colorimetric and fluorometric substrates. An important advantage of these markers is that they can also be used for *in situ* assays, since histological assay formats are available. More recently, bioluminescent markers, such as luciferase and green fluorescent protein, have become popular because these can be assayed in live cells and whole animals and plants. The *lacZ*, *gusA*, luciferase, and green fluorescent protein reporter genes are discussed in more detail in Box 15.1.

---

depends on the host cell's propensity to break down extrachromosomal DNA. Linear DNA is degraded very quickly in mammalian cells, so high-quality supercoiled plasmid vectors are used. Even covalently closed circular DNA tends to remain stable for only one or two days in most animal cells, but this is sufficient for the various transient expression assays. Some cell types, however, are renowned for their ability to maintain exogenous DNA for longer periods. In the human embryonic kidney cell line 293, for example, supercoiled plasmid DNA can remain stable for up to 80 hours (Gorman *et al.* 1990).

Runaway polyomavirus replicons facilitate the accumulation of large amounts of protein in a short time

Transient transformation can also be achieved using replicon vectors that contain origins of replication derived from certain viruses of the polyomavirus family, such as simian virus 40 (SV40) and the murine polyomavirus. These viruses cause lytic infections, i.e. the viral genome replicates to a very high copy number, resulting in cell lysis and the release of thousands of progeny virions. During the infection cycle, viral gene products accumulate at high levels, so there has been considerable interest in exploiting this strategy to produce recombinant proteins.

SV40 was the first animal virus to be characterized in detail at the molecular level, and for this reason it was also the first to be developed as a vector. The productive host range of the virus is limited to certain simian cells. The development of SV40 replicon vectors is discussed in some detail below, but first it is necessary to understand a little of the molecular biology of the virus itself.

SV40 has a small icosahedral capsid and a circular double-stranded DNA genome of approximately 5 kb. The genome has two transcription units, known as the early and late regions, which face in opposite directions. Both transcripts produce multiple products by alternative splicing (Fig. 12.5). The early region produces regulatory proteins, while the late region produces components of the viral capsid. Transcription is controlled by a complex regulatory element located between the early and late regions, and this includes early and late promoters, an enhancer, and the origin of replication. During the first stage of the SV40 infection cycle, the early transcript produces two proteins, known as the large T and small t tumor antigens. The function of the T-antigen is particularly important as this protein binds to the viral origin of replication and is absolutely required for genome replication. All vectors based on SV40 must therefore be supplied with functional T-antigen, or they cannot replicate, even in permissive

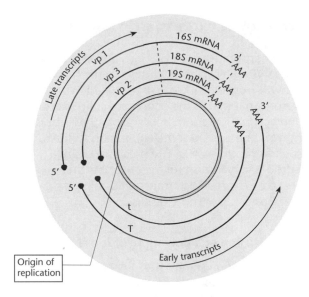

Fig. 12.5 Transcripts and transcript processing of SV40. Intron sequences which are spliced out of the transcripts are shown by purple lines.

cells. The T-antigen also acts as an oncoprotein, interacting with the host's cell cycle machinery and causing uncontrolled cell proliferation.

The first SV40 vectors were viral vectors, and were used to introduce foreign genes into animal cells by transduction. The small size of the viral genome made *in vitro* manipulation straightforward. Either the late region (Goff & Berg 1976) or the early region (Gething & Sambrook 1981) could be replaced with foreign DNA. However, since both these regions are essential for the infection cycle, their functions had to be provided in *trans* initially by a co-introduced helper virus. The use of early replacement vectors was considerably simplified by the development of the COS cell line, a derivative of the African green monkey cell line CV-1 containing an integrated partial copy of the SV40 genome. The integrated fragment included the entire T-antigen coding sequence and provided this protein in *trans* to any SV40 recombinant in which the early region had been replaced with foreign DNA (Gluzman 1981). For example, using this system, Gething & Sambrook (1981) made recombinant viruses that expressed influenza virus hemagglutinin in COS cells.

The major problem with these initial SV40 vectors was that the capacity of the viral capsid allowed a maximum of only about 2.5 kb of foreign DNA to be incorporated. The discovery that plasmids carrying the SV40 origin of replication behaved in the same manner as the virus itself, i.e. replicating to a high copy number in permissive monkey cells, was a

significant breakthrough. Since these SV40 replicons were not packaged into viral capsids, there were no size constraints on the foreign DNA. Many laboratories developed plasmid vectors on this principle (Myers & Tjian 1980, Lusky & Botchan 1981, Mellon *et al.* 1981). In general, these vectors consisted of a small SV40 DNA fragment (containing the origin of replication) cloned in an *E. coli* plasmid vector. Some vectors also contained a T-antigen coding region and could be used in any permissive cell line, while others contained the origin alone and could only replicate in COS cells. Permanent cell lines are not established when SV40 replicons are transfected into COS cells because the massive vector replication eventually causes cell death. However, even though only a low proportion of cells are transfected, the high copy number (10^5 genomes per cell) is compensatory, allowing the transient expression of cloned genes and the harvesting of large amounts of recombinant protein. As an example, one of the pcDNA3.1 series of vectors marketed by Invitrogen Inc. is shown in Fig. 12.6. This version of the vector contains the SV40 origin for high-level expression in COS cells, a *neo* selection cassette, ColE1 and f1 origins for manipulation in bacteria, and an expression cassette driven by the human cytomegalovirus promoter incorporating two epitope tags to facilitate protein purification and an EK site to remove the epitope tags after purification. These components are described in more detail in Box 12.2.

BK and BPV replicons facilitate episomal replication, but the plasmids tend to be structurally unstable

As an alternative to integration, stable transformation can be achieved using a recombinant vector that is maintained as an episomal replicon. While viruses such as SV40 use a lytic infection strategy, others such as human BK virus and Epstein–Barr virus cause latent infections, where the viral genome is maintained as a low to moderate copy-number replicon that does not interfere with host-cell growth. Plasmids that contain such latent origins behave in a similar manner to the parental virus, except they are not packaged in a viral capsid. Such vectors are advantageous because the DNA does not need to integrate in order to be stably maintained, thus stable transformation occurs with an efficiency equal to that normally achieved with transient transformation. Furthermore, while the expression of integrated transgenes is often affected by the surrounding DNA,

Box 12.2 Construct design for high-level transgene expression in animal cells

Many of the expression systems discussed in this chapter are used in experiments where the production of recombinant protein is the ultimate aim. In such cases, it is appropriate for transgene expression to be maximized. Although the different expression systems vary in their total potential yield, in terms of vector design and experimental methodology, the following considerations should apply when high-level expression is required.

The use of a strong and constitutive promoter

Very active promoters provide the highest levels of transgene expression. In viral vectors, transgenes are often expressed under the control of the strongest endogenous promoters, e.g. the baculovirus polyhedrin promoter, the adenoviral E1 promoter and the vaccinia virus p7.5 promoter. Certain viruses contain strong promoters and enhancers that function in a wide range of cell types, and several of these have been subverted for use in plasmid vectors. The elements most commonly used in mammalian cells are the SV40 early promoter and enhancer (Mulligan & Berg 1981b), the Rous sarcoma virus long-terminal-repeat promoter and enhancer (Gorman *et al.* 1982a), and the human cytomegalovirus immediate early promoter (Boshart *et al.* 1985). Although these function widely, they are not necessarily active in all mammalian cells, e.g. the SV40 promoter functions poorly in the human embryonic kidney line 293 (Gorman *et al.* 1989).

The inclusion of an intron

The presence of an intron in a eukaryotic expression unit usually enhances expression. Evidence for the positive effect of introns accumulated in the early years of cDNA expression in animal cells, although there are also many studies in which efficient gene expression was obtained in the absence of an intron (see Kaufman 1990a,b). Nevertheless, most mammalian expression vectors in current use incorporate a heterologous intron, such as the SV40 small t-antigen intron or the human growth-hormone intron, or modified hybrid introns that match the consensus splice donor and acceptor-site sequences. Note that introns may not be used in some expression systems, such as vaccinia virus. The presence of an intron is very important in constructs that are to be expressed in transgenic animals (see Box 13.2 for more discussion).

The inclusion of a polyadenylation signal

Polyadenylation signals (terminators) are required in eukaryotic genes to generate a defined 3′ end to the mRNA. In most cases, this defined end is extended by the addition of several hundred adenosine residues to generate a poly(A) (polyadenylate) tail. This tail is required for the export of mRNA into the cytoplasm, and also increases its stability. In the absence of such a site, the level of recombinant protein produced in transformed cells can fall by as much as 90% (Kaufman 1990a,b). Poly(A) sites from the SV40 early transcription unit or mouse β-globin gene are often incorporated into mammalian expression vectors.

The removal of unnecessary untranslated sequence

Eukaryotic mRNAs comprise a coding region (which actually encodes the gene product) bracketed by untranslated regions (UTRs) of variable length. Both the 5′ and 3′ UTRs can influence gene expression in a number of ways (Kozak 1999). For example, the 5′ UTR may contain one or more AUG codons upstream of the authentic translational start site, and these are often detrimental to translational initiation. The 3′ UTR may contain regulatory elements that control mRNA stability (e.g. AU-rich sequences that reduce stability have been identified (see Shaw & Kamen 1986)). Furthermore, both the 5′ and 3′ UTRs may be rich in secondary structure, which prevents efficient translation. In animal systems, UTR sequences are generally

continued

Box 12.2 *continued*

removed from transgene constructs to maximize expression.

Optimization of the transgene for translational efficiency

The sequence around the translational initiation site should conform to Kozak's consensus, which is defined as 5′-CCRCCAUGG-3′ (Kozak 1986, 1999). Of greatest importance is the purine at the −3 position (identified as R) and the guanidine at position +4. The adenosine of the AUG initiation codon (underlined) is defined as position +1, and the immediately preceding base is defined as position −1. The expression of foreign genes in animals can also be inefficient in some cases due to suboptimal codon choice, which reflects the fact that different organisms prefer to use different codons to specify the same amino acid. If a transgene contains a codon that is commonly used in the source organism but rarely used in the host, translation may pause at that codon due to the scarcity of the corresponding transfer RNA (tRNA). This will reduce the rate of protein synthesis and may even lead to truncation of the protein or frame-shifting. It may therefore be beneficial to "codon-optimize" transgenes for the expression host.

The incorporation of a targeting signal

If the goal of an expression study is to recover large amounts of a functional *eukaryotic* protein, it is necessary to consider whether that protein needs to be post-translationally modified in order to function correctly. For example, many proteins intended for therapeutic use require authentic glycosylation patterns not only for correct function, but also to prevent an immune response in the patient. Since specific types of modification occur in particular cell compartments, it is necessary to consider strategies for targeting the recombinant protein to the correct compartment to ensure that it is appropriately modified. Proteins that need to be glycosylated, for example, must be targeted to the secretory pathway using a *signal peptide*. Many mammalian expression vectors are available for this purpose, and they incorporate heterologous signal peptides. The figure below shows the Invitrogen vector pSecTag2, which incorporates a sequence encoding the murine immunoglobulin light-chain signal peptide for high-efficiency targeting to the secretory pathway. Note also that the C terminus of the recombinant protein is expressed as a fusion to two different epitope tags to facilitate protein purification (see p. 87).

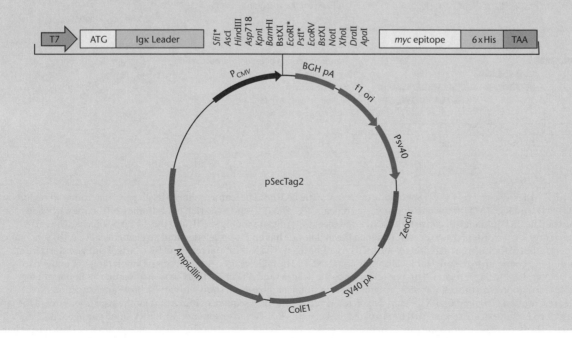

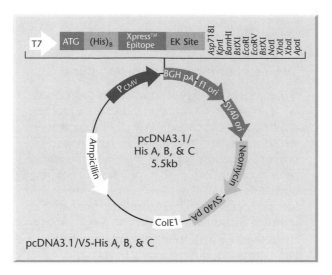

Fig. 12.6 The transient expression vector pcDNA3.1 HisAB&C, which allows donor DNA fragments to be cloned in three reading frames and the recombinant protein to be expressed with integral epitope tags.

a problem known as the position effect (see Box 13.2), these episomal vectors are not subject to such influences.

The human BK polyomavirus infects many cell types and is maintained with a copy number of about 500 genomes per cell. Plasmid vectors containing the BK origin replicate in the same manner as the virus when the BK T-antigen is provided in *trans*.

However, by incorporating a selectable marker such as *neo* into the vector, it is possible to increase the concentration of antibiotics in the medium progressively and select for cells with a higher vector copy number (Fig. 12.7). Cell lines have been propagated for over a year under such conditions (de Benedetti & Rhoads 1991) and stable lines with up to 9000 copies of the genome have been produced. Any transgene

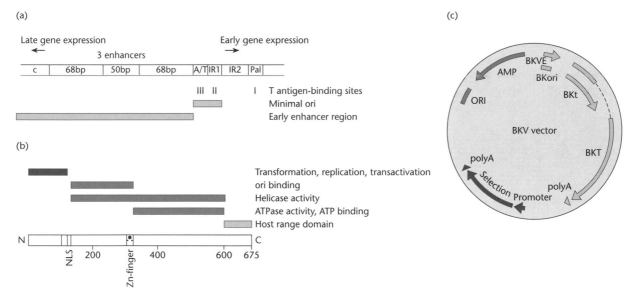

Fig. 12.7 BKV-derived vector. (a) Schematic overview of the BKV *ori*. The features shared by all polyomavirinae *ori* regions are an AT-rich sequence (A/T) at the late side, inverted repeats (IR1 and IR2) and a GC-rich palindrome (Pal) on the early side of the ori. The BKV minimal ori consists of an inverted repeat, T antigen-binding site II, and the 20-bp A/T block. The early enhancer region consists of three repeats, of which the middle one has an 18-bp deletion and an element called c. (b) Schematic representation of the different activities of the BKV large T antigen. The N-terminal region of BKV large T antigen starts from the initiation codon up to the nuclear localization signal (NLS). The middle part contains regions for binding of T antigen to BKV DNA and a zinc finger region. This region localizes the helicase and ATPase activity as well as sequences important for complex formation with the cellular p53 protein. The C-terminal part has been shown to serve a host range function. (c) Representation of a common BKV-derived expression vector. The BKV sequences important for the typical features of the virus, as described in the text, are represented by purple edged bars. BKVE, BKV early promoter; BKt, BKV small t antigen; BKT, BKV large T antigen. Dark purple bars indicate the sequences necessary for selection after transfection in eukaryotic cells. Dark purple bars represent sequences essential for amplification and selection in bacteria.

incorporated into the vector is expressed at very high levels, leading to the recovery of large amounts of recombinant protein (Sabbioni *et al.* 1995). SV40 itself can also be used as a stable episomal vector if the availability of T-antigen is rationed to prevent runaway replication. This has been achieved using a conditional promoter, and by mutating the coding region to render the protein temperature sensitive (Gerard & Gluzman 1985, Rio *et al.* 1985).

The first virus to be developed as an episomal replicon was bovine papillomavirus (BPV). The papillomaviruses are distantly related to the polyomaviruses, and cause papillomas (warts) in a range of mammals. BPV has been exploited as a stable expression vector because it can infect mouse cells without yielding progeny virions. Instead, the viral genome is maintained as an episomal replicon, with a copy number of about 100. The molecular biology of BPV is considerably more complex that that of SV40, but the early part of the infection cycle is similar, involving the production of a T-antigen that:

(a) is required for viral replication and (b) causes oncogenic transformation of the host cell. The early functions of BPV are carried on a 5.5-kb section of the genome, which is called the 69% transforming fragment (BPV$_{69T}$). This was cloned in the *E. coli* plasmid pBR322, and was shown to be sufficient to establish and maintain episomal replication, as well as induce cell proliferation (Sarver *et al.* 1981a). Initially, the ability of the virus to cause uncontrolled cell proliferation was used to identify transformed cells, but this limited the range of cell types that could be used. The incorporation of selectable markers such as *neo* allows transformants to be selected for resistance to G418, permitting the use of a wider range of cell types (Law *et al.* 1983). BPV$_{69T}$ replicons have been used to express numerous proteins, including rat preproinsulin (Sarver *et al.* 1981b) and human β-globin (Di Maio *et al.* 1982). Generally, the plasmids are maintained episomally, but the copy number varies from 10 to over 200 vector molecules per cell (Fig. 12.8). Some investigators have reported

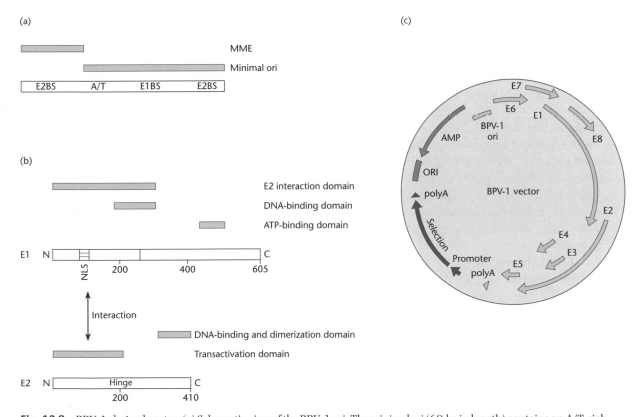

Fig. 12.8 BPV-1-derived vector. (a) Schematic view of the BPV-1 ori. The minimal ori (60 bp in length) contains an A/T-rich region, the E1-binding site (BS) including an 18-bp DS element and an E2-binding site of 12 bp. The minichromosome maintenance element (MME) is composed of multiple binding sites (12) for the transcriptional activator E2. (b) Functional domains of E1 and E2. E1 has several enzymatic functions such as an ATPase activity, a DNA-binding domain and an E2 interaction domain. E2 consists of a transactivation domain, which interacts with E1, and a DNA-binding domain. Both domains are linked with a hinge region. (c) Schematic representation of a BPV-1 expression vector (69%). BPV-1 sequences consisting of the BPV-1 ori and the eight E genes are represented by dark purple bars. Dark purple bars indicate sequences for eukaryotic selection, whereas purple edged bars indicate bacterial replication and selection sequences.

the long-term maintenance of such episomal transformants (Fukunaga *et al.* 1984) while others found a tendency for the construct to integrate into the genome (Ostrowski *et al.* 1983, Sambrook *et al.* 1985). Recombination within the vector or between vectors is also a fairly common observation resulting in unpredictable spontaneous deletions and plasmid oligomerization. Recombination events and copy number appear to be affected by multiple factors, including the host-cell type, the incorporated transgene, and the structure of the vector itself (see Mecsas & Sugden 1987).

Replicons based on Epstein–Barr virus facilitate long-term transgene stability

Unlike BPV replicons, vectors based on Epstein–Barr virus (EBV) replicate very stably in mammalian cells. EBV is a herpesvirus (also see discussion of herpes simplex virus later in the chapter), with a large double-stranded DNA genome (approximately 170 kb), which predominantly infects primate and canine cells. It is also naturally lymphotrophic, infecting B-cells in humans and causing infectious mononucleosis. In cultured lymphocytes, the virus becomes established as an episomal replicon with about 1000 copies per cell (Miller 1985). Although the virus itself only infects lymphocytes, the genome is maintained in a wide range of primate cells if introduced by transfection. Only two relatively small regions of the genome are required for episomal maintenance – the latent origin (*oriP*) and a gene encoding a *trans*-acting regulator called Epstein Barr nuclear antigen 1 (EBNA1) (Yates *et al.* 1984, Reisman *et al.* 1985). These sequences have formed the basis of a series of latent EBV-based plasmid expression vectors, which are maintained at a copy number of 2–50 copies per cell (Fig. 12.9). The first EBV vectors comprised the *oriP* element cloned in a bacterial plasmid, and could replicate only if EBNA1 was supplied in trans. Yates *et al.* (1984) described the construction of a shuttle vector, pBamC, comprising *oriP*, a bacterial origin, and ampicillin resistance gene derived from pBR322, and the neomycin phosphotransferase marker for selection in animal cells. A derivative vector, pHEBO, contained a hygromycin resistance marker instead of *neo* (Sugden *et al.* 1985). The presence of a selectable marker in such vectors is important, because if selection pressure is not applied, EBV replicons are lost passively from the cell population at about 5% per cell generation. Yates *et al.* (1985) added the EBNA-1 gene to pHEBO, producing a con-

struct called p201 that was capable of replicating independently. Similar constructs have been developed in other laboratories (Lupton & Levine 1985). EBV replicons have been used to express a wide range of proteins in mammalian cell lines, including the epidermal growth factor receptor (Young *et al.* 1988), the tumor necrosis factor receptor (Heller *et al.* 1990), and an Na+K+ ATPase (Canfield *et al.* 1990). Generally, such studies have resulted in high-level and long-term gene expression (reviewed by Margolskee 1992, Sclimenti & Calos 1998). It has been suggested that the stability of EBV replicons may reflect the fact that replication is limited to once per cell cycle, unlike BPV replicons that replicate continuously throughout S-phase (DuBridge & Calos 1988, Gilbert & Cohen 1987). This, together with the large genome size, allows EBV-derived vectors to carry large DNA fragments, including mammalian cDNAs and genes. EBV replicons have therefore been used for the construction of episomal cDNA libraries for expression cloning (Margolskee *et al.* 1988), and more recently for the preparation of genomic libraries (Sun *et al.* 1994). The *oriP* element has also been incorporated into yeast artificial chromosome vectors carrying large human genomic DNA fragments. These linear vectors were then circularized *in vitro* and introduced into human cells expressing EBNA1, whereupon they were maintained as episomal replicons (Simpson *et al.* 1996). EBV can also replicate lytically in primate B-cells, and this requires a separate origin (*oriLyt*) and a distinct *trans*-acting regulator called ZEBRA. Vectors have been developed containing both origins, and these are maintained at a moderate copy number in cells latently infected with the virus, but can be amplified up to 400-fold if ZEBRA is supplied by transfecting the cells with the corresponding viral gene under the control of a constitutive promoter (Hammerschmidt & Sugden 1988). Interestingly, while the latent vector remains as a circular replicon, vector DNA isolated from induced cells is present as multicopy concatemers, reflecting the rolling circle mechanism of lytic replication.

DNA can be delivered to animal cells using bacterial vectors

The exploitation of living bacteria for gene transfer is central to the genetic manipulation of plants. As discussed in Chapter 14, *Agrobacterium tumefaciens* and its close relatives have been used for over 20 years to

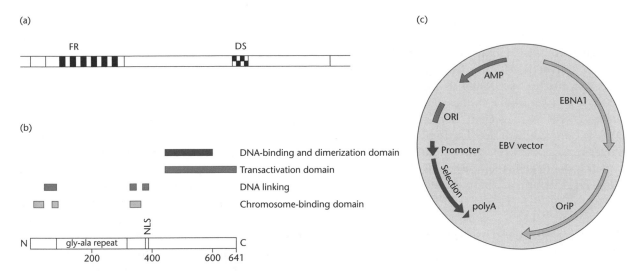

Fig. 12.9 EBV-derived vector. (a) *oriP* consists of two non-contiguous elements, namely the family of repeats (FR) and the dyad symmetry (DS) element, containing 20 and four binding sites for EBNA1, respectively. (b) Representation of EBNA1. The central domain is composed entirely of a repetitive array of glycine and alanine residues. Most of this domain can be deleted without affecting EBNA1 function. The DNA-linking domains seem to contribute to the activation of transcription and replication. The chromosome-binding activity of EBNA1 secures the separation to the daughter cells during mitosis. (c) A basic EBV-derived vector consists of the viral sequences *oriP* and EBNA1 (light purple bars), an expression cassette for eukaryotic selection after transfection (dark purple bars), and sequences necessary for replication and selection in bacteria (mid purple bars).

generate transgenic plants and transformed plant cells, and this remains one of the most popular gene-transfer strategies. More recently, it has been shown that *A. tumefaciens* can transfer DNA to cultured human cells (Kunik *et al.* 2001). As discussed below, however, this example of bactofection is not an isolated observation. Indeed there is growing evidence that highly efficient gene transfer to animals can be achieved using a variety of bacterial species.

The protoplast fusion technique discussed above can be regarded as a highly artificial form of bactofection, but the amount of human intervention required distinguishes the technique from those discussed below, which use living bacteria. *Agrobacterium* species can transfer DNA to plants without any human intervention. The first reports of similarly natural gene transfer between bacterial and animal cells were published in the mid-1990s (e.g. Sizemore *et al.* 1995, reviewed by Higgins & Portnoy 1998). Typically, the bacteria invade the host animal cells and undergo lysis within them, releasing plasmid DNA. In the case of *Salmonella* species, lysis occurs in the phagocytic vesicle, while for other species (e.g. *Listeria monocytogenes* and *Shigella flexneri*) lysis occurs after the bacterium has escaped from the vesicle. The plasmid DNA then finds its way to the nucleus, where it is incorporated into the host cell's genome and expressed. In contrast, *A. tumefaciens* has been shown to transfer DNA to mammalian (and plant) cells without

invading them. In this case, transfer occurs by attachment to the outside of the cell followed by conjugation (the transfer of DNA through a conduit called a pilus, which is assembled by the bacterial cell).

An important principle in the use of live bacteria as *invasive* gene-transfer vehicles is that they must be attenuated. This is because the gene-transfer system exploits the natural ability of the bacteria to infect and subvert the activity of eukaryotic cells. Without attenuation, the bacteria would multiply and destroy the host cells. Attenuation is achieved in several ways. The first is to use auxotrophic mutants, i.e. bacterial strains that are unable to manufacture essential molecules such as amino acids, nucleotides, or components of the cell wall. For example, *aroA* mutants are unable to synthesize aromatic amino acids, and *Salmonella typhimurium* and *Shigella flexneri* strains carrying this mutation have been used for gene transfer (reviewed by Weiss & Chakraborty 2001). Alternatively, the bacteria can be engineered so that they undergo inducible autolysis. There are no auxotrophic strains of *Listeria monocytogenes*, so attenuation has been achieved by induced suicide, i.e. introducing an autolysin-encoding gene that is activated once the bacterium is inside the host cell (Dietrich *et al.* 1998). *In vitro*, lysis can also be induced by treating cells with antibiotics.

Thus far, bacteria-mediated gene transfer has been used not only as a general transfection method for the introduction of DNA into cultured cells, but also as a high-efficiency method for gene transfer *in vivo*. There have been many reports of bacterial gene transfer as a method for the delivery of recombinant DNA vaccines (e.g. see Xiang *et al.* 2000, Woo *et al.* 2001) and comparative studies indicate that bacterial transfer is more efficient than equivalent naked DNA vaccines, which are discussed in Chapter 26 (Zoller & Christ 2001). The potential of bacterial gene transfer in gene therapy has also been explored (Paglia *et al.* 2000).

Viruses are also used as gene-transfer vectors

Virus particles have a natural ability to adsorb to the surface of cells and gain entry, and this can be exploited to deliver recombinant DNA into animal cells. Due to the efficiency with which viruses can deliver their nucleic acid into cells, and the high levels of replication and gene expression it is possible to achieve, viruses have been used as vectors not only for gene expression in cultured cells, but also for gene transfer to living animals. Several classes of viral vector have been developed for use in human gene therapy, and at least eight have been used in clinical trials. Still others have been developed as recombinant vaccines.

Before introducing the individual vector systems, we discuss some general properties of viral transduction vectors. Transgenes may be incorporated into viral vectors either by addition to the whole genome, or by replacing one or more viral genes. This is generally achieved either by ligation (many viruses have been modified to incorporate unique restriction sites) or homologous recombination. If the transgene is added to the genome, or if it replaces one or more genes that are non-essential for the infection cycle in the expression host being used, the vector is described as *helper-independent* because it can propagate independently. However, if the transgene replaces an essential viral gene, this renders the vector *helper-dependent* so that missing functions must be supplied in *trans*. This can be accomplished by co-introducing a helper virus, or transfecting the cells with a helper plasmid, each of which must carry the missing genes. Usually steps are taken to prevent the helper virus completing its own infection cycle, so that only the recombinant vector is packaged. It is also desirable to try and prevent recombination occurring between the helper and the vector, as this can generate wild-type replication-competent viruses as contaminants. An alternative to the co-introduction of helpers is to use a *complementary cell line*, sometimes termed a "packaging line", which is transformed with the appropriate missing genes. For many applications, it is favorable to use vectors from which all viral coding sequences have been deleted. These *amplicons* (also described as fully deleted, gutted, or gutless vectors) contain just the *cis*-acting elements required for packaging and genome replication. The advantage of such vectors is their high capacity for foreign DNA and the fact that, since no viral gene products are made, the vector has no intrinsic cytotoxic effects. The choice of vector depends on the particular properties of the virus and the intended host, whether transient or stable expression is required, and how much DNA needs to be packaged. For example, icosahedral viruses such as adenoviruses and retroviruses package their genomes into preformed capsids, whose volume defines the maximum amount of foreign DNA that can be accommodated. Conversely, rod-shaped viruses such as the baculoviruses form the capsid around the genome, so there are no such size constraints. There is no ideal virus for gene transfer – each has its own advantages and disadvantages – and many researchers are investigating the potential of hybrid vectors which combine useful properties from two or more viruses (Lam & Breakfield 2000).

Adenovirus vectors are useful for short-term transgene expression

Adenoviruses are DNA viruses with a linear, double-stranded genome of approximately 36 kb. The genome of serotype Ad5, from which many adenovirus vectors are derived, is shown in Fig. 12.10. There are six early transcription units, most of which are essential for viral replication, and a major late transcript that encodes components of the capsid. Adenoviruses have been widely used as gene-transfer and expression vectors because they have many advantageous features, including stability, a high capacity for foreign DNA, a wide host range that includes non-dividing cells, and the ability to produce high-titer stocks (up to 10^{11} pfu/ml) (reviewed by Imperiale & Kochanek 2004). They are suitable for transient expression in dividing cells because they do not integrate efficiently into the genome, but prolonged expression can be achieved in post-mitotic cells such as neurons (e.g. see LaSalle *et al.* 1993, Davidson *et al.* 1993). Adenoviruses are particularly attractive as gene-therapy vectors because the virions

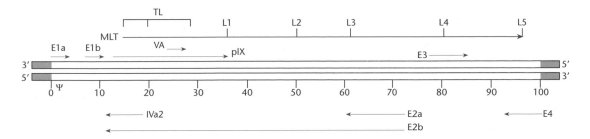

Fig. 12.10   Map of the adenovirus genome, showing the positions of the early transcription units (E), the major late transcript (MLT), the tripartite leader (TL), and other genes (VA, pIX, IVa2). Terminal repeats are shown in purple, ψ is the packaging site.

are taken up efficiently by cells *in vivo* and adenovirus-derived vaccines have been used in humans with no reported side effects. However, the death of a patient following an extreme inflammatory response to adenoviral gene therapy treatment underlines the necessity for rigorous safety testing (Marshall 1999). A number of strategies are being developed to control the activity of the immune system when the vectors are first introduced into the host (reviewed by Benihoud *et al.* 1999, Liu & Muruve 2003).

Most early adenoviral vectors were replication deficient, lacking the essential E1a and E1b genes, and often the non-essential gene E3. These first generation "E1 replacement vectors" had a maximum capacity of about 7 kb and were propagated in the human embryonic kidney line 293. This is transformed with the leftmost 11% of the adenoviral genome, comprising the E1 transcription unit, and hence supplies these functions in *trans* (Graham *et al.* 1977). Although these vectors have been used with great success, they suffer from two particular problems: cytotoxic effects, resulting from low-level expression of the viral gene products, and the tendency for recombination to occur between the vector and the integrated portion of the genome resulting in the recovery of replication-competent viruses. Higher capacity vectors have been developed, which lack the E2 or E4 regions in addition to E1 and E3, providing a maximum cloning capacity of about 10 kb. These must be propagated on complementary cell lines providing multiple functions, and such cell lines have been developed in several laboratories (e.g. Brough *et al.* 1996, Gao *et al.* 1996, Gorziglia *et al.* 1996, Zhou *et al.* 1996). The use of E1/E4 deletions is particularly attractive as the E4 gene is responsible for many of the immunological effects of the virus (Gao *et al.* 1996, Dedieu *et al.* 1997). Unwanted recombination has been addressed through the use of a refined complementary cell line transformed with a specific DNA fragment corresponding

exactly to the E1 genes (Imler *et al.* 1996). An alternative strategy is to insert a large fragment of "stuffer DNA" into the non-essential E3 gene, so that recombination yields a genome too large to be packaged (Bett *et al.* 1993). Gutless adenoviral vectors are favored for *in vivo* gene transfer because they have a large capacity (up to 37 kb) and minimal cytotoxic effects (reviewed by Morsey & Caskey 1999). Therefore, transgene expression persists for longer than can be achieved using first-generation vectors (e.g. see Scheidner *et al.* 1998). Complementary cell lines supplying all adenoviral functions are not available at present, so gutless vectors must be packaged in the presence of a helper virus, which presents a risk of contamination.

Adeno-associated virus vectors integrate into the host-cell genome

Adeno-associated virus (AAV) is not related to adenovirus, but is so-called because it was first discovered as a contaminant in an adenoviral isolate (Atchison *et al.* 1965). AAV is a single-stranded DNA virus, a member of the parvovirus family, and is naturally replication defective such that it requires the presence of another virus (usually adenovirus or herpesvirus) to complete its infection cycle. In adenovirus- or herpesvirus-infected cells, AAV replicates lytically and produces thousands of progeny virions (Buller *et al.* 1981). However, in the absence of these helpers, the AAV DNA integrates into the host cell's genome, where it remains as a latent provirus (Berns *et al.* 1975). In human cells, the provirus integrates predominantly into the same genetic locus on chromosome 19 (Kotin *et al.* 1990). Subsequent infection by adenovirus or herpesvirus can "rescue" the provirus and induce lytic infection.

The dependence of AAV on a heterologous helper virus provides an unusual degree of control over vector replication, making AAV theoretically one

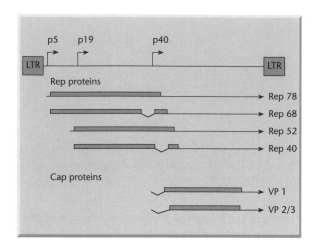

Fig. 12.11 Organization of the adeno-associated virus genome.

of the safest vectors to use for gene therapy. Proviral integration is considered advantageous for increasing the persistence of transgene expression, while at the same time the site specificity of this process theoretically limits the chances of insertional mutagenesis. Other advantages include the wide host range, which encompasses non-dividing cells (reviewed by Muzyczka 1992, Rabinowitz & Samulski 1998).

The AAV genome is small (about 5 kb) and comprises a central region containing *rep* (replicase) and *cap* (capsid) genes flanked by 145-b inverted terminal repeats (Fig. 12.11). In the first AAV vectors, foreign DNA replaced the *cap* region and was expressed from an endogenous AAV promoter (Hermonat & Muzyczka 1984). Heterologous promoters were also used, although in many cases transgene expression was inefficient because the Rep protein inhibited their activity (reviewed by Muzyczka 1992). Rep interference with endogenous promoters is also responsible for many of the cytotoxic effects of the virus. Several groups therefore developed vectors in which both genes were deleted and the transgene was expressed from either an endogenous or heterologous promoter (McLaughlin *et al.* 1988, Samulski *et al.* 1989). From such experiments, it was demonstrated that the repeats are the only elements required for replication, transcription, proviral integration, and rescue. All current AAV vectors are based on this principle (Lehtonen & Tenenbaum 2003). *In vitro* manipulation of AAV is facilitated by cloning the inverted terminal repeats in a plasmid vector and inserting the transgene between them. Traditionally, recombinant viral stocks are produced by transfecting this construct into cells along with a helper plasmid

to supply AAV products, and then infecting the cells with adenovirus to stimulate lytic replication and packaging. This has generally yielded recombinant AAV titers too low to use for human gene therapy, and contaminated with helper AAV and adenovirus. The recent development of transfection-based adenoviral helper plasmids, packaging lines, and the use of affinity chromatography to isolate AAV virions has helped to alleviate such problems (reviewed by Monahan & Samulski 2000, Snyder & Flotte 2002, Flotte 2004).

AAV vectors have been used to introduce genes efficiently into many cell types, including liver (Snyder *et al.* 1997), muscle (Pruchnic *et al.* 2000), and neurons (Davidson *et al.* 2000). However, deletion of the *rep* region abolishes the site specificity of proviral integration, so the vector integrates at essentially random positions which may increase the risk of insertional gene inactivation (Weitzman *et al.* 1994, Yang *et al.* 1997, Young *et al.* 2000). It is also unclear whether the persistence of the vector and prolonged transgene expression is primarily due to vector integration or episomal maintenance of concatemeric dsDNA copies of the genome (see Monahan & Samulski 2000 for discussion). The fact that AAV uses concatemeric replication intermediates has been used to circumvent perhaps the most serious disadvantage of AAV vectors, which is the limited capacity for foreign DNA. This strategy involves cloning a large cDNA as two segments in two separate vectors, which are co-introduced into the same cell. The 5′ portion of the cDNA is cloned in one vector, downstream of a promoter and upstream of a splice donor site. The 3′ portion is cloned in another vector, downstream of a splice acceptor. Concatemerization results in the formation of heterodimers and transcription across the junction yields a mRNA that can be processed to splice out the terminal repeats of the vector. In this way, cDNAs of up to 10 kb can be expressed (Sun *et al.* 2000, Nakai *et al.* 2000).

Baculovirus vectors promote high-level transgene expression in insect cells, but can also infect mammalian cells

Baculoviruses have rod-shaped capsids and large, double-stranded DNA genomes. They productively infect arthropods, particularly insects. One group of baculoviruses, known as the nuclear polyhedrosis viruses, has an unusual infection cycle that involves the production of *nuclear occlusion bodies*. These are proteinaceous particles in which the virions are embedded, allowing the virus to survive harsh environmental conditions such as desiccation (reviewed by Fraser 1992). Baculovirus vectors are used mainly for high-level transient protein expression in insects and insect cells (O'Reilley *et al.* 1992, King & Possee 1992). The occlusion bodies are relevant to vector development because they consist predominantly of a single protein called polyhedrin, which is expressed at very high levels. The nuclear occlusion stage of the infection cycle is non-essential for the productive infection of cell lines, thus the polyhedrin gene can be replaced with foreign DNA, which can be expressed at high levels under the control of the endogenous polyhedrin promoter. Two baculoviruses have been extensively developed as vectors, namely the *Autographa californica* multiple nuclear polyhedrosis virus (AcMNPV) and the *Bombyx mori* nuclear polyhedrosis virus (BmNPV). The former is used for protein expression in insect cell lines, particularly those derived from *Spodoptera frugiperda* (e.g. Sf9, Sf21). The latter infects the silkworm, and has been used for the production of recombinant protein in live silkworm larvae. One limitation of this expression system is that the glycosylation pathway in insects differs from that in mammals, so recombinant mammalian proteins may be incorrectly glycosylated and hence immunogenic (reviewed by Fraser 1992). This has been addressed by using insect cell lines chosen specifically for their ability to carry out mammalian-type post-translational modifications, e.g. those derived from *Estigmene acrea* (Ogonah *et al.* 1996). An innovative approach to this problem is to exploit the indefinite capacity of baculovirus vectors to co-express multiple transgenes, and thus modify the glycosylation process in the host cell line by expressing appropriate glycosylation enzymes along with the transgene of interest. Wagner *et al.* (1996) used this strategy to co-express fowl plague hemagglutinin and β-1,2-N-acetylglucosaminyltransferase, resulting in the synthesis of large amounts of hemagglutinin correctly modified with N-acetylglucosamine residues. Hollister *et al.* (1998) have developed an Sf9 cell line that expresses 1,4-galactosyltransferase under the control of a baculovirus immediate early promoter, such that gene expression is induced by baculovirus infection. These cells were infected with a recombinant baculovirus vector carrying a tissue plasminogen activator transgene, resulting in the production of recombinant protein that was correctly galactosylated.

Polyhedrin gene replacement vectors are the most popular due to the high level of recombinant protein that can be expressed (up to 1 mg per 10^6 cells). The polyhedrin upstream promoter and 5′ untranslated region are important for high-level foreign gene expression and these are included in all polyhedrin replacement vectors (Miller *et al.* 1983, Maeda *et al.* 1985). The highest levels of recombinant protein expression were initially achieved if the transgene was expressed as a fusion, incorporating at least the first 30 amino acids of the polyhedrin protein (Lucklow & Summers 1988). However, this was shown to be due not to the stabilizing effects of the leader sequence on the recombinant protein, but to the presence of regulatory elements that overlapped the translational start site. Mutation of the polyhedrin start codon has allowed these sequences to be incorporated as part of the 5′ untranslated region of the foreign gene cassette, so that native proteins can be expressed (Landford 1988). Replacement of the polyhedrin gene also provides a convenient method to detect recombinant viruses. The occlusion bodies produced by wild-type viruses cause the microscopic viral plaques to appear opalescent if viewed under an oblique light source (OB+), while recombinant plaques appear clear (OB–) (Smith *et al.* 1983). Insertion of the *E. coli lacZ* gene in frame into the polyhedrin coding region allows blue-white screening of recombinants in addition to the OB assay (Pennock *et al.* 1984), and many current baculovirus expression systems employ *lacZ* as a screenable marker to identify recombinants. Substitution of *lacZ* with the gene for green fluorescent protein allows the rapid identification of recombinants by exposing the plaques to UV light (Wilson *et al.* 1997).

The construction of baculovirus expression vectors involves inserting the transgene downstream of the polyhedrin promoter. Since the genome is large, this is usually achieved by homologous recombination using a plasmid vector carrying a baculovirus homology region. A problem with homology-based strategies for introducing foreign DNA into large viral genomes is that recombinants are generated at a low efficiency. For baculovirus, recombinant vectors are recovered at a frequency of 0.5–5% of total virus produced. The proportion of recombinants has been increased by using linear derivatives of the wild-type baculovirus genome containing large deletions, which can be repaired only by homologous recombination with the targeting vector. Compatible targeting vectors span the deletion and provide enough flanking homologous DNA to sponsor recombination between the two elements and generate a viable, recombinant genome. Such approaches

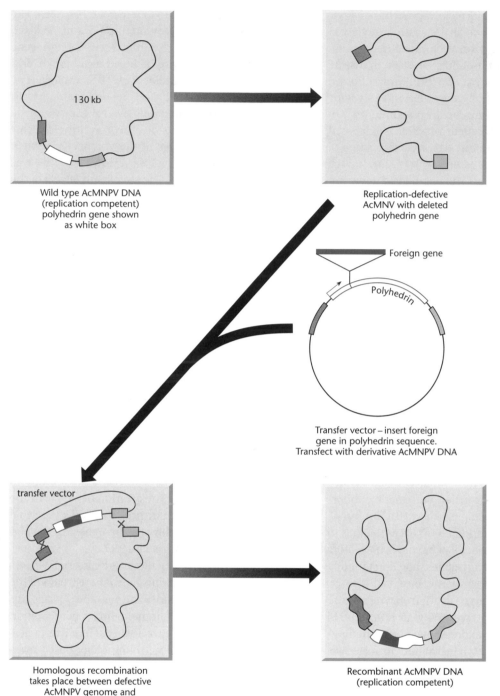

Wild type AcMNPV DNA
(replication competent)
polyhedrin gene shown
as white box

Replication-defective
AcMNV with deleted
polyhedrin gene

Foreign gene

Polyhedrin

Transfer vector – insert foreign
gene in polyhedrin sequence.
Transfect with derivative AcMNPV DNA

transfer vector

Homologous recombination
takes place between defective
AcMNPV genome and
transfer vector

Recombinant AcMNPV DNA
(replication competent)

Fig. 12.12 Procedure for the generation of recombinant baculovirus vectors.

result in the production of up to 90% recombinant plaques (Kitts & Possee 1993) (Fig. 12.12)). Recently, alternative systems have been described in which the baculovirus genome is maintained as a low-copy-number replicon in bacteria or yeast, allowing the powerful genetics of these microbial systems to be exploited. In the bacterial system, marketed by Gibco-BRL under the name "Bac-to-Bac", the baculovirus genome is engineered to contain an

origin of replication from the *E. coli* F plasmid. This hybrid replicon, called a *bacmid*, also contains the target site for the transposon Tn*7*, inserted in-frame within the *lacZ* gene, which is itself downstream of the polyhedrin promoter. The foreign gene is cloned in another plasmid between two Tn*7* repeats, and introduced into the bacmid-containing bacterium, which also contains a third plasmid expressing Tn*7* transposase. Induction of transposase synthesis results in the site-specific transposition of the transgene into the bacmid, generating a recombinant baculovirus genome that can be isolated for transfection into insect cells. Transposition of the transgene into the bacmid interrupts the *lacZ* gene allowing recombinant bacterial colonies to be identified by blue-white screening.

Although baculoviruses productively infect insect cells, they can also be taken up by mammalian cells although without producing progeny virions. The number of mammalian cell lines reported to be transduced by baculoviruses is growing (reviewed by Kost & Condreay 2002) and this suggests that recombinant baculoviruses could be developed as vectors for gene therapy. A number of baculovirus-borne transgenes have been expressed using constitutive promoters, such as the cytomegalovirus immediate early promoter (Hoffman *et al.* 1995) and the Rous sarcoma virus promoter (Boyce & Bucher 1996). More recently, cell lines have been generated that are stably transformed with baculovirus vectors, although it is not clear whether the viral genome has integrated in these cells (Condreay *et al.* 1999).

Baculoviruses are useful not only for the delivery of foreign genes into mammalian cells, but also for the delivery of other viruses. For example, hepatitis C virus does not infect cultured cells, but a hybrid baculovirus containing the entire HCV genome can initiate an HCV infection. Baculoviruses can also be used to improve the production of recombinant viral vectors. As discussed above, vectors based on adenoviruses are generally replication-defective because they lack one or more essential viral gene products, and the highest-capacity amplicons or fully deleted adenoviruses (FD-AdV), contain no viral genes at all, only those *cis*-acting elements required for replication and packaging. Cell lines are not available for the packaging of such vectors so helper viruses are normally required, which leads to contamination of recombinant stocks. Recently, however, a recombinant baculovirus vector has been developed which carries a packaging-deficient copy of the entire adenovirus genome.

Herpesvirus vectors are latent in many cell types and may promote long-term transgene expression

The herpesviruses are large double-stranded DNA viruses that include Epstein–Barr virus (EBV, discussed above) and the herpes simplex viruses (e.g. HSV-I, varicella zoster). Most herpes simplex viruses are transmitted without symptoms (varicella zoster virus is exceptional), and cause prolonged infections. Unlike EBV, which is used as a replicon vector, HSV-I has been developed as a transduction vector (Burton *et al.* 2002, Calderwood *et al.* 2004). Viral replication can occur in many cell types in a wide range of species if the genome is introduced by transfection, but HSV vectors are particularly suitable for gene therapy in the nervous system because the virus is remarkably neurotropic. As with other large viruses, recombinants can be generated in transfected cells by homologous recombination, and such vectors may be replication-competent or helper-dependent (Marconi *et al.* 1996). Alternatively, plasmid-based amplicon vectors can be constructed, which carry only those *cis*-acting elements required for replication and packaging. These require packaging systems to provide the missing functions in *trans* (e.g. Stavropoulos & Strathdee 1998). Therapeutic use of herpesvirus vectors has been limited, but a number of genes have been successfully transferred to neurons *in vivo* (e.g. Boviatsis *et al.* 1994, Lawrence *et al.* 1995). Generally, transgene expression is transient, although prolonged expression has been observed in some neuronal populations (see reviews by Vos *et al.* 1996, Simonato *et al.* 2000). Note that HSV is also transmitted across neuronal synapses during lytic infections, a phenomenon that can be exploited to trace axon pathways (Norgren & Lehman 1998).

Retrovirus vectors integrate efficiently into the host-cell genome

Retroviruses are RNA viruses that replicate via a double-stranded DNA intermediate. The infection cycle involves the precise integration of this intermediate into the genome of the host cell, where it is transcribed to yield daughter genomes that are packaged into virions.

Retroviruses have been developed as vectors for a number of reasons (reviewed by Miller 1992a,b, Blesch 2004). First, certain retroviruses are acutely oncogenic because they carry particular genes that promote host-cell division. Investigation of such viruses has shown that these *viral oncogenes* are in fact gain-of-function derivatives of host genes, *proto-oncogenes*, which are normally involved in the regulation of cell growth. In most cases, the viral oncogenes are found to be expressed as fusions with essential viral genes, rendering the virus replication defective. These *acute transforming retroviruses* therefore demonstrate the natural ability of retroviruses to act as replication-defective gene transfer vectors. Secondly, most retroviruses do not kill the host, but produce progeny virions over an indefinite period. Retroviral vectors can therefore be used to make stably transformed cell lines. Thirdly, viral gene expression is driven by strong promoters, which can be subverted to control the expression of transgenes. In the case of murine mammary tumor virus, transcription from the viral promoter is inducible by glucocorticoids, allowing transgenes controlled by this promoter to be switched on and off (Lee *et al.* 1981, Scheidereit *et al.* 1983). Fourthly, some retroviruses, such as amphotropic strains of murine leukemia virus (MLV), have a broad host range allowing the transduction of many cell types. Finally, retroviruses make efficient and convenient vectors for gene transfer because the genome is small enough for DNA copies to be manipulated *in vitro* in plasmid-cloning vectors, the vectors can be propagated to high titers (up to 10^8 plaque-forming units per ml), and the efficiency of infection *in vitro* can approach 100%.

The major disadvantage of oncoretroviral vectors is that they only productively infect dividing cells, which limits their use for gene-therapy applications (Miller *et al.* 1990, Roe *et al.* 1993). However, lentiviruses such as HIV are more complex retroviruses that have the ability to infect non-dividing cells (Lewis & Emmerman 1994). These were initially developed as vectors for the stable transduction of cells displaying CD4 (Shimada *et al.* 1991, Poznansky *et al.* 1991, Buchschacher & Panganiban 1992, Parolin *et al.* 1994) but recent advances in lentiviral vector design provide improved safety and allow the transduction of multiple cell types (Federico 1999, Sandrin *et al.* 2003, Blesch 2004).

Before discussing the development of retroviral vectors, it is necessary to describe briefly the genome structure, and the molecular biology of the infection cycle (for a comprehensive account, see Weiss *et al.* 1985). A typical retroviral genome map is shown in Fig. 12.13. The infection cycle begins when the viral

envelope interacts with the host-cell's plasma membrane, delivering the particle into the cell. The capsid contains two copies of the RNA genome, as well as reverse transcriptase/integrase. Thus, immediately after infection, the RNA genome is reverse transcribed to produce a cDNA copy. This is a complex process involving two template jumps, with the result that the terminal regions of the RNA genome are duplicated in the DNA as long terminal repeats (LTRs). The DNA intermediate then integrates into the genome at an essentially random site (there may be some preference for actively transcribed regions). The integrated provirus has three genes (*gag*, *pol*, and *env*). The *gag* gene encodes a viral structural protein, *pol* encodes the reverse transcriptase and integrase, and the *env* gene encodes viral envelope proteins. Viral genomic RNA is synthesized by transcription from a single promoter located in the left LTR and ends at a polyadenylation site in the right LTR. Thus, the full-length genomic RNA is shorter than the integrated DNA copy and lacks the duplicated LTR structure. The genomic RNA is capped and polyadenylated, allowing the *gag* gene to be translated (the *pol* gene is also translated by readthrough, producing a Gag-Pol fusion protein that is later processed into several distinct polypeptides). Some of the full-length RNA also undergoes splicing, eliminating the *gag* and *pol* genes and allowing the downstream *env* gene to be translated. Two copies of the full-length RNA genome are incorporated into each capsid, which requires a specific *cis*-acting packaging site termed ψ. The reverse transcriptase/integrase is also packaged.

Retroviral vectors are often replication-defective and self-inactivating

Most retroviral vectors are replication-defective, because removal of the viral genes provides the maximum capacity for foreign DNA (about 8 kb). Only the *cis*-acting sites required for replication and

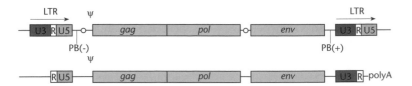

Fig. 12.13 Generic map of an oncoretrovirus genome. Upper figure shows the structure of an integrated provirus, with long terminal repeats (LTRs) comprising three regions U3, R, and U5, enclosing the three open reading frames *gag*, *pol*, and *env*. Lower figure shows the structure of a packaged RNA genome, which lacks the LTR structure and possesses a poly(A) tail. PB represents primer binding sites in the viral replication cycle, and ψ is the packaging signal. The small circles represent splice sites.

packaging are left behind. These include the LTRs (necessary for transcription and polyadenylation of the RNA genome as well as integration), the packaging site ψ which is upstream of the *gag* gene, and "primer-binding sites" which are used during the complex replication process. The inclusion of a small portion of the *gag*-coding region improves packaging efficiency by up to 10-fold (Bender *et al.* 1987). Deleted vectors can be propagated only in the presence of a replication-competent helper virus or a packaging cell line. The former strategy leads to the contamination of the recombinant vector stock with non-defective helper virus. Conversely, packaging lines can be developed where an integrated provirus provides the helper functions but lacks the *cis*-acting sequences required for packaging (Mann *et al.* 1983). Many different retroviruses have been used to develop packaging lines, and since these determine the type of envelope protein inserted into the virion envelope, they govern the host range of the vector (they are said to *pseudotype* the vector). Packaging lines based on amphotropic murine leukemia viruses allow retroviral gene transfer to a wide range of species and cell types, including human cells (e.g. see Cone & Mulligan 1984, Danos & Mulligan 1988). It is still possible for recombination to occur between the vector and the integrated helper provirus, resulting in the production of wild-type contaminants. The most advanced "third-generation" packaging lines limit the extent of homologous sequence between the helper virus and the vector and split up the coding regions so that up to three independent crossover events are required to form a replication-competent virus (e.g. see Markowitz *et al.* 1988).

The simplest strategy for the high-level constitutive expression of single genes in retroviral vectors is to delete all coding sequences and place the foreign gene between the LTR promoter and the viral polyadenylation site. Alternatively, an internal heterologous promoter can be used to drive transgene expression. However, many investigators have reported interference between the heterologous promoter and the LTR promoter (e.g. see Emmerman & Temin 1984, Wu *et al.* 1996). Yu *et al.* (1986) addressed this problem by devising *self-inactivating vectors*, containing deletions in the 3′ LTR which are copied to the 5′ LTR during vector replication, thus inactivating the LTR promoter while leaving internal promoters intact. This strategy also helps to alleviate additional problems associated with the LTR promoter: (i) that adjacent endogenous genes may be activated following integration; and (ii) that the

entire expression cassette may be inactivated by DNA methylation after a variable period of expression in the target cell (Naviaux & Verma 1992).

Since retroviral vectors are used for the production of stably transformed cell lines, it is necessary to co-introduce a selectable marker gene along with the transgene of interest. The expression of two genes can be achieved by arranging the transgene and marker gene in tandem, each under the control of a separate promoter, one of which may be the LTR promoter. This leads to the production of full-length and subgenomic RNAs from the integrated provirus. Alternatively, if the first gene is flanked by splice sites, only a single promoter is necessary because the RNA is spliced in a manner reminiscent of the typical retroviral life cycle, allowing translation of the downstream gene (Cepko *et al.* 1984). Vectors in which the downstream gene is controlled by an internal ribosome entry site have also been used (e.g. Dirks *et al.* 1993, Sugimoto *et al.* 1994). Since the viral replication cycle involves transcription and splicing, an important consideration for vector design is that the foreign DNA must not contain sequences that interfere with these processes. For example, polyadenylation sites downstream of the transgene should be avoided, as these will cause truncation of the RNA, blocking the replication cycle (Miller *et al.* 1983). Retroviruses also remove any introns contained within the transgene during replication (Shimotohno & Temin 1982).

There are special considerations for the construction of lentiviral vectors

Lentiviral vectors are produced in much the same way as MLV and other retroviral vectors, i.e. by replacing essential viral genes with the transgene of interest and using a packaging line to supply the missing viral functions. Most interest has been shown in vectors based on HIV, but systems have also been described that are based on bovine immunodeficiency virus (BIV), equine infectious anemia virus (EIAV), feline immunodeficiency virus (FIV), and simian immunodeficiency virus (SIV). Special considerations for such vectors include the requirement for Tat and Rev functions, and the fact that HIV is one of the few retroviruses known to cause an infectious disease in humans. Therefore, there must be especially stringent precautions to prevent contamination with replication-competent virus when HIV is used as a vector, and this has prompted the development of a series of multicomponent packaging lines

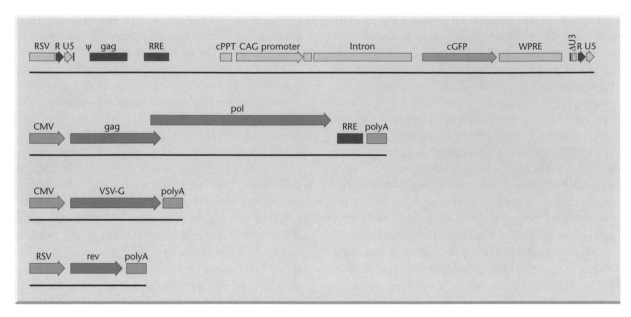

Fig. 12.14 Prototypical plasmids for the production of third-generation HIV-1 lentiviral vectors. In the top panel, the U3 region in the 5'LTR is replaced by an RSV promoter and a deletion has been introduced in the U3 region of the 3'LTR (DU3). The ψ region and a small part of the *gag*-coding sequence necessary for packaging, the Rev response element (RRE), and the central polypurine tract (cPPT) are the only regions left from HIV-1. The transgene, in this case GFP, is driven by a CMV/β-actin promoter (CAG). For mRNA stabilization, a Woodchuck post-transcriptional response element is included (WPRE). The lower three panels show helper plasmids for the production of lentiviral vectors. The first plasmid contains only the coding regions for the HIV *gag* and *pol* genes and the RRE necessary for efficient splicing. The coding region for the accessory genes, regulatory genes, and envelope, the packaging signal (ψ), and the LTRs are removed. The second plasmid codes for the VSV-G envelope. The *rev* gene is expressed on the third plasmid.

in which different viral functions are supplied on different plasmids introduced by transient transfection (Federico 1999, Sandrin *et al.* 2003, Blesch 2004).

The first generation of lentiviral packaging lines involved transfection of 293T cells with three plasmids, one containing a wild-type HIV genome with the *gag*, *pol*, and *env* regions replaced by a transgene; one providing the HIV *gag* and *pol* functions under the control of a constitutive promoter; and one encoding the G-protein from vesicular stomatitis virus (VSV) to replace the HIV *env* function. The combination of these three plasmids resulted in the production of VSV-G pseudotyped lentiviral vectors at a high titer, but some pseudotyped replication-competent HIV were produced by recombination during viral replication and packaging. In the second generation of vectors, the *vpr*, *vpu*, *nef*, and *vif* genes were removed from the first plasmid, since these genes are important for virulence *in vivo* but are not required for productive infection in cell lines. More recently, a third generation of vectors has been produced in which the U3 region of the 5′ UTR has been replaced by the human cytomegalovirus early promoter, therefore making the activity of the LTR independent from the *Tat* gene, which consequently

has been deleted. The U3 region of the 3′ UTR has been modified to include a deletion, making the vector self-inactivating, and several internal modifications have been made including the incorporation of sequences that enhance transgene expression (Fig. 12.14). The *Rev* gene, which is required for *gag* and *pol* expression, has also been removed from the packaging construct and is instead supplied from a fourth plasmid to further reduce the likelihood of replication-competent vectors arising by recombination. Indeed the presence of four plasmids sharing no significant homology makes such recombination events very unlikely even at high production titers.

Sindbis virus and Semliki forest virus vectors replicate in the cytoplasm

The alphaviruses are a family of enveloped viruses with a single-strand positive-sense RNA genome. One of the advantages of using such RNA viruses for gene transfer is that integration into the host genome is guaranteed never to occur. Alphavirus replication takes place in the cytoplasm, and produces a large numbers of daughter genomes, allowing very high-level expression of any transgene (reviewed by Berglung *et al.* 1996). To date, these are the only animal viruses with a replication cycle based solely on RNA to be extensively developed as expression vectors. Sindbis virus (Xiong *et al.* 1989) and Semliki forest virus (SFV) (Liljestrom & Garoff 1993) have been the focus of much of this research. These display a broad host range and cell tropism, and mutants have been isolated with reduced cytopathic effects (Agapov *et al.* 1998, Frolov *et al.* 1999; reviewed by Schlesinger 2001, Rayner *et al.* 2002, Lundstrom 2003).

The wild-type alphavirus comprises two genes, a 5′ gene encoding viral replicase and a 3′ gene encoding a polyprotein from which the capsid structural proteins are autocatalytically derived. Since the genome is made of RNA, it can act immediately as a substrate for protein synthesis. However, because protein synthesis in eukaryotes is dependent on mRNAs possessing a specialized 5′ cap structure, only the replicase gene is initially translated. The replicase protein produces a negative sense complementary strand, which in turn acts as a template for the production of full-length daughter genomes. However, the negative strand also contains an internal promoter which allows the synthesis of a subgenomic positive sense RNA containing the capsid polyprotein gene. This subgenomic RNA is subsequently capped and translated.

A number of different strategies have been used to express recombinant proteins using alphavirus vectors (Berglung *et al.* 1996, Lundstrom 1997). For example, replication-competent vectors have been constructed in which an additional subgenomic promoter is placed either upstream or downstream of the capsid polyprotein gene. If foreign DNA is introduced downstream of this promoter, the replicase protein produces two distinct subgenomic RNAs, one corresponding to the transgene. Such insertion vectors tend to be unstable and have been largely superseded by replacement vectors in which the capsid polyprotein gene is replaced by the transgene. The first 120 b of both the Sindbis and SFV structural polyprotein genes includes a strong enhancer of protein synthesis, which significantly increases the yield of recombinant protein (Sjoberg *et al.* 1994, Frolov & Schlesinger 1994). This is downstream of the translational initiation site, so in many vectors this enhancer region is included so that the foreign gene is expressed as an N-terminal fusion protein. This can result in extremely high levels of recombinant protein synthesis, up to 50% of the total cellular protein.

Both plasmid replicon and viral transduction vectors have been developed from the alphavirus genome. A versatile Sindbis replicon vector, pSinRep5, is currently marketed by Invitrogen (Fig. 12.15). The vector is a plasmid containing bacterial backbone elements, the Sindbis replicase genes and packaging site, and an expression cassette featuring a Sindbis subgenomic promoter, a multiple cloning site, and a polyadenylation site. There is an SP6 promoter upstream of the replicase genes and expression cassette for generating full-length *in vitro* transcripts. There is a second set of restriction sites downstream from the polylinker, allowing the vector to be

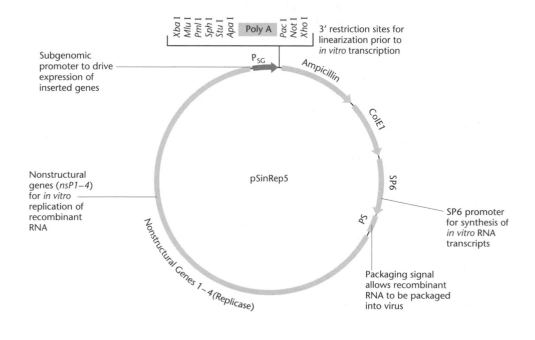

Fig. 12.15 The Sindbis virus-based vector pSinRep5. (Reproduced with permission from Invitrogen.)

linearized prior to *in vitro* transcription. Foreign DNA is cloned in the expression cassette, the vector is linearized and transcribed, and the infectious recombinant Sindbis RNA thus produced is transfected into cells, from which recombinant protein can be recovered. An alternative strategy is to place the entire alphavirus genome under the control of a standard eukaryotic promoter such as SV40, and transfect cells with the DNA. In this case, the DNA expressed in the nucleus and the recombinant vector RNA is exported into the cytoplasm. Such DNA vectors have been described for Sindbis and SFV (Johanning *et al.* 1995, Herweijer *et al.* 1995, Berglund *et al.* 1996). Transduction is a more suitable delivery procedure for gene-therapy applications, and in this case a replicon vector like pSinRep5 is co-introduced with a defective helper plasmid supplying the missing structural proteins. This facilitates one round of replication and packaging and the production of recombinant viral particles that can be isolated from the extracellular fluid. In the original packaging system, the replicon and helper vectors could undergo recombination to produce moderate amounts of contaminating wild-type virus. This problem has been addressed by supplying the structural protein genes on multiple plasmids (Smerdou & Liljestrom 1999) and by the development of complementary cell lines producing the structural proteins (Polo *et al.* 1999).

Vaccinia and other poxvirus vectors are widely used for vaccine delivery

Vaccinia virus is closely related to variola virus, the agent responsible for smallpox. A world-wide vaccination program using vaccinia virus resulted in the elimination of smallpox as an infectious disease. The success of the program raised hopes that recombinant vaccinia viruses, carrying genes from other pathogens, could be used as live vaccines for other infectious diseases.

The poxviruses have a complex structure and a large double-stranded linear DNA genome (up to 300 kb). Unusually for a DNA virus, the poxviruses replicate in the cytoplasm of the infected cell rather than its nucleus. Part of the reason for the large genome and structural complexity is that the virus must encode and package all its own DNA replication and transcription machinery, which most DNA viruses "borrow" from the host cell nucleus.

The unusual replication strategy and large size of the vaccinia genome make the design and construc-

tion of expression vectors more complex than for other viruses (Guo & Bartlett 2004). Since the virus normally packages its own replication and transcription enzymes, recombinant genomes introduced into cells by transfection are non-infectious. Recombinant viruses are therefore generated by homologous recombination, using a targeting plasmid transfected into virus-infected cells. More recently, direct ligation vectors have been developed, and these are transfected into cells containing a helper virus to supply replication and transcription enzymes in *trans* (Merchlinsky *et al.* 1997). Recombinant vectors can be identified by hybridization to the large viral plaques that form on permissive cells. However, the efficiency of this process can be improved by various selection regimes. In one strategy, the transgene is inserted into the viral *Tk* gene and negative selection using the thymidine analog 5-bromodeoxyuridine is carried out to enrich for potential recombinants (Mackett *et al.* 1982). In another, the transgene is inserted into the viral hemagglutinin locus. If chicken erythrocytes are added to the plate of infected cells, wild-type plaques turn red whereas the recombinant plaques remain clear (Shida 1986). Since vaccinia vectors have a high capacity for foreign DNA, selectable markers such as *neo* (Franke *et al.* 1985) or screenable markers such as *lacZ* (Chakrabarti *et al.* 1985) or *gusA* (Carroll & Moss 1995) can be co-introduced with the experimental transgene to identify recombinants.

Transgene expression usually needs to be driven by an endogenous vaccinia promoter, since transcription relies on proteins supplied by the virus. The highest expression levels are provided by late promoters such as P11, allowing the production of up to 1 μg of protein per 10^6 cells, but other promoters such as P7.5 and 4b are used, especially where expression early in the infection cycle is desired (Cochran *et al.* 1985, Bertholet *et al.* 1985). A synthetic late promoter, whose use allows up to 2 μg of protein to be produced per 10^6 cells, has also been developed (Lundstrom 1997). Since the cytoplasm lacks not only host transcription factors but also the nuclear splicing apparatus, vaccinia vectors cannot be used to express genes with introns. Furthermore, the sequence TTTTTNT must be removed from all foreign DNA sequences expressed in vaccinia vectors, since the virus uses this motif as a transcriptional terminator. A useful binary expression system has been developed, in which the transgene is driven by the bacteriophage T7 promoter and the T7 polymerase itself is expressed in a vaccinia vector under the control of a vaccinia promoter (Fuerst *et al.* 1986). Initially, the transgene was placed on a plasmid vector and transfected into cells infected with the recombinant vaccinia virus, but higher expression levels (up to 10% total cellular protein) can be achieved using two vaccinia vectors, one carrying the T7-driven transgene and one expressing the polymerase (Fuerst *et al.* 1987).

In an early demonstration that vaccinia virus could be used to express antigens from other infectious agents, Smith *et al.* (1983a) replaced the vaccinia *Tk* locus with a transgene encoding hepatitis B surface antigen (HBSAg). The transgene was cloned in a plasmid containing the vaccinia thymidine kinase gene, interrupted by one of the vaccinia early promoters. This plasmid was transfected into vaccinia-infected monkey cells, and recombinant vectors carrying the transgene were selected using 5-bromodeoxyuridine. Cells infected with this virus secreted large amounts of HBSAg into the culture medium, and vaccinated rabbits rapidly produced high-titer antibodies to HBSAg. Similarly, vaccinia viruses expressing the influenza hemagglutinin gene were used to immunize hamsters, and induce resistance to influenza (Smith *et al.* 1983a). Recombinant

vaccinia viruses have been constructed expressing a range of important antigens, including HIV and HTLV-III envelope proteins (Hu *et al.* 1986, Chakrabarti *et al.* 1986). In many cases, recombinant vectors have been shown to provide immunity when administered to animals (reviewed by Moss 1996). For example, monkeys infected with recombinant vaccinia and canarypox vectors have shown resistance to SIV and HIV-2 (Hirsch *et al.* 1996, Myagkikh *et al.* 1996). Several recombinant poxviruses have progressed to phase 1 clinical trials, including a vaccine against the Epstein–Barr major membrane antigen (Gu *et al.* 1996). The immunization of wild foxes, using tainted meat, with a recombinant vaccinia virus expressing the rabies virus glycoprotein appears to have eliminated the disease from Belgium (Brochier *et al.* 1991).

Summary of expression systems for animal cells

A variety of gene transfer and expression systems have been discussed in this chapter, and Table 12.5 provides a summary of these systems and their major

Table 12.5 Summary of major expression systems used in animal cells.

System	Host	Major applications
Non-replicating plasmid vectors		
No selection	Many cell lines	Transient assays
Dominant selectable markers	Many cell lines	Stable transformation, long-term expression
DHFR/methotrexate	CHO cells	Stable transformation, high-level expression
Plasmids with viral replicons		
SV40 replicons	COS cells	High-level transient expression
BPV replicons	Various murine	Stable transformation (episomal)
EBV replicons	Various human	Stable transformation (episomal), library construction
Viral transduction vectors		
Adenovirus E1 replacement	293 cells	Transient expression
Adenovirus amplicons	Various mammalian	*In vivo* transfer
Adeno-associated virus	Various mammalian	*In vivo* transfer
Baculovirus	Insects	High-level transient expression
	Various mammalian	*In vivo* transfer
Herpesvirus	Various mammalian	Stable transformation
Oncoretrovirus	Various mammalian and avian	Stable transformation
	ES cells	Transgenic mice (Chapter 13)
Lentivirus	Non-dividing cells, mammalian	*In vivo* transfer
Sindbis, Semliki forest virus	Various mammalian	High-level transient expression
Vaccinia virus	Various mammalian	High-level transient expression

applications. Many factors influence the expression of foreign genes in animal cells, and an understanding of these factors allows transgene expression to be controlled. For the production of recombinant proteins, it is often appropriate to maximize transgene expression, and principles for achieving this are shown in Box 12.2. In other cases, it may be desirable to switch the transgene on and off using inducible expression systems. We explore these considerations in more detail in Chapter 15.

Suggested reading

Blesch A. (2004) Lentiviral and MLV based retroviral vectors for *ex vivo* and *in vivo* gene transfer. *Methods* **33**, 164–72.

Calderwood N.A., White R.E. & Whitehouse A. (2004) Development of herpesvirus-based episomally maintained gene delivery vectors. *Expert Opinion in Biological Theory* **4**, 493–505.

Chou T.W., Biswas S. & Lu S. (2004) Gene delivery using physical methods: an overview. In: Heiser W.C. (ed.) *Gene Delivery to Mammalian Cells. Volume 1 – Nonviral Gene Transfer Techniques*. Humana Press, Towata NJ, pp. 147–65.

Davis M.E. (2002) Non-viral gene delivery systems. *Current Opinion in Biotechnology* **13**, 128–31.

Flotte T.R. (2004) Gene therapy progress and prospects: recombinant adeno-associated virus (rAAV) vectors. *Gene Therapy* **11**, 805–10.

Guo Z.S. & Bartlett D.L. (2004) Vaccinia as a vector for gene delivery. *Expert Opinion in Biological Theory* **4**, 901–17.

Higgins D.E. & Portnoy D.A. (1998) Bacterial delivery of DNA evolves. *Nature Biotechnology* **16**, 138–9.

Imperiale M.J. & Kochanek S. (2004) Adenovirus vectors: biology, design, and production. *Current Topics in Microbiology* **273**, 335–57.

Keown W.A., Campbell C.R. & Kucherlapati R.S. (1990) Methods for introducing DNA into mammalian cells. *Methods in Enzymology* **185**, 527–37.

Liu D., Chiao E.F. & Tian H. (2004) Chemical methods for DNA delivery: an overview. In: Heiser W.C. (ed.) *Gene Delivery to Mammalian Cells. Volume 1 – Nonviral Gene Transfer Techniques*. Humana Press, Towata NJ, pp. 3–23.

Possee R.D. (1997) Baculoviruses as expression vectors. *Current Opinion in Biotechnology* **8**, 569–72.

Schlesinger S. & Dubensky T.W., Jr. (1999) Alphaviruses for gene expression and vaccines. *Current Opinion in Biotechnology* **10**, 434–9.

Van Craenenbroeck K., Vanhoenacker P. & Haegeman G. (2000) Episomal vectors for gene expression in mammalian cells. *European Journal of Biochemistry* **267**, 5665–8.

Weiss S. & Chakraborty T. (2001) Transfer of eukaryotic expression plasmids to mammalian host cells by bacterial carriers. *Current Opinion in Biotechnology* **2**, 467–72.

Yokoyama M. (2002) Gene delivery using temperature-responsive polymeric carriers. *Drug Discovery Today* **7**, 426–32.

CHAPTER 13

Genetic manipulation of animals

Introduction

The genetic manipulation of animals has revolutionized our understanding of biology by making it possible to test gene expression and function at the whole-animal level. Gene-transfer techniques can be used to produce *transgenic animals*, in which every cell carries new genetic information. Similarly, it is possible to introduce specific, pre-selected mutations to the animal genome. The whole animal is the ultimate assay system in which to investigate gene function, particularly for complex biological processes such as development.

Whereas transgenic plants can be regenerated from transformed plant cells or tissue explants, the same cannot be achieved directly in animals. Two fundamental differences between animals and plants make this so. First, animal cells become progressively restricted in their potency as development proceeds. This means that differentiated animal cells are normally unable to dedifferentiate fully and recapitulate the developmental program. Secondly, in most animals, the somatic cells and germ cells (the cells that give rise to gametes) separate at an early developmental stage. Therefore, the only way to achieve germline transformation directly in animals is to introduce DNA into totipotent cells prior to the developmental stage at which the germline forms.

In most cases, this involves introducing DNA into the developing oocyte, egg, or early embryo. In mice, it is also possible to use cultured embryonic stem (ES) cells, which are derived from the preimplantation embryo and can contribute to all the tissues of the developing animal (including the germline) if introduced into a host embryo at the correct developmental stage. These cells are also remarkably amenable to homologous recombination, which allows them to be used for gene targeting, the accurate replacement of a segment of the endogenous genome with a homologous segment of exogenous DNA. Gene targeting can be used to replace endogenous genes with a completely non-functional copy (a null allele) or to make subtle changes, both allowing the function of the endogenous gene to be tested. The same technology can be used for the opposite purpose, i.e. to replace a mutant allele with a functional copy.

While rapid progress has been made with a range of model organisms, particularly *Drosophila*, *Xenopus*, and the mouse, great potential exists for the production of transgenic farm animals with improved or novel traits. The technology for introducing DNA into animals such as chickens, pigs, cattle, and sheep is still in its infancy and is much less efficient than that of mice. Furthermore, it has proven impossible thus far to isolate amenable ES cells from any of our domestic mammals. The desire to generate transgenic livestock has driven research in a different direction, that of nuclear transfer. Although differentiated animal cells are developmentally restricted, their nuclei still contain all the genetic information required to recapitulate the whole of development. Dolly, the first sheep produced following nuclear transfer from a differentiated adult somatic cell to an enucleated egg, has opened the way for animal cloning and the rapid production of élite transgenic herds.

Three major methods have been developed for the production of transgenic mice

The ability to introduce DNA into the germline of mice is one of the greatest achievements of the twentieth century and has paved the way for the transformation of other mammals. Genetically modified mammals have been used not only to study gene function and regulation, but also as bioreactors producing valuable recombinant proteins, e.g. in their milk. Several methods for germline transformation have been developed, all of which require the removal of fertilized eggs or early embryos from donor mothers, brief culture *in vitro*, and then their return to foster

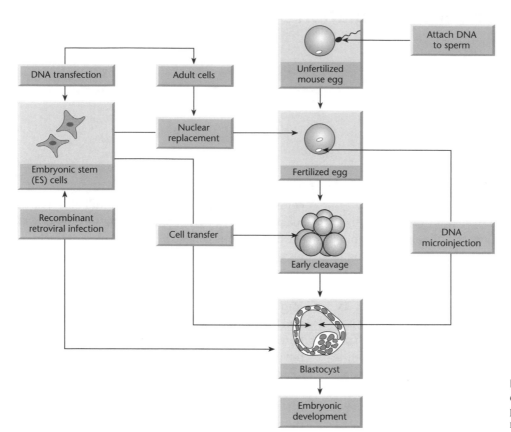

Fig. 13.1 Summary of methods for producing transgenic mammals.

mothers, where development continues to term. These methods are discussed below and summarized in Fig. 13.1. Note that these methods have been developed with nuclear transgenesis in mind, but mitochondria have their own genome. Recently, methods have been developed for the production of mitochondrial transgenics (or transmitochondrial mice), which are considered in Box 13.1.

Pronuclear microinjection involves the direct transfer of DNA into the male pronucleus of the fertilized mouse egg

Direct microinjection of DNA was the first strategy used to generate transgenic mice. Simian virus 40 (SV40) DNA was injected into the blastocoele cavities of preimplantation embryos by Jaenisch & Mintz (1974). The embryos were then implanted into the uteri of foster mothers and allowed to develop. The DNA was taken up by some of the embryonic cells and occasionally contributed to the germline, resulting in transgenic mice containing integrated SV40 DNA in the following generation. Transgenic mice have also been recovered following the injection of viral DNA into the cytoplasm of the fertilized egg (Harbers *et al.* 1981).

The technique that has become established is the injection of DNA into one of the pronuclei of the egg (reviewed by Palmiter & Brinster 1986). The technique is shown in Fig. 13.2. Just after fertilization, the small egg nucleus (female pronucleus) and the large sperm nucleus (male pronucleus) are discrete. Since the male pronucleus is larger, this is usually chosen as the target for injection. About 2 pl of DNA solution is transferred into the nucleus through a fine needle, while the egg is held in position with a suction pipette. The injected embryos are cultured *in vitro* to the morula stage and then transferred to pseudopregnant foster mothers (Gordon & Ruddle 1981). The procedure requires specialized micro-injection equipment and considerable dexterity from the handler. The exogenous DNA may integrate immediately or, less commonly, may remain extrachromosomal for one or more cell divisions. Thus the resulting animal may be transgenic or may be mosaic for transgene insertion. The technique is reliable, although the efficiency varies, so that 5–40% of mice developing from manipulated eggs contain the transgene (Lacy *et al.* 1983). However, once the transgene is transmitted through the germline, it tends to be stably inherited over many generations. The exogenous DNA tends to form head-to-tail

Box 13.1   Mitochondrial transgenesis in mice

Several human genetic diseases are caused by mitochondrial mutations, and to model such diseases it is necessary to introduce DNA into the mitochondrial genome of suitable animals. This cannot be achieved using standard transgenesis techniques and a novel method has been developed in which donor cell lines carrying mitochondrial mutations are enucleated and then fused with recipient mouse zygotes. The resulting *transmitochondrial mice* carry the mutation in every cell and pass the mutation through the female germline, since mitochondria in the sperm only rarely contribute to the embryo.

The procedure for producing transmitochondrial mice is shown in Fig. B13.1. This involves five distinct steps. First, a cell line must be produced carrying mitochondria containing the appropriate transgene or mitochondrial mutation. There are many ways to achieve this, but one of the most efficient is fusion between cells depleted in mitochondrial DNA (mtDNA) and synaptosomes from aged mice which often contain mtDNA deletions and other mutations. The resulting cell lines are called cybrids because their cytoplasm is a hybrid of two cell lines. The next step is to remove the nucleus from the cybrid cells, so that they can act as mitochondrial donors without introducing additional nuclei into the egg. This is generally achieved by centrifugation in the presence of cytochalasin B. After washing the enucleated cybrids, zygotes are collected from donor females and a micromanipulator is used to drill a hole through the zona pellucida and place the cybrids in the perivitelline space. After a brief recovery period in culture, embryos containing cybrids are washed in a medium that promotes fusion and placed between the

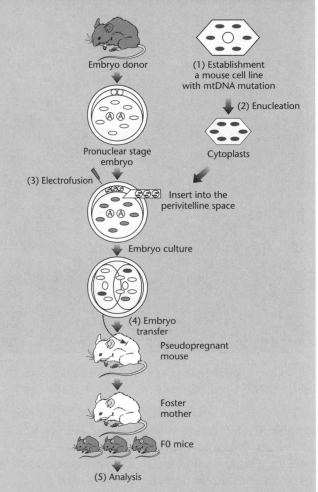

Fig. B13.1 Procedure for generating transmitochondrial mice. See text for details.

electrodes of a fusion chamber. A long AC pulse (225 V, 10–15 s) is used to align the cytoplasts and oocytes, and then a brief DC pulse (2500 V, 20 ms) is used to induce fusion, which occurs within the next hour. After a brief recovery period in culture, the fused embryos are implanted into pseudopregnant foster mothers and raised to term.

arrays prior to integration, and the copy number varies from a few copies to hundreds. The site of integration appears random and may depend on the occurrence of natural chromosome breaks. Extensive deletions and rearrangements of the genomic DNA often accompany transgene integration (Bishop & Smith 1989).

Recombinant retroviruses can be used to transduce early embryos prior to the formation of the germline

As discussed in Chapter 12, recombinant retroviruses provide a natural mechanism for stably introducing DNA into the genome of animal cells. Retroviruses

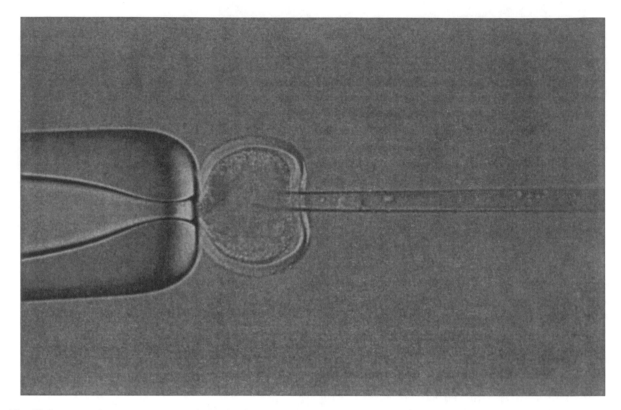

Fig. 13.2 Pronuclear microinjection of a fertilized mouse egg. The two pronuclei are visible, and the egg is held using a suction pipette. The DNA is introduced through a fine glass needle. (Photograph courtesy of Roslin Institute.)

are able to infect early embryos (as well as ES cells, see below), so recombinant retroviral vectors can be used for germline transformation (Robertson *et al.* 1986). An advantage over the microinjection technique is that only a single copy of the retroviral provirus is integrated, and the genomic DNA surrounding the transgenic locus generally remains intact. The infection of preimplantation embryos with a recombinant retrovirus is technically straightforward and, once the infected embryos are implanted in the uterus of a foster mother, can lead to germline transmission of the transgene. However, there are also considerable disadvantages to this method, including the limited amount of foreign DNA that can be carried by the virus, the possible interference of viral regulatory elements with the expression of surrounding genes, and the susceptibility of the virus to *de novo* methylation, resulting in transgene silencing. The founder embryos are always mosaics with respect to transgene integration (reviewed by Jaenisch 1988). Retroviral transduction is therefore not favored as a method for generating fully transgenic animals, but it is useful for generating transgenic sectors of embryos. For example, the analysis of chicken-limb buds infected with recombinant

retroviruses has allowed many of the genes involved in limb development to be functionally characterized (see review by Tickle & Eichele 1994).

Transgenic mice can be produced by the transfection of ES cells followed by the creation of chimeric embryos

ES cells are derived from the inner cell mass of the mouse blastocyst and thus have the potential to contribute to all tissues of the developing embryo (Evans & Kaufman 1981, Martin 1981). The ability of ES cells to contribute to the germline was first demonstrated by Bradley *et al.* (1984) and requires culture conditions that maintain the cells in an undifferentiated state (Joyner 1998). Since these cells can be serially cultured like any other established cell line, DNA can be introduced by transfection or viral transduction and the transformed cells can be selected using standard markers, as discussed in Chapter 12. In contrast, since there is no convenient way to select for *eggs or embryos* that have taken up foreign DNA, each potential transgenic mouse generated by pronuclear microinjection must be tested by Southern blot hybridization or the PCR to confirm transgene integration.

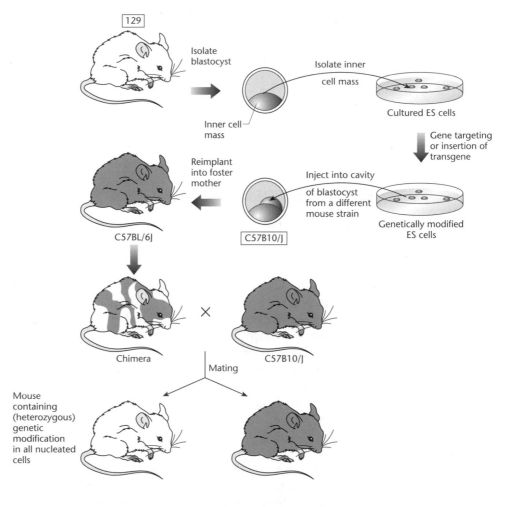

Fig. 13.3 Procedure for generating transgenic mice using transformed ES cells. The ES cells, which are manipulated in culture, contribute potentially to all the tissues of the embryo, including the germline. Germline transmission of the transgene can be confirmed by checking that offspring of chimeric embryos carry ES cell coat color markers.

Transfected ES that survive selection are introduced into the blastocoele of a host embryo at the blastocyst stage, where they mix with the inner cell mass. This creates a true chimeric embryo.[1] The contribution of ES cells to the germline can thus be confirmed using visible markers. Most ES cell lines in common use are derived from mouse strain 129, which has the dominant coat color agouti. A popular strategy is to use host embryos from a mouse strain such as C57BL/6J, which has a recessive black coat color. Colonization of the embryo by vigorous ES cells can be substantial, generating chimeras with patchwork coats of black and agouti cell clones. If the ES cells have contributed to the germline, mating chimeric males with black females will generate heterozygous transgenic offspring with the agouti coat color, confirming germline transmission of the foreign DNA. Most ES cells in use today are derived from male embryos, resulting in a large sex bias towards male chimeras (McMahon & Bradley 1990). This is desirable because male chimeras sire many more offspring than females. The procedure for generating transgenic mice using ES cells is shown in Fig. 13.3.

ES cells can be used for gene targeting in mice

Pronuclear microinjection and retroviral transfer are useful for the *addition* of DNA to the mouse genome. However, in many cases it is desirable to modify the endogenous genome in a specific manner, such as introducing a defined mutation in a selected gene. In yeast, gene targeting by homologous

---

[1] Organisms which are mixtures of genetically distinct cells may be described as either mosaics or chimeras, but the two terms have distinct meanings. A mosaic is a mixture of genetically distinct cells which nevertheless have a common ancestor. This would apply, for example, to two lineages of cells in an animal which have arisen by mutation, or by transgene integration. A chimera on the other hand is a mixture of genetically distinct cells with different ancestors, as occurs when ES cells from one mouse line are injected into the blastocoele of another mouse line.

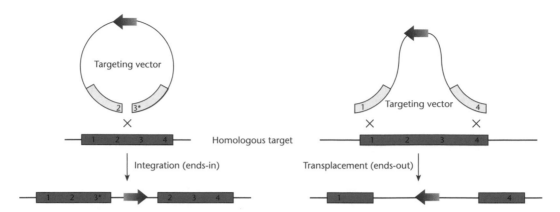

Fig. 13.4 Structure and integration mechanisms of the two major types of targeting vector, (a) insertion and (b) transplacement. Regions of homology between the vector and target are designated by the same number.

recombination occurs with high efficiency (Chapter 11). In contrast, when the first gene-targeting experiments in animal cells were carried out in the 1980s, only a very low frequency of targeted recombination was achieved. These experiments involved the correction of mutations in selectable markers such as *neo*, which had been introduced into cell lines as transgenes by standard methods (e.g. Thomas *et al.* 1986). Smithies *et al.* (1985) were the first to demonstrate targeting of an endogenous gene. They introduced a modified β-globin gene containing the bacterial marker *supF* into a human fibroblast x mouse erythroleukemia cell line and screened large numbers of potential recombinants by re-isolating the modified locus, using *supF* as a cloning tag. This experiment demonstrated that the frequency of homologous recombination was up to 1000-fold lower than that of random integration. Targeting occurs with significantly higher efficiency in certain cell lines, including mouse ES cells. The combination of pluripotency, amenability for *in vitro* manipulation, and capacity for homologous recombination makes ES cells uniquely suitable for the generation of targeted mutant mice, i.e. mice carrying the same mutation in every cell and transmitting it through the germline. Gene targeting in ES cells was first achieved by Thomas & Capecchi (1987), who disrupted the *hprt* gene with the *neo* marker and selected recombinant cells using either G418 or 6-thioguanine, a toxic guanine analog that is only incorporated into DNA if the nucleotide salvage pathway is functional (see p. 224). Doetschman *et al.* (1987) also successfully targeted the *hprt* locus, although they used a mutant recipient cell line and repaired the locus with homologous DNA, subsequently selecting on HAT medium (see p. 225).

Gene-targeting vectors may disrupt genes by insertion or replacement

Targeting vectors are specialized plasmid vectors that promote homologous recombination when introduced into ES cells. This is achieved by the inclusion of a homology region, i.e. a region that is homologous to the target gene, allowing the targeting vector to synapse with the endogenous DNA. Both the size of the homology region and the level of sequence identity have been shown to play an important role in the efficiency of gene targeting (Hasty *et al.* 1991b, Deng & Capecchi 1992, Te Riele *et al.* 1992). Recombination is also more efficient if the vector is linearized prior to transfection.

Most gene-targeting experiments have been used to disrupt endogenous loci, resulting in targeted null alleles (this strategy is often termed "gene knockout"). Two types of targeting vector have been developed for this purpose: insertion vectors and replacement (or transplacement) vectors (Thomas & Capecchi 1987; Fig. 13.4). Insertion vectors are linearized *within* the homology region, resulting in the insertion of the entire vector into the target locus. This type of vector disrupts the target gene but leads to a duplication of the sequences adjacent to the selectable marker. This is not always a desirable configuration, since duplication of the target sequences can lead to a subsequent homologous recombination event that restores the wild-type genotype (Fiering *et al.* 1995). Replacement vectors are designed so that the homology region is co-linear with the target. The vector is linearized *outside* the homology region prior to transfection, resulting in crossover events in which the endogenous DNA is replaced by the incoming DNA. With this type of vector, only sequences within the homology region (not the vector backbone) are inserted, so the homology region itself must be interrupted to achieve gene knockout. Insertion and replacement vectors are thought to be equally efficient, but replacement vectors have been used in the majority of knockout experiments. In both cases, however, it is possible for transcription to occur through the targeted locus, producing low amounts of RNA. This may be spliced in such a way as to remove the targeted exon, resulting in a residual amount of functional protein (e.g. Dorin *et al.* 1994, Dahme *et al.* 1997).

Sophisticated selection strategies have been developed to isolate rare gene-targeting events

The first gene-targeting experiments involved the selectable *Hprt* locus, which is present on the X chromosome, allowing targeted events to be selected in male ES cells without the requirement for homozygosity. The first non-selectable genes to be targeted were *int-2* (also known as *fgf-3*) (Mansour *et al.* 1988) and the oncogene *c-abl* (Schwartzberg *et al.* 1989). In each case, it was necessary to include a selectable marker in the targeting vector to identify transformed cells. In the case of insertion vectors this is placed anywhere on the vector backbone, while in replacement vectors the marker must interrupt the homology region. The *neo* marker has been most commonly used, allowing transformed ES cells to be selected using G418. However, other dominant markers are equally applicable (e.g. see Von Melchner *et al.* 1992), and *Hprt* can be used in combination with *hprt⁻* mutant ES cells (Matzuk *et al.* 1992).

The use of a single marker fails to discriminate between targeted cells and those where the construct has integrated randomly. This problem can be addressed by combined positive-negative selection using *neo* and the herpes simplex virus (HSV) *Tk* gene (Mansour *et al.* 1988). If the *neo* marker is used to interrupt the homology region in a replacement vector, transformed cells can be selected using G418. The HSV *Tk* gene is placed outside the homology region, such that it is inserted by random integration but not by homologous recombination. Therefore, cells that have undergone homologous recombination will survive in the presence of the toxic thymidine analogs ganciclovir or FIAU (1-(2-deoxy-2-fluoro-β-D-arabinofuranosyl)-5-iodouracil), while in those cells containing randomly integrated copies of the *Tk* gene, the analogs will be incorporated into the DNA, resulting in cell death.

A different strategy is to make expression of the *neo* gene dependent on homologous recombination, which is achieved by having no promoter in the vector (promoterless *neo* vector). Under these circumstances, the *neo* gene is not expressed following random integration, but comes under the control of the endogenous gene's promoter following a genuine targeting event (e.g. see Schwartzberg *et al.* 1989, Mansour *et al.* 1993). This method works only if the endogenous gene is normally expressed in ES cells. Another alternative to positive-negative selection, which is used in many laboratories, is simply to screen large numbers of G418-resistant transfected cells by the PCR to identify genuine recombinants (Hogan & Lyons 1988, Zimmer & Gruss 1989). Screening can be carried out relatively quickly without recourse to cloning.

Two rounds of gene targeting allow the introduction of subtle mutations

While gene targeting has often been used to disrupt and hence inactivate specific endogenous genes by introducing large insertions, more refined approaches can be used to generate subtle mutations. The precise effects of minor deletions or point mutations cannot be assessed using the simple targeting strategies discussed above, which necessarily leave the selectable marker and, in some cases, the entire targeting vector integrated at the target site. Furthermore, many investigators have reported that the strong promoters used to drive marker-gene expression can affect the regulation of neighboring genes, in some cases up to 100 kb away from the targeted locus (Pham *et al.* 1996). Two major strategies for the introduction of subtle mutations have been devised, each involving two rounds of homologous recombination. These are the "hit-and-run" strategy, involving a single insertion vector, and the "tag-and-exchange" strategy, involving two replacement vectors (Fig. 13.5). Strategies involving Cre recombinase have also been developed and are discussed on p. 307.

The "hit-and-run" or "in-out" strategy (Hasty *et al.* 1991a, Valancius & Smithies 1991) involves the use of an insertion vector carrying two selectable markers, such as *neo* and *Tk*. The insertion event is positively selected using G418. As discussed above, the use of an insertion vector results in duplication of the homology region in the targeted clone, although in this case the homology region derived from the vector is modified to contain the desired subtle mutation. The success of this strategy relies on a second

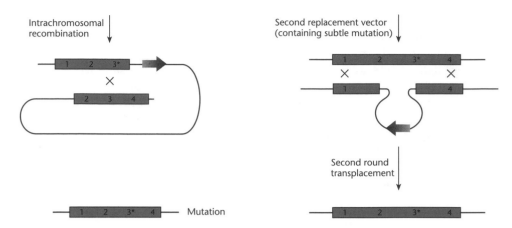

Fig. 13.5 Procedure for the introduction of subtle mutations using (a) an insertion vector and (b) two transplacement vectors. Regions of homology between the vector and target are designated by the same number. An asterisk indicates the position of a point mutation in the vector, which is to be introduced at the target locus.

intrachromosomal homologous recombination event, which replaces the endogenous allele with the mutant and deletes the markers, allowing the second-round recombinants to survive in the presence of ganciclovir. However, the second homologous recombination event occurs at a very low frequency and, in 50% of cases, restores the locus to its original wild-type configuration.

The "tag-and-exchange" strategy also requires two homologous recombination events, but in this case two replacement-type vectors are used. For example, Moore *et al.* (1995) demonstrated the principle using *neo* for positive selection and HSV *Tk* for negative selection. The first "tag" vector was designed to mutate the target gene by inserting a large cassette containing the selectable markers. This event was positively selected with G418. The second "exchange" vector introduced the desired mutation and eliminated the selectable markers, allowing the second-round recombinants to be selected for the absence of *Tk*.

Recent advances in gene-targeting technology

The low efficiency of gene targeting in mammalian somatic cells, particularly human cells, is a significant bottleneck in the use of this technique for applications such as gene therapy. Therefore, many researchers have looked at ways to increase the frequency of targeting events by manipulating the cellular recombination machinery, either by the application of chemical inhibitors or through genetic intervention. One example of the former approach is the application of 1,5-isoquinolinediol, an inhibitor of poly (ADP-ribose) polymerase (PARP). This was

shown to increase the frequency of gene-targeting events in mouse fibroblasts by tenfold (Semionov *et al.* 2003). Targets for genetic manipulation include genes involved in homologous and illegitimate recombination, some of which have been identified through mutations that inhibit gene targeting. Examples include the mouse *Rad51*, whose overexpression in ES cells significantly increased the frequency of gene targeting (Dominguez-Bendala *et al.* 2003) and the human *RAD52* gene, whose overexpression increased the gene-targeting frequency in a human somatic line (Yanez and Porter 2002). Zwaka & Thomson (2003) were the first to demonstrate gene targeting in human ES cells, using a promoterless *neo* vector to target the *HPRT1* and *OCT4* loci. However, highly efficient gene targeting in human cell lines has been demonstrated using adeno-asociated virus vectors, which give targeting frequencies of up to 1% (10,000 times more efficiency than transfection methods) (Russel *et al.* 1998, 2002, Hirata *et al.* 2000, 2002). The introduction of double-stranded DNA breaks at the sites of vector homology within chromosomal DNA can increase targeting efficiencies still further when recombinant AAV vectors are used (Porteus *et al.* 2003, Miller *et al.* 2003).

Applications of genetically modified mice

Applications of transgenic mice

Transgenic mice, i.e. mice containing additional transgenes as opposed to those with targeted mutations, have been used to address many aspects of gene function and regulation. A vast literature has accumulated on this subject, and genes concerning every conceivable biological process have been investigated (e.g. see Houdebine 1997). As well as their use for basic scientific investigation, transgenic mice can be used for more applied purposes, such as models for human disease and the production of valuable pharmaceuticals. Many mouse models for human diseases have been generated by gene knock-out (see below), but gain-of-function models have also been generated by adding transgenes. For example, much information concerning the pathology of prion diseases has arisen from the study of transgenic mice expressing mutant prion transgenes (reviewed by Gabizon & Taraboulos 1997). Transgenic mice expressing oncogenes have been extensively used to study cancer (reviewed by Macleod & Jacks 1999).

For illustrative purposes, we now consider some early experiments that demonstrated how transgenic mice can be used for the analysis of gene function and regulation, but also highlighted some limitations of the transgenic approach. Brinster *et al.* (1981) constructed plasmids in which the promoter of the mouse metallothionein-1 (*MMT*) gene was fused to the coding region of the HSV *Tk* gene. The thymidine kinase (TK) enzyme can be assayed readily and provides a convenient reporter of *MMT* promoter function. The endogenous *MMT* promoter is inducible by glucocorticoid hormones and heavy metals, such as cadmium and zinc, so it was envisaged that the hybrid transgene, *MK* (metallothionein-thymidine kinase), would be similarly regulated. The gene was injected into the male pronucleus of fertilized eggs, which were then incubated *in vitro* in the presence or absence of cadmium ions (Brinster *et al.* 1982). As expected, TK activity was found to be induced by the metal. By making a range of deletions of mouse sequences upstream of the *MMT* promoter sequences, the minimum region necessary for inducibility was localized to a stretch of DNA 40–180 nucleotides upstream of the transcription-initiation site. Additional sequences that potentiate both basal and induced activities extended to at least 600 bp upstream of the transcription-initiation site. The mouse egg was therefore being used in the same way as transfected cell lines, to dissect the activity of a functional promoter (Gorman *et al.* 1982b; see Box 12.1).

The same *MK* fusion gene was injected into embryos, which were raised to transgenic adults (Brinster *et al.* 1981). Most of these mice expressed the *MK* gene and in such mice there were from one to 150 copies of the gene. The reporter activity was inducible by cadmium ions and showed a tissue distribution very similar to that of metallothionein itself (Palmiter *et al.* 1982b). Therefore these experiments showed that DNA sequences necessary for heavy-metal induction and tissue-specific expression could be functionally dissected in both eggs and transgenic mice. For unknown reasons, there was no response to glucocorticoids in either the egg or the transgenic-mouse experiments.

In a dramatic series of experiments, Palmiter *et al.* (1982a) fused the *MMT* promoter to the rat growth-hormone gene. This hybrid gene (*MGH*) was constructed using the same principles as the *MK* fusion. Of 21 mice that developed from microinjected eggs, seven carried the *MGH* fusion gene and six of these grew significantly larger than their littermates. The mice were fed zinc to induce transcription of the *MGH* gene, but this did not appear to be absolutely necessary, since they showed an accelerated growth rate before being placed on the zinc diet. Mice containing high copy numbers of the *MGH* gene (20–40 copies per cell) had very high concentrations of growth hormone in their serum, some 100–800 times above normal. Such mice grew to almost double the weight of littermates at 74 days old (Fig. 13.6).

The similarities between the tissue distribution of normal *MMT* expression and that of the hybrid transgenes encouraged the hope that transgenic

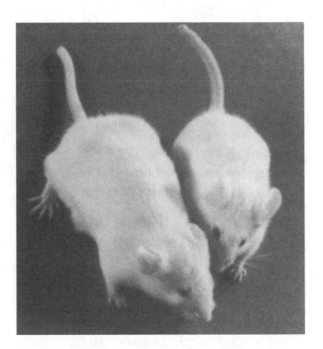

Fig. 13.6 Transgenic mouse containing the mouse metallothionein promoter fused to the rat growth-hormone gene. The photograph shows two male mice at about 10 weeks old. The mouse on the left contains the *MGH* gene and weighs 44 g; his sibling without the gene weighs 29 g. In general, mice that express the gene grow two to three times as fast as controls and reach a size up to twice the normal. (Photograph by courtesy of Dr. R.L. Brinster.)

mice would provide a general assay for functionally dissecting DNA sequences responsible for tissue-specific or developmental regulation of a variety of genes. However, there were also some unexpected findings. For example, independently derived transgenic mice carrying the *MK* transgene showed significant variations in the levels and patterns of transgene expression. Furthermore, while transgenic founders transmitted the construct to their progeny as expected, when reporter activity was assayed in these offspring the amount of expression could be very different from that in the parent. Examples of increased, decreased, or even totally extinguished expression were found. In some, but not all, cases, the changes in expression correlated with changes in methylation of the gene sequences (Palmiter *et al.* 1982b). These results provided the first examples of two complex phenomena, *position effects* (Box 13.2) and *de novo transgene silencing* (Box 15.2), which often affect integrated transgenes.

Box 13.2   Position effects

Independently derived transgenic animals and plants carrying the same expression construct often show variable levels and patterns of transgene expression. In many cases, such variation is dependent on the site of transgene integration, and this phenomenon has been termed the *position effect* (reviewed by Wilson *et al.* 1990). Position effects result from the influence of local regulatory elements on the transgene, as well as the architecture of the surrounding chromatin. For example, an integrated transgene may come under the influence of a local enhancer, resulting in the alteration of its expression profile to match that of the corresponding endogenous gene. The position dependence of the phenomenon has been demonstrated in mice by isolating the entire transgenic locus from such an anomalous line and microinjecting it into the pronuclei of wild-type eggs, resulting in "secondary" transgenic lines with normal transgene expression profiles (Al-Shawi *et al.* 1990). Position effects are also revealed by enhancer-trap constructs, which contain a minimal promoter linked to a reporter gene (O'Kane & Gehring 1987; see Chapter 15).

Unlike the specific influences of nearby regulatory elements, chromatin-mediated position effects are generally non-specific and repressive. They reflect the integration of the transgene into a chromosomal region containing repressed chromatin (heterochromatin). The molecular features of heterochromatin, including its characteristic nucleosome structure, deacetylated histones and, in many cases, hypermethylated DNA,

spread into the transgene, causing it to be inactivated (Huber *et al.* 1996, Pikaart *et al.* 1998). In some cases, variegated transgene expression has been reported due to cell-autonomous variations in the extent of this spreading process (reviewed by Heinkoff 1990). Negative chromosomal position effects can be troublesome in terms of achieving desirable transgene expression levels and patterns; thus a number of different strategies have been used to combat them.

Incorporating dominantly acting transcriptional control elements

Certain regulatory elements are thought to act as master-switches, regulating the expression of genes or gene clusters by helping to establish an open chromatin domain. The locus control region (LCR) of the human β-globin gene cluster is one example (Forrester *et al.* 1987). Transgenic mice carrying a human β-globin transgene driven by its own promoter show a low frequency of expression and, in those mice that do express the transgene, only a low level of the mRNA is produced (e.g. Magram *et al.* 1985, Townes *et al.* 1985). However, inclusion of the LCR in the expression construct confers high-level and position-independent expression (Grosveld *et al.* 1987). There is evidence that LCRs induce chromatin remodeling over large distances. For example, the murine immunoglobulin heavy-chain LCR has been shown to induce histone acetylation in a linked c-*myc* gene (Madisen *et al.* 1998).

continued

Box 13.2 *continued*

This suggests that LCRs could protect against position effects by converting heterochromatin to open euchromatin at the site of transgene integration (Festenstein *et al.* 1996, Milot *et al.* 1996). The interested reader can consult several comprehensive reviews of LCR research (Bonifer 1999, Grosveld 1999, Li *et al.* 1999).

Using boundary elements/matrix attachment regions

Boundary elements (insulators) are sequences that can block the activity of enhancers when placed between the enhancer and a test transgene driven by a minimal promoter. For example, an "A element" with insulator activity is found upstream of the chicken lysozyme gene. This inhibits the activity of a reporter gene when interposed between the promoter and an upstream enhancer, but not when placed elsewhere in the construct (Stief *et al.* 1989). However, by flanking the entire construct with a pair of A elements, the transgene is protected from chromosomal position effects (Stief *et al.* 1989). This protective effect works not only in cell lines, but also in transgenic animals (McKnight *et al.* 1992) and plants (Mlynarova *et al.* 1994). Many boundary elements are associated with matrix-attachment regions (MARs), sequences dispersed throughout the genome that attach to the nuclear matrix, dividing chromosomes into topologically independent loops (reviewed by Spiker & Thompson 1996). It is therefore possible that transgenes flanked by such elements are maintained in an isolated chromatin domain into which heterochromatin cannot spread. However, not all boundary elements are associated with MARs (e.g. see Mirkovitch *et al.* 1984). Similarly, some MARs do not function as boundary elements but as facilitators of gene expression (e.g. Van der Geest & Hall 1997).

Using large genomic transgenes

Conventional transgenes generally comprise complementary DNAs (cDNAs) or intronless "minigenes" expressed under the control of viral promoters or cell-type-specific regulatory elements. Such transgenes are highly sensitive to position effects. Over the last few years, there has been an increasing appreciation that the regulation of eukaryotic gene expression is far more complex and involves much more upstream and downstream DNA than previously thought (reviewed by Bonifer 1999, 2000). The correct, high-level expression of transgenes is favored by the use of genomic constructs that include introns and large amounts of flanking sequence from the source gene (e.g. see Bonifer *et al.* 1990, Lien *et al.* 1997, Nielsen *et al.* 1998). Such constructs are likely to include multiple enhancers, dominant regulatory elements such as LCRs, and boundary elements, which all act together to protect the transgene from position effects.

Dominantly acting transgenes (transgene rescue)

Some conventional transgenes, including β-globin and α-fetoprotein (Chada *et al.* 1986, Kollias *et al.* 1986, Hammer *et al.* 1987) are very sensitive to position effects and *de novo* silencing. Other genes appear to be less sensitive to these phenomena, e.g. immunoglobulin and elastase (Storb *et al.* 1984, Swift *et al.* 1984, Davis & MacDonald 1988). Although the reason for this is not clear, the less sensitive transgenes are assumed to in some way induce or define an open chromatin domain. In some cases, such sequences have been used to protect more susceptible transgenes from negative position effects by introducing the two transgenes simultaneously (e.g. see Clark *et al.* 1992).

Site-specific integration

Site-specific recombination systems (see Chapter 15) can be used to introduce transgenes into a locus known to lack negative position effects, if a target site for the recombinase can be introduced at such a locus. The Cre-*loxP* system has been used to this effect in mammalian cells (see Fukushige & Sauer 1992).

Yeast artificial chromosome (YAC) transgenic mice

Studies of the *MMT* promoter and others have demonstrated the principle that transgenes with minimal flanking sequences tend not to be expressed in the same manner as the corresponding endogenous gene. In many cases, it has also been shown that authentic patterns and levels of protein expression occur only when the intact gene is used, and this can span tens or hundreds of kilobase pairs of DNA (Box 13.2). The transfer of large DNA segments to the mouse genome has been achieved by transformation with yeast artificial chromosome (YAC) vectors. Jakobovits *et al.* (1993) were the first to report transformation of ES cells with a YAC vector, via fusion with yeast spheroplasts. The vector contained the entire human *HPRT* locus, nearly 700 kb in length. The disadvantage of this method is that the endogenous yeast chromosomes were co-introduced with the vector. Alternative strategies involve isolation of the vector DNA by pulsed-field gel electrophoresis (p. 17), followed by introduction of the purified YAC DNA into mouse eggs by pronuclear microinjection or transfection into ES cells. The latter technique is more suitable because microinjection involves shear forces that break the DNA into fragments. YAC transfer to ES cells has been achieved by lipofection, as discussed in Chapter 12. YAC transgenics have been used to study gene regulation, particularly by long-range regulatory elements, such as locus-control regions (reviewed by Lamb & Gerhart 1995). They have also been used to introduce the entire human immunoglobulin locus into mice, for the production of fully humanized antibodies (Mendez *et al.* 1997). It is also possible to introduce chromosomes and chromosome fragments into ES cells using a technique called microcell-mediated fusion. This involves the prolonged mitotic arrest of cultured human cells, using an inhibitor such as colchicine. Eventually, the nucleus breaks up into vesicles containing individual chromosomes, which can be rescued as microcells comprising a nuclear vesicle surrounded by a small amount of cytoplasm and a plasma membrane (Fournier & Ruddle 1977). Transgenic mice have been generated using ES cells that were fused to human microcells, and evidence for germline transmission and expression of the human chromosome was obtained (Tomizuka *et al.* 1997).

Applications of gene targeting

Since the first reports of gene targeting in ES cells, an ever-increasing number of targeted mutant mice have been produced. These have been discussed in several comprehensive reviews (Brandon *et al.* 1995a,b,c, Soriano 1995, Muller 1999) and a number of Internet databases have been established to keep track of the results (see Sikorski & Peters 1997). The phenotypes of homozygous, null mutant mice provide important clues to the normal function of the gene. Some gene knockouts have resulted in surprisingly little phenotypic effect, much less severe than might have been expected. For example, *myoD*, whose expression in transfected fibroblasts causes them to differentiate into muscle cells, and which was therefore a good candidate as a key regulator of myogenesis, is not necessary for development of a viable animal (Rudnicki *et al.* 1992). Similarly, the retinoic acid γ receptor is not necessary for viable mouse development in knockout mice (Lohnes *et al.* 1993), even though this receptor is a necessary component of the pathway for signaling by retinoids and has a pattern of expression quite distinct from other retinoic acid receptors in embryos. Such observations have prompted speculation that genetic redundancy may be common in development, and may include compensatory up-regulation of some members of a gene family when one member is inactivated. An example of this is the up-regulation of *myf-5* in mice lacking *myoD* (Rudnicki *et al.* 1992). Gene knockouts have also been used as mouse models of human diseases such as cystic fibrosis, β-thalassemia, and fragile X syndrome (reviewed by Bedell *et al.* 1997; see Chapter 26).

While most gene-targeting experiments in mice have been used to introduce mutations into genes (either disruptive insertional mutations or subtle changes), the scope of the technique is much wider. The early gene-targeting experiments demonstrated that this approach could also be used to correct mutated genes, with obvious applications in gene therapy. Homologous recombination has also been used to exchange the coding region of one gene for that of another, a strategy described as "gene knock-in". This has been used, for example, to test the ability of the transcription factors Engrailed-1 and Engrailed-2 to compensate for each other's functions. Hanks *et al.* (1995) replaced the coding region of the *engrailed-1* gene with that of *engrailed-2*, and showed that the *engrailed-1* mutant phenotype could be rescued. A more applied use of gene knock-in is the replacement of parts of the murine immunoglob-ulin genes with their human counterparts, resulting in the production of humanized antibodies in trans-genic mice (Moore *et al.* 1995a). The Cre-*loxP* site-specific recombinase system has been used extensively in ES cells to generate mice in which conditional or inducible gene targeting is possible and to produce defined chromosome deletions and *trans*-locations as models for human disease. We shall discuss the many applications of Cre-*loxP* and other site-specific recombinase systems in Chapter 15.

Standard transgenesis methods are more difficult to apply in other mammals and birds

The three major routes for producing transgenic mice have also been used in other mammals and birds, particularly in farm animals. The efficiency of each procedure is much lower than in mice. Pro-nuclear microinjection in mammals such as sheep and cows, for example, typically results in less than 1% of the injected eggs giving rise to transgenic ani-mals. Added to this, the recovery of eggs from donor animals and the reimplantation of transformed eggs into foster mothers is a less efficient procedure and requires, at great expense, a large number of donors and recipients. The eggs themselves are also more difficult to manipulate – they are very delicate and tend to be opaque. It is often necessary to centrifuge the eggs in order to see the pronuclei. In chickens, it is possible to remove eggs just after fertilization and microinject DNA into the cytoplasm of the germinal disc, where the male and female pronuclei are to be found. However, it is not possible to return the manipulated eggs to a surrogate mother, so they must be cultured *in vitro*. Using this procedure, Love *et al.* (1994) obtained seven chicks, equivalent to about 5% of the eggs injected, that survived to sexual maturity. One cockerel transmitted the transgene to a small proportion of his offspring, indicating that he was mosaic for transgene integration.

The use of retroviruses to produce transgenic chickens has been reported by Bosselman *et al.* (1989). These investigators injected a replication-defective recombinant reticuloendotheliosis virus carrying the *neo* gene into laid eggs and found that approximately 8% of male birds carried vector sequences. The transgene was transmitted through the germline in a proportion of these birds and was stably expressed in 20 transgenic lines. As discussed in Chapter 12, oncoretroviral vectors can only infect dividing cells because they can only gain entry to the nucleus when the nuclear membrane breaks down. Since the nuclear envelope breaks down during meiosis as well as mitosis, Chan *et al.* (1998) were able to produce transgenic cattle following the injec-tion of replication-defective retroviral vectors into the perivitelline space of isolated bovine oocytes. Retroviral integration occurred during the second meiotic division, resulting in the production of a number of transgenic offspring. The same technique was later used to generate the first ever transgenic primate, a rhesus monkey named ANDi[2] (Chan *et al.* 2001), but the technique was very inefficient: 224 oocytes were injected to produce one live transgenic monkey; a number of further transgenic fetuses failed to develop to term. As well as their reliance on nuclear division, oncoretroviral vectors are also subject to *de novo* silencing and consequent loss of transgene expression. These problems have been addressed more recently through the development of lentivirus vectors, which can infect non-dividing cells (p. 245). Such vectors have been used to pro-duce transgenic mice, rats, cattle, and pigs (Louis *et al.* 2002, Pfeifer *et al.* 2002, Hoffman *et al.* 2003, 2004; reviewed by Pfeifer 2004).

After nearly two decades of research, it has also proved impossible to derive reliable ES cell lines from any domestic mammal other than mice (although chicken ES cells have been isolated, and have been used to develop transgenic strategies for birds; Pain *et al.* 1999, Prelle *et al.* 1999). As an alternative target, some researchers have sought to isolate primordial germ cells (PGCs), the embryonic cells that give rise to gametes. These can be transfected directly, or cultured as embryonic germ (EG) cells, which are morphologically very similar to ES cells and could provide a route for the direct transforma-tion of the germline (Resnick *et al.* 1992). Chicken PGCs have been isolated from the germinal crescent, infected with a recombinant retrovirus and replaced in the embryo, leading to the development of chimeric birds producing transgenic offspring (Vick *et al.* 1993). PGCs from mice, rabbits, cattle, and pigs have also been isolated and transformed, although it has been difficult to persuade the cells to contribute to the germline once introduced into the host animal (Brinster 2002). A similar technique has been devel-oped more recently, in which male PGCs are allowed

---

[2] The name comes from the phrase inserted DNA written backwards.

to mature into spermatogonia (the immature sperm cells) prior to transfection. These cells can be reintroduced into the testis, where they give rise to transgenic sperm. This method has been used successfully to produce transgenic mice and pigs (Kim *et al.* 1997, Honaramooz *et al.* 2003).

Intracytoplasmic sperm injection uses sperm as passive carriers of recombinant DNA

The injection of sperm heads directly into the cytoplasm of the egg (intracytoplasmic sperm injection (ICSI)) can overcome infertility in humans. It has been shown that sperm heads bind spontaneously to naked plasmid DNA *in vitro*, suggesting that sperm injections could be used to achieve transformation. This was demonstrated by Perry *et al.* (1999), who mixed mouse sperm with plasmid DNA carrying the gene for green fluorescent protein (GFP). These sperm were injected into unfertilized oocytes, and a remarkable 94% of the resulting embryos showed GFP activity. Random transfer of these embryos to pseudopregnant females resulted in development to term, and in about 20% of cases the mice were transgenic. This method could be adaptable to other animals. Rhesus monkey oocytes fertilized in the same manner gave rise to a number of embryos with GFP activity, but this only lasted until the blastula stage, suggesting that there was no stable integration. However, several monkeys developed to term, showing that the procedure was compatible with normal development (Chan *et al.* 2000).

Nuclear transfer technology can be used to clone animals

The failure of traditional transgenesis techniques to yield routine procedures for the genetic modification of mammals other than mice has driven researchers in search of other methods. Over fifty years ago, Briggs & King (1952) established the principle of nuclear transfer in amphibians by transplanting nuclei from the blastula of the frog *Rana pipens* to an enucleated egg, obtaining a number of normal embryos in the process. In *Xenopus laevis*, nuclei from various types of cell in the swimming tadpole can be transplanted to an egg that has been UV-irradiated to destroy the peripheral chromosomes, and similar results are obtained (reviewed by Gurdon 1986, 1991). The important principle here is that, while animal cells become irreversibly committed to their fate as development proceeds, the nuclei of most cells still retain all the genetic information required for the entire developmental program and can, under appropriate circumstances, be reprogrammed by the cytoplasm of the egg to recapitulate development. In all species, it appears that the earlier the developmental stage at which nuclei are isolated, the greater their potential to be reprogrammed. Nuclear transplantation can be used to generate clones of animals with the same genotype by transplanting many somatic nuclei from the same individual into a series of enucleated eggs (King & Briggs 1956). This allows animals with specific and desirable traits to be propagated. If possible in mammals, this would have obvious applications in farming.

Nuclear transfer in mammals has been practiced with success for the last decade, although rabbits and farm animals, such as sheep, pigs, and cows, are far more amenable to the process than mice. In each case, donor nuclei were obtained from the morula or blastocyst-stage embryo and transferred to an egg or oocyte from which the nucleus had been removed with a pipette (Smith & Wilmut 1989, Willadsen 1989, Collas & Robl 1990, McLaughlin *et al.* 1990). The donor nucleus can be introduced by promoting fusion between the egg and a somatic cell. A brief electric pulse is often used to achieve this, as it also activates embryonic development by stimulating the mobilization of calcium ions.

A major advance was made in 1995, when two live lambs, Megan and Morag (Fig. 13.7), were produced by nuclear transfer from cultured embryonic cells (Campbell *et al.* 1996). This demonstrated the principle that mammalian nuclear transfer was possible using a cultured cell line. The same group later reported the birth of Dolly (Fig. 13.8), following nuclear

Fig. 13.7 Megan and Morag, the first sheep produced by nuclear transfer from cultured cells. Reproduced by kind permission of the Roslin Institute, Edinburgh.

Fig. 13.8 Dolly and her lamb Bonnie. Dolly was the first mammal to be generated by nuclear transfer from an adult cell. Reproduced by kind permission of the Roslin Institute, Edinburgh.

transfer from an adult mammary epithelial cell line (Wilmut *et al.* 1997). This was the first mammal to be produced by nuclear transfer from a differentiated adult cell, and aroused much debate among both scientists and the public concerning the possibility of human cloning (see Johnson 1998). It was suggested that a critical factor in the success of the experiment was the quiescent state of the cells in culture, allowing synchronization between the donor and recipient cell cycles (reviewed by Wilmut *et al.* 2002). For the production of Dolly, this was achieved by lowering the level of serum in the culture medium, causing the cells to withdraw from the cell cycle due to lack of growth factors. However, the success rate was very low: only one of 250 transfer experiments produced a viable lamb, a phenomenon that has been blamed on a lack of fundamental understanding of the nuclear reprogramming events that occur following transplantation (Shi *et al.* 2003). Similar transfer experiments have since been carried out in mice, cows, pigs, goats, cats, dogs, rabbits, mules, and rats using variations on the transfer methodology developed by Wilmut and colleagues (Cibelli *et al.* 1998, Wakayama *et al.* 1998, Baguisi *et al.* 1999, Polejaeva *et al.* 2000, Shin *et al.* 2002, Chesne *et al.* 2002, Galli *et al.* 2003, Woods *et al.* 2003, Zhou *et al.* 2003, Lee *et al.* 2005; see reviews by Denning & Priddle 2003, Edwards *et al.* 2003). Despite major efforts, there has been no success thus far in the production of a cloned primate although

various individuals, nations, and religious sects have put forward as yet unsubstantiated claims to have produced the first cloned human being. There are also several ongoing projects looking at the feasibility of cloning rare animals or animals in captivity representing species that are extinct in the wild. For example, a cloned guar (a rare and endangered type of ox native to Asia) was born to a surrogate cow in 2001, although it died the same day from an infection.

The success of nuclear transfer in domestic mammals provides a new route for the production of transgenic animals. This involves the introduction of DNA into cultured cells, which are then used as a source of donor nuclei for nuclear transfer. Such a cell-based strategy has many advantages over traditional techniques, such as microinjection, including the ability to screen transformed cells for high-level transgene expression prior to the nuclear-transfer step. The production of a transgenic mammal by nuclear transfer from a transfected cell line was first achieved by Schnieke *et al.* (1997), who introduced the gene for human factor IX into fetal sheep fibroblasts and transferred the nuclei to enucleated eggs. The resulting sheep, Polly, produces the recombinant protein in her milk and can therefore be used as a bioreactor (Chapter 26). McCreath *et al.* (2000) succeeded in producing a transgenic sheep by nuclear transfer from a somatic cell whose genome had been specifically modified by gene targeting. A foreign gene was introduced into the *COL1A1* locus and was expressed at high levels in the lamb. Nuclear transfer was also used to produce the first "gene knockout" mammals other than mice: Denning *et al.* (2001) produced cloned lambs with targeted disruptions of either the *PRP* gene, which encodes the prion protein, or the *GGTA1* gene, which encodes an enzyme that adds carbohydrate groups to proteins that provoke the human immune system, one of the most significant obstacles to xenotransplantation, the transplantation of organs from animals such as monkey, sheep and pigs to human recipients. In 2002, two groups independently reported targeted disruption of the *GGTA1* gene in pigs, whose organs are similar in size to their human counterparts and are envisaged as the most likely source for xenotransplants (Dai *et al.* 2002, Lai *et al.* 2002). In each of these reports only one allele of the autosomal target gene was disrupted. The first homozygous *GGTA1* knockout pigs were produced by Phelps *et al.* (2003) using heterozygous cells derived from the earlier cloned pigs (Lai *et al.* 2002). The aim of the experiment was to disrupt the second *GGTA1* allele by gene targeting and then select cells lacking the gal-epitope using a fungal toxin. However, on further analysis it was established that the second allele had been produced by a spontaneous point mutation within the *GGTA1* gene, so in the strict sense the pigs were still heterozygous, although for two null alleles.

Gene transfer to *Xenopus* can result in transient expression or germline transformation

Xenopus oocytes can be used as a heterologous expression system

Gurdon *et al.* (1971) first showed that *Xenopus* oocytes synthesized large amounts of globin after they had been microinjected with rabbit globin mRNA. Since then, the *Xenopus* oocyte expression system has been a valuable tool for expressing a very wide range of proteins from plants and animals (Colman 1984). *X. laevis* is an African clawed frog. Oocytes can be obtained in large numbers by removal of the ovary of an adult female. Each fully grown oocyte is a large cell (0.8–1.2 mm diameter) arrested at first meiotic prophase. This large cell has a correspondingly large nucleus (called the *germinal vesicle*), which is located in the darkly pigmented hemisphere of the oocyte.

Due to the large size of the oocytes, mRNA – either natural or synthesized by transcription *in vitro*, using phage-T7 RNA polymerase (Melton 1987) – can be readily introduced into the cytoplasm or nucleus by microinjection. This is achieved using a finely drawn glass capillary as the injection needle, held in a simple micromanipulator. DNA can also be injected. The oocyte nucleus contains a store of the three eukaryotic RNA polymerases, enough to furnish the needs of the developing embryo at least until the 60,000-cell stage. The RNA polymerases are available for the transcription of injected exogenous DNA. Using this system, it has therefore been possible to express complementary DNAs (cDNAs) linked to a heat-shock promoter or to mammalian virus promoters (Ballivet *et al.* 1988, Ymer *et al.* 1989, Swick *et al.* 1992). In addition, vaccinia virus vectors (Chapter 12) can be used for gene expression in the cytoplasm (Yang *et al.* 1991).

An important aspect of the oocyte expression system is that recombinant proteins are usually correctly post-translationally modified and directed to the correct cellular compartment. For example,

oocytes translate a wide variety of mRNAs encoding secretory proteins, modify them, and correctly secrete them (Lane *et al.* 1980, Colman *et al.* 1981). Foreign plasma-membrane proteins are generally targeted to the plasma membrane of the oocyte, where they can be shown to be functional. The first plasma-membrane protein to be expressed in this system was the acetylcholine receptor from the electric organ of the ray, *Torpedo marmorata* (Sumikawa *et al.* 1981). Injected oocytes translated mRNA extracted from the electric organ and assembled functional multi-subunit receptor molecules in the plasma membrane (Barnard *et al.* 1982). Following this work, the oocyte has become a standard heterologous expression system for plasma-membrane proteins, including ion channels, carriers, and receptors. The variety of successfully expressed plasma-membrane proteins is very impressive. However, there are examples of foreign channels and receptors being non-functional in oocytes, either due to lack of coupling to second-messenger systems in the oocyte, incorrect post-translational modification, or other reasons (reviewed in Goldin 1991).

Xenopus oocytes can be used for functional expression cloning

Functional expression cloning using oocytes was first developed by Noma *et al.* (1986), using a strategy outlined in Fig. 13.9. The following example, the cloning of the substance-K receptor, is illustrative. It has been found that oocytes can be made responsive to the mammalian tachykinin neuropeptide,

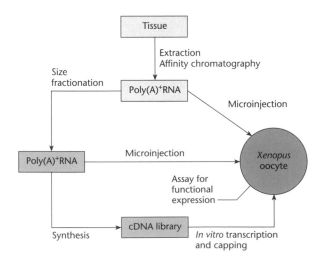

Fig. 13.9 Strategy for functional expression cloning, using *Xenopus* oocytes as a heterologous expression system.

substance K, by injecting an mRNA preparation from bovine stomach into the oocyte cytoplasm. The preparation contains mRNA encoding the substance-K receptor protein, which is evidently expressed as a functional protein and inserted into the oocyte membrane. Masu *et al.* (1987) exploited this property to isolate a cDNA clone encoding the receptor. The principle was to make a cDNA library from stomach mRNA, using a vector in which the cDNA was flanked by a promoter for the SP6 or T7 RNA polymerase. This allowed *in vitro* synthesis of mRNA from the mixture of cloned cDNAs in the library.

The receptor clone was identified by testing for receptor expression following injection of synthetic mRNA into the oocyte cytoplasm. Repeated subdivision of the mixture of cDNAs in the library led to the isolation of a single cloned cDNA. The strategy described above can only be applied to cloning single-subunit proteins, not proteins composed of different subunits or proteins whose function in oocytes requires more than one foreign polypeptide. This limitation was overcome by Lubbert *et al.* (1987), who used a hybrid depletion procedure to clone a serotonin-receptor cDNA.

A prerequisite for using the oocyte in functional expression cloning is a knowledge of the oocyte's own ion channels, carriers, and receptors. Endogenous activity may mask or interfere with the sought-after function (for a review, see Goldin 1991).

Transient gene expression in *Xenopus* embryos is achieved by DNA or mRNA injection

Messenger RNA, synthesized and capped *in vitro*, can be microinjected into dejellied *Xenopus* embryos at the one- or two-cell stage. The mRNA is distributed among the descendants of the injected cells and is expressed during early development. This approach has been exploited very widely for examining the developmental effects resulting from the overexpression of normal or altered gene products (reviewed by Vize & Melton 1991).

DNA can be introduced into *Xenopus* embryos in the same manner. However, unlike the situation in mammals, where the injected DNA integrates rapidly into the genome, exogenous DNA in *Xenopus* persists episomally and undergoes extensive replication (Endean & Smithies 1989). Bendig & Williams (1983) provide a typical example of this process. They injected a recombinant plasmid carrying *Xenopus* globin genes into the egg and showed that the amount of plasmid DNA increased 50- to 100-fold by the gastrula stage. In later development, the amount of DNA per embryo decreased, and most of the persisting DNA co-migrated with high-molecular-weight chromosomal DNA. This difference between mammals and amphibians probably reflects their distinct modes of early development. In mammals, cleavage divisions are slow and asynchronous. Gene expression occurs throughout early development and supplies the embryo with the proteins it requires at a steady rate. Conversely, there is no transcription in the early *Xenopus* embryo and yet the cleavage divisions are rapid and synchronous. DNA replication relies on stored maternal gene products, so there is a stockpile of chromatin assembly proteins and replication enzymes. Exogenous DNA injected into *Xenopus* eggs is therefore assembled immediately into chromatin and undergoes replication in tune with the rapid DNA synthesis already occurring in the nucleus (Leno & Laskey 1991). Etkin *et al.* (1987) have analyzed the replication of a variety of DNAs injected into *Xenopus* embryos. It was found that various plasmids increase to different extents. This was not simply related to the size of the plasmid, but also reflected the presence of specific sequences that inhibited replication. Replication has also been found to depend upon the conformation and number of molecules injected (Marini *et al.* 1989).

Transgenic *Xenopus* embryos can be produced by restriction enzyme-mediated integration

DNA injected into early *Xenopus* embryos is expressed in a mosaic fashion during development, regardless of the promoter used, which limits the use of this system for the analysis of gene expression and function. Some of the DNA does become incorporated into the genome and may be transmitted through the germline (Rusconi & Schaffner 1981). However, integration occurs at a very low frequency and, given the long generation interval of *Xenopus laevis* (12–18 months from egg to adult), this is not an efficient way to generate transgenic frogs.

A simple and efficient process for large-scale transgenesis in *Xenopus* has become available only in the last 10 years (Kroll & Amaya 1996). In this technique, known as restriction-enzyme-mediated integration (REMI), linearized plasmids containing the transgene of interest are mixed with decondensed sperm nuclei and treated with limiting amounts of a restriction enzyme to introduce nicks in the DNA. The nuclei are then transplanted into unfertilized *Xenopus* eggs, where the DNA is repaired, resulting in the integration of plasmid DNA into the genome. This technique allows the production of up to 700 transgenic embryos per person per day, most of which survive at least to the swimming-tadpole stage. The decondensed nuclei are extremely fragile, so careful handling and transplantation within about 30 min are required for a good yield of normal transgenic embryos. In some cases, viable transgenic adults have been derived from the tadpoles and transgenic *X. laevis* lines have been established (Bronchain *et al.* 1999, Marsh-Armstrong *et al.* 1999). A disadvantage of *X. laevis* is that the species is tetraploid. Offield *et al.* (2000) have therefore established transgenic lines of the closely related but diploid species *Xenopus tropicalis*, which also has a shorter generation interval than its tetraploid cousin.

Since *Xenopus* is used worldwide as a developmental model organism, transgenic *Xenopus* technology has been rapidly adopted in many laboratories and is being used to examine (or in many cases re-examine) the roles of developmental genes. Thus far, the sophisticated tools used in transgenic mice have not been applied to *Xenopus*, but this is only a matter of time. Recently, an inducible expression system based on the use of a *Xenopus* heat-shock promoter was described, allowing inducible control of the GFP gene. This system has been used to investigate Wnt signaling in early *Xenopus* development (Wheeler *et al.* 2000). As discussed above, one of the early successes in transgenic mouse methodology was the expression of rat growth hormone, resulting in transgenic mice up to twice the size of their non-transgenic siblings. The role of growth hormone in amphibian metamorphosis has now been examined by expressing *Xenopus* growth hormone in transgenic frogs. The transgenic tadpoles developed at the same rate as control tadpoles, but typically grew to twice the normal size (Huang & Brown 2000). After metamorphosis, the transgenic frogs also grew much more quickly than controls and showed skeletal defects.

Gene transfer to fish is generally carried out by microinjection, but other methods are emerging

Fish transgenesis can be used to study gene function and regulation, e.g. in model species, such as the zebrafish (*Danio rerio*) and medaka (*Oryzias latipes*), and to improve the traits of commercially important species, such as salmon and trout. Gene-transfer technology in fish has lagged behind that in mammals, predominantly due to the lack of suitable regulatory elements to control transgene expression. The first transgenic fish carried transgenes driven by mammalian or viral regulatory elements, and their performance varied considerably. For example, attempts to express growth-hormone genes in trout initially met with little success, and this may have been due to the inability of fish cells to process mammalian introns correctly (Betancourt *et al.* 1993). However, fish are advantageous assay systems for several reasons, including their fecundity, the fact that fertilization and development are external, and the ease with which haploid and uniparental diploid embryos can be produced (Ihssen *et al.* 1990).

Like frogs, the injection of DNA into fish eggs and early embryos leads to extensive replication and expression from unintegrated transgenes, so that fish, like frogs, can be used for transient expression assays (Vielkind 1992). Some of the DNA integrates into the genome, leading to germline transmission and the production of transgenic fish lines (reviewed by Iyengar *et al.* 1996, Zbikowska 2003). There has been recent progress in the development of novel methods to enhance DNA integration in fish, based on the use of purified enzymes from integrating DNA elements (i.e. transposons and retroviruses). The successful transfer of exogenous genes to zebrafish embryos using retroviral vectors was followed by the development of a hybrid technique in which DNA was injected into zebrafish embryos along with purified retroviral integrase, resulting in earlier transgene integration and a significant increase in integration efficiency. This prompted a search for natural fish enzymes that could perform the same function, but unfortunately all the fish transposons that were isolated contained deletions or other mutations that destroyed enzyme activity. This problem was addressed by collecting the sequences of many fish transposons and deriving a theoretical ideal sequence, resulting in a synthetic transposon system known as *Sleeping Beauty*. The use of *Sleeping Beauty* as a gene-transfer vector in zebrafish increases the efficiency of transgene integration over 20-fold, and could well be suitable for other species of fish (Zbikowska 2003). More recently, methods such as electroporation and particle bombardment have also been utilized with success for fish transgenesis, as well as sperm-mediated gene transfer. A novel technique applied in some species is the electroporation of sperm with plasmid DNA followed by sperm-mediated DNA transfer, a procedure which is analogous in some ways to the REMI technique described for amphibians.

Gene transfer to fruit flies involves the microinjection of DNA into the pole plasma

P elements are used to introduce DNA into the *Drosophila* germline

P elements are transposable DNA elements that, under certain circumstances, can be highly mobile in the germline of *D. melanogaster*. The subjugation of these sequences as specialized vector molecules in *Drosophila* was a landmark in *Drosophila* genetics. Through the use of P-element vectors, any DNA sequence can be introduced into the genome of the fly.

P elements cause a syndrome of related genetic phenomena called *P-M hybrid dysgenesis* (Bingham *et al.* 1982, Rubin *et al.* 1982). Dysgenesis occurs when males of a P (paternally contributing) strain are mated with females of an M (maternally contributing) strain, but not when the reciprocal cross is made. The syndrome predominantly affects the germline and induces a high rate of mutation and frequent chromosomal aberrations, resulting in abnormal (dysgenic) hybrid offspring. In extreme cases, there is failure to produce any gametes at all.

Hybrid dysgenesis occurs because P strains contain transposable genetic elements, P elements, which are mobilized in the eggs of M-strain females (eggs that are permissive for P-element transposition are described as M-cytotype). The P elements do not cause dysgenesis in crosses within P strains, because they are not mobilized in P-cytotype eggs. This is because the P element encodes a repressor of its own transposase, which prevents transposition. When a sperm from a P-cytotype male fertilizes the egg of an M-cytotype female, the absence of repressor in the egg results in temporary derepression of the transposase, such that P-element transposition occurs at a high frequency. The high rate of mutation characteristic of the dysgenesis syndrome reflects the insertion of P elements into multiple genetic loci.

Several members of the P-element family have been cloned and characterized (O'Hare & Rubin 1983). The prototype is a 2.9-kb element, while other members of the family appear to have arisen by internal deletion events. The elements are characterized by perfect 31-bp inverted terminal repeats, which are recognized by the transposase. The prototype element contains a single gene, comprising four exons, encoding the transposase (a truncated version of the transposase may act as the repressor). The transposase primary transcript is differentially spliced in germ cells and somatic cells, such that functional transposase is produced only in germ cells. Laski *et al.* (1986) showed this clearly by making a P-element construct in which the differentially spliced third intron was precisely removed. This element showed a high level of somatic transposition activity. Naturally occurring short P elements are generally defective, because they do not encode functional transposase. However, they do possess the inverted terminal repeats and can be activated *in trans* by transposase supplied by a non-defective P element in the same nucleus.

Spradling & Rubin (1982) devised an approach for introducing P-element DNA into *Drosophila* chromosomes. Essentially, a recombinant plasmid comprising a 2.9-kb P element together with some flanking *Drosophila* DNA sequences, cloned in the pBR322 vector, was microinjected into the posterior pole of M-cytotype embryos. The embryos were injected at the syncytial blastoderm stage, when the cytoplasm has not yet become partitioned into individual cells (Fig. 13.10). The posterior pole was chosen because this is where the germline originates, and P-element DNA in this region was expected to be incorporated into the genome in a proportion of the germ cells.

A screen of progeny lines showed that P elements had indeed integrated at a variety of sites in each of the five major chromosomal arms, as revealed by *in situ* hybridization to polytene chromosomes. P element integration occurred by transposition, not by random integration. This was proved by probing Southern blots of restricted DNA and showing that the integrated P element was not accompanied by the flanking *Drosophila* or pBR322 DNA sequences present in the recombinant plasmid (Spradling & Rubin 1982). The injected plasmid DNA must therefore have been expressed at some level *before* integration, so as to provide transposase.

These experiments showed that P elements could transpose with a high efficiency from injected plasmids into diverse sites in the chromosomes of germ cells. At least one of the integrated P elements in each progeny line remained functional, as evidenced by the hypermutability it caused in subsequent crosses to M-cytotype eggs.

Natural P elements have been developed into vectors for gene transfer

Rubin & Spradling (1982) exploited their finding that P elements can be artificially introduced into the *Drosophila* genome. A possible strategy for using the P element as a vector would be to attempt

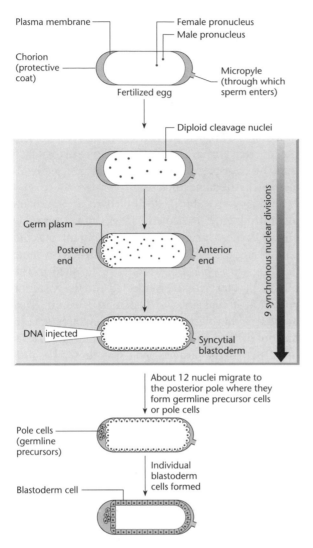

Labels on figure:
Plasma membrane
Female pronucleus
Male pronucleus
Chorion (protective coat)
Micropyle (through which sperm enters)
Fertilized egg
Diploid cleavage nuclei
Germ plasm
Posterior end
Anterior end
9 synchronous nuclear divisions
DNA injected
Syncytial blastoderm
About 12 nuclei migrate to the posterior pole where they form germline precursor cells or pole cells
Pole cells (germline precursors)
Individual blastoderm cells formed
Blastoderm cell

Fig. 13.10 Early embryogenesis of *Drosophila*. DNA injected at the posterior end of the embryo just prior to pole-cell formation is incorporated into germline cells.

P-M dysgenic cross, in which transposase activity was therefore expected to be high. A disadvantage of this approach was that frequent mutations and chromosomal aberrations would also be expected. In the other approach, the plasmid carrying the recombinant P element was co-injected with a plasmid carrying the non-defective 2.9-kb element.

In the first experiments of this kind, embryos homozygous for the *rosy* mutation were microinjected with a P-element vector containing a wild-type *rosy* gene. Both methods for providing complementing transposase were effective. Rosy$^+$ progeny, recognized by their wild-type eye color, were obtained from 20–50% of injected embryos. The chromosomes of these flies contained one or two copies of the integrated *rosy* transgene. The *rosy* gene is a particularly useful genetic marker. It produces a clearly visible phenotype: Rosy$^-$ flies have brown eyes instead of the characteristic red color of Rosy$^+$ flies. The *rosy* gene encodes the enzyme xanthine dehydrogenase, which is involved in the production of a precursor of eye pigments. The *rosy* gene is not cell-autonomous: expression of *rosy* anywhere in the fly, for example in a genetically mosaic fly developing from an injected larva, results in a wild-type eye color. Selectable markers have been used instead of visible markers to identify transformed flies. These include the alcohol dehydrogenase gene *adh* (Goldberg *et al.* 1983) and *neo* (Steller & Pirrotta 1985). Other eye-color markers have been used, including *white* and *vermilion* (Ashburner 1989, Fridell & Searles 1991), as well as alternative visible markers, such as *rough* (which restores normal eye morphology) and *yellow* (which restores normal body pigmentation and is particularly useful for scoring larvae) (Locket *et al.* 1992, Patton *et al.* 1992).

A simple P-element vector is shown in Fig. 13.11 (Rubin & Spradling 1983). It consists of a P element cloned in the bacterial vector pUC8. Most of the P element has been replaced by the *rosy* gene, but the terminal repeats essential for transposition have been retained. The vector includes a polylinker site for inserting foreign sequences. Transposition of the recombinant vector into the genome of injected larvae is brought about by co-injecting a helper P element, which provides transposase *in trans* but which cannot transpose itself because of a deletion in one of its terminal inverted repeats. Such an element is referred to as a *wings-clipped* element (Karess & Rubin 1984). An alternative strategy is to inject purified transposase protein (Kaufman & Rio 1991). The capacity of P-element vectors is large, although

to identify a suitable site in the 2.9-kb P-element sequence where insertion of foreign DNA could be made without disrupting sequences essential for transposition. However, an alternative strategy was favored. A recombinant plasmid was isolated which comprised a short (1.2 kb), internally deleted member of the P-element family together with flanking *Drosophila* sequences, cloned in pBR322. This naturally defective P element did not encode transposase (O'Hare & Rubin 1983). Target DNA was ligated into the defective P element. The aim was to integrate this recombinant P element into the germline of injected embryos by providing transposase function *in trans*. Two approaches for doing this were initially tested. In one approach a plasmid carrying the recombinant P element was injected into embryos derived from a

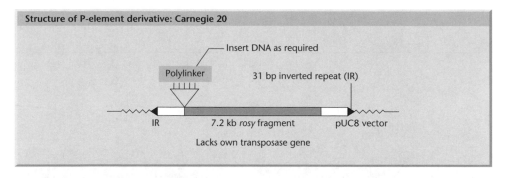

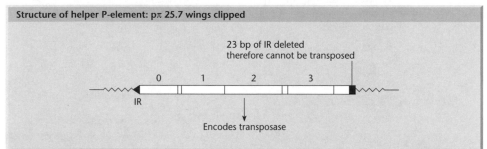

Fig. 13.11 P-element derivatives as a vector system. (See text for details.)

increasing the size of the recombinant element appears to reduce the transposition frequency. Inserts of over 40 kb have been successfully introduced into flies (Haenlin *et al.* 1985) and this has allowed the construction of cosmid libraries using P-element vectors (Speek *et al.* 1988).

Gene targeting in *Drosophila* has been achieved using a combination of homologous and site-specific recombination

The transformation of flies is normally achieved by the injection of P elements, borne on plasmid vectors, into the pole plasm region of early embryos. The direct injection of DNA does not result in gene targeting even if the construct is homologous to an endogenous gene, and this is because the P element functions specifically as an integrating element. Although rare targeting events have been recovered by selection in a *Drosophila* somatic cell line, flies have no cell system equivalent to ES cells for the production of transgenics.

To overcome this problem, gene targeting in *Drosophila* is accomplished by generating the donor molecule *in vivo*, which is achieved using a novel combination of endonuclease digestion and site-specific recombination (Rong 2002). Three P elements are required, initially introduced into three different fly lines and then stacked by two generations of crossing. The first P element contains the gene encoding the yeast site-specific recombinase FLP, which excises DNA sequences between direct

repeats of the *FRT* site as a circular product. The second P element contains the gene encoding the rare-cutting endonuclease I-*Sce*1, which has a 22-bp recognition site not found anywhere in the wild-type *Drosophila* genome. Both genes are under control of a heat-shock promoter. The final element contains the sequence of the donor molecule, which is homologous to the intended target site elsewhere in the genome, together with an adjacent marker, the *white* gene, which produces flies with the wild-type red eye color in a background of flies with white eyes. The donor sequence is interrupted by the recognition site for I-*Sce*1, and the donor sequence and marker are flanked by *FRT* sites (Fig. 13.12a). When all three elements are present in the same fly line, the flies are heat-shocked. The FLP enzyme then excises the donor molecule as a circular element containing a marker gene, homology region, and a single *FRT* site. Simultaneously, the I-*Sce*1 enzyme cleaves the donor molecule within the homology region, producing a linear construct analogous to the insertion type element shown in Fig. 13.4. The donor sequence can then align with its target and undergo homologous recombination, resulting in the targeted disruption of the endogenous gene (Figure 13.12b).

To recover targeting events, the heat-shocked flies are mated to wild types. Most progeny of this cross lack the marker gene because the FLP-excised donor molecule is often lost. However, where targeting has been successful, the marker will be present in the offspring and will display the corresponding

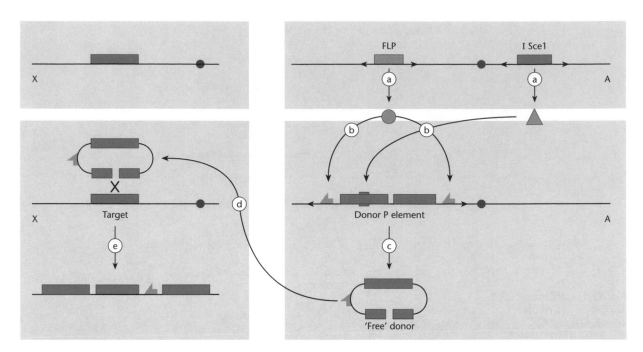

Fig. 13.12 Gene targeting in *Drosophila*. In this example, the target locus is on the X chromosome (X) and is identified by a purple box (top left panel). Flies with three P elements are made by crossing independently transformed lines. In this example, the elements containing the *flp* and I-*sce*1 transgenes are present on the same autosome (A) (top right panel) while the element containing the donor sequence is present on another autosome (A) (bottom right panel). Expression of *flp* and I-*Sce*1 transgenes (a) results in the production of the corresponding enzymes which act on the donor construct (b) producing a linearized free donor molecule (c). In the bottom left panel, this is shown synapsing with the target locus (d) and recombining to generate a duplicated target gene (e) interrupted by the marker and one *FRP* site. Circle is FLP enzyme, arrow is *FRP* site, triangle is I-*Sce*1, purple bar is I-*Sce*1 target site.

phenotype. Each stage of the targeting process can be confirmed in different ways. Successful excision of the donor sequence can be confirmed by FLP expression, since only the intact donor element has the marker gene flanked by two *FRT* sites. However, the marker will survive FLP-mediated excision through random integration as well as homologous recombination. Therefore, genuine targeting events must be confirmed by genetic linkage analysis and molecular analysis, such as PCR or Southern blot hybridization. As with gene targeting in mammals, the efficiency of targeting in *Drosophila* is both locus dependent and related to the length of the homology region. Unlike the situation in mammals, however, there are only a maximum of two donor molecules in the cell (if the cell is in G2) rather than the hundreds or thousands introduced into mammalian cells by transfection. This suggests that *in vivo* donors are far more efficient substrates for homologous recombination than exogenously supplied DNA.

As is the case for gene targeting in mice, the *Drosophila* targeting strategy leads to a duplication of the homology region. However, because *Drosophila* genes can be small compared to those in mammals,

providing enough homology to facilitate efficient recombination may in some cases require the entire gene to be present. Various schemes have been devised to ensure mutation of the target locus where the homology region encompasses the entire gene. The one shown in Fig. 13.13 involves the creation

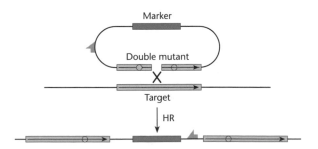

Fig. 13.13 Scheme for generating targeted mutants in *Drosophila*. The targeting construct contains a point mutation either side of the I-*Sce*1 recognition site, so that insertion by homologous recombination generates two copies of the target gene, each containing a different mutation. Spontaneous intrachromosomal recombination to resolve such duplications has not been observed in *Drosophila*, although the presence of two mutations would prevent reversion to wild type should such an event occur.

of two point mutations, one either side of the I-*Sce*1 site. This produces a tandem duplication of the entire target gene separated by the marker, each copy harboring a different mutation.

Suggested reading

Brinster R.L. (2002) Germline stem cell transplantation and transgenesis. *Science* **296**, 2174–6.

Capecchi M.R. (1989) The new mouse genetics – altering the genome by gene targeting. *Trends in Genetics* **5**, 70–6.

Edwards J.L., Schrick F.N., McCracken M.D., van Amstel S.R., Hopkins F.M., Welborn M.G. & Davies C.J. (2003) Cloning adult farm animals: a review of the possibilities and problems associated with somatic cell nuclear transfer. *American Journal of Reproduction & Immunology* **50**, 113–23.

Moraes C.T., Dey R. & Barrientos A. (2001) Trans-mitochondrial technology in animal cells. *Methods in Cell Biology* **65**, 397–412.

Muller U. (1999) Ten years of gene targeting: targeted mouse mutants, from vector design to phenotype analysis. *Mechanical Development* **82**, 3–21.

Palmiter R.D. & Brinster R.L. (1986) Germline transformation of mice. *Annual Review of Genetics* **20**, 465–99.

Pfeifer A. (2004) Lentiviral transgenesis. *Transgenic Research* **13**, 513–22.

Rong Y.S. (2002) Gene targeting by homologous recombination: a powerful addition to the genetic arsenal for *Drosophila* geneticists. *Biochem Biophys Res Communication* **297**, 1–5.

Slack J.M.W. (1996) Developmental biology – high hops of transgenic frogs. *Nature* **383**, 765–6.

Twyman R.M. & Whitelaw C.A.B. (2000) Animal cell genetic engineering. In Spier R.E. (ed.) *Encyclopedia of Cell Technology*. John Wiley & Sons Inc., New York, pp. 737–819.

Zbikowska H.M. (2003) Fish can be first – advances in fish transgenesis for commercial applications. *Transgenic Research* **12**, 379–89.

CHAPTER 14

Gene transfer to plants

Introduction

Plants provide human beings with all manner of useful products: food and animal feed, fibers and structural materials, and small molecules that can be used as dyes, scents, and medicines. Plants have been cultivated for these products since the dawn of history, and for the same length of time people have sought to improve plants by breeding them and selecting the better-performing and most useful varieties. The one limitation of this approach is that breeders are restricted to the existing gene pool in each species or sexually compatible group of species. In order to surmount this barrier, it has been necessary to develop technologies for gene transfer to plants.

During the 1960s and 1970s several attempts to transfer DNA into plant tissues were reported but stable transformation was never confirmed (e.g. Stroun *et al.* 1966, Coe and Straker 1966). The introduction of foreign DNA into a plant followed by stable transmission through the germline was first demonstrated in 1981, when transgenic tobacco plants were generated by transformation using the soil bacterium *Agrobacterium tumefaciens* (Otten *et al.* 1981). In the 25 years following this report, foreign genes have been introduced into well over 100 different plant species either through the use of *A. tumefaciens* or alternative strategies involving direct DNA transfer to plant cells and tissues. In addition, plant viruses have been developed as versatile episomal vectors, allowing high-level transient gene expression. This research has founded an agricultural biotechnology industry in which plants are manipulated to make them resistant to pests and diseases, to improve their tolerance of stress, to improve their nutritional characteristics and even to act as factories producing therapeutic proteins and industrial enzymes (Chapter 26).

A fundamental difference between animals and plants is that organized, differentiated plant tissue shows a high degree of developmental plasticity. Depending on the species, isolated stem segments, leaf disks, or seed-derived callus tissue may be able to regenerate an entire new plant under appropriate culture conditions. For most plant species, some form of tissue culture step is therefore necessary for the successful production of transgenic plants. It should be noted, however, that there is now increasing interest in the use of whole-plant (*in planta*) transformation strategies, in which the need for tissue culture is minimized or eliminated.

Plant tissue culture is required for most transformation procedures

Callus cultures are established under conditions that maintain cells in an undifferentiated state

Tissue culture is the process whereby small pieces of living tissue (*explants*) are isolated from an organism and grown aseptically for indefinite periods on a nutrient medium. For successful plant tissue culture it is best to start with an explant rich in undetermined cells because such cells are capable of rapid proliferation. The usual explants are buds, root tips, nodal stem segments or germinating seeds, and these are placed on suitable culture media where they grow into an undifferentiated mass known as a callus (Fig. 14.1). Since the nutrient media used for plants can also support the growth of microorganisms, the explant is first washed in a disinfectant such as sodium hypochlorite or hydrogen peroxide. Once established, the callus can be propagated indefinitely by subdivision. Usually callus cultures are maintained in the dark because light can induce differentiation of callus cells.

For plant cells to develop into a callus it is essential that the nutrient medium contains the correct balance of plant hormones (phytohormones) to maintain the cells in an undifferentiated state. There are five main classes of plant hormones: auxins, cytokinins, gibberellins, abscisic acid, and ethylene (Fig. 14.2).

Fig. 14.1 Close-up view of a callus culture.

The correct balance of auxins and cytokinins is most important for callus culture, and the exact relative amounts need to be determined empirically for each species and explant type. A low auxin:cytokinin ratio leads to shoot formation whereas a high ratio favors the formation of roots. Requirements for the other hormones vary according to species and explant. Some explants require the presence of gibberellins such as GA_3 for continued growth, whereas abscisic acid tends to be used to encourage specific developmental events, such as somatic embryogenesis. Ethylene is rarely used in tissue culture, although sometimes ethylene produced naturally by cultured cells can inhibit cell growth. Most of the media in common use consist of inorganic salts and trace metals (usually referred to respectively as macroelements and microelements), essential vitamins (thiamine and myoinositol), an organic nitrogen source (usually one or more amino acids), and sucrose as a carbon source. For more complex organic nutrients such as casein hydrolysate, coconut water or yeast extract may also be required. Many plant culture media also include a gelling agent so that plants can grow on the surface of the medium and project roots into the gel as they would naturally into soil.

Callus cultures can be broken up to form cell suspensions, which can be maintained in batches

Depending on the species and culture conditions, callus tissue can become hard and compact or soft and easily breakable. The latter is known as friable callus, and when transferred into liquid medium and agitated, the cell mass breaks up to give a suspension of isolated cells, small clusters of cells, and larger aggregates. Such suspensions can be maintained

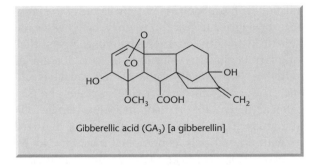

Gibberellic acid (GA_3) [a gibberellin]

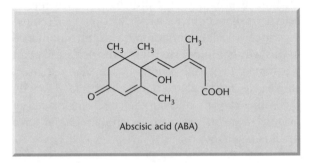

Abscisic acid (ABA)

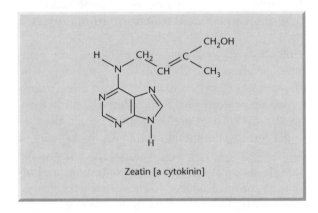

Zeatin [a cytokinin]

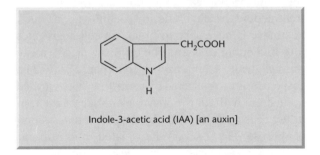

Indole-3-acetic acid (IAA) [an auxin]

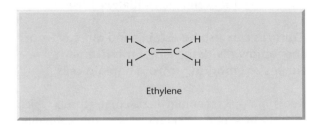

Ethylene

Fig. 14.2 The structures of some chemicals that are plant growth regulators, phytohormones.

indefinitely by subculture but, by virtue of the presence of aggregates, are extremely heterogeneous. Genetic instability adds to this heterogeneity, so that long-term culture results in the accumulation of mutations (*somaclonal variation*) which can adversely affect the vitality and fertility of regenerated plants.

If placed in a suitable medium, isolated single cells from suspension cultures are capable of division. As with animal cells, conditioned medium may be necessary for proliferation to occur. Conditioned medium is prepared by culturing high densities of cells in fresh medium for a few days and then removing the cells by filter sterilization. Medium conditioned in this way contains essential amino acids and plant hormones. Provided conditioned medium is used, single cells can be plated onto solid media in exactly the same way as microorganisms, but instead of forming a colony, plant cells proliferate and form a callus.

Protoplasts are usually derived suspension cells and can be ideal transformation targets

Protoplasts are cells from which the cellulose walls have been removed. They are very useful for genetic manipulation for two reasons: first, several transformation protocols have been developed that work specifically with protoplasts; and second, because under certain conditions, protoplasts from similar or contrasting cell types can be fused to yield somatic hybrids, a process known as protoplast fusion. Protoplasts can be produced from suspension cultures, callus tissue, or intact tissues, e.g. leaf mesophyll cells, by mechanical disruption or, preferably, by treatment with cellulolytic and pectinolytic enzymes. Pectinase is necessary to break up cell aggregates into individual cells and the cellulase digests away the cell wall. After enzyme treatment, protoplast suspensions are collected by centrifugation, washed in medium without the enzyme, and separated from intact cells and cell debris by flotation on a cushion of sucrose (Fig. 14.3). When plated onto nutrient medium, protoplasts will synthesize new cell walls within 5–10 days and then initiate cell division.

Cultures can also be established directly from the rapidly dividing cells of meristematic tissues or embryos, or from haploid cells

Roots and shoots contain meristematic tissue, which is the source of all dividing cells in the elongating roots and stem. Root and shoot tips can be excised

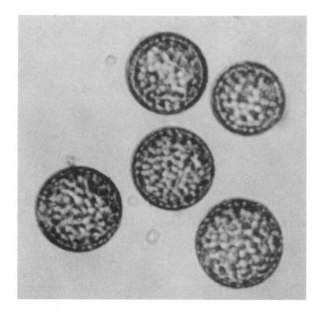

Fig. 14.3 Photomicrograph of tobacco protoplasts.

and cultured directly on solid medium, and will give rise to new organs that can be clonally propagated. Similarly, embryos contain rapidly proliferating cells which can be used as a source of callus. This is the most widely applied strategy to regenerate cereals and other monocotyledonous plants.

Another common source of callus tissue is the male gametophyte, or microspore, which is found in pollen grains. Anther culture is used to provide an environment that stimulates the development of embryos from these cells. During embryogenesis, the haploid tissue may undergo spontaneous or induced chromosome doubling, resulting in so-called dihaploid plants containing two copies of the same haploid genome.

Regeneration of fertile plants can occur through organogenesis or somatic embryogenesis

The developmental plasticity of plant cells means that whole fertile plants can often be regenerated from tissue explants, callus, cell suspensions, or protoplasts by placing them on appropriate media. As discussed above, the maintenance of cells in an undifferentiated state requires the correct balance of phytohormones. However, only cytokinin is required for shoot culture and only auxin for root culture, therefore increasing the level of cytokinins available to the callus induces shoot formation and increasing the auxin level promotes root formation. Ultimately plantlets arise through the development of adventitious roots on shoot buds, or through the development of shoot buds from tissues formed by proliferation at the base of rootlets. The formation of roots and shoots on callus tissue is known as *organogenesis*. The culture conditions required to achieve organogenesis vary from species to species, and have not been determined for every type of callus. As discussed, the adventitious organogenesis of shoots and roots can also occur directly from organized plant tissues such as stem segments, without first passing through a callus stage.

Under certain conditions, cell suspensions or callus tissue of some plant species can be induced to undergo a different development process known as *somatic embryogenesis*. In this process, the cells undergo a pattern of differentiation similar to that seen in zygotes after fertilization, to produce *embryoids*. These structures are embryo-like but differ from normal embryos in being produced from somatic cells and not from the fusion of two germ cells. The embryoids can develop into fertile plants without the need to induce root and shoot formation on artificial media.

The ease with which plant material is manipulated and interconverted in culture provides many opportunities for the development of techniques for gene transfer and the recovery of transgenic plants (Fig. 14.4). DNA can be introduced into most types of plant material – protoplasts, cell suspensions, callus, tissue explants, gametes, seeds, zygotes, embryos, organs, and whole plants – so the ability to recover fertile plants from such material is often the limiting step in plant genetic engineering rather than the DNA transfer process itself. It is also possible to maintain transformed plant cell lines or tissues (e.g. root cultures) producing recombinant proteins or metabolites, in the same way that cultured animal cells can be used as bioreactors for valuable products.

There are four major strategies for gene transfer to plant cells

As is the case for animal cells (Chapter 12), gene transfer to plants can be achieved through four types of mechanism – viral transduction, bacterial gene delivery, and chemical and physical direct DNA transfer. Unlike the situation in animals, where bacterial gene transfer is a relatively new development, *Agrobacterium*-mediated transformation is the most widely used transformation method, particularly for dicotyledonous plants. Physical methods are the next most popular, especially particle bombardment for the transformation of monocotyledonous plants such as cereals. Chemical transfection methods are little used, and are compatible only with protoplasts, which behave in many ways analogously to animal cells. Many of the techniques used to transfect animal cells can therefore be applied to plant protoplasts, e.g. calcium phosphate transfection. All three of the above methods can be used for either transient expression or stable transformation. Another major difference between gene-transfer strategies in animal and plant cells is that no known plant viruses integrate their genetic material into the plant genome as part of the natural infection cycle. Therefore, plant viruses are used as episomal vectors rather than for stable transformation. However, while stable transformation cannot be achieved, plant viruses often cause systemic infections resulting in the rapid production of high levels of recombinant protein throughout the plant, and they can be transmitted through normal infection routes, or by grafting infected scions onto virus-free hosts.

Agrobacterium-mediated transformation

Agrobacterium tumefaciens is a plant pathogen that induces the formation of tumors

Gene transfer from bacteria to plants occurs naturally and is responsible for crown gall disease. This is a plant tumor that can be induced in a wide variety of gymnosperms and dicotyledonous angiosperms (dicots) by inoculation of wound sites with the Gram-negative soil bacterium *A. tumefaciens* (Fig. 14.5). The involvement of bacteria in this disease was established nearly 100 years ago by Smith & Townsend (1907). It was subsequently shown that the crown gall tissue represents true oncogenic transformation, since the undifferentiated callus can be cultivated *in vitro* even if the bacteria are killed with antibiotics, and yet retains its tumorous properties (Fig. 14.6). These properties include the ability to form a tumor when grafted onto a healthy plant, the capacity for unlimited growth as a callus in tissue culture even in the absence of phytohormones necessary for the *in vitro* growth of normal cells, and the synthesis of opines, such as octopine and nopaline, which are unusual amino acid derivatives not found in normal plant tissue (Fig. 14.7).

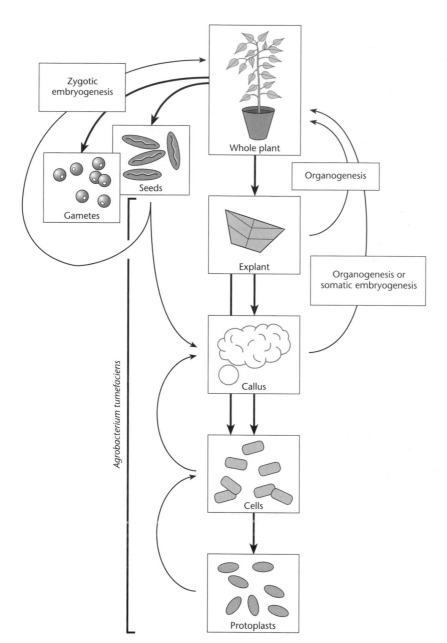

Fig. 14.4 Depending on species, plant tissues are extremely pliable and can be easily interconverted and regenerated in culture. This diagram shows some of the ways in which this flexibility can be exploited to facilitate gene transfer and the creation of transgenic plants.

The metabolism of opines is a central feature of crown gall disease. Opine synthesis is a property conferred upon the plant cell when it is colonized by *A. tumefaciens*. The type of opine produced is determined not by the host plant but by the bacterial strain. In general, the bacterium induces the synthesis of an opine that it can catabolize and use as its sole carbon and nitrogen source. Thus, bacteria that utilize octopine induce tumors that synthesize octopine, and those that utilize nopaline induce tumors that synthesize nopaline (Bomhoff *et al.* 1976, Montaya *et al.* 1977).

The ability to induce tumors is conferred by a Ti-plasmid found only in virulent *Agrobacterium* strains

Since the continued presence of *Agrobacterium* is not required to maintain plant cells in their transformed state, it is clear that some "tumor-inducing principle" is transferred from the bacterium to the plant at the wound site. Zaenen *et al.* (1974) first noted that virulent strains of *A. tumefaciens* harbor large plasmids (140–235 kbp), and experiments involving the transfer of such plasmids between

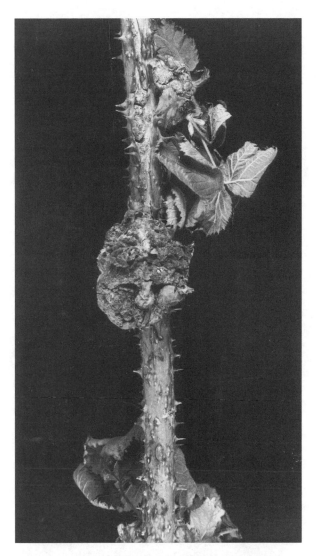

Fig. 14.5 Crown gall on blackberry cane. (Photograph courtesy of Dr. C.M.E. Garrett, East Malling Research Station.)

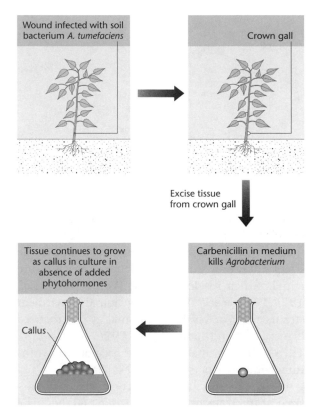

Fig. 14.6 *A. tumefaciens* induces plant tumors, but is not required for the continuous proliferation of those tumors.

Octopine	R = NH_2—C(=NH)—NH—$(CH_2)_3$—
Octopinic acid	R = $NH_2(CH_2)_3$—
Lysopine	R = $NH_2(CH_2)_4$—
Histopine	R =
Nopaline	R = NH_2—C(=NH)—NH—$(CH_2)_3$—
Nopalinic acid	R = $NH_2(CH_2)_3$—

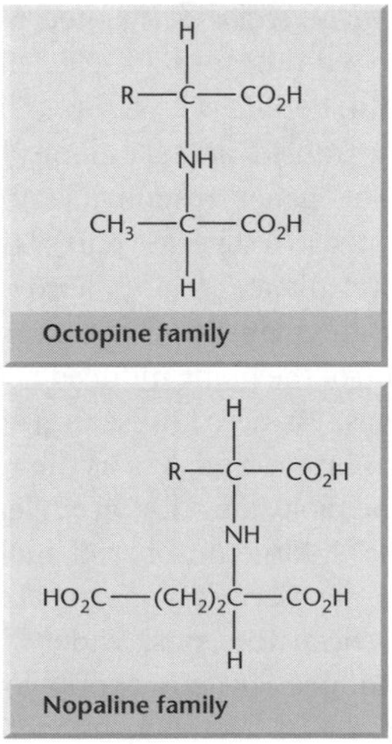

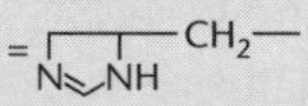

Fig. 14.7 Structures of some opines.

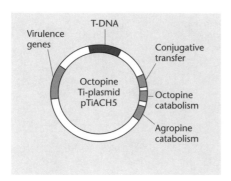
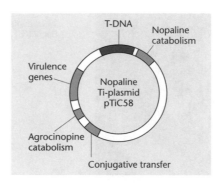

Fig. 14.8 Ti-plasmid gene maps.

various octopine- and nopaline-utilizing strains soon established that virulence and the ability to use and induce the synthesis of opines are plasmid-borne traits. These properties are lost when the bacteria are cured of their resident plasmid (Van Larbeke *et al.* 1974, Watson *et al.* 1975) but acquired by avirulent strains when a virulence plasmid is reintroduced by conjugation (Bomhoff *et al.* 1976, Gordon *et al.* 1979). The plasmids therefore became known as tumor-inducing plasmids (Ti-plasmids).

Ti-plasmids specify the type of opine that is synthesized in the transformed plant tissue and the opine utilized by the bacterium. Plasmids in the octopine group are closely related to each other while those in the nopaline group are considerably more diverse. Between the groups, there are four regions of homology, including the genes directly responsible for tumor formation (Drummond & Chilton 1978, Engler *et al.* 1981; Fig. 14.8). It should be noted that the presence of a plasmid in *A. tumefaciens* does not mean that the strain is virulent. Many strains contain very large cryptic plasmids that do not confer virulence, and in some natural isolates a cryptic plasmid is present together with a Ti-plasmid.

A short segment of DNA, the T-DNA, is transferred to the plant genome

Complete Ti-plasmid DNA is not found in plant tumor cells but a small, specific segment of the plasmid, about 23 kbp in size, is found integrated in the plant nuclear DNA at an apparently random site. This DNA segment is called T-DNA (transferred DNA) and carries genes that confer both unregulated growth and the ability to synthesize opines upon the transformed plant tissue. However, these genes are non-essential for transfer and can be replaced with foreign DNA (see below). The structure and organization of nopaline plasmid T-DNA sequences are usually simple, i.e. there is a single integrated seg-

ment. Conversely, octopine T-DNA comprises two segments, T_L (which carries the genes required for tumor formation) and T_R (which carries the genes for opine synthesis). The two segments are transferred to the plant genome independently and may be present as multiple copies. The significance of this additional complexity is not clear.

In the Ti-plasmid itself, the T-DNA is flanked by 25-bp imperfect direct repeats known as *border sequences*, which are conserved between octopine and nopaline plasmids. The border sequences are not transferred intact to the plant genome, but they are involved in the transfer process. The analysis of junction regions isolated from plant genomic DNA has shown that the integrated T-DNA end points lie internal to the border sequences. The right junction is rather precise, but the left junction can vary by about 100 nucleotides (Yadav *et al.* 1982, Zambryski *et al.* 1982). Deletion of the right-border repeat abolishes T-DNA transfer, but the left-hand border surprisingly appears to be non-essential. Experiments in which the right-border repeat alone has been used have shown that an enhancer, sometimes called the *overdrive sequence*, located external to the repeat is also required for high-efficiency transfer (Shaw *et al.* 1984, Peralta *et al.* 1986). The left-border repeat has little transfer activity alone (Jen & Chilton 1986).

The genes responsible for T-DNA transfer are located in a separate part of the Ti-plasmid called the *vir* (virulence) region. Two of these genes, *virA* and *virG*, are constitutively expressed at a low level and control the plant-induced activation of the other *vir* genes. VirA is a kinase that spans the inner bacterial membrane, and acts as the receptor for certain phenolic molecules that are released by wounded plant cells. A large number of such compounds has been characterized, but one in particular, acetosyringone, has been the most widely used in the laboratory to induce *vir* gene expression (Stachel *et al.* 1985; Fig. 14.9). Notably, phenolic compounds such as

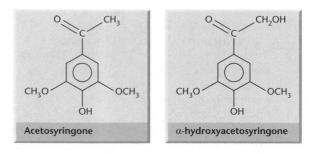

Fig. 14.9 Structures of signal molecules, produced by wounded plant tissue, which activate T-DNA transfer by *A. tumefaciens*.

acetosyringone do not attract bacteria to wounded plant cells. Rather, the bacteria appear to respond to simple molecules such as sugars and amino acids, and the *vir* genes are induced after attachment (Parke *et al.* 1987, Loake *et al.* 1988). Many sugars also synergize the action of the phenolic signals to enhance *vir* gene expression (Shimada *et al.* 1990). Activated VirA transphosphorylates the VirG protein, which is a transcriptional activator of the other *vir* genes. The VirA and VirG proteins show similarities to other two-component regulatory systems common in bacteria (Winans 1992). In addition to *virG*, further genes on the bacterial chromosome also encode transcription factors that regulate *vir* gene expression (reviewed by Kado 1998, Gelvin 2000, 2003).

The induction of *vir* gene expression results in the synthesis of proteins that form a conjugative pilus through which the T-DNA is transferred to the plant cell. The components of the pilus are encoded by genes in the *virB* operon (reviewed by Lai & Kado 2000). DNA transfer itself is initiated by an endonuclease formed by the products of the *virD1* and *virD2* genes. This introduces either single-strand nicks or a double-strand break at the 25-bp borders of the T-DNA, a process enhanced by the VirC12 and VirC2 proteins, which recognize and bind to the overdrive enhancer element. The VirD2 protein remains covalently attached to the processed T-DNA. Recent studies have suggested that the type of T-DNA intermediate produced (single- or double-stranded) depends on the type of Ti-plasmid, with double-stranded T-DNA favored by nopaline plasmids (where the T-DNA is a single element) and single "T-strands" favored by octopine and succinopine plasmids, where the T-DNA is split into noncontiguous sections (Steck 1997). T-strands are coated with VirE2, a single-stranded DNA binding protein. The whole complex, sometimes dubbed the *firecracker*

complex because of its proposed shape, is then transferred through the pilus and into the plant cell. The VirD2 protein has been proposed to protect the T-DNA against nucleases, to target the DNA to the plant cell nucleus, and to integrate it into the plant genome. The protein has two distinct nuclear localization signals, with the C-terminal signal thought to play the major role in targeting the T-DNA (Tinland *et al.* 1992). It has been observed that the nucleus of wounded plant cells often becomes associated with the cytosolic membrane close to the wound site, suggesting that the T-DNA could be transferred directly to the nucleus without extensive exposure to the cytosol (Kahl & Schell 1982). Once in the nucleus, the T-DNA is thought to integrate through a process of illegitimate recombination, perhaps exploiting naturally occurring chromosome breaks (Tinland 1996, Tzfira *et al.* 2004).

The *Agrobacterium* gene-transfer system appears to be a highly adapted form of bacterial conjugation. Many broad-host-range plasmids can transfer from *Agrobacterium* to the plant genome using their own mobilization functions (Buchanan-Wollaston *et al.* 1987) and the *vir* genes encode many components that are common with broad-host-range plasmid conjugation systems (reviewed by Kado 1998). In addition to plants, *Agrobacterium* can transfer DNA to other bacteria, yeast, and filamentous fungi. Recently, a novel insight into the scope of this gene-transfer mechanism was provided by Citovsky and colleagues (Kunik *et al.* 2001) by demonstrating that gene transfer from *Agrobacterium* to cultured human cells was also possible! For the interested reader, T-DNA transfer has been discussed in several comprehensive reviews (Zupan *et al.* 2000, Tzfira & Citofsky 2000, 2002, Gelvin 2003, Valentine 2003).

Disarmed Ti-plasmid derivatives can be used as plant gene-transfer vectors

Genetic maps of T-DNA have been obtained by studying spontaneous and transposon-induced mutants that affect tumor morphology, generating tumors that are larger than normal, or that show "shooty" or "rooty" phenotypes. Although normal tumors can grow on medium lacking auxins and cytokinins, the tumor cells actually contain high levels of these hormones. Ooms *et al.* (1981) therefore proposed that the oncogenes carried on the T-DNA encoded products involved in phytohormone synthesis and that the abnormal morphologies of T-DNA mutants

Table 14.1
Functions of some T-DNA genes in *A. tumefaciens* Ti plasmids.

Gene	Product	Function
ocs	Octopine synthase	Opine synthesis
nos	Nopaline synthase	Opine synthesis
tms1 (*iaaH, auxA*)	Tryptophan-2-mono-oxygenase	Auxin synthesis
tms2 (*iaaM, auxB*)	Indoleacetamide hydrolase	Auxin synthesis
tmr (*ipt, cyt*)	Isopentyl transferase	Cytokinin synthesis
tml	Unknown	Unknown, mutations affect tumor size
frs	Fructopine synthase	Opine synthesis
mas	Mannopine synthase	Opine synthesis
ags	Agropine synthase	Opine synthesis

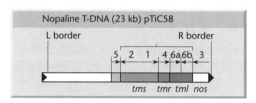

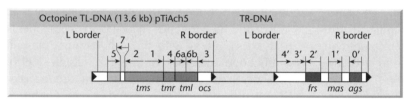

Fig. 14.10 Structure and transcription of T-DNA. The T-regions of nopaline and octopine Ti-plasmids have been aligned to indicate the common DNA sequences. The size and orientation of each transcript (numbered) is indicated by arrows. Genetic loci, defined by deletion and transposon mutagenesis, are shown as follows: *nos*, nopaline synthase; *ocs*, octopine synthase; *tms*, shooty tumor; *tmr*, rooty tumor.

were due to a disturbance in the balance of plant hormones in the callus. The cloning and functional analysis of T-DNA genes has confirmed that those with "shooty" mutant phenotypes encode enzymes for auxin biosynthesis, and those with "rooty" phenotypes are involved in cytokinin production (Weiler & Schroder 1987). Other genes have been identified as encoding enzymes for opine synthesis, while the function of some genes remains unknown (Table 14.1). The transcript maps of T-DNAs from a nopaline plasmid (pTiC58) and an octopine plasmid (pTiAch5) are shown in Fig. 14.10 (Willmitzer *et al.* 1982, 1983, Winter *et al.* 1984).

Interestingly, nucleotide sequencing has revealed that the T-DNA genes have promoter elements and polyadenylation sites that are eukaryotic in nature (De Greve *et al.* 1982a,b, Depicker *et al.* 1982, Bevan *et al.* 1983a,b). This explains how genes from a bacterial plasmid come to be expressed when transferred to the plant nucleus. It is possible that the sequences may have been captured from plants during the evolution of the Ti-plasmid. The ability

of *Agrobacterium* to induce tumors in a wide variety of plants suggested that T-DNA promoters such as those of the *ocs* (octopine synthase) and *nos* (nopaline synthase) genes could be useful for driving transgene expression. These and other promoters used for transgene expression in plants are discussed in Box 14.1.

We have seen that the Ti-plasmid is a natural vector for genetically engineering plant cells because it can transfer its T-DNA from the bacterium to the plant genome. However, wild-type Ti-plasmids are not suitable as general gene vectors because the T-DNA contains oncogenes that cause disorganized growth of the recipient plant cells. To be able to regenerate plants efficiently we must use vectors in which the T-DNA has been *disarmed* by making it non-oncogenic. This is most effectively achieved simply by deleting all of its oncogenes. For example, Zambryski *et al.* (1983) substituted pBR322 sequences for almost all of the T-DNA of pTiC58 leaving only the left- and right-border regions and the *nos* gene. The resulting construct was called pGV3850 (Fig. 14.11). *Agrobacterium*

Box 14.1 Control of transgene expression in plants

Promoters

To achieve high-level and constitutive transgene expression in plants, a very active promoter is required. In dicots, promoters from the *Agrobacterium* nopaline synthase (*nos*), octopine synthase (*ocs*), and mannopine synthase (*mas*) genes have been widely used. These are constitutive and also moderately induced by wounding (An *et al.* 1990, Langridge *et al.* 1989). The most popular promoter for transgene expression in dicots is the 35S RNA promoter from cauliflower mosaic virus (CaMV 35S). This is very active, but can be improved still further by duplicating the enhancer region (Rathus *et al.* 1993). These promoters have a much lower activity in monocots, and duplicating the CaMV 35S enhancer has little effect. Alternative promoters have therefore been sought for transgene expression in cereals (reviewed by McElroy & Brettel 1994). The rice *actin-1* and maize *ubiquitin-1* promoters have been widely used for this purpose (McElroy *et al.* 1995, Christensen & Quail 1996). As well as constitutive promoters, a large number of promoters have been used to direct transgene expression in particular tissues. In monocots, promoters from seed storage-protein genes, such as maize zein, wheat glutenin, and rice glutelins, have been used to target transgene expression to the seeds, which is beneficial for the accumulation of recombinant proteins (Wu *et al.* 1998; reviewed by Bilan *et al.* 1999). Promoters targeting transgene expression to green tissue are also useful (e.g. Graham *et al.* 1997, Datta *et al.* 1998). We discuss inducible expression systems for animals and plants in Chapter 15.

Other components of the expression vector

As discussed for animal cells (Box 12.2), other sequences in the expression vector also influence transgene expression. Generally, the presence of an intron in a plant expression cassette increases the activity of the promoter (Bilan *et al.* 1999). The insertion of a heterologous intron enhances the activity of the CaMV 35S promoter in monocots (e.g. see Mascarenhas *et al.* 1990, Vain *et al.* 1996) and constructs containing the actin or ubiquitin promoters generally include the first intron of the gene (McElroy *et al.* 1991). All transgenes must include a polyadenylation site, which in most cases is derived from the *Agrobacterium nos* gene or the CaMV 35S RNA. Whereas in animals it is conventional to remove untranslated regions from the expression construct, a number of such sequences in plants have been identified as translational enhancers. For example, the 5′ leader sequence of the tobacco mosaic virus RNA, known as the omega sequence, can increase transgene expression up to 80-fold (reviewed by Futterer & Hohn 1996, Gallie 1996). As in animals, the translational start site should conform to Kozak's consensus (Kozak 1999; see Box 12.2 for details) and the transgene should be codon-optimized for the expression host. A good example of the latter is the use of codon-optimized insecticidal toxin genes from *Bacillus thuringiensis* for expression in transgenic crops, leading to dramatically increased expression levels compared with the unmodified genes (Koziel *et al.* 1996). Also, the inclusion of targeting information in the expression cassette may be beneficial for the accumulation of recombinant proteins. For example, recombinant antibodies expressed in plants are much more stable if targeted to the endoplasmic reticulum (ER), since this provides a favorable molecular environment for folding and assembly. Targeting is achieved using an N-terminal signal sequence to direct the ribosome to the ER and a C-terminal tetrapeptide retrieval signal, KDEL, which causes accumulation in the ER lumen (Horvath *et al.* 2000).

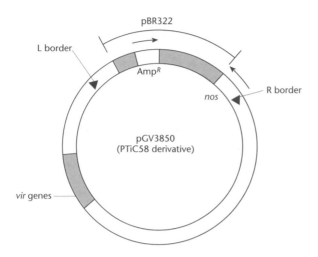

Fig. 14.11 Structure of the Ti-plasmid pGV3850, in which the T-DNA has been disarmed.

carrying this plasmid transferred the modified T-DNA to plant cells. As expected, no tumor cells were produced, but the fact that transfer had taken place was evident when the cells were screened for

nopaline production and found to be positive. Callus tissue could be cultured from these nopaline-positive cells if suitable phytohormones were provided, and fertile adult plants were regenerated by hormone induction of plantlets.

The creation of disarmed T-DNA was an important step forward, but the absence of tumor formation made it necessary to use an alternative method to identify transformed plant cells. In the experiment described above, opine production was exploited as a screenable phenotype, and the *ocs* and *nos* genes have been widely used as screenable markers (reviewed by Dessaux & Petit 1994). However, there are several drawbacks associated with this system, particularly the necessity to carry out enzymatic assays on all potential transformants. To provide a more convenient way to identify transformed plant cells, dominant selectable markers have been inserted into the T-DNA so that transformed plant cells can be selected on the basis of drug or herbicide resistance. The use of selectable markers in plants is discussed in more detail in Box 14.2.

Box 14.2   Selectable markers for plants

Until recently, almost all selectable markers used in plants were dominant selectable markers, providing resistance to either antibiotics or herbicides (see table). Some plants, particularly monocots, are naturally tolerant of kanamycin, and this antibiotic may also interfere with regeneration. In these species, alternative systems, such as hygromycin or phosphinothricin selection, are preferred.

The introduction of markers such as *nptII*, *hpt*, and *dhfr* into the T-DNA of disarmed Ti plasmids provided the first convenient methods to identify transformed plant tissue, and hence opened the way for *Agrobacterium* to be used as a general plant transformation system. The marker and experimental transgene can be cloned in tandem on the same T-DNA. In such cases, it is better for the selectable marker to be placed adjacent to the left-border repeat, since this is transferred to the plant last (Sheng & Citovsky 1996). This strategy reduces the likelihood of obtaining plants

under selection containing the marker alone and not the transgene of interest (see Hellens *et al.* 2000a). Alternatively, cotransformation can be achieved using *Agrobacterium* strains containing two plasmids or by co-inoculating plants with different *Agrobacterium* strains, each containing a single plasmid. Although there is some controversy surrounding the fate of co-introduced T-DNA sequences, it appears that nopaline-type plasmids favor the co-integration of multiple T-DNAs at the same locus, often in an inverted repeat pattern, while octopine-type plasmids favor independent integration sites, which can segregate in progeny plants (Depicker *et al.* 1985, Jones *et al.* 1987, Jorgensen *et al.* 1987, De Block & Debrouwer 1991). In direct transformation methods, the selectable marker can be included either on the same vector as the experimental transgene or on a separate vector, since cotransformation occurs at a high frequency (Schocher *et al.* 1986, Christou & Swain 1990).

continued

Box 14.2 *continued*

Marker	Selection	References
Drug resistance markers		
aad (preferred for chloroplast transformation (see p. 290))	Trimethoprim, streptomycin, spectinomycin, sulfonamides	Svab *et al.* 1990a
ble	Bleomycin	Hille *et al.* 1986
dhfr	Methotrexate	Eichholtz *et al.* 1987
hpt	Hygromycin	Van den Elzen *et al.* 1985
nptII and *aphII*	Kanamycin, neomycin, G418	Pridmore 1987
gat	Gentamicin	Hayford *et al.* 1988
Herbicide resistance markers		
bar and *pat*	Phosphinothricin (bialaphos, glufosinate ammonium, Basta)	De Block *et al.* 1987
csr1-1	Chlorsulfuron	Haughn *et al.* 1988
dhps (*sul*)	Sulfonamides (Asualam)	Guerineau & Mullineaux 1989
epsp	Glyphosate	Shah *et al.* 1986

Recently, public concern that antibiotic- and herbicide-resistance markers could pose a threat to health or the environment has prompted research into alternative innocuous marker systems. One example is the *E. coli* *manA* gene, which encodes mannose phosphate isomerase and confers upon transformed cells the ability to use mannose as a sole carbon source (Negrotto *et al.* 2000). Another is the *A. tumefaciens* *ipt* (isopentyl transferase) gene, located on the T-DNA, which induces cytokinin synthesis and can be used to select plants on the basis of their ability to produce shoots from callus on medium lacking cytokinins (Kunkel *et al.* 1999). Other strategies have also been explored, such as eliminating markers by sexual crossing (Komari *et al.* 1996), transposition (Goldsbrough *et al.* 1993), or site-specific recombination (Dale & Ow 1991, Russel *et al.* 1992, Zubko *et al.* 2000). We consider the use of Cre-*loxP* for marker excision in transgenic organisms in Chapter 15. To verify the successful elimination of particular genes, counterselectable markers are required (e.g. see the discussion of *Tk* as a couterselectable marker for gene targeting in mice, p. 257). In plants, the *A. tumefaciens* T-DNA gene *tms2* has been used as a negative marker. This encodes indoleacetamide hydrolase, an enzyme that converts naphthaleneacetamide (NAM) into the potent auxin naphthaleneacetic acid. In the presence of exogenously applied NAM, transformed callus is unable to produce shoots due to the excess levels of auxin, therefore only callus lacking the gene is able to regenerate into full transgenic plants (Sundaresan *et al.* 1995).

Binary vectors separate the T-DNA and the genes required for T-DNA transfer, allowing transgenes to be cloned in small plasmids

Although disarmed derivatives of wild-type Ti-plasmids can be used for plant transformation, they are not particularly convenient as experimental gene vectors because their large size makes them difficult to manipulate *in vitro*, and there are no unique restriction sites in the T-DNA. Initially, this problem was addressed by the construction of cointegrate vectors. T-DNA isolated from a parent Ti-plasmid was subcloned in a conventional *E. coli* plasmid vector for easy manipulation, producing a so-called *intermediate vector* (Matzke & Chilton 1981). These vectors were incapable of replication in *A. tumefaciens*, and also lacked conjugation functions. Transfer was achieved using a "triparental mating" in which three bacterial strains were mixed together: (a) an *E. coli* strain carrying a helper

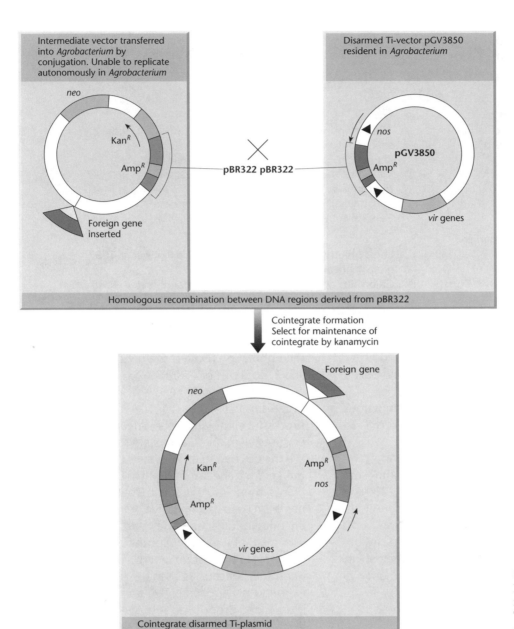

Fig. 14.12
Production of recombinant disarmed Ti plasmid by cointegrate formation.

plasmid able to mobilize the intermediate vector in *trans*; (b) the *E. coli* strain carrying the recombinant intermediate vector; and (c) *A. tumefaciens* carrying the Ti-plasmid. Conjugation between the two *E. coli* strains transferred the helper plasmid to the carrier of the intermediate vector, which was in turn mobilized and transferred to the recipient *Agrobacterium*. Homologous recombination between the T-DNA sequences of the Ti-plasmid and intermediate vector then resulted in the formation of a large cointegrate plasmid, from which the recombinant T-DNA was transferred to the plant genome (Fig. 14.12). In the cointegrate vector system, maintenance of the recombinant T-DNA is dependent

on recombination, which is enhanced if there is an extensive homology region shared by the two plasmids, as in Ti-plasmid pGV3850, which carries a segment of the pBR322 backbone in its T-DNA.

Although intermediate vectors have been widely used, the large cointegrates are not necessary for transformation. The *vir* genes of the Ti-plasmid function in *trans* and can act on any T-DNA sequence present in the same cell. Therefore, the *vir* genes and the disarmed T-DNA containing the transgene can be supplied on separate plasmids, and this is the principle of binary vector systems (Hoekma *et al.* 1983, Bevan 1984). The T-DNA can be subcloned on a small *E. coli* plasmid for ease of manipulation.

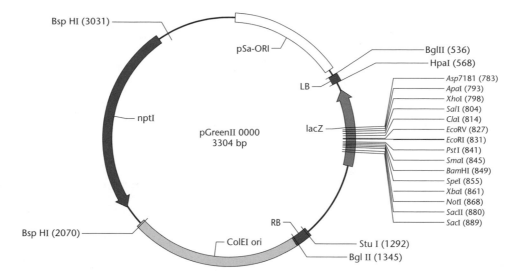

Fig. 14.13 The small and versatile binary vector pGreen, reproduced with permission of Roger Hellens and Phil Mullineaux.

This plasmid, called mini-Ti or micro-Ti, can be introduced into an *Agrobacterium* strain carrying a Ti-plasmid from which the T-DNA has been removed. The *vir* functions are supplied in *trans*, causing transfer of the recombinant T-DNA to the plant genome. The T-DNA plasmid can be introduced into *Agrobacterium* by triparental matings, or by a more simple transformation procedure such as electroporation (Cangelosi *et al.* 1991).

Most contemporary Ti-plasmid transformation systems are based on a binary principle, in which the T-DNA is maintained on a shuttle vector with a broad-host-range origin of replication such as RK2 (which functions in both *A. tumefaciens* and *E. coli*) or separate origins for each species. An independently replicating vector is advantageous because maintenance of the T-DNA is not reliant on recombination, and the binary vector's copy number is not determined by the Ti-plasmid, making the identification of transformants much easier. All the conveniences of bacterial cloning plasmids have been incorporated into binary vectors, such as multiple unique restriction sites in the T-DNA region to facilitate subcloning, the *lacZ* gene for blue-white screening (McBride & Summerfelt 1990) and a λ cos site for preparing cosmid libraries (Lazo *et al.* 1991, Ma *et al.* 1992). A current binary vector, pGreen, is shown in Fig. 14.13 (Hellens *et al.* 2000b). This plasmid is less than 5 kbp in size and has 18 unique restriction sites in the T-DNA, because the T-DNA is entirely synthetic. It has a *lacZ* gene for blue-white selection of recombinants, and a selectable marker that can be used both in bacteria and in the transformed plants. The progressive reduction in size has been made possible by removing essential genes required for replication

in *Agrobacterium* and transferring those genes to the bacterium's genome, or onto a helper plasmid. The pGreen plasmid, for example, contains the Sa origin of replication, which is much smaller than the more traditional Ri and RK2 regions. Furthermore, an essential replicase gene is housed on a second plasmid called pSoup resident within the bacterium. All conjugation functions have also been removed, so this plasmid can only be introduced into *Agrobacterium* by transformation (Hellens *et al.* 2000b).

Agrobacterium-mediated transformation can be achieved using a simple experimental protocol in many dicots

Once the principle of selectable, disarmed T-DNA vectors was established, there followed an explosion in the number of experiments involving DNA transfer to plants. Variations on the simple general protocol of Horsch *et al.* (1985) have been widely used for dicot plants (Fig. 14.14). In the original report, small disks (a few millimeters diameter) were punched from leaves, surface-sterilized, and inoculated in a medium containing *A. tumefaciens* transformed with the recombinant disarmed T-DNA (as cointegrate or binary vector). The foreign DNA contained a chimeric *neo* gene conferring resistance to the antibiotic kanamycin. The disks were cultured for two days and transferred to medium containing kanamycin to select for the transferred *neo* gene, and carbenicillin to kill the *Agrobacterium*. After 2–4 weeks, developing shoots were excised from the callus and transplanted to root-inducing medium. Rooted plantlets were subsequently transplanted to soil, about 4–7 weeks after the inoculation step.

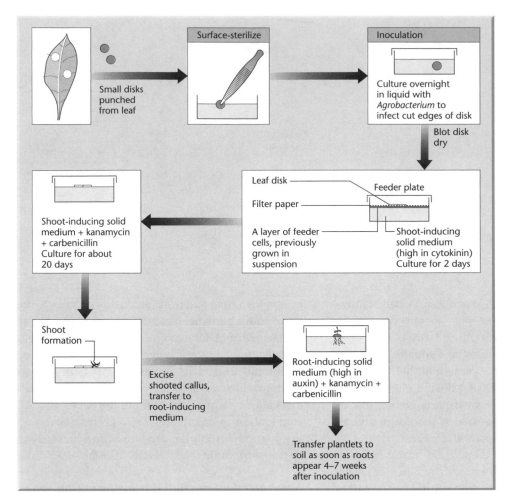

Fig. 14.14 Leaf-disk transformation by *A. tumefaciens*.

This method has the advantage of being simple and relatively rapid. It is superior to previous methods in which transformed plants were regenerated from protoplast-derived callus, the protoplasts having been transformed by co-cultivation with the *Agrobacterium* (De Block *et al.* 1984, Horsch *et al.* 1984). Contemporary protocols for the *Agrobacterium*-mediated transformation of many solanaceous plants are variations on the theme of the leaf disk protocol, although the optimal explant must be determined for each species. Alternative procedures are required for the transformation of monocots, as discussed below.

Monocots were initially recalcitrant to *Agrobacterium*-mediated transformation, but it is now possible to transform certain varieties of many cereals using this method

Until the mid-1990s, most monocotyledonous plants (monocots) were thought to be outside the host range of *Agrobacterium*, prompting research into alternative transformation methods as discussed below. During the 1980s, limited evidence accumulated showing that some monocots might be susceptible to *Agrobacterium* infection (see for example the discussion of agroinfection with maize streak virus DNA on p. 294). However, in most cases there was no convincing evidence for T-DNA integration into the plant genome. In the laboratory, it proved possible to induce tumors in certain monocot species, such as asparagus (Hernalsteens *et al.* 1984) and yam (Schafer *et al.* 1987). In the latter case, an important factor in the success of the experiment was pre-treatment of the *Agrobacterium* suspension with wound exudate from potato tubers. It has been argued that *Agrobacterium* infection of monocots is inefficient because wounded monocot tissues do not produce phenolics such as acetosyringone at sufficient levels to induce *vir* gene expression.

Eventually, however, researchers began to develop modified culture conditions and transformation procedures that worked with at least some monocots. Rice transformation was achieved in the early 1990s but the selection system (based on the *nptII* marker and selection with G418) interfered with regeneration, and only a small number of transgenic plants was produced (Raineri *et al.* 1990, Chan *et al.* 1992, 1993). The use of an alternative marker conferring resistance to hygromycin allowed the regeneration of large numbers of transgenic japonica rice plants (Hiei *et al.* 1994), and the same selection strategy has been used to produce transgenic rice plants representing the other important subspecies, indica and javanica (Rashid *et al.* 1996, Dong *et al.* 1996). More recently, efficient *Agrobacterium*-mediated transformation has become possible for other important cereals, including maize (Ishida *et al.* 1996), wheat (Cheng *et al.* 1997), barley (Tingay *et al.* 1997), and sugarcane (Arencibia *et al.* 1998).

The breakthrough in cereal transformation using *Agrobacterium* reflected the recognition of a number of key factors required for efficient infection and gene transfer to monocots. The use of explants containing a high proportion of actively dividing cells, such as embryos or apical meristems, was found to increase transformation efficiency greatly, probably because DNA synthesis and cell division favor the integration of exogenous DNA. In dicots, cell division is often induced by wounding, whereas wound sites in monocots tend to become lignified. This probably explains why traditional procedures such as the leaf disk method are inefficient in monocots. Hiei *et al.* (1994) showed that the co-cultivation of *Agrobacterium* and rice embryos in the presence of 100 mM acetosyringone was a critical factor for successful transformation. Transformation efficiency is increased further by the use of vectors with enhanced virulence functions. The modification of *Agrobacterium* for increased virulence has been achieved by increasing the expression of *virG* (which in turn boosts the expression of the other *vir* genes) and/or the expression of *virE1*, which is a major limiting factor in T-DNA transfer (reviewed by Sheng & Citovsky 1996), resulting in so-called *supervirulent* bacterial strains such as AGL-1. Komari *et al.* (1996) used a different strategy, in which a portion of the virulence region from the Ti-plasmid of supervirulent strain A281 was transferred to the T-DNA-carrying plasmid to generate a so-called *superbinary vector*. The advantage of the latter technique is that the superbinary vector can be used in any *Agrobacterium* strain.

Binary vectors have been modified to transfer large segments of DNA into the plant genome

A precise upper limit for T-DNA transfer has not been established. It is greater than 50 kbp (Herrera-Estrella *et al.* 1983a,b), but using standard vectors it is difficult to transfer inserts larger than 30 kbp routinely due to instability in the bacterial host. The analysis of very large genes or the transfer of multiple genes (such as those encoding sequentially acting enzymes of a metabolic pathway) can now be achieved thanks to the development of high-capacity binary vectors based on the artificial chromosome type vectors used in *E. coli*. The first to be described was BIBAC2 (Hamilton 1997). This contains an F-plasmid origin of replication and is modeled on the bacterial artificial chromosome (BAC, p. 79). The basic vector transforms tobacco with high efficiency, but the efficiency of transformation drops substantially when large inserts are used. This vector has been used to introduce 150 kbp of human DNA flanked by T-DNA borders into the tobacco genome, although virulence helper plasmids supplying high levels of VirG and VirE in *trans* were critical for successful DNA transfer (Hamilton *et al.* 1996). An alternative vector carrying a P1 origin of replication and modeled on the P1 artificial chromosome (PAC, p. 79) was constructed by Liu *et al.* (1999). This transformation-competent bacterial artificial chromosome (TAC) vector was used to introduce up to 80 kbp of genomic DNA into *Arabidopsis*, and while there was some loss of efficiency with the larger inserts, it was still possible to produce many transgenic plants. Both vectors contain a kanamycin resistance marker for selection in bacteria and *hpt* for hygromycin selection in transgenic plants. For the reasons discussed in Box 14.2, the *hpt* marker gene is placed adjacent to the right border T-DNA repeat. Both vectors also contain the Ri origin for maintenance in *Agrobacterium*, and within the T-DNA region, the *sacB* marker for negative selection in *E. coli*, interrupted by a polylinker for cloning foreign DNA.

One of the most attractive uses of high-capacity binary vectors is for the positional cloning of genes identified by mutation. The ability to introduce large segments of DNA into the plant genome effectively bridges the gap between genetic mapping and sequencing, allowing the position of mutant genes to be narrowed down by complementation. Genomic libraries have been established for several plant species in BIBAC2 and TAC vectors (Hamilton *et al.* 1999, Shibata & Lui 2000) and a number of novel genes have been isolated (e.g. Sawa *et al.* 1999, Kubo & Kakimoto 2001).

Agrobacterium rhizogenes is used to transform plant roots and produce hairy-root cultures

Agrobacterium rhizogenes causes hairy-root disease in plants, and this is induced by root-inducing (Ri) plasmids that are analogous to the Ti-plasmids of

A. tumefaciens. The Ri T-DNA includes genes homologous to the *iaaM* (tryptophan 2-monooxygenase) and *iaaH* (indoleacetamide hydrolase) genes of *A. tumefaciens.* Four other genes present in the Ri T-DNA are named *rol* for *root locus.* Two of these, *rolB* and *rolC,* encode P-glucosidases able to hydrolyze indole- and cytokine-N-glucosides. *A. rhizogenes* therefore appears to alter plant physiology by releasing free hormones from inactive or less active conjugated forms (Estruch *et al.* 1991a,b).

Ri-plasmids are of interest from the point of view of vector development because opine-producing root tissue induced by Ri-plasmids in a variety of dicots can be regenerated into whole plants by manipulation of phytohormones in the culture medium. Ri T-DNA is transmitted sexually by these plants and affects a variety of morphological and physiological traits, but does not in general appear deleterious. The Ri-plasmids therefore appear to be already equivalent to disarmed Ti-plasmids (Tepfer 1984). Transformed roots can also be maintained as hairy-root cultures, which have the potential to produce certain valuable secondary metabolites at higher levels than suspension cultures, and are much more genetically stable (Hamil *et al.* 1987, Signs & Flores 1990). The major limitation for the commercial use of hairy-root cultures is the difficulty involved in scale-up, since each culture comprises a heterogeneous mass of interconnected tissue, with highly uneven distribution (reviewed by Giri & Narassu 2000).

Many of the principles explained in the context of disarmed Ti-plasmids are applicable to Ri-plasmids. A cointegrate vector system has been developed (Jensen *et al.* 1986) and applied to the study of nodulation in transgenic legumes. Van Sluys *et al.* (1987) have exploited the fact that *Agrobacterium* containing both an Ri-plasmid and a disarmed Ti-plasmid can frequently co-transfer both plasmids. The Ri-plasmid induces hairy-root disease in recipient *Arabidopsis* and carrot cells, serving as a transformation marker for the co-transferred recombinant T-DNA, and allowing regeneration of intact plants. No drug resistance marker on the T-DNA is necessary with this plasmid combination.

Direct DNA transfer to plants

Transgenic plants can be regenerated from transformed protoplasts

Protoplast transformation has much in common with the transfection of animal cells (Chapter 12).

The protoplasts must initially be persuaded to take up DNA from their surroundings, after which the DNA integrates stably into the genome in a proportion of these transfected cells. Gene transfer across the protoplast membrane is promoted by a number of chemicals, of which polyethylene glycol has become the most widely used due to the availability of simple transformation protocols (Negrutiu *et al.* 1987). Alternatively, DNA uptake may be induced by electroporation (Shillito *et al.* 1985). As with animal cells, the introduction of a selectable marker gene along with the transgene of interest is required for the identification of stable transformants. This can be achieved using plasmid vectors carrying both the marker and the transgene of interest, but the use of separate vectors also results in a high frequency of co-transformation (Schocher *et al.* 1986). Putative transformants are transferred to selective medium, where surviving protoplasts regenerate their cell walls and commence cell division, producing a callus. Subsequent manipulation of the culture conditions then makes it possible to induce shoot and root development, culminating in the recovery of fertile transgenic plants. The major limitation of protoplast transformation is not the gene-transfer process itself, but the ability of the host species to regenerate from protoplasts. A general observation is that dicots are more amenable than monocots to this process. In species where regeneration is possible, an advantage of the technique is that protoplasts can be cryopreserved and retain their regenerative potential (DiMaio & Shillito 1989).

The first transformation experiments concentrated on species such as tobacco and petunia in which protoplast-to-plant regeneration is well documented. An early example is provided by Meyer *et al.* (1987), who constructed a plasmid vector containing the *nptII* marker gene, and a maize cDNA encoding the enzyme dihydroquercetin 4-reductase, which is involved in anthocyanin pigment biosynthesis. The transgene was driven by the strong and constitutive CaMV 35S promoter. Protoplasts of a mutant, white-colored petunia strain were transformed with the recombinant plasmid by electroporation and then selected on kanamycin-supplemented medium. After a few days, surviving protoplasts had given rise to microcalli, which could be induced to regenerate into whole plants. The flowers produced by these plants were brick red instead of white, showing that the maize cDNA had integrated into the genome and was expressed.

After successful experiments using model dicots, protoplast transformation was attempted in monocots, for which no alternative gene-transfer system was then available. In the first such experiments, involving wheat (Lorz *et al.* 2005) and the Italian ryegrass *Lolium multiflorum* (Potrykus *et al.* 1985a,b), protoplast transformation was achieved and transgenic callus obtained, but it was not possible to recover transgenic plants. The inability of most monocots to regenerate from protoplasts may reflect the loss of competence to respond to tissue culture conditions as the cells differentiate. In cereals and grasses, this has been addressed to a certain extent by using embryogenic suspension cultures as a source of protoplasts. Additionally, since many monocot species are naturally tolerant to kanamycin, the *nptII* marker used in the initial experiments was replaced with alternative markers conferring resistance to hygromycin or phosphinothricin. With these modifications, it has been possible to regenerate transgenic plants representing certain varieties of rice and maize with reasonable efficiency (Shimamoto *et al.* 1988, Datta *et al.* 1990, Omirulleh *et al.* 1993). However, the extended tissue culture step is unfavorable, often resulting in sterility and other phenotypic abnormalities in the regenerated plants.

Protoplast transformation was also the first method developed for gene transfer to the chloroplast genome of higher plants (Golds *et al.* 1993, O'Neill *et al.* 1993). In this context, plastid mutations conferring tolerance to antibiotics through alterations in ribosome structure can be used as an alternative to bacterial antibiotic resistance genes for the selection of plastid transformants (Kavanagh *et al.* 1999).

Particle bombardment can be used to transform a wide range of plant species

An alternative procedure for plant transformation was introduced in 1987, involving the use of a modified shotgun to accelerate small (1–4 μm) metal particles into plant cells at a velocity sufficient to penetrate the cell wall (~250 m/s). In the initial test system, intact onion epidermis was bombarded with tungsten particles coated in tobacco mosaic virus (TMV) RNA. Three days after bombardment, approximately 40% of the onion cells that contained particles also showed evidence of TMV replication (Sanford *et al.* 1987). A plasmid containing the *cat* reporter gene driven by the CaMV35S promoter was then tested to determine whether DNA could be delivered by the same method. Analysis of the epidermal tissue three days after bombardment revealed high levels of transient CAT activity (Klein *et al.* 1987).

The stable transformation of explants from several plant species was achieved soon after these initial experiments. Early reports included the transformation of soybean (Christou *et al.* 1988), tobacco (Klein *et al.* 1988a), and maize (Klein *et al.* 1988b). In each case, the *nptII* gene was used as a selectable marker, and transformation was confirmed by the survival of callus tissue on kanamycin-supplemented medium. The ability to stably transform plant cells by this method offered the exciting possibility of generating transgenic plants representing species that were, at the time, intractable to other transformation procedures. In the first such report, transgenic soybean plants were produced from meristem tissue isolated from immature seeds (McCabe *et al.* 1988). In this experiment, the screenable marker gene *gusA* was introduced by particle bombardment and transgenic plants were recovered in the absence of selection by screening for GUS activity (Box 15.1). Other early successes included cotton, papaya, maize, and tobacco (Finer & McMullen 1991, Fitch *et al.* 1990, Gordon-Kamm *et al.* 1990, Fromm *et al.* 1990, Tomes *et al.* 1990; reviewed by Twyman & Christou 2004). Particle bombardment has also been pivotal in the development of chloroplast transformation technology (see below).

There is no intrinsic limitation to the potential of particle bombardment since DNA delivery is governed entirely by physical parameters (Altpeter *et al.* 2005). Many different types of plant material have been used as transformation targets, including callus, cell suspension cultures, and organized tissues such as immature embryos, meristems, and leaves. The number of species in which transgenic plants can be produced using variants of particle bombardment has therefore increased dramatically over the last 10 years. Notable successes include almost all of the commercially important cereals, i.e. rice (Christou *et al.* 1991), wheat (Vasil *et al.* 1992), oat (Somers *et al.* 1992, Torbert *et al.* 1995), sugarcane (Bower & Birch 1992), and barley (Wan & Lemaux 1994, Hagio *et al.* 1995). Several literature surveys have been published documenting the range of species that have been transformed using this method (Christou 1996, Luthra *et al.* 1997, Twyman & Christou 2004).

The original gunpowder-driven device has been improved and modified resulting in greater control over particle velocity and hence greater reproducibility of transformation conditions. An apparatus based on electric discharge (McCabe & Christou 1993) has been used for the development of variety-independent gene-transfer methods for the more recalcitrant cereals and legumes. Several instruments have been developed where particle acceleration is controlled by pressurized gas. These include a pneumatic apparatus (Iida *et al.* 1990), a "particle inflow gun" using flowing helium (Takeuchi *et al.* 1992, Finer *et al.* 1992) and a device utilizing compressed helium (Sanford *et al.* 1991). Physical parameters such as particle size and acceleration (which affect the depth of penetration and the amount of tissue damage) as well as the amount and conformation of the DNA used to coat the particles, must be optimized for each species and type of explant (Finer *et al.* 1999, Twyman & Christou 2004). However, the nature of the transformation target is probably the most important single variable in the success of gene transfer.

Other direct DNA transfer methods have been developed for intact plant cells

There is a great diversity of approaches for gene transfer to animal cells and many of the same methods have been attempted in plants. Electroporation has been used to transform not only protoplasts (see above) but also walled plant cells, either growing in suspension or as part of intact tissues. In many cases, the target cells have been wounded or pre-treated with enzymes in order to facilitate gene transfer (e.g. D'Halluin *et al.* 1992, Laursen *et al.* 1994). However, immature rice, wheat, and maize embryos can be transformed using electroporation without any form of pre-treatment (Kloti *et al.* 1993, Xu & Li 1994, Sorokin *et al.* 2000). Other transformation methods also involve perforation of the cell, including the use of silicon carbide whiskers (Thompson *et al.* 1995, Nagatani *et al.* 1997), ultrasound (Zhang *et al.* 1991), or a finely focused laser beam (Hoffman 1996). In most of these cases, only transient expression of the introduced DNA has been achieved, although transgenic maize plants have been recovered following whisker-mediated transformation (Frame *et al.* 1994). Finally, microinjection of DNA into plant cells can yield transformed cells or even transgenic plants, although as is the case for animal cells this method is not suitable for large-scale transformation (Crossway *et al.* 1986, Leduc *et al.* 1996, Holm *et al.* 2000).

Direct DNA transfer is also used for chloroplast transformation

So far, we have considered DNA transfer to the plant's nuclear genome. However, the chloroplast is also a useful target for genetic manipulation because thousands of chloroplasts may be present in photosynthetic cells and this can result in levels of transgene expression up to 50 times higher than possible using nuclear transformation. Furthermore, transgenes integrated into chloroplast DNA do not appear to undergo silencing or suffer from position effects that can influence the expression levels of transgenes in the nuclear DNA (see Boxes 13.2 and 15.2). Chloroplast transformation also provides a natural containment method for transgenic plants, since the transgene cannot be transmitted through pollen (reviewed by Maliga 1993).

The first reports of chloroplast transformation were serendipitous, and the integration events were found to be unstable. For example, an early experiment in which tobacco protoplasts were co-cultivated with *Agrobacterium* resulted in the recovery of one transgenic plant line in which the transgene was transmitted maternally. Southern blot analysis of chloroplast DNA showed directly that the foreign DNA had become integrated into the chloroplast genome (De Block *et al.* 1985). However, *Agrobacterium* is not an optimal system for chloroplast transformation because the T-DNA complex is targeted to the nucleus. Therefore, direct DNA transfer has been explored as an alternative strategy. Stable chloroplast transformation was first achieved in the alga *Chlamydomonas reinhardtii*, which has a single large chloroplast occupying most of the volume of the cell (Boynton *et al.* 1988). Particle bombardment was used in this experiment. The principles established using this simple organism were extended to tobacco, allowing the recovery of stable *transplastomic* tobacco plants (Svab *et al.* 1990b). These principles included the use of vectors containing chloroplast homology regions, allowing targeted integration into the chloroplast genome, and use of the selectable marker gene *aad* (encoding aminoglycoside adenyltransferase) which confers resistance to streptomycin and spectinomycin (Zoubenko *et al.* 1994). Efficient chloroplast transformation has been achieved both through particle bombardment (e.g. Staub & Maliga 1992a,b) and PEG mediated transformation (Golds *et al.* 1993, Koop *et al.* 1996). The use of a combined selectable-screenable marker (*aad* linked to the gene for green fluorescent protein) allows the tracking of transplastomic sectors of plant tissue prior to chlorophyll synthesis, so that transformed plants can be rapidly identified (Khan & Maliga 1999). It is now possible to transform the chloroplast genome and then eliminate selectable marker genes after transgene integration (Corneille *et al.* 2001, Iamtham & Day 2000, Klaus *et al.* 2004).

Among crop plants, tobacco (Svab *et al.* 1990a,b), tomato (Ruf *et al.* 2001) and potato (Sidorov *et al.*

1999) chloroplasts have been transformed, as well as rapeseed and other brassicas (Hou *et al.* 2003, Skarjinskaia *et al.* 2003). Most recently, soybean plastid transformation has been achieved (Dufourmantel *et al.* 2004). The major limitations in transforming the chloroplasts of other crop species, especially monocots, include a poor understanding of gene expression in non-green plastids, gene-delivery methods for proplastids, and tissue culture conditions. Thus far, chloroplast transformation by particle bombardment has been achieved only in crops that allow direct organogenesis. Transplastomic plants have been developed with improvements in a number of key agronomic traits (Daniell *et al.* 2004), and there have been many reports of chloroplasts expressing pharmaceutical proteins (summarized by Altpeter *et al.* 2005). These topics are considered in more detail in Chapter 26.

Gene targeting in plants

As discussed earlier in the book, gene targeting is an efficient procedure in bacteria, yeast, and certain animal cells, enabling directed changes to be introduced into the genome by homologous recombination. In plants, homologous recombination is a very inefficient process; only one plant species has been shown to undergo efficient nuclear homologous recombination and that is the moss *Physcomitrella patens*. Among higher plants, low-level gene targeting has been achieved in certain dicots with frequencies ranging from 10^{-3} to 10^{-6} (Lee *et al.* 1990, Ofringa *et al.* 1990, Miao & Lam 1995, Risseeuw *et al.* 1995, 1997, Kempin *et al.* 1997, Reiss *et al.* 2000, Hanin *et al.* 2001) and a transgene has been repaired by homologous recombination in tobacco (Paszkowski *et al.* 1988). However, the most promising results have been achieved using a T-DNA-mediated gene-targeting strategy involving a long homology region in combination with a strong counterselectable marker in rice (Terada *et al.* 2002). Targeting frequencies of up to 1% have been achieved using this system (reviewed by Iida & Tarada 2004, Cotsaftis & Guiderdoni 2005).

In planta transformation minimizes or eliminates the tissue culture steps usually needed for the generation of transgenic plants

Until recently, gene transfer to plants involved the use of cells or explants as transformation targets and an obligatory tissue culture step was needed for the regeneration of whole fertile plants. Experiments using the model dicot *Arabidopsis thaliana* have led the way in the development of so-called *in planta* transformation techniques, where the need for tissue culture is minimized or eliminated altogether. Such methods involve the introduction of DNA, either by *Agrobacterium* or direct transfer, into intact plants. The procedure is carried out at an appropriate time in the plant's life cycle so that the DNA becomes incorporated into cells that will contribute to the germline, directly into the germ cells themselves (often at around the time of fertilization), or into the very early plant embryo. Generally, *in planta* transformation methods have a very low efficiency, so the small size of *Arabidopsis* and its ability to produce over 10,000 seeds per plant is advantageous. This limitation has so far prevented *in planta* techniques from being widely adopted for other plant species.

The first *in planta* transformation system involved imbibing *Arabidopsis* seeds overnight in an *Agrobacterium* culture, followed by germination (Feldman & Marks 1987). A large number of transgenic plants containing T-DNA insertions were recovered but in general this technique has a low reproducibility. A more reliable method has been described by Bechtold *et al.* (1993) in which the bacteria are vacuum infiltrated into *Arabidopsis* flowers. An even simpler technique called floral dip has become widely used (Clough & Bent 1998). This involves simply dipping *Arabidopsis* flowers into a bacterial suspension at the time of fertilization. In both these methods, the transformed plants are chimeric, but give rise to a small number of transgenic progeny (typically about 10 per plant). It has been established that T-DNA is transferred into the ovule during the transformation procedure (Bechtold *et al.* 2000, Ye *et al.* 1999, Desfeux *et al.* 2000). There have been few successful applications of these methods in other plants, although a small number of model species have been transformed as well as radish (Quing *et al.* 2000, Trieu *et al.* 2000, Curtis *et al.* 2001).

Similar approaches using direct DNA transfer have been tried in other species, but germline transformation has not been reproducible. For example, naked DNA has been injected into the floral tillers of rye plants (De La Pena *et al.* 1987) and post-fertilization cotton flowers (Zhou *et al.* 1983) resulting in the recovery of some transgenic plants. Transgenic tobacco has been produced following particle bombardment of pollen (Touraev *et al.* 1997).

An alternative to the direct transformation of germ-line tissue is the introduction of DNA into meristems *in planta* followed by the growth of transgenic shoots. In *Arabidopsis*, this has been achieved simply by severing apical shoots at their bases and inoculating the cut tissue with *Agrobacterium* suspension (Chang *et al.* 1994). Using this procedure, transgenic plants were recovered from the transformed shoots at a frequency of about 5%. In rice, explanted meristem tissue has been transformed using *Agrobacterium* and particle bombardment, resulting in the proliferation of shoots that can be regenerated into transgenic plants (Park *et al.* 1996). Such procedures require only a limited amount of tissue culture.

Plant viruses can be used as episomal expression vectors

As an alternative to stable transformation using *Agrobacterium* or direct DNA transfer, plant viruses can be employed as gene transfer and expression vectors. There are several advantages to the use of viruses. First, viruses are able to adsorb to and introduce their nucleic acid into intact plant cells. However, for many viruses, naked DNA or RNA is also infectious, allowing recombinant vectors to be introduced directly into plants by methods such as leaf rubbing. Second, infected cells yield large amounts of virus, so recombinant viral vectors have the potential for high-level transgene expression. Third, viral infections are often systemic. The virus spreads throughout the plant allowing transgene expression in all cells. Fourth, viral infections are rapid, so large amounts of recombinant protein can be produced in a few weeks. Finally, all known plant viruses replicate episomally, therefore the transgenes they carry are not subject to the position effects that often influence the expression of integrated transgenes (Box 13.2). Since plant viruses neither integrate nor pass through the germline, plants cannot be stably transformed by viral infection and transgenic lines cannot be generated. However, this limitation can also be advantageous in terms of containment.

A complete copy of a viral genome can also be introduced into isolated plant cells or whole plants by *Agrobacterium* or direct DNA transfer. In this manner, it is possible to generate transiently transformed cell lines or transgenic plants carrying an integrated recombinant viral genome. In the case of RNA viruses, transcription of an integrated cDNA copy of the genome yields replication-competent viral RNA, which is amplified episomally, facilitating high-level transgene expression. Transgenic plants are persistently infected by the virus and can produce large amounts of recombinant protein. In the case of DNA viruses, *Agrobacterium*-mediated transient or stable transformation with T-DNA containing a partially duplicated viral genome can lead to the "escape" of intact genomes, which then replicate episomally. The latter process, known as "agroinfection" or "agroinoculation" provides a very sensitive assay for gene transfer. More recently, the *Agrobacterium*-mediated delivery of viral genomes has been enhanced through a process called magnifection, in which amplification of the vector occurs in all infected leaves (Marrilonnet *et al.* 2005).

As well as their use for the expression of whole foreign proteins, certain plant viruses have recently been developed to present short peptides on their surfaces, similar to the phage display technology discussed on p. 147. Epitope-display systems based on cowpea mosaic virus, alfalfa mosaic virus, potato virus X, and tomato bushy stunt virus have been developed as a potential source of vaccines, particularly against animal viruses (reviewed by Lomonossoff & Hamilton 1999, Pogue *et al.* 2002, Yusibov & Rabindran 2004, Khalsa *et al.* 2004, Twyman *et al.* 2005). These systems are discussed in more detail in Chapter 26.

The first plant viral vectors were based on DNA viruses because of their small and simple genomes

The vast majority of plant viruses have RNA genomes. However, the two groups of DNA viruses that are known to infect plants – the caulimoviruses and the geminiviruses – were the first to be developed as vectors because of the ease with which their small, DNA genomes could be manipulated in plasmid vectors.

The type member of the caulimoviruses is cauliflower mosaic virus (CaMV). The 8-kb dsDNA genome of several isolates has been completely sequenced, revealing an unusual structure characterized by the presence of three discontinuities in the duplex. A map of the CaMV genome is shown in Fig. 14.15. There are eight tightly packed genes, expressed as two major transcripts: the 35S RNA (which essentially represents the entire genome) and the 19S RNA (which contains the coding region for gene VI). As discussed earlier in the chapter, the promoter

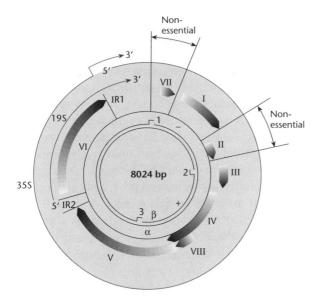

Fig. 14.15 Map of the cauliflower mosaic virus genome. The eight coding regions are shown by the thick gray arrows, and the different reading frames are indicated by the radial positions of the boxes. The thin lines in the center indicate the (plus and minus) DNA strands with the three discontinuities. The major transcripts, 19S and 35S, are shown around the outside.

and terminator sequences for both transcripts have been utilized in plant expression vectors, and the 35S promoter is particularly widely used (Box 14.1).

Only two of the genes in the CaMV genome are non-essential for replication (gene II and gene VII), and since CaMV has an icosahedral capsid, the size of the genome cannot be increased greatly without affecting the efficiency of packaging. The maximum capacity of the CaMV capsid has been defined as 8.3 kb, and with the removal of all non-essential genes, this represents a maximum insert size of less than 1 kb (Daubert *et al.* 1983). This restriction in the capacity for foreign DNA represents a major limitation of CaMV vectors. Thus far it has been possible to express a number of very small transgenes, such as the 240-bp bacterial *dhfr* gene (Brisson *et al.* 1984), the 200-bp murine metallothionein cDNA (Lefebvre *et al.* 1987) and a 500-bp human interferon cDNA (de Zoeten *et al.* 1989). In Chapter 12, we describe how similar limitations were overcome for SV40, a virus that infects primate cells, through the development of replicon vectors and helper viruses or complementary cell lines supplying essential functions in *trans*. Unfortunately, such an approach is not possible with CaMV due to the high level of recombination that occurs, leading to rapid excision of the foreign DNA (Gronenborn & Matzeit 1989).

Geminiviruses are characterized by their twin (geminate) virions, comprising two partially fused icosahedral capsids. The small single-stranded DNA genome is circular, and in some species is divided into two segments called DNA A and DNA B. Interest in geminivirus vector development was stimulated by the discovery that such viruses use a DNA replicative intermediate, suggesting they could be more stable than CaMV, whose RNA-dependent replication cycle is rather error-prone (Stenger *et al.* 1991). Of the three genera of geminiviruses, two have been developed as vectors. The begomoviruses have predominantly bipartite genomes; they are transmitted by the whitefly *Bemisia tabaci* and infect dicots. Species that have been developed as vectors include African cassava mosaic virus (ACMV) and tomato golden mosaic virus (TGMV). The mastreviruses have monopartite genomes; they are transmitted by leafhoppers and predominantly infect monocots. Species that have been developed as vectors include maize streak virus (MSV) and wheat dwarf virus (WDV).

An important additional distinction between these genera is that mastreviruses are not mechanically transmissible. MSV, for example, has never been introduced successfully into plants as native or cloned DNA. Grimsley *et al.* (1987) were able to overcome this problem using *Agrobacterium*, and were the first to demonstrate the principle of agroinfection. They constructed a plasmid containing a tandem dimer of the MSV genome. This dimer was inserted into binary vector, and maize plants were infected with *A. tumefaciens* containing the recombinant T-DNA. Viral symptoms appeared within two weeks of inoculation. Agroinfection has been used to introduce the genomes of a number of different viruses into plants. It can be demonstrated that if the T-DNA contains partially or completely duplicated genomes, single copies of the genome can escape and initiate infections. This may be mediated by homologous recombination or a replicative mechanism (Stenger *et al.* 1991). The study of Grimsley *et al.* (1987) incidentally provided the first evidence that *Agrobacterium* could transfer T-DNA to maize. Agroinfection is a very sensitive assay for transfer to the plant cell because of the amplification inherent in the virus infection and the resulting visible symptoms.

A number of geminiviruses have been developed as expression vectors because of the possibility of achieving high-level recombinant protein expression as a function of viral replication (reviewed by Stanley 1993, Timmermans *et al.* 1994, Palmer & Rybicki 1997). A generally useful strategy is the replacement of the coat protein gene, since this is not required for replication and the strong promoter can be used to drive transgene expression. In the case of begomoviruses, which have bipartite genomes, the coat protein gene is located on DNA A along with all the functions required for DNA replication. Replicons based on DNA A are therefore capable of autonomous replication in protoplasts (e.g. Townsend *et al.* 1986). Geminivirus replicon vectors can facilitate the high-level transient expression of foreign genes in protoplasts. There appears to be no intrinsic limitation to the size of the insert, although larger transgenes tend to reduce the replicon copy number (e.g. Laufs *et al.* 1990, Matzeit *et al.* 1991). Generally, it appears that mastrevirus replicons can achieve a much higher copy number in protoplasts than replicons based on begomoviruses. A WDV shuttle vector capable of replicating in both *E. coli* and plants was shown to achieve a copy number of greater than 3×10^4 in protoplasts derived from cultured maize endosperm cells (Timmermans *et al.* 1992), whereas the typical copy number achieved by TGMV replicons in tobacco protoplasts is less than 1000 (Kanevski *et al.* 1992). This may, however, reflect differences in the respective host cells rather than the intrinsic efficiencies of the vectors themselves.

Geminiviruses are also valuable as expression vectors in whole plants. In the case of the mastreviruses, all viral genes appear to be essential for systemic infection, so coat protein replacement vectors cannot be used in this manner. In contrast, the coat protein genes of ACMV and TGMV are nonessential for systemic infection, but they are required for insect transmission (Briddon *et al.* 1990). Therefore, replicon vectors based on these viruses provide an *in planta* contained transient expression system. Note that viral movement functions are supplied by DNA B, so systemic infections occur only if DNA B is also present in the plant.

In an early study, Ward *et al.* (1988) replaced most of the ACMV AV1 gene with the *cat* reporter gene. In infected tobacco plants, high-level CAT activity was detected for up to four weeks. Interestingly, they found that deletion of the coat protein gene caused a loss of infectivity in plants, but this was restored upon replacement with *cat*, which is approximately the same size as the deleted gene. This and many subsequent reports indicated that, while there may be no intrinsic limit to the size of replicon vectors in protoplasts, systemic infection is dependent on preserving the size of the wild-type DNA A component. A further limitation to this system is that the transmissibility of the recombinant genomes is poor,

probably because they are not packaged. One way in which this can be addressed is to generate transgenic plants in which recombinant viral genomes are produced in every cell. This is achieved by transforming plants with DNA constructs containing a partially duplicated viral genome (Meyer *et al.* 1992). Intact replicons can excise from the delivered transgene in the same way as the MSV genome escapes during agroinfection, and autonomously replicating episomal copies can be detected. Transgenic tobacco plants have also been produced carrying an integrated copy of DNA B (Hayes *et al.* 1988, 1989). In the presence of replication functions supplied by DNA A, the DNA B sequence is rescued from the transgene and can replicate episomally. DNA B can then provide movement functions to DNA A, facilitating the systemic spread of the vector.

Most plant virus expression vectors are based on RNA viruses because they can accept larger transgenes than DNA viruses

Most plant RNA viruses have a filamentous morphology, so the packaging constraints affecting the use of DNA viruses such CaMV should not present a limitation in vector development. However, investigation into the use of RNA viruses as vectors lagged behind research on DNA viruses, awaiting the advent of robust techniques for the manipulation of RNA genomes.

A breakthrough was made in 1984, when a full-length clone corresponding to the genome of brome mosaic virus (BMV) was obtained. Infectious RNA could be produced from this cDNA by *in vitro* transcription (Ahlquist & Janda 1984, Ahlquist *et al.* 1984). The BMV genome comprises three segments: RNA1, RNA2, and RNA3. Only RNA1 and RNA2 are necessary for replication. RNA3, which is dicistronic, encodes the viral coat protein and movement protein. During BMV infection, a subgenomic RNA fragment is synthesized from RNA3, containing the coat protein gene alone. It is therefore possible to replace the coat protein gene with foreign DNA and still generate a productive infection (Ahlquist *et al.* 1987). This was demonstrated by French *et al.* (1986) in an experiment where the coat protein gene was substituted with the *cat* reporter gene. Following the introduction of recombinant RNA3 into barley protoplasts along with the essential RNA1 and RNA2 segments, high-level CAT activity was achieved.

This experiment showed that brome mosaic virus was a potentially useful vector for foreign gene expression. However, to date, BMV has been used

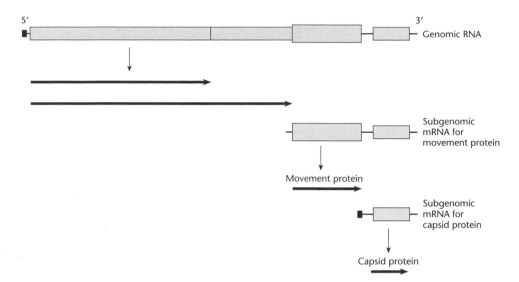

Fig. 14.16 Genome map and expression of tobacco mosaic virus.

solely to study the function of genes from other plant viruses. Following the demonstration that infectious BMV RNA could be produced by *in vitro* transcription, the genomes of many other RNA viruses have been prepared as cDNA copies. Some of these viruses have been extensively developed as vectors for foreign gene expression (see comprehensive reviews by Scholthof *et al.* 1996, Porta & Lomonossoff 2001). Two examples are discussed below.

Tobacco mosaic virus (TMV) is one of the most extensively studied plant viruses and was thus a natural choice for vector development. The virus has a monopartite RNA genome of 6.5 kb. At least four polypeptides are produced, including a movement protein and a coat protein that are translated from subgenomic RNAs (Fig. 14.16). The first use of TMV as a vector was reported by Takamatsu *et al.* (1987). They replaced the coat protein gene with *cat*, and obtained infected plants showing high level CAT activity at the site of infection. However, the recombinant virus was unable to spread throughout the plant because the coat protein is required for systemic infection.

Since there should be no packaging constraints with TMV, Dawson *et al.* (1989) addressed the deficiencies of the TMV replacement vector by generating a replication-competent *addition vector* in which the entire wild-type genome was preserved. Dawson and colleagues added the bacterial *cat* gene, controlled by a duplicated coat protein subgenomic promoter, between the authentic movement and coat protein genes of the TMV genome. In this case, systemic infection occurred in concert with high-level CAT activity, but recombination events in infected plants resulted in deletion of the transgene and the produc-

tion of wild-type TMV RNA. Homologous recombination can be prevented by replacing the TMV coat protein gene with the equivalent sequence from the related *Odontoglossum* ringspot virus (Donson *et al.* 1991). This strategy has been used to produce a range of very stable expression vectors that have been used to synthesize a variety of valuable proteins in plants, such as ribosome-inactivating protein (Kumagai *et al.* 1993) and scFV antibodies (McCormick *et al.* 1999). It has also been possible to produce complete monoclonal antibodies by co-infecting plants with separate TMV vectors expressing the heavy and light immunoglobulin chains (Verch *et al.* 1998).

Potato virus X (PVX) is the type member of the *Potexvirus* family. Like TMV, it has a monopartite RNA genome of approximately 6.5 kb which is packaged in a filamentous particle. The genome map of PVX is shown in Fig. 14.17, and contains genes for replication, viral movement, and the coat protein, the latter expressed from a subgenomic promoter. Reporter genes such as *gusA* and green fluorescent protein have been added to the PVX genome under the control of a duplicated coat protein subgenomic promoter and can be expressed at high levels in infected plants (Chapman *et al.* 1992, Baulcombe *et al.* 1995). As with the early TMV vectors, there is a tendency for the transgene to be lost by homologous recombination, but in the case of PVX, no alternative virus has been identified whose coat protein gene promoter can functionally substitute for the endogenous viral promoter. For this reason, PVX is generally not used for long-term expression, but has been widely employed as a transient expression vector. It has been used for the synthesis of valuable proteins such as antibodies (e.g. Hendy *et al.* 1999,

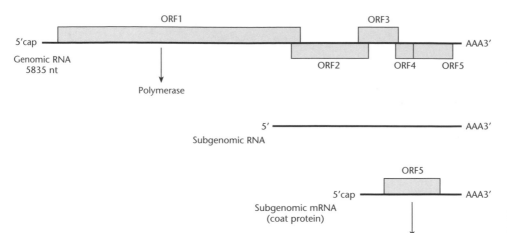

Fig. 14.17 Genome map and expression of potato virus X.

Franconi *et al.* 1999, Ziegler *et al.* 2000) and for the expression of genes that affect plant physiology (e.g. the fungal avirulence gene *avr9*, Hammond-Kosack *et al.* 1995).

The stable transformation of plants with cDNA copies of the PVX genome potentially provides a strategy for extremely high-level transgene expression, because transcripts should be amplified to a high copy number during the viral replication cycle. However, instead of high-level expression, this strategy leads to potent and consistent transgene silencing, as well as resistance to viral infection (English *et al.* 1996, English & Baulcombe 1997). The basis of this phenomenon and its implications for transgene expression in plants are discussed in more detail in Chapter 15. In terms of vector development, however, it is notable that PVX-based vectors are probably most widely used to study virus-induced gene silencing and related phenomena (Dalmay *et al.* 2000) and to deliberately induce silencing of homologous plant genes (reviewed by Baulcombe 1999).

Suggested reading

Cotsaftis O. & Guiderdoni E. (2005) Enhancing gene targeting efficiency in higher plants: rice is on the move. *Transgenic Research* **14**, 1–14.

Daniel H., Khan M.S. & Allison L. (2002) Milestones in chloroplast genetic engineering: an environmentally friendly era in biotechnology. *Trends in Plant Science* **7**, 84–91.

Daniel H., Kumar S. & Dufourmantel N. (2005) Breakthrough in chloroplast genetic engineering of agronomically important crops. *Trends in Biotechnology* **23**, 238–45.

Gamborg O.L. (2002) Plant tissue culture. Biotechnology. Milestones. *In vitro Cell Development Biology – Plant* **38**, 84–92.

Gelvin S.B. (2000) *Agrobacterium* and plant genes involved in T-DNA transfer and integration. *Annual Review of Plant Physiology* **51**, 223–56.

Gelvin S.B. (2003) *Agrobacterium*-mediated plant transformation: the biology behind the "gene-Jockeying" tool. *Microbiology and Molecular Biology* **67**, 16.

Goodwin J.L., Pastori G.M., Davey M.R. & Jones H.D. (2004) Selectable markers: antibiotic and herbicide resistance. *Methods in Molecular Biology* **286**, 191–202.

Hellens R., Mullineux P. & Klee H. (2000) A guide to *Agrobacterium* binary Ti vectors. *Trends in Plant Science* **5**, 446–51.

Maliga, P. (2004) Plastid transformation in higher plants. *Annual Review of Plant Biology* **55**, 289–313.

Marillonnet S., Thoeringer C., Kandzia R., Klimyuk V. & Gleba Y. (2005) Systemic *Agrobacterium tumefaciens*-mediated transfection of viral replicons for efficient transient expression in plants. *Nature Biotechnology*. **23**, 718–23.

Porta C. & Lomonossoff G.P. (1996) Use of viral replicons for the expression of genes in plants. *Molecular Biotechnology* **5**, 209–21.

Porta C. & Lomonossoff G.P. (2002) Viruses as vectors for the expression of foreign sequences in plants. *Biotechnology and Genetic Engineering* **19**, 245–91.

Somers D.A. & Makarevitch I. (2004) Transgene integration in plants: poking or patching holes in promiscuous genomes? *Current Opinion in Biotechnology* **15**, 126–31.

Taylor N.J. & Fauquet C.M. (2002) Microparticle bombardment as a tool in plant science and agricultural biotechnology. *DNA Cell Biology* **21**, 963–77.

Tzfira T. & Citovsky V. (2002) Partners-in-infection: host proteins involved in the transformation of plant cells by *Agrobacterium*. *Trends in Cell Biology* **12**, 121–9.

CHAPTER 15

Advanced transgenic technology

Introduction

Gene transfer experiments in animals and plants have evolved beyond the simple process of introducing additional DNA sequences into target genomes. Now that DNA transfer is becoming routine in many species, attention is shifting to more sophisticated approaches which involve the precise control of transgenes and endogenous genes, both in terms of their structure and their expression. Parallel advances in animal and plant biotechnology have come about through the development of inducible expression systems which facilitate the external regulation of transgenes, and through the exploitation of site-specific recombination systems to make precise insertions and modifications in target sequences. Transgenic technology is particularly advanced in mice, where combinations of gene targeting, site-specific recombination, and inducible transgene expression make it possible to activate and inactivate both transgenes and endogenous genes in a conditional manner.

Other routes to conditional gene silencing have also been explored. These do not involve the direct modification of the target gene but rather the expression of inhibitory genes whose products interfere with the expression of the target. Although many such strategies have been developed, e.g. antisense RNA, ribozymes, interfering antibodies, and dominant negative mutants, one approach in particular has risen to dominate the field. From rather obscure beginnings, the phenomenon of RNA interference (RNAi) has emerged as one of the most exciting recent developments in molecular biology, and may soon make the transition from laboratory to clinic.

Inducible expression systems allow transgene expression to be controlled by physical stimuli or the application of small chemical modulators

In many gene transfer experiments, it is desirable for the introduced transgene to be expressed in a specific manner. In both animals and plants, cell- or tissue-specific promoters are used to restrict transgene expression to certain areas of the organism. In a commercial setting, for example, it is useful to restrict transgene expression in animals to the mammary glands, so that recombinant proteins can be recovered from milk (Wall *et al.* 1999). Similarly, it is useful to restrict transgene expression in plants to the seeds, since this is a stable environment for protein accumulation (Stoger *et al.* 2000). In other cases, it may be necessary to control transgene expression more precisely. For example, if a recombinant protein is toxic, constitutive high-level expression would be lethal and would prevent the recovery of stably transformed cell lines – under these circumstances the experimenter might want to choose the best time to switch the transgene off. Such issues can be addressed through the use of *inducible expression systems*, in which transgene expression is controlled by an external stimulus.

Some naturally occurring inducible promoters can be used to control transgene expression

A number of inducible expression systems have been developed for animals and plants based on promoters from endogenous cellular or viral genes. An early example is the *Drosophila* heat-shock promoter. Most cells respond to elevated temperature by synthesizing *heat-shock proteins*, which include molecular chaperons and other proteins with protective functions (Parsell & Lindquist 1983). The response is controlled at the level of transcription by a heat-labile transcription factor, which binds to heat-responsive promoters in the corresponding genes. The promoter of the *Drosophila hsp70* gene has been widely used to drive transgene expression, both in *Drosophila* itself (Lis *et al.* 1983) and in heterologous systems (e.g. Wurm *et al.* 1986). In transgenic flies, any gene linked to the *hsp70* promoter is more or less inactive at room temperature, but high-level expression in all cells can be induced by heating to 37°C for about 30 min.

The heat-shock promoter is unusual, at least in the context of inducible expression constructs, in that the stimulus that activates it is physical. Most inducible promoters used to control transgene expression respond to chemicals, which must be supplied to the transformed cells or transgenic organisms in order to activate the expression of a linked transgene. In mammals, several promoters are known to be activated by glucocorticoid hormones or synthetic analogs such as dexamethasone. Two of these have been extensively used for inducible transgene expression – the mouse metallothionein promoter (Hager & Palmiter 1981) and the long-terminal-repeat (LTR) promoter of mouse mammary tumor virus (MMTV) (Lee *et al.* 1981). The metallothionein promoter is also induced by interferons and heavy metals, such as cadmium and zinc, allowing the transgene to be activated in transgenic animals by including a source of heavy metals in their drinking water. An example of zinc-induced activation of a rat growth-hormone gene in transgenic mice is discussed on p. 259. A metal-inducible expression system has also been developed for plants, although the components of the system are derived from yeast (Mett *et al.* 1993). An ethanol-inducible system has also been described (Roslan *et al.* 2001).

Endogenous chemically inducible promoters have also been used to control transgene expression in plants. Two systems have been widely employed: the *PR-1a* promoter, which is induced by pathogen infection and by chemical elicitors such as benzothiadiazole (Gorlach *et al.* 1996), and the maize *In2–2* (Inducible gene 2–2) promoter, which is induced by benzenesulfonamide safeners (agrochemicals that are used to increase the tolerance of plants to herbicides; De Veylder *et al.* 1997). The advantage of these systems is that neither inducer is toxic to plants so they can be applied safely in the field as well as under laboratory conditions. Physically inducible systems for plants have also been described: one example is a peroxidase gene promoter isolated from sweet potato (*Ipomoea batatas*), which is inducible by hydrogen peroxide, wounding, or ultraviolet light (Kim *et al.* 2003). A wound-inducible promoter from the tomato hydroxy-3-methylglutaryl CoA reductase 2 (*HMGR2*) gene has been used as the basis of a commercial inducible promoter system called MeGA-PharM (Mechanical Gene Activation Postharvest Manufacturing), which allows transgenes

encoding pharmaceutical proteins in plants to be switched on after the leaves have been harvested and shredded (Ma *et al.* 2003, Fischer *et al.* 2004).

Unfortunately, there are several limitations associated with the use of endogenous inducible promoters which limit their usefulness. First, they tend to be somewhat leaky, i.e. there is a low to moderate level of background transcription in the absence of induction. Second, the level of stimulation achieved by induction (the *induction ratio*) is often quite low, typically less than 10-fold. Third, there are often unwanted side effects caused by the activation of other endogenous genes that respond to the same inductive stimulus. The inducers used with many of these endogenous promoters are also toxic or damaging if contact is prolonged. Finally, the kinetics of induction are generally not ideal. For example, in transgenic animals and plants, there may be differential uptake of the inducer into different cell types and it may be eliminated slowly. For these reasons, there has been great interest in the development of alternative inducible expression systems.

Recombinant inducible systems are built from components that are not found in the host animal or plant

Many of the disadvantages of endogenous inducible promoters can be addressed using recombinant systems, since these can be designed to work more efficiently. All recombinant inducible expression systems are based on a two-component principle in which the transgenic organism is transformed with a transgene encoding a transcription factor whose activity is controlled by the external stimulus or chemical, as well as a transgene regulated by that transcription factor. Because the components are heterologous, there should be no coincidental activation of other endogenous genes in the transgenic organism. For the same reason, the inducer itself should be non-toxic, because it does not interfere with any endogenous processes. Current systems aspire to the ideal in which the inducing agent is taken up rapidly and evenly, but has a short half-life, allowing rapid switching between induced and non-induced states. A range of inducers might be available with different properties. Steps towards this ideal system have been achieved using promoters and transcription factors that are either heterologous in the expression host or completely artificial.

The *lac* and *tet* repressor systems are based on bacterial operons

The first heterologous systems were based on bacterial control circuits (Fig. 15.1). Hu & Davidson (1987) developed an inducible expression system for mammalian cells, incorporating the essential elements of the *lac* repressor control circuit. In *Escherichia coli*, transcription of the *lac* operon is switched off in the absence of lactose by a repressor protein encoded by the gene *lacI*. This protein binds to *operator* sites, the most important of which lies just downstream from the promoter, and thus inhibits transcriptional initiation. In the presence of lactose or a suitable analog, such as isopropyl-β-D-thiogalactoside (IPTG), the Lac repressor undergoes a conformational change that causes it to be released from the operator sites, allowing transcription to commence.

In order to use the *lac* circuit in eukaryotic cells, Hu & Davison (1987) modified the *lacI* gene by adding a eukaryotic initiation codon, and then made a hybrid construct in which this gene was driven by the Rous sarcoma virus (RSV) LTR promoter. The construct was introduced into mouse fibroblasts and a cell line was selected that constitutively expressed the Lac repressor protein. This cell line was transiently transfected with a number of plasmids containing the *cat* reporter gene, driven by a modified simian virus 40 (SV40) promoter. Each of these plasmids also contained a *lac* operator site somewhere within the expression construct. The investigators found that, when the operator sites were placed in the promoter region, transcription from the reporter construct was blocked. However, transcription could be derepressed by supplying the cells with IPTG, resulting in strong chloramphenicol transacetylase (CAT) activity.

A similar system, based on the *E. coli* tetracycline operon, was developed for tobacco plants (Gatz & Quail 1988). The *tet* operon is carried on a bacterial transposon that confers resistance to the antibiotic tetracycline. Similarly to the *lac* system, the *tet* operon is switched off by a repressor protein, encoded by the *tetR* gene, which binds to operator sites around the promoter and blocks transcriptional initiation. Tetracycline itself binds to this repressor protein and causes the conformational change that releases the *tet* operon from repression. Since tetracycline inhibits bacteria at very low concentrations, the *tet* repressor has a very high binding constant for the antibiotic, allowing derepression in the presence of just a few molecules. The *tet* repressor also has very high affinity for its operator sites. Therefore, cell cultures and transgenic tobacco plants expressing TetR were able to inhibit reporter gene expression from a cauliflower mosaic virus (CaMV) 35S promoter surrounded by three *tet* operator sites. This repressed state could be lifted rapidly by the application of tetracycline (Gatz *et al.* 1991).

The *lac* and *tet* repressor systems both show minimal background transcription in the presence of the appropriate repressor protein, and a high induction ratio is therefore possible. In the *lac* system the maximum induction ratio is approximately 50, whereas in the *tet* system up to 500-fold induction has been achieved (Figge *et al.* 1988, Gatz *et al.* 1992).

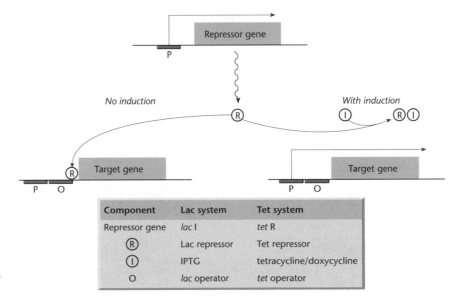

Fig. 15.1 Summary of repression-based inducible expression-control circuits based on the *lac* and *tet* operons of *E. coli*. P = promoter.

Component	Lac system	Tet system
Repressor gene	*lac* I	*tet* R
(R)	Lac repressor	Tet repressor
(I)	IPTG	tetracycline/doxycycline
O	*lac* operator	*tet* operator

Remarkably, the bacterial repressor proteins appear quite capable of interacting with the eukaryotic transcriptional apparatus and functioning as they do in bacteria, despite the many mechanistic differences in transcriptional control between prokaryotes and eukaryotes.

The *tet* activator and reverse activator systems were developed to circumvent some of the limitations of the original *tet* system

A disadvantage of repressor-based systems is that, in order to function effectively, high-level constitutive expression of repressor molecules is required to suppress background transgene activity. However, both the LacI and TetR proteins are cytotoxic at high levels.

To address these problems, TetR and LacI have been converted into activators by generating fusion proteins, in which the repressor is joined to the herpes simplex virus (HSV) VP16 transactivator (Labow *et al.* 1990, Gossen & Bujard 1992). In these systems, only the DNA-binding specificity of the repressor proteins is exploited. The binding of the modified bacterial proteins to operator sites within the transgene leads to transcriptional activation, because the VP16 protein acts positively on the transcriptional apparatus. The operator sites have effectively become enhancers and the inducers (IPTG and tetracycline) have effectively become repressors (Fig. 15.2). The *tet* transactivator (tTA) system has been more widely used than the equivalent *lac* system, because very high levels of IPTG are required to inhibit LacI binding in mammalian cells and this is toxic (Figge *et al.* 1988). Many different proteins have been produced

in mammalian cell lines using the tTA system, particularly cytotoxic proteins, whose constitutive expression would rapidly lead to cell death (Wu & Chiang 1996). In cells, a low background activity has been reported and an induction ratio of approximately 10^5 can be achieved. However, toxic effects of prolonged tetracycline exposure have been reported in transgenic animals, as well as unequal uptake of tetracycline into different organs, resulting in fluctuating basal transcription levels and cell-specific effects (reviewed by Saez *et al.* 1997).

A further modification to the *tet* system has led to marked improvements. A mutated form of the tTA protein has been generated whose DNA-binding activity is *dependent on* rather than abolished by tetracycline (Gossen *et al.* 1995). This protein is called reverse tTA (rtTA) and becomes an activator in the *presence* of tetracycline. In this system, the antibiotic is once again an inducer, but there is no requirement for prolonged exposure (Fig. 15.3). An early example of the use of this system is described by Bohl *et al.* (1997). Myoblasts were transformed with the rtTA system using a retroviral vector in which erythropoietin cDNA was placed under the control of a tetracycline-inducible promoter. These cells were implanted into mice, and erythropoietin secretion could be controlled by feeding the mice doxycycline, a derivative of tetracycline with a shorter half-life. An important finding was that long-term control of the secretion of this hormone was possible, with significant implications for the use of inducible expression systems for gene therapy. Unfortunately, the mutations that reverse the effect of doxycycline also reduce its binding activity, so up to 10 times more antibiotic is required to activate rtTA than inhibit tTA. More recently, new versions of rtTA

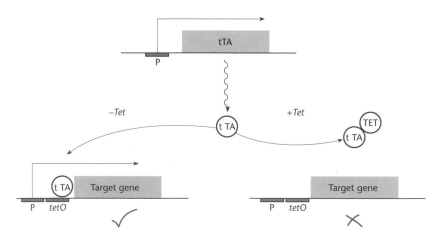

Fig. 15.2 The *tet* transactivator (tTA) system.

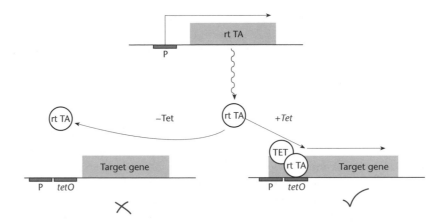

Fig. 15.3 The reverse *tet* transactivator (rtTA) system.

have been developed with mutations at a different site that does not affect doxycycline binding (Urlinger *et al.* 2000, Lamartina *et al.* 2002). Another limitation of the original rtTA system is that the regulator displays a significant residual binding to the *tet* operator in the absence of doxycycline. This has been addressed by the development of Tet-repressible transcriptional repressors or suppressors tTR and tTS, which out-compete rtTA for *tet* operator sites in the absence of doxycycline, but are unable to bind when the antibiotic is present. They do not, however, physically interact with rtTA (Zhu *et al.* 2001). A reverse-transactivator *lac* system has also been developed, with an induction ratio of approximately 10^4 (Baim *et al.* 1991). Another system, based on the pristinamycin (Pip) operon, is available in standard and reverse formats induced by antibiotics of the streptogramin family (Fussenegger *et al.* 2000). The availability of different systems should allow multiple transgenes to be controlled independently by different inducers (Corbel & Ross 2002).

Steroid hormones also make suitable heterologous inducers

Steroid hormones are lipophilic molecules that penetrate cells rapidly and are eliminated within a few hours. The use of heterologous steroids for inducible transgene expression is advantageous because, in addition to their favorable kinetics, such molecules should not activate endogenous signaling pathways in the expression host and should therefore have limited toxicity.

Ecdysone is a steroid hormone found only in insects and is responsible for the extensive morphological changes that occur during molting. As with other steroid-like signaling molecules, the hormone acts through a heterodimeric transcription factor of the nuclear receptor family. In *Drosophila*, this receptor comprises the products of the genes *ecdysone receptor* (*ecr*) and *ultraspiracle* (*usp*). The hormone and its signaling pathway are not found in mammalian cells. Therefore, transgenes including an ecdysone response element in the promoter can be induced by exogenously supplied ecdysone or its analog, muristerone A, in cells or transgenic animals expressing the components of the *Drosophila* receptor. The unmodified *Drosophila* system has a poor induction ratio, but this can be improved using chimeric receptors and mammalian components (Yao *et al.* 1992, 1993). In a significant improvement, No and colleagues were able to achieve an induction ratio of 10^4 by generating a hybrid system in which the ecdysone receptor gene was expressed as a fusion with the HSV VP16 transactivator, and the ultraspiracle protein was replaced with a mammalian homolog, the retinoid X receptor. Background activity was reduced to near zero by altering the DNA sequence recognized by the hybrid receptor (No *et al.* 1996).

The ecdysone system has also been employed successfully in plants. This is due to the identification and development of non-steroidal agonists of the ecdysone receptor which are safe for field use. For example, Martinez *et al.* (1999) developed a hybrid system consisting of the tobacco budworm ecdysone receptor ligand-binding domain fused to the glucocorticoid receptor DNA-binding domain and the VP16 transactivation domain. The receptor responds to tebufenozide (an insecticide better known by its trade name CONFIRM). Similarly, Padidam *et al.* (2003) have developed a system that is based on the spruce budworm ecdysone receptor ligand-binding domain, and responds to another common insecticide, methoxyfenozide (INTREPID). Another system based on the European corn borer ecdysone receptor also responds to this insecticide (Unger *et al.* 2002).

The glucocorticoid receptor has also been developed as a heterologous system in plants (Schena *et al.* 1991, Aoyama & Chua 1997, Moore *et al.* 1997). The system described by Aoyama & Chua comprises the glucocorticoid-receptor steroid-binding domain fused to the DNA-binding domain of the yeast transcription factor GAL4 and the VP16 transactivation domain. In this system, a CaMV 35S promoter modified to contain six GAL4-recognition sites is used to drive transgene expression. Genes placed under the control of this promoter can be induced 100-fold in the presence of dexamethasone. In a recent application, this system was used to express a viral replicase in a plant also expressing human γ-interferon. The intrinsic amplification stimulated gene expression more than 300-fold over the unamplified system (Mori *et al.* 2001). Another interesting development is a dual system which is activated by dexamethasone and repressed by tetracycline (Bohner *et al.* 1999).

Chemically induced dimerization exploits the ability of a divalent ligand to bind two proteins simultaneously

A further strategy for inducible transgene regulation has been developed, exploiting essentially the same principles as the yeast two-hybrid system (p. 459). This technique, termed *chemically induced dimerization* (CID), involves the use of a synthetic divalent ligand to simultaneously bind and hence bring together separate DNA-binding and transactivation domains to generate a functional transcription factor. The initial system utilized the immunosuppressant drug FK-506. This binds with high specificity to an immunophilin protein called FKBP12, forming a complex that suppresses the immune system by inhibiting the maturation of T cells (reviewed by Schreiber 1991). For transgene induction, an artificial homodimer of FK-506 was created, which could simultaneously bind to two immunophilin domains. Therefore, by expressing fusion proteins in which the GAL4 DNA-binding domain and the VP16 transactivator were each joined to an immunophilin domain, the synthetic homodimer could recruit a functional hybrid transcription factor capable of activating any transgene carrying GAL4 recognition elements (Belshaw *et al.* 1996).

Since this homodimer can also recruit non-productive combinations (e.g. two GAL4 fusions), an improved system has been developed using an artificial heterodimer specific for two different immunophilins (Belshaw *et al.* 1996). In this case, FK-506 was linked to cyclosporin A, a drug that binds specifically to a distinct target, cyclophilin. This heterodimer was shown to assemble effectively a transcription factor comprising an FKBP12-GAL4 fusion and a cyclophilin-VP16 fusion, resulting in strong and specific activation of a reporter gene in mammalian cells (Fig. 15.4). A more versatile system has been developed that exploits the ability of another immunosupressant drug, rapamycin, to mediate the heterodimerization of FKBP12 and a kinase known as FRAP (Rivera *et al.* 1996). In this system, FKBP12 and FRAP are each expressed as fusions with the components of a functional transcription factor. In the absence of rapamycin there is no interaction between these fusions, but when the drug is supplied they assemble into a hybrid transcription factor that can activate transgene expression. Transgenic mice containing a growth-hormone gene controlled by a CID-regulated promoter showed no expression in the absence of the inducer, but high levels of human growth hormone 24 h after induction with rapamycin (Magari *et al.* 1997). The advantage of this system is that rapamycin is rapidly taken up by cells *in vivo*, and it decays rapidly. The major disadvantage of immunosuppressant drugs as chemical inducers of dimerization is their pharmacological side effects. Various analogs of rapamycin (rapalogs) have therefore been developed that do not have significant pharmacological effects. Indeed, the best systems use rapalogs that no longer bind to endogenous FKBP and FRAP, but only interact with modified derivatives developed especially for the CID system. Other advantages of rapalogs include their prolonged bioavailability following oral administration and their ability to cross the blood–brain barrier. In the most popular version of the system (Fig. 15.5) three copies of FKBP are joined to a synthetic DNA-binding domain comprising a pair of zinc fingers and a homeodomain, which binds a novel response element, whereas the FKBP-rapamycin binding domain of FRAP is joined to the p65 activation domain subunit of NF-κB. Generally, both hybrid components are expressed from a single transcription unit using an internal ribosome entry site. The target promoter contains 12 copies of the target site followed by a TATA box (Auicchio *et al.* 2002 a,b, Chong *et al.* 2002).

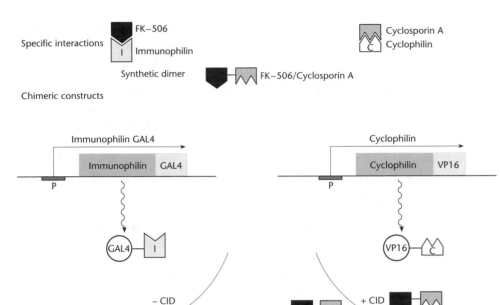

Fig. 15.4 Overview of chemically induced dimerization between the synthetic FK-506/cyclosporin A conjugate and fusion proteins containing immunophilin and cyclosporin domains. Dimerization assembles a functional transcription factor that can activate a promoter with GAL4 response elements.

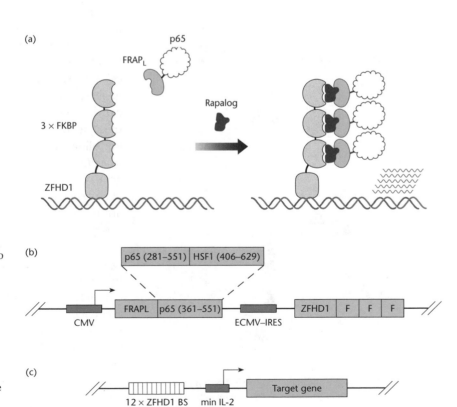

Fig. 15.5 (a) A frequently used configuration of the rapamycin system, in which three tandem copies of the human FKBP are fused to a synthetic DNA-binding domain and FRAP is fused to the DNA activation domain of p65. Rapamycin facilitates interaction between FKBP and FRAP, leading to the assembly of functional transcription factor. (b) Detailed structure of the construct encoding the three copies of FKBP and the synthetic DNA-binding domain ZFHD1 (zinc finger homeodomain 1). ECMV-IRES is the internal ribosome entry site from encephalomyocarditis virus. (c) Structure of a typical target gene showing 12 tandem copies of the ZFHD1-binding site.

Not all inducible expression systems are transcriptional switches

The inducible expression systems discussed above are all regulated at the level of transcription, such that there is often a significant delay between induction and response and between removal of induction and return to the basal state. Where a rapid response to induction is critical, inducible systems that operate at the post-translational level can be utilized. For example, the mammalian estrogen receptor exists in an inert state in the absence of estrogen, because the hormone-binding domain interacts with heat-shock protein 90 (Hsp90) to form an inactive complex. When estrogen is present, it binds to its receptor, causing a conformational change that releases the receptor from Hsp90. The receptor is then free to dimerize and interact with DNA (Fig. 15.6). In principle, any protein expressed as a fusion with the estrogen-binding domain will similarly interact with Hsp90 and form an inactive complex (Picard 1994). A recombinant protein can thus be expressed at high levels in an inactive state, but can be activated by feeding cells or transgenic animals with estrogen or an analog, such as Tamoxifen, which does not induce endogenous estrogen-responsive genes (Littlewood *et al.* 1995). A similar system has been devised using

a mutant-form progesterone receptor, which can no longer bind progesterone but can be induced with the antiprogestin RU486 (Garcia *et al.* 1992, Vegeto *et al.* 1992). An induction ratio of up to 3500 has been demonstrated in transgenic mice and, importantly, the inductive response occurs when the drug is supplied at a dose more than 100-fold below that required for it to function as an antiprogestin (Wang *et al.* 1997a,b).

Site-specific recombination allows precise manipulation of the genome in organisms where gene targeting is inefficient

Until recently, there was no generally applicable method for the precise *in vivo* manipulation of DNA sequences in animal and plant genomes. In mice, gene targeting by homologous recombination allows specific changes to be introduced into pre-selected genes, but it had proved impossible to extend the technique to other animals or to plants. Furthermore, even gene targeting is limited by the fact that the targeted gene is modified in the germline; thus all cells in the mouse are similarly affected from the beginning of development and throughout its entire lifetime.

Over the last 10 years, general methods have become available that allow *in vivo* transgene manipulation in any animal or plant species. Importantly, by using such methods in concert with inducible or cell-type-specific expression systems, it is possible to generate transgenic organisms in which transgenes can be conditionally modified. In mice, the use of these methods in combination with gene targeting allows the production of conditional mutants (*conditional knockouts*), in which an endogenous gene is inactivated specifically in certain cell types or at a particular stage of development. These methods are based on a specialized genetic process, termed *site-specific recombination*.

Site-specific recombination differs from homologous recombination in several important respects. In terms of gene manipulation, the most important differences between these processes concern the availability of the recombinase and the size and specificity of its target sequence. Homologous recombination is a ubiquitous process that relies on endogenous recombinase enzymes present in every cell, whereas site-specific recombination systems are very specialized and different systems are found

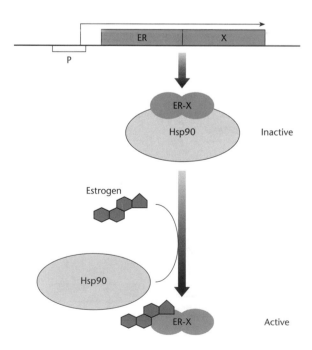

Fig. 15.6 The estrogen-inducible expression system, which works at the post-translational level – see text for details.

in different organisms. Homologous recombination occurs between DNA sequences with long regions of homology but no particular sequence specificity, whereas site-specific recombination occurs at short, specific recognition sites. This means that target sites for site-specific recombination can be introduced easily and unobtrusively into transgenes, but recombination will only occur in a heterologous cell if a source of recombinase is also supplied. The power of site-specific recombination as a tool for genome manipulation thus relies on the ability of the experimenter to supply the recombinase enzyme on a conditional basis.

A number of different site-specific recombination systems have been identified and several have been studied in detail (reviewed by Craig 1988, Sadowski 1993). Some recombinases, such as bacteriophage λ integrase, require various accessory proteins for efficient recombination. However, the simplest systems require only the recombinase and its target sequence. Of these, the most extensively used are Cre recombinase from bacteriophage P1 (Lewanodski & Martin 1997) and FLP recombinase (flippase) from the 2 μm plasmid of the yeast *Saccharomyces cerevisiae* (Buchholz *et al.* 1998). These have been shown to function in many heterologous eukaryotic systems including mammalian cells and transgenic animals and plants (reviewed by Sauer 1994, Ow 1996, Metzger & Feil 1999). Both recombinases recognize 34 bp sites (termed *loxP* and *FRP*, respectively) comprising a pair of 13 bp inverted repeats surrounding an 8 bp central element. *FRP* possesses an additional copy of the 13 bp repeat sequence, although this has been shown to be non-essential for recombination. Cre recombinase has been used most extensively in mammals, because it works optimally at 37°C. The optimal temperature for FLP recombinase is 30°C (Buchholz *et al.* 1996). However, the greatest advantages are seen if multiple systems can be used simultaneously, as has been proposed as a strategy to develop marker-free transgenic plants (Srivastava & Ow 2004).

Site-specific recombination can be used to delete unwanted transgenes

The reaction catalyzed by Cre recombinase is shown in Fig. 15.7. If two *loxP* sites are arranged as direct repeats, the recombinase will delete any intervening DNA, leaving a single *loxP* site remaining in the genome. If the *loxP* sites are arranged as inverted

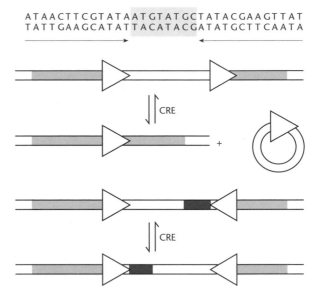

Fig. 15.7 Structure of the *loxP* site and reactions catalyzed by Cre recombinase when paired *loxP* sites, shown as arrows, are arranged in different orientations.

repeats, the intervening DNA segment is inverted. Both reactions are reversible. However, when the *loxP* sites are arranged in the same orientation, excision is favored over reintegration, because the excised DNA fragment is rapidly degraded.

The ability of flanking *loxP* sites to delineate any sequence of interest for site-specific deletion has numerous applications. The most obvious of these is the deletion of unwanted sequences, such as marker genes. This approach has been used, for example, as a simplified strategy to generate targeted mutant mice containing point mutations. Recall from Chapter 13 that traditional strategies for generating subtle mutants in mice involve two rounds of homologous recombination in embryonic stem cells (ES cells) (p. 258). Such strategies are very inefficient, because homologous recombination is a rare event. However, in the Cre recombinase-based approach, a second round of homologous recombination is unnecessary (Kilby *et al.* 1993). As shown in Fig. 15.8, gene targeting is used to replace the wild-type allele of a given endogenous gene with an allele containing a point mutation, and simultaneously to introduce markers, such as *neo* and *Tk*, for positive and negative selection. The positive–negative selection markers within the homology region are flanked by *loxP* sites. A second negative marker (e.g. the gene for diphtheria toxin) is included outside the homology region to select against random integration

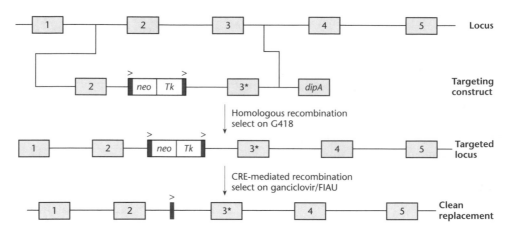

Fig. 15.8 Gene targeting followed by marker excision, catalyzed by Cre recombinase. Initially, positive and negative markers (*neo* and *Tk*), flanked by *loxP* sites, are introduced by homologous recombination (a second negative marker, in this case encoding diphtheria toxin (DIPA), is included outside the homology region to eliminate random integration events). Following selection for *neo* on G418, Cre recombinase is used to excise both markers, leaving a single *loxP* site remaining in the genome. The excision event can be identified by selection for the absence of *Tk*, using ganciclovir or FIAU. Asterisk indicates mutation.

events. Cells that have lost the diphtheria-toxin gene and survive selection for *neo* are likely to represent authentic targeting events. Such cells are then transfected with a plasmid expressing Cre recombinase, which catalyzes the excision of the remaining markers, leaving a clean point mutation and no evidence of tinkering except for a single *loxP* site remaining in one intron. Negative selection using ganciclovir or 1-(2-deoxy-2-fluoro-β-D-arabinofuranosyl)-5 iodouracil (FIAU) identifies cells that have lost the markers by selection against *Tk*.

Similar strategies can be used to remove marker genes from plants, as first demonstrated by Dale & Ow (1991). These investigators used *Agrobacterium* to transform tobacco-leaf explants with a CaMV 35S-luciferase reporter construct. The transfer DNA (T-DNA) also contained a selectable marker for hygromycin resistance, flanked by *loxP* sites. Transgenic plants were regenerated under hygromycin selection and leaf explants from these plants were then transformed with a second construct, in which Cre recombinase was driven by the CaMV 35S promoter. This construct also contained the *nptII* gene and the second-round transgenic plants were selected on kanamycin. Ten of the 11 plants tested were found to be hygromycin-sensitive, even though they continued to express luciferase, showing that the original marker had been excised. Since the *cre*/*nptII* construct was introduced separately, it was not linked to the original T-DNA and segregated in future generations, leaving "clean" transgenic plants containing the luciferase transgene alone.

Site-specific recombination can be used to activate transgene expression or switch between alternative transgenes

While commonly used as a method to inactivate transgenes by deletion, site-specific recombination can also activate transgenes or switch between the expression of two transgenes (Fig. 15.9). In one method, termed *recombinase-activated gene expression* (RAGE), a blocking sequence, such as a polyadenylation site, is placed between the transgene and its promoter, such that the transgene cannot be expressed. If this blocking sequence is flanked by *loxP* sites, Cre recombinase can be used to excise the sequence and activate the transgene.

This strategy was first used in transgenic mice to study the effect of SV40 T-antigen expression in development (Pichel *et al.* 1993). In this case, Cre recombinase was expressed under the control of a developmentally regulated promoter. Essentially the same strategy was used in transgenic tobacco plants to activate a reporter gene in seeds (Odell *et al.* 1994). In this case, Cre recombinase was expressed under the control of a seed-specific promoter. An important feature of both these experiments was the use of two separate transgenic lines, one expressing Cre recombinase in a regulated manner and one containing the target gene. Crosses between these lines brought both transgenes together in the hybrid progeny, resulting in the conditional activation of the transgene based on the expression profile of Cre. This is an extremely versatile and widely used

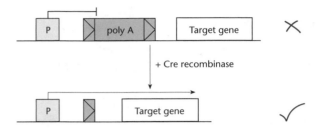

Fig. 15.9 Overview of the recombinase-activated gene expression (RAGE) strategy. A polyadenylation signal is inserted between the promoter and target gene, blocking its expression. However, if this signal is flanked by *loxP* sites, Cre recombinase can be used to excise the block, bringing the promoter and gene into juxtaposition and thus activating gene expression.

strategy, because it allows "mix and match" between different Cre transgenic and "responder" lines. We return to this subject below.

Site-specific recombination can facilitate precise transgene integration

Site-specific integration of transgenes can occur if the genome contains a recombinase recognition site. This may be introduced by random integration or (in mice) by gene targeting. Using an unmodified Cre-*loxP* system, transgene integration occurs at a low efficiency, because, as discussed above, the equilibrium of the reaction is shifted in favor of excision. Initial attempts to overcome this problem by providing transient Cre activity had limited success (see Sauer & Henderson 1990, Baubonis & Saur 1993). However, high-efficiency Cre-mediated integration has been achieved in plants (Albert *et al.* 1995) and mammalian cells (Feng *et al.* 1999) using mutated or inverted *loxP* sites. Site-specific transgene integration into mammalian cells has also been achieved using FLP recombinase (O'Gorman *et al.* 1991).

With recent pressure to develop marker-free transgenic animals and plants in the biotechnology industry, this area of research has benefited from a large amount of funding. Particular progress has been made with commercially important crops such as rice, where targeted integration is now possible at efficiencies that yield hundreds of clones per experiment (Srivastava & Ow 2002, Srivastava *et al.* 2004). Transgene integration by site-specific recombination has many advantages over the random integration that is normally achieved by illegitimate recombination. For example, if a region of the genome can be identified that is not subject to nega-

tive position effects (Box 13.2), transgenic lines with a *loxP* site at this position can be used for the stable and high-level expression of any transgene (e.g. Fukushige & Sauer 1992). Also, due to the precise nature of site-specific recombination, transgenic loci generated by this method are likely to be less complex than loci generated by random integration.

Site-specific recombination can facilitate chromosome engineering

Site-specific recombination between widely separated target sites or target sites on different chromosomes can be used to generate large deletions, translocations, and other types of chromosome mutation. Chromosome engineering by site-specific recombination was first reported by Golic (1991), using FLP recombinase in *Drosophila*, but similar experiments have now been carried out in plants and mice. Precise intra-chromosomal deletions can be generated in mice by two rounds of gene targeting, introducing *loxP* sites at distant sites, followed by Cre-mediated recombination (Ramirez-Solis *et al.* 1995, Li *et al.* 1996). In plants, where gene targeting is very inefficient, an ingenious scheme has been developed where *loxP* sites are arranged in tandem on a transformation construct, one inside a *Ds* transposon and one outside. The transposon is placed between a marker gene and its promoter. When this construct is introduced into tobacco plants containing the autonomous transposon *Ac* to provide a source of transposase, the *Ds* element can excise from the transgene, as revealed by marker-gene expression. In most heterologous plants, *Ac-Ds* elements reintegrate at a position that is linked to the original site. Although the site of reintegration cannot be controlled, this nevertheless defines a large chromosomal segment that can be excised by Cre recombinase (Medberry *et al.* 1995, Osbourne *et al.* 1995). Translocations are more difficult to engineer, because interchromosomal site-specific recombination is inefficient, and inventive selection strategies are required to identify the desired products (e.g. see Qin *et al.* 1994, Smith *et al.* 1995, Van Deursen *et al.* 1995).

Inducible site-specific recombination allows the production of conditional mutants and externally regulated transgene excision

In mice, gene targeting and site-specific recombination can be used in a powerful combined approach

Box 15.1 Visible marker genes

Reporter genes are widely used for *in vitro* assays of promoter activity (Box 12.1). However, reporters that can be used as cytological or histological markers are more versatile, because they allow gene expression profiles to be determined in intact cells and whole organisms.

β-galactosidase and β-glucuronidase

The *E. coli lacZ* gene encodes β-galactosidase, an enzyme that hydrolyzes β-D-galactopyranosides, such as lactose, as well as various synthetic analogs. Like CAT, β-galactosidase activity can be assayed *in vitro*, although with the advantage that the assays are non-radioactive. For example, cell lysates can be assayed spectrophotometrically using the chromogenic substrate ONPG*, which yields a soluble yellow compound (Norton & Coffin 1985). Alternatively, a more sensitive fluorometric assay may be preferred, using the substrate MUG*. For histological staining, the substrate Xgal* yields an insoluble blue precipitate that marks cells brightly. The *lacZ* gene was first expressed in mammalian cells by Hall *et al.* (1983) to confirm transfection. For these experiments, the gene was linked to the SV40 early promoter and the mouse mammary tumor virus (MMTV) LTR promoter. Fusions between the *hsp70* promoter and *lacZ* were also constructed and shown to drive heat-shock-inducible β-galactosidase expression in *Drosophila* (Lis *et al.* 1983). One disadvantage of *lacZ* as a marker is that certain mammalian cells, and many plants, show a high level of endogenous β-galactosidase activity, which can obscure the analysis of chimeric genes (Helmer *et al.* 1984). The *E. coli gusA* gene, which encodes the enzyme β-glucuronidase (GUS), is an alternative (Jefferson *et al.* 1986). This marker is preferred in plants, due to the minimal background activity of the endogenous enzyme (Jefferson *et al.* 1987a), but has also been used successfully in animals (e.g. Jefferson *et al.* 1987b). Similar *in vitro* and histological assay formats to those described for β-galactosidase

are also available for GUS, e.g. a histochemical substrate, X-gluc*, which yields an insoluble blue precipitate.

Luciferase

CAT, GUS, and β-galactosidase are stable proteins, which persist in the cells that express them. One problem with stable reporter proteins is that, while they provide useful markers for gene activation, they are less useful for assaying transcriptional repression or rapid changes in gene activity. Luciferase was introduced as a novel reporter gene in 1986, for use in both plants (Ow *et al.* 1986) and animals (De Wet *et al.* 1987). The original marker gene, *luc*, was isolated from the North American firefly *Photinus pyralis* and encoded a single polypeptide of 550 amino acids. The enzyme catalyzes the oxidation of luciferin, in a reaction requiring oxygen, ATP, and the presence of magnesium ions. When excess substrate is supplied, a flash of light is emitted that is proportional to the amount of enzyme present. This can be detected using a luminometer, a scintillation counter as a luminometer or even photographic film (Wood & DeLuca 1987). Important advantages of the luciferase system include its very high sensitivity (more than 100-fold more sensitive than *lacZ*) and the rapid decay of light emission. Luciferase has therefore been used to analyze the activity of genes with oscillating expression profiles, such as the *Drosophila period* gene (Brandes *et al.* 1996). The amenability of the luciferase system has been expanded by the isolation of alternative luciferases from other organisms, which bioluminesce in different colors (e.g. see Thompson *et al.* 1990). A bacterial luciferase gene, *luxA*, has also been used as a marker in transgenic plants (Koncz *et al.* 1987).

Green fluorescent protein

The most recent addition to the growing family of reporters is green fluorescent protein (GFP), from the jellyfish *Aequoria victoria*. Over

continued

Box 15.1 *continued*

the last 5 years, this remarkable molecule has emerged as one of the most versatile tools in molecular and cellular biology and is being used to investigate an increasing variety of biological processes in bacteria, yeast, animals, and plants (reviewed by Tsien 1998, Haseloff *et al.* 1999, Ikawa *et al.* 1999, Naylor 1999). GFP is a bioluminescent marker that causes cells to emit bright green fluorescence when exposed to blue or ultraviolet light. However, unlike luciferase, GFP has no substrate requirements and can therefore be used as a vital marker to assay cellular processes in real time. Other advantages of the molecule include the fact that it is non-toxic, it does not interfere with normal cellular activity and it is stable even under harsh conditions (Ward & Bokman 1982).

GFP was first used as a heterologous marker in *Caenorhabditis elegans* (Chalfie *et al.* 1994). However, early experiments with GFP expression in a variety of other organisms, including *Drosophila* (Wang & Hazelrigg 1994), mammalian cell lines (Marshall *et al.* 1995), and plants (Haseloff & Amos 1995, Hu & Chen 1995, Sheen *et al.* 1995), identified a number of difficulties in the heterologous expression of the *gfp* gene. Modifications have been necessary for robust GFP expression in some plants (Chiu *et al.* 1996). In *Arabidopsis*, for example, the original *gfp* gene is expressed very poorly due to aberrant splicing. This problem was addressed by removing a cryptic splice site recognized in this plant (Haseloff *et al.* 1997). The original *gfp* gene has been extensively modified to alter various properties of the protein, such as the excitation and emission wavelengths, to increase the signal strength and to reduce

photobleaching (e.g. Heim & Tsein 1996, Zolotukhin *et al.* 1996, Cormack *et al.* 1997). As a result, many variations of the protein are now available such as cyan fluorescent protein (CFP) and yellow fluorescent protein (YFP), which can be used for dual labeling (e.g. Tsien & Miyawaki 1998; reviewed by Ellenberg *et al.* 1999). Fluorescent proteins of other colors are also available, many from coral reef organisms. For example, the proteins DsRed, AmCyn, and ZsYellow are all coral-derived fluorescent proteins available from Clontech. A mutant form of red fluorescent protein from *Anthoza* (Matz *et al.* 1999) changes from green to red fluorescence over time, allowing it to be used to characterize temporal gene expression patterns (Terskikh *et al.* 2000).

GFP is particularly useful for generating fusion proteins, providing a tag to localize recombinant proteins in the cell. This facilitates the investigation of intracellular protein trafficking, and even the transport of proteins between cells. An early example of this application was the use of GFP to monitor the movement of ribonucleprotein particles during oogenesis in *Drosophila* (Wang & Hazelrigg 1994). Kohler *et al.* (1997) have used GFP to study the exchange of molecules between plant organelles, while Wacker *et al.* (1997) have investigated the transport of a GFP-tagged protein along the secretory pathway. The use of GFP to study the real-time dynamics of a systemic viral infection in plants was described by Padgett *et al.* (1996).

* Abbreviations: ONPG: O-nitrophenyl-β-D-galactopyranoside; MUG: 4-methylumbelliferyl-β-D-galactoside; Xgal: 5-bromo-4-chloro-3-indolyl-β-D-galactopyranoside; X-gluc: 5-bromo-4-chloro-3-indolyl-β-D-glucuronic acid.

---

to generate conditional knockout mutants. Essentially, targeting vectors are designed so that part of a selected endogenous gene becomes flanked by *loxP* sites, or *floxed*. The usual strategy is to insert the *loxP* sites into introns flanking an essential exon, since this generally does not interfere with the normal expression of the gene. Cre recombinase is then supplied under the control of a cell-type-specific, developmentally regulated or inducible

promoter, causing the gene segment defined by the *loxP* sites to be deleted in cells or at the developmental stage specified by the experimenter. This addresses a major limitation of traditional gene-knockout techniques, i.e. that, if the mutation has an embryonic lethal phenotype, only its earliest effects can be investigated.

The general methodology for such experiments, as we discussed earlier, is to cross two lines of transgenic mice, one carrying the floxed target gene and the other carrying the conditional *cre* transgene. As the number of reports of such experiments has increased, more and more transgenic mouse lines are becoming available, with Cre expressed under the control of different conditional promoters. For example, a mouse line with Cre expressed specifically in the lens of the eye was generated by Lasko *et al.* (1992). Lines are also available with Cre expressed specifically in the mammary gland (Wagner *et al.* 1997) and developing sperm (O'Gorman *et al.* 1997). Lines in which Cre is expressed in germ cells or in early development are known as "deleter" lines and are used to remove marker genes and generate Cre-mediated constitutive gene knockouts.

In the first examples of the conditional knock-out approach, Gu *et al.* (1994) generated a Cre transgenic line expressing the recombinase under the control of the *lck* promoter, such that it was expressed only in T cells. This strain was crossed to targeted mice in which part of the DNA polymerase β gene was floxed, leading to T-cell-specific excision of an essential exon. Kuhn *et al.* (1995) mutated the same gene, but they used the metallothionein promoter to express Cre recombinase, allowing induction of site-specific recombination with interferon. Although successful, this experiment highlighted many of the inadequacies of inducible promoters. There was pronounced variation in the efficiency of excision in different tissues, probably reflecting differential uptake of the inducer. Furthermore, high-level background activity of Cre was observed in the spleen, resulting in excision of the gene segment in the absence of induction, probably caused by the presence of endogenous interferons. The tTA system has been used to bring Cre expression under the control of tetracycline administration, although a high level of background activity was also observed in this experiment, resulting in excision of the target gene prior to induction (St Ogne *et al.* 1996). Tighter control has been possible using post-translational induction. For example, Cre has been expressed as a fusion with the ligand-binding domain of the estrogen receptor (Fiel *et al.* 1996). When this transgene was crossed into an appropriate responder strain, the background excision was minimal and Cre was strongly induced by Tamoxifen. Several strains of Cre mice are now available, in which Tamoxifen- or RU486-induced site-specific recombination has been shown to be highly efficient (e.g. Brocard *et al.* 1997, Wang *et al.* 1997a, Schwenk *et al.* 1998).

As well as creating conditional mutants, inducible site-specific recombination can be used to control transgene excision externally. A number of reports have been published recently in which Cre or FLP recombinase has been expressed under inducible control in plants using one of the systems discussed at the beginning of the chapter. This allows the recombinase gene to be maintained in an inactive state until the stage of development at which the marker needs to be removed. Furthermore, this strategy is suitable for plants that are propagated vegetatively. The typical strategy, which is to cross plants containing a floxed transgene with plants containing a conditionally expressed *cre* gene, is not possible in plants that do not reproduce by sexual crossing. However, by placing the *cre* gene under inducible control, both the floxed transgene and the *cre* gene can be introduced at the same time (Lyznik *et al.* 1995, Sugita *et al.* 2000, Zuo *et al.* 2001, Hoff *et al.* 2001, Zhang *et al.* 2003).

Many strategies for gene inactivation do not require the direct modification of the target gene

Traditional gene transfer strategies add new genetic information to the genome, resulting in a gain-of-function phenotype conferred by the transgene. Gene targeting and site-specific recombination now provide us with the ability to disrupt or delete specific parts of the mouse genome, allowing loss-of-function phenotypes to be studied, but this approach cannot be used routinely in other animals or in plants. A range of alternative, more widely applicable transgenic strategies have therefore been developed for gene inhibition. These strategies involve the introduction of new genetic information into the genome, but, instead of conferring a gain of function, the transgene interferes with the expression of an endogenous gene, at either the RNA or the protein level. The actual target gene is not affected. The resulting loss-of-function effects are termed *functional knockouts* or *phenocopies*.

Antisense RNA blocks the activity of mRNA in a stoichiometric manner

Antisense RNA has the opposite sense to mRNA. The presence of complementary sense and antisense RNA molecules in the same cell can lead to the formation of a stable duplex, which may interfere with gene expression at the level of transcription, RNA processing, or possibly translation (Green *et al.* 1986). Antisense RNA is used as a natural mechanism to regulate gene expression in a number of prokaryote systems (Simons & Kleckner 1988) and, to a lesser extent, in eukaryotes (e.g. Kimelman & Kirchner 1989, Lee *et al.* 1993, Savage & Fallon 1995).

Transient inhibition of particular genes can be achieved by directly introducing antisense RNA or antisense oligonucleotides into cells. However, the transformation of cells with antisense transgenes (in which the transgene is inverted with respect to the promoter) allows the stable production of antisense RNA and thus the long-term inhibition of gene expression. This principle was established in transgenic animals and plants at about the same time. Katsuki *et al.* (1988) constructed an expression cassette in which the mouse myelin basic protein (MBP) cDNA was inverted with respect to the promoter, thus producing antisense RNA directed against the endogenous gene. In some of the transgenic mice, there was up to an 80% reduction in the levels of MBP, resulting in the absence of myelin from many axons and generating a phenocopy of the myelin-depleted "shiverer" mutation. Smith *et al.* (1988) generated transgenic tomato plants carrying an antisense construct targeting the endogenous polygalacturonase (*pg*) gene. The product of this gene is an enzyme that causes softening and leads to over-ripening. The levels of *pg* mRNA in transgenic plants were reduced to 6% of the normal levels and the fruit had a longer shelf-life and showed resistance to bruising.

Antisense constructs have been widely used in transgenic animals and plants for gene inhibition. However, the efficiency of the technique varies widely and the effects can, in some cases, be non-specific. In some experiments, it has been possible to shut down endogenous gene activity almost completely, as demonstrated by Erickson *et al.* (1993), who used an inverted cDNA to generate antisense RNA against the mouse *wnt-1* gene and reduced endogenous mRNA levels to 2% of normal. Conversely, Munir *et al.* (1990) designed a construct to generate antisense RNA corresponding to the first exon and intron of the mouse *Hprt* gene, and observed no reduction in endogenous mRNA levels at all, even though the presence of antisense RNA was confirmed. The level of inhibition apparently does not depend on the size of the antisense RNA or the part of the endogenous gene to which it is complementary. For example,

Moxham *et al.* (1993) achieved a 95% reduction in the level of $G_{\alpha i2}$ protein through the expression of antisense RNA corresponding to only 39 bp of the gene's 5′ untranslated region.

Conditional gene silencing can be achieved by placing antisense constructs under the control of an inducible promoter. The expression of antisense c-*myc* under the control of the MMTV LTR promoter resulted in the normal growth of transformed cells in the absence of induction, but almost complete growth inhibition in the presence of dexamethasone (Sklar *et al.* 1991). Experiments in which the tTA system was used to control antisense expression in plants have also been reported (e.g. Kumar *et al.* 1995).

Ribozymes are catalytic molecules that destroy targeted mRNAs

Ribozymes are catalytic RNA molecules that carry out site-specific cleavage and (in some cases) ligation reactions on RNA substrates. The incorporation of ribozyme catalytic centers into antisense RNA allows the ribozyme to be targeted to particular mRNA molecules, which are then cleaved and degraded (reviewed by Rossi 1995, James & Gibson 1998). An important potential advantage of ribozymes over antisense RNA is their catalytic activity: ribozymes are recycled after the cleavage reaction and can therefore inactivate many mRNA molecules. Conversely, antisense inhibition relies on stoichiometric binding between sense and antisense RNA molecules.

The use of ribozyme constructs for specific gene inhibition in higher eukaryotes was established in *Drosophila*. In the first such report, Heinrich *et al.* (1983) injected *Drosophila* eggs with a P-element vector containing a ribozyme construct targeted against the *white* gene. They recovered transgenic flies with reduced eye pigmentation, indicating that expression of the endogenous gene had been inhibited. A ribozyme construct has also been expressed under the control of a heat-shock promoter in *Drosophila* (Zhao & Pick 1983). In this case, the target was the developmental regulatory gene *fushi tarazu* (*ftz*). It was possible to generate a series of conditional mutants with *ftz* expression abolished at particular stages of development, simply by increasing the temperature to 37°C.

Ribozymes have also been used in mammalian cell lines, predominantly for the study of oncogenes and in attempts to confer resistance to viruses (reviewed by Welch *et al.* 1998). There has been intensive research into ribozyme-mediated inhibition of HIV, and remarkable success has been achieved using retroviral vectors, particularly vectors carrying multiple ribozymes (reviewed by Welch *et al.* 1998, Muotri *et al.* 1999). So far, there have been relatively few reports of ribozyme expression in transgenic mice. Larsson *et al.* (1994) produced mice expressing three different ribozymes targeted against β_2-macroglobulin mRNA, and succeeded in reducing endogenous RNA levels by 90%. Tissue-specific expression of ribozymes has also been reported. A ribozyme targeted against glucokinase mRNA was expressed in transgenic mice under the control of the insulin promoter, resulting in specific inhibition of the endogenous gene in the pancreas (Efrat *et al.* 1994). Recently, retroviral delivery of anti-neuregulin ribozyme constructs into chicken embryos has been reported (Zhao & Lemke 1998). Inhibition of neuregulin expression resulted in embryonic lethality, generating a very close phenocopy of the equivalent homozygous null mutation in mice.

Cosuppression is the inhibition of an endogenous gene by the presence of a homologous sense transgene

Cosuppression refers to the ability of a sense transgene to suppress the expression of a homologous endogenous gene. This surprising phenomenon was first demonstrated in plants, in a series of experiments designed to increase the levels of an endogenous protein by introducing extra copies of the corresponding gene. In an attempt to increase the amount of pigment synthesized by petunia flowers, Napoli *et al.* (1990) produced transgenic petunia plants carrying multiple copies of the chalcone synthase (*chs*) gene. This encodes an enzyme that converts coumaroyl-CoA and 3-malonyl-CoA into chalcone, a precursor of anthocyanin pigments. The presence of multiple transgene copies was expected to increase the level of enzyme and result in deeper pigmentation. However, in about 50% of the plants recovered from the experiment, exactly the opposite effect was observed, i.e. the flowers were either pure white or variegated with purple and white sectors. Similar findings were reported by Van der Krol *et al.* (1988) using a transgene encoding another pigment biosynthesis enzyme, dihydroflavonol-4-reductase. In both cases, it appeared that integration of multiple copies of the transgene led to the suppression of some or all of the transgenes and the cosuppression of homologous endogenous genes.

While troublesome in terms of generating plant lines with high transgene expression levels, cosuppression can also be exploited as a tool for specific gene inactivation. There have been many reports of this nature. For example, transgenic tomatoes have been produced containing a partial copy of the *pg* gene in the sense orientation (Smith *et al.* 1990). As with the antisense *pg* transgenic plants generated previously by the same group (see above), strong inhibition of the endogenous gene was achieved, resulting in fruit with a prolonged shelf-life. Cosuppression has also been demonstrated in animals (Pal-Bhadra *et al.* 1997, Bahramian & Zabl 1999, Dernberg *et al.* 2000) and is related to a similar phenomenon called *quelling*, which has been described in fungi (reviewed by Selker 1997, 1999).

The mechanism of cosuppression in plants is complex and can involve silencing at either the transcriptional or post-transcriptional levels (for details, see Box 15.2). One of the most remarkable aspects of post-transcriptional gene silencing (PTGS) is that it is a systemic phenomenon, suggesting that a diffusible signal is involved. This can be demonstrated by grafting a non-silenced transgenic scion onto a silenced transgenic host. The silencing effect is able to spread into the graft, and the systemic effect works even if the two transgenic tissues are separated by up to 30 cm of wild-type stem (Palauqui *et al.* 1997, Voinnet *et al.* 1998).

PTGS in plants can be induced not only by integrated transgenes but also by RNA viruses, as long as there is a region of homology between the virus genome and an integrated gene. For example, the virus may carry a sequence that is homologous to an endogenous gene or to a transgene integrated into the host genome. The effect also works if the plant is transformed with a cDNA construct corresponding to part of the virus genome, as demonstrated by Angell & Baulcombe (1997). The rationale behind this experiment was to transform plants with a cDNA construct corresponding to a chimeric potato virus X (PVX) genome containing the *gusA* reporter gene. Expression of the transgene was expected to generate very high levels of β-glucuronidase (GUS) activity, because, after transcription of the transgene, the resulting viral RNA would be amplified by the virus's own replication system. However, disappointingly, all of the transgenic plants produced extremely low levels of viral RNA and GUS activity. The plants also showed an absence of PVX symptoms and were resistant to PVX infection. The virus-induced silencing effect only worked using replication-competent vectors, suggesting that the double-stranded RNA (dsRNA) intermediate involved in viral replication was the trigger for silencing (see Box 15.2).

Such is the efficiency with which PVX RNA can silence homologous genes in the plant genome that PVX vectors have been used very successfully to generate functional knockouts in plants (reviewed by Baulcombe 1999). For example, Burton *et al.* (2000) described the infection of tobacco plants with PVX vectors containing a cDNA sequence putatively encoding a cellulose synthase. The inoculated plants showed a dwarf phenotype, and levels of cellulose in affected leaves were reduced by 25%. On the basis of this evidence, the investigators concluded that the cDNA did indeed encode such an enzyme and was capable of cosuppressing the endogenous cellulose synthase gene.

Box 15.2   Gene silencing

As discussed in the main text, there are several forms of gene silencing in eukaryotes, which act in a sequence-specific manner to inhibit the activity of particular genes or transgenes. These forms of silencing can occur at either the transcriptional or post-transcriptional levels. In the former case, no mRNA is produced from the affected gene, while in the latter case transcription is not only permitted, but is actually necessary for silencing to occur. Transcriptional silencing reflects the structure of chromatin, which can form an open configuration (euchromatin) that is permissive for gene expression or a closed configuration (heterochromatin) that represses gene expression. Post-transcriptional silencing (now more commonly referred to as RNA silencing) reflects the activity of particular protein complexes that target mRNAs with a specific sequence for destruction. Both forms of silencing have arisen as mechanisms of defense against invasive nucleic acids (viruses, transposons etc.), and both have been subjugated as mechanisms for gene regulation. There is extensive cross-talk between these silencing pathways.

Position-dependent silencing and context-dependent silencing

These forms of transcriptional silencing can affect single-copy transgenes and are not, therefore, homology-dependent. Position-dependent silencing occurs where a transgene integrates into a genomic region containing heterochromatin. The repressive chromatin structure and DNA methylation can spread into the transgenic locus from the flanking genomic DNA (Matzke & Matzke 1998); therefore silencing results from a negative position effect (position effects are discussed in more detail in Box 12.1). Single-copy transgenes may also be silenced, even if they integrate into a genomic region that lacks negative position effects. For example, integrated retrovirus vectors often undergo *de novo* silencing associated with increased levels of DNA methylation (Jahner *et al.* 1982) and, indeed, this methylation may spread outwards into flanking host DNA and inactivate linked genes (Jahner & Jaenisch 1985). Many unrelated transgenes in animals and plants have been subject to this type of silencing, suggesting that a specific sequence is not responsible. It is possible that eukaryotic genomes possess mechanisms for scanning and identifying foreign DNA sequences, perhaps based on their unusual sequence context, and then inactivating them by methylation (Kumpatla *et al.* 1998). Prokaryotic DNA may be recognized in this manner, since prokaryotic sequences act as a strong trigger for *de novo* methylation, e.g. in transgenic mice (Clark *et al.* 1997).

RNA silencing

All forms of RNA silencing (i.e. PTGS and VIGS in plants, cosuppression, RNAi, quelling in fungi) appear to depend in some way or other on the presence of double-stranded RNA. The biogenesis of dsRNA can occur in many ways,

continued

Box 15.2   *continued*

e.g. through the deliberate introduction or expression of complementary RNA molecules in the cell in RNAi, the production of aberrant dsRNA from complex transgenes in PTGS, or the production of viral replication intermediates in VIGS. Once the dsRNA has formed, it becomes the substrate of a nuclease called Dicer, which reduces it to short duplexes, 21–25 bp in length with overhangs. These duplexes are known as small interfering RNAs (siRNAs). They assemble with several proteins, including one of the Argonaute family, into an endonucleolytic complex known as the RNA induced silencing complex (RISC), which uses one strand of the siRNA to target complementary mRNAs and cleave them. The RISC is efficient, resulting in potent silencing (Tijsterman & Plasterk 2004). In many organisms (but not mammals) there is amplification of the siRNA by RNA-dependent RNA polymerase to increase the potency of the silencing effect. The siRNA molecules can also move between cells, explaining why RNAi and other RNA silencing phenomena are systemic.

RNA silencing is a form of genomic defense

It is likely that RNA silencing evolved as a defence against "invasive" nucleic acids (Yoder *et al.* 1997, Jones *et al.* 1998, Jensen *et al.* 1999, Li *et al.* 1999). This has been supported by the recent isolation of mutants in several organisms that show deficiencies in PTGS or RNAi. Animals impaired for RNAi show increased rates of transposon mobilization, whereas plants impaired for PTGS are more susceptible to viral infection. Interestingly, similar gene products have been identified in diverse organisms, providing further evidence for a link between PTGS and RNAi. A comprehensive discussion of this exciting new area of research is outside the scope of this book, but the interested reader can consult several excellent reviews on the subject (Plasterk & Ketting 2000, Hammond *et al.* 2001).

RNA silencing and gene regulation

Small RNA molecules are also involved in gene regulation, and the structure of these endogenous regulators is very similar to that of the siRNAs produced during RNA interference experiments. These so-called microRNAs (miRNAs) are produced not through the cleavage of longer dsRNA precursors, but are the direct transcripts of small miRNA genes that generate hairpin RNA precursors. The processed miRNAs assemble into a complex called the miRNP (micro-ribonucleoprotein) complex, which binds to the 3' untranslated regions of target mRNAs and regulates their translation. Once considered a peculiarity, the discovery of hundreds of miRNA genes in eukaryotic genomes (Lim 2003) now suggests miRNA may be a major form of gene regulation. Researchers are focusing on identifying the targets of miRNAs and the processes they help to regulate. While siRNAs and miRNAs were once distinguished by their mechanism (mRNA cleavage or translational repression), the extensive overlap between the two pathways now means they are classified by origin (one from dsRNA produced either inside or outside the cell, one from transcription of an endogenous gene). The chosen pathway appears to depend on the type of complex involved – siRNAs are processed by Dicer and assemble into RISCs which cause cleavage, while at least in animals miRNAs are also processed by another enzyme, Drosha, and assemble into miRNPs, which have a regulatory role. The choice between these complexes appears to depend on how perfect the match is between the complementary strands. Perfect duplexes generally behave as siRNAs whereas duplexes with mismatches or bubbles behave as miRNAs (Bartel 2004, Kim 2005).

Cross-talk between RNA silencing and transcriptional silencing

The RNA silencing apparatus also has significant interactions with chromatin-

continued

Box 15.2 *continued*

modifying proteins, including histone-modifying enzymes and DNA methyltransferases. Recent evidence has accumulated that the RNAi machinery can set epigenetic marks in plants and yeast (Wassenegger 2005) as well as in *Drosophila* and mammalian cells (Matzke & Birchler 2005). Although the mechanisms involved are complex and yet to be elucidated fully, it appears that repetitive regions of the genome such as tandem repeats of transposons (and perhaps integrated transgenes) can generate long dsRNA molecules which are processed into siRNA-like structures known as rasiRNAs (repeat-associated siRNAs). These are cleaved by Dicer and form complexes known as RITS (RNA-induced initiation of transcriptional silencing), which mediate histone modification and DNA methylation (Xie *et al.* 2004).

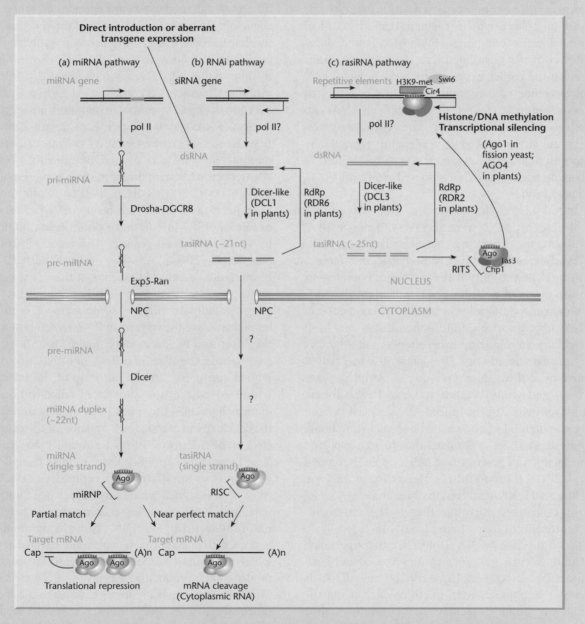

Fig. B15.1 Current model of small RNA pathways which lead to transcriptional and post-transcriptional silencing.

RNA interference is a potent form of silencing caused by the direct introduction of double-stranded RNA into the cell

RNA interference (RNAi) is a sequence-specific gene silencing phenomenon caused by the presence of double-stranded RNA. The process was discovered by researchers working on gene silencing in the nematode worm *C. elegans*, when they observed that either sense *or* antisense RNA could suppress the expression of a homologous gene. It was Fire *et al.* (1998) who first showed that the deliberate introduction of both sense and antisense RNA into worms *at the same time* caused a striking and specific inhibitory effect, which was approximately 10-fold more efficient than either single RNA strand alone. They correctly postulated that the silencing effects seen in the earlier experiments were in fact due to the presence of contaminating dsRNA. Although discovered in *C. elegans*, the principle of RNAi can be traced back even further, since both cosuppression and virus-induced gene silencing in plants are thought to involve the production of dsRNA (due to aberrant transgene expression and viral replication, respectively).

Initial investigation showed that only a few molecules of dsRNA were necessary to induce RNAi in *C. elegans*, suggesting that like ribozymes, RNAi is catalytic rather than stoichiometric. Interference can be achieved only if the dsRNA is homologous to the exons of a target gene, indicating that it is a post-transcriptional process. The phenomenon of RNAi appears to be quite general, and has been used for gene-silencing experiments in many other organisms, including *Drosophila* (Kennerdell & Carthew 2000), mice (Wianny & Zernicka-Goetz 2000), and plants (Waterhouse *et al.* 1998, Chuang & Meyerowitz 2000). Indeed, RNAi is fast becoming the method of choice for large-scale functional analysis in these organisms due to its simplicity, specificity, and potency (see for example Hammond *et al.* 2001, Hannon 2002). We return to the topic of large-scale RNAi screens in Chapter 19.

In *C. elegans* microinjection is the most consistently effective way to induce RNAi. Typically, *in vitro* synthesized dsRNA is injected into the germline of adult worms, and progeny are screened for RNAi-induced phenocopies. However, because RNAi is a systemic phenomenon, microinjection is not the only way it can be achieved. Technically simpler ways to induce RNAi include adding dsRNA to the worms' liquid medium, or even feeding the worms on bacteria that have been engineered to express dsRNA (Maeda *et al.* 2001, Fraser *et al.* 2000). More recently, transgenic strategies to achieve RNAi have become popular. The use of a construct containing adjacent sense and antisense transgenes producing hairpin RNA (e.g. Chuang & Meyerowitz 2000, Tavernarakis *et al.* 2000) or a single transgene with dual opposing promoters (Wang *et al.* 2000) provides a stable source of dsRNA and hence the potential for permanent gene inactivation.

Until 2001, RNAi was not possible in mammals due to an unrelated phenomenon called the interferon response, which shuts down protein synthesis in the presence of dsRNA molecules greater than 30 bp in length and masks any specific effects of gene silencing. However, as more was learned about the mechanism of RNAi, a way around this problem was envisaged. As shown in Box 15.2, RNAi is mediated by an enzyme called Dicer, which chops the dsRNA into short fragments, 21 or 22 bp in length with 2-nt overhangs; these are known as short interfering RNAs (siRNAs). Two groups independently showed that chemically synthesized siRNAs transfected into mammalian cells were capable of inducing specific RNAi effects without inducing the interferon response (Elbashir *et al.* 2001, Caplen *et al.* 2001), although as more experiments have been performed, some examples of the interferon response being induced by RNAi have been reported (Bridge *et al.* 2003, Sledz *et al.* 2003). Another problem with RNAi in mammalian cells is that part of the RNAi pathway appears to be missing. In other organisms, but not in mammals, there is some form of intrinsic amplification of the triggering RNA which prolongs the effect and makes it more potent. In *C. elegans*, for example, the effect of dsRNA injected into adult worms can persist in the offspring of the injected worm! Because of the absence of amplification in mammalian cells, there has been much interest in the development of transgenic systems for the expression of siRNA. Conventional strategies for transgene expression cannot be used because the siRNA genes are so short. Instead, expression cassettes have been designed which are transcribed by RNA polymerase III, since this enzyme transcribes the naturally occurring short RNA genes in mammalian genomes. In one approach, a plasmid is constructed which contains two pol III transcription units in tandem, each producing one of the siRNA strands. The separate strands are thought to assemble spontaneously into the siRNA duplex *in vivo*. A second approach is very similar, but the pol III transcription units are supplied on separate vectors. The individual RNA strands assemble in the same manner. In a third strategy, the plasmid produces a hairpin RNA which assembles into siRNA by self-pairing. Either RNA polymerase II or III can be used to produce hairpin siRNAs because the transgene is longer. Because of these developments, many RNAi experiments have now been performed in mammalian cells (reviewed by Tuschl & Borkhardt 2002, Mittal 2004) and the first reports are beginning to appear concerning the germline transmission of siRNA transgenes in mice and rats, in some cases mirroring the equivalent mutant phenotype (Carmell *et al.* 2003, Hasuwa *et al.* 2002, Kunath *et al.* 2003). Further experiments have shown how siRNA transfected *in vivo* into mouse organs also results in a knockdown phenotype (McCaffrey *et al.* 2002). The medical applications of RNAi are potentially very exciting, and are discussed in Chapter 26.

RNAi is also being applied in plant biotechnology, and recent reports discuss several experiments in which crop plants have been improved using RNAi transgenes. Examples include using RNAi to overcome genetic redundancy in polyploids (Lawrence & Pikaard 2003), modifying plant height in rice by interfering with gibberellin metabolism (Sakamoto *et al.* 2003), changing the glutelin content of rice grains (Kusaba *et al.* 2003) and the oil content of cotton seeds (Liu *et al.* 2002), and controlling the development of leaves (Palatnik *et al.* 2003). The most interesting development is the production of coffee plants that make decaffeinated coffee by using RNAi to suppress caffeine biosynthesis (Ogita *et al.* 2003).

Gene inhibition is also possible at the protein level

Intracellular antibodies and aptamers bind to expressed proteins and inhibit their assembly or activity

Antibodies bind with great specificity to particular target antigens and have therefore been exploited in many different ways as selective biochemical agents. Examples discussed in this book include the immunological screening of cDNA expression libraries (Chapter 6), the isolation of recombinant proteins by immunoaffinity chromatography (Chapter 23), and the development of antibody-based protein chips (Chapter 20). Similarly, oligonucleotides that bind to proteins – known as aptamers – can be used as specific capture agents. The microinjection of antibodies into cells has been widely used to

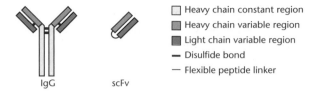

Fig. 15.10 Comparison of a normal immunoglobulin molecule with a single-chain Fv fragment.

block the activity of proteins, but the limitation of this approach is that the inhibitory effect is transient (reviewed by Morgan & Roth 1988). Specific inhibitory effects can also be achieved by microinjecting cells with RNA from hybridoma cell lines (Valle *et al.* 1982, Burke & Warren 1984). Such experiments provided the first evidence that non-lymphoid cells can synthesize and assemble functional antibodies.

To achieve long-term inhibition of specific proteins, cells can be transformed with cDNA constructs that allow the expression of intracellular antibodies (sometimes termed *intrabodies*) (Richardson & Marasco 1995). An important consideration here is that antibodies are large multimeric proteins with, in addition to antigen binding, various effector functions that are non-essential for intracellular protein inhibition. The strategy for expressing intracellular antibodies has been radically simplified using modified antibody forms, such as single-chain Fv (scFv) fragments (Fig. 15.10). These comprise the antigen-binding variable domains of the immunoglobulin heavy and light chains, linked by a flexible peptide arm. Such fragments retain the specificity of the parent monoclonal antibody, but are encoded by a single, relatively small transgene. Further modifications to the expression construct allow the antibody to be targeted to particular intracellular compartments, such as the nucleus, mitochondria, or cytosol. It should be noted, however, that antibodies are normally folded and assembled in the endoplasmic reticulum (ER) and Golgi apparatus and are generally less stable in cell compartments outside the secretory pathway.

Due to their long half-life in the ER, intracellular antibodies have been particularly useful for the inhibition of cell-surface receptors, which pass through this compartment *en route* to the plasma membrane. For example, the cell-surface presentation of functional interleukin-2 (IL2) receptors was completely abolished in Jurkat cells stably expressing an anti-IL2Ra scFv fragment in the ER, rendering these cells insensitive to exogenously applied IL2 (Richardson *et al.* 1995). More recently, the same result has been achieved using lentivirus vectors expressing the scFv fragment, demonstrating how intracellular antibodies can be valuable for gene therapy (Richardson *et al.* 1998). Intracellular antibodies have also been used to abolish the activity of oncogenes (Beerli *et al.* 1994, Cochet *et al.* 1998, Caron de Fromentel *et al.* 1999) and to confer virus resistance by inhibiting replication (reviewed by Rondon & Marasco 1997). Functional antibodies, both full-sized immunoglobulins and fragments, can also be expressed in plants. Hiatt *et al.* (1989) were the first to demonstrate the expression of plant recombinant antibodies, dubbed *plantibodies*, and subsequent experiments have shown that this strategy can be used, as in animal cells, to combat viral diseases by targeting specific viral proteins (Conrad & Fiedler 1998). Antibodies expressed in plants have also been used to interfere with physiological processes in the plant, e.g. antibodies against abscisic acid have been used to disrupt signaling by this hormone in tobacco (Artsaenko *et al.* 1995). There also some examples of expressed aptamers (known as *intramers*) being used to suppress protein activity (e.g. Good *et al.* 1997, Konopka *et al.* 2000, Thomas *et al.* 1997, Shi *et al.* 1999a). As is the case with siRNAs, the expression of intramers usually requires a pol III expression cassette (Famulok & Verma 2002).

Active proteins can be inhibited by dominant-negative mutants in multimeric assemblies

In diploid organisms, most loss-of-function mutations generate recessive or semidominant (dosage-related) phenotypes, because the remaining wild-type copy of the gene provides enough gene product for normal or near-normal activity. However, some loss-of-function mutations are fully dominant over the wild-type allele, because the mutant gene product interferes with the activity of the wild-type protein. Such mutants are known as *dominant negatives*, and principally affect proteins that form dimers or larger multimeric complexes.

The deliberate overexpression of dominant-negative transgenes can be used to swamp a cell with mutant forms of a particular protein, causing all functional molecules to be mopped up into inactive complexes. The microinjection of DNA constructs or *in vitro*-synthesized dominant-negative RNA into *Xenopus* embryos has been widely used to examine the functions of cell-surface receptors in development, since many of these are dimeric (e.g. see Amaya *et al.* 1991, Hemmati-Brivanlou & Melton 1992).

Dominant-negative proteins stably expressed in mammalian cells have been used predominantly to study the control of cell growth and proliferation. A dominant-negative ethylene receptor from *Arabidopsis* has been shown to confer ethylene insensitivity in transgenic tomato and petunia. The effects of transgene expression included delayed fruit ripening and flower senescence (Wilkinson *et al.* 1997).

Suggested reading

Albanese C., Hulit J., Sakamaki T. *et al.* (2002) Recent advances in inducible expression in transgenic mice. *Seminal Cell Development & Biology* **13**, 129–41.

Dykxhoorn D.M. & Lieberman J. (2005) The silent revolution: RNA interference as basic biology, research tool, and therapeutic. *Annual Review Medicine* **56**, 401–23.

Hannon G.J. (2002) RNA interference. *Nature* **418**, 244–51.

Gossen M. & Bujard H. (2002) Studying gene function in eukaryotes by conditional gene inactivation. *Annual Review of Genetics* **36**, 153–73.

Gunsalus K.C. & Piano F. (2005) RNAi as a tool to study cell biology: building the genome–phenome bridge. *Current Opinion in Cell Biology* **17**, 3–8.

Metzger D. & Feil R. (1999) Engineering the mouse genome by site-specific recombination. *Current Opinion in Biotechnology* **10**, 470–6.

Muller U. (1999) Ten years of gene targeting: targeted mouse mutants, from vector design to phenotype analysis. *Mechanical Development* **82**, 3–21.

Padidam M. (2003) Chemically regulated gene expression in plants. *Current Opinion in Plant Biology* **6**, 169–77.

Ristevski S. (2005) Making better transgenic models – Conditional, temporal, and spatial approaches. *Molecular Biotechnology* **29**, 153–63.

Saez E., No D. & West A. (1997) Inducible gene expression in mammalian cells and transgenic mice. *Current Opinion in Biotechnology* **8**, 608–16.

Srivastava V. & Ow D.W. (2004) Marker-free site-specific gene integration in plans. *Trends in Biotechnology* **22**, 627–9.

Tijsterman M. & Plasterk R.H.A. (2004) Dicers at RISC: the mechanism of RNAi. *Cell* **117**, 1–3.

Tuschl T. & Borkhardt A. (2002) Small interfering RNAs: a revolutionary tool for the analysis of gene function and gene therapy. *Molecular Interventions* **2**, 158–67.

Voinnet O. (2002) RNA silencing: small RNAs as ubiquitous regulators of gene expression. *Current Opinion in Plant Biology* **5**, 444–51.

Wang R.H., Zhou X.F. & Wang X.Z. (2003) Chemically regulated expression systems and their applications in transgenic plants. *Transgenic Research* **12**, 529–40.

Part III

Genome Analysis, Genomics, and Beyond

CHAPTER 16

The organization and structure of genomes

Introduction

Across the range of cellular organisms there is an enormous diversity in gene structure. For example, bacterial genomes consist almost entirely of genes whereas in higher eukaryotes genes can be small islands in a large sea of non-coding DNA. Even the genes themselves can be structurally more complex as one moves up the evolutionary tree. At the whole-genome level, there are key differences between the genomes of bacteria, viruses, and organelles on the one hand and the nuclear genomes of eukaryotes on the other. Within the eukaryotes there are major differences in the types of sequences found, the amounts of DNA, and the number of chromosomes.

The genomes of cellular organisms vary in size over five orders of magnitude

Because the different cells within a single organism can be of different ploidy, e.g. germ cells are usu-ally haploid and somatic cells diploid, genome sizes always relate to the haploid genome. The size of the haploid genome also is known as the C-value. Measured C-values range from 3.5×10^3 bp for the smallest viruses, e.g. coliphage MS2, to 10^{11} bp for some amphibians and plants (Fig. 16.1). The largest viral genomes are $1-2 \times 10^5$ bp and are just a little smaller than the smallest cellular genomes, those of some mycoplasmas (5×10^5 bp). Simple unicellular eukaryotes have a genome size ($1-2 \times 10^7$ bp) that is not much larger than that of the largest bacterial genomes. Primitive multicellular organisms such as nematodes have a genome size about four times larger. Not surprisingly, an examination of the genome sizes of a wide range of organisms has shown that the *minimum* C-value found in a particular phylum is related to the structural and organizational complexity of the members of that phylum. Thus the minimum genome size is greater in organisms that evolutionarily are more complex (Fig. 16.2).

A particularly interesting aspect of the data shown

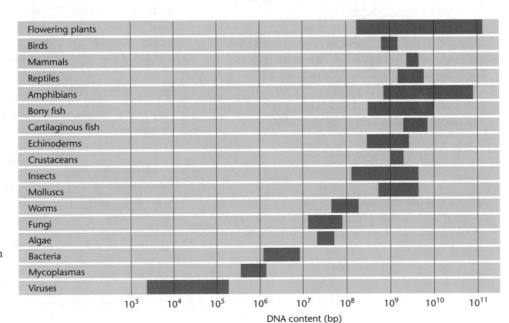

Fig. 16.1 The DNA content of the haploid genome of a range of phyla. The range of values within a phylum is indicated by the shaded area. (Redrawn from Lewin 1994 by permission of Oxford University Press.)

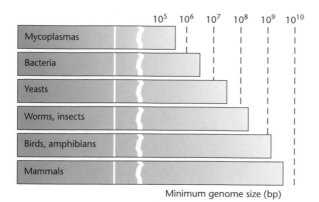

10^5 10^6 10^7 10^8 10^9 10^{10}

Mycoplasmas

Bacteria

Yeasts

Worms, insects

Birds, amphibians

Mammals

Minimum genome size (bp)

Fig. 16.2 The minimum genome size found in a range of organisms. (Redrawn from Lewin 1994 by permission of Oxford University Press.)

in Fig. 16.1 is the range of genome sizes found within each phylum. Within some phyla, e.g. mammals, there is only a twofold difference between the largest and smallest C-value. Within others, e.g. insects and plants, there is a 10- to 100-fold variation in size. Is there really a 100-fold variation in the number of genes needed to specify different flowering plants? Are some plants really more organizationally complex than humans as these data imply? The resolution of this apparent C-value paradox was provided by the analysis of genome complexity using reassociation kinetics (see Box 16.1). Using this technique it was shown that genomes consist of unique sequences of DNA and repeated sequences and that the proportions

---

Box 16.1 Genome complexity can be analyzed using reassociation kinetics

When double-stranded DNA in solution is heated, it denatures ("melts") releasing the complementary single strands. If the solution is cooled quickly the DNA remains in a single-stranded state. However, if the solution is cooled slowly reassociation will occur. In practice, the optimum temperature for reassociation is 25°C below the melting temperature (T_m), that is, the temperature required to dissociate 50% of the duplex. Also, the incubation time and the DNA concentration must be sufficient to permit an adequate number of collisions so that the DNA can reassociate. The size of the DNA fragments affects the rate of reassociation and is conveniently controlled if the DNA is sheared to small fragments.

The reassociation of a pair of complementary sequences results from their collision and therefore the rate depends on their concentration. As two strands are involved the process follows second-order kinetics. Thus, if C is the concentration of DNA that is single stranded at time t, then

$$\frac{dC}{dt} = -kC^2$$

where k is the reassociation rate constant. If C_0 is the initial concentration of single-stranded DNA at time $t = 0$, integrating the above equation gives

$$\frac{C}{C_0} = \frac{1}{1 + k \cdot C_0 t}.$$

When the reassociation is half complete, C/C_0 = 0.5 and the above equation simplifies to

$$C_0 t_{1/2} = \frac{1}{k}.$$

Thus the greater the $C_0 t_{1/2}$ value, the slower the reaction time at a given DNA concentration. More important, for a given DNA concentration the half-period for reassociation is proportional to the number of different types of fragments (sequences) present and thus to the genome size (Britten & Kohne 1968). This can best be seen from the data in Table B16.1. Because the rate of reassociation depends on the concentration of complementary sequences, the $C_0 t_{1/2}$ for organism B will be 200 times greater than for organism A.

Experimentally it has been shown that the rate of reassociation is indeed dependent on genome size (Fig. B16.1). However, this proportionality is only true in the absence of

Table B16.1 Comparison of sequence copy number for two organisms with different genome sizes.

	Organism A	Organism B
Starting DNA concentration (C_0)	10 pg ml^{-1}	10 pg ml^{-1}
Genome size	0.01 pg	2 pg
No. of copies of genome per ml	1000	5
Relative concentration (A vs. B)	200	1

continued

Box 16.1 *continued*

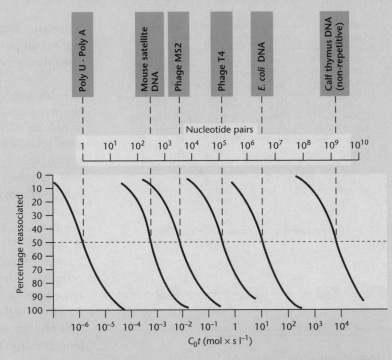

Fig. B16.1 Reassociation of double-stranded nucleic acids from various sources. (Redrawn from Lewin 1994 by permission of Oxford University Press.)

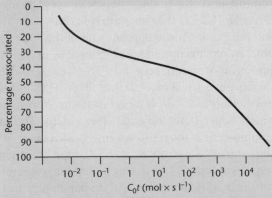

Fig. B16.2 The kinetics of reassociation of calf thymus DNA. Compare the shape of the curve with those shown in Fig. B16.1.

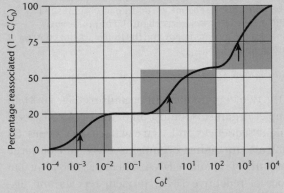

	Fast component	Intermediate component	Slow component
Per cent of genome	25	30	45
$C_0t_{1/2}$	0.0013	1.9	630
Complexity (bp)	340	6.0×10^5	3.0×10^8
Repetition frequency	500000	350	1

Fig. B16.3 The reassociation kinetics of a eukaryotic DNA sample showing the presence of two types of repeated DNA. The arrows indicate the $C_0t_{1/2}$ values for the three components. (Redrawn from Lewin 1994 by permission of Oxford University Press.)

repeated sequences. When the reassociation of calf thymus DNA was first studied, kinetic analysis indicated the presence of two components (Fig. B16.2). About 40% of the DNA had a $C_0t_{1/2}$ of 0.03, whereas the remaining 60% had a $C_0t_{1/2}$ of 3000. Thus the concentration of DNA sequences that reassociate rapidly is 100,000 times the concentration of those sequences that reassociate slowly. If the slow fraction is made up of unique sequences, each of which occurs only once in the calf genome, then the sequences of the rapid fraction must be repeated 100,000 times, on average. Thus the $C_0t_{1/2}$ value can be used to determine the sequence complexity of a DNA preparation. A comparative analysis of DNA from different sources has shown that repetitive DNA occurs widely in eukaryotes (Davidson & Britten 1973) and that different types of repeat are present. In the example shown in Fig. B16.3 a fast-renaturing and an intermediate-renaturing component can be recognized and are present in different copy numbers (500,000 and 350, respectively) relative to the slow component, which is unique or non-repetitive DNA.

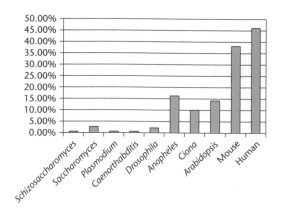

Fig. 16.3 The proportion of repeated DNA in different organisms.

Fig. 16.4 The two types of repeated DNA found in eukaryotes: dispersed repeats (upper part of figure) and tandem repeats (lower part of figure). The purple boxes represent the repeated DNA and the thin black lines the rest of the genome.

of the two vary in different organisms (Fig. 16.3). In simpler organisms almost all of the DNA consists of unique sequences and, as we will see later, genes. By contrast, in higher organisms there can be large amounts of repetitive DNA. There are two principal types of repetitive DNA: tandem repeats and dispersed repeats (Fig. 16.4).

In general, the length of the non-repetitive DNA component tends to increase as we go up the evolutionary tree and reaches a maximum of 2×10^9 bp in mammals. The fact that many plants and animals have a much higher C-value reflects the presence of large amounts of repetitive DNA. Analysis of mRNA hybridization to DNA shows that most of it anneals to non-repetitive DNA, i.e. most genes are present in non-repetitive DNA. Thus genetic complexity is proportional to the content of non-repetitive DNA and not to genome size. As will become clear in the next chapter, repetitive DNA confounds the assembly of a complete sequence of a genome. Therefore, if one wishes to sequence the genome of a representative of a particular phylum it would make sense to select the one with the lowest content of repetitive DNA. In this context, reference to Fig. 16.1 shows why the first two plant genomes sequenced were those of *Arabidopsis* (125 Mb) and rice (430 Mb).

Increases in genome complexity sometimes are accompanied by increases in the complexity of gene structure

In some genes the coding sequence is interrupted by the presence of non-coding (untranslated) sequences known as *introns*. Such genes are known as *split genes* and the parts of these genes that are translated are known as *exons*. Split genes are rare in prokaryotes (Edgell *et al.* 2000, Martinez-Arbaca & Toro 2000) although they are commoner in archaebacteria than eubacteria. Split genes are much commoner in eukaryotes but the number of such genes, and the number and size of introns per gene, increase with genome complexity (Fig. 16.5). For example, the genome of the budding yeast *Saccharomyces cerevisiae* has over 6000 open reading frames but only 330 introns (Lopez & Seraphin 2000). Those genes that are split tend to have just one intron and the longest intron is only 1 kb in size. Introns are much commoner in the fission yeast *Schizosaccharomyces cerevisiae* (43% of the genes are split) but the introns still are small. At the opposite end of the scale, one chicken collagen gene has over 50 exons, the human dystrophin gene has 78 introns, and the *Dscam* gene in *Drosophila* has over 100 introns. Furthermore, in these organisms the introns are much larger than the exons (Fig. 16.6). The dystrophin gene is the most extreme known example of this: the gene has a size of 2.5 Mb but the coding sequence is only 14 kb in length. The longest human intron is 480 kb and this is similar in size to the smallest bacterial genomes!

In genes that are related by evolution the exons are of similar size although the genes themselves may differ greatly in length. This means that the introns must be in the same position but can be of different sizes (Fig. 16.6). Furthermore, if a split gene has been cloned it is possible to sub-clone either the exon or intron sequences. If these sub-clones are used as probes in genomic Southern blots, one can determine if these same sequences are present elsewhere in the genome. Often the exon sequences of one gene are found to be related to sequences in one or more other genes. Some examples of such *gene families* are given in Table 16.1. Multiple copies of an exon also may be found in several apparently unrelated genes. Exons that are shared by several unrelated genes are likely to encode polypeptide regions (domains) that endow the disparate proteins with related properties, e.g. ATP or DNA binding. Some genes appear to be mosaics that were constructed by patching together

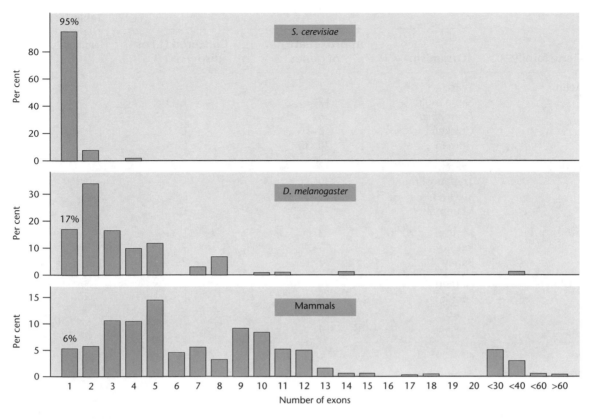

Fig. 16.5 The number of exons in three representative eukaryotes. Uninterrupted genes have only one exon and are totalled in the left-hand column. (Redrawn from Lewin 1994 by permission of Oxford University Press.)

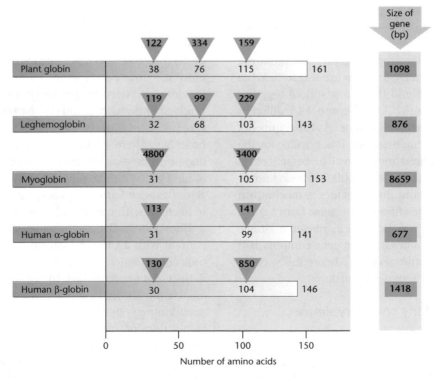

Fig. 16.6 The placement of introns in different members of the globin superfamily. The size of the introns in base pairs is indicated inside the inverted triangles. Note that the size of each polypeptide and the location of the different introns are relatively consistent.

Table 16.1 Some examples of multigene families.

Gene family	Organism	Approximate no. of genes	Clustered (L) or dispersed (D)
Actin	Yeast	1	–
	Slime mold	17	L, D
	Drosophila	6	D
	Chicken	8–10	D
	Human	20–30	D
Tubulin	Yeast	3	D
	Trypanosome	30	L
	Sea urchin	15	L, D
	Mammals	25	D
α-Amylase	Mouse	3	L
	Rat	9	?
	Barley	7	?
β-Globin	Human	6	L
	Lemur	4	L
	Mouse	7	L
	Chicken	4	L

Table 16.1 Some examples of multigene families.

copies of individual exons recruited from different genes, a phenomenon known as *exon shuffling*. It should be noted that although there is a strong correspondence between protein domains and exon structure, particularly in metazoa, the two are not always linked.

There is a second degree of complexity emanating from split genes. After split genes are transcribed the introns are excised and the exons spliced together by a complex of snRNA and some 145 different proteins known as the splicesome. As the number of introns within a gene increases it is possible for the pre-mRNA (unspliced messenger) to be spliced in different ways (Thanaraj *et al.* 2004). This is known as alternative splicing and provides a mechanism for producing a wide variety of proteins from a small number of genes (see Box 16.2 for some simple examples). The *Drosophila Dscam* gene contains 108 exons and alternative splicing theoretically could generate 38,016 different proteins!

Viruses and bacteria have very simple genomes

The genomes of viruses and prokaryotes are very simple structures, although those of viruses show remarkable diversity (for a review see Dimmock *et al.* 2001). Most viruses have a single linear or circular genome but a few, such as reoviruses, bacteriophage φ6, and some plant viruses, have segmented RNA genomes. For a long time it was believed that all eubacterial genomes consisted of a single circular chromosome. However, linear chromosomes have been found in *Borrelia* sp., *Streptomyces* sp., and *Rhodococcus fascians*, and mapping suggests that *Coxiella burnetii* also has a linear genome. Two chromosomes have been found in a number of bacteria including *Rhodobacter spheroides*, *Brucella melitensis*, *Leptospira interrogans*, and *Agrobacterium tumefaciens* (Cole & Saint Girons 1994). In the case of *Agrobacterium*, there is one circular chromosome and one non-homologous linear chromosome (Goodner *et al.* 1999). Linear plasmids have been found in *Borrelia* sp. and *Streptomyces* sp. as well as a number of bacteria with circular chromosomes (Hinnebush & Tilley 1993). *Borrelia* has a very complex plasmid content with 12 linear molecules and nine circular molecules (Casjens *et al.* 2000).

Bacterial genomes lack the centromeres found in eukaryotic chromosomes although there may be a partitioning system based on membrane adherence. Duplication of the genomes is initiated at an origin of replication and may proceed unidirectionally or bidirectionally. The structure of the origin of replication, the *oriC* locus, has been extensively studied in a range of bacteria and found to consist essentially of the same group of genes in a nearly identical order (Cole & Saint Girons 1994). The *oriC* locus is defined

Box 16.2   Examples of alternative splicing

Animals produce antibodies (immunoglobulins) as part of their defense against microbial infection. These immunoglobulins are produced in two forms: a soluble antibody that is secreted and a membrane-bound antibody that helps to identify the producing cell to other cells of the immune system. Both forms of the antibody are produced from the same pre-mRNA molecule by alternative splicing (Fig. B16.4).

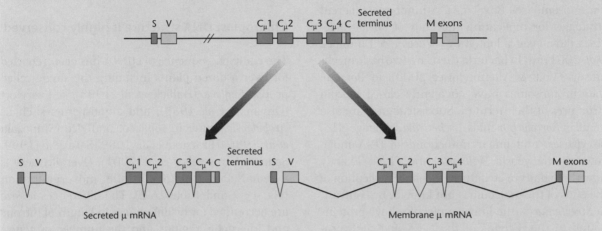

Fig. B16.4 Alternative splicing in immunoglobulin synthesis. The top part of the figure shows the structure of a gene for the μ heavy chain of an IgM immunoglobulin. The exons are shown as boxes and the introns as lines. A pre-mRNA produced from this gene can be spliced such that the transmembrane anchor region (M exons) either is excluded (secreted antibody) or included (membrane-bound antibody).

We consider three genes involved in sex determination in *Drosophila*: *Sxl* (sex lethal), *tra* (transformer), and *dsx* (double sex). Each of these genes produces a pre-mRNA that has two possible splicing patterns depending on whether the fly is male or female (Fig. B16.5). In male flies, splicing results in the production of inactive *Sxl* and *tra* gene products. The *dsx* gene product is functional and inactivates the female-specific genes. In female flies, splicing produces functional *Sxl* and *tra* gene products and these interact with the *dsx* gene to alter its splicing pattern such that inactivation of female-specific genes does not occur.

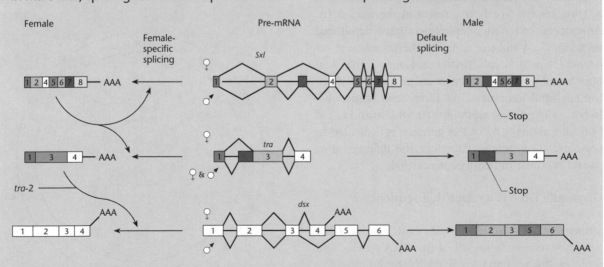

Fig. B16.5 Sex determination in *Drosophila* mediated by alternative splicing. The pre-mRNAs are shown in the middle of the figure with the female splicing pattern on top and the male splicing pattern on the bottom. Note that the inclusion of two exons, number 3 in *Sxl* and number 2 in *tra*, produce messengers that have stop codons resulting in the production of truncated, inactive proteins.

as a region harboring the *dna*A (DNA initiation) or *gyr*B (B subunit of DNA gyrase) genes linked to a ribosomal RNA operon.

Many bacterial and viral genomes are circular or can adopt a circular conformation for the purposes of replication. However, those viral and bacterial genomes which retain a linear configuration need a special mechanism to replicate the ends of the chromosome (see Box 16.3). A number of different strategies for replicating the ends of linear molecules have been adopted by viruses (see Dimmock *et al.* 2001) but in bacteria there are two basic mechanisms (Volff & Altenbuchner 2000). In *Borrelia*, the chromosomes have covalently closed hairpin structures at their termini. Such structures are also found in *Borrelia* plasmids, *Escherichia coli* phage N15, poxviruses, and linear mitochondrial DNA molecules in the yeasts *Williopsis* and *Pichia*. Exactly how these hairpin structures facilitate replication of the ends of the molecule is not known. By contrast, in *Streptomyces*, the linear molecules have proteins bound to the 5′ ends of the DNA and such proteins are also found in adenoviruses, and a number of bacteriophages and fungal and plant mitochondrial plasmids. These terminal proteins probably are involved in the completion of replication. In addition, *Streptomyces* linear replicons have palindromic sequences and inverted repeats at their termini.

The bacterial genomes that have been completely sequenced have sizes ranging from 0.6 to 9.1 Mb. The difference in size between the smallest and the largest is not a result of introns for these are rare in prokaryotes (Edgell *et al.* 2000; Martinez-Abarca & Toro 2000). Nor is it a result of repeated DNA. Analysis of the kinetics of reassociation of denatured bacterial DNA did not indicate the presence of repeated DNA in *E. coli* (Britten & Kohne 1968) and only small amounts have been detected in all of the bacterial genomes that have been sequenced. In both *Mycoplasma genitalium* (0.58 Mb) and *E. coli* (4.6 Mb) about 90% of the genome is dedicated to protein-coding genes. Therefore the differences in size reflect the number of genes carried.

Organelle DNA is a repetitive sequence

Mitochondria and chloroplasts both possess DNA genomes that code for all of the RNA species and some of the proteins involved in the functions of the organelle. In some lower eukaryotes the mitochondrial (mt) DNA is linear but more usually organelle genomes take the form of a single circular molecule of DNA. Because each organelle contains several copies of the genome and because there are multiple organelles per cell, organelle DNA constitutes a repetitive sequence. Whereas chloroplast (ct) DNA falls in the range 120–200 kb, mtDNA varies enormously in size. In animals it is relatively small, usually less than 20 kb, but in plants it can be as big as 2000 kb.

Chloroplast DNA structure is highly conserved

The complete sequence of ctDNA has been reported for over a dozen plants including the single-celled protist *Euglena* (Hallick *et al.* 1993), a liverwort (Ohyama *et al.* 1986), and angiosperms such as *Arabidopsis*, spinach, tobacco, and rice (Shinozaki *et al.* 1986, Hiratsuka *et al.* 1989, Sato *et al.* 1999, Schmitz-Linneweber *et al.* 2001). Overall, there is a remarkable similarity in size and organization (Fig. 16.7 and Table 16.2). The differences in size are accounted for by differences in length of introns and intergenic regions and the number of genes. A general feature of ctDNA is a 10–24 kb sequence that is present in two identical copies as an inverted repeat. The proportion of the genome that is represented by introns can be very high, e.g. in *Euglena* it is 38%.

The chloroplast genome encodes 70–90 proteins, including those involved in photosynthesis, four rRNA genes, and 30 or more tRNA genes. Chloroplast mRNAs are translated with the standard genetic code (cf. mitochondrial mRNA). However, editing

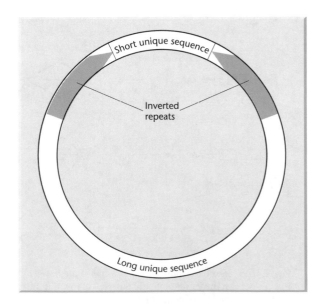

Fig. 16.7 Generalized structure of ctDNA.

Table 16.2 Key features of chloroplast DNA.

Feature	*Arabidopsis*	Spinach	Maize
Inverted repeats	26,264 bp	25,073 bp	22,748 bp
Short unique sequence	17,780 bp	17,860 bp	12,536 bp
Long unique sequence	84,170 bp	82,719 bp	82,355 bp
Length of total genome	154,478 bp	150,725 bp	140,387 bp
Number of genes	128	108	104
rRNA genes	4	4	4
tRNA genes	37	30	30
Protein-encoding genes	87	74	70

events cause the primary structures of several transcripts to deviate from the corresponding genomic sequences by C to U transitions with a strong bias for changes at the second codon position (Maier *et al.* 1995). This editing makes it difficult to convert chloroplast nucleotide sequences into amino acid sequences for the corresponding gene products.

Astasia longa is a colorless heterotrophic flagellate which is closely related to *Euglena gracilis*. It contains a plastid DNA that is 73 kb in length, about half the size of the ctDNA from *Euglena*. Sequencing of this plastid DNA has shown that all chloroplast genes for photosynthesis-related proteins, except that encoding the large subunit of ribulose-1, 5-bisphosphate carboxylase, are missing (Gockel & Hachtel 2000).

Mitochondrial genome architecture varies enormously, particularly in plants and protists

Mitochondria are found in almost every eukaryotic organism and their primary function is the generation of ATP by oxidative phosphorylation. They also contain DNA (mtDNA) that encodes a very limited number of biochemical functions. In marked contrast to this genetic conservatism, the mitochondrial genome is characterized by a bewildering diversity of structures and mechanisms of gene expression (for review, see Burger *et al.* 2003). Although most mtDNA consists of a single DNA molecule of uniform length, molecules of varying length are found in some protists (Fig. 16.8). For example, multiple, circular molecules are found in the fungus *Spizellomyces punctatus* and in trypansosomes. Another protist, *Amoebidium parasiticum*, has mtDNA consisting of several distinct types of linear molecules with terminal and sub-terminal repeat motifs. Terminal repeat motifs also are a feature of the mtDNA of *Tetrahymena*.

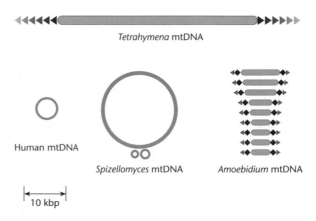

Fig. 16.8 Some unusual mtDNA structures found in protists compared with human mtDNA.

The size of mtDNA in most eukaryotic phyla ranges from 15–60 kb but there are some notable exceptions. The mtDNA of the malarial parasite (*Plasmodium* spp.) is only 6 kb long while that of rice (*Oryza sativa*) is 490 kb and in cucurbit plants it may have a size of 2 Mb. The number of genes encoded by mtDNA ranges from five in *Plasmodium* to 100 in certain flagellates with the average across eukaryotes being 40–50. Very small mtDNA genomes obviously have a restricted coding capacity but as the genome size increases the correlation between size and gene number quickly breaks down. Rather, the differences in mtDNA genome size mostly reflect the size of the intergenic regions and their content of tandem repeats. The mitochondrial genome of most animals is very compact and has very little intergenic space and no introns.

Introns are a common feature of mtDNA but a number of unusual gene structures also has been described. For example, some genes are split into as many as eight modules that are scrambled through the genome and located on both strands of the DNA. Separate transcription of these subgenic modules yields discrete pieces of rRNA that are held together via base pairing of complementary sequences.

The organization of nuclear DNA in eukaryotes

The gross anatomy of chromosomes is revealed by Giemsa staining

In certain eukaryotes, a variety of treatments will cause chromosomes in dividing cells to appear as a series of light- and dark-staining bands (Fig. 16.9). In G-banding, for example, the chromosomes are subjected to controlled digestion with trypsin before Giemsa staining, which reveals alternating positively (dark G-bands) and negatively (R-bands or pale G-bands) staining regions. As many as 2000 light and dark bands can be seen along some mammalian chromosomes. An identical banding pattern (Q-banding) can be seen if the Giemsa stain is replaced with a fluorescent dye such as quinacrine, which intercalates between the bases of DNA. A structural basis for metaphase bands has been proposed that is based on the differential size and packing of DNA loops and matrix-attachment sites in G- vs. R-bands (Saitoh & Laemmli 1994). Bands are classified according to their relative location on the short arm (p) or the long arm (q) of specific chromosomes; e.g. 12q1 means band 1 on the long arm of chromosome 12. If the chromosome DNA is treated with acid and then alkali prior to Giemsa staining, then only the centromeric region stains and this is referred to as *heterochromatin*. The unstained parts of the chromosome are called *euchromatin*. Because Giemsa stain shows preferential binding to DNA rich in AT base pairs, the

dark G-bands should be AT-rich and the light G-bands GC-rich. This has been confirmed by other methods (Zoubak *et al.* 1996, Saccone *et al.* 1999).

CpG islands represent a different form of compositional heterogeneity. They were originally identified as short regions of mammalian DNA which contained many sites for the restriction endonuclease *Hpa*II and for this reason originally were called *Hpa*II Tiny Fragment (HTF) islands. This island DNA is found in short regions of 1–2 kb, which together account for approximately 2% of the mammalian genome. It has distinctive properties when compared with the DNA in the rest of the genome. It is unmethylated, GC-rich, and does not show any suppression of the dinucleotide CpG. By contrast, bulk genomic DNA has a GC content which is much lower, is methylated at CpG, and the CpG dinucleotide is present at a much lower frequency than would be predicted from its base composition. CpG islands have been found at the 5′ ends of all housekeeping genes and of a large proportion of genes with a tissue-restricted pattern of expression (Craig & Bickmore 1994).

Telomeres play a critical role in the maintenance of chromosomal integrity

The ends of eukaryotic chromosomes are also the ends of linear duplex DNA and as such require a special structure to ensure that they are maintained (see Box 16.3). The reason for this is connected with the way in which double-stranded DNA is replicated (Fig. 16.10). DNA synthesis occurs in a 5′ → 3′ direction and is initiated by extension of an RNA primer. After removal of this primer there is no way of completing the 5′ end of the molecule. In the absence of a method for completing the ends, chromosomes would become shorter after each cell division. Telomeres are specialized nucleic acid sequences whose role is to protect the ends of chromosomes. In most eukaryotes the telomere consists of a short

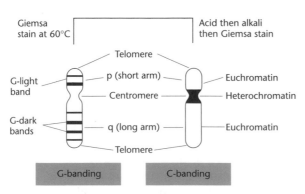

Fig. 16.9 Banding patterns revealed on chromosomes by different staining methods. Note that intercalating fluorescent dyes produce the same pattern as Giemsa stain at 60°C. In the C-banding technique some heterochromatin may be detected at the telomeres.

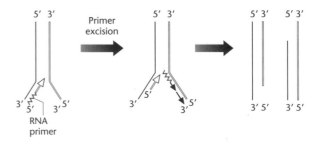

Fig. 16.10 Formation of two daughter molecules with complementary single-stranded 3′ tails after primer excision.

Box 16.3   Telomerase, immortality, and cancer

Telomeres are found at the ends of chromosomes and their role is to protect the ends of chromosomes. They also stop chromosomes from fusing to each other. Telomeres consist of repeating units of TTAGGG that can be up to 15,000 bp in length. The enzyme telomerase is a ribonucleoprotein enzyme whose RNA component binds to the single-stranded end of the telomere. An associated reverse transcriptase activity is able to maintain the length and structure of telomeres by the mechanism shown in Fig. 16.11.

Telomerase is found in fetal tissues, adult germ cells, and cancer cells. In normal somatic cells the activity of telomerase is very low and each time the cells divide some of the telomere (25–200 bp) is lost. When the telomere becomes too short the chromosome no longer divides and the host cell dies by a process known as apoptosis. Thus, normal somatic cells are mortal and in tissue culture they will undergo 50–60 divisions (Hayflick limit) before they senesce. In contrast to mammals, indeterminately growing multicellular organisms, such as fish and crustaceae, maintain unlimited growth potential throughout their entire life and retain telomerase activity (for a review see Krupp *et al.* 2000).

Cancer cells can divide indefinitely in tissue culture and thus are immortal. Telomerase has been found in cancer cells at activities 10- to 20-fold greater than in normal cells. This presence of telomerase confers a selective growth advantage on cancer cells and allows them to grow uncontrollably. Telomerase is an ideal target for chemotherapy because this enzyme is active in most tumors but inactive in most normal cells.

If recombinant DNA technology is used to express telomerase in human somatic cells maintained in culture, senescence is avoided and the cells become immortal. This immortalization usually is accompanied by an increased expression of the *c-myc* oncogene to the levels seen in many cancer cell lines.

Chromosomal rearrangements involving telomeres are emerging as an important cause of human genetic diseases (Knight & Flint 2000). Telomere-specific clones have been used in combination with fluorescence *in situ* hybridization (FISH, see p. 353) to detect abnormalities not found by conventional cytogenetic analysis.

Table 16.3   Length of the telomere repeat in different eukaryotic species.

Species	Length of telomere repeat
S. cerevisiae (yeast)	300 bp
Mouse	50 kb
Human	10 kb
Arabidopsis	2–5 kb
Cereals	12–15 kb
Tobacco	60–160 kb

As can be seen from Fig. 16.11, the ends of telomeres are not blunt-ended but have 3′ single-stranded overhangs of 12 or more nucleotides. The enzyme telomerase, also called telomere terminal transferase, is a ribonucleoprotein enzyme whose RNA component binds to the single-stranded end of the telomere. An associated reverse transcriptase activity is able to maintain the length and structure of the telomeres by the mechanism shown in Fig. 16.11. For a detailed review of the mechanism of telomere maintenance the reader should consult the paper by Shore (2001).

Tandemly repeated sequences can be detected in two ways

Repeated sequences were first discovered 30 years ago during studies on the behavior of DNA in centrifugal fields. When DNA is centrifuged to equilibrium

repeat of TTAGGG many hundreds of units long but in *S. cerevisiae* the repeat unit is TG_{1-3}. These repeats vary considerably in length between species (Table 16.3) but each species maintains a fixed average telomere length in its germline.

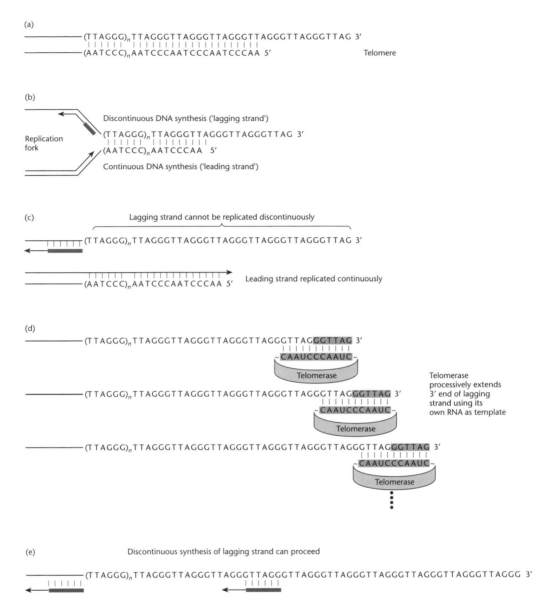

Fig. 16.11 Diagram of telomere replication and the role of telomerase. (a) The hexameric repeat sequence found at telomere ends in human chromosomes. (b) and (c) A replication fork attempting to replicate the telomere. The Okazaki fragment (purple arrow) allows replication of all but the most terminal portion of the lagging strand. (d) and (e) Telomerase, carrying its own RNA template (5′-CUAACCCUAAC-3′), extends the lagging strand at the end of the chromosome and allows replication. (Redrawn with permission from Nussbaum *et al.* 2001. W.B. Saunders Publishing, 2001, a division of Harcourt.)

in solutions of CsCl, it forms a band at the position corresponding to its own buoyant density. This in turn depends on its percentage G + C content:

$$\rho\,(\text{density}) = 1.660 + 0.00098\,(\%\,\text{GC})\ \text{g cm}^{-3}.$$

When eukaryotic DNA is centrifuged in this way the bulk of the DNA forms a single, rather broad band centered on the buoyant density which corresponds to the average G : C content of the genome. Frequently one or more minor or *satellite* bands are seen (Fig. 16.12). The behavior of satellite DNA

on density gradient centrifugation frequently is anomalous. When the base composition of a satellite is determined by chemical means it often is different to that predicted from its buoyant density. One reason is that it is methylated, which changes its buoyant density.

Once isolated, satellite DNA can be radioactively labeled *in vitro* and used as a probe to determine where on the chromosome it will hybridize. In this technique, known as *in situ hybridization*, the chromosomal DNA is denatured by treating cells that have been squashed on a cover slip. The

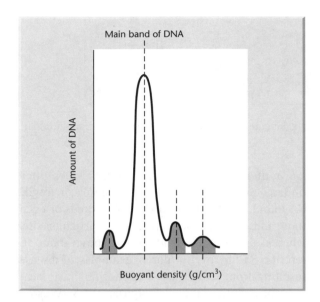

Fig. 16.12 Detection of three satellite DNA bands (dark shading) on equilibrium density gradient centrifugation of total DNA.

localization of the sites of hybridization is determined by autoradiography. Using this technique, most of the labeled satellite DNA is found to hybridize to the heterochromatin present around the centromeres and telomeres. Because RNA that is homologous to satellites is found only rarely the heterochromatic DNA most probably is non-coding.

When satellite DNA is subjected to restriction endonuclease digestion only one or a few distinct low-molecular-weight bands are observed following electrophoresis. These distinct bands are a tell-tale sign of tandemly repeated sequences. The reason (Fig. 16.13) is that if a site for a particular restriction endonuclease occurs in each repeat of a repetitious tandem array, then the array is digested to unit-sized fragments by that enzyme. After elution of the DNA band from the gel it can be used for sequence analysis either directly or after cloning. However, the sequence obtained is a consensus sequence and not necessarily the sequence of any particular repeat unit because sequence divergence can and does occur very readily. Note that if such sequence divergence occurs within a restriction endonuclease cleavage site in the repeated units then digestion with the enzyme produces multimers of the repeat unit ("higher order repeats") (Fig. 16.13).

Tandemly repeated sequences can be subdivided on the basis of size

The amount of satellite DNA and its sequence varies widely between species and can be highly

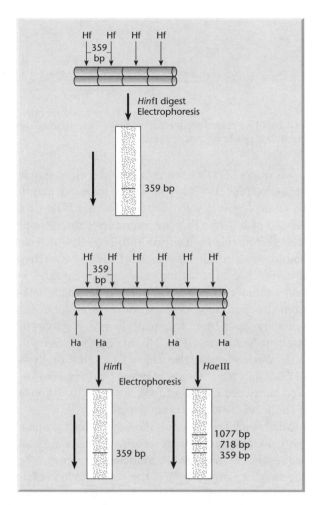

Fig. 16.13 Digestion of purified satellite DNA by restriction endonucleases. The basic repeat unit is 359 bp long and contains one endonuclease *Hin*fI (Hf) site. Digestion with *Hin*fI converts most of the satellite DNA to a set of 359 bp long fragments. These are abundant enough to be seen as a band against the smear of other genomic fragments after gel electrophoresis and staining with ethidium bromide. Digestion of the DNA with endonuclease *Hae*III (Ha) yields a ladder of fragments that are multiples of 359 bp in length. (Redrawn with permission from Singer & Berg 1990.)

polymorphic within a species. Thus, 1–3% of the rat genome is accounted for by centromeric satellite DNA whereas it is 8% in the mouse and 23% in the cow. The length of the satellite repeat unit varies from the $d(AT)_n:d(TA)_n$ structure found in the land crab to complete gene clusters like that encoding the different species of ribosomal RNA (Fig. 16.14). *Minisatellites* are sequences found throughout the genome whose repeat units are 14–500 bp in length. They demonstrate intraspecies polymorphism and because of this they were used (Monckton & Jeffreys 1993, Alford & Caskey 1994) in the earliest versions of DNA profiling (often referred to as DNA "fingerprinting").

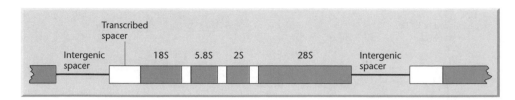

Fig. 16.14 Detailed architecture of a rDNA repeat unit. Transcribed regions are shown in boxes. Coding sequences are shown in purple and spacer regions in white.

Microsatellite DNA families, also known as simple sequence repeats (SSRs) or simple sequence length polymorphisms (SSLPs), are small arrays of tandem repeats of 1–13 bp that are interspersed throughout the genome (for review, see Toth *et al.* 2000). Two features of microsatellites are worth noting. First they are inherited in a Mendelian manner (Fig. 16.15) and so they can be used for checking genetic relationships. Secondly, they are one of the mechanisms whereby restriction fragment length polymorphisms are generated (Fig. 16.16). The properties of microsatellites make them ideal for forensic analysis of materials from scenes of crime (see Box 16.4) as well as investigations of food authenticity (Woolfe & Primrose 2004) and animal theft.

Microsatellites comprise 3% of the human genome with an average density of one SSR per 2 kb of sequence. Of these SSRs, dinucleotide repeats are the most common (0.5% of the genome) closely followed by runs of $(dA \cdot dT)_n$ at 0.3% of the genome. Trinucleotide repeats are much rarer. Within the dinucleotide repeats there is a heavy bias towards dAC·dTG and dAT·dTA repeats with dGC·dCG being extremely rare. The significance of these repeats in normal genes is not known but they can be the locus

for a number of inherited disorders when they undergo unstable expansion. For example, in fragile X syndrome patients can exhibit hundreds or even thousands of the CGG triplet at a particular site, whereas unaffected individuals only have about 30 repeats. So far, over a dozen examples of disease resulting from trinucleotide expansion have been described (Sutherland & Richards 1995, Warren 1996, Mitas 1997, Bowater & Wells 2000). Similar trinucleotide repeats have been discovered in bacteria (Hancock 1996) and yeast (Dujon 1996, Mar Alba *et al.* 1999, Richard *et al.* 1999) following complete genome sequencing. As in the human case, perfect trinucleotide repeats in yeast are subject to polymorphic size variation while imperfect ones are not. Some pathogenic bacteria use length variation in simple repeats to change the antigens on their surfaces so that they can evade host immune attack.

Tandem repetition of DNA sequences also occurs within coding regions. For example, linked groups of identical or near-identical genes sometimes are repeated in tandem. These are the gene families described earlier (p. 326). However, tandem repetition also occurs within a single gene; for example, the *Drosophila* "glue" protein gene contains 19 direct

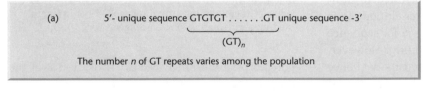

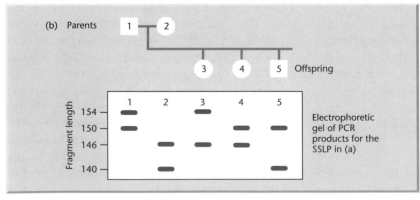

Fig. 16.15 The use of SSLPs in inheritance studies. (a) Structure of an SSLP. (b) Schematic representation of an SSLP in a five-member family. The two alleles carried by the father are different from those carried by the mother. The children inherit one allele of the SSLP from each parent. (Redrawn with permission from Dogget 1992, courtesy of University Science Books.)

Box 16.4 The forensic application of SSLPs

The existence of SSLPs is of great utility in paternity testing and criminal investigations, since they allow ready comparison of DNA samples in the absence of detailed genetic information by the generation of a DNA profile or fingerprint. In principle, a multilocus DNA fingerprint can be generated either by the simultaneous application of several probes, each one specific for a particular locus, or by applying a single DNA probe that simultaneously detects several loci. When DNA profiling was first developed (Jeffreys *et al.* 1985a), multilocus probes were used and these were derived from a tandemly repeated sequence within an intron of the myoglobin gene. These probes can hybridize to other autosomal loci – hence their utility. The first criminal court case to use DNA fingerprinting was in Bristol, UK, in 1987, when a link was shown between a burglary and a rape. In the following year, DNA-fingerprinting evidence was used in the USA. It is worth noting that DNA evidence has been used to prove innocence as well as guilt (Gill & Werrett 1987).

In criminal cases, a major disadvantage of multilocus probes is the complexity of the DNA fingerprint provided. Showing innocence is easy, but proving identity is fraught with problems. The issue boils down to calculations of the probability that two profiles match by chance as opposed to having come from the same person (Lewontin & Hartl 1991). For this reason forensic scientists have moved to the use of single-locus probes and an example is shown in Fig. B16.6. The latest variation of the technique targets 13 SSLP loci and makes a determination of sex (X or Y). For each locus there are multiple alleles (Table B16.2) and the frequency of each allele is known for different ethnic groups. Consequently it is possible to show that the probability of two individuals sharing the same profile is less than one in a billion (thousand million).

Another advantage of single-locus probes is that it is possible to convert the DNA profile into a numerical format. This enables a database to be established and all new profiles

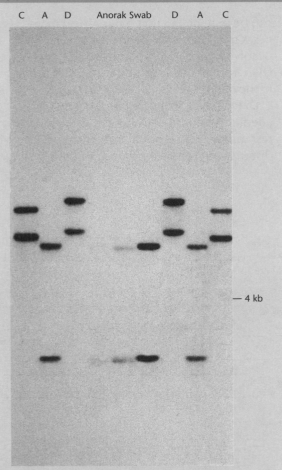

Fig. B16.6 Use of a single-locus probe to determine the identity of a rapist. Semen was extracted from an anorak and a vaginal swab. The victim's profile is in track D and that of two suspects in tracks A and C. The profile matches individual A. (Photo courtesy of Dr. P. Gill.)

Table B16.2 Numbers of alleles for the different SSLPs used in human DNA profiling.

Locus	Number of alleles
D3S1358	8
VWA	11
FGA	14
D8S1179	12
D21S11	22
D18S51	21
D5S818	10
D13S317	8
D7S820	10
D16S539	9
THO1	7
TPOX	8
CSF1PO	10

continued

Box 16.4   *continued*

can be matched to that database. In many cases the police have been able to re-open unsolved murder or rape cases and match DNA from the scene of the crime with profiles in the database.

Detection of SSLPs requires that an adequate amount of DNA be present in the test sample. This is not a problem in paternity disputes but can be an issue in forensic testing.

With the advent of single-locus probes, the amount of DNA required is much less of an issue, since the test loci in the sample can be amplified by the PCR. As a result, it now is possible to type DNA from a face-mask worn by a bank robber, a cigarette-butt discarded at the scene of a crime or the back of a stamp on an envelope used to send a "poison-pen" or blackmail letter.

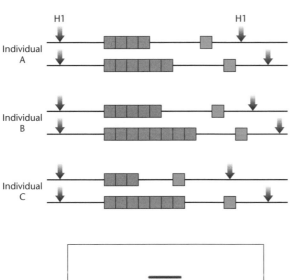

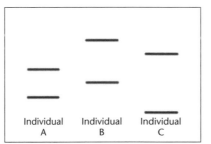

Fig. 16.16   Restriction fragment length polymorphisms caused by a variable number of tandem repeats between the two *Hin*fI restriction sites. The upper part of the diagram shows the DNA structure for three different individuals. The lower part of the diagram shows the pattern obtained on electrophoresis of *Hin*fI cut DNA from the three individuals after hybridization with a probe complementary to the sequence shown in pink.

tandem repeats of a sequence 21 base pairs long that encodes seven amino acids. The repeats are not perfect but show divergence from a consensus sequence. Another example is the gene for α2(1) collagen found in chicken, mouse, and humans. The gene comprises 52 exons with introns varying in length from 80 to 2000 bp. However, all the exon sequences are multi-

ples of 9 bp and most of them are 54 or 108 bp long. This accounts for the observed primary sequence of collagen, which has glycine in every third position and a very high content of proline and lysine.

Dispersed repeated sequences are composed of multiple copies of two types of transposable elements

Dispersed throughout eukaryotic genomes are multiple copies of different transposable elements and hence these elements are dispersed repeated sequences. In studies on reassociation kinetics (Box 16.1) these transposable elements are characterized as moderately repetitive DNA. Eukaryotic transposable elements are divided into two classes (Fig. 16.17) based on the properties of their transposition intermediate. In class 1 elements the transposition intermediate is formed from mRNA

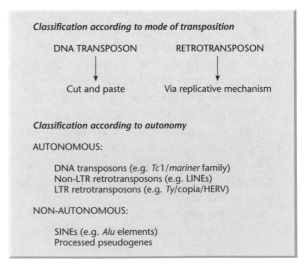

Fig. 16.17   Classification of transposable elements found in eukaryotes.

encoded by the element. Since these elements resemble retroviruses they are usually referred to as retrotransposons. In class 2 elements the transposition intermediate is the transposon DNA. Both classes of transposon can be subdivided into autonomous and non-autonomous elements. Autonomous elements encode the gene products required for transposition. Non-autonomous elements have no significant coding capacity but retain the DNA sequences necessary for transposition. Integration of transposons generally results in the duplication of a short genomic sequence at the site of insertion but the size and sequence of these duplications vary among the various transposon families.

Retrotransposons can be divided into two groups on the basis of transposition mechanism and structure

The LTR group of retrotransposons has long terminal repeats (LTRs) in direct orientation at both ends (Fig. 16.18). If they are autonomous they contain at least two genes: the *gag* gene encodes a capsid-like protein and the *pol* gene encodes a polyprotein with protease, reverse transcriptase, RNase H, and integrase activities. The non-autonomous

elements lack most or all of the *gag* and *pol* genes and can be variable in size. Examples of LTR retrotransposons are the *Ty1–Ty5* elements of yeast and the *copia* and *gypsy* elements originally found in *Drosophila*. The *Ty1/copia* and *Ty3/gypsy* families have been found in all animal, plant, and fungal species that have been examined but their organization and distribution show extensive variation. There is no relationship between the total copy number and the host genome size, and the copy number can vary widely between closely related species within a genus (Kumar 1996). Up to 7% of the human genome is composed of endogenous retroviruses (HERVs) that resemble LTR retrotransposons (Fig. 16.18). Most of these HERVs have accumulated nonsense mutations and no longer encode functional retroviral proteins (Smit 1999).

Non-LTR retrotransposons are divided into long interspersed nuclear elements (LINEs) and short interspersed nuclear elements (SINEs). LINEs encode three proteins: *ORF1*, a *gag*-like protein; an endonuclease; and reverse transcriptase. Both LINEs and SINEs terminate by a simple sequence repeat, usually poly(A). A well-studied group of SINEs is the *Alu* family that is found in Old World primates (Batzer & Deininger 2002). This family is named for an *Alu*I

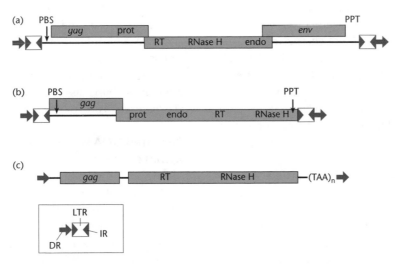

Fig. 16.18 Major types of retroelements. Overall organization of (a) a retrovirus, the avian leukosis virus; (b) a retrotransposon, the yeast *Ty1* element; and (c) a non-LTR (long terminal repeat) retrotransposon, the *Drosophila* I factor. Open reading frames are depicted by purple boxes. The *gag* gene encodes structural proteins of the virion core, including a nucleic-acid-binding protein; the *env* gene encodes a structural envelope protein, necessary for cell-to-cell movement; prot, protease involved in cleavage of primary translation products; RT, reverse transcriptase; RNase H, ribonuclease; endo, endonuclease necessary for integration in the host genome; LTR, long terminal repeats containing signals for initiation and termination of transcription, and bordered by short inverted repeats (IRs) typically terminating in 5′–TG . . . CA–3′; PBS, primer binding site, complementary to the 3′ end of a host tRNA, and used for synthesis of the first (−) DNA strand; PPT, polypurine tract used for synthesis of the second (+) DNA strand; DR, short direct repeats of the host target DNA, created upon insertion. (Reprinted from Grandbastien 1992 by permission of Elsevier Science.)

Fig. 16.19 Structure of a typical 282-bp *Alu* element flanked by direct repeats (purple arrows). The *Alu* element itself consists of an inexact duplication of two monomer units separated by a mid A-rich region. The right monomer contains an additional sequence of approximately 30 bp (dark green) that is absent from the left monomer. The consensus sequence usually is followed by an A-rich region resembling a poly(A) tail.

restriction endonuclease site typical of the sequence. *Alu* elements are 280-nucleotide sequences that lack introns (Fig. 16.19) and there are at least 1 million of them in the human genome. They are transcribed but not translated because they lack an open reading frame.

DNA transposons are simpler than retrotransposons

DNA transposons have terminal inverted repeats and target site duplications of conserved length. Autonomous DNA transposons have a single gene that encodes a transposase responsible for cutting and pasting the transposon to new locations. Non-autonomous DNA transposons usually are derived from autonomous transposons by deletion. The best-known DNA transposons are the *Tc1/mariner* super-family and the non-autonomous MITES (miniature inverted-repeat transposable elements).

Helitrons are a new group of class 2 transposons that have been found in plants, nematodes, flies, fish, and fungi (Kapitonov & Jurka 2001, Poulter *et al.* 2003). They replicate via a rolling-circle mechanism (see p. 193) and encode a replicase and a helicase. Some helitrons, called *helentrons*, also encode an endonuclease similar to that found in LINEs.

Transposon activity is highly variable across eukaryotes

Eukaryotic genomes contain substantially differing amounts of repetitive DNA (Table 16.4) because of differential propagation and deletion of transposons. The distribution of transposable elements in the human genome is quite different to that in the fruit fly, nematode, and *Arabidopsis* genomes. First, the euchromatic portion of the human genome has a much higher density of transposable element copies. Second, the human genome is filled with copies of ancient transposons whereas the transposons in other organisms, particularly the fruit fly, are of

more recent origin. This probably reflects the efficiency of housecleaning through genomic deletion. Thirdly, two repeat families (LINE1 and *Alu*) account for 60% of all interspersed repeat sequences in the human genome but the other genomes studied contain many different transposon families. Similarly DNA transposons represent only 6% of all interspersed repeats in humans but 25, 49, and 87% in the fruit fly, *Arabidopsis*, and the nematode, respectively.

Genome analyses have shown that even closely related lineages can experience radically different rates of transposition (Eichler & Sankoff 2003). For example, retrotransposition among great-ape species has slowed to a crawl when compared with that of Old World monkeys and differential rates of SINE and LINE retrotransposition are believed to be responsible for the increased genome size of primates compared with other mammals. The wide range of genome sizes found in cereals (420–16,000 Mb) is due to a high level of transposition activity with transposons inserting into transposons. So extensive is transposon amplification that counterbalancing deletion mechanisms (e.g. illegitimate recombination) are required to prevent genetic obesity.

Repeated DNA is non-randomly distributed within genomes

Replication of transposons depends on selecting a favorable chromosomal site for integration of their genomic DNA. Different retroelements meet this challenge by targeting distinctive chromosomal regions (Bushman 2003). Within mouse and human DNA, L1 LINEs preferentially associate with gene-poor AT-rich regions whereas *Alu* SINEs accumulate within GC-rich (gene-rich) areas. Similarly, among cereal genomes, LTR retrotransposons are found mostly in intergenic regions whereas MITEs are found within low-copy genic sequences. Among other eukaryotes where repeats constitute less than 10% of the genome, repeats accumulate within heterochromatic

Organism	Species	% repeat DNA in genome
Human	*Homo sapiens*	46
Mouse	*Mus musculus*	38
Mosquito	*Anopheles gambiae*	16
Mustard weed	*Arabidopsis thaliana*	14
Sea squirt	*Ciona intestinales*	10
Pufferfish	*Takifugu rubripes*	<10
Baker's yeast	*Saccharomyces cerevisiae*	2.4
Fruitfly	*Drosophila melanogaster*	2
Fission yeast	*Schizosaccharomyces pombe*	0.35
Nematode	*Caenorhabditis elegans*	<1

Table 16.4
Frequency of repeated DNA in different genomes.

regions and are found infrequently in euchromatin. Within an organism there can be tremendous variation in the distribution of repeat sequences. Some regions of the human genome are extraordinarily dense in repeats whereas others are nearly devoid of repeats. For example, a 525-kb region on chromosome Xp11 has an overall transposable element density of 89% and includes a 200 kb segment with a density of 98%. By contrast, the density in the four homeobox clusters is less than 2%.

Eukaryotic genomes are very plastic

Genomic DNA is often thought of as the stable template of heredity, largely dormant and unchanging, apart from the occasional point mutation. Nothing could be further from the truth. DNA is dynamic and is constantly subjected to rearrangements, insertions, and deletions as a result of the activity of transposable elements. For example, LINE insertions have been found to be responsible for certain cases of hemophilia, thalassemia, Duchenne muscular dystrophy, and chronic granulomatous disease. A particularly striking example is a LINE insertion in the *APC* gene of adenocarcinoma cells from a colon cancer patient, but not in the surrounding normal tissue (Miki *et al.* 1992). In addition to duplicating themselves, LINEs can carry with them genomic flanking sequences that are downstream from their 3'-untranslated regions and could have a role in exon shuffling (see Box 16.5). SINEs may be hot-spots for recombination and a number of examples have been identified in human genetic disorders (the reader interested in this topic should consult the review of Luning Prak & Kazazian 2000).

In plants there are a number of classic studies showing that transposons are responsible for reshaping genomes. For example, bursts of retrotransposon activity have resulted in a doubling of the maize genome (SanMiguel *et al.* 1998). Particularly interesting is the observation of Kalendar *et al.* (2000) that the copy number of an LTR transposon (*BARE-1*) varies almost threefold among nearby wild barley populations subjected to different levels of water stress.

Pseudogenes are derived from repeated DNA

Pseudogenes are sequences of genomic DNA with such similarity to normal genes that they are regarded as non-functional copies or close relatives of genes. They are formed in two ways. Classical *duplicated pseudogenes* are formed when genes that are tandemly duplicated accumulate mutations such that one of the genes becomes non-functional. These mutations may prevent transcription and/or translation. *Processed pseudogenes* are formed by the accumulation of mutations in a gene that has been retrotransposed to a new location. They are characterized by an absence of introns that are present in the parental gene indicating that cDNA synthesis was involved in their formation.

Segmental duplications are very large, low-copy-number repeats

Segmental duplications are 1–200 kb blocks of genomic DNA with >90% identity that are found at two or several locations in the genome. In mammalian DNA, the pericentromeric and subtelomeric regions of chromosomes are filled with segmental duplications but the amount varies between species. For example, segmental duplications in the mouse, rat, and in humans account for 1.5%, 3%, and

Box 16.5 Exon shuffling mediated by a LINE

Moran *et al.* (1999) were able to demonstrate in cultured cells the transduction of a gene by a long interspersed nuclear element (LINE). A neomycin (*neo*) reporter gene was placed downstream from a LINE sequence and both the promoter and the initiation codon of the *neo* gene were replaced with a 3′ splice site. The only way that this modified *neo* gene could be expressed after transduction was if it was inserted into an actively transcribed gene and spliced onto the transcript derived from that gene. Such events were readily detected and characterized as authentic LINE insertions in several genes.

The above result suggests a simple mechanism by which exons can be shuffled in the human genome (see Fig. B16.7). Transcription of a LINE element within a gene *X* fails to terminate at its own weak polyadenylation (poly(A)) signal but terminates instead at the poly(A) signal of the gene. The reverse transcriptase and endonuclease encoded by this transcript bind to the poly(A) tail and insert a cDNA copy into gene *Y*. This step results in the transduction of gene *X* into gene *Y* thereby creating a new gene.

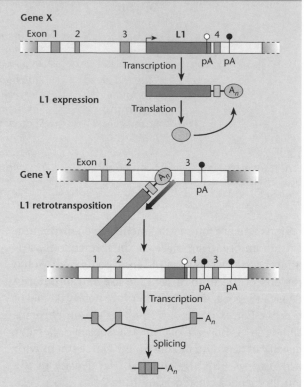

Fig. B16.7 Possible mechanism for LINE involvement in exon shuffling. L1, LINE element; pA, polyadenylation signal; A_n, poly(A) tail. Reverse transcriptase and endonuclease are shown in gray. (Reprinted from Eickbush 1999 by permission of the American Association for the Advancement of Science.)

5.5% respectively of total DNA. Segmental duplications are much less common in the *Saccharomyces*, nematode, and fruitfly genomes. By contrast, 24 segmental duplications account for 58% of the genome of *Arabidopsis* but only one of them is in the centromeric region.

Segmental duplications can be divided into two categories: interchromosomal and intrachromosomal duplications. The former are defined as segments that are duplicated among non-homologous chromosomes. For example, a 9.5 kb genomic segment from the human adrenoleukodystrophy locus from Xq28 has been duplicated to regions near the centromeres of chromosomes 2, 10, 16, and 22. Intrachromosomal duplications occur within a particular chromosome or chromosomal arm. This category includes several duplicated segments, also known as low copy repeat sequences, which mediate recurrent chromosomal structural rearrangements associated with genetic disease in humans. For example, on chromosome 17 there are three copies of a 200 kb repeat separated by 5 Mb of sequence and two copies of a 24 kb repeat separated by 1.5 Mb of sequence. These sequences are so similar that they can undergo recombination (for a review of this topic the reader should consult Eichler 2001).

The human Y chromosome has an unusual structure

The Y chromosome has attracted special interest from geneticists and evolutionary biologists because of its distinctive role in mammalian sex determination. It, and the other sex chromosome (X), originated a few hundred million years ago from the same ancestral autosome. The two then diverged in sequence such that today only relatively short regions at either end of the Y chromosome are homologous to the corresponding regions of the X chromosome. The remaining 95% of the modern-day Y chromosome is male-specific and is designated as MSY (male-specific region of Y). The MSY region is a mosaic of heterochromatic sequences and three classes of euchromatic sequences: X-transposed, X-degenerate, and ampliconic (Skaletsky *et al.* 2003).

About 15% of the MSY consists of X-transposed sequences and they are still 99% identical to

their X-chromosome counterparts. These sequences are dominated by a high proportion of dispersed repetitive sequences and contain only two genes. A further 20% of the MSY consists of X-degenerate sequences that are more distantly related to the X chromosome and has higher gene content than X-transposed sequences (Fig. 16.20). The remainder of MSY consists of a web of Y-specific repetitive sequences (amplicons) that make up a series of

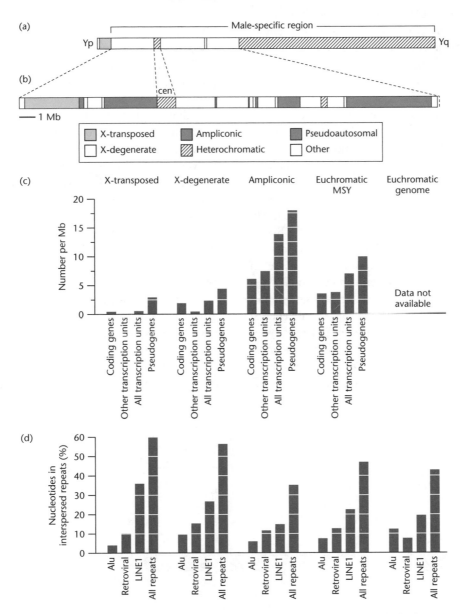

Fig. 16.20 The male-specific region of the Y chromosome. (a) Schematic representation of the whole chromosome, including the pseudoautosomal and heterochromatic regions. (b) Enlarged view of a 24-Mb portion of the MSY, extending from the proximal boundary of the Yp pseudoautosomal region to the proximal boundary of the large heterochromatic region of Yq. Shown are three classes of euchromatic sequences, as well as heterochromatic sequences. A 1-Mb bar indicates the scale of the diagram. (c), (d), Gene, pseudogene, and interspersed repeat content of three euchromatic sequence classes. (c) Densities (numbers per Mb) of coding genes, non-coding transcription units, total transcription units, and pseudogenes. (d) Percentages of nucleotides contained in *Alu*, retroviral, LINE1, and total interspersed repeats. Redrawn from Skaletsky *et al.*, with permission from *Nature*.

palindromes. These palindromes come in a range of sizes, up to 3 Mb in length, with the two arms of the palindrome having 99.9% identity. The ampliconic DNA has the highest gene content and a very high pseudogene content compared with the rest of the MSY (Fig. 16.20).

Centromeres are filled with tandem repeats and retroelements

During mitosis, the key chromosomal element responsible for directing operations is the centromere and its associated kinetochore complex. Most organisms have monocentric chromosomes. That is, the centromere is located at a single point on the chromosome. A few species, such as the sedge *Luzula* and the nematode *C. elegans*, have holocentric chromosomes in which the microtubules attach throughout the length of the chromosome.

In contrast to telomeres (see p. 334), centromeres are not specified by highly conserved sequences. In the budding yeast *S. cerevisiae* the centromere is precisely defined and only 125 bp is needed to mediate spindle attachment. By contrast, in the fission yeast *Schizosaccharomyces pombe* the centromere is 40–120 kb in size and in humans it has a size of 3 Mb. In *S. pombe*, the three centromeres consist of a core sequence flanked by inverted repeats, which in turn are flanked by tandem and dispersed repeats. Centromeres of *Drosophila* consist of different types of satellite DNA, containing embedded transposable elements along with islands of unique sequence DNA. A key component of mammalian centromeres

is alphoid DNA, which consists of a 171 bp motif repeated in a tandem head-to-tail fashion and then organized into higher order repeat arrays. In plants, *Arabidopsis* has a similar centromere structure to humans but the tandem repeats are 178 bp in length while cereal centromeres contain up to 200 copies of a *gypsy*-like retroelement called *cereba*.

Summary of structural elements of eukaryotic chromosomes

The different structural elements that have been discussed in the previous section are summarized in Fig. 16.21. A key feature of eukaryotic DNA is that as the genome size increases so too does the amount of repeated DNA, and hence the density of coding sequences must decrease. This has been confirmed by the data on gene density derived from the different DNA sequencing projects (Table 16.5). However, no generalizations can be made about the frequency of the different types of repeated DNA (tandem repeats, transposons, pseudogenes, segmental duplications) in different organisms or even within the chromosomes of the same organism. By contrast, as a general rule of thumb, the average number of introns per gene, and their size, increase as one moves up the evolutionary tree.

From a practical point of view, repeated sequences are a major source of problems. During genomic cloning experiments, recombination can occur between repeats and this can cause scrambling of DNA sequences. Even if no scrambling occurs, mistakes can occur during the synthesis phase of

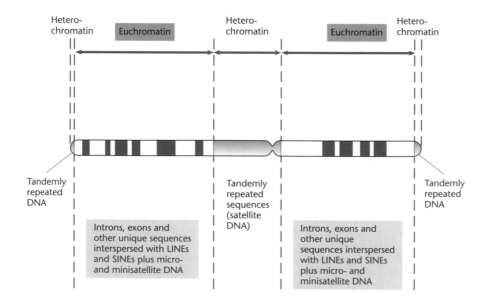

Fig. 16.21 The location of repeated sequences within a typical human chromosome.

Table 16.5 Average gene density in different organisms.

Organism	Gene density (genes/Mb)
Bacteria	800–1100
Saccharomyces cerevisiae	446
Plasmodium (malarial parasite)	221
Caenorhabditis elegans (nematode)	196
Arabidopsis thaliana	175
Fugu (puffer fish)	150
Ciona (sea squirt)	97
Drosophila	71
Human	20

DNA sequencing and the commonest is polymerase slippage on microsatellites. Finally, during data assembly repeats can lead to incorrect assembly of genomic fragments.

Suggested reading

Shapiro J.A. & van Steinberg R. (2005) Why repetitive DNA is essential to genome function. *Biological Reviews of the Cambridge Philosophical Society* **80**, 227–50.

Eichler E.E. & Sankoff D. (2003) Structural dynamics of eukaryotic chromosome evolution. *Science* **301**, 793–7.

These two papers present detailed analyses of the role of repeated DNA in the evolution of genomes.

Burger G., Gray M.W. & Lang B.F. (2003) Mitochondrial genomes: anything goes. *Trends in Genetics* **19**, 709–16.

This paper describes the amazing diversity of mitochondrial genomes that is found, particularly in the eukaryotic protists.

Feschotte C., Jiang N. & Wessler S.R. (2002) Plant transposable elements: where genetics meets genomics. *Nature Reviews Genetics* **3**, 329–41.

A refreshing look at plant transposon elements.

Jobling M.A. & Gill P. (2004) Encoded evidence: DNA in forensic analysis. *Nature Reviews Genetics* **5**, 739–51.

An excellent review of the principles and the issues of DNA profiling.

Kwok P.-Y. & Chen X. (2003) Detection of single nucleotide polymorphisms. *Current Issues in Molecular Biology* **5**, 43–60.

This review covers the infinite ways of detecting SNPs.

Sharpless N.E. & DePinho R.A. (2004) Telomeres, stem cells, senescence, and cancer. *Journal of Clinical Investigation* **113**, 160–8.

This review provides a detailed analysis of the significance of telomerase in cancer.

Skaletsky H., Kuroda-Kawaguchi T., Minx P.J., *et al.* (2003) The male-specific region of the Y chromosome is a mosaic of discrete sequence classes. *Nature* **423**, 825–37.

This paper provides the first clear insight to the structure of the Y chromosome and is destined to become a classic.

Stankiewicz P. & Lupski J.R. (2002) Genomic architecture, rearrangements and genomic disorders. *Trends in Genetics* **18**, 74–82.

Although focused on the role of structural rearrangements on genetic disease this review is well worth reading.

Useful websites

Each year, the January 1 issue of *Nucleic Acids Research* has short reviews of many of the different molecular biology and genomics databases and has links to the associated websites. These reviews can be accessed via http://www.nar.oupjournals.org/content/

CHAPTER 17

Mapping and sequencing genomes

Introduction

Genetics is the study of the inheritance of traits from one generation to another. As such, it examines the phenotypes of the offspring of sexual crosses. Useful as these data may be, they cannot provide an explanation for the biological basis for a phenotype for that requires biochemical information. In the first half of this book we described the different techniques of gene manipulation and gene analysis, and their application has greatly facilitated our understanding of the biology of different phenotypes. For example, cloned genes can be sequenced and this provides data on the amino acid sequence of the gene product. A search of databases can provide information on gene function and this function can be confirmed by overexpressing the gene product, purifying it, and then undertaking biochemical characterization. Finally, the gene can be subjected to site-directed mutagenesis and the phenotype of the variants determined by reintroducing them to their natural host cells.

Although the techniques of gene manipulation are exceedingly powerful they often are not sufficient to understand phenotypes. For example, before many genes can be cloned they need to be mapped close to a convenient marker. While this may be easy in an organism such as *Drosophila* where many mutants are available, it is much more difficult in humans or in organisms whose genetics have been poorly studied. What are required are methods to generate high-density chromosome maps quickly and easily. There is another reason why such high-density maps are required. Biologists who once were content to investigate simple biochemical pathways or phenomena now want to know how all the components of an organism are produced and interact and how they evolved. Also, they want to investigate the "big" themes in biology. For example, how did speech and memory evolve, what changes occurred as the primates evolved, etc. Investigating these issues requires knowledge of the entire genome sequence of the organisms of interest and whole-genome sequencing is greatly facilitated if there are high-density maps.

Why do high-density maps facilitate whole-genome sequencing? The explanation is simple. The largest stretch of DNA that can be sequenced by the Sanger method (p. 126) in a single pass is 600–800 nucleotides and many of the genomes to be sequenced are over 1 million times larger. Therefore, it is necessary to break each genome into much smaller pieces for sequencing and then join the pieces together again. However, joining the pieces together requires markers because two fragments can only be shown to overlap if they have unique sequences (i.e. no repeated sequences) in common. For most organisms of interest the density of genetic markers is woefully inadequate and this has led to the development of physical markers.

The first physical map of an organism made use of restriction fragment length polymorphisms (RFLPs)

Botstein *et al.* (1980) were the first to recognize that DNA probes could be used to identify polymorphic sequences and their initial target was restriction fragment length polymorphisms (RFLPs). These RFLPs are generated in various ways:

1 loss of an endonuclease cleavage site, most probably as a result of a base change (Fig. 17.1);
2 by insertion or deletion of a stretch of DNA;
3 by variation in the number of repeats of a microsatellite (p. 336 *et seq.*).

To generate an RFLP map the probes must be highly informative. This means that the locus must not only be polymorphic, it must be *very* polymorphic. If enough individuals are studied, any randomly selected probe will eventually discover a polymorphism. However, a polymorphism in which one allele exists in 99.9% of the population and the other in 0.1% is

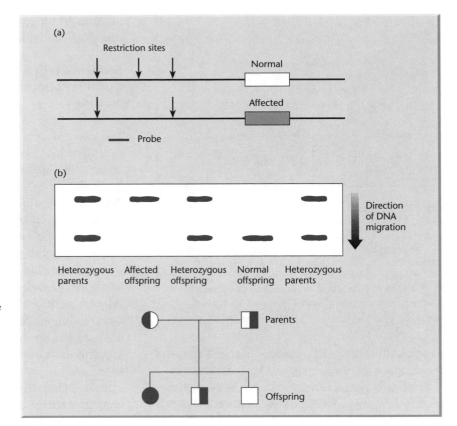

Fig. 17.1 Example of a RFLP and its use for gene mapping. (a) A polymorphic restriction site is present in the DNA close to the gene of interest. In the example shown, the polymorphic site is present in normal individuals but absent in affected individuals. (b) Use of the probe shown in Southern blotting experiments with DNA from parents and progeny for the detection of affected offspring.

of little utility because it seldom will be informative. Thus, as a general rule, the RFLPs used to construct the genetic map should have two, or perhaps three, alleles with equivalent frequencies.

The first RFLP map of an entire genome (Fig. 17.2) was that described for the human genome by Donis-Keller *et al.* (1987). They tested 1680 clones from a phage library of human genomic DNA to see whether they detected RFLPs by hybridization to Southern blots of DNA from five unrelated individuals. DNA from each individual was digested with 6–9 restriction enzymes. Over 500 probes were identified that detected variable banding patterns indicative of polymorphism. From this collection, a subset of 180 probes detecting the highest degree of polymorphism was selected for inheritance studies in 21 three-generation human families (Fig. 17.3). Additional probes were generated from chromosome-specific libraries such that ultimately 393 RFLPs were selected. The various loci were arranged into linkage groups representing the 23 human chromosomes by a combination of mathematical linkage analysis and physical location of selected clones. The latter was achieved by hybridizing probes to panels of rodent–human hybrid cells containing varying human chromosomal complements (see p. 359). RFLP maps have not been restricted to the human genome. For

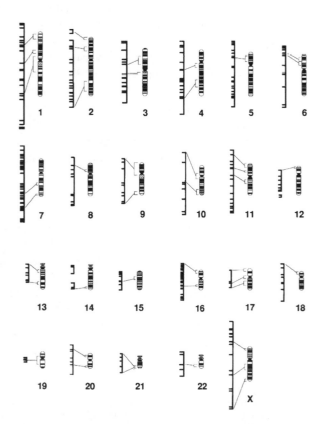

Fig. 17.2 The first RFLP genetic linkage map of the entire human genome. (Reproduced from Donis-Keller *et al.* 1987, with permission from Elsevier Science.)

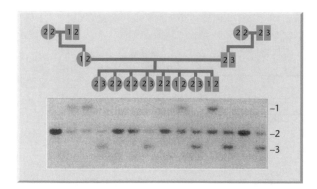

Fig. 17.3 Inheritance of a RFLP in three generations of a family. The RFLP probe detects a single locus on human chromosome 5. In the family shown, three alleles are detected by Southern blotting after digestion with *Taq*1. For each of the parents it can be inferred which allele was inherited from the grandmother and which from the grandfather. For each child the grandparental origin of the two alleles can then be inferred. (Redrawn from Donis-Keller *et al.* 1987, with permission from Elsevier Science.)

example, RFLP maps have been published for most of the major crops (see for example Moore *et al.* 1995).

The human genome map produced by Donis-Keller *et al.* (1987) was a landmark publication. However, it identified RFLP loci with an average spacing of 10 centimorgans (cM). That is, the loci had a 10% chance of recombining at meiosis. Given that the human genome is 4000 cM in length, the distance between the RFLPs is 10 Mb on average. This is too great to be of use for gene isolation. However, if the methodology of Donis-Keller *et al.* (1987) was used to construct a 1 cM map, then 100 times the effort would be required! This is because 10 times as many probes would be required and 10 times more families studied. The solution has been to use

more informative polymorphic markers and other mapping techniques.

Sequence tags are more convenient markers than RFLPs because they do not use Southern blotting

The concept of sequence-tagged sites (STSs) was developed by Olson *et al.* (1986) in an attempt to systematize landmarking of the human genome. Basically, an STS is a short region of DNA about 200–300 bases long whose exact sequence is found nowhere else in the genome. Two or more clones containing the same STS must overlap and the overlap must include the STS.

Any clone that can be sequenced may be used as an STS provided it contains a unique sequence. A better method to develop STS markers is to create a chromosome-specific library in phage M13. Random M13 clones are selected and 200–400 bases sequenced. The sequence data generated are compared with all known repeated sequences to help identify regions likely to be unique. Two PCR primer sequences are selected from the unique regions which are separated by 100–300 bp and whose melting temperatures are similar (Fig. 17.4). Once identified, the primers are synthesized and used to PCR amplify genomic DNA from the target organism and the amplification products analyzed by agarose gel electrophoresis. A functional STS marker will amplify a single target region of the genome and produce a single band on an electrophoretic gel at a position corresponding to the size of the target region (Fig. 17.5). Alternatively, an STS marker can be used as a hybridization probe.

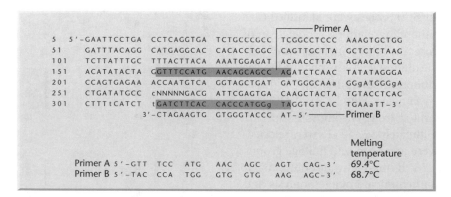

Fig. 17.4 Example of a sequence-tagged site (STS). The STS developed from the sequence shown is 171 bases long. It starts at base 162 and runs through base 332. Primer A is 21 bases long and lies on the sequenced strand. Primer B is also 21 bases long and is complementary to the shaded sequence towards the 3′ end of the sequenced strand. Note that the melting temperatures of the two primers are almost equal. (Reproduced with permission from Dogget 1992, courtesy of University Science Books.)

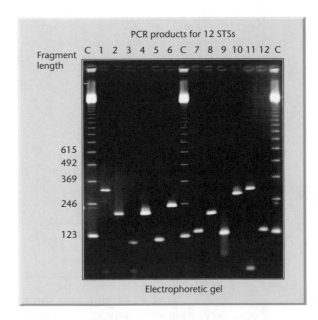

Fig. 17.5 Confirmation that an STS is a unique sequence on the genome. Note that the 12 STSs from chromosome 16 shown above appear as single bands after amplification and hybridization to a chromosome 16 genomic library. (Reproduced with permission from Dogget 1992, courtesy of University Science Books.)

In organisms with large amounts of repetitive DNA the generation of an appropriate sequence, and confirmation that it is an STS, can be time consuming. Adams *et al.* (1991) suggested an alternative approach. The principle of their method is based on the observation that spliced mRNA contains sequences that are largely free of repetitive DNA. Thus partial cDNA sequences, termed expressed sequence tags (ESTs), can serve the same purpose as the random genomic STSs but have the added advantage of pointing directly to an expressed gene. In a test of this concept, partial DNA sequencing was conducted on 600 randomly selected human cDNA clones to generate ESTs. Of the sequences generated, 337 represented new genes, including 48 with similarity to genes from other organisms, and 36 matched previously sequenced human nuclear genes. Forty-six ESTs were mapped to chromosomes.

In practice, there are a number of operational considerations associated with the use of ESTs. First, they need to be very short to ensure that the two ends of the sequence are contiguous in the genome, i.e. are not separated by an intron. Secondly, large genes may be represented by multiple ESTs which may correspond to different portions of a transcript or various alternatively spliced transcripts. For example, one of the major databases holds over 1300 different EST sequences for a single gene product, serum albumin. While this may or may not be a problem in constructing a physical map, it is problematical in the construction of a genetic map.

If it is desirable to select a single representative sequence from each unique gene, then this is accomplished by focusing on 3′ untranslated regions (3′ UTRs) of mRNAs. This can be achieved using oligo(dT) primers if the mRNA has a poly(A) tail. Two advantages of using the 3′ UTRs are that they rarely contain introns and they usually display less sequence conservation than do coding regions (Makalowski *et al.* 1996). The former feature leads to PCR product sizes that are small enough to amplify. The latter feature makes it easier to discriminate among gene family members that are very similar in their coding regions.

Single nucleotide polymorphisms (SNPs) are the most favored physical marker

Single nucleotide polymorphisms (SNPs, pronounced "snips") are single base-pair positions in genomic DNA at which different sequence alternatives (alleles) exist in a population (Fig. 17.6). In highly outbred populations, such as humans, polymorphisms are considered to be SNPs only if the least abundant allele has a frequency of 1% or more. This is to distinguish SNPs from very rare mutations. In practice, the term SNP is typically used more loosely than required by the above definition. For example, single base variants in cDNAs (cSNPs) are usually classed as SNPs because most of these will reflect underlying genomic DNA variants although they could result from RNA editing. Single base-pair insertions or deletions (indels) also are considered to be SNPs by some workers. A special subset of SNPs is one where the base change alters the sensitivity of a sequence to cleavage by a restriction endonuclease. These are known as restriction fragment length polymorphisms (RFLPs, see p. 347) or "snip-SNPs". Another important subset is one in which the two alleles can be distinguished by the presence or absence of a particular phenotype. Good examples here are emphysema in humans, caused by a C > T change in the gene for α_1-antitrypsin, and sickle cell anemia. In the latter case, an A > T change results in the replacement of a glutamine residue in β-globin by a valine residue but also destroys an *Mst*II site thereby generating an RFLP.

SNPs probably are the most important sequence markers for physical mapping of genomes. The reason for this is that they have the potential to

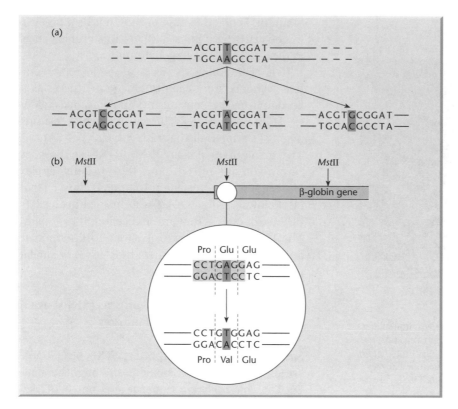

Fig. 17.6 Examples of single nucleotide polymorphisms. (a) Three possible SNP variants of a T/A base pair. (b) A single base change in the β-globin gene that destroys a restriction site for endonuclease *Mst*II (CC/TNAGG) thereby generating a restriction fragment length polymorphism (RFLP). This base change results in amino acid residue 6 being changed from glutamate to valine and is the cause of sickle cell disease.

provide the greatest density of markers. For example, their frequency is estimated to be 1 per 1000 base pairs which, in the case of the human genome, means 3 million in total. Originally the Human SNP Consortium planned to map 300,000 of them (Masood 1999) but the latest version of the SNP database (dbSNP) contains 10 million (www.ncbi.nlm.nih.gov/SNP/).

Given their importance, many different methods for detecting SNPs have been developed (Table 17.1; for reviews see Kristensen *et al.* 2001, Twyman 2004, Kwok & Chen 2003). SNP detection can be

Table 17.1 Different methods for detecting SNPs.

Approach	Examples	Reference
Enzymatic	Restriction fragment length polymorphisms (RFLP)	See text (p. 347)
	Amplified fragment length polymorphisms (AFLP)	See text (p. 352)
	Cleavase fragment length polymorphisms (CFLP)	Sander *et al.* (1999)
	Ligation-based assays	Favis *et al.* (2000)
	Randomly amplified polymorphic DNA (RAPD)	See text (p. 351)
	Direct termination PCR (DT-PCR)	Chen & Hebert (1998)
Electrophoretic	Single-strand conformation polymorphism (SSCP)	Nataraj *et al.* (1999)
	Heteroduplex analysis (CGGE, TGGE, DGGE)	See text (p. 351)
	DNA sequencing	See chapter 8
Solid phase	Oligonucleotide arrays	See text (p. 136)
Chromatographic	Denaturing high performance liquid chromatography (DHPLC)	See text (p. 351)
Physical	Differential sequencing with mass spectrometry	Graber *et al.* (1999)
	Fluorescence exchange-based methods	Marras *et al.* (1999)
In silico	Examining EST data	See text (p. 349)

divided into two different activities: scanning DNA sequences for previously unknown polymorphisms to facilitate mapping, and screening (genotyping) of individuals for known polymorphisms (see p. 355). Scanning for new SNPs can be further divided into the global or random approach and the regional or targeted approach. The latter relies almost totally on direct sequencing or on denaturing high performance liquid chromatography (dHPLC).

The technique of dHPLC takes advantage of the fact that denaturation of ds DNA is highly dependent on its sequence. Even a single nucleotide difference between two DNA molecules can alter their melting characteristics sufficiently for them to be distinguished. Several techniques for the detection of mutations are based on the differences in mobility of partially denatured molecules and include constant gradient gel electrophoresis (CGGE), denaturing gradient gel electrophoresis (DGGE), and temperature gradient gel electrophoresis (TGGE). Instead of electrophoresis as a detection step, the partially heat denatured DNA can be analyzed using HPLC matrices (Arnold *et al.* 1999, Cho *et al.* 1999). The advantage of dHPLC is that analysis is rapid (~5 minutes per sample) and can be automated through the use of an autosampler.

Polymorphic DNA can be detected in the absence of sequence information

Polymorphic DNA can be detected by amplification in the absence of the target DNA sequence information used to generate STSs. Williams *et al.* (1990) have described a simple process, distinct from the standard PCR, which is based on the amplification of genomic DNA with *single* primers of arbitrary nucleotide sequence. The nucleotide sequence of each primer was chosen within the constraints that the primer was nine or 10 nucleotides in length, between 50 and 80% G + C in composition and contained no palindromes. Not all the sequences amplified in this way are polymorphic but those that are (randomly amplified polymorphic DNA, RAPD) are easily identified. RAPDs are widely used by plant molecular biologists (Reiter *et al.* 1992, Tingey & Del Tufo 1993) to construct maps because they provide very large numbers of markers and are very easy to detect by agarose gel electrophoresis. However, they have two disadvantages. First, the amplification of a specific sequence is sensitive to PCR conditions, including template concentration, and hence it can be difficult to correlate results

obtained by different research groups. For this reason, RAPDs may be converted to STSs after isolation (Kurata *et al.* 1994). A second limitation of the RAPD method is that usually it cannot distinguish heterozygotes from one of the two homozygous genotypes. Nevertheless, Postlethwait *et al.* (1994) have used RAPDs to develop a genetic linkage map of the zebrafish (*Danio rerio*).

A different method for detecting polymorphisms, which is not subject to the problems exhibited by RAPDs, has been described by Konieczny and Ausubel (1993). In this method, STSs are derived from genes that have already been mapped and sequenced. Where possible the primers used are chosen such that the PCR products include introns to maximize the possibility of finding polymorphisms. The primary PCR products are subjected to digestion with a panel of restriction endonucleases until a polymorphism is detected. Such markers are called cleaved amplified polymorphic sequences (CAPS). The way in which CAPS are detected is shown in Fig. 17.7. Note that whereas RFLPs are well suited to mapping newly cloned DNA sequences, they are not convenient to use for mapping genes, such as plant genes, which

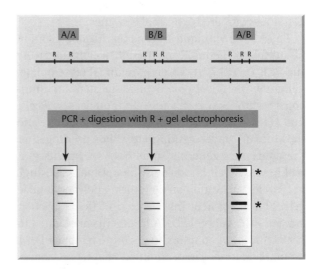

Fig. 17.7 Generation and visualization of CAPS markers. Unique-sequence primers are used to amplify a mapped DNA sequence for two different homozygous strains (A/A and B/B) and from the heterozygote A/B. The amplified fragments from strains A/A and B/B contain two and three recognition sites, respectively, for endonuclease R. In the case of the heterozygote A/B, two different PCR products will be obtained, one of which is cleaved twice and the other three times. After fractionation by agarose gel electrophoresis the PCR products from the three strains give readily distinguishable patterns. The asterisks indicate bands that will appear as doublets. (Redrawn with permission from Konieczny & Ausubel 1993.)

are first identified by mutation. CAPS are much more useful in this respect.

AFLPs resemble RFLPs and can be detected in the absence of sequence information

Amplified fragment length polymorphism (AFLP) is a diagnostic fingerprinting technique that detects genomic restriction fragments and in that respect resembles the RFLP technique (Voss *et al.* 1995). The major difference is that PCR amplification rather than Southern blotting is used for the detection of restriction fragments. The resemblance to the RFLP technique was the basis for choosing the name AFLP. However, the name AFLP should not be used as an acronym because the technique detects *presence* or *absence* of restriction fragments and *not* length differences. The AFLP approach is particularly powerful because it requires no previous sequence characterization of the target genome. For this reason it has been widely adopted by plant geneticists. It also has been used with bacterial and viral genomes (Voss *et al.* 1995). It has not proved useful in mapping animal genomes because it is dependent on the presence of high rates of substitutional variation in the DNA; RFLPs are much more common in plant genomes compared to animal genomes.

The AFLP technique is based on the amplification of subsets of genomic restriction fragments using the PCR (Fig. 17.8). To prepare an AFLP template, genomic DNA is isolated and digested simultaneously with two restriction endonucleases, *Eco*RI and *Mse*I. The former has a 6-bp recognition site and the latter a 4-bp recognition site. When used together these enzymes generate small DNA fragments that will amplify well and are in the optimal size range (<1 kb) for separation on denaturing polyacrylamide gels. Following heat inactivation of the restriction enzymes the genomic DNA fragments are ligated to *Eco*RI and *Mse*I adaptors to generate template DNA for amplification. These common adapter sequences flanking variable genomic DNA sequences serve as primer binding sites on the restriction fragments. Using this strategy it is possible to amplify many DNA fragments without having prior sequence knowledge.

The PCR is performed in two consecutive reactions. In the first pre-amplification reaction, genomic fragments are amplified with AFLP primers each having one selective nucleotide (see Fig. 17.8). The PCR products of the pre-amplification reaction are diluted and used as a template for the selective amplification

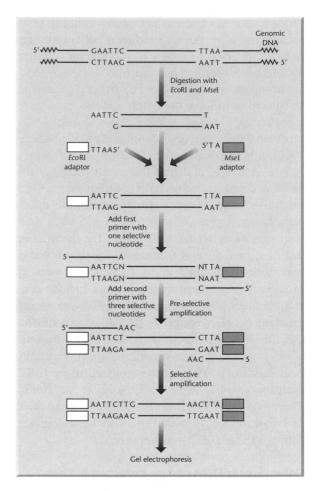

Fig. 17.8   Principle of the amplified fragment length polymorphism (AFLP) method (see text for details).

using two new AFLP primers which have two or three selective nucleotides. In addition, the *Eco*RI selective primer is radiolabeled. After the selective amplification the PCR products are separated on a gel and the resulting DNA fingerprint detected by autoradiography.

The AFLP technique will generate fingerprints of any DNA regardless of the origin or complexity. The number of amplified fragments is controlled by the cleavage frequency of the rare cutter enzyme and the number of selective bases. In addition, the number of amplified bands may be controlled by the nature of the selective bases. Selective extension with rare di- or trinucleotides will result in a reduction of the number of amplified fragments.

The AFLP technique is not simply a fingerprinting technique; rather, it is an enabling technology that can bridge the gap between genetic and physical maps. Most AFLP fragments correspond to unique positions on the genome and hence can be exploited as landmarks. In higher plants AFLPs may be the most effective way of generating high-density maps. The AFLP markers also can be used to detect corresponding genomic clones. Finally, the technique can be used for fingerprinting cloned DNA segments. By using no or few selective nucleotides, restriction fragment fingerprints will be produced which subsequently can be used to line up individual clones and make contigs.

Physical markers can be placed on a cytogenetic map using *in situ* hybridization

There are a number of different kinds of genome maps, e.g. cytogenetic, linkage, physical, etc. The classic cytogenetic map gives visual reality to other maps and to the chromosome itself. Because it does not rely on the cloning of DNA fragments it avoids the pitfalls that this procedure can introduce, particularly with YACs (see p. 213). Genetic linkage mapping allows the localization of inherited markers relative to each other. As with cytogenetic maps, linkage maps examine chromosomes as they are in cells. Although the methodology used to construct cytogenetic and linkage maps can lead to errors, they nevertheless are used as gold standards against which the physical maps are judged. The importance of *in situ* hybridization is that it enables this comparison to be made. Providing hybridization of repeated sequences is suppressed and provided no DNA chimeras are present, a cloned fragment or restriction fragment should anneal to a single location on the cytogenetic map. Furthermore, the physical map order should match that found by *in situ* hybridization. Where genetic markers have been located on the cytogenetic map by *in situ* hybridization they also can be positioned on the physical map.

Originally, *in situ* hybridization of unique sequences utilized radioactively labeled probes and it was a technique which required a great deal of technical dexterity. Today, the methods used are all derivatives of the fluorescence *in situ* hybridization (FISH) method developed by Pinkel *et al.* (1986). In this technique, the DNA probe is labeled by addition of a reporter molecule. The probe is hybridized to a preparation of metaphase chromosomes which has been air-dried on a microscope slide and in which the DNA has been denatured with formamide. Following hybridization, and washing to remove excess probe, the chromosome preparation is incubated in a solution containing a fluorescently labeled affinity molecule which binds to the reporter on the hybri-

dized probe. The preparation is then examined with a fluorescence microscope. If large DNA probes are used, they will contain many repetitive sequences which will bind indiscriminately to the target. This non-specific binding can be eliminated by competitive suppression hybridization. Before the main hybridization the probe is mixed with an aqueous solution of unlabeled total genomic DNA. This saturates the repetitive elements in the probe so that they no longer interfere with the *in situ* hybridization of the unique sequences.

Conventional FISH has a resolving power of ~1 Mb. If a higher resolution is necessary then less condensed chromosomes need to be used as the target. Highly elongated metaphase chromosomes have been prepared by mechanically stretching them by cytocentrifugation. This results in chromosomes that are 5–20 times their normal length. Laan *et al.* (1995) have shown that these stretched chromosomes are excellent for fast and reliable ordering of clones that are separated by at least 200 kb. They also can be used to establish the centromere–telomere orientation of a clone. The disadvantages of the method are that it cannot be used to generate reliable measurements of the distances between signals nor can it be used to localize unknown sequences on a chromosome. This is because the stretching of individual chromosomes is highly variable.

In interphase nuclei the chromatin is less condensed than in metaphase chromosomes and hence provides a good target for high resolution FISH (Trask *et al.* 1989, Yokota *et al.* 1995). The resolution of FISH can be further improved by loosening the organization of the interphase chromatin using concentrated salt, alkali, or detergent treatment of the cell preparations (Parra & Windle 1993). These techniques, which are commonly referred to as fiber-FISH, provide a resolution that permits the detection of a probe to a single fiber. Theoretically, the resolution of fiber-FISH is the same as the resolving power of the light microscope, i.e. 0.34 mm. This is equivalent to 1 kb and has been achieved by Florijn *et al.* (1995).

Padlock probes allow different alleles to be examined simultaneously

Another variation on multicolor FISH is the use of padlock probes (Nilsson *et al.* 1997). Padlock probes are oligonucleotides that can be ligated to form a circle if they bind to a sequence of exact complementarity. The lateral arms of the oligonucleotide

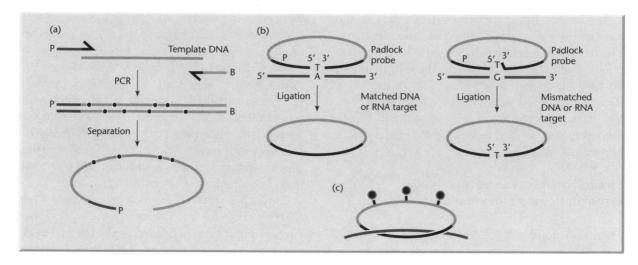

Fig. 17.9 The use of padlock probes to detect SNPs. (a) Synthesis of padlock probes by PCR. The 5′ ends of the PCR primers (dark purple and light purple) define the two target complementary sequences of the padlock probe to be constructed. One primer has a 5′ phosphate (P) to permit ligation whereas the other has a 5′ biotin (B) which is used to remove this strand by capture on a solid support. Black dots denote labeled nucleotides incorporated during PCR. (b) Hybridization of a padlock probe (black and gray) to a target DNA sequence (purple). The padlock probe can be converted to a circle by a ligase (left) but ligation is inhibited if the ends of the probe are mismatched to their target and the probe remains linear (right). (c) Direct detection of a reacted probe catenated to a target sequence. (Reprinted from Baner *et al.* 2001 by permission of Elsevier Science.)

twist around the DNA target forming a double helix and their termini are designed to juxtapose so that they may be ligated enzymatically. Nilsson *et al.* (1997) used two different oligonucleotide probes, each corresponding to a different sequence variant of a centromeric alpha-satellite repeat, and differing by only a single base pair. The closure of the two alternative padlocked probes occurred only when there was perfect sequence recognition (Fig. 17.9). By labeling the two probes with different fluorescent dyes it was possible to monitor the two sequence variants simultaneously. Antson *et al.* (2000) have extended the use of padlock probes to the detection of SNPs in single-copy sequences although methods for signal amplification were required, and more recently the method has been adapted for high-throughput genotyping (see Box 17.1).

Physical mapping is limited by the cloning process

The starting point for the construction of a physical map is the fragmentation of genomic DNA followed by cloning of all the resulting fragments. Any fragments that share a unique marker, e.g. an STS, must overlap and can be joined to form a contig. Although this map-building process is conceptually simple it is not without its problems. For example, regions recalcitrant to cloning lead to uncloseable gaps. Distortions also are created when cloned DNA under-

goes rearrangement or deletion or when unlinked fragments get co-ligated during the cloning process. As noted earlier, such distortions can be detected by *in situ* hybridization but it would be more satisfactory if they could be avoided altogether. The only way that this can be achieved is to avoid the cloning process by mapping DNA directly. Three methods of doing this have been developed: optical mapping, radiation hybrid mapping, and HAPPY mapping.

Optical mapping is undertaken on single DNA molecules

Optical mapping involves the imaging of single DNA molecules during restriction enzyme digestion. In the original method described by Schwartz *et al.* (1993), fluid flow is used to stretch out fluorescently stained DNA molecules dissolved in molten agarose along with a restriction enzyme. When the agarose gels the molecules are fixed in place and cutting of the DNA is triggered by the diffusion of magnesium ions into the gelled mixture. Fluorescence microscopy is used to visualize the cleavage sites, which appear as gaps between bright condensed pools of DNA on the fragment end flanking the cut site.

Since the original description of optical mapping many different methodological improvements have been introduced (Aston *et al.* 1999), two of which deserve mention here. First, in the original method

Box 17.1 High-throughput genotyping using padlock probes

The major genetic difference between the individuals in an outbreeding population is SNPs. While the majority of the SNPs are of no biological consequence, a fraction of the substitutions have functional significance and are the basis for the diversity observed. For example, in humans there are about 3.2 million SNP differences between any two individuals of which a small proportion will be responsible for the variation in physical features, disease susceptibility, drug response etc. Not susrprisingly, drug companies and the medical profession are very interested in being able to determine the SNP profile of each individual that they treat so that they can optimize drug selection (personalized medicine). This in turn demands the

availability of methods for large-scale genotyping.

Most of the methods used for SNP detection involve the use of the PCR and each PCR reaction requires the use of two primers. If one tries to analyze 1000 or 10,000 SNP loci simultaneously then this will involve the use of the same number of primer pairs. Unfortunately, once more than a few primer pairs are involved, cross-reactions become an unmanageable problem. To circumvent this problem, Hardenbol *et al.* (2003) developed a method based on padlock probes that requires only a single pair of primers for PCR amplification.

The probes used to detect each SNP consist of seven segments (Fig. B17.1a). At

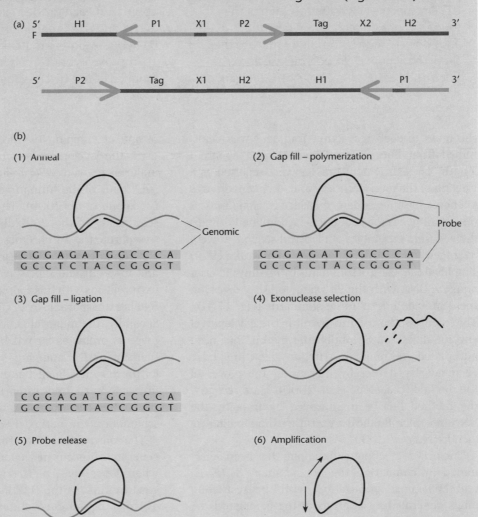

Fig. B17.1 The use of molecular inversion probes for high-throughput genotyping. Redrawn from Hardenbohl *et al.*, with permission from *Nature*.

continued

Box 17.1 *continued*

the termini of the probes there are two regions of homology (H1, H2) that are unique for each SNP and these are identified by a unique detection tag sequence. Each probe also has two common primer binding sites (P1, P2) and two common cleavage sites (X1, X2). The way in which these probes are used is shown in Fig. B17.1b. A mixture of genomic DNA, 1000 or more probes and thermostable ligase and DNA polymerase is prepared in quadruplicate and incubated at a temperature that favors annealing. The sequences homologous to each SNP hybridize to their complementary sequences in the genome, creating a circular conformation with a single nucleotide gap between the termini of each probe. Unlabeled dATP, dCTP, dGTP, and dTTP respectively is added to each of the four reactions. In reactions where the added nucleotide is complementary to the single-base gap, DNA polymerase adds the nucleotide and DNA ligase closes the gap.

This results in a covalently closed circular molecule that encircles the genomic strand (i.e. a padlock) to which it is hybridized. Exonuclease is added and this removes any probe that has not been circularized.

Before the padlock probes can be amplified they need to be released from the genomic DNA. This is done by adding the enzyme uracil-*N*-glycosylase, which depurinates the uracil residues incorporated in the cleavage site X1. Upon heating, strand scission occurs at the abasic site and this releases the probe from the genomic DNA. In the process, the probe has been molecularly inverted and the primer binding sites now are at the ends of the molecule thereby enabling PCR amplification. After the amplification step, the presence or absence of each probe in the dATP, dCTP, dGTP, and dTTP reaction mixes is determined by hybridization to matching microarrays. To improve hybridization, the probes are reduced in size by cleavage at X2.

the mass of each restriction fragment was determined from fluorescence intensity and apparent length. These fragment masses were reported as a fraction of the total clone size and later converted to kilobases by independent measure of clone masses (i.e. cloning vector sequence). Additionally, maps derived from ensembles of identical sequences were averaged to construct final maps. Lai *et al.* (1999) simplified the sizing of fragments by mixing bacteriophage λ DNA with the test sample. After digestion these provide internal size standards (Fig. 17.10). They can also be used to monitor the efficiency of enzyme digestion. Secondly, the method has been automated and map-construction algorithms have been developed thereby increasing the power of the method (Giacolone *et al.* 2000). More recently, the method has been improved by trapping the DNA in a microfluidic device rather than in agarose (Jendrejeck *et al.* 2003).

The accuracy of optical mapping has been determined by comparing the optical maps of *E. coli* and *Deinococcus radiodurans* with the restriction maps generated by analysis of the complete DNA sequences of these genomes. In both cases the error rate was less than 1% (Lin *et al.* 1999a). The power

of optical mapping has been demonstrated by its use in the development of maps for the genome of the malarial parasite (*Plasmodium falciparum*) and the *DAZ* locus of the human genome (Lai *et al.* 1999, Giacolone *et al.* 2000). The original plan for the sequencing of the 24.6 Mb malarial genome was to separate the 14 chromosomes by pulsed-field gel electrophoresis (PFGE) (see p. 17) and then construct and map chromosome-specific libraries. There are two problems with this approach. First, chromosomes 5–9 are unseparable by PFGE and migrate as a blob. Secondly, the malarial genome is AT-rich and this presents problems for reliable library construction. By using optical mapping, ordered restriction maps for the enzymes *Bam*HI and *Nhe*I were derived from unfractionated *P. falciparum* DNA and assembled into 14 contigs corresponding to the chromosomes. A schematic of the method used is shown in Fig. 17.10.

The construction of accurate physical maps and the generation of accurate sequence data are very difficult when clones have a high content of repetitive DNA. A good example is the *DAZ* locus which maps to the q arm of the human Y chromosome. Prior to optical mapping, our knowledge of the structure of the *DAZ* locus was sketchy but by optical mapping of 16

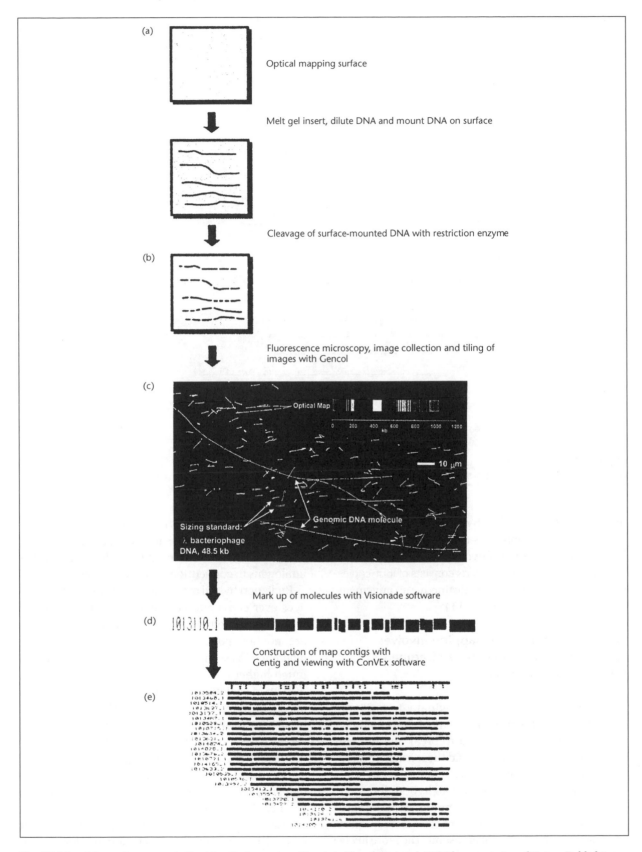

Fig. 17.10 Schematic representation of optical mapping. (Reprinted from Z. Lai *et al.* 1999 by permission of Nature Publishing Group, New York.)

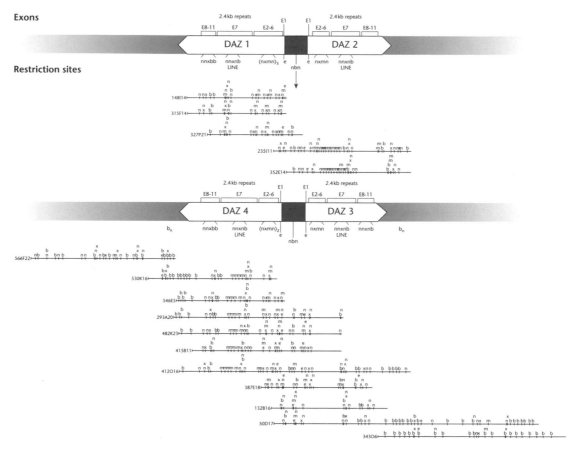

Fig. 17.11 Map of the *DAZ* region of the human Y chromosome generated by optical mapping. Sites for restriction enzymes are represented as follows: x, *Xho*I; b, *Bam*HI; n, *Nhe*I; e, *Eag*I; m, *Mlu*I. (Redrawn with permission from Giacolone *et al.* 2000.)

different BAC clones that hybridize to the locus the structure became apparent (Giacolone *et al.* 2000). Rather than consisting of two copies of the *DAZ* gene, as originally thought, the locus consists of four copies of the gene, each with its own distinctive arrangement of repetitive elements (Fig. 17.11).

Radiation hybrid (RH) mapping involves screening of randomly broken fragments of DNA for specific markers

RH mapping was developed originally to facilitate the mapping of the human genome. In this method, a high dose of X-rays is used to break the human chromosome into fragments and these fragments are recovered in rodent cells using somatic cell hybrids (see Box 17.2). The rodent–human hybrid clones are isolated and examined for the presence or absence of specific human DNA markers. The farther apart two markers are on the chromosome, the more likely a given dose of radiation will break the chromosome between them, placing the markers on two separate

chromosomal fragments. By estimating the frequency of breakage, and thus the distance, between markers it is possible to determine their order in a manner analogous to conventional meiotic mapping.

Radiation hybrid mapping has a number of advantages over conventional genetic (meiotic) mapping. First, chromosome breakage is random and there are no hot-spots, interference, or gender-specific differences as seen with recombination. Secondly, a much higher resolution can be achieved, e.g. 100–500 kb in radiation mapping as opposed to 1–3 Mb in genetic mapping, and the resolution can be varied by varying the radiation dose. Finally, it is not necessary to use polymorphic markers; monomorphic markers such as STSs can be used as well.

The possible outcomes in a radiation hybrid mapping experiment are shown schematically in Fig. 17.12. In this figure, *P* is the *probability of retention*, i.e. the probability that a DNA segment is present in an RH clone or the proportion of clones containing a specific DNA segment. The value of *P* is a function of the radiation dose and the cell lines used

Box 17.2 Somatic cell hybrids

Somatic cell hybrids are cell lines that contain a full complement of chromosomes of one species but only one or a limited number of chromosomes from the second. They are formed by fusing the cells of the two different species by applying conditions that select against the two donor cells. One of the parents may be sensitive to a particular drug and the other might be a mutant requiring special conditions for growth, e.g. thymidine kinase negative cells do not grow in hypoxanthine, aminopterin and thymidine (HAT) medium. In the presence of the drug and HAT medium, only hybrid cells containing a functional thymidine kinase gene can grow. If hybrid cells are grown under non-selective conditions after the initial selection, chromosomes from one of the parents tend to be lost more or less at random. In the case of human–rodent fusions, which are the most common, the human chromosomes are preferentially lost. Eventually only one or a few chromosomes from one parent remain. In this way rodent cell lines containing one or two human chromosomes have been constructed and these can be used to isolate the individual human chromosomes by FACS.

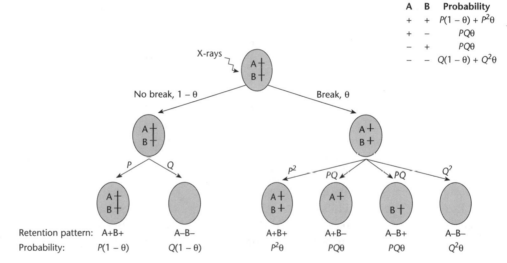

Fig. 17.12
Probability of different outcomes in a radiation hybrid experiment.

and usually ranges from 10–50% with maximum mapping power occurring when $P = 50\%$. It is assumed that P is constant over small regions or the entire chromosome and that segments are lost or retained independently of each other.

The *breakage probability* (θ) is the probability that two markers are separated by one or more breaks. The closer the linkage between two markers, the closer to zero is the value of θ whereas for unlinked markers the value is 1. Note that breakage probability is a function of the radiation dose and the cell lines used.

In radiation hybrid mapping distances are expressed in Rays or centiRays, where 1 cR equals a 1% frequency of breakage. Distances are calculated from the expression:

$$\text{Map distance (D)} = -\log_e(1 - \theta)\text{Rays}$$

and assume that the number of breaks is determined by a Poisson distribution with mean D. This is analogous to Haldane's map function for genetic mapping. In the above equation, if θ is 0.1 then the distance is 11 cR and if it is 0.3 the distance is 36 cR. Distances determined by radiation hybrid mapping can be converted to physical distances if the physical distance between any two markers is known. For example, the q arm of human chromosome 11 is 86 Mb and 1618 cR. Therefore, 1 cR is equal to 53 kb. However, the mapping distance is dependent on the radiation dose applied as shown in Table 17.2.

Although radiation hybrid mapping was originally developed for the human genome it has been

Table 17.2 Variation in mapping distance with radiation dose.

Radiation dose (rad)	Physical distance of 1 cR (kb)
3,000	300
9,000	50–70
10,000	25
50,000	4

extended to a number of other vertebrate genomes including the mouse, rat, dog, pig, cow, horse, baboon, and zebra fish (for references see Geisler *et al.* 1999). Once a radiation hybrid panel is available, one can map any physical marker (STS, EST, simple sequence length polymorphism (SSLP), etc.) by looking for linkage to existing markers on the map.

HAPPY mapping is a more versatile variation on RH mapping

A number of operational issues have restricted the utility of RH mapping. First, it takes considerable effort to generate an RH panel, even for mammalian genomes, and complications arise because the biological activity of donor fragments in the host cell lines biases their segregation. Second, the presence of a host genome in the hybrids precludes the use of generic markers such as AFLPs and RAPDs. Finally, the technique is not applicable to plants. HAPPY mapping represents an easy and generic alternative to RH mapping and has been used to map the human genome (Dear *et al.* 1998), animal genomes (Konfortev *et al.* 2000, Piper *et al.* 1998) and the *Arabidopsis* genome (Thangavelu *et al.* 2003).

The principle of HAPPY mapping is shown in Fig. 17.13 and involves breaking genomic DNA randomly by irradiation or shearing followed by an optional size fractionation step. Markers then are segregated by diluting the resulting fragments to give aliquots containing one haploid genome equivalent (hence "HAPPY" mapping).

Markers are detected using the PCR and linked markers tend to be found together in an aliquot. The map order of markers, and the distance between them, are deduced from the frequency with which they co-segregate. The advantages of this method are that it is fast and accurate and is not subject to the distortions inherent in cloning, meiotic recombina-

tion, or hybrid cell formation. The technique has been used with particular success with a number of protist genomes (Piper *et al.* 1998, Konfortov *et al.* 2000) whose high AT content causes problems in cloning and sequencing.

It should be clear from the above discussion that the useful resolution and range of a HAPPY mapping panel depends upon the mean size of the DNA fragments and this can be controlled by size fractionation. As a rough approximation, a panel will be able to resolve the order of markers over distances as small as 0.1 times the mean fragment size. More closely spaced markers will co-segregate almost completely and will not be resolved. Conversely, the technique will detect the linkage between markers which are separated by up to 0.8 times the mean fragment size. More widely spaced markers do not significantly co-segregate. Therefore, by controlling the degree of fragmentation of the DNA, panels may be created of any desired range or resolution.

It is essential that the different mapping methods are integrated

Each of the mapping methods described above has its advantages and disadvantages and no one method is ideal. The size and complexity of the genome being analyzed can greatly influence the methodologies employed in map construction. Nor is it uncommon for different research groups to use different mapping methods even when working on the same genome. Ultimately, the maps generated by the different methods need to be integrated. This is happening. For example, the BAC clones used to construct the human cytogenetic map (Cheung *et al.* 2001) have STSs that reference the radiation hybrid and linkage maps. Chen *et al.* (2001) have devised a mapping method for rapid assembly and ordering of BAC clones on a radiation hybrid panel using STSs and PCR. Similarly, integrated maps are being constructed in other species, e.g. the rat (Bihoreau *et al.* 2001), maize (Yim *et al.* 2002), and rice (Chen *et al.* 2002).

Many genomic mapping projects involve ordering two classes of objects relative to one another. These are *breakpoints* and *markers* (Table 17.3). Breakpoints, so called because they represent subdivisions of the genome, are defined by a specific experimental resource. Markers consist of unique sites in the genome and should be independent of any particular experimental resource. Although both types of object are essential for map construction, the

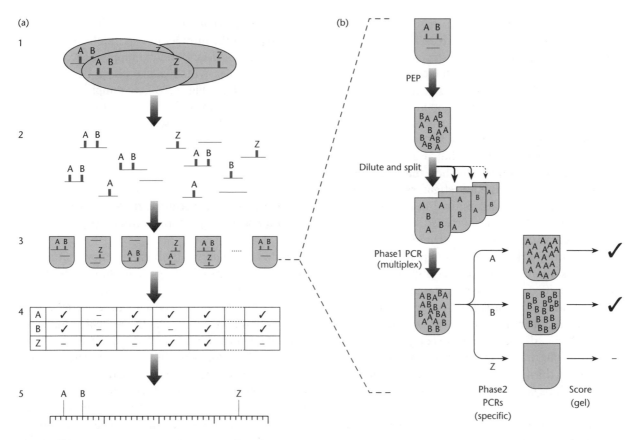

Fig. 17.13 Principle of HAPPY mapping and marker typing. (a) Overview. DNA carrying STS markers (A, B, Z) is extracted from cells (1) and broken randomly to give a pool of fragments (2). These are dispensed at limiting dilution into a series of aliquots – the mapping panel (3). The panel is screened by PCR to produce a table (4) showing the marker content of each aliquot. Linked markers (A, B) are found to co-segregate; remote markers (B, Z) do not. Co-segregation frequencies reflect marker-to-marker distances, allowing a map (5) to be computed. (b) Expanded view of marker screening using a three-step PCR. The protocol is illustrated for one aliquot of the mapping panel. All DNA in the sample is first preamplified >100-fold using primer extension PCR (PEP). This material is diluted and split into subfractions for multiple rounds of screening. One subfraction is amplified in a multiplex PCR for many markers (Phase 1). The products of this reaction are then diluted and split again, and screened for individual markers (A, B, Z) in turn, using hemi-nested primers (Phase 2). Results are scored on gels, determining the marker content of the aliquot.

Table 17.3 Categorization of map objects. (Reproduced with permission from Cox *et al.* 1994, © American Association for the Advancement of Science.)

Mapping method	Experimental resource	Breakpoints	Markers
Meiotic	Pedigrees	Recombination sites	DNA polymorphisms
Radiation hybrid	Hybrid cell lines	Radiation-induced chromosome breaks	STSs
In situ hybridization	Chromosomes	Cytological landmarks	DNA probes
STS content	Library of clones	End points of clones	STSs
Clone-based fingerprinting	Library of clones	End points of clones	Genomic restriction sites

map itself should be defined in terms of markers, especially those based on DNA sequence. One reason for this is that markers are permanent and easily shared. They can be readily stored as DNA sequence information and distributed in this fashion. By contrast, breakpoints are defined by experimental resources that tend to be transient and cumbersome to distribute. The most important reason to use sequence-based markers is that they can be easily screened against any DNA source. Thus they can be used to integrate maps constructed by diverse methods and investigators. Such integration is crucial to the assembly and assessment of maps.

The breakpoints divide the genome into "bins" corresponding to the regions between breakpoints. In assessing mapping progress a first step is to determine the number of "bins" that are occupied by markers and the distribution of these markers within each bin. Although some investigators report only the total number of markers used to construct the map, it is the number of occupied bins that provides the measure of progress. The distribution of the number of markers per bin is important because the goal is to have markers evenly, or at least randomly, spread rather than clustered.

The second step in assessing progress is to identify those occupied bins that are ordered relative to one another and to estimate the confidence in the ordering. Note that assignment of markers to bins can proceed throughout a mapping project but the ordering of bins is only possible as a project matures. Thus ordering is a good indication of the degree of completion. Finally, the distance in kilobases between ordered markers in a map needs to be measured.

As noted in Table 17.3, different experimental resources are used to construct the different kinds of maps. If consistency is to be achieved then it follows that different groups must use *identical* experimental material. Thus a key part of mapping is the construction of genomic libraries and cell lines by one research group for distribution to everyone who needs them. This has huge cost and logistical problems. Nevertheless, such libraries and cell lines are being made available. For example, Osoegawa *et al.* (2000, 2001) have constructed BAC libraries of the human and mouse genomes that are the universal reference material. Similarly, the BACs used for the cytogenetic mapping of the human genome (Cheung *et al.* 2001) are available from various stock centers.

In the context of resources, the work of the Centre d'Etude du Polymorphisme Humain (CEPH) deserves special mention. This organization was set up as a reference source for studies on human genetics. The need for such a reference source stems from the fact that directed matings in humans are not acceptable and because the human breeding cycle is too long to be experimentally useful. Consequently, CEPH maintains cell lines from three-generation human families, consisting in most cases of four grandparents, two parents and an average of eight children (Dausset

et al. 1990). Originally cell lines from 40 families were kept but the number now is much larger. Such families are ideal for genetic mapping because it is possible to infer which allele was inherited from which parent (see Fig. 17.3 for example). The CEPH distributes DNAs from these families to collaborating investigators around the world.

Sequencing genomes

High-throughput sequencing is an essential prerequisite for genome sequencing

As noted in the previous chapter, genomes range in size from millions of base pairs to thousands of millions. Given that a single Sanger sequencing reaction allows 500–600 bases to be sequenced, it is clear that automation is essential. The theoretical sequencing capacity of an automated DNA sequencer is easy to calculate. For a four-dye slab gel system, the capacity is the number of sequencing reactions that can be loaded on each gel, times the number of bases read from each sample, times the number of gels that can be run at once, times the number of days this can be carried out per year. For a 24-channel sequencer the capacity calculated in this way is 2.7 million bases per year. To use just one such instrument to sequence the human genome would require over 1000 years for single pass coverage and this clearly is not a practicable proposition.

To meet the demands of large-scale sequencing, 96-channel instruments have become commonplace and at least one 384-channel instrument has been developed (Shibata *et al.* 2000). By switching from slab gels to capillary systems, the electrophoresis run time is greatly reduced and nine runs can be achieved per 24 h period. Various other improvements to the biochemistry of the sequencing reactions and the chemistry of the gel matrix mean that the read length can be extended from the usual 500–600 bases to 800 bases. As a consequence it now is possible to generate 1–6 million bases of sequence per machine per month (Meldrum 2000b, Elkin *et al.* 2001) and Amersham Biosciences claim over 1 million bases in 8 hours or 2.8 million bases in 24 hours for their MegaBACE 4000 instrument. By way of comparison, 10 years ago the best that could be achieved was only (!) 40,000 bases per month (Fleischmann *et al.* 1995). Those laboratories engaged in sequencing large genomes have large numbers of sequencing machines and can generate millions of base sequences per day, e.g. in excess of 18 million (Elkin *et al.* 2001).

To achieve the levels of sequence data quoted above it is not sufficient to have sequencing instruments with a high capacity. There are many manipulative steps required before samples are loaded on gels in preparation for electrophoresis. For example, DNA has to be isolated, fragmented, and then cloned or amplified. The DNA then has to be re-isolated and subjected to the various sequencing reactions described earlier. Each of these procedures is labor intensive. Not surprisingly, there now are machines that can automate every one of them and details of these machines have been provided by Meldrum (2000a).

There are two different strategies for sequencing genomes

Two different strategies have been developed for the sequencing of whole genomes: the "clone-by-clone" approach and "whole-genome shotgun sequencing". The relative merits of the two strategies have generated much debate, particularly in relation to the sequencing of the human genome. An understanding of this debate, and the terminology used (Box 17.3), is necessary to appreciate some of the landmark genome sequencing publications. More recently, the debate has cooled with the adoption of hybrid strategies for the sequencing of genomes.

As noted earlier, Sanger sequencing has an accurate read length of 500–800 bases. One approach to sequencing a genome would be a shotgun approach in which a random selection of sequencing reads would be collected from a larger target DNA sequence. With sufficient oversampling (coverage) it should be possible to infer the complete genome sequence by piecing together the individual sequence reads. In practice, two problems will arise. First, the presence of dispersed repeated sequences will confound the sequence assembly. Second, assembly of sequences into contigs will be possible but there will be gaps between contigs that will need closing by other means. The numbers of these problems should increase linearly with the size of the genome to be analyzed. Consequently, it was assumed that targets the size of cosmids represented the limit of the shotgun approach. Thus, whole genomes would be sequenced by first developing a set of overlapping cosmids that had been ordered by physical mapping (Fig. 17.14). This is the "clone-by-clone" or "map-based" approach and it was used successfully to generate the complete genome sequence of *Saccharomyces cerevisiae* (Goffeau *et al.* 1996) and the nematode *Caenorhabditis elegans* (*C. elegans* Sequencing Consortium 1998).

The view that cosmids represented the size limit for shotgun sequencing was destroyed when Fleischmann *et al.* (1995) determined the 1,830,137 bp sequence of the bacterium *Hemophilus influenzae* without prior mapping. The starting point was the preparation of genomic DNA, which then was mechanically sheared and size fractionated. Fragments between 1.6 and 2.0 kb in size were selected,

---

Box 17.3   Some definitions used in genome sequencing projects

Contig   An overlapping series of clones or sequence reads that corresponds to a contiguous segment of the source genome

Coverage   The average number of times a genomic segment is represented in a collection of clones or sequence reads. Coverage is synonymous with redundancy

Minimal tiling path   A minimal set of overlapping clones that together provides complete coverage across a genomic region

Sequence-ready map   An overlapping bacterial clone map with sufficiently redundant clone coverage to allow for the rational selection of clones for sequencing

Finished sequence   The complete sequence of a clone or genome with a defined level of accuracy and contiguity

Full-shotgun sequence   A type of pre-finished sequence with sufficient coverage to make it ready for sequence finishing

Prefinished sequence   Sequence derived from a preliminary assembly during a shotgun-sequencing project

Working draft sequence   A type of pre-finished sequence with sufficient coverage (8- to 10-fold) to make it ready for sequence finishing

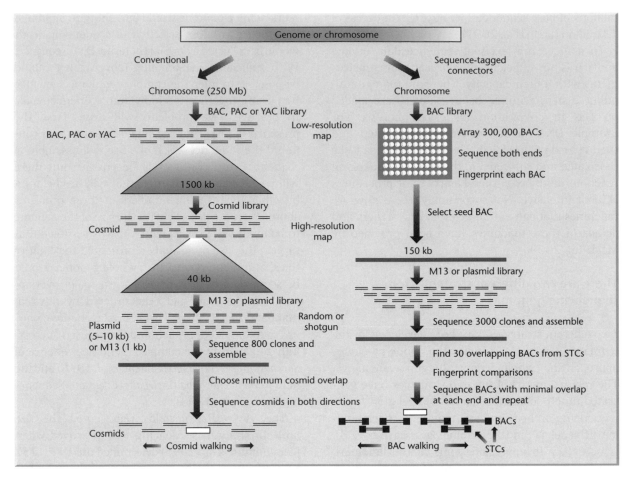

Fig. 17.14 Comparison of the conventional sequence approach with that proposed by Venter *et al.* (1996) (see text for full details). BAC, bacterial artificial chromosone: STC, sequence-tagged connector. (Redrawn from Venter *et al.* 1996 with permission from *Nature*, © Macmillan Magazines Ltd.)

this narrow range being chosen to minimize variation in the growth of clones. In addition, with a maximum size of only 2.0 kb the number of complete genes on a DNA fragment is minimized thereby reducing the chance of their loss through expression of deleterious gene products. The selected fragments were ligated to a sequencing vector and again size fractionated to minimize contamination from double-insert chimeras or free vector. Finally, all cloning was undertaken in host cells deficient in all recombination and restriction functions to prevent deletion and rearrangement of inserts.

Many other bacterial genomes have been sequenced using the same basic method as was used for the *H. influenzae* genome (see Fig. 18.1). However, it should be realized that much of the success of the method stems from the fact that bacterial genomes contain virtually no repeated DNA (see p. 328) to complicate contig assembly. Nevertheless, the success of the method led Venter *et al.* (1996) to propose

a method for sequencing the entire human genome without first constructing a physical map. Rather than using yeast artificial chromosomes (YACs) and cosmids it would use bacterial artificial chromosomes (BACs) (Fig. 17.14), which can accept inserts up to 350 kb in length. A BAC library is prepared which has an average insert size of 150 kb and a 15-fold coverage of the genome in question. The individual clones making up the library are arrayed in microtiter wells for ease of manipulation. Starting at the vector-insert points, both ends of each BAC clone are then sequenced to generate 500 bases from each end. These BAC end sequences will be scattered approximately every 5 kb across the genome and make up 10% of the sequence. These "sequence-tagged connectors" (STCs) will allow any one BAC clone to be connected to about 30 others. In effect, STCs are simply very long STSs (see p. 348).

Each BAC clone would be fingerprinted (see Box 17.4) using one restriction enzyme to provide the

Box 17.4 The principle of restriction enzyme fingerprinting

The principle of restriction enzyme fingerprinting was originally developed for the nematode *Caenorhabditis elegans* (Coulson *et al.* 1986) and yeast (Olson *et al.* 1986). In its original format the starting material was a genomic library made in cosmids. Each cosmid clone was digested with a restriction endonuclease with a hexanucleotide recognition sequence and which leaves staggered ends, e.g. *Hind*III. The ends of the fragments were labeled by end-filling with reverse transcriptase in the presence of a radioactive nucleoside triphosphate. The *Hind*III was destroyed by heating and the fragments cleaved again with a restriction endonuclease with a tetranucleotide recognition sequence, e.g. *Sau*3AI. The fragments were separated on a high resolution gel and detected by autoradiography, the output being a clone fingerprint (Fig. B17.2). Note that the fingerprint is not an order of restriction

sites; rather, it is a series of clusters of bands based on the probability of overlap of clones.

Although restriction enzyme fingerprinting was applied to genomes such as those from *Drosophila* (Siden-Kiamos *et al.* 1990), *Arabidopsis* (Hauge *et al.* 1991), and the human genome (Bellané-Chantelot *et al.* 1992, Trask *et al.* 1992), the method generally fell out of favor except for microbial genomes (for reviews see Cole & Saint Girons 1994, Fonstein & Haselkorn 1995). The reasons for this were twofold. First, cosmids with an average insert size of 40 kb have relatively few sites for any single restriction endonuclease. This means that to generate the necessary numbers of fragments to allow statistically significant matching of overlaps it is necessary to use double digests and radioactive labeling methods. This methodology is not reproducible, nor is it amenable to large-scale mapping efforts (Marra *et al.* 1997, Little 2001). Secondly, the problems of detecting

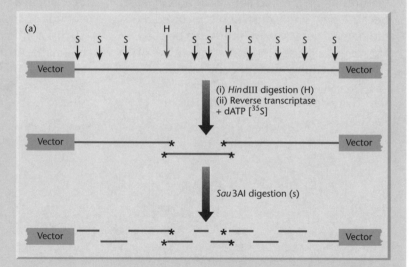

Fig. B17.2 The principle of restriction-fragment fingerprinting. (a) The generation of labeled restriction fragments (see text for details). (b) Pattern generated from four different clones. Note the considerable band sharing between clones 1, 2 and 3 indicating that they are contiguous whereas clone 4 is not contiguous and has few bands in common with the other three. (c) The contig map produced from data shown in (b). (Adapted and redrawn with permission from Coulson *et al.* 1986.)

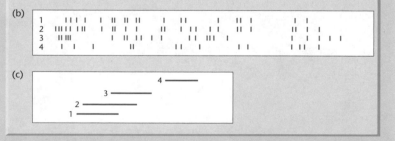

continued

Box 17.4   *continued*

overlaps from fingerprint patterns grow exponentially as the size of the genome being mapped increases.

Two developments have led to a resurgence in restriction enzyme fingerprinting. The first of these was the development of a high-throughput method of fingerprinting using large-insert clones (Marra *et al.* 1997). In this method bacterial artificial chromosome (BAC) and P1-derived artificial chromosome (PAC) clones replace the cosmids used by Coulson *et al.* (1986). Because these clones have insert sizes of 150 kb they yield many easily detectable fragments on digestion with a single restriction enzyme (Fig. B17.3). By measuring the relative mobilities of the fragments it is possible to develop a fingerprint of each clone and to identify other clones that share a large proportion of fragments with the same relative mobilities. In this way it is possible to infer the overlap of clones and construct contigs. Two other features of the method are worth noting. First, because only single digestion is involved, non-radioactive detection methods can be used. However, because of the small quantities of DNA obtainable with low-copy-number vectors, such as BACs and PACs, high-sensitivity detection is required. This is

achieved by using fluorescent dyes and highly sensitive fluorescence imagers. Secondly, the method enables the size of each restriction fragment and each clone to be estimated. This is particularly useful when designing sequencing strategies.

The second development that led to the resurgence of restriction fragment mapping was the creation of powerful software called fingerprinted contigs (FPC) to carry out the massive task of comparing the fingerprints of different clones and determining which ones overlap (Soderlund *et al.* 2000). The FPC software is used in conjunction with the methodology of Marra *et al.* (1997). To determine if two clones overlap, the number of shared bands is counted, where two bands are considered "shared" if they have the same size within a tolerance. The probability that the number of shared bands is a coincidence is computed and, if this score is below a user specified cutoff, the clones are considered to overlap. In a test of this software the largest contig constructed consisted of 9534 clones. Such a contig could not be constructed manually. An additional benefit of the FPC software is that it can accommodate the use of physical markers such as sequence-tagged sites (STSs), etc.

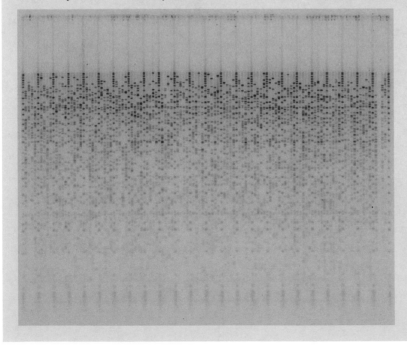

Fig. B17.3 A typical agarose-mapping gel showing human P1-derived artificial chromosomes (PACs) digested with *Hind*III. Clones are present in triplicate to verify stability during propagation and to check for the possibility of cross-contamination. DNA size standards are present in every fifth lane. (Photograph kindly supplied by Dr. M. Marra.)

insert size and detect artifactual clones by comparing the fingerprints with those of overlapping clones. A seed BAC of interest is sequenced and checked against the database of STCs to identify the 30 or so overlapping BAC clones. The two BAC clones showing internal consistency among the fingerprints and minimal overlap at either end then would be sequenced. In this way the entire human genome could be sequenced with just over 20,000 BAC clones. This proposal was not universally accepted but subsequently a modification of it was used to sequence chromosome 2 (19.6 Mb) of *Arabidopsis thaliana* (Lin *et al.* 1999b) and the entire human genome (International Human Genome Sequencing Consortium 2001).

Whereas Venter *et al.* (1996) proposed a hierarchical shotgun approach in which end sequences (STCs) are used to provide long-range continuity across the genome, Weber & Myers (1997) proposed a total shotgun approach to the sequencing of the human genome. In this approach, DNA would be sheared and size-selected before being cloned in *E. coli*. Cloned inserts would fall into two classes:

long inserts of size 5–20 kb and short inserts of size 0.4–1.2 kb. Sequencing read lengths would be of sufficient magnitude so that the two sequence reads from the ends of the short inserts overlap. Both ends of the long inserts also would be sequenced and, because their spacing and orientation are known, they can be used to create a scaffold on which the short sequences can be assembled. This approach was not well received (Green 1997, Marshall & Pennisi 1998) but Myers *et al.* (2000) were able to show the validity of the approach by applying it to the 120 Mb euchromatic portion of the *Drosophila melanogaster* genome. Venter *et al.* (2001) then applied the method to the sequencing of the human genome. However, in both these cases, end sequences also were determined for 50 kb inserts in BACs to provide additional scaffolding information. Also, in the case of the human sequence, any sequence-tagged sites (STSs) that were detected helped to locate the fragment in the context of the overall genome STS map. An indication of the workload associated with sequencing the human genome in this way can be obtained by consideration of the data in Table 17.4.

Table 17.4 Sequencing statistics for the shotgun sequencing of the human genome as undertaken by Venter *et al.* (2001). Note that the DNA sequenced was derived from five different individuals (A, B, C, D, and F).

	Individual	Number of reads for different insert libraries				Total number of base pairs
		2 kbp	10 kbp	50 kbp	Total	
No. of sequencing reads	A	0	0	2,767,357	2,767,357	1,502,674,851
	B	11,736,757	7,467,755	66,930	19,271,442	10,464,393,006
	C	853,819	881,290	0	1,735,109	942,164,187
	D	952,523	1,046,815	0	1,999,338	1,085,640,534
	F	0	1,498,607	0	1,498,607	813,743,601
	Total	13,543,099	10,894,467	2,834,287	27,271,853	14,808,616,179
Fold sequence coverage (2.9 Gb genome)	A	0	0	0.52	0.52	
	B	2.20	1.40	0.01	3.61	
	C	0.16	1.17	0	0.32	
	D	0.18	0.20	0	0.37	
	F	0	0.28	0	0.28	
	Total	2.54	2.04	0.53	5.11	
Fold clone coverage	A	0	0	18.39	18.39	
	B	2.96	11.26	0.44	14.67	
	C	0.22	1.33	0	1.54	
	D	0.24	1.58	0	1.82	
	F	0	2.26	0	2.26	
	Total	3.42	16.43	18.84	38.68	
Insert size (mean)	Average	1951 bp	10,800 bp	50,715 bp		
Insert size (SD)	Average	6.10%	8.10%	14.90%		

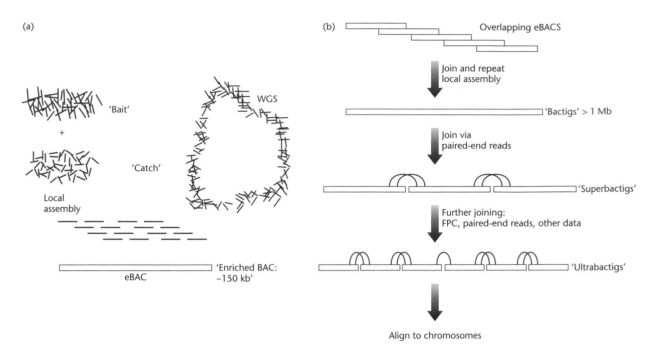

Fig. 17.15   The new combined genome sequencing strategy. (a) Formation of eBACs. Modest sequence coverage (~1.8-fold) from a BAC is used as bait to catch whole genome sequence reads from the same region of the genome. These reads, and their mate pairs, are assembled using Phrap to form an eBAC. This stringent local assembly retains 95% of the catch. (b) Creation of higher order structures. Multiple eBACs are assembled into bactigs based on sequence overlaps. The bactigs are joined into superbactigs by large clone mate-pair information (at least two links), extended into ultrabactigs using additional information (single links, FPC contigs, synteny, markers), and ultimately aligned to genome-mapping data (radiation hybrid and physical maps) to form the complete assembly. Reproduced with permission from *Nature*.

A combination of shotgun sequencing and physical mapping now is the favored method for sequencing large genomes

There is no doubt that assembling a complete genome sequence is relatively easy if one has a collection of sequence-ready clones that already are ordered on a physical map. However, constructing the map is very laborious and takes much longer than sequencing. By contrast, whole-genome shotgun sequencing is very fast but sequence assembly is very prone to errors. In sequencing the human genome, Venter *et al.* (2001) made extensive use of the physical maps provided by the International Human Genome Sequencing Consortium (2001). The draft mouse genome sequence was produced by shotgun sequencing but under-represented duplicated regions because of problems in assembling sequences (Mouse Genome Sequencing Consortium 2002).

To overcome the limitations of the two basic approaches to genome sequencing, the Rat Genome Sequencing Project Consortium (2004) adopted a combined approach involving shotgun sequencing

and BAC sequencing (Fig. 17.15). In the combined approach, shotgun sequence data are progressively melded with light sequence coverage of individual BACs (*BAC skims*) to yield intermediate products called *enriched BACs* (eBACs). eBACs covering the whole genome then are joined into longer structures called *bactigs*. These bactigs are joined to form larger structures called superbactigs and these superbactigs link up to form ultrabactigs. Other data such as STCs, DNA fingerprints, and physical markers are used as appropriate to facilitate the sequence assembly process.

Gaps in sequences occur with all genome-sequencing methodologies and need to be closed

No matter what sequencing strategy is used there always are gaps at the assembly stage. These gaps fall into two categories: sequence gaps where a template exists and physical gaps where no template occurs. Sequence gaps can be closed using a primer-directed walking strategy as shown in Fig. 17.16. Physical gaps are much harder to close. In the case of

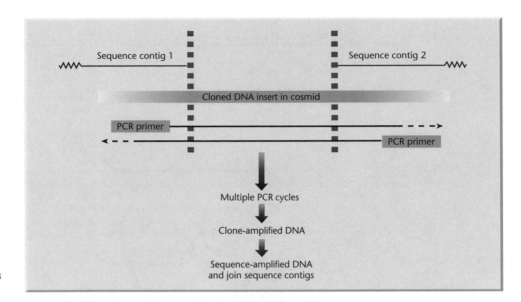

Fig. 17.16 Linking DNA sequence contigs by walking.

the shotgun sequencing of the *H. influenzae* genome, a number of techniques were used. For example, oligonucleotide primers were designed and synthesized from the end of each contig. These primers were used in hybridization reactions based on the premise that labeled oligonucleotides homologous to the ends of adjacent contigs should hybridize to common DNA restriction fragments (Fig. 17.17a). Links were also made by searching each contig end against a peptide database. If the ends of two contigs matched the same database sequence, then the two contigs were tentatively assumed to be adjacent (Fig. 17.17b). Finally, two λ libraries were constructed from genomic *H. influenzae* DNA and were probed with the oligonucleotides designed from the ends of each contig. Positive plaques then were used to prepare templates and the sequence was determined from each end of the λ clone insert. These sequence fragments were searched against a database of all contigs and two contigs that matched the sequence from the opposite ends of the same λ clone were ordered (Fig. 17.17c). The λ clone then provided the template for closure of the sequence gap.

Another method for closing gaps is to use representational difference analysis (Lisitsyn *et al.* 1993). In this technique one undertakes a subtractive hybridization of the library DNA from the total genomic DNA. In this way, Frohme *et al.* (2001) were able to close 11 out of 13 gaps in the sequence of the bacterium *Xylella fastidiosa*. Although this method is useful for isolating sequences that fall within gaps, any sequences isolated will not be useful if the sequences cannot be assembled into a contig that is anchored on at least one end of the gap.

Direct cloning of DNA that is missing in libraries is possible using transformation-associated recombination (TAR). As originally described (Larionov *et al.* 1996), one starts with a YAC containing *Alu* sequences. This vector is cleaved to generate two fragments, one consisting of *Alu*–telomere and the other of *Alu*–centromere–telomere. If these fragments are mixed with high-molecular-weight human DNA containing *Alu* sequences, recombination occurs during transformation to generate new YACs containing large human DNA inserts. Essentially, the *Alu* sequences act as "hooks" and Noskov *et al.* (2003) have shown that the minimal length of sequence homology required is 60 bp. The hooks need not be *Alu* sequences but could be sequences derived from the ends of contigs that are used to trap DNA spanning sequence gaps.

Some of the gaps in complete sequences arise because of the difficulty of cloning DNA containing centromeres and telomeres. The absence of telomeres is easily explained: the absence of restriction sites in the $(TTAGGG)_n$ repeat of telomeres means that they are unlikely to be inserted into cloning vectors. The solution to this problem is to use half-YACs. These are circular yeast vectors containing a single telomere that have a unique restriction site at the end of the telomere (Riethman *et al.* 1989). On cleavage with the appropriate restriction enzyme, a linear molecule containing a single telomere at one end (half-YAC) is generated. This molecule is incapable of replicating in a linear form in yeast unless another telomere is added. Using such half-YACs, Riethman *et al.* (2001) were able to link 32 telomere regions to the draft sequence of the human genome.

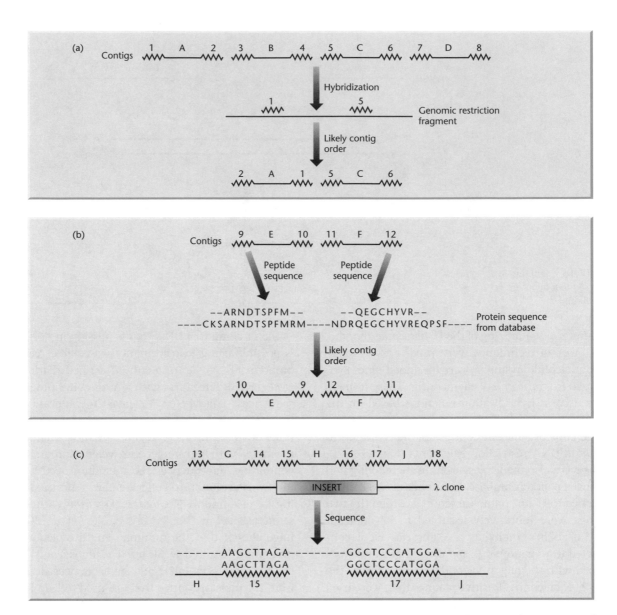

Fig. 17.17 The three methods for closing the physical gaps in sequencing by the method of Fleischmann *et al.* (1995). In each case the sequences at the ends of contigs are shown as wavy lines and individual sequences are given separate numbers. Contigs are denoted by large capital letters. In (b) the individual amino acids are represented by the standard single letter code. See text for a detailed description of each method.

The quality of genome-sequence data needs to be determined

When the techniques of gene manipulation were developed in the late 1970s and early 1980s, cloning was used to identify genes and then gene sequencing was used to characterize the gene product. With the development of high-throughput genome sequencing there has been a fundamental shift in the way biological research is undertaken. Today, the sequencing of genomes not only drives the discovery of genes but also is used to understand the biology of organisms and their evolution. However, if genome-sequence data is going to be used in this way then it is essential that it is complete and that it is accurate. In 1997, the leading scientists planning the sequencing of the human genome established standards, known as the Bermuda standards, for sequence fidelity (http://www.gene.ucl.ac.uk/hugo/bermuda2.htm). These standards stated that the finished sequence should be 99.99% accurate and that the sequence should be contiguous, i.e. there should be no gaps. As noted earlier, every eukaryotic genome has regions that

Table 17.5 Sequencing statistics for some eukaryotic genomes.

Organism	Year	Millions of bases sequenced	Total coverage (%)	Coverage of euchromatin (%)
Saccharomyces cerevisiae	1996	12	93	100
Caenorhabditis elegans	1998	97	99	100
Drosophila melanogaster	2000	116	64	97
Arabidopsis thaliana	2000	115	92	100
Human chromosome 21	2000	34	75	100
Human chromosome 22	1999	34	70	97
Human genome (consortium)	2001	2693	84	90
Human genome (Venter *et al.*)	2001	2654	83	88–93

are difficult to clone and/or sequence. The extent of these regions varies widely in different species and each sequencing consortium has made a pragmatic decision as to when they have sufficient level of coverage for publication (Table 17.5). Further work then is undertaken to eliminate the gaps. In the case of the human genome, the finished sequence was available for all of the 24 chromosomes by mid-2004.

All published genome sequences are derived by assembling thousands or millions of sequence reads and generating a consensus base call for each position in the assembly. The accuracy of a sequence is the measure of how likely the base pairs in a consensus sequence are to be the correct base call. For 99.99% accuracy this means only one incorrect base per 10,000 bp and in the case of the human genome sequence this equates to a total of 300,000 base pair errors in a background of 3 million SNPs! Accuracy usually is determined from error probability assessments generated by DNA base-calling software (PHRED scores, p. 131) and by examining discrepancies between overlapping clone sequences. A retrospective quality assessment of the human genome sequence showed that most of the consortium participants met the desired accuracy standard and identified the commonest sources of errors (Schmutz *et al.* 2004).

Suggested reading

Breen M. *et al.* (2001) Chromosome-specific single locus FISH probes allow anchorage of an 1800-marker integrated radiation-hybrid/linkage map of the domestic dog genome to all chromosomes. *Genome Research* **11**, 1784–95. *This paper is an excellent example of how different mapping methods can be combined to develop a comprehensive physical and genetic map.*

Donis-Keller H. *et al.* (1987) A genetic linkage map of the human genome. *Cell* **51**, 319–37. *This is a classic paper that presented the first physical map of the human genome. It describes the identification and mapping of a large number of RFLPs.*

Fauth C. & Speicher M.R. (2001) Classifying by colors: FISH-based genome analysis. *Cytogenetics and Cell Genetics* **93**, 1–10. *This review details the different applications of FISH and the simultaneous use of fluorochromes.*

Fleischmann R.D. *et al.* (1995) Whole-genome random sequencing and assembly of *Haemophilus influenzae* Rd. *Science* **269**, 496–512. *This is a classic paper detailing the first complete sequence of a bacterial genome and details the principles involved in shotgun sequencing very large pieces of DNA.*

Green E.D. (2001) Strategies for the systematic sequencing of complex genomes. *Nature Reviews Genetics* **2**, 573–83. *An excellent review that covers sequencing strategy in more detail than possible in this chapter.*

Venter J.C. *et al.* (2001) The sequence of the human genome. *Science* **291**, 1304–51.
International Human Genome Sequencing Consortium (2001) Initial sequencing and analysis of the human genome. *Nature* **409**, 860–933.
These two papers and the articles that accompany them give details of how two large sequencing groups undertook the massive task of sequencing the human genome. As well as describing the methodology that was used, they contain a wealth of interesting information.

International Human Genome Sequencing Consortium (2004) Finished: the euchromatic sequence of the human genome. *Nature* **431**, 931–45.
This paper details the conversion of the "working draft" of the human genome sequence into the finished sequence.

Weber J.L. & Myers E.W. (1997) Human whole-genome shotgun sequencing. *Genome Research* **5**, 401–9.

Green P. (1997) Against whole-genome shotgun. *Genome Research* **5**, 410–17.

These two papers need to be read together as they give a good overview of the arguments for and against shotgun sequencing of large genomes. The first paper deals in detail with potential problems of repeated sequences, etc.

Weier H.U. (2001) DNA fiber mapping techniques for the assembly of high-resolution physical maps. *Journal of Histochemistry and Cytochemistry* **49**, 939–48. *This review summarizes different approaches to DNA fiber mapping and recent achievements in mapping ESTs and DNA replication sites.*

Meyers B.C., Scalabrin S. & Morgante M. (2004) Mapping and sequencing complex genomes: let's get physical! *Nature Reviews Genetics* **5**, 578–88.

An excellent review that supplements the material in this chapter, particularly in the areas of clone fingerprinting and contig assembly.

Margulies M., Egholm M., Altman W.E. *et al.* (2005) Genome sequencing in microfabricated high-density picolitre reactors. *Nature* **10**, 1038.

This paper documents the application of pyrosequencing to whole-genome sequencing as opposed to the resequencing of short stretches of DNA.

Chan E.Y. (2005) Advances in sequencing technology. *Mutation Research* **573**, 13–40.

A review focusing on SNP discovery and personalized medicine.

Ambrust E.V., Berges J.A., Bowler C. *et al.* (2004) The genome of the diatom *Thalassiosira pseudonana*: ecology, evolution and metabolism. *Science* **306**, 79–86.

This article describes the use of a combination of sequencing and optical mapping to sequence a 34 Mb genome.

Useful websites

http://www.ncbi.nim.nih.gov/
This is the website of the US National Center for Biotechnology Information. It contains a wealth of information relating to genome sequencing and contains links to many other useful sites.

http://www.ebi.ac.uk/
This is the website of the European Bioinformatics Institute. It contains a wealth of information on tools for sequencing and analyzing genomes. It also provides an up-to-date, standardized, and comprehensively annotated view of the genomic sequence of organisms whose genomes have been completely sequenced.

http://nar.oupjournals.org/
Every year, the first issue of *Nucleic Acids Research* (published in January) has short reviews of all the molecular biology databases. All these reviews can be downloaded free from this website.

CHAPTER 18

Comparative genomics

Introduction

Comparative genomics is the study of the differences and similarities in genome structure and organization in different organisms. For example, how are the differences between humans and other organisms reflected in our genomes? How similar are the number and types of proteins in humans, fruit flies, worms, plants, yeasts, and bacteria? Essentially, comparative genomics is no more than the application of the bioinformatics methods described in Chapter 9 to the analysis of whole-genome sequences with the objective of identifying biological principles, i.e. biology *in silico*. In a sense this statement greatly underplays the real value of comparative genomics for, as the reader will soon see, it is an extremely powerful technique and provides biological insights that could not be achieved in any other way.

There are two drivers for comparative genetics. One is a desire to have a much more detailed understanding of the process of evolution at the gross level (the origin of the major classes of organism) and at a local level (what makes related species unique). The second driver is the need to translate DNA sequence data into proteins of known function. The rationale here is that DNA sequences encoding important cellular functions are more likely to be conserved between species than sequences encoding dispensable functions or non-coding sequences. Until recently it was thought that the ideal species for comparison are those whose form, physiology, and behavior are as similar as possible but whose genomes have evolved sufficiently that non-functional sequences have had time to diverge. More recently, Bofelli *et al.* (2004) have shown that by comparing genomes that are very distantly related, e.g. mammals and fish, it is possible to identify conserved sequences that, presumably, have a significant function.

The formation of orthologs and paralogs are key steps in gene evolution

In order to compare genome organization in different organisms it is necessary to distinguish between *orthologs* and *paralogs*. Orthologs are homologous genes in different organisms that encode proteins with the same function and which have evolved by direct vertical descent. Paralogs are homologous genes within an organism encoding proteins with related but non-identical functions. Implicit in these definitions is that orthologs evolve simply by the gradual accumulation of mutations, whereas paralogs arise by gene duplication followed by mutation accumulation. Good examples of paralogs are the protein superfamilies described in Chapter 16 (see Fig. 16.6 and Table 16.1).

There are many biochemical activities that are common to most or all living organisms, e.g. the citric acid cycle, the generation of ATP, the synthesis of nucleotides, DNA replication, etc. It might be thought that in each case the key proteins would be orthologs. Indeed, "universal protein families" shared by all archae, eubacteria, and eukaryotes have been described (Kyrpides *et al.* 1999). However, there is increasing evidence that functional equivalence of proteins requires neither sequence similarity nor even common three-dimensional folds (Galperin *et al.* 1998, Huynen *et al.* 1999). The existence of two or more distinct sets of orthologs that are responsible for the same function in different organisms is called non-orthologous gene displacement. Now that close to 200 different genomes have been sequenced it is clear that gene displacement occurs within most essential genes. That is, there are at least two biochemical solutions to each cellular requirement. Only about 60 genes have been identified where gene displacement has not been observed (as yet) and most of these encode components of the transcription and translation systems (Koonin 2003).

Protein evolution occurs by exon shuffling

Analysis of protein sequences and three-dimensional structures has revealed that many proteins are composed of discrete domains. These so-called mosaic proteins are particularly abundant in the metazoa. The majority of mosaic proteins are extracellular or constitute the extracellular parts of membrane-bound proteins and thus they may have played an important part in the evolution of multicellularity. The individual domains of a mosaic protein are often involved in specific functions which contribute to its overall activity. These domains are evolutionarily mobile which means that they have spread during evolution and now occur in otherwise unrelated proteins (Doolittle 1995). Mobile domains are characterized by their ability to fold independently. This is an essential characteristic because it prevents misfolding when they are inserted into a new protein

environment. To date, over 60 mobile domains have been identified.

A survey of the genes that encode mosaic proteins reveals a strong correlation between domain organization and intron–exon structure (Kolkman & Stemmer 2001); i.e. each domain tends to be encoded by one or a combination of exons and new combinations of exons are created by recombination within the intervening sequences. This process yields rearranged genes with altered function and is known as *exon shuffling*. Because the average intron is much longer than the average exon and the recombination frequency is proportional to DNA length, the vast majority of crossovers occur in non-coding sequences. The large number of transposable elements and repetitive sequences in introns will facilitate exon shuffling by promoting mismatching and recombination of non-homologous genes. An example of exon shuffling is described in Box 18.1.

Box 18.1   Hemostatic proteins as an example of exon shuffling

The process of blood coagulation and fibrinolysis involves a complex cascade of enzymatic reactions in which inactive zymogens are converted into active enzymes. These zymogens belong to the family of serine proteases and their activation is accompanied by proteolysis of a limited number of peptide bonds. Comparison of the amino acid sequences of the hemostatic proteases with

those of archetypal serine proteases such as trypsin shows that the former have large N-terminal extensions (Fig. B18.1). These extensions consist of a number of discrete domains with functions such as substrate recognition, binding of co-factors, etc. and the different domains show a strong correlation with the exon structure of the encoding genes.

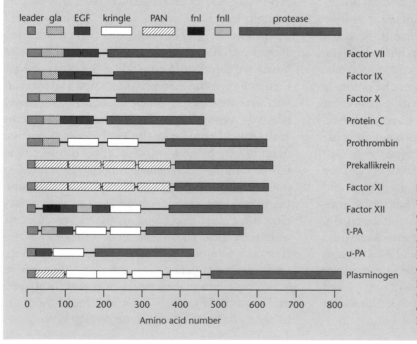

Fig. B18.1 Domain structures of the regulatory proteases of blood coagulation and fibrinolysis. The different domains: gray, serine protease domain; dark purple, EGF-like domain; dotted, Gla domain; cross-hatch, PAN domain; light purple, fibronectin type II domain (fn2); black, fibronectin type I domain (fn1). (Adapted from Kolkman & Stemmer 2001.)

Although mosaic proteins are most common in the metazoa, they are found in unicellular organisms. Because a large number of microbial genomes have been sequenced, including representatives from the three primary kingdoms (Archaea, Eubacteria, and Eukarya), it is possible to determine the evolutionary mobility of domains. With this in mind, Wolf *et al.* (2000a) searched the genomes of 15 bacteria, four archaea, and one eukaryote for genes encoding proteins consisting of domains from the different kingdoms. They found 37 examples of proteins consisting of a "native" domain and a horizontally acquired "alien" domain. In several instances the genome contained the gene for the mosaic protein as well as a sequence encoding a stand-alone version of the alien domain, but more usually the stand-alone counterpart was missing.

Comparative genomics of bacteria

By mid-2004 the website of the National Center for Biotechnology Information listed 173 bacteria (19 Archaea and 154 Eubacteria) whose genomes had been sequenced (http://www.ncbi.nlm.nih. gov/genomes/MICROBES/Complete.html). Simple analysis of the sequence data reveals two features of note. First, the genome sizes vary from 0.49 Mb (*Nanoarchaeum equitans*) to 9.1 Mb (*Bradyrhizobium japonicum* and two species of *Streptomyces*), i.e. a more than 18-fold difference. Secondly, the gene density is remarkably similar across all species and is about 1 gene per kilobase of DNA. This means that large prokaryotic genomes contain many more genes than smaller ones. By contrast, the human genome contains only twice as many genes as *Drosophila*. So how can we account for the size diversity of prokaryotes?

When the different genomes are arranged in size order (Fig. 18.1) some interesting features emerge. First, the archaebacteria exhibit a very much smaller range of genome sizes. This could be an artifact of the small number of genomes examined but more probably reflects the fact that most of them occupy a specialized environment and have little need for metabolic diversity. The exception is *Methanosarcina acetivorans*. This bacterium is known to thrive in a broad range of environments and at 5.8 Mb has the largest archaeal genome (Galagan *et al.* 2002). Second, the smallest eubacterial genomes are found in those organisms that normally are found associated with animals or humans, e.g. mycoplasmas, rickettsias, chlamydiae, etc. Those organisms that can occupy a greater number of niches have a larger genome size. Not surprisingly, there is a good correlation between genome size and metabolic and functional diversity as demonstrated by the size of the genomes of *Bacillus* and *Streptomyces* (formation of spores, antibiotic synthesis), rhizobia (symbiotic nitrogen fixation), and *Pseudomonas* (degradation of a wide range of aromatic compounds).

The minimal gene set consistent with independent existence can be determined using comparative genomics

The genome of *N. equitans* is the smallest sequenced to date (Waters *et al.* 2003) but this organism is an obligate symbiont. This begs the question, what is the minimal genome that is consistent with a free-living cellular organism? In reality, this is a nonsensical question unless one specifies a defined set of environmental conditions. Conceivably, the absolute minimal set of genes will correspond to the most favorable conditions possible in which all essential nutrients are provided and there are no environmental stress factors. If one ignores functionally important RNA molecules and non-coding sequences, the problem is one of defining the minimal protein set.

The first attempt at identifying the minimal protein set was made by compiling a list of orthologous proteins in *Hemophilus influenzae* and *M. genitalium* (Mushegian & Koonin 1996). The expectation was that this list would predominantly contain proteins integral for cell survival as both bacteria are essentially parasites and thus should have shed auxiliary genes. Altogether 244 orthologs were identified but this list is unlikely to be complete because of the occurrence of non-orthologous gene displacements. Some of these gene displacements can be inferred because both organisms appear to have key metabolic pathways that are incomplete. In this way, Mushegian (1999) extended the minimal protein set to 256 genes.

The problem with the above approach is that if one is too strict in defining the degree of similarity between two proteins required to constitute orthologs then the minimal protein set is greatly underestimated. A variation of the above method is to identify orthologous groups, i.e. clusters of genes that include orthologs and, additionally, those paralogs where there has been selective gene loss following gene duplication. When this approach was taken with four eubacteria, one archaebacterium, and one yeast, 816 clusters of orthologous groups (COGs) were

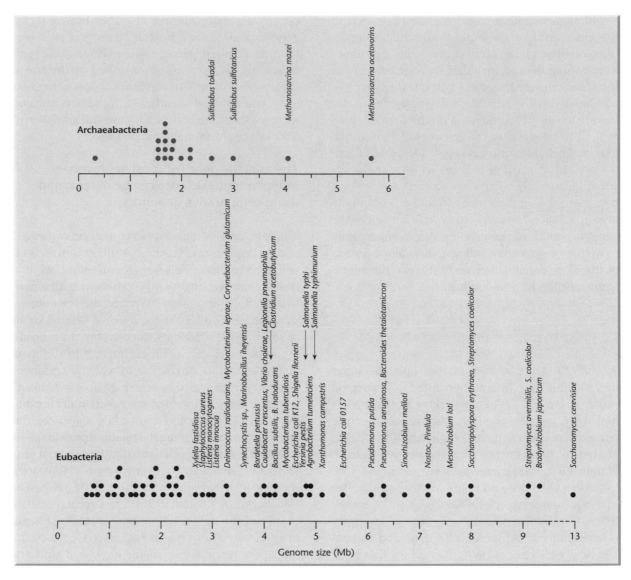

Fig. 18.1 Genome sizes of archaebacteria, some eubacteria, and one eukaryote whose genomes have been completely sequenced.

identified. Of these, 327 contained representatives of all three kingdoms (Mushegian 1999). Based on this set of 327 proteins it was possible to reconstruct all the key biosynthetic pathways. When the analysis was repeated with sequence data from an additional three archaebacteria and 12 eubacteria, the minimal protein set was slightly reduced to 322 COGs.

Larger microbial genomes have more paralogs than smaller genomes

Comparison of the *P. aeruginosa* (6.3 Mb) and *E. coli* (4.5 Mb) genomes indicates that the large genome of

P. aeruginosa is the result of greater genetic complexity rather than differences in genome organization. Distributions of open-reading frame (ORF) sizes and inter-ORF spacings are nearly identical in the two genomes. If the larger genome of *P. aeruginosa* arose by recent gene duplication one would expect it to have a similar number of paralogous groups compared to the other large bacterial genomes and a larger number of ORFs in each group. In fact, the number of ORFs in the paralogous groups in *Pseudomonas* is similar to the other genomes. Thus selection for environmental versatility (Box 18.2) has favored genetic capability through the development of numerous small paralogous gene families

Box 18.2 Correlation of genome sequence data with the biology of bacteria

Pseudomonas aeruginosa

Pseudomonas aeruginosa is a bacterium that is extremely versatile both ecologically and metabolically. It grows in a wide variety of habitats including soil, water, plant surfaces, biofilms, and both in and on animals including humans. A major problem with *P. aeruginosa* is its resistance to many disinfectants and antibiotics. Pseudomonads are characterized by a limited ability to grow on carbohydrates but a remarkable ability to metabolize many other compounds including an astonishing variety of aromatics.

Analysis of the genome of *P. aeruginosa* (Stover *et al*. 2000) reveals a general lack of sugar transporters and an incomplete glycolytic pathway, both of which explain the poor ability to grow on sugars. By contrast, it has large numbers of transporters for a wide range of metabolites and a substantial number of genes for metabolic pathways not found in many other bacteria such as *E. coli*. As might be expected for an organism with great metabolic versatility, a high proportion of the genes (>8%) are involved in gene regulation. The organism also has the most complex chemosensory system of all the complete bacterial genomes with four loci that encode probable chemotaxis signal-transduction pathways. Finally, sequencing revealed the presence of a large number of undescribed drug efflux systems which probably account for the inherent resistance of the organism to many antibacterial substances.

Caulobacter crescentus

Caulobacter crescentus is a bacterium that is found in oligotrophic (very low nutrient) environments and is not capable of growing in rich media. Not surprisingly, genome sequencing (Nierman *et al*. 2001) has shown that the bacterium possesses a large number of genes for responding to environmental substrates. For example, 2.5% of the genome is devoted to motility, there are two chemotaxis systems and over 16 chemoreceptors. It also has 65 members of the family of outer membrane proteins that catalyze energy-dependent transport across the membrane. By contrast, the metabolically versatile *P. aeruginosa* has 32 and other bacteria fewer than 10.

The bacterium also has an obligatory life cycle involving asymmetric cell division and differentiation (Fig. B18.2). Thus it comes as no surprise that genome sequencing reveals a

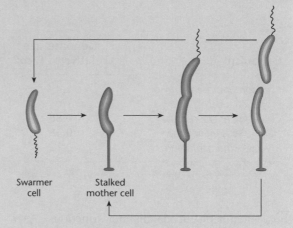

Fig. B18.2 The life cycle of *Caulobacter crescentus*.

very high number of two-component signal-transduction proteins, e.g. 34 histidine protein kinase (HPK) genes, 44 response regulator (RR) genes, and 27 hybrid (HPK/RR) genes. In addition, the frequency of the GAnTC target site for DNA methylation was much less than would be expected if it occurred at random.

Deinococcus radiodurans

This bacterium is remarkable for its ability to survive extremely high doses of ionizing radiation. For example, it can grow in the presence of chronic radiation (6 kilorads/hour) and withstand acute exposures to 1500 kilorads. The organism also is resistant to dessication, oxidizing agents, and ultraviolet radiation. These properties could be the result of one or more of prevention, tolerance, and repair. Genome sequencing (White *et al*. 1999, Makarova *et al*. 2001) has shown that systems for the prevention and tolerance of DNA damage are present but that the key mechanism of resistance is an extremely efficient DNA repair system. Although all of the DNA repair genes identified in *D. radiodurans* have functional homologs in other prokaryotes, no other species has the same high degree of gene redundancy. The bacterium also has multiple genes for proteins involved in exporting oxidation products of nucleotides. Another important component may be the presence of DNA repeat elements scattered throughout the genome. These repeats satisfy several expected requirements for involvement in recombinational repair, including that they are intergenic, they are ubiquitous, and they occur at a frequency that is comparable to the number of double-stranded DNA breaks that can be tolerated.

Organism	Genome size relative to *E. coli*	Percentage of proteins belonging to paralogs
Pseudomonas aeruginosa	1.4	75
Escherichia coli	1	50
Caulobacter crescentus	0.88	48
Hemophilus influenzae	0.38	35
Mycoplasma genitalium	0.12	26

Table 18.1
Relationship between paralogs and genome size.

whose members encode distinct functions. As a general rule, one would expect that as the size of the prokaryotic genome increases then the number of paralogs also would increase, and this is what has been observed (Table 18.1). Furthermore, the biochemical bias in these paralogs reflects the biology of the host organism (Box 18.2).

Analysis of all the prokaryotic genomes sequenced to date has revealed two intriguing observations. First, almost half the ORFs identified are of unknown biological function. This suggests that a number of novel biochemical pathways remain to be identified. Secondly, approximately 25% of all ORFs identified are unique and have no significant sequence similarity to any other available protein sequence. Although this might be an artifact of the small number of bacterial species studied by whole-genome analysis, it does support the observation of incredible biological diversity between bacteria. More importantly, it indicates that there are large numbers of new protein families yet to be discovered, e.g. over 1000 proteins in each of *Bacillus subtilis*, *E. coli*, and *Deinococcus radiodurans*!

Because the DNA and protein sequence databases are updated daily it pays to revisit them from time to time to determine if homologs to previously unidentified proteins have been found. It also pays to re-examine sequence data as new and more sophisticated bioinformatics tools are being developed. The benefits of this can be seen from the work of Robinson *et al.* (1994). They re-examined 18 Mb of prokaryotic DNA sequence and uncovered more than 450 genes that had escaped detection. A more specific example is that of Dandekar *et al.* (2000) who re-examined the sequence data for *Mycoplasma pneumoniae*. They identified an additional 12 ORFs and eliminated one identified previously and found an additional three RNA genes. They also shortened eight protein reading frames and extended 16 others.

Horizontal gene transfer may be a significant evolutionary force but is not easy to detect

Horizontal, or lateral, gene transfer is the occurrence of genetic exchange between different evolutionary lineages. It is generally recognized that horizontal gene transfer has occurred but there is considerable debate about the extent of its occurrence. For example, Gogarten *et al.* (2002) believe that it occurs much more than has hitherto been recognized whereas Kurland *et al.* (2003) feel that it has had little influence on genome phylogeny. Now that so many microbial genomes have been sequenced it might be thought that detecting lateral gene transfer would be easy but there are doubts about the validity of some of the methods used to detect it. Basically, two methods are used: the detection of sequences with unusual nucleotide composition and the detection of a gene, or genes, for a function that is totally absent in all closely related species. For example, analysis of the genomes of two bacterial thermophiles indicated that 20–25% of their genes were more similar to genes in archaeabacteria than those of eubacteria (Aravind *et al.* 1998, Nelson *et al.* 1999). These archaeal-like genes occurred in clusters in the genome and had a markedly different nucleotide composition and could have arisen by horizontal gene transfer.

Garcia-Vallve *et al.* (2000) have developed a statistical procedure for predicting whether genes of a complete genome have been acquired by horizontal gene transfer. This procedure is based on analysis of G + C content, codon usage, amino acid usage, and gene position. When it was applied to 24 sequenced genomes it suggested that 1.5–14.5% of genes had been horizontally transferred and that most of these genes were present in only one or two lineages. However, Koski *et al.* (2001) have urged caution in the use of codon bias and base composition to predict horizontal gene transfer. They compared the ORFs of *E. coli* and *Salmonella typhi*, two closely related bacteria that are estimated to have diverged 100 million years ago. They found that many *E. coli* genes of normal composition have no counterpart in *S. typhi*. Conversely, many genes in *E. coli* have an atypical composition and not only are also found in *S. typhi*, but are found at the same position in the genome, i.e. they are *positional* orthologs.

Karlin (2001) has defined genes as "putative aliens" if their codon usage difference from the average gene exceeds a high threshold and codon usage differences from ribosomal protein genes and chaperone genes also are high. Using this method, in preference to variations in G + C content, he noted that stretches of DNA with anomalous codon usage were frequently associated with pathogenicity islands. These are large stretches of DNA (35–200 kb) that encode several virulence factors and are present in all pathogenic isolates of a species and usually absent from non-pathogenic isolates. Of particular relevance is that they encode an integrase, are flanked by direct repeats, and insert into the chromosome adjacent to tRNA genes (Hacker *et al.* 1997). In this respect, pathogenicity islands resemble temperate phages and could have been acquired by new hosts by transduction (Boyd *et al.* 2001). Alternatively, spread could have been achieved by conjugative transposons. There are many other putative examples of horizontal gene transfer (see Gogarten *et al.* 2002, for a list) but the evidence that transmission occurred in this way is much scantier than for pathogenicity islands with the possible exception of RNA polymerase (Iyer *et al.* 2004).

The comparative genomics of closely related bacteria gives useful insights into microbial evolution

Now that so many microbial genomes have been sequenced it is possible to undertake comparative genomic studies between closely related bacteria or distantly related bacteria. Both kinds of studies are valuable because they reveal different kinds of information. Studies on distantly related bacteria are covered in the next section and here we cover only studies on bacteria that are phylogenetically close.

The most detailed comparative analysis of related bacteria has been undertaken on three genera of the Enterobacteriaceae: *Escherichia*, *Shigella*, and *Salmonella* (Chaudhuri *et al.* 2004). Initially a comparison was made between one laboratory strain of *E. coli* and two O157 enteropathogenic isolates

(Hayashi *et al.* 2001, Perna *et al.* 2001) and later this was supplemented by inclusion of a uropathogenic strain (Welch *et al.* 2002). These studies showed that the genomic backbone is homologous but the homology is punctuated by hundreds of lineage-specific islands of introgressed DNA scattered throughout the genome. Also, the pathogenic strains are 590–800 kb larger than the laboratory strain and this size difference is caused entirely by variations in the amount of island DNA. Many of these islands are at the same relative backbone position in the different pathogens but the island sequences are unrelated. A more surprising finding was that only 39% of the proteins that each strain encodes are common to all of the strains. Furthermore, the pathogen genomes are as different from each other as each pathogen is from the benign strain. A later analysis of the genome of *Shigella flexneri*, a major cause of dysentery, indicated that this bacterium has the same genome structure as *E. coli* and even should be considered as a distinct strain of *E. coli* rather than as belonging to a different genus (Wei *et al.* 2003).

As noted earlier, distinctive codon usage is considered to be an indicator of horizontal gene transfer. Analysis of the different *E. coli* genomes showed that the islands had distinctly different codon usage and a 3–4.5 fold higher use of certain rare codons. Of the approximately 2000 genes that were found in islands in the pathogens only about 10% of them were shared. However, many of these shared genes are related to genes associated with bacteriophages or insertion sequences suggesting that they may have been involved in horizontal gene transfer. Many of the other, non-shared island genes encode known pathogenicity determinants. When different uropathogenic strains are compared, e.g. ones responsible for cystitis, pyelonephritis, and urosepsis, many of their island genes are unique to one strain too. These results suggest that both pathogenic and non-pathogenic strains of *E. coli* have evolved through a complex process. The ancestral backbone genes that define *E. coli* have undergone slow accumulation of vertically acquired sequence changes but the remainder of the genes may have been introduced by numerous occurrences of horizontal gene transfer.

Salmonella species are considered to be close relatives of *E. coli* and two serovars (*S. typhi* and *S. typhimurium*) have been completely sequenced (McClelland *et al.* 2001, Parkhill *et al.* 2001a) and compared to the *E. coli* genome (Fig. 18.2), with which they share extensive synteny. As would be expected, the relationship between *S. typhi* and

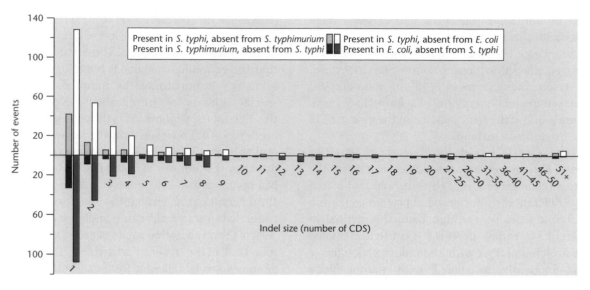

Fig. 18.2 Distribution of insertions and deletions in *S. typhi* relative to *E. coli* and *S. typhimurium*. The graph shows number of insertion–deletion events plotted against the size of the inserted or deleted element (shown as number of genes), clearly indicating that most of the events involve a small number of genes. Values above the lines represent genes present in *S. typhi*; values below the line represent genes absent in *S. typhi*. Dark bars show the comparison with *S. typhimurium*; light bars with *E. coli*. (Redrawn with permission from Parkhill *et al.* 2001b.)

S. typhimurium is very much closer than between *S. typhi* and *E. coli*, although there still are significant differences. There are 601 genes (13.1%) that are unique to *S. typhi* compared with *S. typhimurium* and 479 genes (10.9%) unique to *S. typhimurium* relative to *S. typhi*. By contrast, there are 1505 genes (32.7%) unique to *S. typhi* relative to *E. coli* and 1220 genes (28.4%) unique to *E. coli* relative to *S. typhi*. Another difference between *S. typhi* and *S. typhimurium* is the presence of 204 pseudogenes in the former and only 39 in the latter. In most cases these pseudogenes are relatively recent because they are caused by a single frameshift or stop codon. It is worth noting that complete sequencing of closely related genomes facilitates the detection of pseudogenes. This is because a frame-shift or premature stop codon is only recognizable if the gene is colinear with a functional homologous gene in another genome. One biological difference between the two *Salmonella* serovars is that *S. typhi* only infects humans, whereas *S. typhimurium* can infect a wide range of mammals. This may be related to differences in pseudogene content because many of the pseudogenes in *S. typhi* are in housekeeping functions and virulence components.

The bacterium *Bacillus anthracis* is of much current interest as it is the causative agent of anthrax and has been used as a bioterrorism agent. It has long been considered to be closely related to *B. cereus*, which can cause food poisoning, and *B. thuringiensis*, which is pathogenic for certain insects. A comparative genomic analysis of these three strains has shown that while they differ in their chromosomal backbone the major differences in pathogenicity are due to plasmid-borne genes (Radnedge *et al.* 2003, Rasko *et al.* 2004, Hoffmaster *et al.* 2004). Originally it was thought that *B. cereus* lacked the plasmids pXO1 and pXO2 that respectively encode the lethal toxin complex and the poly-gamma-glutamic acid capsule, both of which are key virulence factors. However, similar plasmids have been found in non-pathogenic *B. cereus* strains and only differ from the corresponding ones from *B. anthracis* by the lack of a pathogenicity island containing various toxin genes.

There have been a number of genomic comparisons made between different species of *Mycobacterium*. Of these, the most interesting is that between *M. tuberculosis* and *M. leprae*, the causative organisms of tuberculosis and leprosy (Table 18.2). Of the 1604 ORFs in *M. leprae*, 1439 had homologs in *M. tuberculosis*. Most of the 1116 pseudogenes were translationally inert but also had functional counterparts in *M. tuberculosis*. Even so, there has still been a massive gene decay in the leprosy bacillus. Genes that have been lost include those for part of the oxidative respiratory chain and most of the microaerophilic and anaerobic ones plus numerous catabolic systems. These losses probably account for the inability of microbiologists to culture *M. leprae* outside of animals. At the genome organization level, 65 segments

Table 18.2
Comparison of the
genomes of two
Mycobacterium spp.
(Reproduced from
Cole *et al.* 2001.)

Feature	Mycobacterium leprae	Mycobacterium tuberculosis
Genome size	3,268,203	4,411,532
G + C (%)	57.79	65.61
Protein coding (%)	49.5	90.8
Protein coding genes (No.)	1604	3959
Pseudogenes (No.)	1116	6
Gene density (bp per gene)	2037	1114
Average gene length (bp)	1011	1012
Average unknown gene length (bp)	338	653

showed synteny but differ in their relative order and distribution. These breaks in synteny generally correspond to dispersed repeats, tRNA genes, or gene-poor regions, and repeat sequences occur at the junctions of discontinuity. These data suggest that genome rearrangements are the result of multiple recombination events between related repetitive sequences.

Comparative analysis of phylogenetically diverse bacteria enables common structural themes to be uncovered

Certain structural themes start to emerge as more and more bacterial genomes are sequenced and comparisons made between these sequences. One such theme is the presence of pathogenicity islands in pathogens and their absence from non-pathogens. Another is that chromosomal inversions in closely related bacteria are most likely to occur around the origin or terminus of replication (Eisen *et al.* 2000, Suyama & Bork 2001). Finally, many genomes are littered with prophages and prophage remnants but the exact significance of these is not known.

The systematic comparison of gene order in bacterial and archaeal genomes has shown that there is very little conservation of gene order between phylogenetically distant genomes. A corollary of this is that whenever statistically significant conservation of gene order is observed then it could be indicative of organization of the genes into operons. Wolf *et al.* (2001) undertook a comparison of gene order in all the sequenced prokaryotic genomes and found a number of potential operons. Most of these operons encode proteins that physically interact, e.g. ribosomal proteins and ABC-type transporter cassettes. More important, this analysis enabled functions to be assigned to genes based on predictions of operon function (Chapter 23).

Comparative genomics can be used to analyze physiological phenomena

The bacterium *Deinococcus radiodurans* is characterized by its ability to survive extremely high doses of ionizing radiation. Although the complete genome has been sequenced this has not been sufficient to provide a convincing explanation for the observed physiological phenotype (see Box 18.2, p. 377). Part of the problem is that there are no other organisms that exhibit the same degree of radiation resistance with which to make comparisons. However, Makarova *et al.* (2003) have made more progress with understanding the basis for hyperthermophily. In this context, hyperthermophily is the ability to grow at temperatures exceeding 75°C whereas thermophily is the ability to grow in the range 55–75°C. Complete genome sequences were available for 11 hyperthermophiles including eight archaea from six distinct lineages and three bacteria from diverse phyla. Sequences also were available for 14 thermophiles. Initially a search was made for COGs which met the following criteria:

1　The COGs must encode proteins and be found in at least three hyperthermophiles.
2　The number of hyperthermophiles with a particular COG should be greater than the number of mesophiles.
3　More than 50% of the organisms with a particular COG should be thermophiles.

Altogether, 290 COGs met the above search criteria but most of them were found only in archaeal hyper-

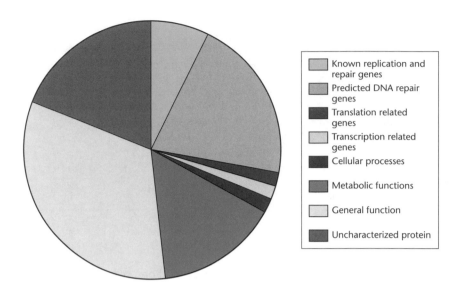

Fig. 18.3 Functions of the 58 COGs associated with hyperthermophily. (Figure reproduced from Makarova *et al.*, 2003, *Trends in Genetics* **19**, 172–6, with permission from Elsevier.)

thermophiles. Therefore the search was refined so that at least one eubacterial hyperthermophile had to encode each COG. In this way 58 COGs were identified as being associated with the hyperthermophilic phenotype. These COGs encode a variety of different cellular functions (Fig. 18.3) and include previously uncharacterized protein families.

Comparative genomics of organelles

Mitochondrial genomes exhibit an amazing structural diversity

Mitochondria are ubiquitous in eukaryotes and play a key role in the generation of ATP through the coupling of electron transport and oxidative phosphorylation. Although the function of mitochondria is highly conserved the structure of the mitochondrial genome exhibits remarkable variation in conformation and size (Fig. 18.4; see Burger *et al.* 2003 for review). Whereas the mtDNAs of animals and fungi are relatively small (15–20 kb), those of plants are very large (200–2000 kb). Plant mitochondria rival the eukaryotic nucleus, and especially the plant nucleus, in terms of the C-value paradox they present: i.e. larger plant mitochondrial genomes do not appear to contain more genes than smaller ones but simply have more spacer DNA. Plant mitochondria also have a large amount of DNA derived from the chloroplast, the nucleus, viruses, and other unknown sources. This process probably is facilitated by the existence of an active, transmembrane potential-dependent mechanism of DNA uptake (Koulintchenko

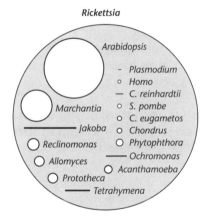

Fig. 18.4 Size and gene content of mitochondrial genomes compared with an α-Proteobacterial (*Rickettsia*) genome. Circles and lines represent circular and linear genome shapes, repectively. (Reprinted from Gray *et al.* 1999 by permission of the American Association for the Advancement of Science.)

et al. 2003). The C-value paradox extends to plant–animal comparisons, where the *Arabidopsis* mtDNA is 20 times larger than human mtDNA but has less than twice the number of genes (Fig. 18.5). Even within a single genus, in this case different species of the yeast *Schizosaccharomyces*, there can be a four-fold variation in the amount of non-coding DNA (Bullerwell *et al.* 2003).

As a result of the steady accumulation of sequence data it now is evident that mtDNAs come in two basic types. These have been designated as "ancestral" and "derived" (Gray *et al.* 1999) and their characteristics are summarized in Table 18.3. It is generally believed that mitochondria are the direct descendants of a bacterial endosymbiont that

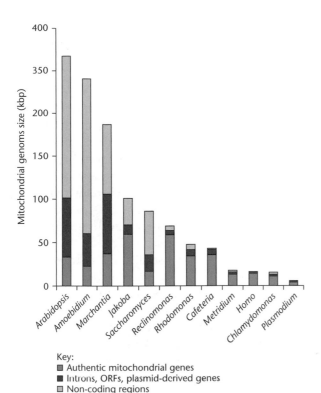

Key:
■ Authentic mitochondrial genes
■ Introns, ORFs, plasmid-derived genes
□ Non-coding regions

Fig. 18.5 Mitochondrial genome size and coding content across eukaryotes: length of coding regions of authentic mitochondrial genes, introns, intronic ORFs, phage-like reverse transcriptases, and DNA polymerases, and intergenic regions. Species names are: *Reclinomonas americana* (jakobid flagellate); *Rhodomonas salina* (cryptophyte alga); *Marchantia polymorpha* (liverwort, bryophyte); *Cafeteria roenbergensis* (stramenopile flagellate); *Arabidopsis thaliana* (flowering plant, angiosperm); *Homo sapiens* (vertebrate animal); *Metridium senile* (cnidarian animal); *Saccharomyces cerevisiae* (ascomycete fungus); and *Plasmodium falciparum* (apicomplexan protist). *Amoebidium parasiticum* (ichthyosporean protist); *Jakoba libera* (jakobid flagellate); and *Chlamydomonas reinhardtii* (green alga, chlorophyte). Reproduced from Burger *et al.*, 2003, with permission from Elsevier.

became established in a nucleus-containing cell and an ancestral mitochondrial genome is one that has retained clear vestiges of this eubacterial ancestry. The prototypal ancestral mtDNA is that of *Reclinomonas americana*, a heterotrophic flagellated protozoon. The mtDNA of this organism contains 97 genes including all the protein-coding genes found in all other sequenced mtDNAs. Derived mitochondrial genomes are ones that depart radically from the ancestral pattern. In animals and many protists this is accompanied by a substantial reduction in overall size and gene content. In plants, and particularly angiosperms, there has been extensive gene loss but size has increased as a result of frequent duplication of DNA and the capture of sequences from the chloroplast and nucleus (Marienfeld *et al.* 1999).

If mitochondria are derived from a bacterium, what is the closest relative of that bacterium that exists today? The current view is that it is *Rickettsia prowazekii*, the causative agent of epidemic typhus. This organism favors an intracellular lifestyle that could have initiated the endosymbiotic evolution of the mitochondrion. The genome of *R. prowazekii* has been sequenced and the functional profile of its genes shows similarities to mitochondria (Andersson *et al.* 1998). The structure, organization, and gene content of the bacterium most resemble those of the mtDNA of *Reclinomonas americana*.

Gene transfer has occurred between mtDNA and nuclear DNA

The principal function of the mitochondrion is the generation of ATP via oxidative phosphorylation. At least 21 genes encode proteins critical for oxidative phosphorylation and one would expect all of these

Table 18.3 Properties of ancestral and derived mtDNAs. (Reprinted from Marienfeld *et al.* 1999 by permission of Elsevier Science.)

Ancestral mtDNA	Derived mtDNA
1 Many extra genes compared with animal mtDNA	**1** Extensive gene loss
2 rRNA genes that encode eubacteria-like 23S, 16S, and 5S rRNAs	**2** Marked divergence in rDNA and rRNA structure
3 Complete, or almost complete, set of tRNA genes	**3** Accelerated rate of sequence divergence in both protein-coding and rRNA genes
4 Tight packing of genetic information with few or no introns	**4** Highly biased use of codons including, in some cases, elimination of certain codons
5 Eubacterial-like gene clusters	**5** Introduction of non-standard codon assignments
6 Use standard genetic code	

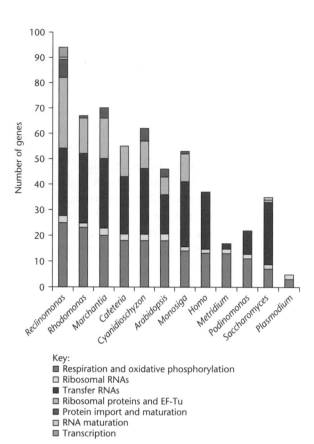

Key:
- ■ Respiration and oxidative phosphorylation
- □ Ribosomal RNAs
- ■ Transfer RNAs
- ▫ Ribosomal proteins and EF-Tu
- ■ Protein import and maturation
- □ RNA maturation
- ▪ Transcription

Fig. 18.6 Mitochondrial gene classes and their representation across eukaryotes. Species names are: *Reclinomonas americana* (jakobid flagellate); *Rhodomonas salina* (cryptophyte alga); *Marchantia polymorpha* (liverwort, bryophyte); *Cafeteria roenbergensis* (stramenopile flagellate); *Cyanidioschyzon merolae* (red alga); *Arabidopsis thaliana* (flowering plant, angiosperm); *Monosiga brevicollis* (choanozoan flagellate); *Homo sapiens* (vertebrate animal); *Metridium senile* (cnidarian animal); *Pedinomonas minor* (green alga, chlorophyte); *Saccharomyces cerevisiae* (ascomycete fungus); and *Plasmodium falciparum* (apicomplexan protist). Reproduced from Burger *et al.*, 2003, with permission from Elsevier.

genes to be located in the mtDNA. Similarly, an mtDNA location would be expected for the genes encoding the 14 ribosomal proteins that are required to translate mtRNA. However, sequence data indicate that many mitochondrial genomes lack a number of key genes (Fig. 18.6) and the missing genes can be found in the nucleus. Functional transfer of mitochondrial genes to the nucleus has stopped in animals, hence their consistency in size. Part of the reason for this is that further transfer is blocked by changes in the mitochondrial genetic code. However, this gene transfer continues to occur in plants and protists because there is no genetic code barrier to transfer. Note that it is not just intact genes that are

transferred for Woischnik & Moraes (2002) found human mitochondrial pseudogenes in the nuclear genome. Many of these pseudogenes comprised parts of two adjacent mitochondrial genes.

In the case of the mitochondrial *cox2* gene, transfer to the nucleus is still on-going in the case of the legumes (Palmer *et al.* 2000). Analysis of 25 different legumes identified some genera in which the *cox2* gene was located in the mitochondrion, some in which it was nuclear, and some where it was present in both genomes. In most cases where two copies of the gene are present, only one gene is transcriptionally active, although at least one genus was found in which both genes are transcribed.

Adams *et al.* (2000a) studied the distribution of the *rps10* gene in 277 angiosperms and identified 26 cases where the gene has been lost from the mtDNA. In 16 of these loss lineages, the nuclear gene was characterized in detail. To be active in the nucleus, a gene acquired from mtDNA must be inserted into the nuclear genome in such a way that a mature translatable mRNA can be produced. Moreover, the resulting protein is made in the cytoplasm and must be targeted to and imported into mitochondria. What emerged was that in some cases pre-existing copies of other nuclear genes have been parasitized with the *rps10* coding sequence. In several instances a mitochondrial targeting sequence has been co-opted to provide entry for the RPS10 protein back into the mitochondrion but different nuclear genes provide this sequence in different plants. In other cases, the RPS10 protein is imported despite the absence of an obvious targeting sequence. These results, and similar findings for other mitochondrial genes (Adams *et al.* 2001), provide confirmation that nuclear transfer is on-going and has happened on many separate occasions in the past. Nor is nuclear transfer confined to mitochondrial genes for Millen *et al.* (2001) have made similar observations with chloroplast genes. Henze & Martin (2000) have reviewed the mechanisms whereby this transfer can occur.

Horizontal gene transfer has been detected in mitochondrial genomes

In the previous section we discussed intracellular horizontal evolution whereby genes moved between the mitochondrion and the nucleus. However, cross-species acquisition of DNA by plant mitochondrial genomes also has been detected. The first example detected was that of a homing group I intron (Palmer *et al.* 2000). These introns encode site-specific endonucleases with relatively long target sites that catalyze their efficient spread from intron-containing alleles to intron-lacking alleles of the same gene in genetic crosses. This intron has been detected in the mitochondrial *cox1* gene of 48 angiosperms out of 281 tested. Based on sequence data for the intron and the host genome, it appears that this intron has been independently acquired by cross-species horizontal transfer to the host plants on many separate occasions. What is not clear are the identities of the donor and recipient in each individual case. By contrast with this group I intron, the 23 other introns in angiosperm mtDNA belong to group II and all are transmitted in a strictly vertical manner.

More recently, Bergthorsson *et al.* (2003) have reported widespread horizontal transfer of mitochondrial genes between distantly related angiosperms, including between monocotyledonous and dicotyledenous plants. The genomic consequences of these mtDNA-to-mtDNA transfers include gene duplication, recapture of genes previously lost through transfer to the nucleus, and a chimeric (half-monocot, half-dicot) ribosomal protein gene.

Comparative genomics of eukaryotes

The minimal eukaryotic genome is smaller than many bacterial genomes

In determining the minimal genome we are seeking to answer a number of different questions. What is the minimal size of the genome of a free-living unicellular eukaryote and how does it compare with the minimal bacterial genome? That is, what are the fundamental genetic differences between a eukaryotic and a prokaryotic cell? Next, what additional genetic information does it require for multicellular coordination? In animals, what are the minimum sizes for a vertebrate genome and a mammalian genome? Finally, what is the minimum size of genome for a flowering plant? Given that many eukaryotic genomes contain large amounts of noncoding DNA these questions have to be answered by considering both genome size and the number of proteins that are encoded.

The smallest eukaryotic genome that has been sequenced is that of the obligate intracellular parasite *Encephalitozoon cuniculi* (Katinka *et al.* 2001). This has a genome size of only 2.9 Mb although its close relative *E. intestinalis* may have a genome that is even smaller (2.3 Mb). Genome compaction in these organisms is achieved by a reduction in the length of intergenic spacers and a shortness of most putative proteins relative to their orthologs in other eukaryotes. Even so, *E. cuniculi* has approximately 2000 ORFs, which is 7–8 times the number in the minimal bacterial genome. The genome of the yeast *Schizosaccharomyces pombe* has about 4800 ORFs (Wood *et al.* 2002) but is unlikely to represent the minimal free-living eukaryotic genome unless the *E. cuniculi* genome has lost many more essential genes than those metabolic and biosynthetic ones already recognized. The multicellular fungus *Neurospora crassa* has approximately 10,000 ORFs (Galagan *et al.* 2003), about 25% fewer than the fruitfly *Drosophila melanogaster* (Adams *et al.* 2000b). Many of these genes do not have homologs in either *Saccharomyces cerevisiae* or *S. pombe* (Borkovich *et al.* 2004) but exactly how many of them are essential for multicellular existence remains to be seen.

Comparative genomics can be used to identify genes and regulatory elements

As noted in Chapter 9 accurately identifying genes in a complete genome sequence can be very difficult and identifying regulatory elements can be even harder still. A powerful method for finding functional elements such as genes and regulatory regions is to align orthologous genomic sequences from different species and search out regions of sequence conservation. The rationale for this approach is that mutations in functional DNA will be deleterious and thus counter-selected thereby resulting in a reduced rate of evolution of functional elements. The two most important factors affecting the results of a comparative analysis are the amount of divergence being captured and the phylogenetic scope of the aligned sequences (Cooper & Sidow 2003). The amount of divergence affects the power and resolution of the analyses. The scope, which is defined as the narrowest taxonomic group that encompasses all analyzed sequences, affects the applicability of conclusions and the generality of the results. For example, a dipteran scope that includes *Drosophila* (fruitfly) and *Anopheles* (mosquito) can be used to find elements that were present in their common ancestor as well as ones present before the diversification of hexapods, arthropods, and metazoa (Fig. 18.7).

An example of a comparative analysis with a narrow scope is the genomic comparison of *S. cerevisiae* with three other species of *Saccharomyces* (Kellis *et al.*

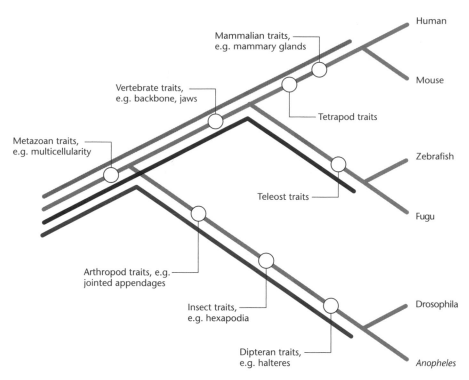

Fig. 18.7   The importance of scope and the impact of shared ancestry on comparative sequence analysis. The tree describing the relationships among six actively studied genomes is drawn in light purple (not to scale). Each colored line indicates the phylogenetic scope that applies to each pair of species at the terminal nodes: gray line, placental mammal scope; black line, teleost scope; dark purple line, dipteran scope. Overlaps of the colored lines indicate shared ancestry and capture traits shared by the indicated scopes and, by implication, shared functional elements. Open circles and associated text show various traits that exemplify the major animal clades and the branch of the tree on which they arose. Reproduced from Cooper *et al.* (2003), with permission from Elsevier.

2003). The gene analysis resulted in a major revision of the *S. cerevisiae* gene catalog that affected 15% of all genes, reduced the total count by about 500 genes, and identified 43 new small ORFs (50–99 amino acids). This latter finding is particularly significant since small ORFs can only be considered putative genes in the absence of function or conservation in different species. A comparative analysis with a more divergent scope is that between the pufferfish (*Fugu rubripes*) and human genomes (Aparicio *et al.* 2002). This identified almost 1000 putative genes that had not been identified in the two published reports on the human genome sequence.

The direct identification of regulatory elements is very difficult since they are short (6–15 bp), tolerate some degree of sequence variation, and follow few known rules. Computational analysis of single genome sequences has been used successfully to identify regulatory elements such as promoters associated with known sets of genes. However, this approach is of relatively little value in identifying other regulatory elements involved in gene expression (enhancers, silencers) and chromatin organiza-

tion (insulators, matrix attachment regions). As the examples below show, comparative analysis is much more useful in this respect. Comparisons within a narrow scope are particularly useful as they permit almost the entire genome to be scanned for regulatory regions. In this way Kellis *et al.* (2003) were able to recognize an additional nine regulatory protein motifs in addition to the 42 that were already known.

Enhancers are regulatory elements that upregulate gene expression by sequence-specific positioning of transcriptional activators. Enhancers can function independently of position and orientation although they generally are located within hundreds of kilobases of their target genes. Using comparative analysis, Spitz *et al.* (2003) discovered a cluster of enhancer elements that are conserved between mammals and *Fugu*. These enhancers coordinate expression between *Hoxd* genes and nearby genes that are evolutionarily unrelated.

Silencers are elements that are capable of repressing transcription. Many are found near their corresponding promoter but there are other types. Sequencing of the chicken CD4 gene showed that it

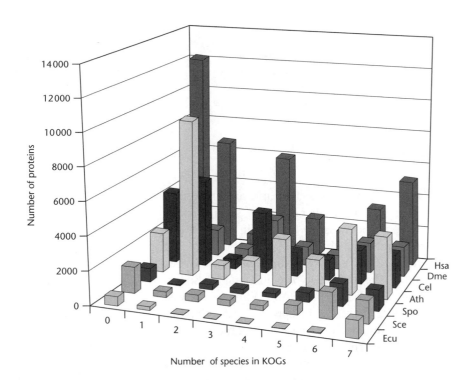

Fig. 18.8
Assignment of proteins from each of the seven analyzed eukaryotic genomes to KOGs with different numbers of species and to LSEs. 0, Proteins without detectable homologs (singletons); 1, LSEs. Species abbreviations: Ath, *Arabidopsis thaliana*; Cel, *Caenorhabditis elegans*; Dme, *Drosophila melanogaster*; Ecu, *Encephalitozoon cuniculi*; Hsa, *Homo sapiens*; Sce, *Saccharomyces cerevisisae*; Spo, *Schizosaccharomyces pombe*.

is similar to the mammalian CD4 gene and has a functional human silencer (Koskinen *et al.* 2002). This level of distant conservation suggests that this silencer has a fundamental role in controlling gene expression.

Insulator elements are barriers that separate domains within chromatin and confine the actions of regulatory elements to their appropriate targets. They can block the action of enhancers as well as prevent the spread of chromatin condensation from nearby regions. Farrell *et al.* (2002) discovered conserved genomic regions that flank the β-globin loci in mouse and man. These regions contain binding sites for CTCF, a protein known to be important for enhancer-blocking insulator activity.

Matrix attachment regions (MARs) are regions of DNA that are involved in the binding to the nuclear matrix. Glazko *et al.* (2003) aligned intergenic sequences from mouse and man and identified conserved segments. Further analysis showed that 11% of these had sequence motifs characteristic of MARs and that many of them precede the 5′ ends of genes. This latter observation suggests a role in regulating transcription.

Comparative genomics gives insight into the evolution of key proteins

Koonin *et al.* (2004) have undertaken a comprehensive evolutionary classification of the proteins encoded

in seven completely sequenced eukaryotic genomes: three animals (man, nematode, and fruitfly), one plant (*Arabidopsis*), a budding yeast, a fission yeast, and the microsporidian *E. cuniculi*. In particular, they looked for eukaryotic clusters of orthologous groups (KOGs) and the results are shown in Fig. 18.8. The fraction of proteins assigned to KOGs tends to decrease with increasing genome size, except for the obligate parasite *E. cuniculi*. By contrast, lineage-specific expansions of paralogous groups show the opposite trend with the largest numbers being in the higher eukaryotes. Only a minority of KOGs have readily detectable prokaryotic counterparts, indicating the extent of innovation linked to the origin of eukaryotes.

A total of 131 KOGs were represented by a single gene in each of the seven genomes. Since these KOGs are present in the minimal genome of *E. cuniculi* they must encode core biological functions. Nearly all of them encode subunits of known multiprotein complexes and many of them are involved in rRNA processing, ribosome assembly, intron splicing, transcription, and protein assembly and trafficking.

The evolution of species can be analyzed at the genome level

The yeasts *Saccharomyces paradoxus*, *S. mikatae*, and *S. bayanus* are estimated to have separated from *S. cerevisiae* 5–20 million years ago. The genomes of all four have been sequenced and Kellis *et al.* (2003)

Species	Reciprocal Translocations	Inversions	Segmental Duplications
S. paradoxus	0	4	3
S. mikatae	4	13	0
S. bayanus	5	3	0

Table 18.4
Genomic rearrangements in three yeast species when compared with *S. cerevisiae*.

have undertaken a comparative analysis. They found a high level of "genomic churning" in the vicinity of the telomeres and gene families in these regions showed significant changes in number, order, and orientation. Only a few rearrangements were seen outside of the telomeric regions and these are summarized in Table 18.4. All 20 inversions were flanked by tRNA genes in opposite transcriptional orientation and usually these were of the same iso-acceptor type. The role of tRNA genes in genomic inversion has not been noted before. Of the nine translocations, seven occurred between Ty elements and two between highly similar pairs of ribosomal genes.

At the gene level, five genes were unique to *S. paradoxus*, eight genes unique to *S. mikatae*, and 19 unique to *S. bayanus*. Most of them encoded functions involved in sugar metabolism or gene regulation. The majority (86%) of these unique genes were located near a telomere or a Ty element, locations that are consistent with rapid genome evolution. One gene was identified that appears to be evolving very rapidly and across the four species showed 32% nucleotide identity and 13% amino acid identity. Functionally it appears to be involved in sporulation, which in yeast is a stage in sexual reproduction. In this regard, it is consistent with the observation that many of the best-studied examples of positive selection in other organisms are genes related to gamete function. One gene also was identified that showed perfect 100% conservation at the amino acid and the nucleotide level. The latter observation is very unusual given the redundancy of the genetic code and suggests that the gene might encode an antisense RNA.

Analysis of dipteran insect genomes permits analysis of evolution in multicellular organisms

The fruit fly *Drosophila melanogaster* and the malaria mosquito *Anopheles gambiae* are both highly adapted,

Table 18.5   Genome statistics for the mosquito and fruit fly.

	Anopheles	Drosophila
Genome size	278 Mb	165 Mb
Total exon length	10 Mb	13.6 Mb
Total intron length	22.6 Mb	12.9 Mb
Average introns per gene	3.5	4.7
Average amino acids per protein	548	649

successful dipteran species that diverged about 250 million years ago. They share a similar body plan and a considerable number of other features but differ in terms of ecology, morphology, and life style. For example, *Drosophila* feeds on decaying fruit while *Anopheles* feeds on the blood of specific hosts. A number of obvious differences can be seen at the whole-genome level (Table 18.5) but these give little insight into the evolutionary process.

When the two genomes are compared at the protein level (Zdobnov *et al.* 2002) five classes of protein can be recognized (Fig. 18.9). A total of 6089 orthologs were identified in the two species and their average sequence identity was 56%. By contrast there is 61% sequence identity of orthologs between the pufferfish and humans, which diverged 450 million years ago. This indicates that insect proteins diverge at a higher rate than vertebrate proteins. This could be because insects have a much shorter life cycle and may experience different selective pressures. When the orthologs are classified according to gene ontology it is not surprising to find that the proteins involved in immunity show the greatest divergence and structural proteins are the most conserved.

The "many-to-many" orthologs shown in Fig. 18.9 represent groups of genes in which gene duplication has occurred in one or both species after divergence,

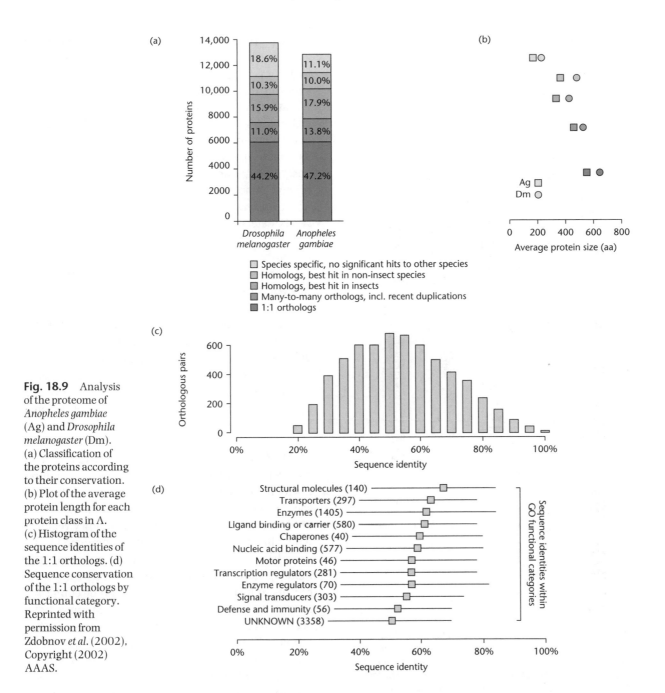

Fig. 18.9 Analysis of the proteome of *Anopheles gambiae* (Ag) and *Drosophila melanogaster* (Dm). (a) Classification of the proteins according to their conservation. (b) Plot of the average protein length for each protein class in Λ. (c) Histogram of the sequence identities of the 1:1 orthologs. (d) Sequence conservation of the 1:1 orthologs by functional category. Reprinted with permission from Zdobnov *et al.* (2002), Copyright (2002) AAAS.

i.e. paralogy. These, and the homologs, probably represent adaptations to environment and life strategies leading to changes in cellular and phenotypic features. For example, four *Anopheles* paralogs without a counterpart in *Drosophila* are similar to the human gene encoding leukotriene B4 12-hydroxy dehydrogenase, an enzyme that can inactivate the proinflammatory leukotriene B4. The anopheline mosquito may have acquired this gene to facilitate the taking of a blood meal. A total of 579 orthologs were restricted to *Anopheles* and *Drosophila* and did

not even share domains with proteins identified in the other organisms whose genomes have been sequenced. Most of those that could be annotated encoded specific odorant and taste receptors, cuticle proteins, pheromone and pheromone-binding proteins, and insect-specific defense molecules.

The dynamics of gene evolution can be analyzed by comparing the intron and exon structure of the 1:1 orthologs. For example, equivalent introns in *Drosophila* have only half the length of those in *Anopheles* whereas exon lengths and intron

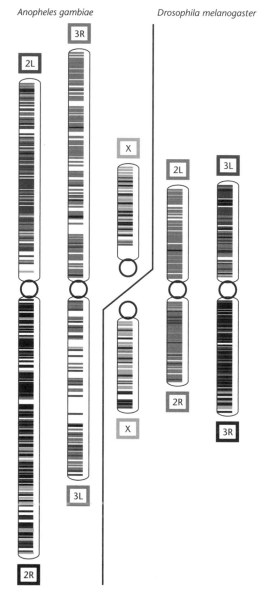

Anopheles gambiae   *Drosophila melanogaster*

Fig. 18.10 Homology of chromosomal arms in insects. Each chromosomal arm is marked by a color shown around its name (pairs of chromosomes with significant homology, such as *Dm*2L/*Ag*3R, use the same color). Coloring inside the schematic chromosome arms denotes microsynteny matches to a region in the other species; the color shown is the color of the chromosome containing the matching region in the other species. Reprinted with permission from Zdobnov *et al.* (2002). Copyright (2002) AAAS. Full-color version available at www.blackwellpublishing.com/primrose

frequencies are roughly similar. Approximately 55% of *Anopheles* introns in 1:1 orthologs have equivalent positions in *Drosophila* but almost 10,000 introns have been lost or gained between the two species. The rate of gain or loss of introns has been calculated to be one per gene per 125 million years.

Given that the two diptera being studied are estimated to have diverged 250 million years ago one would expect that, in addition to changes in exon/intron structure, there would be significant variations in genome structure. Indeed, the gene order of the 1:1 orthologs has only been retained over very small distances and this is referred to as microsynteny. However, at the macro level, chromosomal arms exhibit significant remnants of homology between the two species and major inter-arm transfers and intra-arm shuffling of gene order can be detected (Fig. 18.10).

A number of mammalian genomes have been sequenced and the data is facilitating analysis of evolution

The genomes of humans, the mouse, and the rat have been completely sequenced and good progress is being made with the genome of the chimpanzee (International Human Genome Sequencing Consortium 2001, Venter *et al.* 2001, Mouse Genome Sequencing Consortium 2002, Rat Genome Sequencing Project Consortium 2004, The International Chimpanzee Chromosome 22 Consortium 2004). Figure 18.11 shows an analysis of the three completely sequenced genomes. About 1 billion nucleotides (40% of rat genome) align in all three species and this "ancestral core" contains 94–95% of the known coding exons and regulatory regions, which in turn represent 1–2% of the genome. A further 30% of the rat genome aligns only with the mouse genome and consists largely of rodent-specific repeats. A further 15% of the rat genome comprises rat-specific repeats. More genomic changes have been detected in the rodent lineages than in the human. These include approximately 250 large rearrangements between a hypothetical rodent ancestor and human, approximately 50 between this ancestor and rat, and a similar number between the ancestor and the mouse.

The rat, mouse, and human genomes encode similar numbers of genes and the majority have persisted without deletion or duplication since the last common ancestor. About 90% of the genes have strict orthologs in all three genomes but, compared with humans, the rodents have expanded gene families for functions associated with reproduction, immunity, olfaction, and metabolism of xenobiotics. These features are not surprising given what we know about rodent biology! Almost all the human genes known

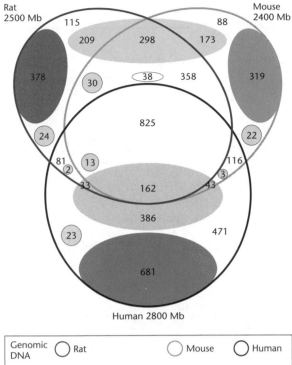

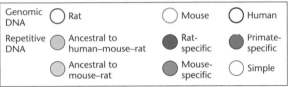

Human 2800 Mb

Genomic DNA	◯ Rat	◯ Mouse	◯ Human
Repetitive DNA	◯ Ancestral to human–mouse–rat	◯ Rat-specific	◯ Primate-specific
	◯ Ancestral to mouse–rat	◯ Mouse-specific	◯ Simple

Fig. 18.11 Aligning portions and origins of sequences in rat, mouse, and human genomes. Each outlined ellipse is a genome, and the overlapping areas indicate the amount of sequence that aligns in all three species (rat, mouse, and human) or in only two species. Non-overlapping regions represent sequence that does not align. Types of repeats classified by ancestry: those that predate the human–rodent divergence, those that arose on the rodent lineage before the rat–mouse divergence, species-specific, those that are rat-specific, mouse-specific, human-specific and simple, each indicated as shown in the key and placed to illustrate the approximate amount of each type in each alignment category. Uncolored areas are non-repetitive DNA – the bulk is assumed to be ancestral to the human–rodent divergence. Numbers of nucleotides (in Mb) are given for each sector (type of sequence and alignment category). Reproduced with permission from *Nature*.

to be associated with disease have orthologs in the rat and mouse genomes but there is one surprising finding. Many SNPs causing disease in man are found in mice but these mice are phenotypically normal.

The comparison of the human genome with that of the chimpanzee is perhaps the most interesting of all the genomic comparisons that can be made, as the chimpanzee is our closest living relative. In particular, comparative analysis should help to uncover

the genetic basis of cognitive function, bipedalism, and speech development. At the time of writing the complete chimpanzee genome was not available but the 33.3 Mb sequence of chromosome 22 had been completed (The International Chimpanzee Chromosome 22 Consortium 2004). Nearly 1.5% of the chimpanzee genome had single base substitutions when compared with its human equivalent (chromosome 21) in addition to approximately 68,000 insertions or deletions. These differences are sufficient to generate changes in most of the 231 coding sequences. In addition, different expansion of particular subfamilies of retrotransposons was observed between the different lineages, suggesting different impacts of retrotransposition on human and chimpanzee evolution. The full impact of these changes remains to be deciphered.

Comparative genomics can be used to uncover the molecular mechanisms that generate new gene structures

The comparative analyses described in the previous sections indicate that there is a general process of new gene origination. This raises the question of the origin of these new genes. Several molecular mechanisms are known to be involved in the creation of new gene structures (Fig. 18.12) and can operate singly or in combination (Long *et al.* 2003). A good example is *jingwei*, the first identified gene that has originated recently (2 My) in the evolutionary timescale (Fig. 18.13). This gene arose in the common ancestor of two *Drosophila* species. The starting point was the *yellow emperor* gene that duplicated to give the *yellow emperor* and *yande* genes. Whereas *yellow emperor* maintained its original functions, *yande* underwent modification. In particular, mRNA of the alcohol dehydrogenase gene retroposed into the third intron of *yande* as a fused exon and recombined with the first three *yande* exons. This formed *jingwei*, a gene that is translated into a chimeric protein.

Once created, new genes such as *jingwei* may become modified beyond recognition. Examples of this kind of change include domains involved in protein–protein interactions such as von Willebrand A, fibronectin type III, immunoglobulin, and SH3 modules (Ponting *et al.* 2000). These domains show extensive proliferation in higher eukaryotes but have only a distant relationship to homologs in prokaryotes and lower eukaryotes.

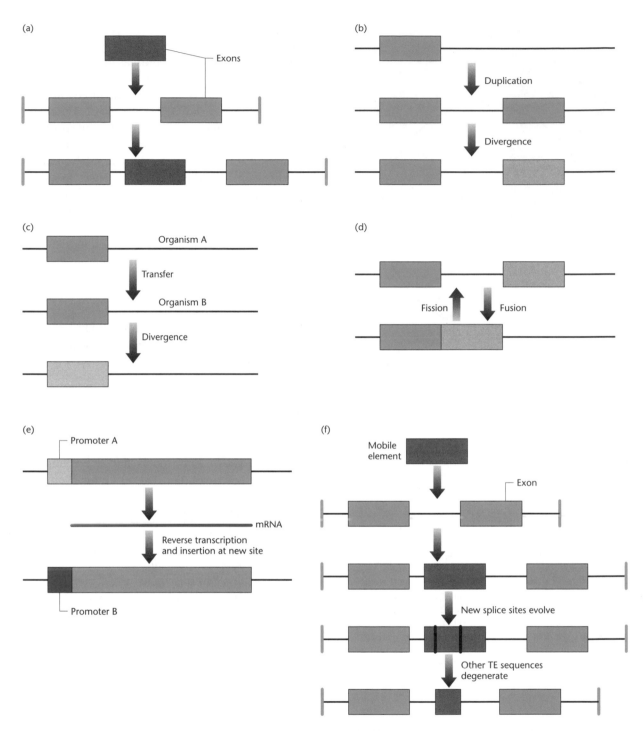

Fig. 18.12 Mechanisms whereby new genes arise. (a) exon capture (exon shuffling); (b) duplication of a gene followed by sequence divergence of the duplicate; (c) divergence of a gene following transfer to a new host; (d) fusion of two separate genes or separation of two fused activities; (e) movement of a gene sequence via an mRNA intermediate followed by coupling to a promoter; (f) capture of a transposable element (TE) followed by degeneration of the TE sequences. Real examples of these mechanisms can be found in the review of Long *et al.* (2003).

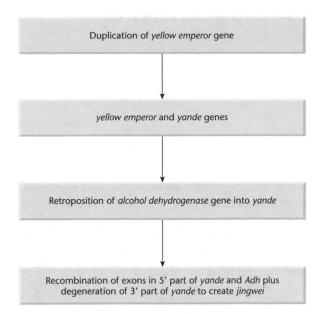

Fig. 18.13 Genomic events leading to the formation of the new gene *jingwei*.

Suggested reading

Kellis M., Patterson N., Endrizzi M., *et al.* (2003) Sequencing and comparison of yeast species to identify genes and regulatory elements. *Nature* **423**, 241–54.
This is rapidly becoming a classic paper on the use of comparative genomics to decipher genome sequences but it also provides insights to the genomic changes that exist between species.

Koonin E.V. (2003) Comparative genomics, minimal gene-sets and the last universal common ancestor. *Nature Reviews Microbiology* **1**, 127–36.
Koonin E.V., Federova N.D., Jackson J.D., *et al.* (2003) A comprehensive evolutionary classification of proteins encoded in complete eukaryotic genomes. *Genome Biology* **5**, R7.

Eugene Koonin probably knows more than anyone about extracting evolutionary information from sequence databases. The two papers cited above are but a tiny sample of his analyses.

Koonin E.V. (2005) Virology: Gulliver among the Lilliputians. *Current Biology* **15**, R167–9.
An analysis of the genome of a virus that is much bigger than many parasitic bacteria.

Long M., Betran E., Thornton K. & Wang W. (2003) The origin of new genes: glimpses from the young and old. *Nature Reviews Genetics* **4**, 865–75.
This is one of the few reviews that attempt to discuss where new genes come from.

Paterson A.H., Bowers J.E., Chapman B.A., *et al.* (2004) Comparative genome analysis of monocots and dicots, towards characterization of angiosperm diversity. *Current Opinion in Biotechnology* **15**, 120–5.
Pedulla M.L., Ford M.E., Houtz J.M., *et al.* (2003) Origins of highly mosaic mycobacteriophage genomes. *Cell* **113**, 171–82.
These two papers cover topics not discussed in this chapter, the comparative genomics of plants and viruses respectively, and are well worth reading.

Each year, the January 1 issue of *Nucleic Acids Research* is devoted to short reviews of the different molecular biology and genomics databases. A considerable number of these databases are for the purposes of comparative genomics and all are linked to relevant websites. An example is given below.

Useful website

http://colibase.bham.ac.uk
This is the website for *coli*BASE, an online database for the comparative genomics of *E. coli* and its close relatives. Now that a number of different strains of *E. coli* have been completely sequenced it is clear that there is much more genomic heterogeneity than expected.

CHAPTER 19

Large-scale mutagenesis and interference

Introduction

One of the most powerful ways to determine the function of a gene is to mutate it and study the resulting phenotype. In this respect, the link between gene and function can be approached from two directions. Traditional "forward genetics" involves random mutagenesis followed by screening to recover mutants showing impairment for a particular biological process. The essence of forward genetics is that one starts by identifying a mutant phenotype and then works towards the gene by mapping and cloning. "Reverse genetics" is the opposite approach, where one starts with a cloned gene whose function is unknown. This gene is mutated deliberately and the effect analyzed. In the post-genomics era, both forward and reverse genetics are being developed as high-throughput tools for the functional mapping of genomes (Hamer *et al.* 2001, Ramachandran & Sundaresan 2001, Nagy *et al.* 2003, Friedman & Perrimon 2004).

There are three basic strategies, which we will explore in turn. The first is the systematic approach of deliberately mutating every single gene in the genome, one at a time, and generating banks of specific mutant strains. This can only yield a comprehensive mutant library if the entire genome sequence is available. The second is a random approach in which genes are mutated indiscriminately. Individual mutations are then cataloged by obtaining flanking sequence tags, and genes are annotated by matching the tags to entries in sequence databases. This method can be applied to any species, even if there is little or no existing sequence information. Moreover, in species with complete or well-advanced genome projects, random mutagenesis may uncover genes that have been missed by other annotation methods. Each of these strategies has further advantages and disadvantages. For example, although the systematic approach provides exhaustive genome coverage, it is a labor-intensive process and depends on pre-existing sequence information. The random mutagenesis approach is rapid and relatively inexpensive,

but there is no control over the distribution of mutations so saturation may be difficult to achieve. The third approach encompasses a group of techniques that generate functional *phenocopies* of mutant alleles, i.e. the likeness of a mutation without actually altering the DNA sequence of the organism. The most widely used method in this category is RNA interference (Friedman & Perrimon 2004).

Genome-wide gene targeting is the systematic approach to large-scale mutagenesis

Mutations can be introduced into predefined genes *in vivo* through a process termed *gene targeting*, which involves homologous recombination (Chapter 13). Where the aim is to inactivate the target gene completely and generate a null allele, the term *gene knockout* is often used. Homologous recombination occurs to a greater or lesser degree in all organisms, but the efficiency varies considerably. In bacteria and yeast, and also in the moss *Physcomitrella patens*, the process is highly efficient and gene transfer with a suitable targeting vector results in homologous recombination more than 90% of the time. Since microbial genomes also contain fewer genes than those of higher eukaryotes, these species are ideally suited to a functional genomics strategy based on systematic gene knockouts.

Homologous recombination in higher eukaryotes occurs at a much lower efficiency. Even if a homologous target is available in the genome, DNA introduced into most animal and plant cells is 100,000 times more likely to integrate randomly than recombine with its target. Until very recently, the only higher eukaryote species that was amenable to gene-targeting technology was the mouse (reviewed by Muller 1999), and this is due to the special properties of embryonic stem (ES) cells. As discussed in Chapter 13, ES cells can be cultured like any established cell line, but they are derived from the very early mouse embryo and are therefore *pluripotent*. This means that if the cells are injected into a mouse blastocyst they can colonize the embryo and contribute to all its tissues, including the germline. The other important property of ES cells is that they have an unusual propensity for homologous recombination. Although random integration still occurs 1000-fold more frequently, PCR-based screening or appropriate selection strategies can be use to identify correctly targeted cells. These can be injected into mouse blastocysts to give rise to genetic chimeras. If colonization of the germline has occurred in these animals, their offspring will carry the targeted mutation in every cell. More recently, gene targeting has also been achieved in *Drosophila* using a combination of homologous and site-specific recombination (Rong & Golic 2000), in sheep and pigs by gene targeting in somatic cells followed by nuclear transfer to enucleated eggs (McCreath *et al.* 2000, Dai *et al.* 2002, Lai *et al.* 2002, Phelps *et al.* 2003), and in human ES cells (Zwaka & Thomson 2003). Gene targeting has also been achieved in plants, albeit at a low efficiency in most species (e.g. Kempin *et al.* 1997). The exceptions are the moss *Physcomitrella patens*, and rice (Chapter 14).

The only organism in which systematic gene targeting has been achieved is the yeast *Saccharomyces cerevisiae*

The genome of *S. cerevisiae* contains about 6000 open-reading frames, which by comparison to higher eukaryotes is a small number. Since the yeast genome has been completely sequenced, several systematic gene knockout projects have been initiated, one by a consortium of European research organizations named the European Functional Analysis Network (EUROFAN) (Dujon 1998) and another by a consortium of US and European laboratories named the *Saccharomyces* Gene Deletion Project consortium (Winzeler *et al.* 1999). The EUROFAN project involves the use of PCR-generated targeting cassettes in which a selectable marker is placed between ~50 bp elements corresponding to the flanking sequences of each yeast gene. Targeting with such constructs results in the replacement of the entire endogenous coding region with the marker, thus generating a null allele (Baudin *et al.* 1993). The *Saccharomyces* Gene Deletion Project consortium has generated similar targeting cassettes corresponding to about 96% of yeast genes (Winzeler *et al.* 1999, Giaever *et al.* 2002). Each contains, in addition to the selectable marker and yeast flanking sequences, two unique 20-bp "barcodes" placed just inside the yeast homology

Fig. 19.1   Barcoding strategy for yeast deletion strains. Dark purple boxes represent yeast homology regions, which recombine with the endogenous gene (crosses). Gray boxes represent unique oligonucleotide barcodes for unambiguous strain identification. Pale purple represents selectable marker gene. (See main text for details.)

region at each end (Fig. 19.1). These provide a means to detect the presence of specific strains in a population rapidly by hybridization of PCR products to an oligonucleotide chip, known as a barcode chip (Shoemaker *et al.* 1996). Thus the growth properties of potentially all the targeted yeast strains can be assayed in parallel (Giaever *et al.* 1999, 2002).

The advantage of these systematic approaches is that strains can be maintained as a central resource and then distributed to laboratories worldwide for functional analysis. Each mutant strain can be subjected to many different assays in parallel to determine the function of the missing gene product rapidly. For example, this approach has been used to screen for genes required for centromeric adhesion (Marston *et al.* 2004). Specialized screening may also be carried out e.g. to assay for sensitivity or resistance to particular drugs (Bianchi *et al.* 1999, Giaever *et al.* 2004, Baetz *et al.* 2004, Lum *et al.* 2004). In one report, the yeast deletion library was used to establish synthetic genetic arrays in which double-mutant phenotypes could be studied systematically (Tong *et al.* 2001).

It is unlikely that systematic gene targeting will be achieved in higher eukaryotes in the foreseeable future

Despite the benefits of gene targeting, in particular the precision with which specific mutant alleles can be designed, there has been no coordinated program for systematic gene knockouts in any other model organism. However, due to the efforts of individual researchers, several thousand independently derived targeted mouse strains have already been produced, and several excellent and comprehensive databases are available on the Internet which list these mouse strains (Box 19.1). It is therefore possible that a mouse knockout project could evolve over the next few years with the aim of completing the knockout catalog and providing a comprehensive genome-wide data set.

Box 19.1   Internet resources for gene targeting

Websites providing information on systematic knockout projects in yeast

http://mips.gsf.de/proj/eurofan/index.html
The EUROFAN website (Dujon 1998)
http://sequence-www.stanford.edu/group/
yeast_deletion_project The *Saccharomyces*
Gene Deletion Project website (Winzeler
et al. 1999)

Websites containing information and resources for transgenic and gene targeted mice

http://jaxmice.jax.org/index.shtml The
Jackson Laboratory website, describing
over 2500 strains of targeted mutant mice
http://tbase.jax.org TBASE, a comprehensive
transgenic and targeted mutant database
run by the Jackson Laboratory
http://www.bioscience.org/knockout/
knochome.htm Frontiers in Science gene
knockout database
http://biomednet.com/db/mkmd BioMedNet
Mouse Knockout Database

Since gene targeting has also been achieved recently in *Drosophila*, it is possible that this organism may be the next to benefit from a systematic gene-targeting program. However, the effort required to knock out the estimated 13,000 genes in the *Drosophila* genome may be too great given the more complex process of the gene-targeting methodology compared to yeast (p. 272).

Genome-wide random mutagenesis is a strategy applicable to all organisms

Saturation mutagenesis has been used for many years to identify mutations affecting specific biological processes. Essentially, the idea behind such an experiment is to mutagenize a population of whatever species is under study and recover enough mutants to stand a reasonable chance that each gene in the genome has been "hit" at least once. This population can then be screened to identify mutants in a particular function. For example, large-scale screens have been carried out in the past to look for replication mutants in bacteria, cell-cycle mutants in yeast, and more recently, developmental mutants in *Drosophila* and the zebrafish. The difference between these traditional studies and the new science of functional genomics is that, in the former, the majority of the mutants were discarded. Researchers focused on a particular area and ignored mutants affecting other processes because they were not interested in them. In functional genomics, *all* mutations are interesting, and the idea is to catalog them, generate

a sequence signature from each affected gene, and use these signatures to annotate full-length genomic and cDNA sequences housed in databases.

Insertional mutagenesis leaves a DNA tag in the interrupted gene, which facilitates cloning and gene identification

By far the most popular mutagenesis strategy in functional genomics is *insertional mutagenesis*, where a piece of DNA is randomly inserted into the genome causing gene disruption and loss of function. The DNA may constitute a *transposable element*, i.e. a sequence that can jump from site to site in the genome when supplied with the necessary enzyme (*transposase*), or it may be a foreign DNA sequence that is introduced into the cell. The main advantage of this strategy over traditional forms of mutagenesis is that the interrupted gene becomes "tagged" with the insertion element, hence the strategy is sometimes termed *signature-tagged mutagenesis* (STM). Simple hybridization- or PCR-based techniques can be used to obtain the flanking DNA. The sequence of this flanking DNA can then be used to interrogate sequence databases, allowing the tagged gene to be associated with its "parent" genomic clone or cDNA. If the insertion also generates a mutant phenotype, the gene in the database can then be ascribed a tentative function. Another advantage of insertional mutagenesis is that the insertion element can be modified into an entrapment vector to provide information about the gene it interrupts (Box 19.2)

Box 19.2   Gene traps and other advanced gene tagging vectors

Gene tagging is the use of an insertional mutagen to mark interrupted genes with a unique DNA sequence. This DNA sequence subsequently can be used as a target for hybridization or as an annealing site for PCR primers, allowing flanking sequences to be isolated. One simple way to achieve this is by inverse PCR, in which primers annealing at each edge of an insertion element are designed to point outwards. In this way, if a genomic fragment is circularized, the PCR can be used to amplify flanking regions (see Fig. B19.1). By careful design, however, simple gene tagging vectors can be modified in a number of ways to expedite cloning and provide more information about the interrupted genes. Some of these refinements are considered below.

Plasmid rescue vector

The insertion element in this type of vector contains the origin of replication and antibiotic resistance marker from a bacterial plasmid (Perucho *et al.* 1980). Genomic DNA from a tagged organism is digested with a restriction enzyme that does not cut in the insert, and the resulting linear fragments are self-ligated to form circles. The complex mixture of circles is then used to transform bacteria, which are grown under antibiotic selection. The circle containing the origin of replication and resistance gene is propagated as a plasmid while all the other circles are lost. In this way, the genomic sequences flanking the insert can be isolated and selectively amplified in a single step. Although more time-consuming than the direct amplification of flanking sequences by PCR, "rescued" plasmids can be maintained as a permanent resource library.

Gene trap vector

In this vector, the insertion element contains a visible marker gene such as *lacZ* (encoding β-galactosidase) or *gusA* (encoding β-glucuronidase) downstream of a splice acceptor site (Gossler *et al.* 1989, Friedrich &

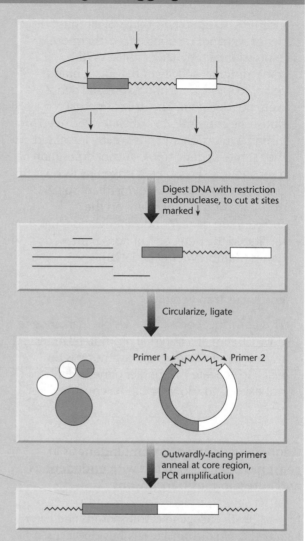

Digest DNA with restriction endonuclease, to cut at sites marked ↓

Circularize, ligate

Primer 1   Primer 2

Outwardly-facing primers anneal at core region, PCR amplification

Fig. B19.1   Inverse PCR. The core region is indicated by the wavy line. Restriction sites are marked with arrows, and the left and right regions which flank the core region are represented by closed and open boxes. Primers are designed to hybridize with core sequences and are extended in the directions shown. PCR amplification generates a linear fragment containing left and right flanking sequences.

Soriano 1991, Skarnes *et al.* 1992, Wurst *et al.* 1995). The marker gene is therefore activated only if the element inserts within the transcription unit of a gene and generates a transcriptional fusion. This strategy selects for insertions into genes and is very useful in animals and plants with large amounts of non-genic DNA (Evans *et al.* 1997, Springer 2000). Early gene trap vectors depended on in-frame

Box 19.2   *continued*

insertion, so up to two-thirds of all "hits" on genes were not recognized. Furthermore, expression of the marker relied on the transcriptional activity of the surrounding gene, so inserts into non-expressed genes were not detected. The use of internal ribosome entry sites has obviated the need for in-frame insertion and has greatly increased the hit rate of gene traps. The incorporation of a second marker, which is driven by its own promoter but carries a downstream splice donor making it dependent on the surrounding gene for polyadenylation, has facilitated the detection of non-expressed genes (Zambrowicz *et al.* 1998).

Enhancer trap vector

This construct comprises a visible marker gene downstream of a minimal promoter. Under normal circumstances the promoter is too weak to activate the marker gene and it is not expressed. However, if the construct

integrates in the vicinity of an endogenous enhancer, the marker is activated and reports the expression profile driven by the enhancer (O'Kane & Gehring 1987). The enhancer is often a long way from the gene so enhancer trapping is not a convenient method for cloning novel genes. However, it can be exploited in other ways, e.g. to drive the expression of a toxin and thus ablate a specific group of cells. This technique has been widely used in *Drosophila* (O'Kane & Moffat 1992).

Activation tagging

In this technique, the insertion element carries a strong outward-facing promoter. If the element integrates adjacent to an endogenous gene, that gene will be activated by the promoter. Unlike other insertion vectors, which cause loss of function by interrupting a gene, an activation tag causes gain of function through overexpression or ectopic expression (Kakimoto 1996).

Genome-wide insertional mutagenesis in yeast has been carried out with endogenous and heterologous transposons

In yeast, both endogenous transposons and heterologous (bacterial) transposons have been used for saturation mutagenesis. The endogenous retrotransposon *Ty* has been used as an insertional mutagen, and libraries of mutants have been generated carrying the *Ty* element as a "genetic footprint" (Smith *et al.* 1995, 1996). Several copies of the *Ty* element are normally present in the yeast genome, so an element modified to carry a unique DNA signature was used as a mutagen to enable unambiguous identification of interrupted genes. A PCR-based strategy was developed in which a primer annealing to the modified element was used in combination with a number of gene-specific primers to identify insertions at particular loci (Fig. 19.2). This strategy, which has been widely adopted in other functional genomics programs (see below), allows highly parallel analysis of large populations of yeast cells, increasing the likelihood that an insertion will be detected in a given gene. A disadvantage of *Ty*-based functional genomics

is the tendency for the element to insert at the 5′ end of genes transcribed by RNA polymerase III, i.e. mainly tRNA genes (Ji *et al.* 1993). Protein-encoding genes, which represent the vast majority of the transcriptome, are transcribed by RNA polymerase II.

Genome-wide insertional mutagenesis in yeast has also been performed with an *Escherichia coli* Tn*3* transposon modified to make a very sophisticated gene trap vector (Ross-Macdonald *et al.* 1997, 1999). Transposons from *E. coli* are not mobilized in yeast, but mutagenesis can be carried out in the surrogate environment of a bacterial cell transformed with a yeast genomic construct. From there, the genomic DNA is reintroduced into yeast cells, where it recombines with the endogenous genome, replacing the existing sequence. Ross-Macdonald and colleagues have generated a bank of *E. coli* strains housing genomic DNA clones covering the entire yeast genome. Mutant clones were then isolated from these strains and reintroduced into diploid yeast cells, generating a library of 12,000 mutant yeast strains. There are several advantages to this strategy including the fact that stable insertions are generated, allowing the production of mutant lines as a

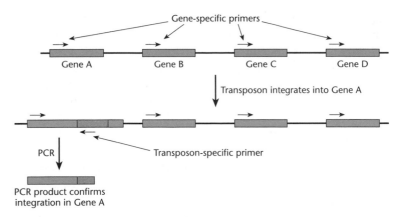

Fig. 19.2 Principle of target-selected random mutagenesis. Random insertional mutagenesis is carried out on a population. Mutants for any particular gene can then be identified by PCR amplification using one gene-specific primer and one primer specific for the insertional mutagen. For organisms with completely sequenced genomes, inserts can be identified in any gene. For organisms with incomplete genome sequences, the target gene has to be isolated and sequenced first so that gene-specific primers can be designed. In large populations, pools of cells/seeds/embryos, etc. can be screened and then deconvoluted to identify individual mutant strains.

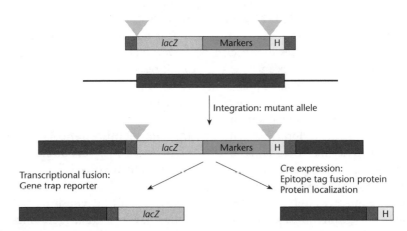

Fig. 19.3 Multifunctional *Escherichia coli* Tn*3* cassette used for random mutagenesis in yeast. The cassette comprises Tn*3* components (black), *lacZ* (gray), selectable markers (light purple) and an epitope tag (very light purple, H). The *lacZ* gene and markers are flanked by *loxP* sites (triangles). Integration generates a mutant allele which may or may not reveal a mutant phenotype. The presence of the *lacZ* gene at the 5′ end of the construct allows transcriptional fusions to be generated, so the insert can be used as a reporter construct to reveal the normal expression profile of the interrupted gene. If Cre recombinase is provided, the *lacZ* gene and markers are deleted, leaving the endogenous gene joined to the epitope tag, allowing protein localization to be studied.

permanent resource, and that *E. coli* transposons are not subject to the insertional bias of yeast *Ty* elements, and therefore provide a better random coverage of the genome. Even so, there is still a certain degree of site selection, resulting in the preferential recovery of mutant alleles for some genes and the omission of others. The modified Tn*3* transposon comprised a reporter gene downstream of a splice acceptor bracketed by subterminal *loxP* sites, allowing excision of most of the insert following expression of Cre recombinase. Outside the *loxP* sites, the transposon also carries an in-frame epitope tag. Thus,

following Cre-mediated excision of the transposon, the endogenous gene becomes fused to the epitope tag, allowing the characterization and localization of proteins (Fig. 19.3).

Genome-wide insertional mutagenesis in vertebrates has been facilitated by the development of artificial transposon systems

One of the key functional objectives of the mouse genome project is the development of a comprehensive insertional mutagenesis library in ES cells (Denny

& Justice 2000). Tagging by insertional mutagenesis has a long history in mouse genetics. It has been demonstrated serendipitously in transgenic mice in cases where the transgene has by chance integrated into an endogenous gene and inactivated it (e.g. see Rijkers *et al.* 1994). Also, it is notable that approximately 5% of all naturally occurring recessive mutations in the mouse are caused by a particular family of retroviruses when they integrate into or adjacent to an endogenous gene. Indeed, the cloning of several mouse genes has been facilitated by virtue of their linkage to a proviral sequence (e.g. Bowes *et al.* 1993).

Gene trapping in ES cells is being carried out by several groups. The concept of ES gene trap libraries was first put into practice by Hicks *et al.* (1997), who analyzed 400 inserts and identified 63 genes and anonymous cDNAs on the basis of flanking sequence tags. The German Gene Trap Consortium (GGTC) is aiming to produce and characterize 20,000 gene trap ES lines using a retroviral gene trap vector. By the year 2000, more than 12,000 lines had been produced and flanking sequences had been determined to generate a database of gene trap sequence tags (GSTs) (Wiles *et al.* 2000). A similar project was carried out by the US biotechnology company Lexicon Genetics. The result was a library of gene trap ES cell lines called OmniBank, represented by over 50,000 "OmniBank sequence tags" corresponding to both expressed and non-expressed genes. Both these organizations have websites providing more information on the programs and a searchable database to identify tagged genes (Box 19.3).

Retrovirus-based vectors have been used to create *cell-based* mutant libraries in vertebrates, but direct application in whole animals has been more challenging because of the lack of active retrotransposons in vertebrate genomes (Ivics & Izsvak 2005). For example, vertebrate genomes contain millions of *mariner*-type transposons, all of which appear to be non-functional due to the accumulation of mutations. This problem has been addressed in one case by creating a new transposon through the derivation of a functional consensus sequence from the sequences of many non-functional copies. The strategy was based on the assumption that different mutations would occur in different copies of the element, and that a consensus sequence should represent an active, archetypal element. By looking at endogenous fish *mariner*-type retrotransposons in this manner, the *Sleeping Beauty* transposon system was created (Ivics *et al.* 1997). *Sleeping Beauty* has been used for large-scale mutagenesis studies in fish (Zbikowska 2003) as well as in mammals (Ivics *et al.* 1997, Clarke *et al.* 2004). Similarly, *Frog Prince* is a reconstructed amphibian retrotransposon based on endogenous non-functional elements in the frog *Rana pipens*. It has been shown to work in a wide range of vertebrate cells (Miskey *et al.* 2003). In another approach, an endogenous retrotransposon with low activity has been "enhanced" to increase its transposition frequency. Han & Boeke (2004) have modified a mouse LINE1 retrotransposon to remove an adenosine-rich tract within one of the open reading frames without altering the coding sense, thereby increasing the transposition frequency over 200-fold. In a third approach, Dewannieux *et al.* (2004) screened through the 1000 or so IAP (intracisternal A particle) retrotransposons in the mouse genome to identify those with intact open reading frames and, by implication, the highest activity. Neither of these approaches has yet matched the impressive 10% transposition efficiency of *Sleeping Beauty* (determined by the introduction of an antibiotic resistance cassette into cultured cells) although further modification of the transposons could enhance them further.

Insertional mutagenesis in plants can be achieved using *Agrobacterium* T-DNA or plant transposons

Insertional mutagenesis is the most appropriate option for functional genomics in plants because of the very low efficiency with which homologous recombination occurs in higher plant species. A number of different systems have been developed, and essentially these fall into two groups: those involving the use of T-DNA from *Agrobacterium tumefaciens* (Azpiroz-Leehan *et al.* 1997, Krysan *et al.* 1999), and those involving the use of plant transposons (Walbot 2000, Hamer *et al.* 2001, Ramachandran & Sundaresan 2001). In each case, if a large enough population of tagged lines is available, there is a good chance of finding a plant carrying an insert within any gene of interest. Mutations that are homozygous lethal can be maintained in the population as heterozygous plants.

T-DNA mutagenesis requires gene transfer by *A. tumefaciens*

A. tumefaciens is the soil bacterium used most extensively to generate transgenic plants (Chapter 14).

Box 19.3   Internet resources for genome-wide random mutagenesis projects

Insertional mutagenesis in yeast

http://ygac.med.yale.edu/mtn/insertion_
   libraries.stm

Insertional and ethylnitrosourea mutagenesis in the mouse

For a more extensive list of mouse resources see Brekers & Harabe de Angelis (2001) and Sanford *et al.* (2001).

http://lena.jax.org/~jcs/Delbank.html
   DELBank project information

http://www.lexgen.com/omnibank/
   omnibank.htm OmniBank gene trap ES cell
   library, information and resources

http://baygenomics.ucsf.edu/ BayGenomics
   gene trap project

http://socrates.berkeley.edu/~skarnes/
   resource.html A useful resource for mouse
   gene trap insertions

http://tikus.gsf.de German Gene Trap
   Consortium ES cell library, information and
   resources

http://www.gsf.de/ieg/groups/
   enu-mouse.html German Human Genome
   Project mouse ENU mutagenesis screen

http://www.mgu.har.mrc.ac.uk/mutabase
   UK/French consortium mouse ENU
   mutagenesis program

Insertional mutagenesis in plants

http://www.biotech.wisc.edu/
   NewServicesAndResearch/Arabidopsis/

default.htm *Arabidopsis* Knockout Facility
   at the University of Wisconsin

http://mbclserver.rutgers.edu/~dooner/
   PGRPpage.html Maize "*Ac* gene engine"
   project to generate evenly distributed *Ac*
   inserts in the maize genome

http://www.zmdb.iastate.edu/zmdb/
   sitemap.html Maize Gene Discovery and
   RescueMu project

http://mtm.cshl.org Maize Targeted
   Mutagenesis database

http://nasc.nott.ac.uk/ima.html Information
   and database of *Ds* insertion lines and
   flanking sequences (Parinov *et al.* 1999)
   with instructions for obtaining seeds

http://www.jic.bbsrc.ac.uk/sainsbury-
   lab/jonathan-jones/SINS-database/sins.htm
   Information and database of sequenced
   insertion sites (SINS) from *dSpm* lines
   (Tissier *et al.* 1999) with instructions for
   obtaining seeds

http://signal.salk.edu/cgi-bin/tdnaexpress
   The Salk Institute Genome Analysis
   Laboratory (SIGnAL)

http://nasc.nott.ac.uk Nottingham *Arabidopsis*
   Stock Centre

http://arabidopsis.org/abrc *Arabidopsis*
   Biological Resource Center

Insertional mutagenesis in *Drosophila*

http://www.fruitfly.org/p_disrupt/index.html
   Berkeley *Drosophila* gene disruption project

Plant transformation involves the transfer of a small segment of DNA, called the T-DNA, from a plasmid harbored by the bacterium into the plant genome. This T-DNA can act as an insertional mutagen, and has thus far been used for genome-wide mutagenesis programs in *Arabidopsis thaliana* and rice (Walden 2002). Since T-DNA is not a transposon, it has no ability to "jump" following integration, and so has the advantage of generating stable insertions in primary transformants. T-DNA integration is also believed to favor genes, but within the "gene space" of the genome integration is essentially random.

This is not the case for most plant transposons (see below). One disadvantage of T-DNA, however, is its tendency to generate complex, multicopy integration patterns, sometimes involving the deletion or rearrangement of surrounding genomic DNA. This can complicate subsequent analysis, especially if the PCR is used to confirm integrations within specific genes.

A number of research groups have invested resources in generating banks of *Arabidopsis* T-DNA insertion lines (Feldmann & Marks 1987, Bouchez *et al.* 1993, Campisi *et al.* 1999, Krysan *et al.* 1999,

Weigel *et al.* 2000). This reflects the availability of simple and convenient techniques for *Agrobacterium*-mediated transformation of *Arabidopsis*, facilitating rapid saturation of the genome (Bechtold & Pelletier 1998, Clough & Bent 1998). Furthermore, *Arabidopsis* is a gene-dense plant with small introns and little intergenic space; about 80% of the genome is thought to represent genes.

Currently, over 130,000 T-DNA tagged *Arabidopsis* lines are made available by the University of Wisconsin *Arabidopsis* Knockout Facility,[1] which maintains a searchable database of mutant lines (Box 19.3). These lines comprise two populations, one generated by the insertion of a simple T-DNA construct (Krysan *et al.* 1999), and one generated by the insertion of an activation tag (Weigel *et al.* 2000). Such populations can be used for comprehensive reverse genetics screens (McKinney *et al.* 1995, Winkler *et al.* 1998, Krysan *et al.* 1999). DNA from the tagged lines is maintained as a series of hierarchical pools. These can be screened in several rounds of PCR using gene-specific primers supplied by the customer and T-DNA-specific primers supplied by the facility. Indeed, the system is very similar to that used in yeast (see Fig. 19.2). If a "hit" is achieved, corresponding seeds can be ordered and the customer can then grow plants with a particular gene disrupted.

Modified T-DNA vectors have been used in *Arabidopsis* not only as activation tags (Weigel *et al.* 2000) but also as gene and promoter traps (Feldmann 1991, Lindsey *et al.* 1993, Babiychuk *et al.* 1997, Campisi *et al.* 1999). A gene trap T-DNA vector has also been used by Jeon *et al.* (2000) in a genome-wide screen of rice. These investigators produced over 22,000 primary transformants carrying the T-DNA insertion, more than half of which contained multiple T-DNA copies. Over 5000 tagged lines were analyzed for reporter gene expression in leaves and roots, 7000 lines were analyzed for expression in flowers, and 2000 for expression in seeds. Overall, about 2% of the lines showed marker gene activity, in some cases ubiquitous but in many cases restricted to highly specific cell types or tissues. Rice lines containing T-DNA insertions have been generated and the flanking sequences have been characterized by Sha *et al.* (2004).

Transposon mutagenesis in plants can be achieved using endogenous or heterologous transposons

Transposons have been widely used for insertional mutagenesis in plants and this has led to the discovery of many new genes (reviewed by Gierl & Saedler 1992). Several transposons have been used for genome-scale mutagenesis projects, including *Activator* (*Ac*), *Suppressor-mutator* (*Spm*), and *Mutator* (*Mu*) from maize, *Tam3* from *Antirrhinum majus*, and *Tph1* from *Petunia*. Unlike T-DNA, transposons tend to generate simple, single-copy insertions. However, as is the case for *Ty* elements in yeast, most plant transposons show pronounced "target-site preference", which can make it difficult to achieve whole-genome saturation. Also, while T-DNA generates stable inserts, additional crossing steps are required to stabilize those generated by plant transposons. This is because transposons have the intrinsic ability to mobilize unless their source of transposase is removed. Control is generally achieved by the use of "two-component" transposon systems, comprising an autonomous (self-mobilizing) element and a non-autonomous derivative. For example, the maize transposon *Activator* (*Ac*) is autonomous because it encodes its own transposase, but shorter derivatives of *Ac* called *Dissociation* (*Ds*) lack the transposase gene. However, *Ds* elements can transpose if transposase is provided by *Ac*. Thus, where *Ac* and *Ds* are present in the same genome, both elements can be mobile. However, if *Ac* is removed by crossing, progeny plants can be recovered with stable *Ds* insertions.

The properties of transposons vary, making different transposon families suitable for different applications. Although maize *Ac/Ds* has been widely used for genome-scale mutagenesis, it actually demonstrates a phenomenon called "local transposition", i.e. it jumps preferentially to linked sites (as does the *Sleeping Beauty* element discussed above). This can make saturation difficult, particularly if

---

[1] Some confusion in terminology can arise here. A *gene knockout* was originally defined as a null mutation produced by gene targeting (homologous recombination). More recently, the term has been used to describe any sort of induced null mutation, including those generated by (random) insertional mutagenesis. Hence, the *Arabidopsis* Knockout Facility maintains a collection of null mutants generated not by gene targeting but by random T-DNA insertion. Even worse, it is becoming common for mutations generated by random insertion to be called "targeted mutations" even though this description is strikingly inaccurate. Hence, the Maize Targeted Mutagenesis Project concerns a population of random insertional mutants generated by the transposon *Mutator*. The term "target-selected mutagenesis" is more accurate, referring to the fact that researchers can identify a mutation in a particular target gene, from a randomly mutagenized population, using a PCR-based assay on pooled DNA.

there are a small number of founder lines, but it can also be an advantage for generating multiple mutant alleles in one gene, or for generating mutations in several clustered genes in a local genomic region. If necessary, selection systems can be devised to select against closely linked transpositions (Parinov *et al.* 1999, Tissier *et al.* 1999). This problem can also be circumvented by generating a population of maize plants with *Ac/Ds* elements spaced at regular 10–20 cM intervals throughout the genome. Such a project is indeed in progress at the Waksman Institute, Rutgers University, and can be seen at the following website: http://mbclserver.rutgers.edu/~dooner/PGRPpage.html.

Another maize transposon, *Mutator* (*Mu*), does not show preferential local transposition, and is therefore a potentially better global mutagen than *Ac*. However, it does preferentially insert into transcription units, making it an excellent tool for gene disruption. Several genome-wide mutagenesis projects have therefore been established in maize using *Mu*, including the Trait Utility System for Corn (TUSC) developed by Pioneer Hi-Bred International, and the Maize Targeted Mutagenesis (MTM) project. In each case, PCR primers facing away from the transposon are used in combination with a gene-specific primer to identify insertions into specific genes, with DNA pooled from maize plants in the field as the template. These resources have been used successfully by a number of investigators (e.g. Bensen *et al.* 1995, Das & Martiensen 1995, Hu *et al.* 1998). The Maize Gene Discovery Project uses a modified *Mu* transposon called *RescueMu*, which can be used for plasmid rescue from whole-genomic DNA. The rescued plasmids have been used to generate DNA libraries containing the *Mu* insertion sites. The PCR is carried out on these plasmids to identify insertions in genes of interest. In all the facilities, seeds corresponding to each insertion can be supplied. A genome-wide study using *RescueMu* has been published recently (Fernandez *et al.* 2004). Websites with further information on maize genome projects, search facilities, and databases of tagged genes are listed in Box 19.3.

Some transposons, such as the *Drosophila* P-element (see below) can only function in their host species. Others, including *Ac* and another maize transposon called *Suppressor-mutator* (*Spm*) are more promiscuous and these can be used in a range of heterologous plants (Osbourne & Baker 1995). The transposons must initially be introduced into the foreign genome either as a T-DNA or as a conventional transgene delivered by a method such as particle bombardment. Once integrated into the genome, however, normal transposition may then occur. Functional genomics programs using *Ac* have been initiated in *Arabidopsis* (Ito *et al.* 1999, Seki *et al.* 1999, Kuromori *et al.* 2004) and tomato (Meissner *et al.* 2000). *Spm* has been used for several large-scale *Arabidopsis* mutagenesis projects (e.g. Tissier *et al.* 1999, Speulman *et al.* 1999). Where *Ac* has been used in *Arabidopsis*, the investigators have exploited local transposition to saturate genomic regions surrounding the original integration site. In both studies (Ito *et al.* 1999, Seki *et al.* 1999), a cDNA scanning strategy was used to isolate ESTs from this region, leading to rapid gene annotation.

The genome-wide transposon mutagenesis projects in *Arabidopsis* have produced a large number of mutant lines. As for the T-DNA insertion lines discussed above, many of the transposon lines are now maintained at a central resource, in this case the Nottingham *Arabidopsis* Stock Centre in the UK, and the *Arabidopsis* Biological Resource Center in America. This includes 960 DNA pools from *Spm* insertion lines corresponding to 48,000 inserts (Tissier *et al.* 1999) and about 2600 DNA pools from multiple *Spm* insertion lines, representing up to 65,000 insertions (Speulman *et al.* 1999). The flanking sequences have been determined and analyzed by BLAST searches, and the results are available on searchable databases, allowing the rapid identification of interrupted genes (Seki *et al.* 1999, Parinov *et al.* 1999, Speulman *et al.* 1999, Tissier *et al.* 1999). Websites for these resources are listed in Box 19.3.

Insertional mutagenesis in invertebrates

Genome-wide insertional mutagenesis programs have been initiated in several other animals. For example, the *Tc1* transposon has been used to generate a frozen bank of insertion mutants of the nematode *C. elegans* yielding 5000 lines and 16 newly identified genes (Zwall *et al.* 1993). P-elements (Chapter 13) have been developed as insertional mutagens (Spradling & Rubin 1982) and have been used to clone and characterize many *Drosophila* genes (reviewed by Cooley *et al.* 1988). Currently, a genome-wide mutagenesis program is ongoing at The Berkeley *Drosophila* Genome Project, with the aim of generating a comprehensive library of mutant fly strains (Spradling *et al.* 1995, 1999). In the initial phase of the program, about 4000 mutagenized lines were examined for P-element insertions. Redundant strains (i.e. allelic mutations) were eliminated, leaving 1045 unique inserts identifying over 1000 genes with homozygous lethal phenotypes. This corresponds to approximately 25% of all "essential" genes on the autosomal chromosomes. As discussed above, saturation of the genome has proven difficult because P-elements, like many transposons, show a pronounced insertional bias (Liao *et al.* 2000). Also, many interrupted genes do not reveal a mutant phenotype, perhaps because the element has inserted into a non-essential region, such as an intron. With the recent completion of the *Drosophila* genome sequence (Adams *et al.* 2000), Spradling and colleagues have begun to collect sequence signatures from the flanks of each P-element insertion and identify genes by comparing these signatures to the genomic sequence. This removes the dependency on a mutant phenotype. At the same time, an additional set of P-element vectors is being used, which function as activation tags (Rorth 1996, Rorth *et al.* 1998). Hopefully, many of the genes that do not reveal informative loss of function phenotypes will show gain of function phenotypes in this screen. More recently, Thibault *et al.* (2004) have shown that the combined use of two different transposon systems with distinct site preferences (P-elements and *piggyback* transposons), can accelerate genome saturation – they achieved 53% saturation with 29,000 inserts. The Berkeley *Drosophila* Genome Project maintains websites for both the P-element mutagenesis program and the P-element gene misexpression program (Box 19.3). Many of the disrupted lines are also available from *Drosophila* resources, such as the Bloomingdale Stock Center: http://flybase.bio.indiana.edu/stocks.

Chemical mutagenesis is more efficient than transposon mutagenesis, and generates point mutations

While insertional mutagenesis is likely to remain the most popular approach to functional genomics because of the way interrupted genes are marked with a DNA tag, the use of chemical mutagens is more efficient, and also generates point mutations, which can be more informative under certain circumstances. In the mouse, for example, models of human disease phenotypes are much more likely to arise by point mutation than by the insertion of a large DNA cassette. Furthermore, insertional mutagenesis tends to generate null alleles (complete loss

of gene function) which in many cases is lethal, while point mutations often have less severe effects.

The alkylating agent ethylnitrosurea (ENU) is the most powerful mutagen available for mice (Russell *et al.* 1979; reviewed by Justice *et al.* 1999). Large-scale mutagenesis screens have been carried out successfully in other species using this chemical (e.g. Mullins *et al.* 1994 used ENU in zebrafish) but most of the screens carried out in mice have been limited in their scope (e.g. Bode *et al.* 1988, Shedlovsky *et al.* 1988, Rinchik *et al.* 1990). Two groups of researchers have carried out genome-wide ENU mutagenesis screens in mice. As part of the German Human Genome Project, Balling and colleagues have screened 14,000 ENU-mutagenized mouse lines for dominant and recessive mutations affecting a large number of clinically important phenotypes (Hrabe de Angelis & Balling 1998, Hrabe de Angelis *et al.* 2000). Categories included allergy, immunology, clinical chemistry, nociception (response to pain), and dysmorphology (abnormal structure). In the initial study, 182 mutants were cataloged and many more are still undergoing analysis. Simultaneously, a consortium of UK and French researchers reported a similar large-scale experiment (Nolan *et al.* 2000). In this case, 26,000 mice were screened for dominant mutations and 500 were recovered. The mice were tested for visible developmental defects from birth to weaning, and then subjected to a battery of functional, behavioral, and biochemical tests over the next eight weeks. In both programs, mutations were mapped by interspecific backcrossing using a genome-wide panel of microsatellite polymorphisms (Balling 2001).

Libraries of knock-down phenocopies can be created by RNA interference

A phenocopy has the same appearance as a mutant phenotype, but there are no changes to the DNA sequence. While phenotypes are caused by mutations, phenocopies are generated by interfering with gene expression. For example, antisense RNA can be used to inactivate the messenger RNA corresponding to a particular gene, or antibodies can be used to inactivate the protein. In each case, there is a loss of gene function while the gene itself remains intact (Chapter 15). Over the last five years, RNA interference has become the method of choice for large-scale functional studies because of its simplicity and the potency of its effects. The basis of RNA interference is discussed in Chapter 15, while in this chapter we consider how the technique has been applied in large-scale studies.

RNA interference has been used to generate comprehensive knock-down libraries in *Caenorhabditis elegans*

The genome sequence of the nematode worm *Caenorhabditis elegans* was published by the *C. elegans* Genome Sequencing Consortium in 1998. Of the higher (multicellular) eukaryotes, *C. elegans* has the smallest genome and the most convenient biological properties for high-throughput handling – it is small, hermaphrodite, and can be viably stored as frozen stocks. Therefore the worm is a very attractive target for functional genomic studies. Coincidentally, *C. elegans* is the model organism in which the phenomenon of RNA interference was first documented (Fire *et al.* 1998). Over the last few years, several large-scale RNAi screens have been carried out in *C. elegans*. In many cases, the phenocopies of the RNAi functional knockouts are equivalent to genuine knockout phenotypes produced by traditional mutagenesis strategies.

Microinjection is currently the most consistently effective way to induce RNAi. Typically, *in vitro* synthesized dsRNA is injected into the germline of adult worms, and progeny are screened for RNAi-induced phenocopies. Due to the laborious nature of the microinjection procedure, carrying out individual injections corresponding to each of the 15,000 genes in the *C. elegans* genome is a challenging task, but Gonczy *et al.* (2000) described just such an approach for the functional analysis of chromosome III. This group synthesized over 2200 individual dsRNA molecules (corresponding to over 95% of the genes on the chromosome) and then carried out systematic microinjections followed by screening for RNAi-induced phenocopies that affected the first two cleavage divisions of development. Remarkably, they obtained a hit rate of over 6% (133 genes were found to be involved in cleavage). This group of genes included all seven chromosome III genes that had previously been shown through traditional mutagenesis to affect cell division, providing important validation of the procedure.

Sugimoto and colleagues (Maeda *et al.* 2001) used the alternative and simpler approach of adding dsRNA to the worms' medium to test the function of approximately 2500 genes represented in the *C. elegans* EST database. They found that nearly one-third of the genes revealed an RNAi phenocopy that could be easily scored under the dissecting microscope. Ahringer and colleagues adopted another approach, that of feeding worms on bacteria expressing dsRNA (Fraser *et al.* 2000). In a pilot study, they generated a library of bacterial strains expressing over 2400 different dsRNAs, representing just under 90% of the genes on chromosome I. Worms were then fed on these bacterial strains, and RNAi-induced phenocopies were sought in the progeny. Of the genes tested 14% revealed observable phenocopies, and in 90% of the genes that had previously been characterized by mutagenesis, the mutant phenotypes and RNAi phenocopies were concordant. More recently, this project has been expanded to cover the entire *C. elegans* genome, and a library has been made publicly available covering 86% of the genes (Kamarth & Ahringer 2003, Kamarth *et al.* 2003). This feeding library has been used for the systematic functional analysis of different classes of genes, such as G-protein-coupled receptors (Keating *et al.* 2003) and genes involved in apoptotic DNA degradation (Parrish *et al.* 2003). Interestingly, it has proven more difficult to generate phenocopies of neuronal gene mutations in *C. elegans* by RNAi, one specific example of the general problem of variable penetrance associated with this technique. Recently, however, a mutation that sensitizes worms to the effect of RNAi has been identified which may alleviate this problem (Simmer *et al.* 2003). Another problem with RNAi is variable specificity, or "off-target effects". This is likely to reflect the design of siRNA constructs, and may be resolved as researchers gain more experience. At the current time, the problem of off-target effects is generally tackled by designing more than one RNAi construct for each target gene (Jackson *et al.* 2003).

The first genome-wide RNAi screens in other organisms have been carried out

RNAi is not restricted to the worm, and it has been used for mid-scale screens in a number of other organisms, including the fruit fly, plants, the mouse, and human cells. An example of such a screen carried out in *Drosophila* was reported by Lum *et al.* (2003). These investigators discovered several new components of the Hedgehog signaling pathway using RNAi constructs representing about 40% of the genome. The first genome-wide screen in *Drosophila* was reported recently (Boutros *et al.* 2004). This involved constructs representing 91% of the genome and sought to identify genes required for cell growth and viability. Duplicate screens were performed to eliminate false positives and nearly 450 dsRNAs generated a recognizable phenotype in both screens, most of which represented genes that had not been identified in previous mutagenesis experiments. Small-scale screens in mammalian cells using synthetic siRNAs or short hairpin RNA (shRNA) transgenes have been reported (e.g. Aza-Blanc *et al.* 2003). More recently, several large-scale RNAi screens have been carried out in mammalian cells. In one such study, which sought to identify novel regulators of the transcription factor NF-κB, 16,000 shRNA constructs were synthesized, two for each of the target genes to improve the hit rate (Zheng *et al.* 2004). In a second study, 15,000 shRNA retrovirus vectors were used to identify regulators of proteasome function (Paddison *et al.* 2004). In a third study, Berns *et al.* (2004) synthesized 8000 retroviral shRNA vectors to study regulators of p53. In each of the latter studies, a barcode strategy was used to allow pooling of the vectors and increase the throughput of the technique.

Suggested reading

Anderson K.V. (2000) Finding the genes that direct mammalian development: ENU mutagenesis in the mouse. *Trends in Genetics* **16**, 99–102. *A commentary on the use of ENU for large-scale mutagenesis studies in the mouse, focusing on development.*

Barstead R. (2001) Genome-wide RNAi. *Current Opinion in Chemical Biology* **5**, 63–6.

Kim S.K. (2001) *C. elegans*: mining the functional genomic landscape. *Nature Reviews Genetics* **2**, 681–9.

Sugimoto, A. (2004) High-throughput RNAi in *Caenorhabditis elegans*: genome-wide screens and functional genomics. *Differentiation* **72**, 81–91.
Current accounts of functional genomics in C. elegans using RNA interference and other methods.

Beckers J. & Hrabe de Angelis M. (2001) Large-scale mutational analysis for the annotation of the mouse genome. *Current Opinion in Chemical Biology* **6**, 17–23. *An excellent summary of mutational analysis in the mouse, covering transgenics, gene traps, knockouts, and ENU mutagenesis.*

Cecconi F. & Meyer B.I. (2000) Gene trap: a way to identify novel genes and unravel their biological function. *FEBS Letters* **480**, 63–71.

Stanford W.L., Cohn J.B. & Cordes S.P. (2001) Gene-trap mutagenesis: past, present and beyond. *Nature Reviews Genetics* **2**, 756–68.
Two good summaries of insertional mutagenesis projects in the mouse, using gene-trap vectors.

Coelho P.S.R., Kumar A. & Snyder M. (2000) Genome-wide mutant collections: toolboxes for functional genomics. *Current Opinion in Microbiology* **3**, 309–15. *An excellent and very accessible review concentrating on gene knockout and random mutagenesis projects in yeast and other microbes, with comparison to similar projects in higher eukaryotes.*

Hamer L., DeZwaan T.M., Montenegro-Chamorro M.V., Frank S.A. & Hamer J.E. (2001) Recent advances in large-scale transposon mutagenesis. *Current Opinion in Chemical Biology* **5**, 67–73. *A summary of recent transposon mutagenesis projects, concentrating on yeast and plants.*

Maes T., De Keukeleire P. & Gerats T. (1999) Plant tagnology. *Trends in Plant Sciences* **4**, 90–6.

Parinov S. & Sundaresan V. (2000) Functional genomics in *Arabidopsis*: large-scale insertional mutagenesis complements the genome sequencing project. *Current Opinion in Biotechnology* **11**, 157–61.
These reviews provide broad coverage of transposon and T-DNA mutagenesis studies in plants.

Friedman A. & Perrimon N. (2004) Genome-wide high-throughput screens in functional genomics. *Current Opinion in Genetics & Development* **14**, 460–76.

Nagy A., Perrimon N., Sandmeyer S. & Plasterk R. (2003) Tailoring the genome: the power of genetic approaches. *Nature Genetics* **33**: 276–84.

Rijkers T., Peetz A. & Ruther U. (1994) Insertional mutagenesis in transgenic mice. *Transgenic Research* **3**, 203–15.

CHAPTER 20

Analysis of the transcriptome

Introduction

Important insights into gene function can be gained by *expression profiling*, i.e. determining where and when particular genes are expressed. For example, some genes are switched on (induced) or switched off (repressed) by external chemical signals reaching the cell surface. In multicellular organisms, many genes are expressed in particular cell types or at certain developmental stages. Furthermore, mutating one gene can alter the expression of others. All this information helps to link genes into functional networks, and genes can be used as *markers* to define particular cellular states.[1]

In the past, genes and their expression profiles have been studied on an individual basis. Therefore, defining functional networks in the cell has been rather like completing a large and complex jigsaw puzzle. More recently, technological advances have made it possible to study the expression profiles of thousands of genes simultaneously, culminating in *global expression profiling*, where every single gene in the genome is monitored in one experiment. This can be carried out at the RNA level (by direct sequence sampling or through the use of DNA arrays, as discussed in this chapter) or at the protein level (as discussed in the next chapter). Global expression profiling produces a holistic view of the cell's activity. Complex aspects of biological change, including differentiation, response to stress, and the

onset of disease, can thus be studied at the genomic level. Instead of defining cell states using single markers, it is now possible to use clustering algorithms to group data obtained over many different experiments and identify groups of co-regulated genes. This provides a new way to define cellular phenotypes, which can help to reveal novel drug targets and develop more effective pharmaceuticals. Furthermore, anonymous genes can be functionally annotated on the basis of their expression profiles, since two or more genes that are co-expressed over a range of experimental conditions are likely to be involved in the same general function.

Traditional approaches to expression profiling allow genes to be studied singly or in small groups

Since the 1970s, techniques have been available to monitor gene expression on an individual basis, and many of these are discussed in Chapter 2. The principles behind each technique are summarized in Box 20.1. Analysis at the RNA level invariably relies on the specificity of nucleic acid hybridization. The target RNA is either directly recognized by a labeled complementary nucleic acid probe, or is first converted into cDNA and then hybridized to a specific pair of primers which facilitate amplification by the PCR.

Unfortunately, none of these traditional techniques is particularly suited to global expression profiling. This is because the experimental design is optimized for single gene analysis, i.e. each experiment works on the principle that a single nucleic acid probe (or primer combination) is used to identify a single target. Although it is possible to modify at least some of the techniques for the parallel analysis of multiple genes (*multiplexing*), the procedure becomes increasingly technically demanding and laborious as more genes are assayed simultaneously.

---

[1] In this context, a marker is a gene whose expression defines a particular cellular phenotype. For example, a neuronal marker is a gene expressed only in neurons and a cancer marker is a gene expressed only in tumors. The term *marker* is also used in a variety of alternative ways, e.g. to describe landmarks on physical and genetic maps, and to describe genes that confer selectable or scorable phenotypes on transformed cells and transgenic animals and plants. Standard sized proteins or nucleic acids used as references in electrophoresis experiments are also known as markers.

Box 20.1   Gene-by-gene techniques for expression analysis

The techniques described below are some of the most widely applied in molecular biology. For a more detailed discussion of these procedures and an extensive list of original references, the interested reader should consult Sambrook & Russell (2000).

Northern blot and RNA dot blot

In these similar, hybridization-based techniques, RNA from a complex source is transferred to a membrane and immobilized, either without prior fractionation (dot blot) or after fractionation by electrophoresis (northern blot). A labeled probe (DNA, antisense RNA, or an oligonucleotide) is then hybridized to the immobilized RNA. The dot blot can indicate the presence or absence of a particular transcript, and allows rough quantification of the amount of RNA if several samples are compared. In addition, the northern blot allows size determination and can reveal the presence of homologous transcripts of different sizes, such as alternative splice variants. In both cases, the probe is applied to the membrane in great excess to the target, and hybridization is carried out to saturation so that the signal intensity reflects the abundance of the immobilized target. Disadvantages of these techniques include their low sensitivity and the large amount of input RNA required. It is difficult to detect rare transcripts using this method.

Reverse northern blot

In this technique, individual target cDNAs or genomic DNA fragments are immobilized on a membrane and hybridized with a complex probe, i.e. a probe prepared from a heterogeneous RNA source. When carried out

to saturation, this technique is often used simply to identify or confirm the presence of genes in large genomic clones (see Chapter 6). However, if non-saturating hybridization is carried out (i.e. with the immobilized target in great excess) the signal intensity reflects the abundance of hybridizing molecules in the probe. DNA array hybridization is based on this principle (see main text).

Nuclease protection

A labeled antisense RNA probe is hybridized in solution with a complex RNA population. The probe is present in excess, and the mixture is then treated with a selective nuclease such as RNaseA, which digests single-stranded RNA but not double-stranded RNA. Hybridization between the probe and target RNA protects the probe from degradation, and simultaneously allows the signal to be visualized on a sequencing gel. This technique is sensitive and, since probe-target binding is stoichiometric, allows quantification of the target molecule.

RT-PCR (reverse transcriptase polymerase chain reaction)

In this technique a population of mRNA molecules is reverse transcribed to generate an equivalent population of cDNAs. These are then amplified by PCR using primers specific for a particular gene or genes. With appropriate controls, RT-PCR can be semi- or fully quantitative, and the amplification is such that very rare target molecules can be detected and quantified. Many experiments can be carried out in parallel, meaning that RT-PCR is easily adaptable for multiplex analysis.

For global analysis, it has been necessary to develop novel technologies with a high degree of automation, which allow thousands or tens of thousands of genes to be assayed simultaneously with minimal labor. We discuss the development and application

of such technologies in the following sections, but it should be borne in mind that the principles of molecular recognition, which underlie the simpler methods such as northern blots, remain largely unchanged.

The transcriptome is the collection of all messenger RNAs in the cell

The full complement of mRNA molecules produced by the genome has been termed the *transcriptome*, and methods for studying the transcriptome are grouped under the term *transcriptomics* (Velculescu *et al.* 1997). Taking human beings as an example, it has been shown that only 3% of the genome is represented by genes, suggesting that the transcriptome is much simpler than the genome. This is not the case, however, because the transcriptome is much more than just the transcribed portion of the genome. The complexity of the transcriptome is increased by processes such as alternative splicing and RNA editing, so that each gene can potentially give rise to many transcripts, each of which may have a unique expression profile. In extreme cases, where a gene has many introns and undergoes extensive differential processing, one gene may potentially produce thousands or even millions of distinct transcripts. An example is the *Drosophila* gene *Dscam* (the homolog of the human Down syndrome cell adhesion molecule), which can be alternatively spliced to generate nearly 40,000 different mature transcripts (twice the number of genes in the *Drosophila* genome). Each of these transcripts potentially encodes a distinct receptor that may play a unique role in axon guidance (Schmucker *et al.* 2000). Other examples of this phenomenon are discussed by Graveley (2001).

Complex as the transcriptome is, it is never seen as a complete system *in vivo*. This is because all genes are not expressed simultaneously, in the same tissues, at the same levels. Cells transcribe a basic set of *housekeeping genes* whose activity is required at all times for elementary functions, but other *luxury genes* are expressed in a regulated manner, e.g. as part of the developmental program or in response to an external stimulus. Similarly, post-transcriptional events such as splicing are also regulated processes. Researchers use phrases such as "human brain transcriptome" or "yeast meiotic transcriptome" to emphasize this. A typical human cell is thought to express, on average, about 15,000–20,000 different mRNAs, some of which have housekeeping functions and some of which are more specialized. A proportion of these will be splice variants of the same primary transcript. Some of the mRNAs will be very abundant, some moderately so, and others very rare. For a truly global perspective of RNA expression in the cell, all of these transcripts must be quantified at the same time. This requires a highly parallel assay format which is both sensitive and selective.

There are two major types of strategy currently used for global RNA expression analysis, which we will discuss in turn:

- The direct sampling of sequences from source RNA populations or cDNA libraries, or from sequence databases derived therefrom (Lorkowski & Cullen 2004). See Box 20.2 for Internet resources.
- Hybridization analysis with comprehensive, non-redundant collections of DNA sequences immobilized on a solid support. These are known as DNA arrays.

Although such analysis is often called *transcriptional profiling* it is important to emphasize that one is not really looking at the level of transcription, but at the steady-state mRNA level, which also takes into account the rate of RNA turnover. Furthermore, most of the transcriptional profiling techniques discussed below do not measure absolute RNA levels, but rather compare relative levels within and/or between samples.

Box 20.2　Sequence sampling resources on the Internet

http://www.ncbi.nlm.nih.gov/UniGene/
　ddd.cgi?ORG=Hs Digital Differential Display
http://cgap.nci.nih.gov/CGAP/Tissues/
　XProfiler

SAGE websites

http://www.ncbi.nlm.nih.gov/SAGE
http://www.genzymemolecularoncology.com/
　sage/

http://www.sagenet.org
http://www-dsv.cea.fr/thema/get/sade.html
http://www.prevent.m-u-
　tokyo.ac.jp/SAGE.html
http://www.urmc.rochester.edu/smd/crc/
　swindex.html
http://sciencepark.mdanderson.org/ggeg
http://genome-www.stanford.edu/
　cgi-bin/SGD/SAGE/querySAGE

Steady-state mRNA levels can be quantified directly by sequence sampling

The first large-scale gene expression studies involved the sampling of ESTs from cDNA libraries

Historically, the first global gene-expression studies were based on the large-scale sequencing of random clones from cDNA libraries (Okubo *et al.* 1992). A cDNA library that has not been "normalized" is representative of the mRNAs in the source population used to prepare it. Some mRNAs (and corresponding cDNAs) are likely to be highly abundant, and some extremely rare. If 5000 clones are picked randomly from the library and partial sequences obtained, abundant transcripts would be more frequently represented among the sequences than rare transcripts. Statistical analysis of these results would allow relative expression levels to be determined, and comparisons of libraries prepared from different sources (e.g. disease vs. normal, induced vs. uninduced) should allow the identification of differentially expressed genes. The limitation of this approach is the expense involved in producing cDNA libraries and carrying out the large-scale sequencing projects required to make the data statistically significant (Audic & Claverie 1997). However, as the amount of EST data continues to increase for model organisms, it is now a viable approach to interrogate this data to see how often particular genes are represented, giving a digital representation of gene expression. This approach is not particularly sensitive and depends on the availability of EST data from appropriate sources, but it demonstrates the principle that direct sampling of sequences *in silico* can be used to derive expression data. Indeed, there have been several reports of differentially expressed genes identified using EST sequence sampling (e.g. see Claudio *et al.* 1998, Vasmatzis *et al.* 1998).

Serial analysis of gene expression uses concatemerized sequence tags to identify each gene

The major problem with EST sampling is the large amount of sequencing required. This had been addressed by Veculescu *et al.* (1995) in a technique called serial analysis of gene expression (SAGE). Essentially, this involves the generation of very short ESTs (9–14 nt), known as SAGE tags, which are joined into long concatemers that are cloned and

sequenced. The size of the SAGE tag approaches the lower limit for the unambiguous identification of specific genes (Adams 1996). If we consider a random sequence of 9 nt, there are 4^9 possible combinations of the four bases, or 262,144 sequences. This is approximately eight-fold the number of genes in the mammalian genome. However, an 8-nt SAGE tag would provide only about 65,000 variations, which is only about twice the number of mammalian genes. The concatemerized tags are sequenced and the sequence is analyzed to resolve the individual tags; the representation of each tag provides a guide to the relative abundances of the different mRNAs. Compared to the random sequencing of cDNA libraries, SAGE is up to 50 times more efficient, because each concatemer represents the presence of many cDNAs. The major advantage of SAGE, however, is that the data obtained are digital representations of absolute expression levels, which allows direct comparison between new experiments and existing databases. Thus as the amount of SAGE data grows it becomes increasingly possible to carry out computer searches to identify differentially expressed genes (e.g. see Veculescu *et al.* 1999).

The SAGE method as originally described is shown in Fig. 20.1. PolyA$^+$ RNA is reverse transcribed using a biotinylated oligo-dT primer and the cDNA is digested with a restriction enzyme that cleaves very frequently, such as *Nla*III (this recognizes the 4-bp sequence CATG and would be expected to cut, on average, every 250 bp). The 3′ end of each cDNA is then captured by affinity to streptavidin, resulting in a representative pool of cDNA ends that can be used to generate the SAGE tags. Note that, by selecting 3′ ends, there is generally more sequence diversity. Although many genes are members of multigene families with conserved sequences, these tend to diverge in the 3′ untranslated region even when the coding sequences are very similar. The pool is then split into two groups, each of which is ligated to a different linker. The linkers contain recognition sites for a type IIs restriction enzyme such as *Fok*I, which has the unusual property of cutting outside the recognition site, but a specific number of base pairs downstream. Cleavage with such an enzyme therefore generates the SAGE tag attached to part of the linker. The tags are ligated "tail to tail" to generate dimers called "ditags", and then amplified by the PCR using the linkers as primer annealing sites. The amplified products are then cleaved with *Nla*III to remove the linkers, and the ditags are concatemerized. Concatemers are then cloned by standard

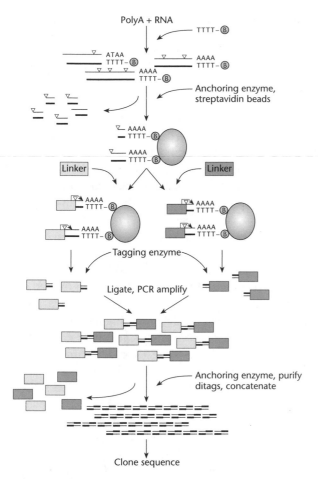

Fig. 20.1 Principle of serial analysis of gene expression (SAGE). (Adapted from Veculescu *et al.* 1997.) ▽ Anchoring enzyme (*Nla* III); ↰ Tagging enzyme (*Fok* I); ⊛ Biotin.

to characterize the transcriptome when yeast are grown on alternative carbon sources (Kal *et al.* 1999) and investigate common characteristics of highly expressed genes (Jansen *et al.* 2000). In the malaria parasite *Plasmodium falciparum*, SAGE has been used to study the transcriptome at different stages of the life cycle (Munasingh *et al.* 2001, Patankar *et al.* 2001). In plant biology, SAGE has been used to study rice development and *Arabidopsis* stress responses (Matsumura *et al.* 1999, Lee *et al.* 2003). A recent study has used SAGE to characterize genes associated with developmental arrest and longevity in *C. elegans*: over 150 genes were identified in this investigation, including genes encoding histones and a novel telomere-associated protein (Jones *et al.* 2001). However, SAGE has been most widely used in the context of human diseases, e.g. to analyze changes in mRNA levels associated with cancer (Polyak *et al.* 1997, Zhang *et al.* 1997) and other diseases (e.g. de Waard *et al.* 1999, Ryo *et al.* 1999). The use of SAGE in human studies has been recently reviewed (Tuteja & Tuteja 2004a,b).

Massively parallel signature sequencing involves the parallel analysis of millions of DNA-tagged microbeads

In a novel approach to global expression analysis, Brenner and colleagues (Brenner *et al.* 2000) have described massively parallel signature sequencing (MPSS). This is a hybrid of microarray technology and sequence sampling, in which millions of DNA-tagged microbeads are aligned in a flow cell and analyzed by fluorescence-based sequencing. The principle of the technique is that cDNA clones can be sequenced by sequential rounds of cleavage with a type IIs restriction enzyme, followed by adapter annealing, with each adapter able to "decode" the sequence of bases left in the overhang of the restriction cleavage site. Decoding is achieved through the use of conjugated labels, which can be analyzed by flow cytometry. The highly parallel nature of the technique results from the ability to simultaneously analyze the fluorescent signal of thousands of microbeads in a flow cell, and provides the same degree of throughput as large-scale cDNA sequencing.

The method is complex and is shown in Fig. 20.2. Initially, a comprehensive population of 32-mer oligonucleotides is synthesized, which are used as tags. At the same time, an equivalent population of complementary "anti-tags" is synthesized, and covalently attached to plastic microbeads. The

methods, and the resulting plasmids are sequenced to reveal the composition of the concatemer. Accurate sequencing is essential in SAGE because even single nucleotide errors (*miscalls*) could result in the incorrect identification of a tag and false expression data for a particular gene. A similar error rate in standard cDNA sequencing would be irrelevant, because of the length of the sequence. More recently, the SAGE method has been adapted so it can be used with small amounts of starting material (Datson *et al.* 1999, Peters *et al.* 1999, Neilson *et al.* 2000, Bosch *et al.* 2000, Ye *et al.* 2000).

The SAGE method of expression profiling, with various modifications to increase the likelihood of identifying genes unambiguously, has been applied to many different systems. In microbial biology, SAGE has been used to study the yeast transcriptome and divide the genome into functional expression domains (Veculescu *et al.* 1997). It has been used

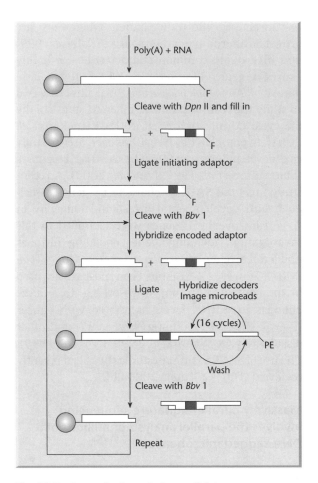

Fig. 20.2 Principle of massively parallel signature sequencing (MPSS) technique. PE = fluorescent label. (Adapted from Brenner 2000.)

32-mer tags are then mixed as a 100-fold excess with a population of cDNAs, and ligated to form conjugates. In the original method, about 10^7 different 32-mers were mixed with a population of about 5×10^4 cDNAs to generate approximately 5×10^{11} different conjugates. One percent of this mixture was taken, ensuring that each cDNA was likely to be represented and attached to a different tag, i.e. only 1% of the available tags were used. In the next stage of the procedure, amplified tagged cDNAs were end labeled with a fluorescent probe and attached to microbeads bearing complementary anti-tags. Since only 1% of the tags were used, only 1% of the anti-tags were recognized and 99% of the beads were discarded; this was achieved by FACS. Next, the cDNA was cleaved by the restriction enzyme *Dpn*II to remove the fluorescent label and generate a cohesive site to which an initiating primer could anneal. The initiating primer recognizes the *Dpn*II overhang, and carries the recognition site for a type IIs enzyme,

*Bbv*1, which cleaves a specific number of bases downstream of the recognition site, therefore chewing a small fragment from the cDNA. The resulting four-base overhang is dependent on the cDNA sequence not the restriction enzyme. In the next stage, the cleaved cDNA is annealed to a set of 16 encoded adaptors. Each adaptor recognizes a specific 4-nt overhang and carries a unique sequence at the other end which is recognized by a fluorescent-labeled decoding oligonucleotide. Scanning of the bead after each round of hybridization therefore reveals the 4-bp overhang. Importantly, the encoding adaptor also carries a *Bbv*1 site, allowing repetition of the process. Therefore, a series of 4-bp calls can be followed on the same microbead in a flow cell, generating a sequence signature for each cDNA.

The accuracy of the method was determined by carrying out MPSS analysis on early and late log phase yeast cells and comparing the signatures obtained with the sequences in public databases: over 90% were represented. Expression analysis was carried out on human THP-1 cells. Over 1.5 million MPSS signatures were obtained from induced THP-1 cells, while nearly 1850 cDNA clones were conventionally sequenced. For most of the genes analyzed, the expression levels revealed by cDNA sequence sampling and MPSS analysis were very similar, although there were discrepancies for a small number of genes which remain unexplained. Developments of this technique which allow the detection of unlabeled cDNA sequences (Steemers *et al.* 2000) and which increase sensitivity by using gold nanoparticles rather than fluorescent tags (Taton *et al.* 2000) have been reported. Although more expensive than SAGE, the MPSS technique has been applied in several studies, including whole-transcriptome analysis in *Arabidopsis* (Meyers *et al.* 2004) and a comparison of human cell lines (Jongeneel *et al.* 2003).

DNA microarray technology allows the parallel analysis of thousands of genes on a convenient miniature device

Microarray hybridization has emerged as the method of choice for high-throughput RNA expression analysis (Shoemaker & Linsley 2002, Whitchurch 2002, Venktasubbarao 2004). DNA microarrays comprise a series of DNA elements arranged as spots (*features*) in a grid pattern on a miniature solid support. These arrayed targets are hybridized (*interrogated*) with a complex probe, i.e. a probe comprising many different sequences, which is prepared from an RNA population from a particular cell type or tissue.[2] The composition of the probe reflects the abundances of individual transcripts in the source RNA population. If an excess of target DNA is provided and hybridization occurs when the kinetics are linear, the intensity of the hybridization signal for each feature represents the relative level of the corresponding transcript in the probe. These conditions are generally met because, depending on the type of array, each individual feature comprises 10^6–10^9 molecules, only a small proportion of which will be "occupied" during any hybridization reaction even in the case of abundant RNAs. The use of arrays allows simultaneous measurement of the relative levels of many transcripts. However, one intrinsic limitation is that microarrays are "closed systems" in that only the sequences represented on the arrays can be measured. In contrast, sequence sampling techniques are "open systems" because any sequence in the transcriptome can be sampled.

Before discussing how DNA microarrays are used for expression profiling, it is necessary to provide some background on the development of array technology and methodology. There are two major types of DNA array used in expression analysis: spotted DNA arrays and printed oligonucleotide chips. Their principal features are compared in Table 20.1.

Spotted DNA arrays are produced by printing DNA samples on treated microscope slides

A spotted DNA array is made by transferring (*spotting*) actual DNA clones (or more usually PCR products derived therefrom) individually onto a solid support where they are immobilized. The technology arose directly from conventional hybridization analysis, and the first high-density cDNA arrays, now described as *macroarrays*, were essentially the same as the gridded reference libraries discussed on p. 114. Cloned cDNAs stored in a matrix format in microtiter plates were transferred to nitrocellulose or nylon membranes in a precise grid pattern, allowing rapid identification of the clones corresponding to positive hybridization signals. For expression analysis, complete libraries could be hybridized with complex probes, generating a "fingerprint" specific to a particular RNA source (Gress *et al.* 1992, Zhao *et al.* 1995). Early examples of the use of macroarrays for expression analysis include studies of differential gene expression in the mouse thymus and human muscle (Nguyen *et al.* 1995, Pietu *et al.* 1996).

Nylon macroarrays are generally about 10–20 cm^2 in size, and the feature density is low, with typically 1–2 mm between targets (i.e. 10–100 targets per square cm). This has some advantages: the arrays are easy to manufacture (and are therefore relatively inexpensive), and they are also simple to use because standard hybridization procedures are applicable. For this reason, macroarrays are still manufactured by a number of commercial suppliers[3] and the technology for in-house array production is readily available, involving simple robotic devices or even hand-held arrayers. The principal disadvantages of macroarrays are: (i) the low feature density limits the number of sequences that can be interrogated simultaneously; and (ii) hybridization must be carried out in a large volume using a radioactive probe, the results being obtained by autoradiography or preferably using a phosphorimager. Although radioactive probes are sensitive, comparative gene expression analysis (e.g. mutant vs. wild type or stimulated vs. non-stimulated tissue) requires the preparation of duplicate arrays, or the sequential probing, stripping, and reprobing of the same array with two different probes. Both these strategies can generate inter-experimental variation that can give misleading results. Also, the large volume of solution required to cover the membrane limits the probe concentration, reducing the efficiency of the

---

[2] In order to maintain continuity in this book, we define a "probe" as a labeled population of nucleic acid molecules in solution, and a "target" as an unlabeled population of nucleic acid molecules usually immobilized on a solid support. These definitions are generally followed by researchers using spotted arrays. However, care should be exercised when reading literature concerning the use of Affymetrix GeneChips, since exactly the opposite convention is followed. Each feature on an Affymetrix GeneChip is termed a *probe cell* or simply a *probe*, and the labeled nucleic acids in solution, which hybridize to the features, are described as *targets*. We deliberately ignore this nomenclature for the sake of clarity.

[3] Although nylon macroarrays are sold by many biotechnology companies, they tend to be called microarrays in the accompanying literature. The distinction between a macroarray and a microarray is not clear-cut. The term *microarray* was initially coined to describe the high-density arrays printed on small glass chips, which contrasted sharply with the original macroarrays printed on large nylon membranes. Confusion arises now that nylon arrays can be manufactured with a size and feature density similar to that of the glass arrays. A convenient cut-off point for a microarray might be an overall size of 1–2 cm^2 and a spacing between spots of 0.5 mm, but this is purely arbitrary.

Table 20.1 Properties of different types of DNA array for expression analysis.

Property	Spotted nylon macroarrays	Spotted glass microarrays	Affymetrix GeneChips
Target composition	dsDNA fragments (genomic or cDNA clones, or PCR products derived from them)		Single-stranded oligonucleotides
Target source	Maintained clone sets, either annotated or anonymous. Must be derived from source RNA or purchased from licensed vendors		Sequences derived from public and/or private databases. Chemically synthesized
Target size	Typically 100–300 bp		Typically 20–25 nt
Array format	Individual features represent non-redundant clones; hybridization sensitivity high		Single clones represented by sets of ~20 non-overlapping oligos to reduce false positives
Density (features per cm^2)	1–10*	>5000	64,000 for available chips, but experimental versions up to 1,000,000
Manufacture	Robotized or manual spotting	Robotized spotting	On-chip photolithographic synthesis
Substrate	Nylon	Glass	Glass or silicon
Probe labeling	Radioactive or enzymatic	Dual fluorescent	Fluorescent
Hybridization	High volume (up to 50 ml*), ~65°C	Very low volume (10 μl), ~65°C	Low volume (200 μl), 40°C
Data acquisition	Autoradiography or phosphorimager for isotopic probes, flatbed scanner for enzymatic probes	Confocal scanning	Confocal scanning
Cost of prefabricated arrays	Low	Moderate	High
In-house manufacture	Inexpensive	Expensive, but prices falling	Not currently available

* Note that nylon microarrays are also available: these have a density of up to 5000 features per cm^2, and require only 100–200 μl of hybridization solution.

hybridization reaction. However, extensive miniaturization of nylon arrays has been difficult because the resolution of the signal provided by radioactive probes is poor. Fluorescent probes have a higher resolution but cannot be used on nylon membranes because the substrate has a high level of autofluorescence, generating a low signal-to-noise ratio. It has been possible to produce nylon microarrays with about 200 μm between features (up to 5000 targets per cm^2) but their analysis requires expensive high-resolution imaging devices (Bertucci *et al.* 1999). An

alternative system, which uses enzymatic rather than radioactive probes, gives a high-resolution signal that can be detected with a low-cost scanning apparatus, but with some loss of sensitivity (Chen *et al.* 1998).

A breakthrough in spotted array technology came with the development of microarrays on glass chips (Schena *et al.* 1996, 1998). Glass is an inert substrate and must be coated before DNA will adhere. Usually, the negatively charged phosphate groups of DNA are exploited for immobilization on positively

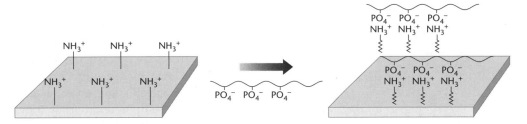

Fig. 20.3 DNA binding by electrostatic interactions. Binding of DNA molecules to amine-derivatized surfaces by ionic interaction between positively charged amino groups and the negatively charged phosphate groups.

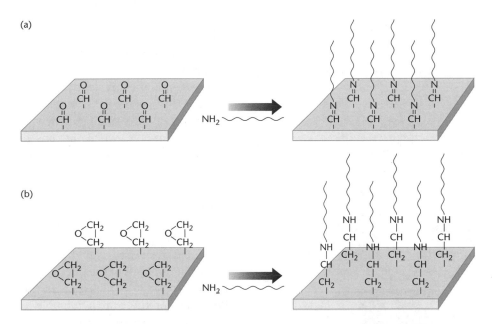

Fig. 20.4 (a) DNA binding by Schiff base reaction. Binding of DNA molecules to aldehyde-derivatized surfaces. An amino crosslinker is used to covalently attach DNA to the aldehyde-derivatized surface. (b) Binding of DNA molecules to epoxy-derivatized slides. An amino crosslinker is used to covalently attach DNA to the epoxy-derivatized surface.

charged surface groups provided e.g. by poly-L-lysine; the DNA must then be cross-linked to the surface (Fig. 20.3). Alternatively, amino groups can be attached to the DNA and immobilized on aldehydes or epoxy-derivatized surfaces (Fig. 20.4). Since glass is non-porous and has very little autofluorescence, fluorescent probes can be used and they can be applied in very small hybridization volumes. The greater resolution afforded by fluorescent probes allows feature density to be increased significantly compared to nylon macroarrays, and the small hybridization volume improves the kinetics of the reaction. Together, these advantages mean that more features can be assayed simultaneously with the same amount of probe without loss of sensitivity. Thus, glass arrays can routinely be manufactured

with up to 5000 features per square cm. The major advantage of fluorescent probes, however, is that different fluorophores can be used to label different RNA populations. These can be simultaneously hybridized to the same array, allowing differential gene expression between samples to be monitored directly (Shalon *et al.* 1996). The most common strategy is to use Cy3, which fluoresces bright red, and Cy5, which fluoresces bright green, to label different probes. If a particular cDNA is present only in the Cy3-labeled population, the spot on the array appears red. If another cDNA is present only in the Cy5-labeled population, the spot on the array appears green. cDNAs that are equally represented in both populations contain equivalent proportions of each label and appear yellow. In this way, it is

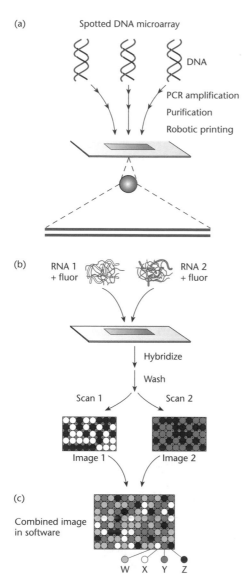

(a)   Spotted DNA microarray

DNA

PCR amplification
Purification
Robotic printing

(b)   RNA 1   RNA 2
      + fluor + fluor

Hybridize
Wash

Scan 1   Scan 2

Image 1   Image 2

(c)
Combined image
in software

W   X   Y   Z

Fig. 20.5 Spotted DNA arrays. (a) Principle of array manufacture. Each robotically printed feature corresponds to one gene or cDNA. (b) Hybridization of glass arrays with differentially labeled RNA probes followed by scanning on separate channels to detect fluorescence. (c) Combined image showing four types of signal: W – genes represented equally in both RNA populations; X – genes represented only in RNA 1; Y – genes represented only in RNA 2; Z – genes represented in neither RNA population. ((a) and (b) redrawn with permission from Harrington *et al.* 2000 by permission of Elsevier Science.)

easy to identify potentially interesting differentially expressed genes (Fig. 20.5).

Due to the many advantages of glass microarrays, including their amenability for automated spotting, this format has emerged as the most popular type of spotted array for expression profiling. However, until recently, a major disadvantage was the cost of production, with the result that the technology

was beyond the reach of all but the best-funded laboratories. Researchers were faced with the initial choice of purchasing ready-made arrays from a commercial source or investing in the resources required for in-house array manufacture. Prefabricated glass arrays are designed for single use, so a simple series of experiments with an appropriate number of replicates can carry a hefty price tag. Unfortunately, the cost of a commercially available precision robot for array manufacture is even greater: $100,000 or more even for those with the simplest specifications. Additionally, clone sets usually need to be purchased to provide the features for a homemade array (Box 20.3).

Genomic clones tend to be used to derive features for bacterial arrays because the lack of introns makes them essentially equivalent to cDNAs. This is also true in the case of the yeast *Saccharomyces cerevisiae*, where introns are small and few in number. In higher eukaryotes, where introns are larger and more common, it is much more convenient to use cDNAs instead of genomic clones. However, full-length cDNA clones are neither necessary nor particularly desirable because of the prevalence of large gene families with conserved sequences. The use of partial cDNA sequences that exploit the differences between related clones avoids cross-hybridization. Most of the cDNA sequence information that exists in databases is in the form of ESTs (Box 9.1), and these are a valuable resource for the manufacture of spotted arrays.

It is beyond the scope of most laboratories to prepare comprehensive clone sets *de novo* for array manufacture. Only large-scale sequencing projects can provide the materials and data required to make comprehensive arrays, and this is the domain of biotechnology and pharmaceutical companies, consortia of academic laboratories, and collaborations between academic institutes and industry. Typically, such organizations make their clone sets available commercially through licensed vendors (for a selected list, see Box 20.3). An example is the UniGene collection of human (and mouse) clustered, sequence-verified ESTs. This collection is available from Incyte Genomics and Research Genetics. More recently, the financial barriers to the general use of microarrays have begun to fall. This reflects a number of factors, including competition between companies producing prefabricated arrays, increasing numbers of universities investing in microarray core facilities, the availability of protocols that allow robots for array manufacture to be built in the laboratory for under

Box 20.3 Selected sources of clone sets for the manufacture of spotted arrays

Company	Website	Resources
Research Genetics	http://www.resgen.com	Human UniGene collection
		Mouse UniGene collection
		Rat cDNA clone collection
		Genome-wide *Caenorhabditis elegans* partial ORF primers
		Drosophila cDNA collection
		Genome wide *Saccharomyces cerevisiae* ORF primers
		Genome wide *S. cerevisiae* intergenic primers
Incyte Genomics	http://www.incyte.com	Human UniGene collection
		Mouse UniGene collection
		8000 *Arabidopsis thaliana* cDNA clones
		Candida albicans complete ORF collection
Genosys Biotech	http://www.genosys.com	*Escherichia coli* complete ORF collection
		Bacillus subtilis complete ORF collection
		Partial clone collections for several other bacteria

Fig. 20.6 A homemade microarraying robot, which can be constructed for approximately $30,000. (Reprinted from Thompson *et al.* 2001 by permission of Elsevier Science.)

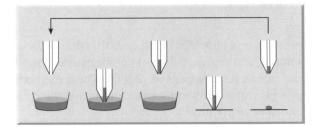

Fig. 20.7 Principle of array manufacture by capillary contact printing.

$50,000 (Fig. 20.6), and the development of novel printing technologies (see below). Instructions for building arraying robots using simple and readily available components are available on the Internet (Box 20.3) and have been discussed in several recent articles (e.g. see Bowtell 1999, Cheung *et al.* 1999, Duggan *et al.* 1999, Thompson *et al.* 2001).

There are numerous printing technologies for spotted arrays

The original method for producing spotted arrays was contact printing, which involves the use of a capillary spotting pin (or quill) that draws up a defined amount of liquid from wells in a microtiter

plate (Fig. 20.7). The pin is then placed in contact with the array surface, and this causes some liquid to be deposited. The pin is thoroughly washed and dried in an automated cycle before returning to the microtiter dish for the next sample (Duggan *et al.* 1999, Xiang & Chen 2000). The speed at which arrays can be produced is increased by using multiplex print heads that deposit samples in a block. A number of alternative "non-contact" printing methods are also available. The *pin and ring system*, devised at Genetic Microsystems and currently marketed by Affymetrix, is popular (Fig. 20.8). The "ring" is inserted into the well of a microtiter plate and draws up a certain amount of liquid. The "pin" then extends through the ring and carries a smaller droplet of solution down onto the array surface. Non-contact printing technologies have also been developed for microarray fabrication and include piezoelectric devices similar to those found in inkjet

Fig. 20.8 A non-contact printing method for array manufacture. This is the pin and ring system developed by Genetic Microsystems and currently marketed by Affymetrix. (Courtesy of Affymetrix.)

printers, and bubblejet printheads that deposit DNA samples on the substrate as a bubble extended from the nozzle (Okamoto *et al.* 2000). These methods provide a more uniform spot size, reducing the variation between features.

Oligonucleotide chips are manufactured by *in situ* oligonucleotide synthesis

The alternative to a spotted DNA array is a high-density prefabricated oligonucleotide chip (Lockhardt *et al.* 1996, Lipshutz *et al.* 1999). These are similar to DNA arrays in that they consist of gridded DNA targets that are interrogated by hybridization. However, while DNA arrays consist of double-stranded clones or PCR products that may be up to several hundred base pairs in length, oligo chips contain single-stranded targets ranging from 25–70 nt. Dual hybridization is not used for expression profiling on oligo chips. Instead, probes for chip hybridization are made from cleaved, biotinylated cRNA (RNA that has been transcribed *in vitro* from cDNA). Comparative expression analysis is carried out by hybridization of alternative cRNA samples to identical chips, followed by comparison of signal intensities (Fig. 20.9). Oligo chips can be made in the same way as spotted DNA arrays, by robotically transferring chemically synthesized oligonucleotides

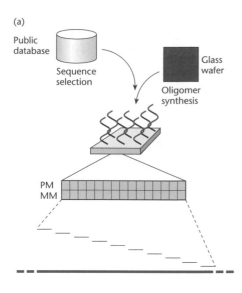

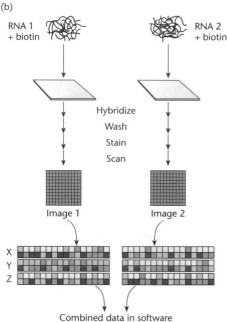

Fig. 20.9 Oligonucleide chips. (a) Principle of chip manufacture. Note that sequences are obtained from public or private databases and sythesized on the chip. Each gene is represented by 20 non-overlapping oligonucleotides, each with a perfect match (PM) and mismatch (MM) feature. (b) Principle of chip hybridization using biotin-labeled cRNA probes. (Redrawn with permission from Harrington *et al.* 2000.)

from microtiter dishes to a solid support, where they are immobilized (e.g. see Yershov *et al.* 1996). However, the maximum array density is increased almost tenfold if the oligos are printed directly onto the glass surface.

Direct "on-chip" synthesis of high-density oligonucleotide arrays was developed by Steve Fodor

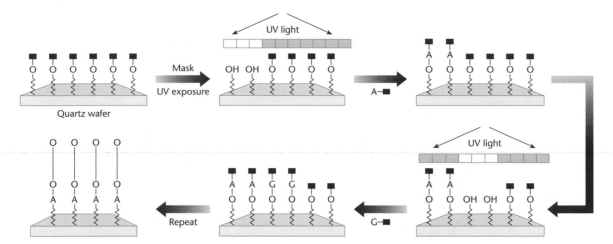

Fig. 20.10 GeneChip fabrication by *in situ* synthesis by photoactivation and deprotection of nucleic acids. Photomasks are used to pattern UV light at localized regions to selectively synthesize a patterned array. This method was developed by Affymetrix.

and colleagues at the US biotechnology company Affymetrix Inc., using a light-directed printing technology known as photolithography (Fodor *et al.* 1991, 1993, Pease *et al.* 1994). The procedure is complex but essentially involves the use of a glass or silicon wafer that is hydroxylated and silanized so that DNA can be covalently attached to the surface in a simple chemical reaction. However, the covalent binding sites are blocked by a photolabile protecting group. A chromium mask is then applied to the surface of the chip which determines which areas are exposed to light. Under illumination, the protecting groups in these areas are destroyed, allowing the addition of a single nucleotide, which is also blocked with a photolabile protecting group. If this process is repeated using a series of different masks and different nucleotides, a precise grid can be generated containing millions of defined and precisely arrayed oligos (Fig. 20.10). The process is highly accurate and allows the production of the densest arrays currently available (up to 64,000 features over an area slightly larger than 1 cm^2 in commercially available GeneChips, but experimental chips with a density of >10^6 targets per cm^2 have been produced). However, the use of physical masks makes the GeneChip very expensive.

More recently, alternative technologies have been developed which reduce the capital costs involved in chip manufacture by employing virtual masks (Fig. 20.11). For example, NimbleGen Inc. has developed a technology using UV light reflected from a miniature array of aluminum mirrors to focus selectively on the appropriate areas of the chip and destroy the photolabile groups. This is known as the

maskless array synthesizer (MAS) (Singh-Gasson *et al.* 1999). Other companies, such as Xeotron Inc. and Febit GmbH have developed similar maskless technologies based on digital micromirrors (Gao *et al.* 2001). Finally, CombiMatrix Inc. has developed a system using individually addressable microelectrode arrays to synthesize many different oligonucleotides *in situ* in parallel reactions (Tesfu *et al.* 2004).

Spotted arrays and oligo chips have similar sensitivities

Although subject to some debate, it is believed that spotted microarrays and high-density oligo chips perform equally well in terms of sensitivity (see discussion by Granjeaud *et al.* 1999). An important difference between the two types of array, however, is that the manufacture of oligo chips relies entirely on pre-existing sequence information. Conversely, spotted DNA arrays can be generated using anonymous (i.e. non-annotated) clones from uncharacterized cDNA libraries, and can therefore be used for *de novo* gene discovery. Oligonucleotide chips can be designed on the basis of genome sequence data or a collection of cDNA or EST sequences. The advantage of this is that chips can be devised *in silico*, i.e. using computer databases as a source of information, with no need to maintain physical DNA clone sets. Oligonucleotide arrays are therefore highly advantageous for expression profiling in those organisms with complete or near-complete genome sequences, or comprehensive EST collections, but less useful for other organisms. In contrast, it is quite possible to

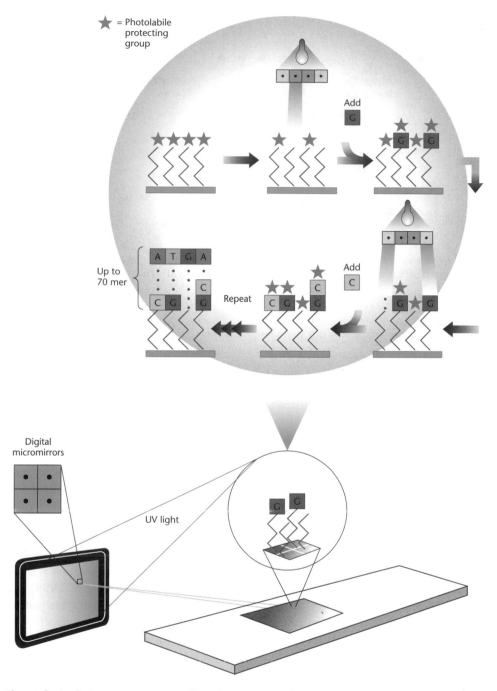

Fig. 20.11 The synthesis of microarrays using NimbleGen Systems MAS technology is very similar to traditional oligonucleotide synthesis with some important exceptions. Unlike conventional oligo synthesis, arrays are synthesized on glass slides rather than controlled pore glass supports. Another key difference is that the deprotection steps are performed by photodeprotection rather than by acid deprotection. The illustration here depicts digital micromirrors reflecting a pattern of UV light, which deprotects the nascent oligonucleotide and allows addition of the next base.

generate cDNA arrays for largely uncharacterized organisms, although positive hybridization signals then have to be further characterized by sequencing the corresponding clone.

The specificity of hybridization to spotted ÐNA arrays is relatively high, due to the length of each target. As discussed above, distinction between members of gene families can be achieved by selecting the least conserved region of the cDNA. Hybridization specificity is more of a problem when the target size is smaller, and this has been addressed in the case of Affymetrix GeneChips by the use of hybridization controls and redundant targets (Lipshutz *et al.* 1999). Each mRNA is represented by up to 20, non-overlapping oligonucleotides, such that the likelihood of obtaining a false positive result is greatly reduced. The chips are designed to contain both "perfect match" (PM) and "mismatch" (MM) oligos for each specific target (Fig. 20.9). The perfect match is expected to hybridize along its whole length, while the mismatch contains a mismatching base at a central site, thus acting as a control for cross-hybridization. The signal obtained from the MM control is subtracted from that of the corresponding PM to reveal the actual level of specific hybridization. Distinction between members of gene families is achieved by designing oligos matching the least conserved regions of the gene.

As transcriptomics technology matures, standardization of data processing and presentation become important challenges

The raw data from DNA array experiments are monochrome images of hybridized arrays. Where dual fluorescence has been used, images are obtained on two channels, rendered in false color, and combined. These visual data must be normalized (corrected for background) and quantified, and the algorithms used for this purpose are often provided with the image-recording apparatus. Background may be generated by non-specific hybridization, autofluorescence, dust, and other contaminants or poor hybridization technique. The background may not be constant over the entire array, so local background values must be obtained. The final data provide a measurement of *relative* expression levels. These factors, in combination, make it very difficult to compare results across different experiments (Quackenbush 2002).

The normalized data from a microarray experiment are generally presented as a *gene expression matrix*, which shows the normalized signal intensities for each feature over a range of experimental conditions. With dual fluorescence, two measurements are taken, one from each channel. In other cases, readings may be taken from a number of identical arrays representing e.g. a series of developmental time points or a cell culture exposed to different concentrations of a drug. It is important to have control features on each array so that the data can be normalized for variation across arrays. The data can then be grouped according to similar expression profiles using a clustering algorithm (reviewed by Raychaudhuri *et al.* 2001, Noordewier & Warren 2001, Quackenbush 2001). This involves convert-

ing the gene expression matrix into a distance matrix showing the pairwise differences between the expression levels of each possible combination of genes. The data are then clustered to generate a tree-like graph called a dendrogram. In *hierarchical clustering* methods, the two most similar genes are clustered first and these define a new merged data point. The analysis is repeated until all the genes are clustered together. Other popular methods include *k-means clustering*, in which the expected number of clusters is specified at the outset, and the generation of *Kohonen self-organizing maps*, a similar process refined by the use of neural nets. These algorithms can take a long time to run if the data set is very large. Run times can be limited by employing *feature reduction* strategies, such as the elimination or merging of redundant and uninformative genes or expression profiles. Several bioinformatics tools are available over the Internet to carry out clustering analysis of microarray expression data, such as the EPCLUST program, which is part of the Expression Profiler suite (http://ep.ebi.ac.uk/EP/).

Because of the many different ways in which microarray data can be obtained, processed, and presented, there has been an international effort to develop a set of rules and conventions for the standardization of microarray data presentation. The conventions have been devised by the MAGE (microarray and gene expression) group and are discussed in detail on the Microarray Gene Expression Database (MGED) (http://www.mged.org) (see Brazma *et al.* 2001, Quackenbush 2004). The MIAME (minimum information about a microarray experiment) standards include six properties of microarray experiments that can be used as descriptors to ensure that experiments can be repeated accurately. These properties are: overall experimental design, array layout, probe source and labeling method, hybridization procedures and parameters, measurement and normalization procedure, and details of any controls. Standard formats for data transfer have also been proposed, including a microarray and gene expression object model and markup language (MAGE-OM, MAGE-ML). One of the primary aims of MAGE is to persuade all scientific journals to require that microarray data are deposited in one of three databases which adhere to the MIAME conventions: ArrayExpress (Brazma *et al.* 2003), Gene Expression Omnibus (GEO) (Edgar *et al.* 2002), and the Center for Information Biology Gene Expression Database (CIBEX) (Ikeo *et al.* 2003). More information about MIAME and MAGE can be found in open letters to journal editors (Ball *et al.* 2002, 2004).

Expression profiling with DNA arrays has permeated almost every area of biology

Much of the early literature about DNA microarrays concerned methodology development and proof-of-principle studies. Indeed, between the years 1995 and 1999, the number of published papers describing the theory and practice of array hybridization far outweighed the number of papers reporting actual experiments! In the last five years, there has been an exponential increase in the number of array-based experiments and the applications are extremely diverse, covering many different organisms and ranging from basic studies of biological processes to clinical applications and pharmacology. Hundreds of papers reporting microarray data appear each month, and at least 100 review articles describing microarray-based analysis of different organisms and systems were published in 2005. Comprehensive coverage of this burgeoning field would require an entire book of its own! If any trends at all can be resolved from this mass of data it is the proliferation of studies in which microarrays have been used to study the expression of entire microbial transcriptomes, and the increasing use of microarrays to investigate different human diseases. The discussion below relates to these two important topics. Further applications of expression profiling, e.g. in agriculture, biotechnology, and developmental biology, are considered in Chapter 26.

Global profiling of microbial gene expression

The first genome-wide expression profiling experiments were reported in 1997. Spotted arrays were manufactured containing PCR-amplified open reading frames representing most of the 6200 genes in the *S. cerevisiae* genome (De Risi *et al.* 1997, Lashkari *et al.* 1997). These investigators analyzed the transcriptional profile of yeast cells shifted from fermentation (anaerobic) to aerobic metabolism, and as they were subjected to a variety of environmental manipulations, including heat shock. In each case, about 5% of the interrogated genes showed highly significant changes in expression induced by the experimental conditions, when unstimulated yeast cells were used as a source of control RNA. Genome-wide expression profiling with arrays has also been carried out for a number of complex biological processes in yeast, such as sporulation (Chu *et al.* 1998), the cell cycle (Spellman *et al.* 1998), and response to glucose (Gleade *et al.* 2003). These studies have

allowed tentative functions to be assigned to a number of previously uncharacterized genes, based on their informative expression patterns. For example, in the study by Spellman and colleagues, 800 cell cycle-regulated genes were identified, about 400 of which were inducible by cyclins. Genes have also been identified whose expression is dependent on the ploidy (number of chromosome sets) of the cell (Galitski *et al.* 1999). Furthermore, transcriptional profiling of yeast cells exposed to drugs has allowed novel drug targets to be identified (e.g. see Marton *et al.* 1998, Lockhart 1998).

Affymetrix GeneChips have also been manufactured representing all the ORFs in the yeast genome. Wodicka and colleagues reported the first use of GeneChips for transcriptional profiling in yeast, when they compared yeast grown on minimal and rich media (Wodicka *et al.* 1997). The Affymetrix yeast GeneChip has also been used to profile yeast cells exposed to alkylating agents (chemicals that cause damage to DNA). Forty-two genes were found to be induced by DNA damage, and for almost all of these genes, the results of the chip experiment were confirmed by traditional gene-by-gene northern blot hybridization (Jelinsky & Samson 1999). Genome-wide transcriptional analysis of the mitotic cell cycle has also been carried out (Cho *et al.* 1998) as well as a comprehensive analysis of the meiotic transcriptome (Primig *et al.* 2000).

Bacteria have smaller genomes than yeast, which should make transcriptional profiling using DNA arrays a simpler process. An array containing all 4290 genes of the *E. coli* genome was produced by Tao *et al.* (1999) and interrogated using RNA from bacteria growing on glucose-rich and minimal medium. Over 200 genes were shown to be induced on minimal medium, including a number of previously identified stress-response genes. About 120 genes were induced by growth on rich medium, many of these involved in protein synthesis. Note that array hybridization in bacteria is complicated by the difficulty in selectively labeling mRNA. In eukaryotes, mRNA has a polyadenylated tail which can be used to selectively prime first-strand cDNA synthesis, generating a labeled probe devoid of rRNA and tRNA. Bacterial mRNA generally lacks a polyadenylate tail, so hybridization is carried out with total RNA. Nevertheless, the presence of rRNA does not appear to interfere with the sensitivity of the hybridization (Richmond *et al.* 1999). Richmond and colleagues looked at the genome-wide transcriptional profile of *E. coli* after exposure to heat shock and the lactose analog IPTG. They used both glass microarrays with a fluorescent probe and nylon arrays with a radioactive probe, finding that the former produced more reliable and consistent results. Expression profiling in bacteria has also been used to identify potential new drug targets. For example, Wilson *et al.* (1999) exposed *Mycobacterium tuberculosis* to isoniazid, a drug commonly used to treat tuberculosis. RNA extracted from treated and untreated bacteria was used to interrogate a genome-wide DNA array. As well as identifying genes involved in the biochemical pathway representing the drug's known mode of action, a number of other genes were induced which could be exploited in the development of novel therapeutics. The effect of BCG (Bacille Calmette-Guérin) vaccines on *M. tuberculosis* has also been investigated by DNA array (Behr *et al.* 1999).

Over the last few years, there has been a shift in emphasis in genome-wide array hybridization experiments. The aim of the expression profiling studies discussed above has been, essentially, to assemble a list of genes that are specifically induced or repressed under particular conditions. With the development of more sophisticated data analysis tools, more recent experiments have tracked transcriptional changes over tens or hundreds of different conditions. Clustering these data allows subtle changes in gene expression patterns to be revealed. As an example we consider a series of experiments carried out by Hughes *et al.* (2000) in which yeast cells were exposed to drugs such as itraconazole, which inhibit sterol biosynthesis. This treatment resulted in significant changes in the expression of hundreds of genes, suggesting that the drug had many specific targets in the yeast cell. However, by looking at the data generated in this experiment in concert with the expression profiles revealed under 300 other conditions (including various mutants, chemical treatments, and physiological parameters) it became apparent that most of the effects were non-specific, and that the only genes specifically affected by itraconazole were those involved in the sterol biosynthetic pathway. The large-scale use of microarrays in this series of experiments resulted in a compendium expression database, which allowed expression profiles over multiple conditions to be compared. In this way, it was possible not only to compile lists of co-regulated genes, but also to assign functions to orphan reading frames and identify drug targets. Thus an anonymous transcript known as *YER044c* was shown to be co-regulated with

other sterol biosynthetic genes, strongly indicating a role in sterol metabolism, and expression of the *erg2p* gene was shown to be influenced by the anesthetic drug dyclonine, suggesting a candidate homologous target in humans (Hughes *et al.* 2000). Microarray experiments have also been carried out using panels of mutants to test gene functions under a similar set of conditions (e.g. Giaever *et al.* 2002, Dassgupt *et al.* 2003).

Applications of expression profiling in human disease

Arrays have been widely used to investigate transcriptional profiles associated with human disease, and to identify novel disease markers and potential new drug targets. Many investigators have used arrays to profile transcriptional changes associated with cancer and this area of research has been recently reviewed (Marx 2000, Guo *et al.* 2003, Russo *et al.* 2003, Albertson & Pinkel 2003, MacGreggor 2003, Mischel *et al.* 2004, Bertucci *et al.* 2004). In one of the earliest studies, De Risi and colleagues used cDNA arrays to investigate the ability of human chromosome 6 to suppress the tumorigenic phenotype of the melanoma cell line UACC-903 (De Risi *et al.* 1996). A number of novel tumor-suppressor genes were identified. Spotted cDNA arrays have also been used to investigate global gene expression in rheumatoid arthritis and inflammatory bowel disease (Heller *et al.* 1997), insulin resistance (Aitman *et al.* 1999), and asthma (Syed *et al.* 1999). In some cases, the investigation of global gene expression profiles has led to the discovery of novel links between biological processes. For example, Iyer *et al.* (1999) investigated the transcriptional profile of serum-starved cells following the addition of fresh serum, using a human cDNA array containing approximately 8600 genes. While many of the genes induced at early time points were well-characterized proliferation-response genes, the investigators also found that a large number of genes induced at later time points were known to be involved in the wound response, an example being *FGF7*. A functional link between serum starvation and wounding had not previously been identified. Affymetrix produce a range of different chips for human, mouse, and plant genomes in addition to yeast and bacteria, thus human GeneChips have also been used for disease profiling. For example, the HUGeneF1 chip contains features representing nearly 7000 human genes. This was interrogated using RNA from human foreskin fibroblasts at several time points after infection with human cytomegalovirus (Zhu *et al.* 1998). One day post-infection, 364 transcripts were shown to have undergone significant changes in expression level and it is likely that some of these may strongly influence the progress of the infection, and could represent useful drug targets.

As array technology has matured, one emerging application with an important impact on medicine is the use of expression profiling for the classification of tumors. Unlike the clustering approaches discussed above, which are unsupervised (i.e. there are no predefined groups), tumor classification is a supervised type of analysis (i.e. the data are placed into categories that have already been defined). Different forms of cancer are generally identified by a histological phenotype, which is subject to visual interpretation and human error. Recently, a number of studies have shown that gene expression profiles can be a useful way to classify tumors, and that such profiles can be defined more rigorously. For example, a systematic study of 60 diverse cancer cell lines held at the National Cancer Institute using an array containing about 10,000 cDNAs showed that each line could be distinguished clearly on the basis of its expression profile (Ross *et al.* 2000). The same cell lines have recently been profiled to determine relationships between RNA levels and drug responses (Scherf *et al.* 2000).

Expression profiling has also been useful for distinguishing very similar types of cancer, an approach called *class prediction* where subtypes of a disease are known, and *class discovery* where they are not. For example, Perou and colleagues have used expression profiling to distinguish different classes of breast cancer (Perou *et al.* 1999, 2000), while Golub *et al.* (1999) used the self-organizing maps algorithm to analyze the transcriptional profiles of a number of leukemia samples, correctly placing them into the two known categories: acute myeloid (AML) and acute lymphoblastic (ALL). An array containing 7000 cDNAs was used in this analysis, and about 50 genes were shown to be differentially expressed. Similarly, Alizadeh *et al.* (2000) used the 18,000-gene "lymphochip" array available from Research Genetics to study non-Hodgkin's lymphoma. Interestingly, this experiment revealed two previously unknown subclasses of the disease with different clinical characteristics. Bittner *et al.* (2000) have used expression profiling for class discovery in cutaneous melanoma, and were also able to distinguish aggressive metastatic melanomas through the analysis of microarray hybridization results. In each of these studies, prediction of the correct type of cancer will help to ensure appropriate treatment is carried out, and the microarray experiments themselves may even reveal novel drug targets.

Suggested reading

Schena M., Shalon D., Davis R.W. & Brown P.O. (1995) Quantitative monitoring of gene expression patterns with a complementary DNA microarray. *Science* **270**, 467–70.

Veculescu V.E., Zhang L., Vogelstein B. & Kinzler K.W. (1995) Serial analysis of gene expression. *Science* **270**, 484–7.

Two seminal papers in the development of high-throughput expression profiling at the mRNA level.

Altman R.B. & Raychaudhuri S. (2001) Whole-genome expression analysis: challenges beyond clustering. *Current Opinion in Structural Biology* **11**, 340–7.

Blohm D.H. & Guiseppe-Elie A. (2001) New developments in microarray technology. *Current Opinion in Biotechnology* **12**, 41–7.

Bowtell D.D.L. (1999) Options available – from start to finish – for obtaining expression data by microarray. *Nature Genetics* **21**, 25–32.

Lander E.S. (1999) Array of hope. *Nature Genetics* **21**, 3–4.

Lipshutz R.J., Fodor S.P.A., Gingeras T.R. & Lockhart D.J. (1999) High density synthetic oligonucleotide arrays. *Nature Genetics* **21**, 20–4.

Lorkowski S. & Cullen P. (2001) High-throughput analysis of mRNA expression: microarrays are not the whole story. *Expert Opinion on Therapeutic Patents* **14**, 377–403.

Quackenbush J. (2002) Microarray data normalization and transformation. *Nature Genetics* **32**, 496–501.

Shoemaker D.D. & Linsley P.S. (2002) Recent developments in DNA microarrays. *Current Opinion in Microbiology* **5**, 334–7.

Venkatasubbarao S. (2004) Microarrays – status and prospects. *Trends in Biotechnology* **12**, 631–7.

A selection of excellent reviews charting the development of array technology for expression profiling, and recent advances in manufacture, methodology, and data analysis.

CHAPTER 21

Proteomics I – Expression analysis and characterization of proteins

Introduction

As discussed in the previous chapter, important insights into gene function can be gained by studying gene expression at the mRNA level. However, analysis of the transcriptome has several disadvantages reflecting the fact that mRNA represents an early stage of gene expression whereas, for most genes, the protein is the final, functional molecule in the cell. One disadvantage is that the abundance of a given transcript may not reflect the abundance of the corresponding protein due to post-transcriptional gene regulation; not all mRNAs are translated at the same rate and some may not be translated at all (Gygi *et al.* 1999). Second, protein diversity is often generated after transcription, sometimes by the synthesis of two or more types of protein from a given mRNA, but more often by post-translational modification (e.g. glycosylation, phosphorylation). Since protein activity is often dependent on post-translational modification, the abundance of a protein *per se* may not represent the level of protein *activity* in the cell. Finally the function of a protein may also depend on its localization, and trafficking between compartments can sometimes be used in a regulatory manner (e.g. to control the activity of transcription factors). Therefore, even the abundance of a correctly modified protein cannot be guaranteed to represent the true intracellular activity of the protein. All these factors must be taken into account in functional analyses and can only be addressed by studying proteins directly.

The entire complement of proteins synthesized by a given cell or organism has been termed the *proteome* (see Wasinger *et al.* 1995). The field of *proteomics*, the global study of proteins, has developed from this concept (Twyman 2004). In one sense, proteomics is the direct equivalent of transcriptomics, i.e. the study of protein expression and abundance. However, because proteins are functional molecules rather than simply information carriers, they are a much richer source of data than nucleic acids. Proteins can be studied in terms of sequence and abundance (like mRNAs) but also relevant are three-dimensional structure, modification, localization, biochemical and physiological function, and interactions with other proteins and other molecules. The next three chapters explore these topics; the current chapter focuses on protein expression analysis and methods for high-throughput protein characterization.

Protein expression analysis is more challenging than mRNA profiling because proteins cannot be amplified like nucleic acids

The global analysis of protein expression presents some technical hurdles that are not found in transcriptional profiling, and these reflect the concepts of target abundance and molecular recognition. Transcriptional profiling using arrays depends on two processes: the ability to selectively amplify particular DNA molecules as targets for hybridization (this may be achieved by molecular cloning, PCR amplification, or direct synthesis in the case of oligonucleotides), and the ability of the probe to hybridize to all targets under similar conditions. Since all DNA and RNA molecules, regardless of sequence, obey similar hybridization kinetics, thousands of DNA sequences arrayed on a solid substrate can be hybridized simultaneously with a heterogeneous probe. For proteins, there is no amplification procedure as can be applied to nucleic acids, and molecular recognition is mediated by antibodies, other proteins, or other ligands, and for each type of protein the reaction kinetics are distinct. Therefore, while there has been considerable interest in the development of protein microarrays as a direct parallel to DNA microarrays, the difficulty with determining a standard set of conditions for molecular recognition has pushed proteomic research in alternative directions. Currently, the most widely used technologies for proteome analysis are protein separation by 2D-electrophoresis and/or multidimensional liquid chromatography followed by annotation using high-throughput mass spectrometry.

There are two major technologies for protein separation in proteomics

Two-dimensional electrophoresis produces a visual display of the proteome

Technology for the global analysis of protein expression was established in 1975, following the development of a two-dimensional polyacrylamide gel electrophoresis (2D-PAGE) procedure that could be applied to complex protein mixtures such as those extracted from whole cells and tissues (Klose 1975, Scheele 1975). The method involved first-dimension isoelectric focusing (i.e. separation of proteins according to charge) followed by second-dimension fractionation according to molecular mass. In the first application of this technology, 1000 proteins from the bacterium *Escherichia coli* were resolved on a single gel (O'Farrell 1975).

The principle of isoelectric focusing (IEF) is that electrophoresis is carried out in a *pH gradient*, allowing each protein to migrate to its isoelectric point, i.e. the point at which its pI value is equivalent to the

surrounding pH and its net charge is zero. Size fractionation is achieved by equilibrating the isoelectric-focusing gel in a solution of the detergent sodium dodecylsulfate (SDS), which binds non-specifically to all proteins and confers a uniform negative charge. The focused proteins are then separated in the perpendicular plane to the first separation (Fig. 21.1). In the original study (O'Farrell 1975), *E. coli* proteins were separated by isoelectric focusing in a tube gel, i.e. a gel cast in a thin plastic tube. When the IEF run was complete, the tube was cracked open and the proteins exposed to SDS by immersion of the gel in an SDS solution. The tube gel was then attached to an SDS-PAGE slab gel, i.e. a flat gel cast between two plates, and the proteins were separated by size. The basic procedure for 2DGE has changed little since this time although the rather cumbersome tube gels have been largely replaced by strip gels, which are easier to handle and give more reproducible separations. Additionally, the pH gradients were initially established using synthetic carrier ampholytes, which are collections of small amphoteric buffering molecules with pI values corresponding to the pH range of the focusing gel. These are mobile and have a tendency to drift towards the cathode of the gel, distorting the separation. They have been largely replaced with immobilized pH gradient gels (IPG gels) in which the buffering groups are attached to the gel matrix (Gorg *et al.* 2000).

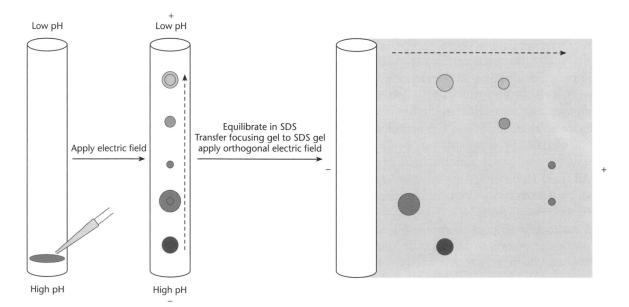

Fig. 21.1 Two-dimensional electrophoresis using a tube gel for isoelectric focusing and a slab gel for SDS-PAGE. The proteins are separated in the first dimension on the basis of charge and in the second dimension on the basis of molecular mass. The circles represent proteins, with shading to indicate protein pI values and diameters representing molecular mass. The dotted line shows the direction of separation.

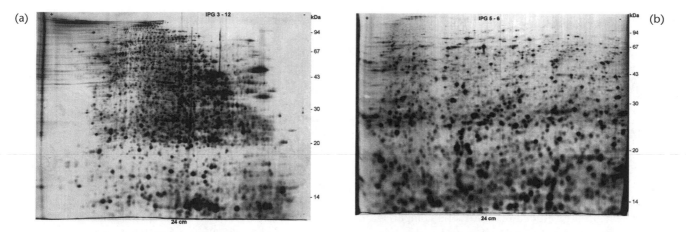

Fig. 21.2 Two-dimensional gel electrophoresis of mouse liver proteins followed by silver staining. (a) Wide immobilized pH gradient (3–12). (b) Narrow immobilized pH gradient (5–6) showing much greater resolution of proteins. (Reproduced with permission from Angelika Görg and *Electrophoresis*.)

After fractionation, the protein gel is stained. There is a wide choice of agents capable of staining proteins non-specifically, including silver nitrate and dyes such as Coomassie brilliant blue, which are commonly used on western blots. For proteomics applications, however, these agents lack the sensitivity to detect scarce proteins, they have a narrow linear range, and silver staining can introduce covalent modifications which interfere with further analysis by mass spectrometry (see below). Therefore, a new series of reagents has been developed with proteomics very much in mind. These are highly sensitive fluorescent stains known as SYPRO dyes, which are sensitive, have a broad linear range, and do not introduce covalent modifications into proteins (Lim *et al.* 1997). The outcome is a unique pattern of dots, each dot representing a protein, providing a fingerprint of the proteins in the cell (Fig. 21.2).

The sensitivity, resolution, and representation of 2D gels need to be improved

It is possible to resolve approximately 2000–3000 proteins on a standard 2D gel, but many of these represent the abundant and superabundant classes, whereas the rarest proteins (often the most interesting) are difficult to detect (Rabilloud 2002). The proteins in a cell differ in abundance over four to six orders of magnitude, with most of the total protein content represented by a relatively small number of abundant or superabundant proteins. In yeast, for example, it is estimated that 50% of the proteome comprises the output of just 100 genes and that the most abundant proteins are present at levels exceed-

ing 1,000,000 copies per cell. Increasing the resolution of a 2DGE separation can increase the chance of detecting rare proteins because the small spots are less likely to be masked by the large spots of abundant proteins. The resolution of 2DGE depends on the separation length in both dimensions, and can thus be increased if very large format gels are used (reviewed by Herbert *et al.* 2001), or if multiple IEF gels are run, each with a narrow pH range (these are known as zoom gels; Fig. 21.2) (Sanchez *et al.* 1997, Rabilloud *et al.* 1997). Alternatively, to increase the resolution of proteins within a particular pH range, gels can be run with non-linear pH gradients preferably with a flattened pH gradient between pH 4 and 7, which accounts for the majority of proteins (Fig. 21.2). Strategies to increase the resolution of IEF gels are often facilitated by pre-fractionation, e.g. to remove proteins outside the pH range of a zoom gel since these can accumulate at the electrodes and distort the focusing of the remaining proteins.

Because proteins are so diverse in terms of their chemical and physical properties, it is virtually impossible to devise a method that leads to the unbiased representation of all proteins on a 2D gel. The most important factor in determining which proteins are represented is the solubilization step, and for general applications the procedure has not changed very much since it was first developed in 1975. The standard lysis buffer includes a chaotropic agent to disrupt hydrogen bonds (urea, or a combination of urea and thiourea), a non-ionic detergent such as NP-40, and a reducing agent. These conditions are not suitable for the solubilization of membrane proteins and this is why membrane proteins are under-represented on standard 2D gels. The recovery of membrane proteins can be increased by choosing stronger detergents, such as CHAPS, and by selectively enriching the initial sample for membrane proteins, e.g. by preparing membrane fractions (Gorg *et al.* 2000).

Multiplexed analysis allows protein expression profiles to be compared on single gels

While the analysis of a single protein sample by 2DGE can be useful, e.g. for cataloging the proteins present in a given cell or tissue, the comparison of gels prepared using related samples (e.g. disease vs. normal tissue) can reveal protein spots that are differentially expressed. Unfortunately, the large number and complex distribution of spots on a typical 2D gel means that comparisons across gels are extremely difficult, and must be carried out using image analysis software. One problem with this is that, due to minor variations in the chemical and physical properties of electrophoretic gels, it is impossible to exactly reproduce the conditions from one experiment to another. Gel-matching software, such as the freely available program MELANIE II (www.expasy.ch/ch2d/melanie), typically works by establishing the positions of unambiguous landmark spots and then stretching, skewing, and rotating parts of the image to match other spots. This may involve calculating values for spot intensities and interspot distances between neighboring spots as variables in the algorithm (reviewed by Pleissner *et al.* 2001). Another useful program is the Java applet CAROL (http://gelmatching.inf.fu-berlin.de/Carol.html) which can be used to compare any two gel images over the Internet.

More recently, the difficulties encountered in running multiple gels and comparing spot patterns have been addressed by the development of techniques for multiplexed proteomics. One example is difference gel electrophoresis, which employs dual fluorescent labels applied separately to different protein samples, which are then loaded onto the same gel (Lilley *et al.* 2001). The principle is similar to DNA microarray experiments using dual labels and the signal intensities from each probe show the relative abundance of each protein in each sample. Different protein samples are labeled on lysine side chains with succinimidyl esters of propyl-Cy3 and methyl-Cy5, two fluorophores that emit light at different wavelengths. The protein samples are mixed prior to separation and loaded onto the 2D gel together. After electrophoresis, the gel is scanned using a CCD camera or fluorescence reader fitted with two different filters and two sets of data are obtained. The images from each filter can be pseudocolored and combined, immediately revealing the spots representing proteins with differential abundance (Fig. 21.3). The use of further labels, e.g. Cy2, can allow even more samples to be run concurrently. Because the samples run together, all differences in gel preparation, running conditions, and local gel structure are eliminated, which considerably simplifies the downstream analysis.

The advantages of DIGE in terms of data analysis are undeniable, but the technique also has its drawbacks in that the fluorescent labels used are less sensitive than both SYPRO dyes and silver staining. This primarily reflects the fact that only a small proportion of the proteins in each sample can be labeled otherwise solubility is lost and the proteins precipitate during electrophoresis. A further consequence of partial labeling is that the bulky fluorescent conjugate retards the proteins during the SDS-PAGE separation so the gels must be post-stained, e.g. with Coomassie brilliant blue, to identify the "true" protein spot to be excised for downstream analysis by mass spectrometry. Registration errors between the labeled and unlabeled protein populations are minimized during isoelectric focusing because the fluorescent conjugates are charge matched.

The sensitivity of standard gels can be combined with the convenience of multiplex fluorescence in a new area of multiplexed proteomics in which the same SYPRO reagent is used to stain and compare protein spots on different gels, but the gels can also be stained with additional reagents that identify specific classes of proteins (Patton & Beecham 2001, Patton 2000). These proteins can be used as landmarks for gel matching but more importantly the technique can be used to identify subsets of proteins in the proteome that share specific functional attributes. A number of stains have been developed by companies such as Molecular Probes Inc. that recognize various structurally or functionally related proteins, e.g. glycoproteins and phosphoproteins, oligo-histidine tagged proteins, calcium-binding proteins, and proteins that have the capability to bind or metabolize particular drugs.

Multidimensional liquid chromatography is more sensitive than 2DGE and is directly compatible with mass spectrometry

Liquid chromatography (LC) is often used either upstream of 2DGE to pre-fractionate samples or

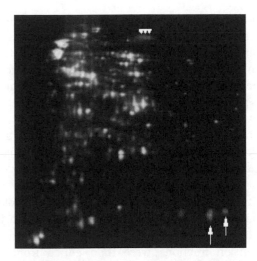

Fig. 21.3 2D DIGE overlay image of Cy3- (green) and Cy5- (red) labeled test-spiked *Erwinia carotovora* proteins. The protein test spikes were three conalbumin isoforms (arrowheads) and two myoglobin isoforms (arrows). Spots that are of equal intensity between the two channels appear yellow in the overlay image. As spike proteins were eight times more abundant in the Cy5 channel, they appear as red spots in the overlay. The gel is oriented with the acidic end to the left. Reprinted from Lilley *et al.* (2002) Two-dimensional gel electrophoresis: recent advances in sample preparation, detection and quantitation. *Current Opinion in Chemical Biology* **6**, 46–50. © 2002, with permission from Elsevier.

downstream to separate tryptic peptides prepared from individual gel spots (see below). However, the flexibility of LC methods in terms of combining different separative principles makes multidimensional chromatography an attractive technology to replace 2DGE all together. Many of the limitations of 2DGE are circumvented by LC systems. For example, HPLC columns allow the loading of large sample volumes, which can be concentrated on the column making low-abundance proteins easier to detect. Many of the proteins that are difficult to analyze by 2DGE (e.g. membrane proteins, very basic proteins) can be separated easily using appropriate resins. Proteins separated in the liquid phase do not need to be stained in order to be detected. Perhaps most importantly, the fact that LC methods can separate peptides as well as proteins, and the ability to couple LC columns directly to the mass spectrometer mean that the entire analytical process from sample preparation to peptide mass profiling can be automated. The disadvantages of LC methods are that the visual aspects of protein separation by 2DGE are lost, including the pI and molecular mass data that can be determined from the positions of spots on the gel (these data can be used in database searches). LC is also a serial analysis technique, while multiple gels

containing related samples can be run at the same time.

The sequential application of different chromatographic techniques exploiting different physical or chemical separative principles can provide sufficient resolution for the analysis of very complex protein or peptide mixtures (Wang & Hanash 2003, Wehr 2002). For example, the sequential use of ion-exchange chromatography (which separates proteins by charge) and RP-HPLC (which separates proteins approximately in a mass-dependent fashion) can achieve the same resolution as 2DGE, with added advantages of automation, increased sensitivity, and better representation of membrane proteins. However, it is necessary to consider the practical limitations of such multidimensional chromatography techniques. The first issue to address is the compatibility of the buffers and solvents used in different steps of each procedure. In the example discussed above, the elution buffer used in the first-dimension ion-exchange step would have to be a suitable solvent for RP-HPLC, and the elution buffer for the second-dimension RP-HPLC step would need to be compatible with the solvents used in the sample preparation stage for mass spectrometry. Otherwise, many of the advantages of speed, resolution, and automation would be lost as the fractions were taken off line to be cleaned up and prepared.

Fortunately, the solvents and buffers described above are indeed compatible, and ion exchange followed by RP-HPLC-MS has been used by several investigators to analyze the proteomes of organisms ranging from yeast to humans. The compatibility of RP-HPLC with the solvents used in mass spectrometry means that HPLC is almost universally used as the final separation method in multidimensional chromatography. Several other profiling methods, such as size exclusion chromatography (Lecchi *et al.* 2003), ion-exchange chromatography (Wang & Hanash 2003), have been used as a first-dimension separation method in combination with HPLC, sometimes with a prior affinity chromatography step resulting in a tri-dimensional separation strategy (Lee & Lee 2004).

Initially, multidimensional chromatography was achieved by a discontinuous process in which fractions were collected from the ion-exchange or gel filtration column and then manually injected into the HPLC column (Gygi *et al.* 1999a). Although labor intensive, the advantage of a discontinuous multidimensional system is the absence of time constraints. The fractions eluting from the first column can be stored offline indefinitely, and fed one-by-one into the HPLC column, which is directly coupled to the mass spectrometer. A further advantage is that large sample volumes can be applied to the first column in order to obtain sufficient amounts of low-abundance proteins for analysis in the second dimension.

The need for manual sample injection can be circumvented by equipping the first column with an automatic fraction collection system and a column-switching valve. Fractions are then collected from the first column across the elution range, and the switching valve can bring the RP-HPLC column in line to receive the fractions sequentially. Alternatively, some researchers have developed apparatus comprising a single ion-exchange column coupled, via an appropriate set of switching valves, to multiple HPLC columns arranged in parallel (Fig. 21.4). In this scheme, fractions emerging from the first column are directed sequentially to the multiple HPLC columns, and the cycle is repeated when the first column has been regenerated (Opitek et al. 1998).

A third strategy for multidimensional chromatography separations is the use of biphasic columns, in which the distal part of the column is filled with reversed-phase resin and the proximal part with another type of matrix. As long as the elution solvents for each type of resin do not interfere with each other, this allows the stepped elution of fractions from the first resin and the gradient elution of second-dimension fractions from the second. This technique was pioneered by Link and colleagues as direct analysis of large protein complexes (DALPC) and was modified by Yates and colleagues into a system called multidimensional protein identification technology (MudPIT) (see Washburn et al. 2001). As shown in Fig. 21.5, peptide mixtures loaded onto the ion-exchange resin were eluted using a stepped gradient of salt, resulting in the release of first-dimension fractions into the reversed-phase resin. Second-dimension fractions were then eluted from the reversed-phase resin into the mass spectrometer using a gradient of acetonitrile. This process, and the subsequent regeneration step, did not interfere with the ion-exchange chromatography step, and after regeneration another fraction was released from the ion-exchange resin by increasing the salt concentration. When this method was applied to the yeast proteome, over 5000 peptides could be assigned to a total of 1484 yeast proteins, representing about

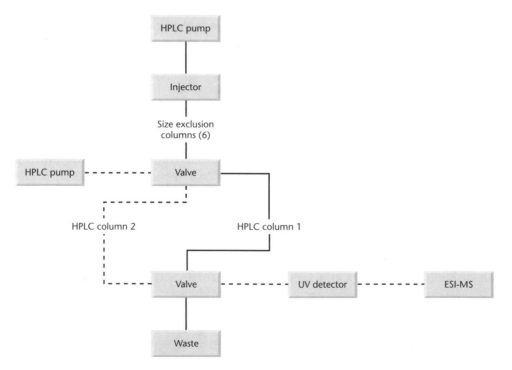

Fig. 21.4 Continuous multidimensional chromatography with column switching. In this example, simplified from Opiteck et al. (1998) two HPLC columns working in parallel receive alternating eluates from a bank of six size exclusion columns in series. After sample injection and separation by size exclusion chromatography, eluate from the size exclusion columns is directed to HPLC column 1 using a four-port valve (thick line). While the peptides are trapped in this column, HPLC column 2 is eluted and the sample is directed to the detector and fraction collector (broken line). After flushing and equilibrating column 2, the valves are reversed allowing column 2 to be loaded with the next fraction from the size exclusion separation, while column 1 is eluted.

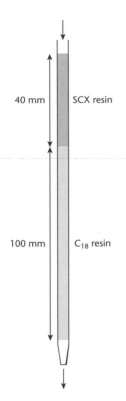

40 mm | SCX resin

100 mm | C₁₈ resin

Fig. 21.5 Continuous multidimensional chromatography using a biphasic column. In this example, simplified from the MudPIT method (Washburn *et al.* 2001) a 140 mm × 0.1 mm fused silica capillary is packed at the distal end with 5-μm C$_{18}$ (reversed phase) particles and at the proximal end with 5-μm strong cation exchange (SCX) particles.

one-quarter of the yeast proteome. The sample appeared to be representative of all classes of proteins, including those usually under-represented in 2DGE experiments.

Mass spectrometry is used for protein characterization

High-throughput protein annotation is achieved by mass spectrometry and correlative database searching

As early as the beginning of the 1980s, it was suggested that a 2DE database should be established to catalog human proteins and identify proteins whose presence or absence was associated with disease (Anderson & Anderson 1982). However, because there was little information available in sequence databases and no high-throughput techniques for protein annotation, such differentially expressed proteins often remained uncharacterized. In the

mid-1980s, the first convenient methods for protein sequencing became available. This began with the development of automated Edman degradation "sequenators", and a few years later a method was devised for blotting 2D gels onto PVDF (polyvinylidine difluoride) membranes, which allowed direct protein sequencing. Although Edman degradation is sensitive and perhaps remains the most convenient method for determining N-terminal sequences, it is insufficient to cope with the vast amounts of data arising from today's 2DE experiments. A breakthrough in high-throughput protein annotation came with improvements in mass spectrometry techniques coupled with the development of algorithms allowing protein databases to be searched on the basis of molecular mass (Yates 2000, Anderson & Mann 2000, Lahm & Langen 2000, Paterson 2000, Mann & Pandey 2001).

Mass spectrometry involves the ionization of target molecules in a vacuum, and accurate measurement of the mass of the resulting ions. A mass spectrometer has three component parts: an ionizer, which converts the anylate into gas phase ions; a mass analyzer which separates the ions according to their mass/charge ratio (m/z); and an ion detector. Generally, large molecules such as proteins and nucleic acids are broken up and degraded by the ionization procedure, but more recently, sensitive instruments that are capable of *soft ionization*, i.e. the ionization of large molecules without significant degradation, have been developed. This allows accurate mass measurements of whole proteins and peptide fragments, data that can be used to search protein databases to identify particular proteins (Fig. 21.6).

Two major strategies are used to characterize proteins (Aebersold & Mann 2003). The first is *peptide mass fingerprinting*, which is often carried out using a mass spectrometer with an electrospray ionization (ESI) or a matrix assisted laser desorption ionization (MALDI) source coupled with a time of flight (TOF) analyzer. A glossary of these terms is provided in Box 21.1. Briefly, protein spots are excised from a 2D gel and digested with a specific endopeptidase, such as trypsin, to generate peptide fragments. These are then analyzed by mass spectrometry to determine their molecular masses, and these data are used to search protein databases. Computer algorithms have been developed by a number of groups for correlating MS-determined peptide masses with virtual peptide masses derived from protein databases (Henzel *et al.* 1993, James *et al.* 1993, Mann *et al.* 1993, Pappin *et al.* 1993, Yates *et al.* 1993). The

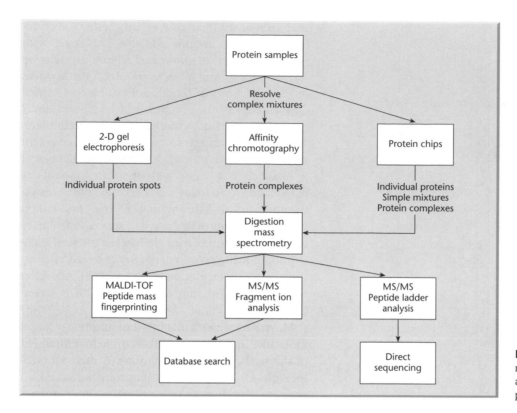

Fig. 21.6 Current routes to protein annotation in proteomics.

searches are useful only if a significant amount of sequence data exists, i.e. in the case of organisms with complete or well-advanced genome projects. In yeast, for example, it has been possible to calculate the masses of the tryptic peptide fragments from every protein, based on the translation of all known open reading frames (ORFs), thus allowing rapid and precise protein annotation. Unfortunately, EST sequences are not useful for peptide mass fingerprinting because the algorithms require correlation between the masses of several peptide fragments from the same protein and a given database entry. ESTs are generally too short to represent a significant number of peptide fragments. The problem is compounded by the propensity for MS data not to agree with stored sequences. This may reflect genuine sequence errors in some cases, but often occurs due to unrecognized post-translational modifications to the protein, non-specific proteolysis, the presence of contaminating proteins, or the absence of a particular sequence from the database. The accuracy of the MS measurement may also vary, although this has been addressed by the development of improved MALDI-MS instruments with ion-focusing mirrors and delayed extraction devices, which help to reduce their spread of kinetic energy and thus increase accuracy and resolution.

Where peptide mass fingerprinting fails to identify any proteins matching those present in a given sample, the fragments of one or more individual peptides may provide important additional information (Lamond & Mann 1997; reviewed by Choudhary *et al.* 2001a,b). The data can be used in two ways. First, the *uninterpreted* fragment ion masses can be used in correlative database searching to identify proteins whose peptides would likely yield similar CID spectra (collision-induced dissociation; see Box 21.1) under the same fragmentation conditions. In probability-based matching, virtual CID spectra are derived from the peptides of all protein sequences in the database and these are compared with the observed data to derive a list of potential matches. In cross-correlation, it is the degree of overlap between the observed and predicted peaks that determines the best potential match. Several algorithms, such as Sequest (Fenyo 2000, Chakravarti *et al.* 2002), are available for database searching with uninterpreted data. Second, the peaks of the mass spectrum can be *interpreted*, either manually or automatically, to derive partial *de novo* peptide sequences that can be used as standard database search queries. The advantage of both these approaches is that correlative searching is not limited to databases of full protein sequences. Both uninterpreted

Box 21.1   Glossary of terms used in mass spectrometry, and useful Internet resources

Mass spectrometry

A technique for accurately determining molecular masses by calculating the mass : charge ratio of ions in a vacuum. A mass spectrometer is an instrument combining a source of ions, a mass analyzer that can separate ions according to their mass : charge ratio, and an ion detector.

Soft ionization

The ionization of large molecules, such as proteins and nucleic acids, without causing significant amounts of fragmentation.

Matrix-assisted laser desorption ionization (MALDI)

A soft ionization method used for peptide mass fingerprinting. The analyte, a mixture of peptide fragments resulting from tryptic digestion of a particular protein, is first mixed with a light-absorbing "matrix compound" such as dihydroxybenzoic acid, in an organic solvent. The solvent is then evaporated to form crystals and these are transferred to a vacuum. The dry crystals are targeted with a laser beam. The laser energy is absorbed and then emitted (desorbed) as heat, resulting in expansion of the matrix and anylate into the gas phase. A high voltage is applied across the sample to ionize it, and the ions are accelerated towards the detector.

Electrospray ionization (ESI)

A soft ionization method used for fragment ion searching. The analyte is dissolved in an appropriate solvent and pushed through a narrow capillary. A potential difference is applied across the capillary such that charged droplets emerge and form a fine spray. Under a stream of heated inert gas, each droplet rapidly evaporates so that the solvent is removed as the analyte enters the mass analyser and the ions are accelerated towards the detector.

Quadrupole

A mass analyzer that determines the mass : charge ratio of an ion by varying the potential difference applied across the ion stream, allowing ions of different mass : charge ratios to be directed towards the detector. A quadrupole comprises four metal rods, pairs of which are electrically connected and carry opposing voltages that can be controlled by the operator. More than one quadrupole may be connected in series, as in triple quadrupole mass spectrometry. Varying the voltage steadily over time allows a mass spectrum to be obtained.

Time of flight (TOF)

A mass analyzer that determines the mass : charge ratio of an ion by measuring the time taken by ions to travel down a flight tube to the detector (Karas & Hillenkamp 1988).

Tandem mass spectrometry (MS/MS)

Mass spectrometry using an instrument with two mass analyzers, either of the same type or a mixture. A number of hybrid quadrupole/TOF instruments have been described (e.g. Krutchinsky et al. 2000, Shevchenko et al. 2000). The mass analyzers may be separated by a collision cell that contains inert gas and causes ions to dissociate. MS/MS is generally used for fragment ion analysis, because particular peptide fragments can be selected using the first mass analyzer, fragmented in the collision cell and the fragments can be separated in the second analyzer.

Collision-induced dissociation (CID)

The use of a collision cell between mass analyzers to excite ions and make them dissociate into fragments.

Mass spectrometry resources on the Internet (peptide mass and fragment fingerprinting tools)

http://www.expasy.ch/tools/#proteome
http://www.seqnet.dl.ac.uk/Bioinformatics/
http://www.narrador.embl-heidelberg.de/GroupPages/PageLink/peptidesearchpage.html
http://prospector.ucsf.edu

fragment ion masses and the short peptide sequences derived from interpreted spectra can be used to search through the less robust but much more abundant EST data. Millions of ESTs have been obtained and deposited in public databases, and when translated they represent a rich source of information about proteins. However, ESTs cannot be searched in peptide mass fingerprinting because that technique relies on the presence of sequences corresponding to several intact peptides. ESTs are generally too short (100–300 bp) to contain more than one intact peptide.

Specialized strategies are used to quantify proteins directly by mass spectrometry

Protein quantitation at the mass spectrometry stage makes it possible to use in-line liquid-phase separation methods such as multidimensional chromatography and capillary electrophoresis (see above). The general approach is to label alternative samples with equivalent reagents, one of which contains a heavy isotope and one of which contains a light isotope. The samples are mixed, separated into fractions, and analyzed by mass spectrometry. The ratio of the two isotopic variants can be determined from the heights of the peaks in the mass spectra and used to identify proteins with differential abundance. Several variants of the approach can be used which are discussed below and summarized in Fig. 21.7.

One of the first developments in quantitative mass spectrometry was a class of reagents known as isotope-coded affinity tags (ICATs) (Gygi *et al.* 1999). These are biotinylated derivatives of iodoacetamide, a reagent that reacts with the cysteine side chains of denatured proteins. Two versions of the reagent are used, one normal or light form and one heavy or deuterated form in which a hydrogen atom is replaced by deuterium. The heavy and light forms are used to label different protein samples and then the proteins are combined and digested with trypsin. The biotin allows cysteine-containing peptides to be isolated from the complex peptide mixture through affinity to streptavidin, therefore considerably simplifying the number of different peptides entering the mass spectrometer (Fig. 21.8).

An alternative to the ICAT labeling of proteins that is not selective for cysteine-containing peptides is to label the peptides *after digestion* (Tao & Aebersold 2003, Goshe & Smith 2003, Sechi & Oda 2003). When trypsin cleaves a protein and generates a peptide with a new C terminus, it introduces an oxygen atom derived from a molecule of water into the carboxyl group of the peptide. This can be used to differentially label peptides derived from alternative

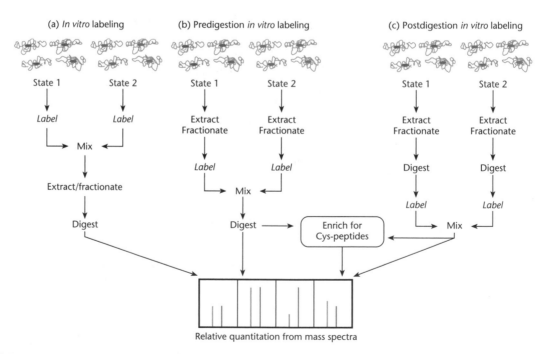

Fig. 21.7 Overview of MS-based strategies for quantitative proteomics. Depending on the point at which the label is introduced, most procedures are classified as (a) *in vivo* labeling, (b) predigestion labeling *in vitro*, or (c) postdigestion labeling *in vitro*. Reprinted from Sechi & Oda (2003) Quantitative proteomics using mass spectrometry. *Current Opinion in Chemical Biology* **7**, 70–7. © 2003 with permission from Elsevier.

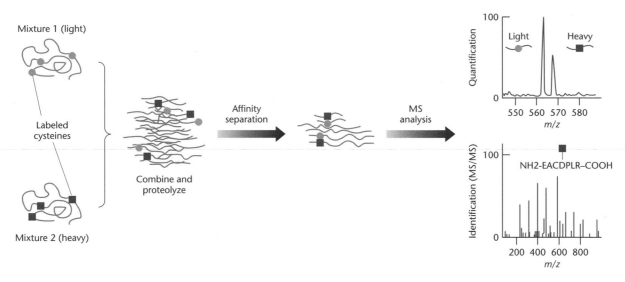

Fig. 21.8   The ICAT reagent strategy for protein quantitation. Two protein mixtures representing two different cell states are treated with the isotopically light (gray) or heavy (purple) ICAT reagents, respectively. The labeled protein mixtures are then combined and proteolyzed; tagged peptides are selectively isolated and analyzed by MS. The relative abundance is determined by the ratio of signal intensities of the tagged peptide pairs. Reprinted from Tao and Aebersold (2003) Advances in quantitative proteomics via stable isotope tagging and mass spectrometry. *Current Opinion in Biotechnology* **14**, 110–18. © 2003 with permission from Elsevier.

protein samples if normal water is used in one buffer and water substituted with heavy oxygen (^{18}O) is used in the other (Fig. 21.9). The abundance of the peptides can then be compared, since they will appear as doublets separated by two mass units.

In another group of methods, cells treated under different conditions are grown in media containing either normal or heavy isotopes of nitrogen, carbon, or hydrogen. A useful approach is the use of labeled amino acids (stable-isotope labeling with amino acids in cell culture, SILAC) (Washburn *et al.* 2001). The cells can then be harvested, combined, and the proteins extracted for separation, tryptic digestion, and analysis by mass spectrometry. Equivalent peptides from each sample will differ in mass by a single mass unit and can easily be identified as doublets in mass spectra. The relative amounts of the two peptides can be determined on the basis of the relative heights of the two peaks. The advantage of this metabolic labeling approach is that the label is introduced early in the experiment, therefore eliminating variation arising from sample preparation and purification losses (Fig. 21.10). However, it can only be used for the analysis of live cells that can be maintained in a controlled environment. It is not useful, for example, for tissue explants, biopsies, body fluids, or cells that are difficult to maintain in culture.

A final strategy involves the use of mass-coded chemical tags, which avoids the need for stable iso-topes. In the MCAT (mass-coded abundance tag) method, proteins from one sample are labeled with O-methylisourea and those from the other sample are not labeled at all. This differs from the other methods discussed above where both samples are labeled but with different isotopes. This method is simple and inexpensive, but less accurate than those involving isotopes.

Protein modifications can also be detected by mass spectrometry

Almost all proteins are modified in some way during or after synthesis, either by cleavage of the polypeptide backbone or chemical modification of specific amino acid side chains. This phenomenon, which is known as post-translational modification (PTM), provides a direct mechanism for the regulation of protein activity and greatly enhances the structural diversity and functionality of proteins by providing a larger repertoire of chemical properties than is possible using the 20 standard amino acids specified by the genetic code. Several hundred different forms of chemical modification have been documented, but here we focus on phosphorylation, a reversible form of modification which occurs in all cells (Yan *et al.* 1998).

As discussed above, MALDI-TOF MS is most often used to analyze intact peptides, and correlative

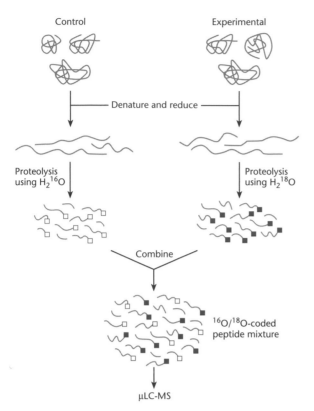

Fig. 21.9 Enzymatic stable isotope coding. For enzymatic labeling, proteins from two distinct proteomes are proteolytically digested in aqueous buffer containing either normal water ($H_2^{16}O$; white squares) or isotopically labeled water ($H_2^{18}O$; purple squares). This encoding strategy effectively labels every C terminus produced during digestion. The samples are combined at the peptide level and then analyzed by microcapillary LC-MS. Reprinted from Goshe and Smith (2003) Stable isotope-coded proteomic mass spectrometry. *Current Opinion in Biotechnology* **14**, 101–9. © 2003 with permission from Elsevier.

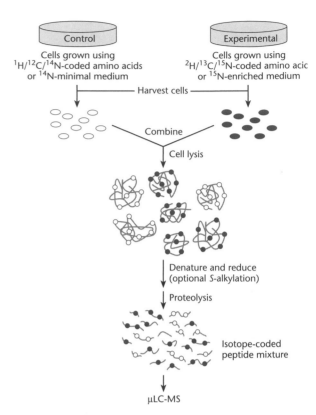

Fig. 21.10 Metabolic stable isotope coding. For metabolic labeling, cells from two distinct cultures are grown on media supplemented with either normal amino acids ($^1H/^{12}C/^{14}N$) or ^{14}N-minimal media (white spheres) or stable isotope amino acids ($^2H/^{13}C/^{15}N$) or ^{15}N-enriched media (purple spheres). These mass tags are incorporated into proteins during translation, thus providing complete proteome coverage. An equivalent number of cells for each sample is combined and processed for microcapillary LC-MS analysis. Reprinted from Goshe and Smith (2003) Stable isotope-coded proteomic mass spectrometry. *Current Opinion in Biotechnology* **14**, 101–9. © 2003 with permission from Elsevier.

database searching (peptide mass fingerprinting) allows the derived masses to be matched against the theoretical peptides of known proteins. Therefore, if the identity of the protein is known or can be deduced from the peptide masses, phosphopeptides can be identified simply by examining the mass spectrum for mass shifts of 80 (the mass of PO_3) compared to predicted masses. Parallel analysis in which the sample has been treated with alkaline phosphatase can also be helpful, since peaks corresponding to phosphopeptides in the untreated sample should be absent from the treated sample (Fig. 21.11).

The analysis of fragment ions serves two purposes in phosphoproteomics. First, phosphopeptides preferentially yield diagnostic, phosphate-specific fragment ions such as $H_2PO_4^-$, PO_3^- and PO_2^-, which have masses of approximately 97, 79, and 63 respectively. Phosphoserine and phosphothreonine are more labile in this respect than phosphotyrosine. The presence of such ions in a mass spectrum therefore indicates the presence of a phosphopeptide in the sample. Secondly, fragmentation along the polypeptide backbone can yield peptide fragments that allow a sequence to be built up *de novo*. This sequence will include the phosphoamino acid, providing a definitive location for the phosphorylated residue.

Since the phosphate group provides a negative charge, phosphate-specific fragment ions are usually obtained by MS/MS using an ESI ion source in negative ion mode. Collision-induced dissociation (CID) is used to generate the fragments and analysis is usually carried out using a triple quadrupole or more sensitive hybrid quadrupole-quadrupole-TOF machine (an ion trap can also be used, although this is less common because the instrumentation is not

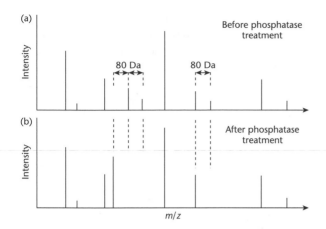

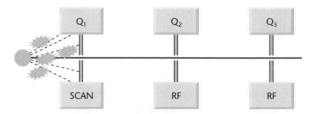

Fig. 21.12 Detection of phosphopeptides using in-source CID in a triple quadrupole mass spectrometer. Excess energy used during ionization causes fragmentation to occur at the ion source. The ion stream is scanned in Q_1 for phosphate reporter ions such as PO_3^- ($m/z = 79$). These pass through the other quadrupoles (running in RF mode) to the detector.

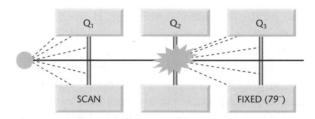

Fig. 21.13 Detection of phosphopeptides using precursor ion scan mode in a triple quadrupole mass spectrometer. The entire ion stream is scanned in Q_1, allowing selected ions through to the collision chamber in Q_2. The fragmented ions then pass through Q_3, which is fixed to detect phosphate reporter ions such as PO_3^- ($m/z = 79$). Only if intact phosphopeptide ions pass through Q_1 will reporter ions pass through Q_3 to the detector.

Fig. 21.11 Phosphopeptide identification by MALDI-TOF MS mapping combined with alkaline phosphatase treatment. (a) The MALDI-TOF MS spectrum of a proteolytic digest. Phosphopeptides are indicated by peaks shifted by multiples of 80 Da ($HPO_3 = 80$ Da) relative to predicted unphosphorylated peptide masses. (b) The disappearance of such peaks upon treatment with a phosphatase confirms their identity as phosphopeptides. Reprinted from McLachlin and Chait (2001) Analysis of phosphorylated proteins and peptides by mass spectrometry. *Current Opinion in Chemical Biology* **5**, 591–602. © 2001 with permission from Elsevier.

widely available). Three general strategies are used. The first is called in-source CID and requires excess energy during ionization to induce multiple collisions and produce the phosphate reporter ions in the emerging ion stream (Fig. 21.12). This method also fragments the peptide backbone to a lesser extent and can therefore provide some peptide sequence information (see below). Normal ionization energy levels are used in precursor ion scanning. In this mode, the first quadrupole (Q_1) is used to scan the ion stream, CID occurs in Q_2 (running in RF mode), and the third analyzer (Q_3 or TOF) is set to detect phosphate reporter ions, such as PO_3^-, induced by collision. Phosphopeptides are thus identified when a precursor ion scanned in Q_1 yields a phosphate fragment that is detected in Q_3 (Fig. 21.13). In neutral loss scan mode, both Q_1 and Q_3 are set to scan the ion stream. Q_1 scans the full mass range, Q_2 is used as the collision cell, and Q_3 scans a parallel range to Q_1 but at an m/z ratio that is $98/z$ lower, with the intention of detecting the neutral loss of H_3PO_4 (Fig. 21.14).

All these methods can be combined with in-line HPLC, either using reversed-phase material alone or in combination with a second-dimension separation matrix, such as a strong cation exchange resin. A good example of this approach was reported by Ficarro *et al.* (2002). They digested yeast protein lysates with trypsin, converted the peptides into

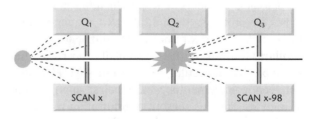

Fig. 21.14 Detection of phosphopeptides using neutral loss scan mode in a triple quadrupole mass spectrometer. The entire ion stream is scanned in Q_1, allowing selected ions (x) through to the collision chamber in Q_2. Q_3 is set to scan the fragmented ions in parallel to Q_1, but at a lower mass range (e.g. x-98, where 98 is the mass of H_3PO_4). Only phosphopeptide ions which lose H_3PO_4 during CID will pass through Q_3 to the detector.

methyl esters and enriched the phosphopeptide pool using IMAC. This was coupled to a nanoflow HPLC column that fed fractions directly into an ESI-mass spectrometer. Over 1000 phosphopeptides were identified in this procedure, and the sequences of 216 peptides were obtained allowing 383 phosphorylation sites to be determined.

Protein microarrays can also be used for expression analysis

2D-electrophoresis is an open system for proteome analysis, rather like direct sequence sampling is an open system for transcriptome analysis (p. 410). The advantage of an open system is that potentially all proteins can be detected, but the disadvantage is that they also have to be characterized, which relies on downstream annotation by mass spectrometry (similarly, directly sampled cDNAs have to be characterized by downstream sequencing). DNA arrays are closed systems in transcriptome analysis, i.e. the data obtained are constrained by the number and nature of sequences immobilized on the array. However, it is not necessary to characterize any of the features on the array by sequencing because the sequences are already known. Similarly, *protein arrays* are emerging as a useful closed system for proteome analysis. These are miniature devices in which proteins, or molecules that recognize proteins, are arrayed on the surface (Templin *et al.* 2002, Schweitzer & Kingsmore 2002, Cutler 2003, Zhu & Snyder 2003).

In concept, protein arrays are no different to DNA arrays, but they suffer from several practical limitations. First, the manufacture of DNA arrays is simplified by the availability of methods, such as the polymerase chain reaction, for amplifying any nucleic acid sequence. No amplification procedure exists for proteins. Second, all DNA sequences are made of the same four nucleotides and hence behave similarly in terms of their chemical properties. The principles of molecular recognition (hybridization between complementary base pairs) apply to all sequences. For this reason, hybridization reactions can be carried out in highly parallel formats using a single complex probe. Conversely, proteins are made of 22 amino acids specified by the genetic code plus many others generated by post-translational modification, so they have diverse chemical properties. For example, some proteins are soluble in water while others are lipophilic. Recognition parameters vary widely so the same reaction conditions could never be used for all proteins. Third, the homogeneity of DNA molecules means that labels are incorporated evenly and labeling does not interfere with hybridization. Binding to solid substrates such as nylon and glass does not interfere with hybridization either. However, the labeling of proteins is much more variable and both labeling and attachment to a substrate could interfere with protein binding, either

by affecting the way the protein folds or by blocking the binding site. Despite these differences, protein arrays have been manufactured in many of the ways discussed above for DNA arrays, including variations on standard contact printing (Mendoza *et al.* 1999, Lueking *et al.* 1999), inkjetting (Etkins 1998), and photolithography (Mooney *et al.* 1996, Jones *et al.* 1998). The concept of the protein array, protein microarray, or protein chip covers a wide range of different applications. Some of these are considered below.

Antibody arrays contain immobilized antibodies or antibody derivatives for the capture of specific proteins

On these devices, antibodies are attached to the array surface, so the protein array or chip can be thought of as a miniaturized solid-state immunoassay. Antibodies interact with specific proteins and are highly discriminatory, so they are suited to the detailed analysis of protein profiles and expression levels. The feasibility of this approach has been demonstrated, for example, using a recombinant Staphylococcal protein A covalently attached to a gold surface. The recombinant protein A has five immunoglobulin G binding domains, allowing antibodies to be attached by the Fc region, therefore exposing the antigen-binding domain (Kanno *et al.* 2000). Antibody arrays have also been generated by using banks of bacterial strains expressing recombinant antibody molecules (de Wildt *et al.* 2000). There are three different formats for this type of assay. The first is a standard immunoassay in which the antibodies are immobilized and are used to capture labeled proteins from solution. Protein expression levels are quantified by measuring the signal (usually fluorescent) which has been incorporated into the proteins. A recent report (Haab *et al.* 2001) describes the use of such an array comprising 115 different antibodies. Another groundbreaking aspect of this report was that two protein samples, each labeled with a different fluorophore, were exposed to the array simultaneously, and differential protein expression could be monitored. The second format is a miniature sandwich assay, in which unlabeled proteins are captured from solution and detected with a second, labeled antibody. Although this format requires two antibodies recognizing distinct epitopes for each protein, it is not necessary to label the target population of proteins, a process that is inefficient and variable. An example of this approach for the detection of human cytokines is provided by Moody *et al.* (2001). The third format involves a tertiary detection system and therefore offers even greater sensitivity. One example is the *immunoRCA technique*, which involves rolling-circle amplification (Schweitzer & Kingsmore 2002, Kingsmore & Patel 2003). The principle of this technique is that a protein, captured by an immobilized antibody, is recognized by a second antibody in a sandwich assay as above, but the second antibody has an oligonucleotide covalently attached to it. In the presence of a circular DNA template, DNA polymerase, and the four dNTPs, rolling-circle amplification of the template occurs resulting in a long concatamer comprising hundreds of copies of the circle, which can be detected using a fluorescent-labeled oligonucleotide probe. Antibody arrays have been reviewed (Lal *et al.* 2002, Peluso *et al.* 2003).

Antigen arrays are used to measure antibodies in solution

On these devices, protein antigens are attached to the array surface. They are used for reverse immunoassays, i.e. detecting antibodies in solution (e.g. see Paweletez *et al.* 2001, Joos *et al.* 2000). The antigens may be proteins or other molecules such as peptides or carbohydrates. Several reports have been published in which arrays of allergens have been used to screen serum samples for IgE reactivity in allergic responses and autoimmune diseases. In some cases it has proven possible not only to confirm the presence of such antibodies but also to carry out quantitative analysis. More recently, antigen arrays have been used to serodiagnose patients with viral infections.

General protein arrays can be used for expression profiling and functional analysis

This type of device can contain any type of protein, and is used to assay protein–protein interactions and protein interactions with other molecules. A range of detection strategies may be used, including labeling of the interacting molecules, or label-independent methods such as surface plasmon resonance. For example, Ge *et al.* (2000) developed a system for studying molecular interactions using a universal protein array (UPA) system where protein samples are transferred from 96-well microtiter plates to nylon membranes. The technology has also been applied to the arraying of cDNA expression libraries,

such that screening can be carried out not only with nucleic acid probes but also with antibodies or other ligands directed at the recombinant proteins (Bussow *et al.* 1998). One of the most impressive demonstrations of the power of protein array technology was provided by MacBeath & Schreiber (2000). They used an array of proteins on a glass slide to screen for ligands, enzyme substrates, and protein–protein interactions. Functional assays can also be carried out on such arrays. For example, Zhu *et al.* (2000) produced an array containing nearly all the protein kinases of the yeast proteome and carried out kinase assays on 17 different substrates. More notably, the same group has also produced a glass microarray containing nearly all the proteins in the yeast proteome (5800 spots) and used this to screen for various functions such as phospholipid binding and interactions with calmodulin (Zhu *et al.* 2001).

Other molecules may be arrayed instead of proteins

These are not protein arrays in the strict sense because they do not consist of arrayed proteins. However, they are considered here because they are analogous to antibody arrays, i.e. they contain specific capture agents that interact with proteins. DNA arrays fall within this class if they are used to analyze DNA–protein interactions (see for example Bulyk *et al.* 2001 and Iyer *et al.* 2001). Aptamers, single-stranded nucleotides that interact specifically with proteins, could also be used in this manner (see review by Mirzabekov & Kolchinsky 2001). Another area of active current research is the development of chips containing artificial recognition sites for proteins. The concept of *molecularly imprinted polymers* (MIPs) depends on the ability to emboss a polymeric substrate with recognizable molecular imprints that mimic the actual recognition molecules. For example, Shi *et al.* (1999) have described a procedure for coating recognition molecules in sugar, which is then overlain by a hexafluoropropylene polymer. They coated a mica surface with streptavidin and then covered it with a disaccharide which molded the shape of the protein. After applying the polymer, the protein and mica were removed leaving a streptavidin MIP. This was subsequently shown to preferentially bind biotin. It is not yet clear whether this technology has the sensitivity or specificity to be applied to proteome-wide expression analysis.

Some biochips bind to particular classes of protein

Instead of specific molecular interactions, these devices use broad-specificity capture agents. As above, they can be termed protein chips but not protein arrays. For example, Ciphergen Biosystems Inc. market a range of ProteinChips with various surface chemistries to bind different classes of proteins. Although relatively non-specific compared to antibodies, complex mixtures of proteins can be simplified and then analyzed by mass spectrometry. An advantageous feature of this system is the ease with which it is integrated with downstream MS analysis, since the ProteinChip itself doubles as a modified MALDI plate (Fung *et al.* 2001, Weinberger *et al.* 2001). After the chip has been washed to remove unbound proteins, it is coated with a matrix solution and analyzed by time-of-flight MS. This allows surface-enhanced laser desorption and ionization (SELDI), which provides more uniform mass spectra than MALDI and allows protein quantification. Other protein chip platforms use surface plasmon resonance to detect and quantify protein binding. This involves measuring changes in the refractive index of the chip surface caused by increases in mass (Malmqvist & Karlsson 1997). Protein chips produced by the US company BIAcore are based on this concept, and other chips combine surface plasmon resonance measurements with mass spectrometry (Nelson *et al.* 2000).

Solution arrays are non-planar microarrays

Recent developments in microfluidic devices indicate that the next generation of miniature protein assay platforms may be solution arrays. These could provide increased sensitivity (due to the kinetics of binding in solution) and higher throughput (because the arrays are constructed in three dimensions).

Examples include flow-through chips containing microchannels (Cheek *et al.* 2001), an array system based on beads labeled with different concentrations of fluorescent dyes (Morgan *et al.* 2004, Edwards *et al.* 2004), a system based on fiber optics (Gunderson *et al.* 2004), and bar-coded nanoparticles (reviewed by Zhou *et al.* 2001, Venkatasubbarao 2004).

Suggested reading

Aebersold R. & Mann M. (2003) Mass spectrometry-based proteomics. *Nature* **422**, 198–207.

Cutler P. (2003) Protein arrays: The current state-of-the-art. *Proteomics* **3**, 3–18.

Lee W.C. & Lee K.H. (2004) Applications of affinity chromatography in proteomics. *Analytical Biochemistry* **324**, 1–10.

Lesney M.S. (2001) Pathways to the proteome: from 2DE to HPLC. *Modern Drug Discovery*, 33–9.

Lilley K.S., Razzaq A. & Dupree P. (2001) Two-dimensional gel electrophoresis: recent advances in sample preparation, detection and quantitation. *Current Opinion in Chemical Biology* **6**, 46–50.

Link A.J. (2002) Multidimensional peptide separations in proteomics. *Trends in Biotechnology* **20**, S8–S13.

Mann M., Hendrickson R.C. & Pandey A. (2001) Analysis of proteins and proteomes by mass spectrometry. *Annual Review of Biochemistry* **70**, 437–73.

Mann M. & Jensen O.N. (2003) Proteomic analysis of post-translational modifications. *Nature Biotechnology* **21**, 255–61.

Nagele E., Vollmer M., Horth P. & Vad C. (2004) 2D-LC/MS techniques for the identification of proteins in highly complex mixtures. *Expert Review Proteomics* **1**, 37–46.

Sechi S. & Oda Y. (2003) Quantitative proteomics using mass spectrometry. *Current Opinion in Chemical Biology* **7**, 70–7.

Wang H. & Hanash S. (2003) Multi-dimensional liquid phase-based separations in proteomics. *Journal of Chromatography B* **787**, 11–18.

Zhu H. & Snyder M. (2003) Protein chip technology. *Current Opinion in Chemical Biology* **7**, 55–63.

CHAPTER 22

Proteomics II – Analysis of protein structures

Introduction

In Chapter 9, we discussed the principles of sequence analysis and showed how, under certain circumstances, the function of a new gene could be assigned by comparison with genes with similar sequences whose functions have already been determined. Although not a foolproof method for functional annotation, this strategy has been successful in the functional classification of 30–40% of the genes discovered in the genome projects. In the *Saccharomyces* Genome Project, for example, 30% of the identified genes had already been characterized by individual experiments, 30% were given tentative functions by sequence analysis alone, while another 30% were anonymous in that they lacked any form of functional annotation. The remaining 10% were questionable open reading frames (ORFs).

One way in which functions can be assigned to anonymous genes (also known as *orphans* or *ORFans* because they do not belong to any known gene family) is to express them in bacteria, purify the encoded protein, and examine its structure. The product of an orphan gene is known as a *hypothetical protein*, because in most cases the protein itself has not been isolated from its natural source and studied directly. By comparing the structure of hypothetical proteins to previously solved protein structures, tentative functions can often be assigned. This is because three-dimensional protein structures are much more strongly conserved in evolutionary terms than primary amino acid sequences, and it is the three-dimensional protein structure (rather than the primary sequence) which actually carries out the biochemical function of the molecule. The determination of protein structure was once regarded as the ultimate stage in any biological study. Such analysis would only be undertaken when there was a clear and thorough understanding of the protein's function, and the desire was to understand the functional implications of its structure and interactions at the atomic level. *Structural proteomics* has turned this idea on its head. The aim of structural proteomics is to express large numbers of genes (including as many orphan genes as possible) and compare the three-dimensional structures of the resulting proteins revealing distant evolutionary relationships between orphan genes and those whose functions are known. This brings the determination of protein structure to the beginning of the investigative process (Norin & Sundstrom 2002, Aloy *et al.* 2002, Sali *et al.* 2003).

Sequence analysis alone is not sufficient to annotate all orphan genes

As the sequencing phase of each genome project nears completion, potentially all the protein-coding sequences in the organism become available. The quickest way to assign a probable function to a new gene or cDNA is to carry out *pairwise comparisons* to all other sequences existing in databases using alignment tools such as BLAST (Altschul *et al.* 1990) or FASTA (Pearson & Lipman 1996) (see Chapter 9). Conserved sequence often indicates conserved function, so if cDNA X is very similar in sequence to cDNA Y, and cDNA Y encodes a protein phosphatase, then it is likely that cDNA X also encodes a protein phosphatase. Although such analysis cannot usually reveal the precise role of a gene in a cellular or whole-organism context, it can often reveal a biochemical function. Further experiments would be necessary for a more in-depth functional understanding e.g. by determining the gene's expression pattern and mutant phenotype. Even so, the knowledge that the gene product is a phosphatase would help in developing experimental strategies to investigate its function more thoroughly.

Powerful though tools such as BLAST and FASTA are, they are only really useful when there is >30% sequence identity between the query sequence and entries in the sequence databases. As the evolutionary relationship between sequences becomes more distant, pairwise comparisons tend to become less reliable. Thus, where there is 20–30% identity, over half of all evolutionary relationships fail to be detected by these standard alignment tools (Brenner *et al.* 1998, Murzin 1998). One way to address this problem is to carry out multiple sequence comparisons, which use the characteristics of sets of related sequences as a search query. PSI-BLAST (Altschul *et al.* 1997; p. 173) is an algorithm that generates profiles of multiple sequences as a search query and can find three times as many homologs as a standard BLAST search (Park *et al.* 1998, Martin 1998). Multiple sequence alignments can also be useful, identifying patterns and motifs that are conserved over great evolutionary distances. Such alignments are collected in secondary sequence databases such as BLOCKS and PRINTS (p. 162). As the power of these strategies increases, however, it becomes more important to screen matches for false positives, which can be generated by low-complexity regions representing common protein structures. For example, amino acid sequences corresponding to transmembrane helices and coiled-coil domains are so common that they are found in many unrelated proteins (Huynen *et al.* 1998). Another potential downfall is pollution in sequence databases. It is known that all databases contain a small but significant proportion of errors, so even a confident hit generated by an anonymous gene or cDNA can be misleading. It is worthwhile considering the fact that annotating a new gene solely on the basis of someone else's annotation of an existing gene can propagate and reinforce any errors that have been made (Bork & Bairoch 1996, Zhang & Smith 1997, Brenner 1999).

Protein structures are more highly conserved than sequences

Sequence comparison is a very powerful method for functional annotation, but even with sophisticated multiple alignment tools, fewer than 50% of newly discovered genes can be matched to previously identified genes with known functions. Although the success rate is likely to improve as more information accumulates in the databases, it is still a significant bottleneck to functional annotation. For example, it was shown that even two years after the complete genome sequence of the yeast *Saccharomyces cerevisiae* became available, the number of orphan ORFs had not been significantly reduced (Fischer & Eisenberg

1999). This is despite the massive increase in available sequence data from other organisms over the same time period, and suggests that a substantial number of gene families exist with no associated functional information in any of the sequenced genomes (Jones 2000). It is here that the determination of three-dimensional protein structure can help. Structure is much more strongly conserved than sequence because of *degeneracy*, a concept that is most often applied to nucleotide sequences but is also relevant to protein structures. In the genetic code, most amino acids are specified by more than one codon, and in some cases up to six codons are used. Therefore, a certain amount of divergence in a nucleotide sequence can occur without altering the primary amino acid sequence of a protein (Fig. 22.1). In the same way, chemically similar amino acids such as leucine, isoleucine, and valine can be regarded as conserved residues rather than mismatches when comparing protein sequences, because they would be expected to influence the folding of the polypeptide backbone in a similar manner. This is why protein sequences are often compared using similarity scores rather than identity scores, because similarity scores take into account the significance of different substitutions (p. 163). However, the level of degeneracy tolerated in terms of converting a primary amino acid sequence into a three-dimensional structure is much greater than this. Indeed, only a small number of residues in a globular protein are actually associated with protein function, and these may vary in their position within a sequence as long as their spatial arrangement is conserved (Kassua & Thornton 1999). Thus, even sequences with very

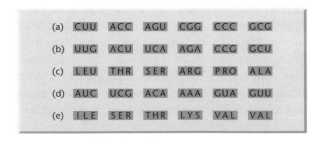

Fig. 22.1 Degeneracy in nucleotide and amino acid sequences. Up to six codons may specify the same amino acid, with the result that nucleotide sequences (a) and (b), which are conserved at only eight out of a possible 18 positions, specify exactly the same peptide (c). Nucleotide sequence (d), which matches sequence (a) at only four positions, encodes a different peptide (e), but one in which the amino acids at each position have very similar chemical properties to those in (c). These two peptides would be likely to fold in similar ways.

(a)

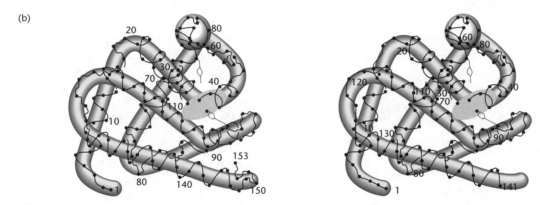

```
  1    V L S P A D K T N V K A A W G K V G A H A G E Y G A E A L E R M F L S F P T T K T Y F P H F - - - -    46
       . | | . . : . . . | . . . | | | | . | . . . . : | . | . | . | . : | . . . | . | . . . | . . |
  1    G L S D G E W Q L V L N V W G K V E A D I P G H G Q E V L I R L F K G H P E T L E K F D K F K H L K    50


 47    - - D L S H G S A Q V K G H G K K V A D A L T N A V A H V D D M P N A L S A L S D L H A H K L R V D    94
       | . . . . | . . : | . | | . . | . . | | . . . . . : . . . . . : | : . . | | . | . : :
 51    S E D E M K A S E D L K K H G A T V L T A L G G I L K K K G H H E A E I K P L A Q S H A T K H K I -    99


 95    P V N F - K L L S H C L L V T L A A H L P A E F T P A V H A S L D K F L A S V S T V L T S K Y R       141
       | | . : : . . : | . | : : . . | . : . . | . : | . . . . . . . . : : : | . | . . . . . . . . : . | . | :
100    P V K Y L E F I S E C I I Q V L Q S K H P G D F G A D A Q G A M N K A L E L F R K D M A S N Y K E L G F Q G    153
```

(b)

Fig. 22.2 (a) Alignment of human α-globin and human myoglobin using the EBI EMBOSS-Align program (http://www.ebi.ac.uk/emboss/align/). Note the sequences are very different (26% identity, 39% similarity). (b) Despite divergence at the sequence level, the three-dimensional structures are remarkably conserved.

few conserved residues can generate similar structures (Finkelstein & Ptytsin 1987). It has been estimated that 20–30% of orphan genes could be annotated immediately by determining the structures of the encoded proteins and comparing these to known structures (Bork & Koonin 1998, Blundell & Mizuguchi 2000).

A hypothetical example of the power of structural proteomics was discussed by Shapiro & Harris (2000) using the proteins hemoglobin and myoglobin. The solved structures of these proteins are very similar, reflecting the fact that they are both oxygen carriers (Aronson *et al.* 1994) (Fig. 22.2). However, the sequences are so divergent that a BLAST search with myoglobin would not elicit a hit on α- or β-globulin, the components of hemoglobin (Fig. 22.2). If the myoglobin sequence were to be obtained for the first time as part of a genome project, how would it be annotated? Sequence comparison might not be appropriate, but structural comparison would clearly

show similarities to hemoglobin, allowing myoglobin to be classified as an oxygen carrier.

Conserved protein structure may imply conserved function but how strong is the likelihood? The frequency with which similar structure corresponds to similar function has been calculated by Koppensteiner *et al.* (2000). These investigators compared the structures and functions of a large number of protein folds and found correlation between structure and function in 66% of cases. This means that, if a hypothetical protein is structurally similar to a characterized protein, there is a 66% chance that the functions are also related. On the other hand, there is a 33% chance that they are not, and the relationship in structure is coincidental. This is partly due to the fact that a number of protein folds are known to serve a variety of functions. These common structures are known as *superfolds* (Orengo *et al.* 1994). For example, the TIM barrel, which consists of eight α-helices and eight β-sheets, is associated with no fewer than 16 different functions (Hegyi & Gernstein 1999). It is found in the majority of enzymes, and if this structure is revealed in a hypothetical protein, it is generally of no practical help in functional annotation.

Structural proteomics has required developments in structural analysis techniques and bioinformatics

Techniques for solving protein structures are notoriously slow and laborious. In order to bring structural biology into the genomics era, it has been necessary to develop novel high-throughput methods and marry these to bioinformatic procedures for structural analysis and comparison. The last few years have seen advances in the two principal techniques for structural determination – X-ray crystallography

and nuclear magnetic resonance spectroscopy, as well as improvements in computer algorithms for structure–structure matching, structure prediction, and model building. At the same time, standard classification schemes for protein structures and databases of structural classification have also been developed.

Protein structures are determined experimentally by X-ray crystallography or nuclear magnetic resonance spectroscopy

The most direct way to determine the three-dimensional structure of a hypothetical protein is to solve it using atomic resolution techniques, the principal methods being X-ray crystallography and nuclear magnetic resonance (NMR) spectroscopy (Box 22.1). However, both these procedures are notoriously slow, involving many laborious preparative

---

Box 22.1   Methods for solving protein structures

X-ray crystallography

The basis of X-ray crystallography is the ability of a precisely oriented protein crystal to scatter incident X-rays onto a detector in a predictable manner. The scattered X-rays can positively or negatively interfere with each other, generating signals called *reflections* (see Fig. B22.1). The nature of the scattering depends on the number of electrons in an atom, and electron density maps can thus be reconstructed using a mathematical function called the *Fourier transform*. The more data used in the Fourier transform, the greater the resolution of the technique, and the more accurate the resulting structural model. Structural determination depends on both the amplitude and phase of the scattering, but only the amplitude can be calculated from the intensities of the diffraction pattern. For a long time, the only way in which this phase problem could be addressed and macromolecular structures completely solved was *multiple isomorphous replacement*, i.e. the use of heavy atom derivatives of the protein crystal. The incorporation of heavy atoms into a protein crystal by soaking alters the diffraction intensity allowing the phase to be calculated.

More recently, the laborious process of finding heavy atom derivatives has been superceded by techniques such as *multiwavelength anomalous diffraction* (MAD). MAD takes advantage of the presence of anomalously scattering atoms in the protein structure. The magnitude of anomalous scattering varies with the wavelength of the

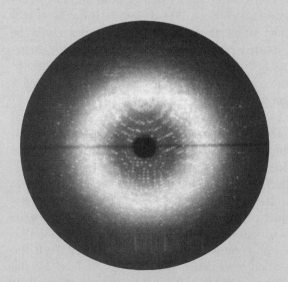

Fig. B22.1   X-ray diffraction image of a protein phosphatase. (Courtesy of Daniela Stock, MRC Laboratory of Molecular Biology, Cambridge.)

Box 22.1 *continued*

incident X-rays, and thus requires precisely tuneable incident X-radiation, which can be found only at *synchrotron radiation sources*. Isomorphous crystals can be produced rapidly by incorporating the amino acid derivative selenomethionine into the expressed protein (Hendrickson *et al.* 1990). Anomalous diffraction data collected at several wavelengths can then be used to determine both the phase and amplitude of the scattering, allowing high-throughput determination of protein structures. Recently, it has been reported that under optimal conditions a protein crystal structure can be solved using MAD analysis in about 30 min (Walsh *et al.* 1999).

Nuclear magnetic resonance spectroscopy

NMR spectroscopy is used to determine the structure of proteins in solution. The basis of NMR is that some atoms, including natural isotopes of nitrogen, phosphorus and hydrogen, are intrinsically magnetic and can switch between magnetic spin states in an applied magnetic field. This is achieved by the absorbance of electromagnetic radiation, generating magnetic resonance spectra (see Fig. B22.2). While the resonance frequency for different atoms is unique, it is also influenced by the surrounding electron density. This means that magnetic resonance frequency is shifted in the context of different chemical groups (a so-called *chemical shift*) allowing the discrimination between e.g. methyl and aromatic groups. The manner in which NMR

decays depends on the structure and spatial configuration of the molecule. For example, the *nuclear Overhauser effect* (NOE) results from the transfer of magnetic energy through space if interacting nuclei are < 0.5 nm apart. NOE spectroscopy (NOESY) shows proximal atoms as symmetrical peaks superimposed over the typical NMR spectrum. In this way, the analysis of NMR spectra can determine the three-dimensional spatial arrangement of atoms, and can allow a set of distance constraints to be built up to determine protein structure. The technique is generally applicable only to small proteins (< 25 kDa) although recent innovations have stretched this limitation to about 100 kDa (see main text).

Circular dichroism spectroscopy

Circular dichroism (CD) describes the optical activity of asymmetric molecules as shown by their differing absorption spectra in left and right circularly polarized light. CD spectrophotometry between 160 and 240 nm allows the characterization of protein secondary structure, because α-helices and β-sheets generate distinct spectra. With the recent application of synchrotron radiation CD (SRCD) to protein structural studies, this technique has emerged as a useful complement to X-ray crystallography and NMR spectroscopy for the analysis of protein structure. The use of the technique in structural genomics has been recently reviewed (Wallace & Janes 2001).

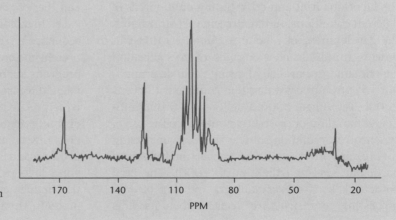

Fig. B22.2 Typical appearance of an NMR spectrum. This data can be built into a series of structural models based on atomic distance constraints.

steps that have to be optimized for each protein analyzed. A second bottleneck is data acquisition, especially for X-ray crystallography, where synchrotron radiation sources are required. A third bottleneck is data processing, i.e. converting the X-ray diffraction patterns and NMR spectra into structures that can be submitted to the protein structural databases. Recent advances in all these areas have made high-throughput structural determination a realistic goal.

Both X-ray crystallography and NMR spectroscopy require milligram amounts of very pure protein. The first challenge in structural genomics is therefore to find suitable expression systems and purification procedures which are widely applicable and produce proteins that fold properly to achieve their native state. Advantages of heterologous expression systems include the ease with which amino acid derivatives can be incorporated, as these are required for procedures such as multi-wavelength anomalous diffraction phasing (Box 22.1), and the ability to include an affinity tag, which greatly simplifies protein purification (reviewed by Stevens 2000a). Otherwise, optimal chromatography procedures must be established for every protein, a significant bottleneck in sample preparation (Heinemann *et al.* 2001, Mittl & Grutter 2001). Various systems have been tested, including *E. coli*, yeast, insect cells, and cell-free systems (reviewed by Edwards *et al.* 2000). *E. coli* has the advantage of low cost, which is an important consideration in a large-scale structural genomics program. However, eukaryotic systems are advantageous where protein folding requires chaperons and where post-translational modification is required to achieve the native structure.

Successful X-ray crystallography is dependent on the production of high-quality protein crystals. Defining conditions to achieve crystallization has long been regarded as an art, and typically involves the laborious trial-and-error testing of hundreds of different conditions. A major recent breakthrough is the development of robotic workstations that can process hundreds of thousands of crystallization experiments in parallel, allowing rapid determination of optimum crystallization conditions (Stevens 2000b, Terwilliger 2000, Mueller *et al.* 2001). Automated handling of crystals can maximize the use of synchrotron radiation sources therefore increasing the number of crystals that can be analyzed in a particular session (Abola *et al.* 2000). The final bottleneck in X-ray crystallography, that of data processing and structure determination itself, is being addressed with the development of powerful new software packages (reviewed by Lamzin & Perrakis 2000). NMR spectroscopy is carried out in solution and is therefore suitable for the analysis of proteins that are recalcitrant to crystallization. Until recently, the technique was applicable only to proteins with a low molecular weight (<25 kDa) but a number of advances in instrumentation have helped to increase this threshold significantly (see review by Riek *et al.* 2000). Following sample preparation, the major bottlenecks in NMR structure determination are data recording and analysis. Data acquisition is becoming quicker with improvements in instrument design, particularly novel probes (Medek *et al.* 2000) and higher frequency magnets (Lin *et al.* 2000) leading to higher signal-to-noise ratios and greater resolution. Data analysis times, typically ranging from weeks to months, are being cut down through the use of integrated software packages for spectral analysis and model building (e.g. Koradi *et al.* 1998, Xu *et al.* 2001).

An important consideration is that increasing the speed at which protein structures are solved should not reduce the quality of the structures generated. With this in mind, a number of computer programs have been developed for structural validation and checking the consistency of structural data, e.g. PROCHECK (Laskowski *et al.* 1993) and SFCHECK (Vaguine *et al.* 1999).

Protein structures can be modeled on related structures

An alternative to solving protein structures directly is to predict how a given amino acid sequence will fold. If the sequence with unknown structure (the *target sequence*) shows >25% identity to another sequence whose structure is known (the *template sequence*), then the structure of the target sequence can be predicted by *comparative modeling* (also called homology modeling). In this method, multiple sequence alignment is carried out with the target protein and one or more templates. Then a computer program such as SWISS-PDBVIEWER is used to align the target structurally on the template. For closely related proteins this is an easy process, but for distantly related proteins with some insertions and deletions, it can be more difficult. The compact core of the protein is always easier to model than the more variable loops, and algorithms are sometimes used that match the loops to databases of loops from other proteins. This is called a "spare parts algorithm".

Where the level of sequence identity is too low for comparative modeling, two broad classes of predictive methods can be used: those in which sequences are compared to known structures to test "goodness of fit" (*fold recognition methods*) and those in which structural information is derived directly from primary sequence data without reference to known structures (*ab initio prediction methods*).

Fold recognition is also known as *threading*, and can be defined as a technique that attempts to determine whether the sequence of an uncharacterized protein is consistent with a known structure by systematically matching its compatibility to a collection of known protein folds (Jones *et al.* 1992, Bryant & Lawrence 1993). Recent algorithms such as GenTHREADER (Jones 1999a) incorporate aspects of both sequence profiling (as used in comparative modeling) and threading, to increase the confidence with which structures are assigned. Fold recognition algorithms are "trained" on fold libraries, such as those found in structural databases like Pfam and BLOCKS (Chapter 9), to establish parameters of recognition. This method is widely used to assign tentative functions to the large sets of anonymous cDNA sequences arising from genome projects (e.g. Fischer & Eisenberg 1997, 1999; Rychlewski *et al.* 1998, 1999). *Ab initio* prediction of tertiary protein structure is still a distant goal, and the algorithms in current use are suitable only for the prediction of secondary structures. However, secondary structure prediction is achieved with reasonable accuracy. For example, the PSIPRED method devised by Jones (1999b) matches solved protein structures in more than 75% of cases. For the prediction of tertiary structures, a mini-threading algorithm (i.e. an *ab initio* method incorporating fold data in the form of fragments from resolved protein structures) has been used to predict the structure of NK-lysin successfully (Jones 1997, reviewed in Jones 2000).

The usefulness of structural modeling depends on the reliability with which the models can accurately predict protein structures. This has been assessed by rigorously comparing several thousand structural predictions with actual solved structures, in a series of critical assessment of structural prediction (CASP) experiments (reviewed by Moult *et al.* 1999, Murzin 2001). *Ab initio* methods are the least reliable and often fail to predict folds that are identified by fold recognition methods. However, fold recognition is limited by the accuracy with which the sequence of the hypothetical protein can be aligned over that of the template for comparison (reviewed by Moult & Melamud 2000). Programs such as Dali (Holm & Sander 1995) and VAST (Gibrat *et al.* 1996), can be used for structural alignment, as recently reviewed by Szustakowski & Weng (2000).

Protein structures can be aligned using algorithms that carry out intramolecular and intermolecular comparisons

Once the tertiary structure of a protein has been determined by X-ray crystallography or NMR spectroscopy, or modeled by comparative modeling techniques, it is deposited in the PDB and can be accessed by other researchers. As discussed at the beginning of the chapter, the key benefit of structural data in proteomics is the ability to compare protein structures and predict functions on the basis of conserved structural features. There are two requirements to fulfil this aim – an objective method for comparing protein structures and a system of structural classification that can be applied to all proteins, so that protein scientists in different parts of the world use the same descriptive language.

Several programs are available, many of which are free over the Internet, that convert PDB files into three-dimensional models (e.g. Rasmol, MolScript, Chime). Furthermore, a large number of algorithms have been written to allow protein structures to be compared. Generally, these work on one of two principles although some of the more recent programs employ elements of both. The first method is intermolecular comparison, where the structures of two proteins are superimposed and the algorithm attempts to minimize the distance between superimposed atoms (Fig. 22.3). The function used to measure the similarity between structures is generally the root mean square deviation (RMSD), which is the square root of the average squared distance between equivalent atoms. The RMSD decreases as protein structures become more similar, and is zero if two identical structures are superimposed. Examples of such algorithms include Comp-3D and ProSup. The second method is intramolecular comparison, where the structures of two proteins are compared side by side, and the algorithm measures the internal distances between equivalent atoms within each structure and identifies alignments in which these internal distances are most closely matched (Fig. 22.4). An example of such an algorithm is Dali. Algorithms that employ both methods include COMPARER and VAST.

Similar methods are used to gauge the accuracy of structural models when the actual structures

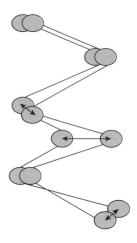

Fig. 22.3 Intermolecular comparison of protein structures, with circles representing Cα atoms of each amino acid residue and lines representing the path of the polypeptide backbone in space. This process involves the superposition of protein structures and the calculation of distances between equivalent atoms in the superimposed structures (shown as bi-directional arrows). These distances are used to calculate the root mean square deviation (RMSD).

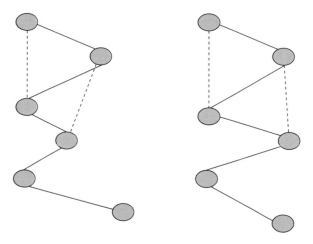

Fig. 22.4 Intramolecular comparison of protein structures, with circles representing Cα atoms of each amino acid residue and lines representing the path of the polypeptide backbone in space. This process involves side-by-side analysis based on comparative distances between equivalent atoms within each structure (shown as color-coded dotted lines).

become available. When alignments are good, as is generally the case with comparative modeling, then very accurate models are possible. RMSDs of less than 1.0Å represent very good predictions, since this is similar to the degrees of difference between two separate experimental determinations of the same protein structure. When the percentage sequence identity between template structures and target sequence exceeds 70% it is reasonable to expect that the model should be accurate to an RMSD of less

than 2–3Å even using completely automated methods. When the percentage identity drops below 40% then getting a good alignment, often with manual intervention, becomes more critical.

The annotation of proteins by structural comparison has been greatly facilitated by standard systems for the structural classification of proteins

Functional annotation on the basis of protein structure requires a rigorous and standardized system for the classification of different structures. One way to do this would be to identify particular structural features, which are known as folds, and group functional protein domains according to which folds are present. This is not as easy as it sounds because there may be a continuous range of intermediate structures between two particular fold types, especially for those folds that appear in many proteins; this is known as the Russian doll effect. It is estimated that there are about 1000 superfamilies of protein folds in total (Brenner *et al.* 1997, Zhang & DeLisi 1998, Wolf *et al.* 2000b) although much higher numbers have been suggested (e.g. see Swindells *et al.* 1998).

Existing classification schemes are generally hierarchical: at each level, proteins are classified according to the particular folds they possess, allowing new proteins to be assigned to the appropriate category. There are various different classification schemes based on pairwise structural comparisons which involve arbitrary thresholds of similarity. Consequently, different schemes can place the same protein in different families or subfamilies according to these cutoff levels. In one study, three structural classification schemes were compared (Hadley & Jones 1999). These are Structural Classification of Proteins (SCOP; Murzin *et al.* 1995), Class, Architecture, Topology and Homologous super-family (CATH; Orengo *et al.* 1997), and Fold classification based on Structure-Structure alignment of Proteins (FSSP; Holm & Sander 1996). SCOP is a manual classification scheme using a number of parameters, including human knowledge of evolutionary relationships. In contrast, CATH and FSSP are automatic systems relying entirely on sequence comparisons and geometric criteria. CATH is semi-automatic, with comparisons carried out using the SSAP algorithm, but the results of comparisons are manually curated. FSSP is fully automated, with structural comparisons carried out using the Dali algorithm. Perhaps because of these differences, only about

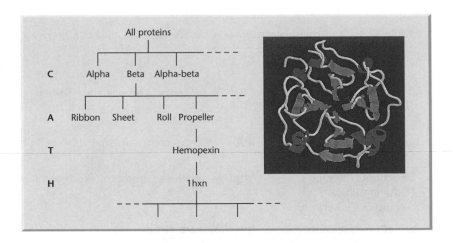

Fig. 22.5 The CATH (Class, Architecture, Topology, Homologous superfamily) hierarchical protein structure classification scheme. This example classification is featured on the CATH homepage (http://www. biochem.ucl.ac.uk/bsm/cath/lex/ cathinfo.html), and shows the single domain hemopexin protein 1hxn (Baker *et al.* 1993). In CATH this protein is classified as C (mainly β) A (propeller) T (hemopexin) H (1hxn). Note that the classification system can branch further after level H.

60% of the proteins investigated showed the same results with all three classification schemes. Over one-third were placed in different groupings by the three schemes (Hadley & Jones 1999). An example CATH classification is shown in Fig. 22.5.

Tentative functions can be assigned based on crude structural features

Where sequence and structural comparisons both fail to provide functional information, structural analysis can often reveal functional characteristics of proteins at a simpler level. For example, scanning the surface of a protein can reveal clefts that are likely to represent ligand-binding sites (Laskowski *et al.* 1996) or domains that probably interact with other proteins (Jones & Thornton 1995, 1997). This allows the low-resolution classification of proteins, e.g. a protein with a large cleft could be tentatively assigned the designation of "enzyme". Higher resolution characterization may be possible by matching the shape of the cleft to a library of small molecular shapes. This can be achieved with the help of drug design software such as DOCK (Briem & Kuntz 1996) and HOOK (Eisen *et al.* 1994), which match potential ligands to binding sites on the protein surface. A number of methods have been published recently that allow the prediction of active sites in enzymes based on the spatial distribution of critical residues in different fold families (Wallace *et al.* 1997, Russell *et al.* 1998). These classifications are not based on homology, since there is no evidence of an evolutionary relationship. The possession of common structural features between two proteins in the absence of a proven evolutionary relationship is known as analogy.

International structural proteomics initiatives have been established to solve protein structures on a large scale

Structural proteomics initiatives have been set up all over the world, some comprising dispersed laboratories working towards a common goal and some focused at particular centralized sites (Stevens *et al.* 2001, Zhang and Kim 2003). In America, the National Institute of General Medical Sciences (NIGMS) has funded nine structural proteomics pilot centers (Terwilliger 2000), and several academic and industrial consortia have been established in America, Europe, and Japan (Terwilliger 2000, Heinemann 2000, Yokoyama *et al.* 2000). The major structural proteomics centers are listed in Table 22.1. While the overall goal of structural proteomics is to provide representative structures for all protein families, various different approaches have been used to select an initial set of target proteins (Brenner 2000, Linial & Yona 2000). In the broadest sense, structural genomics aims to provide total coverage of "fold space" by solving representative structures from every protein superfamily. Once this has been achieved, structural databases will provide a comprehensive resource for the annotation of all future orphan genes.

One current problem is that the Protein Databank is highly redundant, i.e. most of the structures that have been added in the last five years match existing structures. This suggests that some structures are very much under-represented. By deliberately choosing to solve the structures of hypothetical proteins, it is hoped that the Protein Databank will become enriched with proteins containing novel folds. As stated above, due to the degeneracy of the "protein folding code" many different sequences can

Table 22.1 A selection of structural genomics research consortia and their target organisms.

Program	URL	Target organisms
Berkeley Structural Genomics Center	http://www.cchem.berkeley.edu/~shkgrp	*Methanococcus jannaschii* *Pyrococcus horikoshii* *Mycoplasma pneumoniae*
Joint Center for Structural Genomics	http://www.jcsg.org/scripts/prod/home.html	*Thermotoga maritima* *Caenorhabditis elegans*
Structural Proteomics (Ontario Clinical Genomics Center)	http://www.uhnres.utoronto.ca/proteomics/	*Methanobacterium thermoautotrophicum*
Structure 2 Function	http://s2f.carb.nist.gov	*Haemophilus influenzae*
Fold Diversity Project	http://proteome.bnl.gov/	Human disease and pathogen proteins
Mycobacterium tuberculosis Structural Genomics Consortium	http://www.doe-mbi.ucla.edu/TB/	*Mycobacterium tuberculosis*
Northeast Structural Genomics Consortium	http://www.nesg.org/	*Saccharomyces cerevisiae* *Caenorhabditis elegans* *Drosophila melanogaster* (and human homolog)
RIKEN Structural Genomics/Proteomics Initiative	http://www.rsgi.riken.go.jp/	*Thermus thermophilus* HB8 *Arabidopsis thaliana* *Mus musculus*
Southeast Collaboratory for Structural Genomics	http://secsg.org/	*Caenorhabditis elegans* *Pyrococcus furiosus* Human

fold in a similar manner, and the total number of fold types that exists is likely to be relatively small, in the region of 1000 (Brenner *et al.* 1997, Zhang & DeLisi 1998, Wolf *et al.* 2000a). It is estimated that up to 10,000 protein structures will have to be solved to identify all the folds in existence, since many of the solved structures will be related and there will be no way to establish whether a given hypothetical protein contains novel or known folds prior to structural elucidation (Burley *et al.* 1999, Linial & Yona 2000).

Many structural proteomics groups have focused on microbes, which have smaller genomes (and thus smaller proteomes) than higher eukaryotes, but a fundamentally similar basic set of protein structures. Several groups have chosen thermophilic bacteria such as *Methanococcus jannaschii*, *Methanobacterium thermoautotrophicum*, and *Thermotoga maritima* for their pilot studies, on the basis that proteins from these organisms should be easy to express in *E. coli* in

a form suitable for X-ray crystallography and/or NMR spectroscopy (e.g. see Kim *et al.* 1998, Lesley *et al.* 2002). A favored strategy in model eukaryotes is to focus on proteins that are implicated in human diseases because such research is more likely to receive generous funding from pharmaceutical companies looking for novel drug targets (see e.g. Burley *et al.* 1999, Heinemann *et al.* 2000, Schwartz 2000). Overall, the idea has been to choose target structures that maximize the amount of information returned from the structural proteomics programs.

The progress of the structural proteomics projects is difficult to judge at present since the early years have been taken up largely by technology and infrastructure development. The overall aims can be summarized as shown in Fig. 22.6. A common theme emerging from these projects is a "funnel effect" in terms of the number of solved structures compared to the number of proteins chosen for analysis. This is due to the failure of a proportion of the target

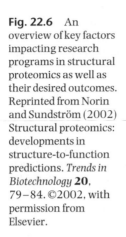

Fig. 22.6 An overview of key factors impacting research programs in structural proteomics as well as their desired outcomes. Reprinted from Norin and Sundström (2002) Structural proteomics: developments in structure-to-function predictions. *Trends in Biotechnology* **20**, 79–84. ©2002, with permission from Elsevier.

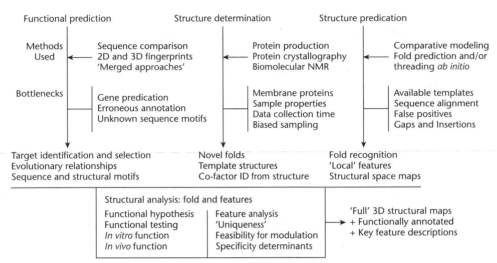

proteins at each stage of the analysis procedure (e.g. in the case of X-ray crystallography, the essential stages are: cloning, expression, solubilization, purification, crystallization, and structural determination). Despite an overall success rate that is probably no higher than 10% at present, a large number of structures have emerged from recently established and semi-automated production pipelines and the PDB is likely to expand quickly over the next few years as these structures are confirmed and deposited.

In principle, the value of the structural proteomics approach has been validated by the functional annotation of many of the initial hypothetical proteins chosen for structural analysis. For example,

of the first 10 proteins analyzed in the *Methanobacterium thermoautotrophicum* project, seven could be assigned a function due to structural similarity with known protein folds or other structural criteria, including the presence of bound ligands in the crystal (Christendat *et al.* 2000; Table 22.2). The presence of a ligand or cofactor can often be helpful, and was instrumental in the functional annotation of *M. jannaschii* hypothetical protein MJ0577, the first structure to be generated in a structural proteomics initiative. In this example the crystal contained ATP, suggesting a role in ATP hydrolysis that was later confirmed by biochemical experiments. Therefore, even when the structure of a protein does

Table 22.2 Functional annotation of the first 10 proteins to be structurally solved in the *Methanobacterium thermoautotrophicum* structural genomics program (see http://www.uhnres.utoronto.ca/proteomics/ for details and updates). Note that only one-third of the proteins contained a novel fold, and two of these could be functionally annotated by other methods (in the case of MTH0152 by the presence of a bound co-factor, and in the case of MTH1048 because of similarity with a functional ortholog). In total, eight of the 10 proteins could be annotated based on structural data alone.

Protein ID	Recognized fold	Predicted function	Reference*
MTH0040	Three helix bundle	Zinc-binding RNA polII subunit	Mackereth *et al.* (2000)
MTH0129	TIM barrel	Orotidine monophosphate decarboxylase	Wu *et al.* (2000)
MTH0150	Nucleotide binding	NAD⁺ binding protein	
MTH0152	(novel)	Nickel binding	
MTH0538	Rossmann fold	ATPase	Cort *et al.* (2000)
MTH1048	(novel)	RNA polII subunit	Yee *et al.* (2000)
MTH1175	Ribonuclease H	(unknown)	
MTH1184	(novel)	(unknown)	
MTH1615	Armadillo repeat	Transcription factor	
MTH1699	Ferredoxin-like	Transcriptional elongation	Kozlov *et al.* (2000)

* Proteins with no explicit references are discussed in Christendat *et al.* (2000).

not match any other in the database, structural analysis may still provide functional information that can be followed up with other experiments. Another interesting example is hypothetical protein TM0423, from *Thermotoga maritima*, which co-purified and co-crystallized with a molecule of Tris buffer. In this case, the position of the buffer suggested that the protein would be able to bind to glycerol, and identified it as a glycerol hydrogenase.

While one goal of structural proteomics is to assign functions to hypothetical proteins on the basis of their relationship to known folds, another is to discover new folds and assemble a comprehensive directory of protein space. It appears that about 35% of the structures emerging from current structural proteomics initiatives contain novel folds, which confirms the hypothesis that protein space is finite and probably comprises at most a few thousand distinct structures. Every time a new fold is discovered, a little bit more of that protein space is filled. Furthermore, many of the new folds can be assigned functions because they bind particular ligands or have other properties, and this reveals new structure–function relationships that can be applied more

widely. Sequence analysis and structural comparisons with these novel folds can identify previously unanticipated evolutionary relationships. At some point in the future, we may reach the stage where there is no such thing as an orphan gene or a hypothetical protein.

Suggested reading

Aloy P., Oliva B., Querol E., Aviles F.X. & Russel, R.B. (2002) Structural similarity to link sequence space: new potential superfamilies and implications for structural genomics. *Protein Science* **11**, 1101–16.

Fiser A. (2004) Protein structure modelling in a proteomics era. *Expert Review Proteomics* **1**, 97–100.

Lan N., Montelione G.T. & Gerstein M. (2003) Ontologies for proteomics: towards a systematic definition of structure and function that scales to the genomic level. *Current Opinion in Chemical Biology* **7**, 44–54.

Sali A., Glaeser R., Earnest T. & Baumeister W. (2003) From words to literature in structural proteomics. *Nature* **422**, 216–25.

Zhang C. & Kim S.-H. (2003) Overview of structural genomics: from structure to function. *Current Opinion in Chemical Biology* **7**, 28–32.

CHAPTER 23

Proteomics III – Protein interactions

Introduction

We have learned in the last five chapters that information about the function of a gene can be gained from the analysis of DNA sequence, genome organization, protein structure, expression profile, and mutant phenotype. However, this information rarely lets us see the whole picture. More often, it provides suggestions or clues that need to be followed up by further experiments. This being the case, how can we rigorously define the function of a gene? At the most fundamental level, gene function reflects the behavior of proteins. It is the proteins that actually carry out cellular activities and interact with the environment, thus gene function can ultimately be broken down into a series of molecular interactions that take place among proteins and between proteins and other molecules. When things go wrong, through mutation or otherwise, it is ultimately due to the failure of these normal interactions. And our efforts to treat diseases, through the use of drugs, ultimately depend on the ability of those drugs to modulate protein interactions in a beneficial manner.

In this final proteomics chapter, we discuss the techniques that are used to study protein interactions, and how these have recently been adapted for high-throughput analysis on a proteomic scale (Drewes & Bouwmeester 2003, Figeys 2003, Physiky *et al.* 2003, Titz *et al.* 2004). The function of an uncharacterized protein is often suggested by its spectrum of inter-

actions. For example, if protein X is uncharacterized but interacts with proteins Y and Z, both of which are part of the RNA splicing machinery, it is likely that protein X is involved in this process also. If this reasoning is applied on a global scale, every protein in the cell can eventually be linked into a functional network (Walhout & Vidal 2001). This functional network has been called the *interactome* (Sanchez *et al.* 1999).

Protein interactions can be inferred by a variety of genetic approaches

A number of classical genetic approaches can be used to show potential interactions between proteins, including screening for suppressor mutations, i.e. mutations in one gene that partially or fully compensate for a mutation in another (Hartman & Roth 1973). In many cases such mutations exist because the proteins encoded by the two genes interact. The primary mutation causes a change in protein structure that prevents the interaction, but the suppressor mutation introduces a complementary change in the second protein that restores it (Fig. 23.1). Such genetic techniques have been widely employed in amenable organisms like *Drosophila* and yeast.

The advantage of genetic screens is that they provide a short cut to functionally significant interactions, sifting through the proteome for those interactions that have a recognizable effect on the overall phenotype.

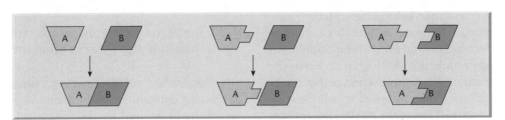

Fig. 23.1   Genetic tests for protein interactions. Suppressor mutations. Two proteins, A and B, normally interact. A mutation in A prevents the interaction, causing a loss of function phenotype, but this can be suppressed by a complementary mutation in B which restores the interaction.

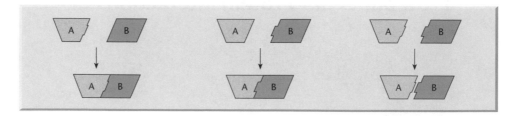

Fig. 23.2 Synthetic lethal effect. The same two proteins can still interact if there is a mutation in either A or B which does not drastically affect the interaction between them. However, if the mutations are combined, protein interaction is abolished and a loss-of-function phenotype is generated.

However, it is important to remember that genetic screens only provide *indirect* evidence for interactions and further direct evidence, at the biochemical level, must also be obtained. One potential problem is that the suppressor mutation may map to the same gene as the primary mutation, since second mutations in the same gene can suppress the primary mutant phenotype by introducing a compensatory conformational change within the same protein. Even if the suppressor maps to a different gene, the two gene products might not actually interact. For example, a mutation that abolishes the activity of an enzyme required for amino acid biosynthesis could be suppressed by a gain of function mutation in a transport protein that increases the uptake of that amino acid from the environment. Furthermore, the mutations do not necessarily have to change the structures of proteins X and Y. An alternative explanation is that protein X and protein Y must be present at the correct ratio, which may mean they function in a stoichiometric complex or may simply indicate that their activities must be balanced to maintain metabolic homeostasis. If the primary mutation changes the quantity of protein X, e.g. by altering the promoter, it could be suppressed by a compensatory change in the amount of protein Y caused by a similar mutation.

Another genetic approach is a screen for enhancer mutations, i.e. those that *worsen* the phenotype generated by a primary mutation (e.g. see Koshland *et al.* 1985, Huffaker *et al.* 1987). One example of this strategy is the synthetic lethal screen, where individual mutations in the genes for proteins X and Y do not prevent interaction and are therefore viable, but simultaneous mutations in both genes prevent the interaction and result in a lethal phenotype (Fig. 23.2). An example of this strategy applied on a large scale has been provided by Tong *et al.* (2001). These investigators established a synthetic genetic array (SGA) system in which a mutation in one yeast gene can be crossed to a set of 5000 viable deletion mutants, allowing synthetic interactions to be mapped in a systematic fashion. This can be used to identify all the proteins involved in the same pathway or complex as a particular query protein. Mutations in different genes that generate similar phenotypes often indicate that the protein products are part of the same complex or the same biochemical or signaling pathway. For pathways, the order of protein function can often be established by epistasis. In this type of experiment, loss-of-function and gain-of-function mutations (with opposite phenotypes) are combined in the same cell or organism. If a loss-of-function mutation in gene X overrides a gain-of-function mutation in gene Y, it suggests that protein X acts downstream of protein Y in the pathway (Fig. 23.3).

One final traditional genetic strategy worth mentioning is the dominant negative approach, where a loss-of-function mutation generates a dominantly interfering version of the protein that quashes the activity of any normally functioning version of the protein in the same cell (Herskowitz 1987). This generally suggests that the protein acts as a multimeric complex and that the nonfunctioning version of the protein interferes with the normal version when they are present in the same complex. Like suppressor mutants, however, these methods provide evidence that two gene products interact but they do not provide definitive proof. There are many other plausible explanations for such genetic effects and candidate protein interactions must be confirmed at the biochemical level.

New methods based on comparative genomics can also infer protein interactions

The availability of complete genome sequences for many different organisms allows comparative genomics to be used for the functional annotation of proteins. Three methods have been developed to infer protein interactions directly from genomic data. These work best in bacteria because more bacterial genome sequences are available for comparison and

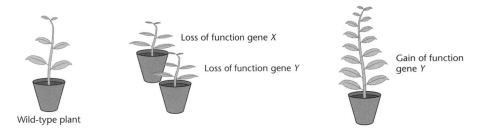

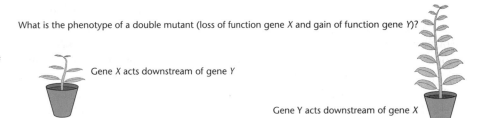

Fig. 23.3 Establishing gene order in a pathway by epistasis. A loss of function mutation in either gene X or Y causes a plant to become stunted. A gain of function mutation in gene Y causes a growth burst. If the phenotype of the double mutant is stunted, then X acts downstream of Y, but if its phenotype is tall, then the converse is true.

because bacterial genomes are often organized into functional units called operons where the encoded proteins tend to have related functions.

The first is called the domain fusion or Rosetta stone method (Fig. 23.4) and is based on the principle that protein domains are structurally and functionally independent units that can operate either as discrete polypeptides or as part of the same polypeptide chain. Therefore, multi-domain proteins in one species may be represented by two or more interacting subunits in another. A well-known example is the *S. cerevisiae* topoisomerase II protein, which has two domains, and which is represented by the two separate subunits GyrA and GyrB in *E. coli*. The domain fusion method can be summarized as follows:

- The sequence of protein X, a single-domain protein from genome 1, is used as a similarity search query on genome 2. This identifies any single-domain proteins related to protein X and also any multi-domain proteins, which we can define as protein X-Y.
- The sequence of protein X-Y can then be used in turn to find the individual gene for protein Y in genome 1. If the protein Y gene in genome 1 was uncharacterized until this point, then the Rosetta stone method successfully provides the first functional annotation for protein Y. It also provides indirect evidence that protein Y interacts with protein X.
- The sequence of protein X-Y may also identify further domain fusions, such as protein Y-Z. This links three proteins into a functional group and possibly identifies an interacting complex.

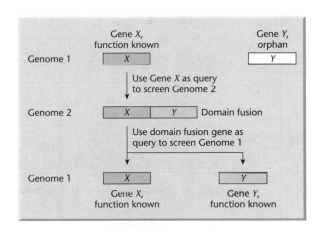

Fig. 23.4 Principle of the domain fusion (Rosetta stone) method of functional annotation. The sequence of gene X, of known function from genome 1, is used as a search query to identify orthologs in genome 2. The search may reveal single-domain orthologs of gene X, but may also reveal domain fusion genes such as XY. As part of the same protein, domains X and Y are likely to be functionally related. The sequence of domain Y can then be used to identify single-domain orthologs in genome 1. Thus gene Y, formally an orphan with no known function, becomes annotated because of its association with gene X.

As well as revealing previously unknown interactions between different protein families, iterative screening of multiple genomes can link many different proteins into an interaction map, based on gene fusion and gene fragmentation events that have occurred over an evolutionary timescale. Several experiments involving functional annotation in bacteria and eukaryotes by this method have been reported (e.g. see Marcotte *et al.* 1999, Shirasu *et al.* 1999) and it has proven possible to construct protein interaction maps on this basis (Enright *et al.* 1999).

The second comparative genomic method is based on the knowledge that bacterial genes are often organized into operons and that such genes are often functionally related even if their sequences are diverse (Jacob & Monod 1961). Therefore, if two genes are neighbors in a series of bacterial genomes, it suggests they are functionally related and that their products may interact (Dandekar *et al.* 1998, Overbeek *et al.* 1999). Caution is required in expanding this conservation of gene position principle to all bacterial genomes, however, as it is becoming evident that genes whose functions are apparently unrelated may also be organized into operons. Furthermore, while there is some evidence for functionally related gene neighbors in eukaryotes, the value of conserved gene position as a predictive tool remains to be established.

The final method is based on phylogenetic profiling and exploits the evolutionary conservation of genes involved in the same function. For example, the conservation of three or four uncharacterized genes in 20 aerobic bacteria and their absence in 20 anaerobes might indicate that the products are required for aerobic metabolism (Pellegrini *et al.* 1999). Since proteins usually function as complexes, the loss of one component would render the entire complex non-functional, and would tend to lead to the loss of the other components over evolutionary time since mutations in the corresponding genes would have no further detrimental effect. Such differential genome analysis has been used by Huynen *et al.* (1998) to identify species-specific functions of *Helicobacter pylori*. The use of phylogenetic profiling to assign a function to the yeast hypothetical protein YPL207W is shown as an example in Fig. 23.5.

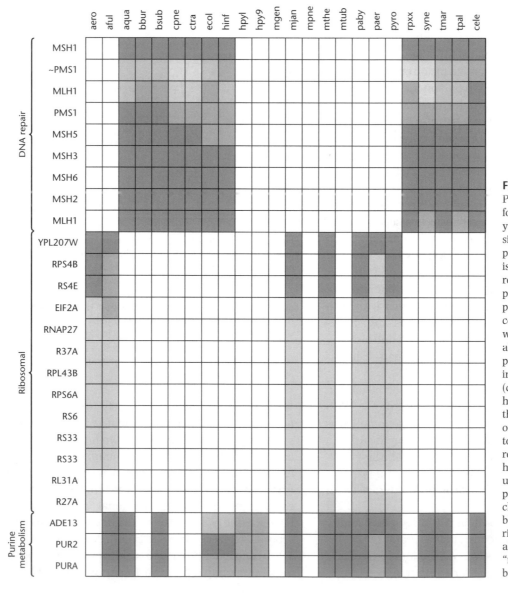

Fig. 23.5
Phylogenetic profiles for three groups of yeast proteins sharing similar co-inheritance patterns. Each row is a graphical representation of a protein phylogenetic profile, with elements colored according to whether a homolog is absent (white box) or present (colored box) in each of 24 genomes (columns). When homology is present, the elements are shaded on a gradient from light to dark purple (darker representing stronger homology). The uncharacterized protein (YPL207W) clusters with the middle block, all of which are ribosomal proteins, allowing the function "ribosomal protein" to be assigned.

Traditional biochemical methods for protein interaction analysis cannot be applied on a large scale

In the pre-genomic era, direct protein interaction studies – like all other biochemical analyses – were carried out largely on an individual basis. There are many different ways in which protein interactions can be demonstrated, some of which are discussed briefly in Box 23.1. The mainstay of traditional protein

interaction analysis is a core of biochemical methods providing direct evidence for interactions. Techniques such as co-immunoprecipitation, affinity chromatography, and cross-linking have been employed for over 25 years to characterize the interactions of individual proteins (reviewed by Phizicky & Fields 1995). Furthermore, as discussed in the previous chapter, there are methods such as X-ray crystallography and nuclear magnetic resonance spectroscopy that can be used to characterize protein interactions at the

Box 23.1   Biochemical methods for detecting and characterizing candidate protein–protein interactions

Co-immunoprecipitation. Interaction between protein X and protein Y is demonstrated by the addition of (usually monoclonal) antibodies against protein X to a cell lysate. Precipitation of the antibody–protein X complex results in the co-precipitation of protein Y.

Affinity chromatography. Interaction between protein X and protein Y is demonstrated by the "capture" of protein X on some kind of affinity matrix, e.g. a Sepharose column, when a cell lysate is passed through. Protein Y also remains attached to the column by virtue of its interaction with protein X, while non-interacting proteins are washed through. Affinity capture may be achieved using antibodies against protein X. Alternatively, protein X may be expressed as a fusion with an epitope tag or a molecule such as glutathione-S-transferase (GST), which binds to glutathione-coated Sepahrose beads. In a related technique, protein X can be immobilized on a membrane in a manner similar to the western blot and used to screen for interacting proteins in a cell lysate. This has been called a far-western blot (Blackwood & Eisenman 1991).

Cross-linking. Interaction between protein X and protein Y is demonstrated where cells or cell lysates are exposed to a cross-linking agent, and immunoprecipitation of protein X results in the co-precipitation of protein Y. Protein Y can be released by cleavage of the cross-link.

Fluorescent resonance energy transfer (FRET). Interaction between protein X and protein Y is

demonstrated where energy is transferred from an excited donor fluorophore to a nearby acceptor fluorophore, a phenomenon called fluorescent resonance energy transfer (FRET). FRET occurs only when the two fluorophores are up to 10 nm apart, and can be detected by the change in the emission wavelength of the acceptor fluorophore. FRET analysis can be carried out if protein X and protein Y are conjugated with fluorophores such as Cy3 and Cy5. Alternatively, they can be expressed as fusions with different fluorescent proteins, e.g. enhanced cyan fluorescent protein (donor) and enhanced yellow fluorescent protein (acceptor), in which case the technique may be called BRET (*bioluminescence resonance energy transfer*). The advantages of FRET/BRET analysis are that the normal physiological conditions inside the cell are maintained (analysis can be carried out *in vivo*) and that transient as well as stable interactions can be detected (see Day 1998, Mahajan *et al.* 1998).

Surface plasmon resonance (SPR) spectroscopy. Surface plasmon resonance is an optical resonance phenomenon occurring when surface plasmon waves become excited at the interface between a metal surface and a liquid. Interaction between protein X (immobilized on the metal surface) and protein Y (free in solution) is demonstrated by a change in the refractive index of the surface layer, which can be detected using a photodiode array. SPR technology is used to detect protein interactions on functional protein chips (p. 439).

atomic level. There are also techniques applied at the level of cell biology, which provide correlative evidence for interaction between specific proteins. For example, studies in which two proteins are localized *in situ* with labeled antibodies can show they co-exist in the same cellular compartment at the same time.

While these traditional biochemical and physical techniques are highly informative on an individual basis, none is particularly suited to high-throughput analysis. They also provide no easy way to link newly identified interacting proteins with their corresponding genes. Since the aim of functional genomics is to determine functions for the many anonymous genes and cDNAs amassing in databases, high-throughput strategies that link genes and proteins into functional networks are essential. This has been achieved by technology development in three areas, which we will discuss in turn:

1   Library-based interaction mapping.
2   High-throughput protein analysis and annotation.
3   Bioinformatics tools and databases of interacting proteins, which provide a platform for organizing and querying the increasing amount of interaction data.

Library-based screening methods allow the large-scale analysis of binary interactions

Library-based methods for protein interaction analysis allow hundreds or thousands of proteins to be screened in parallel, and all experimentally identified proteins are linked to the genes or cDNAs that encode them. Therefore, once interacting proteins have been detected, the corresponding clones can be rapidly isolated and used to interrogate DNA sequence databases (Pelletier & Sidhu 2001). Essentially, there are two broad classes of library: those in which protein interactions are assayed *in vitro* and those where the interaction take place within the environment of a cell. In most library-based interaction screening technologies, only binary interactions (i.e. interactions between pairs of proteins) can be investigated.

In vitro expression libraries are of limited use for interaction screening

In principle, it is possible to use any standard expression library for protein interaction screening. Immunological screening (the use of antibodies as probes) was developed in the 1970s and is essentially a specialized form of interaction analysis (Broome & Gilbert 1976; Chapter 6). There is no reason why other proteins should not also be used as "probes". Indeed, a diverse range of proteins has been used in this manner, with the aim of pulling out interacting partners. One example is provided by MacGregor *et al.* (1990), who sought interacting partners for the transcription factor c-Jun by screening a cDNA expression library with a biotin labeled c-Jun probe.

The traditional clone-based library, however, is not an ideal platform for proteome-wide interaction screening. The studies discussed above involved labor-intensive and technically demanding screening procedures, which would be unsuitable for high-throughput studies. Phage display (Smith 1985; p. 147) is a more suitable alternative. The principle of phage display is the expression of fusion proteins in such a way that a foreign peptide sequence is "displayed" on the bacteriophage surface. Libraries of phage can be produced and screened to identify peptides that interact with a given probe (such as an antibody) which is immobilized on a membrane or in the well of a microtiter plate. Screening is basically a reiterative affinity-purification process, in which non-interacting phages are discarded and bound phages are eluted and used to reinfect *E. coli*. After several rounds of such "panning" the remaining, tightly-bound phage are isolated and the inserts sequenced to identify the interacting peptides (for a review, see Sidhu *et al.* (2000)). The advantages of phage display over other *in vitro* library systems are that libraries of great complexity (up to 10^{12}) can be generated, and several rounds of highly selective screening can be carried out with intrinsic amplification at each step. Screening and panning can be carried out in an array format in microtiter dishes, making the technique amenable to high-throughput processing. One disadvantage of the system, however, is that only small foreign peptides can be incorporated into the coat protein, and this may limit the number of interactions that can be detected. Thus, despite its potential for scale-up, phage display and similar methods in which peptides are displayed on the surface of cells have yet to be exploited for genome-scale interaction analysis.

The yeast two-hybrid system is an *in vivo* interaction screening method

The major disadvantage of all *in vitro* library systems is that interactions occur in an unnatural environment where the protein may be incorrectly folded or partially unfolded. The yeast two-hybrid system, initially described by Fields & Song (1989), is the prototype of a range of related techniques in which protein interactions are assayed *in vivo*. The principle of the system is that proteins often comprise several functionally independent domains, which can function not only when they are covalently linked in the same polypeptide chain, but also when they are brought together through noncovalent interactions. Transcription factors generally contain independent DNA-binding and transactivation domains, and this means that a functional transcription factor can be created if separately expressed DNA-binding and transactivation domains can be persuaded to interact. On this basis, it is possible to use the two-hybrid system to confirm interactions between known proteins and to screen for unknown proteins that interact with a given protein of interest.

It is the latter approach that is most relevant in functional genomics. The general strategy is as follows: protein X is expressed as a fusion (a hybrid) with the DNA-binding domain of a transcription factor to generate a "bait". A library of "prey" is then generated in which each clone is expressed as a fusion protein with the transcription factor's transactivation domain. The final component of the system is a reporter gene that is activated specifically by the two-hybrid transcription factor. Mating between haploid yeast cells carrying the bait construct and those carrying the library of prey results in diploid cells carrying both components. In those cells where the bait interacts with the prey, the transcription factor is assembled and the reporter gene activated, allowing the cells to be isolated and the DNA sequence of the interactor identified. This principle is illustrated in Fig. 23.6.

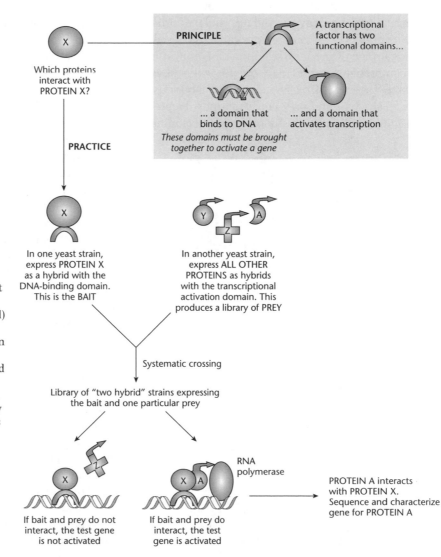

Fig. 23.6 The principle of the yeast two-hybrid system. The bait protein (X) is expressed as a fusion (or hybrid) with the DNA-binding domain of a transcription factor. Yeast expression libraries are then created in which many different proteins are expressed as fusions with the transcription-activating domain of a transcription factor. Mating between bait and prey strains of yeast results in diploid cells expressing both the bait and a candidate prey. Where interactions occur, a functional transcription factor is assembled and a test gene is activated. The interacting protein can be identified by sequencing the expression construct in the corresponding clone in the prey library.

To make the two-hybrid system suitable for genome-wide interaction mapping, comprehensive libraries of baits must be used to screen comprehensive libraries of prey, resulting in huge numbers of combinatorial interactions. Two general strategies have been devised: the matrix approach and the random library method. These are discussed in turn below.

In the matrix approach, defined clones are generated for each bait and prey

Matrix interaction screening involves panels of defined bait and prey, i.e. constructs derived from known open reading frames (ORFs), which are mated systematically in an array format (Fig. 23.7). Since this approach depends on the availability of sequence data corresponding to each protein, it can only be used for pre-defined proteins. However, interacting proteins can be identified immediately on the basis of their array coordinates, and the corresponding cDNAs or genomic clones can be retrieved. The advantage of the matrix approach is that it is fully comprehensive and can provide exhaustive proteome coverage. However, it is also laborious, because each bait and prey construct must be prepared individually by PCR followed by subcloning

in the appropriate expression vector. Haploid yeast cells of opposite mating types are then transformed with the bait and prey constructs, respectively, and arrayed in microtiter plates. Specific pair-wise combinations are generated by mating, and candidate interactions are assayed in the resulting diploid cells.

The first matrix type study to be reported was a small-scale but systematic analysis of interactions among the proteins of the *Drosophila* cell cycle. In this investigation, Finley & Brent (1994) screened a panel of known cyclin-dependent kinases, and revealed a network of 19 protein interactions including many cyclins. More recently, the matrix approach has been used for several proteome-wide interaction screens in yeast. Uetz *et al.* (2000) used a standard matrix approach to screen a proteome-wide library of prey constructs (over 6000 ORFs) with 192 baits. To increase confidence in potential interactors, the screening was carried out twice and only interactions identified in both screens were selected for further analysis. Using this approach, 87 of the baits were shown to be involved in reproducible interactions, with approximately three interactions per bait (a total of 281 interactions all together). The same authors also described a modified matrix assay for high-throughput interaction screening (Fig. 23.8). Instead of generating cell lines expressing specific prey constructs, cells were transformed *en masse* with pools of prey. These were screened with 5300 ORF baits, and 692 interacting protein pairs were identified. About half of these interactions were reproducible.

Pools of clones have also been used in another global study of protein interactions in yeast. Ito and colleagues used pools of 96 baits and 96 prey in 430 combinatorial assays such that over four million potential interactions were tested in parallel (Ito *et al.* 2000). Approximately 850 positive colonies were obtained, and short regions of the bait and prey plasmids were sequenced to derive short sequence signatures termed *interaction sequence tags* (ISTs). This experiment identified 175 interacting protein pairs, only 12 of which were previously known. Four million interactions is approximately 10% of the total number of potential interactions within the yeast proteome (assuming that every protein could interact with every other protein). Scaling this experimental format up to the proteome-wide level, Ito *et al.* (2001) identified 4549 interactions among 3278 proteins, 841 of which demonstrated three or more independent interactions.

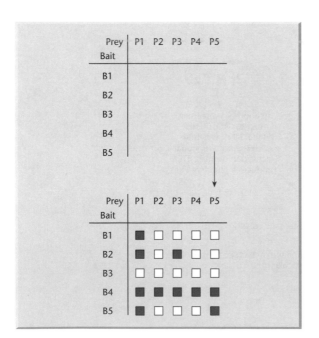

Fig. 23.7 The matrix system for high-throughput two-hybrid screening. Defined panels of bait and prey constructs maintained in haploid yeast strains are exhaustively tested for interactions by systematic mating. (Reprinted from Legrain *et al.* 2001 by permission of Elsevier Science.)

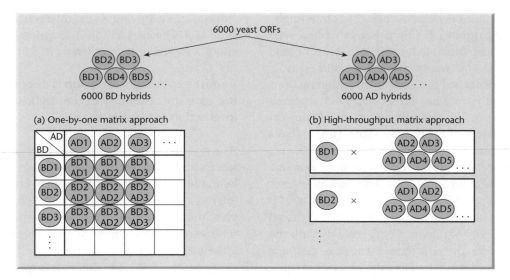

Fig. 23.8 The matrix and pooled matrix strategies. In the pooled matrix system, defined baits are tested against haploid yeast strains carrying pools of potential prey. If an interaction is detected, the pooled strain can be deconvoluted to identify individual interactors. The pooling strategy increases the throughput of the assay, depending on the number of constructs per pool. AD = activation domain; BD = binding domain. (Reprinted from Legrain & Selig 2000 by permission of Elsevier Science.)

A matrix format was also used to screen for interactions within the vaccinia virus proteome. All 266 ORFs were systematically tested against each other by McCraith *et al.* (2000) resulting in about 70,000 individual matings. Thirty-seven interactions were detected, only nine of which were previously known. Most recently, a matrix assay was used in a genome pilot study in the mouse, the first large-scale Y2H screen in a mammalian system (Suzuki *et al.* 2001).

In the random library method, bait and/or prey are represented by random clones from a highly complex expression library

The alternative to matrix format experiments is to generate prey libraries from random genomic fragments or cDNAs (Fig. 23.9). The prey can be screened using defined ORFs as baits, or for comprehensive proteome × proteome analysis, random libraries can be prepared for bait as well as prey. Unlike the matrix method, where all constructs are predefined and candidate interactors can be traced on the basis of their grid positions in the array, interacting clones in the random library must be characterized by sequencing and then they are compared to sequence databases for annotation.

In the first description of such an experiment, Bartel *et al.* (1996) generated random libraries of DNA-binding domain fusions and activation domain fusions from the genome of the *E. coli* bacteriophage

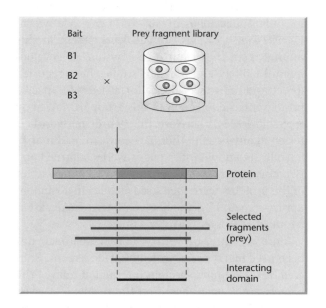

Fig. 23.9 The random library system for high-throughput two-hybrid screening. Defined panels of bait are tested for interactions against a panel of potential prey maintained as a high complexity random library. Each interactor may be defined by a number of overlapping constructs, therefore allowing specific interacting domains to be identified. (Reprinted from Legrain *et al.* 2001 by permission of Elsevier Science.)

T7, which produces 55 proteins. The authors reported 25 interactions between separate proteins, only four of which had been previously described. In several cases, interactions were also found between different domains of the same protein. Interestingly, a significant number of interactions that had been previously demonstrated using biochemical and genetic techniques were not detected in this assay.

The use of defined ORFs to screen for interactions in a random library has been carried out predominantly in specific protein complexes or biological processes. Fromont-Racine *et al.* (1997) used as baits 15 ORFs corresponding to yeast-splicing proteins, and screened a highly complex random genomic library containing more than one million clones. They identified 145 potential prey in a total of 170 interactions. Approximately half of the identified prey were already-known splicing proteins while the other half were uncharacterized. The same genomic library was screened by Flores *et al.* (1999) using 15 of the 17 known components of the yeast RNA polymerase III complex, and in a further study of the spliceosome (Fromont-Racine *et al.* 2000) using 10 baits omitted from the initial screen. The same strategy was employed by Walhout *et al.* (2000) to investigate the interactions of 29 proteins involved in vulval development in the nematode worm *Caenorhabditis elegans*. Vulval development in this animal is an important model system in developmental biology and previous studies have reported direct and indirect evidence for at least 11 specific protein interactions within this group (reviewed by Kornfeld 1997). Therefore, the 29 proteins were first tested against each other in a conventional matrix format: six known interactions were confirmed and two novel interactions were revealed. Twenty-seven of the proteins were then used as baits in a random library screen; 17 of them were shown to take part in a total of 148 potential interactions.

More recently, the ORF × library approach has been used in a series of proteome-wide screens. Rain *et al.* (2001) built a protein interaction map of the bacterium *H. pylori* based on the results of a yeast two-hybrid assay in which 261 *H. pylori* proteins were screened against a highly complex library of genomic fragments. Over 1200 interactions were identified, which allowed nearly half of the genome to be assembled into a protein interaction map. Although no protein complexes have been defined biochemically in this species, homologous proteins in other bacteria have been studied in this manner. Interestingly, only about half of the interactions that are known to occur in *E. coli* were identified when homologous *H. pylori* proteins were used as baits in the two-hybrid system. Giot *et al.* (2004) used the same strategy for a proteome-wide study in *Drosophila*, identifying 4780 interactions, while Li *et al.* (2004) generated an interaction map for the entire *C. elegans* proteome, listing 2135 interactions.

Robust experimental design is necessary to increase the reliability of two-hybrid interaction screening data

As a genomic tool, the yeast two-hybrid system suffers from several intrinsic limitations, including its limitation to the detection of protein–protein binary interactions, the necessity to localize proteins in the yeast nucleus, and its reliance on transcription factors. Some of these limitations have been addressed in derivative systems which are discussed in Box 23.2. Other problems include the high frequency of false positive and false negative results, which in some cases can be remedied by experimental design. False positives occur where the reporter gene is expressed in the absence of any specific interaction between the bait and prey. This may reflect "autoactivation" where the bait or prey can activate the reporter gene unassisted, or a phenomenon known as "sticky prey" where a particular prey protein can interact non-specifically with a series of baits. In large-scale two-hybrid screens, these types of false positive are quite easy to detect and can be eliminated. Other false positives occur through spontaneous mutations and can be more difficult to identify. Typically, researchers using matrix format screens use *reproducibility* as a measure of confidence in their results. For example, as mentioned above, Uetz *et al.* (2000) carried out two independent screens of their matrix and only accepted interactions occurring in both screens. Ito *et al.* (2001) regarded as plausible interactors only those proteins identified by three or more independent hits. In the analysis of vaccinia virus (McCraith *et al.* 2000), each assay was carried out four times and only those interactions occurring in three or four of the assays were accepted. Confidence in random library screens is increased by independent hits from overlapping clones. For example, Fromont-Racine *et al.* (1997) devised a system in which confidence in a prey was assessed according to the number of overlapping fragments interacting with a specific bait, the size of the fragments, and how many times the specific interaction was recorded. False positives can also be caused by genuine, but irrelevant interactions. For example, it is possible that two proteins normally found in separate cell compartments could, by chance, interact when they are both expressed in the yeast nucleus.

Box 23.2   Derivatives of the yeast two-hybrid system

The original yeast two-hybrid system was developed to test or screen for interactions between pairs of proteins. Once established, however, investigators turned their attention towards improvements and enhancements that allowed the detection of different types of interactions. One of the earliest adaptations made it possible to detect interactions between proteins and small peptides, which can be useful to define minimal sets of conserved sequences in interaction partners. Other derivatives allowed higher order complexes to be studied, by expressing the bait with a known interaction partner in the hope of attracting further complex components. A major disadvantage of yeast for the analysis of higher eukaryotic protein interactions is the limited amount of post-translational protein modification that occurs. This was addressed by carrying out two-hybrid screens in a strain of yeast expressing a mammalian kinase, therefore enabling the usual phosphorylation target sites to be occupied. Another variant of the yeast two-hybrid system, known as the yeast one-hybrid system, is useful for the identification of transcription factors. Essentially, this involves the transformation of yeast with a construct comprising a minimal promoter and reporter gene, with several tandem copies of a candidate transcription factor-binding motif placed upstream. A cDNA expression library is then prepared in which all proteins are expressed as transactivation-domain hybrids. These will activate the target gene only if they contain a DNA-binding domain that interacts with the chosen promoter sequence. This system can only identify proteins that bind to DNA autonomously. The one-and-a-half hybrid system is similar, but can detect proteins that bind DNA as heterodimers with a second, accessory protein. The one-two hybrid system can search for both autonomous binders and proteins that bind only as hetetrodimers. Another variant of the two-hybrid system, known as bait and hook or three-hybrid, is

useful for the identification of RNA-binding proteins or protein interactions with small ligands. In this system, one of the components (the hook) comprises the DNA-binding domain of a transcription factor and a sequence-specific RNA-binding protein that attaches to one end of a synthetic RNA molecule. The other end of the RNA molecule contains the sequence for which candidate interactors are sought. A prey library is constructed as normal, with each protein expressed as a fusion to a transactivation domain. Only in cells where the prey interacts with the RNA sequence attached to the hook will the transcription factor be assembled and the reporter gene activated. Another interesting variant is the reverse two-hybrid system, which uses counter-selectable markers to screen for the *loss* of protein interactions. This is used to identify mutations that disrupt specific interaction events, and could conceivably be used to find drugs that disrupt interactions between disease-causing proteins. Various systems using dual baits have also been described, and these can be used to find mutations that block specific interactions between a given prey protein and one of two distinct baits.

As well as derivative systems based on transcriptional activation as described above, a number of related technologies have been developed that do not rely on transcription, and therefore circumvent problems of autoactivation. The split ubiquitin system (ubiquitin-based split protein sensor, USPS) involves fusing a bait protein to the amino-terminal portion of ubiquitin and prey proteins to the separated carboxyl terminal. Interactions reassemble an active ubiquitin molecule, which then causes protein degradation, a process that can be monitored by western blot. The SOS recruitment system (SRS) or CytoTrap involves the fusion of bait proteins to a myrisylation signal, thus targeting them to the plasma membrane, while prey are expressed as fusions to the mammalian signal transduction protein SOS. In yeast strains deficient for CDC25 (the yeast

continued

Box 23.2 *continued*

ortholog of SOS) survival is possible only if SOS is recruited to the membrane, and this can only occur if the bait and prey proteins interact and form a membrane-targeted complex. The RAS recruitment system (RRS) is based on similar principles (i.e. the prey proteins are expressed as RAS fusions) in a yeast strain deficient for RAS.

Further systems have been devised which do not depend on yeast at all. Several bacterial two-hybrid systems have been developed in *E. coli*, one of which is based on the reconstruction of a split adenylyl cyclase enzyme and its activation of the *lac* or *mal* operons. A mammalian two-hybrid system

has been developed in which interactions between bait and prey assemble the enzyme β-lactamase activity, which can be monitored in real time. Similarly, the protein complementation assay involves the reassembly of the enzyme dihydrofolate reductase (DHFR) from two inactive components. This is used to monitor protein interactions in plant cells. Interaction between bait and prey reassembles the enzyme and allows it to bind a fluorescein-conjugated derivative of its normal substrate methotrexate. Free methotrexate is exported from plant cells but the enzyme–substrate complex is retained, allowing fluorescence to be monitored.

False negatives are revealed when known protein interactions are not detected, and when similar studies reveal different sets of interacting proteins with little overlap. For example, as discussed above, Uetz *et al.* (2000) carried out two different types of screen, one involving discrete baits and one involving pools of baits. Although a large number of potential interactors were found in each screen, only 12 were common to both screens. The similar proteome-wide study carried out by Ito *et al.* (2000, 2001) identified nearly 3300 interacting proteins, but of the 841 proteins found to be involved in three or more interactions, only 141 were in common with the set of 692 interacting proteins catalogued by Uetz and colleagues (Hazbun & Fields 2001). Similarly, Fromont-Racine *et al.* (1997) screened their random genomic library with three of the same splicing proteins that Uetz and colleagues used to screen the arrayed yeast prey ORFs. Interestingly, about 10 high-confidence prey were identified by each bait in each screen, but for two of the baits, only two prey were found to be common to both screens. For the remaining bait, there were six overlaps (Uetz & Hughes 2000).

The tendency for similar screens to identify different sets of interacting proteins probably has several causes. First, the selection strategy used in each library may influence the interactions that take place. Second, the interactions may well be extremely complex, i.e. each screen only reveals a subset of the interactions taking place. Third, the matrix and random library methods have been shown to differ in their sensitivity. Direct comparison of the two strategies reveals that random library screening produces more candidate interactors that the matrix

method, i.e. the matrix method suffers from a higher incidence of false negatives. This may be due to incorrect folding of the fusion protein, due perhaps to the presence of the fusion partner, or may reflect undetected PCR errors that occur during clone construction. The problem of false negatives is not so severe in the case of random libraries because each prey clone is represented by an overlapping series of fragments giving much more scope for interacting functional domains to form. This certainly seems to have been an advantage in the case of Rain *et al.* (2001), who used a prey library of protein fragments and succeeded in linking many of the *H. pylori* proteins into a functional network. An extreme example of the difference between matrix and random library screening is shown in the study of Flajolet *et al.* (2000), who investigated interactions among the 10 mature polypeptides produced by hepatitis C virus. Constructs expressing these 10 polypeptides in a matrix format revealed no interactions at all, not even the well-characterized interactions among the capsid proteins. However, a library screening using random genomic fragments revealed all the expected interactions as well as some novel ones. It is likely that the prey constructs generated in the matrix strategy failed to fold properly and therefore could not behave as the normal proteins would *in vivo*. The use of fragment libraries rather than intact ORFs has a further advantage. Where interactions occur between a bait and a series of overlapping prey fragments, the common sequence shared by a number of interacting prey fragments can, in principle, identify the particular domain of the protein that interacts with the bait (see for example Siomi *et al.* 1998, Flores *et al.* 1999).

Systematic analysis of protein complexes can be achieved by affinity purification and mass spectrometry

The limitations of the yeast two-hybrid method are predominantly related to the non-physiological nature of the assay system. As discussed above, protein interactions are not studied in their natural context, but are reconstituted artificially in the yeast nucleus. This does not accurately reflect the conditions under which most interactions occur (leading to false negatives) and brings together proteins that would not usually encounter each other in the cell (leading to false positives). Overall, only about half of the interactions predicted from yeast two-hybrid screens are expected to be true.

Affinity-based methods allow interactions to be studied in their natural context, reducing the appearance of irrelevant interaction data. Unlike the yeast two-hybrid system, they also allow the investigation of higher order interactions, including the purification and categorization of entire complexes. The major bottleneck to large-scale affinity-based interaction analysis, that of identifying the proteins in each complex, has been removed by advances in mass spectrometry that allow the characterization of very low-abundance protein samples (in the low femtomole range). These developments have culminated in the use of affinity purification and mass spectrometry for the systematic analysis of protein complexes, providing a global view of protein interactions known as the complexome.

The first reports of complex analysis by mass spectrometry, such as the analysis of the yeast spliceosome (Neubauer *et al.* 1997, Gottschalk *et al.* 1998, Rigaut *et al.* 1999) and the human spliceosome (Neubauer *et al.* 1998) involved the antibody-based affinity

purification of specific, known components, and thus relied on the availability of suitable antibodies. For whole-complexome analysis, a less selective approach is required for bait preparation and affinity tags have been used instead. These can be attached to any protein of interest and used to capture that protein on a suitable affinity matrix. Two large-scale studies of the yeast complexome were carried out in 2002. In one of these studies (Ho *et al.* 2002), 725 bait proteins selected to represent multiple functional classes were transiently expressed as fusions with the FLAG epitope, a short peptide that can be recognized by a specific antibody. Cell lysates were prepared from each yeast strain and complexes isolated by affinity capture with an anti-FLAG antibody. Over 1500 captured complexes were separated by SDS-PAGE and characterized by tandem mass spectrometry (Chapter 21). When redundant and non-specific interactions were eliminated, this revealed a total of 3617 interactions among 1578 proteins. In the second study (Gavin *et al.* 2002) the tandem affinity purification (TAP) procedure was used. This involves the expression of each bait protein as a fusion to a calmodulin-binding peptide and staphylococcal protein A, with the two elements separated by a protease recognition site (Fig. 23.10). Instead of expressing these constructs transiently, the investigators used gene targeting to replace nearly 2000 yeast genes with a TAP fusion cassette. Yeast cells expressing each bait–TAP fusion cassette were lysed and the cell lysate was passed through an immunoglobulin affinity column to capture the protein A component of the bait fusion. After washing to remove non-specific binding, the bound complexes were selectively eluted by the addition of the protease. Highly selective binding was then carried out in a second round of affinity chromatography using calmodulin as the affinity

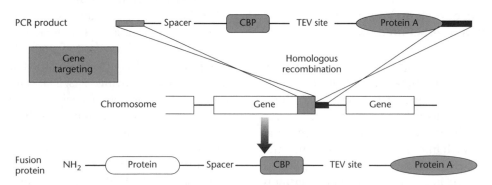

Fig. 23.10 The tandem affinity purification (TAP) cassette, consisting of a PCR-derived gene-specific homology region for targeting each yeast gene, and a generic region comprising a spacer, a calmodulin-binding peptide, a protease cleavage site recognized by tobacco etch virus protease and staphylococcal protein A. The TAP procedure is described in the text. (TEV, tobacco etch virus; CBP, calmodulin-binding protein.) Reprinted from Gavin A-C. *et al.* (2002) Functional organization of the yeast proteome by systematic analysis of protein complexes. *Nature* **415**, 141–7. © 2002 Nature Publishing Group.

matrix in the presence of calcium ions. The proteins retained in this step were eluted by adding the calcium chelating agent EGTA, and were examined by mass spectrometry. Gavin and colleagues found 4111 interactions involving 1440 proteins. Only 10% of the identified complexes had been completely characterized beforehand. In about 30% of the complexes, new components were identified while nearly 60% were entirely novel. Half of the complexes had five or fewer components, and most had fewer than 20. About 10% of complexes had over 30 components and the largest complex contained 83 proteins.

The mass spectrometry approaches were more sensitive than the large-scale yeast two-hybrid experiments when compared to literature benchmarks but still failed to detect about 60% of known interactions, suggesting a high false negative rate. In part, this may reflect the fact that affinity-based methods favor the recovery of stable complexes rather than transient ones. In contrast, the yeast two-hybrid system can detect transient interactions because even short-lived interactions will cause some activation of the reporter gene. The two mass spectrometry studies of the yeast complexome also showed a low degree of overlap, perhaps because of the different experimental approaches. As mentioned on p. 464, the recovery of interacting proteins depends to a large degree on the amount of bait. Gavin *et al.* (2002) used a gene-targeting strategy such that each bait was expressed at roughly physiological levels, whereas Ho *et al.* (2002) overexpressed their baits, which may have had a significant effect on complex architecture. Overall, it appears there is no ideal method for the large-scale collection of interaction data and that interaction maps should be built from a variety of complementary sources.

Protein localization is an important component of interaction data

Knowledge of *protein localization* can provide important evidence either to support or challenge the data from interaction screens. At the very least, showing that two proteins exist in the same cell and in the same subcellular compartment at the same time, indicates that such interactions *could* happen. If this is backed up by FRET analysis or cross-linking studies (p. 457), then the interactions are almost certainly genuine. However, care must be taken to ensure that such experiments are conducted on intact cells with normal levels of gene expression, since both cellular

damage and protein overexpression can result in proteins escaping from their normal compartments and contaminating others.

As well as helping to confirm or dismiss claimed interactions, protein localization data can be useful in their own right. It is in some cases possible to propose a protein's function based solely on its location, e.g. proteins located in the thykaloid membrane of a chloroplast are probably involved in photosynthesis. For these reasons, many investigators have carried out studies of subcellular or organellar proteomes (*organelle proteomics*) and several attempts have been made to catalog protein localization data on an even larger scale.

In one such study, thousands of yeast strains were generated in which a particular gene was replaced with a substitute bearing an epitope sequence (see Davis 2004). Each strain therefore produced one protein labeled with an epitope tag, allowing the protein to be localized using antibodies and fluorescence microscopy. High-throughput imaging was used to determine the localization of nearly 3000 proteins. The results suggested that about half of the yeast proteome is cytosolic, about 25% is nuclear, 10–15% is mitochondrial and 10–15% is found in the secretory pathway. Within the above classifications, about 20% of the proteome was represented by transmembrane proteins. About 1000 proteins of unknown function were included in the analysis and knowledge of their locations may help in the design of further experiments to determine more precise functions. More recently, Huh *et al.* (2003) have carried out a similar study in which proteins were labeled with green fluorescent protein, allowing real-time analysis and the localization of 70% of the yeast proteome into 22 compartment categories.

A pilot experiment has also been performed using mammalian cells, where the cells were grown on a DNA chip containing arrays of expression constructs (Howbrook *et al.* 2003). The array was first coated with a lipophilic transfection reagent, a chemical that promotes DNA uptake, and then immersed in a dish of rapidly growing cells. The cells covered the array, took up the DNA in each area of the array and expressed the corresponding proteins. After a few days, the array was recovered, the cells were fixed *in situ*, and cells in each area were examined by indirect immunofluorescence to determine where the proteins were located. A number of well-characterized proteins were correctly localized, validating the accuracy of the method (e.g. the transcription factor MEFC2 was observed in the nucleus). The major advantage of this method is that the number of proteins investigated simultaneously is limited only by the number of expression constructs that can be fitted on an array. It may therefore be possible to study 5000–10,000 different proteins in parallel. Several companies are developing imaging technology which is compatible with high-throughput localization studies, with the ultimate aim of building up a three-dimensional map of the cell containing localization and interaction data.

Interaction screening produces large data sets which require extensive bioinformatic support

Protein interaction data from a number of international collaborations are being assimilated in databases that can be accessed over the Internet (see Table 23.1). Most of them originated from the large-scale interaction screens described above, and are largely focused on the yeast proteome (e.g. the

Table 23.1 Databases of protein interactions.

Database	Acronym	URL	Content	References
Database of Interacting Proteins	DIP	dip.doe-mbi.ucla.edu	Experimentally determined protein–protein interactions	Xenarios *et al.* (2000, 2001)
Database of Ligand Receptor Partners	DLRP	http://dip.doe-mbi.ucla.edu/dip/DLRP.cgi	Ligand–receptor complexes involved in signal transduction	Xenarios *et al.* (2000, 2001)
Biomoloecular Interaction Network Database	BIND	http://www.blueprint.org/bind/bind.php	Molecular interactions, complexes, and pathways	Bader & Hogue (2000), Bader *et al.* (2001)
Protein-Protein-Interaction and Complex Viewer	MIPS-CYGD	http://mips.gsf.de/proj/yeast/CYGD/interaction/	Protein–protein interactions from large-scale screens	Mewes *et al.* (2000)
Hybrigenics	PIM	www.hybrigenics.fr	Protein interactions in *H. pylori*	Rain *et al.* (2001)
General Repository for Interaction Datasets	GRID	http://biodata.mshri.on.ca/grid	Central repository for yeast protein interactions	
Molecular Interactions Database	MINT	http://cbm.bio.uniroma2.it/mint/	Protein interactions with proteins, nucleic acids, and small molecules	
Curagen *Drosophila* Interactions Database		http://portal.curagen.com/cgi-bin/interaction/flyHome.pl	Protein interactions in *Drosophila*	
Curagen Yeast Interactions Database		http://portal.curagen.com/cgi-bin/interaction/yeastHome.pl	Protein interactions in yeast	Uetz *et al.* (2000)
Saccharomyces Genome Database	SGD	http://www.yeastgenome.org/	Comprehensive structural and functional information, including interactions	

Biomolecular Interaction Network Database, the Database of Interacting Proteins, the Comprehensive Yeast Genome Database, and the Saccharomyces Genome Database). Several tens of thousands of interactions are listed, many of which await further functional validation. These databases have been augmented with additional data from other sources. Importantly, a potentially very large amount of data concerning individual protein interactions is "hidden" in the scientific literature going back many years. It will be a challenge to extract this information and integrate it with that obtained from recent high-

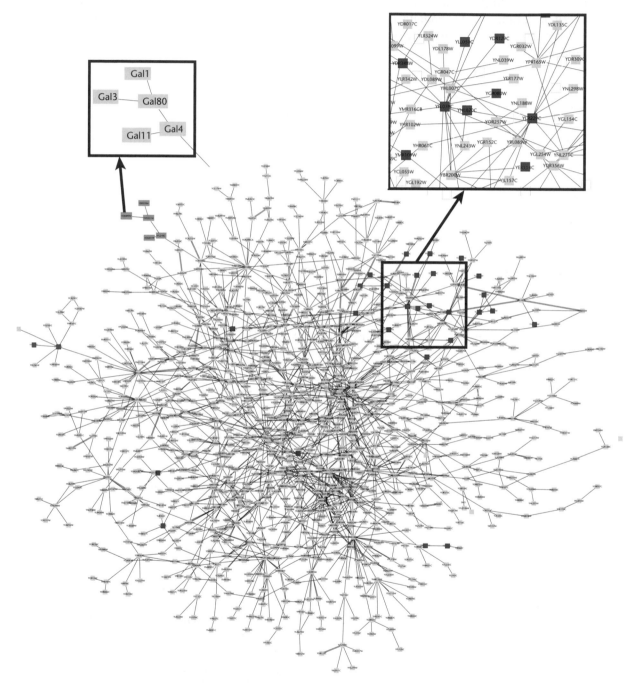

Fig. 23.11   Binary interaction map including 1200 interacting proteins based on published interactions. Inset shows close-up of region highlighted in box. Highlighted in red are cell structure proteins (a single functional class). Proteins in this category can be observed to cluster primarily in one region. Although interacting proteins are not depicted in a way that is consistent with their known cellular location (i.e. those proteins known to be present in the nucleus in the center of the interaction map and those present in plasma membranes in the periphery), signal-transduction pathways (or at least protein contact paths) can be inferred from this diagram. Reprinted from Tucker C.L., Gera J.F. & Uetz P. (2001) Towards an understanding of complex protein networks. *Trends in Cell Biology* **11**, 102–6. ©2001 with permission from Elsevier.

throughput experiments. Several bioinformatics tools have been developed to trawl through the literature databases and identify keywords that indicate protein interactions so that such references can be scrutinized by the human curators of interaction databases (reviewed by Xenarios & Eisenberg 2001).

Another challenge is to find a simple way to present protein interaction data in a readily accessible and understandable way. As discussed above, the yeast proteome is likely to consist of over 6000 basic proteins (not including variations generated by post-translational modifications, which could increase this number substantially). Each protein is thought to interact, on average, with three others. The simplest way to represent interacting components in a system is a chart with interacting proteins joined by lines. Depicting the entire yeast proteome in such a way is likely to yield a map of incredible complexity and intricacy, and it is easy to imagine the informa-

tion becoming lost in the mass of detail. Schwikowski *et al.* (2000) have assimilated binary interaction data for about 2500 yeast proteins and generated an interaction map that included approximately 1200 of them. The map is reproduced in Fig. 23.11 and initially it appears very complex. However, if proteins with particular functions in the cell are highlighted they tend to cluster into regional interaction centers. This can be further simplified to give a functional interaction map, in which basic cellular processes are linked together by virtue of protein interactions (Fig. 23.12). Thus, proteins involved in cell cycle control interact not only with each other, but also with proteins involved in related processes such as cell polarity, cytokinesis, DNA replication, and mitosis. Proteins involved in DNA recombination interact among themselves and also with proteins involved in DNA repair and chromosome maintenance. Figure 23.13 shows the complex interaction map

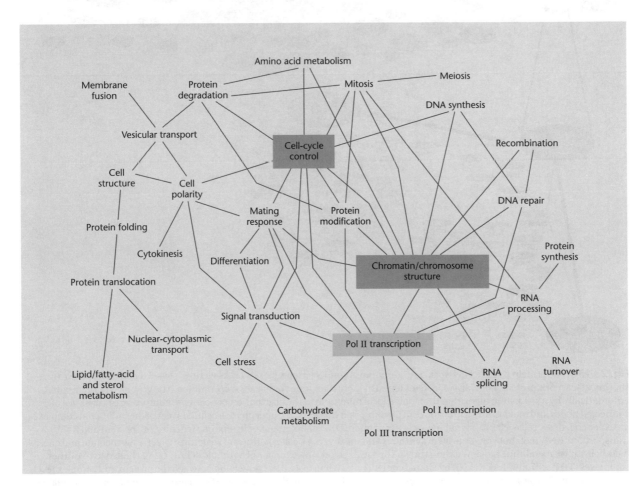

Fig. 23.12 Functional group interaction map derived from the detailed map. Each line indicates that there are 15 or more interactions between proteins of the connected groups. Connections with fewer than 15 interactions are not shown because one or a few interactions occur between almost all groups and often tend to be spurious – that is, based on false positives in two-hybrid screens or other assays. Note that only proteins with known function are included and that about one-third of all yeast proteins belong to several classes. Reprinted from Tucker C.L., Gera J.F. & Uetz P. (2001) Towards an understanding of complex protein networks. *Trends in Cell Biology* **11**, 102–6. ©2001 with permission from Elsevier.

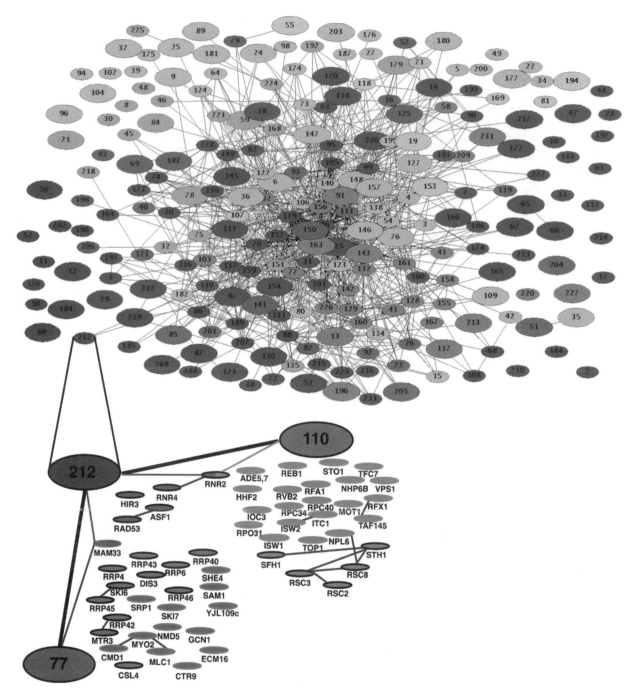

Fig. 23.13 The protein complex network, and grouping of connected complexes. Links were established between complexes sharing at least one protein. For clarity, proteins found in more than nine complexes were omitted. The graphs were generated automatically by a relaxation algorithm that finds a local minimum in the distribution of nodes by minimizing the distance of connected nodes and maximizing the distance of unconnected nodes. In the upper panel, cellular roles of the individual complexes are color coded: ●, cell cycle; ●, signaling; ●, transcription, DNA maintenance, chromatin structure; ●, protein and RNA transport; ●, RNA metabolism; ●, protein synthesis and turnover; ●, cell polarity and structure; violet, intermediate and energy metabolism; ●, membrane biogenesis and traffic. The lower panel is an example of a complex (TAP-C212) linked to two other complexes (TAP-C77 and TAP-C110) by shared components. It illustrates the connection between the protein and complex levels of organization. Purple lines indicate physical interactions as listed in the Yeast Proteome Database. Full-color version available at www.blackwellpublishing.com/primrose

resulting from the yeast–protein complex screen (Gavin *et al.* 2002). This has been simplified by omitting proteins found in more than nine complexes. As shown in the insert, each complex can be inspected for individual proteins, again providing the researcher with multiple levels of detail. As with the binary map, complexes with similar functions tend to share components and interactions, while there are fewer interactions between functionally unrelated complexes.

The existence of such maps is not only a valuable basic resource, but as it grows it will provide a basis to *define* novel interactions. The interaction map provides a benchmark with which to judge the plausibility of newly discovered interactions, and help to eliminate false positives. Statistical analysis of the binary map, for example, shows that nearly three-quarters of all protein interactions occur within the same functional protein group, while most others occur with related functional groups. An unexpected interaction between proteins involved in, for example, Pol I transcription and vesicular transport should be regarded with suspicion and tested by rigorous biochemical and physical assays. Many will be disproved, although some implausible interactions are inevitable.

Suggested reading

Drewes G. & Bouwmeester T. (2003) Global approaches to protein–protein interactions. *Current Opinion in Cell Biology* **15**, 1–7.

Figeys D. (2003) Novel approaches to map protein–protein interactions. *Current Opinion in Biotechnology* **14**, 1–7.

Phizicky E.M., Bastiaens P.I.H., Zhu H., Snyder M. & Fields S. (2003) Protein analysis on a proteomic scale. *Nature* **422**, 208–15.

Phizicky E.M. & Fields S. (1995) Protein–protein interactions: methods for detection and analysis. *Microbiology Review* **59**, 94–123.

Titz B., Schlesner M. & Uetz P. (2004) What do we learn from high-throughout protein interaction data? *Expert Review Proteomics* **1**, 111–21.

Tong A.H.Y., Lesage G., Bader G.D. *et al.* (2004) Global mapping of the yeast genetic interaction network. *Science* **303**, 808–13.

CHAPTER 24

Metabolomics and global biochemical networks

Introduction

The central dogma of molecular biology is that "DNA makes RNA makes protein". One consequence of this dogma is that many scientists have tried to explain biological phenomena solely in terms of gene expression and protein synthesis. Where consideration has been given to small molecules it usually has been at the level of an individual biochemical pathway. However, there are a number of problems with this approach. First, metabolic pathways never exist in isolation but are part of much larger networks. Second, increases in mRNA levels do not always correlate with increases in protein levels (Gygi *et al.* 1999). Third, once translated, a protein may not be functional due to protein–protein or protein–ligand interactions. An example of the failure of the traditional approach is a study of the control of glycolytic flux (ter Kuile & Westerhoff 2001). This showed that flux is rarely regulated by gene expression alone and in one particular case was regulated 30% by gene expression and 70% by metabolism. Thus, in trying to understand biological systems at the level of the intact cell, tissue, organ, or organism we need to construct a global biochemical network that links mRNA, proteins, and metabolites (Brazhnik *et al.* 2002). A simple example of such a network is shown schematically in Fig. 24.1.

The importance of metabolites in global biochemical networks can be deduced from a consideration of a single biochemical step: the conversion of fumarate to aspartate by aspartase (aspartate ammonia lyase). As can be seen from Fig. 24.2, fumarate can participate in seven other reactions and aspartate in nine others. If the level of aspartase is increased significantly, say by cloning, then this could have the effect of pulling fumarate away from the other pathways. At the same time, the increase in the level of aspartate will affect the levels of some or all of the metabolites derived directly from it. For example, more aspartate

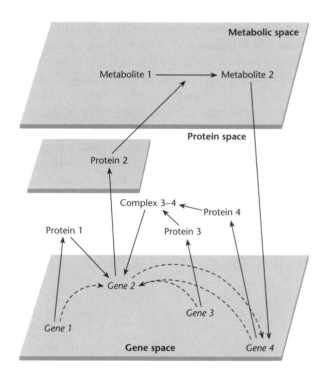

Fig. 24.1 An example of a biochemical network. The molecular constituents are organized into three levels (spaces): mRNAs, proteins, and metabolites. Solid arrows indicate interactions. Three different mechanisms of gene–gene interactions are shown: regulation of gene 2 by the protein product of gene 1; regulation of gene 2 by the complex 3–4 formed by the protein products of genes 3 and 4; and regulation of gene 4 by metabolite 2 which is produced by protein 2. Reproduced from Brazhnik *et al.* (2002), with permission from Elsevier.

may be channeled towards threonine, lysine, methionine, and isoleucine. As the levels of these other amino acids increase, key enzymes will undergo feedback inhibition and the synthesis of many more may be repressed. That is, there will be a major shift in the amounts and activities of many enzymes caused by changes in the amounts of different metabolites. Furthermore, the number of different proteins affected will be related to the magnitude of the increase in the level of the activity of the aspartase.

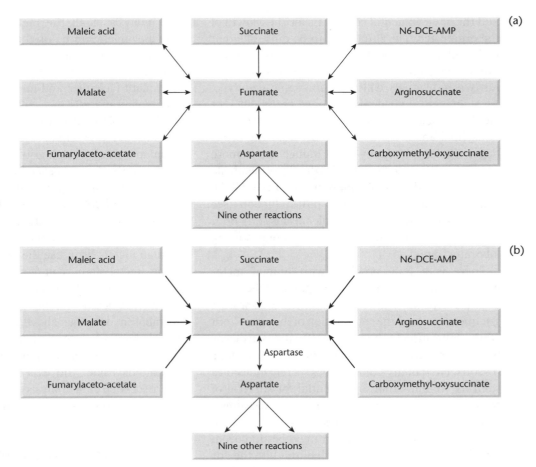

Fig. 24.2 Changes in metabolite flux caused by overexpression of a protein as exemplified by the different biochemical reactions involving fumarate. In (a), the double-headed arrows represent the steady-state levels of the eight different reactions in which fumarate participates. In (b), the overexpression of aspartase reduces the levels of seven of these metabolites and increases the level of aspartate.

There are different levels of metabolite analysis

There is a long history of metabolite analysis but until recently this was focused on (i) metabolite target analysis and (ii) metabolite profiling. The first of these is concerned with the utilization of specialized protocols for the study of difficult analytes, e.g. phytohormones. The second attempts quantitation of a small number of predefined metabolites, e.g. the blood analytes used as biomarkers of human disease. With the development of genomics and its associated 'omics, two new forms of metabolite analysis have been developed: (iii) metabolomics and (iv) metabolic fingerprinting. The goal of metabolomics is the unbiased identification and quantitation of all the metabolites present in a sample taken from an organism. Metabolic fingerprinting focuses on the collection and analysis of data from crude metabolite mixtures to rapidly classify samples without the need to separate individual metabolites. Of these four approaches, metabolomics is the one best suited for the investigation of global biochemical networks because it focuses on the quantification of individual

metabolites without any bias as to choice of target. Metabolite fingerprinting is more suited to the identification of gene function. For example, in a technique known as FANCY (Functional ANalysis by Co-response in Yeast), metabolic fingerprinting is used to compare metabolic changes following perturbation of known genes with those following perturbation of genes of unknown function. If an unknown gene yields a similar result it is assigned a similar function (Raamsdonk *et al.* 2001).

Metabolomics studies in humans are different from those in other organisms

There is a significant difference in the kinds of metabolomics studies that will be undertaken by plant biologists and microbiologists compared with animal biologists. Plant biologists almost certainly will use inbred lines and will either want to compare different lines (cultivar 1 versus cultivar 2 or genetically modified versus non-modified) or the effect of environment (soil, weather, fertilizer, etc.) on a particular line. Microbiologists will have a similar approach. In both cases one controls the genotype and the environmental conditions for the growth of the organisms. The situation can be quite different with animals, particularly humans. For a start, human populations are outbred and so in most cases no two individuals will have the same genotype. Also, different individuals can have vastly different diets even when they are part of the same family or household and some of them may be taking drugs (legal and illegal!). Many drugs induce one or more members of the cytochrome P450 family of metabolic enzymes and the metabolite profile of a cell will be completely different pre- and post-induction (see Nicholson & Wilson 2003 for fuller discussion). Both of these problems can be minimized if one is using laboratory animals.

There is a final problem with animal metabolomics and that is the role of the gut microflora. All animals with an alimentary canal have associated gut microflora and the composition of this microflora can be substantially changed by alterations in diet or by antibiotic treatment. This is important for four reasons. First, many gut microbes secrete molecules such as vitamins that are essential for survival of their host. Second, the microbes may break down dietary components into molecules that can be assimilated as food (e.g. rumen microbes in cows). Third, the body can excrete compounds into the gut where they are transformed and then readsorbed. Finally, many gut organisms behave as commensals but can become pathogens as a result of environmental triggers (Gilmore & Ferretti 2003), and most readers will be familiar with the physiological and biochemical consequences that can ensue.

Because humans are all outbred and controlling their diet is very difficult, most metabolomics studies on humans have focused on the identification of particular metabolites that are indicative of disease (biomarkers) and the toxicity of drugs. The use of biomarkers is well established in clinical chemistry and a good example is the association of glucose in the urine with diabetes. However, there are many diseases such as degenerative changes and cancer whose presence cannot be detected until there are gross physiological changes. By screening all the metabolites in plasma or excreted in urine it might be possible to identify ones that are associated with a particular disease state. Once upon a time, such a "needle in a haystack" approach would have involved screening each and every metabolite individually. In reality, educated guesses would be made about potential biomarkers and only these would be screened, with a negative correlation being the obvious outcome. However, the global approach to the analysis of all the metabolites in a cell, i.e. metabolomics, can be used to make the search for biomarkers much more tractable.

From the above discussion it should be clear that plant biologists and microbiologists use global metabolite analysis in a different way from those studying disease in humans. One consequence has been the development of confusing and inconsistent terminology. Most biologists who work with plants or microbes define *metabolomics* as the unbiased identification and quantitation of all the small molecules (<1000 daltons) in the system being studied and this is the definition used in this chapter. Other workers, particularly those interested in human metabolism, use the term *metabonomics*. To some, metabolomics and metabonomics are one and the same thing whereas to others metabonomics refers specifically to the complexities of animal metabolism described above. A third group consider metabolomics to refer to metabonomics at the level of a single cell rather than a larger system. Leaving definitions aside, the experimental methods and the associated data analysis essentially are the same regardless of the system being studied.

Compromises have to be made in choosing analytical methodology for metabolomics studies

The number of different compounds that can be found in any one cell is vast and these compounds will be representative of many different classes, e.g. amino acids, carbohydrates, lipids, steroids, vitamins, flavonoids, plus a vast range of secondary metabolites and xenobiotics such as terpenoids, macrolides, etc. The physical properties of these different classes of molecule mean that no single method will permit the separation and quantitation of all of them. Of all the methods developed for metabolite detection only two have the capability of resolving large numbers of metabolites: NMR spectroscopy and mass spectrometry (Bundy *et al.* 2002, Fiehn *et al.* 2000, Nicholson *et al.* 2002). Even so, under the best conditions these methods will discern only 500–1000 different molecules and it is not known whether this represents 5% or 50% of the total compounds in a sample. Furthermore, many of the compounds that are resolved by mass spectrometry may remain unidentified.

A major technological challenge encountered in metabolomics is dynamic range (Sumner *et al.* 2003). Dynamic range defines the concentration boundaries of an analytical determination over which the instrumental response as a function of analyte concentration is linear. Most mass spectrometers have a dynamic range of 10^4–10^6 for pure compounds but this range is significantly reduced by the presence of other chemical components. This means that compounds present at high levels can interfere with the detection of compounds present at low levels. This problem is confounded by ion suppression from matrix effects if crude cell extracts are not cleaned up first (Choi *et al.* 2001) by means of gas or liquid chromatography. Gas chromatography separates molecules on the basis of their volatility and double-bond character whereas liquid chromatography is mostly used to separate molecules on the basis of their hydrophobicity/hydrophilicity.

Many different types of mass spectrometer are available and the ones most commonly used in metabolomics are described in Box 24.1. (Also see Box 21.1 which discusses mass spectrometry in

proteomics.) A good workhorse system is gas chromatography coupled to mass spectrometry (GC/MS) but this requires that samples be volatile. This requirement is met by chemical derivatization but adds to the analysis time. Typically, GC/MS is performed with affordable single quadrupole mass analyzers and provides high separation efficiencies that can resolve complex biological mixtures. Newer GC/MS systems incorporating time-of-flight (TOF) mass analyzers give higher mass accuracies and the detectors have higher scan speeds thereby permitting much higher sample throughput.

NMR spectroscopy is a rapid, non-destructive method and can be used for metabolite identification. However, if suffers from two disadvantages when compared to mass spectrometry: low sensitivity (Fig. 24.3) and difficulties in resolving individual metabolites if no chromatographic separation is used prior to detection and identification. For

---

Box 24.1 Basic mass spectrometry in metabolomics

Overview

A mass spectrometer is an instrument that measures the mass to charge ratio (m/z) of individual molecules that have been converted into ions, i.e. electrically charged. The different functional units of a mass spectrometer are shown in Fig. B24.1. The sample to be analyzed enters the vacuum chamber through an inlet and in the case of hyphenated mass spectrometry this will be a connection from

a gas chromatograph (GC/MS), liquid chromatograph (LC/MS) or other separation system. The sample is ionized and volatilized in the ion source and the gas phase ions are sorted in the mass analyzer according to their m/z ratios. The detector collects the sorted ions and the ion flux is converted to a proportional electrical current. Finally, the data system records the magnitude of these electrical signals as a function of m/z and converts the information into a mass spectrum.

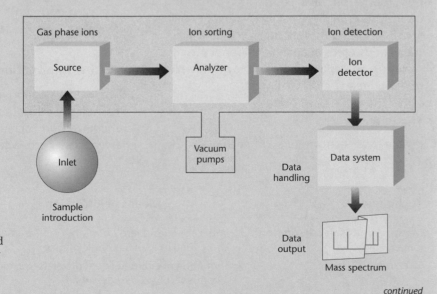

Fig. B24.1 Schematic representation of the key components of a mass spectrometer. (Figure reproduced courtesy of the American Society for Mass Spectrometry.)

continued

Box 24.1 *continued*

The ion source

Samples can be ionized in a number of different ways. One common method (*electron ionization* or *electron impact*) is to bombard the vaporized sample with a beam of energetic electrons. The energy of the electrons generally is much greater than that of the bonds which hold the molecule together. Thus, not only does ionization occur but bonds are broken and fragments are formed giving rise to the ions other than the intact molecule that appear in the mass spectrum. Although both positive and negative ions are generated in the ion source, most of the ions formed are positive and hence the vast majority of measurements are carried out in positive mode.

Electron ionization is a very hard process and it can lead to such extensive fragmentation of the sample that mass and structure are difficult to determine. Consequently, lower energy techniques have been developed based on chemical and desorption ionization. The methods most used in metabolomics are listed in Table B24.1.

The analyzer

The analyzer uses dispersion or filtering to sort ions according to their m/z ratio or a related property. The most widely used analyzers are magnetic sectors, quadrupole mass filters, quadrupole ion traps, Fourier transform ion cyclotron resonance spectrometers (FT-ICR), and time-of-flight mass analyzers. Basically, the various types of analyzer differ in their resolving power and the accuracy with which m/z ratios can be determined.

The detector

In all mass spectrometers, other than FT-ICR instruments, the ions are detected after mass analysis by converting the detector-surface collision energy of the ions into emitted ions, electrons, or photons that are sensed with light or charge detectors.

The data system

Computer-based data systems are used for spectrum acquisition, storage, and presentation. Typically they include software for quantitation, spectral interpretation, and compound identification. The latter is achieved by comparing the spectrum obtained from a sample with the spectra of known compounds that are maintained in online libraries. Spectra are available for only a small proportion of all known compounds and each type of mass spectrometer generates a different spectrum for any one pure compound. That is, there are different spectral libraries for each type of mass spectrometer and some libraries are much more extensive than others.

Tandem mass spectrometry (MS/MS)

This a procedure for identifying compounds in complex mixtures and determining structures of unknown substances. The instrument used has two mass analyzers. Ions with a particular m/z value are selected in the first analyzer, fragmented, and sent to the second analyzer.

Further information

The reader wishing more information should consult the education section of the website of the American Society for Mass Spectrometry (www.asms.org/whatisms/).

Technique	Means of ionization
Matrix assisted laser desorption/ionization (MALDI)	Impact of high-energy photons on a sample embedded in a solid organic matrix
Electrospray	Formation of charged liquid droplets from which ions are desolvated or desorbed

Table B24.1 Some soft ionization methods used in mass spectrometry.

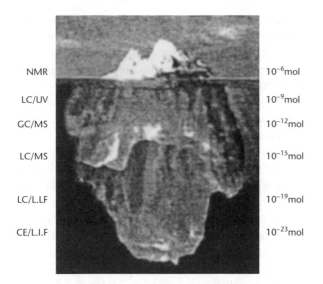

these reasons NMR spectroscopy is more suited to metabolic fingerprinting, e.g. biomarker discovery, than a comprehensive analysis of cellular metabolites. Regardless of what analytical method is used it will be a compromise between speed, selectivity, and sensitivity (Weckwerth & Fiehn 2002, Fiehn & Weckwerth 2003, Sumner *et al.* 2003).

Sample selection and sample handling are crucial stages in metabolomics studies

Ideally, metabolomic data should accurately describe physiological processes as responses to developmental, genetic, or environmental changes but problems of sample selection make this goal difficult to achieve. The first problem is one of compartmentalization. At its simplest, during sample preparation all subcellular compartmentalization is lost. The problem is more complex when one deals with organs since the different tissues within the organ may have different metabolite distributions. A good example of this is the potato tuber shown in Fig. 24.4. Samples were

Fig. 24.3 A comparison of the relative sensitivities of different analytical methods. NMR has rapid analysis times but suffers from lower sensitivity thus allowing visualization of only the more concentrated metabolites (i.e. the tip of the iceberg). GC/MS and LC/MS provide good selectivity and sensitivity. CE/LIF (laser induced fluorescence) provides very high sensitivity but lower selectivity. Reproduced from Sumner *et al.* (2003), with permission from Elsevier.

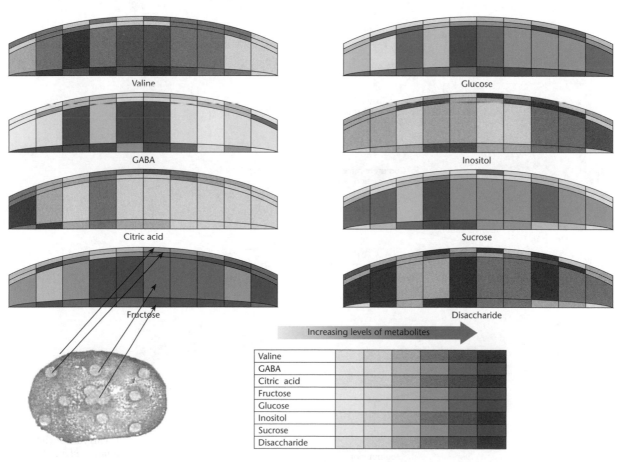

Fig. 24.4 The distribution of eight different metabolites throughout a single potato tuber. (Figure courtesy of Prof. Howard Davies.)

taken from different positions within the tuber and analyzed for eight metabolites. Each metabolite had a complex distribution pattern and no two patterns were the same. The second problem is one of timing, and tomatoes provide a good example. Suppose one wishes to determine the metabolite content of a number of different tomatoes. All of these tomatoes should be analyzed at exactly the same stage in the ripening process but how does one determine that they all are at the same stage?

If any organism is being maintained in a relatively stable environment then its metabolome also will be stable, at least over time periods ranging from a few minutes to a few hours. However, once a tissue sample is taken from the organism then the metabolome of that tissue no longer is in a stable environment and will undergo rapid changes. Therefore it is important that samples are flash frozen as soon as possible and in a consistent manner.

After sampling the tissues need to be homogenized and the small molecules extracted in an unbiased process. However, there are no comprehensive comparisons of extraction techniques in terms of their reproducibility, robustness, and recovery of the different classes of compounds that will be present (Weckwerth & Fiehn 2002). The most commonly used technique is extraction with alcohols or water/alcohol mixtures but the efficiency of this will vary greatly depending on the tissue being sampled; e.g. a lettuce leaf, a sugar-rich fruit, a fatty tissue such as liver, and a hard tissue such as bone. It also is unclear which factors most affect the robustness of protocols, which is the susceptibility to error caused by the use of slightly altered conditions. Such alterations may include subtle differences in extraction times, temperatures, solvent compositions, and staff skills.

If the object of a study simply is the analysis of the metabolome then the process depicted in Fig. 24.5

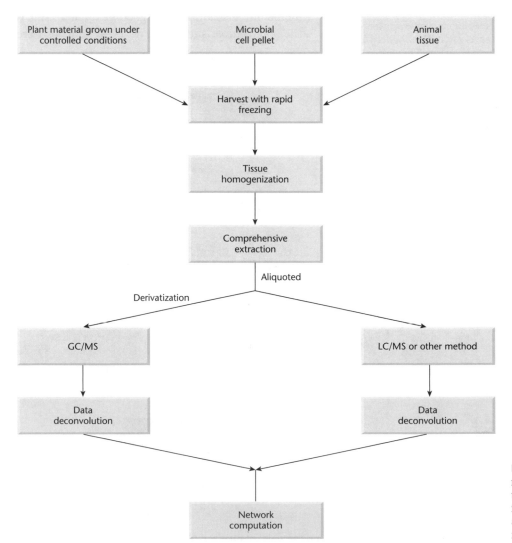

Fig. 24.5 Schematic representation of the steps involved in carrying out an analysis of the metabolites in a sample.

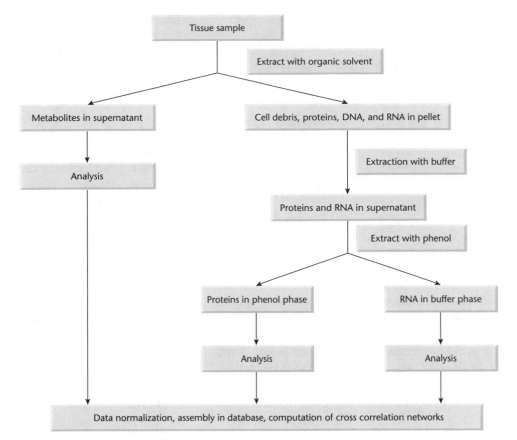

Fig. 24.6 Schematic representation of the methodology required if it is desired to construct a biochemical network linking metabolites, proteins, and mRNA.

will suffice. However, if the objective is to generate a global biochemical network then the transcriptomic, proteomic, and metabolomic analyses must be done on the same sample. A suitable schema for this has been proposed by Fiehn & Weckwerth (2003) and is shown in Fig. 24.6.

Metabolomics produces complex data sets

A single sample analyzed by GC/MS can provide data on up to 500 distinct compounds and this makes simple comparison of data sets impossible. Special methods are needed to reduce the dimensionality of the data and present it in simple visual form. These methods are of two types: supervised and unsupervised. In supervised methods there is an initial calibration step using a training data set, i.e. a set of observations that have been classified by independent means. This approach is particularly applicable to metabolic fingerprinting where profiles are generated for samples known to come from "normal" and "diseased" tissue and then used to classify samples of unknown origin. Another example is the FANCY approach to identifying gene function that was described earlier (p. 473). Here, a fingerprint is

generated for a knockout mutant of a gene of unknown function and compared with the finger prints of knockout mutants in different genes of known function.

Unsupervised data analysis methods require no information other than the original data set and for metabolomics the most popular method is principal component analysis (PCA). In PCA the variance in a set of multivariate data (the levels of different compounds detected by MS) is described in terms of a set of underlying orthogonal variables (principal components). That is, the original metabolite concentrations can be expressed as a particular linear combination of the principal components. PCA is a linear additive model in that each principal component (PC) accounts for a portion of the total variance of the data set. Often, as few as two or three PCs account for over 90% of the total variance and these are used to reduce the dimensionality of the data set. The data then can be visualized graphically when the values for the first two PCs (two-dimensional graphs) or three PCs (three-dimensional graphs) are plotted.

An example of the way in which PCA can be used to analyze metabolomics data is provided by a study on potato tubers that had been engineered to be frost

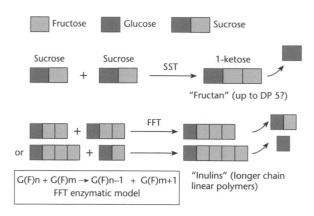

Fig. 24.7 The synthesis of fructans and inulins from sucrose. SST, sucrose:sucrose 1-fructosyltransferase; FFT, fructan:fructan 1-fructosyltransferase. (Figure courtesy of Prof. John Draper.)

resistant. One transgenic line expressed a foreign gene for sucrose:sucrose 1-fructosyltransferase (SST) and the other line expressed SST and a fructan:fructan 1-fructosyltransferase (FFT). These result in the production of fructans and inulins, respectively (Fig. 24.7). Using GC/MS, the relative concentrations of the different metabolites in the two transgenic lines and the parent potato cultivar were determined. Figure 24.8a shows a PCA analysis of these data and it is clear that separation can be achieved using the first two PCs. Furthermore, the different compounds in the samples that are responsible for the separation can be identified. Analysis of the top five of these components showed them to be

fructans. When these components were removed from the data set (Fig. 24.8b) the separation was lost. This confirms that the metabolic differences seen between the transgenic lines and the parent were due to the activity of the transgenes. If the separation had not been lost then it would have suggested that other metabolic changes also were important.

In the potato example quoted above, the objective was to compare lines where the genetic differences were simple and known, i.e. addition of one or two transgenes. However, in many metabolomics experiments the genetic changes might not be known as in a comparison of diseased versus normal tissue. This makes the analysis more difficult. If metabolomic analysis shows that diseased tissue has a higher level of fumarate then fumarate level can be used as a biomarker for that disease. To understand why fumarate levels are increased it is necessary to know the changes in all the eight compounds that can be interconverted to fumarate (Fig. 24.2). If one of these eight compounds is decreased significantly then less fumarate is going down the pathway of which it is the first intermediate. On the other hand, if one of these eight compounds is increased then the re-actions producing it could have increased flux and could be forcing it in the direction of fumarate. That is, one uses metabolic networks to try and establish the chain of causality that leads to the observations. For this to be possible it is essential to have metabolic maps that are comprehensive. This is not always the case (see below).

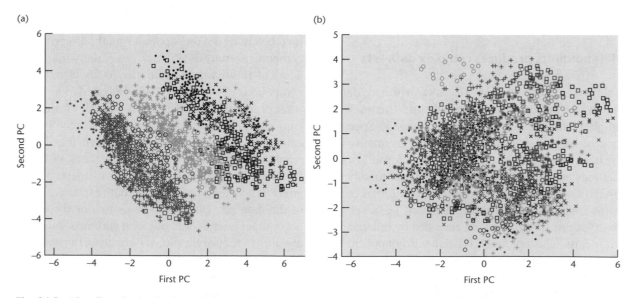

Fig. 24.8 The effect of individual metabolites on the clustering of potato genotypes. (a) Principal component analysis with data for all metabolites included. (b) Principal component analysis after removal of the top five discriminatory metabolites. Black points, data for parent strain; light purple points, data for the line carrying the SST gene; dark purple points, data for the line carrying SST and FFT. (Figure courtesy of Prof. John Draper.)

A good reference database is an essential prerequisite for preparing global biochemical networks but currently is missing

As mentioned above, it is useful to use a biochemical context for visualization and interpretation of metabolomic data. Ideally, this biochemical context should include the known enzyme activities that catalyze each reaction, the proteins with these activities, their biochemical characteristics (k_M values, etc.), and the genes that encode them. That is, we need a detailed biochemical database. A number of such databases exist (for summary, see Sumner *et al.* 2003) but none meet the needs of metabolomics. The one that is most used is KEGG (Kanehisa *et al.* 2002; www.genome.ad.jp/kegg/). It contains information on genes from Genbank (p. 158) and biochemical reactions from the ENZYME database (Bairoch 2000). There are a number of problems associated with the KEGG database. First, it lists the enzymes present in each organism only if an annotated gene sequence for that enzyme activity exists in Genbank. However, many enzymes are known to exist in particular organisms even though the corresponding gene has not been found from analysis of whole-genome sequences. Second, just because a gene can be annotated based on sequence similarity does not necessarily mean that the proposed enzyme activity actually exists in the cell. The presence of the activity needs to be confirmed experimentally. Finally, KEGG takes no account of isozymes but for metabolomics the different tissue distributions and kinetic characteristics could be very important. Despite these reservations, KEGG is a very useful database.

Suggested reading

Sumner L.W., Mendes P. & Dixon R.A. (2003) Plant metabolomics: large-scale phytochemistry in the functional genomics era. *Phytochemistry* **62**, 817–36.
Presents a good overview of the methodologies associated with plant metabolomics.

Nicholson J.K. & Wilson I.D. (2003) Understanding "global" systems biology: metabonomics and the continuum of metabolism. *Nature Reviews Drug Discovery* **2**, 668–76.
Presents an overview of mammalian metabolic conversions and proposes a new probabilistic model to help understand idiosyncratic drug reactions.

Ideker T. & Lauffenburger D. (2003) Building with a scaffold: emerging strategies for high- to low-level cellular modelling. *Trends in Biotechnology* **21**, 255–62.
Describes the use of computational cellular models for the analysis of complex biological systems.

Dunn W.B., Bailey N.J. & Johnson H.F. (2005) Measuring the metabolome: current analytical technologies. *Analyst* **130**, 606–25.
A review that discusses the advantages and disadvantages of mass spectrometry, NMR, and other spectroscopy techniques.

Wang M., Lamers, R.J., Korthout H.A. *et al.* (2005) Metabolomics in the context of systems biology: bridging traditional Chinese medicine and molecular pharmacology. *Phytotherapy Research* **19**, 173–82.
Dunckley T., Coon K.D. & Stephan D.A. (2005) Discovery and development of biomarkers of neurological disease. *Drug Discovery Today* **10**, 326–34.
Two papers that illustrate the different ways that metabolomics is being utilized to solve problems.

Part IV

Applications of Gene Manipulation and Genomics

CHAPTER 25

Applications of genomics: understanding the basis of polygenic disorders and identifying quantitative trait loci

Introduction

Geneticists use the term "complex trait" to describe any phenotype that does not exhibit classical Mendelian recessive or dominant inheritance attributable to a single gene locus. Most, but not all, complex traits can be explained by polygenic inheritance, i.e. these traits require the simultaneous presence of mutations in multiple genes. Polygenic traits may be classified (Lander & Schork 1994) as discrete traits, measured by a specific outcome (e.g. development of diabetes or cleft palate), or quantitative traits measured by a continuous variable (e.g. grain yield, body weight). In general, discrete traits are of particular interest to human geneticists and quantitative traits are of particular interest to plant and animal breeders, although there are significant exceptions. There is another difference. Human populations are outbred whereas plant and animal breeders use inbred populations and the methods used for gene identification reflect these differences.

Despite the differences cited above, the general methodology used to identify genes associated with discrete traits and quantitative traits is the same (Glazier *et al.* 2002). This methodology involves four steps:

1. Establish significant genome-wide evidence for linkage or association of the trait with a particular chromosomal region. Typically, the trait will be localized to a 10–30 cM region of the genome. In humans this equates to 10–30 Mb of DNA with a coding potential of 100–300 genes.
2. Fine mapping is undertaken to reduce the size of the critical region to one that permits sequencing to be undertaken.
3. DNA sequence analysis is undertaken to identify any candidate nucleotide variants.
4. Attempts are made to demonstrate that replacement of the variant nucleotide(s) results in the swapping of one phenotype for another.

This basic methodology forms a recurring theme in the examples described in the sections that follow.

Investigating discrete traits in outbreeding populations (genetic diseases of humans)

Mendelian traits are relatively easy to study but account for only a small proportion of human disease. Most human diseases are polygenic in nature and these include cardiovascular disease, asthma, cancer, diabetes, rheumatoid arthritis, obesity, alcoholism, and schizophrenia. Such complex diseases involve multiple genes, environmental effects, and their interactions. Rather than being caused by specific and relatively rare mutations, complex diseases and traits may result principally from genetic variation that is relatively common in the population. The fact that a large number of genes, many with small effects, are involved in many complex diseases greatly complicates efforts to identify genetic regions involved in the disease process and makes replication of results difficult. The distinction in terminology between Mendelian and complex traits is not meant to imply that complex diseases do not follow the rules of Mendelian inheritance; rather, it is an indication that the inheritance pattern of complex traits is difficult to discern.

There are three main approaches to mapping the genetic variants involved in a disease: functional cloning, the candidate gene strategy, and positional cloning (see Chapter 6). In functional cloning, knowledge of the underlying protein defect leads to localization of the responsible gene. In the candidate gene approach, genes with known or proposed function with the potential to influence the disease phenotype are investigated for a direct role in disease. Positional gene cloning is used when the biochemical nature of the disease is unknown (the norm!). The responsible gene is mapped to the correct location on the chromosome and successive narrowing of the candidate interval eventually results in the identification of the correct gene.

The gene-finding methods described above are used in conjunction with two other analytical methods. These are model-free (or nonparametric) linkage analysis and association (or linkage disequilibrium) mapping. Model-free methods make no assumption about the inheritance pattern, the number of loci involved, or the role of environment. Rather, they depend solely on the principle that two affected relatives will have disease-predisposing alleles in common. In linkage disequilibrium (LD) mapping one looks at co-inheritance in populations of unrelated individuals. There is another difference. Linkage analysis can be used only for coarse mapping (e.g. only 10% recombination will be observed in a region of 10 Mb), whereas LD can be used for fine mapping as resolution is limited only by the spacing of the markers used. Consequently, most effort is being devoted to developing physical maps with a high marker density.

The reader who is not familiar with the methods used in the study of human genetics will have great difficulty in understanding the primary literature. The reason for this is the widespread use of specialist terminology. To assist readers a glossary is provided in Box 25.1.

Box 25.1   Glossary of terms used in human genetics

Ascertainment bias

This is the difference in the likelihood that affected relatives of the cases will be reported to the geneticist as compared with the affected relatives of controls.

Concordance

If two related individuals in a family have the same disease they are said to be concordant for the disorder (cf. discordance).

Discordance

If only one member of a pair of relatives is affected with a disorder then the two relatives are said to be discordant for the disease (cf. concordance).

Familial aggregation

Because relatives share a greater proportion of their genes with one another than with unrelated individuals in the population, a primary characteristic of diseases with complex inheritance is that affected individuals tend to cluster in families. However, familial aggregation of a disease does not necessarily mean that a disease has a genetic basis as other factors could be at work.

Founder effect

If one of the founders of a new population happens to carry a relatively rare allele, that allele will have a far higher frequency than it had in the larger group from which the new population was derived. The founder effect is well illustrated by the Amish in Pennsylvania, the Afrikaners in South Africa, and the French-Canadians in Quebec. An early Afrikaner brought the gene for variegate porphyria and the incidence of this gene in South Africa is 1 in 300 compared with 1 in 100,000 elsewhere.

Genome scan

This is a method whereby DNA of affected individuals is systematically analyzed using hundreds of polymorphic markers in a search for regions that are shared by the two sibs (cf.) more frequently than on a purely random basis. When elevated levels of allele sharing are found at a polymorphic marker it suggests that a locus involved in the disease is located close to the marker. However, the more polymorphic the loci studied the more likely it is that elevated allele sharing occurs by chance alone and hence one looks for high LOD scores.

Index case

See "proband".

Multiplex family

A family with two or more affected members.

continued

Box 25.1 *continued*

Nonparametric (model-free) analysis

This method makes no assumption concerning the number of loci or the role of environment and chance in causing lack of penetrance (q.v.). Instead, it depends solely on the assumption that two affected relatives will have disease-predisposing alleles in common.

Parametric (model-based) linkage analysis

This method of analysis assumes that there is a particular mode of inheritance (autosomal dominant, X-linked, etc.) that explains the inheritance pattern. Therefore one looks for evidence of a genetic locus that recombines with a frequency that is less than the 50% expected with unlinked loci.

Penetrance

In clinical experience, some disorders are not expressed at all even though the individuals in question carry the mutant alleles. Penetrance is the probability that such mutant alleles are phenotypically expressed.

Proband

The member through whom a family with a genetic disorder is first brought to attention (ascertained) is the proband or index case if he or she is affected.

Relative risk

The familial aggregation (q.v.) of a disease can be measured by comparing the frequency of the disease in the relatives of an affected individual with its frequency in the general population. The relative risk ratio is designated by the symbol λ. In practice, one measures λ for a particular class of relative, e.g. sibs, parents.

Sibs

Brothers and sisters are sibs.

Simplex family

A family in which just one member has been diagnosed with a particular disease.

Transmission disequilibrium test

This tests whether any particular alleles at a marker are transmitted more often than they are not transmitted from heterozygous parents to affected offspring. The benefit of this test is that it only requires trios (q.v.).

Trio

An affected child plus both parents.

Model-free (nonparametric) linkage analysis looks at the inheritance of disease genes and selected markers in several generations of the same family

Any kind of genetic marker can be used in linkage mapping and in classical genetics these markers are other phenotypic traits. In practice, it is difficult to detect linkages for loci more than 25 cM apart. Thus, to be useful, markers need to be distributed throughout the genome at a frequency of at least one marker every 10 cM. In humans there are not enough phenotypic traits that have been mapped to give anything like the desired marker density. For this reason, physical markers are very attractive. The first such markers to be described were restriction fragment length polymorphisms (RFLPs, see p. 346) but the ones favored today are the single nucleotide polymorphisms (SNPs, see p. 349) because they occur once every 1000 bp. When large multigeneration pedigrees are available (e.g. the Centre d'Etude du Polymorphisme Humain (CEPH) families, p. 362) linkage analysis is a powerful technique for locating disease genes and has been applied to a number of simple Mendelian traits. The probability of linkage is calculated and expressed as a logarithm$_{10}$ of odds (LOD) score, with a value above 3 being significant (Box 25.2). If linkage to a marker is observed then the chromosomal location of that marker is also the location of the disease gene.

Box 25.2 Logarithm of odds scores
(Adapted from Connor & Ferguson-Smith 1997)

Figure B25.1 shows pedigrees for two families affected by an autosomal dominant disorder. In family A the affected man in the second generation has received the disease allele together with RFLP allele 1 from his father. Similarly, he has received the normal allele and RFLP allele 2 from his mother. If these two loci are on the same chromosome then it follows that he must have one chromosome that carries the disease allele together with RFLP allele 1 and the other carries the normal allele and RFLP allele 2. Consequently, the arrangement of the disease and marker alleles, also known as the *phase,* can be deduced with certainty in this individual. If the loci are linked it will be apparent in the next generation as a tendency for the disease allele to segregate with RFLP allele 1 and the normal allele to segregate with RFLP allele 2. This is indeed the case in family A, where four affected offspring carry RFLP allele 1 and the five unaffected children only carry RFLP allele 2.

If the loci described above are not linked, the probability of such a striking departure from independent assortment occurring by chance in nine offspring is the probability of correctly calling heads or tails for nine consecutive tosses of a coin. That is:

$$(0.5)^9 = 0.002.$$

However, if these two loci are linked such that there is only a 10% chance of crossing over (i.e. a recombination fraction, or θ, of 0.1), the probability of the disease segregating with RFLP allele 1 or the normal allele with RFLP allele 2 is:

$$(0.9)^9 = 0.4.$$

It follows that linkage at 10% recombination is 200 times (0.4/0.002) more likely than no linkage. Similarly, if the disease allele and the RFLP allele are identical, then no recombination could occur and the recombination fraction would be zero. For this family, this is 500 times (1/0.002) more likely than no linkage.

The usual way of representing these probability ratios is as logarithms, referred to as logarithm of odds (LOD) or Z scores. For family A at a recombination fraction of 10% the LOD score is $\log_{10} 200 = 2.3$ and at a recombination fraction of zero it is $\log_{10} 500 = 2.5$.

The figure also shows a two-generation pedigree for family B. In this case, all four affected siblings carry RFLP allele 2 and another four healthy siblings do not. As in family A, this signifies a marked disturbance of independent assortment and suggests linkage between the disease and RFLP allele 1. If this is the case, the youngest child must represent a recombinant because he has inherited RFLP allele 1 from his father but not the disease. However, it could be that the youngest child is non-recombinant and all the other children represent crossovers between the two loci.

Family A

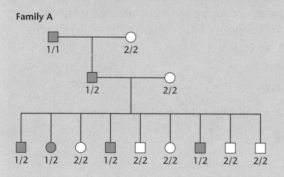

Family B

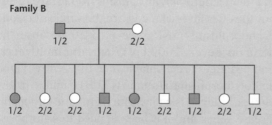

Fig. B25.1 Two families with an autosomal dominant trait showing results of DNA analysis for a marker restriction fragment length polymorphism (RFLP) with alleles 1 and 2.

continued

Box 25.2 *continued*

Although this is much less likely, it cannot be excluded in the absence of phase information from the grandparents. Calculation of the LOD scores for such a family is more complicated because the two possible phases must be taken into account. Whichever phase is considered, at least one recombination event must have taken place and hence the recombination fraction cannot be zero (see Table B25.1).

Analysis of the combined data from the two families shows that the maximum LOD score is 3.3 and this occurs at 10% recombination.

Table B25.1 Logarithm of odds (LOD) scores at values of the recombination fraction from 0 to 40% for the two families shown in Fig. B25.1.

	Recombination fraction (%)				
	0	**10**	**20**	**30**	**40**
Family A	2.7	2.3	1.8	1.3	0.7
Family B	$-\infty$	1.0	0.9	0.6	0.3
Total	$-\infty$	3.3	2.7	1.9	1.0

The ease with which data from phase-known and phase-unknown families can be combined in this way is the reason why the use of LOD scores has become universal for the analysis of linkage data. The maximum value of the LOD score gives a measure of the statistical significance of the result. A value greater than 3 is usually accepted as demonstrating that linkage is present and in most situations it corresponds to the 5% level of significance used in conventional statistical tests. Conversely, if LOD scores below -2 are obtained, this indicates that linkage has been excluded at the corresponding values of the recombination fraction.

The relationship between the recombination fraction and the actual physical distance between the loci depends on several factors. A recombination fraction of 0.1 (10% recombination) corresponds to a map distance of 10 cM. However, with increasing distance between the loci the recombination fraction falls as a result of the occurrence of double crossovers. In humans, 1 cM is equivalent to 1 Mb of DNA on average.

Conventional linkage analysis seldom works for complex diseases. The involvement of many genes and the strong influence of environmental factors mean that large multigeneration pedigrees are seen only rarely. Consequently, analysis is undertaken of families in which both parents and at least two children (sib pairs) have the disease in question. These are known as *nuclear families*. The way this analysis is undertaken is shown in Fig. 25.1. Suppose that we

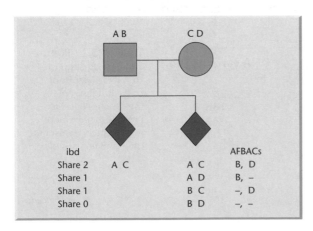

Fig. 25.1 (*left*) Affected sib pair families. A nuclear family pedigree is shown with the father (■) and mother (●) in the first row and the two affected children of either sex (♦) in the second row. Assume for simplicity that we can distinguish all four parental alleles, denoted A, B, C, and D in the genetic region under study, with the parental alleles ordered such that A and C are transmitted from the father and mother, respectively, to the first affected child. Four possible configurations among the two offspring with respect to the alleles inherited from the parents are possible: they can share both parental alleles (AC); they can share an allele from the father (A) but differ in the alleles received from the mother (C and D); they can share an allele from the mother (C) but differ in the alleles received from the father (A and B); or they can share no parental alleles in common. These four configurations are equally likely if there is no influence of the genetic region under consideration on the disease. The parental alleles that are never transmitted to the affected sib pair in each family type are used as a control population in association studies using nuclear family data, the so-called affected family-based control (AFBAC) sample. (Redrawn with permission from Thomson 2001.)

believe that a certain region of the genome is implicated in a disease state and that we can distinguish the four parental chromosomes (A, B, C, D). If the region under test does not carry a gene predisposing to disease then the chances of two affected sibs having two, one, or no parental chromosome regions in common are 25, 50, and 25%, respectively. On the other hand, deviation from this Mendelian random expectation indicates that the affected sibs have chromosome regions that are *identical by descent* (ibd) suggesting the presence of genes predisposing to the disease in question. Physical markers, particularly microsatellites, are ideal for distinguishing the chromosome regions derived from each parent. Not only are micro-

satellites highly polymorphic, but a sufficient number of them have been placed throughout the genome.

The first complex disease to be analyzed using genome-wide linkage scans was type 1 diabetes (Field *et al.* 1994, Hashimoto *et al.* 1994) and this demonstrated linkage to the major histocompatibility complex (see Box 25.3). Since then, a number of other complex diseases have been mapped including bipolar mood disorder (McInnes *et al.* 1996) and Crohn's disease (Hugot *et al.* 1996, Rioux *et al.* 2000). A similar methodology has been used to identify quantitative trait loci (QTLs) controlling adult height (Hirschhorn *et al.* 2001) and human longevity (Geesaman *et al.* 2003).

Box 25.3   The major histocompatibility complex

Higher animals, including humans, are able to distinguish between "self" and "non-self" and to mount a reaction against a very broad spectrum of foreign antigens. This reaction is mediated by the immune response. Genetic factors play a key role in the generation of the normal immune response and, as a result of mutation, in aberrant immune reactions including immunodeficiency and autoimmune disease. A large number of genes play a role in the development and functioning of the immune system but only those of the major histocompatibility complex (MHC) are considered here.

The MHC is composed of a large cluster of genes located on the short arm of chromosome 6. On the basis of structural and functional differences these genes are divided into three classes and each class is highly complex and polymorphic. Two of the three classes correspond to the genes for human leukocyte antigens (HLA) that are cell surface

proteins. These antigens are very important for the normal functioning of the immune system and were first discovered following attempts to transplant tissue between unrelated individuals. A class I antigen consists of two polypeptide units, a polymorphic peptide encoded by the MHC, and an invariant polypeptide encoded by a gene outside the MHC. Class two molecules are heterodimers of α and β subunits, both of which are encoded by the MHC. The class III genes are not HLA genes but include genes for polymorphic serum proteins and membrane receptors.

The HLA system comprises many genes and is highly polymorphic with many antigenic variants having been recognized at the various loci (Table B25.2). Because the HLA alleles are so closely linked they are transmitted together as haplotypes. Each individual has two haplotypes, one on each copy of chromosome 6, and the alleles are co-dominant. Each child receives one

HLA locus	Antigenic variants (no.)	DNA variants (no.)
HLA-A	25	83
HLA-B	53	186
HLA-C	11	42
HLA-DR (β chain only)	20	221
HLA-DQ (α and β chains)	9	49
HLA-DP (α and β chains)	6	88

Table B25.2 Protein and DNA variation at HLA loci. Because of the redundancy of the genetic code it is possible to have more DNA sequence variants than protein variants.

continued

Box 25.3 *continued*

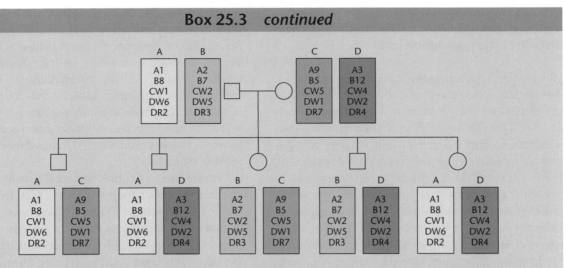

Fig. B25.2 The inheritance of HLA haplotypes. Usually a haplotype is transmitted, as shown in this figure, as a unit. In extrememly rare instances, a parent will transmit a recombinant haplotype to the child.

haplotype from each parent (Fig. B25.2) and there is a 25% chance that two children with the same parents inherit matching HLA haplotypes. Because the success of tissue transplantation is closely linked to the degree of similarity between HLA haplotypes, the favored donor for bone marrow or organ transplantation is a brother or sister who has an identical HLA haplotype.

As more and more information has accumulated about the HLA genes it has become clear that there is an association between specific HLA genes or haplotypes and certain diseases. For example, in one national study, only 9% of the population had the HLA-B27 allele but it was present in 95% of those with the chronic inflammatory disease ankylosing spondylitis. Similarly, 28% of the population carried the HLA-DQ2 allele but it was present in 99% of the population with celiac disease. It is unlikely that HLA genes alone are responsible for specific diseases. Rather, they may contribute to disease predisposition along with other genetic and environmental factors. For example, they probably influence the susceptibility of different individuals to particular infectious agents. They also can play a role in complex diseases as exemplified by type 1 diabetes.

There are two major types of diabetes mellitus: juvenile-onset or insulin dependent (type 1) and adult-onset or insulin-independent (type 2). Type 1 diabetes has a frequency of 0.5% in the Caucasian population and results from an autoimmune destruction of the insulin-producing cells in the pancreas. Genetic factors alone do not cause type 1 diabetes because if one twin of an identical pair develops the disease there is only a 40% chance that the matching twin also will become diabetic. Nevertheless, there is strong evidence for genetic factors and, as noted on p. 490, the first study on model-free analysis of a complex disease linked type 1 diabetes with the MHC locus. Individuals heterozygous for HLA-DR3 or HLA-DR4 are particularly susceptible to diabetes. This fits with the concept of type 1 diabetes being an autoimmune disease, since DR3 and DR4 are found in a locus known to regulate the immune response.

Further insight into the mechanism responsible for type 1 diabetes has come from a molecular analysis of the HLA-DQ genes. The presence of aspartic acid at position 57 of the DQβ chain is closely associated with resistance to type 1 diabetes, whereas other amino acids at this position confer susceptibility. About 95% of patients with type 1 diabetes are homozygous for DQβ genes that do not encode aspartate at position 57. Since position 57 of the β chain is critical for antigen binding and presentation to T cells, changes in this amino acid could play a role in the autoimmune response that destroys the insulin-producing cells.

Linkage disequilibrium (association) studies look at the co-inheritance of markers and the disease at the population level

Whereas linkage analysis is undertaken in families, association studies are undertaken on unrelated cases and controls. If there is a significant association of a marker and a disease state then this may implicate a candidate gene in the etiology of a disease. Alternatively, an association can be caused by LD of marker allele(s) with the gene predisposing to disease. LD implies close physical linkage of the marker and the disease gene. As might be expected, LD is not stable over long time periods because of the effects of meiotic recombination. Thus, the extent of LD decreases in proportion to the number of generations since the LD-generating event. In general, the closer the linkage of two SNPs then the longer the LD will persist in the population but other factors do have an influence, e.g. extent of inbreeding, presence of recombination hotspots, etc. Reich *et al.* (2001) have shown that, in a US population of northern European descent, LD typically extends for about 60 kb. By contrast, LD in a Nigerian population extends for a much shorter distance (5 kb) reflecting the fact that northern Europeans are of more recent evolutionary origin.

So, it should be apparent that the ideal population for LD studies will be one that is isolated, has a narrow population base, and can be sampled not too many generations from the event causing the disease mutation. The Finnish and Costa Rican populations are considered ideal because they are relatively homogenous and show LD over a much wider distance than US populations. This is particularly important because it influences the number of markers that need to be used. In a typical *linkage* analysis one uses markers every 10 Mb (10 cM) but for LD studies one needs many more markers. For a US population of northern European descent the markers would need to be every 20–50 kb on average but this could be extended to every 200–500 kb for Finnish or Costa Rican populations.

Genome-wide LD scans have been undertaken to locate simple Mendelian traits. Lee *et al.* (2001) were able to use such a methodology to localize the critical region for a rare genetic disease (SLSJ cytochrome oxidase deficiency) in a close-knit isolated community. More typically, LD analysis is used to fine-map traits following initial localization to chromosomal regions by linkage analysis as described in the previous section. SNPs are ideal for this purpose because

over 10 million of them from different ethnic groups have been mapped (www.ncbi.nlm.nih.gov/SNP/). For studies on complex diseases, evidence for LD is sought in nuclear families or trios because this avoids possible ethnic mismatching between patients and randomly ascertained controls. In such cases, the parental alleles that are never transmitted to the affected offspring are used as the controls, the so-called affected family-based control (AFBAC) sample.

Two studies on Crohn's disease illustrate how LD can be used in fine mapping (Fig. 25.2). In the first study (Rioux *et al.* 2001), linkage analysis had shown that susceptibility to Crohn's disease mapped to an 18-Mb region of chromosome 5 with a maximal LOD score at marker D5S1984. Using 56 microsatellites, LD was detected between Crohn's disease and two other markers, IRF1p1 and D5S1984, which are 250 kb apart. All the known genes in this region were examined for allelic variants that could confer increased susceptibility to Crohn's disease but no candidate genes were identified. This was a little surprising because this genomic region encodes the cytokine gene cluster that includes many plausible candidate genes for inflammatory disease. Because no obvious candidates had emerged, a detailed SNP map was prepared with the markers spaced every 500 bp. Many of these SNPs showed LD with susceptibility to Crohn's disease, confirming the presence of a gene predisposing to Crohn's disease in the area under study.

In the second study (Hugot *et al.* 2001), linkage analysis had mapped a susceptibility locus for Crohn's disease to chromosome 16. With the aid of 26 microsatellites the locus was mapped to a 5-Mb region between markers D16S541 and D16S2623. LD analysis showed a weak association of Crohn's disease with marker D16S3136, which lies between the other two markers. A 260-kb region around marker D16S3136 was sequenced but only one characterized gene was identified and this did not appear to be a likely Crohn's disease candidate. Sequencing also identified 11 SNPs and three of these showed strong LD with Crohn's disease in 235 affected families, indicating that the susceptibility locus was nearby. By using the GRAIL program and an expressed sequence tag (EST) homology search (p. 171) a number of putatively transcribed regions were identified and one of these (*NOD2*, but now known as *CARD15*) was identified as the susceptibility locus. Further analysis showed that some of the SNPs used in the LD study were the causative mutations.

(a)

Linkage map

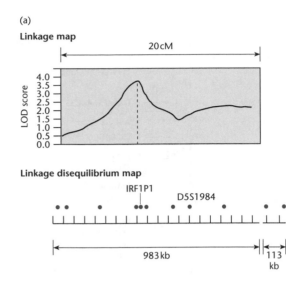

Linkage disequilibrium map

(b)

Linkage map

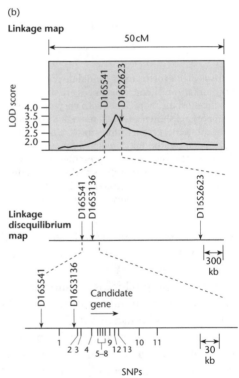

Fig. 25.2 Details of the mapping of two loci associated with Crohn's disease. (a) Mapping of a locus on chromosome 5 by Rioux *et al.* (2001). (b) Mapping of a locus on chromosome 16 by Hugot *et al.* (2001). The numbers along the bottom line correspond to the SNPs used in fine mapping. All the SNPs except 10 and 11 showed tight linkage (see text for further details).

Once a disease locus is identified, all the 'omics can be used to analyze it in detail

In the study of Crohn's disease (CD) described above, Hugot *et al.* (2001) mapped three polymorphisms to

the *NOD2/CARD15* locus. These polymorphisms were R702W (an arg/trp replacement at position 702), G908R (a gly/arg replacement at position 908), and a frameshift mutation (1007fs). Together these mutations represent 81% of the polymorphisms seen at the *NOD2/CARD15* locus and a search for other mutations at this locus identified a further 27 variants (Lesage *et al.* 2002). Once the different mutations had been identified population studies became possible and these have shown different allele frequencies in different CD populations throughout the world. For example, the *NOD2/CARD15* mutations are absent in Japanese CD populations but frequent in European populations. Also, Jewish CD populations have a much higher prevalence of a particular allele than the three most common European mutations combined. More important, mutations at the *NOD2/CARD15* locus account for only 25% of cases of CD and determine only ileal disease (Ahmad *et al.* 2002) whereas mutations in the HLA genes (see Box 25.3) determine overall susceptibility to CD.

The *NOD2/CARD15* locus encodes a protein that is a member of a family of intracellular cytosolic proteins that have a role in response to bacterial antigens. Expression studies have shown that it is synthesized in epithelial cells in the small and large intestine. The highest levels of expression are in the specialized epithelial Paneth cells, which are located in the crypts of the small intestine. Although the function of the Paneth cells is unknown they have been shown to secrete antibacterial substances in response to bacterial cell-wall components.

The structure of the *NOD2/CARD15* protein is shown in Fig. 25.3. The C-terminal portion of the molecule consists of a leucine-rich region (LRR) that is involved in bacterial binding. Approximately

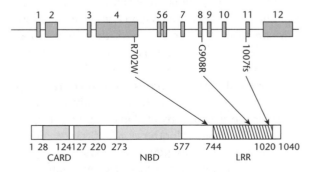

Fig. 25.3 The intron/exon structure of the *NOD2/CARD15* gene and the two-dimensional NOD2/CARD15 protein structure. CARD, caspase activating recruitment domain; NBD, nucleotide binding domain; LRR, leucine-rich region. Reproduced from Russell *et al.*, with permission from Elsevier.

93% of the mutations in *NOD2/CARD15* have been located in the LRR region and they have a diminished ability to activate nuclear factor-κβ (NF-κβ). This suggests that *NOD2/CARD15* acts to protect the intestinal population from bacterial invasion and in CD this protective mechanism malfunctions (Bonen *et al.* 2003).

The murine *NOD2/CARD15* locus has been identified and shown to function in the same way as its human equivalent, e.g. expression is induced by bacterial cell-wall components. A knockout mouse model has been developed but the mice do not develop the intestinal pathology characteristic of CD (Pauleau & Murray 2003). Rather, the knockout mice are more likely to survive bacterial challenge than wildtype mice. This unexpected result shows that the phenotype of complex traits is dependent on the total genetic background of the host in which a mutated allele sits.

The integration of global information about DNA, mRNA, and protein can be used to facilitate disease-gene identification

In the study on Crohn's disease cited in the previous section, the 'omics techniques were used to better understand the biochemical basis of the disease after the gene had been identified. However, in a completely different approach, Mootha *et al.* (2003) used the 'omics to identify the gene that is defective in patients with a particular disease (French-Canadian type Leigh syndrome, LSFC). At the outset it was known that the disease is caused by a deficiency in cytochrome oxidase even though patients do not have mutations in genes for the structural subunits or assembly factors. Genome-wide association studies had shown that the gene maps to chromosome 2p16-21. Using the tools of bioinformatics, 30 genes were identified in the candidate region and there was strong experimental evidence for 15 of them (Fig. 25.4). However, no connection was known between any of these 15 genes and mitochondrial biology.

Functionally related genes tend to be transcriptionally coregulated and this certainly is true in yeast for those genes involved in oxidative phosphorylation. Therefore, Mootha *et al.* (2003) decided to systematically identify genes that exhibited expression patterns resembling those of known mitochondrial genes. The rationale was that any gene that is coregulated with mitochondrial genes might encode polypeptides targeted to this organelle. An examination of the data in publicly available microarray

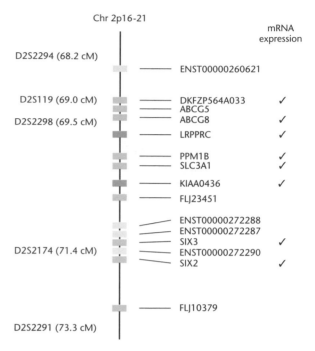

Fig. 25.4 Physical map of the LSFC candidate region. Microsatellite markers and genetic distances are shown to the left of the chromosome map. Genes with varying levels of annotation support are shown with different colors (RefSeq gene, blue; Ensembl gene, green; human mRNA, orange). An additional 15 computationally predicted genes lie within this region but are not shown. Genes represented in mRNA expression sets are indicated with a check to the right of the gene names. Reproduced from Mootha *et al.* (2003). Copyright (2003) National Academy of Sciences, USA.

databases showed that just one of the 15 genes (*LRPPRC*) had an expression pattern similar to other mitochondrial proteins.

If the *LRPPRC* gene product is a mitochondrial protein then it should be found in intact mitochondria. Since nothing was known about the gene product, Mootha *et al.* (2003) adopted a novel proteomics approach. They purified human mitochondria, extracted the proteins from them, digested them with trypsin, and analyzed them by tandem mass spectrometry (MS/MS). All the peptides that were identified then were checked to see if they could have been encoded by any of the 15 genes in the candidate region. A total of 12 peptides matched with sequences in the *LRPPRC* gene and no matches were found with any other genes in the LSFC candidate region.

The above result strongly suggested that defects in the *LRPPRC* gene are responsible for LSFC. To confirm that this is indeed the case, all the *LRPPRC* exons were sequenced in DNA from LSFC patients and normal controls. With one exception, DNA from all the patients was found to have a single base change

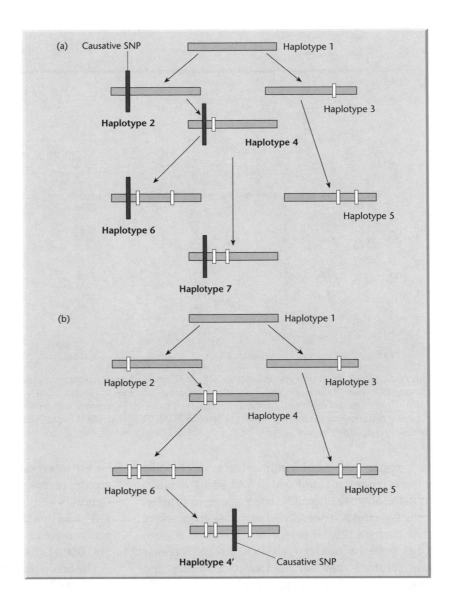

Fig. 25.5 The two ways in which a causative single nucleotide polymorphism (SNP) can become associated with a particular haplotype. In (a) the causative SNP arises early, whereas in (b) it arises late. (Redrawn with permission from Judson *et al.* 2000.)

in exon 9 that would cause a missense change in an amino acid that is conserved in humans, mice, rats, and *Fugu*. One patient was found to be heterozygous for the exon 9 mutation but also was heterozygous for a deletion in exon 35. These results provide definitive genetic proof that *LRPPRC* is the gene that is defective in LSFC and clearly show how the 'omics can be used in gene identification.

The existence of haplotype blocks should simplify linkage disequilibrium analysis

The pattern of SNPs in a stretch of DNA is known as the haplotype. Figure 25.5 shows the evolution of a number of theoretical haplotypes, some of which include an SNP causing disease. From this figure, it is clear that not all SNPs would be predictive of the

disease. Also, not all haplotypes are informative and their detection in an LD association study would complicate interpretation of the data. This is exactly the situation encountered in the two studies on Crohn's disease described above. To understand haplotype structure better, Daly *et al.* (2001) undertook a detailed analysis of 103 SNPs within the 500-kb region on chromosome 5q31 associated with Crohn's disease. Their results showed a picture of discrete haplotype blocks of tens to hundreds of kilobases, each with limited diversity punctuated by apparent sites of recombination (Fig. 25.6). In a corresponding study, Johnson *et al.* (2001) genotyped 122 SNPs in nine genes from 384 individuals and found a limited number of haplotypes (Fig. 25.7).

The existence of haplotype blocks should greatly simplify LD analysis. Rather than using all the SNPs

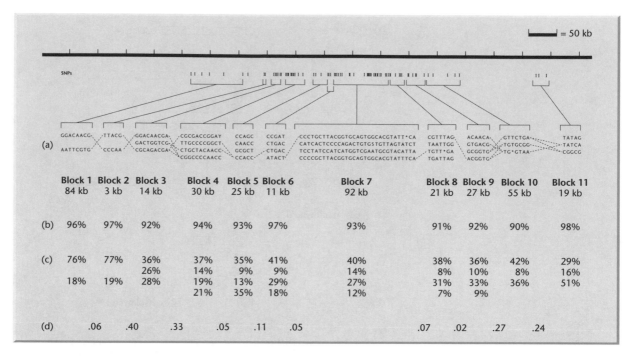

Fig. 25.6 Block-like haplotype diversity at 5q31. (a) Common haplotype patterns in each block of low diversity. Dashed lines indicate locations where more than 2% of all chromosomes are observed to transition from one common haplotype to a different one. (b) Percentage of observed chromosomes that match one of the common patterns exactly. (c) Percentage of each of the common patterns among untransmitted chromosomes. (d) Estimated rate of haplotype exchange between the block. (Reprinted from Daly *et al.* 2001 by permission of Nature Publishing Group, New York.)

in a region, we can identify exactly which SNPs will be redundant and which will be informative in association studies. The latter are referred to as haplotype tag SNPs (htSNPs) and they are markers that capture the haplotype of a genomic region. Thus, once the haplotype blocks in any given region are identified they can be treated as alleles and tested for LD. This not only simplifies the analysis but it reduces the number of SNPs that need to be genotyped. With this in mind, the US National Institutes of Health has funded a haplotype mapping ("HapMap") project whose aim is to catalog the common haplotype blocks in multiple human populations (Couzin 2002).

Although the concept of the HapMap project is very simple, reducing it to practice may be more complex (Cardon & Abecasis 2003). In the first detailed genome-wide analysis of haplotype blocks, Gabriel *et al.* (2002) focused on SNPs with minor allele frequencies >10% in 51 genomic regions. Their samples included individuals from a number of distinct ethnic groups. Two significant observations were made. First, haplotype blocks can be detected with relatively few markers and that within each block, three to five haplotypes can account for 90% of all chromosomes in the population. Second, the haplotype blocks are shorter in populations of African

ancestry (average 11 kb) than in the other ethnic groups (average 22 kb). Based on these results, Gabriel *et al.* (2002) proposed that through careful SNP selection all the common haplotypes in the genome could be identified with no more than 300,000–1,000,000 SNPs.

The above proposal is dependent on block boundaries and haplotype diversity remaining relatively stable as more markers are examined. However, a study of chromosome 19 indicated that only one-third of the chromosome exhibited a block-like structure (Phillips *et al.* 2003). Also, most investigations of haplotype blocks have used a small set of SNPs that are not representative of the variants in the population. Rather, common alleles are over-represented and rare alleles are under-represented. This frequency bias is unlikely to be a problem if common diseases are caused by common variants (Pritchard & Cox 2002) but it will limit the utility of the method for diseases caused by rarer alleles. Finally, the size of the haplotype blocks is influenced by marker density: denser marker panels have identified more short blocks whereas sparser marker panels have identified a smaller number of blocks of greater length. Despite these limitaions, the existence of haplotypes undoubtedly will facilitate the identification of disease genes.

CFLAR

C/T	A/T	A/G*	T/–	G/T	G/A	Freq.
C	A	A	T	G	G	46.00%
.	.	G	.	T	A	44.25%
T	.	G	.	T	A	8.75%
.	T	G	.	.	A	0.50%

CASP10

C/T	C/T	A/G	A/G	C/T	G/A	A/G	T/A	G/A	G/C	G/A	Freq.
C	C	A	A	C	G	A	T	G	G	G	44.00%
T	.	G	.	.	A	.	.	.	C	A	39.00%
T	.	G	.	.	A	.	A	.	C	A	7.00%
T	.	G	.	.	.	G	A	.	C	A	6.25%
.	T	G	G	T	.	.	.	A	.	A	1.75%
.	T	G	G	.	.	.	A	.	C	A	0.50%

GAD2

A/G	C/T	C/A	A/G	A/G	C/A	G/A	A/G	C/A/T	T/G	C/T	G/C*	T/A*	Freq.
A	C	C	A	G	C	G	A	C	T	C	G	T	45.00%
.	T	A	G	A	.	.	.	T	.	.	.	A	28.00%
G	.	.	.	.	.	G	A	G	T	C	.	.	12.75%
.	.	.	.	.	A	.	T	.	.	.	.	A	8.25%
.	.	.	.	.	.	.	.	T	.	.	.	A	1.75%
.	T	A	G	A	.	.	.	T	.	T	C	.	1.00%
.	.	.	.	.	A	.	T	.	.	.	.	.	0.75%
													0.50%

H19

G/C	G/C	G/T	T/C	C/T	C/T	A/T	A/G	C/G	G/C	G/T	G/A	G/A	Freq.
G	G	G	T	C	C	A	A	C	G	G	G	G	34.75%
C	.	T	C	T	.	T	G	G	C	T	A	A	19.00%
.	C	.	C	T	.	T	G	G	C	T	A	A	15.00%
.	.	.	.	.	.	.	.	.	.	.	A	A	10.25%
.	.	T	C	T	.	T	G	G	C	T	A	A	5.75%
.	.	.	.	.	T	.	.	.	.	.	.	.	4.50%
.	C	.	C	T	.	T	G	.	C	T	A	A	1.00%
.	.	.	C	T	.	T	G	G	C	T	A	A	1.00%
.	.	.	.	.	.	.	G	C	T	A	A	.	1.00%
C	.	T	C	T	.	T	G	G	C	T	A	.	0.50%
.	C	.	C	T	.	T	G	.	C	T	A	.	0.50%
.	C	.	C	T	.	T	G	G	C	T	.	A	0.50%
.	.	.	.	.	.	.	.	.	C	.	.	.	0.50%
.	.	.	.	.	.	.	G	C	.	.	A	A	0.50%

INS

A/C	C/T	A/T	C/G	C/T	C/T	C/A	C/T	G/T	G/A	G/A	C/A	C/T	G/A	Freq.
A	C	A	C	C	C	C	C	G	G	G	C	C	G	45.00%
.	.	.	.	.	.	.	.	.	.	.	A	.	.	20.00%
C	T	T	G	.	T	A	T	T	A	.	.	.	.	13.25%
.	.	.	.	.	.	.	.	.	.	.	.	.	A	11.25%
C	.	T	.	T	.	A	.	.	.	A	A	.	.	3.75%
.	.	.	.	.	.	.	.	.	.	.	.	T	.	3.50%
C	.	.	.	.	.	.	.	.	.	.	.	.	.	1.50%
C	.	T	.	T	.	A	.	.	.	.	.	.	.	0.50%
.	T	T	G	.	T	A	T	T	A	.	.	.	.	0.50%

TCF8

G/A	A/G	C/T	T/C	T/C	T/G	C/G	T/C	A/G	T/C	A/G	G/A	T/C	A/G	Freq.
G	A	C	T	T	T	C	T	A	T	A	G	T	A	33.50%
.	.	T	.	C	.	.	.	.	.	G	.	.	.	13.75%
.	.	T	C	C	.	.	.	.	.	.	.	.	.	13.25%
.	.	T	C	C	.	.	C	.	.	.	.	C	.	8.25%
.	.	T	.	C	.	.	.	.	.	.	.	.	.	8.00%
.	.	T	C	C	.	.	.	.	.	.	.	C	.	5.25%
.	.	.	.	C	.	G	.	.	.	A	.	.	.	4.50%
A	.	.	.	.	.	.	.	.	.	.	.	.	.	3.75%
.	G	T	.	C	.	.	.	.	.	G	.	.	.	2.25%
.	.	.	.	.	.	.	.	G	.	.	.	.	.	1.75%
.	.	T	C	C	G	.	.	.	.	.	.	C	.	1.75%
.	G	T	T	C	.	.	.	C	G	.	.	.	.	1.25%
.	.	.	.	.	.	.	.	.	.	.	.	.	G	0.75%
.	.	.	.	.	C	.	.	.	.	.	.	.	.	0.75%
.	.	T	.	C	.	.	.	.	.	.	.	.	.	0.75%

CASP8

T/G	T/C	G/A	G/T	C/G*	G/A	G/A	C/G	C/T	G/C	A/G	A/G	C/A	Freq.
T	T	G	G	C	G	G	C	C	G	A	A	C	39.00%
.	.	A	.	.	A	A	G	.	.	.	.	A	18.25%
.	.	.	.	G	.	.	.	.	.	.	G	.	12.75%
.	C	.	.	.	.	.	.	.	.	.	.	.	9.25%
.	.	A	.	.	A	.	G	.	.	.	.	A	6.50%
.	.	A	.	.	A	.	G	.	.	G	.	A	3.75%
.	.	.	.	.	.	.	.	.	.	.	.	A	2.75%
G	.	A	T	.	.	.	.	.	.	.	.	A	1.75%
.	.	.	.	.	.	.	.	G	T	C	.	A	1.75%
.	.	.	.	.	.	.	.	G	T	.	.	A	1.25%
.	.	A	.	.	A	.	G	.	.	.	.	A	1.00%
G	.	A	T	.	.	.	.	G	.	.	.	A	0.75%

SDF1

G/A	A/G	C/T	G/C	G/A	G/A	C/T	C/T	A/G	G/C	A/T	T/C	+/–	T/C	G/A	+/–*	T/C*	G/A	C/T*	T/C*	C/T	G/G*	Freq.
G	A	C	G	G	G	C	C	A	G	A	T	+	T	G	+	T	G	C	T	C	C	30.75%
.	.	.	.	.	.	.	.	.	.	.	.	.	.	.	–	.	A	T	.	.	.	17.50%
.	.	.	.	.	.	T	.	.	C	.	.	.	.	.	A	–	.	T	.	.	G	15.00%
A	G	.	.	A	.	.	T	G	.	T	C	.	C	.	–	.	T	.	.	.	G	14.25%
.	G	.	C	.	A	.	T	.	.	.	.	.	.	.	–	.	.	T	.	.	.	10.00%
.	.	T	.	.	.	.	.	.	.	.	.	.	.	.	–	.	.	T	.	.	.	5.50%
.	.	.	.	.	.	.	.	.	.	.	.	.	.	.	–	.	.	T	.	.	.	1.50%
.	.	.	.	.	.	.	.	.	.	.	.	.	.	.	.	.	.	C	T	G	.	1.50%
A	G	.	.	A	.	.	T	G	.	T	C	.	.	.	–	.	C	.	T	.	G	0.75%
A	G	.	.	A	.	.	T	G	.	T	C	.	C	.	–	.	C	A	T	.	.	0.50%

Fig. 25.7 Common European haplotypes and their haplotype tag single nucleotide polymorphisms (htSNPs) observed at nine genes. Boxed SNPs represent the htSNPs that can capture the common haplotypes that are segregating in European populations. Dots represent the allele that is found on the most common haplotype. Asterisks indicate SNPs described in dbSNP. (Reprinted from Johnson *et al.* 2001 by permission of Nature Publishing Group, New York.)

Investigating quantitative trait loci (QTLs) in inbred populations

Particular kinds of genetic cross are necessary if QTLs are to be mapped

The basis of all QTL detection, regardless of the species to which it is applied, is the identification of association between genetically determined phenotypes and specific genetic markers. However, there is a special problem in mapping QTLs and other complex trait genes and that is penetrance; i.e. the degree to which the transmission of a gene results in the expression of a trait. For a single gene trait, biological or environmental limitation accounts for penetrance, but in a multigenic trait the genetic context is important. Hence, the consequences of inheriting one gene rely heavily on the co-inheritance of others. For this reason careful thought needs to be given to the analysis methodology.

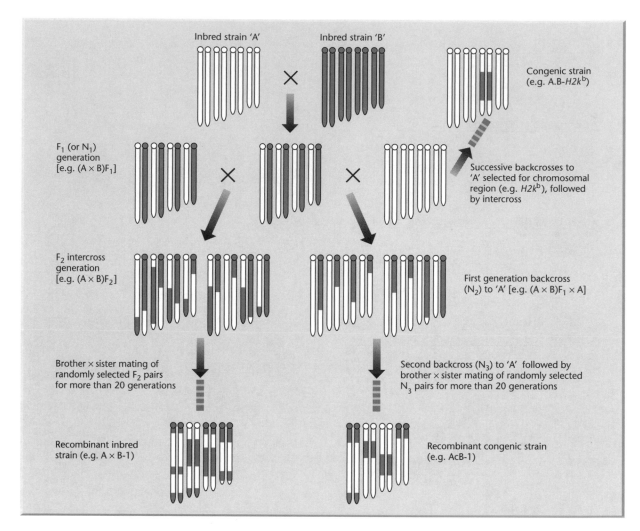

Fig. 25.8 Crosses used in the analysis of complex traits. (Adapted from Frankel 1995 by permission of Elsevier Science.)

In inbreeding populations, i.e. crops and laboratory and domestic animals, the kinds of cross used to dissect QTLs are shown in Fig. 25.8. Populations are typically generated by crosses between inbred strains, usually the first generation backcross (N_2) or intercross (F_2). Higher generation crosses (N_3), panels of recombinant inbred lines (RIL), recombinant congenic strains (RC), or inbred strains themselves may also be used. The chromosomal content in these panels is the heart of the study. They define which alleles are inherited in individuals, so that chromosomal associations can be made, and provide genetic recombination information so that location within a chromosome can be deduced. F_1 hybrids are genetically identical to each other but individuals in subsequent generations are not. Backcross progeny reveal recombination events on only one homolog, the one inherited from the F_1 parent, but intercross progeny reveal such events on both homologs. RILs and RC

strains also harbor recombinations but, unlike backcross and intercross progeny, these are homozygous at all loci as a result of inbreeding. A congenic strain of animal or near isogenic line (NIL) of a plant has only one chromosomal region that distinguishes it from one of its parents. Because they have an unchanging genotype, RILs and congenic strains/NILs offer an elegant way of discriminating between the role of the environment and of genetic factors in the expression of phenotype.

Identifying QTLs involves two challenging steps

The traditional approach for QTL analysis occurs in two stages and both are very resource intensive. The first step consists of a large cross between at least two strains in which hundreds or thousands of progeny are assayed for relevant phenotypes (Fig. 25.9) and

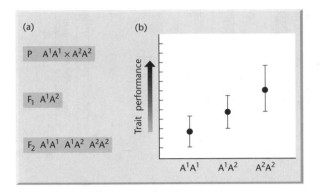

Fig. 25.9 Relationship between the performance for a quantitative trait and genotypes at a marker locus for an F_2 interbreeding population. (a) Genetic composition of the F_2 population sampled. (b) Quantitative performance of different types of F_2 genotypes.

genotyped for polymorphic markers spanning the genome. Because such crosses involve the simultaneous segregation of multiple QTLs only those of large effect are likely to be detected. The second step involves molecular identification of the genetic variants that are responsible for each QTL. This step typically involves studying individual QTLs in isolation by performing 5–10 generations of backcrosses. The objective is to construct congenic strains/NILs with chromosomal segments carrying alternative alleles of the QTL on an otherwise isogenic background. The congenic strains/NILs then are interbred to facilitate fine-structure mapping. A classical example of this approach is provided by studies on sugar content and fruit shape of tomatoes.

One of the major objectives of tomato breeding is to increase the content of total soluble solids (TSS or brix; mainly sugars and amino acids) in fruits to improve taste and processing qualities. TSS in fruits of wild tomatoes (*Lycopersicon pennelli*) can reach up to 15% of the fruit's weight, three times higher than is seen in cultivated tomatoes (*L. esculentum*). To resolve the genetic basis for this variation, a set of 50 NILs was developed from a cross between *L. pennelli* and *L. esculentum*. Each of the NILs contained a single RFLP-defined *L. pennelli* chromosome segment and collectively they covered the genome. Using these NILs it was possible to identify 23 QTLs that increase brix (i.e. polygenic inheritance controls brix). One of these QTLs (Brix9-2-5) was located on a particular NIL that was defined by a 9 cM segment on chromosome 9. When plants of *L. esculentum*, the NIL homozygous for the *L. pennelli* allele of Brix9-2-5 and their F_1 hybrids were compared over a three-

year period, the only trait associated with the NIL was brix (Fridman *et al.* 2000).

To map the Brix9-2-5 QTL, 7000 F_2 progeny of the NIL hybrid (Fig. 25.10) were subjected to RFLP analysis with two markers (CP44 and TG225) and this identified 145 recombinants. Of these recombinants, 28 were localized to the region between the two ends of BAC91A4 and they could be subdivided into six recombination groups. When the brix content of the recombination groups was tested the Brix9-2-5 QTL was found to be associated with an 18 kb segment of the chromosome. This 18-kb region was sequenced and various primer pairs used to amplify different regions. One of these amplicons, about 1 kb in length, was found to cosegregate with the QTL. Further mapping reduced the QTL to a 484 bp region of chromosome 9 and sequence analysis identified this as part of an invertase expressed exclusively in flowers and fruits of tomatoes. Initially the biochemical changes resulting from the Brix9-2-5 allele were not clear but an extension of the QTL analysis to five different tomato species localized the polymorphism to an amino acid change near the catalytic site (Fridman *et al.* 2004).

Two key morphological changes that accompanied tomato domestication were fruit size and shape. Wild tomatoes (*L. pennelli*) have small (~2 g), round berries whereas commercial cultivars have fruit weighing 50–1000 g that comes in a variety of shapes (round, oval, pear-shaped, etc). Genetic crosses between wild and cultivated tomatoes, like those described above, have shown that most of the variation in size and shape is due to fewer than 30 QTLs. One of the key QTLs, fw2.2, accounts for about 30% of the variance in fruit weight and was the first plant QTL to be mapped and cloned (Alpert & Tanksley 1996). The fw2.2 polymorphism has been localized to a gene encoding a 22 kD protein (ORFX) with similarities to a human oncogene (Frary *et al.* 2000). Comparative sequencing of fw2.2 alleles suggested that the modulation of fruit size was attributable to 5′ regulatory variation among the alleles rather than to differences in the structural protein. Further support for this idea was provided by an analysis of transcription in NILs (Cong *et al.* 2002). Large- and small-fruited tomatoes differed in peak transcript levels by one week and this was associated with changes in mitotic activity in the early stages of fruit development. Finally, an artificial gene-dosage series was constructed by generating transgenic plants with 0, 1, 2, 3, or 4 copies of the small-fruited allele of fw2.2 (Liu *et al.* 2003). Analysis of a variety of

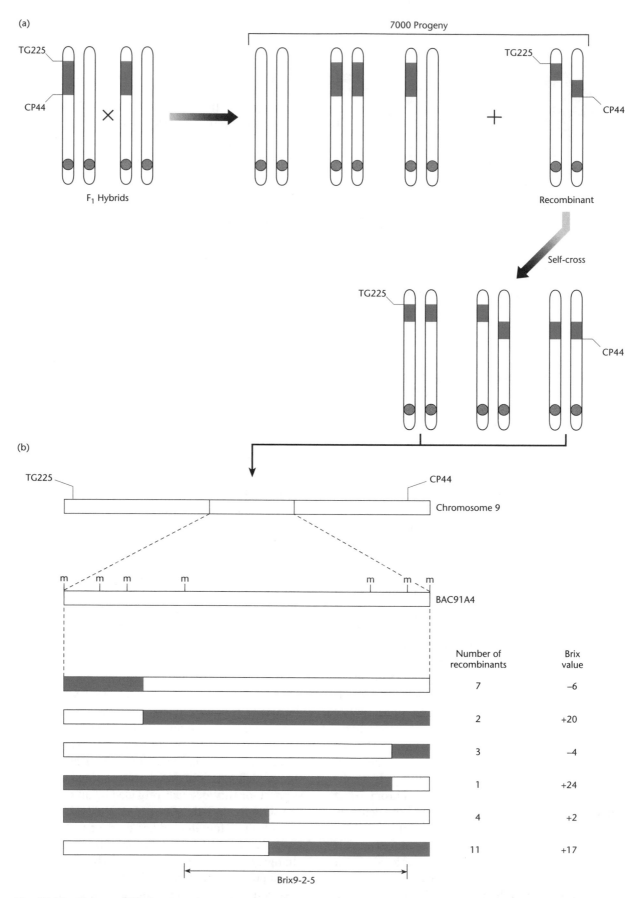

Fig. 25.10 Mapping of the Brix9-2-5 locus. (a) F_1 hybrids are self-crossed and the 7000 progeny tested with RFLP markers CP44 and TG225 to detect those in which recombination has occurred in the region derived from *L. pennelli* (shown in blue). The recombinants are crossed and selection made for homozygous recombinants. (b) The 28 recombinants where crossing over occurred within the region present in BAC91A4 were subjected to fine-scale mapping to identify the crossover more precisely and were tested for brix content.

Table 25.1 Cloned QTLs in plants (adapted from Paran & Zamir 2003).

Plant	QTL	Phenotype	Gene	Variation
Arabidopsis	*EDI*	Flowering time	Cryptochrome photoreceptor	Altered protein function
Arabidopsis	*PHYA*	Hypocotyl elongation	Phytochrome A	Altered protein function
Rice	*Hd1*	Flowering time	Transcription factor	Loss of function
Rice	*Hd6*	Flowering time	Protein kinase	Loss of function
Rice	*Hd3a*	Flowering time	FLOWERING LOCUS T	Unknown
Maize	*tb1*	Plant architecture	Transcription factor	Expression level
Maize	*Dwarf8*	Flowering time	Transcription factor	Unknown
Tomato	*Brix9-2-5*	Sugar content	Invertase	Altered protein function
Tomato	*fw2.2*	Fruit weight	Regulatory gene	Expression level
Tomato	*Ovate*	Fruit shape	Regulatory gene	Loss of function

agronomic factors showed a clear correlation between fruit size and gene dosage and that fw2.2 controls fruit growth in a tissue-specific manner.

Various factors influence the ability to isolate QTLs

At the time of going to press only a very limited number of QTLs had been mapped and sequenced, and Table 25.1 shows the list for plants. Examples from the animal world are given by Glazier *et al.* (2002). In the tomato examples described above, isolation was greatly facilitated by the fact that the selected QTL had a major effect. If the traits being studied were due to a number of combinations of several genes, each of much smaller effect, then analysis would have been much more difficult. Another factor that governs the success of searches for QTLs is the availability of physical markers. Fine-scale mapping was a key part of the identification of the tomato QTLs and such mapping would not be possible without markers such as SNPs, RFLPs, etc.

Fortunately, genome mapping and sequencing projects are in progress for a significant number of crop and domestic animal species (Box 25.4). Finally, from the details of the isolation of the Brix9-2-5 QTL presented above it should be clear that finding QTLs puts a heavy demand on resources. One way of minimizing this problem is to use chromosome substitution strains (Singer *et al.* 2004).

Chromosome substitution strains make the identification of QTLs easier

Chromosome substitution strains (CSSs) are produced as panels from crosses between a donor strain (A)

and a host strain (B) such that strain CSS-*i* carries both copies of chromosome *i* from the donor strain but all other chromosomes from the host. Thus, for an organism with four chromosomes the panel would consist of: 1A1A2B2B3B3B4B4B, 1B1B2A2A3B3B4B4B, 1B1B2B2B3A3A4B4B, 1B1B2B2B3B3B4A4A. There is a significant difference between CSSs and congenic strains/NILs. The former have an entire chromosome substitution whereas the latter have only a portion of the chromosome substituted.

Construction of CSS panels is conceptually simple and occurs in two steps. In the first, AB F$_1$ hybrids are repcatedly backcrossed to the host strain (B) and at each cycle selection is made for progeny carrying a non-recombinant copy of the desired chromosome. Once strains are identified that are heterosomic (A/B) for the desired chromosome on an otherwise isogenic host (B/B) background they are intercrossed to produce progeny homosomic for A/A. In practice, constructing the panel is dependent on detailed physical maps and in the case of the mouse took seven years and involved analysis of 17,000 mice (Singer *et al.* 2004). However, once constructed, a CSS panel is a valuable resource for studying the genetic control of phenotypic variation.

The CSS mouse panel of Singer *et al.* (2004) consists of 22 strains, one for each of the 19 autosomes, the two sex chromosomes, and the mitochondria. This panel of strains was compared with the host and donor strains in terms of sterol levels, amino acid levels, diet-induced obesity, and anxiety. A difference in any of these parameters between the host and donor strains that is seen in one or more of the CSS strains is probably due to a QTL. Overall, evidence was found for 150 QTLs. More convincing proof of the existence of particular QTLs would be their

Box 25.4   Internet resources for domestic animal and crop plant genome projects

The following websites provide current information on the progress of mapping and sequencing the genomes of a variety of domesticated species. Many of the sites also provide links to further resources, including gene mutation and sequence databases and functional analysis tools.

Domestic animals

A useful overview is the document "Coordination of Programs on Domestic Animal Genomics: the Federal Framework" which can be found at www.csrees.usda.gov/nea/animals/pdfs/nstc_progress_rpt.pdf

Dog

http://mendel.berkeley.edu/dog.html Dog Genome Project (University of California, Berkeley, University of Oregon, Fred Hutchinson Cancer Research Center)

Cow

http://www.marc.usda.gov/genome/cattle/cattle.html Meat Animal Research Center

Pig

http://www.marc.usda.gov/genome/swine/swine.html Meat Animal Research Center
http://www.projects.roslin.ac.uk/pigmap/pigmap.html Pig genome resources at the Roslin Institute, including links to the PigMAP linkage map and PIGBASE pig genome database

Sheep

http://www.marc.usda.gov/genome/sheep/sheep.html Meat Animal Research Center
http://www.projects.roslin.ac.uk/sheepmap/front.html UK Sheep Genome Mapping Project, Roslin Institute

Horse

http://www.vgl.ucdavis.edu/~lvmillon/ Veterinary Genetics Laboratory, UC Davis

Chicken

http://www.ri.bbsrc.ac.uk/chickmap/ChickMapHomePage.html ChickMaP project, Roslin Institute
http://poultry.mph.msu.edu/ US Poultry Genome Project, Michigan State University

Crop plants

A useful overview of plant genome databases can be found at www.cbi.pku.edu.cn/mirror/GenomeWeb/plant-gen-db.html

Cotton

http://algodon.tamu.edu/htdocs-cotton/cottondb.html USDA Agricultural research center

Sorghum

http://algodon.tamu.edu/sorghumdb.html USDA Agricultural research center

Barley

http://ukcrop.net/barley.html Scottish Crop Research Institute

Maize

http://www.agron.missouri.edu/ Comprehensive maize genomics and functional genomics database

Wheat and oats

http://ars-genome.cornell.edu/cgi-bin/WebAce/webace?db=graingenes

mapping to specific locations on the implicated chromosome(s). When eight of the putative 150 QTLs were analyzed further it was possible to map them to the substituted chromosome and three of them were located close to known QTLs of similar phenotype. In an extension of this work, Krewson *et al.* (2004) noted that the onset of puberty in mice occurred earlier in the donor strain than in the host strain. Using the CSS panel, this phenotype was linked to chromosomes 6 and 13. When the CSS-6 and CSS-13 strains were crossed with the host strain, the F_1 generation had a timing of puberty onset that was intermediate between the two parents.

The attraction of the CSS mouse panel is that one can use it to identify different QTLs associated with a particular phenotype, e.g. anxiety, and then use the well-tried methods to identify the causative genes. Once these genes have been characterized in the mouse they can be used to probe the human sequence for orthologs. Once the human orthologs have been identified, patient pools can be screened for the presence of SNPs. That is, inbred mouse strains can be used to identify disease genes in outbreeding human populations. Similarly, other quantitative traits could be mapped in the mouse and used to identify orthologs in domestic animals.

The level of gene expression can influence the phenotype of a QTL

Reference to Table 25.1 shows that mutations in coding regions that result in alterations in protein function are responsible for some QTLs. However, other QTLs arise from changes in gene expression. Generally speaking, it is much more difficult to identify such changes because they occur in DNA regions outside the coding region and sometimes at some distance from it. Furthermore, little is known about how variation in DNA sequences might affect the baseline level of gene expression among individuals. The first attempt at quantifying this variation has been undertaken by Morley *et al.* (2004). Using microarrays, they ascertained the baseline expression level of ~8500 active genes in immortalized cells from 14 CEPH families (see p. 362). Then they focused on a subset of genes (~3500) with more variation in expression level between unrelated individuals than between replicate samples within individuals. Attempts then were made to link these genes to chromosomal locations and significant linkage was found for 984 of them. Both *cis*- and *trans*-acting loci were identified although most operate *in trans*.

The significance of the work of Morley *et al.* (2004) is that the level of gene expression is a trait like many others and now is amenable to genetic analysis. Analyzing typical quantitative traits in humans (e.g. blood pressure, levels of serum metabolites) has proved extremely difficult in the past but dissecting the genetic contribution now should be possible.

Understanding responses to drugs (pharmacogenomics)

The perfect therapeutic drug is one that effectively treats a disease and is free of unwanted side-effects. Over the past 25 years many important new classes of drugs have been launched. However, even the most successful and effective of these provide optimal therapy only to a subset of those treated. Some individuals with a particular disease may receive little or no benefit from a drug while others may experience drug-related adverse effects. Such individual variations in response to a drug are responsible for the high failure rates of new drug candidates at the clinical trials stage.

Pharmacogenomics is the study of the association between genomic, genetic, and proteomic data on the one hand and drug response patterns on the other. The objective is to explain inter-patient variability in drug response and to predict the likely response in individuals receiving a particular medicine. As such, pharmacogenomics has the potential to influence the way approved medicines are used as well as have an impact on how clinical trials are designed and interpreted during the drug development process. Relevant information may be derived during the clinical trial recruitment phase, following the treatment phase, or both.

Genetic variation accounts for the different responses of individuals to drugs

There are two fundamental causes of individual responses to drugs. The first of these is variation in the structure of the target molecule. If a drug acts by blocking a particular receptor then it may be that the receptor is not identical in all individuals. A good example is the variation observed in the response of patients with acute promyelocytic leukemia (APL) to all-*trans*-retinoic acid (ATRA). Some patients who contract APL do so because of a balanced translocation between chromosomes 15 and 17 that results in the formation of a chimeric PML-RARα receptor gene. Other patients have a translocation between chromosomes 11 and 17 that results in the formation of a chimeric PLZF-RARα receptor. These chimeras are believed to cause APL by interference with RAR function. Clearly, the RAR receptor is different in the two types of APL and this is reflected in the fact that only the first type responds to ATRA (He *et al.* 1998).

Another good example of receptor variation is the polymorphism exhibited by the β$_2$-adrenergic receptor. Nine naturally occurring polymorphisms have been identified in the coding region of the receptor and one of these (Arg16Gly) has been well studied. Asthmatic patients who are homozygous for the Arg16 form of the receptor are 5.3 times more likely to respond to albuterol (salbutamol) than those who are homozygous for the Gly16 form. Heterozygotes give an intermediate response, being 2.3 times more likely to respond than Gly16 homozygotes (Martinez *et al.* 1997). Also of interest is the ethnic variation in frequency of this polymorphism: in Caucasians, Asians, and those of Afro-Caribbean origin it is 0.61, 0.40, and 0.50, respectively.

The second cause of variation in drug response is differences in pharmacokinetics: differences in the way that a particular drug is adsorbed, distributed, metabolized, and excreted (ADME) by the body. A drug that is not absorbed or that is metabolized too quickly will not be effective. On the other hand, a drug that is poorly metabolized could accumulate and cause adverse effects (Fig. 25.11). Obviously, variations in ADME can have multiple causes and the best studied are polymorphisms in drug transport and drug metabolism.

An important polymorphism associated with drug transport is a variant in the multidrug-resistance gene *MDR-1*. The product of this gene is an ATP-dependent membrane efflux pump whose function is the export of substances from the cell, presumably to prevent accumulation of toxic substances or metabolites. A mutation in exon 26 of the *MDR-1* gene (*C3435T*) correlated with alterations in plasma levels of certain drugs. For example, individuals homozygous for this variant exhibited fourfold higher plasma levels of digoxin after a single oral dose and an increased maximum concentration on chronic dosage (Hoffmeyer *et al.* 2000). Other substrates for this transporter are a number of important drugs that have a narrow therapeutic window; i.e. there is little difference between the drug concentration that gives the desired effect and that which is toxic. Therefore, this polymorphism could have a major impact on the requirement for individual dose adjustments for carriers of this mutation. In this context it should be noted that the mutation is probably very common, given that 40% of the German population are homozygous for it.

Most drugs are altered before excretion, either by modification of functional groups or by conjugation with molecules that enhance water solubility (e.g. sugars). Most functional group modifications are mediated by the cytochrome P450 group of enzymes and the genes that encode them are highly polymorphic (for review see Ingelman-Sundberg *et al.* 1999). Over 70 allelic variants of the *CYP2D6* locus have been described and, of these, at least 15 encode non-functional gene products. Phenotypically, four different types can be recognized (Table 25.2). The significance of these phenotypes is illustrated by the response of patients to the drug nortriptyline. Most patients require 75–150 mg per day in order to reach a steady-state plasma concentration of 50–150 μg per liter. However, poor metabolizers need only 10–20 mg per day, whereas ultrarapid metabolizers need 300–500 mg per day. If the genotype or phenotype of the patient is not known then there is a significant risk of overdosing and toxicity (poor metabolizers) or underdosing (ultrarapid metabolizers). The polymorphism of the *CYP2D6* locus is particularly important because it is implicated in the metabolism of over 100 drugs in addition to nortriptyline.

Pharmacogenomics is being used by the pharmaceutical industry

The pharmaceutical industry is using genomics in two ways (Roses 2004). First, it is using genomics at an early stage in clinical trials to differentiate those

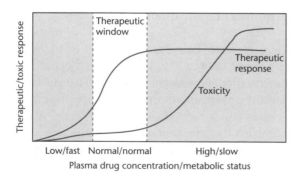

Fig. 25.11 Plasma drug concentrations in different patients after receiving the same doses of a drug metabolized by a polymorphic enzyme.

Table 25.2 The four drug metabolism phenotypes.

Phenotype	Frequency (%)	Cause
Extensive metabolizers	75–85	
Intermediate metabolizers	10–15	Both gene copies encode an enzyme with lower activity than normal
Poor metabolizers	5–10	Homozygous for two low-activity or non-functional alleles
Ultrarapid metabolizers	2–7	Amplification of the *CYP2D6* locus

individuals who respond well to a drug from those who do not respond. This information then can be used to select responders for later clinical trials. This is known as "efficacy pharmacogenomics". The second application is known as "safety pharmacogenomics". This involves identifying unique markers associated with adverse effects and then managing subsequent risk.

Roses (2004) has provided a good example of a clinical trial where efficacy pharmacogenomics is being used for patient selection. The drug in question is an anti-obesity drug. In early clinical trials

patients receiving the drug fell into three classes: non-responders, responders, and hyper-responders (Fig. 25.12). Analysis of the literature suggested 21 genes that might be responsible for the observed effect. Consequently, all the patients were screened for 112 SNPs associated with these 21 genes. One SNP was identified where non-responders were homozygous for one allele and hyper-responders were homozygous for the other allele. As expected, "normal" responders were heterozygous for the alternative alleles. A larger clinical trial is underway to determine if it is possible to exclude non-responders

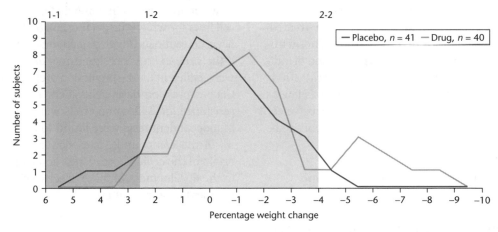

Fig. 25.12 Efficacy pharmacogenetics for an obesity drug. The graph illustrates the results of a small double-blind Phase-IIA efficacy clinical trial of a molecule intended for the treatment of obesity. Weight change in the 40 drug-treated patients (purple line) is compared with 41 placebo-treated patients (blue line) during the two-month double-blind trial. There is an obvious hyper-responder subgroup in the treated patients, with patients losing as much as 9% weight during the trial. Two SNPs from two candidate genes that are related to the proposed mechanism of action of this molecule, and one SNP from another candidate gene that is thought to be implicated in theories of obesity, segregated with the hyper-responders. Each hyper-responder was homozygous for a single allele (labeled 2-2; light-shaded section of the graph) with patients on the left side of the curves being more likely to be 1-1 homozygous (dark purple section). Heterozygous patients (1-2, light purple section) clustered in the middle. In this experiment, treated patients with the 2-2 genotype for any of the three SNPs on average lost ~3.3 kg, whereas treated patients with the 1-1 genotype gained an average of ~1.3 kg. This pattern reassures us that the molecule has efficacy and that subsequent Phase-IIB trials might be enriched by using only patients who are 1-2 heterozygous and 2-2 homozygous, with the exclusion of 1-1 homozygous patients. In addition, although the 1-1 subgroup might be less responsive to this specific treatment, this subgroup could be used in clinical trials of other obesity-drug candidates, which might subsequently allow a drug to be developed that is complementary to the first. *n*, total number of patients in study group. Reproduced from Roses *et al.*, with permission from *Nature*.

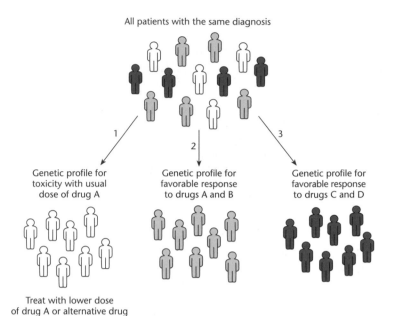

All patients with the same diagnosis

1   2   3

Genetic profile for toxicity with usual dose of drug A

Genetic profile for favorable response to drugs A and B

Genetic profile for favorable response to drugs C and D

Treat with lower dose of drug A or alternative drug

Fig. 25.13 Pharmacogenomics has the potential to subdivide a population of patients with the same empiric diagnosis (e.g. hypertension) into subgroups that have inherited differences in their metabolism of and/or sensitivity to particular drugs. One subset of the population might be at substantially greater risk of serious toxicity (1), whereas other subsets may have receptor polymorphisms or disease pathogenesis polymorphisms that make them more responsive to different treatment options (2 vs. 3).

from trials on the basis of their SNP profile. If so, it will reduce the cost of future clinical trials.

The occurrence of adverse effects to drugs is a major problem for pharmaceutical companies. If the frequency of adverse effects is high then they will be detected at an early stage in clinical development. More often, the frequency is low and significant adverse effects are not seen until post-launch, when tens of thousands of patients have been treated. Serious adverse effects might lead to product withdrawal and hence the accurate identification of individuals at risk would be invaluable. Roses (2004) has calculated that haplotype mapping of as few as 10–20 individuals exhibiting adverse effects (safety pharmacogenomics) could enable "risk" genes to be identified.

Personalized medicine involves matching genotypes to therapy

From the material presented in the early part of the chapter it should be clear that we are beginning to understand the genetic and biochemical causes of common but complex diseases. Initially, this will lead to better classification of the different subtypes of a disease and hence to better diagnoses. This in turn will facilitate selection of the most appropriate therapies. Later, when we know the exact cause of each disease, we should be able to develop drugs that will treat the cause rather than the symptoms as we do at present. Ultimately, genetic analysis of affected individuals will suggest what drugs *could* be used to

treat the disease and a second, pharmacogenomic analysis will determine which drugs *should* be used (Fig. 25.13).

A different way of using pharmacogenomic data has been suggested by a study on the chemosensitivity of different cancer cell lines. Staunton *et al.* (2001) used oligonucleotide chips to study the expression levels of 6817 genes in a panel of 60 human cancer cell lines for which the chemosensitivity profiles had been determined. Their objective was to determine if the gene-expression signatures of untreated cells were sufficient for the prediction of chemosensitivity. Gene-expression-based classifiers of sensitivity or resistance for 232 compounds were generated and in independent tests were found to be predictive for 88 of the compounds, irrespective of the tissue of origin of the cells. These results could open the door to the development of more effective chemotherapy regimes for cancer patients.

Suggested reading

Glazier A.M., Nadeau J.H. & Altman T.J. (2002) Finding genes that underlie complex traits. *Science* **298**, 2345–9.
This review provides an excellent introduction to the material covered in this chapter.

Russell R.K., Nimmo E.R. & Satsangi J. (2004) Molecular genetics of Crohn's disease. *Current Opinion in Genetics and Development* **14**, 264–70.
This paper presents a more detailed account of the material presented on Crohn's disease in this chapter.

Paran I. & Zamir D. (2003) Quantitative traits in plants: beyond the QTL. *Trends in Genetics* **19**, 303–6.
This short review describes not only the tomato work described in this chapter but the excellent work done on understanding the evolution of maize.

Barton N.H. & Keightley P.D. (2002) Understanding quantitative genetic variation. *Nature Reviews Genetics* **3**, 11–21.

Flint J., Vladar W., Shifman S. & Mott R. (2005) Strategies for mapping and cloning quantitative trait genes in rodents. *Nature Reviews Genetics* **6**, 271–86.
Two excellent reviews that describe in detail the work done in a number of classical model systems.

Singer J.B., Hill A.E., Burrage L.C., *et al.* (2004) Genetic dissection of complex traits with chromosome substitution strains of mice. *Science* **304**, 445–8.

Like many papers from Eric Lander and his group, this one is destined to become a classic.

Morley M., Moloney C.M., Weber T.M., *et al.* (2004) Genetic analysis of genome-wide variation in human gene expression. *Nature* **430**, 743–7.
This is another paper that is destined to become a classic.

Roses A.D. (2004) Pharmacogenetics and drug development: the path to safer and more effective drugs. *Nature Reviews Genetics* **5**, 645–56.
This excellent review is just one of a number of reviews on pharmacogenomics that appeared in the September 2004 issue of Nature Reviews Genetics.

Crawford D.C. & Nickerson D.A. (2005) Definition and clinical importance of haplotypes. *Annual Review of Medicine* **56**, 303–20.

CHAPTER 26

Applications of recombinant DNA technology

Introduction

Living organisms have been used by man since the dawn of history – animals and plants are used for food, to provide fiber and building materials, as a source of dyes and other chemicals, and as a source of drugs. Until the 1980s, only natural diversity could be explored although selective breeding and deliberate mutation allowed new strains of microbes and new varieties of plants and animals to be developed with improved traits. Recombinant DNA technology brought about a complete revolution in the way living organisms are exploited. By transferring new DNA sequences into microbes, plants, and animals, or by removing or altering DNA sequences in the endogenous genome, completely new strains or varieties can be created to perform specific tasks.

The first example of this approach with a real practical application was the transformation of *E. coli* with the gene for human insulin and the production of recombinant insulin in cultured bacteria. Since then, over 100 human proteins with medical applications have been expressed in bacteria, and the technology has been extended to yeast, filamentous fungi, animal and plant cells, and transgenic animals and plants. Similarly, proteins derived from human and animal pathogens have been expressed in various host species to provide a reliable source of safe vaccines. However, the production of recombinant proteins is only one example of how gene transfer to microbes, animals, and plants can be used to generate useful products. In other contexts, it is not the protein product of the transgene that is of interest, but the metabolic impact of that protein. Genes can be introduced that extend, curtail, or deviate existing metabolic pathways and direct flux toward particular small molecules, or novel genes can be introduced to create completely new metabolic pathways. This allows bacteria, plants and animal cells to produce novel chemical compounds, drugs, vitamins, specific fats and oils, and even plastics.

In other cases, the aim of genetic engineering is not to produce a valuable substance, protein or otherwise, but to alter the phenotypic properties of the host in some important way. Examples include fish expressing growth hormone to make them grow faster, and crop plants expressing proteins which confer resistance to pests and diseases, tolerance of harsh environments, or improved nutritional composition. While there have been advances in the production of farm animals with improved traits, plant biotechnology has seen the greatest benefits and genetically modified crops are now grown on 75,000,000 hectares of land around the world.

Trait modification through germline manipulation is practiced in animals and plants, but the same technology can be applied to humans in a more limited sense to prevent or cure diseases. This field of medicine, known as gene medicine, includes the concepts of DNA vaccines, using gene transfer to kill cancer cells, and introducing DNA into specific cells to repair or compensate for genetic defects. In the converse approach, animal genomes may be deliberately mutated by gene-transfer techniques to recreate the mutations seen in human inherited diseases or develop disease models that can be used to investigate how diseases occur and test novel drugs.

A comprehensive survey of all the applications of gene manipulation would take several further books (Fig. 26.1). Therefore, in this last chapter we select three topics of interest which show how the technology of gene manipulation can be applied and what benefits can be gained from the genetic modification of microbes, animals, and plants.

Theme 1: Producing useful molecules

Recombinant therapeutic proteins are produced commercially in bacteria, yeast, and mammalian cells

One of the earliest commercial applications of gene manipulation was the production of human

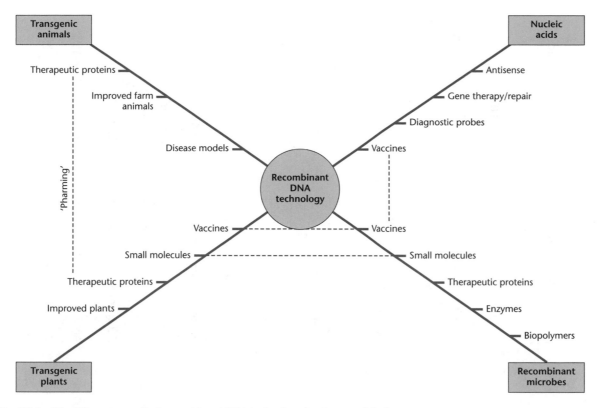

Fig. 26.1 The different ways that recombinant DNA technology has been exploited.

therapeutic proteins in bacteria. Not surprisingly, the first such products were recombinant versions of proteins already used as therapeutics: human growth hormone and insulin. Prior to the advent of genetic engineering, human growth hormone was obtained from pituitary glands removed from cadavers. Not only did this limit the supply of the hormone but in some cases resulted in recipients contracting Creutzfeld–Jacob syndrome. The recombinant approach resulted in unlimited supplies of safe material. This safety aspect has been extended to various blood components, anticoagulants, and growth factors that were originally isolated from blood but now carry the risk of HIV infection. As the methods for cloning genes became more and more sophisticated an increasing number of cytokines, inerferons, and interleukins have been identified and produced in bacteria (Table 26.1).

The first generation of protein drugs were exact copies of the human molecules but protein engineering now is being used to develop second-generation molecules with improved properties, e.g. by replacing certain amino acids or removing particular protein domains. Many of the proteins listed in Table 26.1 are modified in this manner to improve

efficacy or longevity when administered by injection. Another trend is the increasing production of novel therapeutics combining parts of different proteins. Nowhere is this more apparent than in the production of antibody-based drugs, which often combine the binding specificity of antibodies with the activity of other proteins such as cytokines and lymphokines to achieve therapeutic effects on particular target cells.

The early commercial recombinant proteins are typical of protein therapeutics produced in *E. coli* in that they are relatively simple aglycosylated proteins which accumulate as inclusion bodies within the bacterial cell and must be subjected to *in vitro* denaturation and refolding. Because of the importance of glycosylation, the addition of sugar chains to proteins during their synthesis, the *E. coli* system has been superseded in many cases by mammalian cells because the latter can produce more complex proteins and achieve correct folding and glycosylation *in vivo* (Wurm 2004). However, *E. coli* is much cheaper to cultivate than mammalian cells and more recent advances such as low-gene-dosage systems which facilitate the expression of soluble proteins (Simmons *et al.* 1996), and targeting systems which

Table 26.1 Biopharmaceuticals approved in the United States and Europe.

Product	Company	Therapeutic indication	Date approved
Recombinant blood factors			
Factor VIII			
Helixate NexGen (octocog alfa; rh Factor VIII produced in BHK cells)	Bayer	Hemophilia A	2000 (EU)
ReFacto (moroctocog-alfa; B-domain-deleted rh Factor VIII produced in CHO cells)	Genetics Institute/Wyeth Europa	Hemophilia A	1999 (EU), 2000 (US)
Kogenate (rh Factor VIII produced in BHK cells. Also sold as Helixate by Aventis Behring through a license agreement)	Bayer	Hemophilia A	1993 (US), 2000 (EU)
Bioclate (rh Factor VIII produced in CHO cells)	Aventis Behring	Hemophilia A	1993 (US)
Recombinate (rh Factor VIII produced in an animal cell line)	Baxter Healthcare/ Genetics Institute	Hemophilia A	1992 (US)
Other blood factors			
NovoSeven (rh Factor VIIa produced in BHK cells)	Novo Nordisk	Some forms of hemophilia	1996 (EU), 1999 (US)
Benefix (rh Factor IX produced in CHO cells)	Genetics Institute	Hemophilia B	1997 (US, EU)
Recombinant anticoagulants			
Tissue plasminogen activator			
Tenecteplase (also marketed as Metalyse) (TNK-tPA, modified r tPA produced in CHO cells)	Boehringer Ingelheim	Myocardial infarction	2001 (EU)
TNKase (tenecteplase; modified r tPA produced in CHO cells; see Tenecteplase entry above)	Genentech	Myocardial infarction	2000 (US)
Ecokinase (reteplase, r tPA; differs from human tPA in that 3 of its 5 domains have been deleted. Produced in *E. coli*)	Galenus Mannheim	Acute myocardial infarction	1996 (EU)
Rapilysin (reteplase, r tPA; see Ecokinase)	Roche	Acute myocardial infarction	1996 (EU)
Retavase (reteplase, r tPA; see Ecokinase)	Boehringer Mannheim/ Centocor	Acute myocardial infarction	1996 (US)
Activase (alteplase, rh tPA produced in CHO cells)	Genentech	Acute myocardial infarction	1987 (US)
Hirudin			
Refludan (anticoagulant; recombinant hirudin produced in *S. cerevisiae*)	Hoechst Marion Roussel/ Behringwerke AG	Anticoagulation therapy for heparin-associated thrombocytopenia	1997 (EU), 1998 (US)
Revasc (anticoagulant; recombinant hirudin produced in *S. cerevisiae*)	Aventis	Prevention of venous thrombosis	1997 (EU)

Recombinant hormones

Insulin

Product	Company	Indication	Date
Actrapid/Velosulin/Monotard/Insulatard/Protaphane/Mixtard/Actraphane/Ultratard (all contain rhInsulin produced in *S. cerevisiae* formulated as short/intermediate/long acting product)	Novo Nordisk	Diabetes mellitus	2002 (EU)
Novolog (insulin aspart, short-acting rh insulin analog produced in *S. cerevisiae*).	Novo Nordisk	Diabetes mellitus	2001 (US)
Novolog mix 70/30 (contains insulin aspart, short-acting rhInsulin analog as one ingredient (see also Novomix 30))	Novo Nordisk	Diabetes mellitus	2001 (US)
Novomix 30 (contains insulin aspart, short-acting rh insulin analog as one ingredient)	Novo Nordisk	Diabetes mellitus	2000 (EU)
Lantus (insulin glargine, long-acting rh insulin analog produced in *E. coli*)	Aventis	Diabetes mellitus	2000 (EU US)
Optisulin (insulin glargine, long-acting rh insulin analog produced in *E. coli*, see Lantus)	Aventis	Diabetes mellitus	2000 (EU)
NovoRapid (insulin aspart, rh insulin analog)	Novo Nordisk	Diabetes mellitus	1999 (EU)
Liprolog (Bio Lysprol, an insulin analog produced in *E. coli*)	Eli Lilly	Diabetes mellitus	1997 (EU)
Insuman (rh insulin produced in *E. coli*)	Hoechst AG	Diabetes mellitus	1997 (EU)
Humalog (insulin lispro, an insulin analog produced in *E. coli*)	Eli Lilly	Diabetes mellitus	1996 (US, EU)
Novolin (rh insulin)	Novo Nordisk	Diabetes mellitus	1991 (US)
Humulin (rh insulin produced in *E. coli*)	Eli Lilly	Diabetes mellitus	1982 (US)

Human growth hormone

Product	Company	Indication	Date
Somavert (pegvisomant; hGH analog (antagonist) produced in *E. coli*)	Pfizer	Treatment of acromegaly	2003 (US), 2002 (EU)
Nutropin AQ (rh GH produced in *E. coli*)	Schwartz Pharma	Growth failure/Turner's Syndrome	1994 (US), 2001 (EU)
Serostim (r hGH)	Serono Laboratories	Treatment of AIDS-associated catabolism/wasting	1996 (US)
Saizen (r hGH)	Serono Laboratories	hGH deficiency in children	1996 (US)
Genotropin (r hGH produced in *E. coli*)	Pharmacia & Upjohn	hGH deficiency in children	1995 (US)
Norditropin (r hGH)	Novo Nordisk	Treatment of growth failure in children due to inadequate growth hormone secretion	1995 (US)
BioTropin (r hGH)	Savient Pharmaceuticals	hGH deficiency in children	1995 (US)
Nutropin (r hGH produced in *E. coli*)	Genentech	hGH deficiency in children	1994 (US)

Table 26.1 *(cont'd)*

Product	Company	Therapeutic indication	Date approved
Humatrope (r hGH produced in *E. coli*)	Eli Lilly	hGH deficiency in children	1987 (US)
Protropin (r hGH, differs from hGH only by containing an additional N-terminal methionine residue. Produced in *E. coli*)	Genentech	hGH deficiency in children	1985 (US)
Follicle-stimulating hormone			
Follistim (follitropin-β, rh FSH produced in CHO cells)	NV Organon	Infertility	1997 (US)
Puregon (rh FSH produced in CHO cells)	NV Organon	Anovulation and superovulation	1996 (EU)
Gonal F (rh FSH produced in CHO cells)	Ares-Serono	Anovulation and superovulation	1995 (EU), 1997 (US)
Other hormones			
Forsteo (recombinant human parathyroid hormone produced in *E. coli*)	Eli Lilly	Treatment of established osteoporosis in postmenopausal women	2003 (EU)
Forteo (teriparatide; recombinant shortened human parathyroid hormone produced in *E. coli*)	Eli Lilly	Treatment of osteoporosis in some postmenopausal women	2002 (US)
Ovitrelle also termed Ovidrelle (rh choriogonadotrophin produced in CHO cells)	Serono	Used in selected assisted reproductive techniques	2000 (US), 2001 (EU)
Thyrogen (thyrotrophin-α, rh TSH produced in CHO cells)	Genzyme	Detection/treatment of thyroid cancer	1998 (US), 2000 (EU)
Luveris (lutropin alfa; rh luteinizing hormone produced in CHO cells)	Ares-Serono	Some forms of infertility	2000 (EU)
Forcaltonin (r salmon calcitonin produced in *E. coli*)	Unigene	Paget disease	1999 (EU)
Glucagon (rh glucagon produced in *S. cerevisiae*)	Novo Nordisk	Hypoglycemia	1998 (US)
Recombinant hematopoietic growth factors			
Erythropoietin			
Aranesp (darbepoetin alfa; long acting r EPO analog produced in CHO cells)	Amgen	Treatment of anemia	2001 (US, EU)
Nespo (darbepoetin alfa; see also Aranesp; long acting r EPO analog produced in CHO cells)	Dompe Biotec	Treatment of anemia	2001 (EU)
Neorecormon (rh EPO produced in CHO cells)	Roche	Treatment of anemia	1997 (EU)
Procrit (rh EPO produced in a mammalian cell line)	Ortho Biotech	Treatment of anemia	1990 (US)
Epogen (rh EPO produced in a mammalian cell line)	Amgen	Treatment of anemia	1989 (US)
Granulocyte-macrophage colony stimulating factor			
Neulasta (pegfilgrastim, r pegylated GM-CSF (filgrastim). Also marketed in EU as Neupopeg)	Amgen/Dompec Biotech	Chemotherapy induced neutropenia	2002 (US, EU)

Product	Company	Indication	Date
Leukine (r GM-CSF, differs from the native human protein by one amino acid, Leu 23. Produced in *E. coli*)	Immunex (now Amgen)	Autologous bone marrow transplantation	1991 (US)
Neupogen (filgrastim, r GM-CSF differs from human protein by containing an additional N-terminal methionine. Produced in *E. coli*)	Amgen	Chemotherapy-induced neutropenia	1991 (US)

Recombinant interferons and interleukins

Interferon-α

Product	Company	Indication	Date
Pegasys (Peginterferon α-2a, produced in *E. coli*)	Hoffman-La Roche	Hepatitis C	2002 (EU, US)
PegIntron A (PEGylated r IFN-α-2b produced in *E. coli*)	Schering-Plough	Chronic hepatitis C	2000 (EU), 2001 (US)
Viraferon (rIFN-α-2b produced in *E. coli*)	Schering-Plough	Chronic hepatitis B, C	2000 (EU)
ViraferonPeg (PEGylated rIFN-α-2b produced in *E. coli*)	Schering-Plough	Chronic hepatitis C	2000 (EU)
Alfatronol (rh IFN-α-2b produced in *E. coli*)	Schering-Plough	Hepatitis B, C and various cancers	2000 (EU)
Viraferon (rh IFN-α-2b produced in *E. coli*)	Schering-Plough	Hepatitis B, C	2000 (EU)
Intron A (r IFN-α-2b produced in *E. coli*)	Schering-Plough	Cancer, genital warts, hepatitis	1986 (US), 2000 (EU)
Alfatronol (rh IFN-α-2b produced in *E. coli*)	Schering-Plough	Hepatitis B, C and various cancers	2000 (EU)
Rebetron (combination of ribavirin and rh IFN-α-2b produced in *E. coli*)	Schering-Plough	Chronic hepatitis C	1999 (US)
Infergen (r IFN-α, synthetic type I IFN produced in *E. coli*)	Amgen/Yamanouchi Europe	Chronic hepatitis C	1997 (US), 1999 (EU)
Roferon A (rh IFN-α-2b produced in *E. coli*)	Hoffmann-La Roche	Hairy cell leukemia	1986 (US)

Interferon-β

Product	Company	Indication	Date
Rebif (rh IFN-β-1a produced in CHO cells)	Ares-Serono	Relapsing/remitting multiple sclerosis	1998 (EU), 2002 (US)
Avonex (rh IFN-β-1a produced in CHO cells)	Biogen	Relapsing multiple sclerosis	1997 (EU), 1996 (US)
Betaferon (r IFN-β-1b, differs from human protein by C17→S. Produced in *E. coli*)	Schering AG	Multiple sclerosis	1995 (EU)
Betaseron (rIFM-β-1b, differs from human protein by C17→S. Produced in *E. coli*)	Berlex Labs/Chiron	Relapsing/remitting multiple sclerosis	1993 (US)

Others

Product	Company	Indication	Date
Kineret (anakinra; r IL-1 receptor antagonist produced in *E. coli*)	Amgen	Rheumatoid arthritis	2001 (US)
Neumega (r IL-11, lacks N-terminal proline of native molecule. Produced in *E. coli*)	Genetics Institute	Prevention of chemotherapy-induced thrombocytopenia	1997 (US)
Proleukin (r IL-2, differs from human molecule in that it is devoid of an N-terminal alanine and contains C125→S substitution. Produced in *E. coli*)	Chiron	Renal cell carcinoma	1992 (US)
Actimmune (rh IFN-γ-1b produced in *E. coli*)	Genentech	Chronic granulomatous disease	1990 (US)

Table 26.1 (cont'd)

Product	Company	Therapeutic indication	Date approved
Recombinant vaccines			
Hepatitis B			
Ambirix (combination vaccine, containing r HBsAg produced in *S. cerevisiae* as one component)	GlaxoSmithKline	Immunization against hepatitis A and B	2002 (EU)
Pediarix (combination vaccine containing rHBsAg produced in *S. cerevisiae* as one component)	SmithKline Beecham	Immunization of children against various conditions inducing hepatitis B	2002 (US)
HBVAXPRO (r HBsAg produced in *S. cerevisiae*)	Aventis Pharma	Immunization of children & adolescents against hepatitis B	2001 (EU)
Twinrix (adult & pediatric forms in EU. Combination vaccine containing rHBsAg produced in *S. cerevisiae* as one component)	SmithKline Beecham (EU); GlaxoSmithKline (US)	Immunization against hepatitis A and B	1996 (EU) (adult), 1997 (EU) (pediatric), 2001 (US)
Infanrix-Hexa (combination vaccine, containing r HBsAg produced in *S. cerevisiae* as one component)	SmithKline Beecham	Immunization against diphtheria, tetanus, pertussis, *Haemophilus influenzae* type b, hepatitis B and polio	2000 (EU)
Infanrix – Penta (combination vaccine, containing rHBsAg produced in *S. cerevisiae* as one component)	SmithKline Beecham	Immunization against diphtheria, tetanus, pertussis, polio and hepatitis B	2000 (EU)
Hepacare (r S, pre-S & pre-S2 HBsAgs produced in a mammalian (murine) cell line)	Medeva Pharma	Immunization against hepatitis B	2000 (EU)
Hexavac (combination vaccine, containing rHBsAG produced in *S. cerevisiae* as one component)	Aventis Pasteur	Immunization against diphtheria, tetanus, pertussis, hepatitis B, polio and *H. influenzae* type B	2000 (EU)
Procomvax (combination vaccine, containing r HBsAg as one component)	Aventis Pasteur	Immunization against *H. influenzae* type B and hepatitis B	1999 (EU)
Primavax (combination vaccine, containing r HBsAg produced in *S. cerevisiae* as one component)	Aventis Pasteur	Immunization against diphtheria, tetanus, and hepatitis B	1998 (EU)
Infanrix Hep B (combination vaccine containing rHBsAg produced in *S. cerevisiae* as one component)	SmithKline Beecham	Immunization against diphtheria, tetanus, pertussis and hepatitis B	1997 (EU)
Twinrix (adult and pediatric forms; combination (pediatric) vaccine containing r HBsAg produced in *S. cerevisiae* as one component)	SmithKline Beecham	Immunization against hepatitis A and B	1996 (EU) (adult), 1997 (EU)
Comvax (combination vaccine, containing HbsAg produced in *S. cerevisiae*, as one component)	Merck	Vaccination of infants against *H. influenzae* type B and hepatitis B	1996 (US)

Product	Company	Indication	Year
Tritanrix-HB (combination vaccine, containing r HBsAg produced in *S. cerevisiae* as one component)	SmithKline Beecham	Vaccination against hepatitis B, diphtheria, tetanus and pertussis	1996 (EU)
Recombivax (r HBsAg produced in *S. cerevisiae*)	Merck	Hepatitis B prevention	1986 (US)

Other

Product	Company	Indication	Year
Lymerix (r OspA, a lipoprotein found on the surface of *B. burgdorferi*. Produced in *E. coli*)	SmithKline Beecham	Lyme disease vaccine	1998 (US)
Tricelluvax (combination vaccine containing r modified pertussis toxin as one component)	Chiron SpA	Immunization against diphtheria, tetanus and pertussis	1999 (EU)

Monoclonal antibody-based products

Product	Company	Indication	Year
Bexxar (tositumomab; radiolabeled monoclonal antibody directed against CD20, produced in a mammalian cell line)	Corixa/GlaxoSmithKline	Treatment of CD20 positive follicular non-Hodgkin lymphoma	2003 (US)
Xolair (Omalizumab; rIgG1kMab that binds IgE, produced in CHO cells)	Genentech	Asthma	2003 (US)
Humira (adalimumab; r human monoclonal antibody (antiTNF) created using phage display technology and produced in a mammalian cell line)	Abbott Laboratories	Rheumatoid arthritis	2002 (US)
Zevalin (Ibritumomab Tiuxetan; murine monoclonal antibody produced in a CHO cell line, targeted against the CD20 antigen. A radiotherapy agent)	IDEC Pharmaceuticals	Non-Hodgkin lymphoma	2002 (US)
Mabcampath (EU) or Campath (US) (alemtuzumab; a humanized monoclonal antibody directed against CD52 surface antigen of B-lymphocytes)	Millennium & ILEX (EU); Berlex, ILEX & Millennium Pharmaceuticals (US)	Chronic lymphocytic leukemia	2001 (EU, US)
Mylotarg (gemtuzumab zogamicin; a humanized antibody-toxic antibiotic conjugate targeted against CD33 antigen found on leukemic blast cells)	Wyeth	Acute myeloid leukemia	2000 (US)
Herceptin (trastuzumab, humanized antibody directed against human epidermal growth factor receptor 2 (HER2))	Genentech (US); Roche Registration (EU)	Treatment of metastatic breast cancer if tumor overexpresses HER2 protein	1998 (US), 2000 (EU)
Remicade (infliximab, chimeric mAb directed against TNF-α)	Centocor	Treatment of Crohn disease	1998 (US), 1999 (EU)
Synagis (palivizumab, humanized mAb directed against an epitope on the surface of respiratory syncytial virus)	MedImmune (US); Abbott (EU)	Prophylaxis of lower tract respiratory disease caused by syncytial virus in pediatric patients	1998 (US), 1999 (EU)
Zenapax (daclizumab, humanized mAb directed against the α-chain of the IL-2 receptor)	Hoffmann-La Roche	Prevention of acute kidney transplant rejection	1997 (US), 1999 (EU)
Humaspect (Votumumab, human mAb directed against cytokeratin tumor-associated antigen)	Organon Teknika	Detection of carcinoma of the colon or rectum	1998 (EU)
Mabthera (Rituximab, chimeric mAb directed against CD20 surface antigen of B lymphocytes. See also Rituxan)	Hoffmann-La Roche	Non-Hodgkin lymphoma	1998 (EU)

Table 26.1 (cont'd)

Product	Company	Therapeutic indication	Date approved
Simulect (basiliximab, chimeric mAb directed against the α-chain of the IL-2 receptor)	Novartis	Prophylaxis of acute organ rejection in allogeneic renal transplantation	1998 (EU)
LeukoScan (Sulesomab, murine mAb fragment (Fab) directed against NCA 90, a surface granulocyte nonspecific cross-reacting antigen)	Immunomedics	Diagnostic imaging for infection/inflammation in bone of patients with osteomyelitis	1997 (EU)
Rituxan (rituximab chimeric mAb directed against CD20 antigen found on the surface of B lymphocytes)	Genentech/IDEC Pharmaceuticals	Non-Hodgkin lymphoma	1997 (US)
Verluma (Nofetumomab murine mAb fragments (Fab) directed against carcinoma-associated antigen)	Boehringer Ingelheim/NeoRx	Detection of small-cell lung cancer	1996 (US)
Tecnemab KI (murine mAb fragments (Fab/Fab2 mix) directed against HMW-MAA)	Sorin	Diagnosis of cutaneous melanoma lesions	1996 (EU)
ProstaScint (capromab-pentetate murine mAb-directed against the tumor surface antigen PSMA)	Cytogen	Detection/staging/follow-up of prostate adenocarcinoma	1996 (US)
MyoScint (imiciromab-pentetate, murine mAb fragment directed against human cardiac myosin)	Centocor	Myocardial infarction imaging agent	1996 (US)
CEA-scan (arcitumomab, murine mAb fragment (Fab), directed against human carcinoembryonic antigen, CEA)	Immunomedics	Detection of recurrent/metastatic colorectal cancer	1996 (US, EU)
Indimacis 125 (Igovomab, murine mAb fragment (Fab2) directed against the tumor-associated antigen CA 125)	CIS Bio	Diagnosis of ovarian adenocarcinoma	1996 (EU)
ReoPro (abciximab, Fab fragments derived from a chimeric mAb, directed against the platelet surface receptor GPIIb/IIIa)	Centocor	Prevention of blood clots	1994 (US)
OncoScint CR/OV (satumomab pendetide, murine mAb directed against TAG-72, a tumor-associated glycoprotein)	Cytogen	Detection/staging/follow-up of colorectal and ovarian cancers	1992 (US)
Orthoclone OKT3 (Muromomab CD3, murine mAb directed against the T-lymphocyte surface antigen CD3)	Ortho Biotech	Reversal of acute kidney transplant rejection	1986 (US)
Miscellaneous recombinant products			
Bone morphogenetic proteins			
Inductos (dibotermin alfa; r Bone morphogenic protein-2 produced in CHO cells)	Genetics Institute/Wyeth Europa	Treatment of acute tibia fractures	2002 (EU)
Infuse (rh Bone morphogenic protein-2 produced in CHO cells)	Medtronic Sofamor Danek	Promotes fusion of lower spine vertebrae	2002 (US)
Osteogenic protein 1 (rh osteogenic protein 1 BMP-7, produced in CHO cells:	Howmedica (EU); Stryker (US)	Treatment of non-union of tibia	2001 (EU, US)
Galactosidase			
Fabrazyme (rh α-galactosidase produced in CHO cells)		Fabry disease (α-galactosidase A deficiency)	2001 (EU), 2003 (US)

Replagal (rh α-galactosidase produced in a continuous human cell line)	Genzyme TKT Europe	Fabry disease (α-galactosidase A deficiency)	2001 (EU)

Other

Aldurazyme (laronidase; r-alfa-L-iduronidase produced in CHO cells)	Biomarin Pharmaceuticals	Treatment of mucopolysaccharidosis	2003 (US)
Amevive (alefacept; a dimeric fusion protein consisting of the extracellular CD2-binding portion of the human leukocyte functional antigen 3 (LFA-3) linked to the Fc region of human IgG1. Produced in CHO cells)	Biogen	Treatment of adults with moderate to severe chronic plaque psoriasis	2003 (US)
Fasturtec (Elitex in US; rasburicase, rUrate oxidase produced in S. cerevisiae)	Sanofi-Synthelabo	Hyperuricemia	2001 (EU), 2002 (US)
Natrecor (nesiritide, recombinant form of human B-type natriuretic peptide (rhBNP))	Scios	Acutely decompensated congestive heart failure	Aug 2001
Xigris (drotrecogin-α; rh activated protein C produced in a mammalian (human) cell line)	Eli Lilly	Severe sepsis	2001 (US), 2002 (EU)
Enbrel (r TNF receptor–IgG fragment fusion protein produced in CHO cells)	Amgen (US); Wyeth (EU)	Rheumatoid arthritis	1998 (US), 2000 (EU)
Beromun (rh TNF-α, produced in E. coli)	Boehringer Ingelheim	Adjunct to surgery for subsequent tumor removal, to prevent or delay amputation	1999 (EU)
Ontak (r IL-2–diphtheria toxin fusion protein that targets cells displaying a surface IL-2 receptor)	Seragen/Ligand Pharmaceuticals	Cutaneous T-cell lymphoma	1999 (US)
Regranex (rh PDGF produced in S. cerevisiae)	Ortho-McNeil Pharmaceutical (US); Janssen-Cilag (EU)	Lower extremity diabetic neuropathic ulcers	1997 (US),1999 (EU)
Vitravene (fomivirsen, an antisense oligonucleotide)	ISIS pharmaceuticals	Treatment of cytomegalovirus retinitis in AIDS patients	1998 (US)
Cerezyme (r-β-glucocerebrosidase produced in E. coli. Differs from native human enzyme by one amino acid, R495H, and has modified oligosaccharide component)	Genzyme	Treatment of Gaucher disease	1994 (US)
Pulmozyme (dornase-α, r DNase produced in CHO cells)	Genentech	Cystic fibrosis	1993 (US)

Data were collected from several industry sources (http://www.fda.gov, http://www.eudra.org/en_home.htm, http://www.phrma.org).
Products are listed in chronological order, with the most recent approvals first. Where more than one drug in the same category was approved in a single year, they are listed alphabetically. Several products have been approved for multiple indications, but only the first indication for which it was approved is listed here. Products are listed by trade name. Abbreviations: r, recombinant; rh, recombinant human; CHO, Chinese hamster ovary; BHK, baby hamster kidney; mAb, mono-clonal antibody; tPA, tissue plasminogen activator; hGH, human growth hormone; FSH, follicle stimulating hormone; TSH, thyroid stimulating hormone; EPO, erythropoietin; GM-CSF, granulocyte-macrophage colony stimulating factor; IFN, interferon; IL, interleukin; HbsAg, hepatitis B surface antigen; PDGF, platelet derived growth factor; TNFR, tumor necrosis factor receptor.

direct recombinant proteins to the periplasm (Jeong & Lee 2000), have been very successful (see Baneyx & Mujacic 2004 for a review). Mammalian proteins targeted to the bacterial periplasm are more likely to fold correctly because this compartment has the ability to form and isomerize disulfide bonds. An alternative strategy, which has also been successful, is the expression of protein disulfide isomerases in the bacterial cytosol along with the recombinant protein of interest (Besette *et al.* 1999).

Yeast cells grow in a similar manner to bacterial cells and, like bacteria, require simple and relatively inexpensive media for growth. However, they are eukaryotes and are therefore able to fold and assemble complex recombinant proteins much more efficiently than bacteria. The secretion of recombinant proteins from cultured yeast cells allows the formation of disulfide bonds, proteolytic maturation, N- and O-linked glycosylation, and other post-translational modifications that occur either not at all or very inefficiently in bacteria. *Saccharomyces cerevisiae* was the first yeast used for recombinant protein production, and it is used for the commercial production of several approved drugs and vaccines (Table 26.1) including the current hepatitis B virus vaccine (Valenzuela *et al.* 1982).

Unfortunately, as a general production system for recombinant pharmaceuticals, *S. cerevisiae* suffers from a number of limitations including low product yield, poor plasmid stability, difficulties in scaling up production, the hyperglycosylation of recombinant human glycoproteins, and inefficient secretion (Romanos *et al.* 1992, Dominguez *et al.* 1998). Although improved *S. cerevisiae* strains have been described, some of which generate glycan chains compatible with humans (Chiba *et al.* 1998) other yeast species have been developed as alternative production hosts. These include another popular model organism – *Schizosaccharomyces pombe* – as well as the methylotrophic yeasts *Pichia pastoris* and *Hansenula polymorpha*, the dairy yeast *Kluyveromyces lactis*, and others such as *Schwanniomyces occidentalis* and *Yarrowia lipolytica* (Chapter 11). These organisms often outperform *S. cerevisiae* in terms of yield, reduced hyperglycosylation, and secretion efficiency. The methylotrophic yeasts in particular are now emerging as competitive production systems (Cereghino *et al.* 2002, Gerngross 2004).

Filamentous fungi have also been explored as production systems for recombinant therapeutic proteins, reflecting their high capacity for protein secretion (Punt *et al.* 2002, Gerngross 2004).

Heterologous gene expression was first achieved in filamentous fungi using the laboratory organism *Aspergillus nidulans*, but can now also be carried out in a variety of industrially important species, such as *A. niger*, *A. oryzae*, *Trichoderma reesei*, *Acremonium chrysogenum*, and *Penicillium chrysogenum*.

Despite the success of microbes in biotechnology, it is mammalian cells that have dominated the biopharmaceutical industry since the mid-1990s. This is because only mammalian cells can glycosylate human proteins in the correct manner (Andersen & Krummen 2002). There are several principal cell lines of choice: Chinese hamster ovary (CHO) cells, and the murine myeloma cell lines NS0 and SP2/0, BHK and HEK-293 cells, and the human retinal line PER-C6 – these have been used to produce most of the recombinant therapeutic products licensed by the FDA up to June 2006.

Transgenic animals and plants can also be used as bioreactors to produce recombinant proteins

The production of growth hormone in the serum of transgenic mice (Palmiter *et al.* 1982a, see p. 259) provided the first evidence that recombinant proteins could be produced, continuously, in the body fluids of animals. Five years later, several groups reported the secretion of recombinant proteins in mouse milk. In each case, this was achieved by joining the transgene to a mammary-specific promoter, such as that from the casein gene. The first proteins produced in this way were sheep β-lactoglobulin (Simons *et al.* 1987) and human tissue plasminogen activator (tPA) (Gordon *et al.* 1987, Pittius *et al.* 1988). There have been over 100 such reports since these early experiments, and a selection is listed in Table 26.2.

Although proteins can be produced at high concentrations in mouse milk (e.g. 50 ng/ml for tPA) the system is not ideal due to the small volume of milk produced. Therefore, other animals, such as sheep and goats, have been investigated as possible bioreactors. Such animals not only produce large volumes of milk, but the regulatory practices regarding the use of their milk are more acceptable. An early success was Tracy, a transgenic ewe producing extremely high levels (30 g/l) of human α1-antitrypsin in her milk (Wright *et al.* 1991). Artificially inseminated eggs were microinjected with a DNA construct containing an *AAT* gene fused to a β-lactoglobulin promoter. These eggs were implanted

Table 26.2 A selection of recombinant proteins that have been expressed in the milk of transgenic animals.

Coding sequence			Promoter region	
Gene	**Source**	**Transgenic species**	**Gene**	**Source**
α_1 antitrypsin	Mouse	Mice	WAP	Rabbit
α_1 antitrypsin	Human	Mice	β-lactoglobulin	Sheep
α_1 antitrypsin	Human	Sheep	β-lactoglobulin	Sheep
α-glucosidase	Human	Mice	α_{s1}-casein	Bovine
α-lactalbumin	Bovine	Mice	α-lactalbumin	Bovine
α-lactalbumin	Goat	Mice	α-lactalbumin	Goat
α-lactalbumin	Guinea-pig	Mice	α-lactalbumin	Guinea-pig
α-lactalbumin	Human	Rats	α-lactalbumin	Human
α_{s1}-casein	Bovine	Mice	α_{s1}-casein	Bovine
Anti-CD6 antibodies	Mouse/Human	Mice	WAP	Rabbit
Antithrombin III	Human	Goat	J-casein	Goat
β-casein	Bovine	Mice	β-casein	Bovine
β-casein	Bovine	Mice	α-lactalbumin	Bovine
β-casein	Goat	Mice	β-casein	Goat
β-casein	Rat	Mice	β-casein	Rat
β-interferon	Human	Mice	WAP	Mouse
β-lactoglobulin	Sheep	Mice	β-lactoglobulin	Sheep
γ-interferon	Human	Mice	β-lactoglobulin	Sheep
κ-casein	Bovine	Mice	β-casein	Goat
κ-casein	Goat	Mice	β-casein	Goat
CFTR	Human	Mice	β-casein	Goat
EPO	Human	Mice	β-lactoglobulin	Bovine
EPO	Human	Rabbits	β-lactoglobulin	Bovine
Factor VIII	Human	Sheep	β-lactoglobulin	Sheep
Factor IX	Human	Mice	β-lactoglobulin	Sheep
Factor IX	Human	Sheep	β-lactoglobulin	Sheep
Fibrinogen	Human	Mice	WAP	Mouse
FSH	Bovine	Mice	J-casein	Rat
GM-CSF	Human	Mice	α_{s1}-casein	Bovine
Growth homone	Bovine	Mice	WAP	Rat
Growth homone	Human	Mice	J-casein	Rat
Hepatitis B surface antigen	Human	Goat	α_{s1}-casein	Bovine
IGF-1	Human	Rabbits	α_{s1}-casein	Bovine
Interleukin-2	Human	Rabbits	β-casein	Rabbit
Lactoferrin	Human	Mice	α_{s1}-casein	Bovine
Lactoferrin	Human	Cattle	α_{s1}-casein	Bovine
Lysozyme	Human	Mice	α_{s1}-casein	Bovine
Protein C	Human	Mice	WAP	Mouse
Protein C	Human	Pigs	WAP	Mouse
Serum albumin	Human	Mice	β-lactoglobulin	Sheep
Superoxide dismutase	Human	Mice	β-lactoglobulin	Sheep
Superoxide dismutase	Human	Mice	WAP	Mouse
Surfactant protein B	Human	Mice	WAP	Rat
TAP	Human	Mice	WAP	Rat
t-PA	Human	Mice	WAP	Mouse
t-PA	Human	Mice	α_{s1}-casein	Bovine
t-PA	Human	Rabbits	α_{s1}-casein	Bovine
t-PA	Human	Goats	WAP	Mouse
Trophoblastin	Sheep	Mice	α-lactalbumin	Bovine
Urokinase	Human	Mice	α_{s1}-casein	Bovine
WAP	Mouse	Mice	WAP	Mouse
WAP	Rat	Mice	WAP	Rat
WAP	Mouse	Pigs	WAP	Mouse
WAP	Mouse	Sheep	WAP	Mouse

into surrogate mothers, of which 112 gave birth. Four females, including Tracy, and one male were found to have incorporated intact copies of the gene and all five developed normally. Over the lactation period sheep can produce 250–800 liters of milk so the production potential is significant.

Using similar protocols Ebert *et al.* (1991) have demonstrated the production of a variant of human tPA in goat milk. Of 29 offspring, one male and one female contained the transgene. The transgenic female underwent two pregnancies and one out of five offspring was transgenic. Milk collected over her first lactation contained only a few mg of tPA per liter, but improved expression constructs have since resulted in an animal generating several grams per liter of the protein. Recombinant human antithrombin III, which is used to prevent blood clots forming in patients who have undergone heart bypass operations, was the first protein expressed in transgenic animal milk to reach commercial production, and is currently marketed by Genzyme Transgenics Corporation.

The production of foreign proteins in secreted body fluids has the obvious advantage that transgenic animals can be used as a renewable source of the desirable molecule. In addition to milk, other production systems have been investigated including serum (Massoud *et al.* 1991), semen (Dyck *et al.* 1999), and urine (Kerr *et al.* 1998, Ryoo *et al.* 2001, Zbikowska *et al.* 2002). In each case, an important consideration is whether the protein is stable, and whether it folds and assembles correctly. The assembly of complex proteins comprising up to three separate polypeptides has been demonstrated in milk, e.g. fibrinogen (Prunkard *et al.* 1996), collagen (John *et al.* 1999), and various immunoglobulins (e.g. Castilla *et al.* 1998). Other abundantly secreted fluids which are likely to be exploited for recombinant protein expression in the future include the albumen of hens' eggs and silkworm cocoons. Hens' eggs would be commercially attractive but this technology is currently limited by the lack of efficient gene-transfer methods in birds (Harvey *et al.* 2002). There has also been some success with silkworms, using both microinjection (e.g. Nagaraju *et al.* 1996) and infection of silkworm larvae with baculovirus vectors (Yamao *et al.* 1999, Tamura *et al.* 1999, Tomita *et al.* 2003).

Plants are a useful alternative to animals for recombinant protein production because they are inexpensive to grow and scale up from laboratory testing to commercial production. Therefore, there is much interest in using plants as production systems for the synthesis of recombinant proteins (Twyman *et al.* 2003, 2005, Ma *et al.* 2003). There is some concern that therapeutic molecules produced in animal expression systems could be contaminated with small quantities of endogenous viruses or prions, a risk factor that is absent from plants. Furthermore, plants carry out very similar post-translational modification reactions to animal cells with only minor differences in glycosylation patterns, although there is some controversy as to the exact capabilities of plants in this context (Cabanes-Macheteau *et al.* 1999, Seveno *et al.* 2004, Shah *et al.* 2004).

A selection of therapeutic proteins that have been expressed in plants is listed in Table 26.3. The first such report was the expression of human growth hormone, as a fusion with the *Agrobacterium* nopaline synthase gene, in transgenic tobacco and sunflower tissue, although in this study only the transcript and not the protein was detected (Barta *et al.* 1986). Hiatt *et al.* (1989) reported the expression of an antibody in tobacco and this represents a large class of therapeutic molecules although the antibody Hiatt and colleagues chose to express was raised against a plant epitope and had no therapeutic applications. The first human therapeutic protein expressed in plants was human serum albumin (Sijmons *et al.* 1990). Tobacco has been the most frequently used host for recombinant protein expression, although edible crops including cereals, fruits, and vegetables are becoming more popular since recombinant proteins produced in such crops could in principle be administered orally with minimal purification. The expression of human antibodies in plants has particular relevance in this context because the consumption of plant material containing recombinant antibodies could provide passive immunity (i.e. immunity brought about without stimulating the host immune system). Many different types of antibody have been expressed in plants, predominantly tobacco, including full-size immunoglobulins, Fab fragments, and scFvs. For example, a fully humanized antibody against herpes simplex virus-2 has been expressed in soybean (Zeitlin *et al.* 1998). Even secretory IgA antibodies, which have four separate polypeptide components, have been successfully expressed in plants. This experiment involved the generation of four separate transgenic tobacco lines, each expressing a single component, and the sequential crossing of these lines to generate plants in which all four transgenes were stacked (Ma *et al.* 1995). Plants producing recombinant sIgA against the oral pathogen *Streptococcus mutans* have been

Table 26.3 A selection of therapeutic recombinant proteins that have been expressed in genetically modified plants.

(a) Biopharmaceuticals

Protein	Host plant system	Potential application
Adenosine deaminase	Maize	SCID
Angiotensin-1 converting enzyme	Tobacco and tomato (virus infected)	Hypertension
Aprotinin	Maize	Bleeding, pancreatitis
Collagen, human	Tobacco	Tissue repair
Enkephalins, human	*Arabidopsis*, canola seeds	Antihyperanalgesic
Epidermal growth factor, human	Tobacco, transgenic and transient expression	Wound repair, cell proliferation
Erythropoietin, human	Tobacco suspension cells	Anemia
Factor XIII (A-domain)	Tobacco	Bleeding
Glutamic acid decarboxylase (GAD65)	Tobacco	Diabetes
Glucocerebrosidase	Tobacco	Gaucher's disease
Growth hormone, human	Tobacco leaves (chloroplast expression), tobacco seeds	Pituitary dwarfism
Hemoglobin (α- and β-globin), human	Tobacco seeds	Blood substitute
Hirudin, leech	Canola	Anticoagulant
Insulin-like growth factor-1	Tobacco, rice	Diabetes
Interleukin-2	Potato	Antiviral, anticancer
Interleukin-4	Tobacco	
Interleukin-10	Tobacco	Antiviral, anticancer
Interleukin-12	Tobacco	Antiviral, anticancer
Interleukin-18	Tobacco	Antiviral, anticancer
Intrinsic factor	*Arabidopsis* leaves	Diagnostic, supplement
Lactoferrin	Potato tubers, rice, tobacco, viral vectors in tobacco	Antimicrobial
Lysozyme	Rice grains, rice cell culture	Antimicrobial
Pancreatic lipase	Tobacco, maize	Exocrine pancreatic deficiency
Protein C, human	Tobacco	Anticoagulant
Secreted alkaline phosphatase, human	Tobacco (secretion)	Ovarian and testicular cancer
Serum albumin, human	Tobacco and potato leaves, tobacco chloroplasts, potato tubers	Burns, liver cirrhosis
SMAP-29	Tobacco	Antimicrobial
Synthetic elastin	Tobacco, Tobacco chloroplasts	Tissue repair
α-Interferon, human	Rice, turnip	Hepatitis
α-Trichosanthin, viral	Tobacco (virus infected)	HIV

Table 26.3 (*cont'd*)

(b) Recombinant antibodies

Antigen	Antibody format	Production system
B cell lymphoma, murine 38C13	scFv	Virus vectors in tobacco leaves
Carcinoembryonic antigen	scFv, IgGl	Tobacco agroinfiltation
	dAb	Tobacco, agroinfiltration and transgenic
	scFv	Rice, rice cell cultures, wheat, pea
CD-40	scFv-fusion	Tobacco suspension cells
Colon cancer antigen	IgG	Virus vectors in tobacco leaves
Creatine kinase	IgGl, Fab	Tobacco leaves, *Arabidopsis* leaves
	scFv	Tobacco leaves
Rhesus D antigen	IgGl	*Arabidopsis* leaves
Ferritin	scFv	Tobacco leaves
Hepatitis B virus surface antigen	IgG	Tobacco leaves
Herpes simplex virus 2	IgGl	Soybean
HIV antibodies in blood	scFv-fusion	Tobacco leaves, barley grains, potato tubers
Human choriogonadotropin	scFv, dAb, IgG	Tobacco leaves
Human IgG	IgGl	Alfalfa
Interleukin-4	scFv	Tobacco roots
Interleukin-6	scFv	Tobacco roots
Streptococcal surface antigen (I/II)	sIgA	Tobacco leaves
	IgGl	Tobacco leaves
	IgGl	Secretion from tobacco roots
Substance P	VH	Tobacco leaves

generated (Ma *et al.* 1998), and these plant-derived antibodies ("plantibodies") are being commercially developed as the drug CaroRx™, marketed by Planet Biotechnology Inc., which at the time of writing is coming to the end of Phase II clinical trials.

Plants have been explored as cheap, safe, and efficient production systems for subunit vaccines (Table 26.4), with the added advantage that orally administered vaccines can be ingested by eating the plant, therefore limiting the need for processing and purification (reviewed by Mason & Arntzen 1995, Walmsey & Arntzen 2000). The earliest demonstration was the expression of a surface antigen from the bacterium *S. mutans* in tobacco. This bacterium is the cause of tooth decay, and it was envisaged that stimulation of a mucosal immune response would prevent the bacterium colonizing the teeth and therefore protect against cavities (Curtis & Cardineau, 1990 Patent Application).

A number of edible transgenic plants have been generated expressing antigens derived from animal viruses. For example, rabies glycoprotein has been expressed in tomato (McGarvey *et al.* 1995), hepatitis B virus antigen in lettuce (Ehsani *et al.* 1997), and cholera antigen in potato (Arakawa *et al.* 1997). As well as animal virus antigens, autoantigens associated with diabetes have also been produced (Ma *et al.* 1997, Porceddu *et al.* 1999). Plants have also been infected with recombinant viruses expressing various antigen epitopes on their surfaces. Cowpea mosaic virus (CMV) has been extensively developed as a heterologous antigen-presenting system (Porta *et al.* 1994, Lomonossoff & Hamilton 1999). There has been some success in vaccination trials using recombinant CMV vectors expressing epitopes of HIV gp41 (McLain *et al.* 1995, 1996a, 1996b) and canine parvovirus (Dalsgaard *et al.* 1997). The first clinical trials using a plant-derived vaccine were

Table 26.4 Vaccine candidates produced in plants.

Antigen	Production system
For animal diseases	
Bovine herpes virus (type 1) glycoprotein D	Tobacco leaves
Bovine rotavirus	Tobacco leaves, transgenic and virus infected
Canine parvovoris VP2 epitope	Arabidopsis thaliana leaves
	Tobacco leaves (chloroplast expression)
Enterotoxigenic E. coli fimbral subunit FaeG	Tobacco leaves
Foot and mouth disease virus VP1 epitope	Alfalfa leaf
	Arabadopsis thaliana leaves
	Potato tuber
Infectious bronchitis virus (IBV) S1 glycoprotein	Potato tuber
Mannheimia hemolytica A1 leukotoxin 50	White clover leaves
Porcine epidemic diarrhea virus (PEDV) spike protein	Tobacco leaves
	Potato leaves
Rabies virus glycoprotein (G)	Tomato leaves and fruit
Rabies virus glycoprotein (G) and nucleoprotein (N)	Viral vectors in tobacco and spinach leaves
Rinderpest virus hemagglutinin (H) protein	Pigeon pea leaves, tobacco leaves
Transmissible gastroenteritis coronavirus	Arabidopsis thaliana leaves
N-terminal domain of the spike glycoprotein (S)	
	Maize seeds
	Potato tubers
	Tobacco leaves
For human diseases	
Bacillus anthracis protective antigen	Tobacco leaves
Enterotoxigenic E. coli conjugal pilus, A subunit	Tobacco leaves
Enterotoxigenic E. coli heat labile toxin, B subunit	Maize seeds
	Potato tubers
	Tomato fruit and leaves
E. coli O157:H7 intimin	Tobacco leaves
Hepatitis B virus surface antigen	Cherry tomatillo leaves and fruit
	Lettuce leaves
	Lupin callus tissue
	Potato tubers
	Soybean cell suspension cultures
	Tobacco leaves
Hepatitis C virus HVR1 epitope of E2 envelope protein	Tobacco leaves
Hepatitis E virus ORF2 protein	Tomato leaves and fruit
Human cytomegalovirus glycoprotein B	Tobacco seeds
Human immunodeficiency virus (type 1) p24 protein	Virus vectors in tobacco
Human immunodeficiency virus (type 1) p120 protein	Virus vectors in tobacco
Human papillomavirus (type 11) L1 protein	Tobacco leaves, potato tuber
Human papillomavirus (type 16) E7 protein	Virus vectors in tobacco
Human papillomavirus (type 16) L1 protein	Tobacco leaves, potato tubers
Measles virus hemagglutinin	Carrot leaves and taproots
	Tobacco leaves
Norwalk virus capsid protein	Tobacco leaves, potato tubers
Plasmodium falciparum (PfMSP1)	Tobacco leaves
Respiratory syncytial virus fusion protein	Tomato fruit
Rotavirus NSP4 protein	Potato tubers
Rotavirus VP6 protein	Potato tubers
	Tomato suspension cells
	Virus vectors in tobacco leaves
Rotavirus VP7 protein	Potato tubers
Simian/Human immunodeficiency virus (SHIV) 89.6p Tat	Potato tubers
Tetanus toxin, fragment C	Tobacco leaves (chloroplast expression)
V. cholerae cholera toxin, B subunit (as single antigen)	Potato tubers
	Tobacco leaves
	Tobacco leaves (chloroplast expression)
	Tomato leaves and fruit

conducted in 1997, and involved the ingestion of transgenic potatoes expressing the B subunit of the *E. coli* heat labile toxin, which causes diarrhea. This resulted in a successful elicitation of mucosal immunity in test subjects (Tacket *et al.* 1998).

Metabolic engineering allows the directed production of small molecules in bacteria

When the large-scale production of penicillin began in the 1940s, yields were measured in micrograms per liter of culture. Demand for the antibiotic was outstripping supply and higher yielding strains were badly needed. Since nothing was known about the biosynthetic pathway a program of strain improvement was set in place that involved random mutation and screening. The best strain from each cycle of improvement then became the starting point for the next round of selection. In this way the yield of penicillin was steadily increased until it reached the tens of grams per liter that can be achieved today. As each new antibiotic was discovered the same process of strain improvement was applied. In every case, the biochemical and genetic bases of the beneficial mutations were not known. Only when the details of gene regulation and metabolic pathway regulation (allosteric control) had been elucidated could we even begin to understand how antibiotic yields might have been improved.

Based on our current knowledge of metabolic regulation we can predict that the changes in the improved strains described above will involve all of the following:

- Removal of rate-limiting transcriptional and allosteric controls.
- Kinetic enhancement of rate-limiting enzymes.
- Genetic inhibition of competing pathways.
- Enhanced carbon commitment to the primary metabolic pathway from central metabolism.
- Modification of secondary metabolic pathways to enhance energy metabolism and the availability of enzymatic cofactors.
- Enhanced transport of the compound out of the cell.

Today, if one starts with a wild-type strain and wants to turn it into an overproducer then recombinant DNA technology would be used to make these desired changes in a rational way as exemplified below by phenylalanine. There are many other examples of rational metabolic engineering in bacteria and these have been reviewed by Chotani *et al.* (2000).

Phenylalanine is a key raw material for the synthesis of the artificial sweetener aspartame. Phenylalanine can be synthesized chemically but is too expensive if made this way. In the 1980s bacterial strains that overproduced phenylalanine were developed using the traditional mutation and selection method. At the same time a program of rational strain development was instituted at G.D. Searle, the company that owned the patent to aspartame. The starting point for this program was an analysis of the biosynthetic pathway (Fig. 26.2). Removing feedback inhibition at key points was an essential first step. In the case of a phenylalanine producer it is essential to knock out any feedback inhibition of the pathway from chorismate to phenylalanine and this was achieved by selecting strains resistant to phenylalanine analogs. The conversion of E4P and PEP to DAHP is also subject to feedback inhibition but since there are three different enzymes here, each inhibited by a different aromatic end product, all that is necessary for a phenylalanine overproducer is to clone the tryptophan-sensitive enzyme and have it overexpressed. To overcome repression of enzyme synthesis the existing promoters were removed and replaced with one that could be controlled more easily in industrial-scale fermentations.

The above changes removed the natural control circuits. The next step was to remove competing pathways, i.e. the synthesis of tyrosine and tryptophan. This was achieved easily by making a tyrosine and tryptophan double auxotroph. Note that stable (non-reverting) auxotrophs can best be made by deleting part or all of the relevant genes. This is a task that is easy using recombinant DNA technology. Once all the control circuits and competing pathways had been removed, attempts were made to increase the carbon flux through the biosynthetic pathway. Surprisingly, overexpressing all the genes in the pathway did not enhance the yield of phenylalanine. One explanation was that the supply of precursors (E4P and/or PEP) was rate limiting. This was confirmed when cloning transketolase (to enhance E4P levels) and eliminating pyruvate kinase (to enhance PEP levels) enhanced yields.

Metabolic engineering provides new routes to small molecules

Recombinant DNA technology can be used to develop novel routes to small molecules. Good examples are the microbial synthesis of the blue dye indigo (Ensley *et al.* 1983) and the black pigment melanin (Della-Cioppa *et al.* 1990). Neither compound is

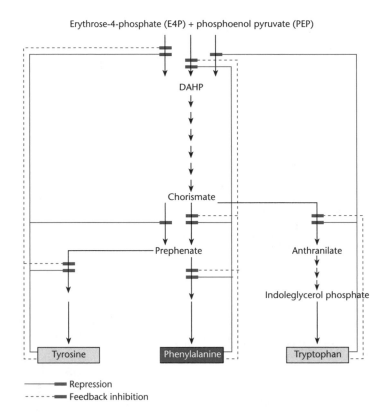

Fig. 26.2 The regulation of the biosynthesis of aromatic amino acids. Note that indoleglycerol phosphate is converted to tryptophan via indole but that the indole normally is not released from the tryptophan synthase complex.

produced in bacteria. The cloning of a single gene from *Pseudomonas putida*, that encoding naphthalene dioxygenase, resulted in the generation of an *E. coli* strain able to synthesize indigo in a medium containing tryptophan (Fig. 26.3). Similarly, cloning a tyrosinase gene in *E. coli* led to the conversion of tyrosine to dopaquinone, which spontaneously converts to melanin in the presence of air. To overproduce these compounds one generates a strain of *E. coli* that overproduces either tryptophan or tyrosine rather than phenylalanine as described above. With both indigo and melanin, yields are improved by increasing the levels of cofactors. Also, in the case of indigo biosynthesis it is necessary to engineer the tryptophan synthase gene. The reason for this is that indole is an intermediate in the biosynthesis of tryptophan (Fig. 26.2). However, normally it is not free in the cytoplasm but remains trapped within the tryptophan synthase complex. By modifying the *trpB* gene, encoding the subunit of tryptophan synthase, it was possible for the indole to be released for conversion by the dioxygenase (Murdock *et al.* 1993).

One disadvantage of the new route to indigo is that one of the intermediates in its synthesis, indoxyl, can undergo an alternative spontaneous oxidation to isatin and indirubin. The latter compound is an

isomer of indigo with similar dyeing properties but is a deep burgundy color instead of blue. To make textile-quality indigo, there must be no indirubin present. Screening soil microorganisms with the capacity to degrade indole resulted in the identification of an enzyme, isatin hydrolase, which can degrade isatin to isatic acid. After cloning the gene for isatin hydrolase in the indigo overproducing strains, the indigo product obtained performed as well as chemically produced material.

A slightly different approach to that above has yielded a new route to vitamin C. The conventional process starts with glucose and comprises one microbiological and four chemical steps (Fig. 26.4). By cloning in *Erwinia* the *Corynebacterium* gene encoding 2,5-diketogluconic acid reductase, the process can be simplified to a single microbiological and a single chemical step (Anderson *et al.* 1985). After observations of unexpectedly low yields of 2-ketogulonic acid in the recombinant strain, it was found that 2-ketogulonic acid was converted to l-idonic acid by an endogenous 2-ketoaldonate reductase. Cloning, deletion mutagenesis, and homologous recombination of the mutated reductase gene into the chromosome were some of the several steps taken to develop an organism capable of accumulating large amounts (120 g/l) of 2-ketogulonic acid (Lazarus *et al.* 1990).

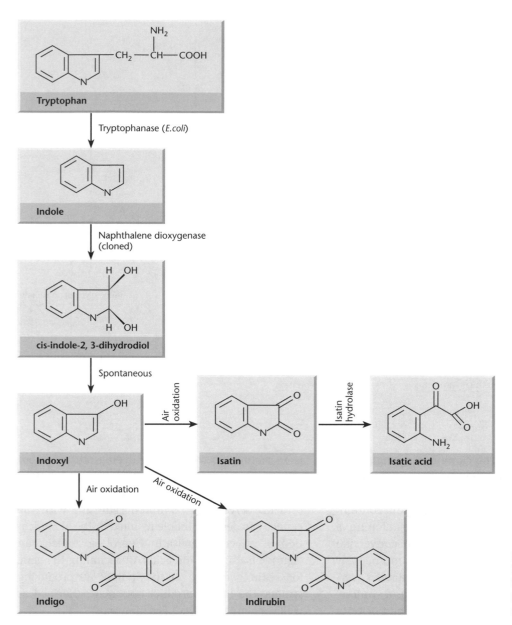

Fig. 26.3 The biosynthesis of indigo in *E. coli* and the formation of alternative end products.

So far, attempts to manufacture vitamin C directly from glucose have been unsuccessful. However, enzymes that can convert 2-ketogulonic acid to ascorbic acid have been identified and the objective now is to clone these activities into *Erwinia* (Chotani *et al.* 2000).

Combinatorial biosynthesis can produce completely novel compounds

A number of widely used antibiotics and immunosuppressants belong to a class of molecules known as polyketides. These molecules, which are synthesized by actinomycetes, have a fairly complex structure (Fig. 26.5). The genes involved in the biosynthesis of polyketides are clustered, thereby facilitating the cloning of all of the genes controlling the pathway. The first cluster to be cloned (the *act* genes) was that for actinorhodin. When parts of the *act* gene cluster were introduced into streptomycetes making related polyketides completely new antibiotics were produced (Hopwood *et al.* 1985). For example, introducing the *actVA* gene from *Streptomyces coelicolor* into a strain that makes medermycin leads to the synthesis of mederrhodin A (Fig. 26.6). This approach has been repeated many times since with other polyketides (for review see Baltz 1998) and is known as combinatorial biosynthesis.

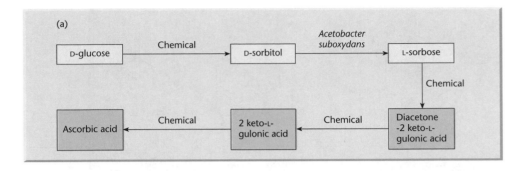

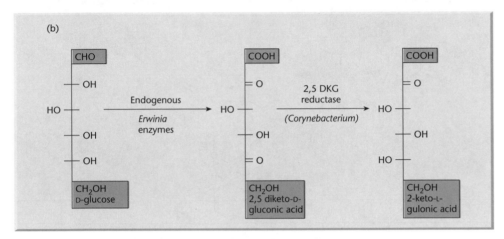

Fig. 26.4 Simplified route to vitamin C (ascorbic acid) developed by cloning in *Erwinia* the *Corynebacterium* gene for 2,5-diketogluconic acid reductase. (a) Classical route to vitamin C. (b) The simplified route to 2-ketogulonic acid, the immediate precursor of vitamin C.

Once a number of polyketide biosynthetic gene clusters had been cloned and sequenced, new insights were gained into the mechanism of synthesis and two enzymic modes of synthesis were discovered. In particular, polyketide synthesis takes place on an enzyme complex in a manner analogous to fatty acid synthesis. Furthermore, there are two types of complex. In type II complexes, the different enzyme activities are encoded by separate subunits. By contrast, in type I synthesis all the different enzyme activities are encoded by a single, very large gene. The polyketide synthases are prime candidates for DNA shuffling and this approach has been widely adopted (Baltz 1998). However, novel polyketides can also be generated simply by changing the *order* of the different activities in type I synthases (McDaniel *et al.* 1999).

Metabolic engineering can also be achieved in plants and plant cells to produce diverse chemical structures

Plants synthesize an incredibly diverse array of useful chemicals. Most are products of *secondary metabolism*, that is, biochemical pathways that are not involved in the synthesis of essential cellular components but more complex molecules that provide additional functions. Examples of these functions are attraction of pollinators and resistance to pests and pathogens. In many cases, these secondary metabolites have specific and potent pharmaceutical properties in humans: well-known examples include caffeine, nicotine, morphine, and cocaine.

Plants have long been exploited as a source of pharmaceutical compounds, and a number of species are cultivated specifically for the purpose of extracting drugs and other valuable molecules. We discussed above how gene transfer to bacteria and yeast can be used to produce novel chemicals, so in theory it would be possible to transfer the necessary components from these useful plants into microbes for large-scale production. However, the secondary metabolic pathways of plants are so extensive and complex, that in most cases such a strategy would prove impossible. Fortunately, advances in plant transformation have made it possible to carry out metabolic engineering in plants themselves (Capell & Christou 2004), and large-scale plant cell cultures can be used in the same manner as microbial cultures for the production of important phytochemicals (reviewed by Verpoorte 1998, Verpoorte *et al.* 2000).

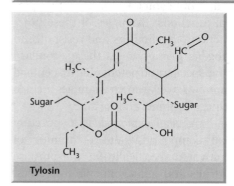

Erythromycin

Daunorubicin

Avermectin

Tylosin

Fig. 26.5 Some examples of polyketides.

Production of vinblastine and vincristine in *Catharanthus* cell cultures is a challenge because of the many steps and control points in the pathway

The secondary metabolic pathways of most plants produce the same basic molecular skeletons, but these are "decorated" with functional groups in a highly specific way, so that particular compounds may be found in only one or a few plant species. Furthermore, such molecules are often produced in extremely low amounts, so extraction and purification can be expensive. For example, the Madagascar periwinkle *Catharanthus roseus* is the source of two potent anti-cancer drugs called vinblastine and vincristine. These terpene indole alkaloids are too complex to synthesize in the laboratory and there are no alternative natural sources. In *C. roseus*, these molecules are produced in such low amounts that over one hectare of plants must be harvested to produce a single gram of each drug, with a commercial value of over $1 million.

It would be much more convenient to produce such drugs in fermentors containing cultured plant cells, and this has been achieved for a number of compounds, two of which (paclitaxel and shikonin)

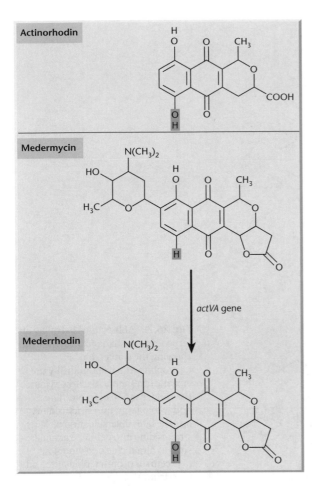

Fig. 26.6 The formation of the new antibiotic mederrhodin from medermycin by the *actVA* gene product.

have reached commercial production (see Verpoorte *et al.* 2000). However, cell suspension cultures often do not produce the downstream products made by the parent plant. This applies to vinblastine and vincristine from *C. roseus*, and also to other important drugs such as morphine, codeine, and hyoscyamine. Part of the reason for this is the complexity of secondary metabolism. The entire pathway is not completed in a single cell, but is often segregated into different cell types, with consequent shuttling of intermediates between cells. Within the cell, different stages of the pathway are also compartmentalized, so that intermediates must be transported between organelles. As in bacteria, knowledge of the target biosynthetic pathway is therefore essential for metabolic engineering, but in plants only a few secondary pathways are understood in sufficient detail. Examples include the phenylpropanoid and flavonoid pathways, which yield anthocyanins (plant pigments) and phytoalexins (antimicrobial compounds), and the terpene indole alkaloid biosynthetic pathway,

which generates important alkaloids such as vinblastine and vincristine. As more plant genomes are sequenced, we are likely to learn much more about such pathways.

All terpene indole alkaloids derive from a single universal precursor called strictosidine. This is formed by the convergence of two pathways, the iridoid pathway (culminating in secologanin) and the terpenoid pathway (culminating in tryptamine). Strictosidine is formed by the condensation of secologanin and tryptamine, catalyzed by the enzyme strictosidine synthase, and is then further modified in later steps to produce the valuable downstream alkaloids (Fig. 26.7). The conversion of tryptophan to tryptamine is a rate-limiting step in the terpenoid pathway, and this has been addressed by overexpressing the enzyme tryptophan decarboxylase in *C. roseus* cell suspension cultures. However, while transformed cultures produced much higher levels of tryptamine, no downstream alkaloids were synthesized (Goddijn *et al.* 1995, Canel *et al.* 1998). The simultaneous overexpression of the next enzyme in the pathway, strictosidine synthase, did increase the levels of the alkaloid ajmalicine, a useful sedative, in some cultures, but did not result in the synthesis of vinblastine or vincristine (Canel *et al.* 1998).

It thus seems that single-step engineering may remove known bottlenecks only to reveal the position of the next. The limited success of single-gene approaches has resulted in the development of alternative strategies for the coordinated regulation of entire pathways using transcription factors. Using the yeast one-hybrid system (p. 463) a transcription factor called ORCA2 has been identified that binds to response elements in the genes for tryptophan decarboxylase, strictosidine synthase, and several other genes encoding enzymes in the same pathway. A related protein, ORCA3, has been identified using insertional vectors that activate genes adjacent to their integration site. By bringing the expression of such transcription factors under the control of the experimenter, entire metabolic pathways could be controlled externally (see review by Memelink *et al.* 2000).

The production of vitamin A in cereals is an example of extending an endogenous metabolic pathway

Although the metabolic pathways for vitamin biosynthesis have been more fully elucidated in microbes than in plants, the engineering of plants

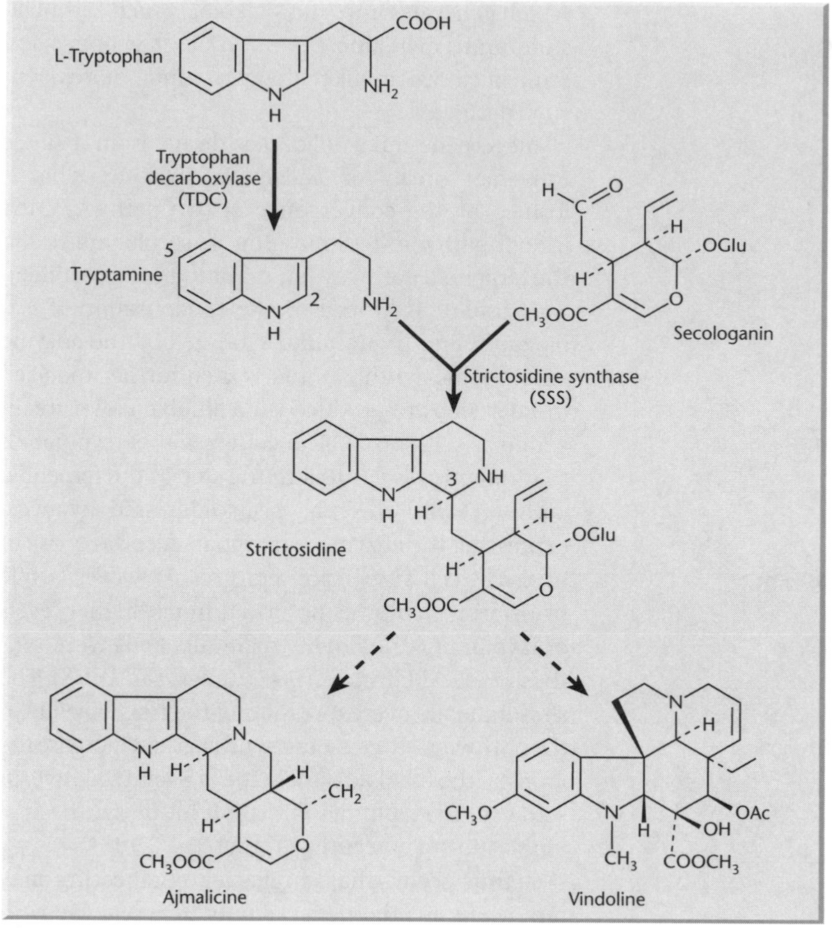

Fig. 26.7 Abbreviated pathway of terpene indole alkaloid biosynthesis, showing the conversion of tryptophan into tryptamine by the enzyme tryptophan decarboxylase, the condensation of tryptamine and secologanin into strictosidine, and the later diversification of strictosidine into valuable akaloids, such as ajmalicine and vindoline (a precursor of both vinblastine and vincristine).

for enhanced vitamin synthesis is now gaining much attention (Herbers 2003). Vitamin A, or 11-*cis*-retinal, is a dietary component required in all human cells but it is particularly important in the eye, where it functions as the lipid prosthetic group of the visual pigment opsin. Vitamin A deficiency is a significant health threat in the developing world, and is the most common (yet preventable) cause of blindness in developing countries. Humans usually obtain vitamin A directly from animal sources, but can synthesize it if provided with its immediate precursor, provitamin A (β-carotene), which is present at high levels in certain fruits and vegetables. The recommended daily allowance of vitamin A is expressed as retinol equivalents, and is equal to about 6 mg of β-carotene per day. There is very little β-carotene in cereal grains, which represent the staple diet for many of the world's poorest people.

The synthesis of carotenes in plants begins with the linkage of two geranylgeranyl diphosphate molecules to form the precursor phytoene (Fig. 26.8). The conversion of phytoene into β-carotene requires three further enzymatic steps. All four steps are absent in cereal endosperm tissue, so cereal grains accumulate geranylgeranyl diphosphate but not the downstream metabolic products in the pathway. The synthesis of β-carotene in cereals therefore represents an example of metabolic pathway *extension*, where novel enzymatic activities must be introduced into the plant and expressed in the endosperm to extend the pathway beyond its endogenous end point. The four enzyme activities in the β-carotene synthesis pathway missing in cereal grains are phytoene synthase, phytoene desaturase, ζ-carotene desaturase, and lycopene β-cyclase (Fig. 26.8). The first major breakthrough was the development of rice grains accumulating phytoene. Burkhardt *et al.* (1997) described rice plants transformed with the daffodil (*Narcissus pseudonarcissus*) phytoene synthase gene, which accumulated high levels of this metabolic intermediate. Further work by the same group (Ye *et al.* 2000) produced transgenic rice plants expressing the daffodil genes encoding phytoene synthase and lycopene β-cyclase, and the *crtI*

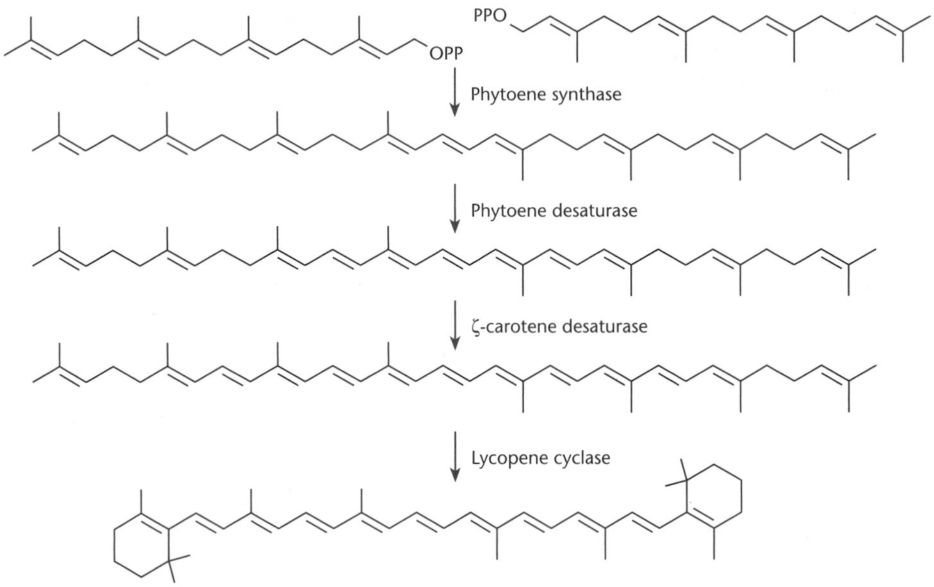

Fig. 26.8 Enzymatic steps and metabolic products in the β-carotene biosynthesis pathway which are missing in cereal grains.

gene from the bacterium *Erwinia uredovora*, which encodes an enzyme with both phytoene desaturase and ζ-carotene desaturase activities. The daffodil genes were expressed under the control of the rice glutelin-1 promoter, which is endosperm specific, while the bacterial gene was controlled by the constitutive cauliflower mosaic virus (CaMV) 35S promoter. This ground-breaking, multi-gene engineering approach resulted in golden colored rice grains containing up to 2 μg/g of β-carotene, in which case a moderate rice meal of 100 g would represent about 10% of the RDA for vitamin A. Interestingly, similar results were achieved in rice plants containing phytoene synthase and the bacterial *crt*I gene but lacking lycopene β-cyclase (Beyer *et al.* 2002). This suggested either that rice grains contain a residual endogenous lycopene β-cyclase activity or that the endogenous enzyme is dormant in wild-type grains but induced in the transgenic grains by the high levels of metabolic intermediates.

The "Golden Rice" project represented not just a technological breakthrough but also a model of humanitarian science that serves as an example for the deployment of other crops addressing food insecurity. From the beginning, the clear aim of the project organizers was to maintain freedom to operate and to provide the technology free of charge to subsistence farmers in developing countries, a feat that required careful negotiation over more than 100 intellectual and technical property rights (Potrykus 2001). Golden Rice fulfils an urgent need;

it complements traditional interventions for vitamin A deficiency and provides a real opportunity to address a significant world health problem. It was developed to benefit the poor and disadvantaged, and will be given away to subsistence farmers with no attached conditions. It requires no additional inputs compared with other rice varieties. To avoid biosafety concerns, Golden Rice lines have been generated with the innocuous metabolic selection marker *mpi*, which allows regenerating plants to grow when mannose is the only carbon source and avoids the use of antibiotics or herbicides for selection (Lucca *et al.* 2001).

While β-carotene synthesis in rice has the greatest potential to address real food security and health problems, experiments in other plants have revealed further useful information about this important metabolic pathway. Transgenic tomatoes have been described expressing *E. uredovora* phytoene synthase (*crt*B) (Fraser *et al.* 2000) and *crt*I (Romer *et al.* 2000) as well as a β-cyclase gene from *Arabidopsis thaliana* (Rosati *et al.* 2000). In the first case, fruit-specific expression of *crt*B was achieved using the tomato polygalacturonase promoter, and the recombinant protein was directed to the chromoplasts using the tomato phytoene synthase-1 transit sequence. Total fruit carotenoids were found to be two- to four-fold higher than in wild-type plants. Romer *et al.* (2000) expressed *crt*I constitutively, under the control of the CaMV 35S promoter. This unexpectedly reduced the total carotene content by about 50% but the levels of β-carotene increased threefold to 520 µg/g dry weight. This probably reflects the existence of complex feedback mechanisms acting at several different levels, a possibility discussed in detail by Giuliano *et al.* (2000). Rosati *et al.* (2000) used the fruit-specific tomato phytoene desaturase promoter to express the *A. thaliana* β-lycopene cyclase gene and increased β-carotene levels in transgenic fruits to 60 µg/g fresh weight. Work is ongoing to determine how the β-carotene pathway is regulated and what steps need to be taken to overcome feedback control.

The enhancement of plants to produce more vitamin E is an example of balancing several metabolic pathways and directing flux in the preferred direction

Vitamin E is actually a group of eight hydrophobic compounds: α-, β-, γ-, and δ-tocopherol, and the unsaturated equivalents α-, β-, γ-, and δ-tocotrienol. Dietary vitamin E is obtained mainly from seeds, and its function in the body is to prevent the oxidation and polymerization of unsaturated fatty acids. Deficiency leads to general wasting, kidney degeneration, and infertility. The α-, β-, γ-, and δ-derivatives differ in the number and position of methyl groups around the chroman ring as shown in Fig. 26.9. The most potent vitamer is RRR-α-tocopherol, but

common dietary sources of natural vitamin E such as soy oil are much richer in γ-tocopherol, which has only 10% of the activity of α-tocopherol, while α-tocopherol itself is only a minor component. Natural vitamin E supplements, which account for 10–15% of the total vitamin E market, are produced mainly from soy oil by chemically converting γ-tocopherol to α-tocopherol (Subramanian *et al.* 2000).

Tocopherol synthesis in plants requires the input from two metabolic pathways. The shikimate pathway generates homogentisic acid, which forms the aromatic ring of the compound, whereas the side chain is derived from phytyldiphosphate, a product of the methylerythritol phosphate (MEP) pathway (Fig. 26.10). These precursors are joined together by the enzyme homogentisic acid prenyltransferase (HPT) to form the intermediate 2-methyl-6-phytylbenzoquinol (MPBQ). MPBQ is the substrate for two enzymes, tocopherol cyclase and MPBQ methyltransferase. The former enzyme produces δ-tocopherol while the latter introduces a second methyl group to form 2,3-dimethyl-5-phytylbenzoquinol. The action of tocopherol cyclase on this intermediate produces γ-tocopherol. Both γ-tocopherol and δ-tocopherol act as substrates for the enzyme γ-tocopherol methyltransferase (γ-TMT), producing α- and β-tocopherol, respectively. The relative abundance of the four tocopherols in different plants is dependent on the activities of the enzymes discussed above.

Recent work carried out using the model plant *A. thaliana* has shown how the levels of vitamin E activity can be increased in the seeds of this plant, either by increasing the total amount of vitamin E or by shifting the vitamer profile towards the most potent form, α-tocopherol. Shintani & Della Penna (1998) expressed the *Synechocystis* PCC6803 and *A. thaliana* genes encoding γ-TMT in *A. thaliana* seeds using the carrot seed-specific DC3 promoter. This resulted in a radical shift from γ- to α-tocopherol (and from δ- to β-tocopherol) showing that nutritional enhancement in plants was possible without altering total vitamin E levels. In contrast, Savidge *et al.* (2002) overexpressed *A. thaliana* HPT, producing twice the level of vitamin E found in normal seeds, while Geiger *et al.* (2001) expressed the *E. coli tyr*A gene, which encodes a dual-function enzyme (chorismate mutase and prephenate dehydrogenase), resulting in up to three times the normal level of vitamin E. A more recent study has shown how the results obtained from experiments in *A. thaliana* can be used to produce soybean crops with enhanced nutritional

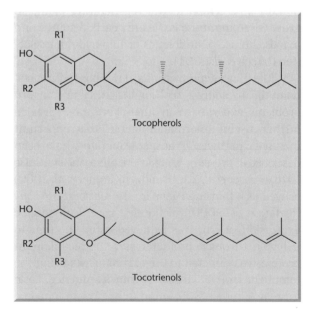

Fig. 26.9 Structure of vitamin E. For α derivatives, R1, 2, and 3 are methylated. For β derivatives, R1 and 3 are methylated. For γ derivatives, R2 and 3 are methylated. For δ derivatives, only R3 is methylated.

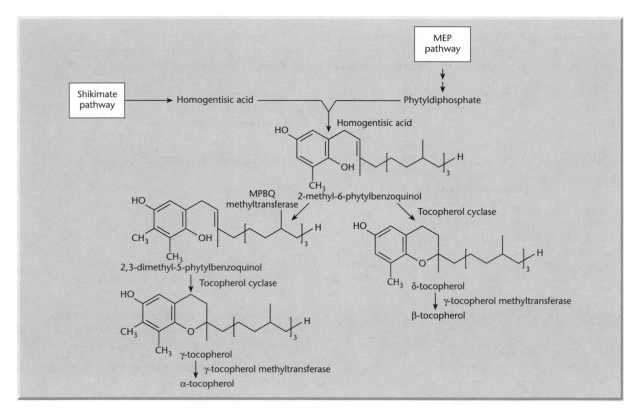

Fig. 26.10 Late steps in tocopherol biosynthesis. MPBQ = 2-methyl-6-phytylbenzoquinol.

properties that have the real potential to address food and nutritional insecurity. As for Golden Rice, the success of this study relied on the simultaneous expression of multiple genes, in this case the *A. thaliana VTE*3 and *VTE*4 genes, encoding MPBQ methyltransferase and γ-tocopherol methyltransferase (Van Eenennaam *et al.* 2003). The transgenic soybeans showed a significant elevation in the total amount of vitamin E activity (five-fold greater than wild-type plants), which was attributable mainly to an eight-fold increase in the levels of α-tocopherol, from its normal 10% of total vitamin E to over 95%. In this case, multiple gene transfer was used not to extend a metabolic pathway, but to regulate the existing steps to increase flux toward desired metabolites.

Theme 2: Improving agronomic traits by genetic modification

For centuries, plants and domestic animals have been bred to select the most desirable traits, either in terms of their appearance, hardiness, or yield of useful products. In conventional breeding, the farmer or breeder relies on chance recombination events to bring together beneficial combinations of genes when particular crosses are performed. Genetic engineering provides a short cut to conventional breeding, by introducing such genes directly into the genome. Genetic engineering also removes the constraints of sexual compatibility when carrying out crosses, since genes from any source (bacteria, animals, and plants) can be transferred into the recipient organism. Nowhere are the benefits of genetic engineering for trait improvement more apparent than in plant biotechnology, given that commercial genetically modified (GM) crops now cover over 75 million hectares of farmland in over 20 countries (Christou & Twyman 2004). In this section we discuss some of the agronomic traits in plants that have been improved by genetic manipulation.

Herbicide resistance is the most widespread trait in commercial transgenic plants

Herbicides are used to kill weeds, and generally affect processes that are unique to plants, e.g. photosynthesis or amino acid biosynthesis (see Table 26.5). Both crops and weeds share these processes, and

Table 26.5 Mode of action of herbicides and method of engineering herbicide-resistant plants.

Herbicide	Pathway inhibited	Target enzyme	Basis of engineered resistance to herbicide
Glyphosate	Aromatic amino acid biosynthesis	5-Enol-pyruvyl shikimate-3-phosphate (EPSP) synthase	Overexpression of plant *EPSP* gene or introduction of bacterial glyphosate-resistant *aroA* gene
Sulfonylurea	Branched-chain amino acid biosynthesis	Acetolactate synthase (ALS)	Introduction of resistant *ALS* gene
Imidazolinones	Branched-chain amino acid biosynthesis	ALS	Introduction of mutant *ALS* gene
Phosphinothricin	Glutamine biosynthesis	Glutamine synthetase	Overexpression of glutamine synthetase or introduction of the *bar* gene, which detoxifies the herbicide
Atrazine	Photosystem II	Q_B	Introduction of mutant gene for Q_B protein or introduction of gene for glutathione-*S*-transferase, which can detoxify atrazines
Bromoxynil	Photosynthesis		Introduction of nitrilase gene, which detoxifies bromoxynil

developing herbicides that are selective for weeds is very difficult. An alternative approach is to modify crop plants so that they become resistant to broad-spectrum herbicides, i.e. incorporating selectivity into the plant itself rather than relying on the selectivity of the chemical. Two strategies to engineer herbicide resistance have been adopted. In the first, the target molecule in the cell either is rendered insensitive or is overproduced. In the second, a pathway that degrades or detoxifies the herbicide is introduced into the plant. An example of each strategy is considered below.

Glyphosate is a non-selective herbicide that inhibits 5-enol-pyruvylshikimate-3-phosphate (EPSP) synthase, a key enzyme in the biosynthesis of aromatic amino acids in plants and bacteria. A glyphosate-tolerant *Petunia hybrida* cell line obtained after selection for glyphosate resistance was found to overproduce EPSP synthase as a result of gene amplification. A gene encoding the enzyme was subsequently isolated and introduced into petunia plants under the control of a CaMV promoter. Transgenic plants expressed increased levels of EPSP synthase in their chloroplasts and were significantly more tolerant to glyphosate (Shah *et al.* 1986). An alternative approach to glyphosate resistance has been to

introduce a gene encoding a mutant EPSP synthase. This mutant enzyme retains its specific activity but has decreased affinity for the herbicide. Transgenic tomato plants expressing this gene under the control of an opine promoter were also glyphosate tolerant (Comai *et al.* 1985). Following on from this early research, several companies have introduced glyphosate tolerance into a range of crop species, with soybean and cotton the first to reach commercialization (Padgette *et al.* 1996, Nida *et al.* 1996). Currently, nearly three-quarters of all transgenic plants in the world are resistant to glyphosate (James 2004).

Phosphinothricin (PPT) is an irreversible inhibitor of glutamine synthetase in plants and bacteria. Bialaphos, produced by *Streptomyces hygroscopicus*, consists of PPT and two alanine residues. When these residues are removed by peptidases the herbicidal component PPT is released. To prevent self-inhibition of growth, bialaphos-producing strains of *S. hygroscopicus* also produce the enzyme phosphinothricin acetyltransferase (PAT), which inactivates PPT by acetylation. The *bar* gene that encodes the acetylase has been introduced into potato, tobacco, and tomato cells using *Agrobacterium*-mediated transformation. The resultant plants were resistant to commercial

(a)

(b)

Fig. 26.11 Evaluation of phosphinothricin resistance in transgenic tobacco plants under field conditions. (a) Untransformed control plants. (b) Transgenic plants. (Photographs courtesy of Dr. J. Botterman and the editor of *Biotechnology*.)

formulations of PPT and bialaphos in the laboratory (De Block *et al.* 1987) and in the field (De Greef *et al.* 1989) (Fig. 26.11). More recently, it has been shown that bialaphos-resistant transgenic rice plants which were inoculated with the fungi causing sheath blight disease and subsequently treated with the herbicide were completely protected from infection (Uchimiya *et al.* 1993). This agronomically important result depends on the observation that bialaphos is toxic to fungi as well as being a herbicide. PPT resistance is widely used in plants as a selectable marker (see p. 284), however, it has also been introduced into a number of different crops for weed control, including sugarcane and rice (Gallo-Meagher & Irvine 1996, Oard *et al.* 1996).

The benefit of herbicide-resistant transgenic crops is the increased yield and seed quality as competing weed species are eliminated. However, there was initial concern that this would come with an associated penalty of increased herbicide use, which could have a serious impact on the environment. Contrary to these predictions, the introduction of herbicide-resistant plants has actually reduced chemical use by up to 80% in many areas, as farmers adopt better weed-control policies and switch to herbicides with

low-use rates. A further risk is that transgenes for herbicide tolerance could spread to weed species, resulting in a new breed of "superweeds" (Kling 1996). It is too early to say whether this will be a problem. Although a range of herbicide-resistant transgenic crops is being tested, only crops resistant to glyphosate or bromoxynil are currently grown on a commercial scale. The benefits and risks of herbicide-resistant crops have been reviewed (Gressel 1999, 2000). Although herbicide resistance is the most common strategy for weed control, several others are being evaluated including enhanced allelopathy, in which crop plants are modified to produce their own weed-inhibiting compounds (reviewed by Duke *et al.* 2002). Typically, the primary crop is the target of modification, but such approaches can also be used to produce aggressive "cover crops" that inhibit weeds before the primary crop is planted. There has been recent interest in the use of suicide transgenes inducible by temperature or photoperiod cues to cause cover crops to self-destruct at the correct time of year (Stanilaus & Cheng 2002). It is also possible to modify microbial pathogens so that they preferentially attack weeds (e.g. Amsellum *et al.* 2002).

Virus-resistant crops can be produced by expressing viral or non-viral transgenes

Major crop losses occur every year as a result of viral infections, e.g. tobacco mosaic virus (TMV) causes losses of over $50 million per annum in the tomato industry. There is a useful phenomenon known as cross-protection in which infection of a plant with one strain of virus protects against superinfection with a second, related strain. The mechanism of cross-protection is not fully understood but it is believed that the viral coat protein is important. Powell-Abel *et al.* (1986) developed transgenic plants which express the TMV coat protein and which greatly reduced disease symptoms following virus infection. Since that observation, the principle of heterologous coat protein expression has been extended to many different plants and viruses (reviewed by Beachy *et al.* 1990). In the case of resistance to TMV, the coat protein must be expressed in the epidermis and in the vascular tissue through which the virus spreads systemically (Clark *et al.* 1990). Transgenic squash containing multiple viral coat protein genes and demonstrating resistance to cucumber mosaic virus, watermelon mosaic virus 2, and zucchini yellow mosaic virus was the first virus-resistant transgenic crop to reach commercial production (Tricoli *et al.* 1995). A related strategy in which viral replicase was expressed in transgenic rice plants protected those plants from several different isolates of rice yellow mottle virus (RYMV) because the replicase was more conserved than the coat protein. In the best-performing lines, viral replication was completely blocked over several generations (Pinto *et al.* 1999).

While the heterologous coat protein approach can be successful, it has been demonstrated that, in many cases, the effect of transgene expression is mediated at the RNA rather than the protein level. This can be proven by generating transgenic plants carrying coat protein genes that cannot be translated to yield functional proteins, as first shown by Lindbo and Dougherty (1992) using the tobacco etch virus coat protein gene. The transgene RNA apparently interferes with viral replication (a phenomenon called RNA-mediated viral resistance, RMVR). This requires homology between the transgene and the target virus, and involves high-level transgene transcription but low-level accumulation of the transcript, features very similar to those of post-transcriptional gene silencing (p. 314).

A different method of minimizing the effects of plant virus infection was developed by Gehrlach *et al.* (1987). They generated plants that expressed the satellite RNA of tobacco ringspot virus and such plants were resistant to infection with tobacco ringspot virus itself. Another potential method of inducing resistance to viruses is the production of antiviral proteins in transgenic plants. American pokeweed produces an antiviral protein called dianthrin that functions as a ribosome-inactivating protein. The cDNA for this protein has been cloned (Lin *et al.* 1991) and expressed in tobacco, providing resistance against African cassava mosaic virus (ACMV) (Hong *et al.* 1996). Interestingly in this experiment, the dianthrin gene was expressed under the control of an ACMV promoter, such that the antiviral protein was expressed only upon viral infection. In this manner, the toxic effects of constitutive transgene expression were avoided.

Several further classes of non-viral gene products can be used to provide specific protection against viruses, including ribosome-inactivating proteins, which block protein synthesis (Moons *et al.* 1997, Tumer *et al.* 1997), ribonucleases, which destroy the virus genome (Watanabe *et al.* 1995), 2′,5′ oligoadenylate synthetases, which interfere with replication (Ogawa *et al.* 1996), and ribozymes, which cleave viral genomes causing them to be degraded by cellular enzymes (de Feyter *et al.* 1996, Kwon *et al.* 1997). Antibodies specific for virion proteins have also been used to protect plants from viruses. In the first demonstration of this approach, Tavladoraki *et al.* (1993) expressed a single chain Fv fragment (scFV) specific for ACMV in transgenic *N. benthamiana*, and demonstrated resistance to viral infection. Other groups have generated transgenic tobacco plants expressing antibodies specific for tobacco mosaic virus, resulting in reduced infectivity. Voss *et al.* (1995) expressed full-size IgG immunoglobulins while Zimmermann *et al.* (1998) expressed scFv fragments. Targeting scFv fragments to the plasma membrane also provides protection against virus infection (Schillberg *et al.* 2001).

Resistance to fungal pathogens is often achieved by manipulating natural plant defense mechanisms

Progress also has been made in developing resistance to fungal pathogens, which are traditionally controlled by appropriate farming practices (e.g. crop rotation) and the application of expensive and environmentally harmful fungicides. A straightforward approach is to engineer plants with antifungal proteins from heterologous species. This was first demonstrated by Broglie *et al.* (1991), who showed that expression of bean chitinase can protect tobacco and oilseed rape from post-emergent damping off caused by *Rhizoctonia solani*. Plants synthesize a wide range of so-called "pathogenesis-related proteins" (PR proteins), such as chitinases and glucanases, which are induced by microbial infection and there are now many examples of such proteins expressed in plants to provide protection against fungi – in some cases multiple PR proteins have been expressed with synergistic effects: tobacco, carrot, and tomato have each been engineered to express both chitinase and glucanase simultaneously, and these plants show greater fungal resistance, indicating that the two enzymes can work synergistically (van den Elzen *et al.* 1993, Zhu *et al.* 1994, Jongedijk *et al.* 1995).

Plants also synthesize anti-fungal peptides called defensins, and other anti-fungal proteins that have been shown to confer resistance to pathogens when overexpressed in transgenic plants (Cary *et al.* 2000, Osusky *et al.* 2000, Li *et al.* 2001); some of these have been shown to provide protection under field trial conditions (Gao *et al.* 2000). As the genes for more of these proteins have been cloned and characterized, the number of transgenic plants constitutively expressing such proteins continues to rise (Punja 2001). For example, tobacco osmotin has been expressed in transgenic potato, providing resistance to *Phytophthora infestans* (Liu *et al.* 1994), and in transgenic rice, providing resistance to *R. solani* (Lin *et al.* 1995). Instead of using a protein to provide direct protection, a metabolic engineering strategy can be utilized. Phytoalexins are alkaloids with anti-fungal activity, and transforming plants with genes encoding the appropriate biosynthetic enzymes can increase their synthesis. Hain *et al.* (1993) generated tomato plants expressing the grapevine gene for stilbene synthase, and these plants demonstrated increased resistance to infection by *Botrytis cinerea*. The grape stilbene synthase (resveratrol synthase) gene has since been expressed in tobacco, tomato, barley, rice, and wheat, and has been shown to confer resistance to a range of fungal pathogens (Stark-Lorenzen *et al.* 1997, Thomzik *et al.* 1997, Leckband & Lorz 1998, Fettig & Hess 1999).

An alternative to the use of anti-fungal proteins or metabolites is to manipulate the hypersensitive response, which is a physiological defense mechanism used by plants to repel attacking pathogens. Resistance occurs only in plants carrying a resistance gene (*R*) that corresponds to an avirulence (*avr*) gene in the pathogen. Elicitors (signaling molecules) released by pathogens are detected by the plant and activate a range of defense responses, including cell death, PR-gene expression, phytoalexin synthesis, and the deposition of cellulose at the site of invasion, forming a physical barrier. Importantly, the hypersensitive response is systemic, so that neighboring cells can pre-empt pathogen invasion. A recently developed strategy is to transfer avirulence genes from the pathogen to the plant, under the control of a pathogen-inducible promoter. This has been demonstrated in tomato plants transformed with the *avr*9 gene from *Clasosporium fulvum*, resulting in resistance to a range of fungal, bacterial, and viral diseases (Keller *et al.* 1999, reviewed by Melchers & Stuiver 2000).

Resistance to blight provides an example of how plants can be protected against bacterial pathogens

Bacterial diseases cause significant losses in crop yields, and many different transgenic strategies have been developed to prevent infection or reduce the severity of symptoms (Herrera-Estrella & Simpson 1995, Mourges *et al.* 1998, Punja 2001). One of the most prevalent bacterial diseases of rice is bacterial blight, which causes losses totaling over $250 million every year in Asia alone. This disease has received a great deal of attention due to the discovery of a resistance gene complex in the related wild species *O. longistaminata*. The trait was introgressed into cultivated rice line IR-24 and was shown to confer resistance to all known isolates of the blight pathogen *Xanthomonas oryzae* pv. *oryzae* in India and the Philippines (Khush *et al.* 1990). Further investigation of the resistance complex resulted in the isolation of a gene, named *Xa*21, encoding a receptor tyrosine kinase (Song *et al.* 1995). The transfer of this gene to susceptible rice varieties resulted in plant lines showing strong resistance to a range of isolates of the pathogen (Wang *et al.* 1996, Tu *et al.* 1998, Zhang *et al.* 1998). The *Xa*21 gene has been stacked with two genes for insect resistance to generate a rice line with resistance to bacterial blight and a range of insect pests, and this is due for commercial release in China in the near future (Huang *et al.* 2002a). As with insect-resistance genes, the widespread use of transgenic plants carrying a single resistance factor could prompt the evolution of new pathogen strains with counteradaptive properties. Therefore, other blight-resistance transgenes are being tested for possible deployment either alone or in combination with *Xa*21. For example, Tang *et al.* (2001) have produced rice plants expressing a ferredoxin-like protein that had previously been shown to delay the hypersensitive response to the pathogen *Pseudomonas syringae* pv. *syringae*. In inoculation tests with *X. oryzae* pv. *oryzae*, all the transgenic plants showed enhanced resistance against the pathogen.

The bacterium *Bacillus thuringiensis* provides the major source of insect-resistant genes

Insect pests represent one of the most serious biotic constraints to crop production. For example, more than one-quarter of all the rice grown in the world is lost to insect pests, at an estimated cost of nearly $50 billion. This is despite an annual expenditure of approximately $1.5 billion on insecticides for this crop alone. Insect-resistant plants are therefore desirable not only because of the potential increased yields, but also because the need for insecticides is eliminated and, following on from this, the undesirable accumulation of such chemicals in the environment is avoided. Typical insecticides are non-selective, so they kill harmless and beneficial insects as well as pests. For these reasons, transgenic plants have been generated expressing toxins that are selective for particular insect species.

Research is being carried out on a wide range of insecticidal proteins from diverse sources. However, all commercially produced insect-resistant transgenic crops express toxin proteins from the Gram-positive bacterium *Bacillus thuringiensis* (Bt) (Peferoen *et al.* 1997, de Maagd *et al.* 1999). Unlike other *Bacillus* species, *B. thuringiensis* produces crystals during sporulation, comprising one or a small number of ~130 kDa protoxins called crystal proteins. These proteins are potent and highly specific insecticides. The specificity reflects interactions between the crystal proteins and receptors in the insect midgut. In susceptible species, ingested crystals dissolve in the alkaline conditions of the gut and the protoxins are activated by gut proteases. The active toxins bind to receptors on midgut epithelial cells, become inserted into the plasma membrane, and form pores that lead to cell death (and eventual insect death) through osmotic lysis. Approximately 150 distinct Bt toxins have been identified and each shows a unique spectrum of activity (van Frankenhuyzen & Nystrom 2002).

Bt toxins have been used as topical insecticides since the 1930s, but never gained widespread use because they are rapidly broken down on exposure to daylight, and thus have to be applied several times during a growing season. Additionally, only insects infesting the exposed surfaces of sprayed plants are killed. These problems have been addressed by the expression of crystal proteins in transgenic plants. Bt genes were initially introduced into tomato (Fischhoff *et al.* 1987) and tobacco (Vaeck *et al.* 1987, Barton *et al.* 1987) and later cotton (Perlak *et al.* 1990) resulting in the production of insecticidal proteins that protected the plants from insect infestation. However, field tests of these plants revealed that higher levels of the toxin in the plant tissue would be required to obtain commercially useful plants (Delannay *et al.* 1989). Attempts to increase the expression of the toxin gene in plants by the use of different promoters, fusion proteins, and leader sequences were not successful. However, examination of the bacterial *cry1Ab* and *cry1Ac* genes indicated that they differed significantly from plant genes in a number of ways (Perlak *et al.* 1991). For example, localized AT rich regions resembling plant introns, potential plant polyadenylation signal sequences, ATTTA sequences which can destabilize mRNA, and rare plant codons were all found. The elimination of undesirable sequences and modifications to bring codon usage into line with the host species resulted in greatly enhanced expression of the insecticidal toxin and strong insect resistance of the transgenic plants in field tests (Koziel *et al.* 1993). By carrying out such enhancements, Perlak and colleagues expressed a modified *cry3A* gene in potato to provide resistance against Colorado beetle (Perlak *et al.* 1993). In 1995, this crop became the first transgenic insect-resistant crop to reach commercial production, as NewLeaf™ potato marketed by Monsanto. The same company also released the first commercial transgenic, insect-resistant varieties of cotton (Bollgard™, expressing *cry1Ac* and protected against tobacco bollworm) and maize (YieldGard™, expressing *cry1Ab* and resistant to the European corn borer). Many other biotechnology companies have now produced Bt-transgenic crop plants resistant to a range of insects (reviewed by Schuler *et al.* 1998, de Maagd *et al.* 1999, Hilder & Boulter 1999, Llewellyn & Higgins 2002). Some of these have been extraordinarily successful: for example, Tu *et al.* (2000) showed that a Bt commercial hybrid variety expressing the *cry1Ab* gene produced a 28% yield increase compared to wild-type plants in field trials in China.

Although Bt-transgenic plants currently dominate the market, there are many alternative insecticidal proteins under investigation. Two types of protein are being studied in particular: proteins that inhibit digestive enzymes in the insect gut (proteinase and amylase inhibitors) and lectins (carbohydrate-binding proteins). Research into these alternatives is driven in part by the fact that some insects are not affected by any of the known Bt crystal proteins. Homopteran insects, mostly sap-sucking pests such as planthoppers, fall into this category, but have been shown to be susceptible to lectins such as *Galanthus nivalis* agglutinin (GNA). This lectin has been expressed in many crops, including potato (Shi *et al.* 1994, Gatehouse *et al.* 1996), rice (Bano-Maqbool & Christou 1999), tomato and tobacco (reviewed by Schuler *et al.* 1998).

Drought resistance provides a good example of how plants can be protected against abiotic stress

After pests and diseases, unfavorable environmental conditions (*abiotic stresses*) represent the next major limitation on crop production. One of the most prevalent of these conditions is drought, and the development of transgenic crops with built-in drought resistance could increase the global yield of food by up to 30%. Many plants respond to drought (prolonged dehydration) and increased salinity by synthesizing small, very soluble molecules such as betaines, sugars, amino acids, and polyamines. These are collectively termed *compatible solutes*, and they increase the osmotic potential within the plant, therefore preventing water loss in the short term and helping to maintain a normal physiological ion balance in the longer term (Yancey *et al*. 1982). Compatible solutes are non-toxic even at high concentrations, so one strategy to provide drought resistance is to make such molecules accumulate in transgenic plants (Chen & Murata 2002, Serraj & Sinclair 2002). For example, several species have been engineered to produce higher levels of glycine betaine but in most cases the levels achieved have fallen short of the 5–40 μmol/g fresh weight observed in plants that naturally accumulate this molecule under salt stress conditions (Sakamoto & Murata 2000, 2001). However, transgenic rice plants expressing BADH (beatine aldehyde dehdrogenase), one of the key enzymes in the glycine betaine synthesis pathway, accumulated the molecule to levels in excess of 5 μmol/g fresh weight (Sakamoto *et al*. 1998). In China, transgenic rice plants expressing BADH are likely to be the first commercially released GM plants developed for abiotic stress tolerance, and will be available for small-scale subsistence farmers and large producers (Huang *et al*. 2002b). As well as compatible solutes, plants also produce osmoprotectant proteins known as *dehydrins* in response to drought and salinity stress. The over-expression of such proteins offers another strategy to generate drought-tolerant crop varieties. Xu *et al*. (1996) expressed the barley dehydrin HVA1 in transgenic rice, and the transgenic plants were shown to be resistant to water deficit and salt stress. More recently, several wheat dehydrins have been expressed in transgenic rice and have also been shown to increase dehydration tolerance (Cheng *et al*. 2001, 2002).

Plants can be engineered to cope with poor soil quality

About 65% of the world's potential arable land consists of marginal soils, which are characterized by extremes of pH, limited nutrient availability (particularly phosphorus and iron), and high levels of toxic metal ions (Marshner 1995). Acidic soils, which account for 40% of the arable land, have low levels of available phosphorus and iron but high levels of aluminum. In an acidic environment, both iron and aluminum sequester phosphorus into insoluble or poorly soluble molecules. Aluminum is also toxic in its own right, its major effect being the inhibition of root development. Alkaline soils account for 25% of arable land. The major problem in alkaline soils is the high level of calcium and magnesium ions, which also sequester phosphorus into insoluble and sparingly soluble molecules. Calcium is an essential signaling molecule in plants and high levels of this metal ion can interfere with normal plant growth and metabolism. Despite these problems, many plants have adapted to grow in marginal soils and some tolerant varieties of crop plants have also been produced by mutation and conventional breeding. A common factor among these tolerant plants is the increased exudation of organic acids, such as citrate, malate, and oxalate, from the roots. These substances concentrate in the rhizosphere and are thought to have a number of protective effects, including the displacement and solubilization of phosphorus and iron, the chelation of aluminum, and the attraction to the rhizosphere of beneficial microorganisms, which may also enhance nutrient availability (Lopez-Bucio *et al*. 2000b). The production of transgenic crop plants engineered to exude higher levels of organic acids is therefore an attractive strategy to increase the use of marginal soils. Several species have been transformed with bacterial or plant citrate synthase genes to increase organic acid production and induce tolerance of poor soils (de la Fuenete & Herrera-Estrella 1997, Koyoma *et al*. 2000, Lopez-Bucio *et al*. 2000a). The analysis of root extracts by HPLC showed that citrate levels were 10 times higher than normal, but more importantly, the levels of citrate recorded in root exudates were four times the normal level. These plants could grow in the presence of aluminum ions at 10 times the concentration sufficient to suppress the growth of non-transformed plants, and there was no evidence of aluminum-induced root damage. The transgenic plants also performed better in alkaline soils. They grew and flowered normally and produced seeds, whereas control plants showed restricted vegetal growth and failed to produce flowers or seeds (Lopez-Bucio *et al*. 2000a). The exudation of phytase from the roots of transgenic plants can be used to release phosphorus locked away in organic compounds, since much of the organic phosphorus in soil is present as phytate (Hayes *et al.* 1999). Initial experiments with *Arabidopsis* plants secreting phytase from the roots showed that the transgenic plants were able to grow much better than control plants on phytate medium. The development of crop plants that secrete phytase and organic acids is therefore an important goal for the future.

One of the most important goals in plant biotechnology is to increase food yields

The amount of usable food obtained from a field of plants can vary tremendously, and much effort has been expended in attempts to increase yields by conventional breeding and optimizing farming practices. Genetic engineering provides a wide range of strategies not only for reducing yield losses (by increasing resistance to pests and diseases and providing tolerance to abiotic stress) but also by increasing the intrinsic yield potential of the plants. In terms of yield enhancement, photosynthesis is perhaps the most obvious target for genetic intervention because it determines the rate of carbon fixation, and therefore the overall size of the organic carbon pool. Strategies for increasing photosynthetic activity include the modification of light-harvesting phytochromes and key photosynthetic enzymes. Progress has been made in crop species by attempting to introduce components of the energy-efficient C_4 photosynthetic pathway into C_3 plants, which lose a proportion of their fixed carbon through photorespiration. The key step in C_4 photosynthesis is the conversion of CO_2 into C_4 organic acids by the enzyme phosphoenolpyruvate carboxylase (PEPC) in mesophyll cells. The maize gene encoding PEPC has been transferred into several C_3 crops, including potato (Ishimaru *et al.* 1998) and rice (Matsuoka *et al.* 1998, Ku *et al.* 1999) in order to increase the overall level of carbon fixation. Transgenic rice plants were also produced expressing pyruvate orthophosphate dikinase (PPDK) and NADP-malic enzyme (Ku *et al.* 2000). Preliminary field trials in China and Korea demonstrated 10–30% and 30–35% yield increases for PEPC and PPDK transgenic rice plants, respectively, which was quite unexpected since only one C_4 enzyme was expressed in each case.

Metabolic approaches to increase overall yields focus on the conversion of sugars, representing the direct products of photosynthesis, to the bioavailable storage carbohydrate starch. Several different strategies have been attempted, including the mani-

pulation of enzyme activity and regulation in source tissues to increase the carbon flux to sink tissues (Paul *et al.* 2001), increasing the efficiency of sugar transport to sink tissues (Rosche *et al.* 2002), and the manipulation of enzyme activity in the sink tissue itself to increase the conversion of photoassimilates into starch. Transgenic approaches to increase sugar to starch conversion have focused primarily on the manipulation of the plastidial starch synthesis pathway. For example, inhibition of the plastidial adenylate kinase in transgenic potatoes resulted in a substantial increase in the level of adenylates and a 60% increase in the level of starch, in combination with a 40% increase in tuber yield (Regierer *et al.* 2002).

Other strategies that have been used include the modification of plant architecture to divert resources to harvestable products. In rice, manipulation of the gibberellin signaling pathway by overexpression of a mutant *Arabidopsis* *GAI* gene was sufficient to induce a dominant semi-dwarfing phenotype (Peng *et al.* 1999, Fu *et al.* 2001). Developmental genes have been expressed in an attempt to increase the number of spikelets (seed-bearing structures) in maize (Cacharrón *et al.* 1999) while other researchers have attempted to modify the way plants respond to day length or seasonal cues to induce early flowering and increase the number of harvestable crops each year.

Theme 3: Using genetic modification to study, prevent, and cure disease

Transgenic animals can be created as models of human disease

Mammals have been used as models for human disease for many years, since they can be exploited to carry out detailed analyses of the molecular basis of disease and to test newly developed therapeutics prior to clinical trials in humans. Before the advent of transgenic animal technology, however, models of inherited diseases (i.e. diseases with a genetic basis) were difficult to come by. They could be obtained as spontaneously occurring mutants, suitable mutant animals identified in mutagenesis screens, and susceptible animal strains obtained by selective breeding. Gene manipulation now offers a range of alternative strategies to create *specific* disease models (see reviews by Smithies 1993, Bedell *et al.* 1997, Petters & Sommer 2000).

Some of the earliest transgenic disease models were mice predisposed to particular forms of cancer because the germ line contained exogenously derived oncogenes (e.g. Sinn *et al.* 1987). This exemplifies so-called gain-of-function diseases, which are caused by a dominantly acting allele and can be modeled simply by adding that allele to the normal genome, e.g. by microinjection into eggs. Other gain-of-function diseases that have been modeled in this way include Gerstmann–Straussler–Scheinker (GSS) syndrome, a neurodegenerative disease caused by a dominantly acting mutated prion protein gene. In one patient suffering from this disease, a mutation was identified in codon 102 of the prion protein gene. Transgenic mice were created carrying this mutant form of the gene in addition to the wild-type locus, and were shown to develop a similar neurodegenerative pathology to their human counterparts (Hsiao *et al.* 1990). Other examples of gain-of-function disease models include Alzheimer's disease, which was modeled by overexpression of the amyloid precursor protein (Quon *et al.* 1991), and the triplet repeat disorder spinocerebellar ataxia type 1 (Burright *et al.* 1995). Simple transgene addition can also be used to model diseases caused by dominant negative alleles, as shown for the premature aging disease, Werner's syndrome (Wang *et al.* 2000).

Recessively inherited diseases are generally caused by loss of function, and these can be modeled by gene knockout. The earliest report of this strategy was a mouse model for HRPT deficiency, generated by disrupting the gene for hypoxanthine-guanine phosphoribosyltransferase (Kuehn *et al.* 1987). A large number of genes has been modeled in this way, including cystic fibrosis (Snouwaert *et al.* 1992, Dorin *et al.* 1992), fragile-X syndrome (Dutch-Belgian Fragile X Consortium 1994), β-thallasemia (Skow *et al.* 1983, Ciavatti *et al.* 1995), and mitochondrial cardiomyopathy (Li *et al.* 2000). Gene targeting has been widely used to model human cancers caused by the inactivation of tumor suppressor genes such as *TP53* and *RB1* (reviewed by Ghebranious & Donehower 1998, Macleod & Jacks 1999).

While the studies above provide models of single gene defects in humans, attention is now shifting towards the modeling of more complex diseases that involve multiple genes. This is a challenging area of research but there have been some encouraging early successes. In some cases, the crossing of different modified mouse lines has led to interesting discoveries. For example, *undulated* mutant mice lack the gene encoding the transcription factor Pax-1, and *Patch* mutant mice are heterozygous for a null allele of the platelet-derived growth factor gene.

Hybrid offspring from a mating between these two strains were shown to model the human birth defect spina bifida occulta (Helwig *et al.* 1995). In other cases, such crosses have pointed the way to possible novel therapies. For example, transgenic mice over-expressing human α-globin and a mutant form of the human β-globin gene that promotes polymerization provide good models of sickle cell anemia (Trudel *et al.* 1991). However, when these mice are crossed to those ectopically expressing human fetal hemoglobin in adulthood, the resulting transgenic hybrids show a remarkable reduction in disease symptoms (Blouin *et al.* 2000). Similarly, crossing transgenic mice overexpressing the anti-apoptotic protein Bcl-2 to rds mutants that show inherited slow retinal degeneration resulted in hybrid offspring in which retinal degeneration was strikingly reduced. This indicates that Bcl-2 could possibly be used in gene therapy to treat the equivalent human retinal degeneration syndrome (Nir *et al.* 2000).

The most complex diseases involve many genes, and transgenic models would be difficult to create. However, it is often the case that such diseases can be reduced to a small number of "major genes" with severe effects, and a larger number of minor genes. Thus, it has been possible to create mouse models of Down syndrome, which in humans is generally caused by the presence of three copies of chromosome 21. Trisomy for the equivalent mouse chromosome 16 is a poor model because the two chromosomes do not contain all the same genes. However, a critical region for Down syndrome has been identified by studying Downs patients with partial duplications of chromosome 21. The generation of YAC transgenic mice carrying this essential region provides a useful model of the disorder (Smith *et al.* 1997) and has identified increased dosage of the *Dyrk1a* (*minibrain*) gene as an important component of the learning defects accompanying the disease. Animal models of Down syndrome have been reviewed (Kola & Hertzog 1998, Reeves *et al.* 2001).

Gene medicine is the use of nucleic acids to prevent, treat, or cure disease

While disease modeling uses gene manipulation to create diseases in model organisms, gene medicine refers to the use of the same technology to ameliorate or even permanently cure diseases in humans. Gene medicine has a wide scope and includes the use of DNA vaccines, the targeted killing of disease cells (e.g. cancer cells), the use of oligonucleotides as

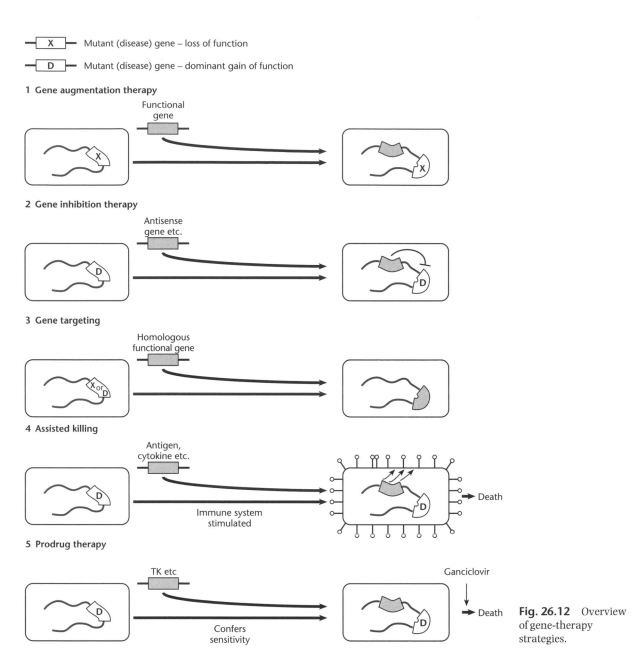

Fig. 26.12 Overview of gene-therapy strategies.

drugs, and the use of gene transfer to correct genetic defects (gene therapy). Gene transfer can be carried out in cultured cells, which are then reintroduced into the patient, or DNA can be transferred to the patient *in vivo*, directly or by using viral vectors. The *ex vivo* approach can be applied only to certain tissues, such as bone marrow, in which the cells are amenable to culture. Gene therapy can be used to treat diseases caused by mutations in the patient's own DNA (inherited disorders, cancers), as well as infectious diseases, and is particularly valuable in cases where no conventional treatment exists, or that treatment is inherently risky. Strategies include (Fig. 26.12):

- gene augmentation therapy (GAT), where DNA is added to the genome with the aim of replacing a missing gene product;
- gene targeting to correct mutant alleles;
- gene inhibition therapy using techniques such as antisense RNA expression or the expression of intracellular antibodies to treat dominantly acting diseases;
- the targeted ablation of specific cells.

Therapeutic gene transfer effectively generates transgenic human cell clones, and for this reason only somatic cells can be used as targets. The prospect of germ-line transgenesis in humans raises serious ethical concerns, and with the rapid advances in technology allowing germ-line transformation and nuclear transfer in numerous mammals, these concerns will need to be addressed in the very near future (Johnson 1998). As an alternative to permanent gene transfer, transient gene therapy can be achieved using oligonucleotides, which can disrupt gene expression at many levels but do not permanently change the genetic material of the cell (Pollock & Gaken 1995).

The tools and techniques for gene therapy are essentially similar to those used for gene transfer to any animal cells. Transfection, direct delivery, or transduction (see Chapter 12) can be used to introduce DNA into cells. Viral vectors are most popular because of their efficiency of gene transfer *in vivo*. However, extreme precautions need to be taken to ensure the safety of such vectors, avoiding potential problems such as the production of infectious viruses by recombination, and pathological effects of viral replication. A number of viral vectors have been developed for gene therapy, including those based on oncoretroviruses, lentiviruses, adenovirus, adeno-associated virus, herpes virus, and a number of hybrid vectors combining advantageous elements of different parental viruses (Robbins *et al.* 1998, Reynolds *et al.* 1999). The risks associated with viral vectors have promoted research into other delivery methods, the most popular of which include direct injection of DNA into tissues (e.g. muscle), the injection of liposome–DNA complexes into the blood, and direct transfer by particle bombardment or other methods. Although inherently much safer than viruses, such procedures show a generally low efficiency (Scheule & Cheng 1996, Tseng & Huang 1998, Kay *et al.* 2001).

DNA vaccines are expression constructs whose products stimulate the immune system

The immune system generates antibodies in response to the recognition of proteins and other large molecules carried by pathogens. With typical vaccines, the functional component of the vaccine introduced into the host is the protein that elicits the immune response. The introduction of DNA into animals does not generate an immune response against the DNA molecule, but, if that DNA is expressed to yield a protein, that protein can stimulate the immune system (Reyes-Sandoval & Ertl 2001). This is the basis of DNA vaccination, as first demonstrated by Ulmer *et al.* (1993) (Fig. 26.13). DNA vaccines generally comprise a bacterial plasmid carrying a gene encoding the appropriate antigen under the control of a strong promoter that is recognized by the host cell. The advantages of this method include its simplicity, its wide applicability, and the ease with which large quantities of the vaccine can be produced. The DNA may be administered by injection, using liposomes or by particle bombardment. In the original demonstration, Ulmer and colleagues introduced DNA corresponding to the influenza virus nucleoprotein and achieved protection against influenza infection. Since then, many DNA vaccines have been used to target viruses (e.g. measles (Cardoso *et al.* 1996); HIV (Wang *et al.* 1993, Fuller *et al.* 1997, Hinkula *et al.* 1997); Ebola virus (Xu *et al.* 1998)), other pathogens (e.g. tuberculosis (Huygen *et al.* 1996)), and even the human cellular prion protein in mice (Krasemann *et al.* 1996).

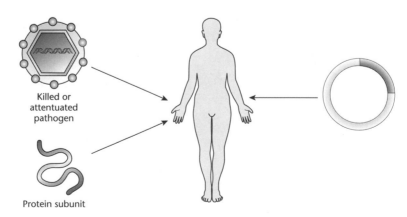

Fig. 26.13 A comparison of conventional vaccination and DNA vaccination.

Killed or attentuated pathogen

Protein subunit

Conventional vaccination Body is presented with antigen

DNA vaccination Body is presented with DNA Antigen expressed inside body

The DNA-vaccination approach has several additional advantages. These include the following:

* Certain bacterial DNA sequences have the innate ability to stimulate the immune system (see Klinman *et al.* 1997, Roman *et al.* 1997).
* Other genes encoding proteins influencing the function of the immune response can be co-introduced along with the vaccine (e.g. Kim *et al.* 1997).
* DNA vaccination can be used to treat diseases that are already established as a chronic infection (e.g. Mancini *et al.* 1996).

In principle, DNA vaccination has much in common with gene therapy (discussed below), since both involve DNA transfer to humans, using a similar selection of methods. However, while the aim of gene therapy is to alleviate disease, by either replacing a lost gene or blocking the expression of a dominantly acting gene, the aim of DNA vaccination is to prevent disease, by causing the expression of an antigen that stimulates the immune system.

Gene augmentation therapy for recessive diseases involves transferring a functional copy of the gene into the genome

The first human genetic engineering experiment was one of *gene marking* rather than gene therapy, and was designed to demonstrate that an exogenous gene could be safely transferred into a patient and that this gene could subsequently be detected in cells removed from the patient. Both objectives were met. Tumor-infiltrating lymphocytes (cells that naturally seek out cancer cells and then kill them by secreting proteins such as tumor necrosis factor, TNF) were isolated from patients with advanced cancer. The cells were then genetically marked with a neomycin-resistance gene and injected back into the same patient (Rosenberg *et al.* 1990).

The first clinical trial using a therapeutic gene-transfer procedure involved a four-year-old female patient, Ashanthi DeSilva, suffering from severe combined immune deficiency, resulting from the absence of the enzyme adenosine deaminase (ADA). This disease fitted many of the ideal criteria for gene-therapy experimentation. The disease was life threatening (therefore making the possibility of unknown treatment-related side-effects ethically acceptable) but the corresponding gene had been cloned and the biochemical basis of the disease was understood. Importantly, since ADA functions in the salvage pathway of nucleotide biosynthesis (p. 225), cells in which the genetic lesion had been corrected had a selective growth advantage over mutant cells, allowing them to be identified and isolated *in vitro*. Conventional treatment for ADA deficiency involves bone marrow transplantation from a matching donor. Essentially the same established procedure could be used for gene therapy, but the bone marrow cells would be derived from the patient herself, and would be genetically modified *ex vivo* (Fig. 26.14). Cells from the patient were subjected to leukopheresis and mononuclear cells were isolated. These were grown in culture under conditions that stimulated T-lymphocyte activation and growth and then transduced with a retroviral vector carrying a normal *ADA* gene as well as the neomycin-resistance gene. Following infusion of these modified cells, both this

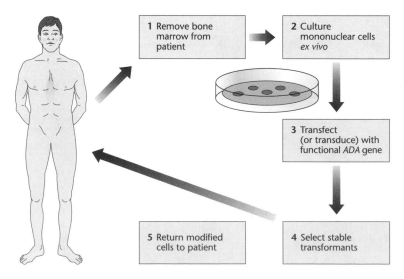

Fig. 26.14 Procedure for *ex vivo* gene therapy, based on the treatment for ADA deficiency.

patient and a second, who began treatment in early 1991, showed an improvement in their clinical condition as well as in a battery of *in vitro* and *in vivo* immune function studies (Anderson 1992). However, the production of recombinant ADA in these patients is transient, so each must undergo regular infusions of recombinant T-lymphocytes. Research is ongoing into procedures for the transformation of bone marrow stem cells, which would provide a permanent supply of corrected cells.

Gene-augmentation therapies for a small number of recessive single-gene diseases are now undergoing clinical trials. We consider cystic fibrosis (CF) as an example (Davies *et al.* 2001). CF is a disorder which predominantly affects the lungs, liver, and pancreas. The disease is caused by the loss of a cAMP-regulated membrane-spanning chlorine channel. This results in an electrolyte imbalance and the accumulation of mucus, often leading to respiratory failure. CF is a recessive disorder, suggesting that the loss of function could be corrected by introducing a functional copy of the gene. Indeed, epithelial cells isolated from CF patients can be restored to normal by transfecting them with the cloned cystic fibrosis transmembrane regulator (*CFTR*) cDNA. Unlike ADA-deficiency, the cells principally affected by CF cannot be cultured and returned to the patient, so *in vivo* delivery strategies must be applied. Targeted delivery of the *CFTR* cDNA to affected cells has been achieved using adenoviral vectors, which have a natural tropism of the epithelial lining of the respiratory system. Recombinant viruses carrying the *CFTR* cDNA have been introduced into patients using an inhaler (Zabner *et al.* 1993, Hay *et al.* 1995, Knowles *et al.* 1995). The *CFTR* cDNA has also been introduced using liposomes (e.g. Caplen *et al.* 1995). While such treatments have resulted in *CFTR* transgene expression in the nasal epithelium, there were neither consistent changes in chloride transport nor reduction in the severity of CF symptoms; they have been largely ineffective.

Gene-therapy strategies for cancer may involve dominant suppression of the overactive gene or targeted killing of the cancer cells

Cancer gene therapy was initially an extension of the early gene-marking experiments. The tumor-infiltrating leukocytes were transformed with a gene for TNF in addition to the neomycin-resistance gene, with the aim of improving the efficiency with which these cells kill tumors by increasing the amount of TNF they secreted. Although TNF is highly toxic to humans at levels as low as 10 μg/kg body weight, there have been no side-effects from the gene therapy and no apparent organ toxicity from secreted TNF (Hwu *et al.* 1993). One alternative strategy is to transform the tumor cells themselves, making them more susceptible to the immune system through the expression of cytokines or a foreign antigen. Another is to transform fibroblasts, which are easier to grow in culture, and then co-inject these together with tumor cells to provoke an immune response against the tumor. A number of such "assisted killing" strategies have been approved for clinical trials (see review by Ockert *et al.* 1999).

Direct intervention to correct cancer-causing genes is also possible. Dominantly acting genes (oncogenes) have been targeted using antisense technology, either with antisense transgenes, oligonucleotides (see Nellen & Lichtenstein 1993, Carter & Lemoine 1993), or ribozymes (Welch *et al.* 1998, Muotri *et al.* 1999). An early report of cancer gene therapy with antisense oligonucleotides was that of Szczylik *et al.* (1991) for the treatment of chronic myeloid leukemia. They used two 18-mers specific for the BCR–ABL gene junction generated by the chromosomal translocation that causes this particular cancer, and showed that colony formation was suppressed in cells removed from cancer patients. Cancers caused by loss of tumor-suppressor gene function have been addressed by replacement strategies in which a functional copy of the appropriate gene is delivered to affected cells (e.g. see Cai *et al.* 1993, Harper *et al.* 1993, Smith *et al.* 1993, Hahn *et al.* 1996). A further strategy, known as prodrug activation therapy, involves the activation of a particular enzyme specifically in cancer cells, which converts a nontoxic "prodrug" into a toxic product, so killing the cancer cells. This can be achieved by driving the expression of a so-called "suicide gene" selectively in cancer cells. An example is the HSV thymidine kinase gene, in combination with the prodrug ganciclovir. Thymidine kinase converts ganciclovir into a nucleotide analog which is incorporated into DNA and blocks replication by inhibiting the DNA polymerase. Activation of the enzyme specifically in cancer cells can be achieved by preferential delivery to dividing cells through the use of oncoretroviruses (e.g. Moolten 1986, Culver *et al.* 1992, Klatzmann *et al.* 1996). Another way is to use transcriptional regulatory elements that are active only in cancer cells (e.g. Harris *et al.* 1994, Su *et al.* 1996).

Suggested reading

Baneyx F. & Mujacic M. (2004) Recombinant protein folding and misfolding in *Escherichia coli*. *Nature Biotechnology* **22**, 1399–408.

Bedell M.A., Jenkins N.A. & Copeland N.G. (1997) Mouse models of human disease. Part II. Recent progress and future directions. *Genes Development* **11**, 11–43.

Capell T. & Christou P. (2004) Progress in plant metabolic engineering. *Current Opinion in Biotechnology* **15**, 148–54.

Christou P. & Twyman R.M. (2004) The potential of genetically enhanced plants to address food insecurity. *Nutrition Research Review* **17**, 23–42.

Dooley K. & Zon L.I. (2000) Zebrafish: a model system for the study of human disease. *Current Opinion in Genetics & Development* **10**, 252–6.

Dyck M.K., Lacroix D., Pothier F. & Sirard M.A. (2003) Making recombinant proteins in animals – different systems, different applications. *Trends in Biotechnology* **21**, 394–9.

Hellwig S., Drossard J., Twyman R.M. & Fischer R. (2004) Plant cell cultures for the production of recombinant diagnostic and therapeutic proteins. *Nature Biotechnology* **22**, 1415–22.

Kay M.A., Glorioso J.C. & Naldini L. (2001) Viral vectors for gene therapy: the art of turning infectious agents into vehicles of therapeutics. *Nature Medicine* **7**, 33–40.

Ma J.K.-C., Drake P.M.W. & Christou P. (2003) The production of recombinant pharmaceutical proteins in plants. *Nature Reviews Genetics* **4**, 794–805.

Macleod K.F. & Jacks T. (1999) Insights into cancer from transgenic mouse models. *Journal of Pathology* **187**, 43–60.

Punt P.J., Van Biezen N., Conesa A., Albers A., Mangnus J. & Van Den Hondel C. (2002) Filamentous fungi as cell factories for heterologous protein production. *Trends in Biotechnology* **20**, 200–6.

Reyes-Sandoval A. & Ertl H.C. (2001) DNA vaccines. *Current Molecular Medicine* **1**, 217–43.

Somia N. & Verma I.M. (2000) Gene therapy: trials and tribulations. *Nature Reviews Genetics* **1**, 91–9.

Tuschl T. & Borkhardt A. (2002) Small interfering RNAs: a revolutionary tool for the analysis of gene function and gene therapy. *Molecular Interventions* **2**, 158–67.

Wurm F.M. (2004) Production of recombinant protein therapeutics in mammalian cells. *Nature Biotechnology* **22**, 1393–8.

References

Aaij C. & Borst P. (1972) The gel electrophoresis of DNA. *Biochim. Biophys. Acta* **269**, 192–200.

Abdallah B., Hassan A., Benoist C., Goula D., Behr J.P., & Demeneix B.A. (1996) A powerful nonviral vector for *in vivo* gene transfer into the adult mammalian brain: polyethylenimine. *Hum. Gene Ther.* **7**, 1947–54.

Abola E., Kuhn P., Earnest T. & Stevens R.C. (2000) Automation of X-ray crystallography. *Nature Struct. Biol.* **7**, 973–7.

Abrahmsen L., Tom J., Burnier J. *et al.* (1991) Engineering subtilisin and its substrates for efficient ligation of peptide bonds in aqueous solution. *Biochemistry* **30**, 4151–9.

Adams K.L. *et al.* (2000a) Repeated, recent and diverse transfers of a mitochondrial gene to the nucleus in flowering plants. *Nature* **408**, 354–7.

Adams K.L., Rosenblueth M., Qui Y.L. & Palmer J.D. (2001) Multiple losses and transfers to the nucleus of two mitochondrial succinate dehydrogenase genes during angiosperm evolution. *Genetics* **158**, 1289–300.

Adams M.D. (1996) Serial analysis of gene expression: ESTs get smaller. *Bioassays* **18**, 261–2.

Adams M.D., Celniker S.E., Holt R.A. *et al.* (2000) The genome sequence of *Drosophila melanogaster. Science* **287**, 2185–95.

Adams M.D., Dubnick M., Kerlavage A.R. *et al.* (1992) Sequence identification of 2,375 human brain genes. *Nature* **355**, 632–4.

Adams M.D., Kelley J.M., Gocayne J.D. *et al.* (1991) Complementary DNA sequencing: expressed sequence tags and human genome project. *Science* **252**, 1651–6.

Adams S.E., Dawson K.M., Gull K., Kingsman S.M. & Kingsman A.J. (1987) The expression of hybrid HIV: Ty virus-like particles in yeast. *Nature* **329**, 68–70.

Aebersold R. & Mann M. (2003) Mass spectrometry-based proteomics. *Nature* **422**, 198–207.

Agapov E.V., Frolov I., Lindenbach B.D. *et al.* (1998) Noncytopathic Sindbis virus RNA vectors for heterologous gene expression. *Proc. Nat. Acad. Sci. USA* **95**, 12989–94.

Ahlquist P. & Janda M. (1984) cDNA cloning and *in vitro* transcription of the complete brome mosaic virus genome. *Mol. Cell. Biol.* **4**, 2876–82.

Ahlquist P., French R. & Bujarski J.J. (1987) Molecular studies of brome mosaic virus using infectious transcripts from cloned cDNA. *Adv. Virus Res.* **32**, 215–42.

Ahlquist P., French R., Janda M. & Loesch-Fries S. (1984) Multicomponent RNA plant virus infection derived from cloned viral cDNA. *Proc. Nat. Acad. Sci. USA* **81**, 7066–70.

Ahmad T., Armuzzi A., Bunce M., *et al.* (2002) The molecular classification of the clinical manifestations of Crohn's disease. *Gastroenterology* **122**, 854–66.

Aihara H. & Miyazaki J. (1998) Gene transfer into muscle by electroporation *in vivo. Nature Biotechnol.* **16**, 867–70.

Aitman T.J. *et al.* (1999) Identification of CD36 (Fat) as an insulin-resistant gene causing defective fatty acid and glucose metabolism in hypertensive rats. *Nature Genet.* **21**, 76–83.

Albert H., Dale E.C., Lee E. & Ow D.W. (1995) Site specific integration of DNA into wild type and mutant lox sites placed in the plant genome. *Plant J.* **7**, 649–59.

Albertson D.G. & Pinkel D. (2003) Genomic microarrays in human genetic disease and cancer. *Hum. Mol. Genet.* **12**, R145–52.

Alford R.L. & Caskey C.T. (1994) DNA analysis in forensics, disease and animal/plant identification. *Curr. Opin. Biotechnol.* **5**, 29–33.

Al-Hasani K., Simpfendorfer K., Wardan H. *et al.* (2003) Development of a novel bacterial artificial chromosome cloning system for functional studies. *Plasmid* **49**, 184–7.

Alizadeh A.A. *et al.* (2000) Distinct types of diffuse large B-cell lymphoma identified by gene expression profiling. *Nature* **403**, 503–11.

Allen N.D., Cran D.G., Barton S.C. *et al.* (1988) Transgenes as probes for active chromosomal domains in mouse development. *Nature* **333**, 852–5.

Aloy P., Oliva B., Querol E., Aviles F.X. & Russel R.B. (2002) Structural similarity to link sequence space: new potential superfamilies and implications for structural genomics. *Protein Sci.* **11**, 1101–16.

Alpert K.B. & Tanksley S.D. (1996) High-resolution mapping and isolation of a yeast artificial chromosome contig containing fw2.2: a major fruit weight quantitative trait locus in tomato. *Proc. Nat. Acad. Sci. USA* **93**, 15503–7.

Al-Shawi R., Kinnaird J., Burke J. & Bishop J.O. (1990) Expression of a foreign gene in a line of transgenic mice is modulated by a chromosomal position effect. *Mol. Cell. Biol.* **10**, 1192–8.

Altpeter F. *et al.* 2005. Particle bombardment and the genetic enhancement of crops: myths and realities. *Mol. Breed.* **15**, 305–27.

Altschul S.F. *et al.* (1997) Gapped BLAST and PSI-BLAST: a new generation of protein database search programs. *Nucl. Acids Res.* **25**, 3389–402.

Altschul S.F., Boguski M.S., Gish W. & Woolton J.C. (1994) Issues in searching molecular databases. *Nature Genet.* **6**, 119–29.

Altschul S.F., Gish W., Miller W., Myers E.W. & Lipmann D.J. (1990) Basic local alignment search tool. *J. Mol. Biol.* **215**, 403–10.

Altshuler D. *et al.* (2000a) The common PPARgamma Pro12Ala polymorphism is associated with decreased risk of type 2 diabetes. *Nature Genet.* **26**, 76–80.

Altshuler D. *et al.* (2000b) An SNP map of the human genome generated by reduced representation shotgun sequencing. *Nature* **407**, 513–16.

Alwine J.C., Kemp D.J., Parker B.A. *et al.* (1979) Detection of specific RNAs or specific fragments of DNA by fractionation in gels and transfer to diazobenzyloxy-menthyl paper. *Methods Enzymol.* **68**, 220–42.

Amaya E., Musci T.J. & Kirschner M.W. (1991) Expression of a dominant negative mutant of the FGF receptor disrupts mesoderm formation in *Xenopus* embryos. *Cell* **66**, 257–70.

Amsterdam A. *et al.* (1999) A large-scale insertional mutagenesis screen in zebrafish. *Genes and Development* **13**, 2713–24.

An G.H., Costa M.A. & Ha S.B. (1990) Nopaline synthase promoter is wound-inducible and auxin-inducible. *Plant Cell* **2**, 225–33.

Anand R., Riley J.H., Butler R., Smith J.C. & Markham A.F. (1990) A 3.5 genome equivalent multi-access YAC library: construction, characterisation, screening and storage. *Nucl. Acids Res.* **18**, 1951–6.

Andersen D.C. & Krummen L. (2002) Recombinant protein expression for therapeutic applications. *Curr. Opin. Biotech.* **13**, 117–23.

Andersen J.S. & Mann M. (2000) Functional genomics by mass spectrometry. *FEBS Lett.* **480**, 25–31.

Anderson N.G. & Anderson L. (1982) The human protein index. *Clinical Chem.* **28**, 739–48.

Anderson S., Marks C.B., Lazarus R. *et al.* (1985) Production of 2-keto-l-gulonate, an intermediate in l-ascorbate synthesis by a genetically modified *Erwinia herbicola. Science* **230**, 144–9.

Anderson W.F. (1992) Human gene therapy. Science **256**, 808–13.

Andersson S.G.E. *et al.* (1998) The genome sequence of *Rickettsia prowazekii* and the origin of mitochondria. *Nature* **396**, 133–40.

Andrulis I.L. & Siminovitch L. (1981) DNA-mediated gene transfer of beta-aspartylhydroxamate resistance into Chinese hamster ovary cells. *Proc. Nat. Acad. Sci. USA* **78**, 5724–8.

Angell S.M. & Baulcombe D.C. (1997) Consistent gene silencing in transgenic plants expressing a replicating potato virus X RNA. *EMBO J.* **16**, 3675–84.

Antson D-O. *et al.* (2000) PCR-generated padlock probes detect single nucleotide variation in genomic DNA. *Nucl. Acids Res.* **28**, E58.

Aoyama T. & Chua N.-H. (1997) A glucocorticoid-mediated transcriptional induction system in transgenic plants. *Plant J.* **11**, 605–12.

Aparicio S., Chapman J., Stupka E., *et al.* (2002) Whole-genome shotgun assembly and analysis of the genome of *Fugu rubripes. Science* **297**, 1301–10.

Apweiler R. *et al.* (2001a) The InterPro database, an integrated documentation resource for protein families, domains and functional sites. *Nucl. Acids Res.* **29**, 37–40.

Apweiler R. *et al.* (2001b) Proteome analysis database: online application of InterPro and CluSTr for the functional classification of proteins in whole genomes. *Nucl. Acids Res.* **29**, 44–8.

Arabidopsis Genome Initiative (2000) Analysis of the genome sequence of the flowering plant *Arabidopsis thaliana. Nature* **408**, 796–813.

Arakawa T., Chong D.K. & Langridge W.H. (1998) Efficacy of a food plant-based oral cholera toxin B subunit vaccine. *Nature Biotechnol.* **16**, 292–7.

Aravind L. *et al.* (1998) Evidence for massive gene exchange between archaeal and bacterial hypothermophiles. *Trends Genet.* **14**, 442–44.

Arber W. & Dussoix D. (1962) Host specificity of DNA produced by *Escherichia coli.* I. Host controlled modification of bacteriophage l. *J. Mol. Biol.* **5**, 18–36.

Arber W. (1965) Host specificity of DNA produced by Escherichia coli. V. The role of methionine in the production of host specificity. *J. Mol. Biol.* **11**, 247–56.

Arencibia A.D., Carmona E.R., Tellez P. *et al.* (1998) An efficient protocol for sugarcane (*Saccharum* spp L) transformation mediated by *Agrobacterium tumefaciens. Transgenic Res.* **7**, 213–22.

Arnold D., Feng L., Kim J. & Heintz N. (1994) A strategy for the analysis of gene expression during neural development. *Proc. Nat. Acad. Sci. USA* **91**, 9970–4.

Arnold N., Gross E., Schwarz-Boeger U., *et al.* (1999) A highly sensitive, fast and economical technique for mutation analysis in hereditary breast and ovarian cancers. *Human Mutation* **14**, 333–9.

Aronson H.E., Royer W.E. Jr & Hendrickson W.A. (1994) Quantification of tertiary structural conservation despite primary sequence drift in the globin fold. *Protein Sci.* **3**, 1706–11.

Artsaenko O., Peisker M., zur Nieden U. *et al.* (1995) Expression of a single chain Fv antibody against abscisic acid creates a wilty phenotype in transgenic tobacco. *Plant J.* **8**, 745–50.

Ashburner M. (1989) *Drosophila: A Laboratory Handbook*. Cold Spring Harbor Laboratory Press, Cold Spring Harbor, New York.

Ashburner M. *et al.* (2000) Gene ontology: tool for the unification of biology. *Nature Genet.* **25**, 25–9.

Aston C., Mishra B. & Schwartz D.C. (1999) Optical mapping and its potential for large-scale sequencing projects. *Trends Biotechnol.* **17**, 297–302.

Atchison R.W., Casto B.C. & Hammond W.M. (1965) Adenovirus-associated defective viral particles. *Science* **149**, 754–6.

Attwood T.K. & Parry-Smith D.J. (1999) *Introduction to Bioinformatics*. Prentice Hall, London.

Auch D. & Reth M. (1990) Exon trap cloning: using PCR to rapidly detect and clone exons from genomic DNA fragments. *Nucl. Acids Res.* **18**, 6743–4.

Audic S. & Claverie J. (1997) The significance of digital gene expression profiles. *Genome Res.* **7**, 986–95.

Auricchio A., Gao G.P., Yu Q.C., Raper S., Rivera V.M., Clackson T. & Wilson J.M. (2002) Constitutive and regulated expression of processed insulin following *in vivo* hepatic gene transfer. *Gene Ther.* **9**, 963–71.

Auricchio A., Rivera V., Clackson T., O'Connor E., Maguire A., Tolentino M., Bennett J. & Wilson J. (2002) Pharmacological regulation of protein expression from adeno-associated viral vectors in the eye. *Mol. Ther.* **6**, 238.

Austin S. & Nordstrom K. (1990) Partition-mediated incompatibility of bacterial plasmids. *Cell* **60**, 351–4.

Axel R., Fiegelson P. & Schutz G. (1976) Analysis of the complexity and diversity of mRNA from chicken oviduct and liver. *Cell* **11**, 247–54.

Aza-Blanc P., Cooper C.L., Wagner K., Batalov S., Deveraux Q.L. & Cooke M.P. (2003) Identification of modulators of TRAIL-induced apoptosis via RNAi-based phenotypic screening. *Mol. Cell* **12**, 627–37.

Azpiroz-Leehan R. & Feldmann K.A. (1997) T-DNA insertion mutagenesis in *A. thaliana*: going back and forth. *Trends Genet.* **13**, 146–52.

Babiychuk E., Fuanghthong M., VanMontagu M., Inze D. & Kushnir S. (1997) Efficient gene tagging in *Arabidopsis thaliana* using a gene trap approach. *Proc. Nat. Acad. Sci. USA* **94**, 12722–7.

Backman K. & Boyer H.W. (1983) Tetracycline resistance determined by pBR322 is mediated by one polypeptide. *Gene* **26**, 197–203.

Bader G.D. & Hogue C.W. (2000) BIND: a data specification for storing and describing biomolecular interactions, molecular complexes and pathways. *Bioinformatics* **16**, 465–77.

Bader G.D., Donaldson I., Wolting C., Ouellette B.F., Pawson T. & Hogue C.W. (2001) BIND: the biomolecular interaction network database. *Nucl. Acids Res.* **29**, 242–5.

Baer R. *et al.* (1984) DNA sequence and expression of the B95.8 Epstein–Barr virus genome. *Nature* **310**, 207–11.

Baetz K., McHardy L., Gable K., Tarling T., Reberioux D., Bryan J., Andersen R.J., Dunn T., Hieter P. & Roberge M. (2004) Yeast genome-wide drug-induced haploinsufficiency screen to determine drug mode of action. *Proc. Nat. Acad. Sci. USA* **101**, 4525–30.

Bagdasarian M., Bagdasarian M.M., Coleman S. & Timmis K.N. (1979) New vector plasmids for gene cloning in *Pseudomonas*. In *Plasmids of Medical, Environmental and Commercial Importance*, eds. Timmis K.N. & Pühler A., pp. 411–22. Elsevier/North-Holland Biomedical Press, Amsterdam.

Bagdasarian M., Lurz R., Rückert B. *et al.* (1981) Specific-purpose plasmid cloning vectors. II. Broad host range, high copy number, RSF1010-derived vectors, and a host-vector system for gene cloning in *Pseudomonas*. *Gene* **16**, 237–47.

Bagdasarian M.M., Amann E., Lurz R., Ruckert B. & Bagdasarian M. (1983) Activity of the hybrid trp-lac (tac) promoter of *Escherichia coli* in *Pseudomonas putida*: construction of broad-host-range, controlled-expression vectors. *Gene* **26**, 273–82.

Bagga J.S. & Wilusz J. (1999) Northwestern screening of expression libraries. *Methods Mol. Biol.* **118**, 245–56.

Baguisi A., Behboodi E., Melican D.T. *et al.* (1999) Production of goats by somatic cell nuclear transfer. *Nature Biotechnol.* **17**, 456–61.

Bahramian M.B. & Zabl H. (1999) Transcriptional and posttranscriptional silencing of rodent alpha1 (I) collagen by a homologous transcriptionally self-silenced transgene. *Mol. Cell. Biol.* **19**, 274–83.

Baim S.B., Labow M.A., Levine A.J. & Shenk T. (1991) A chimeric mammalian transactivator based on the lac repressor that is regulated by temperature and isopropyl beta-d-thiogalactopyranoside. *Proc. Nat. Acad. Sci. USA* **88**, 5072–6.

Bains W. & Smith G.C. (1988) A novel method for nucleic acid sequence determination. *J. Theo. Biol.* **135**, 303–7.

Bairoch A. (2000) The ENZYME database in 2000. *Nucl. Acids Res.* **28**, 304–5.

Balbás P., Soberon X., Merino E. *et al.* (1986) Plasmid vector pBR322 and its special-purpose derivatives – a review. *Gene* **50**, 3–40.

Ball C.A., Brazma A., Causton H. *et al.* (2004) Submission of microarray data to public repositories. *PLoS Biol.* **2**, 1276–7.

Ball C.A., Sherlock G., Parkinson H. *et al.* (2002) An open letter to the scientific journals. *Science* **98**, 539.

Balling R. (2001) ENU mutagenesis: Analyzing gene function in mice. *Ann. Rev. Genome Hum.* **2**, 463–92.

Ballivet M., Nef P., Coutourier S. *et al.* (1988) Electrophysiology of a chick neuronal nicotinic acetylcholine receptor expressed in *Xenopus* oocytes after cDNA injection. *Neuron* **1**, 847–52.

Baltz R.H. (1998) Genetic manipulation of antibiotic-producing *Streptomyces*. *Trends Microbiol.* **6**, 76–83.

Baner J., Nilsson M., Isaksson A., Mendel-Hartvig M., Antson D.O. & Landegren U. (2001) More keys to padlock probes: mechanisms for high-throughput nucleic acid analysis. *Curr. Opin. Biotechnol.* **12**, 11–15.

Baneyx F. & Mujacic M. (2004) Recombinant protein folding and misfolding in *Escherichia coli*. *Nature Biotechnol.* **22**, 1399–408.

Baneyx F. (1999) Recombinant protein expression in *Escherichia coli*. *Curr. Opin. Biotechnol.* **10**, 411–21.

Banfi S., Borsani G., Rossi E. *et al.* (1996) Identification and mapping of human cDNAs homologous to *Drosophila* mutant genes through EST database searching. *Nature Genet.* **13**, 167–74.

Bano-Maqbool S. & Christou P. (1999) Multiple traits of agronomic importance in transgenic indica rice plants: analysis of transgene integration patterns, expression levels and stability. *Mol. Breeding* **5**, 471–80.

Barnard E.A., Houghton M., Miledi R., Richards B.M. & Sumikawa K. (1982) Molecular genetics of the acetylcholine receptor and its insertion and organization in the membrane. *Biol. Cell* **45**, 383.

Barnes W.M. (1980) DNA cloning with single-stranded phage vectors. In *Genetic Engineering*, eds. Setlow J.K. & Hollaender A., Vol. 2, pp. 185–200. Plenum Press, New York.

Barnes W.M. (1994) PCR amplification of up to 35 kbp DNA with high fidelity and high yield from lambda bacteriophage templates. *Proc. Nat. Acad. Sci. USA* **91**, 2216–20.

Barta A., Sommergruber K., Thompson D. *et al.* (1986) The expression of a nopaline synthase-human growth hormone chimaeric gene in transformed tobacco and sunflower callus tissue. *Plant Mol. Biol.* **6**, 347–57.

Bartel, D.P. & Chen, C.Z. (2004) Micromanagers of gene expression: the potentially widespread influence of metazoan microRNAs. *Nat. Rev. Genet.* **5**, 396–400.

Bartel P.L., Roecklein J.A., SenGupta D. & Fields S. (1996) A protein linkage map of *Escherichia coli* bacteriophage T7. *Nature Genet.* **12**, 72–7.

Barth S., Huhn M., Matthey B. *et al.* (2000) Compatible-solute supported periplasmic expression of functional recombinant proteins under stress conditions. *Appl. Environ. Microbiol.* **66**, 1572–9.

Barton K.A., Whitely H.R. & Yang N.S. (1987) *Bacillus thuringiensis* delta endotoxin expressed in transgenic *Nicotiana tabacum* provides resistance to lepidopteran insects. *Plant Physiol.* **85**, 1103–9.

Bass S., Greener R. & Wells J.A. (1990) Hormone phage: an enrichment method for variant proteins with altered binding properties. *Proteins* **8**, 309–14.

Bates P.F. & Swift R.A. (1983) Double *cos* site vectors: simplified cosmid cloning. *Gene* **26**, 137–46.

Bates S., Cashmore A.M. & Wilkins B.M. (1998) IncP plasmids are unusually effective in mediating conjugation of *Escherichia coli* and *Saccharomyces cerevisiae*: involvement of the Tra2 mating system. *J. Bacteriol.* **180**, 6538–43.

Batzer M.A. & Deininger P.L. (2002) ALU repeats and human genomic diversity. *Nature Rev. Genet.* **3**, 370–9.

Baubonis W. & Saur B. (1993) Genomic targeting with purified Cre recombinase. *Nucl. Acids Res.* **21**, 2025–9.

Baudin A., Ozier-Kalogeropoulos O., Denouel A., Lacroute F. & Cullin C. (1993) A simple and efficient method for direct gene deletion in *Saccharomyces cerevisiae*. *Nucl. Acids Res.* **21**, 3329–30.

Baulcombe D.C. (1999) Fast forward genetics based on virus-induced gene silencing. *Curr. Opin. Plant Biol.* **2**, 109–13.

Baulcombe D.C., Chapman S. & SantaCruz S.S. (1995) Jellyfish green fluorescent protein as a reporter for virus-infections. *Plant J.* **7**, 1045–53.

Baxevanis A.D. (2002) The molecular biology database collection: 2002 update. *Nucl. Acids Res.* **30**, 1–12.

Beach L.R. & Palmiter R.D. (1981) Amplification of the metallothionein-I gene in cadmium-resistant mouse cells. *Proc. Nat. Acad. Sci. USA* **78**, 2110–14.

Beachy R., Loesch-Fries S. & Tumer N. (1990) Coat-protein mediated resistance against virus infection. *Ann. Rev. Phytopathol.* **28**, 451–74.

Becher A. & Schweizer H.P. (2000) Integration-proficient *Pseudomonas aeruginosa* vectors for isolation of single-copy chromosomal *lacZ* and *lux* fusions. *Biotechniques* **29**, 948–52.

Bechtold N. & Pelletier G. (1998) In planta *Agrobacterium*-mediated transformation of adult *Arabidosis thaliana* plants by vacuum infiltration. *Methods Mol. Biol.* **82**, 259–66.

Bechtold N., Ellis J. & Pelletier G. (1993) In planta *Agrobacterium*-mediated gene transfer by infiltration of adult *Arabidopsis thaliana* plants. *C. R. Acad. Sci. Paris Life Sci.* **316**, 1194–9.

Bechtold N., Jaudeau B., Jolivet S. *et al.* (2000) The maternal chromosome set is the target of the T-DNA in the *in planta* transformation of *Arabidopsis thaliana*. *Genetics* **155**, 1875–87.

Beck E. & Bremer E. (1980) Nucleotide sequence of the gene *omp*A encoding the outer membrane protein II of *Escherichia coli* K-12. *Nucl. Acids Res.* **8**, 3011–24.

Becker D.M., Fikes J.D. & Guarente L. (1991) A cDNA encoding a human CCAAT-binding protein cloned by functional complementation in yeast. *Proc. Nat. Acad. Sci. USA* **88**, 1968–72.

Beckers J. & Hrabe de Angelis M. (2001) Large-scale mutational analysis for the annotation of the mouse genome. *Curr. Opin. Chem. Biol.* **6**, 17–23.

Bedell M.A., Jenkins N.A. & Copeland N.G. (1997) Mouse models of human disease. Part II. Recent progress and future directions. *Genes Devel.* **11**, 11–43.

Beerli R.R., Wels W. & Hynes N.E. (1994) Intracellular expression of single chain antibodies reverts ErbB-2 transformation. *J. Biol. Chem.* **269**, 23931–6.

Beggs J.D. (1978) Transformation of yeast by a replicating hybrid plasmid. *Nature* **275**, 104–9.

Behr M.A. *et al.* (1999) Comparative genomics of BCG vaccines by whole-genome DNA microarray. *Science* **284**, 1520–3.

Beier M. & Hoheisel J.D. (2000) Production by quantitative photolithographic synthesis of individually quality checked DNA microarrays. *Nucl. Acids Res.* **28**, E11.

Belfort M. & Roberts R.J. (1997) Homing endonucleases: keeping the house in order. *Nucl. Acids Res.* **25**, 3379–88.

Bell G.I., Merryweather J.P., Sanchez-Pescador R. *et al.* (1984) Sequence of a cDNA clone encoding human preproinsulin-like growth factor II. *Nature* **310**, 775–7.

Bellanné-Chantelot C. *et al.* (1992) Mapping the whole human genome by fingerprinting yeast artificial chromosomes. *Cell* **70**, 1059–68.

Belshaw P.J., Ho S.N., Crabtree G.R. & Schreiber S.L. (1996) Controlling protein association and subcellular localization with a synthetic ligand that induces heterodimerization of proteins. *Proc. Nat. Acad. Sci. USA* **93**, 4604–7.

Bender M.A., Palmer T.D., Gelinas R.E. & Miller A.D. (1987) Evidence that the packaging signal of Moloney murine leukemia virus extends into the gag region. *J. Virology* **61**, 1639–46.

Bender W., Spierer P. & Hogness D.S. (1983) Chromosomal walking and jumping to isolate DNA from the Ace and rosy loci and the bithorax complex in *Drosophila melanogaster*. *J Mol. Biol.* **168**, 17–33.

Bendig M.M. & Williams J.G. (1983) Replication and expression of *Xenopus laevis* globin genes injected into fertilized *Xenopus* eggs. *Proc. Nat. Acad. Sci. USA* **80**, 6197–201.

Benhar I. (2001) Biotechnological applications of phage and cell display. *Biotechnol. Advances* **19**, 1–33.

Benihoud K., Yeh P. & Perricaudet M. (1999) Adenovirus vectors for gene delivery. *Curr. Opin. Biotechnol.* **10**, 440–7.

Bensen R.J. *et al.* (1995) Cloning and characterization of the maize *An1* gene. *Plant Cell* **7**, 75–84.

Benson D.A., Karsch-Mizrachi I., Lipman D.J., Ostell J. & Wheeler D.L. (2004) GenBank: update. *Nucl. Acids Res.* **1**, 32.

Bentley D.R., Todd C., Collins C. *et al.* (1992) The development and application of automated gridding for the efficient screening of yeast and bacterial ordered libraries. *Genomics* **12**, 534–41.

Benton W.D. & Davis R.W. (1977) Screening lgt recombinant clones by hybridization to single plaques *in situ*. *Science* **196**, 180–2.

Berger J., Hauber J., Hauber R., Geiger R. & Cullen B.R. (1988) Secreted placental alkaline phosphatase: a powerful new quantitative indicator of gene expression in eukaryotic cells. *Gene* **66**, 1–10.

Berges H., Joseph-Liauzun E. & Fayet O. (1996) Combined effects of the signal sequence and the major chaperone proteins on the export of human cytokines in *Escherichia coli*. *Appl. Environ. Microbiol.* **62**, 55–60.

Berglund P., Quesada-Rolander M., Putkonen P. *et al.* (1997) Outcome of immunization of cynomolgus monkeys with recombinant Semliki Forest virus encoding human immunodeficiency virus type 1 envelope protein and challenge with a high dose of SHIV-4 virus. *AIDS Res. Hum. Retroviruses* **13**, 1487–95.

Berglund P., Tubulekas I. & Liljestrom O. (1996) Alphaviruses as vectors for gene delivery. *Trends Biotechnol.* **14**, 130–4.

Bergthorsson U., Adams K.L., Thomason B. & Palmer J.D. (2003) Widespread horizontal transfer of mitochondrial genes in flowering plants. *Nature* **424**, 197–201.

Berndt C., Meier P. & Wackernagel W. (2003) DNA restriction is a barrier to natural transformation in *Pseudomonas stutzeri* JM300. *Microbiology* **149**, 895–901.

Berns K., Hijmans E.M., Mullenders J., Brummelkamp T.R., Velds A., Heimerikx M., Kerkhoven R.M., Madiredjo M., Nijkamp W., Weigelt B. *et al.* (2004) A large-scale RNAi screen in human cells identifies new components of the p53 pathway. *Nature* **428**, 431–7.

Berns K.I., Pinkerton T.C., Thomas G.F. & Hoggan M.D. (1975) Detection of adeno-associated virus (AAV)-specific nucleotide sequences in DNA isolatd from latently infected Detroit 6 cells. *Virology* **68**, 556–60.

Bertholet C., Drillien R. & Wittek R. (1985) One hundred base pairs of 5′ flanking sequence of vaccinia virus late gene are sufficient to temporally regulate transcription. *Proc. Nat. Acad. Sci. USA* **82**, 2096–100.

Bertucci F. *et al.* (1999) Sensitivity issues in DNA array based expression measurements: advantages of nylon membranes. *Hum. Mol. Genet.* **8**, 1715–22.

Bertucci F., Salas S., Eysteries S. *et al.* (2004) Gene expression profiling of colon cancer by DNA microarrays and correlation with histoclinical parameters. *Oncogene* **23**, 1377–91.

Bessette P.H., Aslund F., Beckwith J. & Georgiou G. (1999) Efficient folding of proteins with multiple disulfide bonds in the *Escherichia coli* cytoplasm. *Proc. Nat. Acad. Sci. USA* **96**, 13703–8.

Betancourt O.H., Attal J., Theron M.C., Puissant C. & Houdebine L.M. (1993) Efficiency of introns from various origins in fish cells. *Mol. Marine Biol. Biotechnol.* **2**, 181–8.

Bett A.J., Prevec L. & Graham F.L. (1993) Packaging capacity and stability of human adenovirus type 5 vectors. *J. Virol.* **67**, 5911–21.

Bevan M. (1984) Binary Agrobacterium vectors for plant transformation. *Nucl. Acids Res.* **12**, 8711–21.

Bevan M., Barnes W. & Chilton M.D. (1983a) Structure and transcription on the nopaline synthase gene region of T-DNA. *Nucl. Acids Res.* **11**, 369–85.

Bevan M., Flavell R.B. & Chilton M.D. (1983b) A chimaeric antibiotic resistance gene as a selectable market for plant cell transformation. *Nature* **304**, 184–7.

Beyer P., Al-Babili S., Ye X., Lucca P., Schaub P., Welsch R. & Potrykus I. (2002) Golden rice: introducing the β-carotene biosynthesis pathway into rice endosperm by genetic engineering to defeat vitamin A deficiency. *J. Nutr.* **132**, 506S–10S.

Bianchi M.M. *et al.* (1999) How to bring orphan genes into functional families. *Yeast* **15**, 513–26.

Bibb M.J., Schottel J.L. & Cohen S.N. (1980) A DNA cloning system for interspecies gene transfer in antibiotic-producing *Streptomyces*. *Nature* **284**, 526–31.

Bibb M.J., Ward J.M. & Hopwood D.A. (1978) Transformation of plasmid DNA into *Streptomyces* at high frequency. *Nature* **274**, 398–400.

Bickley J. & Hopkins D. (1999) Inhibitors and enhancers of PCR. In *Analytical Molecular Biology: Quality and Validation*, eds. Saunders G.C. & Parkes H.C., pp. 81–102. Royal Society of Chemistry, London.

Biery M.C., Stewart F.J., Stellwagen A.E. *et al.* (2000) A simple *in vitro* Tn7-based transposition system with low target site selectivity for genome and gene analysis. *Nucl. Acids Res.* **28**, 1067–77.

Bigey P., Bureau M.F. & Scherman D. (2002) *In vivo* plasmid DNA electrotransfer. *Curr. Opin. Biotechnol.* **13**, 443–7.

Bihoreau M.T. *et al.* (2001) A high-resolution consensus linkage map of the rat, integrating radiation hybrid and genetic maps. *Genomics* **75**, 57–69.

Bilan R., Futterer J. & Sautter C. (1999) Transformation of cereals. *Genet. Eng.* **21**, 113–57.

Bills M.M., Medd J.M., Chappel R.J. & Adler B. (1993) Construction of a shuttle vector for use between *Pasteurella multocida* and *Escherichia coli*. *Plasmid* **30**, 268–73.

Bingham P.M., Kidwell M.G. & Rubin G.M. (1982) The molecular basis of P-M hybrid dysgenesis: the role of the P element, a P-strain-specific transposon family. *Cell* **29**, 995–1004.

Bingle L.E.H. & Thomas C.M. (2001) Regulatory circuits for plasmid survival. *Curr. Opin. Microbiol.* **4**, 194–200.

Birch D.E. (1996) Simplified hot start PCR. *Nature* **381**, 445–6.

Birnboim H.C. & Doly J. (1979) A rapid alkaline extraction procedure for screening recombinant plasmid DNA. *Nucl. Acids Res.* **7**, 1513–23.

Birren B. & Lai E. (1994) Rapid pulsed field separation of DNA molecules up to 250 kb. *Nucl. Acids Res.* **22**, 5366–70.

Birren B.W., Lai E., Clark S.M., Hood L. & Simon M.I. (1988) Optimized conditions for pulsed field gel electrophoretic separations of DNA. *Nucl. Acids Res.* **16**, 7563–82.

Bishop J.O. & Smith P. (1989) Mechanism of chromosomal integration of microinjected DNA. *Mol. Biol. Med.* **6**, 283–98.

Bittner M. *et al.* (2000) Molecular classification of cutaneous malignant melanoma by gene expression profiling. *Nature* **406**, 536–40.

Blackwood E.M. & Eisenman R.N. (1991) Max: a helix-loop-helix zipper protein that forms a sequence-specific DNA-binding complex with c-Fos. *Science* **256**, 1014–18.

Blatny J.M., Brautaset T., Winther-Larsen H.C., Haughan K. & Valla S. (1997) Construction and use of a versatile set of broad-host-range cloning and expression vectors based on the RK2 replicon. *Appl. Environ. Microbiol.* **63**, 370–9.

Blattner F.R., Williams B.G., Blechl A.E. *et al.* (1977) Charon phages: safer derivatives of bacteriophage lambda for DNA cloning. *Science* **196**, 161–9.

Blesch A. (2004) Lentiviral and MLV based retroviral vectors for *ex vivo* and *in vivo* gene transfer. *Methods* **33**, 164–72.

Blobel G. & Dobberstein B. (1975) Transfer of proteins across membranes. I. Presence of proteolytically processed and unprocessed nascent immunoglobulin light chains on membrane-bound ribosomes of murine myeloma. *J. Cell Biol.* **67**, 835–51.

Blochlinger K. & Diggelmann H. (1984) Hygromycin B phosphotransferase as a selectable marker for DNA transfer experiments with higher eucaryotic cells. *Mol. Cell. Biol.* **4**, 2929–31.

Bloom K.S. & Carbon J. (1982) Yeast centromere DNA is a unique and highly ordered structure in chromosomes and small circular minichromosomes. *Cell* **29**, 305–17.

Blouin M.J., Beauchemin H., Wright A. *et al.* (2000) Genetic correction of sickle cell disease: insights using transgenic mouse models. *Nature Med.* **6**, 177–82.

Blum P., Velligan M., Lin N. & Matin A. (1992) DnaK-mediated alterations in human growth hormone protein inclusion bodies. *Biotechnology* **10**, 301–4.

Blundell T.L. & Mizuguchi K. (2000). Structural genomics: an overview. *Prog. Biophys. Mol. Biol.* **73**, 289–95.

Bochmann H., Gehrisch S. & Jaross W. (1999) The gene structure of the human growth factor bound protein GRB2. *Genomics* **56**, 203–7.

Bode V.C., McDonald J.D., Guenet J.L. & Simon D. (1988) hph-1: A mouse mutant with hereditary hyperphenylalaninemia induced by ethylnitrosourea mutagenesis. *Genetics* **118**, 299–305.

Boder E.T. & Wittrup K.D. (1997) Yeast surface display for screening combinatorial polypeptide libraries. *Nature Biotechnol.* **15**, 553–8.

Boeke J.D., Lacroute F. & Fink G.R. (1984) A positive selection for mutants lacking orotidine-5′-phosphate decarboxylase activity in yeast: 5-fluoro-orotic acid resistance. *Mol. Gen. Genet.* **197**, 345–6.

Boeke J.D., Vovis G.F. & Zinder N.D. (1979) Insertion mutant of bacteriophage fl sensitive to *Eco*RI. *Proc. Nat. Acad. Sci. USA* **76**, 2699–702.

Bofelli D., Nobrega M.A. & Rubin E.M. (2004) Comparative genomics at the vertebrate extremes. *Nature Rev. Genet.* **5**, 456–65.

Boguski M.S. (1995) The turning point in genome research. *Trends Biochem. Sci.* **20**, 295–332.

Bohl D., Naffakh N. & Heard J.M. (1997) Long term control of erythropoietin secretion by doxycycline in mice transplanted with engineered primary myoblasts. *Nature Med.* **3**, 299–305.

Bohner S., Lenk I., Rieping M., Herold M. & Gatz C. (1999) Transcriptional activator TGV mediates dexamethasone-inducible and tetracycline-inactivatable gene expression. *Plant J.* **19**, 87–95.

Bolivar F., Rodriguez R.L., Betlach M.C. & Boyer H.W. (1977a) Construction and characterization of new cloning vehicles. I. Ampicillin-resistant derivatives of the plasmid pMB9. *Gene* **2**, 75–93.

Bolivar F., Rodriguez R.L., Greene P.J. *et al.* (1977b) Construction and characterization of new cloning vehicles. II. A multipurpose cloning system. *Gene* **2**, 95–113.

Bolker M., Bohnert H.U., Braun K.H., Gorl J. & Kahmann R. (1995) Tagging pathogenicity genes in *Ustilago maydis* by restriction enzyme-mediated integration (REMI). *Mol. Gen. Genet.* **248**, 547–52.

Bomhoff G.H., Klapwijk F.M., Kester M.C.M. *et al.* (1976) Octopine and nopaline synthesis and breakdown genetically controlled by a plasmid of *Agrobacterium tumefaciens*. *Mol. Gen. Genet.* **145**, 177–81.

Bonen D.K., Ogura Y., Nicolae D.L., *et al.* (2003) Crohn's disease-associated NOD2 variants share a signalling defect in response to lipopolysaccharide and peptidoglycan. *Gastroenterology* **124**, 140–6.

Bonifer C. (1999) Long distance chromatin mechanisms controlling tissue-specific gene locus activation. *Gene* **238**, 277–89.

Bonifer C. (2000) Developmental regulation of eukaryotic gene loci: which cis-regulatory information is required? *Trends Genet.* **16**, 310–15.

Bonifer C., Vidal M., Grosveld F. & Sippel A.E. (1990) Tissue-specific and position independent expression of complete gene domain for chicken lysozyme in transgenic mice. *EMBO J.* **9**, 2843–8.

Bork P. & Bairoch A. (1996) Go hunting in sequence databases but watch out for the traps. *Trends Genet.* **12**, 425–7.

Bork P. & Koonin E.V. (1998) Predicting functions from protein sequences: where are the bottlenecks? *Nature Genet.* **18**, 313–18.

Borkovich K.A., Alex L.A., Yarden O., *et al.* (2004) Lessons from the genome sequence of *Neurospora crassa*: tracing the path from genomic blueprint to multicellular organism. *Microbiol. Mol. Biol. Rev.* **68**, 1–108.

Borodovsky M. & McIninch J. (1993) Recognition of genes in DNA sequence with ambiguities. *Biosystems* **30**, 161–71.

Bosch, I. *et al.* (2000) Identification of differentially expressed genes from limited amounts of RNA. *Nucl. Acids Res.* **28**, e27.

Boshart M., Weber F., Jahn G. *et al.* (1985) A very strong enhancer is located upstream of an immediate early gene of human cytomegalovirus. *Cell* **41**, 521–30.

Bosselman R.A., Hsu R.Y., Boggs T. *et al.* (1989) Germline transmission of exogenous genes in the chicken. *Science* **243**, 533–5.

Bostian K.A., Ellio O., Bussey H. *et al.* (1984) Sequence of the prepro-toxin ds RNA gene of Type 1 killer yeast: multiple processing events produce a two-component toxin. *Cell* **36**, 741–51.

Botstein D. & Davis R.W. (1982) Principles and practice of recombinant DNA research with yeast. In *The Molecular Biology of the Yeast Saccharomyces*, eds. Strathern J.N., Jones E.W. & Botstein D. & Fink G.R. (1988) Yeast: an experimental organism for modern biology. *Science* **240**, 1439–43.

Botstein D., White R.L., Skolnick M. & Davis R.W. (1980) Construction of a genetic linkage map in man using restriction fragment length polymorphisms. *Am. J. Hum. Genet.* **32**, 314–31.

Bouchez D., Camilleri C. & Caboche M. (1993) A binary vector based on Basta resistance for in planta transformation of *Arabidopsis thaliana*. *Comtes Rendus de l'Academy des Sciences Paris III* **316**, 1188–93.

Boussif O., Lezoualch F., Zanta M.A., Mergny M.D., Scherman D., Demeneix B. & Behr J.P. (1995) A versatile vector for gene and oligonucleotide transfer into cells in culture and *in vivo*: polyethylenimine. *Proc. Natl. Acad. Sci. USA* **92**, 7297–301.

Boutros M., Kiger A.A., Armknecht S., Kerr K., Hild M., Koch B., Haas S.A., Consortium H.F., Paro R. & Perrimon N. (2004) Genome-wide RNAi analysis of growth and viability in *Drosophila* cells. *Science* **303**, 832–5.

Boviatsis E.J., Park J.S., Sena-Esteves M. *et al.* (1994) Long-term survival of rats harboring brain neoplasms treated with ganciclovir and a herpes simplex virus vector that retains an intact thymidine kinase gene. *Cancer Res.* **54**, 5745–51.

Bowater R.P. & Wells R.D. (2000) The intrinsically unstable life of DNA triplet repeats associated with human hereditary disorders. *Prog. Nucl. Acid Res. Mol. Biol.* **66**, 159–202.

Bower R. & Birch R.G. (1992) Transgenic sugarcane plants via microprojectile bombardment. *Plant J.* **2**, 409–16.

Bowes C. *et al.* (1993) Localization of a retroviral element within the rd gene coding for the b-subunit of cGMP phosphodiesterase. *Proc. Nat. Acad. Sci. USA* **90**, 2955–9.

Bowtell D.D.L. (1999) Options available – from start to finish – for obtaining expression data by microarray. *Nature Genet.* **21**, 25–32.

Boyce F.M. & Bucher N.L.R. (1996) Baculovirus-mediated gene transfer into mammalian cells. *Proc. Nat. Acad. Sci. USA* **93**, 2348–52.

Boyd E.F., Davis B.M. & Hochhut B. (2001) Bacterio-phage–bacteriophage interactions in the evolution of pathogenic bacteria. *Trends Microbiol.* **9**, 137–44.

Boyko W.L. & Ganschow R.E. (1982) Rapid identification of *Escherichia coli* transformed by pBR322 carrying inserts at the *Pst*I site. *Anal. Biochem.* **122**, 85–8.

Boynton J.E., Gillham N.W., Harris E.H. *et al.* (1988) Chloroplast transformation in *Chlamydomonas* with high-velocity microprojectiles. *Science* **240**, 1534–8.

Bradley A., Evans M., Kaufman M.H. & Robertson E. (1984) Formation of germ line chimeras from embryo-derived teratocarcinoma cell lines. *Nature* **309**, 255–6.

Brady G., Funk A., Mattern J., Schutz G. & Brown R. (1985) Use of gene transfer and a novel cosmid rescue strategy to isolate transforming sequences. *EMBO J.* **4**, 2583–8.

Brandes C., Plautz J.D., Stanewsky R. *et al.* (1996) Novel features of *Drosophila* period transcription revealed by real-time luciferase reporting. *Neuron* **16**, 687–92.

Brandon E.P., Idzerda R.L. & McKnight G.S. (1995a) Targeting the mouse genome: a compendium of knockouts. Part I. *Curr. Biol.* **5**, 625–34.

Brandon E.P., Idzerda R.L. & McKnight G.S. (1995b) Targeting the mouse genome: a compendium of knockouts. Part II. *Curr. Biol.* **5**, 758–65.

Brandon E.P., Idzerda R.L. & McKnight G.S. (1995c) Targeting the mouse genome: a compendium of knockouts. Part III. *Curr. Biol.* **5**, 873–81.

Braslavsky, I., Hebert B., Kartalov E. & Quake S.R. (2003) Sequence information can be obtained from single DNA molecules. *Proc. Nat. Acad. Sci.* **100**, 3969–74.

Brazhnik P., de la Fuente A. & Mendes P. (2002) Gene networks: how to put the function in genomics. *Trends Biotechnol.* **20**, 467–72.

Brazma A., Hingamp P., Quackenbush J. *et al.* (2001) Minimum information about a microarray experiment (MIAME) – toward standards for microarray data. *Nature Genet.* **29**, 365–71.

Brazma A., Parkinson H., Sarkans U., Shojatalab M., Vilo J. *et al.* (2003) ArrayExpress – A public repository for microarray gene expression data at the EBI. *Nucl. Acids Res.* **31**, 68–71.

Brenner S. *et al.* (2000) Gene expression analysis by massively parallel signature sequencing (MPSS) on microbead arrays. *Nature Biotechnol.* **18**, 630–4.

Brenner S., Chothia C. & Hubbard T. (1997) Population statistics of protein structures. *Curr. Opin. Struct. Biol.* **7**, 369–76.

Brenner S.E. (1995) Network sequence retrieval. *Trends Genet.* **11**, 247–8.

Brenner S.E. (1999) Errors in genome annotation. *Trends Genet.* **15**, 132–3.

Brenner S.E. (2000) Target selection for structural genomics. *Nature Struct. Biol.* **7**, 967–9.

Brenner S.E. (2001) A tour of structural genomics. *Nature Rev. Genet.* **2**, 801–9.

Brenner S.E., Chothia C. & Hubbard T.J. (1998) Assessing sequence comparison methods with reliable structurally identified distant evolutionary relationships. *Proc. Nat. Acad. Sci. USA* **95**, 6073–8.

Breter H.J., Knoop M.-T. & Kirchen H. (1987) The mapping of chromosomes in *Saccharomyces cerevisiae*. I. A cosmid vector designed to establish, by cloning cdc mutants, numerous start loci for chromosome walking in the yeast genome. *Gene* **53**, 181–90.

Briddon R.W., Pinner M.S., Stanley J. & Markham P.G. (1990) Geminivirus coat protein gene replacement alters insect specificity. *Virology* **177**, 85–94.

Bridge A.J., Pebernard S., Ducraux A., Nicoulaz A.L. & Iggo R. (2003) Induction of an interferon response by RNAi vectors in mammalian cells. *Nature Genet.* **34**, 263–4.

Briem H. & Kuntz I.D. (1996) Molecular similarity based on DOCK-generated fingerprints. *J. Med. Chem.* **39**, 3401–8.

Briggs R. & King T.J. (1952) Transplantation of living nuclei from blastula cells into enucleated frog's eggs. *Proc. Nat. Acad. Sci. USA* **38**, 455–63.

Brinster R.L. (2002) Germline stem cell transplantation and transgenesis. *Science* **296**, 2174–6.

Brinster R.L., Chen H.Y., Trumbauer M. *et al.* (1981) Somatic expression of Herpes thymidine kinase in mice following injection of a fusion gene into eggs. *Cell* **27**, 223–31.

Brinster R.L., Chen H.Y., Warren R., Sarthy A. & Palmiter R.D. (1982) Regulation of metallothionein-thymidine kinase fusion plasmids injected into mouse eggs. *Nature* **296**, 39–42.

Brisson N., Paszkowski J., Penswick J.R. *et al.* (1984) Expression of a bacterial gene in plants by using a viral vector. *Nature* **310**, 511–14.

Britten R.J. & Kohne D.E. (1968) Repeated sequences in DNA. *Science* **161**, 529–40.

Broach, J.R. Pringle, J.R. & Jones, E.W. (1991) *The Molecular and Cellular Biology of the Yeast Saccharomyces.* Cold Spring Harbor Laboratory Press, Cold Spring Harbor, New York.

Brocard J., Warot X., Wendling O. *et al.* (1997) Spatio-temporally controlled site-specific somatic mutagenesis in the mouse. *Proc. Nat. Acad. Sci. USA* **94**, 14559–63.

Brochier B., Kieny M.P., Costy F. *et al.* (1991) Large-scale eradication of rabies using recombinant vaccinia-rabies vaccine. *Nature* **354**, 520–2.

Broglie K., Chet I., Holliday M. *et al.* (1991) Transgenic plants with enhanced resistance of the fungal pathogen *Rhizoctonia solanii*. *Science* **254**, 1194–7.

Bron S. & Luxen E. (1985) Segregational instability of pUB110-derived recombinant plasmids in *Bacillus subtilis. Plasmid* **14**, 235–44.

Bron S., Bosma P., Van Belkum M. & Luxen E. (1988) Stability function in the *Bacillus subtilis* plasmid pTA1060. *Plasmid* **18**, 8–16.

Bron S., Peijnenburg A., Peeters B., Haima P. & Venema G. (1989) Cloning and plasmid (in)stability in *Bacillus subtilis*. In *Genetic Transformation and Expression*, eds. Butler O.O., Harwood C.R. & Moseley B.E.B., pp. 205–19. Intercept, Andover.

Broome S. & Gilbert W. (1978) Immunological screening method to detect specific translation products. *Proc. Nat. Acad. Sci. USA* **75**, 2746–9.

Brosius J. (1984) Toxicity of an overproduced foreign gene product in *Escherichia coli* and its use in plasmid vectors for the selection of transcription terminators. *Gene* **27**, 161–72.

Brough D.E., Lizonova A., Hsu C., Kulesa V.A. & Kovesdi I. (1996) A gene transfer vector-cell line system for complete functional complementation of adenovirus early regions E1 and E4. *J. Virol.* **70**, 6497–501.

Bryan P.N. (2000) Protein engineering of subtilisin. *Biochim. Biophys. Acta* **1543**, 203–22.

Bryant S.H. & Lawrence C.E. (1993) An empirical energy function for threading protein-sequence through the folding motif. *Proteins* **16**, 92–112.

Buchanan-Wollaston V., Passiatore J.E. & Channon F. (1987) The mob and ori T functions of a bacterial plasmid promote its transfer to plants. *Nature* **328**, 172–5.

Buchholz F., Angrand P.-O. & Stewart A.F. (1998) Improved properties of FLP recombinase evolved by cycling mutagenesis. *Nature Biotechnol.* **16**, 657–62.

Buchholz F., Ringrose L., Angrand P.O., Rossi F. & Stewart A.F. (1996) Different thermostabilities of FLP and Cre recombinases: implications for applied site-specific recombination. *Nucl. Acids Res.* **24**, 4256–62.

Buchschacher G.L.J. & Panganiban A.T. (1992) Human immunodeficiency virus vectors for inducible expression of foreign genes. *J. Virol.* **66**, 2731–9.

Buckler A.J. *et al.* (1991) Exon amplification: a strategy to isolate mammalian genes based on RNA splicing. *Proc. Nat. Acad. Sci. USA* **88**, 4005–9.

Buetow K.H., Edmonson M.N. & Cassidy A.B. (1999) Reliable identification of large numbers of candidate SNPs from public EST data. *Nature Genet.* **21**, 323–5.

Buller R.M., Janik J.E., Sebring E.D. & Rose J.A. (1981) Herpes simplex virus types 1 and 2 completely help adenovirus associated virus replication. *J. Virol.* **40**, 241–7.

Bullerwell C.E., Leigh J., Forget L. & Lang B.F. (2003) A comparison of three fission yeast mitochondrial genomes. *Nucl. Acids Res.* **31**, 759–68.

Bult C.J. *et al.* (1996) Complete genome sequence of the methanogenic archaeon, *Methanococcus jannaschii*. *Science* **273**, 1058–73.

Bulyk M.L. *et al.* (2001) Exploring the DNA binding specificities of zinc fingers with DNA microarrays. *Proc. Nat. Acad. Sci. USA* **98**, 7158–63.

Bundock P. & Hooykaas P.J.J. (1996) Integration of *Agrobacterium tumefaciens* T-DNA in the *Saccharomyces cerevisiae* genome by illegitimate recombination. *Proc. Nat. Acad. Sci. USA* **93**, 15272–5.

Bundock P., Den Dulk-Ras A., Beijersbergen A. & Hooykaas P.J.J. (1995) Trans-kingdom T-DNA transfer from *Agrobacterium tumefaciens* to *Saccharomyces cerevisiae*. *EMBO J.* **14**, 3206–4.

Bundy J.B., Spurgeon D.J., Svendsen C. *et al.* (2002) Earthworm species of the genus *Eisenia* can be phenotypically differentiated by metabolic profiling. *FEBS Lett.* **521**, 115–20.

Burge C. & Karlin S. (1997) Prediction of complete gene structures in human genomic DNA. *J. Mol. Biol.* **268**, 78–94.

Burger G., Gray M.W. & Lang B.F. (2003) Mitochondrial genomes: anything goes. *Trends Genet.* **19**, 709–16.

Burke B. & Warren G. (1984) Microinjection of messenger RNA for an anti-Golgi antibody inhibits intracellular transport of a viral membrane protein. *Cell* **36**, 847–56.

Burke D.T., Carle G.F. & Olson M.V. (1987) Cloning of large segments of exogenous DNA into yeast by means of artificial chromosome vectors. *Science* **236**, 806–13.

Burkhardt P.K., Beyer P., Wunn J. *et al.* (1997) Transgenic rice (*Oryza sativa*) endosperm expressing daffodil (*Narcissus pseudonarcissus*) phytoene synthase accumulates phytoene, a key intermediate of provitamin A biosynthesis. *Plant J.* **11**, 1071–8.

Burkholder J.K., Decker J. & Yang N.S. (1993) Rapid transgene expression in lymphocyte and macrophage primary cultures after particle bombardment-mediated gene transfer. *J. Immunol. Methods* **165**, 149–56.

Burley S.K. *et al.* (1999) Structural genomics: beyond the human genome project. *Nature Genet.* **23**, 151–7.

Burnette W.N. (1981) Western blotting: electrophoretic transfer of proteins from sodium dodecyl sulphate-polyacrylamide gels to unmodified nitrocellulose and radiographic detection with antibody and radioiodinated protein A. *Anal. Biochem.* **112**, 195–203.

Burright E.N., Clark H.B., Servadio A. *et al.* (1995) SCA1 transgenic mice – a model for neurodegeneration caused by CAG trinucleotide expansion. *Cell* **82**, 937–48.

Burset M. & Guigó R. (1996) Evaluation of gene structure prediction programs. *Genomics* **34**, 353–67.

Burton E.A., Fink D.J. & Glorioso J.C. (2002) Gene delivery using herpes simplex virus vectors. *DNA Cell Biol.* **21**, 915–36.

Burton R.A., Gibeaut D.M., Bacic A. *et al.* (2000) Virus-induced silencing of a plant cellulose synthase gene. *Plant Cell* **12**, 691–705.

Bushman F.D. (2003) Targeting survival: integration site selection by retroviruses and LTR-retrotransposons. *Cell* **115**, 135–8.

Bussow K. *et al.* (1998) A method for global protein expression and antibody screening on high-density filters of an arrayed cDNA library. *Nucl. Acids Res.* **26**, 5007–8.

C. elegans Sequencing Consortium (1998) Genome sequence of the nematode *C. elegans*: a platform for investigating biology. *Science* **282**, 2012–18.

Cabanes-Macheteau M., Fitchette-Laine A.C., Loutelier-Bourhis C. *et al.* (1999) N-glycosylation of a mouse IgG expressed in transgenic tobacco plants. *Glycobiology* **9**, 365–72.

Cacharron J., Saedler H. & Theissen G. (1999) Expression of MADS box genes *ZMM8* and *ZMM14* during inflorescence development of *Zea mays* discriminates between the upper and the lower floret of each spiklet. *Development Genes & Evolution* **209**, 411–20.

Cai D.W., Mukhopadhyay T., Lui T., Fujiwara T. & Roth J.A. (1993) Stable expression of the wild type p53 gene in human lung cancer cells after retrovirus mediated gene transfer. *Hum. Gene Ther.* **4**, 617–24.

Calderwood N.A., White R.E. & Whitehouse A. (2004) Development of herpesvirus-based episomally maintained gene delivery vectors. *Expert Opin. Biol. Th.* **4**, 493–505.

Caldwell R.C. & Joyce G.F. (1994) Mutagenic PCR. *PCR Methods and Applications* **3**, S136–40.

Caminci P., Kvam C., Kiamura A. *et al.* (1996) High efficiency full length cDNA cloning by biotinylated CAP trapper. *Genomics* **37**, 327–36.

Campbell K.H.S., McWhir J., Richie W.A. & Wilmut I. (1996) Sheep cloned by nuclear transfer from a cultured cell line. *Nature* **380**, 64–7.

Campisi L. *et al.* (1999) Generation of enhancer trap lines in *Arabidopsis* and characterization of expression patterns in the inflorescence. *Plant J.* **17**, 699–707.

Canel C., Lopez-Cardoso M.I., Whitmer S. *et al.* (1998) Effects of over expression of strictosidine synthase and tryptophan decarboxylase on alkaloid production by cell cultures of *Catharanthus roseus*. *Planta* **205**, 414–19.

Canfield V., Emanuael J.R., Spickofsky N., Levenson R. & Margolskee R.F. (1990) Ouabain resistant mutants of the rat Na, K-ATPase a2 isoform identified using an episomal expression vector. *Mol. Cell. Biol.* **10**, 1367–72.

Cangelosi G.A., Best E.A., Martinetti G. & Nester E.W. (1991) Genetic analysis of *Agrobacterium*. *Methods Enzymol.* **204**, 384–97.

Canosi U., Iglesias A. & Trautner T.A. (1981) Plasmid transformation in *Bacillus subtilis*: effects of insertion of *Bacillus subtilis* DNA into plasmid pC194. *Mol. Gen. Genet.* **181**, 434–40.

Canosi U., Morelli G. & Trautner T.A. (1978) The relationship between molecular structure and transformation efficiency of some *S. aureus* plasmids isolated from *B. subtilis*. *Mol. Gen. Genet.* **166**, 259–67.

Capecchi M.R. (1980) High efficiency transformation by direct microinjection of DNA into cultured mammalian cells. *Cell* **22**, 479–88.

Capecchi M.R. (2000) Choose your target. *Nature Genet.* 26, 159–61.

Capell T. & Christou P. (2004) Progress in plant metabolic engineering. *Curr. Opin. Biotechnol.* **15**, 148–54.

Caplen N.J., Alton E.W.F.W., Middleton P.G. *et al.* (1995) Liposome-mediated *CFTR* gene transfer to the nasal epithelium of patients with cystic fibrosis. *Nature Med.* **1**, 39–46.

Caplen N.J., Parrish S., Imani F., Fire A. & Morgan R.A. (2001) Specific inhibition of gene expression by small double-stranded RNAs in invertebrate and vertebrate systems. *Proc. Nat. Acad. Sci. USA.* **98**, 9742–7.

Cardon L.R. & Abecasis G.R. (2003) Using haplotype blocks to map human complex trait loci. *Trends Genet.* **19**, 135–40.

Cardoso A.I., Blixenkrone-Moller M., Fayolle J. *et al.* (1996) Immunization with plasmid DNA encoding for the measles virus hemagglutinin and nucleoprotein leads to humoral and cell-mediated immunity. *Virology* **225**, 293–9.

Carmell M.A., Zhang L., Conklin D.S., Hannon G.J. & Rosenquist T.A. (2003) Germline transmission of RNAi in mice. *Nature Struct. Biol.* **10**, 91–2.

Caron de Fromentel C., Gruek N., Venot C. *et al.* (1999) Restoration of transcriptional activity of p53 mutants in human tumour cells by intracellular expression of anti-p53 single chain Fv fragments. *Oncogene* **18**, 551–7.

Carroll M.W. & Moss B. (1995) E. coli b-glucuronidase (GUS) as a marker for recombinant vaccinia viruses. *Biotechniques* **19**, 352–5.

Carter G. & Lemoine N.R. (1993) Antisense technology for cancer therapy: does it make sense? *Br. J. Cancer* **67**, 869–76.

Carter P., Bedouelle H. & Winter G. (1985) Improved oligonucleotide site-directed mutagenesis using M13 vectors. *Nucl. Acids Res.* **13**, 4431–43.

Cartier M., Chang M. & Stanners C. (1987) Use of the *Escherichia coli* gene for asparagine synthetase as a selective marker in a shuttle vector capable of dominant transfection and amplification in animal cells. *Mol. Cell. Biol.* **7**, 1623–8.

Cary J.W., Rajasekaran K., Jaynes J.M. & Cleveland T.E. (2000) Transgenic expression of a gene encoding a synthetic antimicrobial peptide results in inhibition of fungal growth *in vitro* and *in planta*. *Plant Sci.* **154**, 171–81.

Casjens S. *et al.* (2000) A bacterial genome in flux: the twelve linear and nine circular extrachromosomal DNAs in an infectious isolate of the Lyme disease spirochete *Borrelia burgdorferi*. *Mol. Microbiol.* **35**, 490–516.

Castanie M.P., Berges H., Oreglia J., Prere M.F. & Fayet O. (1997) A set of pBR322-compatible plasmids allowing the testing of chaperone-assisted folding of proteins over-expressed in *Escherichia coli*. *Anal. Biochem.* **254**, 150–2.

Castilla J., Pintado B., Sola I., Sanchez-Morgado J.M. & Enjuanes L. (1998) Engineering passive immunity in transgenic mice secreting virus-neutralizing antibodies in milk. *Nature Biotechnol.* **16**, 349–54.

Cepko C.L., Roberts B.E. & Mulligan R.C. (1984) Construction and applications of a highly transmissible murine retrovirus shuttle vector. *Cell* **37**, 1053–62.

Cereghino G.P.L., Cereghinou J.L., Ilgen C. & Cregg J.M. (2002) Production of recombinant proteins in fermenter cultures of the yeast *Pichia pastoris. Curr. Opin. Biotechnol.* **13**, 329–32.

Cereghino J.L. & Cregg J.M. (1999) Applications of yeast in biotechnology: protein production and genetic analysis. *Curr. Opin. Biotechnol.* **10**, 422–7.

Cereghino J.L. & Cregg J.M. (2000) Heterologous protein expression in the methylotrophic yeast *Pichia pastoris. FEMS Microbiol. Rev.* **24**, 45–66.

Cereghino J.L., Helinski D.R. & Toukdarian A.E. (1994) Isolation and characterization of DNA-binding mutants of a plasmid replication initiation protein utilizing an *in vivo* binding assay. *Plasmid* **31**, 89–99.

Cesarini G., Helmer-Citterich M. & Castagnoli L. (1991) Control of Col E1 plasmid replication by antisense RNA. *Trends Genet.* **7**, 230–5.

Cesarini G., Muesing M.A. & Polisky B. (1982) Control of Col E1 DNA replication: the *rop* gene product negatively affects transcription from the replication primer promoter. *Proc. Nat. Acad. Sci. USA* **79**, 6313–17.

Chada K., Magram J. & Constantini F. (1986) An embryonic pattern of expression of a human fetal globin gene in transgenic mice. *Nature* **319**, 685–9.

Chakrabarti S., Breaching K. & Moss B. (1985) Vaccinia virus expression vector: coexpression of β galactosidase provides visual screening of recombinant plaques. *Mol. Cell. Biol.* **5**, 3403–9.

Chakrabarti S., Robert-Guroff M., Wong-Staal F., Gallo R.C. & Moss B. (1986) Expression of the HTLV-III envelope gene by a recombinant vaccinia virus. *Nature* **320**, 535–7.

Chakravarti D.N., Chakravarti B. & Moutsatsos I. (2002) Informatic tools for proteome profiling. *Comp. Proteomics* **32**, S4–S15.

Chalfie M., Tu Y., Euskirchen G., Ward W.W. & Prasher D.C. (1994) Green fluorescent protein as a marker for gene expression. *Science* **263**, 802–5.

Champoux J.J. (1995) Roles of ribonuclease H in reverse transcription. In *Reverse Transcriptase*, eds. Skalka, A.M. & Goff, S.P., pp. 103–17. Cold Spring Harbor Press, Cold Spring Harbor, New York.

Chan A.W.S., Chong K.Y., Martinovich C., Simerly C. & Shatten G. (2001) Transgenic monkeys produced by retroviral gene transfer into mature oocytes. *Science* **291**, 309–12.

Chan A.W.S., Homan E.J., Ballou L.U., Burns J.C. & Brennel R.D. (1998) Transgenic cattle produced by reverse transcribed gene transfer in oocytes. *Proc. Nat. Acad. Sci. USA* **95**, 14028–33.

Chan A.W.S., Luetjen C.M., Dominko T. *et al.* (2000) Foreign DNA transmission by ICSI: injection of spermatozoa bound with exogenous DNA results in embryonic GFP expression and live Rhesus monkey births. *Mol. Hum. Reprod.* **6**, 26–33.

Chan M.-T., Chang H.-H., Ho S.-L., Tong W.-F. & Yu S.-M. (1993) *Agrobacterium*-mediated production of transgenic rice plants expressing a chimeric α-amylase promoter/β-glucuronidase gene. *Plant Mol. Biol.* **22**, 491–506.

Chan M.-T., Lee T.-M. & Chang H.-H. (1992) Transformation of indica rice (*Oryza sativa* L.) mediated by *Agrobacterium tumefaciens. Plant Cell Physiol.* **33**, 577–83.

Chang A.C.Y., Nunberg J.H., Kaufman R.K. *et al.* (1978) Phenotypic expression in *E. coli* of a DNA sequence coding for mouse dihydrofolate reductase. *Nature* **275**, 617–24.

Chang L.M.S. & Bollum F.J. (1971) Enzymatic synthesis of oligodeoxynucleotides. *Biochemistry* **10**, 536–42.

Chang S. & Cohen S.N. (1979) High-frequency transformation of *Bacillus subtilis* protoplasts by plasmid DNA. *Mol. Gen. Genet.* **168**, 111–15.

Chang S.S., Park S.K., Kim B.C. *et al.* (1994) Stable genetic transformation of *Arabidopsis thaliana* by *Agrobacterium* inoculation in planta. *Plant J.* **5**, 551–8.

Chapman A.B., Costello M.A., Lee R. & Ringold G.M. (1983) Amplification and hormone-regulated expression of a mouse mammary tumor virus-Ecogpt fusion plasmid in mouse 3T6 cells. *Mol. Cell. Biol.* **3**, 1421–9.

Chapman S., Kavanagh T. & Baulcombe D. (1992) Potato virus-X as a vector for gene-expression in plants. *Plant J.* **2**, 549–57.

Charpentier E., Gerbaud G. & Courvalin P. (1999) Conjugative mobilization of the rolling-circle plasmid pIP823 from *Listeria monocytogenes* BM4293 among Gram-positive and Gram-negative bacteria. *J. Bacteriol.* **181**, 3368–74.

Chater K.F. & Hopwood D.A. (1983) *Streptomyces* genetics. In *Biology of the Actinomycetes*, eds. Goodfellow M. Mordarski M. & Williams S.T., pp. 229–85. Academic Press, London.

Chatoo B.B., Sherman F., Azubalis D.A. *et al.* (1979) Selection of lys2 mutants of the yeast *Saccharomyces cerevisiae* by the utilisation of α-aminoadipate. *Genetics* **93**, 51–65.

Chatterjee P.K. & Coren J.S. (1997) Isolating large nested deletions in bacterial and P1 artificial chromosomes by *in vivo* P1 packaging of products of Cre-catalysed recombination between the endogenous and a transported *loxP* site. *Nucl. Acids Res.* **25**, 2205–12.

Chaudhuri R.R., Khan A.M. & Pallen M.J. (2004) *coli*BASE: an online database for *Escherichia coli*, *Shigella* and *Salmonella* comparative genomics. *Nucl. Acids Res.* **32**, D296–9.

Chee M., Yang R., Hubbell E. *et al.* (1996) Accessing genetic information with high-density DNA arrays. *Science* **274**, 610–14.

Cheek B.J. *et al.* (2001) Chemiluminescence detection for hybridization assays on the flow-thru chip, a three-dimensional microchannel biochip. *Anal. Chem.* **73**, 5777–83.

Chen C. & Okayama H. (1987) High efficiency transformation of mammalian cells by plasmid DNA. *Mol. Cell. Biol.* **7**, 2745–51.

Chen C. & Okayama H. (1988) Calcium phosphate-mediated gene transfer: a highly efficient transfection system for stably transforming cells with plasmid DNA. *Biotechniques* **6**, 632.

Chen C.Y., Oppermann H. & Hitzeman R.A. (1984) Homologous versus heterologous gene expression in the yeast *Saccharomyces cerevisiae*. *Nucl. Acids Res.* **12**, 8951–70.

Chen J. & Hebert P.D.N. (1998) Directed termination PCR: a one step approach to mutation detection. *Nucl. Acids Res.* **26**, 1546–7.

Chen J.J. *et al.* (1998) Profiling expression patterns and isolating differentially expressed genes by cDNA microarray system with colorimetry detection. *Genomics* **51**, 313–24.

Chen K. & Arnold F.H. (1991) Enzyme engineering for non-aqueous solvents: random mutagenesis to enhance activity of subtilisin E in polar organic media. *Biotechnology* **9**, 1073–7.

Chen K. & Arnold F.H. (1993) Tuning the activity of an enzyme for unusual environments: sequential random mutagenesis of subtilisin E for catalysis in dimethylformamide. *Proc. Nat. Acad. Sci. USA* **90**, 5618–22.

Chen M., Presting G., Barbazuk W.B., *et al.* (2002) An integrated physical and genetic map of the rice genome. *Plant Cell* **14**, 537–45.

Chen T.H.H. & Murata N. (2002) Enhancement of tolerance of abiotic stress by metabolic engineering of betaines and other compatible solutes. *Curr. Opin. Plant Biol.* **5**, 250–7.

Chen Y.Z. *et al.* (2001) A bac-based sts-content map spanning a 35-mb region of human chromosome 1p35–p36. *Genomics* **74**, 55–70.

Chen Z. (1996) Simple modifications to increase specificity of the 5′-RACE procedure. *Trends Genet.* **12**, 87–8.

Cheng M., Fry J.E., Pang S.Z. *et al.* (1997) Genetic transformation of wheat mediated by *Agrobacterium tumefaciens*. *Plant Physiol.* **115**, 971–80.

Cheng S., Chang S.Y., Gravitt P. & Respess R. (1994b) Long PCR. *Nature* **369**, 684–5.

Cheng S., Fockler C., Barmes W.M. & Higuchi R. (1994a) Effective amplification of long targets from cloned inserts and human genome DNA. *Proc. Nat. Acad. Sci. USA* **91**, 5695.

Cheng Z.Q., Targolli J., Huang X. & Wu R. (2002) Wheat LEA genes, *PMA80* and *PMA1959*, enhance dehydration tolerance of transgenic rice (*Oryza sativa* L.). *Mol. Breeding* **10**, 71–82.

Cheng Z.Q., Targolli J., Su J., He C.K., Li F. & Wu R. (2001) Transgenic approaches for generating rice tolerant of dehydration stress. In Khush G.S., Brar D.S. & Hardy B. (eds.) *Rice Genetics IV*. Science Publishers, New Delhi, pp. 433–8.

Chesne P., Adenot P.G., Viglietta C., Baratte M., Boulanger L. & Renard J.P. (2002) Cloned rabbits produced by nuclear transfer from adult somatic cells. *Nature Biotechnol.* **20**, 366–9.

Cheung V.G. & Nelson S.F. (1996) Whole genome amplification using a degenerate oligonucelotide primer allows hundrends of genotypes to be performed on less than one nanogram of genomic DNA. *Proc. Nat. Acad. Sci. USA* **93**, 14676–9.

Cheung V.G. *et al.* (2001) Integration of cytogenetic landmarks into the draft sequence of the human genome. *Nature* **409**, 953–8.

Cheung V.G., Morley M., Aguilar F., Massimi A., Kucherlapati R. & Childs G. (1999) Making and reading microarrays. *Nature Genet.* **21**, 15–19.

Chiang T.-R. & McConlogue L. (1988) Amplification of heterologous ornithine decarboxylase in Chinese hamster ovary cells. *Mol. Cell. Biol.* **8**, 764–9.

Chiba Y., Suzuki M., Yoshida S. *et al.* (1998) Production of human compatible high mannose-type (Man$_5$GlcNAc$_2$) sugar chains in *Saccharomyces cerevisiae*. *J. Biol. Chem.* **273**, 26298–304.

Chin J.W., Cropp T.A., Anderson J.C., *et al.* (2003) An expanded eukaryotic genetic code. *Science* **301**, 964–7.

Chissoe S.L., Marra M.A., Hillier L., Brinkman R., Wilson R.K. & Waterson R.H. (1997) Representation of cloned genomic sequences in two sequencing vectors: correlation of DNA sequence and subclone distribution. *Nucl. Acids Res.* **25**, 2960–6.

Chiu W., Niwa Y., Zeng W. *et al.* (1996) Engineered GFP as a vital reporter in plants. *Curr. Biol.* **3**, 325–30.

Cho R.J., Campbell M.J., Winzeler E.A. *et al.* (1998) A genome-wide transcriptional analysis of the mitotic cell cycle. *Mol. Cell* **2**, 65–73.

Cho R.J., Mindrinos M., Richards D.R., *et al.* (1999) Genome-wide mapping with biallelic markers in *Arabidopsis thaliana*. *Nature Genet.* **23**, 203–7.

Choi B.K., Hercules D.M. & Gusev A.I. (2001) Effect of liquid chromatography separation of complex matrices on liquid chromatography-tandem mass spectrometry signal suppression. *J. Chromatography A* **907**, 337–42.

Chong H., Ruchatz A., Clackson T., Rivera V.M. & Vile R.G. (2002) A system for small-molecule control of conditionally replication-competent adenoviral vectors. *Mol. Ther.* **5**, 195–203.

Chong S. & 12 others (1997) Single-column purification of free recombinant proteins using a self-cleavable affinity tag derived from a protein splicing element. *Gene* **192**, 271–81.

Chong S., Montello G.E., Zhang A. *et al.* (1998) Utilizing the C-terminal cleavage activity of a protein splicing element to purify recombinant protein in a single chromatographic step. *Nucl. Acids Res.* **26**, 5109–15.

Chotani G., Dodge T., Hsu A. *et al.* (2000) The commercial production of chemicals using pathway engineering. *Biochim. Biophys. Acta* **1543**, 434–55.

Choudhary J.S., Blackstock W.P., Creasy D.M. & Cottrell J.S. (2001a) Interrogating the human genome using uninterpreted mass spectrometry data. *Proteomics* **1**, 651–67.

Choudhary J.S., Blackstock W.P., Creasy D.M. & Cottrell J.S. (2001b) Matching peptide mass spectra to EST and genomic DNA databases. *Trends Biotechnol.* **19**, S17–S22.

Christendat D. *et al.* (2000) Structural proteomics of an archaeon. *Nature Struct. Biol.* **7**, 903–9.

Christensen A.H. & Quail P.H. (1996) Ubiquitin promoter-based vectors for high-level expression of selectable and/or screenable marker genes in monocotyledonous plants. *Transgenic Res.* **5**, 213–18.

Christou P. (1996) *Particle Bombardment for Genetic Engineering of Plants*. RG Landes Co, Austin TX.

Christou P. & Swain W.F. (1990) Cotransformation frequencies of foreign genes in soybean cell cultures. *Theor. Appl. Genet.* **90**, 97–104.

Christou P. & Twyman R.M. (2004) The potential of genetically enhanced plants to address food insecurity. *Nutrition Res. Rev.* **17**, 23–42.

Christou P., Ford T.L. & Kofron M. (1991) Production of transgenic rice (*Oryza sativa* L.) plants from agronomically important indica and japonica varieties via electric discharge particle acceleration of exogenous DNA into immature zygotic embryos. *Biotechnology* **9**, 957–62.

Christou P., McCabe D.E. & Swain W.F. (1988) Stable transformation of soybean callus by DNA-coated gold particles. *Plant Physiol.* **87**, 671–4.

Chu G. & Sharp P.A. (1981) SV40 DNA transfection of cells in suspension: analysis of the efficiency of transcription and translation of T-antigen. *Gene* **13**, 197–202.

Chu G., Vollrath D. & Davis R. (1986) Separation of large DNA molecules by contour clamped homogenous electric fields. *Science* **234**, 1582–5.

Chu S., DeRisi J., Eisen M. *et al.* (1998) The transcriptional program of sporulation in budding yeast. *Science* **282**, 699–705.

Chuang C.F. & Meyerowitz E.M. (2000) Specific and heritable genetic interference by double-stranded RNA in *Arabidopsis thaliana*. *Proc. Nat. Acad. Sci. USA* **97**, 4985–90.

Church D.M. & Buckler A.J. (1999) Gene identification by exon amplification. *Methods Enzymol.* **303**, 83–99.

Ciavattia D.J., Ryan T.M., Farmer S.C. & Townes T.M. (1995) Mouse model of human beta-zero thalassemia: targeted deletion of the mouse beta major- and beta minor-globin genes in embryonic stem cells. *Proc. Nat. Acad. Sci. USA* **92**, 9259–63.

Cibelli J.B., Stice S.L., Golueke P.J. *et al.* (1998) Cloned transgenic calves produced from nonquiescent fetal fibroblasts. *Science* **280**, 1256–8.

Cirino P.C., Mayer K.M. & Umeno D. (2003) Generating mutant libraries using error-prone PCR. *Methods Mol. Biol.* **231**, 3–9.

Clark A.J., Cowper A., Wallace R., Wright G. & Simmons J.P. (1992) Rescuing transgene expression by co-integration. *Biotechnology* **10**, 1450–4.

Clark A.J., Harold G. & Yull F.E. (1997) Mammalian cDNA and prokaryotic reporter sequences silence adjacent transgenes in transgenic mice. *Nucl. Acids Res.* **25**, 1009–14.

Clark K.J., Geurts A.M., Bell J.B. & Hackett P.B. (2004) Transposon vectors for gene-trap insertional mutagenesis in vertebrates. *Genesis* **39**, 225–33.

Clark W., Register J., Nejidat A. *et al.* (1990) Tissue-specific expression of the TMV coat protein in transgenic tobacco plants affects the level of coat protein-mediated virus protection. *Virology* **179**, 640–7.

Clarke L. & Carbon J. (1976) A colony bank containing synthetic Col E1 hybrid plasmids representative of the entire *E. coli* genome. *Cell* **9**, 91–9.

Clarke L. & Carbon J. (1980) Isolation of a yeast centromere and construction of functional small circular chromosomes. *Nature* **287**, 504–9.

Claudio J.O. *et al.* (1998) Identification of sequence tagged transcripts differentially expressed within the human hematopoeitic hierarchy. *Genomics* **50**, 44–52.

Claverie J.M. (1997) Computational methods for the identification of genes in vertebrate genomic sequences. *Hum. Mol. Gen.* **6**, 1735–44.

Claverys J-P. & Martin B. (2003) Bacterial "competence" genes: signatures of active transformation or only remnants? *Trends in Microbiology* **11**, 161–5.

Clough S.J. & Bent A. (1998) Floral dip: a simplified method for *Agrobacterium*-mediated transformation of *Arabidopsis thaliana*. *Plant J.* **16**, 735–43.

Cocchia M., Kouprina N., Kim, S-J. *et al.* (2000) Recovery and potential utility of YACs as circular YACs/BACs. *Nucl. Acids Res.* **28**, e81.

Cochet O., Kenigsberg M., Delumeau I. *et al.* (1998) Intracellular expression of an antibody fragment neutralizing p21 ras promotes tumor regression. *Cancer Res.* **58**, 1170–6.

Cochran M.A., Puckett C. & Moss B. (1985) *In vitro* mutagenesis of the promoter region for a vaccinia virus gene: evidence for tandem early and late regulatory signals. *J. Virol.* **54**, 30–7.

Cockett M.I., Bebbington C.R. & Yarranton G.T. (1990) High level expression of tissue inhibitor of metalloproteinases in Chinese hamster ovary cells using glutamine synthetase gene amplification. *Biotechnology* **8**, 662–7.

Coco W.M. (2003) RACHITT: Gene shuffling by random chimeragenesis on transient templates. *Methods Mol. Biol.* **231**, 111–27.

Coco W.M., Levinson W.E., Crist M.J., *et al.* (2001) DNA shuffling method for generating highly recombined genes and evolved enzymes. *Nature Biotechnol.* **19**, 354–9.

Coe E.H. & Sarkar K.R. (1966) Preparation of nucleic acids and a genetic transformation attempt in maize. *Crop Sci.* **6**, 432–4.

Coelho P.S.R., Kumar A. & Snyder M. (2000) Genome-wide mutant collections: toolboxes for functional genomics. *Curr. Opin. Microbiol.* **3**, 309–15.

Cohen S.N., Chang A.C.Y. & Hsu L. (1972) Nonchromosomal antibiotic resistance in bacteria: genetic transformation of *Escherichia coli* by R-factor DNA. *Proc. Nat. Acad. Sci. USA* **69**, 2110–14.

Cohn B. (1979) Cosmids: a type of plasmid gene-cloning vector that is packageable *in vitro* in bacteriophage λ heads. *Proc. Nat. Acad. Sci. USA* **75**, 4242–6.

Colbère-Garapin F., Horodniceanu F., Kourilsky P. & Garapin A.C. (1981) A new dominant hybrid selective marker for higher eukaryotic cells. *J. Mol. Biol.* **150**, 1–14.

Cole S.T. & Saint Girons I. (1994) Bacterial genomes. *FEMS Microbiol. Rev.* **14**, 139–60.

Cole S.T. *et al.* (2001) Massive gene decay in the leprosy bacillus. *Nature* **409**, 1007–11.

Coleclough C. & Erlitz E.L. (1985) Use of primer restriction end adapters in a novel cDNA cloning strategy. *Gene* **34**, 305–14.

Collart F.R. & Huberman E. (1987) Amplification of the IMP dehydrogenase gene in Chinese hamster cells resistant to mycophenolic acid. *Mol. Cell. Biol.* **7**, 3328–31.

Collas P. & Robl J.M. (1990) Relationship between nuclear remodeling and development in nuclear transplant rabbit embryos. *Biol. Reprod.* **45**, 455–65.

Collins F.S., Drumm M.L., Cole J.L. *et al.* (1987) Construction of a general human chromosome jumping library, with application to cystic fibrosis. *Science* **235**, 1046–9.

Collins J. & Brüning H.J. (1978) Plasmids usable as gene-cloning vectors in an *in vitro* packaging by coliphage λ: "cosmids". *Gene* **4**, 85–107.

Colman A. (1984) Translation of eukaryotic messenger RNA in *Xenopus* oocytes. In: *Transcription and Translation – A Practical Approach*, eds Hames B.D. & Higgens S.J., pp. 271–302. IRL Press, Oxford.

Colman A., Lane C., Craig R. *et al.* (1981) The influence of topology and glycosylation on the fate of heterologous secretory proteins made in *Xenopus* oocytes. *Eur. J. Biochem.* **113**, 339–48.

Comai L., Facciotti D., Hiatt W.R. *et al.* (1985) Expression in plants of a mutant *aroA* gene from *Salmonella typhimurium* confers tolerance to glyphosate. *Nature* **317**, 741–4.

Condreay J.P., Witherspoon S.M., Clay W.C. & Kost T.A. (1999) Transient and stable gene expression in mammalian cells transduced with a recombinant baculovirus vector. *Proc. Nat. Acad. Sci. USA* **96**, 127–32.

Cone R.D. & Mulligan R.C. (1984) High-efficiency gene transfer into mammalian cells: generation of helper-free recombinant retrovirus with broad mammalian host range. *Proc. Nat. Acad. Sci. USA* **81**, 6349–53.

Cong B., Liu J. & Tanksley S.D. (2002) Natural alleles at a tomato fruit size quantitative locus differ by heterochronic regulatory mutations. *Proc. Nat. Acad. Sci. USA* **99**, 13606–11.

Conrad B., Savchenko R.S., Breves R. & Hofeweister J. (1996) A T7 promoter-specific, inducible protein expression system for *Bacillus subtilis*. *Mol. Gen. Genet.* **250**, 230–6.

Conrad M. & Topal M.D. (1989) DNA and spermidine provide a switch mechanism to regulate the activity of restriction enzyme NaeI. *Proc. Nat. Acad. Sci. USA* **86**, 9707–11.

Conrad U. & Fielder U. (1998) Compartment-specific accumulation of recombinant immunoglobulins in plant cells: an essential tool for antibody production and immunomodulation of physiological functions and pathogen activity. *Plant Mol. Biol.* **38**, 101–9.

Contente S. & Dubnau D. (1979) Characterization of plasmid transformation in *Bacillus subtilis*: kinetic properties and the effect of DNA conformation. *Mol. Gen. Genet.* **167**, 251–8.

Cooke N.E., Coit D., Weiner R.I., Baxter J.D. & Martial J.A. (1980) Structure of cloned DNA complementary to rat prolactin messenger RNA. *J. Biol. Chem.* **255**, 6502–10.

Cooley L., Berg C. & Spradling A. (1988) Controlling P element insertional mutagenesis. *Trends Genet.* **4**, 254–8.

Cooper G.M. & Sidow A. (2003) Genomic regulatory regions: insights from comparative sequence analysis. *Curr. Opin. Genet. Develop.* **13**, 604–10.

Corbel S. & Ross F.M.V. (2002) Latest developments and *in vivo* use of the Tet system: *ex vivo* and *in vivo* delivery of tetracycline-regulated genes. *Curr. Opin. Biotechnol.* **13**, 448–52.

Cormack B.P., Bertram G., Egerton M. *et al.* (1997) Yeast-enhanced green fluorescent protein (yEGFP): a reporter of gene expression in *Candida albicans*. *Microbiology* **143**, 303–11.

Corneille S., Lutz K., Svab Z. & Maliga P. (2001) Efficient elimination of selectable marker genes from the plastid genome by the Cre-lox site-specific recombination system. *Plant J.* **27**, 171–8.

Cornelis P. (2000) Expressing genes in different *Escherichia coli* compartments. *Curr. Opin. Biotechnol.* **11**, 450–4.

Cosloy S.D. & Oishi M. (1973) Genetic transformation in *Escherichia coli* K12. *Proc. Nat. Acad. Sci. USA* **70**, 84–7.

Cotsaftis O. & Guiderdoni E. (2005) Enhancing gene targeting efficiency in higher plants: rice is on the move. *Transgenic Res.* **14**, 1–14.

Coulson A., Sulston J., Brenner S. & Karn L. (1986) Toward a physical map of the genome of the nematode *Caenorhabditis elegans*. *Proc. Nat. Acad. Sci. USA* **83**, 7821–5.

Courtney M., Jallat S., Terrier L.-H. *et al.* (1985) Synthesis in *E. coli* of alpha1-antitrypsin variants of therapeutic potential for emphysema and thrombosis. *Nature* **313**, 149–51.

Couzin J. (2002) New mapping project splits the community. *Science* **296**, 1391–3.

Craig J.M. & Bickmore W.A. (1994) The distribution of CpG islands in mammalian chromosomes. *Nature Genet.* **7**, 376–81.

Craig N.L. (1988) The mechanism of conservative site-specific recombination. *Ann. Rev. Genet.* **22**, 77–105.

Cross S.H. & Bird A.P. (1995) CpG islands and genes. *Curr. Opin. Gen. Develop.* **5**, 309–14.

Crossway A., Oakes J.V., Irvine J.M., Ward B., Knauf V.C. & Shewmaker C.K. 1986. Integration of foreign DNA following microinjection of tobacco mesophyll protoplasts. *Mol. Gen. Genet.* **202**, 179–85.

Cullen B.R. & Malim M.H. (1992) Secreted placental alkaline phosphatase as a eukaryotic reporter gene. *Methods Enzymol.* **216**, 362–8.

Culver K.W., Ram Z., Wallbridge S. *et al.* (1992) *In vivo* gene transfer with retroviral vector-producer cells for treatment of experimental brain tumors. *Science* **256**, 1150–2.

Cunningham T.P., Montelaro R.C. & Rushlow K.E. (1993) Lentivirus envelope sequences and proviral genomes are stabilized in *Escherichia coli* when cloned in low-copy-number plasmid vectors. *Gene* **124**, 93–8.

Curtis R.I. & Cardineau C.A. (1990) Oral immunisation by transgenic plants. World Patent Application, WO 90/02484.

Curtis I.S. & Nam H.G. (2001) Transgenic radish (*Raphanus sativus* L. var. *longipinnatus* Bailey) by floral-dip method – plant development and surfactant are important in optimizing transformation efficiency. *Trans. Res.* **10**, 363–71.

Cutler P. (2003) Protein arrays: the current state-of-the-art. *Proteomics* **3**, 3–18.

Cwirla S.E., Peters E.A., Barrett R.W. & Dower W.J. (1990) Peptides on phage: a vast library of peptides for identifying ligands. *Proc. Nat. Acad. Sci. USA* **87**, 309–14.

D'Halluin K., Bossut M., Bonne E. *et al.* (1992) Transformation of sugar beet (*Beta vulgaris* L.) and evaluation of herbicide resistance in transgenic plants. *Biotechnology* **10**, 309–14.

Dahme M., Bartsch U., Martini R. *et al.* (1997) Disruption of the mouse L1 gene leads to malformations of the nervous system. *Nature Genet.* **17**, 346–9.

Dai Y., Vaught T.D., Boone J. *et al.*(2002) Targeted disruption of the alpha1,3-galactosyltransferase gene in cloned pigs. *Nature Biotechnol.* **20**, 251–5.

Dalbadie-McFarland G., Cohen L.W., Riggs A.D. *et al.* (1982) Oligonucleotide-directed mutagenesis as a general and powerful method for studies of protein functions. *Proc. Nat. Acad. Sci. USA* **79**, 6409–13.

Dale E.C. & Ow D.W. (1991) Gene transfer with subsequent removal of the selection gene from the host genome. *Proc. Nat. Acad. Sci. USA* **23**, 10558–62.

Dalmay T., Hamilton A., Mueller E. & Baulcombe D.C. (2000) Potato virus X amplicons in *Arabidopsis* mediate genetic and epigenetic gene silencing. *Plant Cell* **12**, 369–79.

Dalsgaard K., Uttenthal A., Jones T.D. *et al.* (1997) Plant-derived vaccine protects target animals against a virus disease. *Nature Biotechnol.* **15**, 248–52.

Daly M.J. *et al.* (2001) High-resolution haplotype structure in the human genome. *Nature Genet.* **29**, 229–32.

Dandekar T. *et al.* (2000) Re-annotating the *Mycoplasma pneumoniae* genome sequence: adding value, function and reading frames. *Nucl. Acids Res.* **28**, 3278–88.

Dandekar T., Snel B., Huynen M. & Bork P. (1998) Conservation of gene order: a fingerprint of proteins that physically interact. *Trends Biochem. Sci.* **23**, 324–8.

Dang C. & Jayasena S.D. (1996) Oligonucleotide inhibitors of *Taq* DNA polymerase facilitate detection of low copy number targets by PCR. *J. Mol. Biol.* **264**, 268–78.

Daniell H., Ruiz O.N. & Dhingra A. 2004. Chloroplast genetic engineering to improve agronomic traits. *Methods Mol. Biol.* **286**, 111–37.

Danos O. & Mulligan R.C. (1988) Safe and efficient generation of recombinant retroviruses with amphotropic and ecotropic host ranges. *Proc. Nat. Acad. Sci. USA* **85**, 6460–4.

Das L. & Martienssen R. (1995) Site-selected transposon mutagenesis at the hcf106 locus in maize. *Plant Cell* **7**, 287–94.

Dasgupta N., Wolfgang M.C., Goodman A.L., Arora S.K., Jyot J., Lory S. & Ramphal R. (2003) A four-tiered transcriptional regulatory circuit controls flagellar biogenesis in *Pseudomonas aeruginosa*. *Mol. Microbiol.* **50**, 809–24.

Datson N.A., van de Vosse E., Dauwerse H.G., Bout M., van Ommen G-J.B. & den Dunnen J.T. (1996) Scanning for genes in large genomic regions: cosmid-based exon trapping of multiple exons in a single product. *Nucl. Acids Res.* **24**, 1105–11.

Datson N.A., van der Perk-de Jong J., van den Berg M.P., de Kloet E.R. & Vreugdenhil E. (1999) MicroSAGE: a modified procedure for serial analysis of gene expression in limited amounts of tissue. *Nucl. Acids Res.* **27**, 1300–7.

Datta K., Vasquez A., Tu J. *et al.* (1998) Constitutive and tissue-specific differential expression of the cryIA (b) gene in transgenic rice plants conferring resistance to rice insect pest. *Theor. Appl. Genet.* **97**, 20–30.

Datta S.K., Peterhans A., Datta K. & Potrykus I. (1990) Genetically-engineered fertile Indica rice recovered from protoplasts. *Biotechnology* **8**, 571–4.

Daubert S., Shepherd R.J. & Gardner R.C. (1983) Insertional mutagenesis of the cauliflower mosaic-virus genome. *Gene* **25**, 201–8.

Dausset J., Cann H., Cohen D., Lathrop M., Lalouel J.M. & White R.L. (1990) Collaborative genetic mapping of the human genome. *Genomics* **6**, 575–7.

Davidson B.L., Allen E.D., Kozarsky K.F., Wilson J.M. & Roessler B.J. (1993) A model system for *in vivo* gene transfer into the central nervous system using an adenoviral vector. *Nature Genet.* **3**, 219–23.

Davidson B.L., Stein C.S., Heth J.A. *et al.* (2000) Recombinant adeno-associated virus type 2, 4, 5 vectors: transduction of variant cell types and regions in the mammalian central nervous system. *Proc. Nat. Acad. Sci. USA* **97**, 3428–32.

Davidson E.H. & Britten R.J. (1973) Organization, transcription and regulation in the animal genome. *Quart. Rev. Biol.* **48**, 565–613.

Davies J.C., Geddes D.M. & Alton E.W.F.W. (2001) Gene therapy for cystic fibrosis. J. Gene Med. **3**, 409–17.

Davis B. & MacDonald R.J. (1988) Limited transcription of rat elastase I transgene repeats in transgenic mice. *Genes Devel.* **2**, 13–22.

Davis G.D., Elisee C., Newham D.M. & Harrison R.G. (1999) New fusion protein systems designed to give soluble expression in *Escherichia coli. Biotechnol. Bioeng.* **65**, 382–8.

Davis T.N. (2004) Protein localization in proteomics. *Cur. Opin. Chem. Biol.* **8**, 49–53.

Davison J. (2002) Genetic tools for pseudomonads, rhizobia and other Gram-negative bacteria. *Biotechniques* **32**, 386–401.

Davison J., Chevalier N. & Brunel F. (1989) Bacteriophage T7 RNA polymerase-controlled specific gene expression in *Pseudomonas. Gene* **83**, 371–5.

Dawson W.O., Lewandowski D.J., Hilf M.E. *et al.* (1989) A tobacco mosaic virus-hybrid expresses and loses an added gene. *Virology* **172**, 285–92.

Day R.N. (1998) Visualization of Pit-1 transcription factor interactions in the living cell nucleus by fluorescence resonance energy transfer microscopy. *Mol. Endrocrinol.* **12**, 1410–19.

Dayhoff M.O. ed. (1978) *Atlas of Protein Sequence and Structure, vol. 5*. National Biomedical Research Foundation, Washington DC.

Dayhoff M.O., Eck R.V., Chang M.A. & Sochard M.R. (1965) *Atlas of Protein Sequence and Structure. Vol. 1*. National Biomedical Research Foundation, Silver Spring, MD.

De Benedetti A. & Rhoads R.E. (1991) A novel BK virus-based episomal vector for expression of foreign genes in mammalian cells. *Nucl. Acids Res.* **19**, 1925–31.

De Block M., Botterman J., Vandewiele M. *et al.* (1987) Engineering herbicide resistance in plants by expression of a detoxifying enzyme. *EMBO J.* **6**, 2513–18.

De Block M., Herrera-Estrella L., Van Montagu M., Schell J. & Zambryski P. (1984) Expression of foreign genes in regenerated plants and their progeny. *EMBO J.* **3**, 1681–9.

De Block M., Schell J. & Van Montagu M. (1985) Chloroplast transformation by *Agrobacterium tumefaciens. EMBO J.* **4**, 1367–72.

De Boer H.A. & Hui A.S. (1990) Sequences within ribosome binding site affecting messenger RNA translatability and method to direct ribosomes to single messenger RNA species. *Methods Enzymol.* **185**, 103–14.

De Boer H.A., Hui A., Comstock L.J., Wong E. & Vasser M. (1983b) Portable Shine–Dalgarno regions: a system for a systematic study of defined alterations of nucleotide sequences within *E. coli* ribosome binding sites. *DNA* **2**, 231–41.

De Grado M., Castan P. & Berenguer J. (1999) A high-transformation-efficiency cloning vector for *Thermus thermophilus. Plasmid* **42**, 241–5.

De Greef W., Delon R., De Block M., Leemans J. & Botterman J. (1989) Evaluation of herbicide resistance in transgenic crops under field conditions. *Biotechnology* **7**, 61–4.

De Greve H., Leemans J., Hernalsteens J.P. *et al.* (1982a) Regeneration of normal fertile plants that express octopine synthase from tobacco crown galls after deletion of tumor-controlling functions. *Nature* **300**, 752–5.

De Greve H., Phaese P., Seurwick J. *et al.* (1982b) Nucleotide sequence and transcript map of the *Agrobacterium tumefaciens* Ti plasmid-encoded octopine synthase gene. *J. Mol. Appl. Genet.* **1**, 499–511.

De Groot M.J.A., Bundock P., Hooykaas P.J.J. & Beijersbergen A.G.M. (1998) *Agrobacterium tumefaciens*-mediated transformation of filamentous fungi. *Nature Biotechnol.* **16**, 839–42.

De la Fuente J.M. & Herrera-Estrella L. (1999) Advances in the understanding of aluminum toxicity and the development of aluminum-tolerant transgenic plants. *Advances in Agronomy* **66**, 103–20.

De La Pena A., Lorz H. & Schell J. (1987) Transgenic plants obtained by injecting DNA into young floral tillers. *Nature* **325**, 274–6.

De Maagd R.A., Bosch D. & Stiekema W. (1999) *Bacillus thuringiensis* toxin-mediated insect resistance in plants. *Trends Plant Sci.* **4**, 9–13.

De Risi J. *et al.* (1996) Use of a cDNA microarray to analyse gene expression patterns in human cancer. *Nature Genet.* **14**, 457–60.

De Risi J.L., Iyer V.R. & Brown P.O. (1997) Exploring the metabolic and genetic control of gene expression on a genomic scale. *Science* **278**, 680–6.

De Ruyter P.G.G.A., Kuipers O.P. & De Vos W.M. (1996) Controlled gene expression systems for *Lactococcus lactis* with the food-grade inducer nisin. *Appl. Environ. Microbiol.* **62**, 3662–7.

De Saint-Vincent B.R., Delbruck S., Eckhart W. *et al.* (1981) The cloning and reintroduction into animal cells of a functional CAD gene, a dominant amplifiable genetic marker. *Cell* **27**, 267–77.

De Veylder L., Van Montagu M. & Inze D. (1997) Herbicide safenerinducible gene expression in *Arabidopsis thaliana. Plant Cell Physiol.* **38**, 568–77.

De Vos W.M., Kleerebezem M. & Kuipers O.P. (1997) Expression systems for industrial Gram-positive bacteria with low guanine and cytosine content. *Curr. Opin. Biotechnol.* **8**, 547–53.

De Vos W.M., Venema G., Canosi U. & Trautner T.A. (1981) Plasmid transformation in *Bacillus subtilis*: fate of plasmid DNA. *Mol. Gen. Genet.* **181**, 424–33.

de Waard V., van den Berg B.M.M., Veken J., Schultz-Heienbrok R., Pannekoek H. & van Zonneveld A.J. (1999) Serial analysis of gene expression to assess the endothelial cell response to an atherogenic stimulus. *Gene* **226**, 1–8.

de Wet J.R., Fukushima H., Dewji N.N. *et al.* (1984) Chromogenic immunodetection of human serum albumin and α-l-glucosidase clones in a human hepatoma cDNA expression library. *DNA* **3**, 437–47.

De Wet J.R., Wood K.V., De Luca M., Helinski D.R. & Subramani S. (1987) Firefly luciferase gene: structure and expression in mammalian cells. *Mol. Cell. Biol.* **7**, 725–37.

de Wildt R.M.T., Mundy C.R., Gorick B.D. & Tomlinson I.M. (2000) Antibody arrays for high-throughput screening of antibody-antigen interactions. *Nature Biotechnol.* **18**, 989–94.

De Zoeten G.A., Penswick J.R., Horisberger M.A. *et al.* (1989) The expression, localization, and effect of a human interferon in plants. *Virology* **172**, 213–22.

Dean F.B., Hosono S., Fang L. *et al.* (2002) Comprehensive human genome amplification using multiple displacement amplification. *Proc. Nat. Acad. Sci. USA* **99**, 5261–6.

Dear P.H., Bankier A.T. & Piper M.B. (1998) A high-resolution metric HAPPY map of human chromosome 14. *Genomics* **48**, 232–41.

Dedieu J.F., Vigne E., Torrent C. *et al.* (1997) Long term gene delivery into the liver of immunocompetent mice with E1/E4 defective adenoviruses. *J. Virol.* **71**, 4626–37.

Del Solar G., Giraldo R., Ruiz-Echevarria M.J., Espinosa M. & Diaz-Orejas R. (1998) Replication and control of circular plasmids. *Microbiol. Mol. Biol. Rev.* **62**, 434–64.

Delannay X., La Valle B.J., Proksch R.K. *et al.* (1989) Field performance of transgenic tomato plants expressing the *Bacillus thuringiensis* var. kurstaki insect control protein. *Biotechnology* **7**, 1265–9.

Della-Cioppa G., Garger S.J., Sverlow G.G., Turpen T.J. & Grill L.K. (1990) Melanin production in *Escherichia coli* from a cloned tyrosinase gene. *Biotechnology* **8**, 634–8.

den Dunnen J.T. (1999) Cosmid-based exon trapping. *Methods Enzymol.* **303**, 100–10.

Deng C. & Capecchi M.R. (1992) Re-examination of gene targeting frequency as a function of the extent of homology between the targeting vector and the target locus. *Mol. Cell. Biol.* **12**, 3365–71.

Deng G. & Wu R. (1981) An improved procedure for utilizing terminal transferase to add homopolymers to the 3′ termini of DNA. *Nucl. Acids Res.* **9**, 4173–88.

Denhardt D.T. (1966) A membrane-filter technique for the detection of complementary DNA. *Biochem. Biophys. Res. Commun.* **23**, 641–6.

Denning C. & Priddle H. (2003) New frontiers in gene targeting and cloning: success, application and challenges in domestic animals and human embryonic stem cells. *Reproduction* **126**, 1–11.

Denning C., Burl S., Ainslie A. *et al.* (2001) Deletion of the alpha(1,3)galactosyl transferase (GGTA1) gene and the prion protein (PrP) gene in sheep. *Nature Biotechnol.* 559–62.

Denny P. & Justice M.J. (2000) Mouse as the measure of man? *Trends Genet.* **16**, 283–7.

Depicker A., Herman L., Jacobs A., Schell J. & Van Montagu M. (1985) Frequencies of simultaneous transformation with different T-DNAs and their relevance to the *Agrobacterium*/plant cell interaction. *Mol. Gen. Genet.* **201**, 477–84.

Depicker A., Stachel S., Dhaese P., Zambryski P. & Goodman H.M. (1982) Nopaline synthase: transcript mapping and DNA sequence. *J. Mol. Appl. Genet.* 1, 561–74.

Deretic V., Chandrasekharappa S., Gill J.F., Chaterjee D.K. & Chakrabarty A.M. (1987) A set of cassettes and improved vectors for genetic and biochemical characterization of *Pseudomonas* genes. *Gene* **57**, 61–72.

Dernberg A.F., Zalevsky J., Colaiacovo M.P. & Villeneuve A.M. (2000) Transgene-mediated cosuppression in the *C. elegans* germ line. *Genes Devel.* **14**, 1578–83.

Desfeux C., Clough S.J. & Bent A.F. (2000) Female reproductive tissues are the primary target of *Agrobacterium*-mediated transformation by the *Arabidopsis* floral-dip method. *Plant Physiol.* **123**, 895–904.

Deshayes A., Herrera-Estrella L. & Caboche M. (1985) Liposome-mediated transformation of tobacco mesophyll protoplasts by an *Escherichia coli* plasmid. *EMBO J.* **4**, 2731–7.

Dessaux Y. & Petit A. (1994) Opines as screenable markers for plant transformation. In *Plant Molecular Biology Manual*, 2nd edn, eds. Gelvin S.B. & Schilperoort R.A., Section 3, pp. 1–12. Kluwer Academic Publishers.

Devlin J.J., Panganiban L.C. & Devlin P.E. (1990) Random peptide libraries: a source of specific protein binding molecules. *Science* **249**, 404–6.

Devlin P.E., Drummond R.J., Toy P. *et al.* (1988) Alteration of amino-terminal codons of human granulocyte-colony-stimulating factor increases expression levels and allows efficient processing by methionine amino-peptidase in *Escherichia coli*. *Gene* **65**, 13–22.

Dewannieux M. *et al.* (2004) Identification of autonomous IAP LTR retrotransposons mobile in mammalian cells. *Nature Genet.* **36**, 534–9.

Di Maio D., Triesman R. & Maniatis T. (1982) A bovine papillomavirus vector which propagates as an episome in both mouse and bacterial cells. *Proc. Nat. Acad. Sci. USA* **79**, 4030–4.

Dieffenbach C.W. & Dveler G.S. (eds.) (1995) *PCR Primer: A Laboratory Manual.* Cold Spring Harbor Laboratory Press, Cold Spring Harbor, New York.

Dietrich G., Bubert A., Gentschev I., Sokolovic Z. *et al.* (1998) Delivery of antigen-encoding plasmid DNA into the cytosol of macrophages by attenuated suicide *Listeria monocytogenes. Nature Biotechnol.* **16**, 181–5.

DiMaio J.J. & Shillito R.D. (1989) Cryopreservation technology for plant cell cultures. *J. Tissue Cult. Methods* **12**, 163–9.

Dimmock N.J., Easton A.J. & Leppard K.N. (2001) *Introduction to Modern Virology*, 5th edn. Blackwell Science, Oxford.

Dirks W., Wirth M. & Hauser H. (1993) Dicistronic transcription units for gene expression in mammalian cells. *Gene* **129**, 247–9.

Dittmar M.T., Simmons G., Donaldson Y. *et al.* (1997) Biological characterization of human immuno-deficiency virus type 1 clones derived from different organs of an AIDS patient by long-range PCR. *J. Virol.* **71**, 5140–7.

Dobson M.J., Tuite M.F., Roberts N.A. *et al.* (1982) Conservation of high efficiency promoter sequences in *Saccharomyces cerevisiae. Nucl. Acids Res.* **10**, 2625–37.

Doetschman T., Gregg R.G., Maeda N. *et al.* (1987) Targeted correction of a mutant HPRT gene in mouse embryonic stem cells. *Nature* **330**, 57657–8.

Dogget N. (1992) *The Human Genome Project. Los Alamos Science No. 20.* University Science Books, Sausalito, CA.

Dominguez A., Ferminan E., Sanchez M. *et al.* (1998) Non-conventional yeasts as hosts for heterologous protein production. *Int. Microbiol.* **1**, 131–42.

Dominguez-Bendala J., Priddle H., Clarke A. & McWhir J. (2003) Elevated expression of exogenous Rad51 leads to identical increases in gene-targeting frequency in murine embryonic stem (ES) cells with both functional and dysfunctional p53 genes. *Exp. Cell Res.* **286**, 298–307.

Dong J., Teng W., Buchholz W.G. & Hall T.C. (1996) *Agrobacterium*-mediated transformation of Javanica rice. *Planta* **199**, 612–17.

Donis-Keller H. *et al.* (1987) A genetic linkage map of the human genome. *Cell* **51**, 319–37.

Donson J., Kearney C.M., Hilf M.E. & Dawson W.O. (1991) Systemic expression of a bacterial gene by a tobacco mosaic virus-based vector. *Proc. Nat. Acad. Sci. USA* **88**, 7204–8.

Doolittle W.F. (1995) The multiplicity of domains in proteins. *Ann. Rev. Biochem.* **64**, 287–314.

Dorin J.R., Dickinson P., Alton E.W. *et al.* (1992) Cystic-fibrosis in the mouse by targeted insertional mutagenesis. *Nature* **359**, 211–15.

Dorin J.R., Stevenson B.J., Fleming S. *et al.* (1994) Long term survival of the exon 10 insertional cystic fibrosis mutant mouse is a consequence of low-level residual wild type Cftr gene expression. *Mammalian Genome* **5**, 465–72.

Dovichi N.J. & Zhang J.-Z. (2001) DNA sequencing by capillary array electrophoresis. *Methods Mol. Biol.* **167**, 225–39.

Dower W.J., Miller J.F. & Ragsdale C.W. (1988) High efficiency transformation of *E. coli* by high-voltage electroporation. *Nucl. Acids Res.* **16**, 6127–45.

Drewes G. & Bouwmeester T. (2003) Global approaches to protein–protein interactions. *Curr. Opin. Cell. Biol.* **15**, 1–7.

Drmanac R., Labat I., Brukner I. & Crkvenjakov R. (1989) Sequencing of megabase plus DNA by hybridization: theory of the method. *Genomics* **4**, 114–28.

Drmanac S. *et al.* (1998) Accurate sequencing by hybridization for DNA diagnostics and individual genomics. *Nature Biotechnol.* **16**, 54–8.

Drummond M.H. & Chilton M.-D. (1978) Tumor-inducing (Ti) plasmids of Agrobacterium share extensive regions of DNA homology. *J. Bacteriol.* **136**, 1178–83.

Dubendorff J.W. & Studier F.W. (1991) Controlling basal expression in an inducible T7 expression system by blocking the target T7 promoter with *lac* repressor. *J. Mol. Biol.* **219**, 45–59.

DuBridge R.B. & Calos M.P. (1988) Recombinant shuttle vectors for the study of mutation in mammalian cells. *Mutagenesis* **3**, 1–9.

Dueschle U., Kammerer W., Genz R. & Bujard H. (1986) Promoters of *Escherichia coli*: a hierarchy of *in vivo* strength indicates alternate structure. *EMBO J.* **5**, 2987–94.

Dufourmantel N., Pelissier B., Garc F., Peltier G., Ferullo J.M. & Tissot G. (2004) Generation of fertile transplastomic soybean. *Plant Mol. Biol.* **55**, 479–89.

Dugaiczyk A., Boyer H.W. & Goodman H.M. (1975) Ligation of *Eco*RI endonuclease-generated DNA fragments into linear and circular structures. *J. Mol. Biol.* **96**, 171–84.

Duggan D.J., Bittner M., Chen Y., Meltzer P. & Trent J.M. (1999) Expression profiling using cDNA microarrays. *Nature Genet.* **21**, 10–14.

Duguid J.R., Rohwer R.G. & Seed B. (1988) Isolation of cDNAs of scrapie-modulated RNAs by subtractive hybridization of a cDNA library. *Proc. Nat. Acad. Sci. USA* **85**, 5738–42.

Dujardin N., Van Der Smissen P. & Preat V. (2001) Topical gene transfer into rat skin using electroporation. *Pharm. Res.* **18**, 61–6.

Dujon B. (1996) The yeast genome project: what did we learn? *Trends Genet.* **12**, 263–70.

Duke S.O., Scheffler B.E., Dayan F.E. & Dyer W.E. (2002) Genetic engineering crops for improved weed management traits. *Crop Biotechnol. ACS Symposium Series* **829**, 52–66.

Dunn J.J. & Studier F.W. (1983) Complete nucleotide sequence of bacteriophage T7 DNA and the locations of T7 genetic elements. *J. Mol. Biol.* **166**, 477–535.

Dussoix D. & Arber W. (1962) Host specificity of DNA produced by *Escherichia coli*. II. Control over acceptance of DNA from infecting phage λ. *J. Mol. Biol.* **5**, 37–49.

Dutch-Belgian Fragile X Consortium (1994) Fmr1 knockout mice: a model to study fragile X mental retardation. *Cell* **78**, 23–33.

Duttweiler H.M. & Gross D.S. (1998) Bacterial growth medium that significantly increases the yield of recombinant plasmid. *Biotechniques* **24**, 438–44.

Duyk G.M., Kim S.W., Myers R.M. & Cox D.R. (1990) Exon trapping: a genetic screen to identify candidate transcribed sequences in cloned mammalian genomics DNA. *Proc. Nat. Acad. Sci. USA* **87**, 8995–9.

Dworkin M.B. & Dawid I.B. (1980) Use of a cloned library for the study of abundant poly(A)+ RNA during *Xenopus laevis* development. *Dev. Biol.* **76**, 449–64.

Dyck M.K., Gagne D., Ouellet M. *et al.* (1999) Seminal vesicle production and secretion of growth hormone into seminal fluid. *Nature Biotechnol.* **17**, 1087–90.

Ebert K.M., Selgrath J.P., Ditullio P. *et al.* (1991) Transgenic production of a variant of human tissue-type plasminogen activator in goat milk: generation of transgenic goats and analysis of expression. *Biotechnology* **9**, 835–8.

Eckert K.A. & Kunkel T.A. (1990) High fidelity DNA synthesis by the *Thermus aquaticus* DNA polymerase. *Nucl. Acids Res.* **18**, 3739–44.

Eckhardt T. (1978) A rapid method for the identification of plasmid deoxyribonucleic acid in bacteria. *Plasmid* **1**, 584–8.

Eddy S. (1998) Multiple alignment and sequence searches. *Bioinformatics* **5**, 15–18.

Eddy S. (1999) Noncoding RNA genes. *Curr. Opin. Gen. Develop.* **9**, 695–9.

Edery I., Chu L.L., Soneberg N. & Pelletier J. (1995) An efficient strategy to isolate full length cDNAs based on an mRNA cap retention procedure (CAPture). *Mol. Cell Biol.* **15**, 3363–71.

Edgar R., Domrachev M. & Lash A.E. (2002) Gene Expression Omnibus: NCBI gene expression and hybridization array data repository. *Nucl. Acids Res.* **30**, 207–10.

Edgell D.R., Belfort M. & Shub D.A. (2000) Barriers to intron promiscuity in bacteria. *J. Bacteriol.* **182**, 5281–9.

Edwards A.M. *et al.* (2000) Protein production: feeding the crystallographers and NMR spectroscopists. *Nature Struct. Biol.* **7** (Suppl), 970–2.

Edwards J.L., Schrick F.N., McCracken M.D., van Amstel S.R., Hopkins F.M., Welborn M.G. & Davies C.J. (2003) Cloning adult farm animals: a review of the possibilities and problems associated with somatic cell nuclear transfer. *Am. J. Reprod. Immunol.* **50**, 113–23.

Edwards B.S. *et al.* (2004) Flow cytometry for high-throughput, highcontent screening. *Curr. Opin. Chem. Biol.* **8**, 392–8.

Efrat S., Lieser M., Wu Y. *et al.* (1994) Ribozyme-mediated attenuation of pancreatic b-cell glucokinase expression in transgenic mice results in impaired glucose-induced insulin secretion. *Proc. Nat. Acad. Sci. USA* **91**, 2051–5.

Efstratiadis A., Kafatos F.C., Maxam A.M. & Maniatis T. (1976) Enzymatic *in vitro* synthesis of globin genes. *Cell* **7**, 279–88.

Eguchi Y., Itoh T. & Tomizawa J. (1991) Antisense RNA. *Ann. Rev. Biochem.* **60**, 631–52.

Ehrlich S.D. (1977) Replication and expression of plasmids from *Staphylococcus aureus* in *Bacillus subtilis*. *Proc. Nat. Acad. Sci. USA* **74**, 1680–2.

Ehrlich S.D. (1978) DNA cloning in *Bacillus subtilis*. *Proc. Nat. Acad. Sci. USA* **75**, 1433–6.

Ehrlich S.D., Noirot P., Petit M.A. *et al.* (1986) Structural instability of *Bacillus subtilis* plasmids. In *Genetic Engineering*, eds. Setlow J.K. & Hollaender A., Vol. 8, pp. 71–83. Plenum, New York.

Ehsani P., Khabiri A. & Domansky N.N. (1997) Polypeptides of hepatitis B surface antigen in transgenic plants. *Proc. Nat. Acad. Sci. USA* **190**, 107–11.

Eichholtz D.A., Rogers S.G., Horsch R.B. *et al.* (1987) Expression of mouse dihydrofolate reductase gene confers methotrexate resistance in transgenic petunia plants. *Somat. Cell Mol. Genet.* **13**, 67–76.

Eichler E.E. & Sankoff D. (2003) Structural dynamics of eukaryotic chromosome evolution. *Science* **301**, 793–7.

Eichler E.E. (2001) Recent duplication, domain accretion and the dynamic mutation of the human genome. *Trends Genet.* **17**, 661–9.

Eickbush T. (1999) Exon shuffling in retrospect. *Science* **283**, 1465–7.

Eisen J.A., Heidelberg J.F., White O. & Salzberg S.L. (2000) Evidence for symmetric chromosomal inversions around the replication origin in bacteria. *Genome Biology* **1**, Research, 0011.1–0011.9.

Eisen M.B., Wiley D.C., Karplus M. & Hubbard R.E. (1994) HOOK: a program for finding novel molecular architectures that satisfy the chemical and steric requirements of a macromolecule binding site. *Proteins* **19**, 199–221.

Elbashir S.M., Lendeckel W. & Tuschl T. (2001) RNA interference is mediated by 21- and 22-nucleotide RNAs. *Genes Dev.* **15**, 188–200.

Elkin C.J. *et al.* (2001) High-throughput plasmid purification for capillary sequencing. *Genome Res.* **11**, 1269–74.

Ellenberg J., Lippincott-Schwartz J. & Presley J.F. (1999) Dual colour imaging with GFP variants. *Trends Cell Biol.* **9**, 52–6.

Emmerman M. & Temin H.M. (1984) Genes with promoters in retrovirus vectors can be independently suppressed by an epigenetic mechanism. *Cell* **39**, 449–67.

Endean D. & Smithies O. (1989) Replication of plasmid DNA in fertilized *Xenopus* eggs is sensitive to both the topology and size of the injected template. *Chromosoma* **97**, 307–14.

Engler G., Depicker A., Maenhaut R. *et al.* (1981) Physical mapping of DNA base sequence homologies between an octopine and a nopaline Ti plasmid of *Agrobacterium tumefaciens*. *J. Mol. Biol.* **152**, 183–208.

English J.J. & Baulcombe D.C. (1997) The influence of small changes in transgene transcription on homology-dependent virus resistance and gene silencing. *Plant J.* **12**, 1311–18.

English J.J., Mueller E. & Baulcombe D.C. (1996) Suppression of virus accumulation in transgenic plants exhibiting silencing of nuclear genes. *Plant Cell* **8**, 179–88.

Enright A.J., Iliopoulos I., Kyrpides N.C. & Ouzounis C.A. (1999) Protein interaction maps for complete genomes based on gene fusion events. *Nature* **402**, 86–90.

Ensley B.D., Ratzkin B.J., Osslund T.D. *et al.* (1983) Expression of naphthalene oxidation genes in *Escherichia coli* results in biosynthesis of indigo. *Science* **222**, 167–9.

Erickson R.P., Lai L.W. & Grimes J. (1993) Creating a conditional mutation of Wnt-1 by antisense transgenesis provides evidence that Wnt-1 is not essential for spermatogenesis. *Devel. Biol.* **14**, 274–81.

Erlich H.A. (ed.) (1989) *PCR Technology: Principles and Applications for DNA Amplification.* Stockton Press.

Erlich H.A., Gelfand D.H. & Sakai R.K. (1988) Specific DNA amplification. *Nature* **331**, 461–2.

Ernst J.F. (1988) Codon usage and gene expression. *Trends Biotechnol.* **6**, 196–9.

Eskin B. & Linn S. (1972) The deoxyribonucleic modification and restriction enzymes of *Escherichia coli* B. II. Purification, subunit structure, and catalytic properties of the restriction endonuclease. *J. Biol. Chem.* **247**, 6183–91.

Estruch J.J., Chriqui D., Grossmann K., Schell J. & Spena A. (1991a) The plant oncogene rolC is responsible for the release of cytokinins from glucoside conjugates. *EMBO J.* **10**, 2889–95.

Estruch J.J., Schell J. & Spena A. (1991b) The protein encoded by the rolB plant oncogene hydrolyzes indole glucosides. *EMBO J.* **10**, 3125–8.

Etchegaray J.-P. & Inouye M. (1999) Translational enhancement by an element downstream of the initiation codon in *Escherichia coli. J. Biol. Chem.* **274**, 10079–85.

Etkin L.D., Pearman B. & Ansah-Yiadom R. (1987) Replication of injected DNA templates in *Xenopus* embryos. *Exp. Cell Res.* **169**, 468–77.

Etkins R.P. (1998) Ligand assays: from electrophoresis to miniaturised microarrays. *Clinical Chem.* **44**, 2015–30.

Evans G.A., Lewis K. & Rothenberg B.E. (1989) High efficiency vectors for cosmid microcloning and genomic analysis. *Gene* **79**, 9–20.

Evans M.J. & Kaufman M.H. (1981) Establishment in culture of pluripotential cells from mouse embryos. *Nature* **292**, 154–6.

Evans M.J., Carlton M.B.L. & Russ A.P. (1997) Gene trapping and functional genomics. *Trends Genet.* **13**, 370–4.

Evans P.D., Cook S.N., Riggs P.D. & Noren C.J. (1995) LITMUS: multipurpose cloning vectors with a novel system for bidirectional *in vitro* transcription. *Biotechniques* **19**, 130–5.

Ewing B. & Green P. (1998) Base-calling of automated sequencer traces using Phred II: error probabilities. *Genome Res.* **8**, 186–94.

Ewing B., Hillier L., Wendl M.C. & Green P. (1998) Base-calling of automated sequencer traces using Phred I: accuracy assessment. *Genome Res.* **8**, 175–85.

Famulok M. & Verma S. (2002) *In vivo*-applied functional RNAs as tools in proteomics and

Farrell C.M., West A.G. & Felsenfeld G. (2002) Conserved CTCF insulator elements flank the mouse and human β–globin loci. *Mol. Cell. Biol.* **22**, 3820–31.

Favis R., Day J.P., Gerry N.P., *et al.* (2000) Universal DNA array detection of small insertions and deletions in BRCA1 and BRCA2. *Nature Biotechnol.* **18**, 561–4.

Federico M. (1999) Lentiviruses as gene delivery vectors. *Curr. Opin. Biotechnol.* **10**, 448–53.

Feldmann K.A. & Marks M.D. (1987) *Agrobacterium*-mediated transformation of germinating seeds of *Arabidopsis thaliana* – a non-tissue culture approach. *Mol. Gen. Genet.* **208**, 1–9.

Feldmann K.A. (1991) T-DNA insertion mutagenesis in *Arabidopsis*: mutational spectrum. *Plant J.* **1**, 71–82.

Felgner J.H., Kumar R., Sridhar C.N. *et al.* (1994) Enhanced gene delivery and mechanism studies with a novel series of cationic lipid formulations. *J. Biol. Chem.* **269**, 2550–61.

Felgner P.L., Gadek T.R., Holm M. *et al.* (1987) Lipofection: a highly efficient lipid-mediated DNA-transfection procedure. *Proc. Nat. Acad. Sci. USA* **84**, 7413–17.

Feng Y.Q., Seibler J., Alami R. *et al.* (1999) Site-specific chromosomal integration in mammalian cells: highly efficient CRE recombinase-mediated cassette exchange. *J. Mol. Biol.* **292**, 779–85.

Fenyo D. (2000) Identifying the proteome: software tools. *Curr. Opin. Biotechnol.* **11**, 391–5.

Fernandes J., Dong Q.F., Schneider B., Morrow D.J., Nan G.L., Brendel V. & Walbot V. (2004) Genome-wide mutagenesis of *Zea mays* L. using RescueMu transposons. *Genome Biol.* **5**, R82.

Ferretti L. & Sgaramella V. (1981) Temperature dependence of the joining by T4 DNA ligase of termini produced by type II restriction endonucleases. *Nucl. Acids Res.* **9**, 85–93.

Festenstein R., Tolaini M., Carbella P. *et al.* (1996) Locus control region function and heterochromatin-induced position effect variegation. *Science* **271**, 1123–5.

Fettig S. & Hess D. (1999) Expression of a chimeric stilbene synthase gene in transgenic wheat lines. *Transgenic Res.* **8**, 179–89.

Ficarro S.B., McCleland M.L., Stukenberg P.T., Burke D.J., Ross M.M., Shabanowitz J., Hunt D.F. & White F.M. (2002) Phosphoproteome analysis by mass spectrometry and its application to *Saccharomyces cerevisiae. Nature Biotechnol.* **20**, 301–5.

Fickett J.W. & Hatzigeorgiou A.G. (1997) Eukaryotic promoter recognition. *Genome Res.* **7**, 861–78.

Fickett J.W. & Tung C.S. (1992) Assessment of protein coding measures. *Nucl. Acids Res.* **20**, 6441–50.

Fickett J.W. (1996) Finding genes by computer: the state of the art. *Trends Genet.* **12**, 316–20.

Fiehn O. & Weckwerth W. (2003) Deciphering metabolic networks. *Euro. J. Biochem.* **270**, 579–88.

Fiehn O., Kopka J., Dormann P. *et al.* (2000) Metabolite profiling for plant functional genomics. *Nature Biotechnol.* **18**, 1157–61.

Fiel R., Brocard J., Mascrez B. *et al.* (1996) Ligand-activated site-specific recombination in mice. *Proc. Nat. Acad. Sci. USA* **93**, 10887–90.

Field L.L., Tobias R. & Magnus T. (1994) A locus on chromosome 15q26 (IDDM3) produces susceptibility to insulin-dependent diabetes mellitus. *Nature Genet.* **8**, 189–94.

Fields S. & Song O. (1989) A novel genetic system to detect protein–protein interactions. *Nature* **340**, 245–6.

Fiering S., Epner E., Robinson K. *et al.* (1995) Targeted deletion of 5′ HS2 of the murine beta-globin LCR reveals that it is not essential for proper regulation of the beta-globin locus. *Genes Devel.* **9**, 2203–13.

Figeys D. (2003) Novel approaches to map protein–protein interactions. *Curr. Opin. Biotechnol.* **14**, 1–7.

Figge J., Wright C., Collins C.J., Roberts T.M. & Livingston D.M. (1988) Stringent regulation of stably integrated chloramphenicol acetyl transferase genes by *E. coli lac* repressor in monkey cells. *Cell* **52**, 713–22.

Fincham J.R.S. (1989) Transformation in fungi. *Microbiol. Rev.* **53**, 148–70.

Finer J.J. & McMullen M.D. (1991) Transformation of cotton (*Gossypium hirsutum* L.) via particle bombardment. *Plant Cell Rep.* **8**, 586–9.

Finer J.J., Finer K.R. & Ponappa T. 1999. Particle bombardment mediated transformation. *Curr. Top. Microbiol. Immunol.* **240**, 59–80.

Finer J.J., Vain P., Jones M.W. & McMullen M.D. (1992) Development of the particle inflow gun for DNA delivery to plant cells. *Plant Cell Rep.* **11**, 323–8.

Finley R. & Brent R. (1994) Interaction mating reveals binary and ternary interactions between *Drosophila* cell cycle regulators. *Proc. Nat. Acad. Sci. USA* **91**, 12980–4.

Fire A., Xu S., Montgomery M.K., Kostas S.A., Driver S.E. & Mehlo C.C. (1998) Potent and specific genetic interference by double-stranded RNA in *Caenorhabditis elegans*. *Nature* **391**, 806–11.

Firsheim W. & Kim P. (1997) Plasmid partitioning and replication in *Escherichia coli*: is the membrane the key? *Mol. Microbiol.* **23**, 1–10.

Fischer D. & Eisenberg D. (1997) Assigning folds to the proteins encoded by the genome of *Mycoplasma genitalium*. *Proc. Nat. Acad. Sci. USA* **94**, 11929–34.

Fischer D. & Eisenberg D. (1999a) Finding families for genomic ORFans. *Bioinformatics* **15**, 759–62.

Fischer D. & Eisenberg D. (1999b) Predicting structures for genome proteins. *Curr. Opin. Struct. Biol.* **9**, 208–11.

Fischer R., Stoger E., Schillberg S., Christou P. & Twyman R.M. (2004) Plant based production of biopharmaceuticals. *Curr. Opin. Plant Biol.* **7**, 152–8.

Fischhoff D.A., Bowdish K.S., Perlak F.J. *et al.* (1987) Insect tolerant transgenic tomato plants. *Biotechnology* **5**, 807–13.

Fitch M.M.M., Manshardt R.M., Gonsalves D., Slightom J.L. & Sandford J.C. (1990) Stable transformation of papaya via microprojectile bombardment. *PCR* **9**, 189–94.

Flajolet M. *et al.* (2000) A genomic approach of the hepatitis C virus generates a protein interaction map. *Gene* **241**, 369–79.

Fleischmann R.D. *et al.* (1995) Whole-genome random sequencing and assembly of *Haemophilus influenzae* Rd. *Science* **269**, 496–512.

Flores A. *et al.* (1999) A protein–protein interaction map of yeast RNA polymerase III. *Proc. Nat. Acad. Sci. USA* **96**, 7815–20.

Florijn R.J. *et al.* (1995) High resolution DNA fiberFISH genomic DNA mapping and colour barcoding of large genes. *Hum. Mol. Gen.* **4**, 831–6.

Flotte T.R. (2004) Gene therapy progress and prospects: recombinant adeno-associated virus (rAAV) vectors. *Gene Ther.* **11**, 805–10.

Fodor S.P.A. *et al.* (1991) Light-directed spatially addressable parallel chemical synthesis. *Science* **251**, 767–73.

Fodor S.P.A., Rava R.P., Huang X.C., Pease A.C., Holmes C.P. & Adams C.L. (1993) Multiplexed biochemical assays with bioiogical-chips. *Nature* **364**, 555–6.

Fonstein M. & Haselkorn R. (1995) Physical mapping of bacterial genomes. *J. Bacteriol.* **177**, 3361–9.

Foord O.S. & Rose E.A. (1994) Long-distance PCR. *PCR Methods Appl.* **3**, S149–61.

Forman M.D., Stack R.F., Masters P.S., Hauer C.R. & Baxter S.M. (1998) High level, context dependent misincorporation of lysine for arginine in *Saccharomyces cerevisiae* a1 homeodomain expressed in *Escherichia coli*. *Protein Sci.* **7**, 500–3.

Forrester W.C., Takegawa S., Papayannopoulou T., Atamatoyannopoulos G. & Groudine M. (1987) Evidence for a locus activation region: the formation of developmentally stable hypersensitive sites in globin expressing hybrids. *Nucl. Acids Res.* **83**, 10159–77.

Fortna A. & Gardiner K. (2001) Genomic sequence analysis tools: a user's guide. *Trends Genet.* **17**, 158–64.

Fournier R.E. & Ruddle F.H. (1977) Microcell-mediated transfer of murine chromosomes into mouse, Chinese hamster, and human somatic cells. *Proc. Nat. Acad. Sci. USA* **74**, 319–23.

Frame B.R., Drayton P.R., Bagnall S.V., Lewnau C.J., Bullock W.P., Wilson H.M., Dunwell J.M., Thompson J.A. & Wang K. (1994) Production of fertile transgenic maize plants by silicon-carbide whisker-mediated transformation. *Plant J.* **6**, 941–8.

Franconi A., Roggero P., Pirazzi P. *et al.* (1999) Functional expression in bacteria and plants of an scFv antibody fragment against tospoviruses. *Immunotechnology* **4**, 189–201.

Franke C.A., Rice C.M., Strauss J.H. & Hruby D.E. (1985) Neomycin resistance as a dominant selectable marker for selection and isolation of vaccinia virus recombinants. *Mol. Cell. Biol.* **5**, 1918–24.

Frankel W.N. (1995) Taking stock of complex trait genetics in mice. *Trends Genet.* **11**, 471–7.

Franklin F.C.H., Bagdasarian M., Bagdasarian M.M. & Timmis K.N. (1981) Molecular and functional analysis of the TOL plasmid pWWO from *Pseudomonas putida* and cloning of genes for the entire regulated aromatic ring meta cleavage pathway. *Proc. Nat. Acad. Sci. USA* **78**, 7458–62.

Frary A., Nesbitt T.C., Grandillo S., *et al.* (2000) fw2.2: a quantitative trait locus key to the evolution of tomato fruit size. *Science* **289**, 85–8.

Fraser A.G., Kamath R.S., Zipperlen P., Martinez-Campos M., Sohrmann M. & Ahringer J. (2000) Functional genomic analysis of *C. elegans* chromosome I by systematic RNA interference. *Nature* **408**, 325–30.

Fraser C.M. *et al.* (1995) The minimal gene complement of *Mycoplasma genitalium*. *Science* **270**, 397–403.

Fraser M.J. (1992) The baculovirus infected insect cell as a eukaryotic gene expression system. *Curr. Top. Mol. Biol.* 131–72.

French R., Janda M. & Ahlquist P. (1986) Bacterial gene inserted in an engineered RNA virus: efficient expression in monocotyledonous plant cells. *Science* **231**, 1294–7.

Frengen E., Weichenhan D., Zhao B. *et al.* (1999) A modular, positive selection bacterial artificial chromosome vector with multiple cloning sites. *Genomics* **58**, 250–3.

Fridell Y.W.C. & Searles L.L. (1991) Vermilion as a small selectable marker gene for *Drosophila* transformation. *Nucl. Acids Res.* **19**, 5082.

Fridman E., Carrari F., Liu Y.S., *et al.* (2004) Zooming in on a quantitative trait for tomato yield using interspecific introgression. *Science* **305**, 1786–9.

Fridman E., Pleban T. & Zamir D. (2000) A recombination hotspot delimits a wild-species quantitative trait locus for tomato sugar content to 484 bp within an invertase gene. *Proc. Acad. Sci. USA* **97**, 4718–23.

Friedman A. & Perrimon N. (2004) Genome-wide high-throughput screens in functional genomics. *Curr. Opin. Genet. Dev.* **14**, 460–76.

Friedrich G. & Soriano P. (1991) Promoter traps in embryonic stem cells: a genetic screen to identify and mutate developmental genes in mice. *Genes Devel.* **5**, 1513–23.

Frijters A.C.J., Zhang Z. Van Damme M. *et al.* (1997) Construction of a bacterial artificial chromosome library containing large *Eco*RI and *Hind*III genomic fragments of lettuce. *Theor. Appl. Genet.* **94**, 390–9.

Frischauf A.-M., Lehrach H., Poustka A. & Murray N. (1983) Lambda replacement vectors carrying polylinker sequences. *J. Mol. Biol.* **170**, 827–42.

Frohman M.A. & Martin G.R. (1989) Rapid amplification of cDNA ends using nested primers. *Technique* **1**, 165–70.

Frohman M.A., Dush M.K. & Martin G. (1988) Rapid production of full-length cDNAs from rare transcripts: amplification using a single gene-specific oligonucleotide primer. *Proc. Nat. Acad. Sci. USA* **85**, 8998–9002.

Frohme M. *et al.* (2001) Directed gap closure in large-scale sequencing projects. *Genome Res.* **11**, 901–3.

Frolov I. & Schlesinger S. (1994) Translation of Sindbis virus mRNA: effects of sequences downstream of the initiating codon. *J. Virol.* **68**, 8111–17.

Frolov I., Agapov E., Hoffman T.A. Jr *et al.* (1999) Selection of RNA replicons capable of persistent non-cytopathic replication in mammalian cells. *J. Virol.* **73**, 3854–65.

Fromm M.E., Morrish F., Armstrong C. *et al.* (1990) Inheritance and expression of chimeric genes in the progeny of transgenic maize plants. *Biotechnology* **8**, 833–44.

Fromont-Racine M. *et al.* (2000) Genome-wide protein interaction screens reveal functional networks involving Sm-like proteins. *Yeast* **17**, 95–110.

Fromont-Racine M., Rain J.C. & Legrain P. (1997) Toward a functional analysis of the yeast genome through exhaustive two-hybrid screens. *Nature Genet.* **16**, 277–82.

Fu X.D., Sudhakar D., Peng J.R., Richards D.E., Christou P. & Harberd N.P. (2001) Expression of *Arabidopsis GAI* in transgenic rice represses multiple gibberellin responses. *Plant Cell* **13**, 1791–802.

Fuerst T.R., Earl P.L. & Moss B. (1987) Use of hybrid vaccinia virus T and RNA polymerase system for expression of target genes. *Mol. Cell. Biol.* **7**, 2538–44.

Fuerst T.R., Niles E.G., Studier W.F. & Moss B. (1986) Eukaryotic transient-expression system based on recom-binant vaccinia virus that synthesizes bacteriophage T7 RNA polymerase. *Proc. Nat. Acad. Sci. USA* **83**, 8122–6.

Fuh G. & Sidhu S.S. (2000) Efficient phage display of polypeptides fused to the carboxy-terminus of the M13 gene-3 minor coat protein. *FEBS Lett.* **480**, 231–4.

Fuh G., Pisabarro M.T., Li Y. *et al.* (2000) Analysis of PDZ domain ligand interactions using carboxy-terminal phage display. *J. Biol. Chem.* **275**, 21486–91.

Fukunaga R., Sokawa Y. & Nagata S. (1984) Constitutive production of human interferons by mouse cells with bovine papillomavirus as vector. *Proc. Nat. Acad. Sci. USA* **81**, 5086–90.

Fukushige S. & Sauer B. (1992) Targeted genomic integration with a positive selection *lox* recombination vector allows highly reproducible gene expression in mammalian cells. *Proc. Nat. Acad. Sci. USA* **89**, 7905–9.

Fuller D.J., Corb M.M., Barnett S., Steimer K. & Haynes J.R. (1997) Enhancement of immunodeficiency virus-specific immune responses in DNA-immunized rhesus macaques. *Vaccine* **15**, 924–6.

Fung E.T., Thulasiraman V., Weinberger S.R. & Dalmasso E.A. (2001) Protein biochips for differential profiling. *Curr. Opin. Biotechnol.* **12**, 65–9.

Fussenegger M., Morris R.P., Fux C., Rimann M., von Stockar B., Thompson C.J. & Bailey J.E. (2000) Streptogramin-based gene regulation systems for mammalian cells. *Nature Biotechnol.* **18**, 1203–8.

Futterer J. & Hohn T. (1996) Translation in plants – rules and exceptions. *Plant Mol. Biol.* **32**, 159–89.

genomics research. *Trends Biotechnol.* **20**, 462–6.

Gaasterland T. & Oprea M. (2001) Whole-genome analysis: annotations and updates. *Curr. Opin. Struct. Biol.* **11**, 377–81.

Gabizon R. & Taraboulos A. (1997) Of mice and (mad) cows – transgenic mice help to understand prions. *Trends Genet.* **13**, 264–9.

Gabriel S.B., Schaffner S.F., Nguyen H., *et al.* (2002) The structure of haplotype blocks in the human genome. *Science* **296**, 2225–9.

Galagan J.E., Calvo S.E., Borkovich K.A., *et al.* (2003) The genome sequence of the filamentous fungus *Neurospora crassa*. *Nature* **422**, 859–68.

Galagan J.E., Nusbaum C., Roy A., *et al.* (2002) The genome of *M. acetivorans* reveals extensive metabolic and physiological diversity. *Genome Res.* **12**, 532–42.

Galitski T., Saldanha A.J., Styles C.A., Lander E.S. & Fink G.R. (1999) Ploidy regulation of gene expression. *Science* **285**, 251–4.

Galli C., Lagutina I., Crotti G. *et al.* (2003) A cloned horse born to its dam twin – A birth announcement calls for a rethink on the immunological demands of pregnancy. *Nature* **424**, 635.

Gallie D.R. (1996) Translational control of cellular and viral mRNAs. *Plant Mol. Biol.* **32**, 145–58.

Gallie D.R. (1998) Controlling gene expression in transgenics. *Curr. Opin. Plant Biol.* **1**, 166–72.

Gallie D.R., Gay P. & Kado C.I. (1988) Specialized vectors for members of Rhizobiaceae and other Gram-negative bacteria. In *Vectors. A Survey of Molecular Cloning Vectors and Their Uses*, eds. Rodriguez R.L. & Denhardt D.T., pp. 333–42. Butterworths, London.

Gallo-Meagher M. & Irvine J.E. (1996) Herbicide resistant transgenic sugarcane plants containing the *bar* gene. *Crop Sci.* **36**, 1367–74.

Galperin M.Y. (2005) The Molecular Biology Database Collection: 2005 update. *Nucl. Acids Res.* **33**, D5–D24.

Galperin M.Y., Walker D.R. & Koonin E.V. (1998) Analagous enzymes: independent inventions in enzyme evolution. *Genome Res.* **8**, 779–90.

Gao A.-G., Hakimi S.M., Mittanck C.A., Wu Y., Woerner B.M., Stark D.M., Shah D.M., Liang J.H. & Rommens C.M.T. (2000) Fungal pathogen protection in potato by expression of a plant defensin peptide. *Nature Biotechnol.* **18**, 1307–10.

Gao G.P., Yang Y.P. & Wilson J.M. (1996) Biology of adenovirus vectors with E1 and E4 deletions for liver-directed gene therapy. *J. Virol.* **70**, 8934–43.

Gao X. *et al.* (2001) A flexible light-directed DNA chip synthesis gated by deprotection using solution photogenerated acids. *Nucl. Acids Res.* **29**, 4744–50.

Garcia T., Benhamou B., Gofflo D. *et al.* (1992) Switching agonistic, antagonistic and mixed transcriptional responses to 11β-substituted progestins by mutation of the progesterone receptor. *Mol. Endocrinol.* **6**, 2071–8.

Garcia-Vallve S., Romeu A. & Palau J. (2000) Horizontal gene transfer in bacterial and archaeal complete genomes. *Genome Res.* **10**, 1719–25.

Garrick D., Fiering S., Martin D.I. & Whitelaw E. (1998) Repeat-induced gene silencing in mammals. *Nature Genet.* **18**, 56–9.

Gatehouse A.M.R., Down R.E., Powell K.S. *et al.* (1996) Transgenic potato plants with enhanced resistance to the peach-potato aphid *Myzus persicae*. *Entomol. Exp. Appl.* **79**, 295–307.

Gatz C. & Quail P.H. (1988) Tn-10-encoded Tet repressor can regulate an operator-containing plant promoter. *Proc. Nat. Acad. Sci. USA* **85**, 1394–7.

Gatz C., Frohberg C. & Wendenburg R. (1992) Stringent repression and homogeneous de-repression by tetracycline of a modified CaMV 35S promoter in intact transgenic tobacco plants. *Plant J.* **2**, 397–404.

Gatz C., Kaiser A. & Wendenburg R. (1991) Regulation of a modified CaMV 35S promoter by the Tn-10-encoded Tet repressor in transgenic tobacco. *Mol. Gen. Genet.* **227**, 229–37.

Gavin A.C. *et al.* (2002) Functional organization of the yeast proteome by systematic analysis of protein complexes. *Nature* **415**, 141–7.

Ge H. (2000) UPA, a universal protein array system for quantitative detection of protein-protein, protein-DNA, protein-RNA and protein-ligand interactions. *Nucl. Acids Res.* **28**, e3, I–VII.

Geesaman B.J., Benson E., Brewster S.J., *et al.* (2003) Haplotype-based identification of a microsomal transfer protein marker associated with the human lifespan. *Proc. Nat. Acad. Sci. USA* **100**, 14115–20.

Gehrlach W., Llewellyn D. & Haseloff J. (1987) Construction of a plant disease resistance gene from the satellite RNA of tobacco ringspot virus. *Nature* **328**, 802–5.

Geiger M., Badur R., Kunze I. & Sommer S. (2001) Changing the fine chemical content in organisms by genetically modifying the shikimate pathway. WO 02/00901.

Geisler R. *et al.* (1999) A radiation hybrid map of the zebrafish genome. *Nature Genet.* **23**, 86–9.

Gelade R., Van de Velde S., Van Dijck P. *et al.* Multi-level response of the yeast genome to glucose. *Genome Biol.* **4**, 233.

Gelvin S.B. (2000) *Agrobacterium* and plant genes involved in T-DNA transfer and integration. *Ann. Rev. Plant Physiol. Plant Mol. Biol.* **51**, 223–56.

Gelvin S.B. (2003) *Agrobacterium*-mediated plant transformation: the biology behind the gene jockeying tool. *Microbiol. Mol. Biol. Rev.* **67**, 16–37.

Genilloud O., Garrido M.C. & Moreno F. (1984) The transposon Tn5 carries a bleomycin resistance determinant. *Gene* **32**, 225–33.

Gerard G.F. & D'Allesio J.M. (1993) Reverse transcriptase (EC2.7.7.49): the use of cloned Moloney murine leukemia virus reverse transcriptase to synthesize DNA from RNA. *Methods Mol. Biol.* **16**, 73–94.

Gerard R.D. & Gluzman Y. (1985) A new host cell system for regulated simian virus 40 DNA replication. *Mol. Cell. Biol.* **5**, 3231–40.

Gerhold D. & Caskey C.T. (1996) It's the genes! EST access to the human genome content. *Bioessays* **18**, 973–81.

Gerngross T.U. (2004) Advances in the production of human therapeutic proteins in yeasts and filamentous fungi. *Nature Biotechnol.* **22**, 1409–14.

Gershoni J.M. & Palade G.E. (1982) Electrophoretic transfer of proteins from sodium dodecyl sulfate-polyacrylamide gels to a positively charged membrane filter. *Anal. Biochem.* **124**, 396–405.

Gething M.-J. & Sambrook J. (1981) Cell-surface expression of influenza haemogglutinin from a cloned DNA copy of the RNA gene. *Nature* **293**, 620–5.

Ghebranious N. & Donehower L.A. (1998) Mouse models in tumor suppression. *Oncogene* **17**, 3385–400.

Giacolone J. *et al.* (2000) Optical mapping of BAC clones from the human Y chromosome *DAZ* locus. *Genome Res.* **10**, 1421–9.

Giaever G. *et al.* (1999) Genomic profiling of drug sensitivities via induced haploinsufficiency. *Nature Genet.* **21**, 278–83.

Giaever G., Chu A.M., Ni L., Connelly C., Riles L., Veronneau S., Dow S., Lucau-Danila A., Anderson K., Andre B. *et al.* (2002) Functional profiling of the *Saccharomyces cerevisiae* genome. *Nature* **418**, 387–91.

Giaever G., Flaherty P., Kumm J., Proctor M., Nislow C., Jaramillo D.F., Chu A.M., Jordan M.I., Arkin A.P. & Davis R.W. (2004) Chemogenomic profiling: identifying the functional interactions of small molecules in yeast. *Proc. Nat. Acad. Sci. USA* **101**, 793–8.

Gibrat J.F., Madej T. & Bryant S.H. (1996) Surprising similarities in structure comparison. *Curr. Opin. Struct. Biol.* **6**, 377–85.

Gierl A. & Saedler H. (1992) Plant transposable elements and gene tagging. *Plant Mol. Biol.* **19**, 39–49.

Gilbert D.M. & Cohen S.N. (1987) Bovine papillomavirus plasmids replicate randomly in mouse fibroblasts throughout S phase of the cell cycle. *Cell* **50**, 59–68.

Gill P., Ivanov P.L., Kimpton C. *et al.* (1994) Identification of the remains of the Romanov family by DNA analysis. *Nature Genet.* **6**, 130–5.

Gillam S., Astell C.R. & Smith M. (1980) Site-specific mutagenesis using oligodeoxyribonucleotides: isolation of a phenotypically silent φX174 mutant, with a specific nucleotide deletion, at very high efficiency. *Gene* **12**, 129–37.

Gilmore M.S. & Ferretti J.J. (2003) The thin line between gut commensal and pathogen. *Science* **299**, 1999–2002.

Giot L. *et al.* (2004) A protein interaction map of *Drosophila melanogaster*. *Science* **302**, 1727–36.

Girgis S.I., Alevizaki M., Denny P., Ferrier G.J.M. & Legon S. (1988) Generation of DNA probes for peptides with highly degenerate codons using mixed primer PCR. *Nucl. Acids Res.* **16**, 10371.

Giri A. & Narassu M.L. (2000) Transgenic hairy roots: recent trends and applications. *Biotechnol. Adv.* **18**, 1–22.

Giuliano G., Aquilani R. & Dharmapuri S. (2000) Metabolic engineering of plant carotenoids. *Trends Plant Sci.* 406–9.

Glass J. *et al.* (2000) The complete sequence of the mucosal pathogen *Ureaplasma urealyticum*. *Nature* **407**, 757–62.

Glazier A.M., Nadeau J.H. & Altman T.J. (2002) Finding genes that underlie complex traits. *Science* **298**, 2345–9.

Glazko G.V., Koonin E.V., Rogozin I.B. & Shabalina S.A. (2003) A significant fraction of conserved noncoding DNA in human and mouse consists of predicted matrix attachment regions. *Trends Genet.* **19**, 119–24.

Gluzman Y. (1981) SV40-transformed simian cells support the replication of early SV40 mutants. *Cell* **23**, 175–82.

Gockel G. & Hachtel W. (2000) Complete gene map of the plastid genome of the nonphotosynthetic euglenoid flagellate *Astasia longa*. *Protist* **151**, 347–51.

Goddijn O.J.M., Pennings E.J.M., Van der Helm P., Verpoorte R. & Hoge J.H.C. (1995) Overexpression of a tryptophan decarboxylase cDNA in *Catharanthus roseus* crown gall calli results in increased tryptamine levels but not in increased terpenoid indole alkaloid production. *Transgenic Res.* **4**, 315–23.

Goff S.P. & Berg P. (1976) Construction of hybrid viruses containing SV40 and l phage DNA segments and their propagation in cultured monkey cells. *Cell* **9**, 695–705.

Goffeau A. *et al.* (1996) Life with 6000 genes. *Science* **274**, 546–67.

Gogarten J.P., Doolittle W.F. & Lawrence J.G. (2002) Prokaryotic evolution in light of gene transfer. *Mol. Biol. Evol.* **19**, 2226–38.

Goldberg D.A., Posakony J.W. & Maniatis T. (1983) Correct developmental expression of a cloned alcohol dehydrogenase gene transduced into the *Drosophila* germ line. *Cell* **34**, 59–73.

Goldin A.L. (1991) Expression of ion channels in oocytes. In *Methods in Cell Biology*, Vol. 36. *Xenopus laevis: Practical Uses in Cell and Molecular Biology*, eds Kay KB & Peng HB, pp. 487–509. Academic Press, New York.

Golds T.J., Maliga P. & Koop H.U. (1993) Stable plastid transformation in PEG-treated protoplasts of *Nicotiana tabacum*. *Biotechnology* **11**, 95–7.

Goldsbrough A.P., Lastrella C.N. & Yoder J.I. (1993) Transposition mediated repositioning and subsequent elimination of marker genes from transgenic tomato. *Biotechnology* **11**, 1286–92.

Golic K. (1991) Site-specific recombination between homologous chromosomes in *Drosophila*. *Science* **252**, 958–61.

Golub T.R. *et al.* (1999) Molecular classification of cancer: class discovery and class prediction by gene expression monitoring. *Science* **286**, 531–7.

Gonczy P. *et al.* (2000) Functional genomic analysis of cell division in *C. elegans* using RNAi of genes on chromosome III. *Nature* **408**, 331–6.

Gonnet G.H., Cohen M.A. & Benner S.A. (1992) Exhaustive matching of the entire protein-sequence database. *Science* **256**, 1443–5.

Good P.D. *et al.* (1997) Expression of small, therapeutic RNAs in human cell nuclei. *Gene Ther.* 4, 45–54.

Goodner B.W. *et al.* (1999) Combined genetic and physical map of the complex genome of *Agrobacterium tumefaciens*. *J. Bacteriol.* **180**, 3816–22.

Gordon J.W. & Ruddle F.H. (1981) Integration and stable germ line transmission of genes injected into mouse pronuclei. *Science* **214**, 1244–6.

Gordon K., Lee E., Vitale J.A. *et al.* (1987) Production of human tissue plasminogen activator in transgenic mouse milk. *Biotechnology* **5**, 1183–7.

Gordon M.P., Farrand S.K., Sciaky D. *et al.* (1979) In *Molecular Biology of Plants, Symposium, University of Minnesota*, ed. Rubenstein I. Academic Press, London.

Gordon-Kamm W.J., Spencer T.M., Maugano M.L. *et al.* (1990) Transformation of maize cells and regeneration of fertile transgenic plants. *Plant Cell* **2**, 603–18.

Gorg A., Obermaier C., Boguth G., Harder A. *et al.* (2000) The current state of two-dimensional electrophoresis with immobilized pH gradients. *Electrophoresis* **21**, 1037–53.

Gorlach J., Volrath S., Knauf-Beiter G. *et al.* (1996) Benzothiadiazole, a novel class of inducers of systemic acquired resistance, activates gene expression and disease resistance in wheat. *Plant Cell* **8**, 629–43.

Gorman C.M., Gies D., McCray G. & Huang M. (1989) The human cytomegalovirus major immediate early promoter can be *trans*-activated by adenovirus early proteins. *Virology* **171**, 377–85.

Gorman C.M., Gies D.R. & McCray G. (1990) Transient production of proteins using an adenovirus transformed cell line. *DNA Protein Eng. Technol.* **2**, 3–10.

Gorman C.M., Merlino G.T., Willingham M.C., Pastan I. & Howard B. (1982a) The Rous sarcoma virus long terminal repeat is a strong promoter when introduced into a variety of eukaryotic cells by DNA-mediated transfection. *Proc. Nat. Acad. Sci. USA* **79**, 6777–81.

Gorman C.M., Moffat L.F. & Howard B.H. (1982b) Recombinant genome which expresses chloramphenicol acetyl transferase in mammalian cells. *Mol. Cell. Biol.* **2**, 1044–51.

Gormley E.P. & Davies J. (1991) Transfer of plasmid RSF1010 by conjugation from *Escherichia coli* to *Streptomyces lividans* and *Mycobacterium smegmatis*. *J. Bacteriol.* **173**, 6705–8.

Gorziglia M.I., Kadan M.J., Yei S. *et al.* (1996) Elimination of both E1 and E2a from adenovirus vectors further improves prospects for *in vivo* human gene therapy. *J. Virol.* **70**, 4173–8.

Goshe M.B. & Smith R.D. (2003) Stable isotope-coded proteomic mass spectrometry. *Curr. Opin. Biotechnol.* **14**, 101–9.

Gossen M. & Bujard H. (1992) Tight control of gene expression in mammalian cells by tetracycline-responsive promoters. *Proc. Nat. Acad. Sci. USA* **89**, 5547–51.

Gossen M., Freundlieb S., Bender G. *et al.* (1995) Transcriptional activation by tetracyclines in mammalian cells. *Science* **268**, 1766–9.

Gossler A., Joyner A.L., Rossant J. & Skarnes W.C. (1989) Mouse embryonic stem cells and reporter constructs to detect developmentally regulated genes. *Science* **244**, 463–5.

Gottesmann M.E. & Yarmolinsky M.D. (1968) The integration and excision of the bacteriophage lambda genome. *Cold Spring Harbor Symp. Quant. Biol.* **33**, 735–47.

Gottschalk A. *et al.* (1998) A comprehensive biochemical and genetic analysis of the yeast U1 snRNP reveals five novel proteins. *RNA* **4**, 374–93.

Gourse R.L., Gaal T., Aiyar S.E. *et al.* (1998) Strength and regulation without transcription factors: lessons from bacterial rRNA promoters. *Cold Spring Harbor Symp. Quant. Biol.* **63**, 131–9.

Gowland P.C. & Hardmann D.J. (1986) Methods for isolating large bacterial plasmids. *Microbiol. Sci.* **3**, 252–4.

Graber J.H., Smith C.L. & Cantor C. (1999) Differential sequencing with mass spectrometry. *Genetical Anal. Biomol. Eng.* **14**, 215–19.

Grabherr R. & Bayer K. (2002) Impact of targeted vector design on ColE1 plasmid replication. *Trends Biotechnol.* **20**, 257–60.

Graessmann M. & Graessmann A. (1983) Microinjection of tissue culture cells. *Methods Enzymol.* **101**, 482–92.

Graham F.L. & Van der Erb A.J. (1973) A new technique for the assay of infectivity of human adenovirus 5 DNA. *Virology* **52**, 546–50.

Graham F.L., Smiley J., Russell W.C. & Nairn R. (1977) Characteristics of a human cell line transformed by DNA from human adenovirus type 5. *J. Gen. Virol.* **36**, 59–74.

Graham M.W., Craig S. & Waterhouse P.M. (1997) Expression patterns of vascular-specific promoters RolC and Sh in transgenic potatoes and their use in engineering PLRV-resistant plants. *Plant Mol. Biol.* **33**, 729–35.

Grandbastien M-A. (1992) Retroelements in higher plants. *Trends Genet.* **8**, 103–8.

Granjeaud S., Bertucci F. & Jordan B.R. (1999) Expression profiling: DNA arrays in many guises. *BioEssays* **21**, 781–90.

Grant S.R. (1999) Dissecting the mechanisms of post-transcriptional gene silencing: divide and conquer. *Cell* **96**, 303–6.

Graveley B.R. (2001) Alternative splicing: increasing diversity in the proteomic world. *Trends Genet.* **17**, 100–7.

Gray M.W., Burger G. & Lang F. (1999) Mitochondrial evolution. *Science* **283**, 1476–81.

Green P. (1997) Against a whole-genome shotgun. *Genome Res.* **7**, 410–17.

Green P., Pines O. & Inouye M. (1986) The role of anti-sense RNA in gene regulation. *Ann. Rev. Biochem.* **55**, 569–97.

Gregory R.J., Cheng S.H., Rich D.P. *et al.* (1990) Expression and characterization of the cystic fibrosis transmembrane conductance regulator. *Nature* **347**, 382–6.

Gress T.M. *et al.* (1992) Hybridization fingerprinting of high-density cDNA library arrays with cDNA pools derived from whole tissues. *Mammalian Genome* **3**, 609–10.

Gressel J. (1999) Tandem constructs: preventing the rise of superweeds. *Trends Biotechnol.* **17**, 361–6.

Gressel J. (2000) Molecular biology of weed control. *Transgenic Res.* **9**, 355–82.

Gribskov M. & Burgess R.R. (1983) Overexpression and purification of the sigma subunit of *Escherichia coli* RNA polymerase. *Gene* **26**, 109–18.

Grimsley N., Hohn T., Davies J.W. & Hohn B. (1987) *Agrobacterium*-mediated delivery of infectious maize streak virus into maize plants. *Nature* **325**, 177–9.

Gronenborn B. & Messing J. (1978) Methylation of single-stranded DNA *in vitro* introduces new restriction endonuclease cleavage sites. *Nature* **272**, 375–7.

Gronenborn B. & Matzeit V. (1989) Plant gene vectors and genetic transformation: plant viruses as vectors. In: Schell, J. & Vasil, I.K. (eds.) *Cell Culture and Somatic Cell Genetics of Plants*, **Vol. 6**. San Diego: Academic Press Inc., pp. 69–91.

Grosveld F. (1999) Activation by locus control regions? *Curr. Opin. Genet. Devel.* **9**, 152–7.

Grosveld F., Van Assendelft G.B., Greaves D.R. & Kollias G. (1987) Position independent, high level expression of the human β-globin gene in transgenic mice. *Cell* **51**, 975–85.

Grunstein M. & Hogness D.S. (1975) Colony hybridization: a method for the isolation of cloned DNAs that contain a specific gene. *Proc. Nat. Acad. Sci. USA* **72**, 3961–5.

Gruss A. & Ehrlich S.D. (1989) The family of highly interrelated single stranded deoxyribonucleic acid plasmids. *Microbiol. Rev.* **53**, 231–41.

Gryczan T.J., Contente S. & Dubnau D. (1980) Molecular cloning of heterologous chromosomal DNA by recombination between a plasmid vector and a homologous resident plasmid in *Bacillus subtilis*. *Mol. Gen. Genet.* **177**, 459–67.

Gu H., Marth J.D., Orban P.C., Mossmann H. & Rajewsky K. (1994) Deletion of a DNA polymerase beta gene segment in T-cells using cell type-specific gene targeting. *Science* **265**, 103–6.

Gu S.Y., Huang T.M., Ruan L. *et al.* (1996) First EBV vaccine trial in humans using recombinant vaccinia virus expressing the major membrane antigen. *Devel. Biol. Stand.* **84**, 171–7.

Guarente L. (1987) Regulatory proteins in yeast. *Ann. Rev. Genet.* **21**, 425–52.

Gubler U. & Hoffman B.J. (1983) A simple and very efficient method for generating cDNA libraries. *Gene* **26**, 263–9.

Guerineau F. & Mullineaux P. (1989) Nucleotide sequence of the sulfonamide resistance plasmid R46. *Nucl. Acids Res.* **14**, 4370.

Guhathakurta A., Viney I. & Summers D. (1996) Accessory proteins impose site selectivity during ColE1 dimer resolution. *Mol. Microbiol.* **20**, 613–20.

Gumport R.I. & Lehman I.R. (1971) Structure of the DNA ligase adenylate intermediate: lysine (ε-amino) linked AMP. *Proc. Nat. Acad. Sci. USA* **68**, 2559–63.

Gunderson K.L. *et al.* (2004) Decoding randomly ordered DNA arrays. *Genome Res.* **14**, 870–7.

Guo Q.M. (2003) DNA microarray and cancer. *Curr. Opin. Oncol.* **15**, 36–43.

Guo Z.S. & Bartlett D.L. (2004) Vaccinia as a vector for gene delivery. *Expert Opin. Biol. Th.* **4**, 901–17.

Gurdon J.B. (1986) Nuclear transplantation in eggs and oocytes. *J. Cell Sci. Suppl.* **4**, 287–318.

Gurdon J.B. (1991) Nuclear transplantation in *Xenopus*. *Methods Cell Biol.* **36**, 299–309.

Gurdon J.B., Lane C.D., Woodland H.R. & Marbaix G. (1971) Use of frog eggs and oocytes for the study of messenger RNA and its translation in living cells. *Nature* **233**, 177–82.

Guzman L.M., Belin D., Carson M.J. & Beckwith J. (1995) Tight regulation, modulation, and high-level expression by vectors containing the arabinose pBAD promoter. *J. Bacteriol.* **177**, 4121–30.

Gygi S.P., Rist B., Gerber S.A., Turecek F., Gelb M.H. & Aebersold R. (1999) Quantitative analysis of complex protein mixtures using isotope-coded affinity tags. *Nature Biotechnol.* **17**, 994–99.

Gygi S.P., Rochon Y., Franza B.R. & Aebersold R. (1999) Correlation between protein and mRNA abundance in yeast. *Mol. Cell Biol.* **19**, 1720–30.

Haab B.B., Dunham M.J. & Brown P.O. (2001) Protein microarrays for highly parallel detection and quantitation of specific proteins and antibodies in complex solutions. *Genome Biol.* **2**, research0004.1-0004.13.

Hacia J.G. (1999) Resequencing and mutational analysis using oligonucleotide microarrays. *Nature Genet. Suppl.* **21**, 42–7.

Hacia J.G. *et al.* (1996) Detection of heterozygous mutations in BRCA1 using high density oligonucleotide arrays and two-colour fluorescence analysis. *Nature Genet.* **14**, 441–7.

Hacker J., Blum-Oehler G., Muhldorfer I. & Tschape H. (1997) Pathogenicity islands of virulent bacteria: structure, function and impact on microbial evolution. *Mol. Microbiol.* **23**, 1089–97.

Hadley C. & Jones D. (1999) A systematic comparison of protein structure classifications: SCOP, CATH and FSSP. *Structure with Folding and Design* **7**, 1099–112.

Haenlin M., Steller H., Pirrotta V. & Momer E. (1985) A 43 kb cosmid P transposon rescues the fs(1)K10 morphogenetic locus and three adjacent *Drosophila* developmental mutants. *Cell* **40**, 827–37.

Hager L.J. & Palmiter R.D. (1981) Transcriptional regulation of mouse liver metallothionein I gene by glucocorticoids. *Nature* **291**, 340–2.

Hagio T., Hirabayashi T., Machii H. & Tomotsune H. (1995) Production of fertile transgenic barley (*Hordeum vulgare* L.) plants using the hygromycin resistance marker. *Plant Cell Rep.* **14**, 329–34.

Hahn S., Hoar E.T. & Guarente L. (1985) Each of three "TATA elements" specifies a subset of the transcription initiation sites at the CYC-1 promoter of *Saccharomyces cerevisiae*. *Proc. Nat. Acad. Sci. USA* **82**(24), 8562–6.

Haima P., Bron S. & Venema G. (1987) The effect of restriction on shotgun cloning and plasmid stability in *Bacillus subtilis* Marburg. *Mol. Gen Genet.* **209**, 335–42.

Haima P., Bron S. & Venema G. (1988) A quantitative analysis of shotgun cloning in *Bacillus subtilis* protoplasts. *Mol. Gen. Genet.* **213**, 364–9.

Haima P., Bron S. & Venema G. (1990) Novel plasmid marker rescue transformation system for molecular cloning in *Bacillus subtilis* enabling direct selection of recombinants. *Mol. Gen. Genet.* **223**, 185–91.

Hain R., Reif H.J., Krause E. *et al.* (1993) Disease resistance results from foreign phytoalexin expression in a novel plant. *Nature* **361**, 153–6.

Hall C.V., Jacob P.E., Ringold G.M. & Lee F. (1983) Expression and regulation of *Escherichia coli lacZ* gene fusions in mammalian cells. *J. Mol. Appl. Genet.* **2**, 101–9.

Hallick R.B. *et al.* (1993) Complete sequence of *Euglena gracilis* chloroplast DNA. *Nucl. Acids Res.* **21**, 3537–44.

Hamaguchi M. *et al.* (1992) Establishment of a highly sensitive and specific exon-trapping system. *Proc. Nat. Acad. Sci. USA* **89**, 9779–83.

Hamer L., DeZwaan T.M., Montenegro-Chamorro M.V., Frank S.A. & Hamer J.E. (2001) Recent advances in large-scale transposon mutagenesis. *Curr. Opin. Chem. Biol.* **5**, 67–73.

Hamil J.D., Parr A.J., Rhodes M.J.C., Robins R.J. & Walton N.J. (1987) New routes to plant secondary products. *Biotechnology* **5**, 800–4.

Hamilton A.J. & Baulcombe D.C. (1999) A species of small antisense RNA in posttranscriptional gene silencing in plants. *Science* **286**, 950–2.

Hamilton C.M. (1997) A binary-BAC system for plant transformation with high molecular weight DNA. *Gene* **200**, 107–16.

Hamilton C.M., Frary A., Lewis C. & Tanksley S.D. (1996) Stable transfer of high molecular weight DNA into plant chromosomes. *Proc. Nat. Acad. Sci. USA* **93**, 9975–9.

Hamilton C.M., Frary A., Xu Y.M., Tanksley S.D. & Zhang H.B. (1999) Construction of tomato genomic DNA libraries in a binary-BAC (BIBAC) vector. *Plant J.* **18**, 223–9.

Hammer R.E., Krumlauf R., Camper S.A., Brinster R.L. & Tilghman S.M. (1987) Diversity of alpha-fetoprotein gene expression in mice is generated by a combination of separate enhancer elements. *Science* **235**, 53–8.

Hammerschmidt W. & Sugden B. (1988) Identification and characterization of ori$_{lyt}$, a lytic origin of DNA replication of Epstein-Barr virus. *Cell* **55**, 427–33.

Hammond S.M., Bernstein E., Beach D. & Hannon G.J. (2000) An RNA-directed nuclease mediates posttranscriptional gene silencing in *Drosophila* cells. *Nature* **404**, 293–6.

Hammond S.M., Caudy A.A. & Hannon G.J. (2001) Posttranscriptional gene silencing by double-stranded RNA. *Nature Rev. Genet.* **2**, 110–19.

Hammond-Kosack K.E., Staskowicz B.J., Jones J.D.G. & Baulcombe D.C. (1995) Functional expression of a fungal avirulence gene from a modified potato virus X genome. *Mol. Plant-Microbe Interactions* **8**, 181–5.

Han J.S. & Bockc J.D. (2004) A highly active synthetic mammalian retrotransposon. *Nature* **429**, 314–18.

Hanahan D. & Meselson M. (1980) Plasmid screening at high colony density. *Gene* **10**, 63–7.

Hanahan D. (1983) Studies on transformation of *Escherichia coli* with plasmids. *J. Mol. Biol.* **166**, 557–80.

Hanahan D., Jessee J. & Bloom F.R. (1991) Plasmid transformation of *Escherichia coli* and other bacteria. *Methods Enzymol.* **204**, 63–113.

Hancock J.M. (1996) Simple sequences in a "minimal" genome. *Nature Genet.* **14**, 14–15.

Hanin M., Volrath S., Bogucki A., Briker M., Ward E. & Paszkowski J. (2001) Gene targeting in *Arabidopsis*. *Plant J.* **28**, 671–7.

Hanks M., Wurst W., Anson-Cartwright L., Auerbach A.B. & Joyner A.L. (1995) Rescue of the en-1 mutant phenotype by replacement of en-1 with en-2. *Science* **269**, 679–82.

Hannon, G.J. (2002) RNA interference. *Nature* **418**, 244–51.

Harbers K., Jahner D. & Jaenisch R. (1981) Microinjection of cloned retroviral genomes into mouse zygotes: integration and expression in the animal. *Nature* **293**, 540–3.

Harper J.W., Adami G.R., Wei N., Keyomarsi K. & Ellege S.J. (1993) The p21 Cdk-interacting protein Cip1 is a potent inhibitor of G1 cyclin-dependent kinases. *Cell* **75**, 805–16.

Harrington C.A., Rosenow C. & Retief J. (2000) Monitoring gene expression using DNA microarrays. *Curr. Opin. Microbiol.* **3**, 285–91.

Harris J.D., Gutierrez A.A., Hurst H.C., Sikora K. & Lemoine, N.R. (1994) Gene therapy for cancer using tumour-specific prodrug activation. *Gene Ther.* **1**, 170–5.

Hartl D.L. (1996) EST! EST!! EST!!! *Bioessays* **18**, 1021–3.

Hartl D.L., Nurminsky D.I., Jones R.W. & Lozovskaya E.R. (1994) Genome structure and evolution in *Drosophila*-applications of the framework P1 map. *Proc. Nat. Acad. Sci. USA* **91**, 6824–9.

Hartley J.L., Temple G.F. & Brasch M.A. (2000) DNA cloning using *in vitro* site-specific recombination. *Genome Res.* **10**, 1788–95.

Hartman P.E. & Roth J.R. (1973) Mechanisms of suppression. *Advances in Genetics* **17**, 1–105.

Hartman S.C. & Mulligan R.C. (1988) Two dominant-acting selectable markers for gene transfer studies in mammalian cells. *Proc. Nat. Acad. Sci. USA* **85**, 8047–51.

Harvey A.J. *et al.* (2002) Expression of exogenous protein in the egg white of transgenic chickens. *Nat. Biotechnol.* **20**, 396–9.

Haseloff J. & Amos B. (1995) GFP in plants. *Trends Genet.* **11**, 328–9.

Haseloff J., Dormand E.L. & Brand A.H. (1999) Live imaging with green fluorescent protein. *Methods Mol. Biol.* **122**, 241–59.

Haseloff J., Siemering K.R., Prasher D.C. & Hodge S. (1997) Removal of a cryptic intron and subcellular localization of green fluorescent protein are required to mark transgenic *Arabidopsis* plants brightly. *Proc. Nat. Acad. Sci. USA* **94**, 2122–7.

Hashemzadeh-Bonehi L., Mehraein-Ghomi F., Mitsopoulos C., Hennessey E.S. & Broome-Smith J.K. (1998) Importance of using *lac* rather than *ara* promoter vectors for modulating the levels of toxic gene products in *Escherichia coli*. *Mol. Microbiol.* **30**, 676–8.

Hashimoto L. *et al.* (1994) Genetic mapping of a suscep-tibility locus for insulin-dependent diabetes mellitus on chromosome 11q. *Nature* **371**, 161–4.

Hasty P., Ramirez-Solis R., Krumlauf R. & Bradley A. (1991a) Introduction of subtle mutation into the *Hox-2.6* locus in embryonic stem cells. *Nature* **350**, 243–6.

Hasty P., Rivera-Perez J. & Bradley A. (1991b) The length of homology required for gene targeting in embryonic stem cells. *Mol. Cell. Biol.* **11**, 5586–91.

Hasuwa H., Kaseda K., Einarsdottir T. & Okabe M. (2002) Small interfering RNA and gene silencing in transgenic mice and rats. *FEBS Lett.* **532**, 227–30.

Hauge B.M., Hanley S., Giraudat J. & Goodman H.M. (1991) Mapping the *Arabidopsis* genome. In: *Molecular Biology of Plant Development* (eds. G. Jenkins & W. Schurch). Cambridge University Press, Cambridge.

Haughn G.W., Smith J., Mazur B. & Somerville C. (1988) Transformation with mutant *Arabidopsis* acetolactate synthase gene renders tobacco resistant to sulfonylurea herbicides. *Mol. Gen. Genet.* **211**, 266–71.

Hawley D.K. & McClure W.R. (1983) Compilation and analysis of *Escherichia coli* promoter DNA sequences. *Nucl. Acids Res.* **11**, 2237–55.

Hay J.G., McElvaney N.G., Herena J. & Crystal R.G. (1995) Modification of nasal epithelial potential differences of individuals with cystic fibrosis con-sequent to local administration of a normal CFTR cDNA adenovirus gene transfer vector. *Hum. Gene Ther.* **6**, 1487–96.

Hayashi T. *et al.* (2001) Complete genome sequence of enterohemorrhagic *Escherichia coli* O157:H7 and genomic comparison with a laboratory strain K-12. *DNA Research* **8**, 11–22.

Hayes J.E., Richardson A.E. & Simpson R.J. (1999) Phytase and acid phosphatase activities in extracts from roots of temperate pasture grass and legume seedlings. *Australian J. Plant Physiol.* **26**, 801–9.

Hayes R.J., Coutts R.H.A. & Buck K.W. (1989) Stability and expression of bacterial genes in replicating gemini-virus vectors in plants. *Nucl. Acids Res.* **17**, 2391–403.

Hayes R.J., Petty I.T.D., Coutts R.H.A. & Buck K.W. (1988) Gene amplification and expression in plants by a replicating geminivirus vector. *Nature* **334**, 179–82.

Hayes W. (1968) *The Genetics of Bacteria and their Viruses*, 2nd edn. Blackwell Scientific Publications, Oxford.

Hayford M.B., Medford J.I., Hoffmann N.L., Rogers S.G. & Klee H.J. (1988) Development of a plant trans-formation selection system based on expression of genes encoding gentamicin acetyltransferases. *Plant Physiol.* **86**, 1216–22.

Hayman G.T. & Bolen P.L. (1993) Movement of shuttle plasmids from *Escherichia coli* into yeasts other than *Saccharomyces cerevisiae* using trans-kingdom con-jugation. *Plasmid* **30**, 251–7.

Haynes J.R., McCabe D.E., Swain W.F., Widera G. & Fuller J.T. (1996) Particle-mediated nucleic acid immunization. *J. Biotechnol.* **44**, 37–42.

Hazbun T.R. & Fields S. (2001) Networking proteins in yeast. *Proc. Nat. Acad. Sci. USA* **98**, 4277–8.

He L-Z. *et al.* (1998) Distinct interactions of PML-RARa with co-repressors determine differential responses to RA in APL. *Nature Genet.* **18**, 126–35.

Heale S.M., Stateva L.L. & Oliver S.G. (1994) Intro-duction of YACs into intact yeast cells by a pro-cedure which shows low levels of recombinageni-city and co-transformation. *Nucl. Acids Res.* **22**, 5011–15.

Hedgepeth J., Goodman H.M. & Boyer H.W. (1972) DNA nucleotide sequence restricted by the RI endonucle-ase. *Proc. Nat. Acad. Sci. USA* **69**, 3448–52.

Hegyi H. & Gerstein M. (1999) The relationship between protein structure and function: a compre-hensive survey with application to the yeast genome. *J. Mol. Biol.* **288**, 147–64.

Heim R. & Tsein R. (1996) Engineering green fluorescent protein for improved brightness, longer wavelengths and fluorescence energy resonance transfer. *Curr. Biol.* **6**, 178–82.

Heinemann J.A. & Sprague G.F. (1989) Bacterial conjugative plasmids mobilize DNA transfer between bacteria and yeast. *Nature* **340**, 205–9.

Heinemann U. (2000) Structural genomics in Europe: slow start, strong finish? *Nature Struct. Biol.* **7**, 940–2.

Heinemann U. *et al.* (2000) An integrated approach to structural genomics. *Prog. Biophys. Mol. Biol.* **73**, 347–62.

Heinemann U., Illing G. & Oschkinat H. (2001) High-throughput three-dimensional protein structure determination. *Curr. Opin. Biotechnol.* **12**, 348–54.

Heinkoff S. (1990) Position effect variegation after 60 years. *Trends Genet.* **6**, 422–6.

Heinrich J.C., Tabler M. & Louis C. (1983) Attenuation of white gene expression in transgenic *Drosophila melanogaster*: possible role of catalytic antisense RNA. *Devel. Genet.* **14**, 258–65.

Helfman D.M. & Hughes S.H. (1987) Use of antibodies to screen cDNA expression libraries prepared in plasmid vectors. *Methods Enzymol.* **152**, 451–7.

Helfman D.M., Feramisco J.R., Fiddes J.C., Thomas G.P. & Hughes S.II. (1983) Identification of clones that encode chicken tropomyosin by direct immunological screening of a cDNA expression library. *Proc. Nat. Acad. Sci. USA* **80**, 31–5.

Hellens R., Mullineaux P. & Klee H. (2000a) A guide to *Agrobacterium* binary Ti vectors. *Trends Plant Sci.* **5**, 446–51.

Hellens R.P., Edwards E.A., Leyland N.R., Bean S. & Mullineaux P.M. (2000b) pGreen: a versatile and flexible binary Ti vector for *Agrobacterium*-mediated plant transformation. *Plant Mol. Biol.* **42**, 819–32.

Heller R. *et al.* (1997) Discovery and analysis of inflammatory disease-related genes using cDNA microarrays. *Proc. Nat. Acad. Sci. USA* **94**, 2150–5.

Heller R., Song K., Villaret D. *et al.* (1990) Amplified expression of tumor necrosis factor receptor in cells transfected with Epstein-Barr virus shuttle vector cDNA libraries. *J. Biol. Chem.* **264**, 5708–17.

Heller R., Jaroszeski M., Atkin A., Moradpour D., Gilbert R., Wands J. & Nicolau C. (1996) *In vivo* gene electroinjection and expression in rat liver. *FEBS Lett.* **389**, 225–8.

Helmann J.D. (1995) Compilation and analysis of *Bacillus subtilis* sigma A-dependent promoter sequences: evidence for extended contact between RNA polymerase and upstream promoter DNA. *Nucl. Acids Res.* **23**, 2351–60.

Helmer G., Casadaban M., Bevan M., Kayes L. & Chilton M.-D. (1984) A new chimeric gene as a marker for plant transformation: the expression of *Escherichia coli* b-galactosidase in sunflower and tobacco cells. *Biotechnology* **2**, 520–7.

Helwig U., Imai K., Schmahl W. *et al.* (1995) Interaction between undulated and patch leads to an extreme form of spina-bifida in double-mutant mice. *Nature Genet.* **11**, 60–3.

Hemmati-Brivanlou A. & Melton D.A. (1992) A truncated activin receptor inhibits mesoderm induction and formation of axial structures in *Xenopus* embryos. *Nature* **359**, 609–14.

Hendrickson W.A., Horton J.R. & LeMaster D.M. (1990) Selenomethionyl proteins produced for analysis by multiwavelength anomalous diffraction (MAD): a vehicle for direct determination of three-dimensional structure. *EMBO J.* **9**, 1665–72.

Hendy S., Chen Z.C., Barker H. *et al.* (1999) Rapid production of single-chain Fv fragments in plants using a potato virus X episomal vector. *J. Immunol. Method.* **231**, 137–46.

Henikoff S. & Henikoff J.G. (1993) Performance evaluation of amino acid substitution matrices. *Proteins* **17**, 49–61.

Henze K. & Martin W. (2001) How do mitochondrial genes get into the nucleus? *Trends Genet.* **17**, 383–7.

Henzel W.J. *et al.* (1993) Identifying proteins from two-dimensional gels by molecular mass searching of peptide fragments in protein sequence databases. *Proc. Nat. Acad. Sci. USA* **90**, 5011–15.

Herbers K. (2003) Vitamin production in transgenic plants. *J. Plant. Physiol.* **160**, 821–9.

Herbert B.R. *et al.* (2001) What place for polyacrylamide in proteomics? *Trends Biotechnol.* **19** (Suppl. *Trends Guide to Proteomics II*), S3–S9.

Hermonat O.L. & Muzyczka N. (1984) Use of adeno-associated virus as a mammalian DNA cloning vector: transduction of neomycin resistance into mammalian tissue culture cells. *Proc. Nat. Acad. Sci. USA* **81**, 6466–70.

Hernalsteens J.P., Thiatoong L., Schell J. & Vanmontagu M. (1984) An agrobacterium-transformed cell-culture from the monocot *Asparagus officinalis*. *EMBO J.* **3**, 3039–41.

Herrera-Estrella L. & Simpson J. (1995) Genetically-engineered resistance to bacterial and fungal pathogens. *World J. Microbiol. Biotechnol.* **11**, 383–92.

Herrera-Estrella L., DeBlock M., Messens E. *et al.* (1983b) Chimeric genes as dominant selectable markers in plant cells. *EMBO J.* **2**, 987–95.

Herrera-Estrella L., Depicker A., Van Montagu M. & Schell J. (1983a) Expression of chimaeric genes transferred into plant cells using a Ti-plasmid-derived vector. *Nature* **303**, 209–13.

Herskowitz I. (1974) Control of gene expression in bacteriophage lambda. *Ann. Rev. Genet.* **7**, 289–324.

Herskowitz I. (1987) Functional analysis of genes by dominant negative mutations. *Nature* **329**, 219–22.

Herweijer H., Latendresse J.S., Williams P. *et al.* (1995) A plasmid-based self-amplifying Sindbis virus vector. *Hum. Gene Ther.* **6**, 1161–7.

Heyman J.A. & 17 others (1999) Genome-scale cloning and expression of individual open reading frames using topoisomerase I-mediated ligation. *Genome Res.* **9**, 383–92.

Hiatt A.H., Cafferkey R. & Bowdish K. (1989) Production of antibodies in transgenic plants. *Nature* **342**, 76–8.

Hicks G.G., Shi E.G., Li X.M., Li C.H., Pawlak M. & Ruley H.E. (1997) Functional genomics in mice by tagged sequence mutagenesis. *Nature Genet.* **16**, 338–44.

Hicks J.B., Strathern J.N., Klar A.J.S. & Dellaporta S.L. (1982) Cloning by complementation in yeast: the mating type genes. In *Genetic Engineering*, eds. Setlow J.K. & Hollaender A., pp. 219–48. Plenum Press, New York.

Hiei Y., Ohta S., Komari T. & Kumashiro T. (1994) Efficient transformation of rice (*Oryza sativa* L) mediated by *Agrobacterium* and sequence analysis of the boundaries of the T-DNA. *Plant J.* **6**, 241–82.

Higgins D.E. & Portnoy D.A. (1998) Bacterial delivery of DNA evolves. *Nature Biotechnol.* **16**, 138–9.

Higo K.E., Otaka E. & Osawa S. (1982) Purification and characterization of 30S ribosomal proteins from *Bacillus subtilis*: correlation to *Escherichia coli* 30S proteins. *Mol. Gen. Genet.* **185**, 239–44.

Higuchi R., Dollinger G., Walsh P.S. & Griffith R. (1992) Simultaneous amplification and detection of specific DNA sequences. *Biotechnology* **10**, 413–17.

Higuchi R., Fockler C., Dollinger G. & Watson R. (1993) Kinetic PCR analysis: real-time monitoring of DNA amplification reactions. *Biotechnology* **11**, 1026–30.

Higuchi R., Krummel B. & Saiki R.K. (1988) A general method of *in vitro* preparation and specific mutagenesis of DNA fragments: study of protein and DNA interactions. *Nucl. Acids Res.* **16**, 7351–67.

Hilder V.A. & Boulter D. (1999) Genetic engineering of crop plants for insect resistance – a critical review. *Crop Protection* **18**, 177–91.

Hille J., Verhweggen F., Roelvink P. *et al.* (1986) Bleomycin resistance: a new dominant selectable marker for plant cell transformation. *Plant Mol. Biol.* **7**, 171–6.

Himmelreich R., Hilbert H., Plagens H., Pirkl E., Li B-C. & Herrmann R. (1996) Complete sequence analysis of the genome of the bacterium *Mycoplasma pneumoniae*. *Nucl. Acids Res.* **24**, 4420–49.

Hinkula J., Lundholm P. & Wahren B. (1997) Nucleic acid vaccination with HIV regulatory genes: a combination of HIV-1 genes in separate plasmids induces strong immune responses. *Vaccine* **15**, 874–8.

Hinnebusch J. & Tilly K. (1993) Linear plasmids and chromosomes in bacteria. *Mol. Microbiol.* **10**, 917–22.

Hinnen A., Hicks J.B. & Fink G.R. (1978) Transformation of yeast. *Proc. Nat. Acad. Sci. USA* **75**, 1929–33.

Hiraga K. & Arnold F.H. (2003) General method for sequence-independent site-directed chimeragenesis. *J. Mol. Biol.* **330**, 287–96.

Hirata R., Chamberlain J., Dong R. & Russell D.W. (2002) Targeted transgene insertion into human chromosomes by adeno-associated virus vectors. *Nature Biotechnol.* **20**, 735–8.

Hirata R.K. & Russell D.W. (2000) Design and packaging of adeno-associated virus gene targeting vectors. *J. Virol.* **74**, 4612–20.

Hiratsuka J. *et al.* (1989) The complete sequence of the rice (*Oryza sativa*) chloroplast genome: intermolecular recombination between distinct tRNA genes accounts for a major plastid DNA inversion during the evolution of the cereals. *Mol. Gen. Genet.* **217**, 185–94.

Hirayama T., Ishida C., Kuromori T. *et al.* (1997) Functional cloning of a cDNA encoding Mei2-like protein from *Arabidopsis thaliana* using a fission yeast pheromone receptor deficient mutant. *FEBS Lett.* **413**, 16–20.

Hirsch V.M., Fuerst T.R., Sutter G. *et al.* (1996) Patterns of viral replication correlate with outcome in simian immunodeficiency virus (SIV)-infected macaques: effect of prior immunization with a trivalent SIV vaccine in modified vaccinia virus Ankara. *J. Virol.* **70**, 3741–52.

Hirschhorn J.N. *et al.* (2001) Genome-wide linkage analysis of stature in multiple populations reveals several regions with evidence of linkage to adult height. *Am. J. Hum. Genet.* **69**, 106–16.

Ho Y. *et al.* (2002) Systematic identification of protein complexes in *Saccharomyces cerevisiae* by mass spectrometry. *Nature* **415**, 180–3.

Hoang T.T., Kutchma A.J., Becher A. & Schweizer H.P. (2000) Integration-proficient plasmids for *Pseudomonas aeruginosa*: site-specific integration and use for engineering of reporter and expression strains. *Plasmid* **43**, 59–72.

Hoekma A., Hirsch P.R., Hooykass P.J.J. & Schiperoort R.A. (1983) A binary plant vector strategy based on separation of *vir*- and T-regions of the *Agrobacterium tumefaciens* Ti-plasmid. *Nature* **303**, 179–83.

Hoekstra W.P.M., Bergmans J.E.N. & Zuidweg E.M. (1980) Role of *rec*BC nuclease in *Escherichia coli* transformation. *J. Bacteriol.* **143**, 1031–2.

Hoff T.H. *et al.* (2001) A recombinase mediated transcriptional, induction system in transgenic plants. *Plant Mol. Biol.* **45**, 41–9.

Hoffman C., Sandig V., Jennigs G. *et al.* (1995) Efficient gene transfer into human hepatocytes by baculovirus vectors. *Proc. Nat. Acad. Sci. USA* **92**, 10099–103.

Hoffman F. (1996) Laser microbeams for the manipulation of plant cells and subcellular structures. *Plant Sci.* **113**, 1–11.

Hoffmaster A.R., Ravel J., Rasko D.A., *et al.* (2004) Identification of anthrax toxin genes in a *Bacillus cereus* associated with an illness resembling inhalation anthrax. *Proc. Nat. Acad. Sci. USA* **101**, 8449–54.

Hoffmeyer S. *et al.* (2000) Functional polymorphisms of the human multidrug-resistance gene: multiple sequence variations and correlation of one allele with P-glycoprotein expression and activity *in vivo*. *Proc. Nat. Acad. Sci. USA* **97**, 3473–8.

Hofmann A., Kessler B., Ewerling S., Weppert M., Vogg B., Ludwig H., Stojkovic M., Boelhauve M., Brem G.,

Wolf E. & Pfeifer A. (2003) Efficient transgenesis in farm animals by lentiviral vectors. *EMBO Rep.* **4**, 1054–60.

Hofmann A., Zakhartchenko V., Weppert M., Sebald H., Wenigerkind H., Brem G., Wolf E. & Pfeifer A. (2004) Generation of transgenic cattle by lentiviral gene transfer into oocytes. *Biol. Reprod.* in press.

Hogan B. & Lyons K. (1988) Gene targeting: getting nearer the mark. *Nature* **336**, 304–5.

Hoheisel J.D., Maier E., Mott E. *et al.* (1993) High resolution cosmid and P1 maps spanning the 14 Mb genome of the fission yeast *S. pombe. Cell* **73**, 109–20.

Hohn B. & Murray K. (1977) Packaging recombinant DNA molecules into bacteriophage particles *in vitro. Proc. Nat. Acad. Sci. USA* **74**, 3259–63.

Hohn B. (1975) DNA as substrate for packaging into bacteriophage lambda, *in vitro. J. Mol. Biol.* **98**, 93–106.

Holland P.M., Abramson R.D., Watson R. & Gelfand D.H. (1991) Detection of specific polymerase chain reaction product by utilizing the 5′–3′ exonuclease activity of *Thermus aquaticus* DNA polymerase. *Proc. Nat. Acad. Sci. USA* **88**, 7276–80.

Hollister J.R., Shaper J.H. & Jarvis D.L. (1998) Stable expression of mammalian β1,4-galactosyltransferase extends the N-glycosylation pathway in insect cells. *Glycobiology* **8**, 473–80.

Holm L. & Sander C. (1995) Dali: a network tool for protein structure comparison. *Trends Biochem. Sci.* **20**, 478–80.

Holm P.B., Olsen O., Scnorf M., Brinch-Pedersen H. & Knudsen S. (2000) Transformation of barley by micro-injection into isolated zygote protoplasts. *Transgenic Res.* **9**, 21–32.

Holmes D.L. & Stellwagen N.C. (1990) The electric field dependence of DNA mobilities in agarose gels: a reinvestigation. *Electrophoresis* **11**, 5–15.

Hols P., Baulard A., Garmyn D. *et al.* (1992) Isolation and characterization of genetic expression and secretion signals from *Enterococcus faecalis* through the use of broad-host-range alpha-amylase probe vectors. *Gene* **118**, 21–30.

Honaramooz A. *et al.* (2003) Use of adeno-associated virus for transfection of male-germ cells for transplantation in pigs. *Theriogenology* **59**, 536.

Hong Y.G., Saunders K., Hartley M.R. & Stanley J. (1996) Resistance to geminivirus infection by virus-induced expression of dianthin in transgenic plants. *Virology* **220**, 119–27.

Hopwood D.A. (1999) Forty years of genetics with *Streptomyces*: from *in vivo* through *in vitro* to *in silico. Microbiology* **145**, 2183–202.

Hopwood D.A., Malpartida F., Kieser H.M. *et al.* (1985) Production of "hybrid" antibiotics by genetic engineering. *Nature* **314**, 642–4.

Horland P., Flick J., Johnston M. & Sclafani R.A. (1989) Galactose as a gratuitous inducer of *GAL* gene expression in yeasts growing on glucose. *Gene* **83**, 57–64.

Horsch R.B., Fraley R.T., Rogers S.G. *et al.* (1984) Inheritance of functional genes in plants. *Science* **223**, 496–8.

Horsch R.B., Fry J.E., Hoffmann N.L. *et al.* (1985) A simple and general method for transferring genes into plants. *Science* **227**, 1229–31.

Horvath H., Huang J.T., Wong O. *et al.* (2000) The production of recombinant proteins in transgenic barley grains. *Proc. Nat. Acad. Sci. USA* **97**, 1914–19.

Hosono S., Faruqi A.F., Dean F.B. *et al.* (2003) Unbiased whole-genome amplification directly from clinical samples. *Genome Res.* **13**, 954–64.

Hou B.K., Zhou Y.H., Wan L.H., Zhang Z.L., Shen G.F., Chen Z.H. & Hu Z.M. (2003) Chloroplast transformation in oilseed rape. *Transgenic Res.* **12**, 111–14.

Houard S., Heinderyckx M. & Bollen A. (2002) Engineering of non-conventional yeasts for efficient synthesis of macromolecules: the methylotrophic genera. *Biochimie* **84**, 1089–93.

Houdebine L.M. (1997) *Transgenic Animals: Generation and Use.* Harwood Academic Publishers, Switzerland.

Howbrook D.N., van der Valk A.M., O'Shaughnessy M.C., Sarker D.K., Baker S.C. *et al.* (2003) Developments in microarray technologies. *Drug Discovery Today* **8**, 642–51.

Hrabe de Angelis M. & Balling R. (1998) Large-scale ENU screens in the mouse: genetics meets genomics. *Mutation Res.* **400**, 25–32.

Hrabe de Angelis M. *et al.* (2000) Genome-wide, large-scale production of mutant mice by ENU mutagenesis. *Nature Genet.* **25**, 444–7.

Hsiao C.L. & Carbon J. (1981) Characterization of a yeast replication origin (ars2) and construction of stable minichromosomes containing cloned yeast centromere DNA (CEN 3). *Gene* **15**, 157–66.

Hsiao K.K., Scott M., Foster D. *et al.* (1990) Spontaneous neurodegeneration in transgenic mice with mutant prion protein. *Science* **250**, 1587–90.

Hu G.S., Yalpani N., Briggs S.P. & Johal G.S. (1998) A porphyrin pathway impairment is responsible for the phenotype of a dominant disease lesion mimic mutant of maize. *Plant Cell* **10**, 1095–105.

Hu M.C.-T. & Davidson N. (1987) The inducible lac operator-repressor system is functional in mammalian cells. *Cell* **48**, 555–6.

Hu S.-L., Kosowski S.P. & Dalrymple J.M. (1986) Expression of AIDS virus envelope gene by a recombinant vaccinia virus. *Nature* **320**, 537–40.

Hu W. & Chen C. (1995) Expression of *Aequorea* green fluorescent protein in plant cells. *FEBS Lett.* **369**, 331–4.

Huang H.C. & Brown D.D. (2000) Overexpression of *Xenopus laevis* growth hormone stimulates growth of tadpoles and frogs. *Proc. Nat. Acad. Sci. USA* **97**, 190–4.

Huang J., Pray C. & Rozelle S. (2002a) Enhancing the crops to feed the poor. *Nature* **418**, 678–83.

Huang J., Rozelle S.D., Pray C.E. & Wang Q. (2002b) Plant biotechnology in China. *Science* **295**, 674–7.

Huang R. & Reusch R.N. (1995) Genetic competence in *Escherichia coli* requires poly-beta-hydroxybutyrate/calcium polyphosphate membrane complexes and certain divalent cations. *J. Bacteriol.* **177**, 486–90.

Huang R.Y. & Kowalski D. (1993) A DNA unwinding element and an ARS consensus comprise a replication origin within a yeast chromosome. *EMBO J.* **12**, 4521–31.

Hubank M. & Schatz D.G. (1994) Identifying differences in mRNA expression by representational difference analysis of cDNA. *Nucl. Acids Res.* **22**, 5640–8.

Huber M.C., Kruger G. & Bonifer C. (1996) Genomic position effects lead to an inefficient reorganization of nucleosomes in the 5′-regulatory region of the chicken lysozyme locus in transgenic mice. *Nucl. Acids Res.* **24**, 1443–53.

Huffaker T.C., Hoyt M.A. & Botstein D. (1987) Genetic analysis of the yeast cytoskeleton. *Ann. Rev. Gen.* **21**, 259–84.

Hughes T.R. *et al.* (2000) Functional discovery via a compendium of expression profiles. *Cell* **102**, 109–26.

Hugot J.P. *et al.* (1996) Mapping of a susceptibility locus for Crohn's disease on chromosome 16. *Nature* **379**, 772–3.

Hugot J.-P. *et al.* (2001) Association of NOD2 leucine-rich repeat variants with susceptibility to Crohn's disease. *Nature* **411**, 599–603.

Huh W.K., Falvo J.V., Gerke L.C., Carroll A.S., Howson R.W., Weissman J.S. & O'Shea E.K. (2003) Global analysis of protein localization in budding yeast. *Nature* **425**, 686–91.

Hui A., Hayflick J., Dinkelspiel K. & De Boer H.A. (1984) Mutagenesis of the three bases preceding the start codon of the β-galactosidase mRNA and its effect on translation in *Escherichia coli*. *EMBO J.* **3**, 623–9.

Humphries P., Old R., Coggins L.W. *et al.* (1978) Recombinant plasmids containing *Xenopus laevis* structural genes derived from complementary DNA. *Nucl. Acids Res.* **5**, 905–24.

Huygen K., Content J., Montgomery D.L. *et al.* (1996) Immunogenicity and protective efficacy of a tuberculosis DNA vaccine. *Nature Med.* **8**, 893–8.

Huynen M.A., Dandekar T. & Bork P. (1999) Variation and evolution of the citric acid cycle: a genomic perspective. *Trends Microbiol.* **7**, 281–91.

Hwu P., Yannelli J., Kriegler M. *et al.* (1993) Functional and molecular characterization of tumor-infiltrating lymphocytes transduced with tumor necrosis factor-alpha cDNA for the gene therapy of cancer in humans. *J. Immunol.* **150**, 4101–15.

Iamtham S. & Day A. (2000) Removal of antibiotic resistance genes from transgenic tobacco plastids. *Nature Biotechnol.* **18**, 1172–6.

Ida S. & Terada R. (2004) A tale of two integrations, trans-gene and T-DNA: gene targeting by homologous recombination in rice. *Curr. Opin. Biotechnol.* **15**, 132–8.

Ihssen P.E., McKay L.R., McMillan I. & Phillips R.B. (1990) Ploidy manipulation and gynogenesis in fishes: cytogenetic and fisheries applications. *Trans. Am. Fish Soc.* **119**, 698–717.

Iida A., Morikawa H. & Yamada Y. (1990) Stable transformation of cultured tobacco cells by DNA-coated gold particles accelerated by gas pressure driven particle gun. *Appl. Microbiol. Biotechnol.* **33**, 560–3.

Ikawa M., Yamada S., Nakanishi T. & Okabe M. (1999) Green fluorescent protein as a vital marker in mammals. *Curr. Top. Devel. Biol.* **44**, 1–20.

Ikemura T. (1981a) Correlation between the abundance of *Escherichia coli* transfer RNAs and the occurrence of the respective codons in its protein genes. *J. Mol. Biol.* **146**, 1–21.

Ikemura T. (1981b) Correlation between the abundance of *Escherichia coli* transfer RNAs and the occurrence of the respective codons in its protein genes: a proposal for a synonymous codon choice that is optimal for the *E. coli* translational system. *J. Mol. Biol.* **151**, 389–409.

Ikeo K., Ishi-i J., Tamura T., Gojobori T. & Tateno Y. (2003) CIBEX: Center for information biology gene expression database. *C. R. Biol.* **326**, 1079–82.

Imler J.L. *et al.* (1996) Novel complementation cell lines derived from human lung carcinoma A549 cells support the growth of E1-deleted adenovirus vectors. *Gene Ther.* **3**, 75–84.

Imperiale M.J. & Kochanek S. (2004) Adenovirus vectors: Biology, design, and production. *Curr. Top. Microbiol.* **273**, 335–57.

Ingelman-Sundberg M., Oscarson M. & McLellan R.A. (1999) Polymorphic human cytochrome P450 enzymes: an opportunity for individualized drug treatment. *Trends Pharmacol. Sci.* **20**, 342–9.

Ingmer H. & Cohen S.N. (1993) Excess intracellular concentration of pSC101 RepA protein interferes with both plasmid DNA replication and partitioning. *J. Bacteriol.* **175**, 7834–41.

Innis M.A., Gelfand D.H., Sninsky J.J. & White T.J. (eds.) (1990) *PCR Protocols: A Guide to Methods and Applications.* Academic Press, New York.

Inoue H., Nojima H. & Okayama H. (1990) High efficiency transformation of *Escherichia coli* with plasmids. *Gene* **96**, 23–8.

International Human Genome Sequencing Consortium (2001) Initial sequencing and analysis of the human genome. *Nature* **409**, 860–933.

Ioannou P.A., Amemiya C.T., Garnes J. *et al.* (1994) A new bacteriophage P1-derived vector for the propagation of large human DNA fragments. *Nature Genet.* **6**, 84–9.

Ioshikhes I.P. & Zhang M.Q. (2000) Large-scale human promoter mapping using CpG islands. *Nature Genet.* **26**, 61–3.

Ish-Horowicz D. & Burke J.F. (1981) Rapid and efficient cosmid cloning. *Nucl. Acids Res.* **9**, 2989–98.

Ishida Y., Saito H., Ohta S. *et al.* (1996) High efficiency transformation of maize (*Zea mays* L) mediated by *Agrobacterium tumefaciens*. *Nature Biotechnol.* **14**, 745–50.

Ishimaru K., Okhawa Y., Ishige T., Tobias D.J. & Ohsugi R. (1998) Elevated pyruvate orthophosphate dikinase (PPDK) activity alters carbon metabolism in C3 transgenic potatoes with a C4 maize PPDK gene. *Physiologia Plantarum* **103**, 340–6.

Ito T. *et al.* (2000) Toward a protein–protein interaction map of the budding yeast: a comprehensive system to examine two-hybrid interactions in all possible combinations between the yeast proteins. *Proc. Nat. Acad. Sci. USA* **97**, 1143–7.

Ito T., Chiba T., Ozawa R., Yoshida M., Hattori M. & Sakaki Y. (2001) A comprehensive two-hybrid analysis to explore the yeast protein interactome. *Proc. Nat. Acad. Sci. USA* **98**, 4569–74.

Ito T., Seki M., Hayashida N., Shibata D. & Shinozaki K. (1999) Regional insertional mutagenesis of genes on *Arabidopsis thaliana* chromosome V using the *Ac/Ds* transposon in combination with a cDNA scanning method. *Plant J.* **17**, 433–44.

Ivics Z. & Izsvak Z. (2005) Whole lotta jumpin' goin' on: new transposon tools for vertebrate functional genomics. *Trends Genet.* in press.

Ivics Z. *et al.* (1997) Molecular reconstruction of *Sleeping Beauty*, a Tc1-like transposon from fish, and its transposition in human cells. *Cell* **91**, 501–10.

Iyengar A., Muller F. & Maclean N. (1996) Regulation and expression of transgenes in fish – a review. *Transgenic Res.* **5**, 147–66.

Iyer L.M., Koonin E.V. & Aravind L. (2004) Evolution of bacterial RNA polymerase: implications for large-scale bacterial phylogeny, domain accretion, and horizontal gene transfer. *Gene* **335**, 73–88.

Iyer V.R. *et al.* (1999) The transcriptional program in the response of human fibroblasts to serum. *Science* **283**, 83–7.

Iyer V.R. *et al.* (2001) Genomic binding sites of the yeast cell cycle transcription factors SBF and MBF. *Nature* **409**, 533–8.

Izumi M., Miyazawa H., Kamakura T. *et al.* (1991) Blasticidin S-resistance gene (*bsr*): a novel selectable marker for mammalian cells. *Exp. Cell Res.* **197**, 229–33.

Jackson A.L., Bartz S.R., Schelter J., Kobayashi S.V., Burchard J., Mao M., Li B., Cavet G. & Linsley P.S. (2003) Expression profiling reveals off-target gene regulation by RNAi. *Nature Biotechnol.* **21**, 635–7.

Jackson D.A., Symons R.H. & Berg P. (1972) Biochemical method for inserting new genetic information into DNA of Simian virus 40: circular SV40 DNA molecules containing lambda phage genes and the galactose operon of *Escherichia coli*. *Proc. Nat. Acad. Sci. USA* **69**, 2904–9.

Jacob F. & Monod, J. (1961) Genetic regulatory mechanisms in the synthesis of proteins. *J. Mol. Biol.* **3**, 318–56.

Jaenisch R. & Mintz B. (1974) Simian virus 40 DNA sequences in DNA of healthy adult mice derived from preimplantation blastocysts injected with viral DNA. *Proc. Nat. Acad. Sci. USA* **71**, 1250–4.

Jaenisch R. (1988) Transgenic animals. *Science* **240**, 1468–74.

Jahner D. & Jaenisch R. (1985) Retrovirus-induced *de novo* methylation of flanking host sequences correlates with gene inactivity. *Nature* **315**, 594–7.

Jahner D., Stuhlmann H., Stewart C.L. *et al.* (1982) *De novo* methylation and expression of retroviral genomes during mouse embryogenesis. *Nature* **298**, 623–8.

Jakobovits A., Moore A.L., Green L.L. *et al.* (1993) Germ line transmission and expression of a human-derived yeast artificial chromosome. *Nature* **362**, 255–8.

James H.A. & Gibson I. (1998) The therapeutic potential of ribozymes. *Blood* **91**, 371–82.

James P., Quadroni M., Carafoli E. & Gonnet G. (1993) Protein identification by mass profile fingerprinting. *Biochem. Biophys. Res. Comm.* **195**, 58–64.

James, C. (2004) *Preview: Global Status of Commercialized Biotech/GM Crops: 2004.* ISAAA Briefs No. 32. ISAAA: Ithaca, NY.

Jannière L., Bruand C. & Ehrlich S.D. (1990) Structurally stable *Bacillus subtilis* cloning vectors. *Gene* **81**, 53–61.

Jansen R. & Gerstein M. (2000) Analysis of the yeast transcriptome with structural and functional categories: characterizing highly expressed proteins. *Nucl. Acids Res.* **28**, 1481–8.

Jefferson R.A., Burgess S.M. & Hirsh D. (1986) β-Glucuronidase from *Escherichia coli* as a gene-fusion marker. *Proc. Nat. Acad. Sci. USA* **83**, 8447–51.

Jefferson R.A., Kavanagh T.A. & Bevan M.W. (1987a) GUS fusions: β-glucuronidase as a sensitive and versatile gene fusion marker in higher plants. *EMBO J.* **6**, 3901–7.

Jefferson R.A., Klass M., Wolf N. & Hirsh D. (1987b) Expression of chimeric genes in *Caenorhabditis elegans*. *J. Mol. Biol.* **193**, 41–6.

Jeffreys A.J., Wilson V. & Thein S.L. (1985a) Individual-specific "fingerprints" of human DNA. *Nature* **316**, 76–9.

Jelinsky S. & Samson L. (1999) Global response of *Saccharomyces cerevisiae* to an alkylating agent. *Proc. Nat. Acad. Sci. USA* **96**, 1486–91.

Jen G.C. & Chilton M.D. (1986) The right border region of pTiT37 T-DNA is intrinsically more active than the left border region in promoting T-DNA transformation. *Proc. Nat. Acad. Sci. USA* **83**, 3895–9.

Jendrejack R.M., Dimalanta E.T., Schwartz D.C., *et al.* (2003) DNA dynamics in a microchannel. *Physical Rev. Lett.* **91**, 38–102.

Jensen J.S., Marcker K.A., Otten L. & Schell J. (1986) Nodule-specific expression of a chimaeric soybean leghaemoglobin gene in transgenic *Lotus corniculatus*. *Nature* **321**, 669–74.

Jensen S., Gassama M.P. & Heidmann T. (1999) Taming of transposable elements by homology-dependent gene silencing. *Nature Genet.* **21**, 209–12.

Jeon J.-S., Lee S., Jung K.-H. *et al.* (2000) T-DNA insertional mutagenesis for functional genomics in rice. *Plant J.* **22**, 561–70.

Jeong K.J. & Lee S.Y. (2000) Secretory production of human leptin in *Escherichia coli. Biotechnol. Bioeng.* **67**, 398–407.

Jespers L.S., Messens J.H., De Keyser A. *et al.* (1995) Surface expression and ligand-based selection of cDNAs fused to filamentous phage gene VI. *Bio/Technology* **13**, 378–82.

Ji H., Moore D.P., Blomberg M.A., Braiterman L.T., Voytas D.F., Natsoulis G. & Boeke J.D. (1993) Hotspots for unselected Ty1 transposition events on yeast chromosome III are near tRNA genes and LTR sequences. *Cell* **73**, 1007–18.

Johanning F.W., Conry R.M., LoBuglio A.F. *et al.* (1995) A Sindbis virus mRNA polynucleotide vector achieves prolonged and high level heterologous gene expression *in vivo. Nucl. Acids Res.* **23**, 1495–501.

John D.C.A., Watson R., Kind A.J. *et al.* (1999) Expression of an engineered form of recombinant procollagen in mouse milk. *Nature Biotechnol.* **17**, 385–9.

Johnson G.C.L. *et al.* (2001) Haplotype tagging for the identification of common disease genes. *Nature Genet.* **29**, 233–7.

Johnson M. (1998) Cloning humans? *Bioessays* **19**, 737–9.

Johnston M. (1987) A model fungal gene regulatory mechanism: the GAL genes of *Saccharomyces cerevisiae. Microbiol. Rev.* **51**, 458–76.

Jones A.L., Thomas C.L. & Maule A.J. (1998) *De novo* methylation and cosuppression induced by a cytoplasmically-replicating plant RNA virus. *EMBO J.* **17**, 6385–93.

Jones D.T. (1997) Successful *ab initio* prediction of the tertiary structure of NK-lysin using multiple sequences and recognized supersecondary structural motifs. *Proteins* **S1**, 185–91.

Jones D.T. (1999a) GenTHREADER: an efficient and reliable protein fold recognition method for genomic sequences. *J. Mol. Biol.* **287**, 797–815.

Jones D.T. (1999b) Protein secondary structure prediction based on position-specific scoring matrices. *J. Mol. Biol.* **292**, 195–202.

Jones D.T. (2000) Protein structure prediction in the post-genomic era. *Curr. Opin. Struct. Biol.* **10**, 371–9.

Jones D.T., Taylor W.R. & Thornton J.M. (1992) A new approach to protein fold recognition. *Nature* **358**, 86–9.

Jones I.M., Primrose S.B., Robinson A. & Ellwood D.C. (1980) Maintenance of some Col E1-type plasmids in chemostat culture. *Mol. Gen. Genet.* **180**, 579–84.

Jones J.D.G., Gilbert D.E., Grady K.L. & Jorgensen R.A. (1987) T-DNA structure and gene expression in petunia plants transformed by *Agrobacterium tumefaciens* C58 derivatives. *Mol. Gen. Genet.* **207**, 478–85.

Jones K. & Murray K. (1975) A procedure for detection of heterologous DNA sequences in lambdoid phage by *in situ* hybridization. *J. Mol. Biol.* **51**, 393–409.

Jones S. & Thornton J.M. (1995) Protein–protein interactions: a review of protein dimer structures. *Prog. Biophysics Mol. Biol.* **63**, 31–65.

Jones S. & Thornton J.M. (1997) Prediction of protein–protein interaction sites using patch analysis. *J. Mol. Biol.* **272**, 133–43.

Jones S.J.M. *et al.* (2001) Changes in gene expression associated with developmental arrest and longevity in *Caenorhabditis elegans. Genome Res.* **11**, 1346–52.

Jongedijk E., Tigelaar H., Van Roekel J.S.C., Bres-Vloemans S.A., Dekker I., Van den Elzen P.J.M., Cornelissen B.J.C. & Melchers L.S. (1995) Synergistic activity of chitinases and β-1,3-glucanases enhances fungal resistance in transgenic tomato plants. *Euphytica* **85**, 173–80.

Jongeneel C.V., Iseli C., Stevenson B.J. *et al.* (2003) Comprehensive sampling of gene expression in human cell lines with massively parallel signature sequencing. *Proc. Nat. Acad. Sci. USA* **100**, 4702–5.

Joos T.O. *et al.* (2000) A microarray enzyme linked immuno-sorbent assay for autoimmune diagnostics. *Electrophoresis* **21**, 2641–50.

Jordan M. & Wurm F. (2004) Transfection of adherent and suspended cells by calcium phosphate. *Methods* **33**, 136–43.

Jorgensen R.A., Snyder C. & Jones J.D.G. (1987) T-DNA is organized predominantly in inverted repeat structures in plants transformed with *Agrobacterium tumefaciens* C58 derivatives. *Mol. Gen. Genet.* **207**, 471–7.

Joyner A.L. (ed.) (1998) *Gene Targeting: A Practical Approach*, 2nd edn. Oxford University Press, Oxford.

Julius D., Blair L.C., Brake A.J., Sprague G.F. & Thorner J. (1983) Yeast alpha-factor is processed from a larger precursor polypeptide: the essential role of a membrane-bound dipeptidyl amino-peptidase. *Cell* **32**, 839–52.

Julius D., Scheckman R. & Thorner J. (1984) Glycosylation and processing of prepro-alpha-factor through the yeast secretory pathway. *Cell* **36**, 309–18.

Justice M.J., Noveroske J.K., Weber J.S., Zheng B.H. & Bradley A. (1999) Mouse ENU mutagenesis. *Hum. Mol. Genet.* **8**, 1955–63.

Kado C.I. (1998) *Agrobacterium*-mediated horizontal gene transfer. *Genet. Eng.* **20**, 1–24.

Kahl G. & Schell J.S. (eds.) (1982) *Molecular Biology of Plant Tumours*, pp. 211–67. Academic Press, New York.

Kakimoto T. (1996) CKI1, a histidine kinase homolog implicated in cytokinin signal transduction. *Science* **274**, 982–5.

Kal A.J., van Zonneveld A.J., Benes V. *et al.* (1999) Dynamics of gene expression revealed by comparison of serial analysis of gene expression transcript profiles from yeast grown on two different carbon sources. *Mol. Biol. Cell.* **10**, 1859–72.

Kalendar R., Tanskanen J., Immonen S., *et al.* (2000) Genome evolution of wild barley (*Hordeum spontaneum*) by *BARE-1* retrotransposon dynamics in response to sharp microclimatic divergence. *Proc. Nat. Acad. Sci. USA* **97**, 6603–7.

Kamath R.S. & Ahringer J. (2003) Genome-wide RNAi screening in *Caenorhabditis elegans*. *Methods* **30**, 313–21.

Kamath R.S., Fraser A.G., Dong Y., Poulin G., Durbin R., Gotta M., Kanapin A., Le Bot N., Moreno S., Sohrmann M. *et al.* (2003) Systematic functional analysis of the *Caenorhabditis elegans* genome using RNAi. *Nature* **421**, 231–7.

Kanalas J.J. & Suttle D.P. (1984) Amplification of the UMP synthase gene and enzyme overproduction in pyrazofurin-resistant rat hepatoma cells: molecular cloning of a cDNA for UMP synthase. *J. Biol. Chem.* **259**, 1848–53.

Kane S.E., Troen B.R., Gal S. *et al.* (1988) Use of a cloned multidrug resistance gene for amplification and overproduction of major excreted protein, a transformation regulated secreted acid protease. *Mol. Cell. Biol.* **8**, 3316–21.

Kanehisa M. (1998) Databases of biological information. *Bioinformatics* **5**, 24–6.

Kanehisa M., Goto S., Kawashima S. & Nakaya A. (2002) The KEGG databases at GenomeNet. *Nucl. Acids Res.* **30**, 42–6.

Kanevski I.F., Thakur S., Cosowsky L. *et al.* (1992) Tobacco lines with high copy number of replicating recombinant geminivirus vectors after biolistic DNA delivery. *Plant J.* **2**, 457–63.

Kanno S. *et al.* (2000) Assembling of engineered IgG binding protein on gold surface for highly oriented antibody immobilization. *J. Biotechnol.* **76**, 207–14.

Kao C.M., Katz L. & Khosia C. (1994) Engineered biosynthesis of a complete macrolide in a heterologous host. *Science* **265**, 509–12.

Kapitonov V.V. & Jurka J. (2001) Rolling-circle transposons in eukaryotes. *Proc. Nat. Acad. Sci. USA* **98**, 8714–19.

Kapust R.B. & Waugh, D.S. (1999) *Escherichia coli* maltose-binding protein is uncommonly effective at promoting the solubility of polypeptides to which it is fused. *Protein Sci.* **8**, 1668–74.

Karess R.E. & Rubin G.M. (1984) Analysis of P transposable element functions in *Drosophila*. *Cell* **38**, 135–46.

Karlin S. (2001) Detecting anomalous gene clusters and pathogenicity islands in diverse bacterial genomes. *Trends Microbiol.* **9**, 335–43.

Karn J., Matthews H.W., Gait M.J. & Brenner S. (1984) A new selective cloning vector, λ2001, with sites for *Xba*I, *Bam*HI, *Hind*III, *Eco*RI, *Sst*I and *Xho*I. *Gene* **32**, 217–24.

Kassua A. & Thornton J.M. (1999) Three-dimensional structure analysis of PROSITE patterns. *J. Mol. Biol.* **286**, 1673–91.

Katagiri F., Lam E. & Chua N.H. (1989) Two tobacco DNA-binding proteins with homology to the nuclear factor CREB. *Nature* **340**, 727–30.

Katakura Y., Ametani A., Totsuka M., Nagafuchi S. & Kaminogawa S. (1999) Accelerated secretion of mutant beta-lactoglobulin in *Saccharomyces cerevisiae* resulting from a single amino acid substitution. *Biochim. Biophys. Acta* **1432**, 302–12.

Katinka M.D., Duprat S., Cornillot E., *et al.* (2001) Genome sequence and gene compaction of the eukaryote parasite *Encephalitozoon cuniculi*. *Nature* **414**, 450–3.

Katsuki M., Sato M., Kimura M. *et al.* (1988) Conversion of normal behaviour to shiverer by myelin basic protein antisense cDNA in transgenic mice. *Science* **241**, 593–5.

Kaufman P.D. & Rio D.C. (1991) Germline transformation of *Drosophila melanogaster* by purified P element transposase. *Nucl. Acids Res.* **19**, 6336.

Kaufman R.J. (1990a) Strategies for obtaining high level expression in mammalian cells. *Technique* **2**, 221–36.

Kaufman R.J. (1990b) Vectors used for expression in mammalian cells. *Methods Enzymol.* **185**, 487–511.

Kaufman R.J., Murtha P., Ingolia D.E., Yeung C.-Y. & Kellems R.E. (1986) Selection and amplification of heterologous genes encoding adenosine deaminase in mammalian cells. *Proc. Nat. Acad. Sci. USA* **83**, 3136–40.

Kaufman R.J., Wasley L.C. & Dorner A.J. (1988) Synthesis, processing, and secretion of recombinant human factor VIII expressed in mammalian cells. *J. Biol. Chem.* **263**, 6352–62.

Kaufman R.J., Wasley L.C., Spiliotes A.T. *et al.* (1985) Coamplification and coexpression of human tissue-type plasminogen activator and murine dihydrofolate reductase sequences in Chinese hamster ovary cells. *Mol. Cell. Biol.* **5**, 1730–59.

Kavanagh T.A., Thanh N.D., Lao N.T., McGrath N., Peter S.O., Horvath E.M., Dix P.J. & Medgyesy P. (1999) Homeologous plastid DNA transformation in tobacco is mediated by multiple recombination events. *Genetics* **152**, 1111–22.

Kawasaki E.S. (1990) Amplification of RNA. In *PCR Protocols: A Guide to Methods and Applications*, eds. Innis M.A., Gelfand D.H., Sninsky J.J. & White T.J., pp. 21–7. Academic Press, New York.

Kay M.A., Glorioso J.C. & Naldini L. (2001) Viral vectors for gene therapy: the art of turning infectious agents into vehicles of therapeutics. *Nature Med.* **7**, 33–40.

Keating C.D., Kriek N., Daniels M., Ashcroft N.R., Hopper N.A. Siney E.J., Holden-Dye L. & Burke J.F. (2003) Whole-genome analysis of 60 G protein-coupled receptors in *Caenorhabditis elegans* by gene knockout with RNAi. *Curr. Biol.* **13**, 1715–20.

Keggins K.M., Lovett P.S. & Duvall E.J. (1978) Molecular cloning of genetically active fragments of *Bacillus* DNA in *Bacillus subtilis* and properties of the vector plasmid pUB110. *Proc Nat. Acad. Sci. USA* **75**, 1423–7.

Keilty S. & Rosenberg M. (1987) Constitutive function of a positively regulated promotor reveals new sequences essential for activity. *J. Biol. Chem.* **262**, 6389–95.

Keller H., Pamboukdjian N., Ponchet M. *et al.* (1999) Pathogen-induced elicitin production in transgenic tobacco generates a hypersensitive response and nonspecific disease resistance. *Plant Cell* **11**, 223–35.

Kellis M., Patterson N., Endrizzi M., *et al.* (2003) Sequencing and comparison of yeast species to identify genes and regulatory elements. *Nature* **423**, 241–54.

Kelly T.J. & Smith H.O. (1970) A restriction enzyme from *Hemophilus influenzae*. II. Base sequence of the recognition site. *J. Mol. Biol.* **51**, 393–409.

Kempin S.A., Liljegren S.J., Block L.M., Rounsley S.D., Yanofsky M.F. & Lam E. (1997) Targeted disruption in *Arabidopsis*. *Nature* **389**, 802–3.

Kennerdell J.R. & Carthew R.W. (2000) Heritable gene silencing in *Drosophila* using double-stranded RNA. *Nature Biotechnol.* **17**, 896–8.

Kerr D.E., Liang F., Bondioli K.R. *et al.* (1998) The bladder as a bioreactor: urothelium production and secretion of growth hormone into urine. *Nature Biotechnol.* **16**, 75–9.

Khalsa G., Mason H.S. & Arntzen C.J. (2004) Plant-derived vaccines: progress and constraints. In Fischer R. & Schillberg S. (eds.) *Molecular Farming: Plant-made Pharmaceuticals and Technical Proteins*. John Wiley & Sons Inc., NY.

Khan M.S. & Maliga P. (1999) Fluorescent antibiotic resistance marker for tracking plastid transformation in higher plants. *Nature Biotechnol.* **17**, 910–15.

Khush G.S., Bacalangco E. & Ogawa T. (1990) A new gene for resistance to bacterial blight from *O. longistaminata*. *Rice Genetics Newsletter* **7**, 121–2.

Kieser T. & Hopwood D.A. (1991) Genetic manipulation of *Streptomyces*: integrating vectors and gene replacement. *Methods Enzymol.* **204**, 430–58.

Kieser T., Bibb M.J., Buttner M.J. *et al.* (eds.) (2000) *Practical Streptomyces Genetics*. John Innes Foundation. Norwich, UK.

Kilby N.J., Smith M.R. & Murray J.A. (1993) Site specific recombinases: tools for genome engineering. *Trends Genet.* **9**, 413–21.

Kim J.H. *et al.* (1997) Development of a positive method for male stem cell-mediated gene transfer in mouse and pig. *Mol. Reprod. Dev.* **46**, 515–26.

Kim J.J., Bagarazzi M.L., Trivedi N. *et al.* (1997) Engineering of *in vivo* immune responses to DNA immunization via codelivery of costimulatory molecule genes. *Nature Biotechnol.* **15**, 641–6.

Kim K.K., Hung L.W., Yokota H., Kim R. & Kim S.H. (1998) Crystal structures of eukaryotic translation initiation factor 5A from *Methanococcus jannaschii* at 1.8 A resolution. *Proc. Nat. Acad. Sci. USA* **95**, 10419–24.

Kim K.Y., Kwon S.Y., Lee H.S., Hur Y., Bang J.W. & Kwak S.S. (2003) A novel oxidative stress-inducible

peroxidase promoter from sweet potato: molecular cloning and characterization in transgenic tobacco plants and cultured cells. *Plant Mol Biol.* **51**: 831–8.

Kim L., Mogk A. & Schumann W. (1996) A xylose-inducible *Bacillus subtilis* integration vector and its application. *Gene* **181**, 71–6.

Kim S.K. & Wold B.J. (1985) Stable reduction of thymidine kinase activity in cells expressing high levels of anti-sense RNA. *Cell* **42**, 129–38.

Kim U.-J. *et al.* (1996) Construction and characterization of a human bacterial artificial chromosome library. *Genomics* **34**, 213–18.

Kim V.N. (2005) MicroRNA biogenesis: coordinated cropping and dicing. *Nat. Rev. Mol. Cell Biol.* **6**, 376–85.

Kimelman D. & Kirchner M. (1989) An antisense mRNA directs the covalent modification of the transcripts encoding fibroblast growth factor in *Xenopus* oocytes. *Cell* **59**, 687–96.

King L.A. & Possee R.D. (1992) *The Baculovirus Expression System: A Laboratory Guide*. Chapman & Hall, London.

King T.J. & Briggs R. (1956) Serial transplantation of embryonic nuclei. *Cold Spring Harbor Symp. Quant. Biol.* **21**, 271–90.

Kingsmore S.F. & Patel D.D. (2003) Multiplexed protein profiling on antibody-based microarrays by rolling circle amplification. *Curr. Opin. Biotechnol.* **14**, 74–81.

Kitts P.A. & Possee R.D. (1993) A method for producing recombinant baculovirus expression vectors at high frequency. *Biotechniques* **14**, 810–17.

Klatzmann D., Herson S., Cherin P. *et al.* (1996) Gene therapy for metastatic malignant melanoma: evaluation of tolerance to intratumoral injection of cells producing recombinant retroviruses carrying the herpes simplex virus type 1 thymidine kinase gene, to be followed by ganciclovir administration. *Hum. Gene Ther.* **7**, 155–67.

Klaus, S.M.J., Huang F.-C., Golds T.J. & Koop H.-U. 2004. Generation of marker-free plastid transformants using a transiently cointegrated selection gene. *Nature Biotechnol.* **22**, 225–9.

Klein T.M. & Fitzpatrick-McElligott (1993) Particle bombardment: a universal approach for gene transfer to cells and tissues. *Curr. Opin. Biotechnol.* **4**, 583–90.

Klein T.M., Fromm M.E., Weissinger A. *et al.* (1988a) Transfer of foreign genes into intact maize cells with high velocity microprojectiles. *Proc. Nat. Acad. Sci. USA* **85**, 4305–9.

Klein T.M., Harper E.C., Svab Z. *et al.* (1988b) Stable genetic transformation of intact *Nicotiana* cells by the particle bombardment process. *Proc. Nat. Acad. Sci. USA* **85**, 8502–5.

Klein T.M., Wolf E.D., Wu R. & Sanford J.C. (1987) High-velocity micro-projectiles for delivering nucleic acids into living cells. *Nature* **327**, 70–3.

Klenow H. & Henningsen I. (1970) Selective elimination of the exonuclease activity of the deoxyribonucleic acid polymerase from *E. coli* B by limited proteolysis. *Proc. Nat. Acad. Sci. USA* **65**, 168–75.

Kling J. (1996) Could transgenic supercrops one day breed superweeds? *Science* **274**, 180–1.

Klinman D.M., Yamshchikov G. & Ishigatsubo Y. (1997) Contribution of CpG motifs to the immunogenicity of DNA vaccines. *J. Immunol.* **158**, 3635–9.

Klose J. (1975) Protein mapping by combined isoelectric focussing and electrophoresis of mouse tissues: a novel approach to testing for induced point mutations in mammals. *Humangenetik* **26**, 231–43.

Kloti A., Iglesias V.A., Wunn J. *et al.* (1993) Gene transfer by electroporation into intact scutellum cells of wheat embryos. *Plant Cell Rep.* **12**, 671–5.

Knight S.J. & Flint J. (2000) Perfect endings: a review of subtelomeric probes and their use in clinical diagnosis. *J. Med. Genet.* **37**, 401–9.

Knowles M.R., Hohneker K.W., Zhou Z. *et al.* (1995) A controlled study of adenoviral-vector-mediated gene transfer in the nasal epithelium of patients with cystic fibrosis. *N. Engl. J. Med.* **333**, 823–31.

Kohler R.H., Cao J., Zipfel W.R., Webb W.W. & Hanson M. (1997) Exchange of protein molecules through connections between higher plant plastids. *Science* **276**, 2039–42.

Kok M., Rekik M., Witholt B. & Harayama S. (1994) Conversion of pBR322-based plasmids into broad-host-range vectors by using the Tn3 transposition mechanism. *J. Bacteriol.* **176**, 6566–71.

Kola I. & Hertzog P.J. (1998) Down syndrome and mouse models. *Curr. Opin. Genet. Devel.* **8**, 316–21.

Kolkman J.A. & Stemmer W.P.C. (2001) Directed evolution of proteins by exon shuffling. *Nature Biotechnol.* **19**, 423–8.

Kollias G., Wrighton N., Hurst J. & Grosveld F. (1986) Regulated expression of human ^{A}gamma-, beta-, and hybrid gamma beta-globin genes in transgenic mice: manipulation of the developmental expression patterns. *Cell* **46**, 89–94.

Komari T., Hiei Y., Saito Y., Murai N. & Kumashiro T. (1996) Vectors carrying two separate T-DNAs for co-transformation of higher plants mediated by *Agrobacterium tumefaciens* and segregation of transformants free from selection markers. *Plant J.* **10**, 165–74.

Koncz C., Olsson O., Langridge W.H.R., Schell J. & Szalay A.A. (1987) Expression and assembly of functional bacterial luciferase in plants. *Proc. Nat. Acad. Sci. USA* **84**, 131–5.

Konfortov B.A. *et al.* (2000) A high-resolution HAPPY map of *Dictyostelium discoideum* chromosome 6. *Genome Res.* **10**, 1737–42.

Konieczny A. & Ausubel F.A. (1993) A procedure for mapping *Arabidopsis* mutations using co-dominant ecotype-specific PCR-based markers. *Plant J.* **4**, 403–10.

Konopka, K. *et al.* (2000) Rev-binding aptamer and CMV promoter act as decoys to inhibit HIV replication. *Gene* **255**, 235–44.

Koonin E.V. (2003) Comparative genomics, minimal gene sets and the last universal common ancester. *Nature Rev. Microbiol.* **1**, 127–36.

Koonin E.V., Fedorova N.D., Jackson J., *et al.* (2004) A comprehensive evolutionary classification of proteins encoded in complete eukaryotic genomes. *Genome Biol.* **5**, R7.

Koop H.U., Steinmuller K., Wagner H. *et al.* (1996) Integration of foreign sequences into the tobacco plastome via polyethylene glycol-mediated protoplast transformation. *Planta* **199**, 193–201.

Koppensteiner W.A., Lackner P., Wiederstein M. & Sippl M.J. (2000) Characterization of novel proteins based on known protein structures. *J. Mol. Biol.* **296**, 1139–52.

Koradi R., Billeter M., Engeli M., Güntert P. & Wüthrich K. (1998) Automated peak picking and peak integration in macromolecular NMR spectra using AUTOPSY. *J. Mag. Res.* **135**, 288–97.

Kornfeld K. (1997) Vulval development in *Caenorhabditis elegans*. *Trends Genet.* **13**, 55–61.

Koshland D., Kent J.C. & Hartwell L.H. (1985) Genetic analysis of the mitotic transmission of minichromosomes. *Cell* **40**, 393–403.

Koski L.B., Morton R.A. & Golding G.B. (2001) Codon bias and base composition are poor indicators of horizontally transferred genes. *Mol Biol. Evol.* **18**, 404–12.

Koskinen R., Salomonsen J., Tregaskes C.A., *et al.* (2002) The chicken CD4 gene has remained conserved in evolution. *Immunogenetics* **54**, 520–5.

Kost T.A. & Condreay J.P. (2002) Recombinant baculoviruses as mammalian cell gene-delivery vectors. *Trends Biotechnol.* **20**, 173–80.

Kotewicz M.L., Sampson C.M., D'Alessio J.M. & Gerard G.F. (1988) Isolation of cloned Moloney murine leukemia virus reverse transcriptase lacking ribonuclease H activity. *Nucl. Acids Res.* **16**, 265–77.

Kotin R.M., Siniscalco M., Samulski R.J. *et al.* (1990) Site specific integration by adeno-associated virus. *Proc. Nat. Acad. Sci. USA* **87**, 2211–15.

Koulintchenko M., Konstantinov Y. & Dietrich A. (2003) Plant mitochondria actively import DNA via the permeability transition pore complex. *EMBO J.* **22**, 1245–54.

Kouprina N. & Larionov V. (2003) Exploiting the yeast *Saccharomyces cerevisiae* for the study of the organization and evolution of complex genomes. *FEMS Microbiol. Rev.* **27**, 629–49.

Kouprina N., Ebersole T., Koriabine M., *et al.* (2003a) Cloning of human centromeres by transformation-associated recombination in yeast and generation of functional human artificial chromosomes. *Nucl. Acids Res.* **31**, 922–34.

Kouprina N., Eldarov M., Moyzis R., Resnick M. & Larionov V. (1994) A model system to assess the integrity of mammalian YACs during transformation and propagation in yeast. *Genomics* **21**, 7–17.

Kouprina N., Leem S.H., Solomon G., *et al.* (2003b) Segments missing from the draft human genome sequence can be isolated by TAR cloning in yeast. *EMBO Rep.* **4**, 253–62.

Kovach M.E., Elzer P.H., Hill D.S. *et al.* (1995) Four new derivatives of the broad-host-range cloning vector pBBR1MCS, carrying different antibiotic resistance cassettes. *Gene* **166**, 175–6.

Koyama H., Kawamura A., Kihara T., Hara T., Takita E. & Shibata D. (2000) Overexpression of mitrochondrial citrate synthase in *Arabidopsis thaliana* improved growth on phosphorus-limited soil. *Plant Cell Physiol.* **9**, 1030–7.

Kozak M. (1986) Point mutations define a sequence flanking the AUG initiator codon that modulates translation by eukaryotic ribosomes. *Cell* **44**, 283–92.

Kozak M. (1999) Initiation of translation in prokaryotes and eukaryotes. *Gene* **234**, 187–208.

Kozal M. *et al.* (1996) Extensive polymorphisms observed in HIV-1 cladeB protease gene using high-density oligonucleotide arrays. *Nature Med.* **2**, 753–9.

Koziel M.G., Beland G.L., Bowman C. *et al.* (1993) Field performance of elite transgenic maize plants expressing an insecticidal protein derived from *Bacillus thuringiensis*. *Biotechnol.* **11**, 194–200.

Koziel M.G., Carozzi N.B. & Desai N. (1996) Optimizing expression of transgenes with an emphasis on post-transcriptional events. *Plant Mol. Biol.* **32**, 393–405.

Kramer B., Kramer W. & Fritz H.-J. (1984) Different base/base mismatches are corrected with different efficiencies by the methyl-directed DNA mismatch-repair system of *E. coli*. *Cell* **38**, 879–88.

Kramer R.A., Cameron J.R. & Davis R.W. (1976) Isolation of bacteriophage λ containing yeast ribosomal RNA genes: screening by *in situ* RNA hybridization to plaques. *Cell* **8**, 227–32.

Krasemann S., Groschup M., Hunsmann G. & Bodemer W. (1996) Induction of antibodies against human prion proteins (PrP) by DNA-mediated immunization of PrP$^{0/0}$ mice. *J. Immunol. Methods* **199**, 109–18.

Krewson T.D., Supelak P.J., Hill A.E., *et al.* (2004) Chromosomes 6 and 13 harbor genes that regulate pubertal timing in mouse chromosome substitution strains. *Endocrinology* **145**, 4447–51.

Krishnan B.R., Jamry I., Berg D.E., Berg C.M. & Chaplin D.D. (1995) Construction of a genomic DNA "feature map" by sequencing from nested deletions: application to the HLA class 1 region. *Nucl. Acids Res.* **23**, 117–22.

Kristensen V.N., Kelefiotis D., Kristensen T. & Borresen-Dale A-L. (2001) High-throughput methods for detection of genetic variation. *Biotechniques* **30**, 318–32.

Kroll K.L. & Amaya E. (1996) Transgenic *Xenopus* embryos from sperm nuclear transplantations reveal FGF signaling requirements during gastrulation. *Development* **122**, 3173–83.

Kruger S. (1988) *Eco*RII can be activated to cleave refractory DNA recognition sites. *Nucl. Acids Res.* **16**, 3997–4008.

Krupp G., Bonatz G. & Parwaresch R. (2000) Telomerase, immortality and cancer. *Biotechnol. Ann. Rev.* **6**, 103–40.

Krysan P.J., Young J.C. & Sussman M.R. (1999) T-DNA as an insertional mutagen in *Arabidopsis*. *Plant Cell* **11**, 2283–90.

Ku M.S.B., Agarie S., Nomura M., Fukayama H., Tsuchida H., Ono K., Hirose S., Toki S., Miyao M. & Matsuoka M. (1999) High level expression of maize phosphoenol pyruvate carboxylase in transgenic rice plants. *Nature Biotechnol.* **17**, 76–80.

Kubo M. & Kakimoto T. (2001) The *CYTOKININ HYPERSENSITIVE* genes of *Arabidopsis* negatively regulate the cytokinin signaling pathway for cell division and chloroplast development. *Plant J.* **23**, 385–94.

Kudla B. & Nicolas A. (1992) A multisite integrative cassette for the yeast *Saccharomyces cerevisiae*. *Gene* **119**, 49–56.

Kuehn M.R., Bradley A., Robertson E.J. & Evans M.J. (1987) A potential model for Lesch–Nyhan syndrome through introduction of HPRT mutations in mice. *Nature* **326**, 295–8.

Kuhn R., Schwenk F., Auget M. & Rajewsky K. (1995) Inducible gene targeting in mice. *Science* **269**, 1427–9.

Kuipers O.P., Beerthuyzen M.M., De Ruyter P.G.G.A., Luesink E.J. & De Vos W.M. (1995) Autoregulation of nisin biosynthesis in *Lactococcus lactis* by signal transduction. *J. Biol. Chem.* **270**, 27299–304.

Kukowska-Latallo, J.F., Bielinska, A.U., Johnson, J., Spindler, R., Tomalia, D.A. & Baker, J.R. Jr. (1996) Efficient transfer of genetic material into mammalian cells using Starburst polyamidoamine dendrimers. *Proc. Natl. Acad. Sci. USA* **93**, 4897–902.

Kumagai M.H., Turpen T.H., Weinzettl N. *et al.* (1993) Rapid, high-level expression of biologically-active alpha-trichosanthin in transfected plants by an RNA viral vector. *Proc. Nat. Acad. Sci. USA* **90**, 427–30.

Kumar A. (1996) The adventures of the Ty1-*copia* group of retrotransposons in plants. *Trends Genet.* **12**, 41–3.

Kumar A. *et al.* (1995) Potato plants expressing antisense and sense *S*-adenosylmethionine decarboxylase (SAMDC) transgenes show altered levels of polyamines and ethylene: antisense plants display abnormal phenotypes. *Plant J.* **9**, 147–58.

Kumpatla S.P., Chandrasekharah M.B., Iyer L.M., Li G. & Hall T.C. (1998) Genome intruder scanning and modulation systems and transgene silencing. *Trends Plant Sci.* **3**, 97–104.

Kunath, T. *et al.* (2003) Transgenic RNA interference in ES cell-derived embryos recapitulates a genetic null phenotype. *Nature Biotechnol.* **21**, 559–61.

Kunik T., Tzfira T., Kapulnik Y. *et al.* (2001) Genetic transformation of HeLa cells by *Agrobacterium*. *Proc. Nat. Acad. Sci. USA* **98**, 1871–6.

Kunkel L.M. (1986) Analysis of deletions in DNA from patients with Becker and Duchenne muscular dystrophy. *Nature* **322**, 73–7.

Kunkel T., Niu Q.W., Chan Y.S. & Chua N.H. (1999) Inducible isopentenyl transferase as a high-efficiency marker for plant transformation. *Nature Biotechnol.* **17**, 916–19.

Kunkel T.A. (1985) Rapid and efficient site-specific mutagenesis without phenotypic selection. *Proc. Nat. Acad. Sci. USA* **82**, 488–92.

Kuo C.-L. & Campbell J.L. (1983) Cloning of *Saccharomyces cerevisiae* DNA replication genes: isolation of the CDC8 gene and two genes that compensate for the cdc8-1 mutation. *Mol. Cell. Biol.* **3**, 1730–7.

Kurata N. *et al.* (1994) A 300 kilobase interval genetic map of rice including 883 expressed sequences. *Nature Genet.* **8**, 365–72.

Kurisawa M., Yokoyama M. & Okano T. (2000a) Gene expression control by temperature with thermo-responsive polymeric gene carriers. *J. Control Release* **69**, 127–37.

Kurisawa M., Yokoyama M. & Okano T. (2000b) Transfection efficiency increases by incorporating hydrophobic monomer units into polymeric gene carriers. *J. Control Release* **68**, 1–8.

Kurjan J. & Herskowitz I. (1982) Structure of a yeast pheromone (MF alpha): a putative alpha factor precursor contains four tandem copies of mature alpha factor. *Cell* **30**, 933–43.

Kurland C.G. (1987) Strategies for efficiency and accuracy in gene expression. 1. The major codon preference: a growth optimization strategy. *Trends Biochem. Sci.* **12**, 126–8.

Kurland C.G., Canback B. & Berg O.G. (2003) Horizontal gene transfer: a critical view. *Proc. Nat. Acad. Sci. USA* **100**, 9658–62.

Kuromori T., Hirayama T., Kiyosue Y., *et al.* (2004) A collection of 11,800 single-copy Ds transposon insertion lines in *Arabidopsis*. *Plant J.* **37**, 897–905.

Kurtz D.T. & Nicodemus C.F. (1981) Cloning of a$_{2m}$ globulin cDNA using a high efficiency technique for the cloning of trace messenger RNAs. *Gene* **13**, 145–52.

Kusaba M., Miyahara K., Iida S., Fukuoka H., Takano T., Sassa H., Nishimura M. & Nishio T. (2003) Low glutelin content 1: a dominant mutation that suppresses the glutelin multigene family via RNA silencing in rice. *Plant Cell* **15**, 1455–67.

Kwok P.-Y. & Chen X. (2003) Detection of single nucleotide polymorphisms. *Curr. Issues Mol. Biol.* **5**, 43–60.

Kyrpides N., Overbeek R. & Ouzounis C. (1999) Universal protein families and the functional content of the last common ancestor. *J. Mol. Evol.* **49**, 413–23.

La Vallie E.R., Di Blasio E.A., Kovacic S. *et al.* (1993) A thioredoxin gene fusion expression system that circumvents inclusion body formation in the *E. coli* cytoplasm. *Biotechnology* **11**, 187–93.

Laan M., Kallioniemi O.-P., Hellsten E., Alitalo K., Peltonen L. & Palotie A. (1995) Mechanically stretched chromosomes as targets for high-resolution FISH mapping. *Genome Res.* **5**, 13–20.

Labow M.A., Baim S.B., Shenk T. & Levine A. (1990) Conversion of the lac repressor into an allosterically regulated transcriptional activator for mammalian cells. *Mol. Cell Biol.* **10**, 3343–56.

Lacks S.A., Lopez P., Greenberg B. & Espinosa M. (1986) Identification and analysis of genes for tetracycline resistance and replication functions in the broad-host-range plasmid pLS1. *J. Mol. Biol.* **192**, 753–5.

Lacy E., Roberts S., Evans E.P., Burtenshaw M.D. & Constantini F.D. (1983) A foreign β-globin gene in transgenic mice: integration at abnormal chromosomal positions and expression in inappropriate tissues. *Cell* **34**, 343–58.

Lahm H.W. & Langen H. (2000) Mass spectrometry: a tool for the identification of proteins separated by gels. *Electrophoresis* **21**, 2105–14.

Lai L., Kolber-Simonds D., Park K.W. *et al.* (2002) Production of alpha-1,3-galactosyltransferase knockout pigs by nuclear transfer cloning. *Science* **295**, 1089–92.

Lai Z. *et al.* (1999) A shotgun optical map of the entire *Plasmodium falciparum* genome. *Nature Genet.* **23**, 309–13.

Lal S.P., Christopherson R.I. & dos Remedios C.G. (2002) Antibody arrays: an embryonic but rapidly growing technology. *Drug Discovery Today* **7**, S143–9.

Lam P.Y.P. & Breakefield X.O. (2000) Hybrid vector designs to control the delivery, fate and expression of transgenes. *J. Gene Med.* **2**, 395–408.

Lamartina S., Roscilli G., Rinaudo C.D., Sporeno E., Silvi L., Hillen W., Bujard H., Cortese R., Ciliberto G. & Toniatti C. (2002) Stringent control of gene expression *in vivo* by using novel doxycycline-dependent trans-activators. *Hum. Gene Ther.* **13**, 199–210.

Lamb B.T. & Gerhart J.D. (1995) YAC transgenics and the study of genetics and human disease. *Curr. Opin. Genet. Devel.* **5**, 342–8.

Lamond A.I. & Mann M. (1997) Cell biology and the genome projects: a concerted strategy for characterizing multi-protein complexes by using mass spectrometry. *Trends Cell Biol.* **7**, 139–42.

Lamont P.J., Davis M.B. & Wood N.W. (1997) Identification and sizing of the GAA trinucleotide repeat expansion of Friedreich's ataxia in 56 patients – clinical and genetic correlates. *Brain* **120**, 673–80.

Lamzin V.S. & Perrakis A. (2000) Current state of automated crystallographic data analysis. *Nature Struct. Biol.* **7**, 978–81.

Land H., Grey M., Hanser H., Lindenmaier W. & Schutz G. (1981) 5′-Terminal sequences of eucaryotic mRNA can be cloned with a high efficiency. *Nucl. Acids Res.* **9**, 2251–66.

Landegran U. (1996) The challengers to PCR. A proliferation of chain reactions. *Curr. Opin. Biotechnol.* **7**, 95–7.

Lander E.S. & Schork N.J. (1994) Genetic dissection of complex traits. *Science* **265**, 2037–48.

Lander E.S. (1999) Array of hope. *Nature Genet.* **21**, 3–4.

Landford R.E. (1988) Expression of simian virus 40 T antigen in insect cells using a baculovirus expression vector. *Virology* **167**, 72–81.

Lane C.D., Colman A., Mohun T. *et al.* (1980) The *Xenopus* oocyte as a surrogate secretory system: the specificity of protein export. *Eur. J. Biochem.* **111**, 225–35.

Lane D., Prentki P. & Chandler M. (1992) Use of gel retardation to analyze protein–nucleic acid interactions. *Microbiol. Rev.* **56**, 509–28.

Langley K.E. & Zabin I. (1976) Beta-galactosidase alpha-complementation: properties of the complemented enzyme and mechanism of the complementation reaction. *Biochemistry* **15**, 4866–75.

Langley K.E., Villarejo M.R., Fowler A.V., Zamenhof P.J. & Zabin I. (1975) Molecular basis of beta-galactosidase alpha-complementation. *Proc. Nat. Acad. Sci. USA* **72**, 1254–7.

Langridge W.H.R., Fitzgerald K.J., Koncz C., Schell J. & Szalay A.A. (1989) Dual promoter of *Agrobacterium tumefaciens* mannopine synthase genes is regulated by plant growth hormones. *Proc. Nat. Acad. Sci. USA* **86**, 3219–23.

Lapeyre B. & Amalric F. (1985) A powerful method for the preparation of cDNA libraries: isolation of cDNA encoding a 100-kDa nucleolar protein. *Gene* **37**, 215–20.

Larin Z., Monaco A.P. & Lehrach H. (1991) Yeast artificial chromosome libraries containing large inserts from mouse and human DNA. *Proc. Nat. Acad. Sci. USA* **88**, 4123–7.

Larsson S., Hotchkiss G., Andang M. *et al.* (1994) Reduced b2-macroglobulin mRNA levels in transgenic mice expressing a designed hammerhead ribozyme. *Nucl. Acids Res.* **22**, 2242–8.

LaSalle G.L., Robert J.J., Berrard S. *et al.* (1993) An adenovirus vector for gene transfer into neurons and glia in the brain. *Science* **259**, 988–90.

Lashkari D.A. *et al.* (1997) Yeast microarrays for genome wide parallel genetic and gene expression analysis. *Proc. Nat. Acad. Sci. USA* **94**, 13057–62.

Lasken R.S. & Egholm M. (2003) Whole-genome amplification: abundant supplies of DNA from precious samples or clinical specimens. *Trends Biotechnol.* **21**, 531–5.

Laski F.A., Rio D.C. & Rubin G.M. (1986) Tissue specificity of *Drosophila* P element transposition is regulated at the level of mRNA splicing. *Cell* **44**, 7–19.

Lasko M., Sauer B., Mosinger B. *et al.* (1992) Targeted oncogene activation by site-specific recombination in transgenic mice. *Proc. Nat. Acad. Sci. USA* **89**, 6232–6.

Laskowski R.A., Luscombe N.M., Swindells M.B. & Thornton J.M. (1996) Protein clefts in molecular recognition and function. *Protein Sci.* **5**, 2438–52.

Laskowski R.A., MacArthur M.W., Moss D.S. & Thornton J.M. (1993) PROCHECK: a program to check the stereochemical quality of protein structures. *J. Applied Crystallography* **26**, 283–91.

Lathe R. (1985) Synthetic oligonucleotide probes deduced from amino acid sequence data. *J. Mol. Biol.* **183**, 1–12.

Laufs J., Wirtz U., Kammann M. *et al.* (1990) Wheat dwarf virus *Ac/Ds* vectors: expression and excision of transposable elements introduced into various cereals by a viral replicon. *Proc. Nat. Acad. Sci. USA* **87**, 7752–6.

Laursen C.M., Krzyzek R.A., Flick C.E., Anderson P.C. & Spencer T.M. (1994) Production of fertile transgenic maize by electroporation of suspension culture cells. *Plant Mol. Biol.* **24**, 51–61.

Law M.-F., Byrne J. & Hawley P.M. (1983) A stable bovine papillomavirus hybrid plasmid that expresses a dominant selective trait. *Mol. Cell. Biol.* **3**, 2110–15.

Lawrence M.S., Ho D.Y., Dash R. & Sapolsky R.M. (1995) Herpes simplex virus vectors overexpressing the glucose transporter gene protect against seizure-induced neuron loss. *Proc. Nat. Acad. Sci. USA* **92**, 7247–51.

Lawrence R.J. & Pikaard C.S. (2003) Transgene-induced RNA interference: a strategy for overcoming gene redundancy in polyploids to generate loss-of-function mutations. *Plant J.* **36**, 114–21.

Lay Thein S. & Wallace R.B. (1986) In *Human Genetic Diseases. A Practical Approach*, ed. Davies I.E., pp. 33–50. IRL Press, Oxford.

Lazarus R.A., Seymour J.L., Stafford R.K. *et al.* (1990) A biocatalytic approach to vitamin C production: metabolic pathway engineering of *Erwinia herbicola*. In *Biocatalysis*, ed. Abramowitz D., pp. 136–55. Van Nostrand Reinhold, New York.

Lazo G.R., Stein P.A. & Ludwig R.A. (1991) A DNA transformation competent *Arabidopsis* genomic library in *Agrobacterium*. *Biotechnology* **9**, 963–7.

Le Loir Y., Gruss A., Ehrlich S.D. & Langella P. (1998) A nine-residue synthetic propeptide enhances secretion efficiency of heterologous proteins in *Lactococcus lactis*. *J. Bacteriol.* **180**, 1895–903.

Le Y. & Dobson M.J. (1997) Stabilization of yeast artificial chromosome clones in a *rad54-3* recombination-deficient host strain. *Nucl. Acids Res.* **25**, 1248–53.

Lecchi P., Gupte A.R., Perez R.E., Stockert L.V. & Abramson F.P. (2003) Size-exclusion chromatography in multidimensional separation schemes for proteome analysis. *J. Biochem. Biophys. Methods* **56**, 141–52.

Leckband G. & Lörz H. (1998) Transformation and expression of a stilbene synthase gene of *Vitis vinifera* L. in barley and wheat for increased fungal resistance. *Theoret. App. Genet.* **96**, 1004–12.

Leder P., Tiemeier D. & Enquist L. (1977) EK2 derivatives of bacteriophage lambda useful in the cloning of

DNA from higher organisms: the lgt WES system. *Science* **196**, 175–7.

Lederberg S. (1957) Suppression of the multiplication of heterologous bacteriophages in lysogenic bacteria. *Virology* **3**, 496–513.

Leduc N., Matthys-Rochon E., Rougier M. *et al.* (1996) Isolated maize zygotes mimic *in vivo* early development and express microinjected genes when cultured *in vitro. Devel. Biol.* **10**, 190–203.

Lee B.C. *et al.* (2005) Dogs cloned from adult somatic cells. *Nature* **436**, 641.

Lee C.C., Wu X., Gibbs R.A. *et al.* (1988) Generation of cDNA probes directed by amino acid sequence: cloning of urate oxidase. *Science* **239**, 1288–91.

Lee F., Mulligan R., Berg P. & Ringold G. (1981) Glucocorticoids regulate expression of dihydrofolate reductase cDNA in mouse mammary tumour virus chimaeric plasmids. *Nature* **294**, 228–32.

Lee J.Y. & Lee D.H. (2003) Use of serial analysis of gene expression technology to reveal changes in gene expression in *Arabidopsis* pollen undergoing cold stress. *Plant Physiol.* **132**, 517–29.

Lee K.Y., Lund P., Lowe K. & Dunsmuir P. (1990) Homologous recombination in plant cells after *Agrobacterium*-mediated transformation. *Plant Cell* **2**, 415–25.

Lee R.C., Feinbaum R.L. & Ambros V. (1993) The *C. elegans* heterochromatic gene *lin-4* encodes small RNAs with antisense complementarity to *lin-14. Cell* **75**, 843–54.

Lee S.Y., Choi J.H. & Xu Z. (2003) Microbial cell-surface display. *Trends Biotechnol.* **21**, 45–52.

Lee W.-C. & Lee K.H. (2004) Applications of affinity chromatography in proteomics. *Anal. Biochem.* **324**, 1–10.

Leemans J., Langenakens J., De Greve H. *et al.* (1982b) Broad-host-range cloning vectors derived from the W-plasmid Sa. *Gene* **19**, 361–4.

Leenhouts K.J., Tolner B., Bron S. *et al.* (1991) Nucleotide sequence and characterization of the broad-host-range lactococcal plasmid pWV01. *Plasmid* **26**, 55–66.

Lefebvre D.D., Miki B.L. & Laliberte J.F. (1987) Mammalian metallothionein functions in plants. *Biotechnology* **5**, 1053–6.

Legrain P. & Selig L. (2000) Genome-wide protein interaction maps using two-hybrid systems. *FEBS Lett.* **480**, 32–6.

Legrain P., Wojcik J. & Gauthier J.M. (2001) Protein–protein interaction maps: a lead towards cellular functions. *Trends Genet.* **17**, 346–52.

Lehmann M., Pasamontes L., Lassen S.F. & Wyss M. (2000) The consensus concept for thermostability engineering of proteins. *Biochim. Biophys. Acta* **1543**, 408–15.

Lehtonen E., & Tenenbaum L. (2003) Adeno-associated viral vectors. *Int. Rev. Neurobiol.* **55**, 65–98.

Leno G.H. & Laskey R.A. (1991) The nuclear membrane determines the timing of replication in *Xenopus* egg extracts. *J. Cell Biol.* **112**, 557–66.

Lerner M.R. (1994) Tools for investigating functional interactions between ligands and G-protein-coupled receptors. *Trends Neurosci.* **17**, 142–6.

Lesage S., Zouali H., Cezard J.P., *et al.* (2002) CARD15/NOD2 mutational analysis and genotype-phenotype correlation in 612 patients with inflammatory bowel disease. *Am. J. Hum. Genet.* **70**, 845–57.

Lesley S.A. *et al.* (2002) Structural genomics of the *Thermotoga maritima* proteome implemented in a high-throughput structure determination pipeline. Proc. Nat. Acad. Sci. USA **99**, 11664–9.

Lester S.C., LeVan S.K., Steglich C. & DeMars R. (1980) Expression of human genes for adenine phosphoribosyltransferase and hypoxanthine-guanine phosphoribosyltransferase after genetic transformation of mouse cells with purified human DNA. *Somat. Cell Genet.* **6**, 241–59.

Leuking A. *et al.* (1999) Protein microarrays for gene expression and antibody screening. *Analytical Biochem.* **270**, 103–11.

Levine M., Rubin G.M. & Tjian R. (1984) Human DNA sequences homologous to a protein coding region conserved between homeotic genes of *Drosophila. Cell* **38**, 667–73.

Lewanodski M. & Martin G.R. (1997) Cre-mediated chromosome loss in mice. *Nature Genet.* **17**, 223–5.

Lewin B. (1994) *Genes* V. Cell Press, Cambridge, MA/Oxford University Press, New York.

Lewis S., Ashburner M. & Reese M.G. (2000) Annotating eukaryote genomes. *Curr. Opin. Struct. Biol.* **10**, 349–54.

Lewitter F. (1998) Text-based database searching. *Bioinformatics* **5**, 3–5.

Lewontin R.C. & Hartl D.L. (1991) Population genetics in forensic DNA typing. *Science* **254**, 1745–50.

Li H., Wang J., Wilhelmsson H. *et al.* (2000) Genetic modification of survival in tissue-specific knockout mice with mitochondrial cardiomyopathy. *Proc. Nat. Acad. Sci. USA* **97**, 3467–72.

Li H.W., Lucy A.P., Guo H.S. *et al.* (1999) Strong host resistance targeted against a viral suppressor of the plant gene silencing defense mechanism. *EMBO J.* **18**, 2683–91.

Li Q., Harju S. & Peterson K.R. (1999) Locus control regions coming of age at a decade plus. *Trends Genet.* **15**, 403–8.

Li Q., Lawrence C.B., Xing H.-Y., Babbitt R.A., Bass W.T., Maiti I.B. & Everett N.P. (2001) Enhanced disease resistance conferred by expression of an antimicrobial magainin analog in transgenic tobacco. *Planta* **212**, 635–9.

Li S. *et al.* (2004) A map of the ineractome network of the metazoan *C. elegans. Science* **303**, 540–3.

Li Y., Mitaxov V. & Waksman G. (1999a) Structure-based design of *Taq* DNA polymerases with improved properties of dideoxynucleotide incorporation. *Proc. Nat. Acad. Sci. USA* **96**, 9491–6.

Li Z.W., Stark G., Gotz J. *et al.* (1996) Generation of mice with a 200-kb amyloid precursor protein gene deletion by Cre-recombinase-mediated site-specific recombination in embryonic stem cells. *Proc. Nat. Acad. Sci. USA* **93**, 6158–62.

Liang P. & Pardee A.B. (1992) Differential display analysis of eukaryotic messenger RNA by means of the polymerase chain reaction. *Science* **257**, 967–71.

Liang P., Averboukh L., Keyomarsi K., Saeger R. & Pardee A.B. (1992) Differential display and cloning of messenger RNAs from human breast cancer verses mammary epithelial cells. *Cancer Res.* **52**, 6966–8.

Liao G., Rehm E.J. & Rubin G.M. (2000) Insertion site preferences of the P transposable element in *Drosophila melanogaster*. *Proc. Nat. Acad. Sci. USA* **97**, 3347–51.

Lien L.L., Lee Y. & Orkin S.H. (1997) Regulation of the myeloid-cell-expressed human gp91-*phox* gene as studied by transfer of yeast artificial chromosome clones into embryonic stem cells: suppression of a variegated cellular pattern of expression requires a full complement of distant *cis* elements. *Mol. Cell. Biol.* **17**, 2279–90.

Lilie H., Schwarz E. & Rudolph R. (1998) Advances in refolding of proteins produced in *E. coli. Curr. Opin. Biotechnol.* **9**, 497–501.

Liljestrom P. & Garoff H. (1993) A new generation of animal cell expression vectors based on the Semliki Forest virus replicon. *Biotechnology* **9**, 1356–61.

Lilley K.S., Razzaq A. & Dupree P. (2001) Two-dimensional gel electrophoresis: recent advances in sample preparation, detection and quantitation. *Curr. Opin. Chem. Biol.* **6**, 46–50.

Lim L.P., Glasner M.E., Yekta S., Burge C.B. & Bartel D.P. (2003) Vertebrate microRNA genes. *Science* **299**, 1540.

Lim M. *et al.* (1997) A luminescent europium complex for the sensitive detection of proteins and nucleic acids immobilized on membrane supports. *Anal. Biochem.* **245**, 184–95.

Limbach P.A., Crain P.F. & McCloskey J.A. (1995) Characterization of oligonucleotides and nucleic acids by mass spectrometry. *Curr. Opin. Biotechnol.* **6**, 96–102.

Lin Cereghino G.P., Lin Cereghino J., Ilgen C. & Cregg J.M. (2002) Production of recombinant proteins in fermenter cultures of the yeast *Pichia pastoris. Curr. Opin. Biotechnol.* **13**, 329–32.

Lin Cereghino J. & Cregg J.M. (2001) Heterologous protein expression in the methylotrophic yeast *Pichia pastoris. FEMS Microbiol. Rev.* **24**, 45–66.

Lin J. *et al.* (1999a) Whole-genome shotgun optical mapping of *Deinococcus radiodurans. Science* **285**, 1558–62.

Lin Q., Chen Z., Antoniw J. & White R. (1991) Isolation and characterization of a cDNA clone encoding the anti-viral protein from *Phytolacca americana. Plant Mol. Biol.* **17**, 609–14.

Lin W., Anuratha C.S., Datta K. *et al.* (1995) Genetic engineering of rice for resistance to sheath blight. *Biotechnology* **3**, 686–91.

Lin X. *et al.* (1999b) Sequence analysis of chromosome 2 of the plant *Arabidopsis thaliana. Nature* **402**, 761–8.

Lindbo J.A. & Dougherty W.G. (1992) Untranslatable transcripts of the tobacco etch virus coat protein gene sequence can interfere with tobacco etch virus replication in transgenic plants and protoplasts. *Virology* **189**, 725–33.

Lindsey K., Wei W., Clarke M.C. *et al.* (1993) Tagging genomic sequences that direct transgene expression by activation of a promoter trap in plants. *Transgenic Res.* **2**, 33–47.

Ling L.L., Ma N.S-F., Smith D.R., Miller D.D. & Moir D.T. (1993) Reduced occurrence of chimeric YACs in recombination-deficient hosts. *Nucl. Acids Res.* **21**, 6045–6.

Linial M. & Yona G. (2000) Methodologies for target selection in structural genomics. *Prog. Biophys. Mol. Biol.* **73**, 297–320.

Link A.J., Mock M.L. & Tirrell D.A. (2003) Non-canonical amino acids in protein engineering. *Curr. Opin. Biotechnol.* **14**, 603–9.

Lipshutz R.J., Fodor S.P.A., Gingeras T.R. & Lockhart D.J. (1999) High density synthetic oligonucleotide arrays. *Nature Genet.* **21**, 20–4.

Lipshutz R.J., Taverner F., Hennessy K., Hartzell G. & Davis R. (1994) DNA sequence confidence estimation. *Genomics* **19**, 417–24.

Lis J.T., Simon J.A. & Sutton C.A. (1983) New heat shock puffs and β-galactosidase activity resulting from transformation of *Drosophila* with an hsp70-*lacZ* hybrid gene. *Cell* **35**, 403–10.

Lisitsyn N., Lisitsyn N. & Wigler M. (1993) Cloning the difference between two complex genomes. *Science* **259**, 946–51.

Lisser S. & Margalit, H. (1993) Compilation of *E. coli* mRNA promoter sequences. *Nucl. Acids Res.* **21**, 1507–16.

Little P. (2001) The end of all human DNA maps. *Nature Genet.* **27**, 229–30.

Littlewood T.D., Hancock D.C., Danielian P.S., Parker M.G. & Evan G.I. (1995) A modified oestrogen receptor ligand-binding domain as an improved switch for the regulation of heterologous proteins. *Nucl. Acids Res.* **23**, 1686–90.

Liu D., Raghothama K.G., Hasegawa P.M. & Bressan R.A. (1994) Osmotin overexpression in potato delays development of disease symptoms. *Proc. Nat. Acad. Sci. USA* **91**, 1888–92.

Liu H. & Rashidbaigi A. (1990) Comparison of various competent cell preparation methods for high efficiency DNA transformation. *Biotechniques* **8**, 21–5.

Liu J., Cong B. & Tanksley S.D. (2003) Generation and analysis of an artificial gene dosage series in tomato to study the mechanisms by which the cloned quantitative trait locus fw2.2 controls fruit size. *Plant Physiol.* **132**, 292–9.

Liu Q. & Muruve D.A. (2003) Molecular basis of the inflammatory response to adenovirus vectors. *Gene Ther.* **10**, 935–40.

Liu Q., Singh S.P. & Green A.G. (2002) High-stearic and high-oleic cottonseed oils produced by hairpin RNA-mediated posttranscriptional gene silencing. *Plant Physiol.* **129**, 1732–43.

Liu Y.-G., Shirano Y., Fukaki H. *et al.* (1999) Complementation of plant mutants with large genomic DNA fragments by a transformation-competent artificial chromosome vector accelerates positional cloning. *Proc. Nat. Acad. Sci. USA* **96**, 6535–40.

Liu D. & Knapp J.E. (2001) Hydrodynamics-based gene delivery. *Curr. Opin.Mol. Ther.* **3**, 192–7.

Livak K.J., Flood S.J., Marmaro J., Giusti W. & Deetz K. (1995) Oligonucleotides with fluorescent dyes at opposite ends provide a quenched probe system useful for detecting PCR product and nucleic acid hybridization. *PCR Methods Appl.* **4**, 357–62.

Llewellyn D.J. & Higgins T.J.V. (2002) Transgenic crop plants with increased tolerance to insect pests. In Oksman-Caldentey K-M. & Barz W.H. (eds.) *Plant Biotechnology and Transgenic Plants.* Marcel Dekker NY, pp. 571–95.

Lo D.C., McAllister A.K. & Katz L.C. (1994) Neuronal transfection in brain slices using particle-mediated gene transfer. *Neuron* **13**, 1263–8.

Loake G.J., Ashby A.M. & Shaw C.H. (1988) Attraction of *Agrobacterium tumefaciens* C58C1 towards sugars involves a highly sensitive chemotaxis system. *J. Gen. Microbiol.* **134**, 1427–32.

Lobban P.E. & Kaiser A.D. (1973) Enzymatic end-to-end joining of DNA molecules. *J. Mol. Biol.* **78**, 453–71.

Locket T.J., Lewy D., Holmes P., Medveezky K. & Saint R. (1992) The rough (ro+) gene is a dominant marker in germ line transformation of *Drosophila melanogaster. Gene* **114**, 187–93.

Lockhart D.J. (1998) Mutant yeast on drugs. *Nature Medicine* **4**, 1235–6.

Lockhart D.J., Dong H., Byrne M.C. *et al.* (1996) Expression monitoring by hybridization to high-density oligonucleotide arrays. *Nature Biotechnol.* **14**, 1675–80.

Loenen W.A.M. & Brammar W.J. (1980) A bacteriophage lambda vector for cloning large DNA fragments made with several restriction enzymes. *Gene* **10**, 249–59.

Lohnes D., Kastner A., Dierich M. *et al.* (1993) Function of retinoic acid receptor gamma in the mouse. *Cell* **73**, 643–58.

Lois C., Hong E.J., Pease S., Brown E.J. & Baltimore D. (2002) Germline transmission and tissue-specific expression of transgenes delivered by lentiviral vectors. *Science* **295**, 868–72.

Lokman B.C., Heerikshuizen M., Van den Broek A. *et al.* (1997) Regulation of the *Lactobacillus pentosus* xylAB operon. *J. Bacteriol.* **179**, 5391–7.

Lomonossoff G.P. & Hamilton W.D.O. (1999) Cowpea mosaic virus-based vaccines. *Curr. Top. Microbiol. Immunol.* **240**, 177–89.

Long M., Betran E., Thornton K. & Wang W. (2003) The origin of new genes: glimpses from the young and old. *Nature Rev. Genet.* **4**, 865–75.

Lopata M.A., Cleveland D.W. & Sollner-Webb B. (1984) High level expression of a chloramphenicol acetyltransferase gene by DEAE-dextran mediated DNA transfection coupled with a dimethyl sulfoxide or glycerol shock treatment. *Nucl. Acids Res.* **12**, 5707–17.

Lopez-Bucio J., Martýnez de la Vega O., Guevara-Garcýa A. & Herrera-Estrella L. (2000a) Enhanced phosphorus uptake in transgenic tobacco plants that overproduce citrate. *Nature Biotechnol.* **18**, 450–3.

Lopez-Bucio J., Nieto-Jacobo M.F., Ramýrez-Rodrýguez V. & Herrera-Estrella L. (2000b) Organic acid metabolism in plants: from adaptive physiology to transgenic varieties for cultivation in extreme soils. *Plant Sci.* **160**, 1–13.

Lorenz M.G. & Wackernagel W. (1994) Bacterial gene transfer by natural genetic transformation in the environment. *Microbiol. Rev.* **58**, 563–602.

Lorkowski S. & Cullen P. (2004) High-throughput analysis of mRNA expression: microarrays are not the whole story. *Expert Opin. Therapeut. Patents* **14**, 377–403.

Lory S. (1998) Secretion of proteins and assembly of bacterial surface organelles: shared pathways of extracellular protein targeting. *Curr. Opin. Microbiol.* **1**, 27–35.

Lorz H., Baaker B. & Schell J. (1985) Gene transfer to cereal cells mediated by protoplast transformation. *Mol. Gen. Genet.* **199**, 178–82.

Love J., Gribbin C., Mather C. & Sang H. (1994) Transgenic birds by DNA microinjection. *Biotechnology* **12**, 60–3.

Lovett M. (1994) Fishing for complements: finding genes by direct selection. *Trends Genet.* **10**, 352–7.

Lovett M., Kere J. & Hinton L.M. (1991) Direct selection: a method for the isolation of cDNAs encoded by large genomic regions. *Proc. Nat. Acad. Sci. USA* **88**, 9628–33.

Lowman H.B., Bass S.H., Simpson N. & Wells J.A. (1991) Selecting high-affinity binding proteins by monovalent phage display. *Biochemistry* **30**, 10832–8.

Lu S., Lyngholm L., Yang G. *et al.* (1994) Tagged mutations at the Tox 1 locus of *Cochliobolus heterostrophus* by restriction enzyme-mediated integration. *Proc. Nat. Acad. Sci. USA* **91**, 12649–53.

Lu Z., DiBlasio-Smith E.A., Grant K.L. *et al.* (1996) Histidine patch thioredoxins: mutant forms of thioredoxins with metal chelating affinity that provide for convenient purifications of thioredoxin fusion proteins. *J. Biol. Chem.* **271**, 5059–65.

Lubbert H., Hoffman B.J., Snutch T.P. *et al.* (1987) cDNA cloning of a serotonin 5-HT1C receptor by electrophysiological assays of messenger RNA-injected *Xenopus* oocytes. *Proc. Nat. Acad. Sci. USA* **84**, 4332–6.

Lucca P., Hurrell R. & Potrykus I. (2001) Genetic engineering approaches to improve the bioavailability and the level of iron in rice grains. *Theor. Appl. Genet.* **102**, 392–7.

Lucklow V.A. & Summers M.D. (1988) Signals important for high level expression of foreign genes in

Autographa californica nuclear polyhedrosis virus expression vectors. *Virology* **167**, 56–71.

Lum L., Yao S., Mozer B., Rovescalli A., Von Kessler D., Nirenberg M. & Beachy P.A. (2003) Identification of hedgehog pathway components by RNAi in *Drosophila* cultured cells. *Science* **299**, 2039–45.

Lum P.Y., Armour C.D., Stepaniants S.B., Cavet G., Wolf M.K., Butler J.S., Hinshaw J.C., Garnier P., Prestwich G.D., Leonardson A. *et al.*: Discovering modes of action for therapeutic compounds using a genome-wide screen of yeast heterozygotes. *Cell* **116**, 121–37.

Lundstrom K. (1997) Alphaviruses as expression vectors. *Curr. Opin. Biotechnol.* **8**, 578–82.

Lundstrom K. (2003) Semliki Forest virus vectors for gene therapy. *Expert Opin. Biol. Th.* **3**, 771–7.

Luning Prak E.T. & Kazazian H.H. (2000) Mobile elements and the human genome. *Nature Rev. Genet.* **1**, 134–44.

Luo Y. & Wasserfallen A. (2001). Gene transfer systems and their applications in *Archaea. Systematic and Applied Microbiology* **24**, 15–25.

Lupton S. & Levine A.J. (1985) Mapping genetic elements of Epstein-Barr virus that facilitate extrachromosomal persistence of Epstein-Barr virus-derived plasmids in human cells. *Mol. Cell. Biol.* **5**, 2533–42.

Luria S.E. (1953) Host-induced modifications of viruses. *Cold Spring Harbor Symp. Quant. Biol.* **18**, 237–44.

Lusky M. & Botchan M. (1981) Inhibitory effect of specific pBR322 DNA sequences upon SV40 replication in simian cells. *Nature* **293**, 79–81.

Luthmann H. & Magnusson G. (1983) High efficiency polyoma DNA transfection of chloroquine-treated cells. *Nucl. Acids Res.* **11**, 1295–308.

Luthra R., Varsha R.K.D., Srivastava A.K. & Kumar S. (1997). Microprojectile mediated plant transformation: a bibliographic search. *Euphytica* **95**, 269–94.

Lutz S., Ostermeier M. & Benkovic S.J. (2001a) Rapid generation of incremental truncation libraries for protein engineering using α–phosphothioate nucleotides. *Nucl. Acids Res.* **29**, e16.

Lutz S., Ostermeier M., Moore G.L., *et al.* (2001b) Creating multiple-crossover DNA libraries independent of sequence identity. *Proc. Nat. Acad. Sci. USA* **98**, 11248–53.

Lydiate D.J., Malpartida F. & Hopwood D.A. (1985) The *Streptomyces* plasmid SCP2*: its functional analysis and development into useful cloning vectors. *Gene* **35**, 223–35.

Lysov Y.P., Khorlin A.A., Khrapko K.R., Shick V.V., Florentiev V.L. & Mirzabekov A.D. (1988) DNA sequencing by hybridization with oligonucleotides: a novel method. *Proc. Nat. Acad. Sci. USA* **303**, 1508–11.

Lyznik, L.A. *et al.* (1995) Heat-inducible expression of FLP gene in maize cells. *Plant J.* **8**, 177–86.

Ma H., Kunes S., Schatz P.J. & Botstein D. (1987) Plasmid construction by homologous recombination in yeast. *Gene* **58**, 201–16.

Ma J., Yanofsky M.F., Klee H.J., Bowman B.L. & Meyerowitz E.M. (1992) Vectors for plant transformation and cosmid libraries. *Gene* **117**, 161–7.

Ma J.K., Hiatt A., Hein M. *et al.* (1995) Generation and assembly of secretory antibodies in plants. *Science* **268**, 716–19.

Ma J.K., Hikmat B.Y., Wycoff K. *et al.* (1998) Characterization of a recombinant plant monoclonal secretory antibody and preventive immunotherapy in humans. *Nature Med.* **4**, 601–6.

Ma J.K.-C., Drake P.M.W. & Christou P. (2003) The production of recombinant pharmaceutical proteins in plants. *Nat. Rev. Genet.* **4**, 794–805.

Ma S.W., Zhao D.L., Yin Z.Q. *et al.* (1997) Transgenic plants expressing autoantigens fed to mice to induce oral immune tolerance. *Nature Med.* **3**, 793–6.

MacBeath G. & Schreiber S.L. (2000) Printing proteins as microarrays for high-throughput function determination. *Science* **289**, 1760–3.

Macgregor P.F. (2003) Gene expression in cancer: the application of microarrays. *Expert Rev. Mol. Diagn.* **3**, 185–200.

MacGregor P.F., Abate C. & Curran T. (1990) Direct cloning of leucine zipper proteins: Jun binds cooperatively to CRE with CRE-BP1. *Oncogene* **5**, 451–8.

Mackett, M., Smith, G.L. & Moss, B. (1982) Vaccinia virus: a selectable eukaryotic cloning and expression vector. *Proc. Nat. Acad. Sci. USA* **79**, 7415–19.

Macleod K.F. & Jacks T. (1999) Insights into cancer from transgenic mouse models. *J. Pathol.* **187**, 43–60.

Madisen L., Krumm A., Hebbes T.R. & Groudine M. (1998) The immunoglobulin heavy chain locus control region increases histone acetylation along linked c-*myc* genes. *Mol. Cell Biol.* **18**, 6281–92.

Madison J.T., Everett G.A. & Kung H.K. (1966) On nucleotide sequence of yeast tyrosine transfer RNA. *Cold Spring Harb. Sym.* **31**, 409–10.

Maeda I., Kohara Y., Yamamoto M. & Sugimoto A. (2001) Large-scale analysis of gene function in *Caenorhabditis elegans* by high-throughput RNAi. *Curr. Biol.* **11**, 171–6.

Maeda S., Kawai T., Obinata M. *et al.* (1985) Production of human α-interferon in silkworm using a baculovirus vector. *Nature* **315**, 592–4.

Maes T., De Keukeleire P. & Gerats T. (1999) Plant tagnology. *Trends Plant Sci.* **4**, 90–6.

Magari S.R., Rivera V.M., Luliucci J.D., Gilman M. & Cerasoil F. Jr (1997) Pharmacologic control of a humanized gene therapy system implanted into nude mice. *J. Clin. Invest.* **100**, 2865–72.

Magin-Lachmann C., Kotzamanis G., D'Aiuto L. *et al.* (2003) Retrofitting BACs with G418 resistance, luciferase, and *oriP* and *EBNA-1* – new vectors for *in vitro* and *in vivo* delivery. *BMC Biotechnol.* **3**, 2–10.

Magram J., Chada K. & Costantini F. (1985) Developmental regulation of a cloned adult β-globin gene in transgenic mice. *Nature* **315**, 338–40.

Maguin E., Prevost H., Ehrlich S.D. & Gruss A. (1996) Efficient insertional mutagenesis in lactococci and other Gram-positive bacteria. *J. Bacteriol.* **178**, 931–5.

Mahajan N.P., Linder K., Berry G., Gordon G.W., Heim R. & Herman B. (1998) *Bcl-2* and *bax* interactions in mitochondria probed with green fluorescent protein and fluorescence resonance energy transfer. *Nature Biotechnol.* **16**, 547–52.

Maier R.M., Neckermann K., Igloi G.L. & Kossel H. (1995) Complete sequence of the maize chloroplast genome: gene content, hotspots of divergence and fine tuning of genetic information by transcript editing. *J. Mol. Biol.* **251**, 614–28.

Makalowski W., Zhang J. & Boguski M.S. (1996) Comparative analysis of 1196 orthologous mouse and human full-length mRNA and protein sequences. *Genome Res.* **6**, 846–57.

Makarova K.S. *et al.* (2001) Genome of the extremely radiation-resistant bacterium *Deinococcus radiodurans* viewed from the perspective of comparative genomics. *Microbiol. Mol. Biol. Rev.* **65**, 44–79.

Makarova K.S., Wolf Y.I. & Koonin E.V. (2003) Potential genomic determinants of hyperthermophily. *Trends Genet.* **19**, 172–6.

Makarova K.S., Wolf Y.I. & Koonin E.V. (2003) Potential genomic determinants of hyperthermophily. *Trends Genet.* **19**, 172–6.

Maliga P. (1993) Towards plastid transformation in flowering plants. *Trends Biotechnol.* **11**, 101–7.

Malmqvist M. & Karlsson R. (1997) Biomolecular interaction analysis: affinity biosensor technologies for functional analysis of proteins. *Curr. Opin. Chem. Biol.* **1**, 378–83.

Malpartida R. & Hopwood D.A. (1984) Molecular cloning of the whole biosynthetic pathway of a *Streptomyces* antibiotic and its expression in a heterologous host. *Nature* **309**, 462–4.

Mancini M., Hadchouel M., Davis H.L. *et al.* (1996) DNA-mediated immunization in a transgenic mouse model of the hepatitis B surface antigen chronic carrier state. *Proc. Nat. Acad. Sci. USA* **93**, 12496–501.

Mandel M. & Higa A. (1970) Calcium-dependent bacteriophage DNA infection. *J. Mol. Biol.* **53**, 159–62.

Maniatis T., Hardison R.C., Lacy E. *et al.* (1978) The isolation of structural genes from libraries of eucaryotic DNA. *Cell* **15**, 687–701.

Mann M. & Pandey A. (2001) Use of mass spectrometry-derived data to annotate nucleotide and protein sequence databases. *Trends Biochem. Sci.* **26**, 54–61.

Mann M., Hojrup P. & Roepstorff P. (1993) Use of mass spectrometric molecular weight information to identify protein in sequence databases. *Biological Mass Spectrom.* **22**, 338–45.

Mann R., Mulligan R.C. & Blatimore D. (1983) Construction of a retrovirus packaging mutant and its use to produce helper-free defective retrovirus. *Cell* **33**, 153–9.

Mansour S.L., Goddard J.M. & Capecchi M.R. (1993) Mice homozygous for a targeted disruption of the proto-oncogene *int-2* have developmental defects in the tail and inner ear. *Development* **117**, 13–28.

Mansour S.L., Thomas K.R. & Capecchi M.R. (1988) Disruption of the proto-oncogene *int-2* in mouse embryo-derived stem cells: a general strategy for targeting mutations to non-selectable genes. *Nature* **336**, 348–52.

Mantei N., Boll W. & Weissman C. (1979) Rabbit beta-globin mRNA production in mouse L cells transformed with cloned rabbit beta-globin chromosomal DNA. *Nature* **281**, 40–6.

Mar Alba M., Santibanez-Koref M.F. & Hancock J.M. (1999) Amino acid reiterations in yeast are over-represented in particular classes of proteins and show evidence of a slippage-like mutational process. *J. Mol. Evol.* **49**, 789–97.

Marconi P., Krisky D., Oligino T. *et al.* (1996) Replication-defective herpes simplex virus vectors for gene transfer *in vivo. Proc. Nat. Acad. Sci. USA* **93**, 11319–20.

Marcotte E.M., Pellegrini M., Thompson M.J., Yeates T.O. & Eisenberg D. (1999) A combined algorithm for genome-wide prediction of protein function. *Nature* **402**, 83–6.

Margolskee R.F. (1992) Epstein-Barr virus-based expression vectors. *Curr. Top. Microbiol. Immunol.* **185**, 67–95.

Margolskee R.F., Kavathas P. & Berg P. (1988) Epstein-Barr virus shuttle vector for stable episomal replication of cDNA expression libraries in humans. *Mol. Cell. Biol.* **8**, 2337–47.

Maricnfeld J., Unseld M. & Brennicke A. (1999) The mitochondrial genome of *Arabidopsis* is composed of both native and immigrant information. *Trends Plant Sci.* **4**, 495–502.

Marillonnet S., Thoeringer C., Kandzia R., Klimyuk V. & Gleba Y. (2005) Systemic *Agrobacterium tumefaciens*-mediated transfection of viral replicons for efficient transient expression in plants. *Nat. Biotechnol.* **23**, 718–23.

Marini N., Hiiyanna K.T. & Benbow R.M. (1989) Differential replication of circular DNA molecules co-injected into early *Xenopus laevis* embryos. *Nucl. Acids Res.* **17**, 5793–808.

Marinus M.G., Carraway M., Frey A.Z., Brown L. & Arraj J.A. (1983) Insertion mutations in the *dam* gene of *Escherichia coli* K-12. *Mol. Gen. Genet.* **192**, 288–9.

Markowitz D., Goff S. & Bank A. (1988) A safe packaging line for gene transfer: separating viral genes on two different plasmids. *J. Virol.* **62**, 1120–4.

Marra M.A. *et al.* (1997) High throughput fingerprint analysis of large-insert clones. *Genome Res.* **7**, 1072–84.

Marras S.A.E., Kramer F.R. & Tyagi S. (1999) Multiplex detection of single nucleotide variations using molecular beacons. *Genetical Anal. Biomol. Eng.* **14**, 151–6.

Marshall E. & Pennisi E. (1998) Hubris and the human genome. *Science* **280**, 994–5.

Marshall E. (1999) Gene therapy death prompts review of adenovirus vector. *Science* **286**, 2244–5.

Marshall J., Molloy R., Moss G.W., Howe J.R. & Hughes T.E. (1995) The jellyfish green fluorescent protein: a new tool for studying ion channel expression and function. *Neuron* **14**, 211–15.

Marsh-Armstrong N., Huang H., Berry D.L. & Brown D.D. (1999) Germ line transmission of transgenes in *Xenopus laevis*. *Proc. Nat. Acad. Sci. USA* **96**, 14389–93.

Marshner H. (1995) *Mineral Nutrition of Higher Plants.* London, Academic Press.

Marston A.L., Tham W.H., Shah H. & Amon A. (2004) A genome-wide screen identifies genes required for centromeric cohesion. *Science* **303**, 1367–70.

Martin A.C. *et al.* (1998) Protein folds and functions. *Structure* **6**, 875–84.

Martin G.R. (1981) Isolation of a pluripotent cell line from early mouse embryos cultured in medium conditioned by teratocarcinoma stem cells. *Proc. Nat. Acad. Sci. USA* **78**, 7634–8.

Martinez A., Sparks C., Hart C.A., Thompson J. & Jepson I. (1999) Ecdysone agonist inducible transcription in transgenic tobacco plants. *Plant J.* **19**, 97–106.

Martinez F.D. *et al.* (1997) Association between genetic polymorphisms of the β_2-adrenoreceptor and response to albuterol in children with and without a history of wheezing. *J. Clin. Invest.* **100**, 3184–8.

Martinez-Abarca F. & Toro N. (2000) Group II introns in the bacterial world. *Mol. Microbiol.* **38**, 917–26.

Marton M.J. (1998) Drug target validation and identification of secondary drug target effects using DNA microarrays. *Nature Medicine* **4**, 1293–301.

Maruyama K. & Sugano S. (1994) Oligo-capping: a simple method to replace the cap structure of eukaryotic mRNAs with oligoribonucleotides. *Gene* **138**, 171–4.

Marx J. (2000) Medicine: DNA arrays reveal cancer in its many forms. *Science* **289**, 1670–2.

Mascarenhas D., Mettler I.J., Pierce D.A. & Lowe H.W. (1990) Intron-mediated enhancement of heterologous gene expression in maize. *Plant Mol. Biol.* **15**, 913–20.

Mason H.S. & Arntzen C.J. (1995) Transgenic plants as vaccine production systems. *Trends Biotechnol.* **13**, 388–92.

Masood E. (1999) A consortium plans free SNP map of human genome. *Nature* **398**, 545–6.

Massoud M., Bischoff R., Dalemans W. *et al.* (1991) Expression of active recombinant human alpha l-antitrypsin in transgenic rabbits. *J. Biotechnol.* **18**, 193–204.

Masu Y., Nakayama K., Tamaki H. *et al.* (1987) cDNA cloning of bovine substance-K receptor through oocyte expression system. *Nature* **329**, 836–8.

Matsumoto K., Yoshimatsu T. & Oshima Y. (1983) Recessive mutations conferring resistance to carbon catabolite repression of galactokinase synthesis in *Saccharomyces cerevisiae*. *J. Bacteriol.* **153**, 1405–14.

Matsumura H., Nirasawa S. & Terauchi R. (1999) Technical advance: transcript profiling in rice (*Oryza sativa L.*) seedlings using serial analysis of gene expression (SAGE). *Plant J.* **20**, 719–26.

Matsuoka M., Nomura M., Agarie S., Tokutomi M. & Ku M.S.B. (1998) Evolution of C4 photosynthetic genes and overexpression of maize C4 genes in rice. *J. Plant Res.* **111**, 333–7.

Matz M.V., Fradkov A.F., Labas Y.A. *et al.* (1999) Fluorescent proteins from nonbioluminescent *Anthozoa* species. *Nature Biotechnol.* **17**, 969–73.

Matzeit V., Schaefer S., Kammann M. *et al.* (1991) Wheat dwarf virus vectors replicate and express foreign genes in cells of monocotyledonous plants. *Plant Cell* **3**, 247–58.

Matzke A.J.M. & Chilton M.-D. (1981) Site-specific insertion of genes into T-DNA of the *Agrobacterium* tumour-inducing plasmid: an approach to genetic engineering of higher plant cells. *J. Mol. Appl. Genet.* **1**, 39–49.

Matzke A.J.M. & Matzke M.A. (1998) Position effects and epigenetic silencing of plant transgenes. *Curr. Opin. Plant Biol.* **1**, 142–8.

Matzke, M.A. & Birchler, J.A. (2005). RNAi-mediated pathways in the nucleus. *Nat. Rev. Genet.* **6**, 24–35.

Matzuk M.M., Finegold M.J., Su J.G., Hsueh A.J. & Bradley A. (1992) Alpha inhibin is a tumour-suppressor gene with gonadal specificity in mice. *Nature* **360**, 313–19.

Maxam A.M. & Gilbert W. (1977) A new method for sequencing DNA. *Proc. Nat. Acad. Sci. USA* **74**, 560–4.

Mazodier P. & Davies J. (1991) Gene transfer between distantly related bacteria. *Ann. Rev. Genet.* **25**, 147–71.

Mazodier P., Petter R. & Thompson C. (1989) Intergeneric conjugation between *Escherichia coli* and *Streptomyces* species. *J. Bacteriol.* **171**, 3583–5.

McBride K.E. & Summerfelt K.R. (1990) Improved binary vectors for *Agrobacterium*-mediated plant transformation. *Plant Mol. Biol.* **14**, 269–76.

McCabe D. & Christou P. (1993) Direct DNA transfer using electric discharge particle acceleration (Accell® technology). *Plant Cell Tissue Organ Cult.* **33**, 227–36.

McCabe D.E., Swain W.F., Martinell B.J. & Christou P. (1988) Stable transformation of soybean (*Glycine max*) by particle acceleration. *Biotechnology* **6**, 923–6.

McCaffrey, A.P. *et al.* (2002) RNA interference in adult mice. *Nature* **418**, 38–9.

McClelland M. *et al.* (2001) Complete genome sequence of *Salmonella enterica* serovar Typhimurium LT2. *Nature* **413**, 852–6.

McClelland M., Mathieu-Daude F. & Welsh J. (1995) RNA fingerprinting and differential display using arbitrarily-primed PCR. *Trends Genet.* **11**, 242–6.

McCormick A.A., Kumagai M.H., Hanley K. *et al.* (1999) Rapid production of specific vaccines for lymphoma by expression of the tumor-derived single-chain Fv epitopes in tobacco plants. *Proc. Nat. Acad. Sci. USA* **96**, 703–8.

McCraith S., Hotzam T., Moss B. & Fields S. (2000) Genome-wide analysis of vaccinia virus protein–protein interactions. *Proc. Nat. Acad. Sci. USA* **97**, 4879–84.

McCreath K.J., Howcroft J., Campbell K.H.S. *et al.* (2000) Production of gene targeted sheep by nuclear transfer from cultured somatic cells. *Nature* **405**, 1066–9.

McCutchan J.H. & Pagano J.S. (1968) Enhancement of the infectivity of simian virus 40 deoxyribonucleic acid with diethylaminoethyl dextran. *J. Nat. Cancer Inst.* **41**, 351–7.

McDaniel R., Thamchaipenet A., Gustafsson C. *et al.* (1999) Multiple genetic modifications of the erythromycin polyketide synthase to produce a library of novel "unnatural" natural products. *Proc. Nat. Acad. Sci. USA* **96**, 1846–51.

McEachern M.J., Bott M.A., Tooker P.A. & Helinski D.R. (1989) Negative control of plasmid R6K replication: possible role of intermolecular coupling of replication origins. *Proc. Nat. Acad. Sci. USA* **86**, 7942–6.

McElroy D. & Brettel R.I.S. (1994) Foreign gene expression in transgenic cereals. *Trends Biotechnol.* **12**, 62–8.

McElroy D., Blowers A.D., Jenes B. & Wu R. (1991) Construction of expression vectors based on the rice actin-1 (Act-1), 5′ region for use in monocot transformation. *Mol. Gen. Genet.* **231**, 150–60.

McElroy D., Chamberlain D.A., Moon E. & Wilson K.J. (1995) Development of *gusA* reporter gene constructs for cereal transformation – availability of plant transformation vectors from the CAMBIA molecular genetic resource center. *Mol. Breeding* **1**, 27–37.

McGarvey P.B., Hammond J., Dienelt M.M. *et al.* (1995) Expression of the rabies virus glycoprotein in transgenic tomatoes. *Biotechnology* **13**, 1484–7.

McInnes L.A. *et al.* (1996) A complete genome screen for genes predisposing to severe bipolar disorder in two Costa Rican pedigrees. *Proc. Nat. Acad. Sci. USA* **93**, 13060–5.

McKinney E.C. *et al.* (1995) Sequence based identification of T-DNA insertion mutations in *Arabidopsis*: actin mutants *act2-1* and *act4-1*. *Plant J.* **8**, 613–22.

McKnight R.A., Shamay A., Sankaran L., Wall R.J. & Hennighausen L. (1992) Matrix-attachment regions can impart position-independent regulation of a tissue-specific gene in transgenic mice. *Proc. Nat. Acad. Sci. USA* **89**, 6943–7.

McLain L., Durrani Z., Dimmock N.J. *et al.* (1996b) A plant virus HIV-1 chimaera stimulates antibody that neutralizes HIV-I. In *Vaccines 96*, eds. Brown F., Burton D.R., Collier J., Mekalanos J. & Norrby E. Cold Spring Harbor Laboratory Press, Cold Spring Harbor, New York.

McLain L., Durrani Z., Wisniewski L.A. *et al.* (1996a) Stimulation of neutralizing antibodies to human immunodeficiency virus type I in three strains of mice immunized with a 22 amino acid peptide of gp41 expressed on the surface of a plant virus. *Vaccine* **14**, 799–810.

McLain L., Porta C., Lomonossoff G.P., Durrani Z. & Dimmock, N.J. (1995) Human immunodeficiency virus type I neutralizing antibodies raised to a gp41 peptide expressed on the surface of a plant virus. *AIDS Res. Hum. Retroviruses* **11**, 327–34.

McLaughlin J.A., Davies L. & Seamark R.F. (1990) *In vitro* embryo culture in the production of identical Merino lambs by nuclear transplantation. *Reprod. Fert. Devel.* **2**, 619–22.

McLaughlin S.K., Collis P., Hermonat P.L. & Muzyczka N. (1988) Adeno-associated virus general transduction vectors: analysis of proviral structures. *J. Virol.* **62**, 1963–73.

McMahon A.P. & Bradley A. (1990) The *Wnt-1* (*int-1*) proto-oncogene is required for development for a large region of the mouse brain. *Cell* **62**, 1073–85.

McPherson D.T. (1988) Codon preference reflects mistranslational constructs: a proposal. *Nucl. Acids Res.* **16**, 4111–20.

Mead D.A., Pey N.K., Herrnstadt C., Marcil R.A. & Smith L.M. (1991) A universal method for the direct cloning of PCR amplified nucleic acid. *Biotechnology* **9**, 657–63.

Mecsas J. & Sugden B. (1987) Replication of plasmids derived from bovine papilloma virus type 1 and Epstein-Barr virus in cells in culture. *Ann. Rev. Cell Biol.* **3**, 87–108.

Medberry S.L., Dale E.C., Qin M. & Ow D.W. (1995) Intrachromosomal rearrangements generated by site-specific recombination. *Nucl. Acids Res.* **23**, 485–90.

Medek A., Olejniczak E.T., Meadows R.P. & Fesik S.W. (2000) An approach for high-throughput structure determination of proteins by NMR spectroscopy. *J. Biomol. NMR* **18**, 229–38.

Meissner R., Chague V., Zhu Q., Emmanuel E., Elkind Y. & Levy A. (2000) A high throughput system for transposon tagging and promoter trapping in tomato. *Plant J.* **22**, 265–74.

Melchers L.S. & Stuiver M.H. (2000) Novel genes for disease-resistance breeding. *Curr. Opin. Plant Biol.* **3**, 147–52.

Meldrum D. (2000a) Automation for genomics. I. Preparation for sequencing. *Genome Res.* **10**, 1081–92.

Meldrum D. (2000b) Automation for genomics. II. Sequencers, microarrays, and future trends. *Genome Research* **10**, 1288–303.

Mellon P., Parker V., Gluzman Y. & Maniatis T. (1981) Identification of DNA sequences required for transcription of the human α-globin gene in a new SV40 host-vector system. *Cell* **27**, 279–88.

Melton D.A. (1987) Translation of messenger RNA in injected frog oocytes. *Methods Enzymol.* **152**, 288–96.

Memelink J., Menke F.L.H., Van der Fits L. & Kijne J.W. (2000) Transcriptional regulators to modify secondary metabolism. In *Metabolic Engineering of Plant Secondary Metabolism*, eds. Verpoorte R. & Alfermann A.W., pp. 111–25. Kluwer Academic Press, The Netherlands.

Mendez M.J., Green L.L., Corvalan J.R. *et al.* (1997) Functional transplant of megabase human immunoglobulin loci recapitulates human antibody response in mice. *Nature Genet.* **15**, 146–56.

Mendoza L.G., McQuary P., Mongan A., Gangadharan R., Brignac S. & Eggers M. (1999) High-throughput micro-array-based enzyme-linked immunosorbent assay (ELISA). *Biotechniques* **27**, 778–88.

Merchlinsky M., Eckert D., Smith E. & Zauderer M. (1997) Construction and characterization of vaccinia direct ligation vectors. *Virology* **238**, 444–51.

Mermod N., Ramos J.L., Lehrbach P.R. & Timmis K.N. (1986) Vector for regulated expression of cloned genes in a wide range of Gram-negative bacteria. *J. Bacteriol.* **167**, 447–54.

Meselson M. & Yuan R. (1968) DNA restriction enzyme from *E. coli. Nature* **217**, 1110–14.

Messing J., Crea R. & Seeburg P.H. (1981) A system for shotgun DNA sequencing. *Nucl. Acids Res.* **9**, 309–21.

Metzger D. & Feil R. (1999) Engineering the mouse genome by site-specific recombination. *Curr. Opin. Biotechnol.* **10**, 470–6.

Mewes H.W. *et al.* (2000) MIPS: a database for genomes and protein sequences. *Nucl. Acids Res.* **28**, 37–40.

Meyer P., Heidmann I. & Niedenhof I. (1992) The use of African cassava mosaic virus as a vector system for plants. *Gene* **110**, 213–17.

Meyer P., Heidmann I., Forkmann G. & Saedler H. (1987) A new petunia flower colour generated by transformation of a mutant with a maize gene. *Nature* **330**, 677–8.

Meyers B.C., Tej S.S., Vu T.H., Haudenschild C.D., Agrawal V., Edberg S.B., Ghazal H. & Decola S. (2004) The use of MPSS for whole-genome transcriptional analysis in *Arabidopsis. Genome Res.* **14**, 1641–53.

Miao Z.H. & Lam E. (1995) Targeted disruption of the TGA3 locus in *Arabidopsis thaliana. Plant J.* **7**, 359–65.

Michel B., Niaudet B. & Ehrlich S.D. (1982) Intramolecular recombination during plasmid transformation of *Bacillus subtilis* competent cells. *EMBO J.* **1**, 1565–71.

Michel B., Palla E., Niaudet B. & Ehrlich S.D. (1980) DNA cloning in *Bacillus subtilis*. III. Efficiency of random-segment cloning and insertional inactivation vectors. *Gene* **12**, 147–54.

Michelson A.M. & Orkin S.H. (1982) Characterization of the homoploymer tailing reaction catalyzed by terminal deoxynucleotidyl transferase: implications for the cloning of cDNA. *J. Biol. Chem.* **256**, 1473–82.

Mierendorf R.C., Percy C. & Young R.A. (1987) Gene isolation by screening λgt11 libraries with antibodies. *Methods Enzymol.* **152**, 458–69.

Miki Y. *et al.* (1992) Disruption of the APC gene by a retrotransposal insertion of L1 sequence in a colon cancer. *Cancer Res.* **52**, 643–5.

Millen R.S. *et al.* (2001) Many parallel losses of *infA* from chloroplast DNA during angiosperm evolution with multiple independent transfers to the nucleus. *Plant Cell* **13**, 645–58.

Miller A.D. (1992a) Human gene therapy comes of age. *Nature* **357**, 455–60.

Miller A.D. (1992b) Retroviral vectors. *Curr. Top. Microbiol. Immunol.* **185**, 1–24.

Miller A.D., Jolly D.J., Friedmann T. & Verma I.M. (1983) A transmissible retrovirus expressing human hypoxanthine phosphoribosyltransferase (HPRT): gene transfer into cells obtained from humans deficient in HPRT. *Proc. Nat. Acad. Sci. USA* **80**, 4709–13.

Miller C.A., Beaucage S.L. & Cohen S.N. (1990) Role of DNA superhelicity in partitioning of the pSC101 plasmid. *Cell* **62**, 127–33.

Miller D.G., Adam M.A. & Miller A.D. (1990) Gene transfer by retrovirus vectors occurs only in cells that are actively replicating at the time of infection. *Mol. Cell Biol.* **10**, 4239–42.

Miller D.G., Petek L.M. & Russell D.W. (2003) Human gene targeting by adeno-associated virus vectors is enhanced by DNA double-strand breaks. *Mol. Cell Biol.* **23**, 3550–7.

Miller G. (1985) Epstein-Barr virus. In *Virology*, ed. Fields B.N., pp. 563–89. Raven Press, New York.

Miller L.K., Miller D.W. & Adang M.J. (1983) An insect virus for genetic engineering: developing baculovirus polyhedrin substitution vectors. In *Genetic Engineering in Eukaryotes*, eds. Lurgin P.F. & Kleinhofs A. Plenum Press, New York.

Milman G. & Herzberg M. (1981) Efficient DNA transfection and rapid assay for thymidine kinase activity and viral antigenic determinants. *Somat. Cell Genet.* **7**, 161–70.

Milot E., Strouboulis J., Trimborn T. *et al.* (1996) Heterochromatin effects on the frequency and duration of LCR-mediated gene transcription. *Cell* **87**, 105–14.

Mirkovitch J., Mirault M.E. & Laemmli U.K. (1984) Organization of the higher-order chromatin loop: specific DNA attachment sites on nuclear scaffold. *Cell* **39**, 223–32.

Miroux B. & Walker J.E. (1996) Over-production of protein in *Escherichia coli*: mutant hosts that allow synthesis of some membrane proteins and globular proteins at high levels. *J. Mol. Biol.* **260**, 289–98.

Mirzabekov A. & Kolchinsky A. (2001) Emerging array-based technologies in proteomics. *Curr. Opin. Chem. Biol.* **6**, 70–5.

Mischel P.S., Cloughesy T.F. & Nelson S.F. (2004) DNA-microarray analysis of brain cancer: Molecular classification for therapy. *Nature Rev. Neurosci.* **5**, 782–92.

Miskey, C. et al. (2003) The Frog Prince: a reconstructed transposon from Rana pipiens with high transpositional activity in vertebrate cells. Nucl. Acids Res. **31**, 6873–81.

Mitas M. (1997) Trinucleotide repeats associated with human disease. Nucl. Acids Res. **25**, 2245–53.

Mitragotri, S., Blankschtein, D. & Langer, R. (1996) Transdermal drug delivery using low-frequency sonophoresis. Pharm. Res. **13**, 411–20.

Mittal V. (2004) Improving the efficiency of RNA interference in mammals. Nat. Rev. Genet. **5**, 355–65.

Mittl P.R.E. & Grutter M.G. (2001) Structural genomics: opportunities and challenges. Curr. Opin. Chem. Biol. **5**, 402–8.

Mlynarova L., Loonen A., Heldens J. et al. (1994) Reduced position effect in mature transgenic plants conferred by the chicken lysozyme matrix-attachment region. Plant Cell **6**, 417–26.

Monaco A.P. & Larin Z. (1994) YACs, BACs, and MACs: artificial chromosomes as research tools. Trends Biotechnol. **12**, 280–6.

Monahan P.E. & Samulski R.J. (2000) Adeno-associated virus vectors for gene therapy: more pros than cons? Mol. Med. Today **6**, 433–40.

Monckton D.G. & Jeffreys A.J. (1993) DNA profiling. Curr. Opin. Biotechnol. **4**, 660–4.

Montaya A.L., Chilton M.-D., Gordon M.P., Sciaky D. & Nester E.W. (1977) Octopine and nopaline metabolism in Agrobacterium tumefaciens and crown gall tumor cells: role of plasmid genes. J. Bacteriol. **129**, 101–7.

Moody M.D., Van Arsdell S.W., Murphy K.P., Orencole S.F. & Burns C. (2001) Array-based ELISAs for high-throughput analysis of human cytokines. Biotechniques **31**, 186–94.

Moolten F.L. (1986) Tumour chemosensitivity conferred by inserted herpes thymidine kinase genes: paradigm for a prospective cancer control strategy. Cancer Res. **46**, 5276–81.

Mooney J.F., Hunt A.J., McIntosh J.R., Liberko C.A., Walba D.M. & Rogers C.T. (1996) Patterning of functional antibodies and other proteins by photolithography of silane monolayers. Proc. Nat. Acad. Sci. USA **93**, 12287–91.

Moons A., Gielen J., Vandekerckhove J., Van der Straeten D., Gheysen G. & Van Montagu M. (1997) An abscisic acid and salt stress responsive rice cDNA from a novel plant gene family. Planta **202**, 443–54.

Moore G., Devos K.M., Wang Z. & Gale M.D. (1995) Grasses line up and form a circle. Curr. Biol. **5**, 737–9.

Moore I., Baroux C., Gaelweiter L., Grosskopt D., Mader P., Schell J. & Palme K. (1997) A transactivation system for regulating expression of transgenes in whole plants. J. Exp. Bot. **48S**, 51.

Moore R.C., Redhead N.J., Selfridge J. et al. (1995) Double replacement gene targeting for the production of a series of mouse strains with different prion protein gene alterations. Biotechnology **13**, 999–104.

Mootha V.K., Lepage P., Miller K., et al. (2003) Identification of a gene causing human cytochrome c oxidase deficiency by integrative genomics. Proc. Nat. Acad. Sci. USA **100**, 605–10.

Moran J.V., DeBerardinis R.J. & Kazazian H. (1999) Exon shuffling by L1 retrotransposition. Science **283**, 1530–4.

Morgan D.O. & Roth R.A. (1988) Analysis of intracellular protein function by antibody injection. Immunol. Today **9**, 84–8.

Morgan E. et al. (2004) Cytometric bead array: a multiplexed assay platform with applications in various areas of biology. Clin. Immunol. **110**, 252–66.

Mori M., Fugihara N., Mise K. & Furusawa I. (2001) Inducible high-level mRNA amplification system by viral replicase in transgenic plants. Plant J. **27**, 79–86.

Morsey M.A. & Caskey C.T. (1999) Expanded-capacity adenoviral vectors – the helper-dependent vectors. Mol. Med. Today **5**, 18–24.

Moss B. (1996) Genetically engineered poxviruses for recombinant gene expression, vaccination and safety. Proc. Nat. Acad. Sci. USA **93**, 11341–8.

Mottes M., Grandi G., Sgaramella V. et al. (1979) Different specific activities of the monomeric and oligomeric forms of plasmid DNA in transformation of B. subtilis and E. coli. Mol. Gen. Genet. **174**, 281–6.

Moult J. & Melamud E. (2000) From fold to function. Curr. Opin. Struct. Biol. **10**, 384–9.

Moult J., Hubbard T., Fidelis K. & Pedersen J.T. (1999) Critical assessment of methods of protein structure prediction (CASP): round III. Proteins **S3**, 2–6.

Mourgues F., Brisset M.-N. & Chevreau E. (1998) Strategies to improve plant resistance to bacterial diseases through genetic engineering. Trends Biotechnol. **16**, 203–10.

Mouse Genome Sequencing Consortium (2002) Initial sequencing and comparative analysis of the mouse genome. Nature **420**, 520–62.

Moxham C.M., Hod Y. & Malbon C. (1993) Induction of G_{ia2}-specific antisense RNA in vivo inhibits neonatal growth. Science **260**, 991–5.

Mueller U. et al. (2001) Development of a technology for automation and miniaturisation of protein crystallisation. J. Biotechnol. **85**, 7–14.

Muesing M., Tamm J., Shepard H.M. & Polisky B. (1981) A single base pair alteration is responsible for the DNA overproduction phenotype of a plasmid copy-number mutant. Cell **24**, 235–42.

Muller S., Sanda I.S., Kamp-Hansen P. & Dalboge H. (1998) Comparison of expression systems in the yeasts Saccharomyces cerevisiae, Hansenula polymorpha, Kluyveromyces lactis, Schizosaccharomyces pombe and Yarrowia lipolytica: cloning of two novel promoters from Y. lipolytica. Yeast **14**, 1267–83.

Muller U. (1999) Ten years of gene targeting: targeted mouse mutants, from vector design to phenotype analysis. Mechanisms of Development **82**, 3–21.

Muller U. (1999) Ten years of gene targeting: targeted mouse mutants, from vector design to phenotype analysis. *Mech. Devel.* **82**, 3–21.

Mulligan R.C. & Berg P. (1980) Expression of a bacterial gene in mammalian cells. *Science* **209**, 1422–7.

Mulligan R.C. & Berg P. (1981b) Selection for animal cells that express the *Escherichia coli* gene coding for xanthine-guanine phosphoribosyl-transferase. *Proc. Nat. Acad. Sci. USA* **78**, 2072–6.

Mullins M.C., Hammerschmidt M., Haffter P. & Nusslein-Volhard C. (1994) Large-scale mutagenesis in the zebrafish: in search of genes controlling development in a vertebrate. *Curr. Biol.* **4**, 189–201.

Munasinghe A., Patankar S., Cook B.P. *et al.* (2001) Serial analysis of gene expression (SAGE) in *Plasmodium falciparum*: application of the technique to AT-rich genomes. *Mol. Biochem. Parasitol.* **113**, 23–34.

Munir M., Rossiter B. & Caskey C. (1990) Antisense RNA production in transgenic mice. *Somat. Cell Mol. Genet.* **16**, 383–94.

Muotri A.R., Da Veiga Pereira L., Dos Reis Vasques L. & Menck C.F.M. (1999) Ribozymes and the anti-gene therapy: how a catalytic RNA can be used to inhibit gene function. *Gene* **237**, 303–10.

Murakami H., Hohsaka T. & Sisido M. (2002) Random insertion and deletion of arbitrary number of bases for codon-based random mutation of DNAs. *Nature Biotechnol.* **20**, 76–81.

Murakami H., Hohsaka T. & Sisido M. (2003) Random insertion and deletion mutagenesis. *Methods Mol. Biol.* **231**, 53–64.

Murdock D., Ensley B.D., Serdar C. & Thalen M. (1993) Construction of metabolic operons catalyzing the *de novo* biosynthesis of indigo in *Escherichia coli*. *Biotechnology* **11**, 381–6.

Murphy W.J., Pevzner P.A. & O'Brien S.J. (2004) Mammalian phylogenomics comes of age. *Trends Genet.* **20**, 631–9.

Murray K. & Murray N.E. (1975) Phage lambda receptor chromosomes for DNA fragments made with restriction endonuclease III of *Haemophilus influenzae* and restriction endonuclease I of *Escherichia coli*. *J. Mol. Biol.* **98**, 551–64.

Murray M.J., Kaufman R.J., Latt S.A. & Weinberg R.A. (1983) Construction and use of a dominant, selectable marker: a Harvey sarcoma virus-dihydrofolate reductase chimera. *Mol. Cell. Biol.* **3**, 32–43.

Murray N.E. & Kelley W.S. (1979) Characterization of l*pol*A transducing phages: effective expression of the *E. coli pol A* gene. *Mol. Gen. Genet.* **175**, 77–87.

Murray N.E. (1983) Phage lambda and molecular cloning. In *The Bacteriophage Lambda*, eds. Hendrix R.W., Roberts J.W., Stahl F.W. & Weisberg R.A., Lambda II (Monograph No. 13), Vol. 2. Cold Spring Harbor Laboratory, Cold Spring Harbor, New York.

Murray N.E., Batten P.L. & Murray K. (1973b) Restriction of bacteriophage lambda by *Escherichia coli* K. *J. Mol. Biol.* **81**, 395–407.

Murray N.E., Manduca de Ritis P. & Foster L.A. (1973a) DNA targets for the *Escherichia coli* K restriction system analysed genetically in recombinants between phages phi-80 and lambda. *Mol. Gen. Genet.* **120**, 261–81.

Murzin A. (1998) How divergent evolution goes into proteins. *Curr. Opin. Struct. Biol.* **8**, 380–7.

Murzin A. (2001) Progress in protein structure prediction. *Nature Struct. Biol.* **8**, 110–12.

Murzin A., Brenner S., Hubbard T. & Chothia C. (1995) SCOP – A structural classification of proteins database for the investigation of sequences and structures. *J. Mol. Biol.* **247**, 536–40.

Mushegian A. (1999) The minimal genome concept. *Curr. Opin. Genet. Develop.* **9**, 709–14.

Mushegian A.R. & Koonin E.V. (1996) A minimal gene set for cellular life derived by comparison of complete bacterial genomes. *Proc. Nat. Acad. Sci. USA* **93**, 10268–73.

Muzyczka N. (1992) Use of adeno-associated virus as a general transduction vector for mammalian cells. *Curr. Top. Microbiol. Immunol.* **158**, 97–129.

Myagkikh M., Alpanah S., Markham P.D. *et al.* (1996) Multiple immunizations with attenuated poxvirus HIV type 2 recombinants and subunit boosts required for protection of rhesus macaques. *AIDS Res. Hum. Retroviruses* **12**, 985–92.

Myers E.W. *et al.* (2000) A whole-genome assembly of *Drosophila*. *Science* **287**, 2196–204.

Myers R.M. & Tjian R. (1980) Construction and analysis of simian virus 40 origins defective in tumor antigen binding and DNA replication. *Proc. Nat. Acad. Sci. USA* **77**, 6491–5.

Nagaraju J., Kanda T., Yukuhiro K. *et al.* (1996) Attempt at transgenesis of the silkworm (*Bombyx mon* L.) by egg-injection of foreign DNA. *Appl. Entomol. Zool.* **31**, 487–96.

Nagasaki T. *et al.* (2000) Synthesis of a novel water-soluble polyazobenzene dendrimer and photoregulation of affinity toward DNA. *Mol. Cryst. Liq. Cryst.* **345**, 227–32.

Nagatani N., Honda H., Shimada T. & Kobayashi T. (1997) DNA delivery into rice cells and transformation using silicon carbide whiskers. *Biotechnol. Techniques* **11**, 471–3.

Nagy A., Perrimon N., Sandmeyer S., Plasterk R. (2003) Tailoring the genome: the power of genetic approaches. *Nature Genet.* **33**, 276–84.

Nakai H., Storm T.A. & Kay M.A. (2000) Increasing the size of rAAV-mediated expression cassettes *in vivo* by intermolecular joining of two complementary vectors. *Nature Biotechnol.* **18**, 527–32.

Napoli C., Lemieux C.H. & Jorgensen R. (1990) Introduction of a chimeric chalcone synthese gene into petunia results in reversible co-suppression of homologous genes *in trans*. *Plant Cell* **2**, 279–89.

Nasmyth K.A. & Reed S.I. (1980) Isolation of genes by complementation in yeast: molecular cloning of a cell-cycle gene. *Proc. Nat. Acad. Sci. USA* **77**, 2119–23.

Nataraj A.J., Olivos-Glander I., Kusukawa N. & Highsmith W.E. (1999) Single-strand conformation polymorphism and heteroduplex analysis for gel-based mutation detection. *Electrophoresis* **20**, 1177–85.

Naviaux R.K. & Verma I.M. (1992) Retroviral vectors for persistent expression *in vivo*. *Curr. Biol.* **3**, 540–7.

Naylor L.H. (1999) Reporter gene technology – the future looks bright. *Biochem. Pharmacol.* **58**, 749–57.

Negrotto D., Jolley M., Beer S., Wenck A.R. & Hansen G. (2000) The use of phosphomannose isomerase as a selectable marker to recover transgenic maize plants (*Zea mays* L.) via *Agrobacterium* transformation. *Plant Cell Rep.* **19**, 798–803.

Negrutiu I., Shillito R., Potrykus I., Biasini G. & Sala F. (1987) Hybrid genes in the analysis of transformation conditions. I. Setting up a simple method for direct gene transfer in plant protoplasts. *Plant Mol. Biol.* **8**, 363–73.

Neilson L. *et al.* (2000) Molecular phenotype of the human oocyte by PCR-SAGE. *Genomics* **63**, 13–24.

Nellen W. & Lichtenstein C. (1993) What makes an mRNA anti-sense-itive? *Trends Biochem. Sci.* **18**, 419–23.

Nelson K.E. *et al.* (1999) Genome sequencing of *Thermotoga maritima*: evidence for lateral gene transfer between bacteria and archaea. *Nature* **399**, 323–9.

Ness J.E., Del Cardayre S.B., Minshull J. & Stemmer W.P. (2000) Molecular breeding: the natural approach to protein design. *Adv. Protein Chem.* **55**, 261–92.

Ness J.E., Welch M., Giver L. *et al.* (1999) DNA shuffling of subgenomic sequences of subtilisin. *Nature Biotechnol.* **17**, 893–6.

Neubauer G. *et al.* (1998) Mass spectrometry and EST-database searching allows characterization of the multi-protein spliceosome complex. *Nature Genet.* **20**, 46–50.

Neubauer G., Gottschalk A., Fabrizio P., Séraphin B., Lührmann R. & Mann M. (1997) Identification of the proteins of the yeast U1 small nuclear ribonucleoprotein complex by mass spectrometry. *Proc. Nat. Acad. Sci. USA* **94**, 385–90.

Newman J.R. & Fuqua C. (1999) Broad-host-range expression vectors that carry the L-arabinose-inducible *Escherichia coli araBAD* promoter and the *araC* regulator. *Gene* **227**, 197–203.

Neylon C. (2004) Chemical and biochemical strategies for the randomisation of protein encoding DNA sequences: library construction methods for directed evolution. *Nucl. Acids Res.* **32**, 1448–59.

Nguyen C. *et al.* (1995) Differential gene expression in the murine thymus assayed by quantitative hybridisation of arrayed cDNA clones. *Genomics* **29**, 207–16.

Niaudet B., Goze A. & Ehrlich S.D. (1982) Insertional mutagenesis in *Bacillus subtilis*: mechanism and use in gene cloning. *Gene* **19**, 277–84.

Nicholson J.K. & Wilson I.D. (2003) Understanding "global" systems biology: metabonomics and the continuum of metabolism. *Nature Rev. Drug Disc.* **2**, 668–76.

Nicholson J.K., Connelly J., Lindon J.C. & Names E. (2002) Metabonomics: a platform for studying drug toxicity and gene function. *Nature Rev. Drug Disc.* **1**, 153–61.

Nida D.L., Kolacz K.H., Buehler R.E. *et al.* (1996) Glyphosate-tolerant cotton: genetic characterization and protein expression. *J. Agric. Food Chem.* **44**, 1960–6.

Nielsen L.B., Kahn D., Duell T. *et al.* (1998) Apolipoprotein B gene expression in a series of human apolipoprotein B transgenic mice generated with recA-assisted restriction endonuclease cleavage modified bacterial artificial chromosomes: an intestine-specific enhancer element is located between 54 and 62 kilobases 5′ to the structural gene. *J. Biol. Chem.* **273**, 21800–7.

Nierman W.C. *et al.* (2001) Complete genome sequence of *Caulobacter crescentus*. *Proc. Nat. Acad. Sci. USA* **98**, 4136–41.

Nilsson M., Krejci K., Koch J., Kwiatkowski M., Gustavsson P. & Landegren U. (1997) Padlock probes reveal single-nucleotide differences, parent of origin and *in situ* distribution of centromeric sequences in human chromosomes 13 and 21. *Nature Genet.* **16**, 252–5.

Nir I., Kedzierski W., Chen S. & Travis G.H. (2000) Expression of Bcl-2 protects against photoreceptor degeneration in retinal degeneration slow (rds) mice. *J. Neurosci.* **20**, 2150–4.

Nitsche E.M., Moquin A., Adams P.S. *et al.* (1996) Differential display RT-PCR of total RNA from human foreskin fibroblasts for investigation of androgen-dependent gene expression. *Am. J. Med. Genet.* **63**, 231–8.

No D., Yao T.-P. & Evans R.M. (1996) Ecdysone-inducible gene expression in mammalian cells and transgenic mice. *Proc. Nat. Acad. Sci. USA* **93**, 3346–51.

Nolan P.M. *et al.* (2000) A systematic, genome-wide, phenotype-driven mutagenesis programme for gene function studies in the mouse. *Nature Genet.* **25**, 440–3.

Noma Y., Sideras P., Natto T. *et al.* (1986) Cloning of cDNA encoding the murine IgG1 induction factor by a novel strategy using SP6 promoter. *Nature* **319**, 640–6.

Noordewier M.O. & Warren P.V. (2001) Gene expression microarrays and the integration of biological knowledge. *Trends Biotechnol.* **19**, 412–15.

Nordstrom T. *et al.* (2000b) Direct analysis of single-nucleotide polymorphism on double-stranded DNA by pyrosequencing. *Biotechnol. Applied Biochem.* **31**, 107–12.

Nordstrom T., Nourizad K., Ronaghi M. & Nyren P. (2000a) Methods enabling pyrosequencing on double-stranded DNA. *Anal. Biochem.* **282**, 186–93.

Norgren R.B. Jr & Lehman M.N. (1998) Herpes simplex virus as a transneuronal tracer. *Neurosci. Biobehav. Rev.* **22**, 695–708.

Norin M. & Sundstrom M. (2002) Structural proteomics: developments in structure-to-function predictions. *Trends Biotechnol.* **20**, 79–84.

Norrander J., Kempe T. & Messing J. (1983) Construction of improved M13 vectors using oligodeoxynucleotide-directed mutagenesis. *Gene* **27**, 101–6.

Norton P.A. & Coffin J.M. (1985) Bacterial β-galactosidase as a marker of Rous sarcoma virus gene expression and replication. *Mol. Cell Biol.* **5**, 281–90.

Noskov V.N., Kouprina N., Leem S-H., *et al.* (2003) A general cloning system to selectively isolate any eukaryotic or prokaryotic genomic region in yeast. *BMC Genomics* **4**, 16–27.

Nugent M.E., Primrose S.B. & Tacon W.C.A. (1983) The stability of recombinant DNA. *Devel. Ind. Microbiol.* **24**, 271–85.

Nunberg J.H., Wright D.K., Cole G.E. *et al.* (1989) Identification of the thymidine kinase gene of feline herpesvirus: use of degenerate oligonucleotides in the polymerase chain reaction to isolate herpesvirus gene homologs. *J. Virol.* **63**, 3240–9.

Nur I., Szyf M., Razin A. *et al.* (1985) Procaryotic and eucaryotic traits of DNA methylation in spiroplasmas (mycoplasmas). *J. Bacteriol.* **164**, 19–24.

Nussbaum R.L., McInnes R.R. & Willard H.F. (2001) *Genetics in Medicine*, 6th edn. W.B. Saunders, Philadelphia.

O'Brien J. & Lummis S.C.R. (2004) Biolistic and diolistic transfection: using the gene gun to deliver DNA and lipophilic dyes into mammalian cells. *Methods* **33**, 121–5.

O'Farrell P.H. (1975) High-resolution two-dimensional electrophoresis of proteins. *J. Biol. Chem.* **250**, 4007–21.

O'Gorman S., Dagenais N.A., Qian M. & Marchuk Y. (1997) Protamine-Cre recombinase transgenes efficiently recombine target sequences in the male germ line of mice, but not in embryonic stem cells. *Proc. Nat. Acad. Sci. USA* **94**, 14602–7.

O'Gorman S., Fox D.T. & Wahl G.M. (1991) Recombinase mediated gene activation and site-specific integration in mammalian cells. *Science* **251**, 1351–5.

O'Hare K. & Rubin G.M. (1983) Structures of P transposable elements and their sites of insertion and excision in the *Drosophila melanogaster* genome. *Cell* **34**, 25–35.

O'Hare K., Benoist C. & Breathnach R. (1981) Transformation of mouse fibroblasts to methotrexate resistance. *Proc. Nat. Acad. Sci. USA* **78**, 1527–31.

O'Kane C.J. & Gehring W.J. (1987) Detection *in situ* of genetic regulatory elements in Drosophila. *Proc. Nat. Acad. Sci. USA* **84**, 9123–7.

O'Kane C.J. & Moffat K.G. (1992) Selective cell ablation and genetic surgery. *Curr. Opin. Genet. Devel.* **2**, 602–7.

O'Maille P.E., Bakhtina M. & Tsai M.D. (2002) Structure-based combinatorial protein engineering (SCOPE). *J. Mol. Biol.* **321**, 677–91.

O'Neill C., Horvath G.V., Horvath E., Dix P.J. & Medgyesy P. (1993) Chloroplast transformation in plants: polyethylene glycol (PEG) treatment of protoplasts is an alternative to biolistic delivery systems. *Plant J.* **3**, 729–38.

O'Reilley D.R., Miller L.K. & Luckow V.A. (1992) *Baculovirus Expression Vectors: A Laboratory Manual.* W.H. Freeman, San Francisco.

O'Shaughnessy A., Gnoj L., Scobie K. *et al.* (2004) A resource for large-scale RNA-interference-based screens in mammals. *Nature* **428**, 427–31.

O'Sullivan D.J., Walker S.A., West S.G. & Klaenhammer T.R. (1996) Development of an expression strategy using a lytic phage to trigger explosive plasmid amplification and gene expression. *Biotechnology* **14**, 82–7.

Oard J.H., Linscombe S.D., Braverman M.P. *et al.* (1996) Development, field evaluation, and agronomic performance of transgenic herbicide resistant rice. *Mol. Breeding* **2**, 359–68.

Ockert D., Schmitz M., Hampl M. & Rieber E.P. (1999) Advances in cancer immunotherapy. *Immunol. Today* **20**, 63–5.

Odell J.T., Hoopes J.L. & Vermerris W. (1994) Seed-specific gene activation mediated by the Cre/lox site specific recombination system. *Plant Physiol.* **106**, 447–58.

Offield M.F., Hirsch N. & Grainger R.M. (2000) The development of *Xenopus tropicalis* transgenic lines and their use in studying lens development timing in living embryos. *Development* **127**, 1789–97.

Ofringa R., de Groot M.J., Haagsman H.J., Does M.P., van den Elzen P.J. & Hooykaas P.J. (1990) Extrachromosomal homologous recombination and gene targeting in plant cells after *Agrobacterium*-mediated transformation. *EMBO J.* **9**, 3077–84.

Ogawa T., Hori T. & Ishida I. (1996) Virus-induced cell death in plants expressing the mammalian 2′; 5′ oligoadenylate system. *Nature Biotechnol.* **14**, 1566–9.

Ogita S., Uefuji H., Yamaguchi Y., Koizumi N. & Sano H. (2003) Producing decaffeinated coffee plants. *Nature* **423**, 823.

Ogonah O.W., Freedman R.B., Jenkins N., Patel K. & Rooney B.C. (1996) Isolation and characterization of an insect cell line able to perform complex N-linked glycosylation on recombinant proteins. *Nature Biotechnol.* **14**, 197–202.

Oh S.H. & Chater K.F. (1997) Denaturation of circular or linear DNA facilitates targeted integrative transformation of *Streptomyces coelicolor* A3 (2): possible relevance to other organisms. *J. Bacteriol.* **179**, 122–7.

Ohyama K. *et al.* (1986) Chloroplast gene organization deduced from complete squence of liverwort *Marchantia polymorpha* chloroplast DNA. *Nature* **322**, 572–4.

Okamato K. & Beach D. (1994) Cyclin G is a transcription target of the p53 tumor suppressor protein. *EMBO J.* **13**, 4816–22.

Okamoto T., Suzuki T. & Yamamoto N. (2000) Microarray fabrication with covalent attachment of DNA using bubble jet technology. *Nature Biotechnol.* **18**, 438–41.

Okayama H. & Berg P. (1982) High-efficiency cloning of full-length cDNA. *Mol. Cell. Biol.* **2**, 161–70.

Okubo K., Hori N., Matoba R. *et al.* (1992) Large-scale cDNA sequencing for analysis of quantitative and qualitative aspects of gene expression. *Nature Genet.* **2**, 173–9.

Okubo K. & Matsubara K. (1997) Complementary DNA sequence (EST) collections and the expression information of the human genome. *FEBS Lett.* **403**, 225–9.

Old J.M., Ward R.H.T., Petrov M. *et al.* (1982) First trimester diagnosis for haemoglobinopathies: a report of 3 cases. *Lancet* **ii**, 1413–16.

Old R., Murray K. & Roizes G. (1975) Recognition sequence of restriction endonuclease III from *Haemophilus influenzae. J. Mol. Biol.* **92**, 331–9.

Oliver S.G. (1996) From DNA sequence to biological function. *Nature* **379**, 597–600.

Olivera B.M., Hall Z.W. & Lehman I.R. (1968) Enzymatic joining of polynucleotides. V. A DNA adenylate intermediate in the polynucleotide joining reaction. *Proc. Nat. Acad. Sci. USA* **61**, 237–44.

Olson M.V. *et al.* (1986) A random-clone strategy for restriction mapping in yeast. *Proc. Nat. Acad. Sci. USA* **83**, 7826–30.

Olszewska E. & Jones K. (1988) Vacuum blotting enhances nucleic acid transfer. *Trends Genet.* **4**, 92–4.

Omirulleh S., Abraham M., Golovkin M. *et al.* (1993) Activity of a chimeric promoter with doubled CaMV 35S enhancer element in protoplast-derived cells and transgenic plants in maize. *Plant Mol. Biol.* **21**, 415–28.

Ooms G., Hooykaas P.J.J., Moolenaar G. & Schilperoort R.A. (1981) Crown gall plant tumours of abnormal morphology induced by *Agrobacterim tumefaciens* carrying mutated octopine Ti plasmids: analysis of T-DNA functions. *Gene* **14**, 33–50.

Opiteck G.J., Ramirez S.M., Jorgenson J.W. & Moseley M.A. (1998) Comprehensive two-dimensional high-performance liquid chromatography for the isolation of overexpressed proteins and proteome mapping. *Anal. Biochem.* **258**, 349–61.

Orengo C., Michie A., Jones S., Jones D., Swindells M. & Thornton J. (1997) CATH – A hierarchic classification of protein domain structures. *Structure* **5**, 1093–108.

Orengo C.A., Jones D.T. & Thornton J.M. (1994) Protein superfamilies and domain superfolds. *Nature* **372**, 631–4.

Orkin S.H., Daddona P.E., Shewach D.S. *et al.* (1983) Molecular cloning of human adenosine deaminase gene sequences. *J. Biol. Chem.* **258**, 2753–6.

Orrantia E. & Chang P.L. (1990) Intracellular distribution of DNA internalized through calcium phosphate precipitation. *Exp. Cell Res.* **190**, 170–4.

Orr-Weaver T.L., Szostak J.W. & Rothstein R.L. (1981) Yeast transformation: a model system for the study of recombination. *Proc. Nat. Acad. Sci. USA* **78**, 6354–8.

Osbourne B.I. & Baker B. (1995) Movers and shakers: maize transposons as tools for analyzing other plant genomes. *Curr. Opin. Cell Biol.* **7**, 406–13.

Osbourne B.I., Wirtz U. & Baker B. (1995) A system for insertional mutagenesis and chromosomal rearrangement using the *Ds* transposon and Cre-*lox. Plant J.* **7**, 687–701.

Osoegawa K. *et al.* (2000) Bacterial artificial chromosome libraries for mouse sequencing and functional analysis. *Genome Res.* **10**, 116–28.

Osoegawa K. *et al.* (2001) A bacterial artificial chromosome library for sequencing the complete human genome. *Genome Res.* **11**, 483–96.

Ostermeier M., Shim J.H. & Benkovic S.J. (1999) A combinatorial approach to hybrid enzymes independent of DNA homology. *Nature Biotechnol.* **17**, 1205–9.

Ostrowski M.C., Richard-Foy H., Wolford R.G., Berard D.S. & Hager G.L. (1983) Glucocorticoid regulation of transcription at an amplified episomal promoter. *Mol. Cell. Biol.* **3**, 2045–57.

Osusky M., Zhou G., Osuska L., Hancock R.E., Kay W.W. & Misra S. (2000) Transgenic plants expressing cationic peptide chimeras exhibit broad-spectrum resistance to phytopathogens. *Nature Biotechnol.* **18**, 1162–6.

Otten L., DeGreve H., Hernalsteens J.P., Van Montagu M., Schieder O. *et al.* (1981) Mendelian transmission of genes introduced into plants by the Ti plasmids of *Agrobacterium tumefaciens. Mol. Gen. Genet.* **183**, 209–13.

Ouimet M-C. & Marczynski G.T. (2000) Transcription reporters that shuttle cloned DNA between high-copy *Escherichia coli* plasmids and low-copy broad-host-range plasmids. *Plasmid* **44**, 152–62.

Overbeek R., Fonstein M., D'Souza M., Pusch G.D. & Maltsev N. (1999) The use of gene clusters to infer functional coupling. *Proc. Nat. Acad. Sci., USA* **96**, 2896–901.

Ow D.W. (1996) Recombinase-directed chromosome engineering in plants. *Curr. Opin. Biotechnol.* **7**, 181–6.

Ow D.W., Wood K.V., DeLuca M. *et al.* (1986) Transient and stable expression of the firefly luciferase gene in plant cells and transgenic plants. *Science* **234**, 856–9.

Paddison P.J., Silva J.M., Conklin D.S., Schlabach M., Li M., Aruleba S., Balija V.,

Padgett H.S., Epel B.L., Kahn T.W. *et al.* (1996) Distribution of tombavirus movement in infected cells and implications for cell-to-cell spread of infection. *Plant J.* **10**, 1079–88.

Padgette S.R., Kolacz K.H., Delannay X. *et al.* (1996) Development, identification, and characterization of a glyphosate-tolerant soybean line. *Crop Sci.* **35**, 1451–61.

Padidam M., Gore M., Lu D.L. & Smirnova O. (2003) Chemical-inducible, ecdysone receptor-based gene expression system for plants. *Transgenic Res.* **12**, in press.

Pagano, J.S. & Vaheri, A. (1965) Enhancement of infectivity of poliovirus RNA with diethylaminoethyl-dextran (DEAE-D). *Arch. Gesamte Virusforsch.* **17**, 456–64.

Paglia P., Terrazzini N., Schulze K., Guzman C.A. & Colombo M.P. (2000) *In vivo* correction of genetic defects of monocyte/macrophages using attenuated *Salmonella* as oral vectors for targeted gene delivery. *Gene Therapy* **7**, 1725–30.

Pain B., Chenevier P. & Samarut J. (1999) Chicken embryonic stem cells and transgenic strategies. *Cells Tissues Organs* **165**, 212–19.

Palatnik J.F., Allen E., Wu X., Schommer C., Schwab R., Carrington J.C. & Weigel D. (2003) Control of leaf morphogenesis by microRNAs. *Nature* **425**, 257–63.

Palauqui J.C., Elmayan T., Pollien J.M. & Vaucheret H. (1997) Systemic acquired silencing: transgene-specific post-transcriptional silencing is transmitted by grafting from silenced stocks to non-silenced scions. *EMBO J.* **16**, 4738–45.

Pal-Bhadra M., Bhadra U. & Birchler J.A. (1997) Cosuppression in *Drosophila*: gene silencing of alcohol dehydrogenase by *white-Adh* transgenes is polycomb-dependent. *Cell* **90**, 479–90.

Palmer J.D. *et al.* (2000) Dynamic evolution of plant mitochondrial genomes: mobile genes and introns and highly variable mutation rates. *Proc Nat. Acad. Sci. USA* **97**, 6960–6.

Palmer K.E. & Rybicki E.P. (1997) The use of geminiviruses in biotechnology and plant molecular biology, with particular focus on Mastreviruses. *Plant Sci.* **129**, 115–30.

Palmiter R.D. & Brinster R.L. (1986) Germ-line transformation of mice. *Ann. Rev. Genet.* **20**, 465–99.

Palmiter R.D., Brinster R.L., Hammer R.E. *et al.* (1982a) Dramatic growth of mice that develop from eggs micro-injected with metallothionein-growth hormone fusion genes. *Nature* **300**, 611–15.

Palmiter R.D., Chen H.Y. & Brinster R.L. (1982b) Differential regulation of metallothionein-thymidine kinase fusion genes in transgenic mice and their offspring. *Cell* **29**, 701–10.

Palzkill T.G. & Newlon C.S. (1988) A yeast replication origin consists of multiple copies of a small conserved sequence. *Cell* **53**, 441–50.

Pansegrau W., Lanka E., Barth P. *et al.* (1994) Complete nucleotide sequence of Birmingham IncP alpha plasmids: compilation and comparative analysis. *J. Mol. Biol.* **239**, 623–63.

Pappin D.J.C., Horjup P. & Bleasby A.J. (1993) Rapid identification of proteins by peptide mass fingerprinting. *Current Biology* **3**, 327–32.

Paran I. & Zamir D. (2003) Quantitative traits in plants: beyond the QTL. *Trends Genet.* **19**, 303–6.

Parekh R.N., Shaw M.R. & Wittrup K.D. (1996) An integrating vector for tunable, high copy, stable integration into the dispersed Ty sites of *Saccharomyces cerevisiae*. *Biotechnol. Progress* **12**, 16–21.

Parimoo S., Patanjali S.R., Kolluri R., Xu H., Wei H. & Weissman S.M. (1995) cDNA selection and other approaches in positional cloning. *Anal. Biochem.* **228**, 1–17.

Parimoo S., Patanjali S.R., Shukle H., Chaplin D.D. & Weisman S.M. (1991) cDNA selection: efficient PCR approach for the selection of cDNAs encoded in large chromosomal DNA fragments. *Proc. Nat. Acad. Sci. USA* **88**, 9623–7.

Parinov S., Sevugan M., Ye D. *et al.* (1999) Analysis of flanking sequences from *dissociation* insertion lines: a database for reverse genetics in *Arabidopsis*. *Plant Cell* **11**, 2263–70.

Park J. *et al.* (1998) Sequence comparisons using multiple sequences detect three times as many remote homologues as pairwise methods. *J. Mol. Biol.* **284**, 1201–10.

Park S.H., Pinson S.R.M. & Smith R.H. (1996) T-DNA integration into genomic DNA of rice following *Agrobacterium* inoculation of isolated shoot apices. *Plant Mol. Biol.* **32**, 1135–48.

Parke D., Ornston L.N. & Nester E.W. (1987) Chemotaxis to plant phenolic inducers of virulence genes is constitutively expressed in the absence of the Ti plasmid in *Agrobacterium tumefaciens*. *J. Bacteriol.* **169**, 5336–8.

Parkhill J. *et al.* (2001a) Complete genome sequence of a multiple drug resistant *Salmonella enterica* serovar Typhi CT18. *Nature* **413**, 848–52.

Parkhill J. *et al.* (2001b) Genome sequence of *Yersinia pestis*, the causative agent of plague. *Nature* **413**, 523–7.

Parmley S.E. & Smith P.G. (1988) Antibody-selectable filamentous fd phage vectors: affinity purification of target genes. *Gene* **73**, 305–18.

Parolin C., Dorfman T., Palu G., Gottlinger H. & Sodroski J. (1994) Analysis in human immunodeficiency virus type 1 vectors of *cis*-acting sequences that affect gene transfer into human lymphocytes. *J. Virol.* **68**, 3888–95.

Parra I. & Windle B. (1993) High resolution visual mapping of stretched DNA by fluorescent hybridization. *Nature Genet.* **5**, 17–21.

Parrish J.Z. & Xue D. (2003) Functional genomic analysis of apoptotic DNA degradation in *C. elegans*. *Mol. Cell* **11**, 987–96.

Parrish S., Fleenor J., Xu S., Mello C. & Fire A. (2000) Functional anatomy of a dsRNA trigger: differential requirement for the two trigger strands in RNA interference. *Mol. Cell* **6**, 1077–87.

Parsell D.A. & Lindquist S. (1983) The function of heat-shock protein in stress tolerance – degradation and reactivation of damaged proteins. *Ann. Rev. Genet.* **27**, 437–96.

Paszkowski J., Baur M., Bogucki A. & Potrykus I. (1988) Gene targeting in plants. *EMBO J.* **7**, 4021–6.

Patankar S., Munasinghe A., Shoaibi A., Cummings L.M. & Wirth D.F. (2001) Serial analysis of gene expression in *Plasmodium falciparum* reveals the global expression profile of erythrocytic stages and the presence of anti-sense transcripts in the malarial parasite. *Mol. Biol. Cell.* **12**, 3114–25.

Paterson A.H., Lander E.S., Hewitt J.D., Peterson S., Lincoln S.E. & Tanksley S.D. (2000) Resolution of quantitative traits into Mendelian factors by using a complete linkage map of restriction fragment length polymorphisms. *Nature* **335**, 721–6.

Patton J.S., Gomes X.V. & Geyer P.K. (1992) Position-independent germline transformation in *Drosophila* using a cuticle pigmentation gene as a selectable marker. *Nucl. Acids Res.* **20**, 5859–60.

Patton W. (2000) A thousand points of light; the application of fluorescence detection technologies to two-dimensional gel electrophoresis and proteomics. *Electrophoresis* **21**, 1123–44.

Patton W.F. & Beecham J.M. (2001) Rainbow's end: the quest for multiplexed fluorescence quantitative analysis in proteomics. *Curr. Opin. Chem. Biol.* **6**, 63–9.

Patzer E.J., Nakamura G.R., Hershberg R.D. *et al.* (1986) Cell culture derived recombinant HBsAg is highly immunogenic and protects chimpanzees from infection with hepatitis B virus. *Biotechnology* **4**, 630–6.

Paul M., Pellny T. & Goddijn O. (2001) Enhancing photosynthesis with sugar signals. *Trends Plant Sci.* **6**, 197–200.

Pauleau A.L. & Murray P.J. (2003) Role of NOD2 in the response of macrophages to toll-like receptor agonists. *Mol. Cell Biol.* **23**, 7531–9.

Pavlov A.R., Pavlova N.V., Kozyavkin S.A. & Slesarev A.I. (2004) Recent developments in the optimization of thermostable DNA polymerases for efficient applications. *Trends Biotechnol.* **22**, 253–60.

Paweletez C.P. *et al.* (2001) Reverse phase protein microarrays which capture disease progression show activation of pro-survival pathways at the cancer invasion front. *Oncogene* **20**, 1981–9.

Pearson W.R. & Lipman D.J. (1988) Improved tools for biological sequence comparison. *Proc. Nat. Acad. Sci. USA* **85**, 2444–8.

Pease A.C., Solas D., Sullivan E.J., Cronin M.T., Holmes C.P. & Fodor S.P.A. (1994) Light-generated oligonucleotide arrays for rapid DNA sequence analysis. *Proc. Nat. Acad. Sci. USA* **91**, 5022–6.

Peden K.W.C. (1983) Revised sequence of the tetracycline-resistance gene of pBR322. *Gene* **22**, 277–80.

Peeters B.P.H., De Boer J.H., Bron S. & Venema G. (1988) Structural plasmid instability in *Bacillus subtilis*: effect of direct and inverted repeats. *Mol. Gen. Gent.* **212**, 450–8.

Peferoen M. (1997) Progress and prospects for field use of *Bt* genes in crops. *Trends Biotechnol.* **15**, 173–7.

Peijnenburg A.A.C.M., Bron S. & Venema G. (1987) Structural plasmid instability in recombination- and repair-deficient strains of *Bacillus subtilis*. *Plasmid* **17**, 167–70.

Pelletier J. & Sidhu S. (2001) Mapping protein–protein interactions with combinatorial biology methods. *Curr. Opin. Biotechnol.* **12**, 340–7.

Peluso, P. *et al.* (2003) Optimizing antibody immobilization strategies for the construction of protein microarrays. *Anal. Biochem.* **312**, 113–24.

Peng J.R., Richards D.E., Hartley N.M., Murphy G.P., Devos K.M., Flintham J.E., Beales J., Fish L.J., Worland A.J., Pelica F., Sudhakar D., Christou P., Snape J.W., Gale M.D. & Harberd N.P. (1999) "Green revolution" genes encode mutant gibberellin response modulators. *Nature* **400**, 256–61.

Pennock G.D., Shoemaker C. & Miller L.K. (1984) Strong and regulated expression of *E. coli* β-galactosidase in insect cells using a baculovirus vector. *Mol. Cell. Biol.* **4**, 399–406.

Peralta E.G., Hellmiss R. & Ream W. (1986) *Overdrive*, a T-DNA transmission enhancer on the *A. tumefaciens* tumour-inducing plasmid. *EMBO J.* **5**, 1137–42.

Perlak F.J., Deaton R.W., Armstrong T.A. *et al.* (1990) Insect resistant cotton plants. *Biotechnology* **8**, 939–43.

Perlak F.J., Fuchs R.L., Dean D.A., McPherson S.L. & Fischhoff D.A. (1991) Modification of the coding sequence enhances plant expression of insect control protein genes. *Proc. Nat. Acad. Sci. USA* **88**, 3324–8.

Perlak F.J., Stone T.B., Muskopf Y.M. *et al.* (1993) Genetically improved potatoes: protection from damage by Colorado potato beetles. *Plant Mol. Biol.* **22**, 313–21.

Perna N.T. *et al.* (2001) Genome sequence of enterohaemorrhagic *Escherichia coli* O157:H7. *Nature* **409**, 529–33.

Perou C.M. *et al.* (1999) Distinctive gene expression patterns in human mammary epithelial cells and breast cancers. *Proc. Nat. Acad. Sci. USA* **96**, 9212–17.

Perou C.M. *et al.* (2000) Molecular portraits of human breast tumours. *Nature* **406**, 747–52.

Perry A.C.F., Wakayama T., Kishikawa H. *et al.* (1999) Mammalian transgenesis by intracytoplasmic sperm injection. *Science* **284**, 1180–3.

Perucho M., Hanahan D., Lipsich L. & Wigler M. (1980a) Isolation of the chicken thymidine kinase gene by plasmid rescue. *Nature* **285**, 207–10.

Perucho M., Hanahan D. & Wigler M. (1980b) Genetic and physical linkage of exogenous sequences in transformed cells. *Cell* **22**, 309–17.

Peschel A., Ottenwalder B. & Gotz F. (1996) Inducible production and cellular location of the epidermin biosynthetic enzyme EpiB and improved staphylococcal expression system. *FEMS Microbiol. Lett.* **137**, 279–84.

Peters D.G. *et al.* (1999) Comprehensive transcript analysis in small quantities of mRNA by SAGE-lite. *Nucl. Acids Res.* **27**, e39.

Petrusyte M., Bitinaite J., Menkevicius S. *et al.* (1988) Restriction endonucleases of a new type. *Gene* **74**, 89–91.

Petters R.M. & Sommer J.R. (2000) Transgenic animals and models of human disease. *Transgenic Res.* **9**, 347–51.

Pfeifer A. (2004) Lentiviral transgenesis. *Transgenic Res.* (in press).

Pfeifer A., Ikawa M., Dayn Y. & Verma I.M. (2002) Transgenesis by lentiviral vectors: lack of gene silencing in mammalian embryonic stem cells and preimplantation embryos. *Proc. Nat. Acad. Sci. USA* **99**, 2140–5.

Pham C.T.N., McIvor D.M., Hug B.A., Heusel J.W. & Ley T.J. (1996) Long range disruption of gene expression by a selectable marker cassette. *Proc. Nat. Acad. Sci. USA* **93**, 13090–5.

Phelps C.J., Koike C., Vaught T.D. *et al.* (2003) Production of alpha 1,3-galactosyltransferase-deficient pigs. *Science* **299**, 411–14.

Phillips M.S., Lawrence R., Sachidanandam R., *et al.* (2003) Chromosome-wide distribution of haplotype blocks and the role of recombination hot spots. *Nature Genet.* **33**, 382–7.

Phizicky E., Bastiaens P.I.H., Zhu H., Snyder M. & Fields S. (2003) Protein analysis on a proteomic scale. *Nature* **422**, 208–15.

Phizicky E.M. & Fields S. (1995) Protein–protein interactions: methods for detection and analysis. *Microbiol. Reviews* **59**, 94–123.

Picard D. (1994) Regulation of protein function through expression of chimaeric proteins. *Curr. Opin. Biotechnol.* **5**, 511–15.

Pichel J.G., Lakso M. & Westphal H. (1993) Timing of SV40 oncogene activation by site-specific recombination determines subsequent tumour progression during murine lens development. *Oncogene* **8**, 3333–42.

Pierce J.C., Sauer B. & Sternberg N. (1992) A positive selection vector for cloning high molecular weight DNA by the bacteriophage P1 system: improved cloning efficiency. *Proc. Nat. Acad. Sci. USA* **89**, 2056–60.

Pietu G. *et al.* (1996) Novel gene transcripts preferentially expressed in human muscles revealed by quantitative hybridization of a high-density cDNA array. *Genome Res.* **6**, 492–503.

Pigac J. & Schrempf H. (1995) A simple and rapid method of transformation of *Streptomyces rimosus* R6 and other streptomycetes by electroporation. *Appl. Environ. Microbiol.* **61**, 352–6.

Pikaart M.J., Feng J. & Villepointeau B. (1998) The polyomavirus enhancer activates chromatin accessibility on integration into the *HPRT* gene. *Mol. Cell. Biol.* **12**, 5785–92.

Pinkel D., Straume T. & Gray J.W. (1986) Cytogenetic analysis using quantitative, high-sensitivity, fluorescence hybridization. *Proc. Nat. Acad. Sci. USA* **83**, 2934–8.

Pinto Y.M., Kok R.A. & Baulcombe D.C. (1999) Resistance to rice yellow mottle virus (RYMV) in cultivated African rice varieties containing RYMV transgene. *Nature Biotechnol.* **17**, 702–7.

Piper M.B., Bankier A.T. & Dear P.H. (1998) A HAPPY map of *Cryptosporidium parvum*. *Genome Res.* **8**, 1299–307.

Pirrotta V., Hadfield C. & Pretorius G.H.J. (1983) Microdissection and cloning of the *white* locus and the 3B1–3C2 region of the *Drosophila* X chromosome. *EMBO J.* **2**, 927–34.

Pittius C.W., Hennighausen L., Lee E. *et al.* (1988) A milk protein gene promoter directs expression of human tissue plasminogen activator cDNA to the mammary gland in transgenic mice. *Proc. Nat. Acad. Sci. USA* **85**, 5874–8.

Plasterk R.H.A. & Ketting R.F. (2000) The silence of the genes. *Curr. Opin. Genet. Devel.* **10**, 562–7.

Pleissner K-P., Oswald H. & Wegner S. (2001) Image analysis of two-dimensional gels. In: *Proteomics: From Protein Sequence to Function*, pp. 131–50. BIOS Scientific Publishers, Oxford.

Pogue G.P. Lindbo J.A., Garger S.J. & Fitzmaurice W.P. (2002) Making an ally from an enemy: plant virology and the new agriculture. *Ann. Rev. Phytopathol.* **40**, 45–74.

Polejaeva I.A., Chen S.-H., Vaught T.D. *et al.* (2000) Cloned pigs produced by nuclear transfer from adult somatic cells. *Nature* **407**, 86–90.

Pollack D.J. (2001) *Ureaplasma urealyticum*: an opportunity for combinatorial genomics. *Trends Microbiol.* **9**, 169–75.

Pollock D. & Gaken J. (1995) Antisense oligonucleotides: a survey of recent literature, possible mechanisms of action and therapeutic progress. In *Functional Analysis of the Human Genome*, eds. Farzaneh F. & Cooper D.N., pp. 241–65. BIOS Scientific Publishers, Oxford.

Polo J.M., Belli B.A., Driver D.A. *et al.* (1999) Stable alphavirus packaging cell lines of Sindbis virus and Semliki Forest virus-derived vectors. *Proc. Nat. Acad. Sci. USA* **96**, 4598–603.

Polyak K., Xia Y., Zweier J.L., Kinzler K.W. & Vogelstein B. (1997) A model for p53 induced apoptosis. *Nature* **389**, 300–4.

Ponting C.P. *et al.* (2000) Evolution of domain families. *Advances in Protein Chemistry* **54**, 185–244.

Poquet I., Ehrlich S.D. & Gruss A. (1998) An export-specific reporter designed for Gram-positive bacteria: application to *Lactococcus lactis*. *J. Bacteriol.* **180**, 1904–12.

Porceddu A., Falorni A., Ferradini N. *et al.* (1999) Transgenic plants expressing human glutamic acid decarboxylase (GAD65), a major autoantigen in insulin-dependent diabetes mellitus. *Mol. Breeding* **5**, 553–60.

Porta C. & Lomonossoff G.P. (2001) Viruses as vectors for the expression of foreign sequences in plants. *Biotechnol. Genet. Eng. Rev.* **19** (in press).

Porta C., Spall V.E., Loveland J. *et al.* (1994) Development of cowpea mosaic virus as a high yielding system for the presentation of foreign peptides. *Virology* **202**, 949–55.

Porteus M.H., Cathomen T., Weitzman M.D. & Baltimore D. (2003) Efficient gene targeting mediated by adeno-associated virus and DNA double-strand breaks. *Mol. Cell Biol.* **23**, 3558–65.

Postlethwait J.H. *et al.* (1994) A genetic linkage map for the zebrafish. *Science* **264**, 699–704.

Potrykus I. (2001) Golden rice and beyond. *Plant Physiol.* **125**, 1157–61.

Potrykus I., Paszkowski J., Saul M.W., Petruska J. & Shillito R.D. (1985a) Molecular and general genetics of a hybrid foreign gene introduced into tobacco by direct gene transfer. *Mol. Gen. Genet.* **199**, 169–77.

Potrykus I., Saul M.W., Petruska J., Paszkowski J. & Shillito R. (1985b) Direct gene transfer to cells of a graminaceous monocot. *Mol. Gen. Genet.* **199**, 183–8.

Potter H. (1988) Electroporation in biology: methods, applications and instrumentation. *Anal. Biochem.* **174**, 361–73.

Potter H., Weir L. & Leder P. (1984) Enhancer-dependent expression of human K immunoglobulin genes introduced into mouse pre-B lymphocytes by electroporation. *Proc. Nat. Acad. Sci. USA* **81**, 7161–5.

Poulter R.T., Goodwin T.J. & Butler M.I. (2003) Vertebrate helentrons and other novel helitrons. *Gene* **313**, 201–12.

Powell-Abel P., Nelson R.S., De B. *et al.* (1986) Delay of disease development in transgenic plants that express the tobacco mosaic virus coat protein gene. *Science* **232**, 738–43.

Poyart C. & Trieu-Cuot P. (1997) A broad-host-range mobilizable shuttle vector for the construction of transcriptional fusions to beta-galactosidase in Gram-positive bacteria. *FEMS Microbiol. Lett.* **156**, 193–8.

Poznansky M., Lever A., Bergeron L., Haseltine W. & Sodroski J. (1991) Gene transfer into human lymphocytes by a defective human immunodeficiency virus type 1 vector. *J. Virol.* **65**, 532–6.

Prelle K., Vassiliev I.M., Vassilieva S.G., Wolf E. & Wobus A.M. (1999) Establishment of pluripotent cell lines from vertebrate species – present status and future prospects. *Cells Tissues Organs* **165**, 220–36.

Pridmore R.D. (1987) New and versatile cloning vectors with kanamycin resistance marker gene. *Gene* **56**, 309–12.

Primig M. *et al.* (2000) The core meiotic transcriptome in budding yeasts. *Nature Genet.* **26**, 415–23.

Primrose S.B. & Ehrlich S.D. (1981) Isolation of plasmid deletion mutants and a study of their instability. *Plasmid* **6**, 193–201.

Primrose S.B., Derbyshire P., Jones I.M., Nugent M.E. & Tacon W.C.A. (1983) Hereditary instability of recombinant DNA molecules. In *Bioactive Microbial Products 2: Development and Production*, eds. Nisbet L.J. & Winstanley D.J., pp. 63–77. Academic Press, London.

Pritchard J.K. & Cox N.J. (2002) The allelic architechture of human disease genes: common disease-common variant . . . or not? *Hum. Mol. Genet.* **11**, 2417–23.

Probst F.J., Fridell R.A., Raphael Y. *et al.* (1998) Correction of deafness in Shaker-2 mice by an unconventional myosin in a BAC transgene. *Science* **280**, 1444–7.

Projan S.J. & Archer G.L. (1989) Mobilization of the relaxable *Staphylococcus aureus* plasmid pC221 by the conjugative plasmid pGO1 involves three pC221 loci. *J. Bacteriol.* **171**, 1841–5.

Pruchnic R., Cao B., Peterson Z.Q. *et al.* (2000) The use of adeno-associated virus to circumvent the maturation-dependent viral transduction of myofibres. *Hum. Gene Ther.* **11**, 521–36.

Prunkard D., Cottingham I., Garner I. *et al.* (1996) High-level expression of recombinant human fibrinogen in the milk of transgenic mice. *Nature Biotechnol.* **14**, 867–71.

Ptashne M. (1967a) Isolation of the λ phage repressor. *Proc. Nat. Acad. Sci. USA* **57**, 306–13.

Ptashne M. (1967b) Specific binding of the λ phage repressor to lDNA. *Nature* **214**, 232–4.

Ptashne M. (1992) *A Genetic Switch*, 2nd edn. Blackwell Science, Oxford.

Punja Z.K. (2001) Genetic engineering of plants to enhance resistance to fungal pathogens – a review of progress and future prospects. *Canadian J. Plant Path.* **23**, 216–35.

Punt P.J., Van Biezen N., Conesa A., Albers A., Mangnus J. & Van Den Hondel C. (2002) Filamentous fungi as cell factories for heterologous protein production. *Trends Biotechnol.* **20**, 200–6.

Putney S.D., Herlihy W.C. & Schimmel P. (1983) A new troponin T and cDNA clones for 13 different muscle proteins, found by shotgun sequencing. *Nature* **302**, 718–21.

Puyet A., Sandoval M., López P. *et al.* (1987) A simple medium for rapid regeneration of *Bacillus subtilis* protoplasts transformed with plasmid DNA. *FEBS Microbiol. Lett.* **40**, 1–5.

Qian Z. & Wilusz J. (1993) Cloning cDNA encoding an RNA-binding protein by screening expression libraries using a northwestern strategy. *Ann. Biochem.* **212**, 547–54.

Qin M., Bayley C., Stockton T. & Ow D.W. (1994) Cre recombinase-mediated site-specific recombination between plant chromosomes. *Proc. Nat. Acad. Sci. USA* **91**, 1706–10.

Qing, C.M., Fan, L., Lei, Y. *et al.* (2000) Transformation of Pakchoi (*Brassica rapa* L. ssp. *chinensis*) by *Agrobacterium* infiltration. *Mol. Breed.* **6**, 67–72.

Quackenbush J. (2001) Computational analysis of microarray data. *Nature Rev Genet.* **2**, 418–27.

Quackenbush J. (2002) Microarray data normalization and transformation. *Nature Genet.* **32**, 496–501.

Quackenbush J. (2004) Data standards for 'omic science. *Nature Biotechnol.* **22**, 613–14.

Quon D. Wang Y., Catalano W.R. *et al.* (1991) Formation of β-amyloid protein deposits in the brains of transgenic mice. *Nature* **352**, 239–41.

Raamsdonk L.M., Teusink B., Broadhurst D. *et al.* (2001) A functional genomics strategy that uses metabolome data to reveal the phenotype of silent mutations. *Nature Biotechnol.* **19**, 45–50.

Rabilloud T. (2002) Two-dimensional gel electrophoresis in proteomics: old, old fashioned, but still it climbs up the mountains. *Proteomics* **2**, 3–10.

Rabinowitz J.E. & Samulski J. (1998) Adeno-associated virus expression systems for gene transfer. *Curr. Opin. Biotechnol.* **9**, 470–5.

Radloff R., Bauer W. & Vinograd J. (1967) A dye-buoyant-density method for the detection and isolation of closed circular duplex DNA: the closed circular DNA in HeLa cells. *Proc. Nat. Acad. Sci. USA* **57**, 1514–21.

Radnedge L., Agron P.G., Hill K.K. *et al.* (2003) Genome differences that distinguish *Bacillus anthracis* from *Bacillus cereus* and *Bacillus thuringiensis*. *App. Environ. Microbiol.* **69**, 2755–64.

Rain J.C. *et al.* (2001) The protein–protein interaction map of *Helicobacter pylori*. *Nature* **409**, 211–15.

Raineri D.M., Bottino P., Gordon M.P. & Nester E.W. (1990) *Agrobacterium*-mediated transformation of rice (*Oryza sativa* L.). *Biotechnology* **9**, 33–8.

Ramachandran S. & Sundaresan V. (2001) Transposons as tools for functional genomics. *Plant Physiol. Biochem.* **39**, 243–52.

Ramakrishna R. & Srinivasan R. (1999) Gene identification in bacterial and organellar genomes using GeneScan. *Computers and Chemistry* **23**, 165–74.

Ramanathan A., Huff E.J., Lamers C.C., *et al.* (2004) An integrative approach for the optical sequencing of single DNA molecules. *Anal. Biochem.* **330**, 227–41.

Ramirez-Solis R., Lui P. & Bradley A. (1995) Chromosome engineering in mice. *Nature* **378**, 720–4.

Rashid H., Yokoi S., Toriyama K. & Hinata K. (1996) Transgenic plant production mediated by *Agrobacterium* in Indica rice. *Plant Cell Rep.* **15**, 727–30.

Rasko D.A., Ravel J., Okstad O.A., *et al.* (2004) The genome of *Bacillus cereus* ATCC 10987 reveals metabolic adaptations and a large plasmid related to *Bacillus anthracis* pXO1. *Nucl. Acids Res.* **32**, 977–88.

Rat Genome Sequencing Project Consortium (2004) Genome sequence of the Brown Norway rat yields insights into mammalian evolution. *Nature* **428**, 493–521.

Rathus C., Bower R. & Birch R.G. (1993) Effects of promoter, intron and enhancer elements on transient gene expression in sugarcane and carrot protoplasts. *Plant Mol. Biol.* **23**, 616–18.

Ratzkin B. & Carbon J. (1977) Functional expression of cloned yeast DNA in *Escherichia coli*. *Proc. Nat. Acad. Sci. USA* **74**, 487–91.

Raychaudhuri S., Sutphin P.D., Chang J.T. & Altman R.B. (2001) Basic microarray analysis: grouping and feature reduction. *Trends Biotechnol.* **19**, 189–93.

Rayner J.O., Dryga S.A. & Kamrud K.I. (2002) Alphavirus vectors and vaccination. *Rev. Med. Virol.* **12**, 279–96.

Reese M. *et al.* (2000) Genome annotation assessment in *Drosophila melanogaster*. *Genome Res.* **10**, 483–501.

Reeves R.H., Baxter L.L. & Richtsmeier J.T. (2001) Too much of a good thing: mechanisms of gene action in Down syndrome. *Trends Genet.* **17**, 83–8.

Regierer B., Fernie A.R., Springer F., Perez-Melis A., Leisse1 A., Koehl K., Willmitzer L., Geigenberger P. & Kossmann J. (2002) Starch content and yield increase as a result of altering adenylate pools in transgenic plants. *Nature Biotechnol.* **12**, 1256–60.

Reich D.E. *et al.* (2001) Linkage disequilibrium in the human genome. *Nature* **411**, 199–204.

Reisman D., Yates J. & Sugden B. (1985) A putative origin of replication of plasmids derived from Epstein-Barr virus composed of two *cis*-acting components. *Mol. Cell. Biol.* **5**, 1822–32.

Reiss B., Schubert I., Kopchen K., Wendeler E., Schell J. & Puchta H. (2000) *RecA* stimulates sister chromatid exchange and the fidelity of double-strand break repair, but not gene targeting, in plants transformed by *Agrobacterium*. *Proc. Nat. Acad. Sci. USA* **97**, 3358–63.

Reiter R.S., Williams J.G.K., Feldmann K.A., Rafalsta A., Tingey S.V. & Scolnick P.A. (1992) Global and local genome mapping in *Arabidopsis thaliana* by using recombinant inbred lines and random amplified polymorphic DNAs. *Proc. Nat. Acad. Sci. USA* **89**, 1477–81.

Ren Z. & Black L.W. (1998) Phage T4 SOC and HOC display of biologically active, full length proteins on the viral capsid. *Gene* **215**, 439–44.

Renart J. & Sandoval I.V. (1984) Western blots. *Methods Enzymol.* **104**, 455–60.

Renault P., Corthier G., Goupil N., Delorme C. & Ehrlich S.D. (1996) Plasmid vectors for Gram-positive bacteria switching from high to low copy number. *Gene* **183**, 175–82.

Resnick J.L., Bixler L.S., Cheng L. & Donovan P.J. (1992) Long-term proliferation of mouse primordial germ cells in culture. *Nature* **359**, 550–1.

Reynolds P.N., Feng M. & Curiel D.T. (1999) Chimeric viral vectors – the best of both worlds. *Mol. Med. Today* **5**, 25–31.

Richard G.F., Hennequin C., Thierry A. & Dujon B. (1999) Trinucleotide repeats and other microsatellites in yeast. *Res. Microbiol.* **150**, 589–602.

Richardson J.H. & Marasco W.A. (1995) Intracellular antibodies: development and therapeutic potential. *Trends Biotechnol.* **13**, 306–10.

Richardson J.H., Hofmann W., Sodroski J.G. & Marasco W.A. (1998) Intrabody-mediated knockout of the high affinity IL-2 recdeptor in primary human T cells using a bicistronic lentivirus vector. *Gene Ther.* **5**, 635–44.

Richardson J.H., Sodroski J.G., Waldmann T.A. & Marasco W.A. (1995) Phenotypic knockout of the high-affinity human interleukin 2 receptor by intracellular single-chain antibodies against the alpha subunit of the receptor. *Proc. Nat. Acad. Sci. USA* **92**, 3137–41.

Richmond C.S., Glasner J.D., Mau R., Jin H. & Blattner F.R. (1999) Genome-wide expression in *Escherichia coli* K-12. *Nucl. Acid Res.* **27**, 3821–35.

Riek R., Pervushin K. & Wüthrich K. (2000) TROSY and CRINEPT: NMR with large molecular and supramolecular structures in solution. *Trends Biochem. Sci.* **25**, 462–8.

Riethman H.C. *et al.* (1989) Cloning human telomeric DNA fragments into *Saccharomyces cerevisiae* using a yeast-artificial-chromosome vector. *Proc. Nat. Acad. Sci. USA* **86**, 6240–4.

Riethman H.C. *et al.* (2001) Integration of telomere sequences with the draft human genome sequence. *Nature* **409**, 948–51.

Rigaut G., Shevchenko A., Rutz B., Wilm M., Mann M. & Seraphin B. (1999) A generic protein purification method for protein complex characterization and proteome exploration. *Nature Biotechnol.* **17**, 1030–2.

Rijkers T., Peetz A. & Ruther U. (1994) Insertional mutagenesis in transgenic mice. *Transgenic Res.* **3**, 203–15.

Rinas U., Tsai L.B., Lyons D. *et al.* (1992) Cysteine to serine substitutions in basic fibroblast growth factor: effect on inclusion body formation and proteolytic susceptibility during *in vitro* refolding. *Biotechnology* **10**, 435–40.

Rinchik E.M., Carpenter D.A. & Selby P.B. (1990) A strategy for fine-structure functional analysis of a 6- to 11-centimorgan region of mouse chromosome 7 by high-efficiency mutagenesis. *Proc. Nat. Acad. Sci. USA* **87**, 896–900.

Ringquist S., Shinedling S., Barrick D. *et al.* (1992) Translation initiation in *Escherichia coli*: sequences within the ribosome-binding site. *Mol. Microbiol.* **6**, 1219–29.

Rio D.C., Clark S.G. & Tjian R. (1985) A mammalian host-vector system that regulates expression and amplification of transfected genes by temperature induction. *Science* **227**, 23–8.

Rioux J.D. (2001) Genetic variation in the 5q31 cytokine gene cluster confers susceptibility to Crohn disease. *Nature Genet.* **29**, 223–8.

Rioux J.D. *et al.* (2000) Genomewide search in Canadian families with inflammatory bowel disease reveals two novel susceptibility loci. *Am. J. Hum. Genet.* **66**, 1863–70.

Risseeuw E., Franke-van Dijk M.E. & Hooykaas P.J. (1997) Gene targeting and instability of *Agrobacterium* T-DNA loci in the plant genome. *Plant J.* **11**, 717–28.

Risseeuw E., Offringa R., Franke-van Dijk M.E. & Hooykaas P.J. (1995) Targeted recombination in plants using *Agrobacterium* coincides with additional rearrangements at the target locus. *Plant J.* **7**, 109–19.

Rivera V.M. *et al.* (1996) A humanized system for pharmacologic control of gene expression. *Nature Med.* **2**, 1028–32.

Robbins P.D., Tahara H. & Ghivizzani S.C. (1998) Viral vectors for gene therapy. *Trends Biotechnol.* **16**, 35–40.

Roberts M.W. & Rabinowitz J.C. (1989) The effect of *Escherichia coli* ribosomal protein S1 on the translational specificity of bacterial ribosomes. *J. Biol. Chem.* **264**, 2228–35.

Robertson E., Bradley A., Kuehn M. & Evans M. (1986) Germ line transformation of genes introduced into cultured pluripotential cells by retroviral vector. *Nature* **323**, 445–8.

Robinson K., Gilbert W. & Church G.M. (1994) Large scale bacterial gene discovery by similarity search. *Nature Genet.* **7**, 205–14.

Rodriguez R.L., West R.W., Heyneker H.L., Bolivar P. & Boyer H.W. (1979) Characterizing wild-type and mutant promoters of the tetracycline resistance gene in pBR313. *Nucl. Acids Res.* **6**, 3267–87.

Roe T., Reynolds T.C., Yu G. & Brown P.O. (1993) Integration of murine leukemia virus DNA depends on mitosis. *EMBO J.* **12**, 2099–108.

Rogers J., Goedert M. & Wilson P.M. (1988) An extra sequence in the lambda EMBL3 polylinker. *Nucl. Acid Res.* **16**, 1633.

Rols M.P., Delteil C., Golzio M., Dumond P., Cros S. & Teissie J. (1998) *In vivo* electrically mediated protein and gene transfer in murine melanoma. *Nature Biotechnol.* **16**, 168–71.

Roman M., Martin-Orozco E., Goodman J.S. *et al.* (1997) Immunostimulatory DNA sequences function as T helper-1-promoting adjuvants. *Nature Med.* **3**, 849–54.

Romanos M.A., Scorer C.A. & Clare J.J. (1992) Foreign gene expression in yeast: a review. *Yeast* **8**, 423–88.

Romer S., Fraser P.D., Kiano J.W., Shipton C.A., Misawa N., Schuch W. & Bramley P.M. (2000) Elevation of the provitamin A content of transgenic tomato plants. *Nature Biotechnol.* **18**, 666–9.

Rommens J.M., Iannuzzi M.C., Kerem B.-S. *et al.* (1989) Identification of the cystic fibrosis gene: chromosome walking and jumping. *Science* **245**, 1059–65.

Ronaghi M. (2001) Pyrosequencing sheds light on DNA sequencing. *Genome Res.* **11**, 3–11.

Ronaghi M. *et al.* (1996) Real-time DNA sequencing using detection of pyrophosphate release. *Anal. Biochem.* **242**, 84–9.

Ronaghi M., Petersson B., Uhlen M. & Nyren P. (1998a) PCR-introduced loop structure as primer in DNA sequencing. *BioTechniques* **25**, 876–84.

Ronaghi M., Uhlen M. & Nyren P. (1998b) A sequencing method based on real-time pyrophosphate. *Science* **281**, 363–5.

Rondon I.J. & Marasco W.A. (1997) Intracellular anti-bodies (intrabodies) for gene therapy of infectious diseases. *Ann. Rev. Microbiol.* **51**, 257–83.

Rong Y.S. & Golic K.G. (2000) Gene targeting by homologous recombination in *Drosophila. Science* **288**, 2103–8.

Rorth P. (1996) A modular misexpression screen in *Drosophila* detecting tissue-specific phenotypes. *Proc. Nat. Acad. Sci. USA* **93**, 12418–22.

Rorth P. *et al.* (1998) Systematic gain-of-function genetics in *Drosophila. Development* **125**, 1049–57.

Rosati C., Simoneau P., Treutter D., Poupard P., Cadot Y., Cadic A. & Duron M. (2003) Engineering of flower color in forsythia by expression of two independently-transformed dihydroflavonol 4-reductase and antho-cyanidin synthase genes of flavonoid pathway. *Mol. Breeding* **12**, 197–208.

Rosche E., Blackmore D., Tegeder M., Richardson T., Schroeder H., Higgins T.J.V., Frommer W.B., Offler C.E. & Patrick J.W. (2002) Seed-specific overexpression of the potato sucrose transporter increases sucrose uptake and growth rates of developing pea cotyledons. *Plant J.* **30**, 165–75.

Rose M.D., Novick P., Thomas J.H., Botstein D. & Fink G.R. (1987) A *Saccharomyces cerevisiae* genomic plasmid bank based on a centromere-containing shuttle vector. *Gene* **60**, 237–43.

Rose P.C. & Hui S.W. (1999) Lipoplex size is the major determinant of *in vitro* lipofection efficiency. *Gene Ther.* **6**, 651–9.

Rosenberg S. *et al.* (1990) Gene transfer into humans – immunotherapy of patients with advanced melonoma, using tumor-infiltrating lymphocytes modified by retroviral gene transduction. *N. Engl. J. Med.* **323**, 570–8.

Roses A.D. (2004) Pharmacogenetics and drug development: the path to safer and more effective drugs. *Nature Rev. Genet.* **5**, 645–56.

Roslan H.A., Salter M.G., Wood C.S., White M.R.H., Croft K.P., Robson F., Coupland G., Doonan J., Laufs P., Tomsett A.B. & Caddick M.X. (2001) Characterization of the ethanol-inducible alc gene-expression system in *Arabidopsis thaliana. Plant J.* **28**, 225–35.

Ross D.T. *et al.* (2000) Systematic variation in gene expression patterns in human cancer cell lines. *Nature Genet.* **24**, 227–35.

Rossi J.J. (1995) Controlled, targeted, intracellular expression of ribozymes: progress and problems. *Trends Biotechnol.* **13**, 301–6.

Ross-MacDonald P. *et al.* (1999) Large-scale analysis of the yeast genome by transposon tagging and gene disruption. *Nature* **402**, 413–18.

Ross-Macdonald P., Sheehan A., Roeder G.S. & Snyder M. (1997) A multipurpose transposon system for analyzing protein production, localization, and function in *Saccharomyces cerevisiae. Proc. Nat. Acad. Sci. USA* **94**, 190–5.

Rothman J.E. & Orci L. (1992) Molecular dissection of the secretory pathway. *Nature* **355**, 409–15.

Rousset M., Casalot L., Rapp-Giles B.J. *et al.* (1998) New shuttle vectors for the introduction of cloned DNA in *Desulfovibrio. Plasmid* **39**, 114–22.

Roy M.J., Wu M.S., Barr L.J., Fuller J.T., Tussey L.G., Speller S. *et al.* (2001) Induction of antigen-specific CD8-T cells, T helper cells, and protective levels of antibody in humans by particle-mediated administration of a hepatitis B virus DNA vaccine. *Vaccine* **19**, 764–78.

Roychoudhury R., Jay E. & Wu R. (1976) Terminal labelling and addition of homopolymer tracts to duplex DNA fragments by terminal deoxynucleotidyl transferase. *Nucl. Acids Res.* **3**, 863–77.

Rubin G.M. & Spradling A. (1983) Vectors for P element-mediated gene transfer in *Drosophila. Nucl. Acids Res.* **11**, 6341–51.

Rubin G.M. & Spradling A.C. (1982) Genetic transformation of *Drosophila* with transposable element vectors. *Science* **218**, 348–53.

Rubin G.M. *et al.* (2000) Comparative genomics of the eukaryotes. *Science* **287**, 2204–15.

Rubinstein J.L., Brice A.E., Ciaranello R.D. *et al.* (1990) Subtractive hybridization system using single stranded phagemids with directional inserts. *Nucl. Acids Res.* **18**, 4833–42.

Rudnicki M.A., Braun B., Hinuma S. & Jaenisch R. (1992) Inactivation of *myoD* in mice leads to up-regulation of the myogenic HLH gene *myf*-5 and results in apparently normal muscle development. *Cell* **71**, 383–90.

Rudolph H. & Hinnen A. (1987) The yeast PHO5 promoter: phosphate control elements and sequences mediating mRNA start-site selection. *Proc. Nat. Acad. Sci. USA* **84**, 1340–4.

Ruf S., Hermann M., Berger I. J., Carrer H. & Bock R. (2001) Stable genetic transformation of tomato plastids and expression of a foreign protein in fruit. *Nature Biotechnol.* **19**, 870–5.

Rusconi S. & Schaffner W. (1981) Transformation of frog embryos with a rabbit β-globin gene. *Proc. Nat. Acad. Sci. USA* **78**, 5051–5.

Russel S.H., Hoopes J.L. & Odell J.T. (1992) Directed excision of a transgene from the plant genome. *Mol. Gen. Genet.* **234**, 49–59.

Russell D.W. & Zinder N.D. (1987) Hemimethylation prevents DNA replication in *E. coli. Cell* **50**, 1071–9.

Russell R.B., Saqi M.A., Bates P.A., Sayle R.A. & Sternberg M.J. (1998) Recognition of analogous and homologous protein folds: assessment of prediction success and associated alignment accuracy using empirical substitution matrices. *Protein Engineering* **11**, 1–9.

Russell W.L., Kelly P.R., Hunsicker P.R., Bangham J.W., Maddux S.C. & Phipps E.L. (1979) Specific-locus test shows ethylnitrosourea to be the most potent mutagen in the mouse. *Proc. Nat. Acad. Sci. USA* **76**, 5918–22.

Russell D.W. & Hirata R.K. (1998) Human gene targeting by viral vectors. *Nature Genet.* **18**, 325–30.

Russell D.W., Hirata R.K. & Inoue N. (2002) Validation of AAV-mediated gene targeting. *Nature Biotechnol.* **20**, 658.

Russo G., Zegar C. & Giordano A. (2003) Advantages and limitations of microarray technology in human cancer. *Oncogene* **22**, 6497–507.

Ruysscharet J.-M., El Ouahabi A., Willeaume V. *et al.* (1994) A novel cationic ampiphile for transfection of mammalian cells. *Biochem. Biophys. Res. Commun.* **203**, 1622–8.

Ruzzo A., Andreoni F. & Magnani M. (1998) Structure of the human hexokinase type I gene and nucleotide sequence of the 5′ flanking region. *Biochem. J.* **331**, 607–13.

Rychlewski L., Zhang B. & Godzik A. (1998) Fold and function predictions for *Mycoplasma genitalium* proteins. *Folding Design* **3**, 229–38. [URL: http://cape6.scripps.edu/leszek/genome/cgi-bin/genome.pl?mp]

Rychlewski L., Zhang B.H. & Godzik A. (1999) Functional insights from structural predictions: analysis of the *Escherichia coli* genome. *Protein Sci.* **8**, 614–24.

Ryle A.P., Sanger F., Smith L.F. & Kitai R. (1955) The disulphide bonds of insulin. *Biochem. J.* **60**, 541–56.

Ryo A. *et al.* (1999) Serial analysis of gene expression in HIV-1 infected T cell lines. *FEBS Lett.* **462**, 182–6.

Ryoo, Z.Y. *et al.* (2001) Expression of recombinant human granulocyte macrophage-colony stimulating factor (hGM-CSF) in mouse urine. *Transgenic Res.* **10**, 193–200.

Sabbioni M., Negrini P., Rimessi R., Manservigi R. & Barbatibrodano G. (1995) A BK virus episomal vector for constitutive high expression of exogenous cDNAs in human cells. *Arch. Virol.* **140**, 335–9.

Saccone S. *et al.* (1999) Identification of the gene-richest bands in human prometaphase chromosomes. *Chromosome Research* **7**, 379–86.

Sadowski P.D. (1993) Site-specific genetic recombination: hops, flips and flops. *FASEB J.* **7**, 760–7.

Sadowski P.D. (2003) The Flp double-cross system in a simple efficient procedure for cloning DNA fragments. *BMC Biotechnol.* **3**, 9.

Saez E., No D., West A. & Evans R.M. (1997) Inducible gene expression in mammalian cells and transgenic mice. *Curr. Opin. Biotechnol.* **8**, 608–16.

Sager R., Anisowicz A., Neveu M., Liang P. & Sotiropoulou G. (1993) Identification by differential display of alpha 6 integrin as a candidate tumor suppressor gene. *FASEB J.* **7**, 964–70.

Sagerstrom C.G., Sun B.I. & Sive H.L. (1997) Subtractive cloning: past, present and future. *Ann. Rev. Biochem.* **66**, 751–83.

Saint C.P., Alexander S. & McClure N.C. (1995) pTIM3, a plasmid delivery vector for a transposon-based inducible marker gene system in Gram-negative bacteria. *Plasmid* **34**, 165–74.

Saitoh Y. & Laemmli U.K. (1994) Metaphase chromosome structure: bands arise from a differential folding path of the highly AT-rich scaffold. *Cell* **76**, 609–22.

Sakamoto A. & Murata N. (2000) Genetic engineering of glycinebetaine synthesis in plants: current status and implications for enhancement of stress tolerance. *J. Experiment. Botany* **51**, 81–8.

Sakamoto A. & Murata N. (2001) The use of bacterial choline oxidase, a glycine betaine-synthesizing enzyme, to create stress-resistant transgenic plants. *Plant Physiol.* **125**, 180–8.

Sakamoto A., Murata A. & Murata N. (1998) Metabolic engineering of rice leading to biosynthesis of glycinebetaine and tolerance to salt and cold. *Plant Mol. Biol.* **38**, 1011–19.

Sakamoto T., Morinaka Y., Ishiyama K., Kobayashi M., Itoh H., Kayano T., Iwahori S., Matsuoka M. & Tanaka H. (2003) Genetic manipulation of gibberellin metabolism in transgenic rice. *Nat Biotechnol.* **21**, 909–13.

Sali A., Glaeser R., Earnest T. & Baumeister W. (2003) From words to literature in structural proteomics. *Nature* **422**, 216–25.

Salyers A.A., Shoemaker N.B., Stevens A.M. & Li L.-Y. (1995) Conjugative transposons. an unusual and diverse set of integrated gene transfer elements. *Microbiol. Rev.* **59**, 579–90.

Salzberg S.L., Pertea M., Delcher A.L., Gardner M.J. & Tettelin H. (1999) Interpolated Markov models for eukaryotic gene-finding. *Genomics* **59**, 24–31.

Sambrook J. & Russell D.W. (2001) *Molecular Cloning: A Laboratory Manual*, 3rd edn. CSHL Press, Cold Spring Harbor, NY.

Sambrook J., Rodgers L., White J. & Getling M.J. (1985) Lines of BPV-transformed murine cells that constitutively express influenza virus heamagglutinin. *EMBO J.* **4**, 91–103.

Samulski R.J., Chang L.-S. & Shenk T. (1989) Helper-free stocks of recombinant adeno-associated viruses: normal integration does not require viral gene expression. *J. Virol.* **63**, 3822–8.

Sanchez C. *et al.* (1999) Grasping at molecular interactions and genetic networks in *Drosophila melanogaster* using FlyNets, an Internet database. *Nucl. Acids Res.* **27**, 89–94.

Sanchez J.C. *et al.* (1997) Improved and simplified in-gel sample application using reswelling of dry immobilized pH gradients. *Electrophoresis* **18**, 324–7.

Sanchez O., Navarro R.E. & Aguirre J. (1998) Increased transformation frequency and tagging of developmental genes in *Aspergillus nidulans* by restriction enzyme-mediated integration (REMI). *Mol. Gen. Genet.* **258**, 89–94.

Sander T., Olson S., Hall J., *et al.* (1999) Comparison of detection platforms and post-polymerase chain reaction DNA purification methods for use in conjunction with Cleavase fragment length polymorphism analysis. *Electrophoresis* **20**, 1131–40.

Sandrin V., Russell S.J. & Cosset F.L. (2003) Targeting retroviral and lentiviral vectors. *Curr. Top. Microbiol.* **281**, 137–78.

Sanford J.C., Devit M.J., Russell J.A. *et al.* (1991) An improved, helium-driven biolistic device. *Technique* **3**, 3–16.

Sanford J.C., Klein T.M., Wolf E.D. & Allen N. (1987) Delivery of substances into cells and tissues using a particle bombardment process. *J. Particle Sci. Techniques* **6**, 559–63.

Sanger F., Air G.M., Barrell B.G. *et al.* (1977a) Nucleotide sequence of bacteriophage φX174DNA. *Nature* **265**, 687–95.

Sanger F., Coulson A.R., Hong G.-F., Hill D.F. & Petersen G.B. (1982) Nucleotide sequence of bacteriophage lambda DNA. *J. Mol. Biol.* **162**, 729–73.

Sanger F., Nicklen S. & Coulson A.R. (1977b) DNA sequencing with chain terminating inhibitors. *Proc. Nat. Acad. Sci. USA* **74**, 5463–7.

SanMiguel P., Gaut B.S., Tikhonov A., *et al.* (1998) The palaeontology of intergene retrotransposons of maize. *Nature Genet.* **20**, 43–5.

Santini C., Brennan D., Mennuni C. *et al.* (1998) Efficient display of an HCV cDNA expression library as C-terminal fusion to the capsid protein D of bacteriophage lambda. *J. Mol. Biol.* **282**, 125–35.

Santoro S., Anderson J.C., Lakshman V. & Schultz P.G. (2003) An archaebacterial-derived glutamyl-tRNA synthetase and tRNA pair for unnatural amino acid mutagenesis of proteins in *Escherichia coli*. *Nucl. Acids Res.* **31**, 6700–9.

Sarver N., Gruss P., Law M.-F., Khoury G. & Howley P.M. (1981a) Bovine papilloma virus deoxyribonucleic acid: a novel eucaryotic cloning vector. *Mol. Cell. Biol.* **1**, 486–96.

Sarver N., Gruss P., Law M.-F., Khoury G. & Howley P.M. (1981b) Rat insulin gene covalently linked to bovine papilloma virus DNA is expressed in transformed mouse cells. In *Development Biology Using Purified Genes*, eds. Brown D. & Fox C.R., ICN-UCLA Symposia on Molecular and Cellular Biology, Vol. 23. Academic Press, New York.

Sato S. *et al.* (1999) Complete structure of the chloroplast genome of *Arabidopsis thaliana*. *DNA Res.* **6**, 283–90.

Sauer B. & Henderson N. (1990) Targeted insertion of exogenous DNA into the eukaryotic genome by the Cre recombinase. *New Biol.* **2**, 441–9.

Sauer B. (1994) Site-specific recombination: developments and applications. *Curr. Opin. Biotechnol.* **5**, 521–7.

Savage M.O. & Fallon J.F. (1995) *fgf2* messenger RNA and its antisense message are expressed in a developmentally specific manner in the chick limb bud and mesonephros. *Devel. Dynamics* **200**, 343–53.

Savidge B., Weiss J.D., Wong Y.H.H., Lassner M.W., Mitsky T.A., Shewmaker C.K., Post-Beittenmiller D. & Valentin H.E. (2002) Isolation and characterization of homogentisate phytyltransferase genes from *Synechocystis* sp. PCC 6803 and *Arabidopsis*. *Plant Physiol.* **129**, 321–32.

Sawa S., Watanabe K., Goto K. *et al.* (1999) *FILAMENTOUS FLOWER*, a meristem and organ identity gene of *Arabidopsis*, encodes a protein with zinc finger and HMG-related domains. *Genes Devel.* **13**, 1079–88.

Sayers J.R. & Eckstein F. (1991) A single-strand specific endonuclease activity copurifies with overexpressed T5 D15 exonuclease. *Nucl. Acids Res.* **19**, 4127–32.

Schaefer B.C. (1995) Revolutions in RACE: new strategies for polymerase chain reaction cloning of full length cDNA ends. *Ann. Biochem.* **277**, 255–73.

Schaefer-Ridder M., Wang Y. & Hofschneider P.H. (1982) Liposomes as gene carriers – efficient transformation of mouse L-cells by thymidine kinase gene. *Science* **215**, 166–8.

Schafer W., Gorz A. & Kahl G. (1987) T-DNA integration and expression in a monocot crop plant after induction of *Agrobacterium*. *Nature* **327**, 529–31.

Schaffner W. (1980) Direct transfer of cloned genes from bacteria to mammalian cells. *Proc. Nat. Acad. Sci. USA* **77**, 2163–9.

Scheele G.A. (1975) Two-dimensional gel analysis of soluble proteins: characterization of guinea pig exocrine pancreatic proteins. *J. Biol. Chem.* **250**, 5375–85.

Scheidereit C., Greisse S., Westphal H.M. & Beato M. (1983) The glucocorticoid receptor binds to defined nucleotide sequences near the promoter of mouse mammary tumour virus. *Nature* **304**, 749–52.

Scheidner G., Morral N., Parks R.J. *et al.* (1998) Genomic DNA transfer with high capacity adenovirus vector results in improved *in vivo* gene expression and decreased toxicity. *Nature Genet.* **18**, 180–3.

Schein C.H. & Noteborn M.H.M. (1988) Formation of soluble recombinant proteins in *Escherichia coli* is favored by lower growth temperature. *Biotechnology* **6**, 291–4.

Schein C.H. (1991) Optimizing protein folding to the native state in bacteria. *Curr. Opin. Biotechnol.* **2**, 746–50.

Scheller R.H., Dickerson R.E., Boyer H.W., Riggs A.D. & Itakura K. (1977) Chemical synthesis of restriction enzyme recognition sites useful for cloning. *Science* **196**, 177–80.

Schena M. *et al.* (1996) Parallel human genome analysis: microarray-based expression monitoring of 1000 genes. *Proc. Nat. Acad. Sci. USA* **93**, 10614–19.

Schena M., Heller R.A., Theriault T.P. *et al.* (1998) Microarrays: biotechnology's discovery platform for functional genomics. *Trends Biotechnol.* **16**, 301–6.

Schena M., Lloyd A.M. & Davis R.W. (1991) A steroid-inducible gene expression system for plant cells. *Proc. Nat. Acad. Sci. USA* **88**, 10421–5.

Schena M., Shalon D., Davis R.W. & Brown P.O. (1995) Quantitative monitoring of gene expression patterns with a complementary DNA microarray. *Science* **270**, 467–70.

Scherf U. *et al.* (2000) A gene expression database for the molecular pharmacology of cancer. *Nature Genet.* **24**, 236–44.

Scherzinger E., Bagdasarian M.M., Scholz P. *et al.* (1984) Replication of the broad host range plasmid RSF1010: requirement for three plasmid-encoded proteins. *Proc. Nat. Acad. Sci. USA* **81**, 654–8.

Scheule R.K. & Cheng S.H. (1996) Liposome delivery systems. In *Gene Therapy*, eds. Lemoine N.R. & Cooper D.N., pp. 93–112. BIOS Scientific Publishers, Oxford, UK.

Schiestl R.H. & Petes T.D. (1991) Integration of DNA fragments by illegitimate recombination in *Saccharomyces cerevisiae*. *Proc. Nat. Acad. Sci. USA* **88**, 7585–9.

Schillberg J., Zimmermann J., Findlay K. & Fischer R. (2001) Plasma membrane display of anti-viral single chain Fv fragments confers resistance to tobacco mosaic virus. *Mol. Breeding* in press.

Schimke R.T., Kaufman R.J., Alt F.W. & Kellems R.F. (1978) Gene amplification and drug resistance in cultured murine cells. *Science* **202**, 1051–5.

Schlesinger S. (2001) Alphavirus vectors: development and potential therapeutic applications. *Expert Opin. Biol. Th.* **1**, 177–91.

Schmitz-Linneweber C. *et al.* (2001) The plastid chromosome of spinach (*Spinacia oleracea*): complete nucleotide sequence and gene organization. *Plant Mol. Biol.* **45**, 307–15.

Schmucker D. *et al.* (2000) *Drosophila* Dscam is an axon guidance receptor exhibiting extraordinary molecular diversity. *Cell* **101**, 671–84.

Schmutz J., Wheeler J., Grimwood J., *et al.* (2004) Quality assessment of the human genome sequence. *Nature* **429**, 365–8.

Schnieke A.E., Kind A.J., Ritchie W.A. *et al.* (1997) Human factor IX transgenic sheep produced by transfer of nuclei from transfected fetal fibroblasts. *Science* **278**, 2130–3.

Schocher R.J., Shillito R.D., Saul M.W., Paszkowski S.J. & Potrykus I. (1986) Co-transformation of unlinked foreign genes into plants by direct gene transfer. *Biotechnology* **4**, 1093–6.

Scholthof H.B., Scholthof K.B.G. & Jackson A.O. (1996) Plant virus gene vectors for transient expression of foreign proteins in plants. *Ann. Rev. Phytopathol.* **34**, 299–323.

Scholz P., Haring V., Wittman-Leibold B. *et al.* (1989) Complete nucleotide sequence and gene organization of the broad-host-range plasmid RSF1010. *Gene* **75**, 271–88.

Schreiber S.L. (1991) Chemistry and biology of the immunophillins and their immunosuppressive ligands. *Science* **251**, 283–7.

Schuler T.H., Poppy G.M., Kerry B.R. & Denholm I. (1998) Insect resistant transgenic plants. *Trends Biotechnol.* **16**, 168–75.

Schultz L.D., Hofmann K.J., Mylin L.M. *et al.* (1987) Regulated over-production of the *GAL4* gene product greatly increases expression from galactose-inducible promoters on multi-copy expression vectors in yeast. *Gene* **61**, 123–33.

Schwartz D.C. & Cantor C.R. (1984) Separation of yeast chromosomal-sized DNAs by pulsed field gradient gel electrophoresis. *Cell* **37**, 67–75.

Schwartz D.C. & Koval M. (1989) Conformational dynamics of individual DNA molecules during gel electrophoresis. *Nature* **338**, 520–2.

Schwartz D.C., Li X., Hernandez L.I., Ramnarian S.P., Huff E.J. & Wang Y-K. (1993) Ordered restriction maps of *Sacharomyces cerevisiae* chromosomes constructed by optical mapping. *Science* **202**, 110–14.

Schwartz I. (2000) Microbial genomics: from sequence to function. *Emerging Infectious Diseases* **6**, 493–5.

Schwartzberg P.L., Goff S.P. & Robertson E.J. (1989) Germ line transmission of a c-*abl* mutation produced by targeted gene disruption in ES cells. *Science* **246**, 799–803.

Schwecke T. & 15 others. (1995) The biosynthetic gene cluster for the polyketide immunosuppressant rapamycin. *Proc. Nat. Acad. Sci. USA* **92**, 7839–43.

Schweitzer B. & Kingsmore S.F. (2002) Measuring proteins on microarrays. *Curr. Opin. Biotechnol.* **13**, 14–19.

Schweizer H.P. (2001) Vectors to express foreign genes and techniques to monitor gene expression in pseudomonads. *Curr. Opin. Biotechnol.* **12**, 439–45.

Schweizer H.P. (2003) Application of the *Saccharomyces cerevisiae* Flp-FRT system in bacterial genetics. *Mol. Microbiol. Biotechnol.* **5**, 67–77.

Schwenk F., Kuhn R., Angrand P.O., Rajewsky K. & Stewart A.F. (1998) Temporally and spatially regulated somatic mutagenesis in mice. *Nucl. Acids Res.* **26**, 1427–32.

Sclimenti C.R. & Calos M.P. (1998) Epstein-Barr virus vectors for gene expression and transfer. *Curr. Opin. Biotechnol.* **9**, 476–9.

Scott J.K. & Smith G.P. (1990) Searching for peptide ligands with an epitope library. *Science* **249**, 386–90.

Sechi S. & Oda Y. (2003) Quantitative proteomics using mass spectrometry. *Curr. Opin. Chem. Biol.* **7**, 70–7.

Seelke R., Kline B., Aleff R., Porter R.D. & Shields M.S. (1987) Mutations in the *recD* gene of *Escherichia coli* that raise the copy number of certain plasmids. *J. Bacteriol.* **169**, 4841–4.

Seki M., Ito T., Shibata D. & Shinozaki K. (1999) Regional insertional mutagenesis of specific genes on the *CIC5F11/CIC2B9* locus of *Arabidopsis thaliana* chromosome 5 using the *Ac/Ds* transposon in combination with the cDNA scanning method. *Plant Cell Physiol.* **40**, 624–39.

Selker E.U. (1997) Epigenetic phenomena in filamentous fungi: useful paradigms or repeat-induced confusion? *Trends Genet.* **13**, 296–301.

Selker E.U. (1999) Gene silencing: repeats that count. *Cell* **97**, 157–60.

Semionov A., Cournoyer D. & Chow T.Y. (2003) 1,5-isoquinolinediol increases the frequency of gene targeting by homologous recombination in mouse fibroblasts. *Biochem. Cell Biol.* **81**, 17–24.

Serraj R. & Sinclair T.R. (2002) Osmolyte accumulation: can it really help increase crop yield under drought conditions? *Plant Cell & Environ.* **25**, 333–41.

Seveno M., Bardor M., Paccalet T., Gomord V., Lerouge P. & Faye L. (2004) Glycoprotein sialylation in plants? *Nature Biotechnol.* **22**, 1351–2.

Sgaramella V. (1972) Enzymatic oligomerization of bacteriophage P22 DNA and of linear simian virus 40 DNA. *Proc. Nat. Acad. Sci. USA* **69**, 3389–93.

Sha Y., Li S., Pei Z., *et al.* (2004) Generation and flanking sequence analysis of a rice T-DNA tagged population. *Theor. Appl. Genet.* **108**, 306–14.

Shah D.M., Horsch R.B., Klee H.J. *et al.* (1986) Engineering herbicide tolerance in transgenic plants. *Science* **233**, 478–81.

Shah M.M., Fujiyama K., Flynn C.R. & Joshi L. (2004) Glycoprotein sialylation in plants? Reply. *Nature Biotechnol.* **22**, 1352–3.

Shalon D., Smith S.J. & Brown P.O. (1996) A DNA microarray system for analysing complex DNA samples using two-colour fluorescent probe hybridization. *Genome Res.* **6**, 639–45.

Shapiro L. & Harris T. (2000) Finding function through structural genomics. *Curr. Opin. Biotechnol.* **11**, 31–5.

Sharp P.M. & Bulmer M. (1988) Selective differences among translation termination codons. *Gene* **63**, 141–5.

Shaw C.H., Watson M.D., Carter G.H. & Shaw C.H. (1984) The right hand copy of the nopaline Ti-plasmid 25 bp repeat is required for tumour formation. *Nucl. Acids Res.* **12**, 6031–41.

Shaw G. & Kamen R. (1986) A conserved AU sequence from the 3′-untranslated region of GM-CSF messenger RNA mediates selective messenger RNA degradation. *Cell* **46**, 659–67.

Shedlovsky A., King T.R. & Dove W.F. (1988) Saturation germ line mutagenesis of the murine *t* region including a lethal allele at the *quaking* locus. *Proc. Nat. Acad. Sci. USA* **85**, 180–4.

Sheen J., Hwang S., Niwan Y., Kobayashi H. & Galbraith D.W. (1995) Green fluorescent protein as a new vital marker in plant cells. *Plant J.* **8**, 777–84.

Sheng O.J. & Citovsky V. (1996) *Agrobacterium* plant cell DNA transport: have virulence proteins will travel. *Plant Cell* **8**, 1699–710.

Sheng Y., Mancino V. & Birren B. (1995) Transformation of *Escherichia coli* with large DNA molecules by electroporation. *Nucl. Acids Res.* **23**, 1990–6.

Shepherd N.S. & Smoller D. (1994) The O1 vector system for the preparation and screening of genomic libraries. *Genet. Eng.* **16**, 213–28.

Shepherd N.S., Pfronger B.D., Coulby J.N. *et al.* (1994) Preparation and screening of an arrayed human genomic library generated with the P1 cloning system. *Proc. Nat. Acad. Sci. USA* **91**, 2629–33.

Shi H. *et al.* (1999) Template-imprinted nanosaturated surfaces for protein recognition. *Nature* **398**, 593–7.

Shi H. *et al.* (1999a) RNA aptamers as effective protein antagonists in a multicellular organism. *Proc. Nat. Acad. Sci. USA* **96**, 10033–8.

Shi W., Zakhartchenko V. & Wolf E. (2003) Epigenetic reprogramming in mammalian nuclear transfer. *Differentiation* **71**, 91–113.

Shi Y., Wang M.B., Powell K.S. *et al.* (1994) Use of the rice sucrose synthase-1 promoter to direct phloem-specific expression of beta-glucuronidase and snowdrop lectin genes in transgenic tobacco plants. *J. Exp. Botany* **45**, 623–31.

Shibata D. & Lui Y.-G. (2000) *Agrobacterium*-mediated plant transformation with large DNA fragments. *Trends Plant Sci.* **5**, 354–7.

Shibata K. *et al.* (2000) RIKEN integrated sequence analysis (RISA) system: 384-format sequencing pipeline with 384 multicapillary sequencer. *Genome Res.* **10**, 1757–71.

Shida H. (1986) Nucleotide sequence of the vaccinia virus hemagglutinin gene. *Virology* **150**, 451–62.

Shillito R.D., Saul M.W., Pazkowski J., Muller M. & Potrykus I. (1985) High efficiency direct gene transfer to plants. *Biotechnology* **3**, 1099–103.

Shimada N., Toyoda-Yamamoto A., Nagamine J. *et al.* (1990) Control of expression of *Agrobacterium vir* genes by synergistic actions of phenolic signal molecules and monosaccharides. *Proc. Nat. Acad. Sci. USA* **87**, 6684–8.

Shimada T., Fujii H., Mitsuya A. & Nienhuis W. (1991) Targeted and highly efficient gene transfer into CD4+ cells by a recombinant human immunodeficiency virus retroviral vector. *J. Clin. Invest.* **88**, 1043–7.

Shimamoto K., Terada R., Izawa T. & Fujimoto H. (1988) Fertile transgenic rice plants regenerated from transformed protoplasts. *Nature* **338**, 274–6.

Shimotohno K. & Temin H.M. (1982) Loss of intervening sequences in genomic mouse α-globin DNA inserted in an infectious retrovirus vector. *Nature* **299**, 265–8.

Shin T., Kraemer D., Pryor J., Liu L., Rugila J., Howe L., Buck S., Murphy K., Lyons L. & Westhusin M. (2002) A cat cloned by nuclear transplantation. *Nature* **415**, 859.

Shinozaki K. *et al.* (1986) The complete nucleotide sequence of the tobacco chloroplast genome: its gene organization and expression. *EMBO J.* **5**, 2043–9.

Shintani D. & DellaPenna D. (1998) Elevating the vitamin E content of plants through metabolic engineering. *Science* **282**, 2098–100.

Shirasu K., Lahaye T., Tan M-W., Zhou F., Azevedo C. & Schulze-Lefert P. (1999) A novel class of eukaryotic zinc-binding proteins is required for disease resistance signaling in barley and development in *C. elegans*. *Cell* **99**, 355–66.

Shizuya H., Birren B., Kim U.-J. *et al.* (1992) Cloning and stable maintenance of 300-kilobase-pair fragments of human DNA in *Escherichia coli* using an F-factor-based vector. *Proc. Nat. Acad. Sci. USA* **89**, 8794–7.

Shoemaker D.D., Lashkari D.A., Morris D., Mittmann M. & Davis R.W. (1996) Quantitative phenotypic analysis of yeast deletion mutants using a highly parallel molecular bar-coding strategy. *Nature Genet.* **14**, 450–6.

Shoemaker D.D. & Linsley P.S. (2002) Recent developments in DNA microarrays. *Curr. Opin. Microbiol.* **5**, 334–7.

Shore D. (2001) Telomeric chromatin: replicating and wrapping up chromosome ends. *Curr. Opin. Genet. Develop.* **11**, 189–98.

Short J.M., Fernandez J.M., Sorge J.A. & Huse W.D. (1988) λZAP: a bacteriophage lambda expression vector with *in vivo* excision properties. *Nucl. Acids Res.* **16**, 7583–600.

Shuman S. (1994) Novel approach to molecular cloning and polynucleotide synthesis using vaccinia DNA topoisomerase. *J. Biol. Chem.* **269**, 32678–84.

Shusta E.V., Raines R.T., Pluckthun A. & Wittrup K.D. (1998) Increasing the secretory capacity of *Saccharomyces cerevisiae* for production of single-chain antibody fragments. *Nature Biotechnol.* **16**, 773–7.

Shusta E.V., Van Antwerp J. & Wittrup K.D. (1999) Biosynthetic polypeptide libraries. *Curr. Opin. Biotechnol.* **10**, 117–22.

Siden-Kiamos I., Saunders R.D., Spanos L., *et al.* (1990) Towards a physical map of the *Drosophila melanogaster* genome: mapping of cosmid clones within defined genomic divisions. *Nucl. Acids Res.* **18**, 6261–70.

Sidhu S.S. (2000) Phage display in pharmaceutical biotechnology. *Curr. Opin. Biotechnol.* **11**, 610–16.

Sidhu S.S., Lowman H.B., Cunningham B.C. & Wells J.A. (2000) Phage display for selection of novel binding peptides. *Methods Enzymol.* **328**, 333–63.

Sidorov V.A., Kasten D., Pang S.Z., Hajdukiewicz P.T., Staub J.M. & Nehra N.S. (1999) Stable chloroplast transformation in potato: use of green fluorescent protein as a plastid marker. *Plant J.* **19**, 209–16.

Siegele D.A. & Hu J.C. (1997) Gene expression from plasmids containing the *ara*BAD promoter at sub-saturating inducer concentrations represents mixed populations. *Proc. Nat. Acad. Sci. USA* **94**, 8168–72.

Signs M.W. & Flores H.E. (1990) The biosynthetic potential of plant roots. *Bio-Essays* **12**, 282–5.

Sijmons P.C., Dekker B.M.M., Schrammeijer B. *et al.* (1990) Production of correctly processed human serum albumin in transgenic plants. *Biotechnology* **8**, 217–21.

Sikorski R. & Peters R. (1997) Transgenics on the Internet. *Nature Biotechnol.* **15**, 289.

Sikorski R.S., Michaud W., Levin H.L., Boeke J.D. & Hieter P. (1990) Trans-kingdom promiscuity. *Nature* **345**, 581–2.

Simmer F., Moorman C., Van Der Linden A.M., Kuijk E., Van Den Berghe P.V., Kamath R., Fraser A.G., Ahringer J. & Plasterk R.H. (2003) Genome-wide RNAi of *C. elegans* using the hypersensitive rrf-3 strain reveals novel gene functions. *PLoS Biol.* **1**, E12.

Simmons L.C. & Yansura D.G. (1996) Translational level is a critical factor for the secretion of heterologous proteins in *Escherichia coli*. *Nature Biotechnol.* **14**, 629–34.

Simonato M., Manservigi R., Marconi P. & Glorioso J. (2000) Gene transfer into neurones for the molecular analysis of behaviour: focus on herpes simplex vectors. *Trends Neurosci.* **23**, 183–90.

Simons J.P., McClenaghan M. & Clark A.J. (1987) Alteration of the quality of milk by expression of sheep beta-lactoglobulin in transgenic mice. *Nature* **328**, 530–2.

Simons R. & Kleckner N. (1988) Biological regulation by antisense RNA in prokaryotes. *Ann. Rev. Genet.* **22**, 567–600.

Simonsen C.C. & Levinson A.D. (1983) Isolation and expression of an altered mouse dihydrofolate reductase cDNA. *Proc. Nat. Acad. Sci. USA* **80**, 2495–9.

Singer J.B., Hill A.E., Burrage L.C., *et al.* (2004) Genetic dissection of complex traits with chromosome substitution strains of mice. *Science* **304**, 445–8.

Singer M. & Berg P. (1990) *Genes and Genomes*. Blackwell Scientific Publications, Oxford.

Singer-Sam J., Simmer R.L., Keith D.H. *et al.* (1983) Isolation of a cDNA clone for human X-linked 3-phosphoglycerate kinase by use of a mixture of synthetic oligodeoxyribonucleotides as a detection probe. *Proc. Nat. Acad. Sci. USA* **80**, 802–6.

Singh H. (1993) Specific recognition site probes for isolating genes encoding DNA-binding proteins. *Methods Enzymol.* **218**, 551–67.

Singh H., LeBowitz J.H., Bladwin A.S.J. & Sharp P.A. (1988) Molecular cloning of enhancer-binding protein: isolation by screening of an expression library with a recognition site DNA. *Cell* **52**, 415–23.

Singh-Gasson S. *et al.* (1999) Maskless fabrication of light-directed oligonucleotide microarrays using a digital micro-mirror array. *Nature Biotechnol.* **17**, 974–8.

Sinn E., Muller W., Pattengale P. *et al.* (1987) Coexpression of MMTV/v-Ha-*ras* and MMTV/c-*myc* genes in transgenic mice: synergistic action of oncogenes *in vivo*. *Cell* **49**, 465–75.

Siomi M.C. *et al.* (1998) Functional conservation of the transportin nuclear import pathway in divergent organisms. *Mol. Cell Biol.* **18**, 4141–8.

Sizemore C., Wieland B., Gotz F. & Hillen W. (1991) Regulation of the *Staphylococcus xylosus* xylose utilization genes at the molecular level. *J. Bacteriol.* **174**, 3042–8.

Sizemore D.R., Branstrom A.A. & Sadoff J.C. (1995) Attenuated *Shigella* as a DNA delivery vehicle for DNA-mediated immunization. *Science* **270**, 299–302.

Sjoberg E.M., Suomalainen M. & Garoff H. (1994) A significantly improved Semliki Forest virus expression system based on translation enhancer segments from the viral capsid gene. *Biotechnology* **12**, 1127–31.

Skaletsky H., Kurodo-Kawaguchi T., Minx P.J., *et al.* (2003) The male-specific region of the human Y chromosome is a mosaic of discrete sequence classes. *Nature* **423**, 825–37.

Skarjinskaia M., Svab Z. & Maliga P. (2003) Plastid transformation in *Lesquerella fendleri*, an oilseed Brassicacea. *Transgenic Res.* **12**, 115–22.

Skarnes W.C., Auerbach B.A. & Joyner A.L. (1992) A gene trap approach in mouse embryonic stem cells: the *lacZ* reporter is activated by splicing, reflects endogenous gene expression, and is mutagenic in mice. *Genes Devel.* **6**, 903–18.

Sklar M.D., Thompson E., Welsh M.J. *et al.* (1991) Depletion of c-*myc* with specific antisense sequences reverses the transformed phenotype in ras oncogene-transformed NIH 3T3 cells. *Mol. Cell Biol.* **11**, 3699–710.

Skow L.C., Burkhart B.A., Johnson F.M., Popp P.A. & Popp D.M. (1983) A mouse model for beta-thalassemia. *Cell* **34**, 1043–52.

Sledz C.A., Holko M., de Veer M.J., Silverman R.H. & Williams B.R. (2003) Activation of the interferon system by short interfering RNAs. *Nature Cell Biol.* **5**, 834–9.

Slilaty S.N. & Lebel S. (1998) Accurate insertional inactivation of lacZalpha: construction of pTrueBlue and M13 TrueBlue cloning vectors. *Gene* **213**, 83–91.

Smerdou C. & Liljestrom P. (1999) Two helper RNA system for production of recombinant Semliki Forest virus particles. *J. Virol.* **73**, 1092–8.

Smith A.J., De Sousa M.A., Kwabi-Addo B. *et al.* (1995) A site-directed chromosomal translocation induced in embryonic stem cells by Cre-*loxP* recombination. *Nature Genet.* **9**, 376–85.

Smith C., Watson C., Ray J. *et al.* (1988) Antisense RNA inhibition of polygalacturonase gene expression in transgenic tomatoes. *Nature* **334**, 724–6.

Smith C.L., Econome J.G., Schutt A., Klco S. & Cantor C.R. (1987) A physical map of the *Escherichia coli* K12 genome. *Science* **236**, 1448–53.

Smith D.J., Stevens M.E., Sudanagunta S.P. *et al.* (1997) Functional screening of 2 Mb of human chromosome 21q22.2 in transgenic mice implicates minibrain in learning defects associated with Down Syndrome. *Nature Genet.* **16**, 28–36.

Smith E.F. & Townsend C.O. (1907) A plant-tumor of bacterial origin. *Science* **25**, 671–3.

Smith G.E., Summers M.D. & Fraser M.J. (1983) Production of human β-interferon in insect cells infected with a baculovirus expression vetor. *Mol. Cell. Biol.* **3**, 2156–65.

Smith G.L., Mackett M. & Moss B. (1983a) Infectious vaccinia virus recombinants that express hepatitis B virus surface antigen. *Nature* **302**, 490–5.

Smith G.P. (1985) Filamentous fusion phage: novel expression vectors that display cloned antigens on the virion surface. *Science* **228**, 1315–17.

Smith H.B., Larimer F.W. & Hartman F.C. (1990) An engineered change in substrate specificity of ribu-losebisphosphate carboxylase/oxygenase. *J. Biol. Chem.* **265**, 1243–5.

Smith H.O. & Nathans D. (1973) A suggested nomenclature for bacterial host modification and restriction systems and their enzymes. *J. Mol. Biol.* **81**, 419–23.

Smith H.O. & Wilcox K.W. (1970) A restriction enzyme from *Hemophilus influenzae*. I. Purification and general properties. *J. Mol. Biol.* **51**, 379–91.

Smith K., Johnson K., Bryan T. *et al.* (1993) The *APC* gene product in normal and tumor cells. *Proc. Nat. Acad. Sci. USA* **90**, 2846–50.

Smith L.C. & Wilmut I. (1989) Influence of nuclear and cytoplasmic activity on the development *in vivo* of sheep embryos after nuclear transplantation. *Biol. Reprod.* **40**, 1027–35.

Smith P.A., Tripp B.C., DiBlasio-Smith E.A. *et al.* (1998) A plasmid expression system for quantitative *in vivo* biotinylation of thioredoxin fusion proteins in *Escherichia coli. Nucl. Acids Res.* **26**, 1414–20.

Smith S.B., Aldridge P.K. & Callis J.B. (1989) Observation of individual DNA molecules undergoing gel electrophoreis. *Science* **243**, 203–6.

Smith V., Botstein D. & Brown P.O. (1995) Genetic footprinting: a genomic strategy for determining a gene's function given its sequence. *Proc. Nat. Acad. Sci. USA* **92**, 6479–83.

Smith V., Chou K.N., Lashkari D., Botstein D. & Brown P.O. (1996) Functional analysis of the genes of yeast chromosome V by genetic footprinting. *Science* **274**, 2069–74.

Smithies O. (1993) Animal models of human genetic diseases. *Trends Genet.* **9**, 112–16.

Smithies O., Gregg R.G., Boggs S.S., Koralewski M.A. & Kucherlapati R. (1985) Insertion of DNA sequences into the human b-globin locus by homologous recombination. *Nature* **317**, 230–4.

Snouwaert J.N., Brigman K.K., Latour A.M. *et al.* (1992) An animal-model for cystic-fibrosis made by gene targeting. *Science* **257**, 1083–8.

Snyder R.O. & Flotte T.R. (2002) Production of clinical-grade recombinant adeno-associated virus vectors. *Curr. Opin. Biotechnol.* **13**, 418–23.

Snyder R.O., Miao C.H., Patijn G.A. *et al.* (1997) Persistent and therapeutic concentrations of human factor IX in mice after hepatic gene transfer of recombinant AAV vectors. *Nature Genet.* **16**, 270–6.

Soderlund C. *et al.* (2000) Contigs built with fingerprints, markers and FPC V4.7. *Genome Res.* **10**, 1772–87.

Somers D.A., Rines H.W., Gu W., Kaeppler H.F. & Bushnell W.R. (1992) Fertile, transgenic oat plants. *Biotechnology* **10**, 1589–94.

Song W-Y. *et al.* (1995) A receptor kinase-like protein encoded by the rice disease resistance gene, *Xa21*. *Science* **270**, 1804–6.

Sorge J.A. (1988) Bacteriophage lambda cloning vectors. In *Vectors: A Survey of Molecular Cloning Vectors and Their Uses*, eds. Rodriguez R.L. & Denhardt D.T., pp. 43–60. Butterworth Press, Boston, MA.

Soriano P. (1995) Gene targeting in ES cells. *Ann. Rev. Neurosci.* **18**, 1–18.

Sorokin A.P., Ke X., Chen D. & Elliott M.C. (2000) Production of fertile transgenic wheat plants via tissue electroporation. *Plant Sci.* **156**, 227–33.

Sosio M., Giusino F., Cappellano C. *et al.* (2000) Artificial chromosomes for antibiotic-producing actinomycetes. *Nature Biotechnol.* **18**, 343–5.

Southern E.M. (1975) Detection of specific sequences among DNA fragments separated by gel electrophoresis. *J. Mol. Biol.* **98**, 503–17.

Southern E.M. (1979a) Measurement of DNA length by gel electrophoresis. *Anal. Biochem.* **100**, 319–23.

Southern E.M. (1979b) Gel electrophoresis of restriction fragments. *Methods Enzymol.* **68**, 152–76.

Southern E.M. (1988) Analyzing polynucleotide sequences. *International Patent Application PCT GB 89/01114*.

Southern E.M. (1996a) DNA chips: analysing sequence by hybridization to oligonucleotides on a large scale. *Trends Genet.* **12**, 110–15.

Southern E.M., Maskos U. & Elder J.K. (1992) Analyzing and comparing nucleic acid sequences by hybridization to arrays of oligonucleotides: evaluation using experimental models. *Genomics* **13**, 1008–17.

Southern P.J. & Berg P. (1982) Transformation of mammalian cells to antibiotic resistance with a bacterial gene under the control of the SV40 early region promoter. *J. Mol. Appl. Genet.* **1**, 327–41.

Sowers K.R. & Schreier H.J. (1999). Gene transfer systems for the Archaea. *Trends Microbiol.* **7**, 212–19.

Speck M., Raff J.W., Harrison Lavoie K., Little P.F.R. & Glover D.M. (1988) Smart2, a cosmid vector with a phage lambda origin for both systematic chomosome walking and P-element mediated gene transfer in *Drosophila*. *Gene* **64**, 173–7.

Spellman P.T., Sherlock G., Zhang M.Q. *et al.* (1998) Comprehensive identification of cell cycle-regulated gene of the yeast *Saccharomyces cerevisiae* by microarray hybridization. *Mol. Biol. Cell* **9**, 3273–97.

Speulman E., Metz P.L., van Arkel G., te Lintel Hekkert B., Stiekema W.J. & Pereira A. (1999) A two-component enhancer-inhibitor transposon mutagenesis system for functional analysis of the *Arabidopsis* genome. *Plant Cell* **11**, 1853–66.

Spiker S. & Thompson W.F. (1996) Nuclear matrix attachment regions and transgene expression in plants. *Plant Physiol.* **110**, 15–21.

Spitz F., Gonzalez F. & Duboule D. (2003) A global control region defines a chromosomal regulatory landscape containing the *HoxD* cluster. *Cell* **113**, 405–17.

Spradling A.C. & Rubin G.M. (1982) Transposition of cloned P elements into *Drosophila* germ line chromosomes. *Science* **218**, 341–7.

Spradling A.C. *et al.* (1999) The BDGP Gene Disruption Project: single P element insertions mutating 25% of vital *Drosophila* genes. *Genetics* **153**, 135–77.

Spradling A.C., Stern D.M., Kiss I., Roote J., Laverty T. & Rubin G.M. (1995) Gene disruptions using P transposable elements: an integral component of the *Drosophila* genome project. *Proc. Nat. Acad. Sci. USA* **92**, 10824–30.

Sprague K.V., Faulds D.H. & Smith G.R. (1978) A single basepair change creates a *chi* recombinational hotspot in bacteriophage λ. *Proc. Nat. Acad. Sci. USA* **75**, 6182–6.

Spreng S., Dietrich G. & Niewiesk S. *et al.* (2000) Novel bacterial systems for the delivery of recombinant protein or DNA. *FEMS Immunol. Med. Microbiol.* **27**, 299–304.

Springer P.S. (2000) Gene traps: tools for plant development and genomics. *Plant Cell* **12**, 1007–20.

Srivastava V. & Ow D.W. (2004) Marker-free site-specific gene integration in plans. *Trends Biotechnol.* **22**, 627–9.

Srivastava V. & Ow D.W. (2002) Biolistic mediated site-specific integration in rice. *Mol. Breed.* **8**, 345–50.

Srivastava V. *et al.* (2004) Cre-mediated site-specific integration for consistent transgene expression in rice. *Plant Biotech. J.* **2**, 169–79.

St Ogne L., Furth P.A. & Gruss P. (1996) Temporal control of the Cre recombinase in transgenic mice by a tetracycline responsive promoter. *Nucl. Acids Res.* **24**, 3875–7.

Stachel S.E., Messens E., Van Montagu M. & Zambryski P. (1985) Identification of the signal molecules produced by wounded plant cells that activate T-DNA transfer in *Agrobacterium tumefaciens*. *Nature* **318**, 624–9.

Stallcup M.R., Sharrock W.J. & Rabinowitz J.C. (1974) Ribosome and messenger specificity in protein synthesis by bacteria. *Biochem. Biophys. Res. Commun.* **58**, 92–8.

Stanford W.L., Cohn J.B. & Cordes S.P. (2001) Gene-trap mutagenesis: past, present and beyond. *Nature Rev. Genet.* **2**, 756–68.

Stanilaus M.A. & Cheng C.-L. (2002) Genetically engineered self-destruction: an alternative to herbicides for cover crop systems. *Weed Sci.* **50**, 794–801.

Stanley J. (1983) Infectivity of the cloned geminivirus genome requires sequences from both DNAs. *Nature* **305**, 643–5.

Stark-Lorenzen P., Nelke B., Hänbler G., Mühlbach H.P. & Thomzik J.E. (1997) Transfer of a grapevine stilbene synthase gene to rice (*Oryza sativa* L.). *Plant Cell Rep.* **16**, 668–73.

Staub J. & Maliga P. (1992a) Long regions of homologous DNA are incorporated into the tobacco plastid genome by transformation. *Plant Cell* **4**, 39–45.

Staub J. & Maliga P. (1992b) High-frequency plastid transformation in tobacco by selection for a chimeric *aadA* gene. *Proc. Nat. Acad. Sci. USA* **90**, 913–17.

Staudt L.M., Clerc R.G., Singh H. *et al.* (1988) Cloning of a lymphoid-specific cDNA encoding a protein binding the regulatory octamer DNA motif. *Science* **241**, 577–9.

Stavropoulos T.A. & Strathdee C.A. (1998) An enhanced packaging system for helper-dependent herpes simplex virus vectors. *J. Virol.* **72**, 7137–43.

Stearns T., Ma H. & Botstein D. (1990) Manipulating the yeast genome using plasmid vectors. *Methods Enzymol.* **185**, 280–97.

Steck T.R. (1997) Ti plasmid type affects T-DNA processing in *Agrobacterium tumefaciens*. *FEMS Microbiol. Lett.* **147**, 121–5.

Steemers F.J., Ferguson J.A. & Walt D.R. (2000) Screening unlabeled DNA targets with randomly ordered fiber-optic gene arrays. *Nature Biotechnol.* **18**, 91–4.

Stein L. (2001) Genome annotation: from sequence to biology. *Nature Rev. Genet.* **2**, 493–503.

Steller H. & Pirrotta V. (1985) A transposable P vector that confers selectable G418 resistance of *Drosophila* larvae. *EMBO J.* **4**, 167–71.

Stemmer W.P.C. (1993) DNA shuffling by random fragmentation and reassembly: *In vitro* recombination for molecular evolution. *Proc. Nat. Acad. Sci. USA* **91**, 10747–51.

Stemmer W.P.C. (2004) Rapid evolution of a protein *in vitro* by DNA shuffling. *Nature* **370**, 389–91.

Stenger D.C., Revington G.N., Stevenson M.C. & Bisaro D.M. (1991) Replicational release of geminivirus genomes from tandemly repeated copies – evidence for rolling-circle replication of a plant viral-DNA. *Proc. Nat. Acad. Sci. USA* **88**, 8029–33.

Sternberg N. (1990) Bacteriophage P1 cloning system for the isolation, amplification and recovery of DNA fragments as large as 100 kilobase pairs. *Proc. Nat. Acad. Sci. USA* **87**, 103–7.

Sternberg N. (1994) The P1 cloning system – past and future. *Mammalian Genome* **5**, 397–404.

Stevens R.C. (2000a) Design of high-throughput methods of protein production for structural biology. *Structure with Folding and Design* **8**, R177–85.

Stevens R.C. (2000b) High-throughput protein crystallization. *Curr. Opin. Struct. Biol.* **10**, 558–63.

Stevens R.C., Yokohoma S. & Wilson I.A. (2001) Global efforts in structural genomics. *Science* **294**, 89–92.

Stief A., Winter D.M., Stratling W.H. & Sippel A.E. (1989) A nuclear DNA attachment element mediates elevated and position-independent gene activity. *Nature* **341**, 343–5.

Stinchcomb D.T., Struhl K. & Davis R.W. (1979) Isolation and characterization of a yeast chromosomal replicator. *Nature* **282**, 39–43.

Stoger E., Vaquero C., Torres E., Sack M., Nicholson L., Drossard J., Williams S., Keen D., Perrin Y., Christou P. & Fischer R. (2000) Cereal crops as viable production and storage systems for pharmaceutical scFv antibodies. *Plant Mol. Biol.* **42**, 583–90.

Storb U., O'Brien R.L., McMullen M.D., Gollahon K.A. & Brinster R.L. (1984) High expression of cloned immunoglobulin kappa gene in transgenic mice is restricted to B-lymphocytes. *Nature* **310**, 238–48.

Stoss O., Mogk A. & Schumann W. (1997) Integrative vector for constructing single-copy translational fusions between regulatory regions of *Bacillus subtilis* and the *bgsB* reporter gene encoding a heat-stable beta-galactosidase. *FEMS Microbiol. Lett.* **150**, 49–54.

Stover C.K. *et al.* (2000) Complete genome sequence of *Pseudomonas aeruginosa* PAO1, an opportunistic pathogen. *Nature* **406**, 959–64.

Strathdee C.A., Gavish H., Shannon W.R. & Buchwald M. (1992) Cloning of cDNAs for Fanconi's anaemia by functional complementation. *Nature* **356**, 763–7.

Stroun M., Anker P., Charles P. & Ledoux L. (1966) Bacterial doeoxyribonucleic acid in *Lycopersicon esculentum*. *Nature* **212**, 397–9.

Struhl K. (1983) The new yeast genetics. *Nature* **305**, 391–7.

Struhl K., Stinchcomb D.T., Scherer S. & Davis R.W. (1979) High-frequency transformation of yeast: autonomous replication of hybrid DNA molecules. *Proc. Nat. Acad. Sci. USA* **76**, 1035–9.

Studier F.W. (1991) Use of bacteriophage T7 lysozyme to improve an inducible T7 expression system. *J. Mol. Biol.* **219**, 37–44.

Studier F.W., Rosenberg A.H., Dunn J.J. & Dubendorff J.W. (1990) Use of T7 RNA polymerase to direct expression of cloned genes. *Methods Enzymol.* **185**, 60–89.

Su H., Chang J.C., Xu S.M. & Kan Y.W. (1996) Selective killing of AFP-positive hepatocellular carcinoma cells by adeno-associated virus transfer of the herpes simplex virus thymidine kinase gene. *Hum. Gene Ther.* **7**, 463–70.

Subramani S., Mulligan R. & Berg P. (1981) Expression of mouse dihydrofolate reductase complementary deoxyribonucleic acid in simian virus 40 vectors. *Mol. Cell. Biol.* **1**, 854–64.

Subramaniam S., Slater S., Karberg K., Chen R., Valentin H.E. & Wong Y.H.H. (2000) Nucleic acid sequences to proteins involved in tocopherol biosynthesis. WO 01/79472.

Sugden B., Marsh K. & Yates J. (1985) A vector that replicates as a plasmid and can be efficiently selected in B-lymphoblasts transformed by Epstein-Barr virus. *Mol. Cell. Biol.* **5**, 410–13.

Suggs S.V., Wallace R.B., Hirose T., Kawashima E.H. & Itakura K. (1981) Use of synthetic oligonucleotides as hybridization probes. III. Isolation of cloned cDNA sequences for human beta-2-microglobulin. *Proc. Nat. Acad. Sci. USA* **78**, 6613–17.

Sugimoto Y., Aksentijevich I., Gottesman M.M. & Pastan I. (1994) Efficient expression of drug-selectable genes in retroviral vectors under control of an internal ribosome entry site. *Biotechnology* **12**, 694–8.

Sugita K. *et al.* (2000) A transformation vector for the production of marker-free transgenic plants containing a single copy transgene at high frequency. *Plant J.* **22**, 461–9.

Sukchawalit R., Vattanaviboon P., Sallabhan R. & Mongkolsuk S. (1999) Construction and characterization of regulated L-arabinose-inducible broad host range expression vectors in *Xanthomonas. FEMS Microbiol. Lett.* **181**, 217–23.

Sumikawa K., Houghton M., Emtage J., Richards B. & Barnard E. (1981) Active multi-subunit ACh receptor assembled by translation of heterologous mRNA in *Xenopus* oocytes. *Nature* **292**, 862.

Summers D. (1998) Timing, self-control and a sense of direction are the secrets of multicopy plasmid stability. *Mol. Microbiol.* **29**, 1137–45.

Summers D.K. & Sherratt D.J. (1984) Multimerization of high copy number plasmids causes instability: Col E1 encodes a determinant essential for plasmid monomerization and stability. *Cell* **36**, 1097–103.

Sumner L.W., Mendes P. & Dixon R.A. (2003) Plant metabolomics: large-scale phytochemistry in the functional genomics era. *Phytochemistry* **62**, 817–36.

Sun L., Li J. & Xiao X. (2000) Overcoming AAV vector size limitation through viral DNA heterodimerization. *Nature Med.* **6**, 599–602.

Sun T.-Q., Fenstermacher D.A. & Vos J.-M.H. (1994) Human artificial episomal chromosomes for cloning large DNA fragments in human cells. *Nature Genet.* **8**, 33–41.

Sundaresan V., Springer P., Volpe T. *et al.* (1995) Patterns of gene action in plant development revealed by enhancer trap and gene trap transposable elements. *Genes Devel.* **9**, 1797–810.

Sussman D.J. & Milman G. (1984) Short-term, high-efficiency expression of transfected DNA. *Mol. Cell. Biol.* **4**, 1641–3.

Sutcliffe J.G. (1979) Complete nucleotide sequence of the *Escherichia coli* plasmid pBR322. *Cold Spring Harbor Symp. Quant. Biol.* **43**(1), 77–90.

Sutherland G.R. & Richards R.I. (1995) The molecular basis of fragile sites in human chromosomes. *Curr. Opin. Genet. Develop.* **5**, 323–7.

Suyama M. & Bork P. (2001) Evolution of prokaryotic gene order: genome rearrangements in closely related species. *Trends Genet.* **17**, 10–13.

Suzuki H. *et al.* (2001) Protein–protein interaction panel using mouse full-length cDNAs. *Genome Research* **11**, 1758–65.

Suzuki Y., Ishihara D., Sasaki M. *et al.* (2000) Statistical analysis of the 5′-untranslated region of human mRNA using oligo capped cDNA libraries. *Genomics* **64**, 286–97.

Suzuki Y., Yoshimoto-Nakagawa K., Maruyama K., Suyama A. & Sugano S. (1997) Construction and characterization of full length-enriched and a 5′-end-enriched cDNA library. *Gene* **200**, 149–56.

Svab Z., Hajdukiewcz P. & Maliga P. (1990b) Stable transformation of plastids in higher plants. *Proc. Nat. Acad. Sci. USA* **87**, 8526–30.

Svab Z., Harper E.C., Jones J.D.G. & Maliga P. (1990a) Aminoglycoside 3′-adenyltransferase confers resistance to spectinomycin and streptomycin in *Nicotiana tabacum* plants. *Plant Mol. Biol.* **14**, 197–205.

Swick A.G., Janicot M., Chenevalkastelic T., McLenithan J.C. & Lane M.D. (1992) Promoter cDNA-directed heterologous protein expression in *Xenopus laevis* oocytes. *Proc. Nat. Acad. Sci. USA* **89**, 1812–16.

Swift G.H., Hammer R.E., MacDonald R.J. & Brinster R.L. (1984) Tissue specific expression of the rat pancreatic elastase 1 gene in transgenic mice. *Cell* **38**, 639–46.

Swindells M., Orengo C., Jones D., Hutchinson E. & Thornton J. (1998) Contemporary approaches to protein structure classification. *Bioessays* **20**, 884–91.

Syed F. *et al.* (1999) CCR7 (EBI 1) receptor downregulation in asthma: differential gene expression in human CD4+ T lymphocyte. *Quart. J. Med.* **92**, 463–71.

Szczebara F.M., Chandelier C., Villeret C., *et al.* (2003) Total biosynthesis of hydrocortisone from a simple carbon source in yeast. *Nature Biotechnol.* **21**, 143–9.

Szczylik C., Skorski T., Nicolaides N.C. *et al.* (1991) Selective inhibition of leukaemia cell proliferation by BCR-ABL antisense oligonucleotides. *Science* **253**, 262–5.

Szostak J.W. & Blackburn E.H. (1982) Cloning yeast telomeres on linear plasmid vectors. *Cell* **29**, 245–55.

Szustakowski J.D. & Weng Z. (2000) Protein structure alignment using a genetic algorithm. *Proteins* **38**, 428–40.

Szybalska E.H. & Szybalski (1962) Genetics of human cell lines IV. DNA-mediated heritable transformation of a biochemical trait. *Proc. Nat. Acad. Sci. USA* **48**, 2026–31.

Tabor S. & Richardson C.C. (1995) A single residue in DNA polymerases of the *Escherichia coli* DNA polymerase I family is critical for distinguishing between deoxy- and dideoxyribonucleotides. *Proc. Nat. Acad. Sci.* **92**, 6339–43.

Tacket C.O., Reid R.H., Boedeker E.C. *et al.* (1994) Enteral immunisation and challenge of volunteers given enterotoxigenic *E. coli* CFA/II encapsulated in biodegradable microspheres. *Vaccine* **12**, 1270–4.

Tait R.C., Close T.J., Lundquist R.C. *et al.* (1983) Construction and characterization of a versatile broad host range DNA cloning system for Gram-negative bacteria. *Biotechnology* **1**, 269–75.

Takagi H., Morinaga Y., Tsuchiya M., Ikemura H. & Inouye M. (1988) Control of folding of proteins secreted by a high expression secretion vector, pIN-111-ompA: 16-fold increase in production of active subtilisin E in *Escherichia coli. Biotechnology* **6**, 948–50.

Takamatsu N., Ishikawa M., Meshi T. & Okada Y. (1987) Expression of bacterial chloramphenicol acetyltransferase gene in tobacco plants mediated by TMV-RNA. *EMBO J.* **6**, 307–11.

Takeuchi Y., Dotson M. & Keen N.T. (1992) Plant transformation: a simple particle bombardment device based on flowing helium. *Plant Mol. Biol.* **18**, 835–9.

Takumi T. & Lodish H.F. (1994) Rapid cDNA cloning by PCR screening. *Biotechniques* **17**, 443–4.

Takumi T. (1997) Use of PCR for cDNA library screening. *Methods Mol. Biol.* **67**, 339–44.

Tamura T., Thibert C., Royer C. *et al.* (1999) Germline transformation of the silkworm *Bombyx mori* L. using a piggyBac transposon-derived vector. *Nature Biotechnol.* **18**, 81–4.

Tang K.X., Sun X.F., Hu Q.N., Wu A.Z., Lin C.H., Lin H.J., Twyman R.M., Christou P. & Feng T.Y. (2001) Transgenic rice plants expressing the ferredoxin-like protein (API) from sweet pepper show enhanced resistance to *Xanthomonas oryzae* pv. *oryzae. Plant Sci.* **160**, 1035–42.

Tang M.X., Redemann C.T. & Szoka F.C. Jr. (1996) *In vitro* gene delivery by degraded polyaminedoamine dendrimers. *Bioconj. Chem.* **7**, 703–14.

Tao H., Bausch C., Richmond C., Blattner F.R. & Conway T. (1999) Functional genomics: expression analysis of *Escherichia coli* growing on minimal and rich media. *J. Bacteriol.* **181**, 6425–40.

Tao W.A. & Aebersold R. (2003) Advances in quantitative proteomics via stable isotope tagging and mass spectrometry. *Curr. Opin. Biotechnol.* **14**, 110–18.

Taton T.A., Mirkin C.A. & Letsinger R.L. (2000) Scanometric DNA array detection with nanoparticle probes. *Science* **289**, 1757–60.

Tavernarakis N., Wang S.L., Dorokov M., Ryazanov A. & Driscoll M. (2000) Heritable and inducible genetic interference by double-stranded RNA encoded by transgenes. *Nature Genet.* **24**, 180–3.

Tavladoraki P., Benvenuto E., Trinca S. *et al.* (1993) Transgenic plants expressing a functional single-chain Fv antibody are specifically protected from virus attack. *Nature* **366**, 469–72.

Taylor G.R. & Logan W.P. (1995) The polymerase chain reaction: new variations on an old theme. *Curr. Opin. Biotechnol.* **6**, 24–9.

Te Riele H., Maandag E.R. & Berns A. (1992) Highly efficient gene targeting in embryonic stem cells through homologous recombination with isogenic DNA constructs. *Proc. Nat. Acad. Sci. USA* **89**, 5128–32.

Templin M.F., Stoll D., Schrenk M., Traub P.C., Vohringer C.F. & Joos T.O. (2002) Protein microarray technology. *Trends Biotechnol.* **20**, 160–6.

Tepfer D. (1984) Transformation of several species of higher plants by *Agrobacterium rhizogenes*: sexual transmission of the transformed genotype and phenotype. *Cell* **37**, 959–67.

Terada R., Urawa H., Inagaki V., Tsugane K. & Iida S. (2002) Efficient gene targeting by homologous recombination in rice. *Nat. Biotechnol.* **20**, 1030–4.

ter Kuile B.H. & Westerhoff H.V. (2001) Transcriptome meets metabolome: hierarchical and metabolic regulation of the glycolytic pathway. *FEBS Lett.* **500**, 169–71.

Terskikh A., Fradkov A., Ermakova G. *et al.* (2000) "Fluorescent timer": protein that changes colour with time. *Science* **290**, 1585–8.

Terwilliger T.C. (2000) Structural genomics in North America. *Nature Struct. Biol.* **7**, 935–9.

Tesfu, E. *et al.* (2004) Building addressable libraries: the use of electrochemistry for generating reactive Pd(ii) reagents at preselected sites on a chip. *J. Am. Chem. Soc.* **126**, 6212–13.

Thanaraj T.A., Stamm S., Clark F., *et al.* (2004) ASD: the alternative splicing database. *Nucl. Acids Res.* **32**, D64–9.

Thangavelu M., James A.B., Bankier A., *et al.* (2003) HAPPY mapping in a plant genome: reconstruction and analysis of a high-resolution physical map of a 1.9 Mbp region of *Arabidopsis thaliana* chromosome 4. *Plant Biotechnol. J.* **1**, 23–31.

The International Chimpanzee Chromosome 22 Consortium (2004) DNA sequence and comparative analysis of chimpanzee chromosome 22. *Nature* **429**, 382–8.

Theis J.F. & Newlon C.S. (1997) The ARS309 chromosomal replicator of *Saccharomyces cerevisiae* depends on exceptional ARS consensus sequence. *Proc. Nat. Acad. Sci. USA* **77**, 4559–63.

Thelwell N., Millington S., Solinas A. *et al.* (2000) Mode of action and application of Scorpion primers to mutation detection. *Nucl. Acids Res.* **28**, 3752–61.

Thibault S.T. *et al.* (2004) A complementary transposon tool kit for *Drosophila melanogaster* using P and piggyBac. *Nature Genet.* **36**, 283–7.

Thomas J.G., Ayling A. & Baneyx F. (1997) Molecular chaperones, folding catalysts and the recovery of active recombinant proteins from *E. coli*: to fold or to refold. *Appl. Biochem. Biotechnol.* **66**, 197–238.

Thomas K.R. & Capecchi M.R. (1987) Site-directed mutagenesis by gene targeting in mouse embryo-derived stem cells. *Cell* **51**, 503–12.

Thomas K.R., Folger K.R. & Cappechi M.R. (1986) High frequency targeting of genes to specific sites in the mammalian genome. *Cell* **44**, 419–28.

Thomas M. & Davis R.W. (1975) Studies on the cleavage of bacteriophage lambda DNA with *Eco*RI restriction endonuclease. *J. Mol. Biol.* **91**, 315–28.

Thomas M., Cameron J.R. & Davis R.W. (1974) Viable molecular hybrids of bacteriophage lambda and eukaryotic DNA. *Proc. Nat. Acad. Sci. USA* **71**, 4579–83.

Thomas P.S. (1980) Hybridization of denatured RNA and small DNA fragments transferred to nitrocellulose. *Proc. Nat. Acad. Sci. USA* **77**, 5201–5.

Thomas M. *et al.* (1997) Selective targeting and inhibition of yeast RNA polymerase II by RNA aptamers. *J. Biol. Chem.* **272**, 27980–6.

Thompson A., Lucchini S. & Hinton J.C.D. (2001) It's easy to build your own microarrayer! *Trends Microbiol.* **9**, 154–6.

Thompson E.M., Nagata S. & Tsuji F.I. (1990) *Vargula hilgendorfii* luciferase: a secreted reporter enzyme for monitoring gene expression in mammalian cells. *Gene* **96**, 257–62.

Thompson J.A., Drayton P.R., Frame B.R., Wang K. & Dunwell J.M. (1995) Maize transformation utilizing silicon carbide whiskers – a review. *Euphytica* **85**, 75–80.

Thomzik J.E., Stenzel K., Stöcker R., Schreier P.H., Hain R. & Stahl D.J. (1997) Synthesis of a grapevine phytoalexin in transgenic tomatoes (*Lycopersicon esculentum* Mill.) conditions resistance against *Phytophthora infestans. Physiol. Mol. Plant Path.* **51**, 265–78.

Thorsted P.B. *et al.* (1998) Complete sequence of the IncPbeta plasmid R751: implications for evolution and organisation of the IncP backbone. *J. Mol. Biol.* **282**, 969–90.

Tickle C. & Eichele G. (1994) Vertebrate limb development. *Ann. Rev. Cell Biol.* **10**, 121–52.

Tijsterman M. & Plasterk R.H.A. (2004) Dicers at RISC: the mechanism of RNAi. *Cell* **117**, 1–3.

Timmermans M.C.P., Das O.P. & Messing J. (1992) *Trans* replication and high copy numbers of wheat dwarf virus vectors in maize cells. *Nucl. Acids Res.* **20**, 4047–54.

Timmermans M.C.P., Das O.P. & Messing J. (1994) Geminiviruses and their uses as extrachromosomal replicons. *Ann. Rev. Plant Physiol. Plant Mol. Biol.* **45**, 79–112.

Tingay S., McElroy D., Kalla R. *et al.* (1997) *Agrobacterium tumefaciens*-mediated barley transformation. *Plant J.* **11**, 1369–76.

Tingey S.V. & Del Tufo J.P. (1993) Genetic analysis with RAPD markers. *Plant Physiol.* **101**, 349–52.

Tinland B. (1996) The integration of T-DNA into plant genomes. *Trends Plant Sci.* **1**, 179–84.

Tinland B., Koukolikova-Nicola Z., Hall M.N. & Hohn B. (1992) The T-DNA-linked *vir*D2 protein contains two distinct functional nuclear localization signals. *Proc. Nat. Acad. Sci. USA* **89**, 7442–6.

Tissier A.F., Marillonnet S., Klimyuk V. *et al.* (1999) Multiple independent defective *suppressor-mutator* transposon insertions in *Arabidopsis*: a tool for functional genomics. *Plant Cell* **11**, 1841–52.

Titz B., Schlesner M. & Uetz P. (2004) What do we learn from high-throughput protein interaction data? *Expert Rev. Proteomics* **1**, 111–21.

Tjalsma H., Bolhuis A., Jongbloed J.D.H., Bron S. & Van Dijl J.M. (2000) Signal peptide-dependent protein transport in *Bacillus subtilis*: a genome-based survey of the secretome. *Microbiol. Mol. Biol. Rev.* **64**, 515–47.

Tomes D.T., Wessinger A.K., Ross M. *et al.* (1990) Transgenic tobacco plants and their progeny derived from microprojectile bombardment of tobacco leaves. *Plant Mol. Biol.* **14**, 261–8.

Tomita M. *et al.* (2003) Transgenic silkworms produce recombinant human type III procollagen in cocoons. *Nat. Biotechnol.* **21**, 52–6.

Tomizawa J.-I. & Itoh T. (1981) Plasmid ColE1 incompatibility determined by interaction of RNA I with primer transcript. *Proc. Nat. Acad. Sci. USA* **78**, 6096–100.

Tomizawa J.-I. & Itoh T. (1982) The importance of RNA secondary structure in ColE1 primer formation. *Cell* **31**, 575–83.

Tomizuka K., Yoshida H., Uejima H. *et al.* (1997) Functional expression and germline transmission of a human chromosome fragment in chimaeric mice. *Nature Genet.* **16**, 133–43.

Tong A.H., Evangelista M., Parsons A.B., Xu H., Bader G.D., Page N., Robinson M., Raghibizadeh S., Hogue C.W., Bussey H. *et al.* (2001) Systematic genetic analysis with ordered arrays of yeast deletion mutants. *Science* **294**, 2364–8.

Toole J.J., Knopf J.L., Wozney J.M. *et al.* (1984) Molecular cloning of a cDNA encoding human antihaemophilic factor. *Nature* **321**, 342–7.

Torbert K., Rines H.W. & Somers D.A. (1995) Use of paromomycin as a selective agent for oat transformation. *PCR* **14**, 635–40.

Toth G., Gaspari Z. & Jurka J. (2000) Microsatellites in different eukaryotic genomes: survey and analysis. *Genome Res.* **10**, 967–81.

Touraev A., Stoger E., Voronin V. & Heberle-Bors E. (1997) Plant male germ line transformation. *Plant J.* **12**, 949–56.

Towbin H., Staehelin T. & Gordon J. (1979) Electrophoretic transfer of proteins from polyacrylamide gels to nitrocellulose sheets: procedure and some applications. *Proc. Nat. Acad. Sci. USA* **76**, 4350–4.

Townes T.M. *et al.* (1985) Erythroid-specific expression of human β-globin genes in transgenic mice. *EMBO J.* **4**, 1715–23.

Townsend R., Watts J. & Stanley J. (1986) Synthesis of viral-DNA forms in *Nicotiana-plumbaginifolia* protoplasts inoculated with cassava latent virus (CLV) – evidence for the independent replication of one-component of the CLV genome. *Nucl. Acids Res.* **14**, 1253–65.

Traboni C., Cortese R., Cilibert G. & Cesarini G. (1983) A general method to select M13 clones carrying base pair substitution mutants constructed *in vitro. Nucl. Acids Res.* **11**, 4229–39.

Trask B. *et al.* (1992) Fluorescence *in situ* hybridization mapping of human chromosome 19: mapping and verification of cosmid contigs formed by random restriction fingerprinting. *Genomics* **14**, 162–7.

Trask B.J., Pinkel D. & Van den Engh G.J. (1989) The proximity of DNA sequences in interphase cell nuclei is correlated to genomic distance and permits ordering of cosmids spanning 250 kilobase pairs. *Genomics* **5**, 710–17.

Tricoli D.M., Carney K.J., Russell P.F. *et al.* (1995) Field evaluation of transgenic squash containing single or multiple virus coat protein gene constructs for resistance to cucumber mosaic virus, watermelon mosaic virus 2, and zucchini yellow mosaic virus. *Biotechnology* **13**, 1458–65.

Trieu A.T., Burleigh S.H., Kardailsky I.V. *et al.* (2000) Transformation of *Medicago truncatula* via infiltration of seedlings or flowering plants with *Agrobacterium. Plant J.* **22**, 531–41.

Trieu-Cuot P., Carlier C. & Courvalin P. (1988) Conjugative plasmid transfer from *Enterococcus faecalis* to *Escherichia coli. J. Bacteriol.* **170**, 4388–91.

Trieu-Cuot P., Carlier C., Martin P. & Courvalin P. (1987) Plasmid transfer by conjugation from *Escherichia coli* to Gram positive bacteria. *FEMS Microbiol. Lett.* **48**, 289–94.

Troester H., Bub S., Hunziker A. & Trendelenburg M.F. (2000) Stability of DNA repeats in *Escherichia coli* dam mutant strains indicates a Dam methylation-dependent DNA deletion process. *Gene* **258**(1–2), 95–108.

Trudel M., Saadane N., Garel M.C. *et al.* (1991) Towards a transgenic mouse model of sickle cell disease: hemoglobin SAD. *EMBO J.* **10**, 3157–65.

Tseng W.-C. & Huang L. (1998) Liposome-based gene therapy. *Pharmacol. Sci. Technol. Today* **1**, 206–13.

Tsien R.Y. & Miyawaki A. (1998) Seeing the machinery of live cells. *Science* **280**, 1954–5.

Tsien R.Y. (1998) The green fluorescent protein. *Ann. Rev. Biochem.* **67**, 509–44.

Tsuge K., Matsui K. & Itaya M. (2003) One step assembly of multiple DNA fragments with a designed order and orientation in *Bacillus subtilis* plasmid. *Nucl. Acids Res.* **31**, e133.

Tu J., Ona I., Zhang Q., Mew T.W., Khush G.S. & Datta S.K. (1998) Transgenic rice variety "IR72" with *Xa21* is resistant to bacterial blight. *Theoret. App. Genet.* **97**, 31–6.

Tu J., Zhang G., Datta K., Xu C., He Y., Zhang Q., Khush G.S. & Datta S.K. (2000) Field performance of transgenic elite commercial hybrid rice expressing *Bacillus thuringiensis* δ-endotoxin. *Nature Biotechnol.* **18**, 1101–4.

Tucker C.L., Gera J.F. & Uetz P. (2001) Towards an understanding of complex protein networks. *Trends Cell Biol.* **11**, 102–6.

Tuite M.F., Dobson M.J., Roberts N.A. *et al.* (1982) Regulated high efficiency expression of human interferon-alpha in *Saccharomyces cerevisiae*. *EMBO J.* **1**, 603–8.

Tumbula D.L. & Whitman W.B. (1999) Genetics of *Methanococcus*: possibilities for functional genomics in Archaea. *Mol. Microbiol.* **33**, 1–7.

Tumer N., Hwang D. & Bonness M. (1997) C-terminal deletion mutant of pokeweed antiviral protein inhibits viral infection but does not depurinate host ribosomes. *Proc. Nat. Acad. Sci. USA* **94**, 3866–71.

Tuschl T. & Borkhardt A. (2002) Small interfering RNAs. *Mol. Interv.* **2**, 158–67.

Tuteja R. & Tuteja N. (2004a) Serial analysis of gene expression (SAGE): application in cancer research. *Med. Sci. Monitor* **10**, RA132–40.

Tuteja R. & Tuteja N. (2004b) Serial analysis of gene expression: applications in human studies. *J. Biomed. Biotechnol.* **2**, 113–20.

Twigg A.J. & Sherratt D. (1980) *Trans*-complementable copy-number mutants of plasmid ColE1. *Nature* **283**, 216–18.

Twyman R.M. (2004) *Advanced Text: Principles of Proteomics*. BIOS Scientific Publishers, Oxford UK.

Twyman R.M. (2004) SNP discovery and typing technologies for pharmacogenomics. *Curr Top Med Chem* **4**, 1421–9.

Twyman R.M. & Christou P. (2004) Plant transformation technology – particle bombardment. In *Handbook of Plant Biotechnology*, eds. Christou P. & Klee H. John Wiley & Sons Inc., NY, pp. 263–89.

Twyman R.M. & Primrose S.B. (2003) Techniques patents for SNP genotyping. *Pharmacogenomics* **4**, 67–79.

Twyman R.M., Schillberg S. & Fischer R. (2005) The transgenic plant market in the pharmaceutical industry. *Expert Opin. Emerg. Drugs* (in press).

Twyman R.M., Stoger E., Schillberg S., Christou P. & Fischer R. (2003) Molecular farming in plants: host systems and expression technology. *Trends Biotechnol.* **21**, 570–8.

Tyagi S. & Kramer F.R. (1996) Molecular beacons: probes that fluoresce upon hybridisation. *Nature Biotechnol.* **14**, 303–8.

Tyagi S., Bratu D.P. & Kramer F.R. (1998) Multicolor molecular beacons for allele discrimination. *Nature Biotechnol.* **16**, 49–53.

Tyurin M., Starodubtseva L., Kudryavtseva H., Voeykova T. & Livshits V. (1995) Electrotransformation of germinating spores of *Streptomyces* spp. *Biotechnol. Techniques* **9**, 737–40.

Tzfira T., Li J., Lacroix B. & Citovsky V. (2004) *Agrobacterium* T-DNA integration: molecules and models. *Trends Genet.* **20**, 375–83.

Tzfira T. & Citovsky V. (2000) From host recognition to T-DNA integration: the function of bacterial and plant genes in the *Agrobacterium*–plant cell interaction. *Mol. Plant Pathol.* **1**, 201–12.

Tzfira T. & Citovsky V. (2002) Partners-in-infection: host proteins involved in the transformation of plant cells by *Agrobacterium*. *Trends Cell Biol.* **12**, 121–9.

Uberacher E. & Mural R. (1991) Locating protein-coding regions in human DNA sequences by a multiple sensorneural network approach. *Proc. Nat. Acad. Sci. USA* **88**, 11261–5.

Uchimiya H., Iwata M., Nojiri C. *et al.* (1993) Bialaphos treatment of transgenic rice plants expressing a *bar* gene prevents infection by the sheath blight pathogen (*Rhizoctonia solani*). *Biotechnology* **11**, 835–6.

Uetz P. & Hughes R.E. (2000) Systematic and large-scale two-hybrid screens. *Curr. Opin. Microbiol.* **3**, 303–8.

Uetz P. *et al.* (2000) A comprehensive analysis of protein–protein interactions in *Saccharomyces cerevisiae*. *Nature* **403**, 623–7.

Ullrich A., Bell J.R., Chen E.Y. *et al.* (1985) Human insulin receptor and its relationship to the tyrosine kinase family of oncogenes. *Nature* **313**, 756–61.

Ulmer J.B., Donnelly J.J., Parker S.E. *et al.* (1993) Heterologous protection against influenza by injection of DNA encoding a viral protein. *Science* **259**, 1745–9.

Unger E., Cigan A.M., Trimnell M., Xu R.-J., Kendall T., Roth B. & Albertsen M. (2002) A chimeric ecdysone receptor facilitates methoxyfenozidedependent restoration of male fertility in ms45 maize. *Transgenic Res.* **11**, 455–65.

Urlaub G., Kas E., Carothers A.M. & Chasin L.A. (1983) Deletion of the diploid dihydrofolate reductase locus from cultured mammalian cells. *Cell* **33**, 405–12.

Urlinger S., Baron U., Thellmann M., Hasan M.T., Bujard H. & Hillen W. (2000) Exploring the sequence space for tetracycline-dependent transcriptional activators: novel mutations yield expanded range and sensitivity. *Proc. Nat. Acad. Sci. USA* **97**, 7963–8.

Vaeck M., Reynaerts A., Hofte H. *et al.* (1987) Transgenic plants protected from insect attack. *Nature* **328**, 33–7.

Vagner V., Dervyn E. & Ehrlich S.D. (1998) A vector for systematic gene inactivation in *Bacillus subtilis*. *Microbiology* **144**, 3097–104.

Vaguine A.A., Richelle J. & Wodak S.J. (1999) SFCHECK: a unified set of procedures for evaluating the quality of macromolecular structure-factor data and their agreement with the atomic model. *Acta Crystallographica* **55**, 191–205.

Vain P., Finer K.R., Engler D.E., Pratt R.C. & Finer J.J. (1996) Intron-mediated enhancement of gene expression in maize (*Zea mays* L) and bluegrass (*Poa pratensis* L). *Plant Cell Rep.* **15**, 489–94.

Valancius V. & Smithies O. (1991) Testing an 'in-out' targeting procedure for making subtle genomic modifications in mouse embryonic stem cells. *Mol. Cell. Biol.* **11**, 1402–8.

Valdes J.M., Tagle D.A. & Collins F.S. (1994) Island rescue PCR: a rapid and efficient method for isolating transcribed sequences from yeast artificial chromosomes. *Proc. Nat. Acad. Sci. USA* **91**, 5377–81.

Valentine L. (2003) *Agrobacterium tumefaciens* and the plant: the David and Goliath of modern genetics. *Plant Physiol.* **133**, 948–55.

Valenzuela P., Medina A., Rutter W.J., Ammerer G. & Hall B.D. (1982) Synthesis and assembly of hepatitis B virus surface antigen particles in yeast. *Nature* **298**, 347–50.

Valle G., Jones E.A. & Colman A. (1982) Anti-ovalbumin monoclonal antibodies interact with their antigen in internal membranes of *Xenopus* oocytes. *Nature* **300**, 71–4.

Van den Elzen P.J.M., Jongedijk E., Melchers L.S. & Cornelissen B.J.C. (1993) Virus and fungal resistance: from laboratory to field. *Philosophic. Trans. Royal Soc. London Series B – Biologic. Sci.* **342**, 271–8.

Van den Elzen P.J.M., Townsend J., Lee K.Y. & Bedbrook J.R. (1985) A chimeric hygromycin resistance gene as a selectable marker in plant cells. *Plant Mol. Biol.* **5**, 299–302.

Van der Geest A.H.M. & Hall T.C. (1997) The b-phaseolin 5′ matrix attachment region acts as an enhancer facilitator. *Plant Mol. Biol.* **33**, 553–7.

Van der Krol A.R., Lenting P.E., Veenstra J. *et al.* (1988) An antisense chalcone synthase gene in transgenic plants inhibits flower pigmentation. *Nature* **333**, 866–9.

Van der Veen B.A., Potocki-Veronese G., Albenne C., *et al.* (2004) Combinatorial engineering to enhance amylosucrase performance: construction, selection, and screening of variant libraries for increased activity. *FEBS Lett.* **560**, 91–7.

Van Deursen J., Fornerod M., Van Rees B. & Grosveld G. (1995) Cre-mediated site-specific translocation between nonhomologous mouse chromosomes. *Proc. Nat. Acad. Sci. USA* **92**, 7376–80.

Van Dyk T.K., Gatenby A.A. & LaRossa R.A. (1989) Demonstration by genetic suppression of interaction of GroE products with many proteins. *Nature* **342**, 451–3.

Van Eenennaam A.L., Lincoln K., Durrett T.P., Valentin H.E., Shewmaker C.K., Thorne G.M., Jiang J., Basziz S.R., Levering C.K., Aasen E.D. *et al.* (2003) Engineering vitamin E content: From *Arabidopsis* mutant to soy oil. *Plant Cell* **15**, 3007–19.

Van Frankenhuyzen K. & Nystrom C. (2002) The *Bacillus thuringiensis* toxin specificity database. http://www.glfc.cfs.nrcan.gc.ca/bacillus.

Van Larbeke N., Engler G., Holsters M. *et al.* (1974) Large plasmid in *Agrobacterium tumefaciens* essential for crown gall-inducing ability. *Nature* **252**, 169–70.

Van Sluys M.A., Tempe J. & Fedoroff N. (1987) Studies on the introduction and mobility of the maize *Activator* element in *Arabidopsis thaliana* and *Daucus carota*. *EMBO J.* **6**, 3881–9.

Vara J.A., Portela A., Ortin J. & Jiminez A. (1986) Expression in mammalian cells of a gene from *Streptomyces alboniger* conferring puromycin resistance. *Nucl. Acids Res.* **14**, 4617–24.

Vasil V., Castillo A., Fromm M. & Vasil I. (1992) Herbicide resistant fertile transgenic wheat plants obtained by microprojectile bombardment of regenerable embryogenic callus. *Biotechnology* **10**, 667–74.

Vasmatzis G. *et al.* (1998) Discovery of three genes specifically expressed in human prostate by expressed sequence tag database analysis. *Proc. Nat. Acad. Sci. USA* **95**, 300–4.

Veculescu V.E. *et al.* (1997) Characterization of the yeast transcriptome. *Cell* **88**, 243–51.

Veculescu V.E. *et al.* (1999) Analysis of human transcriptomes. *Nature Genet.* **23**, 387–8.

Veculescu V.E., Zhang L., Vogelstein B. & Kinzler K.W. (1995) Serial analysis of gene expression. *Science* **270**, 484–7.

Vegeto E., Allan G.F., Schrader W.T. *et al.* (1992) The mechanism of RU486 antagonism is dependent on the conformation of the carboxy-terminal tail of the human progesterone receptor. *Cell* **69**, 703–13.

Vellanoweth R.L. (1993) Translation and its regulation: *Bacillus subtilis* and other Gram-positive bacteria. In *Biochemistry, Physiology and Molecular Genetics*, eds. Sonenshein A.L., Hoch J.A., Losick B.R. ASM. Washington DC.

Venkatasubbarao S. (2004) Microarrays – status and prospects. *Trends Biotechnol.* **12**, 631–7.

Venter J.C. *et al.* (1998) Shotgun sequencing of the human genome. *Science* **280**, 1540–2.

Venter J.C. *et al.* (2001) The sequence of the human genome. *Science* **291**, 1304–51.

Venter J.C., Smith H.O. & Hood L. (1996) A new strategy for genome sequencing. *Nature* **381**, 364–6.

Verch T., Yusibov V. & Koprowski H. (1998) Expression and assembly of a full-length monoclonal antibody in plants using a plant virus vector. *J. Immunol. Methods* **220**, 69–75.

Verpoorte R. (1998) Exploration of nature's chemodiversity: the role of secondary metabolites as leads for drug development. *Drug Devel. Today* **3**, 232–8.

Verpoorte R., Van der Heijden R. & Memelink J. (2000) Engineering the plant cell factory for secondary metabolite production. *Transgenic Res.* **9**, 323–43.

Vick L., Li Y. & Simkiss K. (1993) Transgenic birds from transformed primordial germ cells. *Proc. R. Soc. London B Biol. Sci.* **251**, 179–82.

Vieira J. & Messing J. (1982) The pUC plasmids, an M13mp7-derived system for insertion mutagenesis and sequencing with synthetic universal primers. *Gene* **19**, 259–68.

Vieira J. & Messing J. (1987) Production of single-stranded plasmid DNA. *Methods Enzymol.* **153**, 3–11.

Vielkind J.R. (1992) Medaka and zebrafish: ideal as transient and stable transgenic systems. In: *Transgenic Fish*, eds. Hew C.L. & Fletcher G.L., pp. 72–91. World Scientific Press.

Vinson C.R., LaMarco K.L., Johnson P.F., Landschulz W.H. & McKnight S.L. (1988) *In situ* detection of sequence-specific binding activity specified by a recombinant bacteriophage. *Genes Devel.* **2**, 801–6.

Vize P.D. & Melton D.A. (1991) Assays for gene function in developing *Xenopus* embryos. *Methods Cell Biol.* **36**, 367–87.

Voinnet O., Vain P., Angell S. & Baulcombe S.C. (1998) Systemic spread of sequence-specific transgene RNA degradation in plants is initiated by localized introduction of ectopic promoterless DNA. *Cell* **95**, 177–87.

Volff J-N. & Altenbuchner J. (2000) A new beginning with new ends: linearisation of circular chromosomes during bacterial evolution. *FEMS Microbiol. Lett.* **186**, 143–50.

Von Melchner H., DeGregori J.V., Rayburn H. *et al.* (1992) Selective disruption of genes expressed in totipotent embryonal stem cells. *Genes Devel.* **6**, 919–27.

Vos J.-M.H., Westphal E.-M. & Banerjee S. (1996) Infectious herpes vectors for gene therapy. In *Gene Therapy*, eds. Lemoine N.R. & Cooper D.N., pp. 127–53. BIOS Scientific Publishers, Oxford, UK.

Voskuil M.I. & Chambliss G.H. (1998) The 16 region of *Bacillus subtilis* and other Gram-positive bacterial promoters. *Nucl. Acids Res.* **26**, 3584–90.

Voss A., Niersbach M., Han R. *et al.* (1995) Reduced virus infectivity in *N. tabacum* secreting a TMV-specific full size antibody. *Mol. Breeding* **1**, 39–50.

Voss H. *et al.* (1995) Efficient low redundancy large-scale sequencing at EMBL. *J. Biotechnol.* **41**, 121–9.

Wacker I., Kaether C., Kromer A. *et al.* (1997) Microtubule-dependent transport of secretory vesicles visualized in real time with a GFP-tagged secretory protein. *J. Cell Sci.* **110**, 1453–63.

Wagner K.U., Wall R.J., St-Ogne L. *et al.* (1997) Cre-mediated gene deletion in the mammary gland. *Nucl. Acids Res.* **25**, 4323–30.

Wagner R., Liedtke S., Kretzschmar E. *et al.* (1996) Elongation of the N-glycans of fowl plague virus hemagglutinin expressed in *Spodoptera frugiperda* (Sf9) cells by coexpression of human beta 1,2-N-acetylglucosaminyltransferase I. *Glycobiology* **6**, 165–75.

Wahl G.M., De Saint Vincent B.R. & DeRose M.L. (1984) Effect of chromosomal position on amplification of transfected genes in animal cells. *Nature* **307**, 516–20.

Wahl G.M., Lewis K.A., Ruiz J.C. *et al.* (1987) Cosmid vectors for rapid genomic walking, restriction mapping, and gene transfer. *Proc. Nat. Acad. Sci. USA* **84**, 2160–4.

Wakayama T., Perry A.C., Zuccotti M., Johnson K.R. & Yanagimachi R. (1998) Full term development of mice from enucleated oocytes injected with cumulus cell nuclei. *Nature* **394**, 369–74.

Walbot V. (2000) Saturation mutagenesis using maize transposons. *Curr. Opin. Plant Biol.* **3**, 103–7.

Walden R. (2002) T-DNA tagging in a genomics era. *Crit. Rev. Plant Sci.* **21**, 143–65.

Walhout A.J. *et al.* (2000) Protein interaction mapping in *C. elegans* using proteins involved in vulval development. *Science* **287**, 116–22.

Walhout A.J.M. & Vidal M. (2001) Protein interaction maps for model organisms. *Nature Rev. Mol. Cell Biol.* **2**, 55–62.

Walker M.D., Karlsson O., Edlund T., Barnett J. & Rutter W.J. (1986) Sequences controlling cell-specific expression of the rat insulin-1 gene. *J. Cell Biochem.* **73**, Suppl. 10.

Wall R.J. (1999) Biotechnology for the production of modified and innovative animal products: transgenic livestock bioreactors. *Livestock Prod. Sci.* **59**, 243–55.

Wallace A.C., Borkakoti N. & Thornton J.M. (1997) TESS: a geometric hashing algorithm for deriving 3D coordinate templates for searching structural databases: application to enzyme active sites. *Protein Sci.* **6**, 2308–23.

Wallace B.A. & Janes R.W. (2001) Synchrotron radiation circular dichroism spectroscopy of proteins: secondary structure, fold recognition and structural genomics. *Curr. Opin. Chem. Biol.* **5**, 567–71.

Wallace R.B., Johnson P.F., Tanaka S. *et al.* (1980) Directed deletion of a yeast transfer RNA intervening sequence. *Science* **209**, 1396–400.

Wallace R.B., Schold M., Johnson M.J., Dembek P. & Itakura K. (1981) Oligonucleotide directed mutagenesis of the human β-globin gene: a general method for producing specific point mutations in cloned DNA. *Nucl. Acids Res.* **9**, 3647–56.

Walmsey A.M. & Arntzen C.J. (2000) Plants for delivery of edible vaccines. *Curr. Opin. Biotechnol.* **11**, 126–9.

Walther D., Bartha G. & Morris M. (2001) Basecalling with LifeTrace. *Genome Res.* **11**, 875–88.

Wan Y. & Lemaux P. (1994) Generation of large numbers of independently transformed fertile barley plants. *Plant Physiol.* **104**, 37–48.

Wang B., Ugen K.E., Srikantan V. *et al.* (1993) Gene inoculation generates immune responses against human immunodeficiency virus type 1. *Proc. Nat. Acad. Sci. USA* **90**, 4156–60.

Wang G.-L., Holsten T.E., Song W.-Y., Wang H.-P. & Ronald P.C. (1995) Construction of a rice bacterial artificial chromosome library and identification of clones linked to the Xa-21 disease resistance locus. *Plant J.* **7**, 525–33.

Wang G.L., Song W.Y., Ruan D.L., Sideris S. & Ronald P.C. (1996) The cloned gene, *Xa-21*, confers resistance to multiple *Xanthomonas oryzae* pv. *oryzae* isolates in transgenic plants. *Mol. Plant–Microbe Inter.* **9**, 850–5.

Wang H. & Hanash S. (2003) Multi-dimensional liquid phase-based separations in proteomics. *J Chromatog. B* **787**, 11–18.

Wang L., Brock A., Heberich B. & Schultz P.G. (2001a) Expanding the genetic code of *Escherichia coli*. *Science* **292**, 498–504.

Wang L., Ogburn C.E., Ware C.B. *et al.* (2000) Cellular Werner phenotypes in mice expressing a putative dominant-negative human WRN gene. *Genetics* **154**, 357–62.

Wang S. & Hazelrigg T. (1994) Implications for *bcd* mRNA localization from spatial distribution of Exu protein in *Drosophila* oogenesis. *Nature* **369**, 400–3.

Wang Y., De Mayo F.J., Tsai S.Y. & O'Malley B.W. (1997a) Ligand-inducible and liver-specific target gene expression in transgenic mice. *Nature Biotechnol.* **15**, 239–43.

Wang Y., Xu J., Pierson T., O'Malley B.W. & Tsai S.Y. (1997b) Positive and negative regulation of gene expression in eukaryotic cells with an inducible transcriptional regulator. *Gene Ther.* **4**, 432–41.

Wang Z., Engler P., Longacre A. & Storb U. (2001) An efficient method for high-fidelity BAC/PAC retrofitting with a selectable marker for mammalian cell transfection. *Genome Res.* **11**, 137–42.

Ward A., Etessami P. & Stanley J. (1988) Expression of a bacterial gene in plants mediated by infectious geminivirus DNA. *EMBO J.* **7**, 1583–7.

Ward W.W. & Bokman S.H. (1982) Reversible denaturation of *Aequorea green* fluorescent protein: physical separation and characterization of the renatured protein. *Biochemistry* **21**, 4535–50.

Warren S.T. (1996) The expanding world of trinucleotide repeats. *Science* **271**, 1374–5.

Washburn M.P., Wolters D. & Yates J.R. III (2001) Large-scale analysis of the yeast proteome by multidimensional protein identification technology. *Nature Biotechnol.* **19**, 242–7.

Wasinger V.C. *et al.* (1995) Progress with gene product mapping of the Mollicutes: *Mycoplasma genitalium*. *Electrophoresis* **16**, 1090–4.

Wassenegger M. (2005) The role of RNAi/RNAi components in heterochromatin formation. *Cell*, in press.

Watanabe Y., Ogawa T., Takahashi H., Ishida I. Takeuchi Y., Yamamoto M. & Okada Y. (1995) Resistance against multiple plant viruses in plants mediated by a double stranded-RNA specific ribonuclease. *FEBS Lett.* **372**, 165–8.

Waterhouse P.M., Graham M.W. & Wang M.B. (1998) Virus resistance and gene silencing in plants can be induced by simultaneous expression of sense and antisense RNA. *Proc. Nat. Acad. Sci. USA* **95**, 13959–64.

Waters E., Hohn M.J., Ahel A., *et al.* (2003) The genome of *Nanoarchaeum equitans*: insights into early archaeal evolution and derived parasitism. *Proc. Nat. Acad. Sci.* **100**, 12984–8.

Watson B., Currier T.C., Gordon M.P., Chilton M.-D. & Nester E.W. (1975) Plasmid requirement for virulence of *Agrobacterium tumefaciens*. *J. Bacteriol.* **123**, 255–64.

Watson J.D. (1972) Origin of concatameric T7 DNA. *Nature New Biol.* **239**, 197–201.

Watson N. (1988) A new revision of the sequence of plasmid pBR322. *Gene* **70**, 399–403.

Weber J.L. & Myers E.W. (1997) Human whole-genome shotgun sequencing. *Genome Res.* **7**, 401–9.

Weckwerth W. & Fiehn O. (2002) Can we discover novel pathways using metabolomic analysis? *Curr. Opin. Biotechnol.* **13**, 156–60.

Wehr T. (2002) Multidimensional liquid chromatography in proteomic studies. *LCGC* **20**, 954–62.

Wei J., Goldberg M.B., Burland V., *et al.* (2003) Complete genome sequence and comparative genomics of *Shigella flexneri* serotype 2a strain 2457T. *Infection and Immunity* **71**, 2775–86.

Weigel D. *et al.* (2000) Activation tagging in *Arabidopsis*. *Plant Physiol.* **122**, 1003–13.

Weiler E.W. & Schroder J. (1987) Hormone genes and crown gall disease. *TIBS* **12**, 271–5.

Weinberger S.R., Dalmasso E.A. & Fung E.T. (2001) Current achievements using ProteinChip Array technology. *Curr. Opin. Chem. Biol.* **6**, 86–91.

Weiss R., Teich N., Varmus H. & Coffin J. (1985) *RNA Tumor Viruses*, 2nd edn. Cold Spring Harbor Laboratory, Cold Spring Harbor, New York.

Weiss S. & Chakraborty T. (2001) Transfer of eukaryotic expression plasmids to mammalian host cells by bacteria carriers. *Curr. Opin. Biotechnol.* **12**, 467–72.

Weitzman M.D., Kyostio S.R., Kotin R.M. & Owens R.A. (1994) Adeno-associated virus Rep proteins mediate complex formation between AAV DNA and ins integration site in human DNA. *Proc. Nat. Acad. Sci. USA* **91**, 5808–12.

Welch P.J., Barber J.R. & Wong-Staal F. (1998) Expression of ribozymes in gene transfer systems to modulate target RNA levels. *Curr. Opin. Biotechnol.* **9**, 486–96.

Welch R.A., Burland V., Plunket G., *et al.* (2002) Extensive mosaic structure revealed by the complete genome sequence of uropathogenic *Escherichia coli. Proc. Nat. Acad. Sci. USA* **99**, 17020–4.

Welsh J., Chada K., Dalal S.S. *et al.* (1992) Arbitrarily primed PCR fingerprinting of RNA. *Nucl. Acids Res.* **20**, 4965–70.

Wensink P.C., Finnegan D.J., Donelson J.E. & Hogness D.S. (1974) A system for mapping DNA sequences in the chromosomes of *Drosophila melanogaster. Cell* **3**, 315–25.

Wheeler G.N., Hamilton F.S. & Hoppler S. (2000) Inducible gene expression in transgenic *Xenopus* embryos. *Curr. Biol.* **10**, 849–52.

Whelan S., Lio P. & Goldman N. (2001) Molecular phylogenetics: state-of-the-art methods for looking into the past. *Trends Genet.* **17**, 262–72.

Whitchurch, A.K. (2002) Gene expression microarrays. *IEEE Potentials* **21**, 30–4.

White O. *et al.* (1999) Genome sequence of the radioresistant bacterium *Deinococcus radiodurans* R1. *Science* **286**, 1571–7.

Wianny F. & Zernicka-Goetz M. (2000) Specific interference with gene function by double stranded RNA in early mouse development. *Nature Cell Biol.* **2**, 70–5.

Wiberg F.C., Sunnerhagen P. & Bjursell G. (1987) Efficient transient and stable expression in mammalian cells of transfected genes using erythrocyte ghost fusion. *Exp. Cell Res.* **173**, 218–31.

Widera G., Gautier F., Lindenmaier W. & Collins J. (1978) The expression of tetracycline resistance after insertion of foreign DNA fragments between the *Eco*RI and *Hind*III sites of the plasmid cloning vector pBR322. *Mol. Gen. Genet.* **163**, 301–5.

Wigler M., Perucho M., Kurtz D. *et al.* (1980) Transformation of mammalian cells with an amplifiable dominant acting gene. *Proc. Nat. Acad. Sci. USA* **77**, 3567–70.

Wigler M., Silverstein S., Lee L.S. *et al.* (1977) Transfer of purified herpes virus thymidine kinase gene to cultured mouse cells. *Cell* **11**, 223–32.

Wigler M., Sweet R., Sim G.K. *et al.* (1979) Transformation of mammalian cells with genes from procaryotes and eucaryotes. *Cell* **16**, 777–85.

Wilcox A.S., Khan A.S., Hopkins J.A. & Sikela J.M. (1991) Use of 3¢ untranslated sequences of human cDNAs for rapid chromosome assignment and conversion to STSs: implications for an expression map of the genome. *Nucl. Acids Res.* **19**, 1837–43.

Wiles M.V. *et al.* (2000) Establishment of a gene-trap sequence tag library to generate mutant mice from embryonic stem cells. *Nature Genet.* **24**, 13–14.

Wilkins B.M., Chilley P.M., Thomas A.T. & Pocklington M.J. (1996) Distribution of restriction enzyme recognition sequences on broad host range plasmid RP4: molecular and evolutionary implications. *J. Mol. Biol.* **258**, 447–56.

Wilkinson J.Q., Lanahan M.B., Clark D.G. *et al.* (1997) A dominant mutant receptor from *Arabidopsis* confers ethylene insensitivity in heterologous plants. *Nature Biotechnol.* **15**, 444–7.

Wilks A.F. (1989) Two putative protein-tyrosine kinases identified by application of the polymerase chain reaction. *Proc. Nat. Acad. Sci. USA* **86**, 1063–7.

Willadsen S.M. (1989) Cloning of sheep and cow embryos. *Genome* **31**, 956–62.

Williams B.G. & Blattner F.R. (1979) Construction and characterization of the hybrid bacteriophage lambda Charon series for DNA cloning. *J. Virol.* **29**, 555–62.

Williams D.C., Van Frank R.M., Muth W.L. & Burnett J.P. (1982) Cytoplasmic inclusion bodies in *Escherichia coli* producing biosynthetic human insulin proteins. *Science* **215**, 687–8.

Williams J.G.K., Kubelik A.R., Livak K.J., Rafalski J.A. & Tingey S.V. (1990) DNA polymorphisms amplified by arbitrary primers are useful as genetic markers. *Nucl. Acids Res.* **18**, 6531–5.

Williams T.M., Moolten D., Burlein J. *et al.* (1991) Identification of a zinc finger protein that inhibits IL-2 gene expression. *Science* **254**, 1791–4.

Willmitzer L., Dhaese P., Schreier P.H. *et al.* (1983) Size, location and polarity of transferred DNA encoded transcripts in nopaline crown gall tumours: common transcripts in octopine and nopaline tumours. *Cell* **32**, 1045–6.

Willmitzer L., Simons G. & Schell J. (1982) The Ti DNA in octopine crown gall tumours codes for seven well-defined polyadenylated transcripts. *EMBO J.* **1**, 139–46.

Wilmut I., Beaujean N., De Sousa P.A., Dinnyes A. King T.J. *et al.* (2002) Somatic cell nuclear transfer. *Nature* **419**, 583–6.

Wilmut I., Schnieke A.E., McWhir J., Kind A.J. & Campbell K.H.S. (1997) Viable offspring derived from fetal and adult mammalian cells. *Nature* **385**, 810–13.

Wilson C., Bellen H.J. & Gehring W.J. (1990) Position effects on eukaryotic gene expression. *Ann. Rev. Cell Biol.* **6**, 679–714.

Wilson L.E., Wilkinson N., Marlow S.A., Possee R.D. & King L.A. (1997) Identification of recombinant baculoviruses using green fluorescent protein as a selectable marker. *Biotechniques* **22**, 674–81.

Wilson M. *et al.* (1999) Exploring drug-induced alterations in gene expression in *Mycobacterium tuberculosis* by micro-array hybridization. *Proc. Nat. Acad. Sci. USA* **96**, 12833–8.

Winkler R.G., Frank M.R., Galbraith D.W., Feyereisen R. & Feldmann K.A. (1998) Systematic reverse genetics of transfer-DNA-tagged lines of *Arabidopsis*: isolation of mutations in the cytochrome P450 gene superfamily. *Plant Physiol.* **118**, 743–9.

Winter G., Griffiths A.D., Hawkins R.E. & Hoogenboom H.R. (1994) Making antibodies by phage display technology. *Ann. Rev. Immunol.* **12**, 433–55.

Winter J.A., Wright R.L. & Gurley W.B. (1984) Map locations of five transcripts homologous to TR-DNA in tobacco and sunflower crown gall tumours. *Nucl. Acids Res.* **12**, 2391–406.

Winzeler E.A. *et al.* (1999) Functional analysis of the *S. cerevisiae* genome by gene deletion and parallel analysis. *Science* **285**, 901–6.

Wodicka L., Dong H., Mittmann M., Ho M-H. & Lockhart D.J. (1997) Genome-wide expression monitoring in *Saccharomyces cerevisiae*. *Nature Biotechnol.* **15**, 1359–67.

Woischnik M. & Moraes C.T. (2002) Pattern of organization of human mitochondrial pseudogenes in the nuclear genome. *Genome Res.* **12**, 885–93.

Wolf Y.I. *et al.* (2001) Genome alignment, evolution of prokaryotic genome organization, and prediction of gene function using genomic context. *Genome Res.* **11**, 356–72.

Wolf Y.I., Grishin N.V. & Koonin E.V. (2000a) Estimating the number of protein folds and families from complete genome data. *J. Mol. Biol.* **299**, 897–905.

Wolf Y.I., Kondrashov A.S. & Koonin E.V. (2000b) Inter-kingdom gene fusions. *Genome Biol.* **1**, research0013.1–13.

Wolff J.A., Malone R., Williams W.P., Chong W., Acsadi G., Jani A. & Felgner P.L. (1990) Direct gene transfer into mouse muscle *in vivo*. *Science* **247**, 1465–8.

Wong C.-H., Chen S.-T., Hennen W.J. *et al.* (1990) Enzymes in organic synthesis: use of subtilisin and a highly stable mutant derived from multiple site-specific mutation. *J. Am. Chem. Soc.* **112**, 945–53.

Wong T.-K. & Neumann (1982) Electric field mediated gene transfer. *Biochem. Biophys. Res. Commun.* **107**, 584–7.

Woo P.C.Y., Wong L.P., Zheng B.J. & Yuen K.Y. (2001) Unique immunogenicity of hepatitis B virus DNA vaccine presented by live attenuated *Salmonella typhimurium*. *Vaccine* **19**, 2945–54.

Woo S.-S., Jiang J., Gill B.S., Paterson A.H. & Wing R.A. (1994) Construction and characterization of a bacterial artificial chromosome library of *Sorghum bicolor*. *Nucl. Acids Res.* **22**, 4922–31.

Woo T.H., Patel B.K., Smythe L.D., *et al.* (1998) Identification of pathogenic *Leptospira* by TaqMan probe in a LightCycler. *Analytical Biochem.* **256**, 132–4.

Wood K.V. & DeLuca M. (1987) Photographic detection of luminescence in *Escherichia coli* containing the gene for firefly luciferase. *Anal. Biochem.* **161**, 501–7.

Wood V., Gwilliam R., Rajandream M.A., *et al.* (2002) The genome sequence of *Schizosaccharomyces pombe*. *Nature* **415**, 871–80.

Wood W.I., Capon D.J., Simonsen C.C. *et al.* (1984) Expression of active human factor VIII from recombinant DNA clones. *Nature* **312**, 330–7.

Woods G.L., White K.L., Vanderwall D.K., Li G.P., Aston K.I., Bunch T.D., Meerdo L.N. & Pate B.J. (2003) A mule cloned from fetal cells by nuclear transfer. *Science* **301**, 1063.

Woolfe M. & Primrose S.B. (2004) Food forensics: using DNA technology to combat misdescription and fraud. *Trends Biotechnol.* **22**, 222–6.

Wright G., Carver A., Cottom D. *et al.* (1991) High-level expression of active human alpha-1-antitrypsin in the milk of transgenic sheep. *Biotechnology* **9**, 830–4.

Wu C.Y., Adachi T., Hatano T. *et al.* (1998) Promoters of rice seed storage protein genes direct endosperm-specific gene expression in transgenic rice. *Plant Cell Physiol.* **39**, 885–9.

Wu R. & Taylor E. (1971) Nucleotide sequence analysis of DNA. II. Complete nucleotide sequence of the cohesive ends of bacteriophage λ DNA. *J. Mol. Biol.* **57**, 491–511.

Wu R., Bahl C.P. & Narang S.A. (1978) Chemical synthesis of oligonucleotides. *Prog. Nucl. Acid Res. Mol. Biol.* **21**, 101–38.

Wu S.-Y. & Chiang C.-M. (1996) Establishment of stable cell lines expressing potentially toxic proteins by tetracycline-regulated and epitope tagging methods. *Biotechniques* **21**, 718–25.

Wu X., Holschen J., Kennedy S.C. & Parker-Ponder K. (1996) Retroviral vector sequences may interact with some internal promoters and influence expression. *Hum. Gene Ther.* **7**, 159–71.

Wurm F.M. (2004) Production of recombinant protein therapeutics in mammalian cells. *Nature Biotechnol.* **22**, 1393–8.

Wurm F.M., Gwinn K.A. & Kingston R.E. (1986) Inducible overproduction of the mouse c-Myc protein in mammalian cells. *Proc. Nat. Acad. Sci. USA* **83**, 5414–18.

Wurst W., Rossant J., Prideaux V. *et al.* (1995) A large-scale gene-trap screen for insertional mutations in developmentally regulated genes in mice. *Genetics* **139**, 889–99.

Wyber J.A., Andrews J. & D'Emanuele A. (1997) The use of sonication for the efficient delivery of plasmid DNA into cells. *Pharm. Res.* **14**, 750–6.

Xenarios I. & Eisenberg D. (2001) Protein interaction databases. *Curr. Opin. Biotechnol.* **12**, 334–9.

Xenarios I. *et al.* (2001) DIP: the database of interacting proteins: update. *Nucl. Acids Res.* **29**, 239–41.

Xenarios I., Rice D.W., Salwinski L., Baron M.K., Marcotte E.M. & Eisenberg D. (2000) DIP: the database of interacting proteins. *Nucl. Acids Res.* **28**, 289–91.

Xiang C.C. & Chen Y. (2000) cDNA microarray technology and its applications. *Biotechnol. Adv.* **18**, 35–46.

Xiang R., Lods H.N., Chao T.H., Ruehlmann J.M. *et al.* (2000) An autologous oral DNA vaccine protects against murine melanoma. *Proc. Nat. Acad. Sci. USA* **97**, 5492–7.

Xiao J.H., Davidson I., Matthes H., Garnier J.M. & Chambon P. (1991) Cloning, expression, and transcriptional properties of the human enhancer factor TEF-1. *Cell* **65**, 551–68.

Xie, Z., Johansen, L.K., Gustafson, A.M., Kasschau, K.D., Lellis, A.D. *et al.* (2004) Genetic and functional diversification of small RNA pathways in plants. *PLoS Biol.* **2**, E104.

Xiong C., Levis R., Shen P. *et al.* (1989) Sindbis virus: an efficient broad host range vector for gene expression in animal cells. *Science* **243**, 1188–91.

Xu D.P., Duan X.L., Wang B.Y., Hong B.M., Ho T.H.D. & Wu R. (1996) Expression of a late embryogenesis abundant protein gene, *HVA1*, from barley confers tolerance to water deficit and salt stress in transgenic rice. *Plant Physiol.* **110**, 249–57.

Xu L., Sanchez A., Yang Z.-Y. *et al.* (1998) Immunization for Ebola virus infection. *Nature Med.* **4**, 37–42.

Xu Y., Jablonsky M.J., Jackson P.L., Braun W. & Krishna N.R. (2001) Automatic assignment of NOESY cross peaks and determination of the protein structure of a new world scorpion neurotoxin using NOAH/DIAMOND. *J. Mag. Res.* **148**, 35–46.

Yadav N.S., Vanderleyden J., Bennet D., Barnes W.M. & Chilton M.-D. (1982) Short direct repeats flank the T-DNA on a nopaline Ti plasmid. *Proc. Nat. Acad. Sci. USA* **79**, 6322–6.

Yamao M., Katayama N., Nakazawa H. *et al.* (1999) Gene targeting in the silkworm by use of a baculovirus. *Genes Devel.* **13**, 511–16.

Yan J.X., Packer N.H., Gooley A.A. & Williams K.L. (1998) Protein phosphorylation: technologies for the identification of phosphoamino acids. *J. Chromatog. A* **808**, 23–41.

Yancey P.H., Clark M.E., Hand S.C., Bowlus R.D. & Somero G.N. (1982) Living with water stress: evolution of osmolyte systems. *Science* **217**, 1214–22.

Yanez R.J. & Porter A.C. (2002) Differential effects of Rad52p overexpression on gene targeting and extrachromosomal recombination in a human cell line. *Nucl. Acids Res.* **30**, 740–8.

Yang C.C., Xiao X., Zhu X. *et al.* (1997) Cellular recombination pathways and viral terminal repeat hairpin structures are sufficient for adeno-associated virus integration *in vivo* and *in vitro*. *J. Virol.* **71**, 9231–47.

Yang X.C., Karschin A., Labarca C. *et al.* (1991) Expression of ion channels and receptors in *Xenopus* oocytes using vaccinia virus. *FASEB J.* **5**, 2209–16.

Yansich-Perron C., Vieira J. & Messing J. (1985) Improved M13 phage cloning vectors and host strains: nucleotide sequences of the M13 mp18 and pUC19 vectors. *Gene* **33**, 103–19.

Yansura D.G. & Henner D.J. (1984) Use of the *Escherichia coli lac* repressor and operator to control gene expression in *Bacillus subtilis*. *Proc. Nat. Acad. Sci. USA* **81**, 439–43.

Yao T.P., Forman B.M., Jiang Z. *et al.* (1993) Functional ecdysone receptor is the product of EcR and ultraspiracle genes. *Nature* **366**, 476–9.

Yao T.P., Segraves W.A., Oro A.E., McKeown M. & Evans R.M. (1992) *Drosophila* ultraspiracle modulates ecdysone receptor function via heterodimer formation. *Cell* **71**, 63–72.

Yates J.L., Warren N. & Sugden B. (1985) Stable replication of plasmids derived from Epstein-Barr virus in various mammalian cells. *Nature* **313**, 812–15.

Yates J.L., Warren N., Reisman D. & Sugden B. (1984) A *cis*-acting element from the Epstein-Barr genome that permits stable replication of recombinant plasmids in latently infected cells. *Proc. Nat. Acad. Sci. USA* **81**, 3806–10.

Yates J.R. III (2000) Mass spectrometry from genomics to proteomics. *Trends Genet.* **16**, 5–8.

Yates J.R. III, Speicher S., Griffin P.R. & Hunkapiller T. (1993) Peptide mass maps: a highly informative approach to protein identification. *Anal. Biochem.* **214**, 397–408.

Ye S.Q., Zhang L.Q., Zheng F., Virgil D. & Kwiterovich P.O. (2000) MiniSAGE: gene expression profiling using serial analysis of gene expression from 1 µg total RNA. *Analytical Biochemistry* **287**, 144–52.

Ye X.D., Al-Babili S., Kloti A. *et al.* (2000) Engineering the provitamin A (beta-carotene) biosynthetic pathway into (carotenoid-free) rice endosperm. *Science* **287**, 303–5.

Ye G.N., Stone D., Pang S.Z., Creely W., Gonzales K. & Hinchee M. (1999) *Arabidopsis* ovule is the target for *Agrobacterium in planta* vacuum infiltration transformation. *Plant J.* **19**, 249–57.

Yeh R.F., Lim L. & Burge C.B. (2001) Computational inference of homologous gene structures in the human genome. *Genome Res.* **11**, 803–16.

Yershov K. *et al.* (1996) DNA analysis and diagnostics on oligonucleotide microchips. *Proc. Nat. Acad. Sci. USA* **93**, 4913–18.

Yim Y-S., Davis G.L., Duru N.A., *et al.* (2002) Characterization of three maize bacterial artificial chromosome libraries towards anchoring of the physical map to the genetic map using high-density bacterial artificial chromosome filter hybridisation. *Plant Physiol.* **130**, 1686–96.

Ymer S., Schofield P.R., Draguhn A. *et al.* (1989) GABA receptor beta-subunit heterogeneity: functional expression of cloned cDNAs. *EMBO J.* **6**, 1665–70.

Yoder J., Walsh C. & Bestor T. (1997) Cytosine methylation and the ecology of intragenomic parasites. *Trends Genet.* **13**, 335–9.

Yokota H., Van Den Engh G., Hearst J.E., Sachs R.K. & Trask B.J. (1995) Evidence of the organization of chromatin in megabase pair-sized loops arranged along a random walk path in the human G0/G1 interphase nucleus. *J. Cell Biol.* **130**, 1239–49.

Yokoyama M. (2002) Gene delivery using temperature-responsive polymeric carriers. *Drug Discovery Today* **7**, 426–32.

Yokoyama M. *et al.* (2001) Influential factors on temperature-controlled gene expression using thermo-responsive polymeric gene carriers. *J. Artifical Org.* **4**, 138–45.

Yokoyama S. *et al.* (2000) Structural genomics projects in Japan. *Prog. Biophys. Mol. Biol.* **73**, 363–76.

Young B.D., Birnie G.D. & Paul J. (1976) Complexity and specificity of polysomal poly (A)⁺ RNA in mouse tissues. *Biochemistry* **15**, 2823–8.

Young R.A. & Davis R.W. (1983) Efficient isolation of genes by using antibody probes. *Proc. Nat. Acad. Sci. USA* **80**, 1194–8.

Young R.A., Bloom B.R., Grosskinsky C.M. *et al.* (1985) Dissection of *Mycobacterium tuberculosis* antigens using recombinant DNA. *Proc. Nat. Acad. Sci. USA* **82**, 2583–7.

Young S.M., McCarty D.M., Degtyareva N. & Samulski R.J. (2000) Roles of adeno-associated virus Rep and human chromosome 19 in site-specific recombination. *J. Virol.* **74**, 3953–66.

Yu S.F., Von Ruden T., Kantoff P.W. *et al.* (1986) Self-inactivating retroviral vectors designed for transfer of whole genes into mammalian cells. *Proc. Nat. Acad. Sci. USA* **83**, 3194–8.

Yuan R., Hamilton D.L. & Burckhardt J. (1980) DNA translocation by the restriction enzyme from *E. coli* K. *Cell* **20**, 237–44.

Yusibov V. & Rabindran S. (2004) Plant viral expression vectors: history and new developments. In: Fischer R, Schillberg S (eds.) *Molecular Farming: Plant-made Pharmaceuticals and Technical Proteins.* John Wiley & Sons Inc., NY.

Zabner J., Couture L.A., Graham S.M., Smith A.E. & Welsh M.J. (1993) Adenovirus-mediated gene transfer transiently corrects the chloride transport defect in nasal epithelia of patients with cystic fibrosis. *Cell* **75**, 1–20.

Zaccolo M., Williams D.M., Brown D.M. & Gheradi E. (1996) An approach to random mutagenesis of DNA using mixtures of triphosphate derivatives of nucleoside analogues. *J. Mol. Biol.* **255**, 589–603.

Zaenen I., Van Larbeke N., Teuchy H., Van Montagu M. & Schell J. (1974) Super-coiled circular DNA in crown-gall inducing *Agrobacterium* strains. *J. Mol. Biol.* **86**, 109–27.

Zambrowicz B.P., Friedrich G.A., Buxton E.C. *et al.* (1998) Disruption and sequence identification of 2000 genes in mouse embryonic stem cells. *Nature* **392**, 608–11.

Zambryski P., Depicker A., Kruger H. & Goodman H. (1982) Tumor induction by *Agrobacterium tumefaciens*: analysis of the boundaries of T-DNA. *J. Mol. Appl. Gent.* **1**, 361–70.

Zambryski P., Joos H., Genetello C. *et al.* (1983) Ti plasmid vector for the introduction of DNA into plant cells without alteration of their normal regeneration capacity. *EMBO J.* **2**, 2143–50.

Zbikowska H.M. (2003) Fish can be first – advances in fish transgenesis for commercial applications. *Transgenic Res.* **12**, 379–9.

Zbikowska H.M. *et al.* (2002) Uromodulin promoter directs high-level expression of biologically active human alpha1-antitrypsin into mouse urine. *Biochem. J.* **365**, 7–11.

Zdobnov E.M., von Mering C., Letunic I., *et al.* (2002) Comparative genome and proteome analysis of *Anopheles gambiae* and *Drosophila melanogaster*. *Science* **298**, 149–59.

Zehetner G. & Lehrach H. (1994) The reference library system – sharing biological material and experimental data. *Nature* **367**, 489–91.

Zeitlin L., Olmsted S.S., Moench T.R. *et al.* (1998) A humanized monoclonal antibody produced in transgenic plants for immonoprotection of the vagina against genital herpes. *Nature Biotechnol.* **16**, 1361–4.

Zhang C. & DeLisi C. (1998) Estimating the number of protein folds. *J. Mol. Biol.* **284**, 1301–5.

Zhang C. & Kim S.-H. (2003) Overview of structural genomics: from structure to function. Curr. Opin. *Chem. Biol.* **7**, 28–32.

Zhang J. & Deutscher M.P. (1992) A uridine-rich sequence required for translation of prokaryotic mRNA. *Proc. Nat. Acad. Sci. USA* **89**, 2605–9.

Zhang J-H., Dawes G. & Stemmer W.P.C. (1997) Directed evolution of a fucosidase from a galactosidase by DNA shuffling and screening. *Proc. Nat. Acad. Sci. USA*, **94**, 4504–9.

Zhang L. *et al.* (1997) Gene expression profiles in normal and cancer cells. *Science* **276**, 1268–72.

Zhang L., Cui X., Schmitt K. *et al.* (1992) Whole genome amplification from a single cell: implications for genetic analysis. *Proc. Nat. Acad. Sci. USA* **89**, 5847–51.

Zhang L.-J., Cheng L.-M., Xu N. *et al.* (1991) Efficient transformation of tobacco by ultrasonication. *Biotechnol.* **9**, 996–7.

Zhang S., Song W.Y., Chen L., Ruan D., Taylor N., Ronald P., Beachy R. & Fauquet C. (1998) Transgenic elite indica rice varieties, resistant to *Xanthomonas* pv. *oryzae*. *Mol. Breed.* **4**, 551–8.

Zhang X. & Smith T.F. (1997) The challenges of genome sequence annotation or "the Devil is in the Details". *Nature Biotechnol.* **15**, 1222–3.

Zhang X. & Studier F.W. (1997) Mechanism of inhibition of bacteriophage T7 RNA polymerase by T7 lysozyme. *J. Mol. Biol.* **269**, 10–27.

Zhang Z., Gildersleeve J., Yang Y-Y., *et al.* (2004) A new strategy for the synthesis of glycoproteins. *Science* **303**, 371–3.

Zhang Z., Smith B.A.C., Wang L., *et al.* (2003) A new strategy for the site-specific modification of proteins *in vivo*. *Biochemistry* **42**, 6735–46.

Zhang, W. *et al.* (2003) Cre/lox-mediated marker gene excision in transgenic maize (*Zea mays* L.) plants. *Theor. Appl. Genet.* **107**, 1157–68.

Zhao H., Giver L., Shao Z., *et al.* (1998) Molecular evolution by staggered extension process (StEP) *in vitro* recombination. *Nature Biotechnol.* **16**, 285–61.

Zhao J. & Lemke G. (1998) Selective disruption of neuregulin-1 function in vertebrate embryos using ribozyme-tRNA transgenes. *Development* **125**, 1899–907.

Zhao J.J.G. & Pick L. (1983) Generating loss-of-function phenotypes of the *fushi tarazu* gene with a targeted ribozyme in *Drosophila*. *Nature* **365**, 448–51.

Zhao N. *et al.* (1995) High-density cDNA filter analysis: a novel approach for large scale quantitative analysis of gene expression. *Gene* **156**, 207–13.

Zheng L., Liu J., Batalov S., Zhou D., Orth A., Ding S. & Schultz P.G. (2004) An approach to genome-wide screens of expressed smallinterfering RNAs in mammalian cells. *Proc. Nat. Acad. Sci. USA* **101**, 135–40.

Zhong Z., Liu J.L.-C., Dinterman L.M. *et al.* (1991) Engineering subtilisin for reaction in dimethylformamide. *J. Am. Chem. Soc.* **113**, 683–4.

Zhou G., Weng J., Zheng Y. *et al.* (1983) Introduction of exogenous DNA into cotton embryos. *Methods Enzymol.* **101**, 433–81.

Zhou H., Roy S., Schulman H. & Natan M.J. (2001) Solution and chip arrays in protein profiling. *Trends Biotechnol* **19** (Suppl., *A Trends Guide to Proteomics II*), S34–9.

Zhou H.S., O'Neal W., Morral N. & Beaudet A.L. (1996) Development of a complementing cell line and a system for construction of adenovirus vectors with E1 and E2a deleted. *J. Virol.* **70**, 7030–8.

Zhou Q., Renard J.P., Le Friec G., Brochard V., Beaujean N., Cherifi Y., Fraichard A. & Cozzi J. (2003) Generation of fertile cloned rats by regulating oocyte activation. *Science* **302**, 1179.

Zhu H. & Dean R.A. (1999) A novel method for increasing the transformation efficiency of *Escherichia coli*-application for bacterial artificial chromosome library construction. *Nucl. Acids Res.* **27**, 910–11.

Zhu H. & Snyder M. (2003) Protein chip technology. *Curr. Opin. Chem. Biol.* **7**, 55–63.

Zhu H. *et al.* (2000) Analysis of yeast protein kinases using protein chips. *Nature Genet.* **26**, 283–9.

Zhu H. *et al.* (2001) Global analysis of protein activities using proteome chips. *Science* **293**, 2101–5.

Zhu H., Cong J.P., Mamtora G., Gingeras T. & Shenk T. (1998) Cellular gene expression altered by human cytomegalovirus: global monitoring with oligonucleotide arrays. *Proc. Nat. Acad. Sci. USA* **95**, 14470–5.

Zhu Z., Huges K.W., Huang L. *et al.* (1994) Expression of human alpha-interferon cDNA in transgenic rice plants. *Plant Cell Tissue Organ Cult.* **36**, 197–204.

Zhu Z., Ma B., Homer R.J., Zheng T. & Elias J.A. (2001) Use of the tetracycline controlled transcriptional silencer (tTS) to eliminate transgene leak in inducible overexpression transgenic mice. *J. Biol. Chem.* **276**, 25222–9.

Ziegler A., Cowan G.H., Torrance L., Ross H.A. & Davies H.V. (2000) Facile assessment of cDNA constructs for expression of functional antibodies in plants using the potato virus X vector. *Mol. Breeding* **6**, 327–35.

Zimmer A. & Gruss P. (1989) Production for chimaeric mice containing embryonic stem cell (ES) cells carrying a homeobox *Hox1.1* allele mutated by homologous recombination. *Nature* **338**, 150–3.

Zimmerman S.B. & Pheiffer B. (1983) Macromolecular crowding allows blunt end ligation by DNA ligases from rat liver or *Escherichia coli*. *Proc. Nat. Acad. Sci. USA* **80**, 5852–6.

Zimmerman U. & Vienken J. (1983) Electric field induced cell to cell fusion. *J. Membr. Biol.* **67**, 165–82.

Zoller M. & Christ O. (2001) Prophylactic tumor vaccination: comparison of effector mechanisms initiated by protein versus DNA vaccination. *J. Immunol.* **166**, 3440–50.

Zoller M.J. & Smith M. (1983) Oligonucleotide-directed mutagenesis of DNA fragments cloned into M13 vectors. *Methods Enzymol.* **100**, 468–500.

Zolotukhin S.M., Potter W.W., Hauswirth J.G. & Muzyczka N. (1996) A humanized green fluorescent protein cDNA adapted for high level expression in mammalian cells. *J. Virol.* **70**, 4646–54.

Zoubak S., Clay O. & Bernardi G. (1996) The gene distribution of the human genome. *Gene* **174**, 95–102.

Zoubenko O.V., Allison L.A., Svab Z. & Maliga P. (1994) Efficient targeting of foreign genes into the tobacco plastid genome. *Nucl. Acids Res.* **22**, 3819–24.

Zubko E., Scutt C. & Meyer P. (2000) Intrachromosomal recombination between *attP* regions as a tool to remove selectable marker genes from tobacco transgenes. *Nature Biotechnol.* **18**, 422–45.

Zuo J. *et al.* (2001) Chemical-regulated, site-specific DNA excision in transgenic plants. *Nat. Biotechnol.* **19**, 157–61.

Zupan J., Muth T.R., Draper O. & Zambryski P. (2000) The transfer of DNA from *Agrobacterium tumefaciens* into plants: a feast of fundamental insights. *Plant J.* **23**, 11–28.

Zwaal R.R., Broeks A., Meurs J.V., Groenen J.T. & Plasterk R.H. (1993) Target-selected gene inactivation in *Caenorhabditis elegans* by using a frozen transposon insertion mutant bank. *Proc. Nat. Acad. Sci. USA* **90**, 7431–5.

Zwaka, T.P. & Thomson, J.A. (2003) Homologous recombination in human embryonic stem cells. *Nature Biotechnol.* **21**, 319–21.

Appendix: The genetic code and single-letter amino acid designations

		Second position				
		U	C	A	G	
First position	U	Phe / Leu	Ser	Tyr / *ochre* / *amber*/Pyr	Cys / *opal*/Sec / Trp	U C A G
	C	Leu	Pro	His / Gln	Arg	U C A G
	A	Ile / Met	Thr	Asn / Lys	Ser / Arg	U C A G
	G	Val	Ala	Asp / Glu	Gly	U C A G

Third position

A	Alanine	N	Asparagine
B	Asparagine or aspartic acid	P	Proline
C	Cysteine	Q	Glutamine
D	Aspartic acid	R	Arginine
E	Glutamic acid	S	Serine
F	Phenylalanine	T	Threonine
G	Glycine	U	Selenocysteine
H	Histidine	V	Valine
I	Isoleucine	W	Tryptophan
J/O*	Pyrrolysine	X	Any amino acid
K	Lysine	Y	Tyrosine
L	Leucine	Z	Glutamine or
M	Methionine		glutamic acid

* Single letter code for pyrrolysine currently unassigned, but only J and O are unused.

Scale for conversion between kilobase pairs of duplex DNA and molecular weight

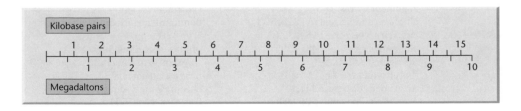

Index